A cameo from the past

A cameo from the past

The prehistory and early history of the Kruger National Park

Dr U. De V. Pienaar

and contributors:
Dr E.J. Carruthers
Y.A. Courtin
Dr J.B. de Vaal
Prof. J.F. Eloff
M. English
Dr S.C.J. Joubert
M. Liebenberg
S. Miller
A. Minnaar
H.H. Mockford
Dr W.H.J. Punt (senior)
J. Verhoef
P.H. Cuénod

Translated by Helena Bryden

Protea Book House
Pretoria
2012

*The author sadly passed away on
1 February 2011 before publication of this book.*

A cameo from the past – Dr U. de V. Pienaar
First edition, first impression in 2012 by Protea Book House

PO Box 35110, Menlo Park, 0102
1067 Burnett Street, Hatfield, Pretoria
8 Minni Street, Clydesdale, Pretoria
protea@intekom.co.za
www.proteaboekhuis.com

TRANSLATOR: Helena Bryden
EDITORS: Iolandi Pool, Danél Hanekom
PROOFREADER: Carmen Hansen-Kruger
COVER DESIGN: Hanli Deysel
TYPOGRAPHY: 10 on 13.5 pt Garamond by Hanli Deysel
PRINTED AND BOUND: in China through Colorcraft Ltd., HK

© 2007 U. de V. Pienaar (Afrikaans text)
© 2012 A. Pienaar (English translation)
ISBN Printed book: 978-1-86919-195-5
ISBN E-book: 978-1-86919-682-0

All rights reserved. No part of this book may be reproduced or transmitted in any form or by any electronic or mechanical means, including photocopying and recording, or by any other information storage or retrieval system, without written permission from the publisher.

CONTENTS

Foreword 7
Introduction 8
Chapter 1 When the Lowveld was wild and desolate 11
Chapter 2 A new order 35
Chapter 3 The first Europeans visit the Lowveld 90
Chapter 4 The Voortrekkers explore the Lowveld 102
Chapter 5 João Albasini (1813–1888) 146
Chapter 6 The bygone days of the Soutpansberg 169
Chapter 7 Pioneers and hunters in the Lowveld 213
Chapter 8 On the trail of the explorers 246
Chapter 9 Prospectors and delving 280
Chapter 10 Transport roads, transport riders and the tsetse fly 301
Chapter 11 The Eastern and Selati railway lines 350
Chapter 12 The demarcation of the border between Mozambique and the ZAR 368
Chapter 13 Game conservation in the Transvaal 1846–1898 and the proclamation of the Sabie Game Reserve 383
Chapter 14 War clouds loom over the Lowveld 1899–1902 403
Chapter 15 Reproclamation of the Sabie Game Reserve and later border changes 434
Chapter 16 Early years: The Sabie and Shingwedzi reserves and the Kruger National Park 1900–1946 444
Chapter 17 Quests for buried treasures 614
Chapter 18 Lonely graves in the bushveld 628
Chapter 19 The wild days of Crooks Corner 647
Chapter 20 The saga of the WNLA in the Kruger National Park 666
Chapter 21 Early supporters and benefactors of the reserves 684
Chapter 22 Memorials and memorial plaques 702
Acknowledgements 721
Historical terrains – National Kruger Park 727
Table with grid references of historical terrains 728
List of maps 734
Index 735

Vision

National parks will be the pride and joy of all South Africans and of the world.

Mission

To develop and manage a system of national parks that represents the biodiversity, landscapes, and associated heritage assets of South Africa for the sustainable use and benefit of all.

A cameo from the past describes the long and sometimes difficult developmental history of SANParks in detail. Despite the good and the bad from the past, the organisation has developed into the leading conservation authority in Africa, responsible for 3 751 113 hectares of protected land in 20 national parks.

SANParks manages a system of parks which represents the indigenous fauna, flora, landscapes and associated cultural heritage of the country. Of all the national parks, most have overnight tourist facilities, with an unrivalled variety of accommodation in arid, coastal, mountain and bushveld habitats.

The parks offer visitors an unparalleled diversity of adventure tourism opportunities including game viewing, bush walks, canoeing and exposure to cultural and historical experiences. Conferences can also be organised in many of the parks. Everyone is free to visit national parks and the more people do so and share and appreciate them, the more secure their future will be.

The national parks of South Africa include: Groenkloof, Kruger, Table Mountain, Marakele, Golden Gate, Camdeboo, Mountain Zebra, Addo Elephant, Garden Route National Park (Tsitsikamma, Knysna, & Wilderness), Bontebok, Agulhas, West Coast, Karoo, Namaqua, Ai-Ais/Richtersveld, Augrabies, Kgalagadi, Mapungubwe, Tankwa Karoo, Pilanesberg and Mokala.

The focus for SANParks in the first decade of democracy has been to make national parks more accessible to tourists to ensure conservation remains a viable contributor to social and economic development in rural areas.

Since 1994, the organisation has managed to transform itself, continue its high research and management standards, expand the land under its protection by 360 000 hectares and has also begun to generate 75 per cent of its operating revenue – a spectacular financial achievement compared to most conservation agencies in the world, including those in developed countries.

Foreword

Nature conservation means different things to people with diverse interests and backgrounds. Some are only interested in conserving the biotic components (fauna and flora) of an area, while others realise that the abiotic components (geology and soil types) are also an inseparable component of nature conservation. What most people do not realise, is that man should also be an integrated component of any nature conservation action.

It can further be stated without a doubt that nature and conservation and national parks go hand in hand. The crucial role which people play in the creation and maintenance of national parks is often underestimated because of short-sighted thinking. The creation of the first national park in the world, the Yellowstone National Park in the USA in 1872, followed in 1898 by the proclamation of the Sabie Reserve, the forerunner of the current Kruger National Park (KNP), stands witness to the fact that mankind is drawn to culturally oriented activities. The cultural process of creating and maintaining national parks is therefore no more than 116 years old, a short time really in the cultural history of mankind. Dr Pienaar, in cooperation with other experts, interpreted the past in order to fill the voids in the way we think, and to make us aware of man's centuries-old cultural-sociological connection with the area now known as the KNP.

The reader will be able to obtain a detailed retrospective view of the distant past (both prehistorical and historical) in the Transvaal Lowveld,[1] as the story unfolds up to 1946. This book reflects the perseverance of key figures and the reader becomes gradually aware of the fact that man did not develop as a separate entity distanced from the developmental history of the KNP. We become more and more aware of the KNP's exceptionally rich historical legacy from the Stone Age, through the San era until today. The influence of man is noticeable everywhere – from archaeological remains, communication systems (or the lack thereof), early settlements, transport riders, explorers and hunters, problems with borders and the tsetse fly, to the creation of management policies and other relevant procedures which are still maintained in the KNP although these have been adapted to a degree.

Nature conservation should be seen as a culturally based action, which has effected impacts by man for man, with the essential sociological results which flows from it. It is this cultural historical interpretation of the KNP which was placed on record by Dr Pienaar and his contributors with great distinction.

This informative publication is an example of academic research which has immediate practical implications and which emphasises how indispensible such research is for the practical management and proper measures by which a natural resource like the KNP should be managed. In this work the reader is able to obtain guidelines for the future and greater insight into legal, administrative, economic, scientific and political contexts which together form the framework for sound management practices.

This book will undoubtedly create a new dimension in the way people will experience and enjoy the Kruger National Park, one of our greatest national assets.

Prof. F.C. Eloff
Chairperson, National Parks Board (now South African National Parks Board)
September 1989

---

1   Transvaal refers here to the area between the Vaal and Limpopo rivers which was divided into four provinces in 1994, namely North West, Gauteng, Limpopo and Mpumalanga. The Lowveld is an ecological region along the eastern borders of Mpumalanga and Limpopo.

Introduction

Dr U. de V. Pienaar

Gezina (née Kruger), President Paul Kruger's wife, who stayed behind in Pretoria due to poor health when he left the country in October 1900.
(ACKNOWLEDGEMENT: TRANSVAAL ARCHIVES)

President Paul Kruger, the father of conservation in southern Africa and founder of the Sabie Reserve, which would eventually become the Kruger National Park, as well as other reserves in the ZAR (South African Republic).

A wise man once said: "A nation without a past is a nation without soul and those that do not honour their past, also do not deserve a future."

Our country at the southern-most tip of Africa certainly has a long and lustrous history that stretches far back into the haze of oblivion.

Educationalists complain that the youth of today show little interest in history and only live for the moment. As a general statement this sounds far-fetched, as the younger generation need not be blamed for this. Thankfully the large conservative part of all the racial groups of South Africa still show a binding national pride while honouring deceased leaders and folk heroes.

The story told here follows a long road with many twists and turns that has not always been strewn with roses. It also would never have been recorded if not for one man's vision. This man, who provided the South African nation with the crown jewel of our national parks, was the State President of the Zuid-Afrikaansche Republiek (South African Republic), Stephanus Johannes Paulus Kruger.

After many years of persuasion and ardent debating in the Volksraad (National Assembly) of the Transvaal, President Kruger succeeded in having the "Sabie Goewernements Wildtuin" (Sabie Government Game Reserve) proclaimed in the Transvaal Lowveld.

In doing this he hoped to ensure that the rich variety of Lowveld fauna and flora between the Sabie and Crocodile rivers would be preserved for future generations. But before he could witness the fruits of his labour, the storm of the Anglo-Boer War broke loose over the Boer republics.

Despite the initial successes of the Boer commandos, they had to retreat before the superior power of the British army. First Bloemfontein and then Pretoria fell into the hands of the occupational forces. Every conceivable measure had to be taken to prevent the president from falling into enemy hands.

On 29 May 1900 he said farewell to his wife Gezina, who was too ill and frail to leave Pretoria. He would never see her again.

After a temporary stay at Waterval-Boven in the eastern Transvaal (Mpumalanga), the Volksraad promulgated a proclamation stating that the president would be granted a six-month leave of absence seeing that his age prevented him from moving around with the commandos and the Executive Council was convinced he could still be of benefit to his country and nation. He would go to Europe to whip up support on behalf of the Republic.

Vice President S.W. Burger would take over the presidency in the meantime. The proclamation was signed by Vice President Burger and Secretary of State F.W. Reitz.

On 11 September 1900 an anguished President Kruger said farewell to his people and left with a small entourage by train for Lourenço Marques (now Maputo). Here they boarded *De Gelderland*, a warship sent by Queen Wilhelmina of the Netherlands, to sail to Europe.

In the summer of 1904 President Kruger moved from Menton to Clarens (Vaud) near Montreux in Switzerland because of his failing health. There a comfortable home named Villa des Prierriers was made available to him.

Shortly before his final sickbed and his death on 14 July 1904, President Kruger sent a last message from the loneliness of his exile to General Louis Botha and his compatriots on 29 June 1904:

Waarde Generaal,
Het is mij een groot voorrecht de ontvangst te kannen erkennen van uw kabelgram van den 25 Mei en uwen brief van den 29n diezelfde maand, waardoor my de groet wordt overgebracht van het congres van 23 tot 25 Mei, te Pretoria gehouden.

By alle treurigheid en leed, die mijn lot zijn, stemde mij deze groet tot dankbaarheid. En van ganscher harte dank ek allen die, tezamen gekomen om te beraadslagen over het heden en de toekomst, gedacht hebben aan hunnen oude Staats President en daardoor getoond hebben het verledene niet te hebben vergeten.

Want wie sich een toekomst scheppen wil, mag het verledene niet uit het oog verliezen.

Daarom: Zoekt in het verledene al het goede en schoone, dat daarin te ontdekken valt, vormt daarnaar uw ideaal en beproeft voor de toekomst dat ideaal te verwezenlijk.

Het is waar: veel van wat was opgebouwd, is thans vernietigd, vernield, gevallen. Doch met eenheid van zin en eenheid van krachten, kan weer worden opgericht, wat thans daarneder ligt.

Het stemt mij eveneens tot dankbaarheid, te zien dat die eenheid, die eendracht bij u regeeren. Vergeet noot de ernstige waarschuwing, die ligt in het woord: "verdeel en heersch", en maak dat dit woord op het Afrikaansche Volk nooit van toepassing zal kunnen zijn.

Dan sullen onze nationaliteit en onze taal blijven en bloeien. Wat ek self nog daarvan zien of beleven sal, ligt in God's hand. Geboren onder de Engelse Vlag, wensch ek niet daaronder te sterven.

Ik heb geleerd te berusten bij de bittere gedachte de oogen te zullen sluiten in den vreemde als een balling, bijna geheel alleen, ver van bloedverwant en vrienden die ik waarschijnlijk nooit zal wederzien, ver van den Afrikaanschen grond die ik wellicht nooit weer betreden zal, ver van het land waaraan ek mijn leven gewijd heb om het te openen voor de beschaving en waar ek een eigen natie zag ontwikkelen.

Maar die bitterheid zal worden verzacht, zoolang ik de overtuiging mag lijven koesteren dat het eenmaal aangevangen werk word voortgezet. Want, dan houden mij staan de hoop en de verwachting dat het einde van dat werk goed zal wezen. – Zoo zij het.

Uit die diepte van mijn hart, groet ek uwen het gansche volk.
(w.g.) S.J.P. Kruger.

For the purpose of this work a summary of the President's main appeal to his nation might be more apt: "If you want to create a future for yourself you cannot lose sight of the past. Therefore: Search in the past for all that was good and noble to create an ideal. Then strive towards fulfilling that ideal."

If one delves into the past of the part of the Lowveld that now bears the President's name, there is indeed much that is good, just and noble on which the future can be built with confidence.

A retrospective of the past of this part of the Lowveld, now the Kruger National Park, opens up a kaleidoscope of images that stretches back to the time when our forefathers settled in the area. Hence evolves a chronicle of improvisation and ingenuity under difficult and even hostile circumstances; of

President Kruger's last view of Africa from the deck of *De Gelderland*, October 1900.
(ACKNOWLEDGEMENT: J. MEINTJIES COLLECTION)

Vice President Schalk Burger and Acting Treasury General P.R. de Villiers, 1902.
(ACKNOWLEDGEMENT: TRANSVAAL ARCHIVES)

INTRODUCTION

Villa des Prierriers 17, Clarens, Switzerland, where President Paul Kruger resided at the time of his death.
(ACKNOWLEDGEMENT: TRANSVAAL ARCHIVES)

daredevilry, perseverance and victory, but also of failure and anguish. There were periods of restlessness and stability, drought and abundance, devastation and recovery, death and new life – all interlaced with the romance and inextinguishable spirit of the Lowveld.

It is this account of the ability of humankind, animals and plants to adapt to varying climate conditions and changing environmental factors that represents the crucible in which the different ecosystems developed.

The history of the Lowveld also embodies the knowledge future generations living in this region will have to use to direct and regulate their sojourn. Without an intensive study of the history of the Lowveld it is not possible for the nature conservator, ecologist or agriculturalist to interpret the current phenomena and make realistic and educated decisions about the management of conservation areas such as the Kruger National Park.

The authors sincerely hope that this account will serve as inspiration to conserve this unique and precious natural inheritance for generations to come.

To enrich visits to the Kruger National Park, a map indicating the known historical sites appears in the back of the book. The map is titled "Historical terrains in the Kruger National Park". Appendix A contains an eight-digit grid map (on 1:50 000 maps of the area) of all these historical sites according to the Universal Transversal Mercator Net system (UTM-system), so sites are described accurately should superficial indications of their locations disappear later on.

Documentation of the history of the area that was proclaimed as the Kruger National Park in May 1926 starts with the settlement of the first humans in the area, namely the San, or Bushmen. After that follows a description of the colonisation of the area and the wars between the different black tribes that moved into the area; the arrival of the first Europeans; the Voortrekker era; the colonisation and early development of the area by hunters, transport riders, farmers, traders, missionaries and prospectors. Then follows an era during which the realisation dawned that the natural resources of the Lowveld were being overexploited and the first attempts were made to establish a conservation ethic, culminating in the proclamation of the Sabie Game Reserve in 1898. Following that, the first years of the reserve after the Anglo-Boer War are described. This part of the history, in which the first warden of the conservation area, Major (later Colonel) J. Stevenson-Hamilton played such a crucial part, is in itself a saga of courage, perseverance, firmness of principle and unyielding love for a cause – a saga that will serve as a guiding principle and inspiration for future generations of nature conservators.

The authors decided to record the history of the Kruger National Park only up to 1946, the year Stevenson-Hamilton retired after 44 years of dedicated service.

At that stage his "Cinderella" – as Stevenson-Hamilton referred to the conserved area in his diary – had already developed into a princess of worldwide renown. What happened to the princess afterwards, and how she developed into a proud and dignified queen among the national parks of the world, is another chronicle for another time.

The editor and contributors wish to invite you to linger a while on the acre of remembrance and so unbury the past.

Chapter 1
When the Lowveld was wild and desolate
1.1 The Stone Age and the San era
Prof. J.F. Eloff

The arrival of the Bantu-speaking groups that settled in the area now known as the Kruger National Park during the nineteenth century heralded the beginning of the end of a long period of habitation by human groups. Hunters and gatherers had lived here for thousands of years, until shortly after the start of the Christian era when the first stockbreeders and agriculturalists moved in from the north. This is discussed in chapter 2.1.

The earliest inhabitants of the area obtained all their food and other provisions from the veld and this in turn had an influence on the natural environment. This exploitation became more intensive later once iron or copper ore was being mined for the production of implements.

The Department of Archaeology at the University of Pretoria carried out research in the Kruger National Park from 1974 until the 1980s on the first people who had lived in and used the area. This work involved a survey of all the sites that had previously been inhabited, after which more intensive research was carried out on selected sites. However, the survey was not so extensive as to rule out the possibility of finding more dwelling places. Nonetheless, hundreds of sites were found that had been inhabited by human groups from the Early Stone Age until the nineteenth century.

The legacy of the prehistoric inhabitants of South Africa can be found all over the country in the form of stone tools, metal objects and other cultural remains, as well as skeletal remains, which give an indication of the physical types that roamed here. For more than a million years, hunter-gatherers, stockbreeders and agriculturalists used their intellectual and technological abilities to survive, reproduce and endure. Their survival depended on the individual's capability, as well as that of the community as a whole, to exploit the environment and adapt to changing circumstances.

Climate shifts, changes in animal and plant life and the availability of essentials demanded that hominids and later humans continuously adapt their way of life, as well as the artefacts essential to their survival.

The early human activities in this area were an integral part of the activities on the whole subcontinent. One can only get an accurate idea of the Park's prehistory when seen against the background of the acceptable division of the prehistoric time periods in South Africa. The period of habitation of South Africa by indigenous tribes, before the settlement of Europeans, is divided into two cultural phases:
1. The Stone Age, typified by communities of hunters and gatherers who manufactured stone tools
2. The Iron Age, during which stockbreeding and agriculture took place, as well as the production of metal objects

In South Africa the Iron Age started after the Christian era when the first communities with the characteristics above moved into the area. This is discussed in Chapter 2. According to archaeologists, the Stone Age lasted more than 500 times longer than the Iron Age. It is estimated that stone tools were being produced more than 1 000 000 years ago in southern Africa. However, no tools dating back to this period have been found in the Kruger National Park yet.

The Stone Age is divided into three lengthy main phases; the Early, Middle and Late Stone Age. Stone tools dating back to all three phases are found in the Kruger National Park, although no material has up to now been found that can be linked to the earliest part of the Early Stone Age.

The three phases of the Stone Age are discussed in broad terms below. As no intensive research has been done as yet on the Stone Age in the Park, this description of the Stone Age is based mainly on results obtained elsewhere in South Africa.

The Early Stone Age
According to archaeologists, the earliest stone tools were produced in South Africa about 1 000 000 years

1. WHEN THE LOWVELD WAS WILD AND DESOLATE

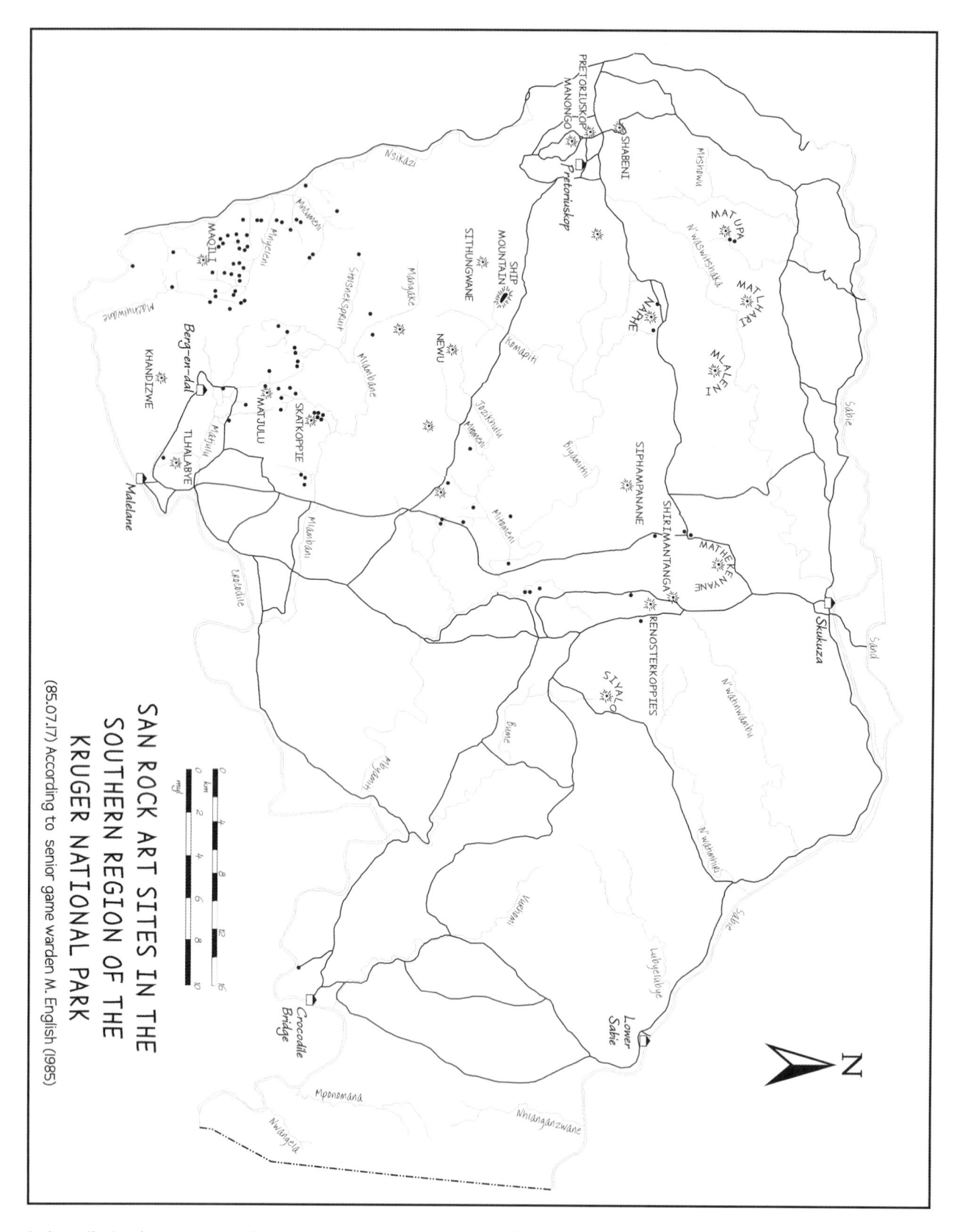

Bushmen (San) rock-painting sites in the southern region of the Kruger National Park (as on 17 July 1985).

ago, while the Early Stone Age ended about 200 000 years ago. In the northern part of the Park, stone tools dating back to this era were found in the Limpopo-Pafuri area, while a concentration of tools also occur along the Makhadzi Spruit on the Lebombo flats south of the Shingwedzi River. In the southern part of the Park, a fair concentration of tools was found along the Timbavati River, which indicates a settlement. A number of tools were also found in the Stolsnek area, but not so many that a settlement could be identified.

Based on information obtained elsewhere in South Africa, the environment during the Early Stone Age would not have differed much from that of today. During times of higher rainfall, which sometimes lasted for thousands of years, riverine vegetation covered larger areas than is now the case.[1] It is interesting that the tools along the Makhadzi occur at a site about 80 to 100 metres from the current course of the spruit, from which it can be deduced that the spruit flowed over a much wider area than it does now. The present tree species probably existed by then, but the animal population differed considerably, as several species have become extinct; for example, more than one elephant species, a giant hartebeest, two horse species, a large type of pig and a giraffe with a short neck, as well as a giant baboon and a giant buffalo. The present species probably already existed, except for the modern-day elephant, bush pig and warthog.

The hominids who exploited the environment during the later part of the Early Stone Age are known as *Homo erectus*. Skeletal remains of *Homo erectus* have also been found at several other sites in Africa, as well as in Europe and the East.

The stone tools they produced were limited in range and included fist-stones, hand-axes and cleavers, as well as chips (flakes) that were sometimes finished off to provide workable edges and were held by hand. All the tools dating back to this period were hand-held, as there are no indications that they were fixed to sticks or any other kind of handle. With these meagre instruments they did all that was necessary: skinning animals; cutting meat; chopping bones; and chopping off branches, to work the wood into clubs or sharpen sticks to serve as spears. These artefacts are found in areas where domestic activities took place.

Part of a fossilised tree trunk on Ntsumanini Ridge, west of Klopperfontein.
(30.07.1984)

However, ancient wooden objects and tools are very rarely found because of their perishable nature.

The fist-stones and axes were produced according to a specific technique developed over thousands of years. The shape and general appearance of these tools are more or less the same everywhere in southern Africa. In fact, further north in Africa, Western Europe and northwest India, the fist-stones and axes that have been found do not differ much from those in southern Africa.

The settlement sites were normally next to rivers, pans or swamps. The hunting parties, which usually consisted of about a dozen people, would have made their homes under trees. There are some indications that they erected grass or branch shelters, but it is not quite certain. It is also not certain that they used fire, because no fireplaces have so far been found in South Africa that can without doubt be linked to *Homo erectus*.

Hunting activities could not have produced much meat, because their weapons – a sharpened stick, a club and possibly a throwing stick and throwing stones – were just not suitable to bring down large or fast-footed smaller animals. Sometimes they did

---

1   Fossil tree trunks have been found in the northern part of the Kruger National Park in the dolerite shales between Klopperfontein and Ntsumanini windmill, which date back to a period when this area must have been much wetter than today. That dinosaurs also roamed in this part of the Park was confirmed when trails ranger A. Louw found the fossil remains of a *Euskelosaurus sp* in the Nyalaland trials area on 27 March 1993, as well as his and ranger B. Pretorius's later find of more dinosaur fossils along Ntsumanini Spruit in March 1996.

manage to kill large animals, as the skeletal remains of larger animal species, even elephant, have been found in deposits at several sites. Skeletal remains found in what used to be a swamp, indicate that they had chased animals into the mud where they could then kill them.

It is reasonably certain that they would have eaten animals that had died of natural causes and also that they would have taken prey off predators. This was common even in recent times. Large animals were slaughtered at the place where they had died, after which parts of the carcass were carried to the camp.

Hunting did not provide their main source of food. For this they were dependent on the gathering of anything edible from the veld. Berries, bulbs and other edible plant material, as well as a number of insect species and insect larvae, were their most important food sources, making them primarily food gatherers rather than hunters.

During the Early Stone Age, technology had developed so slowly that it was almost static, although certain notable changes started taking place. Tools gradually became smaller and better finished, and a larger variety was produced. The most noticeable change by the end of the Early Stone Age was the start of the production of tools that were probably suitable to woody environments. This can be determined from tools that had been made for specific purposes, such as chopping, cutting or scraping. There are even tools that look as though they were used for working wood as one would with a plane. In other words, a certain degree of specialisation started taking place.

The multipurpose fist-stone and hand-axe were gradually produced less and less and eventually disappeared completely. Hunter-gatherer activities probably did not differ much from those of earlier times, except that they took place in a woodier environment. Sites related to this transitional period were found in the Pafuri area of the Kruger National Park, as well at the confluence of the Olifants and Letaba rivers. It is not yet clear to which human type the transitional cultures are connected, but a few findings have been made in southern Africa of a physical type known as *Homo sapiens rhodesiensis*, which dates back to this period. It was possibly they who occupied a few woody areas in the Park. *Homo sapiens rhodesiensis* shows physical similarities to the Neanderthal man of North Africa and Europe.

The Middle Stone Age

A new technological era started in southern Africa about 125 000 years ago. Changes that had started to develop earlier now ran their course. Working tools and equipment that differed markedly from those of the Early Stone Age now became common. The fist-stones and hand-axes were replaced by a group of artefacts of which the most typical was the so-called point. This new era is known as the Middle Stone Age.

Artefacts from the Middle Stone Age, including points and a variety of scrapers, have been found in different places in the Park, but these only consist of a few scattered collections. Collections indicating Middle Stone Age settlements were found at Kostini near the Shingwedzi River, along the lower reaches

Fist-stones and stone-axes dating back to the Early Stone Age which have been found on archaeological sites in the Kruger National Park.
(PHOTOGRAPH: SOUTH AFRICAN NATIONAL PARKS)

of the Tsende (Tsendze) Spruit and slightly west of the Tsende (Tsendze) mouth on the northern bank of the Letaba River, as well as further south in the Stolsnek area. Apart from this, isolated tools were discovered at places where topsoil had been removed by erosion, or by a grader. These meagre findings create the impression that at the time, the Park area was not as popular a habitat as the Mpumalanga and Limpopo plateau. But no definite conclusion about this can be reached up until now with the little information available.

Very little reliable information about the appearance of the Middle Stone Age human is available – less even than those of the Early Stone Age. This is strange, as there are more Middle Stone Age settlements, of which several occur in rock shelters, while the Middle Stone Age was also thousands of years later than the Early Stone Age. From the usable data it can be deduced that the people had a more modern appearance than *Home sapiens rhodesiensis,* but there is as yet no generally accepted species name for this branch of *Homo sapiens.* However, some anthropologists discriminate a physical type, *Homo sapiens afer,* which occurred 100 000 years ago in Africa south of the Sahara and which would have been a primeval forefather of the San (Bushmen). These anthropologists believe that the population of the Middle Stone Age belonged to that group.

The beginning of the Middle Stone Age coincides more or less with the start of the last interglacial and ends during the last glacial period. The last interglacial, a warm period, started about 125 000 years ago and ended about 75 000 years ago, when the last glacial period started. This glacial, a cool period, continued until about 12 000 years ago, when it started getting warmer again at the beginning of the so-called Holocene, in which we are living. The climate during the Middle Stone Age in South Africa was therefore generally warm during the first 50 000 years and generally cold during the next 40 000 years. The Park experienced a warm climate during the interglacial, sometimes warmer than at present, but during the last glacial the average temperature would have been several degrees lower than now and would have resulted in a climate that can be described as temperate. It was considerably cooler on the escarpment. No specific deductions can be made about the rainfall in the Park at the time, but reliable information has been obtained from the escarpment of eastern Mpumalanga and Limpopo, among others, which points towards periods of high and low rainfall during the Middle Stone Age. Such cycles sometimes lasted hundreds and even thousands of years.

The vegetation of the Park would in general not have differed much from that of today. However, in accordance with the changes in rainfall and temperature, some shifting of plant populations would have taken place. During periods of high rainfall, riverine vegetation covered greater areas, and during the last phase of the Middle Stone Age, when it was cooler, a larger part of the Park would have been savannah.

The animal population was largely the same as today. Some archaic species dating back to the Early Stone Age had already disappeared by the start of the Middle Stone Age. The skeletal remains of a giant horse species have been found in the so-called Bushman Rock Shelter in Mpumalanga. In Border Cave, on the border between Swaziland and KwaZulu-Natal, skeletal remains of several extinct species, including a springbok, a horse and a giant buffalo, were found. Apart from these the giant hartebeest, giant warthog and a few other smaller animals represented those Early Stone Age species not

Stone spear points (above) and cutting and scraping tools (below) dating back to the Middle Stone Age, which have been found on archaeological sites in the Kruger National Park.
(PHOTOGRAPH: SOUTH AFRICAN NATIONAL PARKS)

yet extinct. Which of these animals had occurred in the Park during the Middle Stone Age cannot be said at this stage, but it can be accepted that some of the named extinct species would have been present and that, on the whole, the animal species composition would not have differed markedly from that of the present.

The fist-stone and hand-axes of the Early Stone Age were replaced by a larger variety of artefacts, which were almost always made from stone flakes. The flakes were chipped off a suitable stone that had first been shaped according to a standardised method. Although some unfinished flakes were used, others were shaped to fulfil a specific function.

The best-known stone tool dating back to this period is the so-called point. This is a flake with a sharp point and cutting edge used for stabbing and cutting. The point was often glued with resin or some mastic in a cleft stick, or it was tied to a stick with sinew or plant fibres. A good example of this was seen during historical times in Australia, when Aborigines would use some kind of mastic to glue a stone chip in a cleft stick. Fixed to a long stick a point became a spear and fixed to a short handle it served as a cutting tool. Sometimes the chip's sides were shaped to obtain a rough or toothed edge and it was then used to scrape hides or to remove bark from sticks. Some of the scrapers have a clear hollow, deliberately cut out and roughened, with the aim of shaping a stick into a club or stabbing object. A variety of points and scrapers can be found at Middle Stone Age settlements. Split pieces of bone sharpened to a point were also used, but these were not common and are actually more typical of the following Late Stone Age industry. The bow and arrow had probably not been invented yet, while the spear was the typical weapon of the Middle Stone Age.

Apart from stone implements, a number of different objects for general use were probably made from tree bark, skin, sinews, horns and other materials, but these have subsequently disintegrated, so that it is impossible to create a complete picture of this period's tool kit.

As in the Early Stone Age, settlements were close to rivers or pans and not too far away from the type of stone that the people needed for the production of tools. Where settlement sites did not afford ample protection against the elements, they probably erected branch and grass shelters. In Limpopo Province (formerly northern Transvaal) and along the Orange River, half-moon-shaped arrangements of stones dating back to the Middle Stone Age have been found that are probably the remains of shelters. They also used rock overhangs as shelters more often than the people of the Early Stone Age.

Deep middens containing tools and food remains, dating back to the Middle Stone Age, have been found in a number of shelters, indicating long-term habitation of the sites. The midden in Bushman Rock Shelter near the Echo Caves in the vicinity of Ohrigstad in Mpumalanga, is more than five metres deep and was formed over more than 50 000 years of habitation. Stone tools; thousands of pieces of stone chips discarded during the production of tools; skeletal remains of animals; signs of fires made over thousands of years; rotten grass remains, which probably represent sleeping material; and other remains, create a history book about the lives of successive hunter-gatherer communities that lived in the cave. It is a remarkable experience for a researcher to work in such a midden, to remove layer after layer from the top and to consider each as a page from a history book. Each layer reveals its own secrets. In one layer around the ashes of an earlier fireplace, the remains of the tools produced there can be found – probably the handiwork of people who sat around the fire while working. In another layer are the remains of a successful hunt in the form of a concentration of animal bones. In yet another layer are the remains of a fireplace and in the ash are carbonised marula and wild plum pips, as well as the remains of a tortoise. With this careful removal of layer upon layer, the researcher comes in close contact with the people who lived there thousands of years ago. One can visualise the fruit this person collected in the veld and the tortoise he brought home and one cannot help but wonder about the general mood around the fire. Were they singing, talking and laughing? What was the social milieu of the people who sat around that fire and enjoyed their food?

Very little is known about their social and spiritual lives. However, there are findings that suggest that they adorned their bodies. Pieces of haematite and ochre were found at several Middle Stone Age settlements and in a few instances there was evidence that some soft material had been scoured, probably to produce a red powder with which they decorated their bodies. There is a large hole in Swaziland that, according to the type of tools found there, seems to have been dug by Middle Stone Age people to obtain specularite. They presumably used the shiny stone to decorate either their bodies or the clothes they fashioned from skins. If so, it is probably the oldest make-up ever found! There are no indications so far, however, that dyes were used during the Middle Stone Age to create rock paintings. In Europe, the dead of the Middle Stone Age were sometimes sprinkled with red ochre, but no such custom has been observed in South Africa. In fact, none of the meagre remains that have so far been found of the Middle Stone Age people proves without a doubt that they were buried at the sites. Until the opposite has been proved, it is assumed that the dead were not buried at the settlement terrains.

Middle Stone Age people were more successful as hunters than their predecessors, as proven by the skeletal remains of an extinct buffalo, as well as that of rhino, giraffe, leopard, lion and a number of antelope. A spear with a stone point was undoubtedly a more effective weapon to hunt with than the sharpened sticks of the Early Stone Age. Their hunting customs probably did not differ much from those of the San (Bushmen), who lived thousands of years later, except that the Middle Stone Age people had no knowledge of bows and arrows. There are also very few indications of the presence of mice and other micro-mammals in the bone remains in the middens at settlements, and it therefore seems as if these little animals had not been such an important food source for them as for the Bushmen.

The veld was their main food source. Tortoises were very popular, as was a large variety of fruit, berries, roots, bulbs and other plant material.

The stone tools of the Middle Stone Age gradually became smaller and in general more trouble was taken with their finish. In the Stolsnek area in the south of the Kruger National Park, archaeologists have found examples of tools that in terms of typology represent a late phase of the Middle Stone Age. No clear line can be drawn between the end of the Middle Stone Age and the beginning of the Late Stone Age. In Bushman Rock Shelter, the middens of the two periods are separated by a thick layer of stone sherds, which had fallen from the cave roof. This can probably be coupled to an era of low temperatures. Similar phenomena have been observed at a few other places in southern Africa. It would seem that unfavourable climatic conditions forced the inhabitants of these sites to leave, while new inhabitants only arrived thousands of years later. Such a clear separation between the two periods is unusual. The basic criterion to indicate the end of the Middle Stone Age and the beginning of the Late Stone Age is a technological one, in other words a new technique according to which stone tools were made as well as other attributes associated with this technology.

For centuries it has been assumed that modern human behaviour and advanced rational thought first developed in Europe. These suppositions were made after the finding of rock paintings, gravures and primitive sculptures in cave dwellings in France at Lascaux, Chauvet, Niaux, Arcy-sur-Cure, in Spain at Altamira, and in many other places along the Mediterranean coast and in Germany. But recent discoveries in Blombos Cave, near Stilbaai along the South African south coast, have thrown new light on this controversial subject and in the process turned all the old theories topsy-turvy!

A researcher found that a group of people lived unhindered in this extraordinary cave more than 70 000 years ago. They lived during a period with mild climatological conditions (compared to the colder European conditions at the time), died out and disappeared into oblivion.

In 1996 a group of archaeologists made the astounding discovery at Blombos. The researchers were led by Professor Chris Henshilwood, of the Iziko Museum, Cape Town (and later of the Department of Anthropology at the State University of New York and the University of Bergen in Norway). They discovered that this group of cave-dwellers used specialised bone tools, made geometric designs on pieces of dressed red ochre (which had been obtained elsewhere) and made ornaments from sea shells found in the vicinity.

The significance of these discoveries at Blombos Cave was clear: Modern human behaviour, rational thought and sophisticated communication had evolved much earlier than generally surmised and had taken place in Africa – up to 30 000 years earlier than in Europe. Later this spread from Africa to Europe and the rest of the world.

The finds at Blombos also prove that "modern" human behaviour dates back to the Middle Stone Age in Africa, much further ago than has been suspected. As a result, according to information currently available, Africa should be accepted as the cradle of humankind and modern civilisation. It must also be accepted that the "modern Eve", the primeval mother of all people, also came from Africa.

It has recently been confirmed that modern humans are not genetically related to the Neanderthals. A group of Italian researchers of the University of Florence discovered this during an investigation of which the results were published in May 2003. According to Dr Giorgio Bertorelle and his team, Cro-Magnon man, the predecessor of modern man, supplanted the Neanderthals without there being any genetic intermingling. The DNA differences between these two branches of the human family tree are so significant that researchers came to the conclusion that modern man's predecessors had no links with Neanderthal man.

The scale of this work does not allow for more detailed information about these two important findings in connection with the origin of "modern man", but interested readers are referred to the articles and documents by Henshilwood and his contributors and those of Klein and Wong, given in the reference list at the end of this chapter.

The Late Stone Age
Between 30 000 and 40 000 years ago, a vital technological change took place that meant the beginning of the next time period, the Late Stone Age. There are indications that this change was the result of the arrival of immigrants from the north, but there is not enough physical evidence as yet to prove this theory.

The Bushmen, or San as they are referred to today, are the last group of people in southern Africa with a Stone Age culture. As such they represent the end of the Stone Age, both historically and culturally, and of the hunter-gatherer lifestyle, which *Homo erectus* had practised 1 000 000 years ago in southern Africa during the Early Stone Age and which carried on with relatively few changes until the nineteenth century AD.

The people of the Late Stone Age and their culture can be discussed with greater authority than those of the earlier periods, because their weapons and general utensils, as well as their food remains, were better preserved. The rock paintings and engravings, or petroglyphs, that occur over a large part of South Africa are also a rich source of information about the material culture, social customs and even the spiritual life of the Late Stone Age people. In this instance, the information gathered about the San by travellers, ethnologists and other authors is also of exceptional value. Some uncertainties about the Late Stone Age become more understandable if one looks at the material culture, hunting and gathering customs, as well as the general circumstances of these last representatives of the Stone Age.

Although the beginning of the Late Stone Age in South Africa is estimated at being 30 000 to 40 000 years ago, only a few sites containing remains from those early years have been found. Two of these – Heuningneskrans Shelter, next to the road between Ohrigstad and the Strijdom Tunnel, and Border Cave, on the border between southeastern Swaziland and the northwestern part of KwaZulu-Natal – are relatively close to the Kruger National Park. Archaeologists have found that at several sites there was a time lapse between the end of the Middle Stone Age and the Late Stone Age during which the shelters were not occupied. Remains from the Late Stone Age are also more representative of a later phase of the period. It seems, therefore, that certain terrains had been abandoned by the end of the Middle Stone Age, but were shortly afterwards occupied again by people with a Late Stone Age technology, while other sites remained unoccupied for thousands of years. A question arises as to the reason for this. The only logical answer is that the climate, the environment and the animal life in the vicinity of the unoccupied sites were not attractive enough.

These unfavourable conditions must have persisted in South Africa from the end of the Middle Stone Age, between 30 000 and 40 000 years ago, until about 20 000 years ago, and in some cases much longer. The interior of South Africa generally only became densely populated again about 12 000 years ago, but other factors also played a role. Bushman Rock Shelter near the Echo Caves was abandoned after a long period of occupation during the Middle Stone Age and only reoccupied about 12 000 years ago. On the other hand, Heuningneskrans Shelter, just a few kilometres away, was already reoccupied about 30 000 years ago. Further research will hopefully furnish a reason for this.

Deacon (1984a) lists the following typical cultural elements of the Late Stone Age:

- A technique different from that of the Middle Stone Age to fracture stone chips
- The production of small (microlithic) stone tools that were fixed with a mastic to a timber haft
- Specialised hunting equipment, including bows, poisonous arrows, traps and snares
- Fishing equipment, such as specially made hooks
- A variety of equipment to work skins with, for example scrapers, bone needles and bone awls
- Specialised equipment and other necessities for the collection of veld food, for example digging sticks with bored stones, skin or woven carrier bags
- Fibre (thread), ostrich egg shells, which served as water containers, and tortoise shells with different functions
- A series of stone tools that had not been produced during former times and were used for specific tasks, for example, small stone drills and grooved stones for making small beads, reamers to bore stones with and small adzes to work wood with
- Rock paintings and etchings, decorated ostrich shell containers, decorated stone tools and items for personal adornment, for example ostrich shell beads, shells and bone, as well as necklaces and amulets characteristic of formal burial of the dead in graves on which stones were sometimes stacked

These elements are also typical of the Late Stone Age in the Park, although some elements are not common, for example rock engravings[2], while others probably were present before and still exist, but have not been found yet. Archaeological findings all over the country sometimes show significant differences between cultural remains in the different ecological regions and eras, even in a limited area. Characteristic of the Late Stone Age is the greater diversity in human activities and cultural products than during previous periods. The manner of settlement and artefacts of the people who lived along the Letaba River, differ in several aspects from those in the Stolsnek area. Judged against the technology of the hundreds of sites representing the Late Stone Age all over South Africa, widespread changes occurred repeatedly at successive times. The earliest tools are typified by a microlithic technique that was later replaced by non-microlithic artefacts, but eventually there was a return to microlithic technology. Such changes were the results of diversified adaptations to different and even similar environments. Different groups that had made contact also copied each other's ideas and methods. There is also a strong possibility that new settlers introduced new ideas.

Late Stone Age sites are found in many places in the Park, from the Limpopo River in the north to the Crocodile River in the south. Most are open sites near rivers or pans and a few are where there is no open water, for example at Masorini. So many tools

Painting by Samuel Daniell depicting Khoisan warriors armed for an expedition.
(ACKNOWLEDGEMENT: AFRICANA MUSEUM)

---

2 Up to now only found in a cave east of the Nyalaland hiking trail at the upper reaches of the Mashikiri creek. In the Hyena cave ranger S. Mostert discovered engravings (petroglyphs) of animal footprints, amongst others those of the nyala, kudu, wild pig and a strange, snake-like engraving of which the meaning is uncertain.

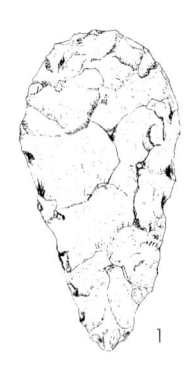

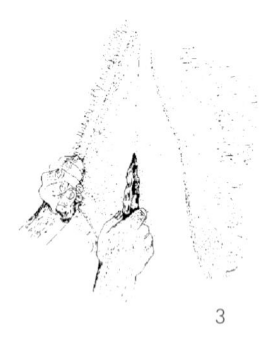

Stone tools of the Early Stone Age. 1: Fist-stone. 2: Axe. 3: A fist-stone being used as a butcher's knife. (Not according to scale.)

have been found at the foot of this hill that it can be assumed that people had lived here for a reasonably long time. Quite a few rock shelters containing tools have also been found, especially in the mountainous southwestern and northern regions. Ranger Mike English has found more than 100 shelters with rock paintings in these two areas. Artefacts dating back to the Later Stone Age can be found at almost all of these shelters, although there are only a few at certain sites.

No human remains that can be linked to the earliest phase of the Late Stone Age with any amount of certainty have up to now been found in the Park. The skeletal remains from the later part of this era show so much similarity with those of the Bushmen that they are in fact classified in the same category. The comprehensive term San, a name bestowed on them by the KhoiKhoi, is used for this group. There is no conclusive answer about the origins of the San. They could have developed here in the south from Middle Stone Age predecessors, or they could have developed further north and then moved southwards. The San are without a doubt the first South Africans. No skeletal remains of the San have so far been found in the Park, but it can be accepted with a reasonable amount of certainty that there are settlements at which graves and skeletons occur.

The early phase of the Late Stone Age, about 30 000 to 12 000 years ago, coincided with the last glacial period, when the world's climate was cooler than it is now. The same is true for southern Africa and it looks as if rainfall during the time was generally lower, and at times significantly lower in certain parts of the country. This is a possible explanation for the low population levels then.

The transition from the last glacial to the interglacial period, or Holocene, in which we are living, lasted about 2 000 years, between 12 000 to 10 000 years ago. This brought on considerable climatic changes in that temperatures started rising, with accompanying changes in rainfall patterns. However, there are climatological indications that the changes in the Transvaal region were not as drastic as further south in the country.

It seems that it became so dry over large parts of the interior a few thousand years after the beginning of the Holocene that the inhabitants moved away.

Very few sites have been found that were inhabited between about 9 000 and 4 500 years ago, except along the coast and on the escarpment. Conditions in the former Transvaal region could not have been as harsh, as there are sites there that date back to this era. It is, however, evident that the two known sites on the escarpment of Mpumalanga, Bushman Rock Shelter and Heuningneskrans, were not inhabited again after they were abandoned about 9 000 years ago. No obvious reason for this has been found so far, but it was probably caused by climatic changes.

In contrast, conditions in the Kruger National Park must have been favourable for human habitation, because material radio-carbon dated to 6 800 years before the present, was found in the bottom layer of a fireplace in a shelter close to Skukuza. These people must therefore have moved in from elsewhere. A layer near the fireplace's surface has been dated to 3 300 years back. This shelter has therefore continuously been occupied for more than 3 000 years, of which more than 2 000 fell within the period that had been so unfavourable in a large part of the country that it became depopulated.

The climate of the area previously known as the Transvaal has probably not changed significantly since the start of the Holocene 12 000 years ago. There were changes in rainfall and temperature and the micro-climate of certain environments could have become unfavourable at times, but the changes were not so significant that they had a hugely negative effect on the hunter-gatherers' chances of survival in a larger area.

Skeletal remains found in archaeological diggings in the Kruger National Park show that there had been few changes in the animal species composition during the Late Stone Age. Shifts in animal populations were sometimes caused by the climatic changes, but only a few of the species that were hunted at the time, have become extinct. A few of the extinct species (types that did not occur during historic times), namely the giant buffalo, giant hartebeest, a giant horse and the southern springbok, disappeared from the Cape provinces about 10 000 years ago. In the Transvaal region between the Vaal and Limpopo rivers no sign of them has been found so far, although some could have occurred here. A tooth of the extinct Bond's springbok was found at Heuningneskrans, but it dated back to an earlier phase of the Late Stone Age, about 20 000 years ago.

The variety of recent species hunted during the Late Stone Age in Mpumalanga (the former eastern Transvaal) is evident from the list Dr Plug compiled of bones identified from the middens in Bushman Rock Shelter. These represent the period during which the shelter was occupied in the Late Stone Age, about 12 000 to 9 000 years ago. This list includes antelope species such as duiker, steenbok, reedbuck, grey rhebok, mountain reedbuck, bushbuck, impala, kudu, sable, roan, blue wildebeest and eland. There are also warthog and zebra remains in most layers, as well as those of smaller animal species and birds. This list obviously does not represent all the species that occurred in the area, but a clear idea of the general composition of the animal population in that environment can nevertheless be formed.

The excavations that have been carried out in the Park so far have not produced enough bones so that a comprehensive list of the species of the Late Stone Age could be drawn up, but it would largely have been the same as that of today. The distribution could have differed somewhat, especially during the earlier phase of the period when the climate was cooler and possibly drier.

Judged by the variety of food remains found at occupied sites, it is apparent that the hunter-gatherers of the Late Stone Age made more efficient use of their environment than their progenitors. This can largely be attributed to improved hunting and gathering techniques, but especially to better weapons and tools.

The bow and arrow was probably introduced during the Late Stone Age, although there is no proof of this. The advantage of this weapon over the spear is obvious, and arrows with poisonous heads were even more efficient. The different types of arrows used by the San – including one with a stone or a bone head; one with a head either directly fixed to the shaft or fixed with a coupling shaft; with or without a feather – gives an indication of which arrows were used early in the Late Stone Age. Remains of these are often found in archaeological layers. No signs have been found in the Park of the types of poison used on the arrowheads. The poisons could have been of plant origin, like the sap of *Acokanthera*, certain *Euphorbia*

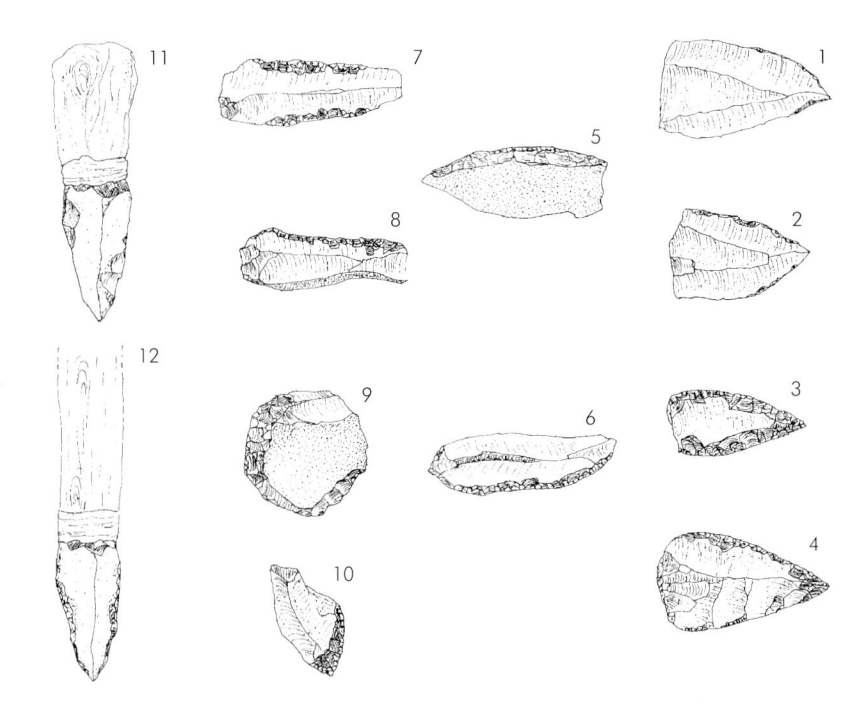

species, *Strophantes komb,* or other plant species that occur in the Park. They could also have used snake venom or poisonous spiders and beetles.

Another technological advance during the Late Stone Age was the production of bone tools. Bone had been used since the earliest times, but during the Late Stone Age more attention was given to the finish. It was also more widely used. Bone tools can be found in most of the Late Stone Age deposits. A good example is those found at Bushman Rock Shelter. Plug (1982) distinguished between three groups of tools dating back to 9 000 to 12 000 years ago:

- Unfinished pieces of bone that show signs of scouring or other signs of usage
- Bone pieces finished like stone tools, including scrapers and small drills, small stabbing tools, burins and cutting tools
- Bone pieces finished by being sanded down to produce neat points, connection shafts, needles (with eyes) and adornments

Points and other bone tools have been found in the Park as well, but not of the same variety as mentioned above.

A characteristic of the stone tools of the Late Stone Age is that they were produced by using a different chipping technique than before. There was also a

Stone tools of the Middle Stone Age. 1–4: Pointed flakes for spearing and cutting purposes. 5–6: Flakes used for cutting purposes. 7–10: Flakes used for scraping skins and for woodworking. 11–12: Pointed flakes inserted, or attached to the ends of short or long hafts. (Not according to scale.)

larger variety with specialised uses. The forms, work edges and user marks point towards specific functions in relation to the working of stones, wood, bone, hides and other materials. Especially during the later part of the Late Stone Age more use was also made of mounted tools, some of which consisted of different small stone chips fixed into a groove. Complete examples of such composite tools have not yet been found in South Africa, but they occur further north in Africa. However, a number of common mounted examples have been found here.

At most sites in the Park, the tool collections dating back to the Late Stone Age generally compare well with those from the rest of the former Transvaal region and other parts of southern Africa that contain microlithic elements and were produced during the Late and Middle Holocene period, from about 8 000 years ago until the historical period. In the northern part of the Park there are sites with tools that could date back to an earlier phase. The Masorini collection has not been studied yet, but it contains typical technological elements of the earliest phase of the Late Stone Age and could belong to the Late Pleistocene. Artefacts found along the lower reaches of the Bangu Spruit may also be older than the other Late and Middle Holocene collections, which are widely distributed throughout the Park.

Space lacks for a complete description of the artefact collection from the Late Stone Age in the Park, but various publications are listed in the bibliography in which detailed descriptions appear of collections from this era that contain similar microlithic and other elements. It is obvious how many tools dating back thousands of years were still used and produced by the San in historical times. Unfortunately there are only a few descriptions by eyewitnesses of how the San made these tools, but the rest of their material culture is well documented. The few changes over such a long period emphasise once again how exceedingly slow the technological development during the Stone Age really was. Development during the Late Stone Age was actually very rapid in comparison with the Early Stone Age, when the standard tools, the fist-stone and hand-axe, had undergone only slight changes over a period spanning hundreds of thousands of years.

The hunter-gatherer pattern also did not change radically over the thousands of years of the Late Stone Age, but changes did take place with respect to the animal species that were hunted.

Excavated animal skeletal remains show that smaller animals were hunted more often during the Holocene than during earlier times. This phenomenon has been noticed at a number of sites in differ-

Stone tools dating back to the Later Stone Age. 1–4: Scrapers. 5–6: Knife blades. 7: Bored stone wedged to the end of a digging stick. 8–9: Grooved stone used to produce beads and ostrich egg beads. 10–11: Lunates. These were inserted in the end, or along both edges of thin timer hafts to form composite arrowheads. 12: Hollow scraper used to smoothen bone or sticks. 13: Stone flake fixed to a wooden shaft to be used as a scraper. (Not according to scale.)

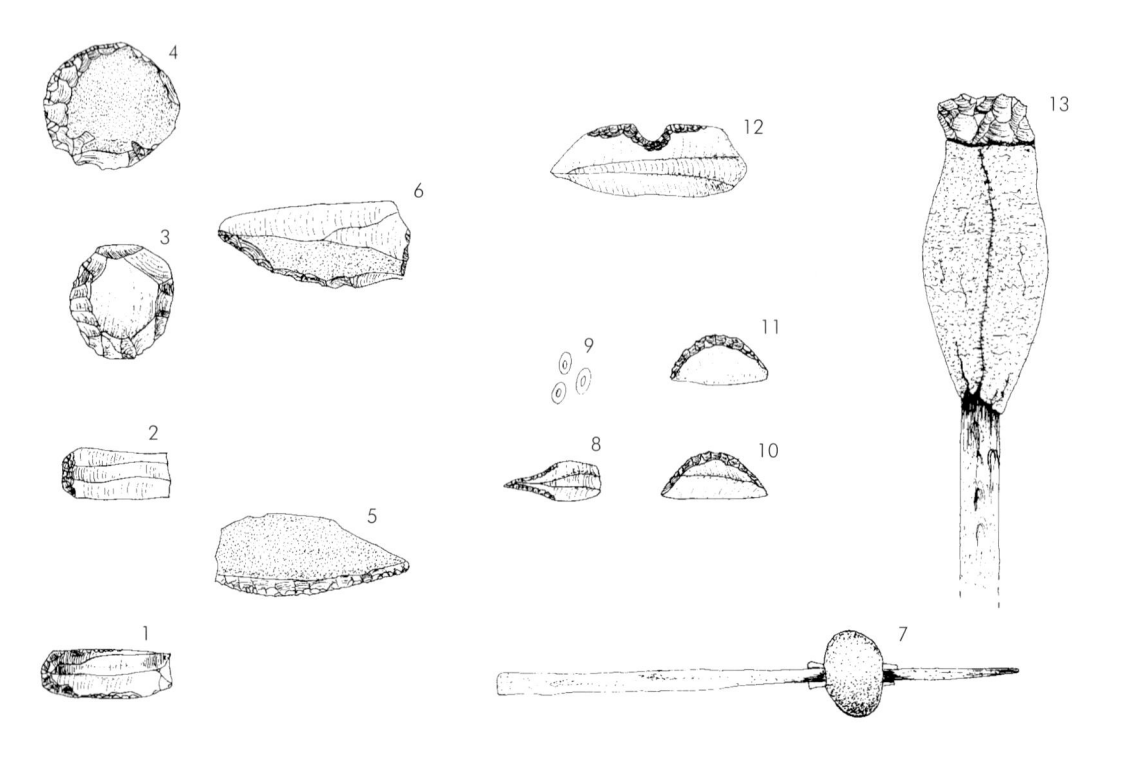

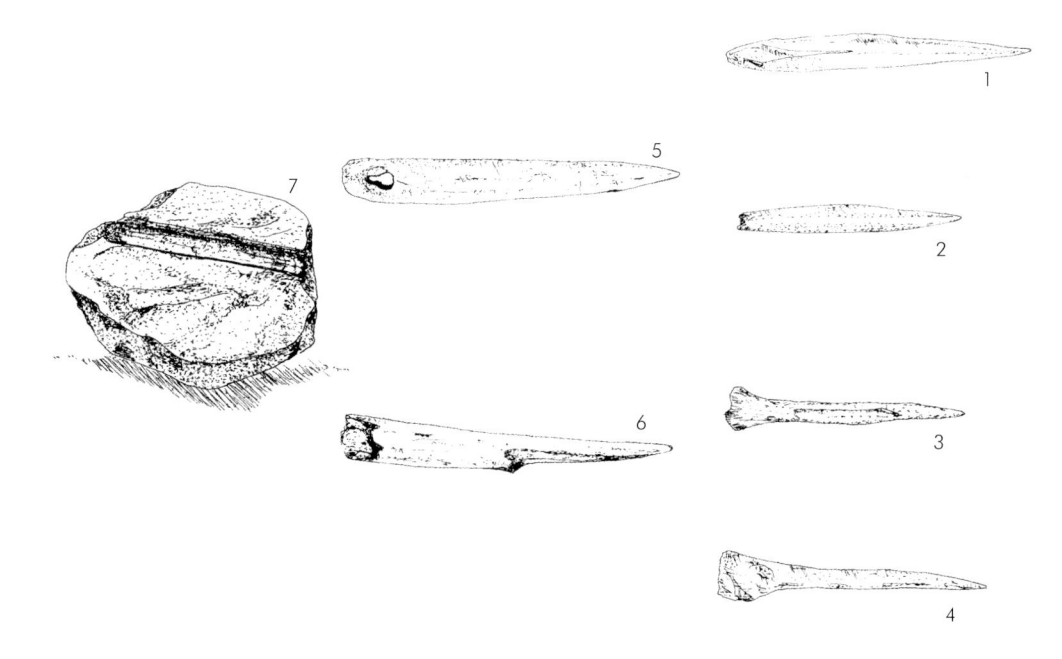

Bone points from the Later Stone Age, used as arrowheads and for the production of skin clothing; 2. smoothened-down point 5. bone needle; 7. grooved stone used to grind strings of beads to a uniform size and shape.

ent parts of South Africa. The analyses by Plug (1981) of skeletal remains from Bushman Rock Shelter show that in the older layers, the remains of mainly grazing herd animals occur, while the single browsers, which are usually active during the night, are well represented in later layers. This change is attributed among others to the fact that more use was made of snares and traps. In the shelter close to Skukuza mentioned earlier, the emphasis fell on the hunting of smaller antelope and other animal species.

A complete picture of the people of the Late Stone Age can only be formed once the settlement sites, which were at the centre of their lives, are studied in detail. Research on the San indicates that they hunted and gathered for only a few hours a day. The lifestyles of their predecessors were not much different and it can be assumed that they also spent the main part of the day in and around their shelters.

Settlements in the Park were normally erected along open water at places where game coming to drink, could be seen. In the southern part, overhanging rocks or hollow cliff shelters were used to observe game on their way to waterholes or rivers. On open sites they erected branch and grass shelters. Inside was a hollow filled with grass, serving as a comfortable place to sleep. Such hollows were also found in rock shelters.

The artefacts lying around at a settlement give a good indication of the activities of the inhabitants. Tools made of skin, wood or plant material perished quickly, although remains of plated grass and fibre strings, as well as leather that had been sewn with a needle, have been found. The most common articles at a site are the following:

- Stone tools, mostly made of chips, sherds and other remains of produced tools. Some of the scrapers, cutting tools, drills and other tools show signs of use.
- Bored stones of different sizes. These were used for a variety of purposes. Rock paintings show that they were used to give weight to a digging stick, among other uses.
- Grooved stones used to sand down ostrich shell beads.
- Flat stones with round hollows (grinding stones or querns) used for grinding and crushing nuts and seed, as well as the powdering of stones to obtain colourants (dyes).
- Bone tools used for a variety of purposes, as mentioned above.
- Ostrich shell and shell beads, in different stages of being manufactured, used for various adornments. These were worn around the neck, legs, arms or other parts of the body, or sewed onto leather bands, carrier bags and clothes. Bone beads and necklaces were used for the same purposes.
- Dyes in the form of pieces of hematite, lumps

of red and yellow ochre[3] or other pieces of rock, used as body paint as well as for rock paintings.

Examples of clothing are rarely found in settlements, but fortunately the rock paintings provide valuable information about this.

The Late Stone Age people buried some of their dead at the sites. However, it is obvious that not everybody was afforded this privilege, as there are fewer graves than one would expect at a site that had been occupied for generations and even thousands of years. Graves provide information not only about people's physical attributes but also about their clothing and adornments, as they were buried wearing these. At Bushman Rock Shelter, for example, the skeletal remains of a baby girl was excavated who had been buried at least 9000 years ago, wrapped in animal skin with a string of ostrich shell beads around her neck.

In a number of graves in the Cape provinces, the bodies of the dead were covered with a thick layer of red ochre. Sometimes, but not always, one or more stones were put on top of the graves. Gravestones with painted figures were found in a few shelters in the south, but these are rare. No graves dating back to the Late Stone Age have been found in the Park so far, but if the distribution patterns of the San are taken into account, chances are good that such graves will be found in future.

Rock paintings and engravings are some of the cultural elements that separate Late Stone Age people from those of earlier periods. Engravings occur in the North West Province (formerly western Transvaal) and further south into the Karoo,[4] but paintings are found over a wider area. More than a hundred shelters with paintings have been found inside the Kruger National Park (compare figure 1 on page 12) and former senior ranger Mike English deserves credit for his enthusiasm in finding and noting down the positions of most of these.[5]

The age of the paintings in the Park can only be speculated about, but it is possible that an estimated age could be determined within the foreseeable future. The specific value of the rock paintings is that they provide first-hand information about the tools, clothing, social life and even rituals of the Late Stone Age people and the animals that had special meaning to them. Therefore it is a special experience for visitors to such sites to see the creations of people who depicted an incident, an observation or an idea with their paint and so left a hint of their culture, centuries ago. If the visitor inspects the area close to the painting, he can also see objects that were used there and can then conjure up a vague image of the group of hunter-gatherers who lived there and depended on the environment to satisfy all their physical, social and spiritual needs.

While the Stone Age people were still practising their rock art and other activities in the area that is now the Kruger National Park, new arrivals, with their cattle and other livestock and their knowledge of food production, metalworking and pottery, arrived. This happened soon after the start of the Christian era and with this another period, the Iron Age, started. The hunter-gatherers still lived in the area before disappearing, probably more than a thousand years ago. Potsherds have been found near the surface in a number of rock shelters and, judging from their decorations and other attributes, they belong to the later part of the first millennium. The potsherds could have been left there by Iron Age people, or the Stone Age inhabitants could have obtained pots from the later inhabitants, which may imply that trading took place between the two groups. In the deposit above the potsherds more stone tools occur. The Stone Age people therefore lived there again, or were still living there after the potsherds had ended up there.

It has not been determined yet when their presence in the Park ended, but it was probably during the second millennium. This signalled the end of the habitation and use of the Park area by hunter-gatherer communities over a period of hundreds of thousands of years.

---

3   In 1986 an archaic mine shaft was discovered at a site just west of the Nwanetsi-Gudzani tourist road and immediately south of Shishangani Spruit, where both red and yellow ochre had been mined by people from the Late Stone Age and probably the Iron Age as well. According to legend, there was another red ochre mine shaft on the northern flank of Ngodzi Hill near the Mooiplaas ranger post. This was mainly used by blacks during historical times.
4   The first clear rock engravings of animal spoor were found in 1986 in Hyena Cave along the upper reaches of Mashakiri Spruit east of Nyasaland trails camp.
5   Colonel James Stevenson-Hamilton compiled a report in 1911 about Bushman rock paintings that he had found in the southwestern corner of the old Sabie Reserve (later the Kruger National Park).

1.2 Bushman (San) rock art in the Kruger National Park

M. ENGLISH

The Bushmen or San, as these diminutive people are more commonly referred to these days, were some of the earliest inhabitants of the area that makes up the Kruger National Park today. They had already occupied the area in significant numbers long before the first black people moved in from the north and east and centuries before the first Europeans set foot in the Lowveld in 1725, and had lived there long before the area was proclaimed as a game reserve in 1898.

The fact that this people once occupied the Lowveld is often regarded with surprise and disbelief by the uninformed, as the original distributional areas of the San are mostly associated with the Drakensberg range and the arid western parts, from Namaqualand in the south and northwards through the Kalahari into Botswana and southern Angola.

The San inhabitants of the Lowveld were, in the early days, nomads who lived off the veld, just as elsewhere. The men hunted with bows and poisoned arrows while the women collected food, including small animals, edible insects, bulbs and rhizomes and other palatable vegetal foods. They did not plant crops or practise stockbreeding and did not stay in one place permanently. They used the numerous shallow caves and rock shelters that are common in both the northern and southern parts of the Park as temporary shelters, as well as the numerous rocky outcrops or inselbergs that occur elsewhere in the area. Archaeologists commonly accept that these early San settlers were responsible for the rock paintings at the shelters.

The area known today as the Kruger National Park, with its numerous rock shelters and rock paintings, as a result of various factors had not been subjected to much human intervention or destruction before 1898, when it was first proclaimed as a game reserve.

In the more than 100 years since, the area has enjoyed absolute protection. It can be accepted that few significant changes have taken place in the composition of the environment since the Late Stone

A typical Bushman (San) shelter in the mountainous southwestern part of the Kruger National Park. The broken clay pot is an indication that the shelter was later inhabited by blacks.
(ACKNOWLEDGEMENT: M. ENGLISH)

Age, except perhaps for the plant physiognomy. The rock shelters and paintings to a large extent still look like the earlier San inhabitants in the area had left them. Climate conditions have definitely undergone cyclic changes through the centuries, but the topography, waterholes, rivers and traditional game paths have probably not changed much either since the period of habitation by the rock-painting artists.

Changes in the density of especially woody plants as a result of changing climatic and fire regimes and later occupation of the area by black tribes would also not have been significant.

It can be accepted that the perennial waterholes in rivers would have dried up during periods of low rainfall, which would have resulted in the migration of animals out of the dry areas. Similar unfavourable environmental conditions would definitely also have led to the movement of the San from such areas, which in turn would have meant that the occupation of certain areas was seasonal, or cyclic in nature.

The same rock shelters that the San occupied had been used for thousands of years by hominids during the different periods of the Stone Age. The San are generally associated with the Late Stone Age. Dating in relation to the Late Stone Age ranges from about 7000 BC to about AD 300 (Cooke, 1969). This does not mean that the rock paintings are that old; there are more accurate methods to determine their age. Many stone tools and other artefacts from the middens in the shelters where the rock paintings occur, have been found that indicate a link with the Late Stone Age.

By the end of the Late Stone Age and the beginning of the Iron Age, the first black settlers also occupied rock shelters in the Park area and gradually started appropriating these shelters. Signs of the utilisation of iron dating as far back as 20 BC have been found in excavations in Zimbabwe. The earliest dating at a terrain south of Limpopo that showed evidence of the use of iron was at a site in Swaziland, which was estimated at AD 400 (Woodhouse, 1971).

A number of old iron furnaces have been found in the Kruger Park area and dating of these sites will start soon.

It is uncertain exactly when the era of cohabitation between the earlier San inhabitants and the later black settlers started in the Lowveld. It is also not known when the black immigrants from the north started their first retaliatory expeditions against the San, who stole their livestock, but these expeditions led to the San leaving the Lowveld for safer areas elsewhere, possibly westwards. It is also possible that they were eradicated by the black tribes.

The only indications of their earlier habitation of the area are their stone tools, other artefacts and, of course, their rock paintings and etchings. Small groups of San that still practise their forefathers' nomadic way of life are found today only in the arid western areas of the Northern Cape, Botswana and Namibia and the southern part of Angola.

Surveys of the Kruger National Park have revealed that the San hardly ever created rock art in the higher-lying rock shelters near the summits of hills or mountains. On the other hand, there are a lot of indications – in the form of potsherds and other artefacts – that these shelters were used by black people. The absence of rock paintings in the higher-lying shelters indicates that the San, and especially the rock artists, preferred the lower-lying shelters, where they probably felt safer.

Stone tools dating back to the Late Stone Age have, however, been found in the vicinity of the higher-lying shelters, indicating that the San living in the area did occupy them at times. They could have been forced to occupy the shelters temporarily when threatened by hostile black settlers and later by the Europeans.

But there are also many indications that the San lived reasonably peacefully alongside the new immigrants in the Lowveld for some time, especially during the earliest period of black occupation.

Some people question if there had been any real contact between the San and the later black immigrants in the Lowveld before the San left the Lowveld for good. Judging by the rock paintings found in the Kruger, it seems there must indeed have been contact. Some of the human figures in certain of the paintings are definitely not San, but tall, thin figures with facial and other characteristics similar to East African black tribes such as the Masai and Watusi. These figures were sometimes painted in detail, with their facial features and curious hairstyles emphasised. This is in sharp contrast with the faceless depictions of other human figures in rock paintings.

A further indication that there was not much conflict between the San and the black immigrants is the lack of battle scenes that occur elsewhere in South Africa.

Up to now, no rock paintings depicting domesticated animals such as cattle, sheep, horses or dogs have been found in the Kruger National Park.

Wilcox proposes three reasons why domesticated animals (more specifically cattle) are not depicted in the rock art of certain regions:
- The people who created the rock paintings had already left the area by the time cattle herders arrived.
- The rock artists were in the area after domesticated animals had been brought in, but did not have the opportunity, or did not use the opportunity, to paint them (possibly because of persecution by the black settlers).
- The artists continued to produce rock paintings and engravings, but for some reason not of domestic animals.

The interest in the San rock paintings in the Kruger National Park dates back to the early years of Colonel Stevenson-Hamilton's service period in the old Sabie Game Reserve.

By 1911 he had already written a report in one of his photo albums about rock paintings he and other rangers had found in the mountainous southern part of the reserve. In January 1912, ranger C.R. de Laporte, of the then Kaapmuiden ranger post, also drew up a report in which he mentioned new rock art sites he had discovered while on patrol, which increased the number of known sites to nine. Unfortunately, these sites were not accurately described or indicated, with the result that their locations are still unknown today.

Interest in the subject was again stimulated years later when Dr U. de V. Pienaar, then chief director of the South African National Parks Board, asked Professor Murray Schoonraad (the then Head of the History of Art Department at the University of Pretoria), to carry out a comprehensive survey of the San rock art sites in the Park and make suggestions on how to best preserve them. At that stage there were only eight known sites, of which six were in the area south of the Sabie River.

Four of the six known terrains were in the Stolsnek ranger section where park ranger Mike English became involved in the survey and subsequently developed a keen interest in San rock art. Since the 1978 survey, English and other members of the field staff have found, documented and mapped another 101 rock art sites in the Kruger National Park. Of these, 95 sites are south of the Sabie River and only six north of it. These new findings increased the total number of known rock art sites from eight to 109 over just eight years.

It has not been possible to conduct a systematic survey of the whole Park, but chances are good that more sites with rock art and even engravings may be found, especially in the mountainous area north of the Punda Maria-Pafuri road. In fact, the first engravings of game spoor were found on a cave floor in this area.

Many of the rock art terrains are in the wilderness areas that are being used for hiking trails, which makes a visit to these areas all the more unique. Many of the sites were in fact found by trail staff along the trail routes.

Most of the rock art sites in the Park show the following attributes:

The large Hyena Cave along the upper reaches of the Makahane Spruit, where there are petroglyphs on the rock floor.
(12.11.1986)

- They are situated in low-lying areas or at the foot of rocky outcrops or hills, often no more than 15 to 20 metres higher than the surrounding area.
- They are close to natural water sources like waterholes in seasonal rivers or spruits, fountains or pans, or in areas where there had been permanent or semipermanent fountains. Good examples are the shelters at Renosterkoppies and the one along the upper reaches of the Mashikiri. At both, the water supply is impermanent or semipermanent and both were probably used by San.
- They have good views of the current game paths or are close to these. It is well known that most of the game paths lead to water or connect water sources. They also follow landscape contour or the more accessible gradients and, because there has been very little change in the topography of the landscape, the current game paths probably follow more or less the same routes they did during the era of San occupation.
- They are camouflaged by vegetation, a feature that probably has not changed much over the years.
- They provide shelter against stormy weather and other natural elements.

One can make the following deductions about the locations of most of the San shelters in the Park:
- Their low position against the hill slopes or close to the foot of rocky outcrops would have made it easy to carry water, wood or food there. It would be effortless to lie in ambush for animals along the game paths. It would even be possible to shoot at the animals with bows and arrows.
- Shelters close to watering places or game paths to water would have ensured a daily concourse of especially impala and waterbuck, with the result that hunters would not have to look for prey very far away, or carry meat over long distances to the shelters.
- Well-camouflaged shelters would provide a good view of the surrounding area, and game movements, or possible enemies, could be watched from these concealed positions.
- The fact that the terrain would provide good shelter against the elements would have contributed to comfort levels and permanent habitation.
- A shelter at a lower level would in the case of an emergency or threat be easier to evacuate than one higher up against a slope or koppie.

Most of the San paintings in the Kruger National Park area depict human figures in different attitudes, wild animals, birds and reptiles.

In certain shelters there are extended friezes with many human and animal figures. In most cases the entire surface of the rock face has not been used to paint on and the artworks are limited to isolated groups or scenes.

Human figures are sometimes depicted in groups that illustrate a social or ceremonial activity. Sometimes human figures are depicted on their own, often separated from animals or other humans in the same panel. The physical attributes of the human figures differ – most are probably San, of typical short stature, with steatopygia (extended bottoms), lordosis (bent postures) and pot bellies.

Other depictions of human figures show similarities to the black tribes of East Africa. These figures are tall and slim with small waists, prominent noses and typical "pageboy" haircuts. Figures showing attributes more typical of southern African tribes also occur, as well as odd stick-like figures.

Rock painting in a Bushman shelter in the mountainous southwestern part of the Kruger National Park. The hairstyle, slim body and prominent shoulders are typical of East African tribes including the Masai and Watusi.
(ACKNOWLEDGEMENT: M. ENGLISH)

Wild animals are usually well proportioned and illustrated in detail – a total of 23 species can be identified. In a hollow cave in a sandstone ridge in the far north of the Park a depiction of an aardvark was found, and on granite rock wall in the mountainous southwestern part there is a painting of an animal that looks like an oribi. In the southern part of the Park the animals most often depicted include reedbuck, giraffe, roan and sable antelope. Eland, impala, kudu, white rhino, lion, leopard, wild dog, klipspringer, hyena, zebra, waterbuck, ostriches and monkeys also often occur in paintings. One painting depicts what looks like a crocodile skin pegged out to dry. A few images of snakes and a tortoise have also been found.

Depictions of elephants have been found at only two sites in the Park – Shantangalani in the north and at a rocky outcrop along the Nwatindlopfu Spruit east of the main Skukuza-Tshokwane road. Depictions of large, organised hunts, which can be seen elsewhere in southern Africa, have not been found in the Park as yet. In truth, there are only a few paintings in which a single figure is hunting game or shooting at game with a bow and arrow.

As far as pigments are concerned, red was used in most of the paintings in the Park, while most paintings were painted using only one colour (monochrome).

A few paintings are dichromatic (bi-coloured) and others polychromatic (where more than two colours were used). Apart from red, colours used include yellow, black, white and grey.

Analyses of the pigments used elsewhere in southern Africa, show the presence of iron-oxides in the red and yellow pigments, namely haematite or ferric oxide (Fe_2O_3) for red, and limonite, or ferric hydroxide ($Fe_4O_3(OH)_6$) for yellow. When heated, haematite takes on different shadings, ranging from red to purple. Charcoal and manganese oxide (pyrolusite, MnO_2) were used as black pigments in rock paintings while shades of white were obtained from bird droppings, kaolin and zinc oxide (ZnO) (Woodhouse, 1971; Willcox, 1978). But in the Park, the only basic element – other than bird droppings and charcoal – used as a pigment in the San paintings found so far, is iron oxide. An old red ochre mine was discovered on the northern slope of Ngodzi Hill, northwest of Letaba, while an old mine shaft was found along the Shishangani Spruit near the Nwanetsi ranger post where notable amounts of red and yellow ochre had been mined over a long period during the Late Stone Age and Early Iron Age. The closest ancient ochre mining site in the southwestern part of the park is about four kilometres southeast of Malelane on a private farm (J.F. Eloff, personal communication). However, it is not possible to determine whether the San artists obtained pigment for their artwork from iron deposits further away from the Park, for example in the vicinity of Consort Gold Mine. It is known, however, that in historical times Australian aborigines walked up to 480 kilometres to obtain red ochre pigment. Such expeditions consisted of 70 to 80 men who had been commissioned to barter valuable pigment for other wares in isolated places (Woodhouse, 1971).

As far as is known, similar trade did not take place

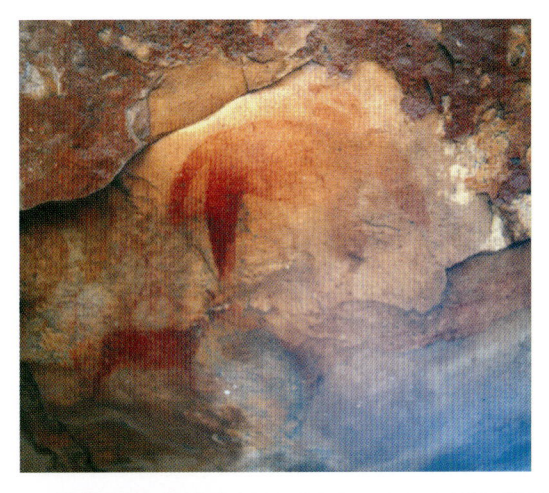

San rock painting depicting an aardvark against the roof of the sandstone arch along the upper reaches of the Mashikiri Spruit. (12.11.1986)

Rock paintings of five elephants in a Bushman (San) shelter which was discovered in 1986 on a granite hill along the Nwatindlopfu Spruit east of the tar road between Skukuza and Tshokwane.
(ACKNOWLEDGEMENT: G. ERASMUS)

in South Africa, but when a well-known mining company decided to exploit a large iron deposit at Bomvu Ridge in Swaziland, they found that an estimated 101 600 000 kilograms of ore had already been removed by ancient miners (Woodhouse, 1971). Archaeological circles regarded this with great interest. A large number of stone tools later identified as ancient mining tools were also found at this terrain, as was a horizontal shaft, 7,7 metres wide and extending into the slope of the ridge for 13,5 metres. The shaft, which had been mined in the haematite deposit, also produced specularite (a kind of schist or slate), which the San rubbed into their hair for cosmetic purposes.

With the aid of carbon dating, the age of this mineshaft has been determined at 28 000 years. According to Woodhouse (1971), indications are that the site was also mined during the Late Stone Age and, according to Inskeep (1978), even as late as the fifth century AD during the Iron Age. This indicates that humans used ochre as a pigment for a very long time.

To date there is no reliable method to determine the age of the rock paintings in the Kruger National Park or anywhere else. Based on paintings elsewhere in South Africa that depict sailing ships, ox-wagons, Voortrekker women wearing bonnets and soldiers in uniform, the age of these paintings can be determined reasonably accurately. Another method is to obtain material that can be carbon-dated from deposits on the floors of shelters or graves that are in some way linked to the paintings on rock faces. Examples of this dating method are two painted gravestones found by John Wymer during an archaeological dig in 1969 in a cave near the mouth of the Klasies River. These stones were found in a stratum containing shells that could be dated to 335 BC (making the gravestones 2322 years old) (Woodhouse, 1971; Philipson, 1977).

Archaeological investigations at San shelters near Renosterkoppies south of Skukuza, indicate that this specific terrain was inhabited on and off over a period of 3000 years by the San and their predecessors.

The binding agent the San used to mix pigments had experts puzzled for years. Willcox (1963) claimed that water had been the most common binding agent, but chemical analyses by Denninger (1971) indicated that blood or blood serum had been used in many instances. A San-Sotho artist who used the blood of a slaughtered ox to mix his red pigments, confirmed these findings. He mixed white pigment with the milk sap of *Asclepia gibba* and black pigment (charcoal) with water (How, 1962; Woodhouse, 1979).

San artists also used animal fat, egg yolk, albumen and the milk sap of different *Euphorbiaceae* as binding agents.

The "brushes" the San used to apply paint to rock walls consisted of feathers inserted into small, hollow pieces of reed. Otherwise they chewed the terminal ends of twigs of specific tree species until a fibrous brush point was obtained. They possibly also used animal tail and mane hair. At some sites it is clear that they painted with their fingers, but this resulted in a considerably cruder art form than the traditional paintings. Examples of finger art have been found at a number of sites in the north of the Park.

The San and their rock art is only one of the many facets of the Kruger National Park. The conservation of historical and prehistoric objects in a national park or other conservation area is just as important as the conservation of the other biotic and abiotic components of the specific ecosystem. Throughout the centuries, San inhabitants documented events and the activities of humans and animals in the area on rock faces. Through the correct interpretation of these rock paintings we obtain insight into the cultural and ritual customs of this people, as well as about the animal populations of the relevant era. Knowledge of historical animal distribution patterns in a specific area

Rock painting in the Kruger National Park depicting a sable or roan cow, a giraffe and a sable or roan bull. This painting also includes some human figures and a large snake.
(ACKNOWLEDGEMENT: M. ENGLISH)

can contribute towards making correct management decisions, for instance when the reintroduction of an animal species in a specific conservation area is considered.

Totally unaware of the future implications of their art, the San artists have contributed not only to our knowledge of the settlement patterns of historical and prehistoric animals in the Lowveld and elsewhere in South Africa, but also to managerial decisions being made about current and future conservation areas.

This is one of the main reasons why it is imperative that all San shelters and art in conservation areas are well documented and mapped, while expert examination of these sites could still provide many clues to unlocking the secrets of this almost extinct tribe at the southernmost point of Africa.

Bushman (San) rock paintings between Nelspruit and White River of what could possibly be a Lichtenstein's hartebeest with a calf and some impala. (From the Walter Battiss Collection.)
(ACKNOWLEDGEMENT: B. WOODHOUSE AND P. MILSTEIN)

Bushman (San) shelter at Mbhandzwe Spruit, Kingfisherspruit section. (17.07.1984)

1.3 Origin of the names San and Bushman

M. English

There are currently several explanations for the meaning and origin of the name "San", which was used by the Hottentots (KhoiKhoi) to describe their contemporaries, the Bushmen.

Lee (1979) quotes Hahn (1881), who claimed that the name "San" or "Sonqua" was adopted by the Dutch settlers at the Cape from the KhoiKhoi name for the San, meaning "native" or "original inhabitant" ("settlers proper").

Later the colonists at the Cape started using the descriptive "Bosjesman" to refer to these people and so the Sonqua or San eventually became known to the world as the Bushmen of South Africa (Lee, 1979). However, Lee (1979) was of the opinion that the name Bushman had a derogatory and racist connotation, therefore the alternative name San. His opinion is not accepted generally, as San actually means "vagabond" in KhoiKhoi.

Smith (1975) noted that the original name "Bosjesmans" had been used by O. Bergh during November 1682 in an official report in which he mentioned "eenige Hottentots, wesende Soomqua alias Bosjesmans" (some Hottentots, which were Soomqua alias Bushmen). Bergh and Hartogh again used the name "Bossiesman" in 1705.

According to Smith, Barrow explained the original name "Bossiesmans" in 1790 as follows:

> ... the central theme of the origin of the name "Bossiesmans" is that the aboriginal "Bossiesmans" skulked in the dwarf bushes (Dutch: *bosjes*) of their arid habitat in the interior, and from this protective cover shot their poisoned arrows at their European foes.

Smith was further of the opinion that the original Dutch names "Bosjesman" and "Boeschenman" had been in use for about a century before any hostile confrontations took place between the San and the European colonists during the last part of the eighteenth century.

Smith also claimed that the correct explanation for the origin of the name "Bosjesman" could be found in the original prefix "bosjes".

> ... long before the early Portuguese rounded the Cape late in the 15th century, one branch of the southwardly migrating pastoral Hottentots deviated westwards along the northern bank of the Gariep (or Grootrivier), ultimately making first contact with the nomadic pigmy "Bossiesmans" somewhere in the northwestern Cape, but below that river. They found that this strange people had curious customs, among which was that they used the aromatic leaves of *Pteronia onobromoides* [bucho], largely to be found in Klein-Namaqualand, in traditional tribal ceremonies. Because of this practice the Hottentots who, unlike the Bantu, appear to have had no special name for these pigmy people, spoke of them as "Sanqua"

Marks against a sandstone wall in a Bushman (San) shelter, caused by sharpening arrowheads, Mbandwe Spruit.
(17.07.1984)

(singular Sab and plural San = bushes; plus qua, the masculine suffix = men of), later far more generally known as *Sonqua,* which was literally and correctly translated by the early expeditioners as "*boschjesmans*" (*bossiesmans*) or simply "*bosjesmans*" i.e. men of the bushes.

The original Dutch names in time became *Boesmans* in Afrikaans and Bushmen in English.

Smith (1975) finally claims that:

... when the Hottentots came into conflict with the Bushmen, they called them Abiqua or Obiqua (= robbermen). The term still survives in the plant vernacular name Abiquaboom or Abiquageelhout (two species of *Tamarix*), as well as in the place names Obiquas Mountains (Tulbach district) and Abiquaspan (Great Namaqualand).

References cited and recommended reading

Bergh, J.S. & Bergh, A.P. 1984. *Stamme en ryke.* Don Nelson, Cape Town.

Bredenkamp, H.C. 1986. The origin of the South African Khoisan communities. In: Cameron, T. & Spies, S.B. (eds). *An Illustrated History of South Africa.* Jonathan Ball, Johannesburg.

Butzer, K.W. 1984. Archaeogeology and Quaternary environment in the interior of Southern Africa. In: Klein, R.G. (ed.). *Southern African Prehistory and Palaeoenvironments.* Balkema, Rotterdam.

Carl, J.D. 1959. *The Prehistory of Southern Africa.* Penguin Books, Harmondsworth.

Cooke, C.K. 1969. *Rock Art of Southern Africa.* Books of Africa, Cape Town.

Deacon, H.J. 1976. *Where Hunters Gathered.* South African Archaeological Society, Claremont.

Deacon, J. 1984a. Later Stone Age people and their descendants in southern Africa. In: Klein, R.G. (ed.). *Southern African Prehistory and Palaeoenvironments.* Balkema, Rotterdam.

Deacon, J. 1984b. *The Later Stone Age of Southernmost Africa.* BAR International Series 213, Oxford.

D'Errico, F., Henshilwood, C.S. & Nillsen, P. 2001. An engrvaved bone fragment from C. 70 000-year old Middle Stone Age levels at Blombos Cave, South Africa: implications for the origin of symbolism and language. *Antiquity* 75:309–318.

Eloff, J.F. 1979. Voorhistoriese mense in die Krugerwildtuin II. *Custos* 8(12):19–26.

Eloff, J.F. 1979. Voorhistoriese mense in die Krugerwildtuin II. *Custos* 9(1):33–38.

Eloff, J.F. 1980. Voorhistoriese mense in die Krugerwildtuin III. *Custos* 9(2):21–25.

English, M. 1980. Rotsskilderinge in die Krugerwildtuin I. *Custos* 9(4):6–13.

English, M. 1980. Rotsskilderinge in die Krugerwildtuin II. *Custos* 9(5):43–45.

English, M. 1980. Rotskuns in die Nasionale Krugerwildtuin. *Custos* 9(9):34–35.

English, M. 1982. The Bushmen and rock art of the Kruger National Park. Unpublished manuscript, Skukuza Archives.

Erasmus, G. 1987. Olifante in oker laat verlede herleef. *Custos* 15(10):16–17.

Grine, E.E. & Henshilwood, C.S. 2002. Additional human remains from Blombos cave, South Africa (1999–2000 excavations). *Journal of Human Evolution* 42:293–302.

Grobbelaar, B.J. 1967. 'n Ondersoek na die verandering van die lewe van die !Kung op tegnologiese en ekonomiese gebiede. Unpublished M.A. thesis.

Hahn, T. 1881. *Tsuni-goam; The Supreme Being of the Khoi-Khoi.* Trübner, London.

Henshilwood, C.S. 1996. A revised chronology for pastrorilism in southernmost Africa: new evidence of sheep at C. 2000 b.p. From Blombos cave. *Antiquity* 70:945–949.

Henshilwood, C.S. 1997. Identifying the collector: Evidence for human processing of the Cape Dune Molerat, *Bathyergus suillus,* from Blombos cave, Southern Cape, South Africa. *Journal of Archaeological Science* 24:659–662.

Henshilwood, C.S. 2001. Blombos cave, Southern Cape, South Africa: Preliminary report on the 1992–1999 excavations of the Middle Stone Age levels. *Journal of Archaeological Science* 28:421–448.

Henshilwood, C.S. & Sealy, J. 1997. Bone artefacts from the Middle Stone Age at Blombos cave, Southern Cape, South Africa. *Current Anthropology* 38:890–895.

Henshilwood, C.S. *et al.* 2001. An early bone tool industry from the Middle Stone Age at Blombos cave, South Africa: implications for the origins of modern human behaviour, symbolism and language. *Journal of Human Evolution* 41:631–678.

Henshilwood, C.S. *et al.* 2002. Emergence of modern human behaviour: Middle Stone Age engravings from Southern Africa. *Science* 295:1278–1280.

Henshilwood, C.S. 2003. The origin of modern human

behaviour: Critique of the models and their test implications. *Current Anthropology* 44(5):627–651.

How, M.W. 1962. *The Mountain Bushmen of Basutoland*. Van Schaik, Pretoria.

Inskeep, R.R. 1978. *The Peopling of Southern Africa*. David Philip, Cape Town.

Klein, R.G. 1977a. The mammalian fauna from the Middle and Later Stone Age (Later Pleistocene) levels of Border Cave. *South African Archaeology Bulletin* 125.

Klein, R.G. 1977b. The ecology of early man in southern Africa. *Science* 197(4299):115–126.

Klein, R.G. 1984a. Later Stone Age faunal samples from Heuningneskrans shelter and Leopard's Hill Cave. *South African Archaeology Bulletin* 140.

Klein, R.G. 1984b. The large mammals of southern Africa: Late Pliocene to Recent. In: Klein, R.G. (ed.). *Southern African Prehistory and Palaeoenvironments*. Balkema, Rotterdam.

Lee, R.B. 1979. *The !Kung San: Men, Women and Work in a Foraging Society*. Cambridge University Press, Cambridge.

Lee, R.B. & De Vore, I. 1976. *Kalahari Hunter-Gatherers*. Harvard University Press, Harvard.

Lewis-Williams, J.D. 1981. *Believing and Seeing*. Academic Press, New York.

Lewis-Williams, J.D. 1983. *The Rock Art of Southern Africa*. Cambridge University Press, Cambridge.

Lowe, C. Van Riet. 1952. Die verspreiding van voorhistoriese rotsgravures en -skilderye in Suid-Afrika. *Archaeological Series No. 7*, Department of Education, Arts & Science.

Malan, B.D. 1965. The classification and distribution of rock art in South Africa. *South African Journal of Science* 61(12):427–430.

Mason, R.J. 1962. *Prehistory of the Transvaal*. University of the Witwatersrand Press, Johannesburg.

Philipson, D.W. 1977. *The Later Prehistory of Eastern and Southern Africa*. Heinemann, London.

Plug, I. 1978. *Die latere Steentydperk van die Boesmanrotsskuiling in Oos-Transvaal*. Unpublished M.A. thesis.

Plug, I. 1981. Some research results on the late Pleistocene and early Holocene deposits of Bushman Rock Shelter, Eastern Transvaal. *South African Archaeology Bulletin* 36.

Plug, I. 1982. Bone tools and shell, bone and ostrich eggshell beads from Bushman Rock Shelter, Eastern Transvaal. *South African Archaeology Bulletin* 37.

Sampson, C.G. 1974. *The Stone Archaeology of Southern Africa*. Academic Press, New York.

Schapera, I. 1930. *The Khoisan Peoples of South Africa*. Routledge and Kegan Paul, London.

Schoonraad, M. 1971. Rock paintings of southern Africa. *Supplement to South African Journal of Science* Special Issue 2, May:62–69.

Schoonraad, M. 1972. Die bewaring en preservering van ons prehistoriese rotskuns. *Historia* 17(4):269–275.

Schoonraad, M. 1980. Sommige petrogliewe van die Magaliesmoot. *Humanites R.S.A.* 6(3):239–269.

Schoonraad, M. & Schoonraad, E. 1965. Rotskuns van Oos-Transvaal. *Outlook* 16(4):10–13.

Schoonraad, M. & Schoonraad, E. 1975. Rotsskilderinge in Oos-Transvaal. In: Barnard, C. (ed.). *Die Transvaalse Laeveld – kamee van 'n kontrei*. Tafelberg, Cape Town.

Silberbauer, G. 1965. *The Bushman Survey Report*. Bechuanaland Government, Gaborone.

Stevenson-Hamilton, J. 1911. Unpublished notes about Bushman rock paintings in the southwestern part of the Sabi Reserve (later Kruger National Park). Skukuza Archives.

Steyn, H.P. 1984. Southern Kalahari San subsistence ecology: a reconstruction. *South African Archaeology Bulletin* 39.

Tobias, P.V. 1986. The last million years in southern Africa. In: Cameron, T. & Spies, S.B. (eds). *An Illustrated History of South Africa*. Jonathan Ball, Johannesburg.

Tyson, P.D. 1986. *Climatic Change and Variability in Southern Africa*. Oxford University Press, Cape Town.

Van Zinderen Bakker, E.M. & Butzer, K.W. 1973. Quaternary environmental changes in southern Africa. *Soil Science* 116.

Volman, T.P. 1984. Early prehistory of southern Africa. In: Klein, R.G. (ed.). *Southern African Prehistory and Palaeoenvironments*. Balkema, Rotterdam.

Willcox, A.R. 1956. *Rock paintings of the Drakensberg, Natal and Griqualand East*. Parrish, London.

Willcox, A.R. 1963. *The Rock Art of South Africa*. Nelson, London.

Willcox, A.R. 1976. *Southern Land: the Prehistory and History of Southern Africa*. Purnell, Cape Town.

Willcox, A.R. 1978. *So-called "Infabulation" in African rock art*. Institute for the Study of Man in Africa, Johannesburg.

Willcox, A.R. & Pager, H.L. 1968. More petroglyphs from the Limpopo Valley, Transvaal. *South African Archaeology Bulletin* 23(90):50–51.

Wong, K. 2005. The morning of the modern mind. *Scientific American* June:66–73.

Woodhouse, H.C. 1971. *Archaeology in Southern Africa*. Purnell, Cape Town.

Woodhouse, H.C. 1979. *The Bushman Art of Southern Africa*. Purnell, Cape Town.

Chapter 2
A new order
2.1 Black invasion and colonisation of the Lowveld
Prof. J.F. Eloff, S. Miller & Dr J.B. de Vaal

About 2000 years ago the San (Bushmen) were still living undisturbed in the area that is now the Kruger National Park. They hunted and gathered food from the veld and obtained everything they needed from their environment. Their rock paintings and artefacts are found together in many shelters in the southern part of the Kruger National Park and here and there further north as well. Collections of artefacts in open terrain alongside rivers, spruits and pans testify to their presence throughout the Park in the past. But they would not stay the undisputed rulers of the Lowveld for long, for small foreign groups were already moving southwards from east Africa down the east coast.

By about AD 400 the first people that were not of San origin settled along the Letaba River. They represented the beginning of an influx of agriculturists and stockbreeders into South Africa and embodied the beginning of the end of the Stone Age. A new cultural phase, known in archaeology as the Iron Age, with agriculture, stockbreeding and metalworking as the most significant cultural elements, would gradually replace the Stone Age and its accompanying hunter-gatherer existence.

The arrival of the new migrants at first did not have a drastic effect on the way of life of the San. There are indications at several places in the Park that the newly arrived immigrants and the San had contact. In certain rock shelters potsherds and even glass beads were found mixed with San tools. The potsherds are similar to those of the Iron Age people who lived in the area after AD 500. The San may have traded meat, skins and ostrich shell beads for pots and other objects. Rock paintings have also been found depicting humans who do not look like the San and use longer bows than they did. There could not have been much reason for conflict in the first 1000 years after the two ethnic groups came into contact with each other, as the Park area provided adequate space to allow peaceful coexistence. The new immigrants only settled in certain localities along the rivers and did not possess enough livestock to provide the San hunters with better prey than the abundant wildlife.

The earliest indication of the presence of the new arrivals in the Park area has been found on the northern bank of the Letaba River. Shortly after them, other groups followed, settling in the vicinity of the Sabie and Crocodile rivers. Their arrival has yielded too little cultural material up to now to allow for a complete reconstruction of their way of life, but it is possible to form a general picture of the way they had lived from information collected at sites in the former Transvaal and KwaZulu-Natal.

They built the walls of their huts using thin poles or sticks and mud, and thatched the roofs with grass, but it is not clear whether the huts had the typical conical-shaped roofs of later rondavels. In settlements the huts were not close together, but were nevertheless grouped into loose units to form a kind of vil-

Dr Pienaar with a painted clay pot on a stone pedestal under a hollow cave shelter on a sandstone hill on the watershed between the Madzaringwe and Mashakiri spruits.
(13.05.1986)

lage or kraal. Their food consisted mainly, and in some instances exclusively, of venison and food foraged from the veld. Some groups kept goats and maybe sheep as well, but they apparently had so few that they seldom slaughtered them. There are no reliable indications that they had cattle. A crop, in this case a type of millet, was found at a terrain near Tzaneen, but nowhere else. The agricultural activities of these people were therefore still very unsophisticated. They were familiar with iron smelting, as pieces of iron slag have been found at several dwelling places, although only in small quantities. Each community probably satisfied only its own needs. None of their completed iron tools has been found so far, but the most essential articles would certainly have been pick-axes, assegai blades and axes.

Between AD 700 and 800, immigrants coming from the north and northwest, and possibly also from an area to the west, started settling in the area that is now the Kruger National Park. Some of their settlements have been found along the Luvuvhu, Letaba, Olifants and Sabie rivers and from there southwards to the Crocodile River. Their way of life was basically the same as that of their east coast predecessors, but their food production and technology were more advanced and their dwelling places indicate both denser and longer habitation. The kraals were larger and could have consisted of 50 or more huts in places. They kept cattle, goats and sheep, as well as dogs, according to information provided by Ina Plug of the Transvaal Museum. They practised stockbreeding on a larger scale than their predecessors, although on a smaller scale than their contemporaries in the former Transvaal and KwaZulu-Natal.[1] While the inhabitants of the Park area were still mainly reliant on hunting, their contemporaries elsewhere relied on cattle and small livestock for meat. Presumably the climate was not favourable for agricultural activities and conditions gradually deteriorated, as by AD 800 some groups' meat was supplied almost exclusively by hunting.

During this time the cultivation of crops also increased. They planted millet, sorghum and several types of beans, although remains of these products have not been found at any of the settlement terrains in the Park. Cultivation methods were undoubtedly similar to those used in Africa in later times. They would chop down shrubs and trees on a suitable piece of land and afterwards burn the wood. Crops were then sown in the ashy soil. When a field became infertile after a few years, they simply cleared a new area, while the former fields, which could rather be described as gardens, were gradually taken over by bush again.

Signs of iron smelting in the form of iron slag are found at most former dwelling places in the Park, but never in large quantities. This proves that each group satisfied only its own needs. The techniques used to smelt iron and produce tools had not been mastered to such an extent that more than what was necessary could be produced. This included pickaxes used in farming, assegai blades and arrowheads for hunting and protection, chopping and cutting tools and awls to make clothing and sleeping karosses. Only the rusted remnants of these iron objects have been found so far, and not enough remains of their smelting ovens to make a faithful reconstruction. An interesting find was pieces of smelted copper at a terrain near Balule that dates back to AD 850. The inhabitants therefore must have produced copper objects. Copper arm and leg rings dating back to the same time have been found at a number of terrains outside the Park. In other words, this was the beginning of the production of and possible trade in copper objects. The copper smelted at Balule probably came from Phalaborwa, where there are indications that copper mining started at that stage.

At the same terrain at Balule, as well as at other terrains near Skukuza and along the Letaba River, other exceptional finds have been made. These are glass beads, which must have come from the East and which therefore indicate trading with the coast. It is of course possible that the beads were brought by the people from the north, and the inhabitants of these terrains may therefore not necessarily have been involved in trading themselves. However, internal trade was already taking place at that stage. Copper was mined at Phalaborwa and salt was harvested at Eiland, close to the western border of the current Park area, and traded, probably for livestock and grain, with other groups. Contact at the coast with strangers bringing wares from the outside world, including glass

---

1 The current Limpopo, Gauteng, North-West, Mpumalanga and KwaZulu-Natal.

beads from India, heralded a new era in the lives of the communities of the interior. Visits to the coast could have taken place as well, as seashells have been found at the inland terrains. It is not clear, however, if these had any connection with trading activities. The contact with the east coast in the eighth and ninth centuries had a definite impact on developments in the northwestern part of the Limpopo Valley and further north as well.

It is interesting that dwelling places inside the Park that can be connected to the first millennium are generally 100 to 200 years younger than those of similar cultures in surrounding areas. This indicates that the inhabitants of these younger communities came from areas that had been occupied by their people for a considerable length of time. The Park area could therefore not have been a popular settlement area for the stockbreeders and agriculturalists of the time. On the other hand, conditions must have been more favourable in some river valleys as settled communities existed of people who owned livestock. As mentioned before, stock farming had deteriorated to such an extent by the nineteenth century that people were much more dependent on venison than their predecessors. Conditions conducive to survival therefore must have deteriorated even further in the Park area. It also seems the area was equally unpopular in terms of agriculture during the following 800 years, until AD 1600. There are few localities in the current Kruger National Park that were inhabited during this period and all of those contained very little cultural material. They could therefore not have been occupied for long periods of time. The only exception is Phalaborwa, just outside the Park's border, where copper mining continued.

In parts of Zimbabwe, the former Transvaal and Natal,[2] the arrival of new groups between AD 800 and 1000 led to economical, technological and political changes. This became the era in the history of South Africa especially typified by the settlement of Bantu-speaking nations.

However, the largest part of the Kruger National Park fell outside the radius of this development until about AD 1600. A development that would eventually involve the people of the Lowveld with trade with the east coast was the coming into being of an influential tribal group in the northwestern Transvaal[3] by AD 800, which would eventually lead to the development of the Great Zimbabwe and Monomotapa empires.

By about AD 800 the Limpopo Valley in the northwestern Transvaal (now Limpopo), southwestern Botswana and southwestern Zimbabwe was occupied by a number of communities whose economic activities and political structures were more advanced than those of the previous period. They practised cattle farming on a large scale and the size of some of the settlements and cattle kraals creates the impression that some groups were more affluent than others. The community structures of smaller groups with relatively uncomplicated ways of life started changing as well, and their economic and political structures became more sophisticated. Signs of trade with the east coast start to appear, including large quantities of glass beads, which had been brought to the interior from the coast to trade for ivory and animal skins.

By the eleventh century this area had developed into the most important economical and political sphere of influence in the interior of southern Africa. The capital of the area was Mapungubwe, situ-

A leadwood pestle found at the ruins of a former black settlement east of Sable Dam.
(OCTOBER 1983)

---

2 The current KwaZulu-Natal.
3 The western part of the current Limpopo Province.

ated south of the Limpopo River across from the current border between Zimbabwe and Botswana. Different trade routes led from here to the east coast – during the twelfth century, significant amounts of gold were taken there and exchanged for glass beads and possibly textiles. In earlier historical documents describing trade along the east coast, ivory is identified as being the most important merchandise coming from the interior, while leopard skins, tortoise shell and slaves are also mentioned. From the twelfth century gold is also highlighted. Sofala (Beira) was at that stage the most important port from where trade took place with the interior, but it is probable that there were other trade centres further south.

Trade along the east coast was controlled by Arab traders, a status quo that would last until the sixteenth century. They used natives as carriers and agents, but also visited the interior themselves.

Mapungubwe was evacuated in the thirteenth century and in the following centuries the Great Zimbabwe and Monomotapa empires would control trade with the east coast. The focus of trade activities therefore moved further away from the current Park area. The earlier trade connections that had brought Mapungubwe into contact with the coast still existed and traders were still using the established trade routes.

Developments in the northwestern Transvaal (Limpopo) and Zimbabwe did not have much influence on the history of the Kruger National Park or the Lowveld, as few people were living in the area until about AD 1600. The area would have served only as a trade route between the interior and the coast. But from the seventeenth century onwards a change started taking place. At that stage most of the surrounding area was already quite densely populated, and groups started moving into unoccupied areas.

The Park environment in some places had by then become more favourable to crop farmers and pastoralists and people from the more densely populated areas started moving there.

By the end of the eighteenth century, the Vhalembethu were already living along the Luvuvhu River. Further south, Sotho-speaking communities were living in the vicinity of Phalaborwa and along the Crocodile River, while Tsonga groups had settled in the Lebombo Mountains further east.

The Vhalembethu lived along the Luvuvhu River up to just north of the Limpopo. They were one of the most southern tribes of the Karanga of Zimbabwe. Tradition has it that they had been living there before the arrival of the Venda, in other words before the start of the eighteenth century. They were subjugated by the Venda, but some Vhalembethu people are still living along the Mutale River, a tributary of the Luvuvhu. During the initial contact between the two groups, the Vhalembethu was treated with disdain by the stronger Venda, with the result that they later tried to conceal their Vhalembethu origins by pretending to be Venda. This is still the case. The Makahane and Thulamela ruins northeast of Punda Maria and other stone structures along the Luvuvhu River are associated with the Vhalembethu.

The Baphalaborwa settled in the vicinity of the current Phalaborwa during the eighteenth century. They concentrated on trading iron wares that they had produced themselves. According to their folklore, they had come from the north and initially settled in the vicinity of Bushbuckridge. From there they moved to their current territory between the Olifants and Letaba rivers, where they soon managed to make a lucrative existence by working iron ore. This is apparently why they named the new locality Phalaborwa, "better than the south", and adopted the name Baphalaborwa, "people of Phalaborwa". They were not the first settlers in that area, as they tell of other people whom they had encountered and driven away. It has also been determined through archaeological findings that people had been smelting iron there in the fifteenth century and had mined copper even earlier. The Baphalaborwa is one of the most eastern tribal groups of the Sotho-speaking nations of the former Transvaal[4] and remains of their earlier settlements can still be seen on Masorini, Shikumbu, Vudogwa and surrounding hills.

The unfavourable environmental conditions for crop and stock farming were not much of an obstacle to them until early in the nineteenth century as the trade in metalwares provided them with a comfortable existence. After that their living conditions started deteriorating as a result of raids from the interior and the introduction of Western wares.

---

4   Limpopo and Mpumalanga.

During the eighteenth century several Sotho-speaking tribes lived in the southern part of the Park area and in Swaziland. Not much is known about their early history, as their encounters with the Swazi and Tsonga at the beginning of the nineteenth century overshadow the memories of the storytellers responsible for relating the tribal history. They originally came from the west, but it is not known when. The Bapai,[5] Mapulana and Bakutswe are the three best-known tribes among the early Sotho-speaking tribes of the southeastern Transvaal[6] and Swaziland. Then there is the Ngomane, who originally lived in the area between the Inkomati and Sabie rivers, bordered by the Lebombo Mountains to the east and the Drakensberg to the west. They claim they were originally Sotho speaking, but lost their Sotho identity as a result of contact with the Tsonga and Swazi. Before the De Cuiper party reached the Crocodile River in 1725 from Delagoa Bay (see chapter 3), they had travelled through densely populated areas, and slightly north of the river they were prevented by a large group of black people from continuing their journey. These people probably belonged to the Ngomane tribe.

De Cuiper also heard of several kraals he would find on the way to Ciremandelle[7] (Phalaborwa). Certain parts of the southern Park area were therefore already inhabited by a reasonable number of people at the start of the eighteenth century.

The area east of the Lebombo Mountains, up to the coast, was for centuries the home of a number of Tsonga tribes. They were mainly crop farmers and had very few, if any, cattle. This is the main reason why they found the area west of the Lebombo Mountains unattractive. There are indications that smaller groups had settled in the mountains before the large nineteenth century influx, but they did not penetrate deep into the Park area. However, some did carry trade wares from the coast deep into the interior.

The people who lived in the Park area by the end of the eighteenth century were members of the tribes mentioned above. There are signs that a few Venda groups inhabited the area, but the evidence is not conclusive. If they had been there, it would have been reasonably far from their tribal area in the vicinity of Soutpansberg.

The settlement of these groups had taken place before the events of the nineteenth century that led to far-reaching changes for many. The following description also has some bearing on the period before that cataclysm. There are many similarities between the lifestyles of the different groups, although clear differences can be detected between cultures if inspected more closely. A general feature of their settlements was the rondavel-type huts grouped in kraals or towns of 20 to 50 huts. The custom of building terraces to live on and on which to build dividing walls was characteristic of the Vhalembethu, while the Baphalaborwa also built their homes on stone terraces they had constructed against the slopes of hills. The Sotho tribes in the south, on the other hand, rarely built with stone and the Tsonga almost never. (Remains of stone walls and terraces occur at quite a few places in the Park area, but many of these date back to the nineteenth century.)

In the area between Lydenburg, Machadodorp and Waterval-Boven there is a large concentration of stone structures, most of which date back to the seventeenth to nineteenth centuries. However, the ruins inside the Park are not part of this complex.

All the tribes made a living by cultivating sorghum, millet and other crops, as well as farming livestock. Cattle and goats were kept, and sometimes a few sheep. The Tsonga had no cattle earlier on, and even in more recent times cattle did not play an important role in their lives. The tribes also kept few cattle. After they settled in an area that was unsuitable for farming but that contained copper and iron, their economy was focused mainly on the production of metal wares they traded for grain, meat, skins and other wares. It is noticeable that iron slag, which points to the working of iron, is not found at as many settlements as earlier. This is a direct result of specialisation in the production of iron and copper, which in turn led to trade between producers and consumers. Similarly, salt, which had been harvested at Eiland, close to the current Park, and at a few other places in the Lowveld, was traded over a wide

---

5 Also called ba-Mbhayi in certain references.
6 Mpumalanga.
7 The name Ciremandelle appears in De Cuiper's diary in Dr W.H.J. Punt's book *The First Europeans in the Kruger National Park 1725*. Punt was convinced that De Cuiper had been referring to Phalaborwa while writing Ciremandelle in his diary, but he did not provide an explanation of the name.

2. A NEW ORDER

Overgrown capture pits along the upper reaches of the Guweni Spruit, which have been used by blacks during the nineteenth century.
(SEPTEMBER 1986)

The huts and a furnace which can be viewed at Masorini are restored structures.
(18.06.1984)

area. As mentioned earlier, people already had been trading salt at Eiland during the first millennium.

Trade over long distances expanded in this part of southern Africa after the Portuguese in the sixteenth century had driven the Arabs away from the existing ports, Sofala and Inhambane, and later started trading from other locations, including Delagoa Bay. Several trade routes went through the Park area during the eighteenth century (Chapter 2.5 gives a comprehensive description of the trade routes). Trade beads are found throughout the Park, from the far north along the Luvuvhu River to Shilowa and Nwanedzi in the east, Masorini in the west, Mooiplaas and Satara in the centre, and Renosterkoppies in the south. All these locations were probably along or near the trade routes.

The inhabitants of the Park before the nineteenth century therefore consisted of a number of tribes that made a living mainly through crop and stock farming. They also hunted and gathered food from the veld to augment their diet. Game was sometimes trapped in pits during big drives and afterwards speared to death. Rows of these pits were found by Colonel Stevenson-Hamilton in the Lebombo Mountains and more recently, remains of this hunting method were also found on the plain along the upper reaches of the Guweni and Mrunsuluku spruits. Even among the Baphalaborwa some tribal members focused on agricultural activities, while others concentrated exclusively on metal production. Members of different tribes sometimes came into contact with each other to trade local or imported wares. History shows that groups of people sometimes peacefully applied for a chief's permission to settle in his area as his subjects. People generally moved about on their own or in groups without any incidents taking place. Even when there was conflict, it did not lead to large-scale disruption, and life carried on relatively peacefully until the first years of the nineteenth century.

Two settlements in the Park area – Masorini and Makahane – deserve a special mention as they illustrate the different lifestyles before the turmoil of the nineteenth century. Masorini represents a community whose existence rested on the production of and trade with iron objects. At Makahane, on the other hand, the stone ruin is similar to some of the later ruins in Zimbabwe that had been the homes of headmen and chiefs, hence people of status in their communities.

Masorini Hill is 11 kilometres from the current Phalaborwa entrance gate to the Park, along the road to Letaba rest camp. Groups of smelters and smiths and their families lived on this hill during the nineteenth century and possibly even earlier. Tradition has it that they were among the ancestors of the Baphalaborwa people who now live near Phalaborwa. Although there is no direct testimony available about Masorini, some of the oldest Baphalaborwa people can still remember the people who used to live there. They did not know the hill by its current name, but as Piene (pronounced Pee-ê-ni).

There are a number of stone wall terraces against the northwestern slope of the hill on which crumbling floors, as well as grinding stones, clay potsherds and other commonly used tools, occur. Furnaces and large amounts of iron slag were found at the foot of the hill. Articles indicating the later presence of whites (possibly after Masorini had been evacuated), were also found. These include the rusted barrel of a muzzle-loader, a few metal buttons from a uniform and articles that provide proof of contact with traders, such as a number of small glass beads and a Portuguese wine bottle.

Students from the Department of Archaeology at the University of Pretoria have carried out excavations and research at the terrain since 1974. Their main aim was to construct as complete a picture as possible of the living terrain. This was done at the request of the Parks Board with the aim to establish an archaeological and ethnological terrain museum at the earlier dwelling on the slope of the hill. Supported by the findings, sketches and plans were drawn up on which the restoration of Masorini was based. To keep the restoration as true as possible to the original, ethnographic research was carried out in the nearby residential areas of the Baphalaborwa on the traditional method of hut-building, as well as on the tools used and the manner in which dwellings were arranged.

Chief Brown Malatji, the minister of agriculture in the former Lebowa Government, gave his active cooperation to the project. Examination of the remains of the huts on the hill did not provide enough information to allow a reliable reconstruction of the different kinds of huts. As a result excavations were then also done at Shikumbu Hill, three kilometres north of Masorini, to obtain additional information about hut-building. Baphalaborwa had lived at Shikumbu as well, except for the northeastern slope, which was occupied by a Karanga group from Zimbabwe who had settled there as subjects of the Baphalaborwa. The last surviving member of this group could provide valuable information about the way of life and iron production on Shikumbu, shortly before his death. Just as at Masorini, iron smelting ovens are also found at the foot of Shikumbu.

The settlement at Masorini is situated against the northwestern slope of the hill and consists of two dwellings. Lower down the hill is a terrace complex on which the remains of burnt huts, rubbish dumps and grinding stones have been found. At the foot of the hill are the remains of large, triangular furnaces, surrounded by large amounts of slag. High up against the slope and about 60 metres from the bottom terraces is another terrace complex. Small ash heaps, grinding stones and hut ruins have been found here as well. Just below these upper terraces are three small forges, each with a large anvil stone and several hammer stones next to it. Footpaths and steps connected the complexes.

Huts, glass beads, grinding stones and ash heaps at the bottom terrace complex indicate the presence of women who carried out domestic activities, while the furnaces and iron slag indicate that men smelted iron from iron ore there. In the upper terrace complex, huts, grinding stones and other objects are like-

Grinding stones found at the ruins of black settlements between Masorini and Vudogwa.
(OCTOBER 1983)

wise signs of female domestic activities. However, the men here did not smelt the iron, but worked the iron with the help of forges and anvils.

It is not clear why the two terrace complexes were separated. There are no large rocks or other visible obstacles that could have prevented the building of terrace walls. Smelters and smiths may have lived apart because of social reasons, as smiths have a higher status than smelters in certain African tribes.

The two groups of craftsmen, the smelters and the smiths, carried out their activities separately but in cooperation with each other. At the foot of the hill were the furnaces to obtain usable iron from the iron ore they had collected at different places nearby. In contrast with other sites, they do not seem to have mined the ore there.

The furnaces were made of clay and looked a bit like a big, triangular pot, about two-thirds of a metre high, with an inner circumference of 60 to 70 centimetres and thick, straight sides. There were long grooves in the three corners through which clay pipes were pushed in order to blow air into the furnace. The inside of the furnace was filled with charcoal. Pieces of iron ore were placed in a clay pot and heated in the furnace for a long period. The eventual product of this process was a mixture of semimelted iron, charcoal and other impurities. This was handed to the smith, who heated it in his own furnace, which was somewhat smaller than that of the smelter, and removed the impurities. A pick axe, for example, was made by heating the iron and hammering it out on a stone anvil. In this manner they manufactured pick axes, axes, adzes, assegai blades, arrowheads, knives and other everyday items or ornaments.

The iron articles were exchanged for grain, meat, skins and other goods. Many glass beads were found in a rock crevice on the hill, and these had probably also been used for trading purposes. Through their trading activities, Masorini (Piene), Shikumbu and the other iron-producing centres in the Baphalaborwa's territory, made a significant contribution to the economy of the community. When these activities came to a halt at the end of the nineteenth century, it also meant the end of an almost 1000-year period during which the Phalaborwa area (which had of course been known by other names before) was known from the east coast to the interior of southern Africa as a place of copper and iron production. During this time, different tribal groups had consecutively lived in and left the area. Some moved to the Soutpansberg area and others to Messina (now Musina) until the Baphalaborwa tribes finally settled there.

The end of Masorini came rather suddenly. The shock waves that came in the wake of the rise of the Zulu empire early in the nineteenth century, struck the Lowveld in the form of raids from the south and the west, the arrival of immigrants from the east and the mutual clashes that flowed from the turmoil. At the same time, cheap Western iron products reduced the demand for local products. And so the rhythmic sounds of smelters' bellows and the knocking of stone hammers on anvils became quiet in the late nineteenth century.

Thousands of years ago Masorini had been occupied by people who had also used the environment, but in a different way than the people who would later live there. These were the hunter-gatherers of the Late Stone Age, whose stone tools are lying on the slopes and at the foot of the hill. Utilisation of the environment therefore changed from an almost unnoticeable exploitation of animals and plants by the Stone Age people to the exploitation of the earth, animals and plants by later miners, metalworkers and farmers. Visitors to Masorini have a unique opportunity to see the cultural remains of both earlier and later inhabitants, to compare them and in this manner form a picture of the changes in the relationship between man and his environment through the centuries.

2.1 BLACK INVASION AND COLONISATION OF THE LOWVELD

The Makahane Ruins on top of Makahane Hill with Makahane Poort in the background.
(MAY 1984)

2.2 The stone-wall culture of the Makahane and Vhalembethu

A stone wall which forms part of the Makahane Ruins on Makahane Hill.
(26.04.1984)

In the northern part of the Kruger National Park, just east of where the Madzaringwe Spruit flows into the Luvuvhu River, is another settlement where stone walls were erected, but in a different building style and with different functions than those at Masorini. This is the ruin of Makahane, situated on a high ridge next to the Luvuvhu River, 48 kilometres northeast of Punda Maria.

Makahane represents a different way of life than that of the iron producers of Masorini. The tribe who lived here and across the Limpopo River was the Vhalembethu, one of a network of tribes in Zimbabwe and the northern Transvaal[8] associated with the building of stone structures in this area.

The building of stone walls was part of a way of life typified by a hierarchical political structure where extraordinary power was vested in the chief. The tradition of sacred leadership among African tribes is a long one. The chief was at the same time ruler, chief justice and high priest of the tribe, as well as the link with the most important ancestral spirits, the person on whom the tribe's welfare depended, the figure around whom the lives of community members revolved, and the symbol of tribal unity.

In certain tribes, including the Shona of Zimbabwe and the Venda of northern Transvaal,[9] the chief was in fact exalted to an almost godlike status. He was a holy man whose slightest acts were praised in song. There was even a "main village language" with unique words to describe the most mundane of his actions, like sleeping or walking. He ate his meals in private and when drinking beer with his advisers, all conversation stopped while he held the drinking pot to his lips. His living quarters in the main kraal (a traditional African village of huts, usually enclosed by a fence) were separated from the rest and were situated either on the highest point or at the back of the kraal, protected by walls so that the chief could move around out of sight of the common people. Very few people had access to this area.

If the development of stone architecture is investigated, one can come to the conclusion that it probably developed out of a desire to emphasise the status of a chief. The prominent spatial position and relative size of the chief's kraal, as well as the neatly stacked stone walls, was indicative of his status. Therefore, this type of architecture is rightly known as "prestige architecture".

The earliest examples of this stone building architecture can be seen at Mapungubwe, where free-standing and terrace walls were erected on the hill where the chief lived. These walls were not very high or impressive, but they already show the architectonic elements of the more impressive structures later erected in Zimbabwe and the northern Transvaal.

During a period spanning the fourteenth to the

---

8   Limpopo.
9   Limpopo.

sixteenth centuries this architectural form reached its pinnacle at Great Zimbabwe, with its high outer walls, passages, friezes, dressed stones and upright stones, which all reflected the status of the chief. This practice was followed at later chiefs' headquarters, such as that at Khami, close to Bulawayo, well into the eighteenth century. Here a cover was even put over a passage that led to the chief's quarters, probably to conceal his movements.

Makahane's dwelling must be seen against the background of the sociopolitical system and the stone architecture described above. Judged by the size and architectural style, it can be deduced that Makahane was not as important as Khami, but nevertheless a figure of authority.

The Makahane Ruins are northeast of Punda Maria on a high ridge along the Luvuvhu River. On top of the ridge is a slightly elevated area, and this is where the stone-walled living quarters were erected.

The walls were built with locally collected, flat sandstone blocks that were stacked on top of each other. The dwelling consists of a circular outer wall with several lower walls that divide the interior into separate units. No bonding agent was used. The flat stones were simply stacked on top of each other in two parallel rows and the space in-between was filled with rubble. For stability, the walls had a broad base, tapering towards the top, and are therefore thicker at the base than at the top. No attempts were made to dig foundations and the walls were built on the ground and rocks. While the builders did try to pack the stones horizontally, no bonds were used where two walls met. There was also no attempt to use stones of the same thickness in one layer. This building style is similar to that of the so-called Acropolis at Great Zimbabwe, although Makahane's walls are much smaller and were built much later.

Because the walls were erected on top of and against the slope of a flat ridge, the bases of the walls on the outside are much lower than on the inside, with the result that the walls look much higher from the outside. In one instance for example, the top of the wall on the outside is about three metres high; while on the inside it is only one meter from ground level. To date no decorations have been found on the walls.

On the living quarters' highest level the walls were erected in such a manner that they formed a courtyard, which can be reached via a passage. In one of the courtyard walls there is a built-in seat with stone steps leading to it and in one of the inner walls there is a neat entrance, also with steps.

There probably used to be a door in this entrance as two V-shaped indentations, which presumably held the wooden doorposts, were created by stacking the stones in an ingenious way. There are several neatly packed platforms on the inside as well as the outside of the walls, but the walls themselves were not packed equally neatly everywhere. For instance, the walls of the section that may have been the meeting place are not constructed with as much care as those in the area with the built-in seat.

In the inner court, right in front of the seat mentioned, stones have been packed flat on the ground in an oval shape. This is Makahane's grave.

To the west of the ruin, on a slightly lower level, is a large terrain that shows signs of earlier habitation. There are no stone ruins here, but signs of habitation in the form of building rubble, glass beads, potsherds and human-made round and elongated holes in the rocky face do occur. There are also signs at the terrain that the Vhalembethu had smelted gold, which they cast in stone hollows. It is not known how they obtained the unrefined gold or gold ore, but it could have come from the well-known Lemba goldsmiths who lived along the western foothills of the Soutpansberg.

In contrast with Mapungubwe, no manufactured gold artefacts were found at Makahane. However, gold artefacts have since been found at the Thulamela ruin on a hill southwest of the old Bobomeni Drift in the Luvuvhu River.[10]

Although the Makahane Ruins are in a dilapidated condition, it is clear that it must have been a chief's stately dwelling, and it is one of only a few sites of its kind that can be linked to a historically-known person and a specific tribe. The history of Makahane, who had been the last chief to inhabit the dwelling, could still be related by his descendants in 1963.

The information about Chief Makahane was obtained from Chief Filemon Makahane, a chief in Venda in Headman Tshivhase's territory, west of the Luvuvhu River.

The gold drop which was found in 1984 at the Makahane Ruins by ranger Leighton Hare.
(PHOTOGRAPH: SOUTH AFRICAN NATIONAL PARKS)

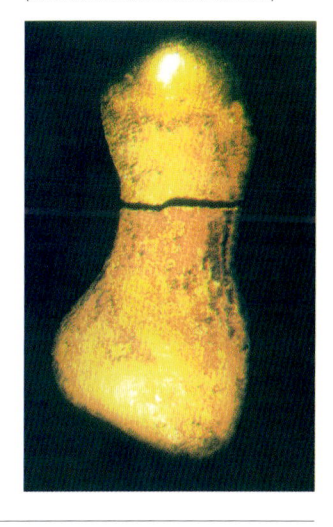

---

10   See chapter 2.3 for more information about the gold smelters of Thulamela.

2. A NEW ORDER

The then already elderly Filemon visited the ruin in the Park accompanied by a number of his advisers in order to relate the history of his forefather to the researchers. From this information it was calculated that Makahane had ruled as a chief of the Vhalembethu during the second half of the eighteenth century from his home next to the Luvuvhu River.

The Vhalembethu had already settled along the Luvuvhu and Mutale rivers and across the Limpopo by the seventeenth century. Makahane's father, the tribe's high chief, lived just north of the river. There were apparently two chiefs who each ruled over his own territory south of the river, of which Makahane was one. The two chiefs did not rule as independent headmen, but were subjects of the high chief north of the river. But because of the vast distance between them and the high chief, the two exerted great power and ruled as almost autonomous leaders. These southernmost members of the Vhalembethu tribe were later conquered by the Venda, probably at the beginning of the nineteenth century.

The Venda settled in the Soutpansberg area only after the Lembethu had already settled south of the Luvuvhu River. According to tradition, the first meeting between the Venda and the Lembethu took place at the Samango River. Venda Chief Ravhura stood on the riverbank and high-handedly shouted at the Lembethu chief: "Who are you?", upon which the latter indignantly yelled back: "I am the elephant bull and somebody you should greet when you come here. I am the chief of the Lembethu and who are you?" Ravhura answered: "I am the elephant's head, the chief of the Nzhelele," and then left without another word.

Highly indignant about the Venda chief's arrogance, the Lembethu chief went home. He stopped to drink water at a spruit near Makuy and his footprints on a rock can still be seen there today.

Chief Makahane is the Vhalembethu about whom the most stories are told and he is particularly known for his cruelty to his subjects and enemies alike. Tradition has it that, if somebody had to be punished for a transgression, the person was lowered by rope down a high cliff on the western side of the ruin. Apparently a large bird[11] had a nest halfway down the cliff. If the person being punished could find a chick in the nest, he had to take it out and show it to the people who had lowered him down. Once they had seen the chick, he would be hoisted up so that he could present it to the chief, who apparently loved the meat of this bird. The chief would then exonerate the culprit as a reward for his find. But if the culprit failed to find a chick, the rope would be cut and he would either fall to his death at the foot of the cliff or land in the river and be devoured by giant crocodiles.

When a head of cattle was slaughtered, the skin would not be stretched on the ground and held in this position with wooden pegs, as was customary. Instead, a couple of people had to bite into the skin and kept it stretched out until dry. When the chief wanted to test the dryness of the skin, he would hit it hard with his walking stick. Anybody who failed to maintain a grip on the skin was thrown over the cliff as well.

As a result of these and other cruelties that cannot be described here, Makahane was so detested by his subjects that they went to his father across the Limpopo to complain about his behaviour. His father was very upset upon hearing the news and after much deliberation made a drastic decision. He called his son Nelombe and said: "Your brother Makahane has lost his mind, as he does not act like a responsible leader towards his subjects any more. He is inhumanly cruel. You must go and kill him – not like a chief, but as an ordinary subject."

Nelombe obeyed and departed for his brother's kraal south of the Limpopo. When he neared the kraal, Nelombe, who was known for his good voice, started singing and dancing. Makahane watched Nelombe and his party approach from afar from his high seat in the stone wall. Because he probably suspected that he had been denounced by his father, he called out: "Look, there comes Nelombe! He is coming to kill me!" Despite his suspicions, he nevertheless called for beer and dancing to celebrate the arrival of his bother in the traditional manner.

A huge fire was made on the dance floor between two baobabs below the chief's living quarters. Mats were spread on the ground around the chief's seat in the wall as well as between his hut and the dance floor for him to walk on. After Makahane had taken

---

11   Possibly the black storks that are still breeding there.

his seat among his excited subjects, Nelombe stepped forward and respectfully knelt in front of him. Makahane's reaction upon this was: "You have come to kill me!"

Nelombe did not answer him, and when he got up the dancing started. Nelombe's dancing was the most agile that they had ever seen and filled them with immense admiration. Even Makahane could not help but watch with interest. Nelombe danced ever closer and closer to Makahane, unsheathed a long knife and in a flash threw Makahane to the ground and slit his throat like that of a goat. Not one of the bystanders lifted so much as a finger to prevent the attack on their chief.

Nelombe took command and ordered that an ox be slaughtered. When the fresh skin was lying on the ground, a number of men ran towards it to spread it open and grip it with their teeth as had been the custom under Makahane's rule, but Nelombe said: "No, the skin must be spread in the usual way with the aid of wooden pegs." With this he ended the rule of the heartless chief and made it clear to the people that they would be able to lead a normal, happy life.

Makahane had given instructions long before his death that he must be buried in front of his seat in the wall. Although he was unloved, his former subjects complied with his request and even made offerings on his grave. When Chief Filemon Makahane visited the ruin in 1963 he first went straight to the grave to make an offering of snuff to his forefather. While kneeling at the grave he greeted his forefather, saying softly and respectfully: "I am here to talk to these people, the white people. They want to know about your doings. Don't be unhappy if I tell them about it, because they want to know it and I must speak the truth."

Makahane's son Mashande, who succeeded his father as chief, immediately left the scene of the murder and settled with a number of followers near the present-day Elim. His son Madadzhe was born there. After they had been living there for a considerable period, the Swiss Mission built the Elim Mission in 1879. In this way Madadzhe, who was by then an old man, came into contact with Christianity.

Shortly afterwards he returned to his grandfather's dwelling and settled there, but it is unclear why he had decided to do so. He probably thought that he would be able to appease the spirits, but he didn't stay there for long. According to legend he left because the spirits had told him that he could not stay as too much blood had been shed there. It is also possible that because of his wish to accept the Christian faith, he wanted to distance himself from the murder scene and the cruelty associated with it. Madadzhe built a new kraal on the banks of the Luvuvhu River, where he was visited and christened by a German missionary, Reverend Wessmann, in 1889. After his christening he took the name August. The old man died a year later.

Tradition has it that when the spirits instructed Madadzhe to leave the ridge of murder, they also said that his daughter Mkonde should act as warden of the place and that she should make regular sacrificial offerings there. After her father's death she built a kraal at the foot of the hill, but because of long, severe droughts and subsequent famines blamed on discord among the spirits, she and her few followers left. And so the last Vhalembethu of Makahane bid a final farewell to the ruins on top of the ridge.

A hollow in a sandstone outcrop at the Makahane Ruins in which gold or other metals could possibly have been poured. (26.04.1984)

2.3 The gold smelters of Thulamela

S. Miller

In 1992, six years after its rediscovery by park ranger Flip Nel in 1986, the archaeological terrain at Thulamela[12] was shown to a group of researchers during a routine visit to the Pafuri area. In July 1992 it was decided to do a test excavation in the main midden with the aim of obtaining material for dating purposes. It was also expected that the exploratory excavation would produce potsherds from which the cultural affinity of the terrain could be determined. However, the excavation produced much more than had been expected and it was a momentous occasion when three gold beads were found, as well as potsherds with droplets of melted gold.

What had been a routine investigation cleared the way for one of the most important archaeological finds in South Africa in the twentieth century: gold artefacts associated with a "prehistoric" southern African terrain that could be studied with the aid of the most modern scientific techniques!

In the years before this discovery, Gold Fields South Africa had made significant financial contributions to the Kruger National Park for the enhancement and dissemination of environmental knowledge. After the discovery of the gold objects at Thulamela, sponsorship was negotiated from this well-known mining company for research and for the development of the terrain.

By July 1993 work could start. The project had been divided right from the beginning between practical field work and research on the one hand and a management action on the other to involve the neighbouring Venda and Tsonga-Shangaan communities in decisions that would previously have been made by the Park's management and scientists. It was a unique approach to civic participation in the general management of a project at the time and probably also the key to the eventual success of Thulamela. The communities were represented by, among others, five Tsonga-Shangaans and five Vendas. Archaeologist Sidney Miller was appointed to manage the direct research and six assistants (three Venda and three Tsonga-Shangaan) were appointed to help him carry out the excavations and rebuild the walls.

Punda Maria, the closest place with modern facilities to the ruins, is 55 kilometres away. To create an infrastructure so that the team could function was therefore first priority, as the project was to stretch over a period of three years. TEBA (The Employment Bureau of Africa), kindly offered the use of its facilities at the Mozambican border post. Fieldwork only started in midsummer 1993–1994, a full six months after the project had been officially launched.

The plan of action was quite simple: The results of the excavations had to be presented to the general public in a readily understandable manner and visitors had to be allowed access to the terrain. Nobody had any idea about the drama and heart-rending scenes awaiting them during the next three years. Neither did anybody have an inkling of the national and international repercussions of this work.

The main objects of the project were:
- To determine the earliest and latest periods of habitation; as well as at which stage the building of the walls was initiated
- To collect enough cultural material for the compilation of a cultural profile of the terrain for comparison with similar terrains in the vicinity
- To draw up a map of the terrain, after which some of the habitation areas would be opened up to determine the settlement pattern and layout

---

12 "Rediscovery" of a site is usually a problematic matter. It is, for instance, acknowledged that Thulamela was known among the Makuleke people as the "place where drums beat". This reference usually indicates important ancestral graves on the terrain and that such a site is deemed holy. Ranger Mike English mentioned to Miller that a former TEBA officer, Harold Mockford, had gold beads in his possession which he used to keep in a bottle. If this is true, then it is quite possible that Mockford had been aware of Thulamela and collected the beads there. Former ranger Harry Kirkman told Dr Pienaar, who was chief research officer at Skukuza at the time, that he had seen the ruins of Thulamela for the first time in 1942 while on a foot patrol along the length of the Luvuvhu River. Former rangers G. Adendorff and D. Lowe had also been aware of the ruins and Dr Pienaar saw them himself while searching for animal carcasses during the anthrax outbreak of 1960. It is, however, true that it was section ranger Flip Nel who pertinently brought the existence of this exceptional cultural feature to the attention of authorities and researchers.

Excavations started in December 1993, but a problem soon emerged. Because of the dilapidated condition of the stone walls, the loose stones were scattered over a wide area, which was earmarked for archaeological excavations. If excavations were to continue in these areas, large quantities of stone would have to be moved. This would mean that information regarding the volume, height and form of the original walls would be lost.

After consultation with the management team, a decision was made to first rebuild the walls with the stones that were still in their "natural" positions. This decision was supported by the fact that visitors to the terrain would be able to form a better idea of what the terrain looked like originally if they could see the walls in their rebuilt state.

This was a new challenge. A great deal of reconstruction had taken place to restore the walls of Great Zimbabwe, but in South Africa this was a totally new concept that had to be investigated. Apart from that, the technique used in Zimbabwe produced unsatisfactory results as the "taste" and "comfort" of the "rebuilders" mostly took precedence over the original's true style and material. After much deliberation and many experiments, the following building method was decided on: All loose stones on top of or lying next to a wall were moved two metres away

Stone walls at Thulamela.

from the wall at right angles. There was also an attempt to determine what effects gravity, vegetation and animals had on the displacement of the different stones from their original to their final positions. In cases where the elements had played a role, the affected stones' original positions were recalculated to counter these influences. This process was followed until the bottom layers of the walls were exposed and the building style of the walls could be determined.

Loose stones were sorted into two heaps; those suitably shaped for the outside layers of the wall and those suitable for filling. The typical Thulamela stone wall consisted of two outer layers, where stones were packed in uniform layers, while the space between the two layers was filled and packed with rubble. After the sorting process had been completed, reconstruction could start. The building style that had been determined from the lower levels of the walls was duplicated in the successive layers and building continued until all the loose stones had been used.[13]

This building technique produced exceptional results, although one aspect is open to criticism: If decorations had been built into the original walls, as are fairly common at the Zimbabwe-type terrains, then the reconstruction process at Thulamela would not reflect this. It was, however, one of the sacrifices that had to be made for the project to succeed.

Between December 1993 and July 1995, about 2000 tons of stone were used for construction. The team handled six tons of stone a day on average, in daytime temperatures that varied between 15 °C and 36 °C in winter and 32 °C and 48 °C in summer … Although there were no serious confrontations between the workers and wild animals, it was definitely a factor that affected the nerves! An important finding was that game had quite an impact on the walls – to such an extent that the reconstructed walls needed regular maintenance. It was therefore only to be expected that the walls of Thulamela would be dilapidated by 1992, if one takes into account the fact that the last inhabitants of the terrain had left 350 years ago.

During the 18 months it took to rebuild the walls, the management team was not idle, as they had much to contemplate.

The practical implications of opening the terrain to the public was one of the major problems, as Kruger Park management was somewhat hesitant to expose the public to possible injuries as a result of contact with wild animals. Reservations, opening hours, travelling times to and from rest camps and the matter of guides and their accommodation, were just some of the issues that had to be considered. The inclusion of Thulamela and other cultural terrains in educational environmental programmes was another issue, as some members of the management team were of the opinion that there is no relation between culture and the environment. A social ecology section was established in the Park about that time. Pressure from this team, as well as from the sponsors, forced the management team to acknowledge that culture and nature cannot be separated and that a holistic approach towards environmental education is essential.

Both the Venda and Tsonga-Shangaan communities were invited to help manage Thulamela. During discussions about educational programmes, community outreach and whether school groups should be allowed to visit the terrain, the question of guardianship repeatedly raised its head. The Tsonga-Shangaan community reasoned that Thulamela belonged to them as they had lived in the area until as recently as 1969,[14] while the Venda community pointed out that the Tsonga-Shangaan had never lived in or on mountains and had also never built with stone. However, they did admit that the Tsonga-Shangaan had been the last inhabitants of the area, but argued that this group had won the area from the Venda only about 200 years ago. This issue would only be resolved to the mutual satisfaction of both parties towards the end of the project.

The most important development during this time was that both communities became truly active participants in the decision-making process. Both groups contributed meaningfully towards the execution of an archaeological project that was their own heritage, as well as towards the acceptance of the concept that scientists do not have exclusive rights to the terrain. This was a sore point for some members of the

---

13 The only modern building technique applied was to reinforce a few of the larger walls with diamond mesh wire to ensure visitors' safety.
14 The Makuleke community was moved in August 1969 from Pafuri to a new terrain at Mthlaveni, southwest of Punda Maria.

management team, as scientific ownership had never been questioned until then. However, the writing was on the wall and especially after the 1994 elections it became clear that communities would not hesitate to make themselves heard in matters that concerned their cultural heritage.

The actual excavations at the sight began almost two years after the start of the project. But the preceding time had not been wasted as a firm foundation had been laid for the execution of the project's archaeological excavation. The first excavations were a series of test trenches in the main midden[15] aimed at obtaining both cultural material and materials such as bone and charcoal that could be dated.

As a result of the very small diameter of the gold beads, a painfully slow technique was used. The excavation trenches were 1,2 metres wide and two metres long, while the depth varied according to the midden's depth. In other words, excavation carried on until either solid rock or sterile ground was reached.

Despite the fact that six people were digging, excavations never exceeded 100 millimetres per day. This represents the same volume of soil that would fill a municipal dustbin. Material in a midden is especially soft and the position of cultural material in such a heap is not of much importance. As such the material could be excavated relatively quickly, but the sorting of the sifted remains was a tedious process.

In the case of Thulamela it was necessary to sift the excavated material through three separate, layered sieves. Not even the finest dust was discarded, but put through a nested graduated screen especially manufactured for the purpose. This screen divided the material into four gradings collected in four different sieves. Only then could the sorting process begin.

The first grading usually removed most of the stones, but sometimes larger potsherds and bones were found among the stones. Such findings were few and far between as most of the large artefacts and cultural material had already been discovered during the physical excavation process. The second grading contained the most cultural material including potsherds, bones, charcoal, metal, wire, bracelets and ankle rings and sometimes pieces of gold wire that had been part of broken bracelets. The sorting of the first two gradings was relatively simple and usually took only a few minutes.

Most time was taken up sorting the third grading in which all material smaller than ten millimetres and larger than one millimetre was handled. Almost all the beads found on the terrain came from this grading. This included imported glass beads, locally manufactured ostrich shell beads and gold beads, as well as small gold droplets, which were the rests of the gold smelting process that had taken place there.

The last grading, which included the dust that had gone through the one-millimetre sieve, was graded only once a week to ensure that the sieves were still working well.

After these excavations, two holes were dug next to the walls to determine when they had been built. These excavations were totally different from the previous ones as there were sizeable stone deposits next to the walls as a result of the walls' gradual collapse. Relatively little cultural material was found in either excavations, and their main contribution was to reveal that the walls had been built on a layer of ash that could be dated. Bone and charcoal samples were submitted to the CSIR's carbon-14 dating laboratory for analysis. However, an exact description of the succession of habitation was not possible because of the position in which the material had been found, the relative reliability of carbon dating and the fact that material had gradually spread over the site.

It could be determined that the site had been inhabited between AD 1200 and 1300 and that only a small part of the terrain had been used. No cultural material was found in the first ash layer from this era, and therefore no particular "character" can be ascribed to the culture of the original inhabitants. However, in the area between the Luvuvhu and Limpopo rivers, not far from Thulamela, two sites have been identified that contain typical Mapungubwe pottery. Although Mapungubwe was not the focal point of the area any more at that stage, it can tentatively be accepted that the first people at Thulamela had affinities with the Mapungubwe culture.

---

15 A midden is a pile on which people who lived at the terrain discarded their daily ash and rubble. Cultural material that landed in it could therefore easily be dislocated when people or animals move over it. In time, the activities of subterrestrial animals like mice, moles, skunks, porcupines and warthogs can also move material out of context.

The first date after that, that can be linked with the stone walls is the late fourteenth century AD, but it cannot be concluded with certainty if the site had been inhabited continuously or if the inhabitants had changed their way of life, or even if different people had moved to Thulamela. What is certain, however, is that the stone walls can be linked to a new pottery style that does not possess the stylistic attributes of the pottery at Mapungubwe, and cannot be linked with any certainty to other cultures such as that of Great Zimbabwe.

The last date obtained points to the site being evacuated in the second half of the seventeenth century. However, the pottery of this last period of habitation was in a classical style that may be linked to the pottery at the Khami site in Zimbabwe. The Khami site is 40 kilometres west of Bulawayo and was one of the smaller cultural focus points after the collapse of Great Zimbabwe.[16]

Apart from the pottery and banded decorations on potsherds, which show affinity to the Khami culture, a number of potsherds show the classic zoomorphic form (animal forms, or forms referencing animals), which is associated mainly with the Khami. Apart from that, a few potsherds were found with some gold droplets attached and showed signs of having been subjected to high temperatures. This served as confirmation that gold had been cast in the main complex at Thulamela.

Potsherds that had been modified to form bobbins were common. These bobbins indicate that the people at Thulamela had spun wild cotton (*Gossypium herbaceum*) and used the yarn to weave fabrics, which they probably used for making clothes.

A large number of animal species could be identified from the bone fragments that were collected, including vlei reedbuck (*Redunca arundinum*), which does not occur in the area any more. Cattle, goat, sheep, chicken and dog bones were also found. The dog bones, which were spread across the midden, could be an indication that these animals had been used as a food source or for hunting purposes. Ivory bracelets found in the midden indicate the utilisation of big game, while the vertebrae and bones of several fish species point toward the utilisation of rivers.

Metals are represented in the form of iron, copper and gold. Remnants of iron bracelets and of spear- or arrowheads were found, as well as small iron adzes, which point to woodworking at the terrain. Especially interesting was an object that looks like a bent spearhead. Early in the twentieth century a similar piece of equipment was documented as being used by the BoLobedu, who used it to carve wooden bowls. These bowl shapes are still found in the former Venda (Limpopo). Copper wire bracelets were common. Part of a molten copper bar was found in the midden, as well as a larger piece of molten copper on the surface next to the midden. Gold was found in the form of wire bracelets, beads and droplets.

A large variety of beads, which had served as ornaments for both men and women, was also found in the midden. Most of these were small, finely finished ostrich eggshell beads and freshwater mussel shell beads, seldom more than five millimetres in diameter. Blue, yellow, green, red, black and white glass beads probably came from the Middle East. Iron, gold and copper beads were also found. The gold beads were particularly small – dimensions of between one and two millimetres were not uncommon.

Two important deductions can be made, based on these finds. Firstly, access to large amounts of meat, clothing made of woven fabrics and rich adornments in the shape of beads and ivory bracelets – these all indicate that the inhabitants of Thulamela would have been considered reasonably well off by their contemporaries. Secondly, some experts assumed that metalworking would not have been allowed in the main complex, but the two pieces of molten copper, the gold droplets and the potsherds with gold droplets, are all indications that in this instance, metalworking did in fact take place in the main complex at Thulamela.

Even at the beginning researchers were relatively certain that Thulamela belongs to the Zimbabwe pattern. It was nevertheless still preferable to document as much of the settlement pattern at Thulamela as possible for comparison with known sites and for

---

16 Khami, which has been relatively extensively excavated by Robinson, exhibits extraordinary cultural characteristics. Commonly used articles which have probably belonged to a Portuguese priest, as well as a stacked Maltese stone cross in the main complex, indicate that Khami had probably been visited by Portuguese travellers during the seventeenth century.

future reference. A map of the terrain, which showed all the stone walls and relevant material such as grinding stones and middens, was drawn up. It clearly indicated the concentration of stone walls where the senior group lived. The smaller groups of stone walls probably indicate where the various "less important" families or followers of the more senior people resided.

Excavations of the main concentration of stone walls provided information about the shape of the huts and the layout of the individual sections inside the main complex. It was this information that indicated where the chief's hut had been, as well as the location of the hut where he would have received visitors under special circumstances. And during the excavation of these units Sidney Miller discovered the graves. All the units were not investigated to the same extent as it is archaeological practice to keep disturbance to a minimum.

No sign of a cattle kraal could be found at Thulamela, although we know that they owned cattle, and that most southern African groups attached great importance to their cattle and the kraal traditionally formed part of the main residential complex.

Grinding stones found inside the stone-walled areas indicate that the inhabitants had used maize (corn) as a food source during the final years of habitation. Maize originally came from North America and was brought to Africa in the seventeenth century by Portuguese traders. It can therefore be assumed that this food source was available to the people of Thulamela for a limited period and only in small quantities.

Interestingly, all the whetstones were found in the main complex. They are made of a particularly fine type of sandstone not found in the vicinity of Thulamela and were used to whet iron tools and weapons. Such metalworking in a main complex of a Zimbabwe-type settlement, where according to experts only royalty lived, is an anomaly about which a considerable amount of research is still necessary.

Metalworking, or at least an interest in metalwork techniques, was further confirmed by a number of tools found in the main complex. Two of these are harpoon heads with complicated grapple hooks that would have tested even a modern-day iron smith's skills to the limit.

Two hoes and a delicate spearhead with two grapple hooks were also among the finds. But the most interesting of the tools found together, was an unshaped hoe.[17]

In the area where the chief would probably have received important guests, the remains of a ceremonial spear were found. A broad, long, iron blade, which had probably never been sharpened, was found with a few pieces of gold sheet. The negative imprints on the sheet, as well as the golden nails still in it, indicated that it had been nailed around the carved end of the wooden handle of the spear. Venda maces were traditionally short spears with carved shafts and there are a few photographs of Venda chiefs with similar spears that were taken in the late nineteenth century.

One of the most beautiful pieces of jewellery discovered in the main complex was a delicately carved ivory pendant on a copper wire band. It was not possible to determine if the necklace once had more pendants, but the piece of ivory by itself testifies to good taste, exceptional craftsmanship and the affluent lifestyle of Thulamela's royalty.

A find that intrigued many people was a piece of Chinese porcelain, which was dated to the sixteenth century. Although small, this piece of blue-on-white ceramic points to the extensive trading contacts of the people of Thulamela.

A fair amount of monoliths were found in and around the main stone-walled complex of Thulamela, and some are of a type of stone not found in the vicinity of Thulamela. It is not clear what the function of the monoliths was, but modern people believe these upright stones may symbolise the ancestral spirits. It would also appear that the distribution of the monoliths around the main kraal indicates where senior members of the community lived. It is possible that the monoliths, as representatives

---

17   At least four sophisticated processes have to be followed to produce iron tools. Iron ore was reduced in a furnace to basic iron, which still contained many impurities in the form of slag. The furnace was then broken down and the iron removed. After that the iron was taken to the smith who heated, hammered and folded it to purify it further. The iron was heated for a third time to produce the unshaped tool. These processes ensured that the right amount of metal was forged into a rod shape for the planned tool. It also ensured that the grain of the hand-forged iron was orientated in the correct direction. The rod was then heated again in the smith's furnace so the smith could shape it into the planned tool.

2. A NEW ORDER

Gold necklaces, a bracelet and pieces of perforated gold foil which have been found in the royal graves at Thulamela.
(ACKNOWLEDGEMENT: SOUTH AFRICAN NATIONAL PARKS)

of the forefathers, had to provide earthly protection to the royals.

In summary: The interpretation of cultural objects is crucial in determining the lifestyle of the people of Thulamela. The grinding stones and hoes give an indication of the limited utilisation of maize and sorghum as food sources, while the monoliths and ceremonial spear reveal something of their spiritual and social behavioural codes. The whetstones back the deduction that metalworking took place in the main complex, while the harpoon heads indicate that these people were willing to hunt hippo and larger fish like catfish. Hippo bones in the midden also back up this deduction.

Thulamela covers about nine hectares, but cultural objects that point to a comfortable lifestyle only occur inside the main stone wall, an area of about half a hectare. It was a case of "give the emperor his due" – which in this instance was evidently everything.

The piece of Chinese ceramic and the double iron bell from West Africa found near the male grave are also of great value. Although these two objects do not represent meaningful international trade, it is nevertheless a sign that the leaders of Thulamela were aware that there is more to the world than what stretches to the horizon. Five hundred years ago, before the Portuguese arrived at the east coast of Africa, merchandise was being transported from these distant areas to Thulamela and back.

Miller suspected that the graves could produce valuable cultural artefacts that could aid the description of the Zimbabwe culture in southern Africa. However, the decision to open the graves was only taken after consultation with representatives of the local communities. Understandably this was not an easy decision for them, as these were the graves of their most revered ancestors and the interference could have unpleasant implications for the spiritual wellbeing of the descendants. After long deliberation, the representatives decided that opening the graves could be to the advantage of the communities and would enable the ancestors to serve and benefit the communities. Clear conditions were set for the handling and the fate of the remains. Respect had to be shown at all times, the research process finished as soon as possible and the remains reburied in the same place.

It was then decided that the forensic archaeological team of the University of Pretoria's medical school would open the graves. Researchers would collect as much information as possible and identify the bones at the terrain, which would limit laboratory examination to a minimum.

The first grave that was opened contained the remains of a woman who had probably died in her forties. In scientific terms she is known as UP 43. Because of the way her body had been arranged in her grave about 350 years ago, Miller and the rest of the team named her Losha, the Venda word for the respectful attitude a woman assumes when she greets a man who is her social senior. To do this, a woman would sit on her haunches, or bow forward and place her hands to the left of her chin.

Losha was 1,73 metres tall, which is taller than most women and men of her time. This opens the door to the possibility that she came from one of the northern African groups. Forensic investigations revealed that she had walked a lot, but never performed any heavy duties. Investigation of her teeth showed that she had had a very high fever at least twice in her lifetime. She was buried in a foetal position, as described above, in a shallow grave underneath a hut floor, with her head pointing to the north. Around her left wrist was a bracelet with 273 gold beads that may have been twisted into three strings. Around her left arm was a gold wire bracelet. No other grave objects were buried with her.

The second grave was that of a man who had died 550 years ago while in his sixties. His skeletal remains were packed into the grave in an orderly manner, with the long arm and leg bones parallel to each other in a northsouth direction. The hip bone was placed on the southern side of the grave against the long bones to form a U-shaped "bowl". The rest of the bones were placed in the "bowl" and the skull on top of the heap. The clean bones indicate that they were packed in the grave after decomposition had taken place. This person may have been buried in another grave and later moved to this grave. Otherwise, the placing of the bones corresponds with the Venda tradition that a chief is sometimes buried long after decomposition.

Forensic data revealed that he had reached a relatively advanced age for the time in which he lived. A leg injury would have caused him to limp, and he probably had a difficult temperament – a deduction made because of the serious osteoporosis in his upper neck vertebrae. A clear cut on the vertebra third from the bottom is similar to that a spear or assegai would have made. Taking his age and leg injury into consideration, it is unlikely that he sustained the back injury in the battlefield. One cultural tradition that could have played a role here was that a chief who could no longer fulfil his duties was ceremonially killed so a new leader could take his place.

Whatever the cause of his death, it is clear that those who buried him had the utmost respect for him, as 70 big gold beads, 16 gold bracelets and almost 1000 ostrich eggshell beads – possibly his own jewellery – had been placed among his remains.

So, after half a millennium in his grave, UP 44, or Nngwe to the team (the Venda word for leopard), again saw the light of day to assist with attempts to understand how the ancestors had lived in Africa. Normal people, just like us, with dreams and problems. In Nngwe's case, serious problems!

During the period that excavations of the graves and research on the remains took place, the management team had time to prepare for the reburial. At this stage the differences between the Tsonga-Shangaan and the Venda about their claims to Thulamela deepened. A senior Tsonga-Shangaan representative declared outright that UP 43 and UP 44 represented his ancestors. To clear up the matter, the management team decided to use a traditional nanga[18] to contact the forefathers so that they could provide a decisive answer.

A well-known female nanga was consulted and she prescribed a ritual that had to be followed to the letter. During sunrise one morning at Thulamela and next to Nngwe's grave, she consulted the forefathers by "throwing the bones" in the presence of all the representatives.

The message was the same three times: "Thulamela is the home of the forefathers of the Venda." The nanga was of Tsonga-Shangaan origin …

The dispute was finally settled and all that still had to be decided was who in Vendaland would be responsible for the reburial. Without any hesitation, the Venda representatives decided on Venda Chief Makahane and his direct family.

As was customary, the funeral consisted of two parts, namely the burial and a feast. The first part began with the presentation of the two containers with the bones. At the foot of Thulamela the researchers turned over their responsibility to the bearers, who were Chief Makahane's relatives. They ascended Thulamela Hill amid soft singing and the rumble of drums. Twice the procession stopped to rest, as tradition prescribed. On top of the hill the bones were removed from their containers and wrapped in cotton. Goat's blood was sprinkled over them and they were placed in children's coffins.

Princess Makahane took off some of her bracelets, placing them in the casket with Losha's bones. In the same manner the chief's stone[19] from Nngwe's grave was placed in the casket with him. This was done in the hope of soothing the dead because their grave objects had been removed.

Losha was then placed in her grave and covered with soil, after which sorghum seeds and maize were strewn over the grave and wet with water and sorghum beer.

The same process was followed with Nngwe, after which Chief Makahane performed the rhino dance to wish the departed a happy journey, as well as to enhance the power and lives of the remaining chiefs.

A couple hundred people attended the occasion, but only a few were allowed to partake in the funeral rituals. The main function of the other mourners was to have a proper feast – a duty that they fulfilled with great enthusiasm and conviction. The Tsonga-Shangaan were responsible for the preparation of the food, which was served in large pots at Pafuri picnic spot. Traditional dancing, in which most of the partygoers participated, followed the meal.

This funeral feast represented the conclusion of three years of work during which people from different cultural backgrounds got to know each other and during which ties of friendship were formed. Africa had revealed a small part of its secrets to us.

---

18 Traditional healer and prophet.
19 A small stone from the stomach of a crocodile that was swallowed by chiefs to confirm their association with the power and awe-inspiring dread of the crocodile.

2.4 The aftermath of the Mfecane in the Lowveld

An examination of the sacrificial altar in a cave near the Mutale mouth.
(01.08.1985)

uMzilikazi, the mightiest figure north of the Vaal River during the Difaqane, as depicted by hunter-explorer William Cornwallis-Harris, who had visited the Ndebele chieftain in 1836.
(ACKNOWLEDGEMENT: AFRICANA MUSEUM)

While the last phase in the history of the Vhalembethu was taking place in the northern Lowveld, the people of Masorini were still occupied with their iron production, and the tribes in what is now the southern part of the Park were planting crops and looking after their livestock, certain developments were taking place in Zululand that would soon affect the lives of all these people living in the Lowveld.

In the history of the Bantu-speaking peoples of South Africa, on occasion some areas would become overpopulated, leading to internal strife, tribes splitting up, the formation of new tribes and migrations from the overpopulated areas. Areas became densely populated after groups settled in fertile territory, where the population increased rapidly due to the favourable conditions. Their patrilocal habitation system and their way of life were so strongly linked to their environment that they would only leave an area under great pressure. An example of this can be seen in northwestern Transvaal[20] in the area located between Brits and Zeerust. As a result of overpopulation, many groups had already left the area during the eighteenth and nineteenth centuries and settled elsewhere as independent tribes, before uMzilikazi's arrival.

A development that had a more direct effect on the Park area and whose effects could be felt across southern Africa, took place in the area that currently consists of Mpumalanga, southern Swaziland and northern KwaZulu-Natal. At the end of the eighteenth century this area was inhabited by a number of small Nguni tribes, but because of an increase in population, the stronger ones started dominating the weaker ones. This led to the development of power blocs, each consisting of a number of tribes, who were locked in a struggle for domination. At that point Shaka, a member of a small Zulu tribe that was part of one of the large, powerful coalitions, appeared on the scene. After a bloody campaign he managed to subjugate all these coalitions – and in the process a large number of people were killed, while many groups fled, spreading devastation and chaos in their wake. This caused a chain reaction that would influence the lives of people over a large part of southern Africa.

This era of disruption and destruction is known as the Mfecane in the Nguni languages and the Difaqane in the Sotho languages. The destruction in the area north of the Tugela River had already started early in the nineteenth century. From there it spread, with the worst turmoil taking place from 1820 to 1835. Whole tribes were eradicated, while others had to seek shelter in inhospitable areas. Many people died of famine and cannibalism was a common occurrence. Leaders and followers who fled Shaka's tyranny invaded occupied areas and subjugated the local tribes so that communities consolidated to form new political units. As a result, new nations emerged out of the turmoil, namely the Swazi, Shangaan-Tsonga[21] and AmaNdebele, as well as the Ngoni and Kololo, who settled north of the Zambezi River. The disruption after 1824 in the interior of the former

---

20 The western part of the current Limpopo Province.
21 In certain sources (also dictionaries) spelled Shangana-Tsonga or Sjangaan-Tsonga.

Transvaal did not have a direct effect on the Lowveld tribes after the arrival of uMzilikazi. Although uMzilikazi's impis attacked the Pedi tribes in the mountainous parts of the eastern Transvaal (Mpumalanga), stole cattle and abducted women and children, they never went further east and their raids were mainly aimed at tribes further west. More refugees arrived in the Lowveld during this period, but most had fled the escarpment because of infighting among the Pedi tribes.

Zwangendaba's trek through Mozambique also does not appear to have affected the people living in the Lowveld. Zwangendaba was the chief of a lower-ranking Nguni-speaking tribe that left their territory in 1822 and fled northwards to escape Shaka's tyranny and extermination campaigns. He and his followers initially moved as quickly as possible and did not deviate much from their planned route to carry out raids. Later they moved in a northwesterly direction and invaded the Monomotapa Kingdom with its impressive stone structures. The high chiefs and chiefs were killed, their homes burned down, and some of their subjects were captured. This brought to an end the centuries-long existence of the Zimbabwe-Monomotapa empire. Zwangendaba's people eventually settled on the far side of Lake Malawi after a trek of thousands of kilometres.

Another Nguni leader who left his territory at more or less the same time as Zwangendaba, was Nxaba. Like Zwangendaba he and his followers also hurried through the southern part of Mozambique and then turned in a northwesterly direction to cross the Zambezi River. There Nxaba died during a fight against the Kololo of Sebetwane. This tribe, which had moved out of the current South Sotho area, left a trail of destruction up to Barotseland, where they eventually settled.

Two groups that had a meaningful effect on the Lowveld tribes were the Swazi and the Shangaan-Tsonga, and their origins can also be traced to the upheavals of the early nineteenth century and Shaka's reign of terror.

The history of the Swazi starts with the Ngwane, a Nguni group consisting of a number of different small tribes and living between the Pongola and Usutu rivers by 1800. Clashes with stronger tribes, such as the Ndwandwe under Chief Zwide Nqumayo, forced the Ngwane tribe under the leadership of Sobhuza I[22] to move northwards, and they eventually settled in the central part of the current Swaziland. Groups of Sotho-speaking people had already settled there, but they were subjected by the Ngwane of Sobhuza and incorporated into the Nguni-speaking nucleus group. Sobhuza allowed the subservient groups' chiefs to continue to rule their tribes and the tribes to live according to their customs, but during times of war they were conscripted into the Ngwane military organisation with its system of age-based regiments. Eventually it became a status symbol to be accepted into the Ngwane regiments and in this manner the Sotho-speaking people eventually adopted the Nguni language and culture.

After the death of Sobhuza, Mswati I, from whom the Swazi nation got their name, succeeded him as high chief. Mswati further developed the regimental system and built military kraals according to Shaka's design. To keep his regiments ready for action, he regularly sent them out on raids against other tribes. The Sotho tribes of the southeastern Transvaal (Mpumalanga) and the southern part of the current Kruger National Park area, the Baphalaborwa, the Pedi tribes on the escarpment, the Lobedu near Duiwelskloof and even the Venda in Soutpansberg were all subjected to Swazi raids. There were also several clashes with the Tsonga in Mozambique. The defenceless tribes living below the escarpment in the Lowveld were regularly raided to establish the rule of the Swazis and to collect tribute. Based on a description by an elderly informant, Colonel Stevenson-Hamilton wrote a striking account of an attack by a group of Swazis on a small kraal (a traditional African village of huts, usually enclosed by a fence) in the current Park area and the way in which this area was turned into a desolated track of land by about 1860. Mswati died in 1865, and he was succeeded 10 years later, in 1875, by Mbandzeni (Umbandine) after a series of intrigues and the murder of the legal heir, Ludvonga. By that time the Swazis were claiming all the inhabitants of the Lowveld south of the Olifants River as their subjects. However, the sparsely populated area's boundaries had not been formally determined and the Sotho

A sketch of Zulu King Shaka, circa 1824, by Lieutenant James King, a trader from Port Natal who had known Shaka.
(ACKNOWLEDGEMENT: *TRAVELS AND ADVENTURES*, SA LIBRARY)

---

22 Also called Somhlolo in certain sources.

groups still living in the area, as well as the Tsonga immigrants from Mozambique, remained more or less autonomous. The Baphalaborwa also retained their independence, but this was slowly eroded by battles with the Pedi and the Lobedu, attacks by cannibals and invasions by Swazi poachers. The raids carried out against them and other tribes disrupted both their iron production and their trade activities. The arrival of European traders and their cheap merchandise carried this disruption a step further.

The Shangaan-Tsonga, more specifically the Shangaan element, also came into being during the Mfecane/Difaqane of the nineteenth century. In the area north of the Tugela River there had been two tribal groups, the Ndwandwe under the leadership of Zwide and the Mthethwa under Shaka, who were locked in a constant struggle for supremacy. The Ndwandwe were defeated by the Mthethwa during a bloody battle and two military leaders, Zwangendaba and Sochangana[23], each fled northwards accompanied by a number of followers in 1821 to escape Shaka's revenge. One of Shaka's prime regiments under the leadership of Songandaba was sent after them, but instead of attacking them he formed an alliance with Sochangana. At the same time, two other tribes led by Nxaba also left the area and migrated northwards. Zwangendaba and Nxaba moved away as fast as they could, but Sochangana, also known as Manukosi or Manukuza, settled for a while along the Tembe River southeast of Swaziland in the Tsongas' territory. Although they were only a small group, thanks to their superior military abilities they nevertheless managed to subjugate the surrounding Tsonga groups and incorporated some of them. Among the Tsonga they were known as the Ngoni.

When Sochangana discovered his new subjects' military prowess, he used them in all his campaigns and pushed them in front of his Zulu impis, much as the Romans used their foreign auxiliary troops in front of their legions. Those on whom this task fell were called *amabuyandhlela* ("those who clean the way").

Sochangana and his followers, who by now had a

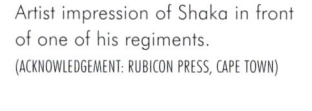

Artist impression of Shaka in front of one of his regiments.
(ACKNOWLEDGEMENT: RUBICON PRESS, CAPE TOWN)

---

23 In certain sources also spelled Soshangana or Soshangane.

Tsonga element, moved northwards from the Tembe River and settled in the Limpopo Valley at Bilini, where they stayed for almost five years. In time, the place where they had settled became known as Kachangana and Sochangana's followers as the Machangana. During the journey of Sochangana and his followers through Swaziland to the north, King Sobhuza I of the Swazi wisely left them in peace. However, Kongwane, the high chief of the Ngomane, marched against them and Sochangana's superior Zulu warriors crushed his army at Sikwameni along the Komati River. Kongwane and his surviving followers fled northwards and settled along the western foothills of the Lebombo Mountains in the present Kruger National Park. The Kongwane, a tributary of the Nwaswitsontso River between Ka-Nwamuriwane Hill and Tshokwane, carries the name of this hot-headed chief of the Ngomane to this day.

In 1828 Shaka sent out his army one last time under the leadership of his faithful lieutenant Naba, to destroy the Ndwandwe who had attempted to flee his rule. Although it seems no real battle had taken place, Shaka was satisfied that Sochangana would not threaten his supremacy any more, as he had moved northwards with his followers. However, this campaign spelt disaster for Shaka, as Shaka's half brothers Dingane and Mhlangana hatched a plot against him before Naba's return. With the help of his induna, Mbopha, they treacherously murdered him on 24 September 1828.

Because Sochangana and his followers were now out of Shaka's reach, he could continue to expand his rule without interference in the area between Delagoa Bay (later Lourenço Marques, still later Maputo) and the Great Sabie River in the north by subjecting the indigenous Tsonga tribes. Still fearing possible attacks by the Zulus, Sochangana left the Limpopo Valley in about 1830 and moved further northwards in the direction of the Great Sabie River and from there to the Zambezi River. Here he clashed with Zwangendaba and Nxaba, the two other refugee Zulu chiefs. After two encounters in which Sochangana triumphed, he temporarily settled close to the Sabie River in eastern Zimbabwe. In 1831 he subjugated the surrounding Shona tribes and conscripted their young men into his army, as he had done with the Tsonga.

From his stronghold next to the Sabie River Sochangana sent out expeditions in all directions. With his main adversaries out of his reach, he was soon able to consolidate the conquered areas between the Zambezi River and Delagoa Bay and in this manner he managed to build an empire that became known as Gaza empire, after his ancestor Gaza.

Sochangana lost many of his warriors as a result of a smallpox epidemic in the mid-1830s. After this, he and his followers moved south and in 1835 they settled in the Limpopo Valley north of its confluence with the Olifants River. Here he and his warriors made contact with Lang Hans van Rensburg's Voortrekker party in July 1836, when they murdered all the whites except for two young children. It was during this southward trek of Sochangana that the first group of Tsonga fled from Mozambique into the Transvaal, apparently out of fear because they had refused Sochangana's request for assistance in his campaign against Zwangendaba and Nxaba in the north. One group of Tsonga refugees settled in the territory of the Bakgaga, a Sotho tribe who lived near the present-day Tzaneen, and another in the territory of the Lobedu of Modjadji. A third group settled in Swaziland, but later also moved into the Lobedu territory.

As was the case with Dingane in Natal and later leaders in Mozambique, Sochangana feared white people and never missed an opportunity to attack them or drive them from his territory.

An attack by Sochangana's warriors on an expedition of Captain William Owen in the vicinity of Delagoa Bay was thwarted, but in October 1833 they attacked the small Portuguese settlement and savagely killed every man, woman and child. Only the young João Albasini escaped the massacre. The clever manner in which he accomplished this is described in chapter 5. In November 1834 Sochangana attacked the Portuguese settlement at Inhambane as well and destroyed it. In 1836 the garrison at Sofala was eliminated in the same way. He also repeatedly attacked the Portuguese forts at Sena and Tete, but with little success.

After this Sochangana did not conduct any major wars and the last years of his reign were typified by the stabilisation of his rule, as he sent his regiments annually as far as the Zambezi River to collect taxes. After his death in 1858 the Gaza empire began to crumble.

Sochangana was succeeded by his son Mawewe, who moved his headquarters to Biyene near Delagoa Bay. Because of his cruel nature he was not very popular among his people. His brother Muzila (Mzila),[24] who had fled to the Transvaal with his followers and found refuge with João Albasini, returned after two years. His return resulted in a long battle for supremacy. Mawewe obtained help from the Swazis, while Muzila was assisted by the Portuguese.

The six-year civil war in Gazaland (Mozambique) came to an end in 1862 after a huge battle along the Lower Sabie River during which Mawewe was soundly defeated by Muzila. The Baronga, whom the Portuguese had supplied with 3000 guns, assisted Muzila during this battle. Mawewe died in about 1886. The descendants of his followers, the amaNdawo, are still living near the Komati and Crocodile rivers.

In about 1876 a deputy chief of the Ngomane, one Ndlabu, whose kraal was situated in the vicinity of the Sabie River, revolted against the rule of high chief Ntuyi of the Ngomane. He fled to Gazaland where he obtained the help of Chief Magude, a deputy chief and subject of high chief Muzila of the Shangaan nation. With Magude's help he defeated Ntuyi in a battle near the present Tenbosch. Ntuyi narrowly escaped with his life, but his son Hoyi was captured and carried off to the land of Muzila. He was held here for a year and then sent back to his people after the death of Muzila. They are still living in the vicinity of Tenbosch along the southern border of the Kruger National Park.

During the years of bloodshed in Mozambique, many Tsonga, including most of the Nwalungu, Loyi and Nkuna families, moved westwards across the Lebombo Mountains and joined the earlier refugees in the Transvaal.[25] An important figure then, who also had a great influence on the history of the Tsonga, was the Portuguese trader João Albasini already mentioned. Even during his stay in Delagoa Bay he provided protection to many Tsonga fugitives. In this way he gained a large following that he built up into a strong military force. His followers recognised him as high chief and gave him the name Juwawa. He later moved to the Transvaal[26] and eventually settled at Goedewensch south of the Soutpansberg. Most of his Tsonga supporters followed him and in later years they became known among the Sotho as the Makwapa.[27] Among the whites they were known as the "knopneuse" (knob noses). During the succession war between Muzila and Mawewe, the stream of refugees to the Transvaal[28] augmented Albasini's ranks and this further enhanced his status as white high chief of the Magwamba.

Muzila maintained good relations with high chief uMzilikazi of the AmaNdebele, but became involved in border disputes with the Swazi. He also clashed with the Portuguese, who were expanding their control over the Gaza area.

These confrontations were continued by Gungunyane,[29] who succeeded Muzila in 1885, but by that time the Gaza state had already disintegrated to such an extent that some tribes began to conclude their own treaties with the Portuguese.

Between 1894 and 1896 the last big battle between Gungunyane and the Portuguese colonial government took place. The Portuguese, who had been the target of the Gazaland high chief's raids for some time, were looking for an excuse to finally destroy his authority and found it in the form of a dispute between two local Tsonga chiefs. The Portuguese supported the claims of deputy chief Umbvesha, while Gungunyane sided with the feudal superior Mahazulu.

At first Mahazulu criss-crossed the Portuguese province of Lourenço Marques and his warriors invaded the outer reaches of the harbour city. From here they were apparently driven back by cannon fire from a British battleship in the bay.

Reinforcements from Portugal were soon on the spoor of the attackers. At Marakwene along the Komati River, Mahazulu's warriors attacked the Portuguese force at daybreak. The Portuguese soldiers were nearly overwhelmed and suffered heavy losses, but managed to resist the attack.

Mahazulu's army retreated northwards to Gaza-

---

24 Spelled Umzila in certain sources.
25 Mpumalanga & Limpopo provinces.
26 Limpopo.
27 Also known as Magwamba.
28 Limpopo.
29 Also known as Gungunyane or Dungunyana.

Artist impression of a ceremony at Tsonga High Chief Gungunyane's village along the Limpopo in Mozambique. Circa 1890.
(ACKNOWLEDGEMENT: TRANSVAAL ARCHIVES)

land, but was pursued by the Portuguese force under the leadership of Captain Frere D'Andrade. Another battle took place at Magule. During this altercation several of Gungunyane's Shangaan regiments were present, but they did not participate in the attack on the whites. Mahazulu's army was defeated once again and had to flee even further north.

After this victory, the Portuguese insisted that Gungunyane should hand over the rebels and when he refused, they prepared an attack on his main kraal. In the meantime, Gungunyane's Shangaan regiments, who had not been overly inspired by their Nguni potentate, absconded. This reduced Gungunyane's army by about 30 000 men and all he had left with which to oppose the Portuguese were his personal bodyguard consisting of Ngoni and a number of faithful Shangaan auxiliaries. They fought bravely against the Portuguese force, but were comprehensively defeated. The Portuguese captured the royal kraal and burned it to the ground.

Gungunyane fled to Tshayimiti, but was attacked again a week or two later by Captain d'Albuquerque and captured. He was first brought to Lourenço Marques (Maputo) and later banned to the Azores, where he died in exile in 1897.

The refugees who had arrived in the Transvaal[30] during the ongoing hostilities between Mawewe and Muzila were all Tsonga. There were no Shangaan among them and that is why descendants of these refugee groups still prefer to be known as Tsonga. They maintain that they are of pure Tsonga extraction, while the later arrivals call themselves Shangaan (Changana) as they are descendants of Sochangana and his Nguni followers, although some of them had close Tsonga relations.

At the end of the nineteenth century and even later the tribes who had taken up arms against Portuguese rule in Mozambique had all fled across the Lebombo Mountains and joined settled groups. Through this, the amalgamation of the Shangaan and Tsonga was carried a step further. The Tsonga did not move to the Transvaal[31] as tribes but only as small groups, which then combined to form new units. They settled mainly in the Gazankulu[32] area and in parts of the current Kruger National Park (compare with the map on page 63).

Soniya, senior wife of the late High Chief Gungunyane of the Ama-Changana, during a visit to Sabie Bridge (Skukuza) in February 1921.
(ACKNOWLEDGEMENT: STEVENSON-HAMILTON COLLECTION, SKUKUZA ARCHIVES)

---

30 The current Mpumalanga and Limpopo provinces.
31 Mpumalanga and Limpopo.
32 The current Mpumalanga.

2. A NEW ORDER

A third tribal group that influenced the history of the Lowveld, although to a lesser extent than the Swazi and Shangaan-Tsonga, was the Pedi living on the Drakensberg escarpment, specifically in Sekhukuneland[33] and vicinity. The Pedi comprises a few tribes who settled here at different times since the beginning of the seventeenth century. One of those tribes was the Rota, who eventually changed their name to Pedi. Eventually this name was used for a whole group of tribes, which each still kept its own tribe name. These tribes eventually developed into an influential, powerful unit through the subjugation of the weak, intermarriage between different chiefs' families and other factors.

The Pedi rule reached its zenith during Thulare's reign at the end of the eighteenth century and the beginning of the nineteenth century. They even established trade relations with the Portuguese in Delagoa Bay[34]. It does not, however, appear as if they were interested in expanding their influence over the distant Lowveld.

After Thulare's death, in the midst of a succession row, uMzilikazi's warriors arrived, captured a number of tribesmen and enslaved them. During this battle and the following confusion, only one of Thulare's sons, Sekwati, managed to survive to follow in his father's footsteps. It was during this time, around 1830, that people were forced by famine to practise

Ruins of a former Pedi settlement at Shilowa Poort of which the inhabitants had been massacred during the 1880s by Tsonga High Chief Gungunyane's warriors.
(OCTOBER 1983)

---

33 Sections of Mpumalanga and Limpopo provinces.
34 Later Lourenço Marques, now Maputo.

cannibalism. Sekwati was succeeded by his son Sekhukhune, after whom the Boers named Sekhukhuneland. His reign had been one of strife and conflict and led to some groups of refugees leaving the area again. However, this did not have much of an effect on the Lowveld tribes.

During Sekhukhune's reign, the Pedi were continuously in conflict with their arch-enemies the Swazi and bloody battles took place on an ongoing basis between Swazi warriors on the attack and local Pedi tribes. In 1864 the Swazi were crushingly defeated during the battle of Maholoholo ("the big battle"). The Pedi army was commanded by Mahlala and two Mapulana chiefs who had joined him – Maripi Moshile (after whom Mariepskop was named) and Chilwane. Trouble with the "Zuid-Afrikaansche Republiek" (South African Republic) during the 1870s, as well as with the British government after the annexation of the Transvaal in 1877, eventually led to an attack on Sekhukhune's stronghold by a couple of British regiments, assisted by a large group of Swazi warriors and a group of Pedi under command of Chief Mampuru. Sekhukhune was defeated and this led to a further decrease in the influence of the Pedi, who had already been divided, although they still manage to maintain a certain degree of unity.

Taking into account the settlement of agriculturalists, pastoralists and metal workers in the area that is now the Kruger National Park, it is clear that it never harboured as dense a human population as the surrounding areas. This can only be ascribed to the relatively unfavourable climate and diseases that affected both humans and their animals. Even so, the archaeological survey of human habitation patterns revealed a large number of sites that had been inhabited during the eighteenth and nineteenth centuries, in some cases for short periods only. Several parts of the Park, namely the Luvuvhu, Phalaborwa and Crocodile River areas and the area slightly north of the Crocodile River, were relatively densely populated during the periods mentioned, while small groups lived in scattered communities near rivers in the rest of the area.

Disturbances during the nineteenth century affected in particular the tribes in the south and those of Phalaborwa, and the rest of the small groups living in this part of the Lowveld were also affected by

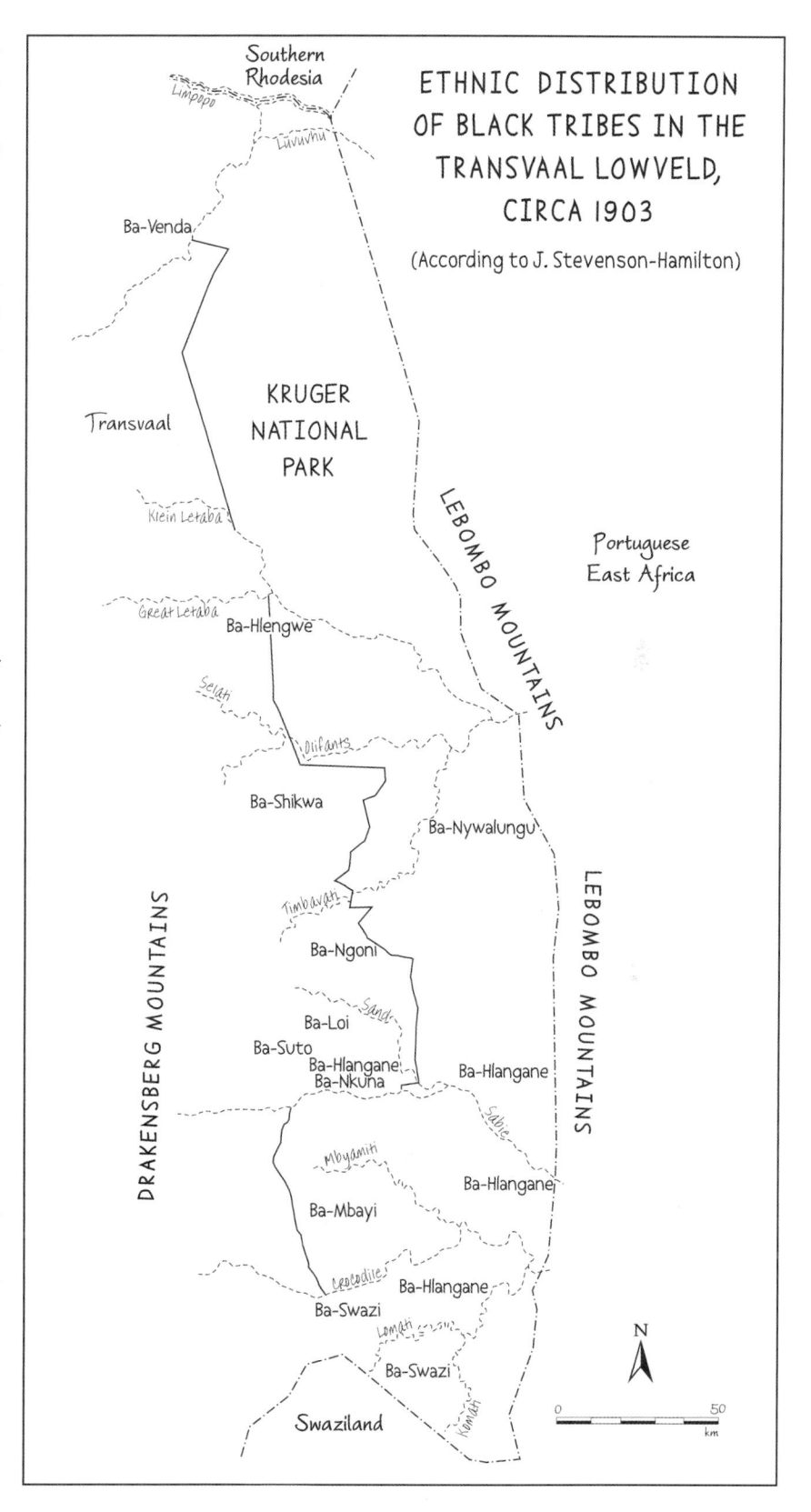

Ethnic distribution of black tribes in the Transvaal Lowveld, circa 1903, according to J. Stevenson-Hamilton.

2. A NEW ORDER

the tidal wave of the Mfecane/Difaqane. The arrivals of refugee groups from Mozambique, which continued until the end of the nineteenth century, also had a marked influence on habitation patterns in the Kruger National Park area.

The last great event in the cultural history of the Kruger National Park was the arrival of the Voortrekkers, traders, transport drivers, hunters, prospectors and fortune seekers during the nineteenth century. When the hunter-gatherers living in the area during the first millennium came into contact with the agriculturalists and metal workers, their way of life did not change dramatically. They did not accept the ways of the immigrants and carried on with their traditional way of life. On the other hand, the immigrants' technology eventually led to the disappearance of the Stone Age people. With the arrival of the whites, foreign elements were introduced in the way of life of the Bantu-speaking tribes. In contrast with their predecessors of the Stone Age, they accepted the new elements into their culture. By doing this, their traditional way of life changed and became more integrated with the new economy and technology. The men went to work in towns and on farms and this changed the self-supporting economy and the social system that had been unique to them.

After the end of the Anglo-Boer War and the re-proclamation of the Sabie Reserve (predecessor of the Kruger National Park) there was a notable black population in the Kruger National Park area consisting mainly of the Ngomane tribe and their Tsonga-Shangaan fellow tribesmen. The latter had emigrated from Mozambique into the Transvaal[35] as a result of the unrest in their traditional territory. In the far-

The aged Chief Chokwane (Tshokwane) (left) and his brother in 1909. The Tshokwane ranger section and picnic spot were named after him.
(ACKNOWLEDGEMENT: STEVENSON-HAMILTON COLLECTION, SKUKUZA ARCHIVES)

---

35 Sections of the current Mpumalanga and Limpopo provinces.

northern parts of the Shingwedzi Reserve, there were also small groups of Venda, remains of the Vhalembethu and Tsonga-speaking Magwamba of chiefs Mhinga, Makuleke, Khubyane, Dzombo, Makhuba and others. Further south the inhabitants were mainly Tsonga-speaking, except for the Sotho-speaking Baphalaborwa in the vicinity of Masorini, Shikumbu and Phalaborwa. The Swazi-speaking Ngomane lived in the southeastern part of the old Sabie Reserve and the ba-Mbhayi (Bapai), who spoke both Swazi and Sotho, lived in the southwestern (Nsikazi) area of the reserve (compare map no. 17).[36] One of Major Stevenson-Hamilton's first tasks after he had been appointed as warden of the Sabie Reserve was to resettle the black people who were not needed for active duty or the management of the reserve, in adjacent areas.

It was this undertaking of Major Stevenson-Hamilton that led to the nickname "Skukuza" given to him by the local blacks, meaning "he who sweeps clean". His headquarters at Sabie Bridge eventually also became known by this name and is now the administrative centre of the Kruger National Park. Most of the Tsonga-Shangaan inhabitants of the Sabie Reserve at the time settled to the west at Acornhoek, Welverdiend and Mhala in what are now the Mpumalanga and Limpopo provinces. The Ngomane settled mainly near Tenbosch and in the southeastern part of the former KaNgwane[37] "homeland" (west and east of Komatipoort).

The ba-Mbhayi settled mainly in the Nsikazi region of KaNgwane[38] and as far west as Waterval-Onder and Airlie. Another settlement area of the Mbhayi-Pedi group was the Bushbuckridge-Mariepskop area of the Drakensberg escarpment.

The events of the past few hundred thousand years related so far, represent only the beginning of a history of a single area where so much has happened that it could fill volume upon volume. The subsequent course the area's history took will be related in the chapters that follow.

References

Bergh, J.S. & Bergh, A.P. 1984. *Stamme en ryke*. Don Nelson, Cape Town.

De Vaal, J.B. 1952. Ysterbewerking in Vendaland. *Yskor Nuus* July:591–596.

De Vaal, J.B. 1975. 'n Voorgeskiedenis. In: Barnard, C. (ed.). *Die Transvaalse Laeveld – kamee van 'n kontrei*. Tafelberg, Cape Town.

De Vaal, J.B. 1984. Ou handelsroetes en wapaaie in Oos- en Noord-Transvaal. *Contree* 16:3–15.

De Vaal, J.B. 1985. Handel langs die vroegste roetes. *Contree* 17:3–14.

Du Buisson, L. 1987. *The White Man Cometh*. Jonathan Ball, Johannesburg.

Du Toit, A.P. 1967. Historiese oorsig van die Phalaborwa van die Laeveld. *Publications of the University of the North*, series A, no.7.

Edgecombe, R. 1986. The Mfecane or Difaqane. In: Cameron, T. & Spies, S.B. (eds). *An Illustrated History of South Africa*. Jonathan Ball, Johannesburg.

Eloff, J.F. 1979. Voorhistoriese mense in die Krugerwildtuin. *Custos* 8(12):19–26.

Eloff, J.F. 1982. Masorini: Geskiedenis kry gestalte. *Custos* 11(2):20–27.

Eloff, J.F. & De Vaal, J.B. 1965. Makahane. *Koedoe* 8:68–74.

Eloff, J.F. & De Vaal, J.B. 1974. Die geheime van Makahane. *Custos* 3(6):21–28.

Evers, T.M. 1974. *Three Iron Age industrial sites in the Eastern Transvaal Lowveld*. M.A. thesis, University of the Witwatersrand, Johannesburg.

Evers, T.M. 1981. The Iron Age in eastern Transvaal. In: Voight, E.A. (ed.). *Guide to Archaeological Sites in the Northern and Eastern Transvaal*. Transvaal Museum, Pretoria.

Ferreira, O.J.O. 2002. *Montanha in Zoutpansberg. 'n Portugese handelsending van Inhambane se besoek aan Schoemansdal, 1855–1856*. Protea Book House, Pretoria.

Flygare, J. 1899. *De Zoutpansbergen en de Bawenda Natie*. De Volkstem Drukkerij, Pretoria.

Fouché, L. 1937. *Mapungubwe. Ancient Bantu Civilization on the Limpopo*. 2 vols. Cambridge University Press, Cambridge.

Grewar, J.F. 1980. *The Vanished Saddleback – The Story of Phalaborwa*. Purnell, Johannesburg.

Gründler, W. 1891. *Geschichte der Bawendamission in Nord-Transvaal*. W. Gründler, Berlin.

---

36 See map on page 63 titled: Terrains of black kraals in the Kruger National Park, 1905–1906.

37 Mpumalanga.

38 Mpumalanga.

Hanisch, E.O.M. 1980. *An archaeological interpretation of certain Iron Age sites in the Limpopo/Shashi Valley.* M.A. thesis, University of Pretoria, Pretoria.

Hartman, J.B. 1972. *Die politieke en judisiële organisasie van die suidelike Changana (Bosbokrand) in die lig van hulle herkoms.* M.A. thesis, University of Pretoria, Pretoria.

Jäckel, M. 1949. *Juwwawa.* Verlag Friedrich Reinhardt A.G., Basel.

Junod, H.A. 1927. *The Life of a South African Tribe.* Macmillan, London.

Kriel, J.D. & Hartman, J.B. 1991. *Gazankulu en sy mense.* Kriel & Hartman, Pretoria.

Kriel, J.D. & Hartman, J.B. 1991. *Khindlimukani Vatsonga: The Cultural Heritage and Development of the Shangan-Tsonga.* Promedia, Pretoria.

Küsel, U.A. 1979. *'n Argeologiese studie van vroeë ystersmelting in Transvaal.* M.A. thesis, University of Pretoria, Pretoria.

Liesegang, G. 1972. New light on Venda traditions. Mahumane's account of 1730. *History in Africa* 4:163–181.

Maggs, T.M. O'C. 1984. The Iron Age south of the Zambezi. In: Klein, R.G. (ed.). *Southern African Prehistory and Paleoenvironments.* A.A. Balkema, Rotterdam.

Maggs, T.M. O'C. 1986. The early history of the Black people in Southern Africa. In: Cameron, T. & Spies, S.B. (eds). *An Illustrated History of South Africa.* Jonathan Ball, Johannesburg.

Martins, F. 1957. *João Albasini ea Colónia de St. Luis.* Agência Geral do Ultramar. Divisão de Publicações e Biblioteca, Lisbon.

Mason, R. 1987. *Origins of the African People of the Johannesburg Area.* Blackshaws Pty. Ltd., Johannesburg.

Matsebula, J.S.M. 1972. *A History of Swaziland.* Longman Southern Africa, Cape Town.

Maylam, P. 1986. *A History of the African People of South Africa: From the Early Iron Age to the 1970s.* David Phillip, Cape Town.

Meyer, A. 1986. *'n Kultuurhistoriese interpretasie van die Ystertydperk in die Nasionale Krugerwildtuin.* D.Phil. thesis, University of Pretoria, Pretoria.

Mönnig, H.O. 1967. *The Pedi.* Van Schaik, Pretoria.

Myburgh, A.C. 1949. The tribes of the Barberton district. *Ethnological Publications* 25. Department of Native Affairs, Union of South Africa.

Omer-Cooper, J.D. 1966. *The Zulu Aftermath.* Longman, London.

Omer-Cooper, J.D. 1987. *History of southern Africa.* James Currey, London.

Plug, I. 1983. *Man, animals and subsistence patterns in the Kruger National Park.* Paper read at a South African Archaeological Society meeting.

Plug, I. 1984. Man and animals in the Kruger National Park. *BAR International Series* 207:228–235.

Plug, I. 1988. *Hunters and herders: an archaeozoological interpretation of prehistoric communities in the Kruger National Park.* D.Phil. thesis, University of Pretoria, Pretoria.

Smit, J. 1996. Op die spoor van vergete geheime. *Custos* November:8–12.

Stayt, H.A. 1968. *The Bavenda.* Oxford University Press, London.

Stubbs, E. 1912. *Historical sketch of the Bawenda.* Manuscript, Parliamentary Library.

Summers, R. 1969. Ancient Mining in Rhodesia and Adjacent Areas. *Museum Memoir* Number 3, National Museums of Rhodesia, Salisbury.

Summers, R. 1971. *Ancient Ruins and Vanished Civilisations of Southern Africa.* T.V. Bulpin, Cape Town.

Thompson, L.C. 1938. The Mu-Tsuku. *South African Journal of Science* 35:396–398.

Thompson, L.C. 1942. The Balemba of Southern Rhodesia. *Native Affairs Annual (Rhodesia)* 19:76–86.

Toscano, F. & Quintinho, J. 1930. *A derrocada do império Vátua e Mousinho de Albuquerque.* Vol. 1&2. Casa Editora Nunes de Carvalho, Lisbon.

Van der Merwe, N.J. & Scully, R.T.K. 1971. The Phalaborwa story: archaeological and ethnographic investigation of a SA Iron Age group. *World Archaeology* 3:178–196.

Van Warmelo, N.J. 1935. A preliminary survey of the Bantu tribes of South Africa. *Ethnological Publication* 5. Government Printer, Pretoria.

Verhoef, J. 1982. *Die oprigting en ingebruikstelling van die Masorini terrein-museum in die NKW.* Thesis submitted for the postgraduate diploma in Museum Science, University of Pretoria.

Verhoef, J. 1993. Terugkeer na Makahane. *Custos* 21(12):28–38.

Verhoef, J. 1994. Krugerwildtuin – die eerste miljoen jaar. *Custos* 22(7):12–18.

Verhoef, J. 1994. Thulamela – 'n proeflopie vir deelname oor kultuurgrense heen. *Custos* 22(7):12–18.

Verhoef, J. & Küsel, M. 1992. Thulamela se goudsmelters. *Custos* 21(5):38–41.

Voigt, E.A. & Plug, I. 1981. *Early Iron Age Herders of the Limpopo Valley.* Transvaal Museum, Pretoria.

Webb, H.S. 1954. The native inhabitants of the southern Lowveld. In: H.S. Webb (ed.). *A Survey of the Resources and Development of the Southern Region of the Eastern Transvaal Lowveld.* Publication of the Lowveld Regional Development Association.

Ziervogel, H.S. 1954. *The Eastern Sotho.* Van Schaik, Pretoria.

2.5 Ancient trade routes through the Lowveld
Dr J.B. de Vaal

When Dr Willem Punt started researching the routes that had been followed by the leading Voortrekkers, Johannes van Rensburg and Louis Trichardt, in the Soutpansberg and the Lowveld, he first studied the trade routes between the ports on the east coast and the interior. It was obvious that the Trekkers with their ox-wagons, in Trichardt's case accompanied by 3000 sheep and 500 cattle, would have followed the most easily traversable routes with adequate, well-distributed watering holes for their livestock. It is therefore not surprising that these routes coincided with the old trade routes.

B.H. Dicke, who had spent most of his life trading in the Lowveld with his farm Modderfontein near Duiwelskloof as base, on his travels collected information from black people not only about the ancient routes, but also the ways in which trading took place. His knowledge of the then Northern Transvaal[39] and the black people there was of the utmost value to Punt. In 1931 Dicke indicated the routes known to him on grid maps, or on the farm maps of 1929. Punt in turn drew maps of the routes relevant to his research.

With the aid of the information gathered by Punt and several nineteenth century maps and information supplied by Dr U. de V. Pienaar, it was possible to indicate the main routes from Delagoa Bay and Inhambane to the interior and from Messina (Musina) to the Rooiberg tin mines, as well as some of the less important routes, on a topographic map of 1936 and to copy it onto tracing-paper.

One of the pioneers in the Lowveld and Northern Transvaal[40] who had used the trade routes, was the Portuguese trader and elephant hunter João Albasini. After 1845 he lived consecutively at Magashulaskraal near the upper reaches of the Sabie River, at Ohrigstad, Lydenburg and Schoemansdal and, from 1857 until his death in 1888, on the farm Goedewensch in the Kleinspelonken. He had a shop at each of these places. Great caravans of black bearers carrying his trade goods on their heads travelled the centuries-old trade pathways.

In 1927 my father bought part of Albasini's former farm Goedewensch (then Goedehoop 362) from his grandson João, who was married to my mother's younger sister. My uncle and aunt built a thatched-roof house close to the remains of the first João's fortress-like homestead. My family lived only about three kilometres east of them across the Luvuvhu River, which was both the western and northern boundary of their farm. The two families used to visit each other on foot. The footpath we used was dilapidated, but clearly a wagon trail, cut out in an easterly direction from the former fort, along the northern bank of the Luvuvhu and through the river at a shallow between two hippo pools full of crocodiles. The road passed almost through our yard, then led further east past Mashao Hill. Kamanjan, an old Shangaan-Tsonga who lived on our farm, was in his youth a subject of the first João, as were his parents. He told me that Albasini had constructed the road in order to travel to Delagoa Bay and the hunting fields. A photograph I took in 1935 clearly shows the road along the riverbank, but it was badly damaged during the construction of the Albasini Dam in 1948.

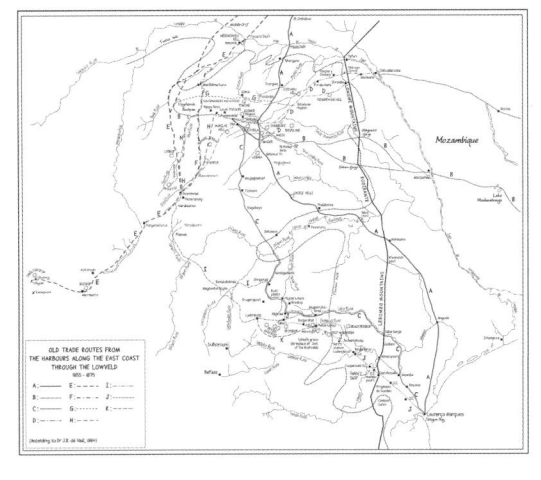

Old trade routes from the east coast harbours through the Lowveld. There is an enlarged version of this map in the back of the book.
(A COPY OF THE ENLARGED MAP IN ENGLISH IS IN THE MAPCASE)

---

39   Limpopo Province.
40   Limpopo and Mpumalanga provinces.

2. A NEW ORDER

It was the historical environment in which I grew up, that motivated me to conduct research on Albasini at the Transvaal Archives during my high school and university years. In this manner my interest in the old trade routes was stimulated. This interest was further piqued by three articles by Punt, published in *Die Volkstem* in 1939, on the topic of the Van Rensburg Voortrekker party's murder. One of the maps with the caption: "Trade routes between Northeastern Transvaal and Inhambane-Delagoa Bay during the time of Trichardt", was a great aid. Punt drew the map based on the information he had obtained from Dicke.

Although it had been stipulated on the map that copyright was reserved, Punt was a good friend and before his death he gave me permission to copy it if need be. Although I repeatedly traversed most of the areas through which the old routes went and in this manner got to know them well, I never followed them on foot like Punt had done. Accompanied by field officials Thys Mostert, Mike English and Gert Erasmus of the Kruger National Park, I was able to visit some of the few remains of the forgotten routes, as well as graves and overnight sites along them. During our holidays, my wife and I, armed with maps, travelled along the old routes both within and outside the Park. My brother, Pieter de Vaal of Louis Trichardt (Makhado), and I also investigated the routes going past Tshivhulane, Mashao, Magor and Rivolla hills.

It is clear that the early traders had made a thorough study of the topography of the areas through which they had to travel. As a result the trade routes usually followed river beds, where water had already cleared the way in a sense, while water for daily use was within reach at the same time. Having to carry water containers would lead to a situation where less trade goods could be transported.

They crossed mountainous areas by using passes and poorts. Elephant paths were also easy routes to follow for the earliest traders from the east coast with their bearers, who had to carry heavy bundles on their heads. According to Pienaar, elephants are the best civil engineers in the animal kingdom, as their footpaths never exceed a gradient of one to three. The trading expeditions were normally undertaken during the dry season when the large rivers were fordable, the climate bearable and the incidence of malaria low.

Another factor that determined the routes to the interior was the locations of the mines where the minerals needed were mined, smelted, cast and worked. For this reason, Phalaborwa and Musina (Messina) for their copper, Tshimbubfe for its iron, Zimbabwe for its gold and Rooiberg for its tin, were all connected to trade routes.

Route A: From Delagoa Bay to Mashonaland

According to a Portuguese map of 1893, there was a trade route that ran from Delagoa Bay[41] northwards and crossed the Komati River at Magude directly east of its confluence with the Nwanedzi River. This path followed the eastern bank of the Nwanedzi to where the river turns westwards in the direction of Nwanedzi Poort, where it flows through the Lebombo Mountains. However, the route carried on northwards and crossed the mountain range, which forms a plateau, at three different places, namely at Mahasane (an Arabic name), Albinduane and Lake Cunque. According to Reverend H. Berthoud's map of the

The old trade route from the Soutpansberg to the east coast ports of Sofala and Inhambane is still clearly visible with a large isiVivan to its right just north of Malonga Fountain.
(01.05.1986)

---

41 Later Lourenço Marques, now Maputo.

Soutpansberg, the route passed east of Pamahoma kraal, via the old farm Voetpad, across the Olifants River and past the copper mine at Phalaborwa in the direction of Witkop or Tsane Hill. The route between the Olifants River and the Lebombo Mountains is indicated on Dr H. Raddatz's nineteenth century map as the "Main footpath to Delagoa Bay". It is also clearly indicated on G.R. von Wielligh's 1890 map of the border between the Zuid-Afrikaansche Republiek (South African Republic, ZAR) and Portuguese East Africa. From Tsane Hill the footpath went northwards past Kasteelkoppies and then, according to Dicke, crossed the Great Letaba River at Lower Letaba Drift, from where it went over Modjadji's Neck in Bolobedo Mountain east of Observation Hill or Shirulurulu. Dicke (1938) also noted the following about this part of the route:

> Some eighteen miles north of the range, a solitary conical mountain draws immediate attention. Its shape reminds one of Vesuvius or Fujiyama ... That cone-like mountain is Levanga (also called Lebanga, Rehanga, or Levaka, according to the different Bantu languages and dialects of the tribes living throughout).
>
> Past Levanga, skirting the Wolowedo range on its eastern point, led the great communication road, the inland short cut between Delagoa Bay and Southern Rhodesia (Mashona and Manicaland).[42]
>
> Along this road, only a native footpath, of course, where men travelled in single file, the Magwamba traders with their bearer columns came from Delagoa Bay to compete with the traders from the mouth of the Limpopo, from Inhambane, and from Sofala. For centuries the native traders of the Thonga tribes, to which the Magwamba belong, had been using this route from Delagoa Bay through the Northern Transvaal lowveld before the inland regions through which they passed had been inhabited.

According to Dicke, the most important trading post of the Magwamba pedlars from Delagoa Bay was situated on the farm Bellevue 75 opposite Levanga Hill.

The footpath led northwards, west of Ha-Magoro Hill, across the farm Schuynshoogte with Tshimbupfe or Ysterberg in the east. It then led through the Luvuvhu River on the farm Weltevreden east of Valdézia and from there via the site of the later mission Beuster, the Phiphidi Falls and Tshivhase, through Thengwe's territory, past Tshongane and through the Limpopo at Malala Drift to Mashonaland and Zimbabwe. Reverend A. Merensky also indicated this footpath on his 1875 map of the Transvaal as a "hunting road", while Berthoud referred to it on his map from Beuster northwards as "a wagon trail to Fort Victoria", therefore also to the Zimbabwe Ruins. Carl Mauch also followed this route in 1871 on his way to Zimbabwe. Mauch spent 28 days with João Albasini on his farm Goedewensch to prepare for his journey on foot and Albasini provided him with Magwamba bearers who had probably been at the Ruins before. He reached the Zimbabwe Ruins on 5 September 1871, after which he made their existence universally known.

During his fateful journey from the Soutpansberg to Delagoa Bay, Louis Trichardt crossed the Lebombo Mountains on 13 March 1838 through the poort of the Little Nwanedzi River (or Mbhatsi). He crossed the Nwanedzi and then followed the old trade route to Delagoa Bay. Albasini described the Delagoa Bay-Mashonaland route to Goedewensch as follows (loosely translated):

> From the fort to a place named Maxuva: from there to Mohamba (Moamba), from there to Pocoana, from there to Clhangano, from there to another place belonging to high chief Maxavella – the place's name is Magemella – and from this place to the Belule River (Olifants River), from there to Palauri (Phalaborwa), from there to the Great Litthave (Letaba) and from there to the Klein Litthave (Letaba) and from there to Macia and from there to Goedewensch, where I live.
>
> Goedewensch is the name of my place and it means *Good Wish* – Litthave is the name of a river, but because there are two rivers with the same name, they are differentiated as Great and Klein (Little) Litthave. Palauri is a region which is called so. Belule or Libelule is a river. Maxavella is the name of a high chief. Clhangano is the name of the place. Pocoana is a high chief known by that

---

42 Now Zimbabwe.

2. A NEW ORDER

Shingwedzi Poort. An old trade route went through the Lebombo Mountains here to the east coast of Mozambique. (01.08.1985)

name. Mohamba is the name of a region; Maxuva is the name of a high chief who is the son of the earlier King Mohamba and a brother of the current king.

According to Dicke, Albasini's settlement was the ending of a branch of the route between Delagoa Bay and Mashonaland. This route had already existed by 1727, when the black trader Mahumane followed it from Delagoa Bay to the land of Inthowelle. All available information points to the fact that Inthowelle was a Venda chief who lived at Dzata by the Nzhelele River (all Venda chiefs are called "Ithovele" as a sign of respect). This route was also followed by the Dutch expedition led by Francois de Cuiper in 1725 from Fort Lijdzaamheid (Fort Rio de Lagoa) during their brief journey into the interior as far as Dawano's kraal at Gomondwane in the current Kruger National Park.

Route B: From Inhambane to the Soutpansberg
According to Dicke (1938), "a regular trade route had developed from Inhambane through the Limpopo, near its junction with the Limvubu and past Tsikundokop". The only available information in connection with this route comes from the previously mentioned Portuguese map of 1893. This indicates the route to the Soutpansberg that Father Joaquim de Santa Rita Montanha and Second Lieutenant Antonio de Souza Teixeira followed in 1855, with their bearers and guides, to establish trade relations with the Boers.

From Inhambane they travelled to Chicuque, Mongo, Guione's kraal, Cambi's kraal, Malamula's kraal, Lake Murraba, Lake Nitemalla, the territories of chiefs Ingoana, Macuacua and Marive, Lake Inhamelanga, Lake Monhembongo, Chiqueta, Madiacune, crossed a drift in the Limpopo, carried on to Chivandana, Maducumne's kraal and Machambo's (Matsambo) kraal, and then went through the Shingwedzi Poort in the Lebombo Mountains to Mocane's kraal, where the route on the map ends.

There were two routes from Shingwedzi Poort that led to the Soutpansberg. One followed the valley of the Great Shingwedzi River, running between the two Shivhulane hills to link up with a route leading from Elim to Pafuri along the southern bank of the Luvuvhu River. According to Dicke the most important depot for the traders from Inhambane was at Shivhulane Hills.

Montanha wrote that he and his party followed another footpath from Shingwedzi Poort. This one apparently joined a footpath that went from Inhambane through Shilowa Poort, after which it followed the banks of the Klein Letaba River. After Montanha had crossed the river, he and his group travelled past Masia's kraal, set against a terraced hill. Their route then went past Kurulen and Mashao Hill to Albasini at Goedewensch.

According to Punt, Van Rensburg had followed this route along the Klein Letaba as early as July 1836. In contrast to Montanha, he had travelled through Shilowa Poort, which is south of Shingwedzi Poort, to Matsambo's kraal on the western bank of the Limpopo, close to where he and his group were murdered by Manukosi's warriors. On Jeppe's map of the Transvaal of 1879 the following caption occurs directly under the name Matsambo: *Battle Ground between Boers and Shoshangaans*. Jeppe probably obtained this information from the Natal surveyor St Vincent Erskine. Erskine had already visited the scene of the murder in 1872 and published his map of it in 1878 with almost the same title as Jeppe's, namely *Battleground between Trek Boers and Shoshangaan Army (Baobab)*.

A. Petermann also gave an indication of the route Montanha had followed to the Soutpansberg on a map of 1870. When compared to the priest's diary and the sources mentioned, it appears that he indicated it incorrectly. When Montanha and a few Boers returned to Inhambane a year later, they followed a partially different route.

Route C: Along the eastern escarpment
According to an unpublished map compiled by Punt in 1953, there was another route from Delagoa Bay that led in a northwesterly direction. Southwest of Pescene the route swerved to the north and crossed the Komati River just north of Moamba. From there it followed the western bank of the Komati and crossed the Lebombo Mountains at Sabie Poort. From there it followed the southern bank of the Sabie, past the current Lower Sabie Rest Camp and Skukuza, and over Magashulaskraal near the hippo pools, where Albasini had built a brick house and started a trading post in 1845/1846. According to Punt's map, the route led from Magashulaskraal to a place south of the current Mac-Mac, first along the southern and then the northern bank of the Sabie River. From there it went to Pilgrim's Rest, Rustplaats and then along the Ohrigstad River to Ohrigstad, which was founded in 1845. According to Dicke, this was also the trade route to Sekhukhuneland. The main route northwards went past Ohrigstad along the Ohrigstad River, through the Olifants River, along the foot of the Drakensberg, through Shikororo's and Magoboya's territories, west past Tzaneen and Duiwelskloof and east past Mokeetsi, over Saliesnek and through the western corner of Mamahila's territory. It joined a footpath that turned off from the Mashonaland route on the farm Grootfontein near the trading post at Bellevue 75 and passed on the western side of the conical Rivola Hill to Elim.

On 3 September 1860, the 30-year-old Portuguese trader and elephant hunter Diocleciano Fernandes das Neves, with his party of 253 bearers and hunters, followed part of this route from Lourenço Marques (now Maputo) to trade with Albasini and at Schoemansdal, hoping to trade ivory on the way and hunt elephant. On 6 September they crossed the Komati River at Moamba and stayed over in the town Gingelim. The night of the seventh they stayed over at Chief Magude's main kraal, which, according to Das Neves, was situated on the western bank of the Sabie River. The night of the eleventh they slept under the trees at chief Iávine's kraal and the next night they camped near a stream with abundant water. On the thirteenth they travelled for three hours before reaching a waterfall in the Sabie, where they rested until the afternoon before resuming their journey. They reached a second waterfall and spent the night on the southern side of the river. Judged by the two waterfalls, they had crossed the Lebombo Mountains through Sabie Poort. On the fourteenth they crossed the river somewhere near Lower Sabie and then travelled northwards. After four hours they reached "a valley through which coursed a stream of excellent water. This valley was very picturesque, a long dense forest of large trees, whose branches were knit together, extending along the south of the valley" (Das Neves, 1879). Dr U. de V. Pienaar believes that this is a description of a patch of trees growing next to the Mlondozi River on the northern side of Muntshe Hill in the Kruger National Park.

Sabie Poort as seen from the west. A centuries-old trade route from the interior to Delagoa Bay crossed the Lebombo here.
(11.11.1986)

2. A NEW ORDER

From here they travelled northwards and reached the route that went from Lourenço Marques (now Maputo) past Mahasane and across the Lebombo Mountains. "In the early morning we continued our journey for a long distance, walking without halting for several hours, reaching a thickly wooded part through which flowed a stream of clear water" (Das Neves, 1879). Pienaar believes this was the Ngotso Spruit.

Here we all bathed, and after breakfast continued our journey until we came to a large tract of level ground, through which flowed an affluent [the Timbavati] of the great river Imbélule [Olifants River] where we camped under majestic trees … The following morning we reached the great river Imbélule, called by the Dutch "Elephant River", because when they discovered it, they found a number of elephants on its banks. I crossed the river on the shoulders of two men, the water reaching up to their chests.

The party passed through the Phalaborwa area and went past a hill that was almost perpendicular. This must have been Kasteelkoppies. Once they reached the Great Letaba River, they walked along it up to Shiluvane. According to Berthoud's map, the footpath ran northeastwards from Shiluvane over Thabina and past the Shiruluru Hills to join the main route, which led northwards.

A very old leadwood tree with a typical Portuguese cross carved deep into its trunk grows next to the river road to the north, just a few kilometres from Letaba Rest Camp in the Kruger National Park. Nobody has yet been able to explain who had carved the cross, but in all probability it was Das Neves on his return journey. Or else it could have been one of the Portuguese soldiers who used to carry Albasini's mail between Lourenço Marques (now Maputo) and Goedewensch.

Route D: From Elim to Pafuri, Chicuala-cuala, Inhambane and Delagoa Bay

Another route to Inhambane and Delagoa Bay led eastwards from Elim through Karringmelk Spruit, over Welgevonden, with Mashao Hill and Tshimbupfe (Ysterberg) on the southern and Shikundu Kop and the Luvuvhu River on the northern side, over Mhinga, Shikokololo (later Punda Maria), Klopperfontein and Baobab Hill to Pafuri. There was a short cut from the trading post at Bellevue 75 in Chief Mavhambe's territory that linked up with this route. According to Punt's map, which was published in 1939 in *Die Volkstem* with the information he had obtained from Dicke, this route led through the Limpopo River at Pafuri and passed south of Chicuala-cuala on the eastern side of the river, from where it led to Matsambo. There the route crossed the footpath that led through Shingwedzi Poort from Inhambane to the Soutpansberg. Further along it followed the Limpopo River to Delagoa Bay.

Shikundu Hill. The trade route between Soutpansberg and the east coast ports Sofala and Inhambane went past here. (01.08.1986)

Dzunwini Hill, a well-known landmark in the northern Kruger National Park. An old trade route to the east coast went past here. (30.07.1984)

2.5 ANCIENT TRADE ROUTES THROUGH THE LOWVELD

According to Berthoud, a footpath turned off this route. This path led eastwards from just south of Klopperfontein past Malonga Fountain, across the Malonga Hills and further into Mozambique to Matiwane, through the Limpopo and the nearby Rio Nuanedzi,[43] to Chicuala-cuala on the eastern bank of the Nuanedzi.

Two elderly black men, James and Samuel Makuleke, told Pienaar that there was also a route from Shikundu Kop over Dotholi and Manangananga to the Dzundwini fountains and from there across the Lebombo flats to Malonga or Shirombe, Matiwane and Chicuala-cuala. This route is indicated on Berthoud's map only up to Dzundwini. Up to that point it was also a wagon trail.

Part of this trade route along the Luvuvhu River was used by Albasini, as well as the hunters from Schoemansdal and the later Soutpansberg hunters, to travel by ox-wagon to Makuleke, close to the confluence of the Luvuvhu and Limpopo rivers. According to Berthoud, the route from Klopperfontein past Malonga Fountain to Chicuala-cuala was also passable by ox-wagon.

In May and June 1836, Van Rensburg and his trek stayed over next to the old trade route on the farm Welgevonden 216 in the Soutpansberg for several weeks on their way to the coast. When they continued their journey they turned off at Mashao Hill and from there followed the route along the Klein Letaba River. To commemorate this, the South African Heritage Resources Agency (formerly the National Monuments Council) erected a memorial plaque at Welgevonden.

When Trichardt left his laager at the Doring River close to the current Elim Hospital at the end of July and in the first half of August 1836 to search for the Van Rensburg party, he followed Van Rensburg's wagon tracks into the Lowveld. According to Punt's 1960 map in *Koedoe*, they travelled from Matsambo northwards along the Limpopo to across the river from Chicuala-cuala, and then along the old trade route past Malonga Fountain, Klopperfontein and Shikokololo on their return journey.

When De Santa Rita Montanha and his party set off on their return journey to Inhambane from Schoemansdal on 23 June 1856, they hired two bearers and guides at Shikundu in exchange for a piece of blue Indian cotton. From there they travelled to Mhinga's kraal where they spent the night. It seems they then had to decide between two routes, because Montanha remarked: "… we followed a bad road with many stones, uphill and downhill over a mountain … By sunset we arrived at a place with water in round holes between rocks where we pitched camp to overnight." This must have been Shantangalani, as the sinkholes in the sandstone ridge fit the description exactly. It is also a day's journey from the route between Punda Maria and Pafuri. From

Malonga Fountain. According to old journey descriptions it used to be a very strong fountain in days gone by.
(JUNE 1959)

Malonga Fountain (in the background) and the direction of a trade route to Sofala and Inhambane.
(21.06.1984)

---

43 This is not the same river as the Nwanedzi in the Kruger National Park. This Nuanedzi (also spelled Nuanetzi) is a river of which the source is north of the Limpopo in Zimbabwe and which flows into the Limpopo south of Chicuala-cuala (see map on page 67).

Chicuala-cuala they travelled more or less in a south-easterly direction and passed Chief Bocota's kraal. They then crossed the Luize River and followed its eastern bank southwards. At a passable drift they crossed the river again and walked southeastwards to the territory of Chief Marive, then going past Pachano's kraal to Ingoana. Here Montanha rejoined the route he had followed the previous year to the town in the Soutpansberg.

Route E: From the Zimbabwe Ruins to the copper mines at Musina (Messina) and the tin mines at Rooiberg.
An old trade route also existed between the Zimbabwe Ruins and the Musina copper mines about which G. Caton-Thomson would write the following:

> … running southwest, leads to the ancient workings of the great copper bearing regions of Messina, whose traffic with Zimbabwe may be indicated by scattered ruins along the line of terminus.

This route must have crossed the Limpopo at Fleures Drift. From there it ran past Msingalele Hill, where iron smelting by the first black smiths had taken place, along the Sand River and then the Brak River. On the farm Longford the route turned south, running past Vetfontein, Vogelstruis, Soutpan and the western point of the Soutpansberg at Vivo. After that it turned eastwards over Mara, where the Buys people later settled. Because of the availability of water the route carried on southwards along the Hout River, past Loskop, Kalkbank, through Moletse's territory and east of Strydomsloop, through Blood River and then over Doornkraal west of the current Polokwane (Pietersburg). Trichardt followed this Hout River route to Soutpan in 1836 when he went looking for Van Rensburg. The route then followed the Sand River, went west past Ysterberg, through Makapane Poort and the current Mokopane (Potgietersrus), along the Mohalakwêna or Nile River past the current Modimolle (Nylstroom) to Buyskop[44] close to Bilabila (the later Warmbaths and now Bela-Bela). According to Punt, there had been a strong fountain at the foot of Buyskop in earlier times that was later blocked by Bushmen (San). The rest of the route is not indicated on any of the old maps I have. However, it would have turned westwards from Bilabila to Smelterskop, where there had been tin smelteries. To make it easier for travellers, it is possible that the route had gone across the flats directly south of the Waterberg, but it also would have had side branches to the various small tin mines in the mountain. Past the Rooiberg and Weynek tin mines the route probably turned south towards Leeupoort. Max Baumann (1919), an engineer, noted the following about an old route in that vicinity:

> One more evidence of the ancient occupation of this district is an ancient road … which starts from the neck, joining Smelters Kop to the Elandsberg and climbs the latter in a northerly direction and at the summit turns east, is lost, or at any rate, has not been further traced. It has been cut out of the side of the hill, and the excavation material filled in on the down hill edge of the road, and it is wide enough to take a wheeled vehicle. The road was not made by any Bantu, for no native has ever been known to pick up a stone out of his path … he will either step over or go round it … and he has never been known to build a path, besides it is much too broad for a native road and apparently leads nowhere. From its state of dilapidation and the vegetation growing in it, it must be very old. It has not been made by the Voortrekkers nor by any Boers, for no boer would climb over a hill when he could go on the flat, round either end of the range as can be easily done. It is, therefore, of earlier age than either Boer or Bantu.
>
> Smelters Kop was a sort of Acropolis, overlooking the Rooiberg Valley – well situated for defence – a place of refuge during times of battle and raids by other tribes. The road may therefore have been built with the object of insuring the safety of the travellers – perhaps heavily laden with ingots – in an easterly direction towards the coast. The road is so exposed that no raiders could approach the road unseen and ambush the travellers.

However, the secret of this old route will not be revealed until it has been followed and fully investigated.

---

44 Also spelled Buiskop on old maps.

Route F: Northwards along the Sand River

Only on *Bartholomew's tourists' map of South Africa* is there a route that turns off from the Brak and Sand rivers route and follows the western bank of the Sand River. Along this route traders could visit Matšhêma, a miniature Zimbabwe on top of a hill on the farm Solvent. This footpath must be very old, because tradition has it that the black Musina copper miners and the inhabitants of Matšhêma were related. There must therefore have been regular communication between them. Excavations there revealed that the inhabitants in ancient times also had trade relations with the coast and the outside world. Glass beads, shells, a piece of white porcelain with a blue dot – probably of Portuguese or Chinese origin – and a piece of green glass were found during the excavations.

From Matšhêma this trade route followed the Sand River to the current Waterpoort. After the path had run through the poort, it turned off in the river bed south of Elim Mountain and joined the Delagoa Bay and Inhambane route. However, the main route continued southwards along the river to Sandrivierspoort, Rhenosterspoort and Doornkraal, where it joined the Hout River route.

Route G: From Soutpan to Tshivhase

According to Berthoud, there was a wagon trail north of the Soutpansberg mountain range between Soutpan and Tshivhase while he was serving as a Swiss missionary at Valdézia from 1881 to 1898. This route crossed the Sand River, Ntambande Spruit and the Nzhelele, and then went past the Dzata Ruins, which had been built by the first Venda immigrants, and Tswime Mountain. The wagon trail joined the Delagoa Bay-Mashonaland route at the Phiphidi Falls and from there it led to Tshivhase. This old wagon trail is now a tarred road to Wyllie's Poort. At the place where the dirt road turns off to Dzata, there must originally also have been a trade path to Vendaland, along which the Venda and possibly also earlier inhabitants living north of the mountain could transport salt, copper, tin, hoes and other trade goods. In fact, metal objects found at several places along the route confirm the theory that this had originally been a trade route linking the three northsouth routes. Its age is not known. H.P. Prinsloo, of the Department of Archaeology at the University of Pretoria, found some potsherds on the farm Klein Afrika at Wyllie's Poort that he had dated. The results indicated that people had lived along this route as early as AD 331.

Routes H, I and J and their branches

In prehistoric times, long before the arrival of the first Europeans, the Drakensberg with its foothills, healthy environment for man and beast, and alluvial gold in the vicinity of Sabie, Pilgrim's Rest, Lydenburg, Machadodorp and Waterval-Boven, must have been fairly densely populated. On the farm Klipbankspruit, south of the Sabie-Lydenburg road in the Drakensberg as well as along the road from Pilgrim's Rest to Lydenburg, there are extensive agricultural terraces. From Lydenburg to Machadodorp and south and north of the main road next to Waterval-Boven more agricultural terraces, as well as large circular stone structures against the slopes, can be seen from the road, especially when the grass is short. The best-preserved circular stone structure, with walls that are still chest high, can be seen at Blouboskraal. Small dams and irrigation channels leading to the terraces have also been discovered.

Baked clay masks were found along the Sterk Spruit southeast of Lydenburg. On the farm Nooitgedacht, which is close to town, and northwest of there, deep mines with horizontal shafts were discovered. When the then owner of the farm, Mr S.P. Coetser, showed me the old workings in 1954, near the shafts there were large circular structures of which the walls had collapsed. He told me that his father had gone down the old shafts and at the bottom he had found a lot of snakes, which had probably fallen in and could not get out again. The only man-made objects he had found inside were round, bored stones and iron wedges. His father believed that gold had been mined there, as had been the case at the abandoned little mines at Sabie, Waterval-Onder and Pilgrim's Rest. This dense population and the mining industry resulted in trade relations with the east coast, and one or more trade routes from Delagoa Bay past Lydenburg into the interior.

There had to have been a trade path leading from Lydenburg in a southeasterly direction, because there is a train stop at the farm Doornkop that is indicated as Voetpad (Footpath) Station on a map of

2. A NEW ORDER

1929, while the adjacent farm's name on the same map is indicated as Kaffervoetpad. These are both names that date back to the Voortrekker era and are coupled to a footpath running across both farms. This reasonably passable footpath down the Drakensberg escarpment must have been used by the Voortrekkers as a wagon trail, because on Loveday's map of 1883 it is indicated as the "Road to Kaap Gold Fields", which joined the "new Road to Delagoa Bay". Further on it joined another former trade route and a later Voortrekker road over Pretoriuskop that crossed the Lebombo Mountains at Matalha Poort and then continued to Delagoa Bay. From Delagoa Bay northwards, this footpath (J) went past Nellmapius's later overnight sites, namely Campos de Corvo, Progresso de Guedes and Castilhopolis, the later Furley's Drift through the Komati, past Tengamanzi by the Crocodile River, Joubertshoop, Pretoriuskop and then to Klipkraal, otherwise over Burger's Hall, Sabie and then to Klipkraal, Pilgrim's Rest and Rustplaats.

Trade path (I) went from Sabie across the Drakensberg to Lydenburg, where it turned in a northwesterly direction to Sekhukhuneland across the Grootdwars River, through Steelpoort Drift, up Magneetshoogte, past Ramakok's Kraal, along the Gompies River, west past Platberg, through Strydpoort and over Marabastad to Doornkraal. From there the route continued along the Hout River, while another went to Rhenosterspoort and later branched in two. One route (H) went west of Matok and Bandelierkop to Schoemansdal and Happy Rest (Muraleni), a terrace where many black people lived years ago. Later school buildings were erected on this terrace. Clay pots dating back to AD 350 were excavated here. From this terrace a footpath led up the mountain, along which hundreds of black women with baskets on their heads had walked to mine ochre (red clay), using iron wedges and round stone hammers. They used the ochre to decorate their clay pots. A bit further on the path led to a small gold mine, which had also been mined by black delvers. From there it led eastwards to connect with the route from Soutpan. Along the footpath on the mountain there is also an *isiVivan* (pile of stones), which had been stacked over many years. The other footpath ran between Rhenosterspoort and Bandelierkop in a northeasterly direction west of Mailaskop to Elim, where it joined the footpaths to Mashonaland and the coast.

From the overnight site at Rustplaas near Ohrigstad, another route led in a northwesterly direction. On the farm Rietfontein it joined one running from Ohrigstad to the farm Goudmyn next to the Steelpoorts River. From there it followed the river and joined the path from Lydenburg.

It was important that there should be a short cut from the Rooiberg tin mines to Delagoa Bay through Lydenburg. According to the topography of the area, this short cut would have turned away from the Rooiberg-Musina route (E) along the Mohalakwêna River in the vicinity of Moorddrif to avoid the mountainous area. It would, according to the old footpaths on Map no. 13, have joined the main route (I) in the vicinity of Sebitiêla's territory, somewhere south of Groothoek near Koornpunt. The correct connection could, however, not be determined.

Trade along the earliest routes

To gather information on the manner in which trade took place along these ancient routes, the researcher is dependent on: (1) archaeological finds at the ports along the east coast and along the old routes in the interior; (2) the first mines, carbon-dating and the metal objects produced there and elsewhere; (3) historical sources and (4) the names of places and metal objects. Before the arrival of the European settlers, there had been great demand for different types of metal and metal articles among the black tribes of

The large isiVivan (stone cairn) along the old trade route just north of Malonga Fountain. (01.05.1986)

the northern and eastern Transvaal[45] interior, as well as among the traders along the east coast. The result was that the most important mines and trading depots were connected to each other and to the east coast harbours by trade routes.

At the Phalaborwa mines, which according to carbon-dating have been mined since AD 700, copper items known as *marale* (singular *lerale*) were cast. These are round rods about 500 millimetres long and 13 millimetres in diameter, each with a cone-shaped head cast onto it at one end. There are up to four small bars, 34 millimetres long, on some of the heads, but examples without any of the small bars also exist.

At Musina black smiths also cast copper bars by using square, rectangular or round wooden shapes to make moulds in wet sand. Rows of holes were made at the bottom of the larger hole with a small stick about as thick as a pencil. Before the melted copper was cast, a ball of clay (or slag) serving as a funnel was placed in the large upper hole, possibly to save copper. After the copper had cooled down, the cast object was removed from the mould and the bars cut off, so that only the copper that had been in the mould and the overflow remained. Turned upside down, the cast looks like a top hat with rows of pegs on it. These casts (*musuku*) were quite common in the Soutpansberg in the past. The fact that several have been found along the old trade routes indicates that they had been used to trade with. Several smaller, single round rods between 50 and 115 mm by 6 mm with cups similar to those of the *marale* cast onto them, have also been found. The cup was probably the funnel hollow in the wet sand.

It is also known that the smiths used to cast copper beads – I have a flat copper cast in my collection, with nodules on it, and in the nodules are little holes. Solid arm and ankle rings were made from copper rods and rings from stretched wire.

Three different types of tin casts have been found: a biscuit-shaped cast with a thin hole in it; a long, triangular rod 508 mm long with sides of 9 mm; and rods, one of which is 406 mm long, 70 mm high and 44 mm broad. On each side of the rod is a protruding point with a series of lateral, proportionally spaced knobs in-between. Another rod is 362 mm long, 25 mm high and 25 mm broad. It does not have any knobs – only two upright points at either side. These rods could easily be transported by traders and worked by smiths.

At Tshimbupfe iron mines, grain transported in skin bags was traded for iron ore, which in turn was transported in the same bags over long distances to the different smelters and smiths. The smiths at Tshimbupfe (Ysterberg) and many other places in the Soutpansberg manufactured small iron axes, battle-axes, arrowheads, knives, razors, sweat scrapers, assegai heads, sickles, hammers and hoes for own use and to trade with. The typical Venda hoe has a long handle with a diamond-shaped blade. The length varies between 432 and 483 mm and the width between 160 and 195 mm. A large, unused hoe in the shape of a spade, with a particularly long handle, was found in the 1940s on the farm Boomplaats near Lydenburg. This hoe had probably been

[top] Iron was smelted in this way at Tshimbubfe (Iron Mountain).
(WITH KIND PERMISSION: H. GROS)

[bottom, left] Five "musuka", a flat copper cast, a rod in the shape of a "lerale" and the remains of a cast with indications that there had been hollow rods on, which were probably used to produce beads. Found next to the old trade routes.
(WITH KIND PERMISSION: DR J.B. DE VAAL)

[bottom, right] A copper "lerale" from the Phalaborwa area.
(WITH KIND PERMISSION: DR J.B. DE VAAL)

---

45 Limpopo and Mpumalanga provinces.

2. A NEW ORDER

[top] A self-made hammer, hoes, an ax, an arrow, some arrowheads and a piece of iron with which the smith probably tried to make an arrowhead. All these objects came from the Soutpansberg.
(WITH KIND PERMISSION: DR J.B. DE VAAL)

[bottom] Kamanjan, who participated in trading along the trade routes between Mashonaland and Delagoa Bay as a young man.
(WITH KIND PERMISSION: DR J.B. DE VAAL)

made by the Roka smiths of Sekhukhuneland. It is 750 mm long and the blade alone is 290 mm long and 170 mm wide.

In prehistoric times gold was mined and worked in the Middelburg, Pilgrim's Rest, Lydenburg and Groblersdal districts, as well as in the Soutpansberg. In the Soutpansberg, the Lemba artisans, who may have Middle Eastern or Semitic origins, monopolised the gold industry. They mined gold and cast it in small rectangular bars. Unfortunately no examples of these have been discovered or described. They used this gold to make various objects, including rings, of which the Voortrekker J.G.S. Bronkhorst saw examples in 1836 at a small gold mine in the Soutpansberg.

Kamanjan, the old Shangaan-Tsonga, had a ring on his finger in 1939, which according to him had been made from gold mined at the Soutpansberg, probably by the Lemba. However, this claim has never been tested. As far as is known, worked gold has been found at only two other localities in the Soutpansberg area – at Mapungubwe in the form of sheet gold, nails and small beads, and at the Matšhêma Ruins in the form of small beads and a piece of wire, 10 mm long and 0,5 mm in diameter.[46] The unworked gold could have had its origin in the gold mines of the Soutpansberg, but it is more probable that traders had brought it from across the Limpopo, after which it was worked by Lemba smiths.

Ore containing 27,5 grams of gold per ton was discovered on the farm Doornpoort in the Groblersdal district, compared to that of the Witwatersrand, which contains only 5 to 7,5 grams per ton.[47] A gold ring weighing 653 grams on which the date 1640 is engraved, was also found on this farm. A similar ring, also with an inscription, was obtained years ago from an unidentified woman by G.J. Joubert of Graskop.

In May 1867 Dr H.T. Wangemann visited Reverend A. Merensky's Berlin-Lutheran mission called Matlala, situated northwest of Polokwane. There was a Lemba kraal about three kilometres from the mission. According to Wangemann (1868) "some Malepa brought some of their so-called gold rings from their former country (Mashonaland) to sell. Merensky bought some of these. They were copper, apparently mixed with gold. They had salvaged this metal in their old fatherland Mashonaland".

Twelve years earlier, in May 1855, a shopkeeper living in the western part of the Witwatersrand showed a Dutchman, K.J. Kok, a few hundred flat copper necklaces he had obtained from the Mashonas on his annual journey to trade ivory. These necklaces contained enough gold to be sent to Durban to be smelted. There is no doubt that these rings were also of Lemba origin, because they used to live in Mashonaland and certainly had been the most important goldsmiths there.

---

46 Smelted or worked gold was later found at two sites in the northern Kruger National Park, namely the Makahane and Thulamela ruins.

47 A ton (2000 pounds) is equal to 1016 kilograms.

Mutual trade

The black tribes who owned mines were involved in a lively trade with other tribes. In the vicinity of Leydsdorp *marale* were traded for wives, mainly by the chiefs. The miners of Musina, like the Tshimbupfe and Ysterberg miners, were not agriculturists. They therefore exchanged copper ore for grain and because they ate meat every day, they and their neighbours across the Limpopo and the surrounding tribes bartered cattle.

The Shangaan-Tsonga of the Soutpansberg and the Lowveld, who were never great iron workers, traded hoes with the Tshimbupfe for grain, cattle and other livestock. Many of the hoes were never used for agricultural purposes, only as trade goods and currency. Former ranger Gus Adendorff found two unused Venda hoes in a cave at Ngirivane near Satara about 38 years ago. The *bogadi* or *lobola* for a strapping young woman was 30 hoes, but if she was slender, it was only 25. It appears as though five hoes were a trade unit, because 25 hoes tied up in bundles of five each were found in a river bed in the northern Transvaal. The following information was provided by Johannes Masini, an old Venda, more than half a century ago: a cow was worth 20 hoes, an ox 11, a goat two, a sheep four and a wife 40. Later, when hard cash was in circulation, a hoe was sold for five shillings.

From 1855 the Soutpansbergers collected taxes from the black inhabitants. Every chief had to pay either five head of cattle, five elephant tusks, 25 copper bars or 20 leopard skins annually. The owner of a hut had to pay either a goat and a sheep, or four hoes. There is no complete record of the tax gains, but apart from cattle, goats, sheep and elephant tusks, Albasini in his role as tax collector collected a total of 1483 hoes between 26 December and 7 April 1863. This gives an indication of the large scale on which iron ore was mined, smelted and worked into hoes and other tools.

Metal and other objects found along the old routes

Iron, copper and tin objects, which have been found at several places in the Soutpansberg, as well as an undamaged coconut, are evidence of traffic and trade along these routes in former times. Tin bars and a *musuku* were found many years ago in a cave at Blouberg and two *musuku* on the farm Spilsby by the Brak River. This is proof of the two-way trade between the tin mines at Rooiberg and the Musina copper mines along the Musina-Brak River-Rooiberg route.

A man working for Mr G.P. du Preez on his farm Vogelstruis found a *musuku* under a marula tree next to the old trade footpath between Musina and Soutpan. Two fine-looking, unused shield-shaped hoes were discovered on the farms Bluebell and Prince's Hill, which are about 25 kilometres apart on the Soutpan-Tshivhase route. According to several old informants, these are of Shona origin and therefore proves that trade had taken place with Mashonaland. On the farm Albert a *musuku* was ploughed up and another was found on the farm Waterpoort. Two

From top to bottom: a Shona, a Pedi and a Venda hoe, found along the old trade routes.
(WITH KIND PERMISSION: DR J.B. DE VAAL)

The gold rhino statue and other gold articles which have been found at Mapungubwe.
(PHOTOGRAPH: PROF. C. VAN RIET-LOWE)

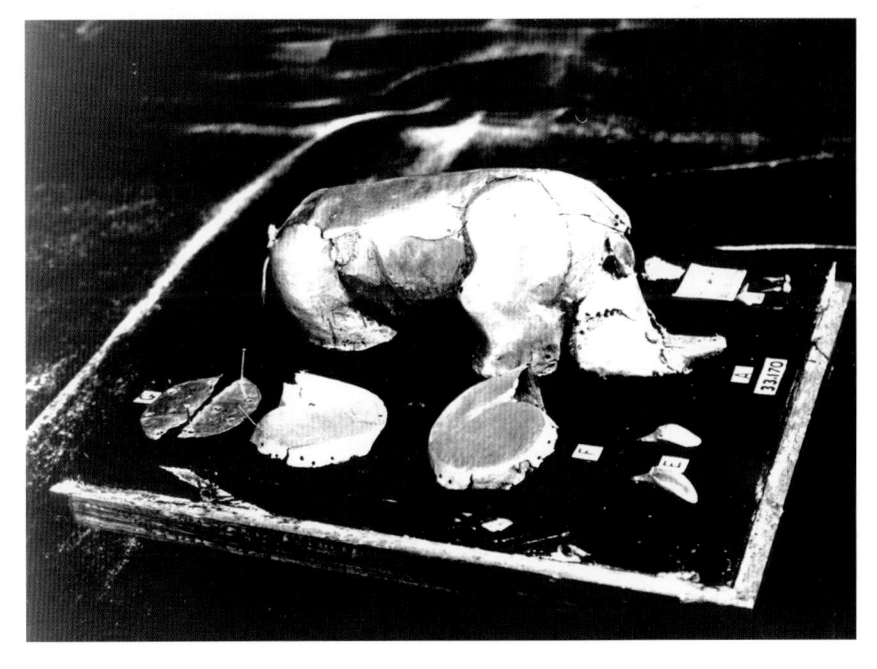

2. A NEW ORDER

Part of a tin rod which was dug up on the farm Madrid.
(WITH KIND PERMISSION: DR J.B. DE VAAL)

undamaged Venda hoes were found in 1941 on Prince's Hill and a *musuku* was found on the neighbouring farm Sandilands. Another *musuku* was found at Rivola Hill, also along an old trade route.

In 1949 a labourer discovered a tin object on M.J. du Plessis's farm Madrid 112[48] at the Levubu settlement east of Makhado (Louis Trichardt) while digging up a tree.

According to Dicke there used to be an old trading post on the farm. Mr Gert Erasmus, an information officer at Skukuza in the Kruger National Park in 1990, told me that his father (also G.J. Erasmus) of the farm Nooitgedacht adjacent to Madrid, had found an undamaged coconut three metres under the ground in a creek while constructing a weir. An old black man told Erasmus junior that the "old people" had owned such nuts, which were holy heirlooms. This proves that the traders also brought coconuts from the east coast on their travels to serve as both food and currency.

Louis Trichardt stayed at his fourth camp site beside the Brak Spruit in the Soutpansberg between 2 March and 31 May 1837, just a few kilometres west of the site of Makhado (earlier the town Louis Trichardt). After that, between June and August of the same year, they stayed at their fifth camp site directly north of Elim next to the Doring River close to one of the old trade routes. At these camp sites he received trade goods as gifts and payment from black chiefs, or traded with the Magwamba and other pedlars. Chief Rossetoe gave him a piece of printed calico, with one half striped and the other flowered. Doors Buys gave Trichardt's wife, Martha, half a length of checked cloth, for which he received a goat and a sheep's tail as a *quid pro quo*. On another occasion he presented her with a length of printed calico.

On 18 April 1837, chiefs Rossetoe and Mashao paid Jan Pretorius and Izak Albach, both members of the Trichardt trek, with a tin bracelet, among others. This must have been made from a solid tin rod because Jan Pretorius "*nam de ring, brak door hem en Albach*", basically meaning that he liked it very much. The payment further consisted of clothing and a blanket woven from wild cotton they had spun themselves, a length of checked cloth eight ell long and a length of white linen five ell long (an ell being six hand breadths or 114.3 cm). Mashao lived along the old trade routes and had undoubtedly got the articles from traders from the coast, except for the blanket.

Chief Moletše sent Trichardt a piece of checked cloth as a gift and Mrs Botha traded articles from Doors Buys for a piece of broken mirror, a small tin box, a powder horn and a length of bafta. Trichardt traded two and a half ell calico from a Magwamba pedlar for one large and one small elephant tusk and one as a present for his employer at Delagoa Bay, on condition that he bring back a keg of brandy. He also paid one elephant bull tusk for some beads and two pieces of clothing.

When Carolus (Karel) Trichardt and an employee named Adonis went to Rossetoe and Mashao to find iron implements, they got eight hoes from Rossetoe. The chief's mother sent Martha Trichardt a piece of checked cloth, as well as a red cashmere ladies' coat or cloak. Chief Rossetoe sent Martha Trichardt a piece of blue chintz and gave Carolus two small pieces of checked cloth. Mashao also bought a length of calico from other pedlars who happened to be present and gave it as a present to Trichardt.

Trichardt once got a six-ell-long piece of bafta and a piece of striped cloth from Kalahari pedlars for his wife, and a blanket and a nightdress for teacher Daniel Pfeffer (Trichardt did not record the price he had paid for these items). Martha Trichardt exchanged a piece of checked cloth for a young elephant bull tusk, and her husband traded a length of cloth for two elephant cow tusks and another four pieces of cloth for another two cow tusks, with the Magwamba pedlar Waai-Waai. Carolus bartered five ell calico for two tusks and old Strijdom two pieces of bafta for half a bull tusk.

During Trichardt's journey from the Soutpansberg

---

48 Madrid is the name of the farm, while 112 is the registered farm number.

to Delagoa Bay, five Magwamba pedlars arrived one evening at the camp site just west of Bandelierkop. He did not want to do business while the light was bad because he wanted to see their wares properly, so the pedlars went to bed. They left during the night, and, judged by their tracks, in a great hurry.

On his trek along the Olifants River, after he had passed through the Strydpoort Mountains at Klynpoort 16 days before, Trichardt bought a piece of tin weighing four or five pounds from a black man at the last turn above Olifantspoort on 10 October 1837.

In the vicinity of Sekororo he exchanged a tusk for a piece of coarse linen and a piece of striped cloth with a Magwamba man. The ancient trade routes were thus not only traffic routes to Trichardt, but also passages along which he could obtain certain necessities.

The same kind of glass beads as those found at Magashulaskraal at the Albasini Ruins and along the old trade routes leading northwards along the eastern escarpment, are also found near Pafuri, Waterpoort, the Matšhêma Ruins, Mapungubwe, Kremetartkop by the Mohalakwêna River, Dzata, the Zimbabwe Ruins and in the Umtali district northeast of the Zimbabwe Ruins. All of these sites are linked to the old trade routes.

The coastal traders

Dr Cyril A. Hromník conducted an intensive study in India of the Indian language, history and culture. According to him, the Indians were the world's first seafarers, first maritime traders and first gold miners. The Indian *sanyatrik* caste was involved in maritime trade by as early as 4000 BC on the ocean named after these seafaring pioneers. Hromník claims that the MaKomati, after whom the Komati River is named, was a Dravida trading caste from southern India who, about 2000 years ago, lived, mined gold and traded ivory in the area through which the river flows. The old goldfields of Swaziland, Barberton, De Kaap, Waterval-Onder and those close to Lydenburg are all in the catchment area of the Komati. The area stretching from St Lucia in the south to the Zambezi River in the north, and from the sea in the east to the Drakensberg in the west, was under the sway of the MaKomati and was known as the Land of the MaKomati. The name MaKomati, in the Portuguese form MaComates, appeared as early as 1589 in a Portuguese document. The *Ma-* is a Bantu prefix that frequently indicates foreigners, for example Magowa for Europeans, while the *-s* in Portuguese is a plural ending. The name MaKomati thus means "the foreigners of Komati". These traders later intermarried with the new black arrivals and formed kingdoms whose leaders were known as *fumos* or *fumus*. These *fumos* lived all along the coast of East Africa as well as along the Zambezi, Sabie, Limpopo and other rivers used by the Asian gold and ivory traders. As a result Delagoa Bay is also known as Mfumo (or Mafumo).

Hromník found remains of extensive Dravidian settlements in the Komati River valley. So far no fewer than nine stone altars and other places of worship have been identified, along with the remains of a rock cloister, a temple, altars, memorial stones called *malekal*[49] and other objects, which shed some light on the religious practices of the ancient MaKomati gold miners. Excavations should cast more light on these findings (personal communication from Dr C.A. Hromník to Dr J.B. de Vaal, 19 September 1985). Komatipoort had thus been a

A smelting furnace at Phalaborwa.
(WITH KIND PERMISSION: DR J.B. DE VAAL)

49 Stone that had been immersed in the ashes of the cremated person and, thereafter, in a special ceremony transferred and planted in a recognised "death place", usually a circle of stones. *Malekal* forms part of the Dravidian (south Indian) mortuary rituals and practices.

natural gate or thoroughfare for trade between Transvaal and Delagoa Bay or Mfumo in ancient times.

A tribe known as the Ama-Lala lives in northern KwaZulu-Natal. They are expert iron smelters and claim that, unlike other Zulus, they are not of Nguni origin. Hromník established that their forefathers were Indians who had intermarried with local black people. At the end of the nineteenth century there were still some Lalas with straight black hair and Asiatic features.

Hromník also sees an Indian influence in the names *Phalaborwa, Lulu* and *lerali*. According to him *Phalaborwa* is derived from the word *phállu* (pl. *phâllâ*), a word in the southwest Indian Konkani language meaning "rod of iron (or another metal)". In Sotho *borwa* means "land of the San" as well as "south". Phalaborwa, therefore, is "the place of iron (or other metal) rods in the land of the San" and indicates a place where Indian-made metal bars, including copper bars, were bought by black people from further north. Until now accepted wisdom had it that black immigrants from the colder south went north and called this place Phalaborwa, meaning "better than in the South", as *phala* means "better" and *borwa* means "south" in Northern Sotho (compare with the explanation in chapter 2.1).

According to Hromník, Lulwe means "the place of iron", as *lu* is "iron" and *we* "place" in Sanskrit, thus Lulwe or Lulukop (Ysterberg). He also claims that the North Sotho word *tšipi* for iron is of Dravida origin, while the Sotho term *lelale* for a blacksmith refers to the Indian-related Lala smiths of northern Zululand.

Dr Willem Punt senior claimed that the Egyptians and the Phoenicians had been the first people to sail along the east coast of Africa. The queen of Shiraz (in modern Iran), for example, sent expeditions along the east coast in 1470 BC to explore the land's southern end,[50] while the Phoenicians from Carthage travelled down the west coast and sailed around the southern tip of Africa in 600 BC. The Indians, however, had been at the east coast long before them.

Arabs settled along the east coast from AD 740 onwards to exchange cloth and beads for the natives' gold and iron. They established a whole series of trading posts from Mogadishu in the north to Sofala (now Beira) in the south. Axelson is of the opinion that the Arabs' quest for gold brought them as far south as Natal. Near Ladybrand, for example, there is a rock painting depicting Bushmen with bows and arrows defending themselves against people wearing turbans and armed with spears and breastplates – probably Arabs or Indians.

In 1953 a retired Lydenburg teacher, Mr C. Bauling, the son of Reverend Bauling of the Nooitgedacht Mission just outside town, related that elderly black people had told his father that the previously mentioned mines on the mission belonged to a brown race with long, brown beards and long, white robes. They rode mules because these, unlike horses, are immune to the tsetse flies and African horse sickness along the road to Delagoa Bay. According to the description they must have been either Arabs or Indians.

During Father Joaquim de Santa Rita Montanha's journey to the Soutpansberg in 1855 he was accompanied by an Arab dealer, Mamoed Amad Saiboe, and his ten black bearers almost all the way to the Limpopo. Five years later, the Portuguese trader D.F. das Neves bought his trade goods from an Arab dealer from India. At the Komati, a 17-year-old black chief told Das Neves that he was the first European whom he had ever seen, because the only "white people" from Lourenço Marques (now Maputo) whom he knew, were Asians. They were probably Kanarese[51] and Banyans who had converted to Christianity. When Carl Mauch was in Lourenço Marques in 1870, he saw many Banyans. They controlled trade and he often saw them in the streets wearing their long robes, always carrying an umbrella or a cane.

In 1890 an elderly Venda told Colonel P.W. Möller of Groblersplaats, Louis Trichardt (now Makhado), that as children they had seen traders with hook noses north of the Soutpansberg. They were dressed in long, white "shirts" and were on their way to the Rooiberg tin mines, and therefore they must have been Arabs or Indians.

One of the San paintings on a projecting rock face on the farm Little Muck, south of the Limpopo and

---

50 King Solomon also sent expeditions along the east coast in the tenth century BC to explore the continent's southern end.
51 Also spelled Canarin in certain references.

west of Musina, depicts a horseman. It looks as if the rider has a turban around his head and a blanket roll or coat tied to the pommel. Might this have been an Indian or Arab trader on his way to Mapungubwe to trade ivory and gold?

The Portuguese along the east coast
Eight years after the Portuguese had sailed around the Cape in 1488, they established a trading post at Sofala, and two years later they annexed Mozambique. They sent a ship to Inhambane only once a year to trade ivory, slaves, amber, honey, butter, rhino horn and hippopotamus and elephant tusks with the natives. At Inhambane they also found pearls.

The Dutch
The Dutch occupied Delagoa Bay from 1721 to 1732. The first commander was Willem van Taak, followed by Jean Michel and, from 1725, the energetic Jan van de Capelle. The Dutch East India Company's instructions to the Governor of the Cape, who in turn relayed them to the commander at Delagoa Bay, was to find an overland route to the goldfields of Monomotapa (Mashonaland), to find the natives' route along which gold was brought to the Portuguese at Delagoa Bay, and to establish a permanent route to the gold, copper and iron mines in the interior. Delagoa Bay had to become the base for the gold trade with Monomotapa.

In 1723 Van de Capelle was able to inform the Governor of the Cape that, according to black informants, very little or no gold could be found in Mozambique; only amber, ivory and wax. He was not aware of the gold the Portuguese had traded at Delagoa Bay. In the same year, blacks arrived from the interior with copper and tin. The colour of both metals was very good, but one of the two tin rods was brittle, while the other one was harder. According to Van de Capelle, the blacks melted tin and copper into an alloy and made rings of it. This alloy had a unique lustre and was more flexible and more attractive than the rings with which the Dutch were trading. Evidently the black craftsmen had known how to produce brass (*messing*), as Dr Percy Wagner would discover at Rooiberg.

Two years later, in 1725, Francois de Cuiper undertook an expedition from Delagoa Bay to the interior to search for the gold of Monomotapa. But because of the hostility of certain black tribes, the expedition had to turn back at what is now Gomondwane in the Kruger National Park, only 48 kilometres from Skukuza. Despite this, they still obtained valuable information in connection with the copper mines of Phalaborwa and the iron mines of Tshimbupfe, as well as the route and distance to Zimbabwe.

At the same time, Van de Capelle obtained gold from people who had come from Paraotte. Their country was seven days' travel from the coast and was probably the northern Transvaal.[52] Apparently one could obtain gold, copper and ivory there. One trader knew Monomotapa (Zimbabwe), as he had visited it before. According to him it was a centre for the gold brought there from the surrounding area as well as further afield, and was ten days' journey from where he lived. In 1732 Van de Capelle exchanged 56 tin bars (20 flat circular ones and 36 round long ones) for 46.7 kilograms of beads. However, the Dutch were forced to leave Delagoa Bay in 1732 as a result of the many fatalities caused by malaria and blackwater fever. In 1759 the Portuguese established a permanent trading post at Delagoa Bay. The settlement Mozambique (15° south) was their headquarters and later a governor general was stationed there, with a governor at Delagoa Bay.

How trade took place along the routes
The old Shangaan-Tsonga Kamanjan said that he had accompanied some Magwamba traders from the Soutpansberg to Delagoa Bay in about 1875, when he was about 20 years old. They first shot elephants with bows and poisoned arrows in the eastern Lowveld. Once they had spotted the elephants, they scattered dust into the air to determine the wind direction. Then they stayed downwind from the animals and stalked them. After an arrow had hit the target, they patiently waited for the animal to die and then hacked out the tusks with a small axe. As soon as they had enough ivory, they used either rawhide straps or strips of bark to tie the tusks to a pole. Two people would carry a tusk on their shoulders, but if the tusk was very heavy a third bearer would walk in the middle.

---

52 Limpopo Province.

They took the ivory and some cash to Delagoa Bay where they bought blankets or shawls (known as *sitossana*), mirrors, tinderboxes, beads and other bric-a-brac in demand in the interior. With heavy bundles on their heads, they returned to the Soutpansberg in groups of 20 or more. From there they went to Tshivhase, Rambuda, Thengwe and across the Limpopo to Mashonaland, following the ancient trade route.

As soon as the Shonas spotted the traders, they climbed the koppies and shouted: "The *vashavi* (traders) have arrived!" News spread quickly and soon the Shonas arrived with their cattle. The traders paid ten shawls or five blankets for a cow. Once they had completed their transactions, they left. Each of the 20 to 50 traders had acquired about five cattle, making a total of 100 to 120 that had to be driven back.

On their return journey they had to be very careful of lions. In the late afternoon some of them looked after the grazing cattle while the rest chopped branches and collected firewood to build a kraal for the cattle and make bonfires during the night. When water was scarce they had to drive the animals as quickly as possible through the dry areas.

Dicke had planned to use this information in an appendix named "Old mineral and trade routes" in his unfinished manuscript *The Northern Transvaal Voortrekkers*, but he died before he could do so. Still, some intriguing information about the way in which trading took place appeared in *The Bush Speaks*, as is clear from the following:

> How was business conducted on such a trade route? The native traders using it did business separately, but provided for protection jointly. Along a route they established fortified depots where always some of them were to be found as there was much traffic on such a route. At the depots the traders rested their caravans and deposited their merchandise and the goods bartered. From the depots their retail pedlars went out in all directions amongst the people with whom they wanted to barter. To the depots the retailers returned to store the ivory, copper, and gold obtained until sufficient bearer loads had accumulated to despatch carriers adequate in number safely to make the journey; the traders had always to be prepared to meet a surprise attack, a hold up, an attempt to rob or to exact tribute. According to the dangers to be met on different stretches between depots, traders would wait at the depots until others would be travelling the same direction with their caravans.
>
> Sometimes punishment had to be meted out to a native clan that had attacked a column or expeditions had to be undertaken to try and recover goods robbed.

In his capacity as Vice Consul of Portugal in the Transvaal, João Albasini reported such a raid to the magistrate of Schoemansdal in 1862. Four black bearers working for the trader Covetti, who lived in Lourenço Marques, had been murdered in the Lowveld and robbed of: 180 pieces of linen; 400 strings of Matombo beads; six bunches of coloured beads; and 12 bunches of blue polished beads.

It would be worthwhile to determine if there are any remains of the fortified depots along the trade routes about which Dicke had written. In 1941 Colonel P.W. Möller, a highly intelligent and observant person, told me that in 1887 he and his brother had gone on a hunting trip to the area between the Klein Letaba and a tributary of the Nsama River (Ntsama or Musana). One day he wounded a mountain reedbuck, which fled up a hill called Medzimoagombi (Matzibanamba or Madzimbanombe). After following the blood spoor, he found the dead buck among the remains of some neatly packed stone ruins. A little north of the ruins and on top of the hill he found a guard or observation tower from which one could see very far over the bushveld in any direction. Because the trade route along the Klein Letaba through the Shingwedzi or Shilowa poorts to Inhambane passed on the southern side of the hill, these ruins, with the observation tower, had probably been one of the fortified depots about which Dicke had written, used by the earliest traders. It seems Albasini had a depot (which has not been found yet) by the Lower Sabie, as well as one at Magashulaskraal, one at Klipkraal and one at Rustplaats. But such depots or rest stops had already been around 120 years before Albasini made use of the routes.

Mahumane, for instance, slept over in huts specially erected for travellers' convenience at two different places along the trade route in 1727 and 1728 during his journeys between Soutpansberg and Delagoa Bay.

The only known organised trade journey and hunting expedition that followed the old trade routes and of which a written record exists, was that of Das Neves in 1860. His group consisted of:

> … one hundred and twenty carriers, with bales of goods for trading with the Dutch; thirty with the merchandise proper for bartering for provender and provisions; three captains or guides for the carriers, seventeen hunters, sixty-eight negroes for transporting the necessary materials for the hunt, five carriers for my personal baggage, four servants, a second and third lieutenant in command, and four carriers for their separate use, in all mustering 253 men.

The early traders paid in beads and cloth for ivory and other products. To maximise their profits they had to keep up to date with the needs and fashions of the black women in the interior. Das Neves had the following to say about this:

> The dress is rather graceful. It consists of a cotton capelana fastened round the waist and scarf thrown gracefully in front, both ends reaching the knees; around their waists they wear many strings of large blue beads and pale missanga (small glass beads).

Das Neves paid the following for an elephant tusk weighing 69 pounds (31.4 kilograms):

> thirty pieces of blue cloth, five pieces of carlagani (Indian cloth), ten capelanas, thirty bunches of missanga and fifteen rows of blue beads, the whole amounting to the value of 66.00 reis (about £15). The tusk was worth in those days 86.700 reis (£19.5.0).

Cowrie shells found at Mapungubwe and the Matshêma Ruins in northern Transvaal[53] during archaeological digs prove that they had also been used as barter there. These shells came from the east coast and were traded in the interior. Archaeologists found some cowrie shells in the Ngorogoro Crater dating as far back as 1500 BC. William Bolts, a representative of an Austrian trading company at the east coast, in a letter dated 16 July 1777, Mfumo (Delagoa Bay), wrote to Andrew Pollet in London:

> Cowries of which there is great plenty to be had on some parts of these coasts and particularly at Iniáca Island is another considerable article of commerce and it merits our attention to encourage the collection of them.

In July 1944 Chief Josia of the Bakoni, who lived in the Leydsdorp district, gave some interesting information about trade in the old days to Dr N.J. van Warmelo, with the assistance of a number of old tribe members at his main kraal. They told him that blacks with bundles on their heads had come from their home Ga Mpfumo (Delagoa Bay) to trade in the interior a long time ago. Their route crossed the Lebombo Mountains at the home of Mphisane, a Portuguese trader, who lived in a poort. Sometimes the Bakoni carried their trade goods to the coast. The import goods included *thaga* (green, red and blue beads), *mabêtlwa* (six-sided blue beads), *maôkolane* (a type of blue cloth) and later guns, gunpowder and lead, which also came from the Portuguese. The *thaga* resemble in colour and shape the Lebedo's *mudala* and the Venda's *vhulungu ha madi* ("beads of water"), which are today only worn by chiefs' wives and are considered holy heirlooms.

Other goods that were probably imported from Delagoa Bay in ancient times were metal suits of armour worn by medieval European knights. While a surveyor was working in Sekhukhuneland in the late 1950s, he once noticed a motionless black man watching him from the shade of a tree. The man was wearing the metal breastplate of a medieval suit of armour. At first he was unfriendly and sullen and refused to identify himself, but the surveyor eventually managed to bribe him with money to reveal that he was one of the chief's secret guards and that the suits of armour were part of their uniform.

Taking all of this information into consideration, it would seem that the Indians, Arabs, Portuguese, Dutch, Austrians, Magwamba and other traders, over a period of many years bartered large blue ring-shaped *matombo* beads, *thaga* beads in different col-

---

53 Limpopo Province.

ours, large blue polished angular *mêthwa* beads, small light-blue *missanga* beads, small beads in different colours, and cowries, known among the Venda as *tshêshêla* ("the laughing shells", so called because the indentations resemble little teeth). Otherwise they traded both finely and coarsely woven linen, flowered calico, a blue cotton fabric known as *maôkolane*, a red fabric known as *moretele*, checked cloth, striped cloth, Indian cotton or *carlagani capelanas* (shawls, *sitossana* and Trichardt's scarves), as well as ladies' capes or coats of red cashmere. Snuffboxes, tinderboxes, straw hats, shoes, men's clothes and socks were all part of the price paid for ivory, rhino horn, rhino and hippo hoof, gold, copper, tin, amber, honey, wax, salt and butter. Guns, ammunition and sometimes even knights' suits of armour were sold to the blacks.

References

Ackermann, D. 1983. Marale van groot argeologiese belang. *Custos* 12(6):30–31.

Albasini, João. 1858–1872. *Correspondencia officialdo consulado Portuguez em a republica Africana do Sui Transvaalsche*. (Official correspondence of the Portuguese Consulate in the *Zuid-Afrikaansche Republiek* Transvaal). Document in possession of the Albasini family of which Dr J.B. de Vaal had a photocopy and an English translation.

Axelson, E.V. 1940. *South-East Africa 1488–1530*. Longmans Green, London.

Axelson, E.V. 1960. *Portuguese in South-East Africa 1600–1700*. University of the Witwatersrand, Johannesburg.

Baumann, M. 1919. Ancient tin mines of the Transvaal. *Metallurgical and Mining Society of South Africa* 19(7):120–132.

Bernhard, E. & Bernhard, F.O. 1969. *The journals of Carl Mauch 1869–1872*. National Archives of Zimbabwe, Salisbury.

Bernhard, F.O. 1971. *Karl Mauch African Explorer*. C. Struik, Cape Town.

Breytenbach, J.H. *Suid-Afrikaanse Argiefstukke, Transvaal no. 2: Notule van die Volksraad van die Suid-Afrikaanse Republiek II, 1851–1853*, Parow, p. 49: E.V.R. 3, Kommissieraadsvergadering, Lydenburg, 27.11.1851, par. 7.

Burke, E.E. (ed.). 1969. *The journals of Carl Mauch*. National Archives, Salisbury.

Das Neves, D.F. 1879. *A Hunting Expedition to the Transvaal*. George Bell & Sons, London.

De Vaal, J.B. 1942. Ysterbewerking deur die Bawenda en die Balemba in Soutpansberg. *Tydskrif vir Wetenskap en Kuns* 3(1):45–50.

De Vaal, J.B. 1942. An ancient mine in the Transvaal. *African Studies* 1(2):151–152.

De Vaal, J.B. 1943. 'n Soutpansbergse Zimbabwe. *South African Journal of Science* 40:303–322.

De Vaal, J.B. 1952. Die ou kopermyne van Messina. *Lantern* 1(5):405–408.

De Vaal, J.B. 1952. Ysterbewerking in Vendaland. *Lantern* 2(2):152–156.

De Vaal, J.B. 1953. Die rol van João Albasini in die geskiedenis van Transvaal. *Archives Year Book for South African History* 16(1):1–155.

De Vaal, J.B. 1958. Die Lemba … die Semiete onder die Bantoe van suidelike Afrika. *Bantoe* 7:51–74.

De Vaal, J.B. 1983. Die Jerusalem-gangers. *Die Transvaler*, 8 April.

De Vaal, J.B. 1984. Ou handelsroetes en wapaaie in Oosen Noord-Transvaal. *Contree* 16:3–15.

De Vaal, J.B. 1985. Handel langs die vroegste roetes. *Contree* 17:3–14.

Dicke, B.H. 1936. *The Bush Speaks: Border Life in the Old Transvaal*. Shuter & Shooter, Pietermaritzburg.

Dicke, B.H. 1938. The Northern Transvaal Voortrekkers. *Archives Year Book for South African History* 4(1):67–170.

Een Hollander (Pseudonym for K.H. Kok). 1898. *Toen en Thans*. Dusseau, Cape Town.

Elton, F. 1872. Journal of an exploration of the Limpopo River. *Journal of the Royal Geographical Society* 42:47.

Evers, J.M. 1974. Iron Age trade in the Eastern Transvaal. *The South African Archaeological Bulletin* 29(113–114):33–37.

Fouché, L. 1937. *Mapungubwe, Ancient Bantu Civilization on the Limpopo*. 2 vols. Cambridge University Press, Cambridge.

Friede, H.M. & Steel, R.H. 1976. Tin mining and smelting in the Transvaal during the Iron Age. *Journal of the South African Institute for Mining and Metallurgy* 76(12):465.

Gieseke, E.D. 1930. Die Eisen-Industrie der Bawenda. *Die Brücke* 8(4):5–9.

Haddon, A.C. 1908. Copper rod currency from the Transvaal. *Man* 8(65):121–122.

Hammerbeck, E.C.J. & Schoeman, J.J. 1976. *Koper in delfstowwe van die Republiek van Suid-Afrika*. Unknown publisher, Pretoria.

Hromník, C.A. 1981. *Indo-Africa*. Juta & Co. Ltd, Cape Town.

Hromník, C.A. 1984. Nkomati means trade and communication. *The Argus*, 10.

Hromník, C.A. 1986. Gold and God in Komatiland. Unpublished manuscript.

Junod, H.A. 1927. *The Life of a South African Tribe* 1 & 2, Macmillan, London.

Le Roux, T.H. (ed.). 1964. *Die dagboek van Louis Trichardt*. Van Schaik, Pretoria.

Liesegang, G. 1977. New light on Venda traditions: Mahumane's account of 1730. *History in Africa* 4:169.

Lindblom, G. 1926. Copper rod currency from Palabora, N. Transvaal. *Man* 26:144–147.

Lowe, C. Van Riet. 1944. Ladybrand may rival Les Eyzies as the mecca of prehistorians. *The Outspan,* 10 March: 27 & 45.

Mason, R. 1962. *Prehistory of the Transvaal*. Witwatersrand University Press, Johannesburg.

Merensky, A. 1888. *Erinnerungen aus dem Missionsleben in Süd-Ost Afrika 1859–1882*. Velhagen & Klasing, Bieleveld.

Meyer, A. 1986. *'n Kultuurhistoriese interpretasie van die Ystertydperk in die Nasionale Krugerwildtuin*. Unpublished D.Phil. thesis, University of Pretoria.

Pienaar, U. de V. 1985. Indicators of a progressive desiccation of the Transvaal Lowveld over the past 100 years, and implications for the water stabilisation programme

in the Kruger National Park. *Koedoe* 28:93–166.
Preller, G.S. (ed.). 1917. *Dagboek van Louis Trichardt. (1836–1838)*. Het Volksblad Drukkerij, Bloemfontein.
Preller, G.S. (ed.). 1940. Stigting en verwoesting van Schoemansdal. *Die Vaderland*, 13 December.
Prinsloo, H.P. 1974. Early iron age site at Klein Afrika near Wylliespoort, Soutpansberg Mountain, South Africa. *South African Journal of Science* 70:271–273.
Punt, W.H.J. 1939. Die Van Rensburgmoord. *Die Volkstem,* 11 and 25 September and 2 October.
Punt, W.H.J. 1953. *Louis Trichardt se laaste skof.* Van Schaik, Pretoria.
Punt, W.H.J. 1960. Waar die Van Rensburgtrek in 1836 vermoor is. *Koedoe* 3:206–237.
Punt, W.H.J. 1975. *The First Europeans in the Kruger National Park 1725.* National Parks Board of Trustees, Pretoria.
Punt, W.H.J. 1975. *The Relationship Between the Pioneer Routes in the Transvaal and Ancient Trade Routes.* South African Archaeological Society, Johannesburg.
Schwellnus, C.M. 1937. Short notes on the Phalaborwa smelting ovens. *South African Journal of Science* 33:911.
Spencer-Smith, T. 1980. Zulu's have Indian ancestors. *Sunday Times*, 22 June.
Stanley, G.H. 1910. Notes on ancient copper workings and smelting in Northern Transvaal. *Proceeding of the University of Durham Philosophical Society* 3(5).
Stanley, G.H. 1929. Primitive metallurgy in South Africa: some products and their significance. *South African Journal of Science*, 26 December:733–748.
Stayt, H.A. 1931. *The BaVenda.* Cass, Oxford.
Suid-Afrikaanse Biografiese Woordeboek vol. 4. 1982. Insert on: Dicke, Bernard Heinrich. Human Sciences Research Council, Pretoria.
Theyl, G.M. 1964. *Records of South-Eastern Africa* 1. Rapport van Jan Van de Capelle 3 Augustus 1723. Cape Town:407–420.
Thompson, L.C. 1949. Ingots of native manufacture. *Nada* 26:13.
Thompson, L.C. 1949. A Native made tin ingot. *Nada* 31:40–41.
Transvaalse Archives (TA). Acquisition A 81, Pater Joaquim de Santa Rita Montanha's journey report.
Transvaal Archives. S.S. 41 State Secretary – Incoming documents: R81/62 supplementary documents, 8 June 1862.
Transvaal Archives. A497 Dr Meyr-Harting-Collection – William Bolts, Mafumo 16 July 1777 to Andrew Pollet, London.
Van Hoepen, E.C.N. 1939. A pre-European Bantu culture in the Lydenburg district, *Argeologiese Navorsinge van die Nasionale Museum* 11(5).
Van Warmelo, N.J. 1940. *The Copper Miners of Messina and the Early History of the Zoutpansberg.* Government Printer, Pretoria.
Van Warmelo, N.J. 1944. The Bakoni of Mametsa. *Ethnological Publication* 15:48.
Wagner, P.A. 1926. Bronze from an ancient smelter in the Waterberg district Transvaal. *South African Journal of Science* 13:899–900.
Wangemann, H.T. 1868. *Ein Reise-Jahr in Süd-Afrika.* Verlag des Missionhauses, Berlin.
Ziervogel, D. 1954. *The Eastern Sotho.* Van Schaik, Pretoria.

Maps

Bartholomew's Tourists' Map of South Africa. 1: 2 500 000, Edinburgh, 1903.

Berthoud, H. 1903. *Map of Zoutpansberg, North Transvaal* 1: 333 000, Berne.

De Vaal, J.B. 1984. Ou Handelsvoetpaaie en wapaaie. *Contree* 16:6. *Graadblaaie* (Farm maps) – Government Printer, Pretoria 1929, in S.P. Engelbrecht Museum, Pretoria.

Jeppe, F. 1879. *Map of the Transvaal and the surrounding territories* 1: 1 850 000, Pretoria, 1879, In: Sandemann, E.F. 1880, *Eight Months in an Ox-wagon*. Griffith & Farran, London.

Jeppe, F. & C.F.W. 1899. *Jeppe's map of the Transvaal or S.A. Republic and surrounding territories.* Pretoria.

Loveday, R.K. 1883. *Map of the Lydenburg Gold Fields, S.A. Republic* (Transvaal). Pretoria, (TA, 1/267).

Merensky, A. 1875. *Original map of the Transvaal or South African Republic including the gold and diamond fields.* 1: 1 850 000, Botshabelo, Berlin.

Petermann, A. 1893. *Originalkarte von Südafrika* 1: 2 000 000, 1870. Provincia de Mocambique. 1: 1 000 000, (TA, 2/5).

Punt, W. 1959. Handelsroetes tussen Noord-oos Transvaal en Inhambane-Delagoabaai ten tyde van Trichardt, volgens B.H. Dicke, *Die Volkstem*, 25 September.

Punt, W.H.J. 1953. Ou wapaaie na Delagoabaai, 4 April, (S.P.E.M.).

Punt, W.H.J. 1953. Louis Trichardt se roetes, kampe en uitspannings in Zoutpansberg. In *Louis Trichardt se laaste skof.* Van Schaik, Pretoria:36.

Punt, W.H.J. 1953. Die Trichardtroete oor die Pietersburgplato. *Louis Trichardt se laaste skof.* Van Schaik, Pretoria:54.

Punt, W.H.J. 1953. Die Trichardtroete deur Pienaarspoort na die Olifantsrivier. *Louis Trichardt se laaste skof.* Van Schaik, Pretoria:67.

Punt, W.H.J. 1953. Die Trichardtroete in die Olifantsriviervallei en oor die Drakensberg. *Louis Trichardt se laaste skof.* Van Schaik, Pretoria:78.

Punt, W.H.J. 1953. Die Trichardtroete deur die Laeveld na Lebomborand. *Louis Trichardt se laaste skof.* Van Schaik, Pretoria:91.

Punt, W.H.J. 1953. Die Trichardtroete van die Lebomborand na Delagoabaai. *Louis Trichardt se laaste skof.* Van Schaik, Pretoria:96.

Punt, W.H.J. 1960. Voortrekkerroetes deur die Krugerwildtuin. *Koedoe* 3:26–237.

Punt, W.H.J. 1960. Terrein van Van Rensburgmoord Julie 1836 soos vasgestel deur die Parkeraadekspedisie 1959 en navorsing van dr. W. Punt (twee kaarte). *Koedoe* 3:217 & 219.

Punt, W.H.J.1964. Frans de Kuiper expedition – 1725. In: *The First Europeans in the Kruger National Park 1725*. National Parks Board, Pretoria:70 & 71.

Raddatz, H. *Map of Zoutpansberg district, Transvaal Goldfields* 1: 500 000 (TA, 3/478).

Sketch Map Of The N.E. Zoutpansberg ZAR, (TA, 2/307).

Von Wielligh, G.R. 1890. Plan aantoonende N.O. Grenzen der Z.A. Republiek tusschen Komatipoort en samenloop van Pafuri en Limpopo. Archive Ref. R5879/87 and R11709/90. *Union of South Africa*, l: 500 000, topographic maps, Government Printer, Pretoria, 1936.

Chapter 3
The first Europeans visit the Lowveld
Dr W.H.J. Punt

There are currently no black tribes living in the Kruger National Park. After the reproclamation of the Sabie Reserve in 1902, those who had been living there were systematically resettled in surrounding areas. However, in previous centuries black tribes occupied the entire Lowveld, as is clear from the accounts of journeys between 1725 and 1838. These black people forged copper and iron, mined gold and traded with the proceeds of hunting, especially ivory. The tribes' economies were simple and no large-scale trading took place between the tribes living at the coast and those in the interior. However, the centuries-old trade routes between various inland sites and the coast do attest to the fact that limited trade did take place between the interior and the settlements along the Mozambican coast, as discussed in chapter 2.5. Although the gold trade between Monomotapa and Sofala (near the current Beira) was a lucrative business, traders did not pay much attention to the coast south of Sofala. The economic development of the Lowveld was also hampered by tribal conflict and wars of conquest.

Another factor that impeded to a great degree the development of trade and colonisation of the area was the two diseases that plagued both man and beast in the Lowveld through the centuries: malaria, transmitted by the *Anopheles* mosquito, and nagana, transmitted by the tsetse fly and deadly to domestic animals.

It is not known exactly when these two diseases first appeared in the Lowveld, but there are indications that this had happened many centuries ago. The threat these diseases posed and the areas in which they occurred varied at different times as a result of climatic cycles and other factors. In 1725, for example, the first Europeans to enter the area found that there were no tsetse flies in the southern part of the Kruger National Park. The fact that the area was densely populated indicates that malaria was also not a problem in the interior at the time. This interpretation is further borne out by the fact that 31 Europeans could travel in the Komati area for 18 days during the winter of 1725 without contracting the disease.

The distribution range of the tsetse fly expanded in the nineteenth century and became the main reason why transport by ox-wagon through the nagana-infested Lowveld to the Mozambican coast was a dangerous and risky undertaking. When Louis Trichardt trekked through the Lowveld from 1836 to 1838, he encountered tsetse flies at several places, especially along the Limpopo and Olifants rivers.

Because of tsetse fly the black tribes in the Lowveld could not keep domestic animals then. The chiefs could nevertheless show the Boers old cattle kraals that had been in use not long before. Between 1725 and 1836, tsetse fly had thus penetrated parts of the area that is now the Kruger National Park and for almost a century nagana made large tracts impassable to trek oxen. Then, in 1896, a widespread rinderpest epidemic in the Lowveld left a trail of bleaching bones in its wake. It almost managed to eradicate the wild hosts of the tsetse fly, including buffalo, bushbuck, warthog, kudu and eland. With the virtual eradication of the game, the tsetse fly also disappeared from the area.

Taking into consideration the factors that negatively affected humans and animals, it is understandable that the Lowveld was impenetrable to Europeans for centuries. In former times it was difficult to reach and economically not too attractive, even to Asian and native traders. The trade routes to the interior were therefore limited in number, as was the extent of most trading activities.

Before Vasco da Gama navigated the Indian Ocean in 1497, Egyptians, Chinese, Persians, Arabs, Indians and Indonesians were already trading along the east coast of Africa. These traders seldom ventured south of Sofala, although the Phoenicians had long since sailed around the southernmost point of Africa. However, this journey was more an exploratory venture than a trade mission.

At the start of the Common Era other nations gradually started moving south from the Red Sea along the east coast. Asian traders, especially, had opened up trade routes to the interior of Africa from the ports they occupied and so the riches of the African continent, such as gold, ivory, slaves, etc., were transported to the East along these routes. Sofala was the southernmost of the larger ports used to export the riches of Monomotapa or Ofira (now Zimbabwe). The present Zimbabwe Ruins were one of the most important centres visited by gold traders from the east coast. South of the Limpopo, in what is now the Lowveld of Mpumalanga, trade was less profitable, although quite a few trade routes originated in the Transvaal, especially from the northern Transvaal (Limpopo) to the coast and Zimbabwe (see chapter 2.5).

According to accounts of the Venda in the Soutpansberg, traders with hook noses wearing long white robes used to walk southwards through the Soutpansberg along the trade routes leading from Zimbabwe. When the Voortrekkers reached the north-eastern Transvaal in 1836, the old trade route over Punda Maria to Sofala and Inhambane was already well known and in general use.

From the Northern Transvaal (Limpopo) more routes led to the Rooiberg tin mines. Tradition has it that these paths existed long before the arrival of the first Europeans. These same routes were later used by the first white inhabitants of the Transvaal, Coenraad de Buys and his sons, as well as Voortrekker leaders Hans van Rensburg, Louis Trichardt and Andries Hendrik Potgieter.

The routes linking what used to be black kraals in the area that is now the Kruger National Park are even older and were used to carry ivory and metalwares (especially iron and copper) from the Lowveld to the Mozambican coast. However, these routes never developed into large and busy trade routes.

The exploration of the interior from the east coast harbours was neglected for centuries and up until the end of 1400, the eastern Transvaal Lowveld (Mpumalanga) remained largely unknown to historians and explorers.

European ships first appeared along the east coast of Africa in 1497. That year Portuguese ships under Vasco da Gama's command rounded the Cape of Good Hope and sailed northwards along the coast to India. Within a few years the Portuguese had occupied most of the ports between Sofala and Malindi, and so the Europeans became the rulers of the East African coast and obtained control over the gold, ivory and slave trade. In 1505 Pedro Anhaya took Sofala from the Arabs. Fort Mozambique was built in 1507 and other Arabic trading posts and ports along the east coast, including Malindi, Mombasa, Zanzibar, Sena and Mogadishu, were systematically taken over by the Portuguese seafarers.

It was noted by Hendriksen (1978) that a Portuguese ship's carpenter, Antonio Fernandes, was the first European to venture into the interior, away from the safety of the east coast ports. In 1514 he undertook an extensive journey from Sofala and reached Zimbabwe, where he even made the acquaintance of the legendary Chief Monomotapa. Although quite a few signs of ancient mining activities have been found in the Kaap Valley and as far west as the so-called "Portuguese Reef" between Sabie and Mount Anderson, no real documentary evidence have been found to date proving that the Portuguese had penetrated the Transvaal Lowveld before 1700. Although Portuguese ships visited the ports of Inhambane and Delagoa Bay on an annual basis to collect ivory and slaves, the black settlers from the interior had brought the trade goods to the coast. According to Hromník (1981 & 1986), the early mine workers in the Transvaal Lowveld were probably traders and miners of

The old Portuguese fort at Mombasa along the east coast of Africa in September 1909.
(ACKNOWLEDGEMENT: STEVENSON-HAMILTON COLLECTION, SKUKUZA ARCHIVES)

Arabic, Persian, Indian or Indonesian origin. Until 1725 the Transvaal Lowveld was a closed book to Europeans, as purposeful and planned expeditions were only undertaken after the Dutch East India Company had annexed Delagoa Bay.

In 1602 the Dutch East India Company was founded in Amsterdam. This powerful trading company, in Dutch the Vereenigde Oos-Indiese Compagnie (VOC), was governed by 17 directors, known as the Lords XVII. It maintained its hegemony for nearly 200 years, gathering enormous riches and exercising great political, cultural and economic influence on countries bordering the Indian Ocean, from the Cape of Good Hope to Japan.

The headquarters and naval base of the Portuguese empire in East Africa was Fort San Sebastião on the small island of Mozambique. In order to obtain a foothold in the gold trade from Monomotapa to the coast, this island had to be conquered. The Lords XVII wasted no time and in 1603 Admiral Stevan van der Hagen, with a fleet of 12 warships, was ordered to capture the island. This attempt failed, as did the later attempts by admirals Paulus van Carden in 1607 and Pieter Verhoef in 1608.

In 1589 Admiral Wybregt van Warwyck took possession of the uninhabited island of Mauritius on behalf of the States General of the Netherlands. In 1638 the VOC built Fort Frederik Hendrik on the island and entered into trade relations with the neighbouring island of Madagascar. The VOC concluded a commercial treaty with the King of Antogil on the east coast of the island and thus obtained a foothold in Africa's east coast.

Meanwhile British and French ships also appeared in the Indian Ocean as competitors of the Dutch and Portuguese for a share of the gold and ivory riches of Africa. Despite all the competition, the Portuguese remained in possession of the harbours between Sofala and the Red Sea. From these harbours overland routes led to Monomotapa, of which the most important route was the one from Sofala to Zimbabwe.

In 1651 the Lords XVII made the very important decision to take possession of Table Bay. As instructed by the VOC, Commander Jan van Riebeeck landed in Table Bay on 6 April 1652 to establish a fort and victualling station. After approximately 50 years the Dutch at long last had a base on the continent of Africa. From here they would be able to obtain a share of the valuable mineral trade of darkest Africa.

The Dutch settlement heralded the beginning of purposeful and carefully planned explorations in search of minerals in southern Africa. In the course of the next two centuries a series of expeditions led to the discovery of the mineral wealth and precious gemstones of our country. Jan van Riebeeck ordered the first expeditions to the interior. He also interrogated the Khoikhoi about the route to the land of gold, Monomotapa. After him, Governor Simon van der Stel undertook a journey in 1685 to the rich copper fields of Namaqualand. In 1689 he bought the Bay of Natal for the Lords XVII with the intention of establishing a VOC trading post closer to Monomotapa (Punt, 1975).

The information they garnered about the east coast harbours convinced the VOC that the unoccupied Delagoa Bay was the nearest bay to the Zimbabwe goldfields, and that it could be reached overland through the Eastern Transvaal. The Lords XVII feared that the British or French might take possession of the Bay of Lourenço Marques before the Dutch could. Thus the proposal to occupy the bay met with the approval of the directors of the VOC. It was an urgent matter and on 25 July 1719 the Lords XVII in Amsterdam decided to order the Political Council in Cape Town to take possession of Delagoa Bay for the VOC.

A few months later the Governor of the Cape Colony, Mauritz de Chavonnes, began arrangements for the expedition to Delagoa Bay. Many volunteers wished to take part in this adventurous undertaking and authorities selected a group of about a hundred enthusiastic soldiers, sailors and artisans.

Two ships, the *Gouda* and *Zeelandia*, were carefully loaded with all the supplies needed to build a fort and plant a vegetable garden. Wood, seed, cannons, weapons, trade goods, beads, medicines, tents, food, equipment etc. were taken on board. Willem van Taak was appointed as commander of the new trading post with Jacob de Bucquoi as surveyor and cartographer.

On 29 March 1721 the two Dutch ships sailed into the bay of Lourenço Marques and dropped anchor in the Rio Esperito Santo. From their ships the settlers viewed the area that would become their home and where they had to establish a new colony that would serve as a trading centre for the gold and ivory trade with Monomotapa.

Delagoa Bay was a beautifully sheltered harbour with high sand dunes, lagoons, low hills, dense bush, palm trees and white sandy beaches. On a hill and along the southern riverbank were native huts. A paradise, but one in which a deadly fever lurked. No trace could be found of previous Portuguese settlements on the mainland. Only on Chefini Island, several kilometres north of the harbour, at the mouth of the Komati River, were a group of deserted huts and the ruins of an old trading post of the Portuguese, long since abandoned.

In the glaring sunlight and scorching heat, Van Taak looked for a suitable spot to build a fort. A high sand ridge north of the river between the two lagoons grabbed his attention. This site was a few metres higher than the surrounding swamp and during high tide it became an island. Later, in about 1800, the Portuguese would build their fort, the Nossa Senhora de Conceiçao de Lourenço Marques on the same spot, while Louis Trichardt outspanned his ox-wagons there on 13 April 1838.

Within a few days Van Taak and his men were hard at work building a wood and earthen fort, searching for water supplies and clearing the bush. Trade negotiations and the purchase of the land were discussed with local chiefs. However, March and April were the worst malaria months in Delagoa Bay. Within a few weeks Van Taak had only a few healthy men left. In spite of all these troubles and with the assistance of the locals, the few healthy men remaining worked so hard that after a few months Fort Rio de Lagoa (later renamed Fort Lijdzaamheid because of the many deaths) was completed, the garden planted and treaties concluded with the local chiefs. Now attention could be paid to the reconnaissance of the interior. After Van Taak's death in 1723, Jean Michel became the provisional leader and Jan van de Capelle his second in command.

With great expectations the VOC officials sailed up the river entering the bay and asked local chiefs about the trade in gold, copper, iron, tin and ivory, as well as routes to the interior. Shortly afterwards, on 9 August 1723, the commander sent Sergeant Jan Christoffel Steffler, a soldier from Magdeburg, and Sergeant Jan Mulder with 18 men on an expedition to the "Blauwe Bergen", or Lebombo Mountains. The soldiers covered a distance of about 64 kilometres (10 hours) westward from Fort Rio de Lagoa to the Lebombo Mountains.

It is accepted by some historians that Steffler's company was the first group of Europeans to enter the Transvaal Lowveld. But this claim was refuted after a study of Steffler and Mulder's account of the expedition. The point they had reached was probably between Nomahasha and Goba in the Lebombo Mountains in Swaziland.

The "Blauwe Bergen" that were the object of the company's journey, were undoubtedly the Lebombo Mountains (or Rooirand) that today form the border between South Africa and Mozambique, as well as between Swaziland and Mozambique. A blue mountain range on the western horizon is clearly visible from the hill in Delagoa Bay. As the crow flies the Lebombo Mountains are about 64 kilometres west of Delagoa Bay and it was this direction in which Steffler travelled.

The end of Steffler's journey can be identified from his description. On the first day he noted that they had crossed the Matola River at its mouth, after which they trekked through the "second" Matola River. Then he and Mulder travelled upstream along the southern bank of the Umbuluzi until they crossed a third river. Here they pitched camp, while Steffler and a few men travelled ahead to explore the route. As a result of this rash action of Steffler his small company was attacked by a group of Swazis on 18 August 1723 while crossing a river. In the ensuing skirmish, both Jan Steffler and Sergeant Eldret Sligting were killed. The rest of the company returned to their camp and managed to reach Fort Rio de Lagoa unharmed.

The expedition achieved little, but they did ascertain that there was coal in the Lebombo Mountains (this is the first reference to the existence of this mineral in South Africa) and that it was used by the Swazis. They also found out that the famous Ysterberg was situated in the northern Transvaal, nine days on foot from the Bay. The expedition, however, did not set foot in the Transvaal despite probably crossing the wooded western border of Swaziland.

3. THE FIRST EUROPEANS VISIT THE LOWVELD

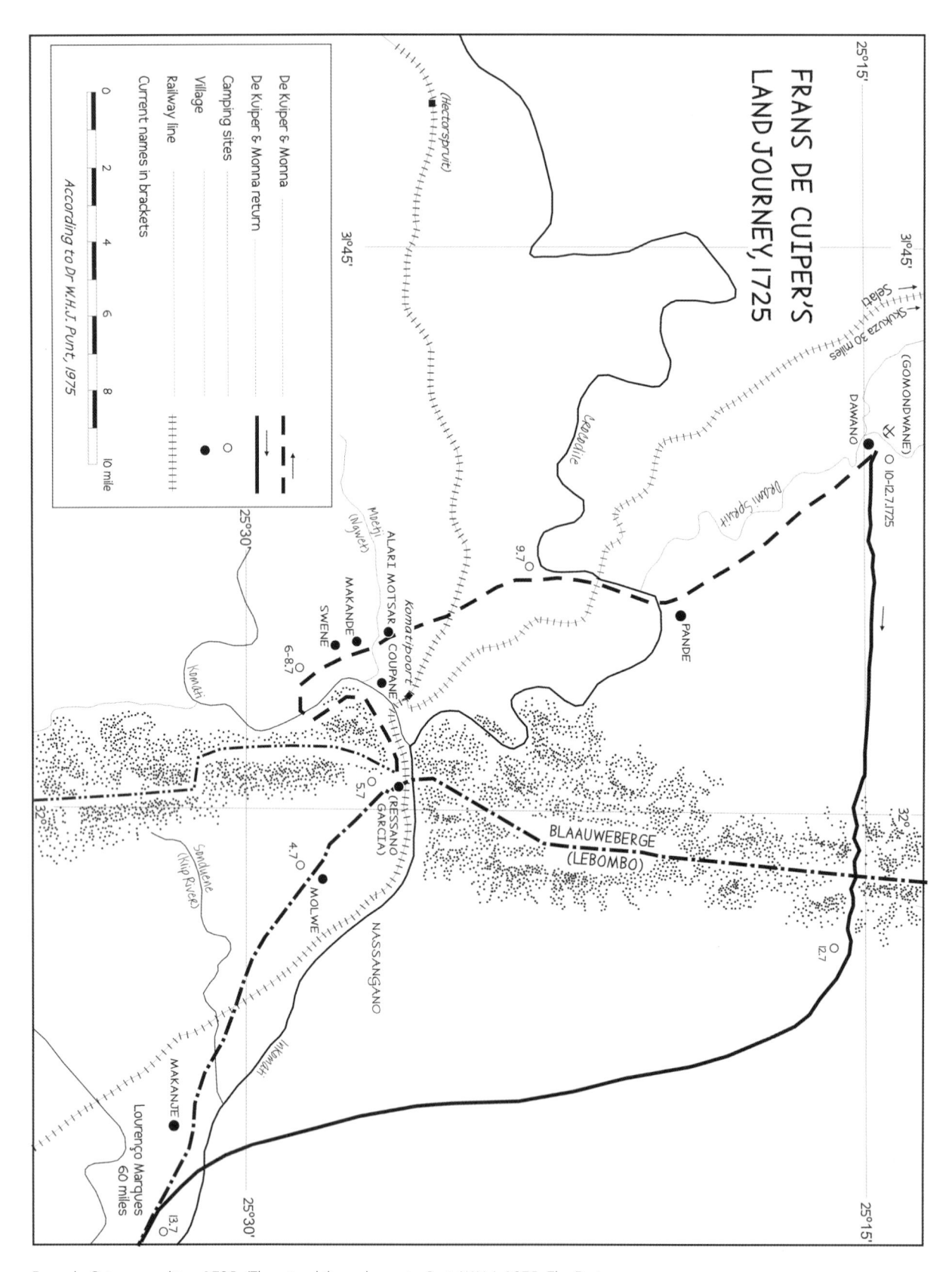

Frans de Cuiper expedition 1725. (There is a bilingual copy in: Punt, W.H.J. 1975. *The First Europeans in the Kruger National Park*. National Parks Board of Trustees, Pretoria.)

In 1725 Jan van de Capelle became provisional commander of the colony and under his strong leadership and inspiring example, the VOC's affairs made rapid progress. The new commander was a patriotic man, inspired with the ambition of reaching the goldfields of Monomotapa and thereby promoting the interests of his company and country. The Dutch had already learnt from the local tribesmen that the road to the north of the Lebombo, or Copper Mountains, went through densely populated areas.

If contact with Monomotapa could be established, it would greatly benefit the Cape Colony. Obviously all export and import trade from the Delagoa Bay colony, a domain of the Cape, would take place via Table Bay where the larger VOC return fleets could take supplies on board.

At the time two age-old routes went from Monomotapa to the coast along which the Portuguese transported their trade goods. The first route was the waterway down the Zambezi and from its mouth northwards to the island of Mozambique. A second route was the Portuguese-controlled footpath westward from Fort Sofala to the gold emporium in Zimbabwe. But during the VOC's occupation of Delagoa Bay there was still no regular trade between the goldfields in the heart of southern Africa and Delagoa Bay, making the prospect a tantalising possibility to the Dutch.

The attitude of the inhabitants of the interior to a European-controlled trade route through their country was an unknown factor, but Van de Capelle thought that their cooperation could be obtained for such a venture. However, the dangerous malaria and blackwater fever, the tsetse fly and the hostile locals would become decisive factors in the VOC's ambitious plans. After due consideration, Van de Capelle decided that an expedition should be undertaken to determine a route from Delagoa Bay to the mines in the interior.

On 18 June 1725 the settlement's political council approved the commander's plan. His instructions were ready on 25 June and the composition of the expedition was announced in the Fort. The costs of the journey were to be defrayed by the government of the Cape of Good Hope.

Van de Capelle selected with great care the soldiers and sailors who would make the adventurous expedition to the interior. The group consisted of a troop (which today would be called a platoon) of 31 men under the leadership of Assistant Francois de Cuiper[1] and Sergeant Johannes Monna. At that time 31 was the standard number of men for a military expedition.

De Cuiper kept a diary (now preserved at the Government Archives in Pretoria) in which he comprehensively described their journey, including their travels through the area that would become the Kruger National Park 250 years later. The handwriting is clear and the fine Dutch easily legible.

There are many interesting details for the geographer, historian, geologist, ethnologist and student of the Dutch and Afrikaans languages and of flora and fauna. His diary is an extremely valuable piece of Africana and, with Louis Trichardt's diary, forms the earliest Dutch description of a journey in the old Transvaal Lowveld (now the Mpumalanga Lowveld) and the adjacent district of Lourenço Marques (Maputo).

All we know about the author of the diary and his men is that they were soldiers in the service of the VOC. The contents of De Cuiper's journal and the instructions of Van de Capelle show that both were educated and cultured men with a good command of the Dutch language. In De Cuiper's journal, Afrikaans words such as "eiers" (eggs) and "trop" (herd) occur, which indicate that the author most likely came from the southern part of the Netherlands.

The 31-man company (including 21 soldiers, five sailors, a drummer and a medical assistant) left Fort Rio de Lagoa on 27 June 1725 with ten pack oxen loaded with provisions, gunpowder, lead, camping gear and other necessities. They also took eight slaughter animals, as well as eight locals to serve as interpreters. They set off in a northwesterly direction from Delagoa Bay to the upper reaches of the Matola River (which they called the Olifants because the Matola area was home to large herds of elephants). Progress across the sandy Mozambican coastal plain was slow and they reached the Komati or Assangno River by the fourth day.

The Dutch named this river the Hottentots River

---

1   In several texts also spelled "de Kuiper" (Punt 1958).

Members of the expedition who investigated De Cuiper's journey in September – October 1957. From left to right: Mr W. Schack (photographer of the National Park Board), Mr P.J. van der Merwe (ranger, Crocodile Bridge), Dr W.H.J. Punt (leader of the expedition) and Mr W. Punt jun. (member).
(OCTOBER 1957)

because the local black people told them that it had its source in the mountains of the Hottentots (Carolina–Machadodorp–Badplaas). The fact that Hottentots (Khoikhoi) used to live in the vicinity of Carolina and Badplaas before 1800 is important historical information. The Voortrekkers had it that the Swazis eradicated this group of Khoikhoi shortly after 1800 and a large number of skeletons were found on a hill near Badplaas in about 1840.

Judging by the description of the river and surrounding area, as well as by the distance covered in four days of marching (about 16 kilometres a day), they reached the Komati at what is now Moamba. At this point the river is an impressively wide, but often shallow stream. Apparently they found many large trees on the way, including one with a trunk 11 metres in diameter (a baobab). De Cuiper's company did not cross the Komati here, but travelled along its southern bank in a westerly direction until they reached a swamp.

The swamp forced the company to travel along a tributary of the Komati until they reached a large, dense stand of aloe next to a spruit, which they named Aloe Spruit. A similar stand of aloe (*Aloe marlothi*) can still be seen today at Chinculo, 24 kilometres east of Ressano Garcia.

Here the Dutch decided to march northwards over the hilly countryside covered in thorn bushes. After a very difficult journey the exhausted group reached a waterhole near what is now Ressano Garcia, where they pitched camp on 4 July 1725.

They decided to stay over for a day to give the pack oxen an opportunity to rest. Johannes Monna and six men were sent ahead to see how far they were from Chief Coupanne's kraal. 5 July 1725 should be recognised as a day of tremendous historical importance, as it was on that day that the first Europeans set foot in the area later known as the Transvaal.

Monna returned in the afternoon and told them that they had a three-hour journey ahead. They would have to cross a low mountain range (the Lebombo between Komatipoort and Ressano Garcia), and had found the place where locals crossed the Komati. The company could not, however, ford the river there themselves as it was too deep and there were very large boulders in the river bed (probably those at the confluence of the Crocodile and Inkomati rivers). When they asked the locals about an alternative drift through the Komati, they were fobbed off and told that the river was unfordable for kilometres upstream.

On 6 July, De Cuiper and his company trekked on, past Coupanne's kraal and along the route explored by Monna. They crossed the river in the vicinity of the current farm Coopersdal. The men carried most of the perishable provisions through the river and they also succeeded in getting all the pack oxen through safely. Further enquiries soon showed that the locals mistrusted them and that Chief Coupanne had given strict instructions to all his subjects that no one was to reveal any information about the trade route to the interior and northern Transvaal.

The next day they once again failed to obtain a guide, and although several locals came to trade cattle for beads, it was impossible to find anyone who would show the route to them, as all the inhabitants were Coupanne's subjects.

By 9 July they had still not obtained a guide, so they decided to advance on their own. After about an hour's journey they reached a large kraal. This was one of the biggest kraals they had yet encountered and consisted of at least a hundred huts. After much cajoling, one Swene, the headman of the kraal, agreed to provide them with a guide, but this guide was less than keen on accompanying them. At the next kraal, which was Alari Motsari's along the Moetji River (now Ngweti Spruit, a tributary of the Crocodile River flowing just south of Komatipoort), they obtained another guide from Chief Coupanne's father. Accompanied by their new guide, they travelled along a well-used path for about an hour before reaching the Crocodile River (then known as Monganje) in the afternoon and pitched camp on the southern bank.

The next day, 10 July 1725, the party crossed the Crocodile River about three kilometres northeast of the current Crocodile Bridge rest camp and so became the first Europeans to set foot in the area now known as the Kruger National Park. The river was about 1.3 metres deep on either side, but no more than 0.6 metres in the middle. The pack oxen could therefore be chased across without any problem, except that the ox carrying the gunpowder slipped on the muddy bottom and toppled into the water with its load. This meant that almost two-thirds of their powder was wet and useless. Chief Coupanne had not yet arrived as promised, but they reached the kraal of Chief Pande, which was close to the north bank of the river.

Here a large crowd gathered around the small group of explorers and started to shout threateningly, while beating their assegais on their shields. De Cuiper was certain they would be attacked and ordered his men to prepare to defend themselves. During the cacophony a reedbuck jumped up in front of them and tried to flee, but was caught by De Cuiper's dogs. The crowd claimed the reedbuck and again encircled the group in a threatening manner. There were at least 500 armed men and when De Cuiper asked them why they were taking such a threatening stance, they answered that they wanted the guides back. To maintain the peace De Cuiper released the guides and the satisfied horde dispersed.

The company of explorers carried on northwards on their own and soon reached a well-travelled path, which they followed northwards. After a while, a black man caught up with them and offered to show them, at a price, the way to the next kraal belonging to Dawano. By four o'clock that afternoon they reached a small river (the Vurhami) near the kraal of Dawano (meaning "I have a child"), where they pitched camp. One of the pack oxen died of an un-

The drift where Francois de Cuiper and his small, brave group of men crossed the Crocodile River on 10 July 1725.
(12.06.1984)

known cause and De Cuiper had it slaughtered. That night it was bitterly cold along the river. On Wednesday, 11 July, they stayed put to rest the oxen, which had been driven very hard. In the meantime they questioned some of the locals about the way to Ciremandelle (Phalaborwa) and Thowelle (Zimbabwe), but each had a different answer, which meant that the company was none the wiser.

The next day the pack oxen were still cripple and they were obliged to stay longer at their campsite. During the day an elderly black man arrived at their camp and intimated that he knew the way to Ciremandelle as well as to Thowelle. According to him, they would reach the river Sabe (Sabie, or "river of fear") after a day's journey. From there it was another day's journey to the kraal of Massawane, another day to Matonie and another two days' journey to Ciremandelle. At Ciremandelle there was too little gold to make trade worthwhile, but abundant copper. At Simangale in the Thowelle area there was ample gold, copper and ivory. He also explained to De Cuiper that the land of Thowelle was at least eight days' journey by foot from Dawano's kraal, therefore at least 380 kilometres. He also told them that many traders went to the gold-producing areas in the land of Tsouke (Zimbabwe) and that there was a river (the Zambezi) that the Portuguese sailed when they came to trade. He had seen the ships and the sailors in their shining armour himself.

De Cuiper was also told about the names of rivers and chiefs between Dawano's kraal and both Ciremandelle and Thowelle. He wrote down the names, but probably became confused about the sequence and locations. Geographers may be interested to know that some of the larger and better-known rivers flowing through the Park already had their Tsonga names in 1725. The old man for instance mentioned the names of the Sabe (Sabie), Matintonde (Nwaswitsontso or "river that flows under the sand and is only visible here and there"), Matibawati (Timbavati or "bitter or brack water"), Moutomme (Ntomeni or "Jackalberry" River or pan) and Imbaloele (Balule or "far-off river", thus the Olifants River).

After they had questioned him for about an hour it became apparent that the old man had lost interest in the conversation, so they gave him a gift as thanks for the information.

It was clear to De Cuiper that they were ill-equipped to make a long and dangerous journey deeper into the interior. The pack oxen were crippled and because of the accident in the Crocodile River most of their gunpowder was useless. The animosity of the black tribesmen also made it impossible to obtain guides or trade food for beads and other wares.

While writing his diary, he heard terrible shouts and suddenly they were surrounded by hundreds of blacks, whistling, shouting and blowing horns. He immediately commanded his men to close ranks and to fire as fast as possible at the bloodthirsty hordes. With their first round of fire they killed six men and seriously wounded about ten. The warriors pelted them with assegais and two of De Cuiper's men were wounded. De Cuiper ordered the men to advance, but when their attackers noticed this they managed to drive the oxen aside and killed all of the animals by stabbing them with assegais. It was clear to the brave group of men that they would have to leave everything behind and retreat across the Lebombo plains. They continued fighting while retreating and killed a large number of attackers. Because of the shortage of gunpowder they were ordered to fire only when the men were within range (60 to 70 metres). They were attacked at about four in the afternoon and were pursued until around seven the evening. When De Cuiper realised that the pursuit had been dropped, he ordered them to continue their flight until they had crossed the Lebombo. That night they marched until about two o'clock and found themselves in mountainous terrain with many valleys and pans. Eventually they reached the eastern coastal plain. During their headlong flight in the dark they could hear dogs barking in all directions. It became so cold that they decided to carry on marching, using the stars as guidance. They kept on walking until about eight o'clock the next morning – Friday 13 July 1725.

At the next kraal they were again confronted by a large number (more than 500) of armed warriors, who threatened to kill them because they had travelled through their territory. After they had shot six of their attackers and wounded several more, the hostile warriors retreated. While the sorely tested group of Dutchmen moved on, a great crowd followed them at a distance. After walking another half an hour, they finally reached the Komati River. The river was wide and about l.5 metres deep and it took them about an hour to ford it (probably near what is now the Incomati Station in Mozambique).

After having safely crossed the river the expedition passed a number of different kraals whose inhabitants were anything but friendly. Eventually they arrived at a kraal on an open plain, where a subject of Mateky cordially welcomed them. When other locals saw this, they came closer and kindly offered the exhausted travellers some food. This they gratefully accepted as it was already five o'clock in the afternoon and they had not eaten since ten o'clock the previous morning. They estimated that they had marched at least 20 Dutch miles (± 56 kilometres) since then.

The small company pitched camp under a large tree and posted guards for the night, as there were still numerous armed and hostile blacks in the vicinity. The next morning they reached the kraal of Chief Samaal Mabotte, where they had rested on their forward journey on 2 July. They then sent a message to the hostile tribesmen still pursuing them, saying that they would be shot if they came any closer. The company marched on until four o'clock in the afternoon when they reached Chief Mambo's kraal, where they were cordially greeted and presented with more than enough food.

They continued their journey early the next morning and crossed the Matola River in the afternoon. After an hour's rest they carried on to Chief Mambeete's kraal. Here they spent the night in the knowledge that they would reach Fort Lijdzaamheid (Fort Rio de Lagoa), and safety, the next day.

On Monday 16 July 1725 they completed the last leg of their journey. They passed Matola's kraal and arrived at Maphumbo's kraal at eight o'clock. Here the locals received them with enthusiasm. The commanding sergeant of the fort, who had been sent by Van de Capelle to accompany them on their return, awaited them. He had brought bread and arrack (palm wine) as refreshment and they cheerfully marched on to reach Fort Lijdzaamheid at ten in the morning, where they were welcomed back by Van de Capelle and those of their comrades who had stayed behind.

The expedition had not achieved its goal of establishing a trade route to the mineral-rich kingdoms of the interior, Zimbabwe in particular. However, their journey was nevertheless of great historical importance and Francois de Cuiper's diary represents the oldest description of an overland journey to the interior of the Transvaal. Remarkably, the whole journal barely makes mention of game, except for elephant and reedbuck along the Crocodile River. (This might be merely by chance or game could have been scarce in the area through which they had travelled, due to the large black settlements they had encountered.)

The diary contains valuable references to climatic conditions, the state of the rivers and water distribution, vegetation and black settlements and customs. It also gives information on geography, the absence of tsetse fly in this part of the Lowveld (note the numerous references to cattle, even at Dawano's kraal) and other aspects of historical importance relevant to this part of southern Africa 250 years ago.

Unfortunately this early exploration of the Lowveld was not followed by subsequent journeys

The Francois de Cuiper commemorative plaque, which was erected by the Historical Monuments Commission at Gomondwane where the VOC expedition had been attacked by Chief Dawano's warriors.

and for more than 100 years no Europeans entered this untouched area.

Interestingly enough, the next big exploration of the Lowveld, by the Voortrekkers led by Louis Trichardt from 1836 to 1838, also yielded a valuable diary – that of Trichardt himself. The first Europeans to set foot in the area that is now the Kruger National Park, in 1725, were members of a Dutch military expedition. Their quest was to establish trade relations with the interior on the instruction of the VOC. A hundred and eleven years would pass before Europeans again entered the area, but then they were citizens of the Transvaal who wanted to reach the coast to trade.

Francois de Cuiper died of fever in 1727, and this was eventually the fate of most of his brave companions. Jan van de Capelle survived the unhealthy climate of Delagoa Bay for six years and in 1730 he was put in charge of the dismantling of Fort Lijdzaamheid. The Governor in Cape Town instructed him to embark all troops and load all material, guns, merchandise, documents and assets on ships bound for Table Bay. Van de Capelle was a purposeful leader who had placed the Delagoa Bay colony on a sound foundation.

However, in 1730 he sent a platoon to ward off a threatened attack on the cattle post of the VOC. Johannes Monna and Sergeant Johannes Mulder from the Cape were in command of the 29 soldiers.

During the clash Sergeant Mulder made a fatal tactical error when, with seven soldiers, he left the ranks of the main body to attack the impis. This broke the formation of the troop, allowing the enemy to penetrate the ranks of the Dutch. The whole company was wiped out during the ensuing battle. The Political Council at the Cape held Van de Capelle responsible for this disaster and when he returned to Cape Town as an assistant merchant in 1731, he was not given a new appointment. He returned to the Netherlands in 1732.

Ultimately malaria and blackwater fever forced the VOC to abandon Delagoa Bay. It is estimated that the VOC lost 1000 men during its ten years of occupation, and Karel Trichardt later described Delagoa Bay as "the place of death".

The VOC played an especially important role in the political development of southern Africa, and its remarkable cultural and historical contribution is often sold short.

The historic expedition by Francois de Cuiper to the Transvaal Lowveld in 1725 was commemorated in July 1975 (exactly 250 years later) by the National Parks Board when Dr W.H.J. Punt unveiled a brass memorial tablet of the Simon van der Stel Foundation. This took place at Gomondwane, where it is thought De Cuiper and his men were attacked by Dawano's warriors on 12 July 1725.

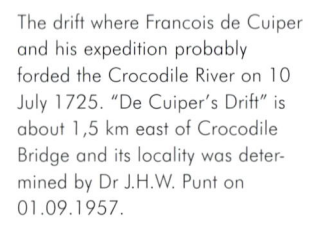

The drift where Francois de Cuiper and his expedition probably forded the Crocodile River on 10 July 1725. "De Cuiper's Drift" is about 1,5 km east of Crocodile Bridge and its locality was determined by Dr J.H.W. Punt on 01.09.1957.

References

Abshire, D.M. & Samuels, M.A. 1969. *Portuguese Africa*. Pall Mall, London.

Anon. 1975. De Cuiper ekspedisie herdenk. *Custos* 4(9): 29–30.

Axelson, E. 1940. *South-East Africa 1488–1530*. Longmans & Co., London.

Axelson, E. 1954. *South African Explorers*. Oxford University Press, Cape Town.

Axelson, E. 1967. *Portugal and the Scramble for Africa, 1875–1891*. University of the Witwatersrand Press, Johannesburg.

Axelson, E. 1987. *Dias and his Successors*. Oxford University Press, Cape Town.

Brown, R. 1904. *The Story of Africa and its Explorers*. Cassel & Co., London.

Coetzee, C.G. 1954. *Die stryd om Delagoabaai en die Suidooskus, 1600–1800*. Unpublished Ph.D. thesis, University of Stellenbosch.

De Kock, W.J. 1957. *Portugese ontdekkers om die Kaap: die Europese aanraking met Suidelike Afrika, 1415–1600*. A.A. Balkema, Cape Town.

Henriksen, T.H. 1978. *Mozambique, a History*. Rex Collins, London.

Hromník, C.A. 1981. *Indo-Africa*. Juta, Johannesburg.

Hromník, C.A. 1984. Nkomati means trade and communication. *The Argus*, Cape Town, 10.

Hromník, C.A. 1986. Gold and God in Komatiland. Unpublished manuscript.

Punt, W.H.J. 1958. Die verkenning van die Krugerwildtuin deur die Hollandse Oos-Indiese Kompanje, 1725. *Koedoe* 1:1–18.

Punt, W.H.J. 1975. *The First Europeans in the Kruger National Park, 1725*. National Parks Board of Trustees, Pretoria.

Theal, G.M. 1897. *History of South Africa* vol. 1:448.

Webb, H.S. 1954. The penetration of the region by Europeans. In: Webb, H.S. (ed.). *The South-Eastern Transvaal Lowveld*. Publication of the Lowveld Regional Development Association.

Chapter 4
The Voortrekkers explore the Lowveld
4.1 The fate of Lang Hans van Rensburg
Dr W.H.J. Punt

The Voortrekker party of Johannes Hendrik Janse van Rensburg (nicknamed Lang [Tall] Hans) left the Eastern Province (now the Eastern Cape) early in 1835 and set off northwards in the direction of the Soutpansberg. His party consisted of 49 whites and a number of coloured servants.

The white members of the trek were Van Rensburg with his wife and four children, Sybrand Bronkhorst with his wife and six children, Gysbert Bronkhorst with his wife and one child, Jacobus de Wet and his wife, Gysbert Bronkhorst Jnr and his wife, Frederik van Wyk with his wife and two children, Hendrik Croucamp with his wife and three children, Petrus Viljoen with his wife and six children, Nicolaas (Klaas Nuweveld) Prinsloo[1] with his wife and eight children and Marthinus Prinsloo, a bachelor (Preller, 1917).

The party travelled with nine wagons, about 500 cattle, a few thousand sheep and a number of horses.

At the Caledon River the Van Rensburg party met up with the trek of Louis Trichardt, who were also on their way to the far northern Transvaal.

After spending a few months at the Caledon, the Van Rensburg trek proceeded towards the Vaal River ahead of the Trichardt trek. In January 1836 both parties encamped on the southern bank of the Vaal River, close to where Standerton is today. From here they journeyed together to form a united front in case of an attack by the bloodthirsty uMzilikazi's[2] looting AmaNdebele[3] warriors, and moved slightly west of where Middelburg is now to the Strydpoort Mountains. Here they camped for about 10 days in April 1836.

In the same month, Van Rensburg decided to move further north after he and Louis Trichardt had a falling-out about ammunition and the road ahead. The Van Rensburg party apparently frequently borrowed lead and supplies from the Trichardt people, but they never returned anything. This inevitably led to a confrontation between the two leaders. Thanks to this row the name Strydpoort (literally Quarrel Pass) was later given to the mountain range where they had made camp.

Afterwards Van Rensburg and his people set off through Pienaarspoort to the Soutpansberg, where they arrived at the beginning of May 1836. From there the trek headed in an easterly direction along the southern slopes of the Soutpansberg and by mid-June they were already temporarily settled near Mashao Hill on the current farm Welgevonden. The late Mr Piet Möller pointed out this camping site on Mr W. van der Merwe's farm to Dr W.H.J. Punt in later years. The group camped here for a few weeks, explored the area and looked for the route to Delagoa Bay or Inhambane.

At the end of May Louis Trichardt and his party arrived at Soutpan and pitched camp there, while the Van Rensburg group was still at Mashao Hill. There are indications that the two leaders were in contact, as Trichardt could later inform the Portuguese in Delagoa Bay that the Van Rensburgs had set out from the Soutpansberg for Delagoa Bay in July. At Soutpan Trichardt was awaiting a patrol of Commandant A.H. Potgieter, who was leading a third trek along the route to the north, to discuss their plans for the future.

The patrol of A.H. Potgieter and J.G.S. Bronkhorst arrived at Trichardt's camp on 24 June 1836 to explore the Soutpan area and further eastwards, with the aim of establishing a new settlement (compare the founding of Zoutpansbergdorp in 1849). Three days later, an 11-man patrol under Potgieter and Bronkhorst's leadership set off in the direction of Sofala (now Beira). In the meantime, Van Rensburg had led his party into the wild eastern bush and then vanished. Potgieter, Bronkhorst and the rest of their patrol eventually returned at the end of July 1836 after having reached Massangena on the Great Save in Mozambique.

By the time of the Potgieter patrol's return, rumours about the murder of Van Rensburg and his

---
1   Also called Klaas Prins (Roos, 1930).
2   Also spelled Msilikazi.
3   Also referred to as Ndebele, Matebeles, or Matabeles.

entire party had already reached the Voortrekkers at Soutpan. After consultation between Potgieter and Trichardt it was decided to send a five-man patrol on horseback on the trail of the Van Rensburg party to determine their fate. The group consisted of Louis Trichardt, his son Karel and three members of the Potgieter party, namely J.G.S. Bronkhorst (whose two brothers and a cousin had been with the Van Rensburg trek), J. Robbertse and A. Swanepoel. Louis Trichardt made brief notes about this extraordinary journey, which started at Soutpan on 30 July 1836, in his diary.

Trichardt's journal describing the patrol reads as follows:

- 31 July: Leave from Matebies's Kraal and ride for 5½ hours until reaching a valley.
- 1 August: Ride for 8½ hours until the second loss (a horse?).
- 2 August: Ditto until reaching "Buffelslaagten" (probably the Shawu Valley in what is now the Kruger National Park), 7 hours.
- 3 August: Ride for 8 hours to the "Sant" River (probably the Shingwedzi in Mozambique).
- 4 August: To "Grootriviers drif" (the Limpopo River), 6½ hours.
- 5 August: To the first Magwamba chief, 6 hours and 10 minutes.
- 6 August: Up to lüKoerie (vicinity of Mapai), 5½ hours.
- 7 August: To "Paddavontijn" (the area east of the Nyandu thicket in Mozambique where there are numerous pans), 5½ hours.
- 8 August (1836): To "Olifantsbos" (Nyandu thicket), 7½ hours; G. Bronkhorst shot an impala there.
- 9 August: To Suur-bier (Sour Beer), 4½ hours; where the third horse was left behind. August 10: To "Matibee tuijn". (Possibly Dzundwini or Shikokololo Fountain.) That day we shot a hartebeest. (Possibly the first reference to the presence of Lichtenstein's hartebeest in the Lowveld.)
- 11 August: To the Surubele, 6½ hours.
- 12 August: To "Sirities tuijn", 5 hours and 20 minutes. Robberts shot a giraffe to ostensibly deliver a letter to Sijbrand, but by 10 o'clock the next day they were hardly finished.
- 13 August: To where the river turns (from the Luvuvhu?), at six they shot a hartebeest and wounded another in the leg, which they could not find.
- 14 August: To "M'assouw" (Mashao Hill), 5½ hours, where I left Bap (a guide?).
- 15 August: To the drift, 8 hours.
- 16 August: Home, 7 hours.
- 17 August: The patrol leaves (i.e. the Potgieter-Bronkhorst party).

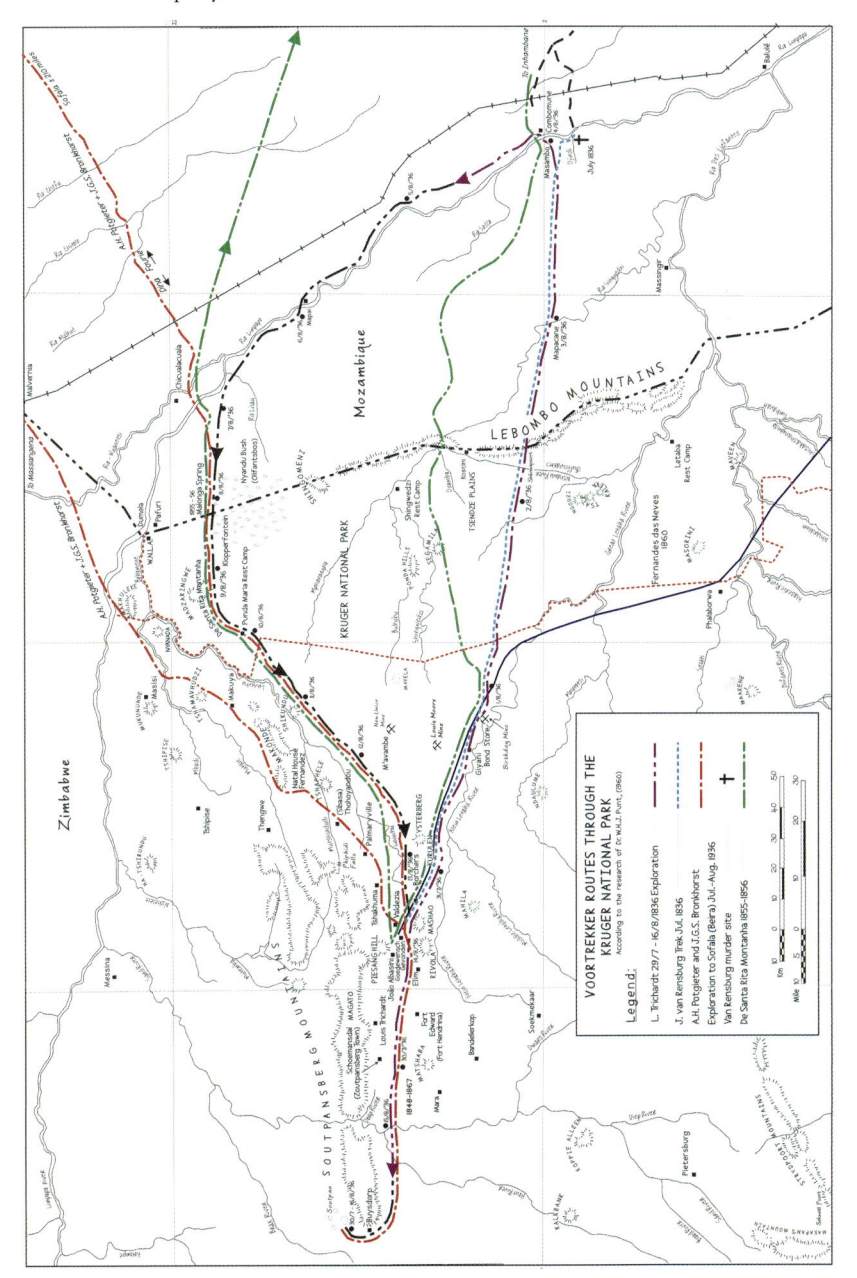

Voortrekker routes through the Kruger National Park.
(THERE IS AN ENLARGED VERSION OF THIS MAP IN THE BACK OF THE BOOK.)

After years of both terrain and documentary research by Punt to solve the mystery of the Van Rensburg massacre, he came to the conclusion that, if the route of Louis Trichardt's search party could be determined accurately, it would provide the best opportunity to determine the place where the massacre had taken place. Louis Trichardt and his party managed to reach Sakana's kraal along the Limpopo and it is known that the Van Rensburg massacre had taken place in the immediate vicinity of Sakana's kraal. It thus had to be determined exactly where the patrol had reached the Limpopo. After intensive investigations, Punt and his co-workers eventually succeeded in mapping the route the patrol had followed. To their surprise, the route indicated that Louis Trichardt had reached the Limpopo in the vicinity of Masambo on 4 August 1836, at a place he had called "Grootriviersdrif". This confirmed documentary information that indicated that the massacre had taken place not far from the confluence of the Olifants and Limpopo rivers.

Before Trichardt's route is discussed any further, it is necessary to take a closer look at the direction in which the Van Rensburg trek headed from Mashao Hill.

At the time, Chief Mashao's[4] hill, which is now on the farm Welgevonden, was a well-known trading post where Portuguese goods were traded with the local black population. Mashao's subjects were therefore well acquainted with the footpaths to the coast. They also knew where Delagoa Bay, Inhambane and Sofala were, from where they obtained trade goods.

Van Rensburg probably received information from Mashao about the ports and trade routes. After that he went to the Rivola Hills, close to Mashao, to look for a way southwards, which is an indication that he had made inquiries about the routes to Delagoa Bay and/or Inhambane. Now the question arises: Which direction did he take to reach the southern coastal port of his choice?

To reach Delagoa Bay from Mashao, the Van Rensburg party could have followed one of four routes:

- The first route went due south along the foot of the Drakensberg and then through Komati Poort to Delagoa Bay.
- A second route was a trade route leading eastwards slightly south of Ysterberg along the left bank of the Klein Letaba River to the Limpopo.
- The third possibility was a trade route via Pafuri to Sofala, which turned southwards on the banks of the Limpopo at Pafuri towards Inhambane and Delagoa Bay.
- A fourth road led northwards from Mashao over Makonde Mountain to the Limpopo and from there along the river via Pafuri towards the mentioned ports.

The Klein Letaba route (2) was the most passable for wagons. Large sections went through open bushveld and, with the exception of the Lebombos, there were no mountains to cross. Route 3 went through a very dense area near Pafuri and would be a major detour for wagons on their way to Delagoa Bay. The longest and most difficult of the roads (4) was the one across Makonde Mountain down the Mutale River to the Limpopo. This road passed through bushy and mountainous terrain, while the Mutale River had several large drifts, making it very difficult to travel by ox-wagon.

The well-known Soutpan (Salt Pan) along the northwestern point of the Soutpansberg where Voortrekker leader Louis Trichardt's party camped between May and the end of August 1836.
(ACKNOWLEDGEMENT: TRANSVAAL ARCHIVES)

4 Also spelled Masjou.

4.1 THE FATE OF LANG HANS VAN RENSBURG

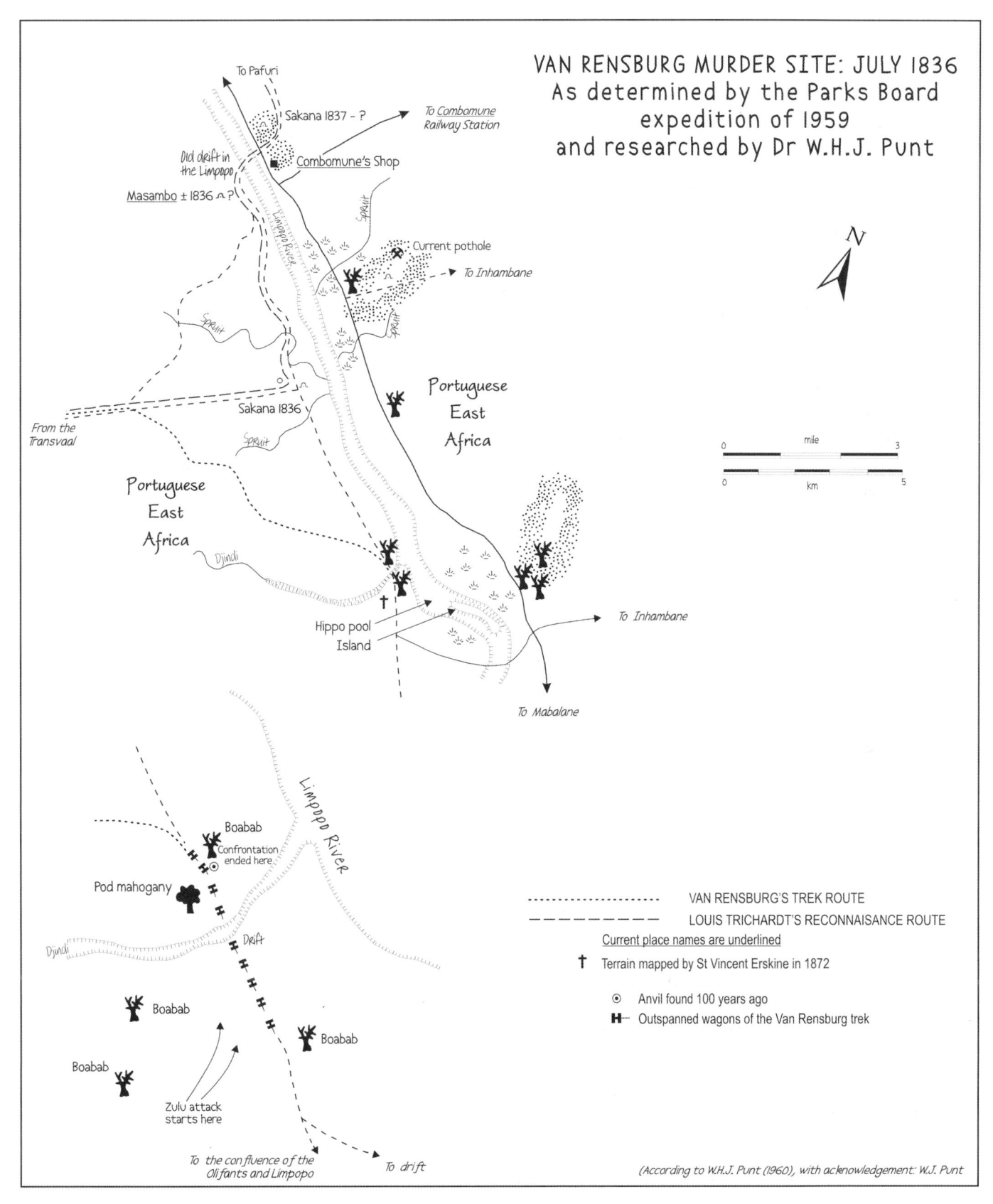

Terrain where the Van Rensburg massacre took place in July 1836.

It is important to determine which route Van Rensburg had taken to reach the Limpopo, as the Voortrekkers were murdered close to where they had reached the river.

An experienced outdoorsman like Van Rensburg would, like any other trekker, have followed the shortest and easiest route. This undoubtedly was the route between Ysterberg and the Klein Letaba to the Limpopo (route 2) (see map number 7).

From the hills at Mashao and Elim there is a magnificent view of the Lowveld to the Lebombo Mountains in the east. On a clear day and by using a pair of binoculars the whole area can be observed, including the roads leading into the Kruger National Park.

Van Rensburg would certainly have planned their route from Mashao Hill to Delagoa Bay/Inhambane from this hilltop. It would have been obvious to him that the route running east along the Klein Letaba's northern bank was relatively open and the area traversable by wagons. It can thus be assumed that his party decided to follow this particular route to the Limpopo.

Months later Jan Pretorius and his group reached the place where the massacre had taken place. On his way back he let Trichardt know that he was 30 shifts away from where the Voortrekkers were camping, thus from where the Kurulen Mission is today to the Limpopo.

Thirty shifts means a distance of 30 times 8.5 kilometres, or about 255 kilometres in total. In reality, the distance between the two localities is about 240 kilometres. The route to the Limpopo along the Klein Letaba through Shilowa Poort in the Lebombo can be completed in 30 shifts. This corresponds with Trichardt's notes about his camping site, which read: "Wij ben 5 of 6 dagen te voet van Sakana weswaarts" (By foot we are five or six days west of Sakana along the Limpopo), taking into account that black messengers could cover a distance of between 40 and 50 kilometres per day.

To determine the date of the murder, it had to be ascertained exactly when Van Rensburg and his party had left Mashao. This could be achieved with the help of Louis Trichardt's diary and J.G.S. Bronkhorst's travel journal. It is known that the Van Rensburg party had been camping near Chief Mashao's kraal for a few weeks at the end of May 1836, before they set off eastwards on their journey to Delagoa Bay (or Inhambane).

Louis Trichardt, who was at Soutpan at the time, was aware of Van Rensburg's itinerary, as is shown in his later letter to the Portuguese authorities at Delagoa Bay. Hans van Rensburg had let his fellow trek leader know that he was not going to wait in the Soutpansberg any longer. The Van Rensburg party was keen to exchange their large stocks of ivory for gunpowder, lead and other trade goods, and left Welgevonden. Shortly afterwards, the patrol of Potgieter and Bronkhorst rode over the abandoned camp site.

If Potgieter and Bronkhorst's patrol had left Soutpan on 27 June, they would have passed Mashao four days later, on about 30 June. The Van Rensburg party had to have left by then, otherwise Bronkhorst would certainly have visited his two brothers who were members of the Van Rensburg party. He would definitely also have mentioned such an encounter in his journal, as he had mentioned their meeting with Louis Trichardt. From this it can be deduced that Potgieter and Bronkhorst did not encounter the Van Rensburg party in the Soutpansberg.

Van Rensburg and his group had therefore probably left several days before the 30th. It could have been on about 25 June 1836, and when Potgieter

A bronze plaque which was erected by the Historical Monuments Commission in Punda Maria rest camp to commemorate the historical Potgieter-Bronkhorst patrol to Sofala in 1836.

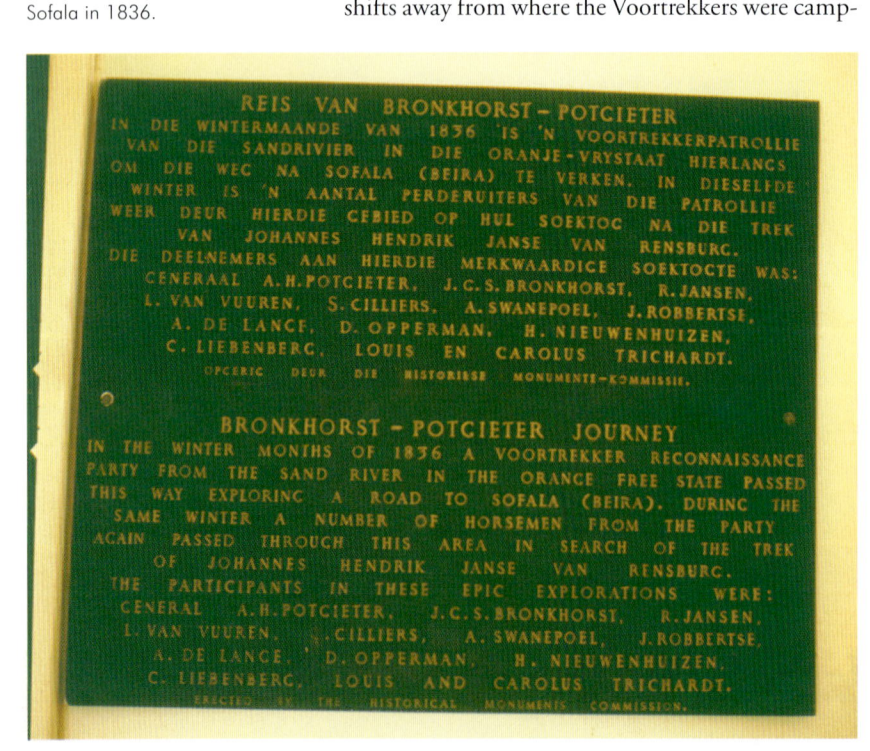

and Bronkhorst arrived at Welgevonden the Van Rensburg trek must already have been far into the wild Lowveld bush.

By the end of July the same patrol again passed Mashao on their return to Soutpan, and again J.G.S. Bronkhorst's journal made no mention of any meeting with the Van Rensburg party.

Judging by Jan Pretorius's letter, it was about 33 shifts from Mashao (Welgevonden) to Sakana. (Welgevonden is three shifts from Kurulen.) If it is accepted that the Van Rensburgs had travelled 33 consecutive shifts from Mashao to Sakana's kraal along the Limpopo, the murder must have taken place in the last week of July 1836.

During June and July, when Hans van Rensburg and his party had trekked through the Lowveld, there was plenty of water and grazing for their livestock (see to Bronkhorst's travel journal). They also did not have to cross any large rivers, except for the Shingwedzi in Mozambique. Game was abundant and few black people lived along the route.

From Mashao the Van Rensburgs travelled along the northern bank of the Klein Letaba until they reached the area now known as the Kruger National Park. They then trekked across the Tsende (Tsendze) Plain towards the Lebombo Mountains.

Opposite the Tsende (Tsendze) Plain, the Lebombo is relatively steep with only two narrow passes through which ox-wagons could comfortably travel, namely Shilowa Poort and Shingwedzi Poort further north. Greone Neck is about 20 kilometres south of Shilowa Poort and it is possible to cross the mountain by wagon there as well, but east of this neck to the Shingwedzi River the terrain is impassable because of dense bush and rocky terrain.

Taking into account the direction Van Rensburg had taken, Shilowa was the obvious choice for a passage through the mountains to the Limpopo as it was also an ancient trade route to Inhambane. It is wide enough for wagons to pass through and, from a few kilometres east of the poort the road is easily traversable to the Shingwedzi.

In Mozambique the Shingwedzi is a large sandy river with isolated waterholes and little or no running water during the winter months. There are a number of good drifts, but on the whole the banks are quite steep and the bed sandy. Van Rensburg had

Shilowa Poort through which the Van Rensburg party trekked in 1836 in the direction of Inhambane.

probably headed towards the well-known, large drift at Mapacane and crossed the river there.

Once through the sandy Shingwedzi, the path went across a flat area with numerous pans, towards the Limpopo. The soil here is clayey, which makes this road almost impassable during the rainy season, as Jan Pretorius would later discover to his dismay.

From the drift in the Shingwedzi at Mapacane there is an ancient footpath leading to the well-known drift through the Limpopo at Chief Masambo's kraal. About eight kilometres before it reaches the kraal, the path splits into three. One path leads northeast towards Masambo Drift, the middle one leads to the chief's kraal and the third goes through the Djindi (a small tributary of the Limpopo) to join the trade route to Inhambane, 1.6 kilometres away, before crossing the Limpopo. At the time of Van Rensburg's arrival, Chief Sakana's kraal was about five kilometres north of the Djindi on the western bank of the Limpopo.

When Van Rensburg was about eight kilometres from the Limpopo he decided to follow the footpath to Djindi, not only because it was the shortest route to the Inhambane trade route, but also because he apparently wanted to avoid Sakana's kraal. At the drift, the Djindi is a small ephemeral river with very high, steep banks. The steep approaches and a very hard limestone bed make it very difficult to cross. This means that it would have been really difficult and time-consuming for the Van Rensburg party to cross the river, with the result they probably outspanned on both sides of the little river.

There is much debate in historical literature about who Chief Sakana really was. Dicke (1944) claimed that Sakana was an unimportant induna at Pafuri. From Trichardt's diary, however, it is clear that Sakana was an important and influential leader along the Limpopo from 1836 until 1838.

Jan Pretorius, who trekked with Trichardt, also reached Sakana's kraal later during his search for the missing Van Rensburg party. His descendants wrote about Chief "Sogana" (Rossouw, 1929). According to them, Sogana is an abbreviation of Sochangana, whose Zulu name was Manukosi. He was one of Shaka's generals who became disloyal and fled to Mozambique in about 1830, where he settled in the Limpopo area and either subjugated or annihilated one tribe after the other.

In 1833 and 1834 he attacked the Portuguese settlements at Delagoa Bay and Inhambane and murdered almost all of the inhabitants. To escape Shaka's avenging warriors, he and his followers moved further and further north along the Limpopo and built kraals at various places along the river. At one stage his kraal was located in the vicinity of Chief Masambo's kraal in 1959 (during the Parks Board expedition). In 1959 a grandson of one of Manukosi's indunas informed Dr Punt that Manukosi had lived on the western bank of the Limpopo in the vicinity of Masambo's kraal at the time the Van Rensburg trek reached the area, and that he had subjected the then Headman Masambo and his people. There is also a significant reference in Louis Trichardt's diary that confirms Sakana and Manukosi was one and the same person.

After Trichardt had informed the Portuguese about the murder of the Van Rensburg trek by letter, the Commandant of Fort Rio de Lagoa replied that the Portuguese and the Zulu King Dingane wanted to send a punitive expedition against Sakana. The Portuguese were of course fully aware of the fact that the Zulu king had a bone to pick with Manukosi over his treachery. It is difficult to accept that the Portuguese and Zulus would collude against any other black chief in Mozambique. In Sakana's case there were, however, reasons for combined military retaliation, as the Portuguese had not forgotten Manukosi's callous attacks on Delagoa Bay and the other ports along the east coast a few years before.

Manukosi's tyranny made him all powerful in the area between the Limpopo and the Indian Ocean. Whoever wanted to travel through his territory, white or black, had to ask for his permission before crossing his empire. Van Rensburg had to do the same as he had to travel from the Limpopo through Manukosi's territory to Inhambane. This meant that Van Rensburg and his people had to come into close contact with Manukosi, which enabled the avaricious chief to see how much livestock the Voortrekkers had.

Further confirmation that Manukosi had been responsible for the murder of the Van Rensburgs came from N.T. Bührmann of Lydenburg. In his notes he mentioned that the "Volksraad" (House of Assembly) of Ohrigstad had decided in 1848 to punish Manukosi for the massacre of the whole trekker party led by Van Rensburg. There is therefore little doubt that Louis Trichardt's Sakana and Jan Pretorius's Sogana was the Zulu Chief Manukosi or Sochangana – the father of the Shangaan nation.

The place, date and way in which the massacre was committed are noted by various sources. According to calculations of the trek distances, Van Rensburg and his party must have arrived at Manukosi's kraal during the last week of July 1836. However, it could have been a little earlier. The arrival of so many whites must have been a huge occasion for the kraal and the large herds of cattle and sheep must have caught Manukosi's eye. Livestock was particularly scarce in his territory because of tsetse fly and Manukosi usually ventured to other areas to steal livestock, as he could only keep goats.

Louis Trichardt's diary is still the most important source of information about the Van Rensburg massacre. With its help, as well as documentary evidence, legends and the examination of the terrain by the Parks Board expedition (Punt, 1960), a fairly accurate scenario of the tragedy could be put together.

On the day of the murder, the Van Rensburgs had set up camp five kilometres south of Sakana's (Manukosi's) kraal along a footpath on the right bank of the Limpopo. The ox-wagons were not drawn into a circle (laager), but spanned out in the path in a long line, one behind the other. Four or five of the wagons stood over a distance of 150 paces north of the Djindi, but the rest had already gone through the drift and were outspanned for 200 paces along the southern side. The camping site was open, near large baobabs. Several of these giants were still standing in 1959 when the Parks Board expedition visited the terrain.

The Limpopo River, which does not hold much water during the winter months, was only about 100 paces east of the wagons. The bank on the western side of the Limpopo is very steep at this specific spot, while there is a large flood plain and swamp on the eastern side.

To this day there is a large hippo pool on the western side and just south of that a long island where a footpath to Inhambane crosses the river. This drift is only about one and a half kilometres from the murder site and the Voortrekkers were probably planning to trek through the Limpopo the next day (see map no. 7).

According to black men whom Louis Trichardt later questioned in the Soutpansberg, Van Rensburg had already explored the road for eight kilometres past the Djindi and the large hippo pool from his second last camping site. Here the Van Rensburg party had apparently shot a few hippos, which they gave to Sakana and his subjects. Despite the chief's warning that Van Rensburg should not travel any further, as a Zulu impi was on its way, the Voortrekkers nevertheless proceeded to the Djindi that day. Van Rensburg intimated that he would not be stopped by a Zulu impi. They apparently arrived late at the camping site that day, otherwise he would certainly have moved all nine wagons through the Djindi. By nightfall, the oxen and small livestock lay down next to the wagons for safety. As evening fell the small Van Rensburg party made their last camp among the giant baobabs in Mozambique.

Later that night they were attacked by a Zulu impi under the leadership of Induna Malitel. The attack was launched from the south and the first wagons were quickly overcome. Despite the surprise of the attack, the Boers nevertheless fought back bravely and retreated towards the wagons further back, but the warriors moved in among the wagons and divided the group into four. They managed to overpower three of the four, but the fourth group fought back so courageously that the attackers were driven back several times. However, by daybreak the Boers' ammunition stock was depleted, and so Trichardt's warning at Strydpoort tragically became true.

Once the Zulu warriors noticed the Voortrekkers' predicament, they chased the cattle in among the defenders. By hiding behind the cattle, they were able to get close enough to kill all the survivors in a brutal manner. Only two children of Frederik van Wyk (Van Rensburg's son-in-law) – a little boy of six and his sister of three or four – survived the attack, as one of the warriors had hidden them under his shield (Roos, 1930).[5] Nobody else, not even the servants, survived the assegais of Manukosi's warriors. About 30 of the warriors were killed, while their leader, Malitel, was wounded, but he recovered and lived for many years near the confluence of the Limpopo and Olifants rivers.

The information about the two children who had survived the tragedy came from Jan Pretorius, as he mentioned in a letter to Trichardt that he had heard white children crying in Sakana's hut. When Pretorius visited the murder scene at the end of 1836, the children had obviously still been alive, but they died shortly afterwards of malaria. When Karel Trichardt arrived at the site in the winter of 1838, he was shown the graves of the two children (Roos, 1930).

The later claim of an author (Bulpin, 1950) that two people (a man and a woman) who were extradited by Swazi King Mswati to the then Zuid-Afrikaansche Republiek (South African Republic) 31 years later, were connected to the two Van Wyk children, is therefore unfounded. This claim is also clearly refuted by the statement of Magistrate W.H. Neethling of Lydenburg that the two hapless people who had been handed over by the Swazis were albino blacks (Preller, 1917).

Manukosi apparently forbade his people to remove anything from the wagons after the massacre and they were set alight later on along with their contents. The livestock was rustled and the goats were herded with those of Sakana, while the rifles that had remained among the bodies were removed. When Karel Trichardt arrived at the scene two years later, he buried the remains of his countrymen.

Van Rensburg and his people had thus become the first white victims of the Zulu supremacy wars,

---

5   One wonders how Karel Trichardt determined this, as there had been no other survivors.

which ended 18 months later with the battle of Blood River in Natal.

In later years, the scene of the murder was forgotten and uncertainty over whether anybody had survived the attack existed for a long time. The few people who had visited the site after the murder did not leave complete descriptions of it. The exact terrain could only be determined again in 1959 by the National Parks Board expedition under the leadership of Dr Punt.

The first expedition in search of the Van Rensburgs was undertaken by Louis Trichardt. According to his notes he had reached Chief Sakana, but did not visit the scene of the murder. On his return to the Soutpansberg he repeatedly questioned black travellers coming from Mozambique whether the Van Rensburg party had been killed on the southern or northern side of the Limpopo, proving that he had not visited the scene of the massacre.

Analysis of the place names and dates that Louis Trichardt recorded during his search for the missing Van Rensburg party enables one to locate the general area of the murder, but not the site itself. The direction of the patrol's route indicates that Trichardt and his men travelled directly towards the Limpopo because of the clear wagon trail left by the Van Rensburgs and the information Chief Mashao had provided.

From Trichardt's diary it can be deduced that he was in a great hurry to reach Van Rensburg. They rode about seven and a half hours per day by horse from Soutpan to the Limpopo, which is an average of 50 kilometres per day. During their return journey the average time decreased to six hours and the distance to an average of 38 kilometres per day. Trichardt and the search party managed to cover the distance between Soutpan and the Limpopo by horse in only six days. On 4 August 1836 the patrol reached Sakana (Manukosi) at "Grootriviersdrif" (Masambo Drift through the Limpopo). Here Trichardt inquired about the Van Rensburgs and found out that the trek had passed Sakana's kraal. The patrol must have had some inkling of what had happened, because they noticed that the blacks had Van Rensburg's binoculars as well as a mirror. At that stage Trichardt realised that it would be dangerous for his patrol of only five men to linger in the area and, after spending the night at Sakana's kraal, the horsemen accepted the offer of guides. The next day, however, they travelled in the opposite direction, away from the murder site northwards along the Limpopo. This decision probably saved Trichardt and his men from the same fate as that of the Van Rensburgs.

On their journey along the Limpopo they met Inkuri (Inkoerie), who lived opposite Mapai on the western bank of the river. (This information was confirmed by descendants of Inkuri who had settled near Schuynshoogte in modern times.)

The guides probably accompanied Trichardt to the Nyandu forest (Trichardt's "Olifantbos" [elephant thicket]). There he discharged them after they had shown him the trade route across Malonga and Klopperfontein through the current Kruger National Park. By then, they were on the same route that J.G.S. Bronkhorst, J. Robbertse and A. Swanepoel had followed a couple of weeks earlier with Potgieter while returning from their reconnaissance trip to Massangena.

Trichardt rode in a westerly direction from the Nyandu forest towards what is now Punda Maria, from where he followed the ancient trade route to Mashao and Soutpan. The daily distances they travelled became shorter and shorter as their horses were ridden very hard. The patrol would lose four horses as a result of either exhaustion or tsetse flies. Judging by the short distance between the "Draaij" and Mashao, it would appear as though the men had even walked some of the time.

However, the patrol must have obtained fresh horses at Mashao, as the travelling times of the next two days increased to seven and eight hours per day again.

Back at Soutpan, Trichardt reported on their journey. Commandant A.H. Potgieter was obviously in a hurry to return to his people in the Free State, as Potgieter, Bronkhorst and the rest of their group were already planning to depart for the south the next day (17 August 1836). Louis Trichardt stayed behind in the Soutpansberg, more or less convinced that the Van Rensburgs had been murdered, but without any tangible evidence of this.

Restless and uncertain, Jan Pretorius, Gert Scheepers, Hendrik Botha and Izak Albagh decided shortly after Trichardt's return to go looking for the Van

Rensburg party themselves. They apparently considered the first patrol's report unconvincing and as some of their family members had been part of the Van Rensburg party, they decided to set off on their trail.

On 18 August, only a day after Potgieter and his men had left, the four men, with their wives, children and livestock in tow, left the Trichardts. They set off eastwards on the Van Rensburg party's trail while the Trichardts and the elderly teacher Daniël Pfeffer remained at Soutpan.

Pretorius and his companions could easily follow the wagon trails left by both the Van Rensburgs and Trichardt's patrol. Legend has it that a hunter by the name of Erasmus could still follow the wagon trail to the Lebombo Mountains 30 years later. The distance between Soutpan and Masambo along the Limpopo River is about 350 kilometres and the small trek could complete this in about 45 shifts, indicating that they reached the river at the end of September 1836. Jan Pretorius and his party met Sakana (Manukosi) and finally came to the conclusion that the Van Rensburgs had been murdered. They also heard the crying of white children in Sakana's huts, and another chief confirmed that Sakana was keeping two white children captive.

This group never reached the actual murder site either and the uncertainty whether the trekkers had been murdered on the eastern or western bank of the Limpopo remained. After having spent some time along the Limpopo, the four families travelled back in their tracks to the Soutpansberg without being molested.

Their return journey was difficult as by this time it was summer and heavy rains were falling over the Lowveld. Summer also meant that there were more tsetse flies and malaria mosquitoes. At last, on 17 January 1837, they reached Ziedie, slightly east of Mashao Hill in the vicinity of the current Kurulen. Their return journey of 30 shifts took two full months to complete, mainly as a result of the fact that they had lost almost all of their trek oxen.

Gert Scheepers died of malaria and was buried in an unknown grave somewhere in the Lowveld. Thus he became possibly the first white to be buried in the area that is now the Kruger National Park.[6] On reaching Ziedie, some of them were very ill with malaria.

From here, Antjie Scheepers secretly sent a San servant with a letter to Louis Trichardt to ask for help, and on 30 January 1837 Louis Trichardt sent his son Pieta with 30 oxen to go to the aid of the sorely tried party. Eventually they all arrived safely back at Trichardt's new camp along the Brak Spruit at the end of February 1837.

The Voortrekkers did not make any further attempts to visit the murder scene from the Soutpansberg, although they kept on making inquiries among the blacks.

The same year Louis Trichardt decided to move to the coast after having waited for almost a year for the "people at the rear" (Potgieter and his party). Louis Trichardt's epic journey started in August 1837 and ended in Delagoa Bay in April 1838 after unbelievable hardships (see chapter 4.3). Here Trichardt undoubtedly again brought the murder of Van Rensburg and his party to the attention of the Portuguese commandant, but no steps were taken to punish Manukosi for this outrage.

Shortly after their arrival in Delagoa Bay in June 1838, Trichardt's eldest son, Karel (Carolus), left on an extended voyage along the east coast on board the sailing ship *Estrella de Damão* (Punt, 1953). From Inhambane Karel undertook a long trade expedition into the interior, during which he undoubtedly would have had to follow the well-known trade route between the port and the densely populated Limpopo area (see chapter 4.4).

He reached the Limpopo at the point where the trade route crosses the river at Sakana's kraal during the winter months of 1838. Karel Trichardt was therefore in all probability the first white person to visit the Van Rensburg murder site. Here he was shown human skeletal remains as well as wagon wheel hoops and naves (Roos, 1930). Karel buried the remains of his murdered compatriots to the best of his ability and also determined that the two captured white children had died of malaria.

After his visit, several other people reached the murder site before 1900. Their findings and descriptions by credible witnesses confirmed the location,

---

6   Punt (1953) was, however, of the opinion that Gert Scheepers had died and been buried during the last days of January 1837 in the vicinity of the current Kurulen Mission, east of Mashao Hill.

4 THE VOORTREKKERS EXPLORE THE LOWVELD

The anvil which was found more than a century ago at the site where the Van Rensburg party had been massacred at Combomune along the Limpopo River in Mozambique. It was brought to South Africa in 1959 by the Punt expedition. It had the following numbers on it: (2) 0-30(18)20.948.

Dr W.H.J. Punt's expedition examine the anvil that they had found at Combomune, and which confirmed the locality of the site where the Van Rensburg party had been massacred.

which was finally determined by the Parks Board expedition of the late Dr Punt in 1959.

The following points serve as evidence that the murder took place near the confluence of the Olifants and Limpopo rivers:

1—Louis Trichardt mentioned in his diary that the Van Rensburg trek had been murdered at the "Groterivier" (Limpopo) near the kraal of Chief Sakana (Manukosi). He also mentioned that his base camp was situated five or six days on foot west of Sakana. (By then he had moved camp to an area near the current town of Louis Trichardt [Makhado], which was about 280 kilometres west of the murder scene at Masambo.)

2—Johannes (Jan) Pretorius mentioned in his letter to Louis Trichardt that the murder had taken place at "Sogana's" kraal by the Limpopo River.

3—Louis Trichardt's son Karel visited the scene of the murder a few years later. His evidence was of great value as he had been a member of the original search party, which reached Sakana in August 1836, and was therefore familiar with the general area. Karel later said that the murder scene was near the confluence of the Crocodile (Limpopo) and Olifants rivers, or rather 1800 paces further north, where they had crossed the river. In 1959 the Parks Board expedition determined that the murder scene was quite a bit further north (in fact about 80 kilometres north of the confluence of the Limpopo and Olifants rivers opposite and west of Combomune town along the Limpopo). It can be deduced from this that Karel had not been at the confluence of the Limpopo and Olifants rivers himself, but had heard from the blacks that it was "close" to the murder scene. According to evidence given to Roos (1930) by Karel, he managed to locate the warrior who had abducted the two children and he confirmed that they were both dead. He also showed the graves to Karel and said that if he opened them he would be able to determine by the long hair that they were in fact the two white children. Karel also found the burnt-out remains of the wagons at the murder scene. The iron had probably been removed to be smelted down. He also found two wheel naves, one of Klaas Prins's (Prinsloo) wagon and another which he could not identify. He buried the skeletal remains at the site.

4—S.W. Burger of Lydenburg recorded that his expedition from Ohrigstad to Inhambane had reached the Limpopo in May 1848. Here they heard that the murder had taken place on the left bank (sic) of the Limpopo. In his account he described the murder scene as being near the "Lempoepoe" (Limpopo) on the eastern side in almost a straight line between Pilgrim's Rest and the port of "Umjanbana" (Inhambane). The Zulu Chief Umzamane, who lived near the Limpopo, told this expedition that his father had been the leader of the Zulu impi who had murdered the Van Rensburg party. During the visit to the site by the expedition consisting of S.W. Burger, Dawid Joubert Jnr and Doors Buys (as guide and interpreter), they also found some of the ill-fated trekkers' ox-wagon wheel hoops still lying at the scene. They were also given two muzzle-loaders; the name J.H.J. van Rensburg was inlaid with ivory on the butt of the one and N. Prinsloo on the butt of the other. Upon their arrival back at Ohrigstad, the Executive Council decided that the rifles had to be handed to the next of kin (Preller, 1917).

5—According to W.H. Brown of Pretoria, the murder site is at the confluence of the Olifants and Limpopo rivers. He declared that a certain Erasmus had told him in 1889 how he had once followed Van Rensburg's wagon trail. Erasmus had been on a hunting trip in 1866 when he joined a group of people who were searching for the trail of the Van Rensburg party. He then travelled with them for a few days. The group was able to follow the wagon tracks as it was still clearly visible in places. Against a certain hill

112

4.1 THE FATE OF LANG HANS VAN RENSBURG

Combomune – "Place of the tragedy".

the tracks left by the wheel hoops were still clearly visible, but on the other side of the hill it disappeared. The trail disappeared completely near the confluence of the Olifants and Limpopo rivers.

6—Senator J. Wannenburg also said that the murder scene was in the vicinity of the confluence of the Olifants and Limpopo rivers after he and Magistrate Kuun had had an interview with an elderly gentleman, Mr Snyman, in their offices in Krugersdorp. The senator declared in 1938 that Mr Snyman had told him that he had spoken to Chief Sakana. During the interview the chief told him that the Van Rensburg trek had been murdered at a place before the confluence of the Crocodile (Limpopo) and Olifants rivers. Also that there were still wagon wheel hoops and other remains of the wagons at the site. Snyman was not certain whether it was on the upper or lower part of the confluence of the two rivers. As far as he could remember, the place where the Van Rensburg wagons had been was on the southern bank (right bank) of the Limpopo River, as the group had been on their way to Inhambane.

7—In 1872 the Natal surveyor St Vincent Erskine undertook an expedition to Mozambique and the Limpopo and accurately mapped the Limpopo's course from Masambo to the confluence. This map was published in 1875 by the Royal Geographical Society of London. On this map, the Djindi River is shown to be on the western (right) bank of the Limpopo with the following caption: "Battleground between Trek Boers and Shoshangans Army (Baobab)". This map is of the utmost importance, as it is the only document on which the murder scene was scientifically indicated a mere 36 years after the murder.

8—In 1917 Gustav Preller asked a missionary, Reverend C. Hoffman, to investigate the Van Rensburg murder. He did this enthusiastically and when Dr Punt questioned him in 1957 about the information he had collected, he stuck to his statement of 1917 to Preller in which he stated that in his opinion it was certain that the Van Rensburgs had trekked from Soutpan to Sakana and that they had been murdered near Sakana's kraal next to the Limpopo. He also stated that Sakana had lived three hours by horse from the confluence of the "Vimpoy" (Vembe, Limpopo, or Crocodile River) with the Olifants and that the Van Rensburgs had been murdered on the right (western) bank of the Limpopo.

9—Another extremely important statement was that of Breggie Pretorius, the wife of Jan Pretorius, who had accompanied her husband to Chief Sakana in 1836. Her second husband, Mr Hans Aucamp of Molteno, later wrote about her experiences. He confirmed that by the time Trichardt arrived in the Soutpansberg, the Van Rensburgs had already left for the Limpopo. Louis Trichardt and a few others decided to follow the Van Rensburgs on horseback and at the confluence of the Olifants and Limpopo he discovered that the Van Rensburgs had been murdered (Rossouw, 1929).

10—N.T. Bührmann of Lydenburg, in documents

113

he had written from 1850 to 1860, stated that the Volksraad in Ohrigstad had decided in 1848 to send a punitive expedition against Manukosi "and so bring the Zulus to task for the murders of the whole trek under Johannes van Rensburg".

The preceding evidence is overwhelmingly in favour of the tradition that the Van Rensburg murder site is near the confluence of the Limpopo and Olifants rivers. After careful consideration of all available information and extensive investigations locally, as well as the questioning of the local population, the National Parks Board expedition under the leadership of the late Dr Punt could determine the scene of the murder in September 1959 beyond any doubt. The expedition ascertained that the murder had taken place close to Combomune ("place of the tragedy") along the Limpopo. Combomune is situated 80 kilometres north of the confluence of the Limpopo and Olifants rivers across from Masambo's kraal, where the old trade route to Inhambane crosses the Limpopo. About 12 kilometres south of Masambo Drift, and along the western (right) bank of the Limpopo, the expedition reached the Djindi River where they found the murder scene, exactly as mapped by St Vincent Erskine in 1872. Here Chief Masambo also handed an old anvil (which according to serial numbers on it dated back to 1820) to the expedition. Legend has it that the anvil belonged to the Van Rensburgs and had been found at the murder scene close to one of the large baobab trees on the left bank of the Djindi and near the western bank of the Limpopo by one Medamo Balois's grandfather, Kohtshane Balois. In this way one of the most provocative secrets of the Voortrekker era could be satisfactorily solved through historical research.

References

Adendorff, G.M. 1984. *Wild Company*. Books of Africa, Cape Town.
Anon. 1959. Voortrekkeraambeeld sleutel tot geheim. *Die Vaderland* 3 October.
Bulpin, T.V. 1950. *Lost Trails on the Lowveld*. Howard B. Timmins, Cape Town.
Bulpin, T.V. 1976. *The Great Trek*. Printpak (Cape) Ltd., Cape Town.
Bulpin, T.V. 1978. *Illustrated Guide to Southern Africa*. Readers Digest Association of S.A. (Pty.) Ltd., Cape Town.
Bulpin, T.V. 1984. *Lost Trails of the Transvaal*. Books of Africa, Cape Town.
Dicke, B.H. 1944. The Northern Transvaal Voortrekkers. *Archives Year Book for South African History* 4(1):67–171.
Erskine, St. V.W. 1869. Journey of exploration to the mouth of the Limpopo. *Journal of the Royal Geographical Society London* 45:45–128.
Erskine, St. V.W. 1875. *Journey to Umzila, King of Gaza, 1872*. Unknown publisher, London.
Krüger, D.W. 1937. Die vestiging van die blanke beskawing in Noordoos-Transvaal. *Die Burger* 9 January.
Muller, C.F.J. 1974. *Die oorsprong van die Groot Trek*. Tafelberg, Cape Town.
Nathan, M. 193T. *The Voortrekkers of South Africa*. Central News Agency Ltd., S.A.
Preller, G.S. 1917. *Dagboek van Louis Trichardt (1836–1838)*. Het Volksblad-Drukkerij, Bloemfontein.
Preller, G.S. 1920. *Voortrekkermense ll. Herinneringe van Karel Trichardt*. Nasionale Pers, Cape Town.
Pretorius, J.C. 1984. Antjie Scheepers: onverskrokke pionier. *S.A. Tydskrif vir Kultuurgeskiedenis* 1(2):4–7.
Punt, W.H.J. 1939. Die Van Rensburg-raaisel I, II, III. *Die Volkstem* 6, 7 & 23 August.
Punt, W.H.J. 1953. *Louis Trichardt se laaste skof*. Van Schaik, Pretoria.
Punt, W.H.J. 1960. Waar die Van Rensburgtrek in 1836 vermoor is. *Koedoe* 3:206–237.
Punt, W.H.J. 1962. 'n Beknopte oorsig van die historiese navorsing in die Nasionale Krugerwildtuin. *Koedoe* 5:123–127.
Roos, J. de V. 1930. Carel Trigardt (1811–1901). *Die Huisgenoot* 24 October.
Rossouw, G.G. 1929. Jan en Breggie Pretorius van die Trichardt-trek. *Die Huisgenoot* 26 July.
Steyn, L.B. 1939. A Voortrekker Mystery solved, I. *The Star* 18 September.
Steyn, L.B. 1939. The trail to Sakana's kraal. II. *The Star* 19 September.
Steyn, L.B. 1939. How Sakana outwitted Trichardt. III. *The Star* 20 September.
Van der Merwe, P.J. 1937. *Die noordwaartse beweging van die Boere voor die Groot Trek*. W.P. van Stockum & Zoon, The Hague.
Van Schalkwyk, S. 1962. Anvil solves old mystery. *South African Panorama* February:26–27.
Venter, C. 1985. *Die Groot Trek*. Don Nelson, Cape Town.

4.2 The expedition of General A.H. Potgieter and J.G.S. Bronkhorst through the Lowveld

Dr U. de V. Pienaar

At the end of May 1836 the trekker party of Voortrekker leader Louis Trichardt arrived at Soutpan, which is situated along the westernmost foothills of the Soutpansberg. Here they camped to await the arrival of Commandant A.H. Potgieter. Johannes Hendrik Janse van Rensburg (Lang [Tall] Hans) and his party had already arrived at Mashao Hill, just east of the current Elim Hospital and Albasini Dam, searching for the route to Inhambane or Delagoa Bay.

Trichardt and Potgieter had agreed to explore the area in the far north with an eye to possible settlement. About a month later, on 24 June 1836, the exploratory patrol of A.H. Potgieter and J.G.S. Bronkhorst finally arrived at the Trichardt camp at Soutpan and stayed over for a few days. It can be assumed that Potgieter used the opportunity to traverse the countryside during that time. He was duly impressed with what he discovered about the area where he would lay out Zoutpansbergdorp (Soutpansbergdorp) in 1849, which would later become Schoemansdal.

This reconnaissance patrol had started on 24 May at the Sand River in the Free State and ended on 2 September 1836, when they returned safely to their laagers at the Sand River after several adventures only to discover that some of their people had been attacked and murdered by the AmaNdebele warriors of uMzilikazi (Silkaats).

Johannes Gerhardus Stephanus Bronkhorst[1] had kept a journal of their expedition. This document, as is the case with Louis Trichardt's diary, contains extremely valuable historical, geographical, climatological, ethnological and natural historical information about this time. As this party on their epic journey were probably the first whites since Francois de Cuiper's group in 1725 to travel through the area now known as the Kruger National Park (with the possible exception of Coenraad de Buys and his sons and the trekker party of Van Rensburg), it is important to examine Bronkhorst's diary in detail. Taking current knowledge into consideration it seems that historians, including Preller (1917), Dicke (1944) and Van der Merwe (1962), were wrong in their deductions that the reconnaissance party had travelled north of the Soutpansberg and crossed the Limpopo quite a distance west of the Bubye mouth.

If Bronkhorst's particularly faithful descriptions of especially the botanical and geographical features along the route are taken into consideration, one can come to the same conclusion as Punt (1960), who maintained that the patrol had in fact travelled along the southern slopes of the Soutpansberg from Mashao Hill (where the Van Rensburg trek camped for a while). They then headed northeast to Makonde Mountain, and after that followed the centuries-old trade route through Makuleke and across the Limpopo and Nuanetsi rivers in the direction of Sofala.

According to Bronkhorst, Potgieter and his group left the Sand River in the Free State on 24 May 1836. The expedition consisted of 11[2] members: Commandant Andries Hendrik Potgieter, Bronkhorst (the chronicler), Roelof Janse, Lourens Janse van Vuuren, Sarel Cilliers, Abraham Swanepoel, J. Robbertse, Adriaan de Lange, Daniël Opperman, H. Nieuwenhuizen and Christian Liebenberg.

From the Sand River they travelled about 12 shifts (a shift is a distance of about 20 to 25 kilometres on horseback and 8.5 to 10 kilometres by ox-wagon) across a grassy plain, where firewood was scarce, to a ridge, arriving on 5 June 1836. They named this ridge Suikerbosrand.

From Suikerbosrand they travelled another four shifts and reached the Olifants River on 9 June. Here they could not find any firewood, but ample sour grass and water. Two shifts further (11 June) brought them to Rhenosterspoort (later renamed Strydpoort because of the argument that had taken place there between Louis Trichardt and Van Rensburg) in the

---

1 After the establishment of Soutpansbergdorp in 1849 by Commandant Potgieter, he appointed J.G.S. Bronkhorst as leader of an expedition that had to negotiate a peace accord in July 1852 with the AmaNdebele chief uMzilikazi at his main kraal, Bulawayo.
2 According to Dr I. van Vuuren (1988), the expedition consisted of 12 members and included Hermanus Potgieter, the Commandant's brother.

The "wild banana" *Ensete ventricosum*, which J.G.S. Bronkhorst saw and described along the southern slope of the Soutpansberg. He was a member of General A.H. Potgieter's patrol to Sofala in July 1836. These examples were photographed in a ravine in the southern slope of Entabeni (upper reaches of the Luvuvhu). (MAY 1986)

Randberg (foothills of the Drakensberg, or Strydpoort Mountain). Here the countryside was uneven with a mixture of sour and sweet grass and thorn trees. From here they rode two and a half days across a grassy plain until they reached a rough plain covered in several types of grass. After another 13 shifts they eventually reached Soutpan at the western point of the Soutpansberg on 24 June. They encountered the first blacks, who were Mantatese (North Sothos or Pedi), at the Rhenoster River. At Soutpan they found the Trichardt party in good health. Karel Trichardt mentioned in his memoirs (Preller, 1920) that Potgieter, Bronkhorst and their party had stayed with them for three days and rested. From Soutpan they travelled further along the southern foothills of the Soutpansberg to Mashao Hill, which they reached four days later (Punt, 1960). Here they discovered that Van Rensburg had already led his party off into the Lowveld. If they had had any contact with them, Bronkhorst would certainly have mentioned it, as two of his brothers were members of the Van Rensburg group.

Bronkhorst recorded that, two shifts away from Soutpan, they encountered trees with thick trunks entwined with roots, which entered the ground lower down. The trees had white bark but no fruit (it was winter). According to this description, these trees could have been either the Lowveld fig *Ficus stuhlmanii* or the knobbly fig *Ficus sansibarica*. Both species occur in the area south of the Soutpansberg, but are rare north of it.

Three shifts further Bronkhorst described a tree "with a fruit like a coconut". "The fruit is hollow on the inside and the content looks like cream of tartar." The colour of the bark and wood reminded him of the spekboom of the Eastern Cape. These trees are very large and one had a circumference of 13 fathoms (23.77 metres). He was definitely describing a baobab. Although these trees are very common north of the Soutpansberg, Van der Merwe (1962) was incorrect in his statement that baobabs are rare in Venda south of the mountain. In fact, they are quite common in the northern part of the Kruger National Park.

Bronkhorst also described another tree from which the natives chopped branches that they then tapped sap from to use as food (drink). He described it as being a large tree resembling an oak. The fruit looked like candles, but they could not eat any as there were no ripe ones. The locals apparently boiled buffalo skins down to a kind of glue, which they then mixed with the sap. This mixture, which they drank, had the appearance of curdled milk. They apparently also used the sap to brew a type of beer.

It is uncertain if this information of Bronkhorst is correct. The tree he described could have been a sausage tree, *Kigelia africana*. Although it is known that the Venda and other tribes use the burnt fruit of this tree to enhance the fermentation process of beer, it is not known whether enough sap can be obtained from chopped-off branches to use as food or drink. What probably happened was that he had found calabashes full of mlala beer in the shade of a sausage tree and misunderstood the people as saying that they obtained the sap from the sausage tree, while they had in fact tapped it from the cut fronds of mlala palms, *Hyphaene natalensis*.

Bronkhorst also described the most common tree in the region, which is the mopane, *Colophospermum mopane*. To him this tree looked like an apricot tree and its fruit like a lentil, although much bigger, and he describes the seed inside the "lentil" as soft and oily, smelling like turpentine. The later settlers in the area indeed ended up calling it the "terpentynboom" (turpentine tree)!

Bronkhorst also mentioned that they encountered thick stands of "banana" trees along the river (at the upper reaches of the Luvuvhu). Both the African wild banana (*Ensete ventricosum*) and the Transvaal wild banana (*Strelitzia caudata*) grow in the ravines of the southern slopes of the Soutpansberg. However, the most conclusive evidence that the group travelled along the southern slopes of the mountain along the old trade route to Sofala comes from Bronkhorst's notes indicating that they had moved past a large, thick bamboo stand that was "half an hour's travel by horse in circumference".

The only place in South Africa where a bamboo bush occurs naturally is east of Sibasa[3] along the old trade route to Sofala. Here a large stand of northern mountain bamboo (*Oxytenanthera abyssinica*) can be found. The Venda hold this place sacred and fashion reed flutes from the stems, which feature in their tribal rituals. How and when the bamboo got there is not known, but it is possible that it had been brought via the old trade routes by local people long before the first whites set foot in the area. This particular bamboo species is indigenous to Zimbabwe and areas as far north as Tanzania.

Nine shifts (225 kilometres) from Trichardt's camp site, the group reached a big running river that was 1780 paces wide and in which water flowed 60 centimetres deep. This could only have been the Limpopo, but the only places where the Limpopo is even nearly that wide are just east of the Bubye mouth and across from the current Manqeva outpost in the Makuleke area of the current Kruger National Park. The latter looks like the most probable and if the route that was indicated by Punt (1960) were to be divided into shifts, then it would correspond to the distance to Soutpan. It thus seems highly unlikely that the party had crossed the Limpopo further to the west, as claimed by Preller, Dicke and Van der Merwe.

On the banks of the Limpopo they found beautiful, large trees. Some of these were so large that one could outspan six or seven wagons in their shade. The leaves were small and green, the trunk smooth and the fruit showed similarities to acorns. This is undoubtedly a description of the nyala tree (*Zanthocercis zambesiaca*), which is a typical feature of the riverine forests along both the Limpopo and the Luvuvhu.

One day's journey further, the group reached another large river and found similar trees. This was the Nuanetsi in the southeastern corner of Zimbabwe. From this river they travelled another six shifts to reach the kraal of the "Knob noses" (Shangaan), who indicated the way to the next kraal about six shifts further on. This was probably Massangena next to the Rio Save in Mozambique (Punt, 1960). They were warmly received by the inhabitants of the kraal, who had been aware of their approach. The people of this kraal could speak Portuguese and had come there to trade beads, linen and other wares with the coastal traders from Sofala (Beira). As currency they offered ivory. The party was also informed that ships were waiting in the port at Sofala for the ivory and other trade goods from the interior.

By that time their horses and oxen were exhausted and Potgieter decided that the party should turn around. They started their return journey on 18 July and were provided with guides who showed them a shorter route. This was probably the trade route from Sofala over Chicuala-cuala and Malonga Fountain to Klopperfontein, Punda Maria (Shikokololo), Ysterberg and Mashao Hill.

The unique holy Venda bamboo clump (*Oxytenanthera abyssenica*) which was discovered by the Potgieter-Bronkhorst patrol along the centuries-old trade route from the Soutpansberg to the east coast in July 1836. It is situated at Tshaulu, about 7,5 km northwest of Shikundo Hill in Venda.
(APRIL 1986)

---

3   The locality of the bamboo bush is at Tshaulu in the Phaswane settlement of Venda Chief Bohwana. It is about eight kilometres northwest of Shikundu Hill and the Venda call this bamboo "Musununu" (Cuénod, 1985).

Bronkhorst mentioned that the Shangaan of this area were a friendly and defenceless people that constantly had to flee attackers from the west and the south. Many were hardly surviving as they had been robbed of all their livestock by uMzilikazi's warriors. At Massangena they also met Doors and Gabriël (wrongly called Karel in the diary), the two coloured sons of Coenraad de Buys. They were there to exchange ammunition for ivory and other trade goods. The blacks referred to the members of the expedition as Magoas (Dutchmen). Bronkhorst made notes of a variety of geographical, geological and climatological features of the countryside through which they travelled and summarised his impressions as follows (loosely translated):

> The climate is fairly warm and there is very little difference between winter and summer [he obviously did not have an inkling of exactly just how hot it can become during summer!]. Lush vegetables grow everywhere. We were there during July and saw all kinds of fruit growing and flowering. In the gardens we found sweet potatoes, maize and all kinds of vegetables. Water to irrigate the lands is not a problem and one can almost say: 'not enough land for the many fountains' [When looking at Bronkhorst's description of the water level of the Limpopo in June and the manuscripts of other explorers and hunters during and after that time, it is clear that water was much more abundant in the Transvaal Lowveld than it is now. This in turn indicates a long, above-average rainfall cycle of 80 to 100 years]. A large city could be built if there were enough people to do so, and each would have their own water. Everything points to it being ideal for a settlement [see the later establishment of Zoutpansbergdorp (Schoemansdal) in 1849]. Wood is plentiful, the cultivatable lands large and extended, so that thousands of families could exist there. At the same time it is also suitable for stock-breeding [another indication that the patrol did not travel through the relatively dry, infertile and tsetse fly-infested area north of the Soutpansberg].
>
> We found a large amount of iron among the Mantatese [North Sotho or Pedi], which they smelted themselves. The iron is of good quality and mixed with steel. They showed us a small ridge in the south [the Ysterberg or possibly Phalaborwa] where the smelting and forging take place and where they make assegais. They exchange these with the 'Knob noses' [Shangaan] for beads and other wares.
>
> The locals also pointed out a mine to us [the gold mines of the Lemba in the Soutpansberg?] where they mine gold from which they make rings. I have seen some examples of these.
>
> At the first tribe we encountered [the Pedi along the Olifants River] we also found good quality tin, which they mine along the Randberg [Drakensberg] and from which they also make rings [the tin mines in the vicinity of Potgietersrus or those at Penge]. They call the tin white iron. Along the Olifants River we saw banks of a silver mineral that looked like leaves [mica]. It is tough and hard to separate. I did not see anything that had been produced from it.
>
> We saw many kinds of game. From Suikerbosrand onwards we encountered elephant all along the way. Along the Vaal and Olifants rivers we saw large numbers of rhino, as well as buffalo, hippo, sable, waterbuck and blesbok the size of hartebeest. We also saw kudu with striped quarters, gemsbok, impala and other familiar game in large numbers.

From Trichardt and Bronkhorst's diaries it would seem that Potgieter and Bronkhorst's patrol returned to Louis Trichardt's laager at Soutpan at the end of June. By this time, the trekkers had already heard rumours about the murder of the Van Rensburg party and they hurriedly nominated a five-man team to go searching for them. The patrol consisted of five horsemen, namely Louis and Karel Trichardt, J.G.S. Bronkhorst,

A flooded Limpopo River, photographed at the site where the Potgieter-Bronkhorst patrol had probably crossed it in July 1836. This wide section of the Limpopo is directly north of the Pafuri ranger post and Manqeva picket.
(MAY 1986)

4.2 THE EXPEDITION OF GENERAL A.H. POTGIETER AND J.G.S. BRONKHORST THROUGH THE LOWVELD

J. Robbertse and A. Swanepoel. The group could find no sign of the Van Rensburg trek, but there was enough evidence to accept that the whole party had been murdered by the warriors of the Zulu renegade Manukosi (Sochangana). They arrived back at Soutpan on 16 August without having any positive news to report.

On 17 August Potgieter, Bronkhorst and the rest of their party set off again on their return journey to their laagers along the Sand River in the Free State. Upon their arrival on 2 September 1836 they found their compatriots in a sad state. Bronkhorst stated that when they were three days south of Trichardt's laager at Soutpan, they sent five men – C. Liebenberg, R. Jansen, A. de Lange, D. Opperman and A. Swanepoel – ahead to fetch fresh pack oxen and horses. On arrival at the first camp along the Sand River, they found a wagon in the river. When Opperman rode closer to investigate, he saw that two wagons had been tied together. In the vicinity of the camp he found a bloody scene. When the rest of the patrol arrived, they discovered the bodies of Liebenberg Snr and his wife, H. Liebenberg, as well as several other bodies they could not identify. They returned to the rest of the patrol the same afternoon with the shocking news about the murder of their companions.

Five patrol members went back to the scene and identified the victims as B. Liebenberg Snr; Johannes du Toit and his wife; H. Liebenberg Jnr and his wife; S. Liebenberg and a little boy; a son of C. Liebenberg; and McDonald the teacher. They could do nothing but bury the dead and afterwards proceeded hastily to their own laager, which they reached three days later. Here they found the survivors as well as some of their livestock.

Conditions here were equally sad. Johannes Bronkhorst's son, G. Bronkhorst, and Christian Liebenberg's son Barend were still missing, while Christiaan Harmse's son had been killed. At the same time they found out exactly how the disaster had happened.

Apparently, a group of AmaNdebele warriors had led Stephanus Erasmus and eight other hunters into an ambush. Erasmus and one of his sons, as well as Pieter Bekker and his son, managed to escape, but two other sons of Erasmus, as well as Johannes Claasen and Karel Kruger, were missing (presumably murdered). They were attacked without any provocation (the warriors were possibly under the impression that they were a band of Griqua robbers). The attackers left with their wagons, as well as the rest of their belongings, including the livestock. Erasmus immediately set off to inform the other laagers that the blacks were on the warpath and were attacking the small, remote laagers and killing the people.

Ten men from the main laager met the band of warriors on 23 August 1836 in an attempt to negotiate a peace accord, but they were attacked immediately and had to beat a hasty retreat to the main laager. Although there were only 35 people at the main laager by the Vaal River, they still managed to drive the attackers away and killed a few of them.

Potgieter immediately decided to trek back across the Vaal River and moved his laager four days' journey closer to Vegkop, between the Wilge and Renos-

Vegkop as seen from Koranna Mountain. The historical battle of Vegkop took place close to this hill which is about 20 km southwest of Heilbron. During this battle, a band of uMzilikazi's warriors under the leadership of field general Khaliphi was comprehensively defeated by a small group of Voortrekkers under Hendrik Potgieter and Sarel Cilliers on 19 October 1836.
(ACKNOWLEDGEMENT: TRANSVAAL ARCHIVES)

The battle of Vegkop in 1836, as depicted by H. Egersdörfer.
(ACKNOWLEDGEMENT: AFRICANA MUSEUM)

ter rivers (south of the current Heilbron), but they were followed by an AmaNdebele army of about 3000 men under the leadership of Khaliphi.[4] The next day Potgieter again rode out with 35 men towards the approaching army to negotiate peace, but they were attacked yet again and had to retreat to the laager. This time they reached the laager just ahead of the charging army and barely had time to clean their rifles. When the attacking army came within 500 paces of the laager they slaughtered two of the trekkers' oxen. After they had consumed the raw meat, they charged the laager with shouts and cries. However, they could not penetrate the trekkers' defences and the small group of trekkers defended themselves so bravely that the enemy was beaten off.

Although more than 1000 assegais were picked up inside the laager after the attack, only two trekkers, Nikolaas Potgieter and Piet Botha, were killed, while 12 others were wounded.

During this battle, which probably took place on 19 October 1836, the trekkers lost 6000 head of cattle and 41 000 sheep and goats. They managed to save their horses, however, as these had been tied up in the laager during the battle.

Three days later they set off after the warriors to get some of their livestock back, but all they managed to find were about 1000 carcasses that had been left behind by the attackers.

With the help of Chief Moroka's friendly Baralong and the larger group of Voortrekkers who were still encamped at Thaba Nchu, the sorely tried trekkers managed to make ends meet. Potgieter and his courageous punitive expedition repeatedly attacked uMzilikazi's AmaNdebele army during January and November 1837. In the process they recovered large numbers of their livestock and destroyed uMzilikazi's strongholds at Thaba Mosega and Gabem, in the Marico district of the western Transvaal (now North-West Province), driving them across the Limpopo. In 1847 Potgieter led another commando from Ohrigstad against uMzilikazi and decisively defeated the AmaNdebele chief at his new settlement north of the Limpopo. This, however, is related in detail in other sources.

References

Bergh, J.S. & Bergh, A.P. 1984. *Stamme en ryke*. Don Nelson, Cape Town.

Bulpin, T.V. 1984. *Lost Trails of the Transvaal*. Books of Africa, Cape Town.

Coates Palgrave, K. 1983. *Trees of Southern Africa*. C. Struik, Cape Town.

Cuénod, P. 1985. Personal communication.

De Jongh, P.S. 1987. *Sarel Cilliers: 'n biografie oor die Voortrekker-figuur*. Perskor, Johannesburg.

Dicke, B.H. 1944. The Northern Transvaal Voortrekkers. *Archives Year Book for South African History* 4(1):67–171.

Du Buisson, L. 1987. *The White Man Cometh*. Jonathan Ball, Johannesburg.

Grobler, H.J. 1919. Storie van 'n vergane Voortrekkerdorp. *Die Brandwag* 24 December:221.

Grobler, H.J. 1938. 'n Voortrekkerdorp wat verrys en verdwyn het. *Die Volkstem* 9 December:14.

Krüger, D.W. 1937. Die vestiging van die blanke beskawing in Noordoos-Transvaal. *Die Burger* 9 January.

Moller-Malan, D. 1953. *Dina Fourie*. Voortrekkerpers, Johannesburg.

Preller, G.S. 1917. *Dagboek van Louis Trichardt (1836–1838)*. Het Volksblad-drukkerij, Bloemfontein.

Preller, G.S. 1920. *Voortrekkermense II. Herinneringe van Karel Trichardt*. Nasionale Pers, Cape Town.

Punt, W.H.J. 1953. *Louis Trichardt se laaste skof*. Van Schaik, Pretoria.

Punt, W.H.J. 1960. Waar die Van Rensburgtrek vermoor is. *Koedoe* 3:206–237.

Van der Merwe, H.J.J.M. 1969. *Scheepsjournael ende Daghregister*. Van Schaik, Pretoria.

Van der Merwe, P.J. 1962. Die Matabeles en die Voortrekkers. *Archives Year Book for South African History* 49(2).

Van der Merwe, P.J. 1962. *Nog verder noord*. Nasionale Boekhandel, Cape Town.

Van Vuuren, I.J. 1988. Personal communication.

---

4   Also spelled Kalipi in certain references.

4.3 Louis Trichardt's trek through the Lowveld

Dr U. de V. Pienaar

The small white population at the southernmost tip of Africa has produced a number of remarkable leaders in the past 300 years. The Great Trek, which was an event of great moment in this part of Africa during the nineteenth century, brought men and women of extraordinary character to the fore. Unfortunately it was not well documented, as only a few of the Voortrekkers kept diaries of their experiences. Louis Trichardt, one of the first leaders of the Great Trek, was one of them.

This valuable document, which was written in eligible, proficient Dutch, is in many ways a unique legacy, as it not only provides insight into the daily lives of the Voortrekkers, but also into Trichardt's character.

Although the diary deals mainly with the Trichardt trek's arduous journey from the Soutpansberg to Delagoa Bay during 1837 and 1838, studies of this document have uncovered valuable information about the early history of the Transvaal and Mozambique. It also contains data of scientific import with regards to the climatic and other environmental conditions, the ethnology of black tribes living in the areas through which he travelled, the social life of the Voortrekkers, the incidence of animal diseases and malaria, and the distribution of tsetse fly, as well as valuable botanical information and data about the occurrence and distribution of game in the 1830s.

Trichardt's prominence stems from the fact that his trek was the first to complete an overland journey from the interior of the Eastern Cape all the way to Delagoa Bay.[1] His 2500 kilometre-long journey through the wilderness lasted three years and resulted in unprecedented suffering for both man and beast. Nevertheless, the small group of 53 in nine ox-wagons, with their large herds of cattle and sheep, managed to reach Delagoa Bay under the efficient and brave leadership of Louis Trichardt without the loss of a single wagon, while only two adults had died during their journey. This is undoubtedly a remarkable achievement, unrivalled in the annals of African expeditions and settlement.

Voortrekker leader Louis Trichardt. This photograph was probably taken in Lourenço Marques (Maputo) shortly before his death in 1838.

Louis Trichardt's ill-fated journey through the wilderness has been documented and discussed in detail by several authors (Preller, 1917; Fuller, 1932; Dicke, 1941; Thom, 1943; Punt, 1953 and others). For the purposes of this work it is sufficient to look at his findings during their trek through the Lowveld and specifically the area that is now the Kruger National Park. However, to obtain adequate insight into this piece of national history, it is important to relate a little about Louis Trichardt the man.

Louis Trichardt was born on 10 August 1783 in the Oudtshoorn district. The Trichardts were of Swedish extraction. The patriarch of the family, Carel Gustavus Triegard (Tregardt), came from the town Kvidinge near Ängelholm and Helsingborg in south-

---

1 The vexed Boer community of the Eastern Cape sent several exploratory commissions to investigate the probability of settlement outside British territory. One of these expeditions was led by Hendrik Scholtz in 1834. This reconnaissance party travelled all the way through the already known Free State, across the plains of southern Transvaal, up to Soutpansberg, where they met the descendants of Coenraad de Buys, whom he had left behind in 1821. They found this area very attractive because of the ample water and game and Scholtz reported this to his compatriots. Louis Trichardt had heard of Scholtz's findings and decided to follow the same route into the interior (Van der Merwe, 1950).

4 THE VOORTREKKERS EXPLORE THE LOWVELD

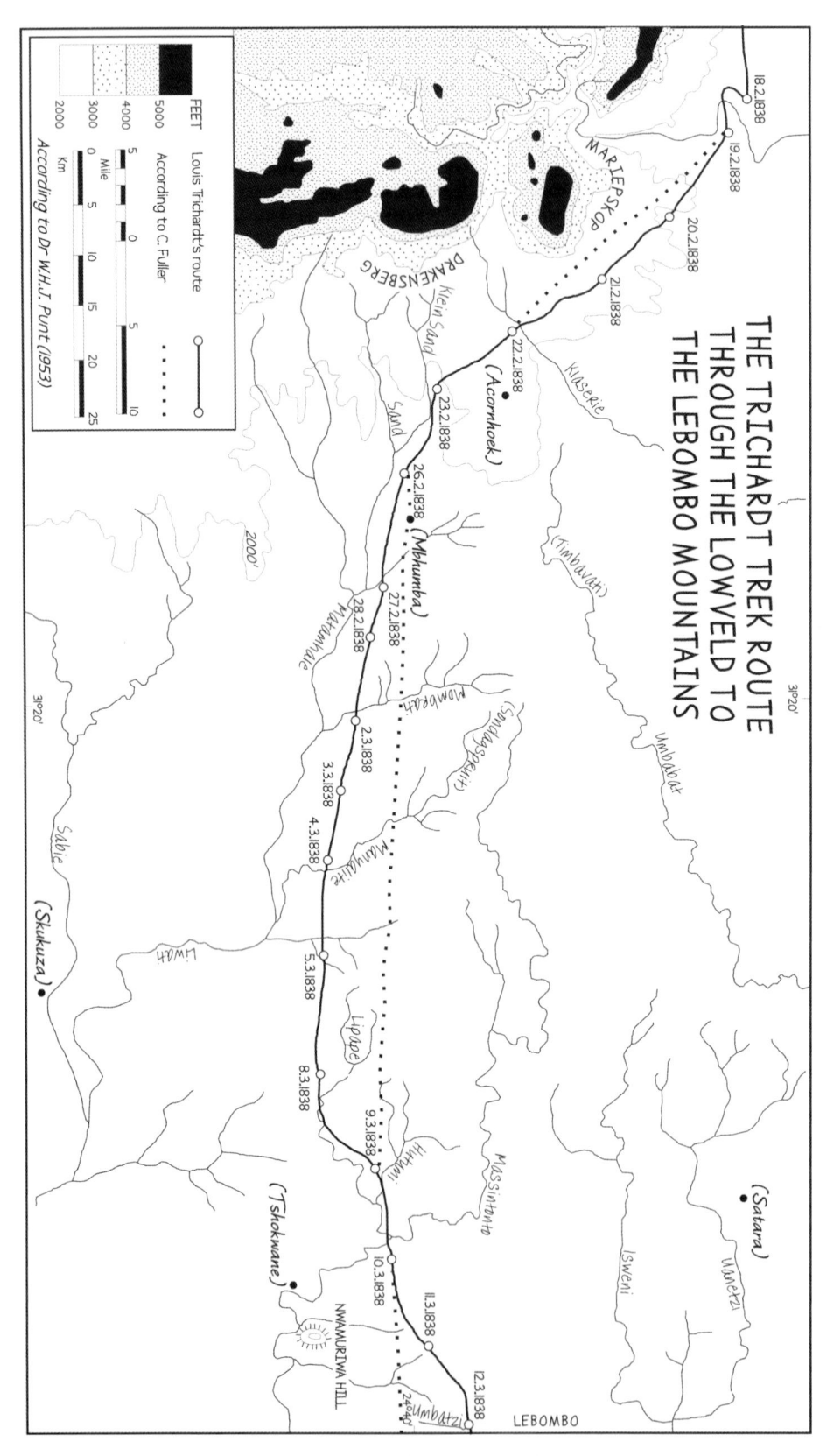

The route followed by the Trichardt Voortrekker party through the Lowveld towards the Lebombo Mountains.

west Sweden. He was a sailor in the service of the Dutch East India Company. In 1744 he was married in Cape Town and in later years moved to Stellenbosch, where he died in 1767 as a "free burgher". His descendants moved eastwards to the Eastern Cape, where they became progressive and prosperous farmers. They were also to play a prominent role in the establishment of Graaff-Reinet in 1786 and in the general activities of the district. It was only natural that they (especially Louis's father, Carel Johannes Trichardt) would become involved in the military and political turmoil at the time centred on Graaff-Reinet with its pro-independence burghers.

One hundred years later, Louis's grandson Colonel Stephanus Trichardt, an officer in the state artillery of the Zuid-Afrikaansche Republic (South African Republic), distinguished himself in the battle of Amajuba during the First Anglo-Boer War. It was during this time that the Transvaal Trichardts heard about a Trichard family in the south of France.

Louis Trichardt's eldest son Karel farmed on Goede Hoop near Standerton in the Transvaal (now Mpumalanga) in the 1880s and he sent his son Jeremiah to the Netherlands in 1885 to further his education. While there, Jeremiah visited his French namesakes at Ramoulins in the Gard *département* in the south of France for two days. After Jeremiah's return to South Africa, Karel decided to change the spelling of their family name. Until then both he and his late father Louis had spelled their surname Tregardt,[2] while other members of the family chose the traditional Dutch spelling Triegaard, or Tregart. After 1890, Karel and his sons accepted the French spelling Trichard, but kept the original "-t" suffix. This is why all the direct descendants of the famous Voortrekker leader spell their surname Trichardt, while Louis himself spelled his surname Tregardt (Punt, 1951).

Louis was 52 when he decided to leave the Cape once and for all and set out northwards in an effort to escape British rule. He was almost 1.8 metres tall and weighed about 82 kilograms, which indicates an athletic build. He must have had a very strong constitution, as no reference could be found anywhere in his diary of him being ill or in bed. From the only existing photograph of him it can be deduced that he had short black hair and bushy eyebrows. Except for neat side-whiskers he was clean shaven with a strong, firm mouth. His brown eyes inspired authority, but were at the same time those of a gentle dreamer. He was always neatly dressed in the fashion of the time and sometimes wore long boots.

His patience, courage and sense of humour made him an ideal leader, even during times of seemingly insurmountable problems. His firm but just handling of the numerous disputes all too common in any community during difficult times, contributed considerably to his being able to keep his people united during the long and arduous journey.

It can be deduced from his diary that Louis was a leader who led by example, keeping his calm rather than exercising an indiscriminate or violent authority. He often had to act as peacemaker. It was he who had to reprimand his quick-tempered son, Karel,

General Piet Joubert (left) and Karel Trichardt in about 1890.
(ACKNOWLEDGEMENT: STATE ARCHIVES, PRETORIA AND MR W.J. PUNT)

---

2   This spelling is currently recognised by historians as the correct one.

when Karel was ready to grab a gun or lay in with his fists. He also had to settle the daily arguments with the stubborn and argumentative Jan Pretorius and had to calm the women who were persistently sniping at each other. He even had to make sure that the young men did not cross the line with Izak Albagh's nubile daughters. Apart from that, he constantly had to give attention to their daily safety, the food and ammunition stocks, the condition of the livestock, the education of the children by the aged Daniël Pfeffer and the spiritual wellbeing of his people. He took weather readings and as such was the first person to record climatological data in the area that would later become the Transvaal. Louis also negotiated with the local blacks to trade provisions, or to find the best route ahead for the wagons.

On Christmas Day 1837, while the morale of the trekker party was at its lowest as they were struggling to find a suitable way down the Drakensberg from Sekororo, a small drama ensued. Three of the boys – Koot Scheepers, Izak Albagh and Frederik Botha – absconded, each with a rifle and a bandolier, with the aim of returning to the Cape by themselves. After four days the three adventurers decided that it would be safer with their parents and headed back to camp. Louis did not want to punish them because they had returned on their own accord, but the widow Scheepers disagreed and gave her son Kootjie a thorough thrashing for his rashness. The other parents followed suit and Karel boxed the young Izak Albagh's ears because he had stolen his best bandolier for the abortive trip back south.

Louis's wisdom in the handling of differences soon became known among the locals, and before long the high chief and other leaders came from near and far to obtain his verdict in connection with personal problems or tribal concerns. His extensive knowledge of the human psyche and an understanding of the black tribes' customs and traditions normally led Louis in his decisions. In this way "Lebese", as he was called by the locals, could nearly always encourage good relations between the different races. For his good advice he was often rewarded with an elephant tusk or some other token of appreciation.

When Louis left his last home in the Cape along the Wit Kei at the foot of Tsomo Mountain in 1835, his party consisted of his own family, including his wife, three sons and a daughter, as well as their elderly Dutch teacher Daniël Pfeffer and a tenant, Hans Strijdom, with his wife and five children.

Hans van Rensburg and his party spent some time with the Trichardt group along the Caledon River before Van Rensburg decided to trek ahead to the Vaal River. However, some members of the Van Rensburg party, including Gert Scheepers, Jan Pretorius, Hendrik Botha and Izak Albagh and their families, decided to stay with Trichardt.

The Trichardt party eventually arrived in the Soutpansberg, where they stayed for 15 months. Then the arrival of two Portuguese guides, Antonio and Lourenço, from Delagoa Bay, caused the Trichardt party to leave their temporary home on 23 August 1837 and set off along the unknown route to Delagoa Bay. Louis, who was almost always called Lewies or Lowies, decided not to follow the same route to the coast that Van Rensburg had taken, because he was aware of the dangers that malaria and tsetse fly posed to his people and animals in the Lowveld. Then of course there was Manukosi and his antagonistic warriors along the Limpopo who would be a deadly threat to the handful of trekkers and their servants.

Taking all of this into consideration, he decided

Mashao Hill, southwest of Elim, where the Trichardt Voortrekker party had been encamped before moving into the Lowveld in 1837. (MAY 1986)

to return to Sekwati's Poort to ask the friendly Pedi Chief Sekwati for the best route to Delagoa Bay.

On 2 October 1837 the trekker party reached the banks of the Olifants River (Balule or "far-off" river) after having travelled south for almost 200 kilometres. Here they waited for more than a week for Chief Sekwati to arrive, only to be told that the route to the Bay was almost impassable for wagons and very dangerous.

Trichardt, however, was not about to throw in the towel and in the middle of October the party set out with courage and determination on what seemed to be an impossible journey across the Drakensberg.

It is well known that it was a mistake to follow the Olifants River in an easterly direction. As a result they had to ford and reford the river 13 times in five days to find suitable terrain over which the wagons could travel. Finally on 2 November, after heated debate, they decided to turn back to a point between the current Penge Mine and the confluence of the Wit (Moetsi) and Olifants rivers. At this point they took the onerous decision to build a road across the Drakensberg at Makeghula[3] (Sekororo Peak), an undertaking that took them almost a month of sweat and blood.

On 30 November they finally outspanned on the eastern summit of the mountain. This place, which they called "green height", was just north of the massive Ngopelle Peak. The most difficult part of the journey was still lying ahead, though, and they searched for days and speculated much about the best route down the mountain. In the end some of the women in the party – Martha Trichardt, Antjie Scheepers and Breggie Pretorius – found a passable route. Everybody, except the know-all Jan Pretorius, was optimistic about the route and immediately started constructing a road down the mountain. Years later Karel Trichardt described their crossing:

> It was so steep that I had to lock the front wheels of the wagon with a chain. We removed the back wheels and bolted them to two wooden crossbars. On top of that there was also a large branch tied to the back to act as a drag-break. We tied the tails of the two hind oxen to the wagon so that they would not go head over heels (Roos, 1930).

The cutting against the mountain summit where the Trichardt party descended with their wagons is still clearly visible today. It is about five kilometres south of the mountain peak Makeghula (1795 metres above sea level) along the edge of the Drakensberg (Fuller, 1932).

Two months later, on 25 January 1838, they were encamped at the foot of the mountain at Marula Spruit (a tributary of the Makhutswi). Louis penned down the general mood of his people in his diary in these simple words: "Today is indeed the happiest day in the whole of our adversity." Although the wagons had made it down the mountain, they were understandably all fairly worse for wear and the trekkers had to stay put for a few days to repair them. During this time they again had problems with Chief Sekororo's people in the form of stock theft. To solve this problem, Trichardt devised a plan whereby they would take a few of Chief Masipana's subjects hostage, promising to eventually release them with the proviso that the theft of the livestock end immediately.

On 5 February the party, with their eight hostages, moved to the "Hoeks" River (Malomanye, a tributary of the Makhutswi on what is now the farm Metz 168). Some of Masipana's warriors followed them intending to free the hostages, but Karel and Pieta drove them away with a few warning shots. The next day the wagons moved on to the "Drai" River (the Kubjaname, which is another tributary of the Makhutswi) on the current farm Enable 159. The ancient trade route to Delagoa Bay across this farm is still visible to this day. On 7 February the party reached the "Klipplaat" River (the Kgogongwe, a tributary of the Molomahlapi). This outspan is situated on the current farm Worcester 164. Here they received a visit from some of the locals and were offered honey.

On 8 February the wagons trekked all the way to the Olifants River (Balule), where they were visited by a delegation from Chief Masipana, who promised to send an elephant tusk in exchange for the release of the hostages. That night more sheep were stolen, despite the fact that they had posted guards.

---

3 Also called Mameghula. This spelling is today once again accepted by historians as the correct one.

Chief Masipana himself arrived the next day with two tusks that he offered to Antonio and Lourenço, the two Portuguese guides, in exchange for the release of the hostages. He was too scared to negotiate with Trichardt himself, however, and left without having achieved his purpose.

At what is now the farm Willows 177, the party had to wait for the flooded Olifants River to subside before they could ford it. On the night of 10 February they had to deal with another stock raid – Karel wounded one of the raiders while the others fled. The next day, 11 February, the water in the drift was still too high for them to cross safely, but they received a delegation from one of the chiefs on the other side of the river, who sent them a calabash of marula beer.

On 12 February the group could at last herd their sheep through the Olifants, an operation that took the whole day. They received help from one of the local chiefs on the other side of the river, who also presented them with an elephant tusk. That night Pieta had to fire a few shots at lions that were after their sheep on the southern side of the river.

The next day they drove the remaining livestock through the river with the help of 60 of Chief Marmanella Omsana's subjects, so that they could pitch camp south of the river that night. This was the seventeenth crossing of this large river by the Trichardt party.

The drift, which is on the current farm Willows 177 at the confluence of the Meetse-Matau Spruit and the Olifants River, corresponds with the legend related by black locals that "Lebese" crossed the Balule (Olifants River) east of the Mabin settlement. This drift was apparently still clearly visible in 1889 about 300 paces east of the corner beacon between Willows 177 and Arthur's Rest 526.

The Voortrekkers camped for another two days on the southern bank of the Olifants River on what is today the farm Dublin 15, while Trichardt inquired among the locals about the presence of tsetse fly and the route to Delagoa Bay. By that time the party had already been in tsetse fly territory for eight days, without the loss of any cattle. Chief Marmanella Omsana provided guides to accompany the Trichardt party further southeastwards.

On Thursday 15 February the wagons reached the "Bergen" River (Blyde River). As a result of earlier heavy rains further up in the Drakensberg the river was flooded, so that neither the wagons nor the stock could get across. In the meantime they negotiated further with the chief about guides and reached a compromise about payment for them.

By 19 February the water level had gone down to such an extent that they could get through. With the aid of 50 black helpers the crossing took place without any problems. This brought the group to the current Moriah 100 where the motorway currently crosses the river.

Several local chiefs paid courtesy visits and requested aid with their problems. Everybody was given a patient hearing or asked to return at a later date. The chiefs gave the "people in front" information about the route and distances to Delagoa Bay and also promised to provide them with guides. At the Blyde River the Voortrekkers were, according to the blacks, at least six days on foot (thus 275 kilometres) from Delagoa Bay. They would therefore still have to complete at least 30 shifts to reach their destination. By that time their small livestock was heavily infested with ticks and the trek was forced to move much slower over the wet terrain.

Their next outspan was next to the Mabeti or Rietspruit on the current farm Antioch 368. Here Trichardt heard Chief Mosalie's request for protection against another chief who wanted to rob her of her position. Both Mosalie and the other chief asked Trichardt for "testimonials" attesting to the fact that they had treated the whites well, as they were certain that other Voortrekkers would follow.

After 1838, hunters had opened up Trichardt's wagon trail on the farms Antioch 368 and Bluebank 236 to such an extent that Dr Punt could easily follow the beaten track. It was particularly clear where stones had been rolled out of the way, after Mr J.I. Nel of the farm Bedford 366 had shown him the passage.

On 21 February the party trekked past Mariepskop to the upper reaches of the Klaserie River, where they found pools. From here the Trichardt group trekked even slower, as he had noticed that the sheep were in a very poor condition as a result of all the ticks. By then they had realised that they would have to reach the coast as soon as possible.

From Klaserie the route led to the "Donsee" outspan along the upper reaches of the Klein Sand (Little Sand) River, which they reached on 23 February. Three days later, they reached the Mehlamhali [4] or Sand River southeast of the current Acornhoek, where they pitched camp. Trichardt received gifts in the form of ivory and food from the chiefs at each new outspan. From Trichardt's conversations with the locals it is obvious that he had been told about the tribal conflicts and battles in Sekwatiland (later to become Lebowa). This knowledge of what was happening in the surrounding areas enhanced Trichardt's status and ensured him of more readily available help and assistance.

At this stage Trichardt noted in his diary that several oxen and cattle were sick, which could have been caused by theileriosis (east coast fever), transmitted by ticks, or could have been the first signs of nagana. From here conditions only worsened and many of the cattle and small livestock died along the way during Trichardt's trek through the Lowveld and across the Mozambican coastal plains to Delagoa Bay.

The area between the Sand River and the Lebombo Mountains was teeming with game at that time and Trichardt often mentioned camels (giraffe). He also mentioned lions, which tried to catch their stock and probably followed their spoor for great distances. There is, however, very little reference to other game species in Trichardt's diary in the area that is now the Kruger National Park. The group had probably been so keen to travel through the area as quickly as possible that there was very little time to hunt. Trichardt probably also had little time to write long reports, although his son Karel later revealed that his father made entries in his diary every night. For ink he used the sap of the wagon bush or little ink bush of which, according to him, there were three or four types. The ink had to be boiled a few times and this is probably the reason why the script has faded so much.

With the help of guides provided by the local chiefs, the trekkers moved eastwards along the northern bank of the Sand River. Where the Mbumba station is now situated on the farm Edinburgh 266, the Trichardt party crossed the spot where the Selati railway line would later be constructed.

Heavy rains hampered their progress, but on 2 March they trekked between the Khokhovela and Manyeleti across the watershed. On their way to the Koemoeri (Kumhuri) outspan along the Manyeleti they trekked across hilly country, which gave Preller the impression that the Trichardts were crossing the Lebombo by that time (Preller, 1917), but it is now known that they only reached the Lebombo Mountains 10 days later.

Two days later they reached the "Ommasloesloetoe" (Mlowati), a tributary of the Sand River in the area that is now the Sabie Sand nature reserve.

On 5 March they camped alongside the "Loäsie" River. This was probably the upper reaches of the Maxipiri River, which means that the trekkers must have followed the Sand River for some distance as Trichardt mentioned a large river to their right.

6 March 1838 is a historic day in the annals of the Kruger National Park, as the Trichardt trek crossed the border of the area that is now the Park on that day. That night the wagons arrived at the upper reaches of the "Liepopa" River (Lipape), where they had to stay for two days to repair one of the wagons. They also used the break to make a few new yoke-pins. While there, the local chief brought them 15 watermelons, three baskets full of peanuts and a rhino horn as gifts.

Trichardt noted that while they were on their way to the Lipape, they had to be preceded by a guide to avoid some pitfalls that had been dug by the local blacks to trap game.

Three days later, on 9 March, they were able to continue their journey and managed to reach the "Outombee" (Vutome River). On the way there,

The bronze plaque which was erected in 1938 by the Historical Monuments Commission at the Tshokwane-Vutome crossing in commemoration of Louis Trichardt and his party's epic journey through the Park area in March 1838.

---

4 Also called Mathimhale or Moathlamohale.

Karel had problems with his oxen and wildly berated and whipped them. When his brother Pieta had a word with him about his behaviour, he immediately wanted to beat Pieta up. Once again, father Louis had to step in to restore the peace. By this time, many of the oxen were sick and Trichardt also noted in his diary that some of the sheep had died.

On Saturday 10 March the party reached the kraal of Chief Halowaan[5] at Ka-Nwamuriwane Hill just east of the confluence of the Vutome and Nwaswitsontso. Pieta reported that he had seen many tsetse flies on the horses and this only added to their fears, as most of their cattle were already sick and had started lagging behind. Louis asked the local Magwambas if they had ever owned any cattle, and they answered in the positive, showing him previously used cattle kraals, still containing dung, close to where they were camping.

The group had to cross the sandstone ridges along the Nwaswitsontso during part of this trek, and the oxen struggled and suffered badly. Although there was quite a bit of bush, it was not very dense and there were also few large trees. Along the river the grass was sweet, but not as tall as along the other rivers next to which they had camped.

It now became a matter of urgency to escape the many tsetse flies along the Nwaswitsontso. They trekked at such a pace that Trichardt did not even allow himself the time to keep his diary up to date. Some authors suggest that he may have been suffering from malaria and could therefore not keep his diary, but this has never been confirmed. They may have camped on the Lebombo flats once more before reaching the "Rooirand" (Lebombo) along the upper reaches of the Lindanda. On 12 March they probably outspanned at the foot of the Lebombo Mountains, just south of Mbhatsi Poort.

On 13 and 14 March they trekked through Mbhatsi Poort and on 15 March they camped close to the kraal of Chief Clensana next to the Wiensaana River (the Uanetsana, a tributary of the Rio Nuanedzi in Mozambique).

Here they shot eight hippos of which they could use very little. The next day they trekked towards the kraal of Chief Clewaan and encountered a large herd of elephants on the way. Trichardt shot at one himself, but he did not mention if he had managed to kill the animal. On their journey along the Uanetsana and later the Rio Nuanedzi he again mentioned game and they shot an impala, a zebra, a blue wildebeest and a number of hippos for the pot.

Because of the poor condition of their livestock, it took them nearly a month to cover the remaining 200 kilometres, more or less, to the coast and they only reached the fort at Delagoa Bay[6] on 13 April 1838. Here they were cordially received and offered accommodation by the governor, Captain Antonio Gamitto.

In this way Louis Trichardt finally achieved his goal of establishing personal contact with the Portuguese at Delagoa Bay after almost eight months of hardship and suffering. But his dream of finding his people a safe place to settle in the interior or along the coast, close enough to trade with the east coast ports, would never materialise. The spectre of death, which had been following their wagon trail since the start of their journey through the unhealthy lowlands,

Mbhatsi Poort in the Lebombo Mountains through which Louis Trichardt trekked to Portuguese East Africa in March 1838.

5   According to Steyn (1932), this chief's real name was Nonyalena.
6   Later Lourenço Marques, now Maputo.

caught up with them at Delagoa Bay. Within eight days of their arrival Mister Daniël Pfeffer was the first to succumb to malaria. Then, one after the other, members of the party fell prey to the feared disease. On 1 May 1838 Louis's beloved wife Martha died and left him a broken man. Five months later, on 25 October 1938, he followed her to the grave and in the end the only adults to survive were Karel and the widows Scheepers, Botha and Pretorius, along with a number of teenagers and children.

The remaining women and children (25 in total) reached Port Natal on 19 July 1839 on board the sailing ship *Mazeppa*, which had been sent by General Andries Pretorius to bring them to safety.

Karel Trichardt only joined the rest of the survivors in May 1840, after he had completed an extended voyage along the east coast of Africa to Abyssinia on his father's orders.

Louis Trichardt and half of his trekker party were buried in Delagoa Bay. In 1964 a monument was erected in the graveyard and it was unveiled in September of the same year by the late Dr W.H.J. Punt, the chairperson of the Trichardt Trekker Society. This monument and garden are still maintained by the Mozambican authorities, a fact that says much about the regard that exists to this day for the brave and inspiring Voortrekker leader. As with so many other human tales, his was inspired by hope and a vision of the future that ended in misery and tears.

The heroic journey of this respected figure in the history of the Afrikaner was not in vain, however. He befriended the black tribes in the Northern Transvaal and established contact with the Portuguese. The Trichardt party's contribution to the knowledge of the geography of the Transvaal was invaluable to their compatriots in Natal in later years. The information that the surviving members of the party, especially Karel, could relate about the climate, grazing conditions, game distribution, black habitation and health-related issues, as well as the geography and topography of the main regions of the Transvaal where they had stayed or through which they had trekked, was of great help to Commandant Hendrik Potgieter when he later had to choose suitable places in which the first Voortrekker communities could settle in the area north of the Vaal River (Mooi River [Potchefstroom], Ohrigstad and Schoemansdal).

Louis Trichardt is rightly regarded as one of the first and most efficient scouts of the area that would later become the Transvaal. He was the only Voortrekker leader to have kept a diary of his journey and this journal is understandably one of the most valuable pieces of Africana today.

The Portuguese community at the coast also valued the information Trichardt relayed to them. It certainly played a major role in the later determination of the Lebombo Mountains as the border between Mozambique and the ZAR.

The tragic final chapter of the epic tale of the Trichardt Voortrekker party will always call into remembrance the major part Louis Trichardt and his party played in the settlement of whites in the Transvaal.

The Trichardt graveyard and memorial in Lourenço Marques (Maputo), which was unveiled on 7 September 1964 by Dr W.H.J. Punt, chairperson of the Louis Trichardt Trekker Society.
(OCT. 1984; ACKNOWLEDGEMENT: DEPARTMENT OF NATURE CONSERVATION, MOZAMBIQUE)

References

Badenhorst, E. 1928. Bedevaart naar 'n Voortrekker-heiligdom. *Ons Vaderland* July.

De Kock, H.C. 1938. *Op Trichardt se spoor oor die Drakensberge.* Van Schaik, Pretoria.

De Kock, H.C. & Punt, W.H.J. 1938. Is Trichardt deur Chueniespoort? *Die Volkstem* September.

De Vaal, P.H.C. 1986. *Die dorp Louis Trichardt.* Kirsten Printers, Louis Trichardt.

Dicke, B.H. 1941. The Northern Transvaal Voortrekkers. *Archives Year Book for South African History* 4(1):67–170.

Fuller, C.F. 1932. *Louis Trichardt's Trek Accross the Drakensberg 1837–1838.* Van Riebeeck Society, Cape Town.

Hofmeyr, N. 1895. Karel Trichardt te Delagoabaai. Letter in *De Volkstem* 3 July & letter in *Land en Volk* 4 July.

Joyce, P. & Evans, P. 1987. *The Golden Escarpment – The story of the Eastern Transvaal.* C. Struik, Cape Town.

Krüger, D.W. 1938. Die weg na die see. *Archives Year Book for South African History* 1(1):31–233.

Muller, C.F.J. 1974. *Die oorsprong van die Groot Trek.* Tafelberg, Cape Town.

Nathan, M. 1937. *The Voortrekkers of South Africa.* Central News Agency (Pty) Ltd., S.A.

Paynter, D. & Nussey, W. 1986. *Kruger – Portrait of a National Park.* Macmillan, Johannesburg.

Preller, G.S. 1917. *Dagboek van Louis Trichardt (1836–1838).* Het Volksblad-drukkerij, Bloemfontein.

Preller, G.S. 1920. *Voortrekkermense II.* Nationale Pers, Cape Town.

Preller, G.S. 1930. Die Trichardt-graftes in Lourenço Marques. *Die Volkstem.*

Preller, G.S. 1930. Die Trichardt-trek. *Ons Vaderland* 12 December.

Punt, W.H.J. 1939. Die Van Rensburg-raaisel, I, II, III. *Die Volkstem* 6, 7 and 23 August.

Punt, W.H.J. 1939. Louis Trichardt is deur Pienaarsnek, *Die Brandwag* 1 December.

Punt, W.H.J. 1941. Op Trichardt se waspore oor die Pietersburg-plato. *Die Huisgenoot* 4 April.

Punt, W.H.J. 1944. Our early Voortrekkers and their links with Mocambique. *The Outspan* June.

Punt, W.H.J. 1944. How I discovered the Trichardt graveyard in Lourenço Marques. *The Outspan* September.

Punt, W.H.J. 1944. Die Trichardt-begraafplaas in Lourenço Marques. *Die Volkstem* August.

Punt, W.H.J. 1945. Trichardt did not pass through Chuenies Poort. *Africana Notes and News* 2(1) March.

Punt, W.H.J. 1951. Louis Trichardt, 1783–1838. *The Outspan* 20 July.

Punt, W.H.J. 1953. *Louis Trichardt se laaste skof.* Van Schaik, Pretoria.

Roos, J. de V. 1930. Herinneringe van Carel Trigardt. *Die Huisgenoot* 24 October.

Steyn, L.B. 1932. Louis Trichardt se worstelstryd deur die wildtuin. *Ons Vaderland* 10 December.

Steyn, L.B. 1939. A Voortrekker mystery solved. *The Star* 18 September.

Thom, H.B. 1943. Enige nuwe besonderhede oor Trigardt en sy mense. *Tydskrif vir Wetenskap en Kuns* July.

Thom, H.B. 1942. Trigardt se nalatenskap. *Die Huisgenoot* 11 December.

Trichardt, S.P.E. 1895. Besoek aan Delagoabaai. Letter in *De Volkstem* 17 July.

Van der Merwe, A.P. 1950. *Die Groot Trek.* Don Nelson, Cape Town.

Venter, C. 1985. *Die Groot Trek.* Don Nelson, Cape Town.

4.4 Karel Trichardt – pioneer and explorer

Dr U. de V. Pienaar

Karel Johannes Trichardt,[1] the eldest son of Voortrekker leader Louis Trichardt, was born on 3 September 1811 on the farm Boschberg in the Somerset East district of the Eastern Cape. His father, Louis, a respected man and field cornet of the Smaldeel ward, also had extensive farming interests.

The frontier Boers had been dissatisfied with British rule of the Cape for a long time, while the Xhosa had been subjecting them to raids from their territory in the Transkei for decades. Because of these and other reasons, a large number of Dutch-speaking colonists decided to move north of latitude 25°S, where they would be beyond the sphere of British influence. Louis Trichardt and his followers were some of the first Voortrekkers to leave their homes in the Eastern Cape to pursue a free and peaceful existence in the uncharted interior.

It appears to have been Trichardt's ideal right from the start to establish contact with the Portuguese at the ports along the Mozambican coast from a base in the interior and establish trade links with them.

The Trichardt family, including Carel Johannes Trichardt, Louis's father, had played an important role in the uprising against British rule by the Graaff-Reinet burghers in 1779. At the time they appealed to the Netherlands for help and in May 1779 Johannes de Freyn, a Dutch ship's captain, one compatriot and 36 blacks were sent from Delagoa Bay (now Maputo) to establish contact with the rebels in Graaff-Reinet. Their brief was to provide moral and military support to the burghers, on behalf of the Dutch authorities in Batavia, against the British administration. After an exceptionally brave attempt, the small group reached the border of the current Eastern Cape. Here they were attacked by hostile inhabitants, however, and only De Freyn and his compatriot managed to escape with their lives, fleeing to Delagoa Bay. Although no contact had been made, the burghers of Graaff-Reinet later heard about the attempt.

It is quite possible that Trichardt made use of De Freyn's geographical information 40 years later on his remarkable journey to the northern Transvaal and Portuguese East Africa.

Louis Trichardt sold one of his farms and abandoned the others before sending his son Karel ahead in 1829 to trek across the border of the Cape Colony with most of his livestock (including 1300 cattle, 8000 sheep and goats, as well as a large number of horses of which 33 belonged to Karel). Karel, who was 18 at the time, sojourned for two years in the vicinity of the Swart Kei River. During this time he visited Xhosa high chief Hintsa with the aim of renting land. The piece of land that he managed to lease lies along the Wit Kei at the foot of Mount Tsomo. In 1830 Karel managed to acquire another 12 000 morgen from the high chief with the right to live there for 99 years.

While living there he married Cornelia Bouwer and they had three children. In the meantime Louis Trichardt had completed his business in the Colony and joined Karel in 1832, accompanied by his wife Martha and their other children.

Judging from this piece of history, Karel was a fearless, intrepid young man who was not only his father's right hand, but also a pillar of strength to him during difficult times.

Although he was not really one for books and could barely sign his name, he readily mastered languages and could speak Portuguese and several African languages. Some historians describe him as a sullen and aggressive person, even as a bully with a violent temper. This is, however, a distorted picture of a man who was destined to become a pioneer and explorer of note and a person whose adventures would be acclaimed even in high circles. His story is closely interwoven with that of the Transvaal Lowveld

---

1   He was christened Carolus Johannes and other sources spell his name Carel, Karolus or Karools. His Portuguese passport gives his names as Carlos João. The Trichardts' surname is given by different sources as Tregardt, Trigardt, Trichard, Trigard, Triegaard or Tregart. Louis and his son Karel initially preferred the original Scandinavian spelling of Tregardt, but after 1895 Karel adopted the original French spelling of Trichard to which he added a 't', which is the spelling used by most of their descendants.

and therefore also with that of the area that is now the Kruger National Park.

Karel was a large, well-built man with exceptional bodily strength, but he rarely showed off this strength or came to blows with anyone. Admittedly he was short tempered, but nevertheless always showed the utmost respect for his father's authority and admonitions. His temper settled down as quickly as it flared up and at heart he was an introverted loner.

Despite the fact that he was not very communicative he made friends easily, as is clear from his later association with the renowned João Albasini, the Portuguese Governor Gamitto, the captain of the *Estrella de Damão* and others. He was a crack shot, a man of the veld and a hunter in a class of his own. Furthermore, he was a practical farmer with an extended knowledge of livestock, as well as a competent artisan and wagon-maker. As is the case with many strong people, Karel also had a soft heart, which was easily touched by human suffering and distress. An example of this is when he interrupted his journey to join his compatriots in Natal to provide a number of starving Zulus with food by shooting several hippos.

His unquenchable thirst for adventure, at least until middle age, impelled him to follow unbeaten tracks into unknown territory where no white man had set foot before. Some of the explorations he undertook by himself can still be described as phenomenal. His adventures overshadow those of the better-known English, French, German and Portuguese explorers of later years, especially when one takes into account that people such as Livingstone, Stanley, Burchell, Baines, James Chapman, Andrew Smith, Andersson, Erskine, Carl Mauch and Major Serpa Pinto were always accompanied by large parties and had the backing of notable sponsors with regards to equipment and finances.

Karel inherited his leadership capabilities from his father. In historical terms he should by now be regarded with the same respect as his father, although he is seldom bestowed this honour as he left neither diary nor journal about his adventurous life of almost 90 years. What is related in this short biography is an attempt, by virtue of his close connection to the prehistory of the current Kruger National Park, to do justice to a pioneer and hero in the history of the Afrikaners whose equal would be difficult to find.

After Louis Trichardt and his family had joined Karel along the Wit Kei at Mount Tsomo, the uprising by Chief Tjali Makomo and his people disrupted their peaceful existence. In the ensuing battles Karel once had a close shave with death. The Trichardts might have stayed on, but in 1835 high chief Hintsa of the Xhosa staged an uprising against British rule. Trichardt decided that the time had come to move north so that they could escape both British rule and the ongoing Xhosa raids. Karel planted a tree on Mount Tsomo to commemorate their decision to move on (Roos, 1930).

According to Karel it was untrue that his father had been declared an outlaw by Sir Harry Smith for allegedly inciting the Xhosa to rebel against the British (Theal, 1908). On the contrary, Sir Benjamin D'Urban had apparently offered him £4000 to stay in the Eastern Cape.

The Trichardt family trekked northwards and reached the "Groot" River (Orange) near its confluence with the Kraai River, which is close to the current town Bethulie, before the end of January 1835. After they had crossed the river they were joined by Izak Albagh[2], a former French soldier with a coloured wife and three daughters, as well as Jan Pretorius, Hendrik Botha, Gert Scheepers and their families. They had all been part of Lang Hans van Rensburg's group. The Van Rensburg party had by then also crossed the river.

According to Karel Trichardt, the Van Rensburgs were a group of less well-off people from the Ghoup (Koup), coarse and dissolute (Roos, 1930). They also had few livestock and the Trichardts did not really associate with them. They were often a day ahead or behind the Trichardt party. The Van Rensburg party had moved away from the Caledon River several months before the Trichardt group finally left the area at the end of 1835, with the result that they reached the Sand and then the Vaal rivers months before the Trichardts. Louis Trichardt and his party had scouted out the journey ahead in the meantime, and by the time they trekked through the Free State the area was not totally unfamiliar to them.

At the end of January 1836 the two parties met

---

2   Also mentioned as Izaak Albach (compare Preller, 1917).

up at the confluence of the Vaal River and Eland Spruit. Here they camped close to each other opposite Suikerbosrand because of the threat posed by uMzilikazi's AmaNdebele force. uMzilikazi lived along the Magaliesberg Mountains at the time.

They also decided to increase their pace across the eastern Highveld to the Olifants River to escape this danger. They trekked a little to the west of where Middelburg is today, through Trichardtspoort (close to Renosterkop), then along the Olifants up to where the river turns west through the Drakensberg east of the current Modimolle (formerly Nylstroom). In this way they had managed to pass the AmaNdebele strongholds unnoticed.

At Strydpoort, close to the current Zebediela, Karel Trichardt had a row with the Van Rensburgs. The latter were constantly busy trading ivory with the locals. The Trichardts refrained from this practice as the weight of the tusks overloaded the wagons, while they did not have a market for the ivory in any case. Karel nevertheless traded one large tusk for a white roan cow from one of the blacks. On his way to Karel to conclude the transaction, the black trader met Klaas Prins(loo), who offered him ten sheep for the tusk. The man refused the offer and delivered the tusk to Karel, after which Klaas Prins(loo) sent two men to take the tusk from Karel. According to Karel, he was considered to be a strong man who would not allow them to take the tusk from him, but he decided that it would be useless to cause trouble with people of their ilk, so he told Klaas to take the tusk seeing that he (Klaas) was penniless and would have no use for it in any case. The Van Rensburgs remained with the Trichardts for a while, but always stayed a shift or two ahead in order to be the first to trade with the local inhabitants.

Near the Soutpansberg, four members of the Van Rensburg group attempted to lift a small barrel of gunpowder from one of the Trichardt wagons. Apparently Karel reached for his rifle and bandolier, but his father stopped him, grabbed a piece of burning wood, walked towards the wagon from which the barrel had been rolled and kicked in the bottom of the barrel with his heel. Then he told the would-be thieves, "Now take it!" With this, the men beat a hasty retreat, upon which Louis remarked, "See how the cowards are running now." When the Van Rensburgs came to beg for powder later on, Louis gave them a few small bags. As a result of the bad blood between the two parties the Van Rensburgs trekked ahead to the Soutpansberg, and the poort through which they trekked became known as Strydpoort in later years. Hendrik Botha, Jan Pretorius, Gert Scheepers, Izak Albagh and their families preferred to remain with the Trichardts.

According to hearsay, the Trichardts met Robert Schoon, a merchant from Grahamstown, who described the correct route to the Soutpansberg to them. It is known that traders from the Cape had been trading on a limited basis with the blacks before Van Rensburg and Trichardt's arrival, but Trichardt did not mention such a meeting in his diary.

Karel later said that they could not catch up with the Van Rensburg party as the Van Rensburgs had been determined to reach Inhambane along the Portuguese coast as soon as possible to trade ivory for gunpowder, liquor and other supplies. The whole Van Rensburg party was eventually massacred by the renegade Zulu Chief Manukosi's (Sochangana) warriors at the end of July 1836 next to the Limpopo, about 80 kilometres north of its confluence with the Olifants River.

The Trichardt trek reached Soutpan along the western foothills of the Soutpansberg at the end of

This memorial sign between Bourke's Luck and Graskop commemorates the epic journeys of Voortrekker leaders Louis Trichardt en Andries Potgieter in their quest to find a way from the interior to the sea.
(06.09.1986)

May 1836. Here they made the acquaintance of Doors and Gabriël de Buys, Coenraad de Buys's sons who had found a home at the Soutpansberg before the Voortrekkers' arrival. Their mother was a black woman, the sister of AmaNdebele high chief uMzilikazi. While there they also heard about Jacob Kruger, who had moved as far north as the Zimbabwe Ruins with his black wife. He died and was buried there after also apparently having visited Inhambane.

The Trichardt party set up camp at Soutpan to await the arrival of a group who were "behind" them. This group, a patrol of 11 men under the leadership of Commandant A.H. Potgieter, arrived at their camp on 24 June. They left again a few days later on an expedition to Sofala (now Beira). Potgieter's patrol, however, only managed to get as far as Massangena along the Great Save in Mozambique and arrived back at Soutpan by the end of July.

By this time, rumours about the Van Rensburg party's massacre had reached the Voortrekkers at Soutpan. After consultation between the two leaders, they decided to send a five-man patrol on horseback into the wilderness to search for them. The patrol consisted of Louis Trichardt, his son Karel and three members of Potgieter's party, namely J.G.S. Bronkhorst, J. Robbertse and A. Swanepoel. The patrol left on 30 July 1836 and reached Sakana's (Manukosi) kraal next to the Limpopo on 4 August. Their stay left them none the wiser about the murder, but they did notice items that had belonged to the Van Rensburgs, including a mirror and a pair of binoculars, among the inhabitants. At the same time they heard rumours that plans were being made to murder them as well. To avoid this fate they outwitted Sakana by returning to Soutpan by a roundabout way, where they arrived on 16 August.

The next day Potgieter and his group left Soutpan to return to their laager next the Sand River in the Free State, with the promise to return to Soutpan as soon as he could organise matters with his followers.

Just a day after Potgieter's departure and not satisfied with the results of the first investigation, the obstinate Jan Pretorius decided to undertake another search for the Van Rensburg party. He, Hendrik Botha, Gert Scheepers, Izak Albagh and their families inspanned their four wagons and drove off into the bush. This journey was almost as calamitous as that of Van Rensburg. Although they managed to reach Sakana's kraal, they also failed to establish without a doubt that the Van Rensburg party had been massacred, although they later claimed that they had heard the crying of white children coming from one of the huts in Sakana's kraal. On their way back they lost almost all of their trek oxen as a result of nagana and Gert Scheepers died in the Lowveld of malaria. After a call for help, which the widow Scheepers had secretly sent with a young San servant to Louis Trichardt, Pieta Trichardt set off on 30 January 1837 with about 30 oxen to the long-suffering group of trekkers to accompany them to safety.

At the end of August, while Pretorius and his party were away, Trichardt moved camp to where Schoemansdal was later founded by Commandant Hendrik Potgieter. This was close to the kraal of the Venda high chief Ramavhoya.[3] Many of Trichardt's cattle died here and on 20 September 1836 he made a note in his diary that the hundredth head of cattle had died (probably of east coast fever).

According to Karel, the Trichardts' laager was attacked by a group of Ramavhoya's warriors between 22 August and 14 September[4] (Roos, 1930). Fortunately, Doors Buys and their own black informers warned them that an attack was imminent, and they were able to repel the attack with great losses to the attackers. The defence consisted of Karel Trichardt and about 30 of their workers, well-armed with assegais, bows and arrows (Preller, 1920). Louis Trichardt and the rest of the party stayed at the laager to protect the women and children. The attack took place at daybreak, but Karel and his men were waiting in ambush and opened fire with four guns. The attack lasted for about three-quarters of an hour before the attackers took to their heels. Apparently, two of the dead had numbered among Manukosi's Nguni warriors who had murdered the Van Rensburg party. Karel later said that he had shot three of the leaders from his horse. Soon after this defeat, Rhamavhoya's brothers executed him and several tribal battles took place between conflicting Venda chiefs in the Soutpansberg in the last few months of 1836.

---

3 Also spelled Rimmabooya or Rammabooia, by other sources.
4 During this period Trichardt did not make any notes in his diary.

During this time, Louis Trichardt befriended Rasethau (Rossetoe) and Mashao (Masjou), two of the local headmen. Rasethau lived in the Soutpansberg close to where Makhado (formerly Louis Trichardt) is today. On 19 December, Louis, Karel and Pieta Trichardt; Gabriël and Doors Buys; and Hans and Willem Strijdom assisted Rasethau in a tribal battle against an antagonistic chief. In the process they confiscated 90 head of cattle, which they divided among themselves.

The Voortrekkers stayed for several months in the vicinity of the Dorps River close to the later Schoemansdal. Karel even dug a furrow to channel water from the river. They built a few wattle-and-daub houses and Mister Pfeffer taught the children. There was abundant game in the area and Trichardt mentioned eland, buffalo, giraffe, blue wildebeest, zebra, warthog and even a rhino. There were also predators, including lion, leopard and hyena, which were hunted by especially Karel and Pieta. During this period Martha Trichardt gave birth to twin boys who died soon after birth and were buried there.

On 17 November the Trekkers moved camp again, this time to Brak Spruit (currently the farm Rondebosch 30), close to the current town of Louis Trichardt. While they were there they received the message about the predicament of Jan Pretorius and his party. This was also their base when they rode out as allies of Chief Rasethau.

On 28 February Trichardt and his people noticed the approaching wagons of Jan Pretorius and his people, on their way from the Lowveld. After an absence of almost seven months Pretorius and other members of the unsuccessful patrol once again joined the Trichardt party.

Two days later the whole group moved further south to a large vlei with reeds and marshes along the Dorps River, just over a kilometre south of the current Makhado. Trichardt called this camp "De Doorns" and they stayed there until 31 May 1837. They once again built a number of wattle-and-daub houses, a school and workshop, and laid out gardens.

Taking the grazing conditions along the western foothills and southern slopes of the Soutpansberg into account, it is clear why the Trichardt party had to move camp time and again. The area in the vicinity of Soutpan, where their original laager was, is sandveld with many elements of the dry Kalahari sandveld to the west of it. The grass cover in the area is of good quality, but could not sustain large numbers of livestock for any length of time. As a result, Pieta Trichardt, who was in charge of the livestock, had to move further and further from camp with the animals. This exposed him to attacks from predators and hostile locals. As the winter of 1836 progressed, the Trichardt trekker community had to move camp to the vicinity of the later Schoemansdal. In the autumn and winter months of 1837 they continued moving further eastwards in search of the Lowveld's better quality sweet grass.

The "De Doorns" camp was close to Chief Rasethau's kraal. As a result of the nearby marshy vlei, several Voortrekkers contracted malaria and died. The widow Scheepers's daughter, Anna, died in March 1837 after a long illness and in April, Hans Strydom's wife Alida also passed away. It was while they were there that they received confirmation from the Magwamba messengers whom they had sent to Mozambique, that the Van Rensburg party had been massacred next to the Limpopo by Manukosi's Zulu warriors.

On 7 March 1837 Trichardt wrote to the governor of Delagoa Bay to inform him of the murder of the Van Rensburgs. He also wanted to know how contact could be established between the trekkers and the harbour so they could trade ivory, skins, livestock and other items for ammunition and provisions. This letter was entrusted to a black messenger. When, after two months, Trichardt had still not heard anything from the Portuguese authorities, he redrafted his first letter on 11 May and this time sent Gabriël Buys and a Magwamba guide by name of Waai-Waai to deliver it. Gabriël could speak Portuguese and could therefore translate the contents of the letter.

During a hunt on 17 May the trekkers had a narrow escape from a herd of elephants, of which they shot a few. On 31 May 1837 the Trichardt party moved eastwards to their fifth camp along the Soutpansberg. This encampment was next to the Doorn River, about 14 kilometres west of Mashao Hill and nine kilometres northwest of Rivola on the current farm Doornspruit 199. At this camp, Chief Mashao became one of Trichardt's regular visitors, while they lost contact with Chief Rasethau. From here they hunted hippo in the Luvuvhu River, three kilometres north of their camp. Karel and Jan Pretorius, in particular, hunted some of the many elephants found in the area. They shot some of the numerous hartebeest (Lichtenstein's hartebeest?), zebra, blue wildebeest, buffalo, eland, kudu, warthog, rhino, reedbuck and other game on an almost daily basis. Lions and "wolves" (hyena) regularly threatened the livestock under Pieta's care. Karel had much to say about hunting lions (Roos, 1930) and he maintained that a lion had to be killed with the first shot, otherwise there would be trouble.

On 1 August 1837 Martha Trichardt gave birth to a little girl but, like her twin brothers, she also died a short while later. At their campsite next to the Doorn River, the trekkers once again built wattle-and-daub houses, as well as a school and a smithy. They laid out gardens, prepared leather straps, made veldskoens, repaired the wagons and carried out a multitude of other chores, while waiting for news from the "agterste mense" (the people at the back, in other words Potgieter and his followers), or from Delagoa Bay. The Trichardt group was unaware of the Potgieter party's problems with the AmaNdebele after the battle of Vegkop. They had, however, received information that uMzilikazi was considering a raid to the north. His army had in fact set off in their direction, but was fortunately halted by a flooded Olifants River.

On 6 August 1837 the Trichardts received news that Gabriël Buys and the Magwamba guide Waai-Waai were on their way back from Delagoa Bay with two soldiers. This was exciting news to the Voortrekkers as they would finally receive a definite answer from the Portuguese about their intended journey to Delagoa Bay.

On the morning of 7 August, Gabriël and the two Portuguese[5] soldiers arrived. In broken English they explained that they had been sent from the fort to accompany the Voortrekkers to Delagoa Bay. After Gabriël had conveyed all the news concerning their journey he left to see his brother Doors, who had been displaced after an altercation with Rasethau and was living in the vicinity of the later Schoemansdal.

After lengthy deliberations, Louis Trichardt decided to undertake the long and dangerous journey to Delagoa Bay as soon as possible. They would not follow the eastern route past Mashao as they feared Manukosi's hostile people living along the Limpopo, as well as the tsetse fly threat along that route. Eventually they agreed to travel back to Sekwati's Poort, from where they would set out for Delagoa Bay. They prepared the wagons, livestock and provisions and on 23 August 1837 the Trichardt party set off from the Soutpansberg to Delagoa Bay.

Karel tried to convince his father to wait for Potgieter and his group, but Louis argued that there would not be enough grazing for all of their livestock in the vicinity of their last encampment, while Manukosi posed a constant threat from the east and uMzilikazi from the southwest, which was too serious to ignore. Furthermore, their ammunition and food supplies were dangerously low and it was imperative that they restock at the harbour as soon as possible. Exploring the coastal area with an eye to possible settlement was also a consideration, as Trichardt knew of Potgieter's interest in making contact with a harbour. In later years Karel said that his father had planned to return via Natal to the Soutpansberg after they had established relations with the Portuguese and had the opportunity to replenish their supplies.

This ideal of the Voortrekker leader would never be fulfilled, but he nevertheless succeeded in reaching Delagoa Bay on 13 April 1838 (read more about this in chapter 4.3). Theirs was an epic journey that took seven and a half months to complete. They crossed the Drakensberg over the most difficult possible terrain and trekked through the wild and inhospitable Lowveld. In Delagoa Bay they were warmly welcomed by Captain Antonio Candido Pedrozo Gamitto, the governor of the settlement, and were shown a place where they could pitch camp. Their guns were temporarily confiscated, but were later returned.

It soon became clear to the Voortrekkers that Delagoa Bay was not a suitable place to settle down because of the unhealthy conditions. It was April, the month in which the malaria mosquito reigned supreme. The elderly Pfeffer was already ill with malaria at the time of their arrival at the Bay and on

---

5   Antonio and Lourenço, or, as the Boers called them, Anthonie and Lourins.

21 April he was the first to succumb to the disease, and thus the first to be buried in a foreign country. Just four days later one of Izak Albagh's daughters died and on 29 April so did Hendrik Botha. In the meantime Martha Trichardt had also fallen ill and despite the tender care of Señora Gamitto in her own home, Louis had to say farewell to his beloved on 1 May 1838. She was buried in a coffin built by her son Karel before her death. After that the following people died at short intervals: Hans Strijdom, Izak Albagh, Jan Pretorius, Izak Albagh's wife, Karel Trichardt's wife Cornelia and her daughters, Jan Pretorius's two sons and the widow Scheepers's three sons, another of Izak Albagh's daughters and Louis's youngest son Jerimias Josias, who was only eight years old. Louis Trichardt himself died on 25 October 1838 at the age of 55 of the feared "yellow fever" (malaria) and a broken heart and was put to rest in the same cemetery as his family and friends.

Of the 46 whites and seven coloured servants who had arrived with Louis at Delagoa Bay, 24 died within 15 months of their arrival. The 25 whites and three coloured servants who had miraculously survived, were left behind in the greatest misery and poverty. They were eventually fetched from Natal by the sailing ship the *Mazeppa* and disembarked on 19 July 1839 in Port Natal. The English missionary Reverend Frances Owen, who had arrived in Delagoa Bay on 8 May 1938 (thus before Louis's death) on board the schooner *Comet*, had witnessed the fate of Trichardt and his people. On his return he wrote a report about the misfortunes of the Trichardts. There was, however, no reaction from the Cape to his report and the captain of the *Comet* also refused to call at Port Natal on the way to the Cape. Thus a valuable opportunity to remove the devastated trekkers from the perils at Delagoa Bay was lost. In his report, Owen also noted that the trekkers had almost no livestock left. Louis had already made a note in his diary on 26 March 1838 that there were only 457 cattle and a few sheep left of his large herds.

In February 1839, almost a year after the Trichardt party's arrival at Delagoa Bay, news of their ordeal reached the Voortrekker leader Andries Pretorius at Pietermaritzburg and plans were made to fetch them. After a number of frustrating delays, Captain Tait set sail with the *Mazeppa* on 25 June from Port Natal to Delagoa Bay to fetch the surviving members of the Trichardt party. Karel Trichardt was the only adult male to have survived the terrible ordeal. The other survivors were the gutsy widow Antjie Scheepers, with one daughter and four sons; Antjie Botha, a widow with two daughters and three sons; the widow Breggie Pretorius and her two sons; Willem Strijdom, a young man with two brothers and two younger sisters; Louw and Izak Albagh jnr.; and Louis Trichardt's two sons Petrus Frederik (Pieta) (20 years old) and the 13-year-old Louis Gustavus, as well as their ten-year-old sister Anna Elizabeth (Annie).

Despite the abominable circumstances in which they found themselves towards the end, the welfare of his people remained Louis Trichardt's priority and he was still trying to negotiate a better future for them at the time of his death.

In June 1838 a Portuguese sailing ship, the *Estrella de Damão*,[6] anchored in Delagoa Bay on its voyage up the east coast of Africa to Djibouti in Somalia.

Karel Trichardt, then a young man of 27 and in the prime of his life, was ordered by his father to accompany the ship on its voyage to search for a healthier area further northwards where the Voortrekkers could settle. To achieve this he was to travel from the ports into the interior.

Karel would need trade goods for this purpose and these were bought from the warehouse in Delagoa Bay – his supplies included black fabric, beads, copper rings and other trinkets, as well as salt. His left his horses and other possessions in the care of Governor Gamitto.

By the third week of June everything was ready and Karel could board the *Estrella de Damão*. The ship's captain befriended Karel, who in turn readily mastered Portuguese, a proficiency that would be a great advantage to him and his compatriots in future. He also distinguished himself as sailor, with the result that the captain promoted him to boatswain, for which he received 25 Spanish *matte* (the equivalent of R6) a month.

The voyage northwards along the coast and back lasted more than a year. The ship called at several ports and countries, including Inhambane, Sofala,

---

6   Some sources refer to it as the *Estrella del Mar*.

Quelimane, the island fortress of Mozambique, Zanzibar, Djibouti in Somalia, Abyssinia (Ethiopia) and Malagasy (Madagascar).

During its voyage the vessel was ravaged by storms and at Cabo Corrientes, just south of Inhambane, they almost foundered, but they nevertheless managed to reach Inhambane harbour safely and anchored there. Karel disembarked there and prepared for a trade expedition into the interior. He hired 30 bearers and they managed to reach the fertile Limpopo valley.

During this journey he once again found himself in the vicinity of Sakana's (Manukosi's) kraal. This time he actually managed to locate the site where the Van Rensburg massacre had taken place, 33 years before the Natal explorer St Vincent Erskine mapped it in 1871. At the site he found the remains of burnt-out ox-wagons as well as human skeletal remains. Karel buried the remains and later mentioned that he had recognised the wheel nave of Klaas Prins(loo)'s wagon, which he had shortened himself.

He also met the warrior who had kidnapped Frederik van Wyk's two children. The man told him that the children had died of malaria and showed him the two small graves. He even offered to open the graves so Karel could see by the long hair that the bodies were those of white people. Karel also supposedly later told Roos that he had brought the rifles of Van Rensburg and Klaas Prins(loo) back with him, but this was probably a misunderstanding on the part of the author, as this discovery was later made by the Burgher expedition from Ohrigstad in 1848 (Roos, 1930).

With only six weeks at his disposal to complete the return journey, Karel could not advance any further into the interior than the vicinity of the present Mozambique-Zimbabwe border. He discovered that the natives living in the area through which he travelled did not own any cattle because of the presence of tsetse flies. Karel did not consider the interior west of Inhambane a suitable area for a Boer settlement, obviously with good reason.

After an absence of three months he was back in Inhambane at the end of September 1838. The ship set sail again five days later.

Sofala (now Beira), which is 480 kilometres north of Inhambane, was to be their next port of call. The *Estrella de Damão* could complete this voyage in three days. Karel disembarked at Sofala during the second week of October 1838 and with the help of 20 bearers and a translator he undertook another trade expedition into the interior. During this two-month journey he walked for 15 days in a westerly direction, covering a distance of 400 kilometres (Preller, 1920).

There had been a trade route between Sofala and Zimbabwe since time immemorial and this was obviously the most suitable route for his safari. By following it Karel reached the Zimbabwe Ruins at the beginning of November 1838 (Preller, 1920). He had therefore observed this mysterious construction 33 years before the German explorer Carl Mauch did so in 1871. Karel examined the ruins, which in his opinion had not been built by blacks, and traded some of his wares for gold dust. It is interesting to note here that Karel Trichardt was not the first non-Portuguese person to have visited the exceptional ruins. This honour probably belongs to Jacob Kruger, who had left the Cape with his black wife long before the start of the Great Trek to settle in Zimbabwe, where he eventually died. When Mauch reached Zimbabwe in May 1871 he found a white compatriot by the name of Adam Renders and his wife Elsie (Lettie) living there. Renders was born in Germany in 1832 and had travelled through America before coming to South Africa.

Renders had accompanied Andries Pretorius and his son Marthinus Wessel to the Transvaal and later settled in the Soutpansberg. From here he undertook hunting expeditions to Mashonaland and in the winter of 1867 he moved to Zimbabwe to settle there. Thus his wife, Elsie Renders, was the first white woman to live at the Zimbabwe Ruins (Punt, 1958).

Adam Renders was hit in the shoulder by a poisoned arrow during a tribal battle in about 1872 and later died of this injury. He was buried about 15 kilometres from the Zimbabwe Ruins. Elsie later returned to the Transvaal and settled in Pietersburg (Polokwane).

During this journey into the interior, Karel Trichardt also explored the highland north of the Zimbabwe Ruins. His comment about what is now the Harare district was that he had found the area healthy, but that the grass was mostly tambookie, which is less appealing to livestock (Preller, 1920). The area nevertheless impressed Trichardt, and Preller speculated that he would have recommended it to his father as a possible place to settle had he still been alive. On his return journey to the coast he travelled in a southeasterly direction. He must therefore have travelled along the Buzi River during the last part of his journey to reach Sofala.

After two and a half months Karel departed again with the merchantman to Quelimane, which is situated 320 kilometres northeast of Sofala on the northern side of the Zambezi delta. Here they anchored at the beginning of January.

It was possibly in this port that Karel nearly lost his life during an encounter with thieves. He later said that he had been busy loading on the wharf until dusk on the day the event took place. The captain had warned him not to work too late as robbers and other unsavoury characters prowled the port area at night. In answer to his question about who would want to kill him, the captain accused the Asian dock workers. Karel did not take much heed of the warning, but while walking to his lodgings that night he saw someone wearing a long coat following him. The only weapon he had on him was a bottle of champagne. When he changed direction to the opposite side of the street, his pursuer did the same. This was enough to spark his anger and he decided the best defence was attack, and hit his pursuer between the eyes with the champagne bottle. The culprit fell down senseless before he could pull his weapon (a dagger with a glass blade). Karel confiscated the weapon and called the police, but the perpetrator managed to escape. The next morning the governor summoned Karel with the news that the footpad had been arrested the previous evening. When the governor asked him if the dagger in Trichardt's possession belonged to him he answered in the affirmative – denial would have been futile since the mark the champagne bottle had left on his forehead was clearly visible. The governor summarily sentenced him to death and ordered the Portuguese soldiers to take him to the beach and shoot him, which they promptly did.

According to Karel, he found the land most suitable for agriculture in Mozambique at a place called "Nekka", between Quelimane and the island fortress of Mozambique. There must have been a homestead before as he found some old orange trees there. He wanted to write to his father to come and settle there, but had no opportunity to do so.

From Quelimane, Karel organised yet another expedition into the interior, this time with 20 bearers and a Portuguese slave called Suzéz as interpreter. He left enough geographical information about this journey in his statement to Odé that his route to the interior can be determined with reasonable accuracy. Karel travelled about 160 kilometres from Quelimane northwest up to the Shire River, which he crossed to reach the northern bank of the Zambezi. By making use of a riverboat to travel between the Portuguese trading posts Sena and Tete along the Zambezi, he managed to reach the "Sandia" rapids (Cahora Bassa). In the time available to him it was, however, impossible for him to have reached the mighty Mosi-oa-Tunya (the Victoria Falls, which were eventually discovered by Livingstone in 1855), despite the claims of certain historians. Karel found the hinterland of Quelimane fertile, with large rice fields that were being tended by locals, but nevertheless unhealthy for habitation by whites.

He returned to Quelimane in early March and once again the *Estrella de Damão* set sail northwards along the coast. After a journey of several days over a distance of more than 600 kilometres, they sailed into the port of the island fortress of Mozambique. At last he had reached the seat of Portuguese authority in Mozambique.

He went to see the Governor General of Mozambique with testimonials from Delagoa Bay, and during this interview he told the Governor General that Portuguese officials had taken a horse immune to African horse sickness from him in Delagoa Bay. This was apparently to serve as tax on goods his father had traded with an English ship (probably the *Comet*). The Governor General answered that he should grab the involved jack-in-office by the neck and take his horse back. This he did as soon as he arrived back in Delagoa Bay!

In Mozambique he sold the proceeds of his trade journeys and the ivory, for which he had obtained authorisation from his father, and concluded his business. The letter giving him permission to do business on his father's behalf had reached Karel in Quelimane.

Karel paid another visit to the interior accompanied by about 12 Portuguese soldiers, but this trip only lasted a few days. The region failed to impress him because of the humidity, heat and the incidence of malaria, which was as bad as in Delagoa Bay.

It was Karel's resolve to return to Delagoa Bay as soon as possible to inform his father of his findings, but it was not possible as the *Estrella de Damão* still had to sail to Abyssinia (Ethiopia). Karel decided to undertake this journey for the sake of his friend Boda, the shipbroker. So, after a month at Mozambique, he boarded the ship in the middle of April 1839.

In later years Karel provided details about this eventful journey along the coast to Djibouti in Somalia (Roos, 1930). When they anchored in the bay of Zanzibar they found a Russian battleship there. The Russians apparently wanted to shoot hippo in the area that later became German East Africa (Tanzania), but were afraid of the local people, who were known to be cannibals. Trichardt let them know that he was not frightened of anyone, cannibals or not, and that he would go and shoot the hippo. While rowing up the river in a boat he was suddenly attacked by a band of blacks who made an unearthly racket. Thanks to his knowledge of Swahili, the language of the east coast, he managed to deduce that they were arguing about who was going to eat which part of his body. He quickly shot four and the rest disappeared like mist before the sun – their hunger appeased for the moment! By that stage cannibals had already killed and devoured 40 Germans in the interior.

Karel also mentioned that he had visited the island of Zanzibar, but did not say whether it was during the forward or return journey. From Zanzibar the *Estrella de Damão* sailed directly to Djibouti and they must have arrived there at the beginning of May 1839.

In his statement to Odé, Karel said that they had managed to sell the whole shipload within two weeks. From this harbour he travelled 270 kilometres inland to Harar, then the capital of the Abyssinian Empire. His description of this city was so vivid that his eldest grandson could still easily recognise it when he was involved in the East African campaign in 1942 (Trichardt, C.J., 1947). On his way to Harar, Trichardt encountered a caravan of 73 riders who were guarding 12 elephants, heavily loaded with trade goods. While in Harar, Trichardt received gifts from the sultan that are in the Trichardt family's possession to this day. After a stay of about two weeks in Djibouti, they weighed anchor at the end of May 1839 and started their long return voyage down the coast. At that point Karel was ill with cholera, but he recovered and recuperated in Mozambique. They stayed in the port of Mozambique for two weeks. This time Karel used the opportunity to visit Majunga Bay in Madagascar (then known as Malagasy). This bay is situated 640 kilometres east of Mozambique on the west coast of Madagascar, but on this occasion they did not land. During his two visits to this fertile island, Trichardt came to the conclusion that of all the places he had seen along the east coast of Africa, Madagascar (which was under French occupation at the time) was, with the exception of the Transvaal, the "crown of the world". He also said that if his father had been alive on his return, they would certainly have settled on Madagascar.

Voortrekker Carolus (Karel) Johannes Trichardt and his second wife, Zacherija Geertruida (née Erasmus). They both died at Middelburg (Transvaal) in 1901.
(ACKNOWLEDGEMENT: STATE ARCHIVES, PRETORIA AND MR W.J. PUNT)

After 14 days in Mozambique they resumed their journey to Delagoa Bay and arrived there at the end of July 1839, after an absence of an entire year. Here he received the news from Governor Gamitto that his father had died in October 1838 and that his surviving compatriots had left for Port Natal on board the *Mazeppa* barely two weeks before. Trichardt visited his friend João Albasini, the energetic young trader who would later play such an important role in the history of Ohrigstad, Lydenburg and Schoemansdal. He also discussed his future trading plans with Albasini and Governor Gamitto.

Karel nominated Antonio José Nobre as his proxy in the city of Mozambique, while Albasini appointed Governor Antonio Gamitto as his first commissioned agent in Delagoa Bay.

It is clear that Karel and João knew each other well as they witnessed each other's authorisation documents. It is possible that Albasini had returned on the same ship as Karel from Mozambique to Delagoa Bay and that they were already business associates at that time.

On 8 August 1839 Trichardt and Albasini were ready to set off on an expedition together. Their destination was most probably Madagascar, especially if one takes into account Trichardt's later declaration that he had visited the island twice (Roos, 1930). This was not a very long journey as they were back in Delagoa Bay after three months. The real reason for their visit to Madagascar is not known, but the establishment of trade relations was probably the most important consideration.

At the end of October 1839 Trichardt was back in Delagoa Bay and once again he drew up a document that awarded power of attorney – this time to Governor Gamitto, as his commissioned agent.

Carel must have undertaken another short journey and he was back in Delagoa Bay by the middle of January 1840.

After returning from his third voyage, Karel was once again staying with the Portuguese governor. He served as harbour master and also served the Portuguese in a military capacity against the Zulus. Governor Gamitto had heard that Dingane's fleeing Zulus were heading in the direction of Delagoa Bay. This Zulu army was defeated decisively at the end of January 1840 at Makondo Hills, about 15 kilometres south of the Pongola River, by Commandant General Andries Pretorius's third punitive commando assisted by Panda's warriors.

Karel was immediately sent out on a reconnaissance trip accompanied by 18 men, as the Governor feared that his small settlement could be attacked and wiped out, as in the days of Zwangendaba (Songandaba) and Manukosi. He navigated up the Maputo River for two days to the confluence of the Usutu and Pongola rivers (Preller, 1920). Dingane and the remains of his army had arrived here, but before they could cross the river they were attacked and almost annihilated by the army of Swazi King Sobhuza I. The Zulu army was overwhelmed and the women, children and livestock (about 40 000 head) were taken as spoils of war by the Swazis.

The next day Karel was summoned by the Swazis, so he crossed the river and was shown the body of the Zulu tyrant Dingane, who had been killed by one of Sobhuza's warriors. Next to Dingane's body lay the bodies of two of his own indunas (headmen), whom he had murdered the night before with the idea that they would precede him if he were to die the next morning.

The Swazi warrior who had speared Dingane was also killed. On Trichardt's question as to why it was necessary, Sobhuza answered: "Now that he has learned how to kill kings, he could do the same to me!"

Karel returned to Delagoa Bay to report to Governor Gamitto that the danger of an attack by the Zulus had been averted. He also informed Gamitto that he was missing his people and therefore wanted to go to Natal. In Delagoa Bay he heard that his brother Pieta had sold their remaining cattle and other valuable possessions to American whalers. The proceeds of this transaction had been £173 – enough to partially pay for their passage with the *Mazeppa* to Port Natal in 1839.

On 27 April 1840 Karel eventually left his friends in Delagoa Bay. Apart from a passport he had two horses (one of his own and one that he bought there and had apparently belonged to Piet Uys previously), eight black helpers and some pack oxen. With this entourage he crossed the Maputo River and travelled southwards along the coast to Port Natal.

Along the way they encountered many Zulus who

4 THE VOORTREKKERS EXPLORE THE LOWVELD

A section of the fort which had been erected during the period 1845–1850 by the Voortrekkers at Ohrigstad. A commemorative hall in honour of General A.H. Potgieter and his fellow Voortrekkers was inaugurated here on 16 December 1952.
(06.09.1986)

had been wounded during the battles with Pretorius's punitive commando and Panda's warriors. Famine prevailed among the local population, and at one place the situation was so dire that Trichardt interrupted his journey to shoot a number of hippos so the meat could be distributed among the starving people.

In the last week of May 1840 Karel arrived at Karel Landman's laager next to the Umlazi River near Port Natal. Here he was offered accommodation at the camp of Jacobus Moolman. Other members of this party included Evert Potgieter, a De Jager family, the Fouries, Groenewalds, Van Rooyens, Oosthuizens, a Scheepers family and others. Oosthuizen and De Jager were later treacherously murdered by locals nearby. At the time of his arrival at the Landmans Karel was already very ill with malaria and had to be nursed for almost two months before he could move on.

After this, he travelled to Pietermaritzburg where he lived with Magistrate Zietsman and recuperated for another six months. During this time he made the acquaintance of Zacherija (Sagriet) Gertruida Erasmus, the youngest daughter of former Commandant Stephanus Petrus Erasmus. They married shortly afterwards.

The Voortrekkers lived peacefully in Natal for another year under the leadership of Commandant General Andries Pretorius before the British annexed it in 1842 and the Boers had to reach for their weapons again. Karel, who had become a member of the Volksraad (Parliament) of the Republic of Natalia, was an active participant in the war, which the Boers lost against the British forces based in Port Natal. After the cessation of hostilities Karel set off on a trade expedition to the Free State in the company of the Dutchman Jan de Vrij. Karel took all his livestock along, hoping to leave them with his brother-in-law Jan Jacobsen near Winburg.

Upon his arrival at Winburg he met an expedition that was about to leave on an expedition to Delagoa Bay. Karel accompanied this expedition under the leadership of Magistrate Vermeulen of Winburg in 1843 as guide and interpreter. Initially the group wanted to travel to the vicinity of the later Ohrigstad, but because the Olifants River was in flood they could not reach their intended destination. After a delay of 14 days they returned without having achieved their purpose. Karel was then joined by his wife, whom he had not seen for seven months, and his father-in-law, who had in the meantime sold his farm and other belongings in Natal.

After staying in the Free State for about six months, Trichardt trekked to the Mooi River (Potchefstroom) in the Transvaal to settle there with his whole family. Commandant A.H. Potgieter was the leader of the Boers there. Shortly after Trichardt's arrival in 1844, Potgieter set off with a commission from Potchefstroom to Delagoa Bay. Karel was asked to accompany the group as guide and interpreter, but he refused the offer, the failure of and lack of appreciation for the 1843 expedition still all too clear a memory. Potgieter's expedition did manage to reach Delagoa Bay and conclude a trade treaty with the Portuguese. Part of Potgieter's group had stayed behind with the wagons on the Drakensberg, while the rest went down the mountain with pack oxen. When Potgieter and his people stayed away longer than expected, the people who had been waiting on top of the mountain feared that he and his group had been killed and named the river where they had gone hunting the "Treur" (Mourning) River. But at the next river they met Potgieter and his group making their return, and named it "Blyde" (Joyful) River.

After Potgieter's return from Delagoa Bay, he and his followers moved from the Mooi River and the Magaliesberg range and established the town Ohrigstad so they could be closer to the Portuguese harbour town, for trading purposes. On the advice of the Dutch trader and shipping magnate Smellekamp, they would also be north of latitude 25°S and therefore beyond the British sphere of influence. Karel Trichardt and his family moved with Potgieter and his followers and settled on the farm Branddraai along the Ohrigstad River.

On 14 October 1845 the Volksraad of Ohrigstad sent an expedition of 26 men under the leadership of Karel Trichardt to Delagoa Bay with the aim of finding a suitable trade route to the Bay. Karel himself, however, said (Preller, 1920) that the group had had only 18 members. This expedition managed to pave the way for a road to Delagoa Bay that would later become known as "De oude wagenweg" (The Old Wagon Route). Against Trichardt's advice, the Volksraad decided that the expedition should take wagonloads of trade goods with them to Delagoa Bay, as a result of which tsetse fly killed almost all of their livestock and horses. The Old Wagon Route soon fell into disuse as in winter there was insufficient water along the way for trek animals.

In January 1847 Karel's old friend João Albasini, who had settled at Magashulaskraal next to the Sabie River, appeared before the Volksraad at Ohrigstad. He informed them that he had found a route to Delagoa Bay that was reasonably free of the feared tsetse fly and along which there was more than enough water. After much arguing and many delays an expedition under the leadership of J. van Rensburg left on 12 July 1847 to investigate and report back on Albasini's suggested trade route. Karel Trichardt played a prominent role in this successful expedition, as well as a following one in July 1848, during which the transport route to Delagoa Bay was clearly demarcated. During the latter journey Willem Pretorius, one of the expedition members, got ill and died of malaria. He was buried by Albasini under a marula tree close to the current Pretoriuskop rest camp, and as a result the nearby hill was named Pretoriuskop. (The late Professor B. Lombaard [1969] determined that the hill's original name had been Mntsobo. He was also convinced that the hill had been named after President M.W. Pretorius, who visited the area in 1865 during a hunting expedition [see explanation in chapter 18]).

Commandant Potgieter had in the meantime led another punitive expedition in 1847 against the AmaNdebele Chief uMzilikazi at his new settlement

Karel Trichardt's house under construction on the farm Rozenkrans between Lydenburg and Pilgrim's Rest, circa 1850–1851.
(ACKNOWLEDGEMENT: TRANSVAAL ARCHIVES)

Lt Col S.P.E. Trichardt, Karel Trichardt's son, in the uniform of the State Artillery (23 Jan. 1874 – Oct. 1901).
(ACKNOWLEDGEMENT: TRANSVAAL ARCHIVES)

in Matabeleland, currently the southwestern part of Zimbabwe. After Potgieter and his people had left Ohrigstad as a result of disputes with J.J. Burger and other problems, the newly elected Volksraad of Lydenburg again sent a commission to Delagoa Bay in May 1850 to establish trade relations, as well as to sign a border treaty. This commission was led by Andries Spies, a Volksraad member, with Cornelis Potgieter, the magistrate of Lydenburg, Flip Coetzer, Hendrik Bucherman and Karel Trichardt as members. At Delagoa Bay the commission had fruitful discussions with Portuguese Governor Joachim Carlos D'Andrade. The two parties managed to reach an agreement and provisional trade and border relations were established. The governor was also cordially invited to pay a courtesy visit to Ohrigstad. As a result Governor D'Andrade attended a commission board meeting on 22 August 1850 in Ohrigstad. He was accompanied by two officers of the garrison at Delagoa Bay, namely F.J. de Mattos and G.G. de Castellão. These talks led to a supplementary agreement to elucidate and explain ambiguous elements in the previous treaty.

With regards to the border issue, the Portuguese suggested that the Boers cede all territory from the last slope of the Drakensberg eastwards. However, this proposal was met with vehement opposition, especially from Karel Trichardt. Although he was not a member of the Volksraad, he nevertheless had great influence with the members as well as the Portuguese governor. He indicated that it would be much more to the Portuguese's benefit if the Boers could settle as closely as possible to Delagoa Bay and suggested the Lebombo Mountain range as a logical border between the two states. Governor D'Andrade declared himself willing to submit this proposal to his government for ratification. Although the proposal was rejected by the Governor General of Mozambique in 1854, it nevertheless served as a basis for successful border negotiations between President M.W. Pretorius of the ZAR (South African Republic) and Consul General Alfredo du Prat of Mozambique in 1869.

In this manner Karel Trichardt played a very important role in negotiations that would eventually lead to the whole of the current Kruger National Park being included in the former ZAR.

Karel Trichardt lived on the farm Branddraai for three years. Because of the high incidence of malaria in the area he moved to a higher-lying property and settled on the farm Rozenkrans (also along the Ohrigstad River), close to the farm Rustplaats belonging to his old friend João Albasini. Here he lived for 20 years and transformed the farm into a veritable Eden. He moved to the farm Goedehoop in the Standerton district in January 1872 to focus more on sheep, cattle and horse breeding. Shortly afterwards he moved to the adjacent farm Trichardtsfontein, which had been laid out by his son Louis. Here he farmed peacefully into his old age.

During a visit by Dr W.H.J. Punt in 1953 to the farm Goedehoop, the owner of the farm, Mr Nothling, showed him a sandstone block in the wall of his house on which the initials of Karel Trichardt, C.J.T., had been engraved. Karel's original homestead on the farm Goedehoop was burnt down during the Anglo-Boer War and with it his famous rifle "Lustig" was also destroyed. His grandson Karel found the barrel after the war, but left it in the ruins.

In 1895 the railway between Lourenço Marques (earlier Delagoa Bay and now Maputo) and Pretoria was completed and ceremoniously inaugurated. As pioneer and last link to the Trichardt trek and their untiring search for a way to the sea, Karel Trichardt, his sons Stephanus,[7] Jeremiah and Piet, some of his grandsons and a few friends were invited on this historic train journey. They arrived in Lourenço Marques on 26 June 1895. During their stay in the harbour city the 84-year-old Karel showed them where his father and the other members of the ill-fated Trichardt Voortrekker party had been buried. This graveyard was carefully reconstructed in later years by Dr Punt. As chairperson of the Louis Trichardt Trekker Society he unveiled a memorial there on 7 September 1964.

During the war against Mphephu in 1898, Karel, who was 87, and his son Stephanus visited the area along the Soutpansberg where the Trichardts had lived from 1836–37. On his recommendation the town of Louis Trichardt was officially proclaimed in February 1899 in memory of his father.

---

7   Commandant S.P.E. Trichardt, then an officer in the State Artillery in Pretoria, who distinguished himself at the battle of Majuba, among others, during the Transvaal War of Independence.

Karel spent his last years with his son Jeremiah in Middelburg, where he died on 25 April 1901, shortly before the British occupation of the eastern Transvaal (now Mpumalanga) during the Anglo-Boer War. The inextinguishable fire and great courage that typified his whole life kept on burning fiercely until shortly before his death. It is said that when the British occupational forces marched into Middelburg, the staunch old man awaited them at the garden gate with his Martini-Henry and ordered them to get out of there!

Indeed a man among men!

References

Anon. 1937. Karl Trichardt. All information assembled. *The Star* 5 January.

Beyers, C. 1938. Report to Col. J. Stevenson-Hamilton about the history of the Delagoa Bay road.

Chase, J.C. 1832. Notice respecting the expedition overland from the colony of Cape of Good Hope to the Portuguese settlement at De la Goa Bay by Messrs. Cowie and Green in 1829. *Grahamstown Journal* August.

Dicke, B.H. 1944. The Northern Transvaal Voortrekkers. *Archives Year Book for South African History* 4(1):67–171.

Du Buisson, L. 1987. *The White Man Cometh*. Jonathan Ball Publishers, Johannesburg.

Du Toit, H. 1896. Herrinneringe van Karel Trichardt. *Onze Courant*, Graaff-Reinet 27 July.

Hofmeyr, N. 1895. Karel Trichardt te Delagoabaai. Letter in *De Volkstem* 3 July.

Krüger, D.W. 1938. Die weg na die see. *Archives Year Book for South African History* 1(1):33–225.

Lombaard, B.V. 1969. Herkoms van die naam Pretoriuskop. *Koedoe* 12:53–57.

Möller-Malan, D. 1953. *Dina Fourie*. Voortrekkerpers, Johannesburg.

Nathan, M. 1937. *The Voortrekkers of South Africa*. Central News Agency (Pty) Ltd, S.A.

Odé, G.A. ca 1894. Die herinneringe van Karel Trichardt opgeteken deur staatsargivaris van die Z.A. Republiek. State Archives, Pretoria R95/94:32.

Preller, G.S. 1917. *Dagboek van Louis Trichardt (1836–1838)*. Het Volksblad Drukkerij, Bloemfontein.

Preller, G.S. 1920. *Herinnering van Karel Trichardt. Voortrekkermense II*. Nasionale Pers, Cape Town.

Preller, G.S. 1934. Internal report to the National Parks Board of Trustees, in connection with the history of the Delagoa Bay transport road.

Pretorius, H.S. 1936. Report to Col. J. Stevenson-Hamilton about Hugo Nellmapius and the Delagoa Bay transport road.

Punt, W.H.J. 1944. Our early Voortrekkers and their links with Mocambique. *The Outspan* June.

Punt, W.H.J. 1948. Op besoek by 'n Voortrekker van Oos-Afrika. *De Nederlandsche Post* January.

Punt, W.H.J. 1950. Personal notes. Skukuza Archives.

Punt, W.H.J. 1953. *Louis Trichardt se laaste sk*of. Van Schaik, Pretoria.

Punt, W.H.J. 1958. Voortrekkerdogter was wit vrou van Zimbabwe. *Fleur* (Magazine Supplement of *Dagbreek*) 24 August.

Roos, J. de V. 1930. Herrinneringe van Carel Trichardt. *Die Huisgenoot* 24 October.

Theal, G.M. 1908. *History of South Africa Since 1795*. I. Swan Sonnenschein, London.

Trichardt, C.J. 1947. Personal communication to Dr W. Punt in Kenia.

Trichardt, S.P.E. 1895. Besoek aan Delagoabaai. Letter in *De Volkstem* 17 July.

Trichardt, S.P.E. 1975. *Geschiedenis, werken en streven van luitenant-kolonel der vroegere Staatartillerie, ZAR, door hemzelve beschreven*. O.J.O. Ferreira (ed.). Human Sciences Research Council, Pretoria.

Van der Merwe, P.J. 1937. *Die noordwaartse beweging van die Boere vóór die Groot Trek (1770–1842)*. W.P. Stockum & Zoon, The Hague.

Chapter 5
João Albasini (1813–1888)
Dr J.B. de Vaal

The ruin next to the Sabie River

Near the western border of the Kruger National Park, about 18 kilometres north of Pretoriuskop rest camp, is the partially rebuilt ruin of a house. It is situated on a level area close to the confluence of the Phabeni Spruit and the Sabie River, with the foothills of the Drakensberg range a few kilometres to the west. The Albasini Ruin, as it is known, is the remains of the Portuguese trader and elephant hunter João Albasini's former home and trading post. The brick building had deteriorated to such an extent that by 1978 it had been reduced to an unstable pile of rubble about a metre wide and ten metres long. Bits of red brick were visible in places, but there was no sign of a wall. It was overgrown with grass, shrubs, a young red ivory, a common hook-thorn and two large fig trees. Another tree that had grown in the middle of the rubble had disappeared completely and only a hole indicated its former position.

The building had already fallen into ruin by 1870. When explorer and geologist Carl Mauch walked from Lourenço Marques (Maputo) to Lydenburg that year, a tree about 4.5 metres high and with a 15 cm thick trunk was growing in the ruin. There was no sign that the land in the vicinity of the former building had once been cultivated. Harry Wolhuter was the first white ranger of the former Sabie Game Reserve to rediscover the ruin. He later said that he had come upon the ruin of a brick building in about 1903. On this occasion he was accompanied by photographer C.A. Yates of Barberton, who took a photograph of the collapsed building with a large fig tree growing inside it.

A very good description of João Albasini's former home was published 22 years later in *The Star* of 20 and 27 June 1925. During the same year Major J. Stevenson-Hamilton, the warden of the Sabie Reserve, made the following significant note in his annual report: "An interesting landmark is the ruin of the house of Albasini, situated on the south bank of the Sabie River, and representing the abode of, so far as is known, the first white man to enter, and reside in the eastern Low Veld. Steps should be taken for the preservation of these interesting remains."

Nothing came of these recommendations and eight years later he again wrote: "The house is now reduced to a mere fragment, but shows signs of having once been of considerable size and having been well built with burnt brick. A few trees and plants put in by Albasini still remain."

A photograph taken during the great drought of 1935 by Dr F.Z. van der Merwe, the medical inspector of Transvaal schools at the time, showed that the ruin had been reduced to a pile of rubble. Five years later, Dr Gustav Preller wrote as follows of two of

João Albasini, the well-known pioneer and trader (1813–1888) in his vice-consular uniform, circa 1860.
(ACKNOWLEDGEMENT: DR J.B. DE VAAL)

Albasini's trading posts (translated): "In the current Kruger National Park there are still two of Albasini's building sites. One is close to the road between Pretoriuskop and Skukuza and another close to Lower Sabie. These abodes had been thatch-roofed brick houses, each with a fortified wall around it to offer protection against hostile natives and wild beasts." The one site is the ruin under discussion, but the ruin near Lower Sabie has not been found yet.

In July 1962 I visited the site at the request of Mr R.J. Labuschagne, accompanied by section ranger P.J. van der Merwe and field ranger Helfas Nkuna. He wanted me to excavate the site, but circumstances did not allow it. In the late seventies Mr R.J. Immink, then the Parks Board's manager of development, felt that the historic ruin should enjoy the attention it deserved before it was too late. This resulted in another visit on 21 September 1978. This time the party consisted of Immink, Dr W.H.J. Punt, Willie de Beer (the Parks Board photographer) and me, from Pretoria; as well as Park head Dr U. de V. Pienaar, P. van Wyk and M. Mostert, from the Kruger National Park. After Immink had dug a trial trench from the western side to the middle of the rubble, it was clear that there was still a wall underneath the rubble and Dr Pienaar asked me to excavate the ruin.

The 80-year-old Merriman (Phutadza) Nkuna, who had been born nearby and grew up there, showed us part of Albasini's irrigation furrow, which was still visible east of the ruin. The group also visited a Spanish (common) reed bush which, according to hearsay, had been planted by Albasini. Another larger reed stand has since been discovered along the southern bank of the Sabie east of the Phabeni mouth.

Excavations started on 30 November 1978, but because of other obligations no more than a week at a time could be spent at the digging. After a year, during which an average of five labourers dug for six weeks and 29 days in total, the site had been opened up to such an extent that by 13 September 1979 the rebuilding of the walls according to Yates's photograph could be started by Parks Board builders.

What was uncovered?
The building is aligned 30° east of magnetic north lengthwise and consisted of four rooms, with external measurements of 15.85 metres by 5.8 metres. The eastern wall has an extension on the northern side measuring 0.37 metres by 0.66 metres. Mr Wim Cooper, the Parks Board's architect at the time, was of the opinion that the extension had been a support structure, erected because the corner was giving problems. The sturdy outer walls were 0.52 metres thick and the inside walls 0.25 metres. On the outside the highest bit of standing wall measured 0,760 metres, while in other places it was only a few bricks high.

The house faced east. On the western side two half-circular steps with a radius of 0.9 metres were uncovered. On the eastern side the rubble was dug out around the two fig trees growing in the wall, to determine if they germinated after the building had been evacuated. It was clear they came up later as their roots were growing on top of and over the wall. Three bricks lying in a curved line between the roots

The Albasini Ruins on Magashulaskraal at Phabeni mouth as photographed by C.A. Yates in 1903.
(ACKNOWLEDGEMENT: DR J.B. DE VAAL)

The clump of Spanish reed which was planted along the Sabie River by João Albasini near his home at Phabeni mouth between 1845 and 1847.
(16.06.1984)

indicated that there had been a semicircular step at the front door as well. The seeds of the Lowveld fig and cluster fig were undoubtedly dropped by birds perching on the wall on both sides of the door opening. These trees were not originally doorposts, as some people had speculated. For this purpose there were many more durable types of wood available, as will be discussed later on. When Mauch visited the ruin in 1870, the two fig trees were not there yet.

On the outside of the southwestern corner, a long column of fallen bricks that were excavated indicated the presence of a chimney. The place where the hearth had been was also exposed and there was still ash on the floor. On the southern side of the building, the position of the bricks indicated the remains of an outside oven. The walls of the house were not built on a stone foundation – instead, the foundation ditches had been dug until a granite gravel layer was reached. The walls were built on top of this using brick and mortar, without any lime. The bottom must have been stable as there were no signs that the walls had cracked due to subsidence. The tree roots, however, caused severe damage in most places.

As far as could be determined the inner and outer walls had not been bonded to each other. It also appeared as though the inner wall between the kitchen and the room next door to it had been built later, as it stands in the middle of the semicircular step and the doorway on the western side.

Rusted nails were found on the doorstep of the northern doorway. These must have held the doorpost in place. Charcoal and wood remains of the doorpost were found on both sides of the wall. Miss S. Kromhout of the Department of Forestry examined these samples by microscope and determined that they were yellow-wood (*Podocarpus latifolius*). This proved that the doorposts, and possibly the window frames as well, had been made of yellow-wood that probably came from the ravines of the Drakensberg, where, according to experts, it still occurs today.

The position of the inner doors between rooms two, three and four could easily be determined, but the position of an inside door leading from room one could not be pinpointed as the only remaining piece of dividing wall was level with the doorstep. A compacted clay floor could also not be found in any of the rooms. The levels on which ash was found in the different rooms (as determined with a levelling instrument), was, however, at the same level as a floor would have been and as such indicated a floor surface. The doorsteps of the two inner doors are also at the same height and thus also indicate where the floor used to be.

Some trapezoidal bricks, which were excavated and

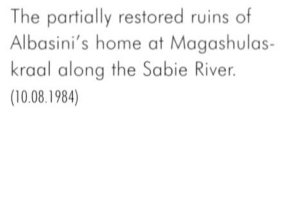

The partially restored ruins of Albasini's home at Magashulaskraal along the Sabie River. (10.08.1984)

stacked on top of each other, show the house had two gables that slanted upwards at an angle of 45 degrees – a single brick tapered on both sides of its one long side, indicating that it served as the finial of the gable. The house had a thatch roof, which later burnt down. This could be deduced from the fact that bits of partially burnt thatch grass were found on the floors. Because of the extreme summer heat in the Lowveld, and if one uses as a guideline a photograph of Albasini's house on Goedewensch (Good Wish) in the Soutpansberg built 11 years later, this house probably had at least one or maybe two verandas.

The brickwork showed good craftsmanship and it was clear that a spirit level had been used throughout. The bond was Flemish with a tendency towards English bond, according to architect Cooper. Because the Portuguese stuck to the Flemish bond while the Voortrekkers built in the English bond, it would seem that the builders were a Voortrekker and a Portuguese – with the Portuguese being Albasini and the Voortrekker Karel Trichardt, as will be shown later.

The bricks used for the building had been moulded and fired about 500 metres northeast of the building site on the western bank of the Phabeni Spruit, as shown me by Phutadza Nkuna. At the site was a knobthorn tree of which the bark had been partially chopped away and which probably served as a marker. Red bricks, similar to those found at the house, were lying around, and three bricks were lying on their sides next to each other. By removing some topsoil, a 5.5 metre-long unbroken row of bricks was uncovered. These were all lying on their sides 15 cm under the surface. These could have been part of an oven's opening or packed like that to dry.

Loose objects found among the rubble
In the bottom layer of rubble up to the floor level, as well as along the outside walls, objects were found that must have belonged to Albasini and hunters that came after him. These consisted of lead granules, lead bullets, a cartridge, percussion caps, home-made nails, pieces of rusted iron plate, a piece of a knife handle, the lower part of a lip drill, a curry comb, broken bottle pieces and sherds of earthenware and porcelain. The objects were analysed as follows:

The ammunition
Lieutenant Colonel P.P.J. van Schalkwyk, then in charge of the Ballistic Section of the Criminal Bureau of the South African Police, agreed that Sergeant F.O. Basson, an ammunition expert, analyse the bullets, cartridge and percussion caps. The cartridge was that of a .577 Snider rifle (14.7 mm), which was manufactured in 1867, according to Barnes (1965). Eleven of the 20 lead bullets matched this calibre rifle while six were fired by a .450 calibre, probably a Martini-Henry. The remaining three were those of either a .475 or a .480 calibre, perhaps an English hunting rifle known as a Black Powder Express. The percussion cap was that of a muzzle-loader.

Because lead bullets were found on the northeastern, northern and northwestern sides of the walls, it can be deduced that hunters had probably used the ruin for target practice later on, or had been firing at someone hiding inside the building.

The nails
Three sizes of home-made nails were found deep in the rubble, and could be identified as follows:
- Large nails of which the longest was 194 mm measuring 10 mm^2 on top, with round heads and pointed tips;
- Medium-sized nails, 100 mm long and shorter, 6 mm^2 on top, with flattened points;
- Smaller nails between 33 and 65 mm long, measuring 5 mm^2.

Because two of the longer nails were found on the inside of the western wall, it can be assumed that the brick line had been fastened to them. They could also have served to hold the roof trusses together, instead of the wooden pegs the Voortrekkers normally used. When the building burned down the nails fell to the floor where they were found. Some of the smallest nails were bent at an angle, which makes it likely that they had been used to fasten planks – possibly those of a door or a window.

The curry comb
As will be shown later, there is documentary proof that Albasini kept horses. The stable must have been close to the front door as the curry comb was found

in a layer of manure. Unfortunately it had been chopped in half, but in one of the six rows of platelets a couple of teeth can still be identified.

The bottle pieces

Not one whole bottle was found in the rubble, only some bottoms and necks with spouts. These were classified according to Edward Fletcher's booklet, *A Bottle Collector's Guide*, as follows:

- Two bottoms of square, semitransparent, green, gin bottles that had been manufactured after 1840. Seen from the side the bottoms are concave, which is typical of bottles from that era.
- The mouths and short necks of four olive-green gin bottles dating back to approximately 1860. These are actually schnapps bottles, which held spiced gin.
- Three oval glass stamps, also olive green, with the words: "Simon Rijnbende & Zonen, Schiedam". These were clearly stamps used on bottles. The gin, which according to Dr Punt had been mixed with tea, was taken as a preventive against malaria. It came from Schiedam in the southern part of the Netherlands, where it was distilled.

The porcelain

Mr E. Bernardi of Bernardi Auctioneers, Pretoria and Mrs M. Venter, then a specialist at the National Cultural Historical and Open Air Museum in Pretoria, provided valuable help with the identification of the 13 different fragments.

- Sherd 1 is a fragment of a Chinese bowl painted in blue. There is a similar bowl in the E.G. Jansen Collection in Pretoria.
- Sherds 2, 3 and 4 are fragments of English ware dating back to the late nineteenth century.
- Sherd 5 is probably a fragment of a willow pattern cup dating back to the late nineteenth century.
- The reconstructed sherd 6 is the rim of a plate painted blue, and of Chinese origin, dating back to 1800–1840. It was part of a set similar to the one to which a large platter in the E.G. Jansen collection belongs.
- Sherd 7 is painted in green and is of European origin, dating back to the late nineteenth century.
- Sherds 8 and 9 are also Chinese, painted blue and part of 6.
- Sherds 10, 11 and 12 are fragments of very fine porcelain that look as though they could be of Chinese origin.

The earthenware

Three pieces of a brown earthenware bottle – the mouth, a piece of the bottom and a piece of the side – were found. A fracture point indicates that the bottle must have had an ear. Ranger A. Espag later found a similar, whole bottle along the lower reaches of the Matjulwana Spruit in the mountainous part of the Malelane section.

The glass beads

A total of 61 glass beads were sifted from the soil that had been removed from inside and directly outside the ruin. Others were picked up from the ground. These were all mounted on white cardboard. With the aid of an own collection and the Mapungubwe beads in the Department of Archaeology of the University of Pretoria, other occurrences of similar beads could be tabulated.

A large, clear, deep-blue, flattened, spherical bead with a diameter of 14 mm and 11 mm long, as well as a similar half measuring 16 mm by 13 mm, could have been part of the Roman Catholic trader João Albasini's rosary. It is striking that most of the other beads are trade beads from previous centuries. Similar examples can be found in places that were also along the ancient trade routes between Delagoa Bay, Inhambane, Sofala and Beira, as was the Albasini trading post. These find-spots are an abandoned village at the foot of Thulamela Hill near Pafuri in the Kruger National Park, Waterpoort in the Soutpansberg district, the Matšhêma Ruins next to the Sand River, the Dzata Ruins in Venda, Mapungubwe and K2 on the farm Greefswald along the Limpopo, Kremetart Mountain near the Mohalakwêna River, on the farm Parma, across the Limpopo at the Zimbabwe Ruins, the Vucha Ruins in the Belingwe Reserve, in Ndanga Cave northeast of the Zimbabwe Ruins, the Mshosho Ruins in the Sabie Reserve and on the farm Headlands near Umtali.

All of these objects are displayed in a small rondavel museum close to the partially rebuilt ruin, together with photographs and a model of the former house and shop, with explanatory notes.

Who was Albasini?

As noted before, João Albasini was the enterprising Portuguese man who survived the malaria-infested, wild Lowveld of 140 years ago, built a large, spacious home along the Sabie River during a time of wattle-and-daub houses, and settled there. But who was he and what was his story?

In the early years of the ZAR (South African Republic) probably no other white man had more influence among the black people of the Northern Transvaal, the eastern Lowveld and Mozambique than the Portuguese-born João Albasini. He was the only white chief of a tribe known as the Magwamba. His entire career was one of great adventure, as trader and elephant hunter first along the east coast, then along the upper reaches of the Sabie River within the boundaries of what is now the Kruger National park, after that in the Voortrekker towns Andries-Ohrigstad, Lydenburg and Zoutpansbergdorp (Soutpansberg, later Schoemansdal), and last on his farm Goedewensch in the Kleinspelonken. There he became the Superintendent of Native Affairs and Vice Consul of Portugal in the ZAR. No wonder that Dr D. Wangemann, the director of the Berlin-Lutheran Mission, had the following to say about him after a visit to Goedewensch in 1884: "This Albasini is one of the most remarkable personalities in South Africa."

João Albasini was born in Lisbon, Portugal on 1 May 1813 and was christened in the Roman Catholic parish or parochial of São Lourenço. His parents were Antonio Augusto Albasini and Maria da Purificacão. Although they were Portuguese subjects and his father worked as a sea captain and ivory trader under the Portuguese flag, they were not of Portuguese origin. Antonio was an Italian farmer from Tyrol who had left his country in 1807 for religious reasons, while his beautiful wife was Spanish. João was Portuguese as far as his birthplace and citizenship were concerned.

The young boy enjoyed reasonable schooling for those days. He studied law, among others, and was a gifted pupil who could speak several European languages. He lived in Lisbon with his parents, a brother and a sister until he was 17. After that he chose the adventurous life of a trader among Africans along the east coast of Africa. In October 1830 João accompanied his father on one of his trade missions to Bahia in Brazil. The following year a trading post in Delagoa Bay (Lourenço Marques, now Maputo) became his responsibility. According to legend the ship in which they had sailed ran ashore along the east coast, but they managed to make it to the beach unscathed.

Albasini built himself a house about 400 paces northeast of the fort near a fountain that was separated from the harbour by a swamp on the south. Apparently he traded from here; in slaves among others. This part of the beach is marked on an 1876 map as Albasini Beach. The fountain, which is now in the Vasco da Gama Garden, was still visible a few years ago.

The living conditions of these first colonists were terrible. Trade was very poor and year after year many fell prey to the deadly malaria. In 1829, only two years before Albasini's arrival in Delagoa Bay, only six out of 40 malaria-sufferers had survived the disease.

After spending some time in Delagoa Bay, probably to school his son in trade, João's father returned to Lisbon. Later João was to hear that both his parents died shortly after his father's return. He never heard from either his brother or sister again. As a young lad of 18 he now had to make his own way in the world. A lesser person might have gone under, but Albasini with his strong personality, sharp intellect and perseverance was determined to make a success of his career. He also made the most of the supply of merchandise his father had left him. Every year during the healthy winter months he sent traders with material, beads, knives, tinderboxes and mirrors from the coast into the interior along the centuries-old trade routes to trade for ivory. With the advent of the rainy season they returned from their long and dangerous journeys with loads of ivory. This he traded with visiting ships for more ammunition, merchandise and provisions.

Albasini is kidnapped

A few years after his arrival in Delagoa Bay in October 1833 the small community suffered a terrible fate. Manukosi (Sochangana), the Zulu who had fled

Shaka's wrath, caused havoc among the small peaceful Thonga tribes with his hundreds of warriors as far as he went. He also attacked the small group of Portuguese at the Bay, forcing them to flee to the nearby island Shefina under the leadership of the officer in charge, Captain Dionisio Antonio Ribeiro. Albasini was apparently one of the refugees.

After Manukosi's warriors had looted and destroyed the fort, they attacked Shefina and took all of the Europeans hostage. The hostages were then brought back to the mainland where they had to dance throughout the night to entertain their captors. While dancing Albasini went down on all fours and started howling like a jackal. This entertained the young warriors to such an extent that they started throwing small pieces of meat at him, which he had to pick up from the ground with his teeth. They soon became bored with this, however, and enlivened matters by tying his hands behind his back and leading him to a door, which he had to knock open with his head while they were beating him with sticks.

After they had untied his hands he was taken to confront a vicious chained dog. "Now the jackal must fight the dog so that we can see which is the strongest," they said. While hesitating and thinking that his last moment had arrived, he was pushed onto the enraged beast. Fortunately he landed on the dog's back and grabbed it by the throat with his right hand and by the ear with the other, while sitting on top of it. The dog was big and strong and Albasini soon realised that he would not be able to control it for long. Upon his plea that they must please prevent their jackal from being bitten to death, he was freed. As Albasini could speak several African languages, knew their customs and could teach them much, his life was spared. His captors massacred the rest of the prisoners in front of him by nailing them to the ground with wooden pegs. After that they took him with them into the interior.

After he had been living with his captors for six months, two black traders arrived from the coast one day. They immediately recognised Albasini and contrived a way to free him. At about eight that same evening the three managed to slip away silently and unnoticed. After a headlong flight through the bush that lasted a day and a night, braving lions and other dangerous animals, they reached safety at daybreak the second morning. By the time they reached Delagoa Bay, a new governor and reinforcements were already in place.

A new beginning

Albasini had nothing left and had to start from scratch. He only managed to get his business off the ground with the help of the Africans who admired him. He did not settle in Delagoa Bay again, but north of it, in the area ruled by Ntimane or Kosine. On 10 September 1836 Albasini was issued with a passport so that he could leave Delagoa Bay, but it is not known why he had decided to go. The area to which he moved had been in the grip of a terrible drought for two years, with the result that many people and animals died of hunger. Being an elephant hunter, he collected ivory and in the process provided many starving tribesmen, who had been plundered by Manukosi not long before, with meat, saving them from a certain death.

Soon Albasini was living like a feudal lord. Tribesmen who had fallen from favour with their chieftains fled to him with their followers. In exchange for food and protection they delivered excellent services as hunters, labourers and bearers of trade wares, including ivory. He selected the most able and taught these men to shoot with muzzle-loaders, after which he sent them out to exploit the rich resources of ivory, rhino horn and skins. He also taught them how to prepare skins and ivory for the market.

Albasini showed military capabilities and organised the marksmen and other able-bodied men – during a time of tribal conflicts and raids he built up a force to be reckoned with. His followers recognised him as their high chief and called him Juwawa, a distortion of his Portuguese name. A Swiss missionary summarised his rule as follows:

> They served him as they would have served a native chief. They gathered to his call, ploughed his fields, were doing all his work, carrying his goods, marching under his orders armed with guns which he had got in Lourenço Marques. They brought to him the first fruits of their crops and a foreleg of each large piece of game they killed. In certain important circumstances Albasini judged cases or ordered his troops about in the costume of a Portuguese officer

... Albasini remained the chief of the Gwamba and the only magistrate they recognised until his death.

Albasini and the Trichardts
Voortrekker leader Louis Trichardt arrived with his trek from Soutpansberg at Delagoa Bay on 13 April 1838. Louis never mentioned Albasini in his diary, so it can be assumed that he was away the four months Trichardt spent there before his death. By 7 September 1838 Albasini was back in Delagoa Bay, serving as a cashier and officer in charge of the garrison Vicente Thomas dos Santos. He was also cashier and director of the Esta Trading Company in Inhambane's trading post or depot in Delagoa Bay, which contained all of the valuable merchandise. On a named date they opened the depot and sealed it again after they had removed the wares that were intended for Inhambane.

In June 1838 Louis Trichardt sent his son Karel to explore the African east coast in order to find a healthier place to settle. It is possible that Karel could have made Albasini's acquaintance at Inhambane while the latter was delivering the goods that had been taken from the warehouse. Dr Punt said he had discovered that Albasini planted a grove of orange trees at Quelimane at the mouth of the Zambezi. The oranges were sold to ships and eaten by the crews to prevent scurvy. This kind of resourcefulness was a characteristic that stood him in good stead throughout his life.

At the end of July 1839, a year and a month after his departure, Karel Trichardt was back in Delagoa Bay and he appointed Antonio J. Nobre as his commissioned agent in Mozambique. In his turn Albasini named the governor of Lourenço Marques, Antonio C.P. Gamitto, as his "first commissioned agent to handle his affairs and look after his business in this settlement of Lourenço Marques". Trichardt and Albasini acted as witnesses for each other, which proves that they must have been old acquaintances and possibly fellow passengers on board the same ship, as well as trading partners along the east coast.

A company to hunt elephants
Albasini was not the only one in Delagoa Bay interested in the ivory trade. On 7 July 1841, ten members of the garrison town had a document drawn up by Notary José Conçalves Martinho according to which they founded an "Entrepreneurs Kompanjie" (entrepreneurial company). They were José Antonio da Silveira, Avelino Chavier de Menezes, Pedro Francisco de Souza, Boda Cassimo, João Albasini, Dionisio Manuel da Silva, Alexandre José de'Ochoa, José de Sequeira as representative of Antonio de Sequeira, Boda Cassimo as representative of Asane Cassimo, and Antonio de Gouveia. These ten would "by common agreement ... shoot elephants in the territories neighbouring this garrison, helping one another in the work in which they participate equally. It will have a duration of three years at the end of which it can be dissolved, but may continue longer if this is decided on."

The sole purpose of the company was "to increase the ivory trade which is today so diminished by the negroes of the interior, whose vagary has made so many inhabitants unhappy, not only of this Garrison, but also of many ports along the coast".

It was further determined that the company would employ 25 Africans to shoot only elephants. The company would be managed by an advisory board, which would meet once a semester to check the books that would be kept by a treasurer and a director. After the profits had been determined and shared among the members, arrangements would be made to cover the expenses of the next semester.

On 12 January 1842 Albasini gave up, among other things, a house and two water tanks he owned in the harbour town, as well as five cows, agricultural tools and four people – probably his slaves. His mind was in all likelihood set on moving deeper into the interior and that is probably why he relinquished some of his fixed assets.

A.H. Potgieter visits Delagoa Bay
In the winter of 1844 something happened to give a totally new direction to Albasini's career. A Boer expedition under the leadership of Chief Commandant Hendrik Potgieter arrived in Delagoa Bay from Potchefstroom. They were there to negotiate trade relations with the Portuguese and to find a place to settle closer to the coast. They reached an agreement with Governor Antonio Joaquim Teixeira according to which the immigrants could move into the interior between 26° and 10° S, up to four days' jour-

By June 1948 the Albasini Ruins at Magashulaskraal were just a heap of rubble.

ney from the coast and ports. Albasini and Potgieter had presumably met before this in the interior, after which Albasini probably provided them with guides. By May the next year, wheels had been set in motion to establish a town closer to the coast and by June-July 1845 the Voortrekkers arrived at a point west of the Drakensberg Mountains, only a few days' journey from Delagoa Bay. Here Potgieter founded his third town, Andries-Ohrigstad. They were convinced that it would be easier to reach Delagoa Bay from there, while they would also be within reach of Albasini and other Portuguese traders.

By the middle of October 1845 the "Raad der Representanten" (board of representatives) again sent an expedition of Boers, under the leadership of Karel Trichardt as field cornet, to Delagoa Bay. They were to clear a road down the mountain and at the same time allocate land for the registration of farms. Despite the fact that they were going to travel in a season unhealthy for both man and animal, and despite Trichardt's advice that they should travel on foot, the party decided to travel by ox-wagon so they could trade ivory and skins at the same time. They managed to reach the harbour, but only after they had lost all of their oxen to the disease transmitted by tsetse fly (nagana), as well as all of their horses as a result of African horse sickness.

Albasini trades Magashulaskraal
In the second half of 1845 Albasini was on his way from Delagoa Bay to Ohrigstad. Although there is no documentary proof of this, it is only logical that the sorely tested Trichardt would have gone to his old friend for assistance in the form of draught animals and bearers. If this was the case, he and his party would also have returned with Albasini to Andries-Ohrigstad. Legend has it that Albasini occasionally used donkeys instead of oxen as trek animals. He travelled on a white mule himself as they are more resistant to African horse sickness and nagana. On 10 December the expedition could report on their journey to Ohrigstad.

During this first journey to the Boers he traded 22 head of cattle for a piece of land from the Kutswe Chief Magashula, who lived along the Sabie River, with the aim of establishing a trading post on the existing trade route between Delagoa Bay and Sekhukhuneland that followed the Sabie River. His bearers would use this trade route when transporting wares between Delagoa Bay, Magashulaskraal and Ohrigstad. It was clear to Albasini that the place was favourably situated for trade with both the blacks and the Boers. Black people referred to this place as eMngomeni and later as Mambatines Drift. The Sotho-speaking Chief Maġashula called it Mokômeng, while Albasini always spoke of Magashulaskraal. The Portuguese in turn called it "Vila Albasini" and later "São Luiz" in honour of the Portuguese king.

During what we assume to have been the journey of Trichardt and his party with Albasini via Magashulaskraal to Ohrigstad, the two probably renewed their friendship and the trade agreement they had closed seven years before, after which they decided to open a shop at Magashulaskraal. After his first trade journey to Ohrigstad, work on the trading post started and, with the help of Albasini's many black followers, they made and fired bricks. They dug an irrigation furrow from Phabeni Spruit to the fertile land east of the building site, where they planted fruit trees, including oranges, papayas and mangoes. They also established a vegetable garden and planted wheat. In 1935 the late Mrs Maria Biccard, one of Albasini's daughters who was by then quite elderly, said that Magashulaskraal had been known for its fine white bread. They probably also planted a variety of vegetables, including runner beans, for why else would they have grown Spanish reed if not for this purpose?

Albasini's partner, Karel Trichardt, was a skilled

artisan. While stopping over in the Soutpansberg he built a forge, made shoe nails and served as a wagon-maker, and as a result was more than able to help Albasini build the shop. His interest in handiwork was later reconfirmed when he bought tools at the English trader Joseph McCabe's auction in Ohrigstad in October 1846. These included a plane, a drill, five chisels, three sickles and a long pit-saw (which was probably purchased with the specific aim of sawing yellow-wood planks in the Drakensberg).

The shop must have been completed by the middle of 1846 as, on 7 July 1846, João Albasini and "Carlos João Trichardt" on the one side and "João Bernardo Juberti" on the other, signed an agreement before Notary Jacob Christovão Zavier Couto and two witnesses. It was agreed that Albasini and Trichardt would open a shop "in the area known as Macazula, where João Bernardo Juberti lives". Johannes Joubert agreed to carry out all tasks assigned to him by Albasini or Trichardt. All expenses, losses and profits incurred would be the owners', while Joubert undertook to trade with everything stored at the post, as one or both instructed him. For his services he would get 50 sheep, 200 pounds of rice, six cans of vinegar, 200 pounds of salt, 40 pounds of coffee and 100 pounds of sugar every year, which would be paid monthly or annually. They also decided to keep 50 head of cattle at Magashulaskraal, with all gains or losses being that of the owners, but Joubert would be able to use them if the owners did not. In exchange for his supervision he would receive half of the calves. He would also keep two horses for Trichardt and Albasini's use. If an animal were to die while being used by the owners, the loss would be theirs, but if it died while being used by Joubert, he would be responsible for the replacement thereof. The owners would also be responsible for anyone injured while on duty.

This document shows that in 1846 Albasini was not the only white person living in the area that is now the Kruger National Park, and that Johannes Joubert could have settled in the vicinity of Ship Mountain before Albasini. The elderly transport rider Org Basson pointed out the ruin of what had been Joubert's home to Dr Punt in 1950. This farm was later called Joubertshoop. At the time the area was just outside the tsetse fly belt, but exactly why Joubert

The possible site along the southern flank of Ship Mountain where Johannes Joubert had built his house before 1845, and therefore the first European homestead in the Lowveld. He was appointed in 1845 to run a trading post for João Albasini and Karel Trichardt at Magashulaskraal. (25.09.1984)

settled there is not known. Punt thought that he had provided the Portuguese traders and their bearers with meat during their travels.

Albasini's followers
Most of Albasini's black followers from Ntimane or Kosine, whom he had protected against Swazi raids, followed him to Magashulaskraal where their services were of inestimable value. Three of the outstanding characters among them who always acted as leaders were Manungu, Monene and Josekhulu[1] (Big Josef).

Manungu
Manungu was of Zulu extraction, robust and strongly built. He had been employed by Albasini at a reasonably young age as "apprenties" (apprentice) and stayed with him until his death. Because he was so reliable and could run fast, his employer always sent him as courier and messenger when there were urgent matters that had to be brought to a close speedily. He was also Albasini's most important elephant hunter, and was usually sent into the bush as leader of a number of armed hunters and bearers during the hunting season to collect ivory. Manungu Hill in the Park was named after him as he used to look after some of Albasini's cattle at the bottom of this hill. Because his encampment was close to the old Voortrekker road between Ohrigstad and Delagoa

---

1   Also spelled Josikhulu in certain documents.

5. JOÃO ALBASINI (1813–1888)

Bay, on a trade route the blacks used, Albasini built a small shop there as well. The fosse of a single hut 10.5 metres west of the shop indicates that it was not a permanent family residence, but merely an outpost. Excavations of the site unearthed nails identical to those that had been found at the Albasini Ruin. This place later became an outspan for transport riders, as proven by the ammunition and pieces of bottle and porcelain dating back to the nineteenth century that were found there.

Josekhulu
Josekhulu looked after a herd of Albasini's cattle at an outpost southeast of Ship Mountain and ran a small shop. According to Org Basson, who used to be a transport rider, this little shop had been situated along the Josekhulu Spruit. He and Dr Punt went to search for it in 1950, but the grass was so tall that they could not find any sign of it.

According to Dr Punt it is no coincidence that this shop was 24 kilometres from Manungu's, thus exactly a day's journey by ox-wagon from one another on the route to Lourenço Marques. In July 1983 senior ranger Thys Mostert found the foundations of some huts, a small rectangular building as well as a grave next to the west bank of the Josekhulu Spruit, about eight kilometres southeast of Ship Mountain. This had been Josekhulu's home and the foundations of a shop and home belonging to a later trader, Thomas Hart, of whom more will be related later. Mostert was able to find the remains due to the fact that it had been a very dry year and the vegetation was therefore sparse.

Josekhulu was never mentioned in any of Albasini's correspondence as he had probably been less important, but his name was nevertheless immortalised in the name of a small creek that flows past Ship Mountain.

Monene
The third mentioned leader among Albasini's followers was Monene, who was of more noble ancestry than Manungu. He was a combatant, strategist and at one stage commander-in-chief of the feared Manukosi's warriors. But after one battle, he had a disagreement with his leader, and the only way he could save his life was to flee to Albasini with his followers, the Maswanganyi tribal group. He was commander of Albasini's army for a long time, first at Magashulaskraal and later in the Soutpansberg, where he would cause endless trouble and have the whole district up in arms.

Albasini's horses and his bread, which was made of home-grown wheat, so impressed the blacks that they sang a song about them that is still being sung today. In 1931 a nursemaid sang it to the children of Reverend Jacques of the Massena Mission Station.

The words of the song are:
Yo ya yo ya amagada hanci,
hanci ya Jiwawa
tata nga vu yi
hi ta dya mapa
wa mapa wa xi lungu
kikigi, kikigi, kikigi, kikigi, kikigi, kikigi

It was translated into English as follows:
We ride horses
the horses of Jiwawa
Father does not come back
we will eat bread,
the bread of the white man
kikigi, kikigi, kikigi, kikigi, kikigi, kikigi

The possible site of Albasini's trading post along the trade route on the southern bank of the Sabie River, between the current Lower Sabie rest camp and Sabie Poort. This historical site was unfortunately destroyed in 1968–1969 during road construction operations between Lower Sabie and Crocodile Bridge.
(27.07.1984)

(The "kikigi" is a sound that mimics the galloping of the horses.)

Trade with the people of Ohrigstad
Albasini had been of great service to the Voortrekkers in those times as his wagons and bearers could transport merchandise from Delagoa Bay across the Lebombo Mountains and through the tsetse fly-infested area to Magashulaskraal. From there the Boers could fetch it with ox-wagons. Occasionally he took his wares all the way to Ohrigstad to trade there. To reach Ohrigstad he travelled by ox-wagon from Magashulaskraal along the southern and then the northern bank of the Sabie River to the farm Klipkraal. Mauch also followed this overgrown path in 1870 and therefore walked along Albasini's old wagon trail. From Klipkraal Albasini followed Potgieter's route of 1844, which is the same one Karel Trichardt followed the following year, west of Graskop, along the Treur River, across the Blyde River and Kaspers Neck to Ohrigstad. His bearers would, however, follow the shorter trade route and rest in turn at Lower Sabie, Magashulaskraal, Klipkraal and Rustplaats along the Ohrigstad River, on the trade route to Ohrigstad. They were exactly 40 kilometres or a day's journey apart and therefore served as overnight stops.

In December 1846 Albasini was once again on his way to Ohrigstad with a full six loads of merchandise. During this visit he offered to show the Boers a passable road to the coast that was free of tsetse flies. Unfortunately for the settlement this journey was impossible just then as the hostile Chief Makkasana had laid siege to the harbour. Another reason why the journey was prevented from taking place was that the people of Ohrigstad had split into two political groups that could not cooperate.

In March 1847 Albasini left Ohrigstad and in May he was back in Lourenço Marques. There he answered a long questionnaire compiled by Acting Governor José Antonio da Silveira about the Voortrekkers and their living conditions. In conclusion he made the following significant remark about possible trade relations with them: "… and I only want to say that it is certain that they are currently unaware in Portugal of what good prospects this has, because if they had known, there would not have been any shortage of people wanting to invest their money here, both to their own advantage as well as to the advantage of the nation. I could have told you much more about the Boers if I had the time, but unfortunately I am in a hurry as I am about to depart for Andries-Ohrigstad."

As a result of the political division among the Boers, Manukosi's hostility, the presence of tsetse flies, malaria, the difficult road and the scarcity of water for their animals in winter along the wagon route to Delagoa Bay, Potgieter and his followers decided to move to the Soutpansberg. The first trekkers arrived there on 3 May 1848. Here they founded a small town called Zoutpansbergdorp (Soutpansberg town) at the foot of the mountain next to a spruit. This name changed in 1855 to Schoemansdal. They hoped to be able to establish trade relations with Inhambane from there.

The members of the Volksraad Party who had stayed behind in Ohrigstad and its vicinity still had to find a suitable route to Delagoa Bay. On 20 April 1848 they decided that an expedition would leave on 20 June 1848 to "clear the way to Delagoa Bay … with the aid of Joas's labourers". By mid-July Karel Trichardt was once again on his way with a commission in an effort to find a better route to Delagoa Bay. It is not known how many people were in the party, but it is known that he was accompanied by J.J. Burger, Willem Pretorius and somebody called Joubert. A youngster by the name of Ngutu led the oxen. Years later an elderly Ngutu said that Pretorius had fallen ill with malaria near Ship Mountain. He was sent back to Ohrigstad accompanied by two Shangaans, who could have been some of Albasini's guides, but he was not destined to leave the Lowveld again and died in the bush near a peaked hill, beneath an old marula tree next to "De oude Wagenweg" (the old wagon trail). At Magashulaskraal Albasini, the only white man in the vicinity, was informed of his death and went to bury him. This hill later became known as Pretoriuskop, and the Mbyamiti Spruit, whose source is in the vicinity, was also indicated on old maps as Pretorius River. Even today it is not sure if these two landmarks were named after Willem Pretorius or President M.W. Pretorius, who visited the area in 1865.

On 21 July Trichardt arrived with his "Joas volk" guides at Delagoa Bay. Thirteen days later, on 3 Au-

João Albasini at his home on his farm Goedewensch near Piesangkop or Luonde in the Spelonken. The Zulu veteran with him is probably one of his headmen, Manungu, circa 1870.
(ACKNOWLEDGEMENT: DR J.B. DE VAAL)

gust, they were on their way back to Ohrigstad, or possibly already in Ohrigstad, when Albasini and one Combrink arrived with two wagons in Delagoa Bay. On their way they had encountered tsetse flies for only about two hours' journey; a distance they covered at night.

The Joubert who had accompanied Trichardt and J.J. Burger could have been David Joubert, a member of the Board of Representatives, or Johannes Bernardus Joubert, Trichardt and Albasini's employee. Johannes Joubert knew the area well and had been to the Bay at least once, and could therefore quite possibly have served as guide. This would explain why Albasini and Combrink only arrived at Delagoa Bay two weeks after the Trichardt commission, as Albasini had to run the shop in Joubert's absence and had to wait for him to take over again. This gap also gave Albasini the opportunity to bury Willem Pretorius. On his return journey Albasini transported a load of merchandise as well as N.T. Bührmann, who had come from Amsterdam to teach at Ohrigstad. Two years earlier, Albasini had transported a number of crates filled with books, which had been sent from the Netherlands to the Boers. They collected these at Magashulaskraal.

Albasini leaves Magashulaskraal
Albasini later wrote the following in a letter about his stay at Magashulaskraal: "I stayed there for two years, but because of changing prospects left that house and moved to the Afrikaans Republic."

It must have been difficult for him to trade at Ohrigstad without a suitable store for his many wares. A house in town, with Magashulaskraal as base for some of his African bearers and elephant hunters, would be more comfortable and lucrative. At the end of 1848 he left his home along the Sabie and built a shop in town, where he and Anthonie Fick were the established traders. Later he bought the farm Rustplaats where he also settled some of his followers and bearers. Karel Trichardt owned the adjoining farm Rozenkrans, where he only built a house in 1874.

Albasini's "changing prospects" were not only material. When he arrived at Ohrigstad during his first visit, there had been a church service in progress. He walked in and, being a Roman Catholic, made the sign of the cross and sat down in the back of the church. After the service one of the Boers, Willem Janse van Rensburg, came to greet him, asked who he was and what the purpose of his visit was. He identified himself in the broken Afrikaans he had probably learnt from his friend Trichardt. Van Rensburg kindly invited him to his house for a cup of coffee. Following Boer tradition, the pretty young daughter of the house, who had clear blue eyes and blonde hair, served the coffee. It was love at first sight for Albasini, and his discovery that a widower was also courting her certainly precipitated his move to Ohrigstad.

The Van Rensburg parents were not happy with the fact that a Portuguese, and a Roman Catholic at that, was interested in their daughter. There was, however, no stopping the two and on 6 March 1850 the young man "Johan Albasienie Litmaat (member)" and the "jongedogter" (young girl) Gertina Petronella Maria Janse van Rensburg were married by Magistrate J. de Clerq, assisted by L. Nel and C. Fourie, two members of the court. Gertina had been born in 1832 in Grahamstown and was therefore 18 years old when she married the 36-year-old João. Gertina's mother was Susanna Elizabeth Oosthuizen. The Albasinis had nine children – three sons and six daughters. João, their third son, died at a young age.

During the course of four years it became increasingly obvious that Ohrigstad was not the best location for a town, as it was unhealthy to both humans and animals. The Volksraad therefore decided to move town to a healthier area about 50 kilometres south. The first people started moving there early in 1850. They named this new town Lydenburg (place of suffering) in memory of their miserable stay in Ohrigstad.

Albasini also moved to Lydenburg and built a shop in town,[2] where he became partners with the Goanese trader Casimiro Simoês. The latter eventually became one of the richest traders in the ZAR (South African Republic). Mariano Luiz de Souza, the progenitor of the De Souzas of Lydenburg, was a clerk in Albasini's shop.

Although Albasini could never master Dutch, by 1848 he understood enough to be able to translate Portuguese letters for the Boer leaders. He also acted as interpreter in court if one of his compatriots was on trial.

Relocation to Soutpansberg

Later on trade in Lydenburg was not as lucrative as it had been a few years earlier, as a large number of people had moved with Commandant Andries Hendrik Potgieter to Soutpansberg. Albasini nevertheless lived and traded in Lydenburg for another three years, until August 1853. By that time he was a wealthy man, as he had bought the government farm Rustplaats between Krugerspos and Ohrigstad for 700 rix-dollars in 1851. With the exeption of the Portuguese and Dutch, no foreigners were allowed to buy land north of the Vaal River. When Albasini left Lydenburg he owned the farm Rustplaats, the adjoining farm Nooitgedacht, Klipkraal in the Drakensberg and four stands in the abandoned Ohrigstad. By August 1853 he was on his way to Soutpansberg with seven wagons and a large black entourage.

As there was no residential area near town where his followers could live, he bought a farm in the eastern Soutpansberg in the Kleinspelonken Range south of the impressive Piesang Kop (Luonde) and called it Goedewensch. The most important family groups who settled there under his leadership were the Maswanganyi, Nwamanungu, Mbangezithe and Ndengeza, collectively called the Magwamba and related to the Thonga of the east coast. Apart from the previously mentioned Manungu and Monene, the other two Magwamba chiefs were Pandeka and Simswane. Their closest neighbours and arch-enemies were the Venda, who lived in the mountains. The small Thonga tribes who had fled from Manukosi 20 years before lived further east. Albasini and his family lived in "Zoutpansbergdorp" for the first four years. Once again he built a shop and did business with the Boers as ivory trader. But the wares for his shop, which were transported from Delagoa Bay by his bearers, were of poorer quality than those brought in by traders from Grahamstown and Natal, and he could not compete with them. He could also not get hold of all the merchandise the Boers needed in Lourenço

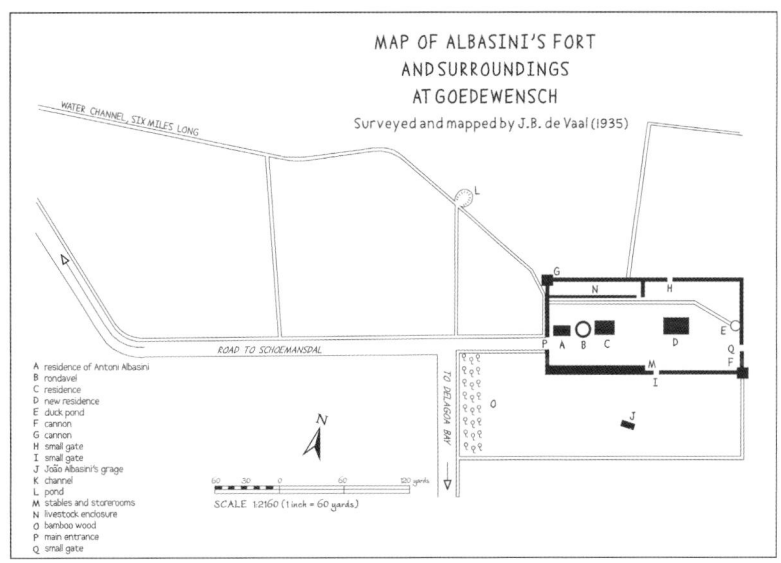

The layout of João Albasini's farm Goedewensch as depicted by Dr J.B. de Vaal in 1935.
(ACKNOWLEDGEMENT: DR J.B. DE VAAL)

MAP OF ALBASINI'S FORT AND SURROUNDINGS AT GOEDEWENSCH
Surveyed and mapped by J.B. de Vaal (1935)

A residence of Antoni Albasini
B rondavel
C residence
D new residence
E duck pond
F cannon
G cannon
H small gate
I small gate
J João Albasini's grave
K channel
L pond
M stables and storerooms
N livestock enclosure
O bamboo wood
P main entrance
Q small gate

SCALE 1:2160 (1 inch = 60 yards)

---

2 According to legend, Albasini also ran a trading and overnight post along the old "harbour road" between Spitskop and Lydenburg during the period 1850–1853. The ruins of this old trading post can still be seen on the old farm Rhenosterhoek, between the Devil's Knuckels and Blystaanhoogte.

Marques, as the Portuguese focussed on trade with the Africans. He was therefore forced to stop trading with the Boers in 1857 and went to live on his farm Goedewensch, from where he operated as elephant hunter, trader among the blacks, farmer and chief of the Magwamba (Knobnoses).

A model farm
At Goedewensch he got off to a good start. Albasini left no stone unturned to improve his farm and transformed it into a veritable Eden. With the large number of workers at his disposal he made it into the most beautiful farm in the district and possibly the whole republic. He built a solid house on a strategically situated hill from where he could keep an eye on all the approaches to his property. He also had a spectacular view of Piesangkop and the picturesque Soutpansberg mountain range. About ten kilometres from the house, on the current farm Beja 224, he constructed a large dam and dug a canal all the way to his house. There the water was divided into different furrows and dams.

To protect the house from attacks he engaged a German by name of Von Marnecke to build a rectangular bulwark around it. The fort was built from brick and plastered with lime, measuring 2.9 metres high and 164.6 metres long by 76.2 metres wide. On all four corners were platforms, each decorated with an orange tree. The bulwark was blown up in the Anglo-Boer War, but judging from the foundations the platforms on the northeastern and southeastern sides had been the highest. Small cannons were placed on these so they could each protect two sides of the fort during times of war. There were also loopholes all along the walls. On the southern side was a large entrance through which a wagon could easily pass. At night the entrance was closed with heavy wooden gates and two armed guards stood guard. Over the entrance was an arch and on top of the arch was a wooden doll that represented a little black boy. On either side of the entrance was a large clay pot. There were also two small gates on the southern and eastern sides of the fort that led to the fields and fruit orchards, while a third on the northern side gave access to the labourers' village.

The last part of the road from Schoemansdal (Zoutpansbergdorp) to the bulwark looked like a street and from the southwest it ran over a distance of 400 paces straight into the main entrance. It was 40 paces wide and paved with stones. On both sides of the road were deep water furrows with lanes of lilac trees growing on the banks. Another road of the same width joined it at an angle. This road led to the hunting grounds and became the road to Delagoa Bay and Inhambane.

There was no shortage of fruit at Goedewensch. On the southeastern side of the bulwark was an orchard with a variety of fruit trees and a coffee plantation, with a hedge of banana trees as protection against cold winds. Albasini's followers lived south of the mountain in large villages, some as far as ten kilometres away. All the able men were well trained and would later have bloody battles with the Venda.

With the exception of Albasini and his family several other people lived on the farm, of whom the Irishman Tom Kelly, the young Mr Reginald Alphons van Nispen and the Van Boeschoten family were the most important. Van Nispen had been a clerk of the court at Potchefstroom before Albasini fetched him to teach his children at Goedewensch. He eventually became a magistrate's clerk at Schoemansdal and after a number of years the magistrate of Soutpansberg.

Although far from town, Goedewensch had many visitors as the road from the south to the hunting grounds ran across it, while all the traders travelling from Delagoa Bay to Schoemansdal usually spent a night at Goedewensch.

Goedewensch, together with Schoemansdal, formed a nucleus of civilisation when the northern corner of the ZAR was still wild and in its formative years. This farm would become the centre of Portuguese interest and black politics in the Soutpansberg. From here Albasini could use his influence as Portuguese Vice Consul in the ZAR and as Native Superintendent of Soutpansberg.

Albasini becomes Vice Consul
In 1858 the Governor General of Mozambique honoured João Albasini by offering him the post of Vice Consul of Portugal in the ZAR. He gladly accepted and applied himself to keep the Portuguese government on the east coast up to date on the political, economic and social conditions in the Republic. He was a particularly systematic person and kept copies

of all his official letters written between 14 April 1858 and 15 October 1872 in a thick file, which he called "Correspondencia oficial do Consulado Portuguêz em a Republica Africana do Sul Transvaalsche" (Official correspondence of the Portuguese Consulate in the South African Republic Transvaal). The file, which consisted of 486 written foolscap pages, was loaned to me in October 1945 by Albasini's daughter, Mrs M.M. Biccard of Pietersburg (Polokwane). She also agreed to it being translated and photocopied by the Transvaal Archives. These documents are a valuable source of information in connection with the earliest relations between the Boers and the Portuguese authorities. It only contains copies of his outward-bound letters to the Portuguese, as the incoming mail was unfortunately burned as trash after his death.

Albasini's appointment as vice consul made the implementation of a postal service between Goedewensch and Lourenço Marques essential. A Portuguese soldier transported the post along the existing trade route to the coast on a monthly basis. An invitation was extended to the Boer government to use the service and an opportunity was created for mutual negotiations.

Somebody who knew him well had the following to say about Albasini's official dress: "On official occasions he wore a dark blue broadcloth suit with moth-eaten gilded braid. His hat looked the same as those that are nowadays worn by Portuguese consuls and governors ... black with gilded braid." A sword in a beautifully adorned scabbard contributed greatly towards his vice-consular dignity. According to a communication by Dr T. da Mota, Consul General of Portugal in the Republic of South Africa in 1980, this was actually an admiral's uniform, which he by rights was not allowed to wear. There is no getting away from the fact, however, that in this dress he cut an imposing figure of great authority, as can be seen in the photo of him. It is also said that he always dressed up in his uniform before opening and reading official letters.

Regulation stipulated that the consulate's office had to be built in a quiet place convenient to traders. Over the main entrance should be the royal coat of arms with the phrase "Consul-General" or "Vice Consul of Portugal". The national flag had to be raised on national holidays, except in countries where the ruling government would not allow it. Albasini's office was a rondavel in the garden, and in front of the door was a shield-shaped coat of arms with the wording "Vice Consulado de Portugal, na Republica Africana Meridional".

One can hardly imagine a more diligent and loyal official than Albasini. He had to deal with a decidedly lax government that treated him very poorly, as he received absolutely no compensation during the 15 years he served as Vice Consul. He had to pay a secretary £100 a year out of his own pocket, with free board and lodging on top of that, provide the postal messengers with free food and accommodation for a week every month and receive honoured Portuguese guests. Taking all of this into account one can get some idea of what the prestigious position must have cost him.

Prospects
As he was still well off Albasini did not mind the lack of remuneration and spared no trouble or expense when it came to establishing and expanding the Portuguese sphere of influence in the Republic. He was a practical person, but also an idealist. In connection with the ivory trade he determined that between 70 000 and 80 000 kilograms of ivory to the value of £120 000 were exported to the British colonies annually. To ensure that this trade took place through Delagoa Bay he negotiated with influential traders such as Ignacio de Paiva Rapozo, who wanted to build a road between Soutpansberg and Delagoa Bay with the cooperation of the ZAR. Steam-driven carriages would be used to transport merchandise along the road. Unfortunately the Republic had too many internal problems at the time and there was no support for such a scheme. Another spoke in the wheel was the hostile Mawewe, a son of the redoubtable Manukosi, who barred the way to the coast for more than a year. As a result nothing came of the plan. It was, however, the forerunner of the Delagoa Bay railway line.

As tax collector
Apart from his duties as Vice Consul, Albasini also played a significant role in the Soutpansberg as poll tax collector. In 1859 Stephanus Schoeman appointed him as Native Superintendent. His duties

required collecting poll taxes from the Vendas east, northeast and southeast of Schoemansdal. He kept regular notes of all income and proceedings with the black tribes in a book he called "my journal" or "tax book". Unlike the file containing his vice-consular letters, this journal could unfortunately not be found. He had to report on the collected revenue on a quarterly basis, or as often as the president and the members of the Executive Council required him to. When the Executive Council held its sessions at Schoemansdal, Albasini, armed with his instructions and his "journal", had to appear before them. He also sent regular reports to the president and members of the Executive Council in Pretoria. When a chief was unwilling to pay his dues, he reported this to the highest authorities and asked for their advice before acting. The president then usually ordered the field cornet and the commandant of the district to punish the chief in question, in cooperation with Albasini. Sometimes it was left to him to apply punitive measures with his Magwamba regiment.

Albasini's regiment was reserved for military service and was therefore known as "Gouvernementsvolk" (the Government's people). It seems that they were exempted from paying taxes as long as they provided their services from time to time. After a battle they were also entitled to a portion of the spoils. At the first sign of danger Albasini would blow a whistle. The guards at the bulwark's main entrance and the armed men, who also had a reed whistle each, would blow in turn and thus the alarm was raised. His whole army, which consisted of more than 2000 trained warriors, could gather at the bulwark within a short period of time, ready for any eventuality. With the exception of a small minority who could shoot and to whom guns were issued during campaigns, most of the men were only armed with shields and assegais.

The poll taxes were normally paid in the form of cattle, goats, sheep, cash, ivory, picks that the Venda smiths had produced, or copper rods. The movables and picks were sold at a public auction after which the cash and ivory were sent to the government. In this way, Albasini collected a good deal to supplement the meagre coffers of the Republic. What he received in compensation is not known, but it seems as though it consisted of a percentage of the collected goods. The chiefs also brought him gifts when they came to pay their taxes.

War in the Soutpansberg

In 1864 war broke out between the Boers and the Vendas, who lived in inaccessible mountain strongholds. There were several reasons for this conflict. After the death of Venda high chief Ramabulana, his two sons Makhado[3] and Davhana became involved in a battle over the succession to the throne. Davhana had to seek refuge and was placed under Albasini's protection by the Executive Council, which of course made Makhado their enemy.

At the same time, after Umzila (or Muzila) had succeeded his father Manukosi, he demanded the extradition of the fugitive general Monene, threatening to halt the Boers' elephant hunts if they did not comply. Monene was captured and the Executive Council ordered his extradition, but before they could do so he escaped from imprisonment in Schoemansdal. In an attempt to recapture him, Field Cornet J.H. du Plessis and Commandant S.M. Venter attacked Venda chiefs who had not been directly involved in Monene's escape. Albasini, in turn, considered it an opportune moment to attack chiefs who had been unwilling to pay taxes and so force them to observe the rules. As a result the whole district found itself at war. It must be added though that the people of Schoemansdal trusted their black riflemen far too much. In the beginning they accompanied the "swartskuts" (black sharpshooters) on hunting expeditions, but later they simply sent them out on their own. Eventually the "swartskuts" had about 200 rifles, which they refused to return. In the meantime certain pedlars were also illegally selling arms and ammunition to the Vendas.

Albasini, his black regiment and the handful of Soutpansbergers were too few to even try to attack the Venda in their mountain refuges and force them into submission. After April 1865, the inhabitants of the town and district had to move into laagers at different places for their own protection. Albasini gave a number of families shelter inside his bulwark at Goedewensch, while the townspeople found refuge in the fort that had been erected in the town centre.

After several government commissions had visited

---

3   Also called Maghato.

the area to investigate matters, Commandant General Paul Kruger eventually arrived with a commando of 400 burghers. The burgher commando and Albasini attacked one of the hostile chiefs, Chief Katse-Katse or "Katlagter" (Babbler), who lived near the town against some steep cliffs, with several thousand black warriors. As the burghers did not have enough ammunition and President Pretorius could not send promised provisions in time, the offensive had to be called off. Schoemansdal was evacuated on 15 July 1867 and its inhabitants moved south and formed laagers about 128 kilometres away at Marabastad and Kalkbank.

Only João Albasini and about 20 less well-off people, who did not have any means of transport, stayed behind at Goedewensch. As a result the district was not completely abandoned by whites. Tradition has it that 200 of Albasini's men were at the fort that night to protect their white chief.

Mrs Biccard, who had been 11 at the time, often told of the indescribable suffering at the farm at the time. Her father had to feed everybody and when their food ran out they had to slaughter some of the best milk cows to survive. When the men ran out of clothing, pants were made for them from blankets, while her mother unpicked the linings of her dresses to give to the women to wear.

Stephanus Schoeman at Soutpansberg
In an attempt to reclaim the burnt-down Schoemansdal, which had been named after him, Stephanus Schoeman arrived with 42 volunteers. They, together with Albasini and his warriors and about 30 Soutpansbergers, attacked several of the weaker tribal chiefs. However, they did not have enough manpower to attempt attacks on Katse-Katse and Makhado in their mountain strongholds. Schoeman was therefore forced to return without having achieved his purpose.

A number of burghers submitted a petition to the Executive Council after the evacuation of Schoemansdal. In it, Albasini was accused of being more an obstacle than a help since the beginning of the campaign against the enemy in the Soutpansberg. They claimed that he never tried to encourage or establish peace with the force he controlled. His poor support of the volunteers was, according to the signatories, proof of this. They claimed that with such a man in their midst they could not possibly expect the blessing of the Almighty in their attempts to safeguard the district. They requested that Albasini be stripped of his power and removed from the area.

It is not known who the petitioners were or how many there were as the original document could not be found in the Transvaal Archives. The petition was published in *De Transvaalsche Argus* – a newspaper of the time – and only a C.B. Janson was mentioned as leader of the petitioners. Nobody in the Soutpansberg knew who this man was.

Although the complaints against Albasini could not be proven, he was discharged as "Superintendent of Native Affairs" and Stephanus Schoeman was appointed as diplomatic agent in Soutpansberg to rule the Magwamba. With the aid of Zulu Chief Umzila's men he was to fight the Venda and force them into submission. But the government had lost sight of the fact that Albasini's followers were bound to him by loyalty and would not be subjected to anybody else's leadership. The result was dissension among the Magwamba and severe conflict between Albasini and Schoeman. Schoeman made scathing attacks on Albasini in his reports, and the latter was eventually summoned to Pretoria to justify his actions. The Executive Council found that Schoeman's accusations were too vague and without any substantial evidence and therefore did not prosecute the former superintendent. Albasini emerged the victor in this battle.

Umzila, who had to help Schoeman, sent about 20 000 Zulu warriors to the Soutpansberg. They attacked Tengwe and Lwamondo, two of the lesser Venda chiefs, but did not even attempt to engage Katse-Katse and Makhado, with the result that they had to withdraw without any success worth mentioning.

Paul Kruger intervenes again
At the end of 1869 Commandant General Paul Kruger was again sent to the Soutpansberg. At Goedewensch he made peace with all of the high chiefs, with the exception of Makhado and Katse-Katse. He also rehabilitated Albasini and reinstated him as legal chief of the Magwamba. This mission more or less restored peace in the district. Although the two most important enemies had not joined the

Col Adolf Schiel, an officer in the State Artillery and Native Commissioner in Soutpansberg during the Venda uprisings which took place between 1889 and 1896. (ACKNOWLEDGEMENT: TRANSVAAL ARCHIVES)

peace negotiations, they did not cause any further trouble.

Borders and colonies
Albasini's greatest aspiration was that the border between the ZAR and the Portuguese territory at the coast be demarcated and he never failed to emphasise the importance of such a step in his correspondence. The matter dragged on for a long time, however, and only in 1869, ten years after he had first brought it up, the Portuguese Consul General Alfredo du Prat was sent to the Republic to discuss the matter with the Boers. He entered into a treaty in the absence of Albasini and without him even knowing about it. The Vice Consul obviously felt very insulted. In his opinion, the Portuguese had been entitled to a much larger area than what had been determined in the treaty.

But what bothered him most was the fact that one of his dreams had been shattered. He had wanted to establish a Portuguese colony in the interior that would be part of the Portuguese territory. The area that he considered most suitable for such a colony was the land that he had traded from Chief Magashula along the Sabie River, thus a large part of what is now the Kruger National Park. On 8 April 1868, before Du Prat's arrival, he had donated this piece of land to the Portuguese government and submitted proposals for the establishment of a colony to the Governor General of Mozambique. The colony was to be called "São Luiz" in honour of the Portuguese king. This is why Magashulaskraal is sometimes indicated as "São Luiz" on old maps. His suggestions were accepted and he was named acting head of the colony in the "Boletim oficial do Governo General da Provincia Mozambique". There were also petitions from Albasini's former clerk, the Goanese-Portuguese Jacob de Couto, as well as from Stephanus Schoeman's secretary, C. von Ludwig, to incorporate Soutpansberg and Lydenburg into the Portuguese domain. Albasini did not support these proposals, however, as he believed such steps would harm the good relations between the two governments.

The date on which Albasini was to move from Soutpansberg to the new Portuguese colony had already been determined when he heard, to his great disappointment, that Magashulaskraal fell in the ZAR, according to the treaty. The Executive Council informed Albasini that they only knew Magashula as a lesser chief and not as a high chief with his own territory. The intended area had been bought years before by the Republic from the legal and acknowledged high chief of the tribe of Sobhuza I, and as such they could not acknowledge any claim by the Portuguese Government to it and considered it to be part of the Republic and to fall within the proclaimed border. This was a disappointment that Albasini could not easily overcome. On 15 August 1870 he wrote an embittered letter to the Governor General of Mozambique: "... and all that had been left to me, was the displeasure to see that all my work, expenses and attempts on behalf of the Kingdom of Portugal over a period of 12 years was in vain. Despite all my efforts, I see that, according to the honourable Mr A. du Prat's treaty, almost all the territory that I believe legally belongs to the King of Portugal has been allocated to the Afrikaans Republic, including the new colony of São Luiz. Thus the harbour of Lourenço Marques has been cut off, because what is the use of the harbour, or the Portuguese settlement Lourenço Marques at the coast, while the interior areas are in the possession of another state?" (Freely translated from the Portuguese.)

Poverty and ill-health
After all the disasters that had befallen him, the once wealthy João Albasini was an impoverished and broken man. The war of 1865 and five years of taking care of a laager on his farm had ruined him financially. He ended up leading a precarious existence without receiving any compensation for the losses he had incurred by attempting to safeguard Soutpansberg for civilisation. His own government also treated him disgracefully.

In an attempt to make ends meet, Albasini and his family moved to the Kimberley diamond fields in 1875 where he recruited labour for the Cape railway and the diamond diggings. He even spent a few years on the farm Kraalkop in the Gatsrand (Back of Beyond) before returning to the Soutpansberg in 1877 – poorer than ever.

Later, after whites had slowly started to occupy the district again, he served the government in the capacity of resident justice of peace, native commissioner and district council member. This provided him with a meagre income. Elephant hunting was not an option any more as the herds had been depleted and the survivors driven away. A few years before his death he suffered a stroke, but when Dr Theodor Wangemann, a seasoned traveller who was an intelligent and keen observer, visited him in 1884, he was on his feet again. Wangemann wrote the following about him and his neat home:

> He acquired so much authority among them (the Magwambas) that thousands react to his command. Now he is an old man. With his long, grey beard, broad-brimmed hat with ostrich feathers on, he appeared picturesque. The Boers appointed him for some time as area commissioner of the northern district and he was succeeded by his son, who, however, does not hold the same measure of authority over the Magwamba. The old man is still feared far and wide by the natives, because wherever he and his death-defying Magwambas arrive there is weeping and moaning.
>
> He organised his home like a nobleman's castle and a citadel. It is surrounded by a stone wall with a kind of bastion on the one side, which would easily withstand any attack by the natives.
>
> Albasini is getting old now, but it looks as if he has at least recovered in spirit after a period of illness. He spoke clearly without mumbling. His words and features depicted benevolence, not revealing anything at all about the strict chief of the natives. His wife provided the Swiss and our own missionaries with valuable assistance. The old man took us to his well-tended garden and coffee plantation. Everything runs smoothly as he has an unlimited native workforce. Lanes of trees lead to the house and provide shade in the garden. A place that has been so densely planted with trees that no ray of sunshine can get through serves as his court of justice (tribunal). He showed us two lines on the ground that separate the two parties, the witnesses and the judges, and which may not be crossed on account of a heavy penalty. He hears the cases, listens to the witnesses, allows the judges to give their opinions and then passes the sentence against which no appeal is allowed.
>
> After a short farewell from his son who lives next door in a house and currently serves as district's commissioner also over Modjadji's area, we said our goodbyes to reach Valdézia, the oldest Swiss missionary station.

About three years after Wangemann's visit, Albasini suffered another stroke. This confined him to his bed for more than a year, until his death on 10 July 1888.

João Albasini's merits

Although he remained loyal to his own government, there is no evidence in any of his letters that he was a spy or in any way disloyal towards the Republic's authorities. He was, however, an astute diplomat, as the following incident illustrates: in August 1860, three months before Stephanus Schoeman was elected president of the ZAR in the Orange Free State in the absence of M.W. Pretorius, Albasini found himself in a very difficult position. Schoeman asked him to buy six cannons for the Republic through the Portuguese harbours. Albasini pretended to be very eager to help. He addressed letters with the same contents to both the Governor General of Mozambique and the governor of Lourenço Marques, and included translations of Schoeman's request. He championed the cause and said he did not doubt

João Albasini's grave in the family graveyard on his farm Goedewensch, just below the wall of Albasini Dam. He died on 10 July 1888.
(08.12.1985)

The grave of João Augusto Albasini II, the first João Albasini's grandson, who was buried with his wife, Susanna Maria Elizabeth, in the family graveyard on Goedewensch. He was the son of Antonio Augusto Albasini, a son of João I, who had been buried elsewhere.
(09.12.1985)

The grave of João (Joe) Albasini III, in front of his great grandfather's grave in the family cemetery on Goedewensch. Joe Albasini was a formidable hunter and in service of the Department of Native Affairs at Sibasa.
(09.12.1985)

that the Republic needed the cannons for use against hostile black tribes, especially to remove them from their mountain strongholds.

After writing the letters he called the Portuguese inhabitants of Soutpansberg together and discussed the matter with them. Taking into account the precarious political state of the country, they decided that it would not be wise to supply the cannons as Schoeman could easily use them against his compatriots. They also expected a clash between the Boers and the British. If the British were to find out that the Boers were fighting with Portuguese cannons, it could lead to an unnecessary misunderstanding between Great Britain and the Portuguese government. A letter containing this decision, completely contradicting the first letter, was marked "confidential" and sent along with the first one to the Governor General only.

Albasini was well aware of the fact that all of Lourenço Marques's official correspondence had to go through the Governor General. He left the governor of Lourenço Marques under the impression that he had to do his best to obtain the cannons. A while later the governor replied that it would be impossible to grant the request as he needed three cannons himself against Mawewe (Manukosi's son). He would, however, contact the Governor General immediately.

But Albasini's confidential letter was already in Mozambique. Early the next year the Governor General asked him to inform Schoeman that it would not be possible to comply with his request. For this bit of diplomacy Albasini was later called a deceiver by Schoeman's biographer in his book *Stormvoël van die Noorde* (Hurricane Bird of the North). But Albasini knew Schoeman well and this would have made it difficult for him to act in any other way.

The honour of being the first Portuguese trader to do business with the people of Ohrigstad and who helped them to find a passable way to the sea belongs to Albasini. He identified with the Boers and their fate by marrying a Voortrekker girl. His long procession of native bearers along the old trade routes into the interior was the forerunner of the Delagoa Bay railway line. Of this he had a clear vision in the form of a road on which steam-powered wagons travelled. In the tales of his followers and their descendants he has taken on a heroic stature with a place of honour in their hearts. And as the secrets of the past are unlocked and the scrolls of history are unfolded, so this staunch foreigner will also grow in stature among the white community of South Africa as the man who had shown their forefathers the way to the sea. What he had envisioned was attempted by President T.F. Burgers, but only Paul Kruger, backed by Johannesburg's millions, could open the way to the sea. It is no coincidence that electric trains between Delagoa Bay (Maputo) and the interior nowadays run only about 200 metres west of his former home and trading post at Magashulaskraal in the Kruger National Park. This is the realisation of the practical dream of a great thinker and idealist.

As indicated before, Albasini also carried out pioneering work in the field of agriculture. On his model farm, Goedewensch, he planted vegetables, grain and a variety of fruit trees. He also planted coffee seeds that had been personally sent to him by President M.W. Pretorius – a first for the Soutpansberg. He harvested enough for own use as well as for that of the neighbours. Today the large coffee plantations nearby at the foot of Mashao Hill and the Albasini Dam in the Luvuvhu River on his former farm are continuations of his pioneering work.

João Albasini did not leave many earthy possessions behind. It was not really necessary, as he left his descendants the fertile Soutpansberg district for which he had surrendered both goods and blood. His remains rest at the place where his rondavel office used to be and the graveyard has been declared a National Monument. Together with that of Louis Trichardt, Andries Hendrik Potgieter, his son Piet, Stephanus Schoeman and other staunch Boer leaders of the North, João Albasini's name will live on. He was possibly the first European to have lived in the Kruger National Park and a pioneer of the eastern Lowveld and the Soutpansberg.

References

Albasini, João. 1858–1872. *Correspondencia Oficial do Consulado Portuguez em a Republica Africana do Sul Transvaalsche*. (Official correspondence of the Portuguese Consulate in the South African Republic).

Albasini, João III. 1982. *João Albasini (1813–1888)*. Published by the Albasini family.

Anon.1925. Juwawa the white chieftain. *The Star*, 20 & 27 June.

Anon. 1948. The historical associations of Sabie: The Long Tom Pass over the Drakensberg. Pamphlet Number 14. Sabie Forestry Museum.

Barnes, F.C. 1965. *Cartridges of the World*. Follet Publishing Company, Chicago.

Beck, H. 1928. Classification and nomenclature of beads and pendants. *Archaeologia LXXXIX*, Oxford.

Breytenbach, J.H. & Pretorius, H.S. Suid-Afrikaanse Argiefstukke, Transvaal No. 1. Notule van die Volksraad van die Suid-Afrikaanse Republiek I, 1844–1850. *Cape Times*, Cape Town.

Breytenbach, J.H. Suid-Afrikaanse Argiefstukke, Transvaal No. 2. Notule van die Volksraad van die Suid-Afrikaanse Republiek II, 1851–1853. *Cape Times*, Parow.

Breytenbach, J.H. Suid-Afrikaanse Argiefstukke, Transvaal No. 3. Notule van die Volksraad van die Suid-Afrikaanse Republiek III, 1854–1858. *Cape Times*, Parow.

Breytenbach, J.H. Suid-Afrikaanse Argiefstukke, Transvaal No. 4. Notule van die Volksraad van die Suid-Afrikaanse Republiek IV, 1859–1863. *Cape Times*, Parow.

Breytenbach, J.H. Suid-Afrikaanse Argiefstukke, Transvaal No. 5. Notule van die Volksraad van die Suid-Afrikaanse Republiek V, 1864–1865. *Cape Times*, Parow.

Breytenbach, J.H. Suid-Afrikaanse Argiefstukke, Transvaal No. 6. Notule van die Volksraad van die Suid-Afrikaanse Republiek VI 1866–1867. *Cape Times*, Parow.

Burke, E.E. (ed.) 1968. *The Journals of Carl Mauch 1869–1872*. National Archives of Rhodesia, Salisbury.

Cachet, F. Lion. 1898. *De worstelstryd der Transvalers aan de volk van Nederland verhaald*. Third edition. Hoveker & Wormser, Amsterdam.

Caton-Thompson, G. 1931. *The Zimbabwe Culture*. Clarendon Press, Oxford.

Corpo Notarial Codice no. 5., Arquivo Historico De Mocambique, Lourenço Marques.

Das Neves, D.F. 1879. *A Hunting Expedition to the Transvaal*. George Bell and Sons, London.

De Vaal, J.B. 1948. João Albasini, wit kaptein van die Magwamba. *Die Huisgenoot*, 4 & 18 July.

De Vaal, J.B. 1953. Die rol van João Albasini in die geskiedenis van die Transvaal. *Archives Year Book for South African History* 16:1–155.

De Vaal, J.B. 1982. *João Albasini (1813–1888)*. University of the Witwatersrand Press, Johannesburg.

De Vaal, J.B. 1983. João Albasini (1813–1888). Camoês Annual Lectures No. 3. Ernest Oppenheimer Institute for Portuguese Studies. University of the Witwatersrand, Johannesburg.

De Vaal, J.B. 1984. Ou handelsvoetpaaie en wapaaie in Oos- en Noord-Transvaal. *Contree* 16:5–15.

Dicke, B.H. 1936. *The Bush Speaks*. Shuter & Shooter, Pietermaritzburg.

Dicke, B.H. 1937. Van moord beskuldig. *Die Huisgenoot*, 18 June.

Engelbrecht, S.P. 1920. *Geschiedenis van de Nederduitsch Hervormde Kerk in Zuid-Afrika I*. J.H. de Bussy, Pretoria.

Ferreira, O.J.O. 1978. *Stormvoël van die Noorde*. Makro Boeke, Pretoria.

Fletcher, E. 1976. *A Bottle Collector's Guide*. Latimer New

Dimensions, Ltd., London.

Fouche, L. 1937. *Mapungubwe, Ancient Bantu Civilization on the Limpopo.* Cambridge University Press.

Grandjean, A. 1889. L'invasion des Zoulou dans le sud-est africain. *Bulletin de la Society Neuchtalose de Geographie.*

Grote Nederlandse Larousse Encyclopedie. 1971. Uitgevery Heideland – Orhis N. v. Hasselt.

Jäckel, M. 1949. *Juwwawa.* Verlag Friedrich Reinhardt A.G., Basel.

Joubert, D.C. 1966. Suid-Afrikaanse Argiefstukke, Transvaal No. 7. Notule van die Volksraad van die Suid-Afrikaanse Republiek VII, 1807–1868. Cape & Transvaal Printers, Cape Town.

Klein, H. 1952. *Land of the Silver Mist.* Howard B. Timmins, Cape Town.

Krynauw, D.W. & Pretorius, H.S. 1949. Transvaalse Argiefstukke, Staatsekretaris, inkomende stukke, 1850–1853. Minerva Printers, Pretoria.

Lombaard, B.V. 1969. Herkoms van die naam Pretoriuskop. *Koedoe* 12:53–57.

Lowe, C. Van Riet 1937. Beads of the water. *Bantu Studies*: 367–372.

Lowe, C. Van Riet 1955. Die glaskrale van Mapungubwe. *Archeological Series* 9. Pretoria.

Martins, F. 1957. *João Albasini ea Colónia de St. Luis.* Agéncia Geral do Ultramar. Divisão de Publicações e Biblioteca, Lisbon.

Paynter, D. & Nussey, W. 1986. *Kruger – Portrait of a National Park.* Macmillan S.A., Johannesburg.

Pienaar, U. de V. & Mostert, M.C. 1983. Op die spoor van die ou transportryers deur die Krugerwildtuin. *Custos* 12(9):27–32 & *Custos* 12(10):12–19.

Potgieter, C. & Theunissen, N.H. 1938. *Kommandant-generaal Hendrik Potgieter.* Afrikaanse Pers, Johannesburg.

Preller, G.S. 1917. *Dagboek van Louis Trichardt.* Het Volksblad Drukkerij, Bloemfontein.

Preller, G.S. 1920. *Voortrekkermense II.* Nasionale Pers, Cape Town.

Preller, G.S. 1940. Stigting en verwoesting van Schoemansdal. *Die Vaderland*, 13 December.

Pretorius, H.S. & Kruger, D.W. 1937. Voortrekker-Argiefstukke 1829–1849. Government Printer, Pretoria.

Punt, W.H.J. 1953. *Louis Trichardt se laaste sk*of. Van Schaik, Pretoria.

Punt, W.H.J. 1962. 'n Beknopte oorsig van die historiese navorsing in die Nasionale Krugerwildtuin. *Koedoe* 5:123–128.

Schofield, J.F. 1938. A preliminary study of the prehistoric beads of the northern Transvaal and Natal. *Transactions of the Royal Society* of *South Africa* 26(4).

Stevenson-Hamilton, J. 1925. Annual Report. (Internal Report to the National Parks Board.)

Stevenson-Hamilton, J. 1933. *The Kruger National Park.* Pretoria.

Toscano, F. & Quintnha, J. 1930. *A derrocada do império Vátua e Mousinho de Albuquerque.* Vol. 1&2. Casa Editora Nunes de Carvalho, Lisbon.

Transvaal Archives, Pretoria. Acquisition A. 81. Description of Pater Joaquim de Santa Rita Montanha's journey to Soutpansberg.

Van der Sleen, W.G.N. 1967. *A Handbook on Beads.* Musée du Verre, Liége.

Wangemann, D. 1886. Ein zweites Reisejahr in Südafrika. Verlag des Missionshauses, Berlin. (According to the *Suid-Afrikaanse Biografiese Woordeboek*, Part I, p. 900, Wangemann's initials were H.T.).

Wolhuter, H. 1948. *Memories of a Game Ranger.* Central New Agency, Johannesburg.

Yates, C.A. 1944. *Juwawa.* In: *Varia.* Unpublished manuscript, Skukuza Archives.

Chapter 6
The bygone days of the Soutpansberg
6.1 Schoemansdal, the former Zoutpansbergdorp (Soutpansberg town) (1848–1867)
Dr J.B. de Vaal

At the end of June 1984, the Transvaal Provincial Administration announced to the media its decision to excavate and restore the abandoned Voortrekker town Schoemansdal as a living museum.

The remains of this town are in Mpumalanga, about 17 kilometres west of Louis Trichardt (Makhado) and four kilometres south of the Soutpansberg Mountains. This was the fourth and last town founded by the bold "Trekker of the North", Andries Hendrik Potgieter.

Before excavations began there was not much to indicate that it had been the site of a prospering Voortrekker town between 1848 and 1867. A neatly plastered, rectangular brick wall, built with funds collected through penny-laying ceremonies in schools across the country, protects a number of Voortrekker graves. One of these graves is that of Commandant General A.H. Potgieter. It is marked by a square, upright granite stone erected in 1938 by the Central Burgher Grave Committee and the Schoemansdal Dingane's Day Festival Committee.

Interestingly, when the former government reburied the remains of Commandant General A.W.J. Pretorius and President T.F. Burgers in the Church Street cemetery in Pretoria, Potgieter's son who lived in Grootspelonken blankly refused that the same be done with his father's remains. Instead, the government erected a monument in the hero's acre of the cemetery in his honour. Hermanus Potgieter, the general's father, was also buried in the cemetery at Schoemansdal. He lived to the ripe old age of 105 and had apparently been in dire financial straits in his last years.

In 1948, during Schoemansdal's centenary celebrations, Reverend T.F. Dreyer of Potchefstroom unveiled a new stone on the grave of Mrs Josina van Warmelo, erected by the church council of the Dutch Reformed Church of Louis Trichardt (Makhado). With a few exceptions, the rest of the graves consist mainly of small, disorderly heaps of stone that do not indicate the real positions of the graves. The late W.J. Grobler (Uncle Koos), a former inhabitant of Schoemansdal, identified some of the graves for Dr Gustav Preller. A map of the cemetery was published in *Die Brandwag* of 24 December 1919. Koos Grobler was born on 11 June 1843 on the farm Koesterfontein in the Rustenburg district. As a little boy of five he and his parents, the pious Douw Gerbrand Grobler and his wife, moved to the Soutpansberg, where he grew up. He and his second wife, Annie, lived on their farm Groblersplaats about six kilometres east of Louis Trichardt (Makhado), where they both eventually died.

About a hundred metres from the cemetery a small dry creek meanders in a westerly direction. This used

Cmdt Gen. Andries Hendrik Potgieter, the founder of Zoutpansbergdorp (later Schoemansdal) in 1848.
(ACKNOWLEDGEMENT: AFRIKAANSE PERS LTD AND THE TRANSVAAL PROVINCIAL MUSEUM SERVICE)

169

The grave of Mrs Josina van Warmelo, the wife of Rev. N.J. van Warmelo, the permanent minister at Schoemansdal between 1864 and 1867. She passed away on 28 January 1865 and was laid to rest in the cemetery at Schoemansdal.
(09.12.1985)

The late H.J. Grobler (Uncle Koos) and Aunt Annie, his second wife (née Venter). Koos Grobler had been a boy of five when he and his parents moved with Gen. Potgieter to Soutpansberg in 1848. He died on his farm Groblersplaats near Louis Trichardt (Makhado) in 1937 and his wife died a few years later. He provided valuable information about Schoemansdal to Dr G. Preller and later to Dr Punt.
(ACKNOWLEDGEMENT: DR J.B. DE VAAL)

to be the Dorp's River. On its northern bank is a ditch that used to be a canal with a fertile piece of land below. This was one of two irrigation canals that had been dug by the people of Schoemansdal. The other one started in the mountain on the farm Versamelhoek and ran past the upper part of the town.

Several rusty nails, beautifully painted willow pattern porcelain sherds, and sanna (flintlock) bullets have been found on the terrain. One of the bullets had a stone inside to preserve lead. Other finds included a small three-toothed bone-handled fork, a bullet mould, a damaged soapstone pipe and glass perfume bottle stoppers. Systematic excavation would certainly produce even more surprising finds.

Reasons for the founding of Schoemansdal
How did it happen that this town was founded, only to be abandoned later? This is a long and interesting story that can only be related here in broad strokes.

In May 1836 the Voortrekker Louis Trichardt and his party had already set up camp at Soutpan (salt pan) along the western foothills of the Soutpansberg range, where they stayed for three months until about 23 August. On 23 June an 11-man patrol under Commandant General A.H. Potgieter arrived at Trichardt's laager en route from Sand River in the Free State. Potgieter's party included the chronicler of their expedition, J.G.S. Bronkhorst. Three days later Potgieter and his party left the Trichardt camp to explore the lay of the land towards Sofala. After they had ridden 16 shifts, they returned to the Soutpansberg following a shorter route. In the last week of July they were back in Trichardt's camp and during that time the two leaders selected a suitable area to lay out a town. This was the site where Schoemansdal would later be built.

On 17 August Potgieter, Bronkhorst and the rest of their patrol left for the Free State to fetch their trek. In September Trichardt moved to his second encampment in the Soutpansberg, which was where he and Potgieter intended to found a town.

Here they waited in vain for Potgieter and his people to join them. Because of a serious shortage of ammunition and provisions, Trichardt decided to visit Delagoa Bay to stock up and then return. This journey ended in tragedy for him as he lost his wife Martha and most of the other members of his party. He also lost his livestock and eventually his own life.

It was with great disappointment that Potgieter heard on his return to the Free State that uMzilikazi's subjects had murdered some of his friends. Although Trichardt was waiting for him, he did not even consider moving northwards at that stage as the AmaNdebele had to be punished first. The battles against them at the Vaal River, Vegkop, Thaba Mosega and Kapain (also called Marikwa) followed in quick succession. After that the Voortrekkers founded Winburg, from where Potgieter left to provide assistance to the Voortrekkers in Natal against the Zulus. At the end of 1838 they laid out the town Mooirivier, which would become Potchefstroom in 1839. This settlement was too far from Delagoa Bay, however, so they decided to move once more and by the middle of 1845 the whips were cracking again. The trekker wagons steadily moved eastwards to a stand on the Drakensberg escarpment. Potgieter had selected this place the previous year while exploring the way to Delagoa Bay in the hopes of negotiating a trade agreement with the Portuguese governor. Here they founded Andries-Ohrigstad.[1]

1   Dr W.H.J. Punt found the ruins of Commandant General A.H. Potgieter's erstwhile home (1845–1848) on his former farm Strydfontein near Ohrigstad in May 1952.

6.1 SCHOEMANSDAL, THE FORMER ZOUTPANSBERGDORP (SOUTPANSBERG TOWN) (1848–1867)

Discord between Potgieter's followers and those of J.J. Burger soon followed, on top of which they were plagued by malaria, tsetse fly, African horse sickness and the ever-looming presence of the hostile high chief Manukosi. The Old Wagon Route down the Drakensberg and across the Lebombo Mountains was a difficult one with too few watering places along it. All of these factors combined, convinced Potgieter that Ohrigstad was not the most suitable place from where Delagoa Bay could be reached and therefore he fixed his gaze on the north, on Inhambane. This port was situated more or less at the same latitude as the Soutpansberg, only slightly further south. In the middle of June 1847 Potgieter left Ohrigstad at the head of a commando of 238 men, travelling to Inhambane via the Soutpansberg in search of a new home and hoping to find an easier, safer route to the coast where they could trade. Only a few days' journey from Inhambane they had to retreat, however, as a large band of Manukosi's warriors blocked their way. The commando had been divided into three for the search, and one of these sections was not able to fend off an attack by a group that outnumbered them. About 60 blacks who had accompanied the reconnaissance were murdered. Back at Ohrigstad, Potgieter decided to follow his dream of 12 years earlier and move to the Soutpansberg.

The Soutpansberg was not unknown to most of the new arrivals as many of them had already undertaken hunting expeditions to the area even before the trek. Jan Valentyn Botha and Piet Potgieter, the commandant's son, were among those who had been on a hunt to the Blouberg area in 1846. During the same year Potgieter had sent two blacks from Ohrigstad to inform people in the area, including Venda Chief Rasethau, of their intended move. In 1847 Johannes Janse van Rensburg returned from an expedition in search of a port that he had led as commandant. He visited several chiefs in the Soutpansberg and received cattle as gifts from chiefs Maraba and Mashaba. He also received elephant tusks and six whips from Venda Chief Ramabulana.

The move to the Soutpansberg

The route from Ohrigstad to the Soutpansberg led to the Steelpoort River, where it turned south and went through Steelpoort Drift, across Magneets-

[top] Sanna bullets, a bullet mould, a three-pronged bone-handled fork, two glass stoppers and a soapstone pipe with part of its stem, which were picked up at the site where Schoemansdal had been. Next to the bullet mould is a paring knife which was picked up in the Makapane's Caves and which had probably also belonged to one of the Voortrekkers.
(ACKNOWLEDGEMENT: DR J.B. DE VAAL)

[bottom] Group photograph of the Burger brothers, circa 1870. At the back from left to right: Johannes Jacobus Burger (mine commissioner, Krugersdorp) and Frederik Burger (farmer from Heidelberg). In front from left to right: Willem Francois Burger (farmer), Jacobus Johannes Burger (chairperson of the Second Volksraad, Lydenburg) and Schalk Willem Burger (later vice president of the ZAR).
(ACKNOWLEDGEMENT: TRANSVAAL ARCHIVES)

hoogte and via Ramakok's Kraal along the Gompies River. While they were in this area, Potgieter received a message that he had to meet his co-leader Pretorius in Potchefstroom. Field Cornet Jan Valentyn Botha was tasked with leading the trek through Strydpoort and then via Marabastad and Doringbult, along the Sand River down to the later farm Klipdam, through Rhenosterspoort, via the Dwars River and past Mashaba's kraal to the site they had selected 12 years earlier. The trek arrived on 3 May 1848, which can be considered the founding date of the town.

At various places along the route some people stayed behind with their families. Gert Koekemoer and P.J.L. Venter stayed at Doringbult; Thobias Legrange, his son Jan, Hendrik Geyser and some others remained at Klipdam; Pieter du Preez and Willem Marais stayed at Weltevreden; and P.P. Hugo, who later became field cornet, stopped at Rietpol. The Zulus in Natal had murdered Pieter du Preez's wife and seven children, while P.P. Hugo's father had lost his life alongside Piet Retief. Pieter du Preez later married the late Hugo's widow and died in 1890 at the age of 83 at Weltevreden, where he was buried. Jan Bosch settled at Rooiwal, where the Berlin Lutheran Mission was later established, while Theunis Botha went to live on Kalkfontein where the Solomondale railway station is today. B.J. Vorster, F. Snyman and others settled even further north.

The ruins of several Voortrekker homes were found in 1935 on the northern bank of the Sand River at Klipdam, as was an upright section of the stone wall of a fort that had measured 40 by 40 square metres. About 200 paces north of the old fortifications was a

Section of a clay wall which had been part of a Voortrekker fort at Klipdam which could still be seen in 1935.
(WITH KIND PERMISSION: DR J.B. DE VAAL)

A view of the Voortrekker cemetery at Klipdam, with Rhenoster Poort hills in the background.
(ACKNOWLEDGEMENT: DR J.B. DE VAAL)

cemetery with more than 40 discernible graves – an indication of the price the pioneers of the North had to pay to make the malaria-infested area habitable.

Some of the graves had slate headstones, each telling a heart-rending story of its own. The couple G.L. and A.J. van Emmenes buried three children there over a period of seven years. Hendrik Josephis was only eight days old when he died on 7 May 1860. A little girl, Aletta Magdalena, was six months old when she passed away on 16 December 1861 and another baby boy, Lourens, was born on 14 December 1866 and died four months later on 13 April 1867. Emmi Adrijana, the daughter of A.P. and E.A. Botha, died on 26 June 1863 at the age of 11 months. Another boy, Aberham Org, whose surname could not be deciphered, died at the age of 14 in 1860, while P.A. Eloff, born on 2 November 1824, died on 6 April 1861. He was 36.

Unaware of the fact that the *Anopheles* mosquito breeds in stagnant pools alongside streams and rivers and that mosquitoes are responsible for transmitting malaria parasites to humans, the Voortrekkers usually built their homes close to the water's edge to have water for household use and irrigation. As a result many of them succumbed to malaria, and possibly also bilharzia. The large cemeteries near their settlements offer ample proof of this.

A new beginning
Immediately after their arrival in the Soutpansberg, the trekkers formed a laager with their wagons, which they fortified by stacking thorn tree branches around it. The men then dug a canal from the Dorp River to irrigate the fertile alluvial land on the northern bank. Most people had brought cuttings, plants and seeds so they could establish small gardens. It was also the right time to sow wheat. What they reaped would be trampled on the threshing floor and later Andries Combrink would mill it.

After the initial sowing and planting they started building houses. The walls of the first few were built using poles with reeds in between, after which they were plastered with clay. The roofs consisted of a bottom layer of common reeds and an upper layer of long thatching grass, which grew well in the vicinity. It was clear that by building houses so close to each other, a town would soon emerge.

At first there was a difference of opinion over the town's name. Potgieter was against the idea of naming it after Field Cornet Botha. On the other hand, it could not be named after him, as his name had already been immortalised in both Potchefstroom and Andries-Ohrigstad. Eventually they compromised and named it Zoutpansbergdorp.

Many of the Magaliesbergers under Pretorius were quite keen to move to the Soutpansberg, but they did not feel free to do so because of the discord between Potgieter and Pretorius. In 1849, Potgieter's secretary C.L. Rabé, a cousin of Pretorius, wrote from the Soutpansberg to Pretorius that he had heard that many people would have liked to have come to the Soutpansberg had it not been for the division, as it would have been an escape for them and their livestock. He added that he knew the district well and could assure him that there was ample room to live a good life in a fertile, healthy environment. There was enough space for another 500 families at least. Why should the stubbornness of a few deny the rest of their compatriots the privilege? He said he knew that rumours had been spread about the unhealthiness of the area, but these were untrue as only four people had died so far and all of them had returned home sick from elephant hunts in the lowland bushveld along the Krokodil River (Limpopo), about four days' travel from there. Nobody in the town and its vicinity had contracted the fever as yet. Some had had relapses from the malaria they had contracted at Ohrigstad, but they had all recovered.

But three years later in 1852, when Dr A. Murray and J.H. Neethling paid a pastoral visit to the Soutpansberg to establish a congregation, malaria was rife. All the inhabitants moved to Kerkbult, about 23 kilometres south of town, as it was healthier there. Four people died of the fever on the day of the move, including Engela Kelly, the mother of Tom Kelly, who later became field cornet and whose descendants are still living in the Soutpansberg district. Potgieter's gunpowder chest served as her coffin. At Kerkbult (Church Hill) people attended church, catechism, confirmations, christenings and weddings under cover of buck sails (tarpaulins) for about three weeks. After that they returned home with their ox-wagons.

It had never been the aim of the Soutpansbergers

to isolate themselves from civilisation or to establish their own republic, and nothing like a Soutpansberg republic ever existed. In 1851 the Volksraad (Assembly of the South African Republic) decided to appoint commandant generals to lead the four districts of the Transvaal. Andries Pretorius was appointed in the Potchefstroom district, J.A. Enslin in Marico, Andries Hendrik Potgieter in the Soutpansberg and W.F. Joubert in Lydenburg. Jan Valentyn Botha was the first field cornet of the Soutpansberg, while the town's administration was also the responsibility of the first magistrate, Abraham Duvenage, and other members of the heemraad (country court). The magistrate and heemraad had occasional meetings with the commandant general.

In 1852 Pretorius signed the well-known Sand River Convention with two British commissioners, Owen and Hogge. This would guarantee the Transvalers the freedom to manage their affairs without interference from Her Majesty's government. Potgieter and his people were dead set against any negotiations with the British and wanted no part in it. The treaty nevertheless needed to be signed by the Volksraad at Rustenburg. Both Boer leaders were present and the two adversaries went into a tent by themselves to negotiate. They stayed in the tent for such a long time that the anxious crowd became worried that the two had come to blows, so they went to investigate. To their utmost surprise and joy they found the two sitting opposite each other at a table with an open Bible between them and their hands clasped together. Thus the two former rivals could part as friends before their deaths, which were around the corner.

Potgieter – the man and his death

Potgieter was a born leader who always had the interests of his people at heart. It is said that he never had to raise his voice at any of his officers when they made a mess of things. Instead he set matters right by addressing them in a kindly manner. He was very religious and, in the absence of a permanent minister, led his people during religious services. His State Bible can still be seen at the Cultural History Museum in Pretoria.

According to Reverend J.H. Neethling he was "a tall man with a venerable face, made to be a commandant", but due to his age and some physical discomfort he had a stoop. His skin was fair, his hair and beard dark brown with a red tinge and his eyes blue-gray. Like all Voortrekker men he wore a broad-brimmed straw hat. He had a wagon-making business in Zoutpansbergdorp, but did not work there himself. Instead, Thobias Legrange and Frans Lottering were the wainwrights while his son Andries served as their assistant. They also made and repaired guns in the workshop.

Potgieter was young at heart and loved children. Every New Year all the town's children gathered in his workshop. There they feasted on delicacies that the townswomen had been baking for days on end. During the festivities the children sang and he played games with them.

Potgieter planned to depart for Makapaan's Poort on 10 August 1852 to found his fifth town, Vredeburg, in honour of the peace he and Pretorius had made at Rustenburg in March. People from both the Magaliesberg and Potchefstroom had promised him that they would move there, but the hostile Pedi high chief Sekwati thwarted this mission, as he now had

This memorial was erected in honour of Voortrekker leader Jacobus Johannes Burger and 50 other Voortrekkers, whose remains had been exhumed in 1942 and reburied on 10 October 1942 in the cemetery at Ohrigstad.
(06.09.1986)

6.1 SCHOEMANSDAL, THE FORMER ZOUTPANSBERGDORP (SOUTPANSBERG TOWN) (1848–1867)

to lead a commando against Sekwati. In his report about the campaign, Potgieter wrote that he became so ill that the war council decided that he should go home, and since his arrival home on 24 September he had been bedridden.

Potgieter died on 16 December 1852 after a hard and dangerous life. The so-called Elephant of the Voortrekkers was laid to rest in the cemetery at Zoutpansbergdorp.

P.J. Potgieter's short-lived rule

Pieter Johannes Potgieter succeeded his father as commandant general when he was only 30 years old. When he took over as leader, there were only two elephant tusks in the public coffers of the Soutpansberg, which he exchanged for 45 pounds (20.45 kg) of gunpowder. During those years large-scale trade in ammunition and rifles with the blacks was common. Some traders known for this practice were Joseph McCabe, Samuel Edwards, Green and Edward Martins.

It was soon clear they were preparing for war against the new settlers. In August 1853 Sekwati's people fired at João Albasini's wagons while he was on his way from Lydenburg, near the former farm Dorenvonteyn of Field Cornet J.V. Botha along the Steelpoort River. When Albasini arrived at Magneetshoogte he sent a large woollen blanket, about four pounds (1.82 kg) of beads and a butcher's knife as a token of peace to Sekwati. These gifts were accompanied with the message that he was a traveller and trader with no ill intentions. Two days later, Sekwati sent his son Tuan and two other tribesmen to Albasini to tell him that he had been wise to reveal the purpose of his journey, as he had given instructions to his subjects to attack any wagons they saw and kill the people.

Shortly after his arrival in Zoutpansbergdorp, where he settled as a trader, a group of people planned to travel to Swartruggens. Albasini warned them not to go under any circumstances and related his narrow escape. His warning fell on deaf ears. The party that set off consisted of Jan Breed, his wife Maria and their three children; Willem Prinsloo with his wife Nellie and their three children; Lourens Bronkhorst; and Flip du Preez, who helped with the wagons. In the Waterberg they pitched camp at a drift in the Nyl River below Mokopane's[2] kraal. Prinsloo

The grave of Cmdt Gen. A.H. Potgieter in the old graveyard at Schoemansdal. He was the founder of this forgotten Voortrekker town.
(09.12.1985)

The monument at Moorddrif, south of Potgietersrust (Mokopane), which was erected in memory of the 28 victims of the Makapaan and Mapela murders in 1853.
(WITH KIND PERMISSION: DR J.B. DE VAAL)

2   An Ndebele chief of the Kekana tribe whose name is usually given as Makapan, Makapane or Makapaan in older references.

175

and Du Preez went to pay Mokopane a visit and were murdered. After that, the rest of the party at the drift was also attacked and cruelly killed. The women were first tied to wagon wheels to witness the slaughter of their husbands and children, and then they were tortured to death. They were decapitated and their heads were impaled on poles lined up in front of the kraal. On the same day Hermanus Potgieter, younger brother of the Commandant General and a rough and unruly character, was killed along with all his people at Chief Mankopane's[3] kraal. A San boy who had been captured but later managed to escape, witnessed how Potgieter was flayed alive, so that his skin could be used as a kaross (skin rug or blanket) by the chief. Willem and Albert Venter were also tortured to death. On one day 28 people lost their lives. From that day on the drift where the murders took place was known as Moorddrif (the ford of murder).

Commandants General Piet Potgieter and M.W. Pretorius rallied the burghers and led a commando against Mokopane. The tribesmen had retreated into a network of caves[4] near the current Mokopane (formerly Potgietersrus). The Boers soon tracked them down, however, and besieged the caves. On 6 November 1854, while standing in the mouth of the cave, Piet Potgieter was shot and killed by a sniper inside. Fearing that his body would be taken and mutilated by Mokopane's people, the brave young Paul Kruger dragged it away with the help of Albasini's chieftain, Manungu. Piet Potgieter was buried on his farm Middelfontein in the Waterberg district, but was later reburied at Potgietersrus, the town named after him.

Stephanus Schoeman succeeds Potgieter

On 19 February 1855, Stephanus Schoeman was elected by the inhabitants of the Soutpansberg and Rhenoster Poort as commandant general during a meeting held in Zoutpansbergdorp. He was sworn in on 1 June by the Volksraad. Schoeman was an excellent organiser, but also rash, short of temper and ambitious. To shore up his position he became engaged to Potgieter's widow Elsje-Maria in June 1855. Immediately after his election he changed the town's name to Schoemansdal and made some improvements. The townspeople had started building a fort as a refuge in times of danger. It measured 50 by 50 square paces and even had bastions to hold cannons. Schoeman had this job, which had started before his election, completed.

By 1855 there was only a water canal in the lower part of town and people living in the upper part of town had no water for irrigation. Under the leadership of Schoeman they decided to dig a second canal from the farm Versamelhoek in the mountain to bring water as close as possible to the fort. The commandant and field cornet were ordered to assist Magistrate A.C. Duvenage and each inhabitant had to provide two black labourers. When the canal was about a kilometre above town, its continued run was determined with water in it.

All the stands now had water and, according to former inhabitants, all the canals along the streets had been finished with brick and lime, which must have afforded a neat and pleasing display. Nicholas Snyman was elected as water warden and was paid ten rix-dollars a month, payable by the inhabitants every three months. Later Pieter Johannes Eloff was elected to the post with payment of one hundred rix-dollars a year. After him J. Haagen, a former clerk of the court, was appointed as waterfiskaal (water bailiff), scout (jailer) and market-master. He also ran the local pound.

Schoeman also had the town surveyed into small erven measuring 70 by 30 paces. Each block consisted of four stands. Inhabitants were given three months from a set date to start building on their erven and the magistrate had to be present to indicate the corner pegs. They also built a jail measuring 3.66 metres wide by 6.10 metres long, with walls two bricks wide and 3.66 metres high. Another decision was to erect a public market where products coming from the interior could be sold, with a public office as close as possible to the market. On 26 June 1855, Magistrate Duvenage could let Schoeman, who was in the Waterberg at the time, know that the town looked

---

3 An Ndebele chief of the Langa tribe to whom many Boers wrongly referred as Mapela, which was the name of his grandfather. During the 1850s he terrorised the whites and harassed the tribes who lived in peace with the Boers. Commandant J.H. Jacobs sent a patrol of 34 men under Field Cornet J.G. Duvenage in December 1855 against him. The commando killed 134 of Mankopane's followers and confiscated large numbers of livestock before returning on 16 December 1855.

4 These caves later became known as Makapan's Caves and are now an International Heritage Site.

totally different from when he had last seen it as most of the erven had been fenced in and built on.

Between 6 July 1855 and 23 June 1856, the Roman Catholic priest mentioned before, Joaquim de Santa Rita Montanha, visited the Soutpansberg. He came from Inhambane at the request of Commandant General Piet Potgieter to negotiate trade relations with the Boers. The priest kept a journal that he updated regularly and this document became a valuable source of information about the town and its people. In the journal he had the following to say about Schoemansdal:

The streets are very long and wide,[5] maybe six fathoms [10.8 m] wide, and run from east to west. By the time I left, there were six of these. The cross streets, of which I think there were also six, ran from north to south and were also very broad. There is running water in furrows or open canals along all the streets, so that everybody has water right in front of their houses. The water comes from a great distance, from the river in the mountain, and runs along all the streets. In some places there are bridges to cross the canals.

I must also say that all the streets are straight and all the inhabitants are obliged to fence their stands. Pigs are not allowed in the streets unless for a very good reason. Nobody is allowed to throw their rubbish on the street. [It is not known what the names of the streets were. One of the townsmen, Jasper Aitchison, gave his address as Kerkstraat (Church Street) 133].

The houses are mostly small, low and unattractive; some have straw roofs; some have walls of dagha [clay], others of wood and dagha, while the more modern ones have brick walls. Some of the houses do not have interior walls and only the place where the bed stands is closed off with a curtain. A few of the houses have glass windows. Everybody building new houses is putting in glass window-panes and interior walls. Almost all the windows open to the outside, while the doors open to the inside. In some poor people's houses, or I do not know what they use it for, there are only holes in the walls instead of windows. These are so small that one can scarcely get one's head through them, so that they look more like openings for doves to enter and leave. Some of these have no shutters and bags are hung on the inside.

According to Montanha, the population of the town came to about 1 800, consisting of about 260 families.[6]

Life in town and the district

The Soutpansberg district, especially just south and southeast of the mountain range, had fertile soil, ample water and a favourable climate – factors that promised a great economic future. The Voortrekkers realised this potential and started utilising it. With enough water for the whole town, they could make gardens for their own use and plant fruit trees. The vegetables and herbs they grew included potatoes, runner beans, red, white and streaky beans, pumpkins, cabbages, turnips, lettuce, chicory, mint, parsley, coriander, onions and garlic.

They planted several kinds of fruit trees, especially yellow peaches. Although there were no grafted trees as yet, the fruit was nevertheless of high quality. According to Montanha, the variety of fruit included apples, grapefruit, lemons, oranges, walnuts, almonds, quinces, apricots, bananas, grapes, chestnuts and "fruit from India and the coast of Africa" (probably mangos, papayas and dates).

Das Neves mentioned that grapevines thrived, but apparently the Boers did not go to a lot of trouble to plant these. Those who had grapevines distilled brandy from the grapes that they sold at a high price. Peach trees were the most common and large amounts of fruit were produced and dried. Peach trees were also planted in rows along all of the streets. Fig trees also did fairly well. He was also of the opinion that there were no better or tastier oranges than those from the Soutpansberg and regretted the fact that so few orange trees had been planted on the stands. Das Neves also wrote that there were about 70 houses in the town during his visit in 1860 (thus substantially fewer than Montanha had mentioned).

---

5   The long streets were called Schoeman Street (closest to the fort), Duvenage Street and Potgieter Street, while the cross streets were called Weeber Street, Market Street, Korte (Short) Street, Burger Street and Jacob Street.
6   The discovery of the old town plan of Schoemansdal, which was probably drawn up in 1855 (Boeyens, 1987), indicate that only 50 erven had been measured out initially.

Almost every family lived on their own stand in their own house. All the erven had long canals that ran at the highest point and in which there was always plenty of running water. As soon as a piece of land had been ploughed and harrowed, the inhabitants opened the sluices so that water could flow down and irrigate the fields. They planted wheat, barley, rye, French beans, peas, maize and millet. Mrs Biccard, Albasini's daughter, later confirmed this observation. The fields below the lower, older canal were apparently particularly fertile.

Tradesmen
There were excellent locksmiths, joiners, smiths, sawyers, wagon-makers, brick-makers and masons in Schoemansdal. Few men could not make a decent pair of shoes. There were also excellent gunsmiths and gunstock-makers. People placed hides in vats with tanning-bark, which they collected in the mountains. They used the tanned hides to make saddles, bridles, harnesses and shoes, mostly for own use. They also sold tobacco pipes made of stone.

Large trees, including yellow-wood, kiaat, Rhodesian teak and matumi, grew in the mountains. These were cut down and sawed into planks using saw pits and long, wide jack saws or pit saws. Das Neves saw some very impressive planks there that were eight meters long, 34 centimetres wide and four centimetres thick. These were sold for 10/- each and exported to the British colonies. He also saw a table with a solid black marble top, almost one and a half metres long and one metre wide, which looked more like ebony than stone. According to Das Neves there was also an abundance of marble in the Soutpansberg, probably on the northern side of the mountain where the Cairo marble quarry was situated later.

Salt
In the bushveld salt was an important and very necessary condiment for domestic use and for the curing of meat, biltong and game hides. The salt pan, after which the mountain range, the district and the town were named, was only about two days' travel by ox-wagon from town, behind the western foothills of the mountain. According to the elderly Annie Grobler, people went to harvest salt in the winter months. During the salt harvesting families from far and near would camp in their wagons around the pan. First a well about two metres deep would be dug at the water's edge, so the salt water could filter into it. Small dams similar in shape to seedbeds, surrounded by low mounds, would be dug next to the well and water was bucketed from the well into these. To prevent a hole forming in the dam during the pouring process, a woven grass or rush mat would be placed over the bed, so the water could be poured onto it. After the water had evaporated, beautiful, white salt remained, which was poured in heaps to dry. Once dry, the salt was stored in canvas bags. Salt was a very expensive commodity and people could sell it for up to 1£ a bag. It could also be traded for clothing and other merchandise. On 6 March 1856 Magistrate A.C. Duvenage wrote to Commandant General Schoeman in Potchefstroom that there was such a large amount of salt in the pan that most of the inhabitants were more than adequately supplied as far as salt was concerned. He also mentioned that they had collected a considerable amount for Schoeman.

Elephant hunting in the Soutpansberg
The main source of income for the Soutpansbergers was elephant hunting, and many elephants were shot in the area that is now the Kruger National Park. Because tsetse fly, African horse sickness and the dense bush caused insurmountable problems for hunters on horseback, most undertook hunting trips on foot. Armed with a supply of rifles, gunpowder, lead bullets, bullet moulds, lead, a coffee kettle, coffee, rusks and eating utensils, the hunters left town in the early hours of the morning. Thus they managed to cover a considerable distance before sunrise. During the earlier years of the town's existence there were elephants in the immediate vicinity of the town. In time, however, their search for elephants took them further and further away, into what is now the Kruger Park, Zimbabwe and Mozambique. Bernard Francois Lotrie, who was one of the most fearless elephant hunters of his time, told how he downed seven elephants within a couple of minutes at the place now called Mara, only 16 kilometres from town.

When the hunters managed to shoot an elephant on their outward journey, they chopped out the tusks, buried them and collected them on their return journey. Although there was a considerable amount of adventure attached to these hunts, they nevertheless were very dangerous. Armed with a primitive muzzle-loader, or the even heavy four-pounder elephant gun, a hunter needed courage and daring.

The dangers to which these hunters were exposed, were plenty. They could be torn apart or trampled by wild animals. Many contracted malaria in summer and ended up in unmarked graves in the wilderness. Johannes Augustus Breedt, who was field cornet of Schoemansdal at one time, told how his brother, who had accompanied him on a hunt, fell ill with malaria near Inhambane. After they had carried him for 12 days in the hope of finding a Portuguese settlement, he died. "I was nearly desperate. I could not even dig a grave, because the soil was too hard and I had no equipment. Eventually I found a place that was almost a natural grave where we buried my brother as best we could."

To prevent people from entering the hunting fields during the unhealthy season, Commandant General Schoeman passed a hunting law in 1855 that stipulated that the hunting season would run from 15 June to 15 October. Transgressors would be fined 500 rix-dollars.

During Schoemansdal's heyday an enormous number of elephants were shot in the Soutpansberg, as well as in the adjacent Mozambique and Zimbabwe. Reverend Piet Huet, who visited the town in 1858, said the following:

> The number of elephants that are shot yearly is almost unbelievable, and those shot by Soutpansberg hunters alone amount to several thousand. Gigantic tusks of 60, 80, 100 pounds and even heavier are brought into town. Portuguese traders as well as traders from Pietermaritzburg come to town on an annual basis to trade these for money and merchandise.

According to a contemporary observer, it was an impressive sight when the staunch old hunters return to town after an absence of between ten days and a month or sometimes even longer. Mrs Maria Biccard could still recall many such scenes from her childhood days at Goedewensch. The hunters' hair and beards would be unkempt and their clothing dirty and torn. Behind the hunters followed the bearers, two or more per tusk, which were tied to poles with rawhide straps. The white tusks were then left in piles in the backyards of the houses while their owners waited for the arrival of the traders. Otherwise they were traded in due course to the local merchants. To amass such large amounts of ivory in a year, they needed between 22 500 to 23 000 pounds of gunpowder, 4000 pounds of lead and 5000 pounds of tin.

Local merchants and ivory traders from afar
There were several general dealers in town. Some of the first ones were Antonie H. Fick, João Albasini, F. Pistorius and Henry Austin. On 3 September 1856 a court case against these four was set down for hearing before Magistrate A.C. Duvenage and four "heemrade" (members of the county court). They were accused of using steelyards (balances with unequal arms, also known as Roman steelyards) that did not tally.

Albasini and his colleagues wanted to know if they could carry on using the steelyards for the moment, as it would be difficult to obtain platform scales and weights in a short time with which the weight of wares could be determined accurately. The court ruled that traders would be allowed to use the steelyards until 31 May 1857. After that, wares would have to be sold according to mass. Merchandise would have to be weighed either on bowl or platform scales, which had to be properly calibrated according to the Republic's laws, while certified weights also had to be used. Non-adherence to the law was punishable by a fine of 500 rix-dollars.

Other traders who did business at Schoemansdal at one time or another included the Goanese Casimiro Simões, August Landsberg, Dietlof Siegfried Maré, Augusto de Carvalho (who died of malaria during one of his visits to the town and was buried in the cemetery), L.M. Nunes, Antonio de Paiva Rapoza, Murphy and Cassimo Gamaal, an Indian who also died there. According to Mrs Biccard, Musgrow had the biggest business in town. Apparently he used to drink a lot and later died in an accident with his horse near Witklip, close to the current Polokwane (Pietersburg).

Casimiro Simões, or old Casmier as he was commonly known, was certainly the most colourful and one of the most prosperous traders. He was one of the first to come from Lourenço Marques with Albasini to trade with the Voortrekkers in Ohrigstad. By the time the Voortrekkers evacuated Ohrigstad and moved to Lydenburg, both men had a shop there. Later Simões moved to Schoemansdal, but he kept his business in Lydenburg going as well.

According to Mrs Biccard, Simões was a Portuguese who had been born in Goa, a Portuguese colony on the west coast of India. As a child she often saw him, and described him as a short, fat, ugly, almost black man who very much resembled a big ape. The children loved him, however, as he always brought each of them a paper bag of sweets whenever he visited Goedewensch. He flourished as a trader because he bought large quantities of the ivory that the hunters brought to town. As most transactions involved the trading of ivory, skins and other trade goods, Simões kept a record of each client's transactions in a voluminous leather-bound ledger that had an alphabetical index in the front. This index contained the names of no less than 141 people, with a double page allocated to each client. On one page he kept a record of the credits the client had earned with his ivory and on the other side the debits the client accumulated by taking wares from the shop. The ledger was titled "Livro Mestre ... Zoutpansberg Septembro 1863 No. II C. Simoens" and was kept in Portuguese. Felazio Cornelio Fernandes, a 35-year-old Goanese, was one of Simões's bookkeepers for a while, but he returned to Goa in 1865. He was succeeded by a young Portuguese, Bras Piedade Pereira, who worked as bookkeeper and shop clerk.

As can be expected, the 141 names in the ledger were spelled according to Portuguese pronunciation and interpretation. Some examples include: Coenraad de Buys, who became Cunrat Bois; Kosie Duvenage, who became Cochi Duwinage; Ferdinand Haenert became Ferdinando Heinert; Christiaan Herbst became Christiaan Herbest; Sarel Eloff became Sarlo Ilof; John Chambers became Jan Kambers; Frederik Potgieter became Frederick Potguiter, etc. Reverend N.J. van Warmelo's name was spelled correctly with "predicante" after it. Nevertheless, this ledger not only provides insight into the trading business, but is also a representative list of the inhabitants of Schoemansdal. Some of the articles Albasini bought from him included sardines, vinegar, tea, saucepans, a tray, mustard, soap, ointment, perfume, snuff, writing paper, ladies' boots, a mantilla, table cloths, shirts, gunpowder and planks.

By virtue of his trade, Simões surely had to transport several wagon loads of ivory to the ports along the east coast every year and return with merchandise for his shops. In March 1860, Merensky and Grützner, two German missionaries on their way from Natal to Lydenburg by ox-wagon to do mission work among the Swazi, encountered Simões with 10 ox-wagons along the road. He had also come from Natal and was on his way to Lydenburg with trade goods.

Although Simões did very well economically, the same could not be said of his personal life. While living in Schoemansdal, he married his second wife. Rumour had it that he had bought her from her father for a large amount of money. To make matters worse, there was a huge age difference between them – she was only 18 while he was about 50. It is therefore understandable that she was discontented.

It was public knowledge that there was a rather intimate relationship between Simões's young wife and Pereira, his clerk. On the night of 18 February 1865 the three had tea together while working on the shop's books. Later that night, Simões's butler Juzeh heard him making strange noises. He got up to investigate, only to find Simões lying swollen and bloated on his bed. Assuming that the man was dead, he set off to Albasini at Goedewensch. Albasini immediately inspanned his cart and horses and rushed to town, only to find his old friend dead in his bed.

Although rumours were rife about the cause of the trader's death, there was no conclusive evidence. He was buried in the cemetery at Schoemansdal at four o'clock the afternoon on Monday, 20 February 1865.

Simões left a large estate. In ivory alone there were 1083 tusks weighing 17 255 pounds (7843 kg). Then there were also 7¼ pounds (3.29 kg) white and 31½ pounds (14.3 kg) black ostrich feathers. These goods were transported in three of Simões's wagons to Pietermaritzburg and from there to Durban, from where it was shipped to London to be sold. After the deduction of costs, a substantial profit of £3 543-3-8 was made. If one considers that the whole estate was worth about £14 000, a fair idea can be obtained of just how lucrative the ivory trade was during the heyday of Schoemansdal.

In the central part of the cemetery in Schoemansdal is a pile of broken bricks, mixed with soil and lime. Old residents claimed that this was the grave of Casimiro Simões, the richest trader in the ZAR (South African Republic) in the 1860s.

Henry Hartley
Apart from the semipermanent merchants who had settled among the Voortrekkers, there were also traders who made the journey from Natal and Grahamstown with provisions and clothing. These they traded for ivory, skins, whips, sjamboks, rawhide straps, rhino horns, etc. One of the first traders to visit the Soutpansberg was Henry Hartley from Grahamstown. The Boers knew him well as he had been to Andries-Ohrigstad on a trade mission before and was an old friend of Commandant General Potgieter from his Cape Colony days.

This trader was not treated as a stranger by the Voortrekkers, but rather as a friend and benefactor. Apart from bringing the most needed wares to the far north, he also brought news. Along his long journey he would meet many acquaintances or hear of others. He usually brought letters as well. Although the post was old, it was nevertheless welcome, even more so because he could add to what had been written. He was often also accompanied by a number of passengers, or other wagons, while men on horseback sometimes rode along for safety reasons.

Demand for ammunition always outstripped supply and, fearing that ordinary citizens would buy it all, Potgieter held the monopoly on it. He would first buy up everything a trader had to offer and then distribute it in equal measure among the townsfolk. Afterwards the pedlar would display his wares on buck sails and the women could shop to their hearts' content. The traders normally paid half in cash and half in trade goods. The price of ivory ranged from 4/6 to £1 a pound, depending on supply and demand.

The most important wares the trader brought were gunpowder, lead, tin, fuses, good rifles for elephant hunting, oil paint, tar, tools for locksmiths, joiners and masons, coffee, sugar, tea, rice, knives, forks, spoons, pocketknives, flowered chintz (in rolls six to nine metres long), moleskin, calico, ready-made men's clothes, hats for both men and women, shoes, dresses, coloured cotton scarves, narrow and wide ribbons and lace, cotton in different colours, hooks and eyes, pins, needles, thimbles, several kinds of medicines with labels prescribing their use, wine, gin, brandy in bottles, etc.

The merchandise the Boers offered was ivory, rhino horns, hippo tusks, timber, different types of rawhide, wild cotton, horned livestock, tanned hides, upper-leather, leather thongs, strops, whips, sjamboks, baskets and other articles made by blacks. One pedlar gave the prices of some of the wares sold as the following: 200 pounds of cotton at £20; 1 pound of ostrich feathers at £2, coffee 1/6 a pound; tea 5/- a pound, Cape brandy 3/- a bottle, beef 3d a pound, mutton 6d a pound, salt 1d a pound, beer 2/6 a bottle, flour 25/- a bag, wheat 15/- a bag, potatoes 10/- a bag, turkeys 6/- each, geese 5/- each, ducks 1/6 each, chickens 1/-, eggs 1/- a dozen, grapes 2/- a basket, oranges 10/- a 100, oxen £5 to £6 each, a cow with a calf £4 and sheep at 15/- each.

The observations of J. Fleetwood Churchill
Unfortunately Hartley himself left no written documentation of his visits to the Soutpansberg. On the other hand, J. Fleetwood Churchill, a traveller who visited the settlement eight years after its founding while accompanying a trader from Durban, gave a lively description in a letter to his brother in England in 1856 about such a mission.

Churchill and the trader left the newly founded Pretoria with two loads of trade goods and arrived after two and a half days' travel at Warmbaths (Bela-Bela). Here they met several groups of people who had come there from the Free State for health reasons. The writer was struck with the immense distance that completely isolated the inhabitants of the Soutpansberg from the outside world. From the last house in the Waterberg they travelled for five days before encountering people again, just after they had crossed Bloedrivier (Blood River) outside the current Polokwane (Pietersburg). These were probably the Venters and Koekemoers at Doringbult. He mentioned that the largest of the six houses was encircled by a clay wall as fortification.

Three hours later they arrived at Rhenoster Poort,

where a number of families and an illiterate Scottish trader, probably John Watt, lived. Here he also noticed a well-built fort that served as protection in times of danger. Under normal circumstances it would have taken them another two days of hard riding to reach Schoemansdal, but it took them much longer as they traded all along the route. On the first day they visited a number of farms where several families lived together for safety reasons. They traded with everybody and obtained a good deal of ivory, as the people had just returned from the hunting fields. Some were already preparing for another hunting expedition. It goes without saying that most discussions revolved around their adventures with elephants, buffalo and lions. On one farm they saw a young man who was recuperating after a lioness had bitten him two months earlier.

With the exception of elephants, the travellers encountered lots of different game along the road from Pretoria. In the vicinity of Bandelierkop Churchill saw eight giraffes, which was the first time he had ever seen these odd-looking animals. In an attempt to follow them he fell off his horse and injured himself so badly that he had to recuperate in a tent at someone's home for a few days. He used Holloway's ointment to treat the bruises. This ointment was apparently highly recommended by the old elephant hunters.

After an eight-day journey from Rhenosterspoort they finally arrived at the Soutpansberg on a Sunday morning. He had the following to say about the town:

> Soutpansberg is a 'Dorp' lying under a 'Berg' of the same name and said to be very sickly in summer. The range of mountains run [sic] East and West. Beyond begins the Elephant hunting grounds. It is said to be the farthest point possible at which the white man can permanently reside. If you ask why, you receive the unanimous reply 'The fever and the fly'. Owing to my accident, I was prevented from ascending the 'Berg' and also from visiting the famous 'Cream of Tarter' [sic] tree, the fruit of which I took home with me in '53. It is inhabited by hunters and traders. The latter of whom generally arrive in April and May and leave in October or November. They thus escape the hot season. The Commandant General of the District also lives here and had just had a very good house built for him. It is both papered and painted but not ceilinged.
>
> There are three or four other very fair houses; better than the one I live in P.M. Berg. The Kirk was nearly complete, barring doors and windows. The Landdrost (Magistrate) officiates. It is two years since they have had a visit from a Minister.
>
> I saw lots of ivory, more than I ever saw before in one place. Our own customers having their share. The largest tusk was 118 lbs Dutch, about 127 lbs English. This is the biggest I have ever seen. Its fellow was 102 lbs. This elephant was shot by two Englishmen. There is a good deal of rivalry amongst the Hunters and Traders as to who shall shoot and buy the heaviest tusks. I expect the 102 lbs will pass through our hands, the other, I am afraid, will go to Grahamstown, being paid for a debt to the only Grahamstown trader now there. More than two-thirds of the ivory now comes into Natal. You will be better able to judge of its position by me telling you it is about 300 miles nearly West of 'Inhambane'. A Portuguese Priest has come from there last year and gone back again this year. It was to visit his two or three countrymen here. The Hunters all go out on foot with six, eight or ten kaffirs along with them carrying a kettle, small bag of coffee and biscuits and a couple of elephant guns from four to eight to ten lb. Their hunts last from ten to thirty days at a time. It is certainly hard work for them especially if they are unfortunate and bring back nothing. The largest quantity shot by one man this year is about 900 lbs weight, in value about £200. This is great luck. It is a curious sight to see a hunter return with his kaffirs after him, carrying teeth. Our Zulus would hardly stand it. Some of the people in 'Soutpansberg' employ Bastards to shoot for them, giving them a share of the proceeds.
>
> My arrival in S. caused some surprise amongst the Traders, who could hardly believe that a Durban Merchant would travel so far. In reckoning up the distance with one or two individuals we made it about 800 miles from Durban. This is under rather than over the mark. What a distance for land carriage. You will ask why not open trade with 'Inhambane'. The very thing the Boers have been trying for two years, but unsuccessfully, on account of the fly and a powerful chief named 'MANICOS' who

will only allow one or two men to pass through his country.

The estimated number of Elephants supposed to be shot this year in this district is nearly 1 000. Fearful slaughter.

After a trader had completed his business, the return journey had to be undertaken before the start of the rainy season. He usually bought so much ivory and other goods from the Voortrekkers that he had to hire one or more wagons and leaders to transport his loads from the Soutpansberg to the coast. The fee paid for a wagon was £15. Churchill and his companion left the Soutpansberg with 5000 pounds of ivory.

Religion and education in Schoemansdal

For most of its 19-year existence Schoemansdal struggled with religion and education. Without a full-time minister, church elder Douw Gerbrand Grobler, Jacobus Nel and Magistrate Abraham Duvenage led church services.

Reference has already been made to the visit by reverends Murray and Neethling in 1852. In 1854 Reverend Dirk van der Hoff visited the town, and in 1858 so did Reverend Piet Huet. Rev Van der Hoff again visited in 1859, in 1861 Reverend Begemann held services (returning for a visit in 1863) and in 1862 so did Reverend G.W. Smits.

In the meantime the Soutpansbergers were busy. Four years after their arrival, the congregation had already collected £150 towards the travelling expenses of a minister who would accept a calling there. They had also started building a church. The builder was G.P. Marnitz. In 1856 Montanha noted in his diary that the building of a church was in progress while he was there. It was seven or eight fathoms long (12.8 or 8.6 m) and more or less four fathoms wide (7.3 m). It had doors on both sides and four windows along the northern and southern sides, and had not been plastered yet. At the end of the year the church council complained as one of the gables was about to collapse.

On 30 June 1864 Reverend N.J. van Warmelo was ordained as minister of the congregation of Soutpansberg by Reverend Smits. This competent and diligent minister arrived in the Transvaal in May 1864 accompanied by his wife Josina (née Van Vollenhoven). They had disembarked in Durban and deacon Venter took them by ox-wagon via Pietermaritzburg and Pretoria to Potchefstroom, where Rev. Van Warmelo was ordained. At the time of their arrival in Schoemansdal there was no parsonage yet, but the bricks and woodwork were ready and waiting. Rev. Van Warmelo was most impressed with the one he had seen in Pretoria. Before building could commence he had to select a stand, and the couple had to give an indication of the number, division and placement of the different rooms. In the meantime they moved into another house.

Prof. S.P. Engelbrecht described Rev. Van Warmelo as follows: "He was somebody with deep and

Dr N.J. van Warmelo, the enthusiastic and respected minister of Schoemansdal's congregation 1864–1867.
(ACKNOWLEDGEMENT: DR J.B. DE VAAL)

The Voortrekker church at Schoemansdal according to an authentic drawing in G.G. Munnik's book, *Kronieke van Noord-Transvaal*.
(ACKNOWLEDGEMENT: DR J.B. DE VAAL)

Mrs Josina van Warmelo, Rev. N.J. van Warmelo's wife. She was a remarkable woman who supported her husband with love and self-sacrifice.
(ACKNOWLEDGEMENT: DR J.B. DE VAAL)

affectionate feelings, somebody who did not seek his own advantage, who understood a way worthy of imitation, to renounce himself and to carry his God-appointed cross. His heart was full of love for others and at the same time sensitive to the love bestowed on him. In each human soul with which he came into contact, he found the heavenly side and that ability allowed him to witness the kingdom of heaven inside the heart. We are especially impressed with the high regard which Rev. Van Warmelo had for the virtues he saw in others. He had respect for the old Voortrekker lineage, the old emigrants who had trekked to the Transvaal because they wanted to be independent. He admired that brave old generation who came to trample the roads to dust, the old population who suffered the heat of the day and the cold of the night. He was grateful for every ray of sunshine in life."

What he had been could largely be attributed to the love and devotion of his wife.

A person with these characteristics would soon gain the esteem of his parish. He was therefore also generally loved in his work circle. The parsonage's doors were always open to everyone and the welcome always cordial. The rough old elephant hunters never left without a cup of coffee and a few encouraging words. This friendliness was always rewarded with vegetables, meat, chickens, eggs and all kinds of treats offered by the women. In 1866 he received a Christmas gift from the church amounting to 60 gold pounds.

At Communion-time Schoemansdal became a hive of activity. People came from near and far with their ox-wagons to praise and thank the Lord for their continued survival in the wild. One can imagine that these weekends must have been very busy times both for Rev. and Mrs Van Warmelo at the parsonage and for the traders. Casimiro Camoês and the German dealer Landsberg kept the inhabitants up to date on events in the rest of the world. On Christmas Day 1864 there were no less than 97 wagons, which all stood alongside and in front of each other on the large church square. All pitched their own tents, because they had arrived on the Wednesday or Thursday and some only left again on the Monday. Rev. Van Warmelo had the following to say about the church square:

> To walk and see this sight, especially from an elevated position, is very picturesque. In the evenings each tent has its own candle or wooden fire – up to as many as 1000 in one night. The families visit each other, conduct their business and do their shopping, all while preparing for Communion. It is usual for people to come and *groeten* (greet) the reverend and his wife at any time and for each and everyone to bring what they can. I find something patriarchal and simple in these visits, which is why I encourage it. If somebody does not come, it is a sign that something is wrong. Because this many people only get together four times a year, one can handle it ... They dress neatly with gloves and bracelets; men find many able wives.

Apart from an extended church council meeting held on the Friday, up to 21 young people confessed their faith. On the Saturday morning they were confirmed as church members and there was a church sermon in the afternoon as preparation for Communion. In the evening, the names of 34 children to be baptised were entered in the church's baptismal register. On the Sunday both Communion and the service after Holy Communion took place. If there had been any deaths, memorial services would be held. As people had travelled far, another church council meeting took place on Monday. With that, the activities for the quarter coinciding with Holy Communion were concluded.

The parsonage at Schoemansdal according to an authentic drawing in G.G. Munnik's book, *Kronieke van Noord-Transvaal*.
(ACKNOWLEDGEMENT: DR J.B. DE VAAL)

6.1 SCHOEMANSDAL, THE FORMER ZOUTPANSBERGDORP (SOUTPANSBERG TOWN) (1848–1867)

During the first few years after the founding of the town, the children's education was neglected as the townspeople were unable to find a permanent teacher. During the day the children had to work and in the evening they were taught by their parents. Often an older brother or sister, a neighbour, or just about anybody who was educated to some degree, served as a teacher. A teacher by name of Jacobus Nel had arrived with the trekkers at Schoemansdal. Koos Grobler said the children had to sit with him in school reciting their lessons out loud – definitely not an ideal teaching method! Jan Helberg also made a small contribution towards their education.

When Reverend Piet Huet visited Schoemansdal in 1858, he taught 50 to 60 children on a daily basis in the church. It was only after the arrival of the Van Warmelos that serious education got off the ground, though. Apart from the catechism lessons, he taught the children in the church on the first couple of mornings of the week. He also started giving singing lessons in which pupils made good progress, as he himself was an excellent pianist and a lover of and expert in classical music. Mrs Van Warmelo contributed by teaching the town's girls needlework, embroidery and knitting every Friday.

The nuptial bliss of the Van Warmelo couple was not to last long, however. On Friday morning, 20 January 1865, she woke up with an aching body, but nevertheless carried on with her chores. By two o'clock that afternoon she felt so bad that she was forced to take to her bed. By then there were signs of dysentery. The doctor was called in and he diagnosed typhoid fever. When her condition deteriorated by the day, two women of the congregation, the widow of church elder D.G. Grobler and a Mrs Van Rensburg, offered their help. She died at 11 o'clock in the morning on Saturday 28 January 1865 and was buried the next day, an event of great interest to the town. Members of the church council carried her coffin in the funeral procession, which slowly made its way to the Voortrekker cemetery, where so many victims of the unhealthy Soutpansberg area were laid to rest.

The following Sunday the whole congregation gathered in the church. As an expression of their sympathy with their minister's terrible loss, all the women were in mourning.

Rev. Van Warmelo often wrote to his mother in the Netherlands of his very happy marriage. According to him, his wife was very capable and always did everything in her power to make life as pleasant as possible for him.

Shortly after this blow he also fell ill, but soon recovered. On the invitation of Rev. Smits he travelled to Rustenburg for a visit where he gradually recovered his strength. Later in 1865 he was back in Schoemansdal and resumed his duties with renewed energy. He realised that he had a huge responsibility towards his congregation, especially the youth. His schooling of the children lasted for almost two years, during which time the number of pupils continuously increased.

On 20 April 1866 he wrote a letter to the president and members of the Executive Council in which he asked them to establish a government school at Schoemansdal. He mentioned that he had been serious about the education of the children from the beginning and wanted to give his best. He was convinced that if the education of the children could be improved, the nation would ultimately benefit, as people would become more dutiful and industrious.

He continued that he had been searching for a suitable teacher for a long time and had finally found one who satisfied all his requirements in the person of Mr C. van Boeschoten. During the seven months Van Boeschoten had been working under his supervision he got to know him as a diligent young man with a keen interest in his subject and for whom it was a pleasure to work with children. He was a teacher in the true sense of the word and would not exchange his position for any other, even a better paid one.

When he was 15, Cornelis van Boeschoten and his parents arrived in South Africa from the Netherlands. They were urged by President M.W. Pretorius to settle in the Soutpansberg district, where they grew coffee. Both his parents, however, died of malaria in 1866. They were survived by five children, of whom Cornelis, who had just turned 21, was the eldest. After teaching in Schoemansdal he worked as a teacher in Pretoria, where he taught, among others, the future poet Jan Celliers. Still later he rose to the position of assistant state secretary of the Republic. Cornelis died in the Netherlands in 1927, at the age of 82.

Cornelis van Boeschoten, Schoemansdal's energetic teacher during the last few years of the town's existence.
(ACKNOWLEDGEMENT: STATE ARCHIVES)

On 26 December 1866 the school commission inspected the standard of education in the church. They evaluated the children's abilities in the subjects of reading, arithmetic, grammar, geography and poetry. The inspection lasted three hours and was considered of great interest from beginning to end, with all concerned coming to the conclusion that the time had been well spent and that education rested on a firm foundation here. They especially praised the young teacher's sound methods. So impressed were they that the chairperson, Rev. Van Warmelo, thanked Mr Van Boeschoten at the end of the proceedings on behalf of the commission and the parents, for his tireless patience and the great interest he showed in his work.

During the inspection something worth mentioning happened. While they were busy with their treatment of arithmetic, which they did with great interest, members of the commission were astonished by Aletta Meyer. This little girl of eight could apparently transpose very high Roman numerals into Arabic numerals in a matter of seconds. They were so impressed, in fact, that they decided to award her a special prize as an incentive. What became of this gifted child is not known, but the words of the poet Thomas Gray in "Elegy written in a country churchyard" are certainly applicable to her:

Full many a flower is born to blush unseen,
And waste its sweetness on the desert air.

Excavations at Schoemansdal, the former Voortrekker town. (09.12.1985)

Prizes in the form of picture books were awarded for progress made, neat handwriting, good behaviour and regular school attendance. The winners were Zybrand Eloff, Francisca Meyer, W.W. Maré and Frederik Eloff. At the close of proceedings Rev. Van Warmelo thanked the parents and urged them to cooperate in making the children's day a pleasant one in the form of a picnic.

Loosely translated, he wrote the following about the picnic:

The second of January will stay in the memory of Soutpansberg for a long time as one of the most pleasant days that people here have spent together. Everybody did their best. We selected a nice spot under some big trees at the foot of the mountain. Ox-wagons were ready and tents pitched and there was food and drink in abundance. It was the nicest cool weather imaginable and spirits were high; a great change from the everyday routine, being able to set aside all problems and forget about misery and to spend the day childlike with children, almost the whole town as one large family. Nobody will easily forget the day on which we came together in seemly merriment.

It appears as though the church was still serving as a school, as Rev. Van Warmelo wrote to his mother in the Netherlands on 26 January 1867 that he was having a big school built. This project kept him very busy. It would be difficult for her to understand why, as in the Netherlands one would employ a carpenter, builder, etc. As long as you could pay, you did not have to be involved at all. In Schoemansdal this was not the case, as the school was built mainly with funding from the church. As minister and chairperson of the school board, he had to keep an eye on things and give advice where necessary.

Rev. Van Warmelo's congregation in Schoemansdal were generous in their support. A year after his wife's death he had to attend a general church meeting and was away for almost two months. During his absence, the congregation built a memorial on his late wife's grave. Again he wrote to his mother: "When I arrived home, the townspeople had a surprise ready for me, something which proves an attentiveness and goodwill, which one would not really expect to find here. They have erected a memorial on my wife's grave, built with bricks and plastered white with the inscription: 'Josina van Warmelo, van Vollenhoven' on the front. A mason was employed for the task and everybody helped to have it ready just in time for my return. They planted a rose hedge around it with a willow in front. I was very touched by this act of the townspeople, most of whom had known her. I sincerely appreciate the fact that they wanted to commemorate and perpetuate her memory."

As mentioned before, the cemetery and the graves went to ruin after the evacuation of Schoemansdal in 1867. Only in 1917, 50 years later, did the Schoemansdal Dingane's Day Commission erect a rectangular brick wall around the disorganised piles of stones. Mrs Van Warmelo's grave, like all the others, was destroyed. Not one grave still had a headstone. Like the Voortrekkers at Klipdam, the Schoemansdallers would have erected headstones on the graves of their loved ones, with at least some information about the deceased on them, however simple. The only conclusion one can reach is that many of the stones were removed while others were smashed to pieces and eventually covered in soil. If a decision were made to excavate the area carefully, no deeper than 15 centimetres, some interesting facts could come to light.

Social life in Schoemansdal
There is not much information available about social life in Schoemansdal, except for what was written by Pater Montanha in 1856 and Mrs Josina van Warmelo 10 years later.

Mrs Van Warmelo mentioned in a letter to her mother in the Netherlands that there were many children in town, but all were shy and scared of the minister and his wife. When naughty, they were threatened with the minister and told that he would give them a hiding. In general they appeared healthy, except that they were almost always dirty as they played outside. Schoemansdal's red soil would be to blame for this!

She regarded the presence of a minister as essential since the population had been spiritually neglected and there were many married couples who were not church members yet. Although they had laws, these were not adhered to and people often did as they pleased. It went badly for several years, as the smallest disagreement could flare up into a huge argument. In the sixties, Schoemansdal actually became notorious for the constant altercations and bickering among its inhabitants. But in the short time they had been there, they could already see an improvement. If this could carry on the town would prosper, in her opinion. She found that it was difficult to get to know the people's better qualities, and therefore their labours were greater and more needed. The young people especially would benefit the most.

People visited each other at all hours. There were no doorbells – some people knocked, but others just walked in. They drank a lot of coffee in Schoemansdal, and wherever you visited you were served coffee, irrespective of the time of day. Montanha also noted this. He determined that they used between 4000 and 5000 pounds (1818 and 2273 kg) of coffee and 10 000 pounds (4545 kg) of sugar, but little tea. For the rest they mainly lived off hunting. Their food consisted of bread, meat, sweet potatoes, vegetables and maize. Except for tea, coffee, sugar and rice, the Van Warmelos didn't have to spend anything on food. Although they had their own chickens, they often received eggs as gifts and from time to time even an ostrich egg.

Nothing but wood was used to make fires. As it was so abundant in the vicinity, blacks carried bundles around and traded them for beads. One could also trade potatoes, maize and pumpkins from them in exchange for blue beads.

During the vegetable season, the rectory received a basket of peas, carrots, figs, cabbages and other soup vegetables on Saturdays. This came in very handy. On Sundays they normally received a whole dinner from Mrs Simões, who lived across the street from them. She was the wife of Casimiro Simões, the Goanese trader mentioned before. She was always neatly dressed and when she paid them a visit she was smartly turned out, wearing a new dress with every visit. She could play a little piano and guitar and had a nice singing voice. Sometimes Mrs Van Warmelo returned her visits and was entertained with song and music. The first time she visited her without her husband, Kobus, Mrs Simões asked her guest if she would like to sing a Dutch song. Josina sang a Christmas song with the following rather woeful refrain about an unrequited love:

Moet ik dan van droefheid smoren,
Hebt jy dan geen liefde voor my,
Kan myn hart u dan niet bekoren?
Ach, ruk my van deez' wereld maar af!
Gy moet my dalen in het graf.

Should sadness smother me,
Do you have no love for me,
Can my heart not enchant you?
Oh, tear me from this earth!
Lower me into my grave.

From time to time she also shared shocking news in her letters. Once three guns were stolen from trader Maré's shop; a father accidentally shot his child at shooting practice; a small girl had burnt her feet, but fortunately they healed well.

Something she found remarkable was the fact that the people married so young. Many girls in their fifteenth or sixteenth year and even younger did so. There was one young couple who married when the bride was not even a full 13 years old. She was the wife of Mr Rood, a Dutchman originally from Amsterdam. Her first child, a dear little boy, was born in April while she only turned 15 in July. She also mentioned that the women aged quickly. Mrs Van Warmelo had met several who looked quite old, only to find out that they were not even 50. They always wore large, colourful bonnets, so one could hardly see their faces. On Sundays they wore their best bonnets, which were decorated with flowers that had been fixed onto the bonnet in a peculiar way. Then they were neatly clothed, wearing gloves, aprons, shawls or cloaks of a variety of cuts and colours. She found it unusual to see these different types of clothing.

Schoemansdal also had to make do without a doctor for most of its existence, relying mostly on tried and tested folk remedies. During the last years there was one doctor, whom Rev. Van Warmelo described as follows:

> We have a doctor who has been living here for a number of years with quite a good reputation. He visited us this afternoon and after he had left, we could not help but laugh at this character. He is somewhat pedantic and during formal occasions, such as visiting the minister, he wears a pair of mildewy, dirty, beaver gloves. He looked so odd that I thought at first that he was wearing a mask. While sitting down he wriggles on the chair and doubles up as if suffering from cramps, so that his body takes on a laughable appearance; definitely a man for Dickens ...

About New Year's Day 1865 Montanha wrote the following:

> The first day of January is a big occasion for these people, a general day of feasting. Everybody puts on their newest and best clothes and gather in groups in the street carrying their guns, singing and playing. When arriving at somebody's house, they all fire into the air and enter the house while making a lot of noise. If they are offered a tipple they stay, otherwise they move on. They look like lunatics and carry on from daybreak until evening.
>
> I also noted that one does not hear any noise or shouting at night, not even the sound of people walking on the floors of their houses. After eight or nine in the evening, one only hears the crying of jackals or barking dogs. Shooting a rifle at night is prohibited.

At the end of 1864 Mrs Van Warmelo wrote that there had been a church service on New Year's Eve – a first for Schoemansdal. Because New Year fell on a Sunday, there was no dancing.

They had lunch with the magistrate and his wife and the food was very good, to her surprise. The hostess, like most Afrikaans women, was a very good confectionery baker. They received a fair amount to take home and therefore had enough treats for some time.

A few elderly people lived in town. They were poor, but had lovely gardens and gladly shared the fruit with the minister and his wife. When she thanked them, they always said: "This is all we can give our minister, who came from such a far-off country with the lady to keep him company."

About weddings Montanha remarked:

> In the meantime I received a number of invitations to attend weddings. Thus as a stranger I could observe their ways of celebrating. The wedding feast started as soon as the couple was married in the pres-

ence of the judge and two witnesses. Immediately after the ceremony they paid the marriage licence. I do not know how much this was, but they certainly paid. The questions posed to the bride and groom were the same as those asked at our marriage ceremonies.

After arriving home, the dance started immediately. There was always somebody playing the fiddle. The most popular dance was the Scottish reel; with two or three pairs dancing. When a man invites a lady to dance, he bows down with both his arms stretched out and usually taps the lady on the knees with his hands. She is not allowed to refuse to accompany him to the middle of the dance floor, no matter what her age. At the end of the dance the man gives the lady a kiss on each cheek. Sometimes they do an English country dance, but very old-fashioned, and many couples participate.

Everybody goes into the dance hall, men, boys and women of all ages, and everybody wants to dance. The men arrive dressed in whatever they like, some even in shirtsleeves. Sometimes I heard a couple of women singing songs, accompanied by the violin, and sometimes the men sing as well.

Stephanus Schoeman and Schoemansdal

After his election as Commandant General of Soutpansberg, Schoeman at first lived in Mooirivier. From there he visited Zoutpansbergdorp and the district in his official capacity. As has already been pointed out, he was focused on having certain proposals regarding the improvement of the town accepted, with the cooperation of the residents, magistrate and council members. All of these were signed as "decisions taken by Commandant General Schoeman".

A while after De Santa Rita Montanha's arrival in the Soutpansberg in 1855, Schoeman consulted him and it was decided that Montanha's delegation would only return to Inhambane in the winter of 1856, accompanied by a group of Boers. He also arranged that the priest marry him and his bride on 20 January 1856 in the Waterberg. His fiancée was the widow of Commandant General P.J. Potgieter, Elsje-Maria Aletta, née Van Heerden. On 31 May he and his bride arrived in Schoemansdal and received a glorious reception. Less than a month later, on 23 June 1856, Montanha returned to Inhambane.

The constitution

During a sitting of the Volksraad at Pienaars River on 19 September 1855, a commission was appointed to draw up a constitution for the ZAR. Schoeman opposed the "Constitutie" as he feared that M.W. Pretorius would become president. He even wrote to somebody that he wanted no part of it. Under his influence, the people of the Soutpansberg rejected the constitution at a public meeting on 29 January 1857. In a voluminous document consisting of 22 lengthy articles, they pulled it apart article by article.

Everything came down to the fact that they did not want to be subjected to the whims of one person, namely the president. At a sitting of the Volksraad in Potchefstroom the reaction to this protest document was so vehement that Schoeman was relieved of his duties as Commandant General. A proclamation was passed that declared the Soutpansberg in a state of blockade and forbade all trade with the town and district. Matters took on such serious proportions that Schoeman challenged Pretorius to a duel to prevent bloodshed and war. Schoemansdal and district completely isolated themselves from the outside world, and Schoeman decided that anybody guilty of sedition would be charged with high treason and executed, imprisoned or banished. They even considered taking up arms and the people of the Soutpansberg counted on help from the Free State. For his part Pretorius declared Schoeman an outlaw and put a price of 1000 rix-dollars on his head. Fortunately no shots were fired. Eventually the two leaders met and after consultation declared that it had all been a huge misunderstanding and made peace.

On 11 February 1858 Schoeman was sworn in as a member of the Executive Council and Commandant General by the Volksraad in Rustenburg. Now Schoemansdal could once again function as part of the Republic. Two months later, on 13 March, Schoeman and his family were back in Schoemansdal. He hoisted the republic's new flag, the Vierkleur (Four Colour) and asked for copies of the constitution from the president. He closed his letter of request with the words: "Your fondest friend and servant." Schoeman expected that everybody whom he had incited against the constitution before, now accept it without question. When they were not pre-

Cmdt Gen. Stephanus Schoeman, the "Stormy Petrel of the North", after whom Schoemansdal was named in 1858.
(ACKNOWLEDGEMENT: STATE ARCHIVES)

pared to do this, he called them rebels and asked Pretorius for 100 or 150 men to call the obstinate group to order by force of arms. Pretorius did not oblige. Although he was delighted with the way matters had turned out, he was certainly not going to associate himself with the capriciousness of the "Stormy Petrel of the North".

However, the kudos for giving shape to Hendrik Potgieter's little town does belong to Stephanus Schoeman and therefore he named it after himself. From the beginning of November 1858 the words "Schoemansdal, Zoutpansberg" appeared as the address on all official letters leaving Soutpansberg.

One of the first matters that needed to be addressed was the need for a punitive expedition against Chief Mankopane, who was held responsible for the murder of 28 people. Schoeman and Commandant Paul Kruger from Rustenburg undertook this together and completed it with great success.

Communication with the outside world is the key to successful administration. On 3 October a decision was made at a public meeting, at which Schoeman was present, to institute a regular postal service between the Soutpansberg and Field Cornet Van Heerden in the Waterberg. Two reliable black men, August and November, would be employed by the town to carry the post to Field Cornet Geyser at Rhenosterspoort. From there, two other men, Adam and Rooibok, would take it to Field Cornet Van Heerden. The carriers would be exempted from doing other work and from paying taxes. The post would leave town on the 25th of each month and Postmaster C. Rabe would receive his salary on the same day. How much this was and how much the carriers were paid was not mentioned.

Unfortunately there was a series of unpleasant incidences in the Soutpansberg. C.J. Rabe, former secretary of Commandant General Potgieter and later of João Albasini, was the Schoemansdal magistrate at the time. He was not a member of the Dutch Reformed Church, but belonged to the Episcopalian Church instead. According to the law he was thus not allowed to serve as magistrate. Shortly after his appointment, Field Cornet Pieter Johannes Eloff's daughter was to marry D.G. Grobler. Eloff did not want Rabe to officiate and instead asked S.A. Jacobs, one of the council members, to do it. On the day of the wedding, Rabe was asked to evacuate the magistrate's chair; a call he blankly refused. As a result the couple had to leave without being married. Shortly afterwards Rabe asked that he be released of his duties and his request was granted.

Eloff proceeded to incite people against Schoeman and Rabe. On 3 February 1859 Schoeman was to move from Schoemansdal to Pretoria. He informed Eloff that he wanted to take the two cannons from the fort with him for self-protection. One of these, "Old Grietjie", had been brought from the Cape by Potgieter. Schoeman did not give a thought to the people of Schoemansdal, who would be left unprotected in the midst of their enemies.

Field Cornet Eloff could not approve of these actions and decided to try keeping the cannons in town for their protection. He was assisted by Johannes Breedt, Hendrik Janse van Rensburg, Cornelis van Wyk, Jan Jacobs and John Chambers. Harsh and ugly words were exchanged during the ensuing altercation. Chambers lost his self-restraint while Schoeman threatened to shoot anybody laying a hand on the cannons. He also informed them that he was in a hurry to leave and would deal with Chambers at a later visit.

A month after Schoeman's departure, a petition signed by 92 townsmen was sent to President M.W. Pretorius on 2 March 1859. In it they informed him that they were firing Schoeman as they did not wish him to rule as their general any longer. As motivation they submitted that he seldom if ever went to church, even if there was a visiting minister. He was also opposed to the pious church council. He did not explain the country's laws well, regarded them as rebels and had asked for help to fight those who did not agree with him, mentioning them by name and threatening to throw them in jail.

The president regarded the cannon incident and the petition in a serious light and arrived on 19 May 1859 with Reverend Dirk van der Hoff and Stephanus Schoeman, both members of the Executive Council. He also brought along people to constitute a Supreme Court. Field Cornet Eloff, Breedt, Van Rensburg, Van Wyk, Jacobs, Grobler and Du Preez were accused of disturbing the peace and misconduct. Eloff did not even attend the court proceedings, as he had apparently been declared insane. The case against the rest was heard on 25 May and they were all found guilty. Each was fined 550 rix-dollars and they also had to pay the court fees. They were later excused on the grounds that they had not been familiar with the country's laws.

Despite the unpleasantness with which President Pretorius had to deal at Schoemansdal, he had the highest praise for the inhabitants of the town. He also mentioned that the black inhabitants had a lot of respect for Commandant General Schoeman.

From 1860 to 1864 Schoeman was involved in the civil war and his most important opponent was Paul Kruger, who later became president. This explains why he had been so anxious to take the cannons with him the previous year. During this turbulent time he was also acting president for a while, but the people of Schoemansdal decided at a national assembly on 6 November 1861 that they would not allow him back in Schoemansdal in this capacity. So as far as the people of the Soutpansberg district were concerned, he had quit the scene.

Reasons for the evacuation of Schoemansdal
The reasons for the evacuation of Schoemansdal and the events that led up to it have already been briefly discussed in chapter 5 about Albasini. It is discussed here in more depth.

During a public meeting held in the Soutpansberg on 19 February 1855, Schoeman stipulated that hunters were not allowed to enter the hunting fields before 15 June. This was to prevent the irresponsible eradication of game and also to prevent the hunters from contracting malaria during the summer months. The meeting accepted it unanimously and a "Wet tot beter regelen der jagt op Olifanten en ander wild in de Z.A. Republiek" (Law for the better regulation of the hunting of elephant and other game in the ZAR) was passed. The legislation was promulgated in the Government Gazette of 22 October 1858. The law stipulated that: the hunting season would only last four months (from 15 June to 15 October); no black man was to be sent on an elephant hunt without being accompanied by a white man; nobody was allowed to take a black marksman or other blacks on an elephant hunt if they had not been registered with a magistrate; nobody was allowed to take more than two black marksmen with them on an elephant hunt; and nobody was allowed to kill more game than he would need, or shoot game only for the sake of the skins.

As the entire existence of the people of Schoemansdal depended on the proceeds of hunting, they did not comply with the law. There was in any case little authority to enforce the law in the first place. At first hunters took gun and ivory bearers with them on their hunts, but as soon as they could shoot well they were sent by themselves to collect ivory, for which they were compensated. New traders also sold rifles and ammunition to the blacks in an irresponsible manner. A.K. Murray was an English trader from Natal who visited the Soutpansberg from July to October 1863. He later said that the ivory trade was the town's most important source of income, but he also mentioned that the supply of ivory was decreasing as elephant herds gradually retreated to the interior. Farmers and traders were sending blacks who had rifles "in their hundreds, if not thousands" on elephant hunts.

At first the ivory arrived in a steady stream, but later the black hunters took to collecting their own ivory with the rifles belonging to their employers, and disappeared with both the ivory and the rifles.[7] When these men realised that the rifles were better weapons against the Boers than their assegais, bows and arrows, the Soutpansbergers found themselves in danger. Many of the loyal hunters were also murdered by hostile tribes and their rifles, gunpowder, lead and ivory taken. This in turn led to punitive expeditions against the guilty chiefs. The whole district soon became embroiled in a civil war. Two successive wars between the Venda and the Shangaan-Tsonga also had a negative impact on the district.

When Manukosi, "the murderer of the east coast", died in 1858, a power struggle developed among his sons for the position of chief. The rightful successor, Muzila,[8] was driven away by his brother Mawewe or Langapuma. Muzila fled to the Soutpansberg to seek protection. The Executive Council placed him and his followers under the custodianship of the Super-

---

7   Some of the four-pounder elephant guns and front-loaders are still found today in hollow tree trunks and rock crevices in the Kruger National Park where the blacks hid them to prevent the Boers from finding them.
8   Also called Umzila or Mzila.

intendent of Native Affairs and Field Cornet of Spelonken, João Albasini. The farm Madrid (then Beja), about ten kilometres east of Albasini's farm Goedewensch, was set aside as a temporary abode for him and his followers. This fuelled Mawewe's anger against Albasini and the Boers. As a result he carried out raids in the Portuguese territory, murdered the Boers' hunters and confiscated their belongings. This not only cut off communication between Delagoa Bay and the Soutpansberg, but prevented trading as well. The hunter and trader Das Neves was one of the people who had a close encounter with Mawewe's marauders in the vicinity of Chicuala-cuala. More is told about this in chapter 8.

When Muzila was eventually reinstated as his father's successor with the aid of chiefs he had befriended, as well as Albasini and the Portuguese at Delagoa Bay, he demanded the extradition of Monene. The latter had stepped out of line while Manukosi's chief field-general and had also found protection with Albasini. To put pressure on the Boers Muzila closed the elephant hunting grounds, inhabited by his subjects, to outside hunters. This did not suit the whites, so they captured Monene and the Executive Council ordered that he be handed to Muzila. But Monene managed to escape from the stocks in Schoemansdal. In their attempt to recapture him, chiefs who offered him protection were attacked by Albasini, Commandant S.M. Venter and Field Cornet J.H. du Plessis and had many of their possessions carried off. Unfortunately innocent chiefs were also attacked, with the result that the district's war with the blacks intensified.

Albasini also made use of the opportunity to launch an attack on Chief Magor in the southeastern part of the district as he had refused to pay his taxes. Despite the fact that they had guaranteed his safety if he came down from his stronghold on top of Magor Hill, he was murdered by Albasini's troops when he arrived at the foot of the mountain. This deed had severe repercussions in the press. Albasini also wanted to get even with Modjadji, the chief of the Lobedo, as she also refused to pay her toll taxes. According to her, her annual contribution was the rain she made as the rain queen of the Soutpansberg.

In 1864, high chief Ramabulana[9] of the Venda died. He had 11 sons, of whom Davhana was the eldest and Makhado[10] the youngest. According to legend, Davhana and his father had never been on good terms, so Ramabulana told him that he would not succeed him as chief after his death. When Ramabulana died the nation had good reason to suspect Davhana of murdering his father by poisoning him.

A power struggle followed that would have far-reaching consequences in the history of the Soutpansberg. Makhado, supported by his uncle Katse-Katse, or Katlagter [Babler], was the other contender. In an ensuing battle Davhana had to flee and leave behind 45 head of cattle, 11 wives and 12 children. He first went to Chiefs Tshivhase and Mphaphuli for protection, but when life became dangerous for him there he fled to Albasini. As many other chiefs had done before him, Davhana found refuge at Goedewensch. He built a kraal about a kilometre from the homestead against a hill facing Schoemansdal. Because of his father Ramabulana's order to obey Albasini, he and his followers tried to live there in peace. The Executive Council approved of this step by Davhana and, as a taxpayer, officially placed him under the authority of Albasini.

As a result of Davhana's escape and the protection afforded him by the Republic, animosity developed between the Boers and Makhado, as well as between the Boers and his supporters, headmen Tshivase, Madzivhandila, Pago and Sehkhupa.

Makhado also accused the whites of killing his father's sister. If they extradited Davhana, however, the relationship would recover.

By the middle of 1864 Davhana was not safe with Albasini any more. He received news that a band of warriors under Funjufunju, or Tromp, who worked as a hunter for Magistrate Jan Vercueil and Michael Buys, were ready to attack him. Davhana had to send his livestock to a safe place and prepare for the battle. He was attacked on the morning of 3 July 1864 and a number of his kraals were burnt down. Davhana was not only able to withstand the attack, but managed to deal the enemy a severe blow. Albasini was highly indignant about the attack by Magistrate Vercueil's hunter on Davhana, as he had been placed under his protection by the Executive Council. He

---

9   Also called Ramapulana.
10   Also called Maghato or Makhato.

asked Vercueil if he had ordered the attack, but did not receive a direct answer. All Vercueil said was that matters concerning Native Affairs fell under the jurisdiction of the commandant and field cornet. In other words, Albasini was complaining to the wrong person. Vercueil did not see any harm in allowing the blacks to settle their own differences, providing it did not harm the whites in turn. Albasini was not satisfied with this answer and informed the Executive Council of the incident, adding that it had led to the same mistrust as before. It was impossible for him to determine what the grounds for the attack were.

No further direct attacks were made on Davhana for a while. But his stay under the protection of Albasini and the mentioned attack were only the first in a series of unpleasant events that would culminate in the war of 1865 and result in the evacuation of Schoemansdal. Makhado was furious. In the beginning of May 1865, Makhado told Albasini to launch an attack on him. Albasini thought this was a cunning way of luring him away from his farm and then creating havoc in his absence. Instead he waited patiently at Goedewensch and laid his plans well. His strategy included hiding his men and leading Makhado into a trap. The battle took place next to the steep northwestern bank of the Luvuvhu River, only a few kilometres east of his bulwark. Excepting the wounded and those who died of their wounds away from the scene, 68 died during the battle, including five headmen. Some of the warriors were trapped above the riverbank and jumped, with the result that 22 drowned in the deep hippo pool below. Albasini did not lose one man.

Conditions were also critical in Schoemansdal. Magistrate Vercueil instructed the town's people to move into the fort. He appealed to district officers Geyser, Venter and Du Plessis for support, but received no cooperation. Albasini had sent an armed contingent to town as protection, but was forced to withdraw it when Makhado threatened to attack Goedewensch.

Vercueil complained to the Executive Council that a divided state could not exist. If the officials could not reach unanimity, the town and district would be doomed. Apart from the arguments among officials, the instructions of the field cornets and superintendents of native affairs overlapped. This caused A.P. Duvenage to twice resign from his superintendency and led to Magistrate Vercueil's remark to the Executive Council that every official imagined himself to be magistrate. When he tried to further the cause of the good and fair, his actions were doomed to failure by the improper steps taken by Commandant Venter and Field Cornets Du Plessis and S.J. van Rensburg, as well as by J. Albasini's selfish behaviour.

Acting Commandant General M.J. Schoeman's commission to Schoemansdal

The government sent a commission of inquiry from Pretoria to investigate the causes of the war and to try to wind up affairs without sending a commando. The guilty parties had to be pointed out and account for their actions. The commission then had to act according to their findings. The commission of inquiry consisted of Commandant General M.J. Schoeman from Rustenburg, Commandant N. Smit from Waterberg and G.J. Verdoorn as secretary. They could commandeer as many privates as they needed to ensure their safety. The group arrived in the Soutpansberg in the middle of July. During a council of war at which the commission members, the district officers and Albasini were present, they unanimously decided that they would not be able to subjugate the enemy without the help of their countrymen. Schoeman sent for Makhado and Katse-Katse, but they refused to come and instead sent Makoebie and Fleur, Makhado's field-general. During the investigation it came to light that Makoebie had murdered two Van der Merwes as well as a number of blacks. Fleur declared that they would not surrender their rifles as they had earned them from the whites over a period of three years. After this they were arrested and placed under Albasini's care at Goedewensch. This was another extremely tactless move, and while the commission was still in Schoemansdal an attack was made on the town to free the two prisoners. The commission therefore had achieved nothing. Through their irresponsible actions they only increased the animosity between Makhado and the people of the Soutpansberg and plunged the district into deeper misery.

Dietlof S.M. Maré was named commandant of the laager at Schoemansdal. As an extra protective measure a fence of thorn branches was planted around the fortifications at a distance of eight paces from the walls. No alcohol was allowed in the fort and only the corporals and the watch were entrusted with the locking and unlocking of the doors. To distinguish herdsmen from the enemy, they were each given a pass.

During the six months that the people of Schoemansdal stayed in the fort they were subjected to the utmost poverty and wretchedness. People with money could still buy food, but the poor had to go hungry or make do with the little they had. With the inhabitants in laagers, the warriors were free to plunder the farms to their hearts' content. They burnt down every house, as well as Reverend Stephanus Hofmeyr's mission at Goedgedacht, and in the process seized 13 rifles and 31 sheep. They also murdered W. Simpson and six servants on various farms.

Albasini achieved two more victories in Kleinspelonken, but lost against Lwamondo. During this battle a young man named Hans Fourie was captured and flayed alive. His skin was later used by Lwamondo as a kaross and his scalp as a cap during special occasions.

After his arrival back in Pretoria, M.J. Schoeman received orders to commandeer burghers to come to the aid of the people of Schoemansdal. However, at Makapaan's Poort he made the commando turn back, claiming that they did not have sufficient ammunition to pass through the poort.

In the Soutpansberg Commandant Geyser and Albasini attacked a kraal behind Piesangkop. On another occasion, when they were supposed to attack Makhado, Albasini let Geyser know that he could not come because of ill health. Instead, he would send his tribe's warriors to the farm Bergvliet at the foot of the mountain directly south of Makhado's kraal. On the appointed day there was, however, no force, and Geyser had to disband his commando. Albasini offered no excuse for breaking his promise. Geyser was extremely upset and let the government know that if the "Gouvernementsvolk voor tuygsdiens" (the government's black troops for military action) remained under Albasini's command any longer he would resign from his position.

President Pretorius intervenes
At this stage President Pretorius and Commandant General S.J.P. Kruger decided to take the matter into their own hands and investigate whether war was really necessary. The other members of the commission were S.T. Prinsloo, D.F.J. Steyn as state attorney and S.J. Meintjies as secretary. They arrived at Schoemansdal at the beginning of December 1865 and immediately assumed their duties. Fleur and Makoebie appeared before the Executive Council, but the state prosecutor soon complained that he could not continue with his investigations as he was of the opinion that some members of the Executive Council were acting as agents for the accused. This suspicion could, however, never be proven.

The slackness of the officials at Soutpansberg was once again demonstrated by the fact that Assistant Commandant General Geyser, Commandant S.M. Venter and Field Cornets Du Plessis and Duvenage, as well as Magistrate Vercueil, all requested to be released from their duties. Only Vercueil and Duvenage were relieved of their posts, but both were appointed as superintendents of native affairs to help Albasini collect taxes. R.A. van Nispen was appointed as magistrate. After Makhado and Katse-Katse had promised to hand over their rifles, Commandant General Kruger gave the townsfolk permission to return to their homes on the second-last day of December. They would be safe as a guard detail had been left behind. By that time 38 homes had already been burnt down in the district.

The firm and tactful actions of Magistrate Van Nispen
At the end of April 1866 Makhado and his subjects had only handed over four of the 200 rifles in their possession. This was considered proof enough of his cunning. In contrast to the division among the officials the previous year, a feeling of solidarity now arose among them, under very difficult circumstances. This could be ascribed partly to the continuous danger and partly to the able and tactful leadership of Magistrate Van Nispen. By submitting regular reports he kept the government informed of conditions in the Soutpansberg. Most of the inhabitants could not hope to be released from their abject poverty unless the hunting laws were revised and they could hunt again. Trade came to a complete standstill, there was no

credit, and the distances to the other districts were too great to allow a sustainable trade in the products of the Soutpansberg. Many traders were suspected of being involved in arms deals with the blacks. From time to time they entered the district without licences, skirted the town and went directly to the tribes, where they would buy ivory collected with rifles obtained from the whites. The whites suffered while the blacks flourished. Conditions were so critical that at one stage Schoemansdal's powder magazine held only 20 pounds of gunpowder despite the constant danger. Van Nispen left no stone unturned in his attempts to bring order to the chaos. He regularly held meetings with the officials in town so difficulties could be smoothed out and mutual trust be established. He submitted recommendations to the government, in particular regarding the contradictory instructions of the superintendents and field cornets. Unfortunately his appointment came too late to allow him to fix matters that had been botched long before.

A.P. Duvenage requests a commando
A.P. Duvenage was delegated to deliver letters to the Executive Council asking for the help of a strong commando. He could not, however, find either the president or any of the members of the Executive Council at their offices. He did find State Secretary Van der Linde, from whom he received not so much as a receipt acknowledging the delivery. After Duvenage had left, a decision was made to conscript a commando to save the Soutpansberg district and suppress the black tribes.

The Soutpansbergers did not know about this decision and held another public meeting with the officials in town. They came to the conclusion that the government was not overly concerned about conditions in the district. When they were informed that a commando had been conscripted, they were not convinced. By now they wanted a direct answer and Duvenage went back to Pretoria.

Yet again the conscripted commando failed to reach the Soutpansberg. The reaction to the call-up was poor with half those called up failing to arrive and the rest unwilling to go, with the result that it had to be disbanded. When the Executive Council looked into the matter, they found that people were less than enthusiastic to lend a hand as the Schoemansdallers had willingly given guns to the blacks and they saw no reason why they should now be shot at with these guns. Kruger was nevertheless instructed to establish a commando again, and failed once more. By this time Albasini, like Van Nispen, had lost his patience and resigned as superintendent and as Vice Consul of Portugal. Neither government accepted his resignation, however.

Maré, Duvenage and Barend Vorster to Pretoria
On 4 January 1867 a public meeting was once again held in Schoemansdal and it was decided to send a delegation to Pretoria for the umpteenth time in a final call for help. Maré, Duvenage and Vorster put their case before the Executive Council a month later. The Executive Council realised that the situation in the Soutpansberg was critical and were finally convinced that the people there needed help urgently. A commando would leave at the end of May. In the meantime the president would endeavour to buy more lead and primers. A High Court would accompany the commando to investigate to what extent the burghers had contributed to the unsafe conditions in Schoemansdal.

Excavations of the foundations of the powder-magazine at Schoemansdal, with the Soutpansberg in the background.
(APRIL 1985)

Another delegation under Stephanus Schoeman

It would seem President Pretorius was very sensitive to the attitude of the commanded burghers to the people of the Soutpansberg, as he decided off his own bat, without consulting the Executive Council, to send another delegation to Schoemansdal. They had to get to the heart of the matter and decide if things could not be set to rights without going to war. Stephanus Schoeman, Hannes Schoeman and S.J. Meintjies arrived in town on 21 May. The next day Katse-Katse and Tshivhase were summoned. Katse-Katse sent a messenger to inform them that he was not willing to surrender the guns, whereupon the commission decided that a clash was inevitable. They also consulted Albasini and he resolved to send an army of about 1500 Magwamba into the field. This would mean that the combined troops, which included 500 coloured men, would number about 3000.

A commando approaches at last

On 6 June Commandant General Paul Kruger reached Kranskop near the Nyl River with only 400 of the burghers who had been commandeered up for duty. From here he asked the president to commandeer as many men as possible. He also requested 2000 pounds of gunpowder, 3000 pounds of lead and 30 000 cartridge cases, as he did not have enough ammunition.

In mid-June the long-awaited commando arrived in Schoemansdal, welcomed by the jubilant cheers of the long-suffering townsfolk. They pitched camp on the western side of the Dorps River. Kruger sent a message to Katse-Katse saying that they did not want to fight and only wanted the rifles. This had no effect. Katse-Katse had so thoroughly entrenched himself directly east of Visierskerf against a steep mountain, at the foot of the cliff, that he feared no whites. His kraals were built on terraces along the rock face and stone walls protected the entire mountain slope. The terrain was so densely overgrown with thorny shrubs and creepers that anybody who wanted to reach the kraals had to crawl on all fours. Trying to use the existing footpaths would mean certain death, as many traps would have been set.

Although the Commandant General knew his commando had little ammunition, he could also recall other battles he had won with much less. In the hope that this would again be the case, he started planning the assault.

According to Sarel Eloff and the future MP Paul Maré, who were young men when they participated in the battle, it was decided to attack Katse-Katse's stronghold from two sides. On a moonlit night a division of about 400 men, mostly members of the Kruger commando, ascended the mountain to fend off any reinforcements from Makhado and others from the east. From there they would also be ready when the rest of the men started shooting from below. Early one morning, the Commandant General, the men from Soutpansberg, which was well represented, and 1500 Magwambas under the leadership of Albasini advanced on Katse-Katse's cliff. Two thousand metres from the cliffs they were met with rifle fire, which the Boers answered with rifles and big guns (a few old cannons known as "Grietjies").

The Schoemansdal men under the leadership of Field Cornet J.C.J. Herbst came closest to the cliff, while Albasini and the Magwambas attacked from the east. They did not see much of the enemy as they were hiding behind rocks in ravines and fought from there with the Boers' elephant guns. Albasini and his Magwambas managed to seize several small areas, but it soon became clear that the battle would not lead to a decisive victory.

The fighting lasted until about four o'clock in the afternoon, when the Commandant General gave the command to withdraw. The burghers who had been on top of the mountain were attacked on their way down, but managed to escape without any loss of blood.

Teacher Cornelis van Boeschoten composed a song about the episode, which Mrs Biccard later managed to recall in parts. Loosely translated it went:

We climbed the mountain at dawn
Our hearts sincere and pure
Climb brothers, climb with courage
Don't forget God, because he is good.

We were not at the cliffs yet
When shots were roaring from the fort
And moments afterwards
Some were praying for mercy.

The Boers did not launch any further attacks. The days were spent playing jukskei and "ride-oxhide" and enjoying other forms of entertainment. Forty-five years after the event, Paul Maré was still convinced that it had been a lily-livered, unmanly affair. He thought the fact that there had been too many people who wanted to play boss might have been to blame. "As young as I was, I could see that there was not much discipline, but we were living in the year 1867."

The Commandant General had been misled by both the government and the delegation from the Soutpansberg about the ammunition. When the three delegates were in Pretoria they assured the Executive Council that every burgher had enough gunpowder and lead. The ammunition President Pretorius would send had been on its way from Natal. But instead of the transport rider travelling via Wakkerstroom as he was supposed to, he went via Harrismith. There it was held for three months as the Transvaal did not have the right to transport ammunition through the Free State. With the permission of the council of war, A.P. Duvenage offered to fetch more ammunition from Pretoria as soon as possible. A day after his departure word came from President Pretorius that there was no ammunition to be found in Pretoria. He and his secretary, Spruyt, were not idle, however. On 8 July they sent a wagon and a six-man guard with 1500 pounds of lead and 12 500 primers from Pretoria to the Soutpansberg. The magistrate of the Waterberg was asked to have it transported the rest of the way with a stronger guard, but unfortunately they could not reach the town in time.

The High Court
In the meantime, the sitting of the much discussed High Court with Magistrate Van Nispen as one of three judges and S.J. Meintjies as acting state prosecutor was also taking place. The purpose of the court was to decide to what extent certain inhabitants could be held accountable for the uprising in the Soutpansberg district. Assistant Commandant General Frederik Geyser, Commandant S.M. Venter, Field Cornet J.H. du Plessis, João Albasini and S.J. van Rensburg were on trial. The accused had to account for the irregularities that had taken place in 1865 during the search for Monene. Geyser, Venter and Albasini were summoned to appear in court on 15 July at ten o'clock. Albasini was accused of murder, child abduction and theft, Geyser and Venter of murder, manslaughter and careless actions, and Geyser of murder, child abduction and careless actions.

Du Plessis was found guilty and fined £500. He also had to return the 300 cattle he had taken from Pago. Venter also had to return a portion of the looted livestock. Vercueil was fired from his position as superintendent because of incitement, released on bail of £1000 and fined 300 rix-dollars.

Du Plessis's sentence caused an enormous upheaval among those present in court. They stormed the bench, forcibly removed Meintjies and the presiding magistrates from court and released the captives. When Meintjies walked out, a member of the public shouted: "Why do you let him walk away upright? He should be crossing the street doubled up." All of this led to divisions among the commando and there was no hope of a united front. On 9 July the war council unanimously decided to inform the High Court that they could not continue with their duties as they did not see their way clear to carrying out the sentences because of the divisions in the laager.

A difficult decision
Because of the scarcity of food and ammunition, high mortality among the cattle and horses, and division among the people in the laager, it was clear to the Commandant General that he could not keep the commando with him. He suggested to the war council that the commando should be sent back under the command of T. Pretorius. He would stay behind with a number of burghers to at least attempt to protect the town until the promised ammunition arrived. They would then launch a second attack on Katse-Katse. But the war council would not agree to this proposal.

An idea that emanated from the large commando laager was that the town should be relocated to a safer place. The inhabitants were still hardly aware of the proposal when a petition was drawn up under the leadership of Rev. Van Warmelo, stating why the town should not be abandoned. They said the value of such a frontier town was that it was the gateway to the Soutpansberg and it would be a disgrace to the Republic to relinquish a town to the blacks without having put up a fight. The about 60 petitioners promised to remain in town, to protect it and stay in control of it if possible. The Commandant General was pleased at their attitude and promised to submit their petition to the war council. The war council was equally impressed and decided not to move the town. The large commando laager would retreat while the responsibility of protecting the town would fall to Jan du Plessis and a handful of volunteers. The townspeople, however, were not prepared to obey the orders of a man who had so recently received such a dishonourable sentence. Even his own supporters discouraged him, causing Du Plessis to declare that it would be impossible for him to protect the town under such circumstances. He started a second petition, which he asked be signed by everyone who agreed with him. This is how people who at first petitioned for the preservation of Schoemansdal, now asked that the war council provide them with wagons so they could move away. The result was that by the time the commando was ready to depart, everybody had to follow. Keeping all of these developments in mind, an order was issued on 12 July that everybody had to pack as fast as possible as the commando would leave on 15 July 1867. Soon the sound of hammer blows could be heard throughout town.

The exodus
Paul Maré later described the withdrawal: "Everything was astir. Many tears were shed. Women cried and children did likewise out of sympathy; and dogs were barking and howling as a result of the unusual commotion. There was no time to kill the tied pigs and their squeals mixed with the anxious noise of a cockerel being caught, either to be slaughtered or to be stuffed into a cage."

In the short time people had at their disposal, they could only pack the most necessary goods. That is why many of Schoemansdal's official documents have not been found, as they were probably left behind or burnt. There was no leeway, not even for Rev. Van Warmelo. That Sunday, instead of delivering his sermon, he had to remove the church's doors and windows and stack them on a wagon. Six months later an eyewitness said that one had to have been an inhabitant of the unfortunate district, to have stayed there for a year and to have experienced the mixed emotions of the people, before one could come close to understanding what they went through.

Johannes Augustus Breedt, a respected elder of the Reformed Church and field cornet of Schoemansdal, remarked in his memoirs:

> ... there had been no immediate danger, but instead of retreating in an orderly fashion, the exodus soon turned panicky and became a headlong flight in which people left behind everything they owned, and therefore were soon doomed to abject poverty and suffering. In a rather senseless manner the poor people threw a few pieces of furniture and clothing onto wagons and left their other possessions for the blacks.
>
> The blacks were also not slow to take advantage of the panic and as the last wagons were leaving town flames started to engulf the houses at the back, as they were setting them alight.
>
> Most people moved to Marabastad in the vicinity of Pietersburg (Polokwane). We also moved our belongings there, or as much of it as I could transport.

Apparently the Magwambas did not want their white chief João Albasini to move away, and when the decision was taken to leave town, he was there. He informed the war council that he and his family would not leave the district, but rather stay on in their fort at Goedewensch. He was therefore ordered to prepare to protect his people and to ask help from high headmen Mswati, uMzilikazi, Madumelane and Muzila, which he did via couriers.

Reaction to the evacuation of Schoemansdal
Five days after the town had been evacuated, Albasini wrote to President Pretorius: "The difficult and bloody work of almost 20 years has been sacrificed, without any significant effort to retain anything ... Only I stayed behind on my farm with about 27 inhabitants and the Magwambas in my area and I will do anything to retain the district for the Republic ..."

He was hurt that he had done so much for the district and had received so much acclamation from both the Executive Council and the Volksraad, only to be summoned to court and accused of murder, child abduction and theft. He had only the highest praise for the Commandant General, saying that Kruger had done everything in his power to retain the town, but that most of the burghers of the "front districts" had refused to stay and Kruger did not have the power to force them to obey. How could he, who had been called the bravest among the brave, catch a rabbit with reluctant dogs?

Commandant General Kruger, who left the town with tears in his eyes, moved with his commando to Moletse Land northwest of the current Polokwane, where he stayed a few days to give the refugees the opportunity to reach the laager at Marabastad. At this camp lower down the Sand River, he informed President Pretorius about the course of events. He talked of moving Schoemansdal to Marabastad or elsewhere and asked him to meet the people who were on their way, lending visible help as the morale in the laager left much to be desired. He was obviously still planning to relieve the town. He also mentioned that 30 burghers had absconded. Unfortunately it was not possible for the president to come, as he was bedridden as a result of all the privations he had suffered on his travels as he tried to find men for the commando. After Kruger had disbanded the commando and left a temporary guard at Makapane Poort, he returned to Pretoria, arriving a month later.

The people were unhappy in the laager at Marabastad. The future MP D.S. Maré vehemently criticised the Commandant General about the evacuation of Schoemansdal and described their position as follows: "We were put on a bare hill, without any provision of accommodation or housing for ourselves, our families, or our possessions. We were left behind in a desperate condition, in which everybody had to make do, exhausted and impoverished by the last years of the war and rumours of war." Rev. Van Warmelo, who found himself at Weltevreden near Rhenosterspoort with Field Cornet Jan du Preez, and his church council members protested to the president and members of the Executive Council over the evacuation of the town. He would never forgive the future president this.

During its September session the Volksraad approved the decision of the war council to disband the High Court. Commandant General Kruger was exonerated of all blame. They felt that the evacuation of Schoemansdal and the fact that it had fallen prey to the "savages" was regrettable and definitely cast a stain on the Republic's army, but it could not be ascribed to cowardly behaviour, intentional dereliction of duties or a lack of military skill on the part of Commandant General S.J.P. Kruger. Many unforeseen circumstances, coupled to divisions in the laager, had led to the incident.

The Volksraad also decided to declare Acting State Prosecutor Meintjies's activities illegal as he had not been sworn in. This decision was slightly modified later in that the cases that had been finalised would be considered binding, but the rest would be regarded as "non-existent and never having happened". Albasini and the other accused who had not been sentenced, were thus exonerated of all blame. And so ended the history of the once prosperous Voortrekker town of Schoemansdal, the former ivory mecca of Transvaal: through unnecessary discord and division.

Today, when one tries to reach the site, one is confronted with a seemingly impenetrable wilderness of Karoo thorn, umbrella-, sickle-bush and buffalo-thorn trees, making it difficult to believe that Commandant General Potgieter had chosen this par-

Row upon row of unidentified Voortrekker graves in Schoemansdal's old cemetery. (09.12.1985)

6. THE BYGONE DAYS OF THE SOUTPANSBERG

ticular spot for his town, and that there had once been a flourishing Voortrekker town with fertile vegetable gardens, fruit orchards and a network of water canals here.

The Dorps River is now a dry donga, overgrown with Karoo thorn and tree pincushions. There can be no doubt that the area had a much wetter climate during the nineteenth century. This is confirmed by all the reports about the former Schoemansdal, as well as the diary of the Voortrekker J.G.S. Bronkhorst. The bush's encroachment on the once open game-rich tree savannah can probably be ascribed in part to the drier climate today. Other factors that might have contributed are years of overgrazing and the absence of regular natural veld fires.

It is also difficult to believe that most of the 122 graves in Schoemansdal's cemetery contain the remains of inhabitants who had died of malaria in a span of only 19 years. Malaria is now rare in this region. As in Ohrigstad, it occurs only in wet years.

Thanks to recent excavations and restoration work future generations will be able to see what this forgotten Voortrekker town once looked like and how the Boers lived in those days. The terrain of the former town has been used since 1917 as a national holiday site where Day of the Covenant (now Day of Reconciliation) and other gatherings take place. In 1984 the executive committee of the former Transvaal Provincial Administration, on the initiative of Administrator W.A. Cruywagen, gave permission that Schoemansdal may function as a provincial museum, and that partial restoration, which would recapture the spirit and look of the Voortrekker settlement, could start. The project was tackled in the 1990s and completed under the guidance of Prof. Hannes Eloff, Dr Udo Küsel (curator of the National Cultural History Museum in Pretoria) and archaeologist and civil engineer Sidney Miller.

Archaeological restorations at Schoemansdal. The foundations of the old fort with its two canon bastions are clearly visible from the air.
(01.05.1986)

References

Acutt, N. 1938. Makapaan se gruweldade, *Die Huisgenoot* 6 May.

Axelson, E. 1967. *Portugal and the Scramble for Africa (1865–1891)*. Witwatersrand University Press, Johannesburg.

Axelson, E. 1967. *Portuguese in South-East Africa 1488–1600*. C. Struik, Cape Town.

Axelson, E. 1969. *Portuguese in South-East Africa 1600–1700*. Witwatersrand University Press, Johannesburg.

Basson, J.J. 1994. *Vestiging en dorpstigting: 'n Kultuurhistoriese studie van Transvaalse pioniersdorpe*. M.A. thesis, University of Pretoria, Pretoria.

Bergh, J.S. (ed.) 1999. *Geskiedenisatlas van Suid-Afrika: Die vier noordelike provinsies*. Van Schaik, Pretoria.

Bergh, J.S. (ed.). 1988. *Herdenkingsjaar 1988: Portugese, Hugenote en Voortrekkers*. De Jager-HAUM, Pretoria.

Boeyens, J.C.A. 1987. Nuwe lig op die uitleg van Schoemansdal. *Suid-Afrikaanse Tydskrif vir Kultuur- en Kunsgeskiedenis* 1(2): 182–186.

Boeyens, J.C.A. 1990. Die konflik tussen die Venda en die blankes in Transvaal, 1864–1869. *Archives Year Book for South African History* 53(2).

Boeyens, J.C.A. 1991. "Zwart Ivoor": Inboekelinge in Zoutpansberg 1848–1869. *Suid-Afrikaanse Historiese Joernaal* 24 May.

Boeyens, J.C.A. 1994. "Black Ivory": the indenture system and slavery in Zoutpansberg 1848–1869. In: E.A. Eldridge & F. Morton (eds). *Slavery in South Africa, Captive Labor on the Dutch Frontier*. University of Natal Press, Pietermaritzburg.

Breytenbach, J.H. et al. *Suid-Afrikaanse Argiefstukke*, Notule van die Volksraad van die Suid-Afrikaanse Republiek, parts I to VII, (1845–1868). Cape Town.

Bruyns, J.A. 1932. 'n Verdwynende Voortrekkerdorp. *Koningsbode* December 1932.

Changuion, L. 1986. *Pietersburg – die eerste eeu 1886–1986*. V&R Printers, Pretoria.

Child, D. 1989. A forgotten South-African town: fever-ridden village in Northern Transvaal was center for ivory trade. *The Settler* 62(2) March/April.

Damangwato. 1868. *To Ophir Direct or the South African Gold Fields*. Unknown publisher, London.

Das Neves, D.F. 1879. *A Hunting Expedition to the Transvaal*. George Bell & Son, London.

De Jager, H.J. 1948. Schoemansdal: Verste voorpos van Boerebeskawing. *Fleur* (Magazine supplement of *Dagbreek*) May.

De Jager, H.J. & De Vaal, P. 1948. *Eeufees-Gedenkboek van Schoemansdal 1848–1948*. Unknown publisher, Louis Trichardt.

De Jong, C. 1993. Skaduwees oor Schoemansdal, 1867. *Pretoriana* 102, March.

De Kock, W.J. 1957. *Portugese ontdekkers om die Kaap: die Europese aanraking met suidelike Afrika*. A.A. Balkema, Cape Town.

De Vaal, J.B. 1938. Die wordingsgeskiedenis van die Soutpansbergse distrik tot 1867. Series in *Die Noord-Transvaler* July 1938 to July 1939.

De Vaal, J.B. 1939. Opvoeding in Schoemansdal. *Historiese Studies* 1(1).

De Vaal, J.B. 1946. Schoemansdal, standplaas van die Voortrekkers. *Die Transvaler* 21 December.

De Vaal, J.B. 1946. Casimiro Simoês. *Die Volkstem* 6 December.

De Vaal, J.B. 1948. Schoemansdal, die verlate Voortrekkerdorp in die Zoutpansberg. *Die Naweek* 10 June.

De Vaal, J.B. 1948. Olifantjag in die dae van Schoemansdal. *Die Naweek* 10 June.

De Vaal, J.B. 1948. Olifantjag in die dae van Schoemansdal. *Die Naweek* 17 June.

De Vaal, J.B. 1948. Ivoorsmouse in Schoemansdal. *Die Naweek* 24 June.

De Vaal, J.B. 1948. Die Makapaan en Mapeiamoorde. *Die Naweek* 1 July.

De Vaal, J.B. 1948. João Albasini, wit kaptein van die Magwamba. *Die Huisgenoot* 4 & 18 June.

De Vaal, J.B. 1953. Die Rol van João Albasini in die Geskiedenis van die Transvaal. *Archives Year Book for South African History* 16(1):1–155.

De Vaal, J.B. 1982. *João Albasini (1813–1888)*. University of the Witwatersrand Press, Johannesburg.

De Vaal, J.B. 1984. Ou handelsvoetpaaie en wapaaie in die Oos-en Noord-Transvaal. *Contree* 16:5–15.

De Vaal, J.B. 1985. Handel langs die vroegste roetes. *Contree* 17:5–14.

De Waal, J.J. 1998. *Die verhouding tussen blankes en die hoofmanne Mokopane en Mankopane in die omgewing van Potgietersrus (1836–1869)*. M.A. thesis, University of South Africa, Pretoria.

De Waal, J.J. 2000. *Schoemansdal: 'n Voortrekkergrensdorp, 1848–1868*. D.Litt. et Phil. thesis, University of South Africa, Pretoria.

Dicke, B.H. 1936. *The Bush Speaks: Border Life in Old Transvaal*. Shuter & Shooter, Pietermaritzburg.

Dixon, W.M. & Grierson, H.J.C. 1967. *The English Parnassus*. Oxford University Press, Oxford.

Du Preez, M.H.C. 1993. *Die wedervaringe van Voortrekker Pieter du Preez (1807–1889)*. Yearbook of the Africana Society of Pretoria.

Elridge, E.A. & Morton, F. (eds). 1994. *Slavery in South Africa: Captive Labour on the Dutch Frontier*. University of Natal Press, Pietermaritzburg.

Engelbrecht, S.P. 1920. *Geschiedenis van de Nederduits Hervormde Kerk in Zuid-Afrika 1*. De Bussy, Pretoria and Amsterdam.

Engelbrecht, S.P. 1927. Schoemansdal. *Die Volkstem* 15 December.

Engelbrecht, S.P. 1936. *Geskiedenis van die Nederduitsch Hervormde Kerk van Afrika*. Unknown publisher, Pretoria.

Engelbrecht, S.P. 1942. *Eeufees-album van die Nederduitsch Hervormde Kerk van Afrika 1842–1942*. Unknown publisher, Pretoria.

Engelbrecht, S.P. 1943, 1944, 1945. Ds. N.J. van Warmelo. *Almanak van die Nederduitsch Hervormde Kerk in Afrika*, 1943:47–74; 1944:89–118; 1945:33–60.

Engelbrecht, S.P. 1952. *Die Nederduitsch Hervormde Gemeente Pietersburg (Zoutpansberg) 1852–1952*. Wallachs, Pretoria.

Ferreira, O.J.O. 1973. Schoemansdal: vergane Voortrekkervoorpos. *Die Taalgenoot* February 1973.

Ferreira, O.J.O. 1977. Schoemansdal: van Voortrekkerpos tot Volksfeesterrein. *Contree* 1, January:5–10.

Ferreira, O.J.O. 1978. *Stormvoël van die Noorde*. Promedia, Pretoria.

Ferreira, O.J.O. 1988. Orde van die Tempel, Orde van Christus en die Portugese ontdekkingstogte. *Suid-Afrikaanse Tydskrif vir Kultuur- en Kunsgeskiedenis* 2(1) January.

Ferreira, O.J.O. 2002. *Montanha in Zoutpansberg. 'n Portugese handelsending van Inhambane se besoek aan Schoemansdal, 1855–1856*. Protea Book House, Pretoria.

Fourie, G. 1948. Die wildernis wen. *Die Brandwag* 23 July.

Grobler, H.J. 1919. Storie van 'n vergane Voortrekkerdorp. *Die Brandwag* 24 December:217–222.

Grobler, H.J. 1929. 'n Vergane Voortrekkerdorp. *Die Vaderland*

6 September.
Grobler, H.J. 1938. 'n Voortrekkerdorp wat verrys en verdwyn het. *Die Volkstem* 9 December:4.
Henriksen, T.H. 1978. *Mozambique: A History*. Rex Collings & David Phillips. London.
Hofmeyer, S. 1890. *Twintig jaren in Zoutpansberg*. J.H. Rose & Co, Cape Town.
Huet, P. 1869. *Het Lot der Zwarten in Transvaal*. J.H. van Pearson, Utrecht.
Joubert, C.J. 1959. *Die soutpan van Soutpansberg*. D.Ed. thesis, University of Pretoria, Pretoria.
Junod, H.A. 1927. *The Life of a South African Tribe I & II*. Macmillan, London.
Kleyn, W.J. 1988. Die lewe van João Albasini. *Militaria* 18(2).
Kriel, C. 1988. Schoemansdal – herlewing van 'n pioniersnedersetting. *Overvaal Musea News* 15(2) September.
Kriel, C. 1992. The influence of the tstetse fly on settlement in Northern Transvaal with special reference to Schoemansdal. *Overvaal Musea News* 18(2) October.
Krüger, D.W. 1937. Die vestiging van die blanke beskawing in Noordoos-Transvaal – uit die geskiedenis van Schoemansdal. *Die Burger* 9 January.
Krüger, D.W. 1945. Andries Ohrigstad. Mislukte Nedersetting se verreikende betekenis. *Die Huisgenoot* 13 July.
Krynauw, D.W. & Pretorius, G.H.S. 1949. *Transvaalse Argiefstukke 1850–1853*. Government Printers, Pretoria.
Le Roux, T.H. (ed.). 1966. *Die dagboek van Louis Trigardt*. Van Schaik, Pretoria.
Mansveldt, S.S. 1942. Schoemansdal – eers 'n bloeiende Voortrekkerdorp, nou 'n volksfeesterrein. *Die Huisgenoot* 20 November.
Macqueen, J. 1862. Journey from Inhambane to Zoutpansberg by Joaquim de Santa Rita Montanha. *Journal of the Royal Geographical Society* 32:63–68.
Maré, P. 1912. Schoemansdal. Het verlaten van die plaats in 1865 (sic). *Die Brandwag* 15 July.
Miller, S.M. 1997. From Schoemansdal to Thulamela: thoughts on reconstruction in the conservation process. *Research by the National Cultural History Museum* 6.
Miller, S.M. & Tempelhoff, J.W.N. 1990. Die romantiek van 'n grensterrein. *Fauna en Flora* 47.
Moerschell, C.J. 1912. *Der Wilde Lotrie*. H. Stürtz, Würzburg.
Möller-Malan, D. 1953. *The Chair of the Ramabulans: A Story of Bantu Life in the Zoutpansberg*. Unknown publisher, Johannesburg.
Möller-Malan, D. 1957. Die donker Soutpansberg: Ramavoyha se "Rooimiere". *Historia* 2(1).
Moolman, J.P.F. 1982. *Die boere se stryd teen die swart stamme in en om die ZAR 1864–1871*. D. Litt. et Phil. thesis, University of South Africa, Pretoria.
Munnik, G.G. (n.d.). *Kronieke van die Noordelike Transvaal*. S.A. Boekwinkel, Pretoria.
Naudé, P.J. 1939. *Die Geskiedenis van Zoutpansberg 1836–1867*. M.A. thesis, University of South Africa, Pretoria.
Nienaber, P.J. 1938. Schoemansdal. *Die Burger* 1 October.
Page, B. 1994. João Albasini (1830–88), a man of many parts. *Lantern* 43(4).
Pelzer, A.N. 1950. *Geskiedenis van die Suid-Afrikaanse Republiek I: Wordingsjare*. A.A. Balkema, Cape Town.
Pont, A.D. 1955. *Nicolaas Jacobus van Warmelo 1835–1892*. Kemink & Zoon N.V., Utrecht.
Potgieter, C. & Theunissen, N.H. 1938. *Kommandant-generaal Hendrik Potgieter*. Afrikaanse Pers, Johannesburg.
Preller, G.S. 1915. *Baanbrekers: 'n hoofstuk uit die voorgeskiedenis van Transvaal*. Volkstem-Drukkerij, Pretoria.
Preller, G.S. 1917. *Dagboek van Louis Trichardt 1836–1838*. Het Volksblad Drukkerij, Bloemfontein.
Preller, G.S. 1921. Schoemansdal en sy kerkhof. *Die Brandwag* 25 January:123–216.
Preller, G.S. 1925. *Voortrekkermense IV*. Nasionale Pers, Cape Town.
Preller, G.S. 1931. *Oorlogsoormag en ander sketse en verhale*. Nasionale Pers, Bloemfontein.
Preller, G.S. 1940. Stigting en verwoesting van Schoemansdal. *Die Vaderland* 13 December.
Pretorius, H.S. & Krüger D.W. 1937. *Voortrekker-argiefstukke 1829–1849*. Government Printers, Pretoria.
Punt, W.H.J. 1953. *Louis Trichardt se laaste skof*. Van Schaik, Pretoria.
Rademeyer, J.I. 1944. Die oorlog teen Magalo (M'pefu). *Historiese Studies* 5(2) June.
Roos, J. De V. 1938. Die ou Ivoorsentrum van Transvaal. *Die Huisgenoot* 3 June.
Smith, A.K. 1970. *The struggle for control of Southern Moçambique, 1720–1835*. D.Phil. thesis, University of California, Los Angeles.
S.P. Engelbrecht-Museum, Pretoria (EMP). *Livor Mestre ... Zoutpansberg Septembro 1863 No. Il C. Simoens*. (Casimiro Simões's ledger).
Stoffberg, D.P. (ed.). 1988. *Verslag van die argeologiese opgrawings van die Schoemansdalse Voortrekkerskans (distrik Louis Trichardt)*. Ethnological Services, SA Defence Force Headquarters, Pretoria.
Suid-Afrikaanse biografiese woordeboek II:5–6. 1972. Insert on: João Albasini. Tafelberg, Cape Town & Johannesburg.
Tempelhoff, J.W.N. 1989. *Die okkupasiestelsel in die distrik Soutpansberg, 1886–1899*. D.Litt. et Phil. thesis, University of South Africa, Pretoria.
Transvaalse Archives (TA), Acquisition: "A.17" Letter of J. Fleetwood Churchill containing a description of his journey to Soutpansberg in 1956. "A.26" Collection of documents, originally from Dietlof S. Maré, Magistrate of Soutpansberg 1865. "A.81" De Santa Rita Montanha's description of his journey from Inhambane to Soutpansberg and back, 1855/56. "TA" State Secretary, Z.A. Republiek (indicated as S.S.). State Secretary, Z.A. Republiek. (Incoming documents, indicated as R.). State Secretary, Z.A. Republiek, (Outgoing documents, indicated as B.B.). Executive Council, Z.A. Republiek (indicated as U.R.). Volksraad Z.A. Republiek (indicatedas E.V.R. en V.R.). State Attorney (indicated as S.P.).
Van Aswegen, H.J. & Verhoef, G. 1982. *Die geskiedenis van Mosambiek*. Butterworth, Durban.
Van Der Merwe, P.J. 1986. Die Matabeles en die Voortrekkers. *Archives Year Book for South African History* 49(2).
Van Pletsen, J.C. 1949. Schoemansdal – Voortrekker outpost of a century ago. *Journal of the Institute of Bankers in South Africa* 46, July.
Van Tonder, B.S.C. 1952. *Die verhouding tussen die Boere in die Zuid-Afrikaansche Republiek en die Portugese van Mosambiek tussen die jare 1836–1869*. M.A. thesis. University of Pretoria, Pretoria.
Wichmann, F.A.F. 1939. Die Nedersetting te Ohrigstad. *Historiese Studies* 1(1):16–25.
Wichmann, F.A.F. 1941. Die wordingsgeskiedenis van die Zuid-Afrikaansche Republiek 1838–1860. *Archives Year Book for South African History* 4(2).
Yates, C.A. 1944. Jawawa . In: *Varia*. Unpublished manuscript, Skukuza Archives.

6.2 Coenraad de Buys – renegade and pioneer of the Soutpansberg

Dr J.B. de Vaal

Upon his arrival in the Soutpansberg in 1836, Louis Trichardt met two coloured brothers, Gabriël and Doris (Doors) de Buys. They lived close to Venda Chief Rasethau; near Trichardt's second encampment and the site of the future Schoemansdal. Gabriël and Doris were the sons of Coenraad de Buys, the well-known pioneer, adventurer, renegade and wanderer. He was the first white person to live in the old Transvaal and patriarch of the remarkable Buys community.

Coenraad de Buys was born in 1761 at Wagenbooms River near Kogmanskloof in the Cape. He was the second son of Johannes de Buys and Christina Scheepers, the widow of Dirk Minnie, and was christened on 24 October 1762 in Cape Town. Shortly afterwards the family settled on their farm Eselsjagt in the Upper Langkloof. In 1769, at the age of seven, he lost his father. The next year his mother married her third husband, Jacob Senekal. After this, Coenraad was fostered by his half-sister Geertruy Minnie and her husband, David Senekal, in the Swellendam district. Judging by his writing he must have had a fair amount of schooling.

Initially he worked as foreman on his foster parents' farm, but after a lengthy court case against his foster father Coenraad moved to his own quit-rent farm, De Brakkerivier, in the Langkloof. In 1784 he moved to Boesmans River, where he joined some family members. This is where he first became involved in criminal activities by illegally trading cattle with the Xhosa. He was also guilty of forging names on a petition and alienated the magistrate of Graaff-Reinet, H.C.D. Maynier, when he and other burghers refused to participate in the Second Frontier War of 1789 against the Xhosa – a war, incidentally, that was to a great degree fuelled by his actions. As a result Maynier and other officials branded him as a troublemaker.

In 1790 three farms – Brandwagt, De Driefonteinen and Brakfontein – were registered in his name and two years later he failed to pay his rent.

De Buys was known to be a staunch patriot and passionately opposed to Dutch rule. When the British came into power in the Cape he was banished from the Colony by Governor Macartney on 14 February 1798 and a reward of 100 rix-dollars was placed on his head. He fled across the Fish River and vowed not to return while the Colony was under British rule.

In 1799 he fruitlessly attempted to reach the Portuguese colonies along the east coast. When he failed, he went to live with the Tamboekies (Thembu). By September 1799 he was back with Xhosa Chief Ngqika (Gaika) along the Tyume River, which is where he met Dr J.T. van der Kemp of the London Missionary Society. Together they settled on land on the other side of the Keiskamma River that Ngqika had made available to them on account of his friendship with De Buys.

In 1803 the Cape was once again under Batavian rule. On 23 June of the same year, De Buys acted as interpreter for General Janssens during negotiations with Ngqika at the Kat River. He did not make a favourable impression on the general and, on Janssens' order, he and a number of other renegades living among the Xhosa had to move to the Langkloof. Here he met J.A. de Mist and company on 31 December 1803. At first he made a better impression on De Mist, with whom he had a long conversation. According to De Mist's personal physician, Dr M.H.C. Lichtenstein, De Buys was about seven feet tall, strong and well proportioned, with a serene attitude, a high forehead, and a certain dignity in his movements. He was humble and reserved with a soft and friendly attitude. These pleasant attributes put paid to all their preconceptions about him.

He answered their questions about various subjects, but when it came to himself and his relations with the Xhosas he gave a sly smile and avoided their questions. They got the impression that he considered their curiosity beneath his dignity, and his craftiness negated the good impression his pleasant looks and manner had made.

De Buys had several liaisons during his wander-

Buyskop in the vicinity of Buysdorp, between Vivo en Mara along the southern foothills of the Soutpansberg, where particularly large specimen of candelabra trees, *Euphorbia ingens* occur. (01.05.1986)

ings. While living with Van der Kemp, who was married to a Xhosa woman, he had an affair with Ngqika's mother, who apparently was a highly intelligent and influential woman.

Between 1782 and 1805 he had a relationship with Maria van der Horst, a coloured woman. After his confirmation, seven of his children were christened in Swellendam between 1807 and 1812. They were Maria Magdalena, Johannes, Georg, Frederik, Eliza, Aletta, Petrus and Elisabeth.

Coenraad officially married Elizabeth on 7 December 1812 in Swellendam. She was from the land of the Makinas behind the Tamboekies and was a sister of uMzilikazi (Moselekatse). He had five sons with her, namely Doris (Doors); Gabriël, born on 17 September 1808; Michael, born on 31 January 1812; Jan; and Baba. Piet and Dorha were the children of other women.

Because Coenraad could apparently not live in peace with his neighbours and because his marriage to a black woman was frowned upon in the Colony, he moved away from the Langkloof, first to the vicinity of the present Beaufort West and later with a number of followers across the Orange River. From the Northern Cape reports reached the Colony about his raids on the livestock of the local tribes. Attempts by the colonial authorities to capture him failed and again they placed a reward on his head – this time it was set at 1000 rix-dollars. In 1815 he and his followers moved to the western Transvaal where he lived among the Huruthse at their old kraal Tshwenyane next to Enzelsberg. They called him Môrô on account of his morning greeting.

After having joined Tswana headman Mokwasele, Setjele's father, he moved to the territory of the Bamangwato. From here he undertook his next journey, which would be along the Limpopo, with three wagons. This must have been in about 1821, when he was 60 years old and his health was deteriorating. On the fifth day of this trek his wife Elizabeth died of malaria. His son Michael later said that his father was grief stricken at their mother's death. In this bereft state of mind he told them to remain in the Soutpansberg area and not to move further inland or back to the Cape. He knew that whites would eventually arrive and trusted that the Lord would look after them. The next morning the children could not find him anywhere as he had apparently set off on his own during the night.

He was never seen again. Reports reached his family that he had made it to the Portuguese Sofala and married a European woman, with whom he had a son and a daughter. Rumour had it that this daughter later married a Portuguese. It is, however, more probable that the weakened Coenraad died of malaria on his way to the east coast. Of his children, Baba froze to death during a snow storm while with the Bamangwato, Dorha died of measles in Bechuanaland (Botswana), and Jan, like his mother, died of malaria.

After his long trek from the Colony, Trichardt met Doris and Gabriël in the Soutpansberg and, as has been mentioned, Gabriël in particular was a great help to him. It was Gabriël who undertook the long journey to Delagoa Bay in May 1837, accompanied by the Magwamba Waai-Waai. He returned in August with the two Portuguese guides Antonio and Lourenço, who would help Trichardt to find Delagoa Bay. Despite Doris's promise to accompany Trichardt, neither he nor Gabriël did so. During their stay in the Soutpansberg, Doris once warned Trichardt and his people of an imminent attack by one of the Venda headmen. In December 1836 Doris and Gabriël also helped Trichardt and his trekkers to stand by their ally, Rasethau, in a battle against a hostile chief.

Shortly after Trichardt's departure from the Soutpansberg, the De Buys people were harried by the Venda, so they fled to Buyskop near what is now Bela-Bela (the former Warmbaths). Here they were besieged once by the AmaNdebele, who planned to let them die of thirst. Gabriël defiantly lifted their last knapsack with water above his head for the enemy to see. Then he poured the water on to the ground and shouted down at them that there was plenty more where that came from. Thanks to this desperate manoeuvre they were left in peace.

Doris also assisted A.H. Potgieter with certain matters, but when he later fell out of favour with him, he turned to the Volksraad Party in Ohrigstad. In 1848 he guided the Burgher expedition from Ohrigstad to the site of the Van Rensburg murder next to the Olifants River. On 8 March 1849, almost a year after Potgieter's move to the Soutpansberg, the Volksraad unanimously appointed him chief and high chief of all the black tribes as far as their territories stretched, except in the direction of the Soutpansberg. He would be accountable to the Volksraad. However, this was a huge mistake as he abused his position by blackmailing people. Soon a warning came from Magistrate Versfeld at Krugerspost that Buys's situation was precarious, followed by a petition by 30 people arguing that his actions were causing turmoil across the country and that he had to be relieved of his position.

In the Soutpansberg Commandant General Schoeman appointed Michael de Buys in 1855 to collect taxes from the blacks. He later declared that he had done this three times and collected in total eight head of cattle, 110 sheep, 400 goats, 120 hoes made by the black smiths and nine elephant tusks.

Michael was very keen to find a missionary to work among his people and Acting President W.C.J. van Rensburg supported him in this. On 13 May 1863 the first missionary couple, Reverend Alexander MacKidd and his wife, arrived at the Buys settlement a few kilometres west of Schoemansdal.

The couple stayed with a sympathetic farmer, Cornelis Lottering, and his wife on their farm Houtrivier, which was 29 kilometres west of town. Lottering also donated a farm to the mission, apparently Goedgedacht, while Rev. MacKidd bought Kranspoort from him. The mission, which still exists today, was laid out on these two farms.

A year after their arrival in the Soutpansberg, Mrs MacKidd passed away and a year later, on 30 April 1865, so did her husband. Rev. Van Warmelo of Schoemansdal delivered the eulogy and the future Reverend Stephanus Hofmeyr, who was also ill at the time, said a few words. Rev. MacKidd's wish to be carried to his grave by the people of the mission was honoured. Stephanus Hofmeyr had been at Goedgedacht since February 1865 to help out with all the work and could therefore take over immediately. He stayed there for 40 years.

Michael de Buys eventually became a strong leader among his father's descendants. About a month before his death on 19 May 1888, at the age of 76, he travelled to Pretoria with a deputation to officially request land for the Buys people. The delegation was cordially received by President Kruger and he granted them 11 000 morgen in perpetuity for the services they had rendered to the Republic through the years. This land became known as Mara and, in contrast with their restless, impetuous ancestor and progenitor, the Buys people are still peacefully living there.

Buysdorp, where the descendants of the renegade and pioneer, Coenraad de Buys, are still living today.
(01.05.1986)

References

Breytenbach, J.H. *Suid-Afrikaanse Argiefstukke*, Notule van die Volksraad van die Suid-Afrikaanse Republiek I. (1844–1850).

De Vaal, J.B. 1953. Die rol van João Albasini in die geskiedenis van Transvaal. *Archives Year Book for South African History* l6(1):1–155.

De Vaal, J.B. 1953. Buys, Coenraad de, *Encyclopaedia of Southern Africa*. Nasionale Boekhandel, Cape Town.

Ferreira, O.J.O. 2002. *Montanha in die Zoutpansberg. 'n Portugese handelsending van Inhambane se besoek aan Schoemansdal, 1855–1856.* Protea Book House, Pretoria.

Hofmeyr, S. 1890. *Twintig jaren in Zoutpansberg*. J.H. Rose 8. Co., Cape Town.

Hoge, T. 1946. Aantekeninge i.v.m. die familie van Coenraad Buys. *Tydskrif vir Volkskunde en Volkstaal* 3(1):17.

Krynauw, D.W. & Pretorius, H.S. 1949. *Transvaalse Argiefstukke*. Government Printer, Pretoria.

Maree, W.L. 1962. *Lig in Soutpansberg*. N.G. Kerkboekhandel, Pretoria.

Millin, S. 1950. *King of the Bastards*. William Heinemann, Melbourne, London, Toronto.

Preller, G.S. 1917. *Dagboek van Louis Trigardt*. Het Volksblad Drukkerij, Bloemfontein.

Preller, G.S. 1918. *Buys en sy bure. Sketse en opstelle*. Van Schaik, Pretoria.

Pretorius, H.S. & Krüger, D.W. 1937. *Voortrekker-argiefstukke 1829–1849*. Government Printer, Pretoria.

Punt, W.H.J. 1953. *Louis Trichardt se laaste skof*. Van Schaik, Pretoria.

Schoeman, A.E. 1933. *Coenraad de Buys*. M.A. thesis. University of Pretoria, Pretoria.

Schoeman, A.E. 1938. *Coenraad de Buys the first Transvaler*. De Bussy, Pretoria.

Suid-Afrikaanse biografiese woordeboek II: 165–167. 1972. Insert on: De Buys, Coenraad. Tafelberg, Cape Town & Johannesburg.

6.3 Dina Fourie's epic journey

Dr J.B. de Vaal

Two of the Voortrekker women who had proven themselves to be heroines, lived in Schoemansdal. They were Helena Lottrie and Dina Fourie. Since the route that Dina and her trekker party followed to Sofala crossed the northern corner of the current Kruger National Park, her gripping adventures, almost superhuman endurance and altruism are described here in brief.

The trekker party consisted of two families with two ox-wagons, namely: Dina; John Chambers, Dina's husband; their four children, Maria (almost five), Jannie (three), Josef or Joof (one), and Susan (three months); her 13-year-old half-brother David; Jacobus Lottrie (also known as Koos Lotring), his wife Hannie and their invalid son Neelsie (four).

Dina Geertruida was the daughter of the Voortrekker Josef Fourie. She was born on 22 October 1829 in Cradock. In 1838, at the age of nine, she left the Cape Colony with her parents and brother Gert. Dina was used to hard work from an early age as it was her and Gert's responsibility to tend and drive their livestock.

During an attack on their laager next to the Boesmans River, her mother sustained serious assegai wounds. These never healed properly and she died on the way from Natal to the Transvaal, leaving the young Dina with the household responsibilities. After her father's second marriage, she went to live with the still unmarried Gert to run his household. After his marriage she moved with him and his family to the farm Klipdam in the Soutpansberg, near the Rhenosterspoort hills. Jacobus Lottrie and his wife Hannie, Rooi (Red) Barend Vorster and his wife Alie also lived there. Jacobus Lottrie was apparently an adventurous hunter who did not hesitate to follow his prey deep into Matabeleland. They were all young people, full of courage and enthusiasm.

Klipdam stayed in the possession of the Fouries as Dina's half sister, Susanna Catharina Fourie, later married Captain Oscar Dahl, a well-known Native Commissioner at the time of the Republic. He built an attractive, fortified, double-storied house on the farm. Because of Sannie's hot temper she became known among the blacks as "Mmasambok". Sannie

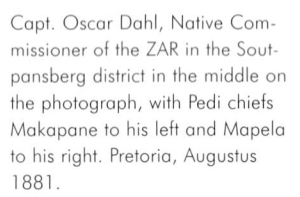

Capt. Oscar Dahl, Native Commissioner of the ZAR in the Soutpansberg district in the middle on the photograph, with Pedi chiefs Makapane to his left and Mapela to his right. Pretoria, Augustus 1881.
(ACKNOWLEDGEMENT: TRANSVAAL ARCHIVES)

was apparently also unusually strong for a woman. When my father travelled from farm to farm in 1919 with a maize threshing machine and a steam engine, he also worked at Klipdam, where he saw how she and a black man loaded grain bags filled with maize onto an ox-wagon.

According to Dina's biographer, Dorothea Möller-Malan, Dina was short and plump with beautiful blue eyes and blonde hair. She apparently had a heart of gold and always had an encouraging word for those who came to her to seek solace. She also never walked past a dog or cat without giving them some attention.

The man she married, John Chambers, was a big, well-built Englishman. Little is known about his past, and it would seem he had either deserted or been released from the Indian army. This adventurer came from England to South Africa and joined the Grahamstown trader Henry Hartley with his own wagon and team of oxen on a trade mission to Ohrigstad.

He later moved with Potgieter to the Soutpansberg and settled there as a merchant. Eventually he became known among the Boers as "Jan Kambers".[1] This is also how his name appeared in general dealer Casimiro Simões's cash book.

Dina and Chambers fell in love and became engaged shortly after their first meeting. After he had built a four-roomed house in town, J. Chambers and D.G. Fourie were married on 22 October 1849 by Magistrate A.C. Duvenhage. Their first child only lived a few weeks and was buried in the town's cemetery.

Like many hunters, Chambers and his good friend Jacobus Lottrie also went hunting in Matabeleland and Banjailand[2] north of the Limpopo. North of the Great Save River he twice saved the life of Chikovele, who was the 16-year-old son of Tsonga Chief Chabane. First he shot and killed a charging elephant that was almost on top of the boy and later he sucked the poison out of a stomach wound inflicted by the poisoned arrow of a hunting rival. For these acts of kindness Chambers received two large elephant tusks from the grateful chief, who assured him that he would never forget his good deeds.

Chambers and the trader Henry Hartley had a healthy respect for Commandant General Hendrik Potgieter, which is why he decided to live with the Boers and to move with Potgieter and his followers to the Soutpansberg. However, he did not hold Potgieter's successor General Stephanus Schoeman in the same regard and demonstrated this in word and deed. When Schoeman wanted to load the two small muzzle-loading cannons known as Grietjies, which were supposed to protect the people of Schoemansdal, on to his wagon on 3 February 1859 on his way to Pretoria, Chambers and Field Cornet Piet Eloff were the ringleaders of the indignant crowd. A month later he was among the 92 petitioners who informed President Pretorius that they had relieved Schoeman of his duties in the Soutpansberg.

When Chambers later heard that Schoeman was on his way back to Schoemansdal, accompanied by the president and members of the High Court, to try the guilty parties in the cannon case, he was filled with dread. He vividly remembered Schoeman's threat to get even. Rather than being delivered into the hands of his enemy, he decided to leave for Portuguese territory. Jacobus Lottrie, against whom there was a case pending as well, joined him. Their exodus probably took place at the middle or end of April 1859, thus at the height of the malaria season.

It was extremely difficult for Dina to move away from all that was familiar, into the unknown. She not only had to leave behind her lovely home, but her family and close friends as well, including Elsje-

The steel fort and fortified outpost built by Capt. Oscar Dahl, Native Commissioner during the Venda rebellions of the 1870s en 1880s at Klipdam along the Sand River, about 27 kilometres northeast of Pietersburg (Polokwane). The fort was eventually moved northwards to a locality between Louis Trichardt (Makhado) and Elim. It was named Fort Hendrina after the wife of Gen. Piet Joubert. After the British forces had occupied it during the Anglo-Boer War, it was renamed Fort Edward.
(ACKNOWLEDGEMENT: TRANSVAAL ARCHIVES)

---

1 Some sources say "Jan Kamer".
2 Southeastern Zimbabwe and the bordering area of Mozambique (i.e. Sabie-Lundi area).

Maria, General Schoeman's wife. Then there was the grave of her first-born child. She stood by Chambers, however, as they saw no other alternative course of action.

According to Malan they took a difficult route; up to Makonde Mountain and from there past Mutale along the Luvuvhu and Limpopo to Chicuala-cuala. This was the same route Potgieter and Bronkhorst had followed in July 1836 during their reconnaissance of the trade route to Sofala. The way was so bumpy that the women and little Maria, carrying the doll her grandmother Chambers had sent her from England, usually walked behind the wagons during the day.

Mawewe, who had temporarily succeeded his father Manukosi, was ill disposed towards the Boers and they always had to be on their guard for his followers. For this reason they decided after they had reached Chicuala-cuala, not to go to Inhambane or Delagoa Bay, but to trek northwards to Sofala, away from him. They also knew from experience that they were travelling through a tsetse fly area, and therefore they camped in open spaces during the day and trekked by night. David led the oxen while John and Jacobus rode next to the wagons with their loaded guns. While their horses were still alive, that is.

For safety's sake they only stayed at Chicuala-cuala for a short while and then moved in the direction of Dumeri on the way to the Chefu River. A heavy, unseasonal rainstorm drenched all of them. As a result, some of the adults suffered malaria relapses. The physically weak Hannie developed ague and became deadly ill. Dina noticed that Hannie was very anxious and scared of dying, but John gently spoke to her, read a passage from the Bible and prayed. After that Hannie became peaceful and declared her faith in her Saviour. A little while later she passed away. Dina and John first wrapped her in a sheet and then in a blanket before they buried her. John also performed the funeral service.

The distance between Chicuala-cuala and Sofala via Massangena is 240 kilometres, which Chambers wanted to cover as speedily as possible. They had noticed tsetse flies on their oxen and knew what would soon follow.

Neelsie, who was entrusted to David's care after his mother's death, wouldn't stop calling for his mother. He became increasingly ill and weak, until he died one night in the midnight hours. He was also buried in a lonely grave along the route to Sofala. His death left a deep impression on everyone, especially the young David and little Maria.

As expected, their oxen soon started dying, one after the other. This eventually forced them to pack the most essential items onto one wagon and leave the other one behind. After a while they had to unload even more of their belongings, as the few remaining oxen could not draw the wagon. Eventually they abandoned the second wagon as well and only packed the most necessary items onto the last living oxen. Jacobus Lottrie was apparently loaded like a pack animal himself.

At that stage John sent a message to his old friend, Chief Chabane, confessing that he was hard pressed and in big trouble. Chabane immediately sent 30 bearers while the women brought curdled milk and sour porridge for little Susan, since Dina could not breastfeed her any more due to malnourishment. The bearers also fetched some of the belongings they had abandoned along the way.

At Chabane's village everybody was very surprised to see the whites, especially the women and children, and they were allocated special huts. After Chief Chabane had welcomed them, they were given unfermented beer, porridge and meat. Chambers told them about his difficulties, while Chabane in turn told him about a leopard that was preying on his livestock. John promised to help. Chabane was very keen on keeping them there a while longer, but reluctantly agreed that they could leave after a week. This interval at least gave Dina the opportunity to wash their clothes and repair a few things. Before they left, John set a snare for the leopard.

Makakikiaan, one of Chabane's indunas, would take them to the next chief, Massurizi, and they would be provided with food by the small villages along the way. The adults walked while the children were carried on litters, all the way to the Chimanimani mountain range. Here Massurizi received a message asking him to accompany and guide them further.

Suddenly David fell ill and was in great pain. Chambers and Lottrie carried him as the bearers believed that he was bewitched. One by one they disappeared into the bush, fearing for their own lives.

Fortunately Massurizi's bearers arrived and could take over. The women also sent along maize meal for porridge. In the meantime David was still suffering unbearably and all Dina had at her disposal was a tin of zinc ointment, which she rubbed into the sore area.

They stayed in two temporary huts near Massurizi's village. As proof that Chabane's bearers had fulfilled their duties, Chambers sent him a copper bracelet with Makakikiaan, while Makakikiaan also received one as a memento. At the farewell Dina and the children could not contain their tears of gratitude.

A traditional healer gave David a root extract, which relieved the pain somewhat, but Dina and John did not want to make small cuts on David's body to rub the medicine in. As his ammunition was running out, John set snares, caught smaller animals, skinned them and put the skins on the painful areas – an old Boer remedy.

Dina got a bottle of brandy from some Portuguese hunters they came across. She heated some of the brandy and soaked pieces of cloth in it to put on the affected areas. John sent a letter to the Portuguese governor at Sofala with the hunters in which he asked for permission to settle there with his family.

David's condition worsened. Apart from the rheumatic fever he seems to have had, he also contracted malaria. Constantly feverish, all he managed to ingest was the white powder of the seeds of the baobab tree, mixed with water. The sour taste partially relieved his burning thirst, which is one of the symptoms of malaria. In his delirious state he spoke about his father, mother and sisters and wanted to know if he was going to die like the others. All Chambers could do was to pray that he be released from his suffering. Fourteen days after the rains, they wrapped his body in a blanket and buried him at Massurizi.

Her beloved half-brother's suffering and eventual death was a severe blow to Dina. Both she and John loved him deeply as he was a wonderfully helpful and obliging person. John could not help but blame himself for David's death and regretted the fact that he had angered Schoeman. Her husband's cry of despair: "Please Lord, help me, I can't carry on!" brought about a change in Dina. She somehow found the strength to support her husband in his time of need.

After David's death, Jacobus Lottrie felt that there was no reason for him to stay with the group any longer, so he told John that he was going to search for the gold mines in Banjailand, which the blacks had told him about. The next morning he was gone and only the Chambers family remained.

John made the children a small round tent from a piece of canvas, which they pitched every night and in which the children could play. Their joy was not to last long, however, as Joof also fell ill without any warning and died soon afterwards. After his burial they resumed their long walk. About 40 kilometres from Sofala they encountered a riverboat on the Búzi River, along which they were walking. The captain handed a letter from the governor to John in which he welcomed them and invited them to stay at the fort. Because of the language barrier Dina could communicate with the governor's wife only by means of gestures. The senhora was very sympathetic and gave them a spacious bedroom with a nice bed, but they struggled to escape the memory of the four graves they had left behind.

After John had informed the governor of Jacobus Lottrie's plans, he became very suspicious and sent an officer and a number of soldiers to fetch Lottrie. The governor then sent him back to the Soutpansberg with an escort. There Lottrie gave Dina's father, Josef Fourie, and other family members the sad news.

Josef Fourie was determined to fetch his daughter, her husband and their remaining children. As soon as he could, he set off by ox-wagon accompanied by Joof, his youngest son, and the family retainer, Sedomi. Jacobus Lottrie had explained the route to him up to Massangena. But Josef would never reach Massangena – he developed pneumonia and died north of Dumeri near the Chefu. He was buried between two thorn trees, a spot he had picked before his death. Joof carved his father's name on the trunk of one of the trees.

John worked for the governor in his office at Sofala and one winter they set off into the interior on a hunting trip. Dina helped with the running of the household while the senhora taught the children and did needlework. Then a grave disaster struck the community at Sofala. There were people suffering from diphtheria on the first sailing boat that returned with the trade winds. Two sailors on shore leave spread the disease, and it was thought that cats brought the deadly bacteria into the fort.

Dina could not witness the suffering of the community living outside the fort without doing something, so she went to nurse the sick with the senhora's permission, leaving her own children in the woman's capable care. She did not sleep inside the fort and received her food through a shutter in the wall. One morning, two weeks later, she did not receive her food, became suspicious and pushed the fort's large doors open. Dina found the barely conscious senhora lying on her back in her room. Next to her lay the body of her small child, who must have died only a little while before as his body was still warm. She found the rest of the children ill and with swollen faces. Not all of them recovered. First they buried the child who had been found next to his mother. After that her own Maria died, then an older boy of Senhor Pegado and then Jannie. With his dying breath he asked if his father was not back yet. Dina's only consolation was that this time her children would be buried in coffins and not wrapped in blankets.

When the message about the death of their children reached John and the governor, they immediately rushed back to Sofala. John found Dina in the little church, praying. Now they only had their baby girl left.

In May the following year, John Chambers undertook a trade expedition up the east coast, as Karel Trichardt had done before him. Dina did not want to stay behind again as she would not be able to stand the loneliness and the longing for her children. With the permission of the ship's captain, Chambers took his wife and daughter along. The captain was nice enough to grant them the use of his cabin.

For the first few days they had good weather and could enjoy the journey while Susan played on the deck. But one night there was a terrible storm and all three became very seasick. John was virtually unconscious, while Susan, the spitting image of her father, was unable to survive the affliction and died. Her small body was sewn up in a piece of canvas with a weight attached to it, as was customary at sea. It sank below the deep waters of the Indian Ocean. Dina packed away Susan's clothes at the bottom of her trunk. After the loss of her only remaining child, Dina became completely withdrawn and refused to talk about her children. John likewise was engulfed in loneliness. All they could do was pray. By then he had had enough of Africa and decided that they should go to live with his parents in England for a few years. To pay for their passage, he wanted to trade the iron from the wagons they had left in the bushveld for ivory from the blacks. Dina secretly hoped that they would have another baby in England. At the end of August they were back in Sofala.

While Dina moved back into the fort, the governor provided John with bearers to accompany him to Massurizi. There he would find help to reach his old friend, Chabane. The journey started off well enough, but as a result of the early rains his clothes were almost never dry. When John reached Chabane he told him of his losses and his hopes for the future. They packed dry branches on top of the abandoned wagons and set the wood alight to remove the iron.

John suddenly became nauseous and at the same time experienced severe pain in his neck and at the back of his head. It was probably malaria affecting his spinal cord. He immediately called Chabane and asked him to send Makakikiaan to fetch Dina from Sofala. When Makakikiaan left for Sofala that afternoon, John was in a critical condition, but still alive.

The guards in the watchtower at Sofala could see a safari approaching in the distance. It was Makakikiaan and his party, but without Chambers. Makakikiaan told the governor about John's illness, but the governor was somewhat sceptical as there was no accompanying letter. Against all advice, Dina nonetheless prepared to leave early the next morning. It was 3 November 1861, just five days short of two years after their arrival at the fort.

Although the governor had provided her with a luxurious litter, Dina preferred to walk most of the way. It was still the rainy season and therefore malaria season as well. To add insult to injury, Mawewe's antagonistic warriors harried them. With the woman entrusted to his care, Makakikiaan and his 80 warriors could not even think of attacking their pestering enemy and teaching them a lesson. They did eventually manage to bribe them with two metres of linen to leave them alone. However, the result of this encounter was that they could not make a fire any more, hiding in dense bush by day and walking at night. Then they still had to find time to hunt for food.

And still their troubles piled up. They encountered Tsongas who were fleeing from the AmaNdebele of uMzilikazi – now they were threatened by the AmaNdebele from the front and by Mawewe's raiders at the back.

They were constantly soaked as the rain would not let up. Eventually Dina had no more dry clothes left as everything was wet and mildewed. To top it all, she fell ill with malaria and was soon unable to walk. Makakikiaan nursed her like a mother would. As there was nothing else to eat, he kept her alive by feeding her marulas, medlars and edible roots. While they rested, he covered her with thorn branches for protection. Once a toothless man-eating lion tried to drag her away in broad daylight, but Makakikiaan saw it in time and they made short work of it.

Later Dina could not remember how long the journey had taken or for how much of the time she was unconscious. She could not even remember why she had a child's doll among her bundles of clothes. By the time they reached Chawane's village she was a little better. The chief awaited her under a large tree. Against all rules of courtesy she immediately asked him where her husband was. While the indunas bowed their heads, Chabane told her that he had died long ago. He led her to John's grave and gave her the two elephant tusks her husband had left her. She received these and placed one on John's grave. The other she handed to Chabane as a token of her gratitude. In her desolation she asked Chabane what she should do. His answer was that he had promised her husband that Makakikiaan would take her to her people. She asked to stay there for another week so she could recuperate, wash her clothes and spend a little more time at the grave of her husband. Chabane agreed, but Makakikiaan could not wait to start the journey as Chabane's people did not have any food for her. The AmaNdebele raiders had taken almost everything they had and destroyed the little that was left. Makakikiaan went to search for wild yams while the women offered her a small calabash of unfermented beer and a handful of dry beans.

When they reached the Great Save River it was in flood. They crossed the river on a raft built using the bark of giant marula trees, while the others sat three-three on a log. There were crocodiles and hippos in the churning flood waters that they had to keep at bay using sticks. Because of the strong current they landed far from the drift on the other bank. Dina's litter, her clothes and Maria's doll washed away during the crossing. All that were left of her meagre possessions were the wet, torn clothes on her body. The bearers built her a temporary shelter near the river so she could dry these rags.

She was so exhausted after this ordeal that her faithful companions had to build a stretcher so they could carry her in her weakened state. Near Chefu they camped one night next to the two thorn trees where her father had been buried. The men read Joof's "letter" on the tree and realised it was the grave of Dina's father, as many of them had got to know him during his hunting trips. Makakikiaan told her that it was her father's grave, but her mind was too clouded to grasp the meaning of his words.

At Dumeri she carved a letter with an assegai point on a piece of bark, in which she asked her people to fetch her at the Limpopo. The couriers delivered it to João Albasini at Goedewensch, as he was the only white man who lived in this eastern corner of the district. He in turn sent it to her brother Gert at Klipdam. Gert and his brother-in-law Niklaas Grobler immediately left for the Limpopo on horseback.

When Dina's bearers reached the Limpopo it was also in flood. Makakikiaan and another tribal elder swam across the flooded river with Dina propped up between them. She often had to struggle for air and was shivering with cold. Although they managed to cross the river unscathed, Dina came down with another bout of malaria immediately afterwards. Hovering between life and death yet again, she lay in a small temporary shelter on the bank of the Limpopo for a full week before they could resume their journey. At the Makonde hills they met two men on horseback. Gert and Niklaas saw a bewildered, unkempt woman. She was emaciated, her hair stringy and filthy and her clothes in tatters. She told her brother Gert that the black men had to accompany them as Makakikiaan had been told to collect a man's jacket and pants or a hat to prove to Chabane that he had delivered her to her people. She also wanted to give him a hatchet.

Their journey from Sofala had taken a full five months to complete. On 1 March 1862 Dina's group finally arrived at Albasini's farm. All she could say was that Makakikiaan had a good heart, that they must reward him well and that he had to say goodbye to her before leaving. Makakikiaan was paid £7.10 while his men shared some money that had been collected among the interested visitors. Albasini slaughtered two cattle for the well-doers and Gert made certain that Makakikiaan got his hatchet, as well as a hat and a jacket to take to Chief Chabane. At their farewell Dina took his hand into hers and held it. This man was her last connection to her husband and children. As weak as she was, she brought his hand to her lips and kissed it. "Makakikiaan, your hand is black, but your heart is white, lily-white ... I say thank you ...". Makakikiaan lifted his assegai up high and shouted "nKosikazi!" (Lady!). His men followed his example, turned around and, with Makakikiaan in the lead, started up a war song while running in file. Now, with the help of the Portuguese, they were free to take on Mawewe's warriors, who had made their journey such a misery.

Dina enjoyed the Albasinis' hospitality for five months. Then, one day, Barend Vorster and his wife Alie came to fetch her with their spring-loaded wagon to take her to her house in Schoemansdal. Piet Eloff had been living in it while she was away. Her best friend Elsje-Maria also came to greet her. Much later, after the civil war had ended, Elsje-Maria asked Dina if she would not come and greet Stephanus Schoeman, who had played such a leading role in the war. Dina agreed. Of what lay buried between them, no one ever said a single word.

On 6 April 1864 Dina married Johan Heinlein, a Swiss bachelor from Zurich. She was 35 and he 42. Despite all the privations she had suffered, they had three sons and a daughter. Johan Heinlein was killed at the age of 75 in the battle at Rooiwal shortly after the outbreak of the Anglo-Boer War. Dina Heinlein of the farm Onverwacht, Spelonken, died on 28 June 1901 in the Pietersburg concentration camp at the age of almost 72. She was buried and her grave can still be seen there. She never failed to express her gratitude to the Almighty who had been so good to her.

During a pastoral visit to Palmietfontein in 1897, Dina related her experiences to her minister, Reverend Coetzee. He wrote these down and his manuscript became the main source on which Dorothea Möller-Malan based her gripping book, *Dina Fourie*. Because the biography consisted of verbal narratives certain dates were incorrect, but some of these could be corrected with the aid of archival and other sources. In the words of the poet Jan Celliers, the following is certainly true of Dina:

Ek sien haar wen, vir man en seun en broeder
Want haar naam
is vrou en Moeder!

Loosely translated, it reads:

I see her win for husband, son and brother
because her name
is wife and Mother!

References

Bulpin, T.V. 1969. *Lost Trails of the Transvaal*, Cape & Transvaal Printers, Cape Town.

Distant, W.L. 1982. *A Naturalist in the Transvaal.* N.H. Porter, London.

Engelbrecht, S.P. 1952. *Die Nederduitsch Hervormde Gemeente Pietersburg (Zoutpansberg) 1852–1952*. Wallachs, Pretoria.

Grobler, H.J. 1938. 'n Voortrekkerdorp wat verrys en verdwyn het. *Die Volkstem* 9 December:14.

Möller-Malan, D. 1953. *Dina Fourie*. Voortrekkerpers, Johannesburg.

Möller-Malan, D. 1957. Die donker Zoutpansberg. *Historia* 2(2) September.

Naudé, P. 1937. Die Geskiedenis van Zoutpansberg 1836–1867. Unpublished M.A thesis, University of South Africa, Pretoria.

Transvaal Archives(TA), State Secretary Z.A. Republic (S.S.), Incoming documents (R). Consulted: R2524/59; R2823/59; R2824/59; R2825/58; R2826/59; R2828/59; R2831/59; R2832/59; R2833/59; R2835/59; R2836/59; R2837/59; R2840/59; R2843/59.

Chapter 7
Pioneers and hunters in the Lowveld
Dr U. de V. Pienaar and co-workers

The first scouts of the South African interior were astounded at both the amount and variety of game they encountered. These great hordes of animals created the impression of there being an endless supply of meat, skins, horns and ivory for hunters and sportsmen. It was also mainly hunting and the search for new hunting grounds that drew the white settlers deeper into the interior. However, looking at old documents it is apparent that this natural resource was not equally abundant or of equal variety everywhere. The density and composition of game concentrations then, as today, correlated with suitable habitat conditions, the specific climatic regime of the time and the presence of large human settlements and the subsequent hunting pressure.

Historical documents show there was little game in what is now the Kruger National Park at the beginning of the nineteenth century, before the "Difaqane" (wars of conquest), when a large and flourishing black population lived in the area between the Drakensberg and the Lebombo mountains. This is reflected in the diary of Francois de Cuiper, who was the first European to visit the Transvaal Lowveld, during an expedition from the Dutch settlement at Delagoa Bay to the interior in 1725. He found large African settlements with extensive irrigated fields, large cattle herds and no tsetse fly. The only game he mentioned were a small herd of elephants next to the Matola River near Delagoa Bay and a single reedbuck next to the Crocodile River. After the wars of conquest in the Lowveld during the first three to four decades of the nineteenth century, game numbers increased rapidly, bush encroachment followed and with that tsetse fly also made a reappearance.

Hunters lucked upon the crest of the wave in the southern part of the subcontinent and a great amount of game skins, horns and ivory was sold at various markets, including those in Grahamstown and Algoa Bay.

At first it did not appear as if the hunters were making a dent in the number of game in certain areas. The well-known hunter and explorer Roualeyn George Gordon-Cumming wrote the following about conditions in 1840:

I beheld the plains and even the hillsides, which stretched away on every side, thickly covered, not with herds, but with one vast mass of springboks; as far as the eye could strain, the landscape was alive with them, until they softened down to a dim mass of living creatures. 'This morning,' remarked a Boer, 'you beheld only one flat covered with springboks, but I give you my word that I have ridden a long day's journey over a succession of flats

Paul Kruger as a young man and brave hunter.
(ACKNOWLEDGEMENT: TRANSVAAL ARCHIVES)

covered with them as far as I could see and as thick as sheep in a fold.'

I.S.C. Cronwright-Schreiner (1925) also wrote that springbok in the Karoo

trekked in such dense masses that they used sometimes to pass right through the streets of the small up-country town. I have known old people who have walked among them and actually now and then touched them with their hands. Men have gone in, armed only with a heavy stick and killed as many as they wished. Native herdsmen have been tramped to death by the buck and droves of Africander sheep carried away, never to be recovered in the surging crowd. So dense is the mass at times and so overpowering the pressure from the millions behind, that if a sloot (gully) is come to, so wide and deep that the buck cannot leap over or go through it, the front ranks are forced in until it is levelled up with their bodies, when the mass marches over and continues its irresistible way ... The Cape Colony has from time to time during recent years been visited by trekbokke though not in such numbers as the old farmers used to describe, and, I have no doubt, truthfully described.

In 1895, however, the upcountry was suffering from a long drought, which was particularly severe in Namaqualand, and the trekbokke began to move well into the Colony ... In the afternoon we gradually left the noise of the hunters behind and drove to quieter quarters, until at length our wish to see large numbers of buck was gratified. On driving over a low neck of land a vast, undisturbed, glittering plain lay before us. Our glance at one sweep took in the great expanse of brown country, bounded in the distance by low koppies, bathed in the wonderful glowing tints of the Karoo; and throughout its whole extent the exquisite antelopes grazed peacefully in the warm afternoon sunshine ... We were three farmers, accustomed to estimate numbers of small stock, and as we had an excellent pair of field glasses we deliberately formed a careful estimate, taking them in sections and checking one another's calculations. We eventually computed the number to be not less than five hundred thousand; half a million springbuck in sight at one moment!

Magistrate W.C. Scully (1913) wrote:

One might as well endeavour to describe the mass of a mile-long sand dune by expressing the sum of its grains in ciphers, as to attempt to give the numbers of antelopes forming the living wave that surged across the desert in 1892 and broke like foam against the western granite range. I have stood on an eminence some twenty feet high, far out on the plains, and seen the absolutely level surface, as wide as the eye could reach, covered with resting springbucks, whilst from over the eastern horizon the rising columns of dust told of fresh hosts advancing ... It is not many years ago since millions of them crossed the mountain range and made for the sea. They dashed into the waves, drank the salt water, and died. Their bodies lay in one continuous pile along the shore for over 30 miles, and the stench drove the Trek-Boers, who were camped near the coast, far inland.

Major Walter McDonald led an official expedition to Delagoa Bay in June 1874 to deliver a consignment of military equipment to Lydenburg (Swart & MacDonald, 1874). With the exception of ten tons of gunpowder, it was a gift from the German government to the Transvaal after the Franco-Prussian War. W.C. Scully (1907) was one of 25 men who trans-

A leadwood tree with bullet holes in its trunk at a place where hunters used to camp south of Ship Mountain. (25.09.1984)

ported the first load of military equipment along the old Delagoa Bay transport route past Skipberg (Ship Mountain) in the current Kruger National Park. He and five other members of the party remained behind with nagana-infected oxen in the Pretoriuskop area while the rest returned to Delagoa Bay with fresh trek animals to fetch the rest of the goods. Scully was a very keen observer and during their stay with the dying oxen he described a stampede of thousands of head of game from a southwesterly direction. According to him the onrush lasted about 20 minutes: "Buffalo, zebra, blue wildebeest, kudu and many other varieties, including troops of giraffe jostled together and rushed wildly on." He specifically mentioned hartebeest as being among the large number of game he had observed. These could of course have been tsessebi that he had wrongly identified, but the possibility that he had seen Lichtenstein's hartebeest, *Sigmoceros lichtensteinii,* cannot be excluded (Milstein, 1986). Here it must be noted that Scully (1913) specifically referred to "tsessaby" (tsessebi), which he had hunted in the same area. He also mentioned the "marvellous richness of animal life on these plains in the early seventies, particularly between the Lebombos and Ship Mountain".

There are many other references in the writings of travellers and hunters of the seventeenth and eighteenth centuries to the richness of animal and plant life in southern Africa, which surpassed those in other parts of the world in both variety and quantity.

But when the economy of the country, especially in these areas, was based mainly on hunting, even such vast numbers could not withstand the onslaught of thousands of hunters, armed with increasingly sophisticated weapons. This downward spiral of the animal population was accelerated by prolonged droughts and catastrophic epidemics, such as the rinderpest epizootic of 1896–97. The game numbers were reduced so drastically that more and more people started calling for better protection of the country's remaining natural heritage. The bluebuck, *Hippotragus leucophaeus*, a relative of the sable and roan antelope and an endemic of the ridged country between the Bot and Gourits rivers, and between the Langeberg range and the sea, was the first to disappear. The last specimen was apparently found in 1779 by Lichtenstein in the vicinity of Swellendam.

Initials chiselled into sandstone on the top of a hill along the Sweni, east of Ngumula Pan. It could be the initials of hunter Adam William Briscoe, with the date 1891. (27.10.1984)

The true quagga, *Equus quagga*, of the Great Karoo, Namaqualand and southern Free State, was the next species to become extinct. Although Gordon Cummings still found large herds in the vicinity of Colesberg in 1843, the last quaggas in the Cape Colony were shot in 1808. By 1878, the last remaining herd in the Free State had also been eradicated. When European zoos asked about sources from which they could supplement their quagga stocks, they discovered to their shock that none of them was to be found in the wild any more. The last of these beautiful animals died in the zoos of London (1872), Berlin (1875) and Amsterdam (1883). The Cape lion, *Panthera leo melanochaitus*, and Burchell's zebra, *Equus burchelli burchelli,* met the same fate by the middle of the nineteenth century (although in the case of Burchell's zebra this has been disputed). Many species were eradicated locally, but managed to survive in other parts of their distribution ranges. Of importance here is Lichtenstein's hartebeest, *Sigmoceros lichtensteinii*, which had been eradicated in the southwestern boundary area of its distribution range north of the Soutpansberg in the Transvaal and the far-northern areas of the current Kruger National Park by the end of the First World War (Milstein, 1986).

Ivory has always had a significant value as trade ware. As colonisation extended east- and northwards, the hunting of elephant, rhino and other game increased in both extent and intensity. Grahamstown became the ivory market of the Cape Colony and most ivory was exported through this channel. In 1824 it was noted that 22 700 kilograms of ivory were traded during the first seven months of the year at the Grahamstown ivory market. In 1825 exports grew to a massive 48 050 kilograms, the tusks of more than 1000 elephants. Large amounts of gunpowder and lead were necessary to shoot this number of elephant and other game. It is mentioned in a document that the magistrate of Grahamstown issued permits for 31 495 kilograms of gunpowder, 14 762 kilograms of lead, 3 699 577 percussion caps, 1069 rifles, 45 revolvers, 2 294 941 bullets and 698 kilograms of dynamite in a period of six months (Pringle, 1982). After the colonisation of Natal, the Orange Free State and the Transvaal, Durban also became an important port for the export of ivory and other game products. Ivory to the value of £31 754 was exported from Durban in 1858 – double the income from any other source in Natal! In a period of 34 years, ivory to the value of £337 109 was exported, with the record year being 1877, when 19 350 kilograms of ivory were shipped from Durban. The ivory and other game products were transported by ox-wagon from far-off places, such as Schoemansdal in the northern Transvaal, by Henry Hartley and other traders. Schoemansdal was an elephant hunter's mecca from 1848 to 1867 and thousands of elephants were shot in the Soutpansberg district and the adjacent Mozambique, Matabeleland and Mashonaland. The missionary Piet Huet (1869) noted that the people of Schoemansdal were almost totally dependent on elephant hunting. He wrote:

> The number of elephants that are shot every year is almost unbelievable and the number shot by Soutpansberg hunters alone amounts to several thousand. Gigantic tusks of 60, 80, 100 pounds and even heavier are brought into town. Portuguese traders as well as traders from Pietermaritzburg come to town annually to trade these for money and merchandise. Some tusks are so heavy that two or three blacks struggle to carry them. Almost the whole of Soutpansberg's existence depends on the hunt. When somebody needs meat, he goes out for two or more days and brings food for all his friends.
>
> During my stay there, one man went out and shot two giraffes, eight hippos and a number of wildebeest and buck. One went out one morning and that same evening brought 11 hartebeest (which are almost the size of a cow) back home. [He might be referring to Lichtenstein's hartebeest.]

Reverend Stefanus Hofmeyr, who did missionary work in the Soutpansberg for 20 years, wrote in 1890:

> In 1865, Zoutpansberg was not a large town, but prospering. Thousands of pounds of ivory are brought in by Boer hunters every year and sent to Natal, and there are more than enough traders. From then on trading started to deteriorate, as elephants as well as other kinds of game were getting scarcer ...
>
> When I first arrived in February 1865 one could still find giraffe, eland, rhino, buffalo, zebra, wildebeest and many other species in abundance. There was also no lack of lions, tigers [leopards], leopards, wolves [hyenas], etc. Now the animals are so scarce that someone I have known since 1865 could easily state what one hears all the time: that the game and their predators have been eradicated.

The heydays of Schoemansdal and the ivory hunters of the time are described in more detail in chapter 6.1.

The livelihoods of the settlers at the frontier (Boer as well as British) and the Voortrekkers during their journeys to the interior ultimately depended on hunting, which also provided them with trade goods (ivory, horns and skins) that could be traded for provisions from the traders who periodically visited the remote areas. Although there certainly were exceptions, the typical Boer hunter only shot enough for personal use and little was wasted, if ever. In their tough pioneer existence they could simply not afford it.

By the middle of the nineteenth century the so-called sports hunters appeared on the scene. These were usually rich men who hunted for pleasure and were responsible for mass killings. In 1860 the largest hunt in living memory was organised by a certain Andrew Hudson Bain of the farm Bainsvlei just east of Bloemfontein. This was in honour of Prince Alfred, the 16-year-old second son of Queen Victoria of Great Britain. The royal party consisted of Sir George Grey, the Governor of the Cape Colony, General Major J.J. Bisset and the prince himself.

The local chief Moroko and his whole tribe were hired to drive large herds of game past the prince and the rest of the hunting party, which consisted of 25 men armed with rifles. They rode on horseback through the large herds and shot indiscriminately, with the result that by the time the hunters had tired of the wanton massacre, more than 1000 head of game were dead or wounded.

As game became scarcer, hunters had to move further and further into the interior in their search for more. After Father De Santa Rita Montanha had arrived back in Mozambique following his visit in 1856 to Schoemansdal, he heard about a Boer hunter, Piet du Preez, who had apparently caused trouble in the territory of a certain chief six days' journey east of Chicuala-cuala in Mozambique.

Jacobus Lottrie, the good friend of John Chambers, of whom more is told in chapter 6.3, also undertook extended hunts to Banjailand[1] in 1859 (the area between the Save and Limpopo rivers in Mozambique). When Carl Mauch, the famous German scientist and explorer, arrived at the Zimbabwe Ruins in 1871, he encountered a Boer hunter, Adam Renders, and his wife, Elsie, who had settled there in 1868 (Punt, 1958).

The courage and perseverance of some of these pioneer hunters in exploring uncharted territory knew no bounds. One of the Colony's frontier farmers, Hendrik Scholtz, scouted out a route from the Eastern Cape through the Free State and southern Transvaal right up to the Soutpansberg in 1834. Here he met the descendants of Coenraad de Buys, who had moved through the area in 1823, and cleared the way for the trekker parties of Hans van Rensburg and Louis Trichardt.

After the founding of the Voortrekker towns Ohrigstad (1845), Krugerspost (1849), Schoemansdal (1848) and Lydenburg (1850), and later the towns of Pilgrim's Rest (1873), Sabie (1895), Barberton (1892), Leydsdorp (1890), Komatipoort (1892) and Nelspruit (1905), it became easier for hunters and other pioneers to explore and exploit the hunting grounds of the Lowveld, especially during the "healthy" winter months.

Thus the area that is now the Kruger National Park was also criss-crossed by hunters and pioneers. Many of these pioneer hunters will stay unknown or will only be remembered by the names they had carved on the trunks of the large baobab trees. Counted among the latter are the Ebersohns (Baobab Hill), Prellers (Malonga), Briscoe[2] 1890 (Ntsumaneni Poort) and H.M. Borter 1890 (Olifantspoort). One man chiselled the initials A.W.B. and the date '91 on to a sandstone outcrop on a hill along the northern bank of the Sweni River near Ngumula Pan. The site of Joubert's grave at Stapelkop Dam, who died there of malaria at the end of the nineteenth century, is also known.

The Stols family from the White River area were well-known hunters and wagon-makers in the two decades before the Anglo-Boer War. They are also sometimes referred to as the Berg (Mountain) Stolse. Jan "Bokwa" Stols' enormous strength was legendary – apparently he could lift a wagon with his shoulders when a wheel had to be changed.

The name "Briscoe 1890" carved into the bark of a baobab tree in Ntsumaneni Poort. (18.06.1984)

The name of an unknown hunter, H.M. Borter and the date July 1890, carved into the trunk of a baobab tree just east of the confluence of the Olifants and Letaba rivers. (17.07.1984)

---

1   Also called Banyan Land.
2   Recent information shows that it was Adam William Briscoe who had carved his name into the trunk of a baobab tree in Ntsumaneni Poort in 1890. He was a hunter and transport rider and a cousin of John Edward Briscoe, who, together with one John Flett Duncan, were the co-owners of the farm Zwartruggens No. 744 between the Sabie and the Olifants rivers (in the vicinity of where Ngotso Dam is today). This farm was expropriated by the State and exchanged in 1927 for three farms in the Soutpansberg district, outside the newly determined boundaries of the Kruger National Park.

   His grandson, Dr Henry W.E. Briscoe was the medical superintendent of the Newcastle hospital. He is by chance also the owner of the historical monument farm Mount Prospect and the farm house of John O'Neill near Laingsnek, where the peace accord was signed between Vice President Paul Kruger of the ZAR and the British negotiators under Sir Henry Evelyn Wood on 23 March 1881. The end of the Transvaal War of Independence was negotiated by President Brand of the Orange Free State after General Piet Joubert had defeated the British forces under Sir George Pomeroy Colley at Majuba on 27 February 1881.

The Briscoe baobab tree in Ntsumaneni Poort in the Bangu area. (18.06.1984)

Gert Frederik Coenraad Stols was a blacksmith, but like the rest of the family he was mad about hunting. In August 1886, when he and his party went hunting along the upper reaches of the Bukweneni Spruit, a tributary of the Nsikazi (or Lozies) River, he contracted malaria and died. He was buried there and the place later became known as Stolsnek.

Tradition has it that another one of the Stols hunters was buried near Pretoriuskop. Apparently he had to find a wounded lion one day and kill it. It seemed the excitement was just too much for him, however, as he died shortly afterwards of a heart attack.

Through the years, Africa has produced its quota of renowned hunters: W.D.M. "Karamojo" Bell; Arthur Newmann; J.H. Hunter and Jim Sutherland of British East Africa; Fredrick Courtenay Selous; Henry Hartley and Johan Colenbrander of the former Matabele- and Mashonaland (Zimbabwe); William Cornwallis Harris; R. Gordon Cumming and William Cotton Oswell of the western Transvaal and Bechuanaland (Botswana); Major P.J. Pretorius of German East Africa and Addo; William Finnaughty of northern Bechuanaland (Botswana) and Barotseland; William Baldwin; H. Anderson Bryden and John G. Millais of Natal, the Free State and Bechuanaland; Petrus Jacobs (regarded by many as the greatest hunter in southern Africa); Jan Viljoen and Marthinus Schwartz of the Kalahari, northern Bechuanaland, Lake Ngami and Matabeleland; and Jan Harm Robbertse of Ovamboland and Angola. Legendary hunters from the Lowveld include: Henry Thomas Glynn; the brothers Bill and Bob Sanderson; Frederick Vaughan-Kirby; E.F. Sandeman; Captain H.F. "Farmer" Francis; Stephanus Schoeman; Abel Erasmus; Harry Wolhuter; Percy FitzPatrick; Miles Robert Bowker; Cecil "Bvekenya" Barnard; Commandant Tom Kelly; Hans Klopper; João Albasini and his great-grandson, Joe; W. Borchers; Piet Möller; Piet Eloff; and Francois Bernard Lotrie.

These hunters are often viewed in conservation circles as senseless exterminators of game and destroyers of nature. This matter should be seen in perspective, however, and their hunting activities should be weighed against the needs of the time. It must be remembered that politicians, including Loveday and Van Wyk, only took notice of the fast disappearing wildlife of the country as a result of the agitation of hunters such as Glynn, Ingle, Vaughan-Kirby, Chapman and others. President Paul Kruger, an enthusiastic hunter in his youth, proclaimed the first nature reserves in the Transvaal, namely the Pongola (1894), Groenkloof (1895) and the Sabie Game Reserve (1898).

It is remarkable that the game populations, especially those in the Sabie Reserve, could withstand the pressure from hunters and the rinderpest epidemic of 1896. This says much about nature's capacity to restore itself. When one takes into account that the first choice for warden of the Sabie Reserve was the well-known hunter Captain H.F. Francis, and that one of the first rangers of the Sabie Reserve, Harry Wolhuter, was also a hunter, their role can be seen in a more positive light.

Following is a number of stories about the lives and times of some of the foremost hunters of the Lowveld.

João Albasini (1813–1888)

The role Albasini played as a pioneer in establishing and maintaining trade relations between the Voortrekkers and the Portuguese is discussed in chapter 5. To achieve this, he did everything he could to find a passable and tsetse fly-free route to the coast. He was also probably the first European to settle in the Transvaal Lowveld when he built a four-roomed house and shop at Magashulaskraal. He used bricks that he had made next to the western bank of the Phabeni Spruit in the current Kruger National Park and baked in an oven. A row of baked bricks, lying on their sides and covered by a 15cm layer of silt, was excavated at the site – these had formed the bottom layer of the oven's mouth.

The few black people who lived in the area before 1845 worked the ground with pick-axes. Albasini dug the first canal from the Phabeni Spruit and planted vegetables and wheat under irrigation. He also planted Spanish (common) reed along the Phabeni and the Sabie River, where only river reed had grown previously, as stakes for his runner beans and for general use. Today these large stands of Spanish reed are evidence of his ambition and knowledge of horticulture. Albasini also planted the first fruit trees in the area, including oranges, papayas, bananas, mangoes and other subtropical fruit, which still thrive there to this day. There is a strong possibility that he planted coffee and sugar cane there as well. Albasini can therefore be considered the founder of irrigation farming in the Lowveld, something without which the economy of the area would have been much poorer today.

At Goedewensch in the Soutpansberg he continued on a much greater scale with his agricultural activities and the cultivation of fruit trees under irrigation. As has been pointed out, he was the first to grow coffee in the Kleinspelonken, and he harvested so much coffee beans that he had enough to share with his neighbours. He was thus also the founder of the extensive coffee plantations that can be found today in the vicinity of Mashao Hill.

Albasini was also a pioneer in the area of administration. His work was particularly neat and systematic – almost faultless, in fact. In one large file, which fortunately still exists, he kept copies of all his outgoing letters as vice consul. In another one, to which he referred in correspondence as "My tax book" and "My journal", he kept records of his activities as Superintendent of Native Affairs. This second book has not been found, but his neat reports on these duties have fortunately been preserved by the Transvaal Archives.

Albasini left no information about his hunting trips or his elephant hunting, unless these notes were made in his journal. According to some elderly blacks, his daughter Maria Biccard and historical sources, he was a formidable elephant hunter and ivory dealer who sometimes went hunting himself or otherwise sent black hunters to do so on his behalf.

After Sochangana (Manukosi) and his warriors had raided the peaceful Tsonga tribes along the east coast, the already hard-hit local people suffered a great drought and resulting famine. Albasini kept many of them alive by shooting elephants – he took the ivory and they got the meat.

In July 1841, when he was a young man of 28, he was one of ten people who started a company to

Six of João Albasini's Magwamba hunters, armed with assegais and six- and four-pounder elephant rifles of the time. This photograph was taken by H. Exton of Pietersburg (Polokwane) during the late 1860s.
(ACKNOWLEDGEMENT: DR J.B. DE VAAL)

7. PIONEERS AND HUNTERS IN THE LOWVELD

The well-known Rÿ Hills east of Bushbuckridge. This used to be a popular hunting area during the nineteenth century. According to legend the area had been inhabited by a cannibalistic tribe after the Difaqane. The entire tribe was later killed by the Swazis. (11.11.1986)

Francois Bernard Rudolph Lotrie (1825–1917), circa 1901, master elephant hunter of Schoemansdal. (ACKNOWLEDGEMENT: C.J. MOERSCHELL)

hunt elephants with the aid of 25 black hunters. He continued this line of business from 1845 onwards at Magashulaskraal. Albasini told W.H. Neethling that, after having moved to Lydenburg in 1850 and establishing himself as a merchant, he had distributed 29 rifles among his followers with which they hunted elephants.

In the Soutpansberg his main source of income was elephant hunting, as was the case with most people there. He continued hunting in the eastern Lowveld. A.K. Murray, the ivory trader from Natal mentioned earlier, visited Schoemansdal from July to October 1863. While in the area he also paid a visit to Albasini at Goedewensch. He said afterwards that Albasini's entire business was built on black hunters, and at any one time several of them would be out in the veld. He had no fewer than 100 rifles in use in the Soutpansberg and, taking into account that each hunter was accompanied by 20 ivory bearers, one can imagine the scale of the undertaking. While Murray was at Goedewensch, three or four hunters and a group of bearers returned from a hunting trip with about 300 pounds of ivory. They walked in single file towards the house and were obviously pleased with the fruits of their labours, which they proudly exhibited in front of everyone.

At New Year Albasini would call together his approximately 700 hunters and the one who had shot the most elephants would receive an ox and be regarded as captain of the hunters the following year. This encouraged a competitive atmosphere, which undoubtedly increased Albasini's annual income substantially.

Francois Bernard Rudolph Lotrie (1825–1917)

As has been mentioned in chapter 6.1, Francois Bernard Lotrie, or Frans Lottering as he was known among his pears, was one of the greatest elephant hunters of his time. But unbeknown to himself, he was much more than just another elephant hunter. The honour of discovering the now world-renowned archaeological site at Mapungubwe Hill on the farm Greefswald along the Limpopo belongs to this Nimrod. In the 1890s, when he was already in his seventies, he lived as a recluse in a rock shelter at the foot of Mapungubwe just over a kilometre from the Limpopo River, the same area where this nature lover had felled so many tuskers in his youth. He often climbed the hill, an act that was forbidden to the blacks, and found several objects there, including a lovely clay pot. He gave this pot to the black man who would eventually lead the Van Graan party there more than 30 years later. Shortly after the Anglo-Boer War this eccentric figure attracted the attention of the German farmer and trader Carl Moerschell, who was farming on Bergfontein near Mara. Lotrie was waiting for his repatriation money at the time.

As he was just about destitute and already in his eighties, Moerschell took pity on the old man and invited him to live as his guest at Bergfontein for several years. He lived in a rondavel near the house and ate at the table of his benefactor. Moerschell often listened with great interest to the stories the old man told at night. Lotrie often talked of Mapungubwe. As he talked he would turn his head towards the horizon with a pensive look in his eyes, as if thinking of times gone by. According to him Mapungubwe was the centuries-old seat of a great ruler. There were peculiar clay pots that differed from the usual ware used by the blacks he knew. He had also picked up other objects, but he never revealed what these were. There was no doubt in his mind that this hill used to be a king's treasury. Unfortunately, Moerschell did not take much notice of these theories, dismissing them as flights of fancy by someone whose knowledge did not reach much beyond the Bible. His reminiscences of his hunting days were regarded with more interest, however, and were published in 1912 in German. The author wrote the following in the front of a copy he gave to Andries van der Walt: "Although I have added some information based on my own knowledge or on historical facts, for explanatory reasons, Lotrie's spirit, which is the Voortrekker spirit, flows through the book: an example and warning to young South Africa." (Loosely translated from the German.)

Like the other hunters, Lotrie undertook most of his hunting trips on foot. He had the following to say about the sanna rifle:

> The handling of the old sanna, as we jokingly used to refer to this heavy old rifle, was more complicated than the modern breech-loaders and it demanded considerable courage and confidence to use them. The four- and six-pounders were most often used, and they were called this depending on whether four or six bullets could be cast from a pound of lead. The handling of the muzzle-loader did not give much trouble to the experienced shot. One had to make fairly certain of your aim. If the first bullet hitting a lion was not fatal, there wouldn't be time for a second round. Even if you could manage to load the rifle, the animal would already be on top of you, inches from your face. I believe the elephant is the most dangerous wild animal. The weak penetration capability of our lead bullets in those times made it essential to shoot from the immediate vicinity of the animal and a bullet through the ear into the brain usually produced the best results. The kick of the old rifles, which were specially made to kill elephants, was, despite a protective pad, very unpleasant in the long run and caused a painful swelling of the shoulder.

Although these hunts went hand in hand with a great deal of adventure, they were still dangerous undertakings thanks to the primitive old muzzle-loaders and demanded all the courage and perseverance a hunter could muster. Lotrie had the following to say about this, in his entertaining manner:

> Hunting elephants is not such a simple matter. From the moment the hunter suspects their presence, no sound should be made to disturb the silence of the wilderness. Felling elephant bulls is the set purpose. They move around singly, in pairs or in herds of up to 30, completely separate from the cows and calves. Once a hunter has discovered fresh spoor, he follows it, all the while paying careful attention to the wind direction. This can often take hours, yes, even days, and can be very stressful and exhausting with the sun beating down, a scarcity of water, or unexpected thunderstorms. If the hunter is riding a well-trained horse, he tries to separate the selected target from the herd. When only a few steps away from his opponent he jumps off the horse and shoots. Immediately he is back in the saddle to reload and to get away from the enraged, charging animal. An outrider or a mounted escort ... keeps a reserve rifle handy and expedites, if necessary, the dispatch of the dying animal.

When Lotrie was hunting on foot, he would get as close as possible to the selected elephant, muzzle-loader in hand. Without fail he would feel as if he were about to go into battle.

> I first make my choice with a practised eye, then dive behind the last bush that separates me from my bull by 10 to 15 paces. They stand and stare at me with raised trunks, until the crack of my shot relieves the general tension ... Now follows shot after shot, as fast as I encounter the charging animals and as fast as I can reload the old sanna. At the first sign of its nearing collapse I leave the animal to one of my helpers while I storm after the fleeing animals, throw powder into the barrel and push the bullet in while running. Completely out of breath I fire at another. After any number of hours the hunt is over. Black people come running from the nearest village, carry the meat to their huts and take up the load of ivory.

Lotrie denied that an elephant would trample a victim to death. One of his labourers once became involved in an altercation with an enraged animal. The elephant apparently grabbed the man with his trunk and threw him to the ground. After that, he gored his unfortunate victim's body with his tusks, knelt on the prostate body and kneed it into a shapeless mass, all the time trumpeting in a tone that sounded like a cry of triumph.

This courageous Nimrod had numerous close shaves while hunting. Once he got a huge fright at Malapiane's kraal north of the Soutpansberg. After several hours' exhausting hunt, he was trudging on, tired and unobservant, only to find himself surrounded by elephants. He did not know what triggered their response, but they started charging from all sides. Totally confused, he started looking around for an escape. There was none, except for a baobab ahead that he could not climb because of its high, smooth trunk. He hotfooted it to the tree and pushed himself into a fold in the trunk as the snuffling giants rushed past. They were so close that he could almost feel their breath.

He had a similar experience next to the Bubye River in Mashonaland, where he and his brother Cornelis were hunting. While he could hear his brother shooting a few hundred paces away, he followed a bull, under the impression that there were no animals behind him. Some instinct urged him to turn around, only to see five or six brutes charging at him. The only possible hiding place was behind a trampled shrub in a small depression just ahead of him. With no other options open to him, he reached it with two or three leaps and fell down flat behind it, waiting for the inevitable. The elephants thudded past, sniffing the air with their trunks. They were so close he could make out their eyelids.

Like several other hunters from Schoemansdal, Lotrie also hunted big game in the eastern Lowveld. During these expeditions Klopperfontein and Malonga were two of his favourite campsites because of the availability of ample, fresh fountain water. These two fountains are situated along the old trade and hunting path via Chicuala-cuala to Inhambane. A few hundred metres east of Malonga Fountain Dr U. de V. Pienaar found a baobab tree with several initials carved into it, among others "B.L. 1860", which probably stands for Bernard Lotrie. The other initials are P.B, H.H. (or H.B. or H.P.), R.C. (or R.O), P. and I.D. 1867. The dates indicate that the inscriptions date back to the Soutpansberg elephant hunting period after Schoemansdal's founding.

Henry Thomas Glynn (1856–1928)

Henry Thomas Glynn was born on 30 November 1856 in Cape Town, one of four sons and four daughters. His father, Henry Glynn, had emigrated from Ireland to the Cape in 1838 and at first settled in Cape Town. His sons were all adventurers who repeatedly undertook expeditions into the interior.

In 1869 Glynn senior, also a keen explorer, hunter and crack shot, moved with his family to the dia-

The initials B L 60 carved on the trunk of a gigantic baobab tree east of Malonga Fountain. They are probably those of Bernard Lotrie (1860).

mond fields of Simonella and Delportshoop in search of his fortune.

They arrived just after the first diamond rush at the place that would later become known as Kimberley. His wife eventually managed to convince him that their children needed more formal schooling. As a result the family moved back to the Cape and bought a large farm near Stellenbosch. Henry Thomas, better known as H.T. among his friends, completed his education at the South African College (later the University of Cape Town).

Then rumours of a large goldfield in the Lydenburg district reached the Cape and Henry's father once again decided to try his luck at the diggings. At the end of 1873 the Glynn family settled at Krugerspost between Ohrigstad and Lydenburg. Just before that an alluvial gold deposit had been discovered at Ponies Krantz (See map no.10) in Pilgrim's Rest. This was close to the older diggings at Geelhoutboom (renamed Mac-Mac by President Thomas Burgers in 1874), Spitzkop and Hendriksdal.

By 1873 there were about 1000 prospectors of various nationalities working in the area. H.T.'s father and his brother Joe bought unrefined gold from the diggers and sold it at a profit. Hunting was their passion, however, and the Glynns spent most of their winters in the hunting fields.

In 1876 all of the inhabitants of Krugerspos moved into an emergency laager, where they had to stay for a year because of the First Sekhukhune Uprising. This problematic period was followed by the Second Sekhukhune War in 1878, which lasted for two years. During this period, the Glynns entertained guests such as Sir Theophilus Shepstone, Colonel (later General) Redvers Buller and Lieutenant (later General) Carrington. Among their other friends were captains Owen, Bowlby, Pennefather and Smythe, who played a major role in the eventual subjugation and capture of the rebel Pedi chief Sekhukhune. The Glynns were also very well disposed towards the Boers and were known and loved far and wide.

In 1880 H.T. bought the farm Grootfontein for R1200 and his father the adjacent Ceylon for R1000. While they were having a brick house built on Grootfontein they rented a house on Spitzkop that belonged to one Van Niekerk.

The name of the town of Sabie, which would eventually be situated on Grootfontein, has an interesting story, especially seeing as no one knew its origins before H.T. rechristened his farm Sabi. Apparently the locals had a superstitious fear of the big river that flowed through Grootfontein. They believed that the spirits of the people who had drowned in the river, had been taken by crocodiles, or whose bodies had been thrown into it after tribal wars, were still wandering in it. Their word for fear is "saba", and so the river became known as the "Saba River" or "river of fear". In Mozambique it is known as the Save, which probably derives from the Tsonga word "chava" (which also means fear), or otherwise the Karanga word "Sava", which means sand.

After the Glynns had settled on Grootfontein they called the farm Sabi, which they later changed to Sabie. H.T. was 23 years old at the time.

The Glynns were not the first Europeans to settle here. Before any permanent settlement took place there had been a hunting camp on Grootfontein and later a trading post on the transport route from Delagoa Bay to Pilgrim's Rest and Lydenburg.

The first whites who lived there on a permanent basis were the three Badenhorst brothers – Dirk, Casper and Hendrik. They arrived in 1844–45 and obtained the first title deeds to Grootfontein in 1846. They then sold the farm for R1500 to Pieter de Villiers, the father-in-law of General Schalk Burger. Glynn in turn bought the farm from De Villiers.

The Glynns' first house can be considered the

Pioneer and hunter, Henry Thomas Glynn (1856–1928).
(ACKNOWLEDGEMENT: INFORMATION AND MARKETING SERVICES, TRANSVAAL PROVINCIAL ADMINISTRATION)

7. PIONEERS AND HUNTERS IN THE LOWVELD

Initials carved into the trunk of a historical old baobab tree just north of the confluence of the Letaba and Olifants rivers. These initials are probably those of Henry and Arthur Glynn and H.F. Francis (1896), all well-known hunters from that era. (18.06.1984)

birth of the town Sabie. The road network was very inadequate during those days and rough wagon trails ran from Spitzkop over the mountain to Lydenburg, and from Spitzkop to Mac-Mac and Pilgrim's Rest through the drift above the Sabie Falls. Transport costs were very high, for instance 6/- for 100 pounds of maize meal. For this reason, H.T. and his father decided in 1881 to build a road from Spitzkop (along the course of the old "Neethling road"), down the escarpment to Magashulaskraal, where João Albasini had had a trading post next to the Sabie River from 1846 to 1848.

This route became very popular for the transport of goods by ox-wagon between Delagoa Bay and the goldfields of Pilgrim's Rest. H.T. and his father also hired geologists and prospectors to look for gold-bearing reefs on Grootfontein. After a long search they eventually found a paying reef. Thanks to Sir Alfred Beit and Lionel Phillips the company Glynn's-Lydenburg was founded with R175 in capital in 1895 and placed under the management of A.L. Neale. The reef, which was named after Glynn, was the largest source of gold in the district and stretched from Eland's Drift through the Sabie to Vaalhoek. H.T. was fortunate to make powerful contacts through Frank Watkins, who was married to his sister Susan and was a member of the Volksraad of the ZAR (South African Republic). Watkins successfully lobbied in the Volksraad for the railway connection between Kaapmuiden and Barberton.

As a break from his trading activities, H.T. and his brother Arthur went on their most extensive exploring and hunting expedition to the Pungwe River, Shiringoma Forest and Gorongoza in Mozambique in 1849.

They shot a large amount of game, especially buffalo, and had many adventures. H.T. was an excellent shot and it is said that he could shoot a rhebok or duiker from the hip with a shotgun without having to aim.

After their return from Portuguese East Africa (Mozambique) in 1896, H.T. decided to build himself a new house, Huntington, near the Sabie Falls. That year he also met Gertrude Gilbertson Dales, of Cawood Castle in York, while travelling to England by boat. They married in 1896 in England and, after touring Europe, returned to Sabie and Huntington.

In later years, H.T., his brother Arthur and their brothers-in-law Gus Stiebel and Colonel Hartley Dales became wealthy mining magnates and developed several more mines, including Glynn's Extension, Serita Mine and Heather Mine. He also played an active role in many other undertakings, including the development of the White River Estate Company with Glen Merriman, Exley Millar, Reverend Maurice Ponsonby and Billy Barnard. To ensure adequate labour for his mining activities, he convinced two Shangaan chiefs from Mozambique to settle on the farms Lunsklip and Rietspruit with their followers.

In time, H.T. became regarded as the laird of Sabie and was known and loved by all. He never said no to a deserving cause and the first school in Sabie, the St Peter's Anglican Church (designed by Sir Herbert Baker), the first hospital and the first bridge across the Sabie River came into being through his efforts.

His greatest love remained hunting and in his book *Game and Gold,* published shortly after his death, he described many of his and his brother Arthur's adventures while on hunting expeditions in the Lowveld. His notes on game distribution, vegetation and climatological conditions during those years are of great value today. Their favourite hunting grounds were around Pretoriuskop, the Nwanedzi along the eastern frontier and Olifantspoort. The initials of H.T. and that of Arthur, above the date 1895, are carved on the trunk of a baobab that grows close to the confluence of the Letaba and Olifants rivers.

H.T. wrote the following of hunting in 1876 and 1877:

Early one morning we saddled up and went in search of the much coveted game – eland. We had ridden some distance when we suddenly espied a troop of about two hundred on a high hill. [This was in the Pretoriuskop area. They shot eight all together and probably wounded many more.]

On another occasion we were out riding near Legogot and saw a large troop of game. We could not make out what they were at first, but we galloped on and found them to be a troop of elands. [They shot six.]

On the Pretoriuskop flats not far from White River in following a troop of brindled gnu, one might come onto a second troop, and a third, then vast herds of zebra and some sassaby, and then more wildebeest and sassaby pouring down from every quarter, until the landscape presented the appearance of a moving mass of game of anything up to about 5000 in number!

The country, extending to Swaziland, contained game of all descriptions and where White River Estates are today, and down the Crocodile River, below the escarpment, all kinds were to be found, except elephants, which were further back in the Lebombo. Black rhinos were killed by us on the opposite side of the Crocodile River, where Nelspruit Station is today, and white rhinos were in the country low down on the Sabie River on the edge of the Matemere [Nwatimhiri] Bush. Giraffe were all over where the Acacia trees were in the low country, and buffaloes were to be found in troops of two and three hundred on the road to Delagoa Bay and all over the country extending to Oliphants River and Lebombo Mountains. I have seen buffalo in the White River country in the open near the kop called Legogot, and roan antelope, sable antelope, waterbuck, sassaby and elands in big troops up to two hundred. Hippos could easily be found in the rivers, and at Pretoriuskop giraffe, blue wildebeest and ostriches were plentiful. No white man lived in these parts then; the Swazis had driven the Basutos [Pedis] out of this country …

Jim Makokel of *Jock of the Bushveld* fame at one time lived on a farm of mine. He was a married man and had a son and daughter, and they were all heavy drinkers …

Elands were all over the country in the early days, and a Boer hunting at Spitzkop, close to where I lived had a fall in an ant-bear hole, the horse rolling over on him and breaking his neck.

Black rhinos were found in the rough country down below White River, Nelspruit and extending across to the Olifants River; there were a few white rhinos as well.

President Paul Kruger when he was hunting down the Spekboom not far from Lydenburg, in the early days, encountered a black rhino. His Sanna exploded, and blew part of his hand and thumb off. He had a very bad time with it … Gangrene set in, and he was advised to kill a goat, and use the contents of the warm stomach on the wound. After three or four goats had been killed, the wound began to heal.

He also gave information that indicates a more water-rich environment at the time:

In hunting in the Transvaal in the low country, extending to the Lebombo Mountains, I came across several small streams that carry some hippo between the Sabie and Olifants rivers. The Wanetsi [Nwanedzi] River is one, and the seacow travel back and forwards from the Portuguese territory.

In 1895 H.T. came to realise that unless something was done to protect the game of the Lowveld, mass extinctions were inevitable. He therefore wrote a letter to President Paul Kruger in which he suggested that a reserve be proclaimed where game could be protected, proposing boundaries he deemed to be practical. President Kruger himself had been agitating since 1884 in the Volksraad for stricter hunting legislation and better game protection. The existing legislation did not have the desired effect of curtailing the systematic eradication of game. It is fitting to mention here that the Voortrekkers did refer to game preservation in their legislation from time to time. For example, in as early as 1837 a hunting ordinance was included in the "Freely Elected" Volksraad's law of nine articles when it convened next to the Vet River in the Free State. In 1846 the Volksraad at Ohrigstad also accepted clauses with regards to game preservation that were so drastic that rifles, ammunition and wagons were confiscated on these grounds. In 1848 Commandant General A.H. Potgieter issued a notice in which foreigners were strictly

A drift through the Olifants River just east of the first Gorge rest camp, which had earlier been used by hunters.
(01.08.1985)

forbidden from hunting in his territory. All of this is proof of some awareness of conservation among the Voortrekkers.

On 22 September 1858 concern was expressed about the fast disappearance of certain species in the Transvaal. The resulting law that was passed clearly stipulated:

> That nobody is allowed to kill more game (by any method) than what he can use for own consumption or can load onto one wagon, or to kill game merely for the sake of the skins.

In 1870 the ZAR took a step forward by implementing closed seasons for certain game species. Elephant and hippo were strictly protected after 1891 and buffalo, eland, giraffe and rhino after 1893. On 13 June 1894 President Kruger proclaimed the first game reserve in Transvaal, namely the Pongola Reserve in the Piet Retief district between the Pongola River, the Swaziland border and the Lebombo Mountains.

On the strength of several different representations from the Lowveld, of which those of H.T. Glynn, Captain J.C. Ingle and others carried much weight, the member of the Volksraad for Krugersdorp, J.L. van Wijk, stated in the Volksraad on 6 September 1895 and again on 17 September that a game reserve in the Lowveld had become a matter of the utmost urgency. His motion was backed by R.K. Loveday of Barberton. After several postponements following the Jameson Raid, the Sabie Reserve (which consisted of the area between the Crocodile and Sabie rivers) was at last proclaimed on 26 March 1898 by President Kruger.

In this way, the pioneer and hunter Henry Thomas Glynn played an important role in the proclamation of an area that would preserve game in the Lowveld for generations to come, eventually becoming the Kruger National Park.

H.T. Glynn died in Sabie in 1928 and was buried next to his father and mother in the private Glynn cemetery on Huntington. He was survived by his wife, G.G. Glynn, and their three children, Billy Glynn, Serita Bucknell and Gerda Green.

J.A. (Abel) Erasmus (1845–1912)

Jacobus Abel Erasmus was born on 8 February 1845 in Weenen, Natal. Abel, as he would become known, was the son of Jacobus Johannes Petrus Erasmus and Maria Margaretha Catherina Jordaan, two prominent figures among the Voortrekkers in Natal. His father died shortly after his birth and when he was eight months old his mother decided to move from Natal with Commandant General A.H. Potgieter and settle in the newly-founded Ohrigstad in the eastern Transvaal.

Here the widow Erasmus eventually married J. de Clercq. Like the other boys of Ohrigstad and district, by the age of 19 the young Abel was knowledgeable about every aspect of the veld. His experiences in this untamed part of the Transvaal schooled the young man in the ways of the wild and prepared

A hunting camp in the Sabie area of the Lowveld during the 1880s.
(ACKNOWLEDGEMENT: TRANSVAAL ARCHIVES)

him for the duties he would later have to carry out on behalf of the ZAR in this remote area. He also attempted to educate himself about farming practices as best he could under the circumstances.

At the age of 19 he married Geertruida Zacharyde Magdalena Kruger, daughter of the well-known pioneer Pieter Ernst Kruger. The couple settled at Krugerspost and made it their permanent home. The young man zealously applied himself to farming and soon his systematic approach and practical knowledge paid dividends. He built dams and dug water canals and irrigation channels. But more than farming kept him busy in his younger years. Gold was discovered in 1873 on the farm Hendriksdal. This led to an immediate rush to obtain options on land that might yield gold. Abel obtained such an option on the farm Geelhoutboom, which was eventually renamed Mac-Mac by President Burgers. He sold this option to J.B. Shires at a considerable profit. After the discovery of gold on his farm Graskop he sold it to President Burgers for £1000.

In those years, money was scarce and land abundant. The profits the young farmer had made from these sales helped him a great deal and made it possible for him to set his aims higher. In February 1876, at the age of 31, he was elected as member of the heemraad (county court) for Lydenburg district. During the same year, high chief Sekhukhune of the Pedi rebelled. Like his father Sekwati he was a subject of the ZAR, but he gradually extended his influence and became increasingly recalcitrant.

One day in 1876 Field Cornet Pieter de Villiers arrived at Henry Glynn's shop in Krugerspos with the unsettling news that a strong Pedi army was on its way. Twenty-five families hastily moved into a laager near Glynn's shop. During the ensuing military operation, De Villiers resigned from his post and Abel took over as field cornet because of his excellent knowledge of the local black languages and his leadership capabilities. On one occasion a band of 5000 warriors attacked the laager defended by only 33 white men and their 25 black helpers. Erasmus and his men managed to fend them off, but the enemy left with 2000 head of cattle.

Later Johannes, one of Sekhukhune's subordinate headmen, was attacked with the aid of a Swazi force of 5000 men under the leadership of Chief Mataffin.

J.A. (Abel) Erasmus (1845–1912), the well-known hunter and native commissioner of the ZAR.
(ACKNOWLEDGEMENT: MR JAN VAN SCHAIK)

Despite discord among the Boer leadership they managed to seize Johannes's stronghold and kill him.

Abel Erasmus, who in the meantime had officially been promoted to field cornet, undertook a punitive patrol with a commando to the Blyde River in an effort to recover the stolen cattle.

In the interim President Burgers led a commando in an attack on Sekhukhune's mountain stronghold. However, the commando would not accept the president as its leader and his army had to retreat to Steel Poort. Here, Captain Von Schlickmann was put in charge of a volunteer corps that attacked the rebellious blacks when necessary. Von Schlickmann was killed during one of the altercations and was replaced by Captain A. Aylward. The Boer commando and volunteers from the goldfields eventually succeeded in quelling the uprising. Through the intercession of Reverend Merensky of the Botshabelo Mission, high chief Sekhukhune petitioned for peace.

The Sekhukhune uprising, the inefficient leadership of President Burgers and various other factors led to the annexation of the Transvaal by Sir Theophilus Shepstone on behalf of Great Britain.

The British administration forced Sekhukhune to take an oath of allegiance to the Crown and he was fined 2000 head of cattle.

The wily Pedi high chief started making trouble

again shortly after the annexation of the Transvaal in April 1877, and soon the whole area was in a state of war. Sir Owen Lanyon, who took over the administration of the Transvaal from Shepstone, and Colonel Hugh Rowlands tried in vain to force Sekhukhune into submission. It was only after the arrival of Sir Garnet Wolseley with a large British force from Natal that a decisive defeat could be inflicted on Sekhukhune, on 28 November 1879. Sekhukhune was taken into custody and jailed in Pretoria. He was, however, released by the Boers after the Transvaal War of Independence in 1881. In 1882 he was murdered in his kraal at Lulukop by his half-brother Mampuru.

While in detention, Sekhukhune had made certain accusations against Abel Erasmus that were believed by the British officials and as a result a warrant of arrest was issued for the Field Cornet. But preliminary investigations showed that the accusations against Erasmus could not be proven. He was found not guilty by Magistrate Vanrenen on 5 July 1880 and released unconditionally. Shortly after the end of the Transvaal War of Independence and when the ZAR had regained its freedom, Abel Erasmus was appointed as commissioner of native affairs for the Lydenburg district. In this capacity he was of great service to the Republic. On several occasions he was appointed as leader of a black auxiliary force during the campaigns against the rebellious Pedi under the leadership of Mampuru, Niabel and Mapog. These rebellions were finally put down in 1883. Later he suppressed Chief Maguba's rebellion with the aid of a Swazi impi.

During his travels as native commissioner to collect taxes from the large black population in the district, Abel Erasmus often had to take obscure routes and mountain passes to reach all the remote hamlets. As a result he got to know the district like the back of his hand. It is no wonder he was often used as guide and interpreter during official expeditions. In 1887 P.D. de Villiers and he were members of Surveyor General G.R. von Wielligh's commission, which had to determine the border between Swaziland and the ZAR. Three years later, in 1890, he was again asked by Von Wielligh to accompany him and Surveyor M.C. Vos during the combined ZAR-Mozambican commission's survey of the border between Transvaal and Mozambique. Erasmus was in charge of the large number of black bearers that had to carry the commission's equipment and provisions. He was also responsible for the daily shooting of game to augment their food supplies. In 1892 he recruited 3000 black men to help build the Eastern railway line between Komatipoort and Nelspruit.

In his dealings with chiefs in the district he was said to act firmly but fairly, and in time he garnered enormous respect among the local population. He was especially strict and ruthless when it came to stopping certain unscrupulous whites from trading in weapons with the blacks. Henry Glynn mentioned in his book *Game and Gold* that Erasmus once said to a traditional healer in his presence: "You are getting very old and will not live much longer," to which the old man answered: "I know that very well, but I am content because I have eaten my share."

In 1896 the Pedis caused problems again, and this time the instigator was Turometzane, Sekhukhune's mother. Superintendent P. Cronjé, Abel Erasmus and Field Cornet Dawid Schoeman made short shrift of the problem by placing Turometzane under arrest and fining her.

Turometzane then turned to the High Court and successfully appealed against her sentence. Her fine was repaid and Abel Erasmus and Field Cornet Dawid Schoeman were each fined £25 for the corporal punishment they had imposed on some of her chiefs.

Through all these troubled days Erasmus never lost the love of hunting he learned as a young boy in the still-unspoiled Lowveld. He had plenty of opportunity to keep his hand in during the many expeditions to the border, and his travels in the district. He was commonly regarded as a crack shot and was soon given the name "Dubula Duzé" (he who shoots from close by) by the blacks. With his luxuriant beard and astride one of his prize horses, he struck an impressive figure.

The large-scale and uncontrolled extermination of game in the Lowveld forced Erasmus to send a petition to the Volksraad on 25 November 1890 in which he proposed that the area between the Sabie and Crocodile rivers be proclaimed a game reserve. G.J. Louw, the special justice of peace of Komatipoort, had done the same on 6 November 1890. Unfortunately these appeals fell on deaf ears.

The proclamation of the Pongola Game Reserve by President Kruger in June 1894 was an occasion of great moment for Erasmus. He was under the impression that this would create an ever-lasting hunter's paradise and in 1898 he actually sent a letter to the provincial secretary in which he requested permission to go and hunt in the reserve. To his utmost disappointment his request was turned down by J.W.B. Gunning, the director of the State Museum in Pretoria, under whose administration the game reserve had been placed.

In later years, Erasmus bought the farm Orinoco in the Lowveld and started breeding an unusual type of Afrikaner cattle that he called the "bruin geelbekke" (brown yellow-muzzles). In 1889 he built a road between Albasini's old home at Magashulaskraal and his farm Orinoco that was eventually lengthened into a communication route to Leydsdorp in the northern Transvaal.

During the Anglo-Boer War, the British irregular unit Steinaecker's Horse built a fort close to Mpisane's kraal on the farm New Forest, not far from where the old pioneer Abel Erasmus kept his cattle on Orinoco.[3] His splendid herd was a very tempting haul for this notorious gang and they lost no time in herding the entire lot to their base at Sabie Bridge. The rightful owner was hopping mad at the loss of his prized herd and made an urgent appeal to General Ben Viljoen to retaliate over this trespass on private property. This led to the attack on Fort Mpisane by the Boer army in August 1901, during which the commanding officer, Captain H.F. Francis, lost his life. Erasmus was instructed to stay on in the district and to carry on as usual with his duties as native commissioner during the war. Later he was also asked to assist with the provision of food supplies to the Boer commandos. Close to the end of the war the old man gave up the struggle and surrendered to the British forces.

After the war, Erasmus spent his retirement on his farm at Krugerspos. Some blamed him for surrendering to the British, and attempts to have him reappointed in his former position did not succeed (Kuit, 1945).

But Erasmus remained a respected figure among the black population of the Lowveld and it was noted that the Swazi's Ndhlovukazi (queen-regent) sent a delegation to commiserate when he died on 30 May 1912 on his farm.

In 1905 he applied for permission to search for a hidden treasure in the Sabie Reserve, but was sent away empty handed by acting warden Major A.A. Fraser.

The well-known mountain pass between Ohrigstad and the Strijdom Tunnel along the Summit Route in Mpumalanga (formerly eastern Transvaal), was named Abel Erasmus Pass on 8 May 1959 in honour of this staunch old pioneer of the Transvaal outposts.

Captain Frederick Vaughan-Kirby – hunter and naturalist

Frederick Vaughan-Kirby was born in 1855 in Ulster, Ireland. He came to South Africa in 1884 and every winter between this date and 1890 he hunted in the Lowveld region of the ZAR. After his return to Europe, his book *In Haunts of Wild Game* was published in England in 1896. Today this work is an invaluable source of information about game distribution and conditions in the Transvaal Lowveld during the last two decades of the nineteenth century. Vaughan-Kirby was an accurate observer with a surprising knowledge of animal behaviour and habitat preferences and was, for his time, remarkably authoritative.

He was clearly passionate about hunting and crisscrossed the area between the Drakensberg (Kahlamba) and the Lebombo Mountains. His most favourite hunting ground was the Satara-Nwanedzi area of the current Kruger National Park. He left very useful notes about that area regarding the species of game to be found there and the distribution of water sources during the six years he had hunted there. He mentioned that the blacks named him "Maqaqamba", but did not explain what it meant. According to locals, the name refers to somebody whose clothes are permanently bespattered with blood (that of game), in other words a hunter.

---

3 According to Milstein (1968) it is in this vicinity where Paul Krantz, the taxidermist of the old State Museum in Pretoria, shot a number of Lichtenstein's hartebeest in 1894 for the museum. A small herd of Lichtenstein's hartebeest could be found on the adjacent farm Dingleydale until 1956, when it was taken over by the S.A. Bantu Trust. This farm had been bought in 1910 by Carel Theodorus Rabie (one of the three original owners of Pretoriuskop).

In the introduction of his book he wrote the following about the Transvaal Lowveld of the late 1880s:

> The game of the country of which I write is still varied and fairly numerous, though of late years the destruction of the South African fauna has been great; and we who have known the country in its earlier days can but look back with feelings of deep regret for what has been, and forward with concern to what may be. It must be borne in mind that many of the animals now to be found only in the 'fly'-infested tracts of the bush-veldt along the course of the Sabi, Oliphants, Limpopo and Singwetsi rivers at one time, well within my own recollection, existed in large numbers in the more broken country amongst the foot-hills of the mountain-range; amongst those which have thus retracted to safer haunts being buffalo, rhinoceros, giraffe, eland and roan antelope.
>
> I can well remember hunting the black rhinoceros in places where now one might walk or ride for hours without turning out as much as a reedbuck. There are still a fair number of koodoo, waterbuck, wildebeeste, Burchell's zebra, and a few sable antelope to be found more or less near to the skirts of civilization, wherever certain natural advantages have proved favourable to their remaining; but they are appreciably diminishing in numbers every year. Nowadays the highest plateaux of the range (the krantz country) afford shelter to many of the smaller antelopes; reedbuck, oribi, and duiker on the open flats and ridges; mountain reedbuck and vaal rhebuck on the hills and krantzes; bushbuck, 'msumbi' (the red or bush duiker), and bush-pigs in the kloofs and small patches of bush; leopards and cheeta still lurk about the kloofs and krantzes; and the 'aardwolf', African foxes, and ant-bear are numerous.
>
> Amongst the foothills (the kloofcountry) we leave behind us the mountain reedbuck; the oribi and vaal rhebuck are met with sparingly on the stretches of open grassland; but here is the true home of the hill leopard, while cheeta are plentiful, and the great spotted hyena not infrequently met with. Hunting dogs and serval are seldom seen; koodoo in the hills, buffalo in the dense kloofs, and bush-pig everywhere; while all the smaller antelopes, with the exception of those mentioned as peculiar to the terracelands, are very numerous. Bush-pig and 'sumbi' especially, swarm in all the dense heavily-wooded kloofs; and klipspringers in the krantzes and kopjes.
>
> Then as we proceed further east we find the long, gaunt, pale-coloured leopard of the Low Country proper, with many intermediate forms; all the smaller antelopes are still met with in numbers, with the exception of the 'msumbi', which is very rare, being only seen in the densely-wooded tracts close to water. Eland, sable antelope, roan antelope, koodoo, waterbuck, sassaby, blue wildebeest, impala, reedbuck, duiker, steenbuck, and rarely the grys-steinbuck, are the antelopes of the Low Country.
>
> Eland and roan antelope are very scarce indeed, a small tract of country between the Oliphants and Limpopo rivers being about the only place where

Capt. Frederick Vaughan-Kirby, hunter and naturalist.
(ACKNOWLEDGEMENT: WILLIAM BLACKWOOD & SONS, LONDON)

small troops can still be found. As lately as five years ago there were elephants on the Timbabati, a herd of over fifty head being encountered by some Boer hunters: they came from the extensive reedbeds at the junction of the Letaba and Oliphants rivers, though they still exist in the dense bush on the Libombo slopes.

Both species of rhinoceros are now practically extinct, the square-mouthed being altogether so, though a few of the prehensile-lipped species remain in the Libombo and the Matamiri bush on the lower Sabi. For many years past the well-known Matamiri bush, lying along the south bank of the Sabi river, has been a favourite resort of *Rhinoceros simus,* but they have become almost extinct even there. This year (1895) I came upon two in that district, a cow and big calf; but they are decidedly rare. The Matamiri bush, however, does not come within the district which I seek to describe, though separated from its southern boundary only by the Sabi river.

Buffalo and hippopotami are to be found; giraffe, ostrich and Burchell's zebra still plentiful; lions everywhere along the courses of the principal rivers, also grey and red foxes and black-backed jackals, spotted hyenas, and cheeta; the bush-pig is very rare in the Low Country, but warthog supplies its place, and is most plentiful; serval, and two smaller species of the Felidae – the 'impaka' and 'imbodhla' of the natives – are found …

The disappearance of the game is not solely attributable to its extermination by the rifle; the advance of civilization has driven it back, and caused it to retire from those parts where it was constantly hunted on horseback, and take refuge in the 'fly'-infested districts, where horses and other domestic animals cannot live. In these secure retreats it is seldom molested: the 'fly' has done more for the preservation of the game than all the game laws ever framed. Fever, in its worst form is rife throughout the summer months, and is another and almost equally important factor in the preservation of the big game from utter annihilation.

Vaughan-Kirby also provided remarkable descriptions of the vegetation of the Lowveld bushveld and plains, a particularly valuable map with place names and indications of game distribution, an appendix with a list of all the Lowveld game species complete with their Latin and local names, and even a list of the smaller mammals and birds of the area. In Vaughan-Kirby's authoritative book there are also many references to water distribution in the area between the Olifants and Sabie rivers at the time of his hunting trips. Some of his descriptions of water conditions are given below, because at most of the places that he noted as having permanent water, there is now no sign of natural water during the dry season. Some of the environmental conditions he described (particularly in the areas Nwanedzi, Sweni, Mavumbye and Gudzani), occurred again in the period 1971 to 1978 with its unusually high rainfall (Pienaar, 1986). Vaughan-Kirby wrote the following:

> Our wagons stand outspanned on the edge of a small clearing in the otherwise low but thick bush on the north side of the Mazimtonti river, and from a comfortable camp, with good water within 400 yards, and near the long stretches of young sweet grass, which has sprung up since the February 'burns'. About 200 yards distant a small stone kopje, rising out of the surrounding bush, forms a remarkable feature of the landscape. It is the lowest of the straggling group comprising the Eland kopjes – thus called for the elands, which but a few years ago were to be found in considerable numbers in their vicinity. [According to his excellent map, they must have camped close to the current Nwaswitsontso hills on the western border and along the upper reaches of the Mtlhowa Spruit, a tributary of the Nwaswitsontso. During winter months there is now no water at all here any more.]
>
> I cannot say whether wildebeeste are able to swim, for I never saw them trying to cross deep water; but once I witnessed a singular incident, which at the time led me to think they could not do so. I was hunting on foot near the junction of the Mjindana [Shinkelengane] and Mabutsha [Mavumbye] rivers, and having hit a good koodoo bull hard, had followed him into an extensive 'gwarra' thicket, where I lost the spoor amongst the numerous tracks of game. Catching sight of an animal standing in the thickets, the nature of which neither I nor my boy could determine, I fired at it. It dashed off,

and we followed the spoor, evidently that of a big wildebeest bull: it led us to the edge of a deep pool, 300 yards long and about 25 yards wide, with 10 feet-high banks. I saw the wildebeest in the water, apparently drowning. [The size of this waterhole is equalled by few of the existing earthen dams in the Park.]

One Sunday morning about 9 pm two boys who had been down to the junction of Mabutsha [Mavumbye] and Mjindana [Shinkelengane], returned with the news that a hippopotamus was in a large hole near that place and distant about ten miles.

The stream narrowed here to a breadth of less than a dozen yards, flowing swiftly over a stony bottom; but a few yards above was an enormous pool – or seacow hole, as they called it – fully a mile in length and as deep for aught I know, and alive with crocodiles. A deadly-looking place with great water-lilies covering its surface scum. [He was probably referring here to the Dumbana waterhole, which in those years was about as big as most of the larger dams in the Kruger National Park today.]

At length we turned our faces westwards towards the Nguanetsi [Nwanedzi] at which river we found Messrs Barber and Bowker camped; and they showed me a very remarkable specimen of an albino reedbuck, a young ram, which one of their party had shot higher up the river.

On August 18 we crossed our old drift on the Manzimtonti and outspanned, and in the afternoon trekked on again along the course of the river by an old native foot-path, and about 5 pm outspanned on the south bank near a large waterhole, and within a mile of the junction – upon the other side – of the Malau and the larger river. [This must have been in the vicinity of Misane Mouth, where a windmill provides water for game today.]

The grass on that side of the river opposite our camp being very dry, we fired it at all points as we went along. It is never advisable to leave too much grass cover round a camp in lion country, and besides, burnt ground is a great assistance in spooring. Torrents of rain fell during the night, accompanied by thunder and most vivid lightning.

I killed a wildebeest bull for bait, which for three days failed to attract them (lions); but seeing some fresh spoor on the banks of a stream, the Lion river, a tributary of the Vimbangwenya [Ngotso] – where lions had been twice to drink …

At length the spoor (of a lion), which hitherto had run parallel with, and about 80 yards distant from, the bank of the stream [Ngotso], turning towards it. Right at the waters edge we held the dogs fast …

On one occasion I was walking along the banks of the Makanbana [close to the current Satara], at a spot where the reeds were very dense, and a lion jumped out of those … [Similar conditions repeated themselves along the Kambana Spruit during the high rainfall years from 1971 to 1978.] My friend F and I were shooting during that season in partnership, and we made our headquarters at my old 'wild-cat camp', near the junction of the Mabutsha [Mavumbye] and Manungu [Gudzani] rivers, tributaries of the Nguanetsi. [This must have been in the vicinity of Shikwembu waterhole in the Mavumbye. This waterhole currently also dries up during the winter months of dry years.]

On the 5th July I did not fire at any of the game seen along the way, my object being only to secure something for bait, at or near Simana kopjes [Ngirivane], as it seemed probable the lions were lying up in some of the dense and extensive patches of cover on the Simana river. About midday we off-saddled close to a considerable pan near the kopjes, containing delightfully cool and – what was indeed a treat in that country, where all the water is brackish, not to say salt – fresh water! [He probably referred here to the Chuhwini Pan, which now only holds water in the winter months in very wet years.]

During the Anglo-Boer War Vaughan-Kirby returned to South Africa and was promoted to the rank of captain in the British forces. In July 1902 he was an unsuccessful candidate for the position of warden of the reproclaimed Sabie Reserve in the Lowveld. His great ideal of becoming involved in the preservation of the country's game resources became a reality in 1911 when he was appointed as the first warden of the Zululand reserves (Hluhluwe and Umfolozi in KwaZulu-Natal). At the time he was already a respected member of the Royal Zoological Society. Here he became known among the blacks as "Mfohioza", and with E. Warren, the director of the Natal Museum in Pietermaritzburg, he became an avid champion for the permanent protection of the last remaining white rhino and the cessation of the veterinary section's campaign of killing game in an effort to eradicate nagana.

In time he acquired a good understanding of the relationship between tsetse flies and their game hosts. Until his retirement in 1928 he contributed greatly to the eventual demise of this wasteful method of containing the spread of tsetse flies. Thus another former hunter helped to protect his adopted country's wildlife from certain eradication.

Edward F. Sandeman – hunter and author
The English traveller and writer Edward F. Sandeman spent eight months in South Africa in 1878, during which he traversed the southern Lowveld for several months (June to August) on hunting trips. After his return home in 1880 he wrote a very interesting book, *Eight months in an ox-waggon*, in which he described his many experiences in South Africa, with the parts about his hunting trips in the Lowveld being of particular interest. Sandeman did not exhibit nearly the same amount of knowledge as Vaughan-Kirby in his observations about wildlife in the Transvaal Lowveld, but there are nevertheless many important references in his work. The enclosed 1879 map by F. Jeppe (*Map of the Transvaal and Surrounding Territories*) is in its own right a remarkable piece of Africana. The map not only contains remarkable detail for those years, but apart from Sandeman's routes it also shows the exploratory routes taken by C. Mauch, E. Mohr, A. Hübner, T. Baines, St V. Erskine, Captain Elton, Colonel Colley, G.P. Moodie, R.T. Hall and E. Cohen.

During his stay in the Lowveld, Sandeman hunted mainly in the vicinity of Pretoriuskop. He and his hunting partners reached the area by ox-wagon from Lydenburg, via the old Harbour Route over Devil's Knuckles, Sabie and Spitzkop. His description of the game species in the Pretoriuskop/Skipberg area provides little more information than what was noted by Henry Glynn and Vaughan-Kirby. However, it is clear from his descriptions that at the time blue wildebeest and zebra were the most abundant game species around Pretoriuskop. One day, Sandeman and his companion, a Mr White, were almost trampled by a herd of zebra and blue wildebeest. They noticed the large cloud of dust of an approaching herd of game between Pretoriuskop and Skipberg, and awaited it excitedly, under the impression that it was a herd of buffalo. When the animals came closer they realised to their disappointment that it was a large herd of zebra and blue wildebeest. He described the scene as follows:

> ... To our intense disgust and disappointment, the vanguard of an enormous troop of quagga, numbering many hundreds, burst upon us ... line after line of quagga passed by, only turning aside sufficiently to avoid upsetting us, but at their tail, instead of buffalo, appeared a troop of wildebeeste.

They did, however, manage to shoot several buffalo along the upper reaches of the Mbyamiti. Here they also found lion prides and unsuccessfully fired at a black rhino with their Martini-Henrys.

Of more importance than his observations about game is Sandeman's description of and notes about the tsetse fly distribution in the Lowveld, with the tsetse fly area's boundary just east of Skipberg (Langkop). He also wrote about water distribution in the area in which they hunted and had the following to say about it:

> Long Kop, or Saddle-back Kop [Skipberg] was now only five or six miles away, and we were told the boundary of the terrible tsetse fly ran along the base of it from north to south. There was a very bad supply of water where we outspanned, and the grass was not especially good, so we determined to trek about three miles to the southeast of Pretoriuskop, where we discovered two deep waterholes [probably the Fayi waterholes] and good pasturage, before making our permanent camp. Within a stone's throw lay the footpath used by the Blacks on their way to and from the diamond or goldfields, and by all going to Delagoa Bay from the Transvaal.
>
> Nearly eight miles we followed the track towards Delagoa Bay, and were soon at the far end of Long Kop [Skipberg]. The bush from this point becomes much thicker, and the general aspect of the country grew wilder and more densely wooded. We went

on the path for another 5 miles before we caught up with Woodward [one of his hunting partners], who waited for us at a place known far and near as Hart's station; once a small but comfortable log-hut, with a small garden around it, but now only a few broken-down walls, and a mass of cinders. The place had a very gruesome appearance, which the tragic tale connected with it helped to foster.

While they were resting at a waterhole in a slab of rock in the Josekhulu Spruit close to Hart's burnt-down wattle-and-daub house, his partner, a Mr Woodward, told him about the tragedy that had taken place there. Thomas Hart had been one of A.H. Nellmapius's "station masters" along the trade route he had built on the instructions of President Thomas Burgers between the goldfields of Pilgrim's Rest and Delagoa Bay. During the Sekhukhune uprising of 1876, the marauding gangs of Chief Maripe, a subject of Sekhukhune, had made the transport route unsafe. All the "station masters" returned to Pilgrim's Rest for their own safety, but Hart refused to leave his station until he received official instruction to do so and spent his time hunting and collecting plants. He also caught and tamed a variety of animals and birds. His pets included small buck, monkeys, snakes, parrots and rollers. Thanks to his pleasant nature he was on good terms with everybody, including the local blacks. But on 10 August 1876 disaster befell the 22-year-old young man in the form of a band of Chief Maripe's raiders.

Woodward described the event to Sandeman as follows:

> One day a party of armed Macatees came up to his door, and in angry terms told him that a few days previously some white men by Pretorius Kop had met a Macatee of their tribe who was in possession of a gun. The whites had taken his gun away from him in the most unjustifiable manner. The boy had returned to the head kraal and told his story. Instantly the present party had set out in pursuit of the offenders, but had been unable to come up with them. However, they were convinced that they were in some manner connected to Hart, and at all events Hart was a white man, even if not connected with the actual robbers of the gun; and they speedily gave him to understand that he must give them his rifle in exchange for the one they had been deprived of. If he did this, they promised to go and leave him in peace. Hart naturally objected to this disposal of his property, and argued long against it, and indeed refused to part with his rifle for any reason. This conversation had taken place while he had been standing in the doorway. He then stepped outside altogether, to try and persuade the whole band to go away and leave him in peace, for he probably saw that some of them were his friends and inclined to be guided by his words. While, however, he was haranguing them, one murderous wretch going behind him put up his rifle, and shot him through the head, blowing off the top of his skull.
>
>   As the band had nothing to gain by remaining, they stripped his little hut of everything of any value, killed all his pets, set fire to the roof and to the outhouses, and then departed in possession of the disputed gun, leaving the murdered man where he had fallen, but stripped of his clothes, which they divided among themselves.

According to Woodward, two white friends of Hart who lived among the Swazi south of the Crocodile River heard about the murder and paid a group of Swazis to bury Hart's remains. This part of the story is incorrect, as it had been noted in an official document that W. Barter, bailiff and inspector of Police at Pilgrim's Rest, received news of Hart's murder on 16 August 1876. During an investigation on 20 August he found Hart's station at Joubertshoop destroyed and deserted, with his remains mutilated by wild animals. After he had buried Hart's skull and other remains, he returned to Pretoriuskop.

When Sandeman and his companions visited the murder scene two years after the murder, parts of the house's walls were still standing, as were the door posts. The horns of several game species that had been hunted by Hart, including kudu, blue wildebeest, tsessebi and sable, were lying around. A roughly constructed bench was standing in front of one of the windows, while there was a dilapidated camp to one side in which he had kept his animals.

While they were looking around, they noticed the first tsetse flies on their eastward journey from Skipberg. The party moved on and Sandeman noted the following important facts about the rest of their hunting trip:

> For some three miles further we kept along the footpath, and then Woodward, under whose direction we moved, turned off to the left and almost at right angles. We took as our guiding beacon a high peak of rock which showed up clear from a low range of broken up hillocks [Makhuthwanini]. As it was, we did not see a living thing, and although we put off making our camp for the night as late as possible, we discovered a small hole of clear water between two high rocks, which lay in the bed of a now dry stream [possibly the Mtlhowa Spruit].
>
> Before we had gone many miles along our new course, the nature of the country changed. The trees gave way to thorns again, and short crisp grass took the place of the long tufts we had hitherto been walking through [compare this to the brackish spots along the lower reaches of the Ntomeni Spruit]. The boys grew more hopeful, and quickened their pace of their own accord. Soon we came across some fresh spoor of impala, going in the same direction as ourselves, and a mile further we came upon a large pool of sweet clear water. [The Ntomeni Spruit has stopped running in the dry season.]
>
> We made our camp for the night by the only good water we had tasted for a couple of months, and there was almost an unlimited supply of it. [According to his description, it looks like the confluence of the Mhlambanyathi and Mbyamiti.] It was discovered more by chance than anything else as none of the boys had any knowledge of our present neighbourhood. Several acres of water were covered thickly with large rugged rocks, piled up one above the other. On the chance of coming across a panther or a lion, we were climbing over them, and in the centre came upon a little open sandy space, on two sides of which, deep down to the foundation of the rocks, were two clear pools of water, one not less than six feet deep and evidently supplied by a spring. [At this specific site there is no sign of these water pools today.]

William (Bill) Sanderson

The eccentric Scottish frontiersman William (Bill) Sanderson and his two brothers, Bob and Tom, lived in the vicinity of Legogote (formerly Lozieskop), where the farm Peebles is today, as well as on the adjacent farm Klipkoppies. This was in the decades just before and after the Anglo-Boer War. Bill Sanderson was one of the many colourful characters who tamed the Lowveld in those years. He was a well-known hunter who was particularly proud of his skills as a marksman. An expert in the local wildlife, he was a good friend of Harry Wolhuter, one of the Sabie Reserve's first rangers. He was also Stevenson-Hamilton's adviser and mentor during his first years as warden of this famous reserve.

The story of Bill Sanderson and his life during and after the Anglo-Boer War is related in chapter 14 ("War clouds over the Lowveld").

Captain H.F. (Farmer) Francis

Captain H.F. (Farmer) Francis made a name for himself in the southern Lowveld in the years before the outbreak of the Anglo-Boer War as a lion hunter, wildlife expert and naturalist par excellence.

He was a brother of W.F. Francis, who had collected the first yellow-billed oxpecker (*Buphagus africanus*) in the Lowveld at Komatipoort on 29 September 1896, before the outbreak of the rinderpest epidemic of 1897 (Stutterheim & Brooke, 1981). Apparently the Francis brothers collected birds for the South African Museum in Cape Town. In 1898 they undertook a collecting trip to the Inhambane district of Mozambique during which they discov-

W. (Bill) Sanderson, pioneer and hunter on his farm near Legogote, 1903.
(ACKNOWLEDGEMENT: STEVENSON-HAMILTON COLLECTION, SKUKUZA ARCHIVES)

ered the interesting tiny greenbul *Phyllastrephus debilis* at Massinga (Clancey, 1971).

During the siege of Ladysmith and the relief of Mafeking (Mafikeng), Francis served in the Imperial Light Horse. His brother, who was also a member of the ILH, was killed in action. Francis later joined the notorious Steinaecker's Horse and was put in charge of the fort built at Mpisane's kraal on the farm New Forest. Because of his knowledge of the Lowveld and its wildlife, Captain Francis was recommended as a suitable candidate to take over the position of warden of the Sabie Reserve after the war. This recommendation was made to the provincial secretary of Transvaal by Tom Casement, the assistant mine commissioner at Barberton, in July 1901. Permission was given on 4 July 1901 that he be appointed as acting warden of the area, and on 2 August 1901 Captain Francis wrote a letter to Tom Casement at Sabie Bridge in which he stated that he would accept a permanent position if the conditions were attractive enough. But before the appointment could be confirmed, news was received that he had been killed on 7 August 1901 during an attack on Fort Mpisane by the Boer forces under the command of General Ben Viljoen.

On 14 August 1901 one of his comrades in arms, Lieutenant G.E. Gray, applied to the civil commissioner to be considered for the position, but without success.

The story of Captain Francis and the battle of Mpisane is related in chapter 11.

H.C.C. (Harry) Wolhuter

Henry Charles Christoffel Wolhuter was born on 14 February 1876 in Beaufort West. His parents later moved to the Transvaal and eventually settled on a farm in the vicinity of Legogote. He undertook his first hunting expedition when he accompanied a family acquaintance to the vicinity of the current Satara at the age of 15. This area and the area further north to the Olifants River were to become his favourite hunting grounds. He was in fact on an extended hunting trip in what is now the Kruger National Park at the time the Anglo-Boer War broke out. Wolhuter also joined the infamous British irregular unit Steinaecker's Horse and served at Komatipoort, Sabie Bridge and other outposts as far north as the Olifants River. He attained the rank of sergeant in the unit and at the end of the war he was recommended for the position of ranger by the warden of the Sabie Reserve, J. Stevenson-Hamilton. He was one of the first rangers appointed by Stevenson-Hamilton on 17 August 1902 (some sources give his date of appointment as being 15 August 1902). Wolhuter was to work for 44 years in this famous reserve until he retired in April 1946, at the same time as his chief. Wolhuter became a living legend when a large lion pounced on him and dragged him from his horse in 1903 at Lindanda. In the ensuing struggle between man and lion, Wolhuter managed to kill the lion with his hunting knife. His career as soldier and ranger is documented in full in chapters 11 and 16.2.

Sir Percy FitzPatrick KCMG (1862–1931)

Sir James FitzPatrick was a well-known politician, author and pioneer of the Lowveld, and also a pioneer of the fruit industry in South Africa. He was born on 24 July 1862 in King William's Town in the then British-Kaffraria (Eastern Cape).

His publications include *The Outspan*, *Through Mashonaland with Pick and Pen*, *The Transvaal from Within*, *Jock of the Bushveld* and *South African Memories*, which was published after his death.

Percy FitzPatrick or "Fitz", as he was commonly known, was the eldest son of James Coleman Fitz-Patrick, a judge of the Cape Colony's High Court,

H.C.C. (Harry) Wolhuter (1876–1864), well-known hunter and ranger on his horse. February 1904.
(ACKNOWLEDGEMENT: STEVENSON-HAMILTON COLLECTION, SKUKUZA ARCHIVES)

and Jenny FitzPatrick, both from Ireland. Two of Judge FitzPatrick's four sons died in action: Tom during the Matabele rebellion and George during the Anglo-Boer War, as a member of the Imperial Light Horse.

Percy FitzPatrick underwent his schooling first at St Gregory's College, Downside, England, and then at St Aidan's College, Grahamstown, and the South African College in Cape Town.

After his father's death in 1880 he left college to support his mother and family. In 1882 he was lured, like so many other young men, to the goldfields at Pilgrim's Rest. He worked as storeman, prospector's assistant and journalist and eventually as transport rider with an ox-wagon between Delagoa Bay, Lydenburg and Barberton (1882–1885).

His adventures as transport rider are related in *The Outspan* (1897) and in his classic work *Jock of the Bushveld*, which was published in 1907.

In the early 1900s he would often recount the antics of his famous dog Jock to his children ("the likkle people") at bedtime. Rudyard Kipling, a good friend of Percy, persuaded him to compile these stories in book form. After he had completed the task, he started searching for an artist who could illustrate the book. Eventually he made contact with the artist Edmund Caldwell in England and brought him to South Africa in 1904 to do the illustrations in the Transvaal Lowveld. At its publication in 1907 the book was a huge success and it was reprinted four times during the first year. It is still one of South Africa's favourite works among old and young, has been translated into several languages and has been reprinted more than 90 times. With this classic, Percy FitzPatrick became one of the most beloved authors of all time. Although he cannot be regarded as one of the great pioneer hunters of the period, he captured the romance of the transport rider era and hunting adventures in the Lowveld and raised it to a level that certainly entitles him to an elevated position in the hunters' gallery of honour.

FitzPatrick and his famous dog, Jock, were inseparable companions. But he admitted that, during all their adventures along the transport road, in the hunting fields and in their gold prospecting days, he always had a third "friend" at hand – his rifle. His hunting rifle was an old, worn-out, mounted police issue – an ex-military .577/.450 Martini-Henry carbine. Despite rough treatment this old firearm, like his dog, never failed him and in tight spots often stood him in great stead. This was also the case with the different hunting rifles (4-, or 10-pounder muzzle-loaders, Westley Richards, Lee-Metfords, Mausers, etc.) of the better-known hunter figures of this romantic era in the Lowveld's history.

In 1886 FitzPatrick married Elizabeth Lilian Cubitt, the daughter of the pioneer John Cubitt, who lost his life during the Sekhukhune wars. Three years later they moved from Barberton, where he had been the editor of the *Barberton Herald,* to the Witwatersrand. In 1891 he led Lord Randolph Churchill's expedition to Rhodesia (Zimbabwe) and also prepared the way for Sir Alfred Beit's expedition to Lobengula's land.

In 1892 he returned to the Rand to serve as head of the information department of Hermann Eckstein & Co. (an affiliate of Wernher, Beit of London, later the well-known Corner House). FitzPatrick entered the political arena and during President Kruger's rule he was a great champion of the political rights of the "outlanders" and became secretary of the so-called Reform Committee, which advocated this cause.

After the failed Jameson Raid at Doornkop, FitzPatrick and the other leaders of the raid were arrested and accused of high treason. Bail was refused and FitzPatrick was sentenced to two years' imprisonment and a fine of £200. However, he was

Sir James Percy FitzPatrick (1862–1931), the well-known politician, author and pioneer of the Lowveld (circa 1920s).
(ACKNOWLEDGEMENT: MRS C. MACKIE-NIVEN, UITENHAGE)

released in May 1896 with the proviso that he refrains from becoming involved in politics for three years. As the political arena was now closed to him, he focused his attention on the gold industry and became Hermann Eckstein's partner and the chairperson of Rand Mines.

One day, Kurt von Veltheim, a political activist, came to see him in his office, all the while playing with his pistol in a sinister manner. Only two days later Von Veltheim killed Woolf Joel in his office!

With the outbreak of the Anglo-Boer War, FitzPatrick was responsible for the founding of the Imperial Light Horse. He was offered a commissioned rank, but both Lord Salisbury (the British prime minister) and Lord Balfour asked him to stay in England as an additional official adviser to the British government on South African affairs.

After the Treaty of Vereeniging, FitzPatrick applied all his energy to the rebuilding of the two Boer republics. Later, in 1910, he played a major role in the establishment of the Union of South Africa as the representative for Transvaal. During his parliamentary career he fought two memorable campaigns in his ward in Pretoria – one in 1906 against Sir Richard Solomon and the other in 1910 against General Louis Botha, the prime minister of the first Union Parliament.

During the First World War, General Jan Smuts sent FitzPatrick on a lecture tour around the country to explain to people the causes of the war, as well as the Union's participation. He fathered the idea that a silence of two minutes should be maintained on Armistice Day. He also bought Delville Wood in France, the scene of the South African Brigade's bravest war action in July 1916, and presented it as a gift to the South African nation.

FitzPatrick had been interested in farming from an early age. When he and a group of friends noticed the potential of the Sunday's River Valley for citrus production, they invested $450 000 in the development of 5000 acres of land under irrigation in the Lower Sunday's River Valley. He travelled to America to study the citrus industry and was responsible for the import of a number of new citrus cultivars into South Africa. He also imported modern equipment for citrus packing and, together with H.E.V. Pickstone, was the driving force behind the founding of the SA Cooperative Citrus Exchange.

FitzPatrick was a pleasant and optimistic man, but his last years were marred by a series of personal tragedies. His eldest son Nugent died in action in France in 1917 and his wife in 1923. His other two sons died within one week of each other over Christmas in 1927 – Alan was killed in an accident in Johannesburg while Oliver died of typhoid in Mexico. This left him with only his daughter, Cecily, who married Jack Niven in 1923.

These tragic events were a severe blow to him. He died on 24 January 1931 on his estate Amanzi near Uitenhage and was buried on The Outlook, which looks out over the Sunday's River Valley. His final resting place was later declared a national monument. There is also a statue of his famous dog, Jock, in Barberton. In 1983 his daughter and her sons sponsored the building of a small private camp in the Kruger National Park, which is known today as the Jock of the Bushveld camp. This lovely camp, close to the old transport routes taken by Jock and his master, was officially handed over to the Parks Board on 13 September 1983 by the donors, Jack and Cecily Mackie-Niven. Built into the camp's main building is a doorpost and door from the shop of Jock's arch enemy, Field Cornet Seedling of Krugerspost. These were donated to the Board by Mr and Mrs J.H. Smith of Krugerspost. A beautiful statue of Jock by the White River artist R. Lawrence was later unveiled in the camp.

Miles Robert Bowker
Miles Robert Bowker was a direct descendant of Miles Bowker, a British 1820 settler who had settled on the

The memorial stone at the Bowkers' hunting campsite at Bowkerkop. (28.04.1984) farm Tharfield on the Kowie side of the Riet River in the Eastern Cape. His cousins Alec Bowker and Charlie White, together with Fred and Harry Barber, the founders of Barberton, often hunted in the Transvaal Lowveld in the two decades preceding the Anglo-Boer War. They were usually accompanied by a certain Miller, who provided the wagons and oxen.

Their favourite campsite was close to two gigantic baobab trees at the foot of a basaltic outcrop just west of the Shipandani valley and three to four kilometres north of the current Mooiplaas ranger station. The hill is now known as Bowker's Kop and in the early 1950s the family erected a marble memorial tablet there with the following inscription:

> Near this baobab tree a party of hunters from the Rand had their camp in 1888. Their names were Miles Robt. Bowker whose name is carved on this tree, Alec Bowker and Charlie White, his cousins, Fred and Harry Barber, the founders of Barberton, and Miller with his ox-wagon who provided the transport. M.R. Bowker. July 10, 1888.

Unfortunately, in the early 1980s elephants damaged one of the two trees so badly that it toppled over. There must surely have been permanent water in the Shipandani Valley at the time of the Bowkers' hunting trips, but today a windmill is used to provide game with water. The Bowkers and Barbers hunted not only in the area north of the Letaba River, but also in the vicinity of Satara.

F. Vaughan-Kirby mentioned in his book *In Haunts of Wild Game* that they also encountered game one winter in the years between 1884 and 1890 along the Nwanetsi Spruit east of the current Satara rest camp, and shot a rare albino reedbuck ram.

The initials A.W.B. '91, chiselled on a rock on a sandstone hill along the Sweni River east of Ngumula Pan, could be those of Alec Bowker. (See also the photograph and caption on page 215, where the possibility is mooted that it may have been Adam Briscoe who had chiselled his initials into the sandstone.)

Another Bowker from the Eastern Cape had a close connection with the Kruger National Park in later years. He was the Member of Parliament for Albany, Thomas B. Bowker, who served as a member of the National Parks Board between 1945 and 1950 and delivered a valuable service in connection with the management of national parks in the then Cape Province. He also donated the funds with which both the concrete dam in the Tshange (Tshanga) Spruit along the southern flank of Tshange (Tshanga) Hill in 1952 and the Tshange (Tshanga) lookout point were built.

S.C.R. (Bvekenya) Barnard (1886–1962)

The professional elephant hunters under discussion here were a very rare group of adventurers. Heat, humidity, fever, biting insects and the threat posed by wild animals and sometimes hostile black tribes were their constant companions in their chosen hunting grounds.

Stephanus Cecil Rutgert Barnard, known among the Tsonga-speaking people of the Lowveld as "Bvekenya" ("the one who swaggers" or "the dandy one"), was the best known of the more recent elephant hunters of the northern Transvaal. Bvekenya was born on 19 September 1886 in Knysna and grew up in the western Transvaal. After serving in the police for a number of years, he decided to become a professional elephant hunter. In 1910 he set off for that fascinating and romantic part of the northeastern Transvaal, Crooks Corner, the strip of land between the Limpopo and Luvuvhu rivers. This area used to be a safe haven from the law for poachers, smugglers and renegades. Its main attraction was the fact that the borders of the former Transvaal, Mozambique and the former Rhodesia (Zimbabwe) all met there. Any pursuing law officials could therefore eas-

S.C.R. "Bvekenya" Barnard (1886–1962), well-known hunter of Crooks Corner and the Ivory Trail. (ACKNOWLEDGEMENT: MR IZAK BARNARD)

The baobab tree (one of two) at Bowker's Kop where the Bowkers had their hunting camp in 1888. (28.04.1984)

ily be given the slip by simply crossing the nearest border. Barnard used this area as his base when he launched poaching trips into neighbouring countries. He lived as a fugitive and was constantly hunted by the police of all three countries.

Elephants were his specialty and he searched far and wide for the best ivory bearers, particularly the legendary elephant bull Ndlulamithi. After having been employed for a time by the recruiting company WNLA, he left the hunting fields in November 1929 and started making an "honest" living. He got married and settled on his farm near Geysdorp in the western Transvaal, where he farmed and provided for his family until his death on 2 June 1962.

Shortly before his death, in February 1962, Barnard granted an interview to R.J. Labuschagne and N.J. van der Merwe of the National Parks Board. A lot of what he had to say was of great interest, especially about conditions in this northeastern corner of the Lowveld, which is now part of the Kruger National Park, and in southern Rhodesia (Zimbabwe) and Mozambique between 1910 and 1929, when he had lived there.

According to Barnard, the buffalo population in the northern Transvaal Lowveld and adjacent area recovered very slowly after the great rinderpest epidemic of 1896–97. Upon his arrival at Crooks Corner in 1910 there were no buffalo in the area between the Limpopo and Luvuvhu rivers. Nor where there any in the adjacent Rhodesia, but he was told by blacks that there was a small herd of four in the vicinity of Massingire and another group of three along the Lundi River.

In 1916 he saw the first two buffalo along the Chefu River in Mozambique. Buffalo, like cattle, are highly susceptible to rinderpest and according to Barnard they had all been killed off by the disease in the Lowveld, as had his father's herd of cattle in the western Transvaal. Out of a total of 300 head of cattle, apparently only three survived. He did believe, however, that at certain places some animals managed to escape the disease. This included the farm Frisgewaagd of a friend of his, where the deadly virus for some inexplicable reason had not claimed any victims. In later years buffalo numbers quickly increased and by the mid-1920s there again were large herds along the Sabie and Lundi in the then Rhodesia.

Nyala were also hit hard by the rinderpest and during his first few years at Crooks Corner he did not encounter a single one. Apparently quite a few survived lower down along the Limpopo, as well as along the Nuanetsi in what was then Rhodesia. Eventually their numbers increased to such an extent that they churned the soil around the waterholes into powder. According to Barnard, nyala made their re-appearance in the Pafuri area in the early 1920s.

As a result of the scarcity of prey, lions and wild dogs were also few and far between after the rinderpest, and some lions became man-eaters.

At the time of Barnard's arrival in 1910, there was still a considerable number of black rhino in this area, as well as in the adjacent British and Portuguese territories, but he only knew of five white rhino, which came to drink in the Luvuvhu River in the Pafuri area. These were apparently shot soon afterwards by black hunters. (In 1870 a white rhino killed a hunter named Jacobs close to a fountain in Rhodesia that afterwards became known as Jacobs' Fountain.) Barnard also claimed that some of the Park's rangers shot and killed a black rhino in the Shingwedzi area in 1928.

There were no elephants in the northern parts of the Kruger National Park in those days, as all of them had been shot by the Boer hunters and their helpers. Barnard therefore shot most of his elephants in the adjoining Rhodesia (Zimbabwe), as far north as the Haroni River. In Portuguese East Africa (Mozambique) he hunted along the lower Limpopo, Chefu and Great Save rivers, as well as on the Banyine Plains. Later he heard that a couple of elephants had been seen in the area along the Shingwedzi River.

Giraffe had also disappeared from the northern parts of the Kruger National Park, as they had long since been driven away by hunters such as Hans Klopper, after whom Klopperfontein was named.

Barnard knew Hans Klopper well. This tall, thin, bearded man lived on his farm Doornspruit near Louis Trichardt (Makhado), where Barnard often visited him.

According to Klopper, he had shot a giraffe in Rhodesia whose spoor measured 11.5 inches (almost 30 cm). He also claimed to have shot a man-eating crocodile that measured almost 22 feet (6.6 m) at the confluence of the Sabie and Lundi rivers.

The numbers of roan and sable antelope were apparently less affected by the rinderpest than those of buffalo, kudu, eland and nyala. According to Barnard there were lots of these animals in the park area and especially in the adjacent southern Rhodesia (Zimbabwe). He mentioned three black names for the roan antelope, namely "shiyanamane", "mtagaisa" and "hangataya".

From 1911 to 1913 the northern Transvaal and adjacent areas were subjected to a great drought. At Christmas 1914 there were a few rainstorms over the Nuanedzi, and the lush regrowth attracted large herds of game of all kinds. Over a distance of about 50 miles (80 kilometres) he encountered large congregations of game, including elephant, hippo (in the rivers), giraffe, zebra, kudu, roan antelope, nyala, bushbuck, oribi (called "nshiki" by the locals), sable, tsessebi and several Lichtenstein's hartebeest. Barnard could not state with certainty that Lichtenstein's hartebeest could be found in the Kruger Park area, but he claimed that they were common in the southeastern part of Rhodesia as far south as Nuanedzi and in the adjacent Mozambique. He gave the impression that these animals could have occurred south of the Limpopo, especially in the area north of the Soutpansberg. According to him the locals called Lichtenstein's hartebeest "mzaas" (mzanza) and tsessebi "kolomane".

In 1916 Barnard caught 12 eland calves in Mozambique with his two dogs, Mac and Brits. They were then suckled by cows he kept in a camp. He hoped to eventually breed them on a farm he planned to buy in the Waterberg, but as a result of the First World War he could not obtain a permit to bring the animals into the Union. At his wits' end, he gave a few of the calves to District Commissioner Forrestall of Chibi, near Fort Victoria in Rhodesia, who was so "grateful" that he immediately issued a new warrant for Barnard's arrest. Later Barnard apparently took the remaining animals, two bulls and three heifers, and the tame donkey with which they used to roam around in his yard, into the Park and released them in the vicinity of Klopperfontein. Some of their descendants are possibly still wandering around the Hlamalala Plains.

During his stay at Pafuri, Barnard never heard of any animals dying of anthrax. He did, however, have a narrow escape in Mozambique when he caused the trigger mechanism of a poisoned spear to go off. The spear had been driven into a long log and hoisted high into a tree above an elephant path. Barnard was carrying a .303 Lee-Metford, given to him by Morty Ash, over his shoulder. The spear hit the rifle and stripped his shirt from his back without leaving a scratch on his body. This saved his life, as the poison the Tsonga make from the creeper *Strophanthus kombe*, is deadly – even to elephants!

Barnard's activities at Crooks Corner and elsewhere had to come to the attention of the park authorities sooner or later. He had the greatest respect for Stevenson-Hamilton as an excellent nature conservator, but did not think much of ranger Kat Coetser stationed at Punda Maria. According to Barnard he was, however, "a good sport" who loved his tipple.

Coetser once asked Barnard to transport Judge F.E.T. Krause and his wife in his tented wagon from Punda Maria to Pafuri. He delivered them safely to their camp, returned, immediately loaded 700 pounds (318 kilograms) of ivory onto the wagon and smuggled this "surplus baggage of Judge Krause" out of the reserve at Punda Maria, right under Coetser's nose.

In his younger days Barnard also befriended Joseph Fourie who, according to him, was the brother of Dina Fourie, whose epic journey is related in chapter 6.3. Barnard concluded his narration with the claim that he had once made an offer to the Parks Board to build a wall along the border of the northern part of the Park to prevent game from wandering into an area where they could be killed by poachers. His reward would be a few tons of Rhodesian teak, jackal berry, kiaat and tamboti, as well as permission to shoot 20 elephants in the Park every year. He was very disappointed when the Board would not accept his offer!

On 19 September 1986, 100 years after his birth, his family erected a memorial in the area where he and many others had experienced such extraordinary things on the so-called Ivory Trail from 1910 to 1929. The story of "Bvekenya" Barnard and his adventures is told in full by T.V. Bulpin in his book *Ivory Trail* (1954) and a short version is included in chapter 19 about the wild days of Crooks Corner.

J.P.J. (Hans) Klopper (1851–1928), after whom Klopperfontein was named.
(ACKNOWLEDGEMENT: THE LATE GUS ADENDORFF)

John Thomas Kelly (1849–1932)

John Thomas Kelly was born on 2 January 1849 in the Voortrekker town of Schoemansdal. His father, Thomas Kelly, and a friend, Terrence William Fitzgerald, had deserted from the British army in Natal and went to Schoemansdal with two English traders to hunt elephant.

Here Fitzgerald married the widow Botes and Kelly her sister, Engela. Shortly after the birth of their son, John Thomas Kelly, both parents died of malaria and were buried in the cemetery at Schoemansdal. The young Tom Kelly was raised by the Fitzgeralds.

It is said that he worked for João Albasini for a while. He later married Maria Johanna (Mita) du Plooy and they had four sons and two daughters. Kelly became interested in hunting at a young age and, like many of the able men of Schoemansdal, he spent the winter months hunting elephant. He became a crack shot and was promoted to field cornet by the Boer leaders during the Venda uprisings of Chief Makhado and his subordinate headmen.

Kelly attained the rank of commandant during the Anglo-Boer War and, with his guerrilla tactics of attack and retreat, became a real thorn in the side to the British. He specifically concentrated on the irregular Bushveldt Carbineers, stationed at Fort Edward near Elim and responsible for terrorising Boer families on farms in the northern Transvaal. Commandant Kelly and Captain Sarel Eloff were involved in the battle at Fort Tuli in November 1899 under Assistant General H.C.J. van Rensburg when Colonel H.C.O. Plumer tried to break through to the besieged Mafeking (now Mafikeng), under Colonel R.S.S. Baden-Powell.

He was also involved in the bombing of armoured trains on the Pietersburg line and an attack on Fort Edward itself. In September 1901 he was betrayed by local blacks and captured at his camp at Dzombo in the current Kruger National Park by a patrol of the Bushveldt Carbineers under Lieutenant H. (Breaker) Morant.

More is told about Kelly's war experiences and capture in chapter 14, "War clouds over the Lowveld". It is said that he was so enamoured with hunting that during the war, before his capture, he found the time to go and shoot a few elephants in the vicinity of the confluence of the Letaba and Olifants rivers.

After Commandant Kelly's return from Ceylon, where he had been held as a prisoner of war, he settled on his farm Boskoppies in the Soutpansberg district, where he died on 27 March 1932.

J.P.J. (Hans) Klopper (1851–1928)

Hans Klopper was another one of the Soutpansberg pioneers who regularly went hunting in the area that is now the northern part of the Kruger National Park. He was usually accompanied by Piet Kelly, a cousin of Tom Kelly, or other hunting partners. His favourite camping place was at the fountain one kilometre northeast of the Shantangalani hills, which was later named Klopperfontein in his honour. There used to be a large fig tree at the fountain under which Hans usually pitched camp. He preferred hunting the giraffe so common in those days, as he apparently had a special liking for the marrow in the long bones. It is said that a pile of giraffe bones was still lying close to his old camping place years after the area had become part of the Shingwedzi Reserve and the first ranger, J.F. Coetser, had settled at Punda Maria in 1919 to control illegal hunting. P.W. Willis confirmed in a letter to Colonel Sandenbergh that he had seen this heap of bones himself in 1903 while visiting the area. The fountain later dried up and in 1955 a concrete reservoir was built here, with two boreholes to provide additional water to game.

During the high-rainfall period of 1971–78 the fountain started running again and maintained its flow until it dried up during the prolonged drought of the 1980s.

Hans Klopper acquired the nickname "Ribada" among the blacks, which means "the one who is always on the move" (to the hunting grounds!). He apparently had another campsite next to the Tsendze, just upstream from the Old Wagon Drift, where he killed all the giraffes in the years preceding the Anglo-Boer War. According to P.W. Willis there was a pile of giraffe bones at this old campsite as well, and he thought the pile of bones at Klopperfontein had looked older than the one by the Tsendze.

According to the blacks to whom Willis had spoken, Klopper arrived in the winter months with his wagons and shot mainly giraffe and eland. During

Hans Klopper's hunting days before 1896 tsetse fly was one of their big problems, while African horse sickness decimated their horses.

After the curtailment of his hunting activities in that area by the proclamation of the Shingwedzi Reserve in 1903, he moved further eastward to Mozambique. In the Skukuza Archives is a written request for permission to travel through the reserve on a hunting trip to Mozambique from Hans Klopper to the provincial secretary, dated 24 May 1921.

In later years he concentrated more on farming at Doornspruit, about 20 kilometres southeast of Louis Trichardt (Makhado) on the road to Elim Hospital, and his old homestead is still standing. He died in 1928 at the age of 77 and was buried in the old family cemetery.

A few other well-known Lowveld hunters

Apart from this group of the most prominent hunters, who had played such an important role in the formative years of the Lowveld, there were other, less well-known figures that also hunted regularly in the area now encompassing the Kruger National Park and in the adjacent Portuguese East Africa (Mozambique). In the years before and just after the Anglo-Boer War there were, among others, Jan (Bokwa), Gert and Gabriël Stols and their family in the White River area; Abel Chapman, the explorer; Commandant Dawid Schoeman of Krugerspost; Captain Ignatius (Naas) Ferreira; P.W. (Pump) Willis and his brother (Clinkers), who were both members of Steinaecker's Horse (P.W. later settled at Acornhoek, where he established a flourishing business concern and traded his gun for a camera); Commandant Piet Möller of Louis Trichardt; and Hermanus Jacobus Grobler, Pieter du Preez[4] and Willem Fitzgerald of Schoemansdal.

In the years after the Anglo-Boer War, W. Borchers, a prominent trader in the Soutpansberg district, and Joe Albasini made names for themselves. Joe Albasini was an official in the Department of Native Affairs at Sibasa and the great-grandson of João Albasini. Joe in particular was a very good shot and downed a lot of elephants and other large game in Mozambique, between the Limpopo and the Great Save rivers. In the early 1920s he was also the owner of Mbaula Ranch between the Great and Klein Letaba rivers.

Comdt Dawid Schoeman and his wife. He had been a well-known hunter in the Transvaal Lowveld in the decades before and after the Anglo-Boer War.
(ACKNOWLEDGEMENT: TRANSVAAL ARCHIVES)

The well-known Klopperfontein between Punda Maria and Pafuri, where Hans Klopper used to have his hunting camp in the era just before and after the Anglo-Boer War.
(27.04.1984)

---

4 More information about Pieter du Preez can be found in Du Preez, M.H.C. (1993). See references.

References

Adendorff, G. 1974. Waar kom die naam Klopperfontein vandaan? *Custos* 3(13):30–31.

Adendorff, G. 1984. *Wild Company.* Books of Africa, Cape Town.

Albasini, João III. 1981. *Joao Albasini: 1813–1888.* A 28-page typescript document distributed by the Albasini family.

Altenroxel, H.S. 1942. *Ich suchte Land in Afrika.* E.A. Seemann Verlag, Leipzig.

Anon. 1965. *Some Lowveld Pioneers.* Publication of the Lowveld, 1820 Settlers Society.

Anon. 1981. *Die verhaal van Henry Thomas Glynn (1856–1928) en van sy gade Gertrude Gilbertson (Dakes) (1876–1970).* Curator, Sabie Forestry Museum.

Batten, H.V. 1913. Letter by himself (15 May 1913) from Musina, to Dr Austin Roberts, Transvaal Museum. Transvaal Museum Archives.

Becker, P. 1985. *The Pathfinders – a Saga of Exploration in Southern Africa.* Harmondworth Viking, London.

Breyer, H.C. (sic) 1916. Letter (23 Augustus 1916) from the Director of the Transvaal Museum to the Administrator of the Transvaal, Pretoria. Transvaal Archives Acquisition T.P.B. 785 (T.A. 3011).

Buchanan, D.T. 1955. *History of Makuleka.* Cyclostyled manuscript. Skukuza Archives.

Buckley, T.E. 1876. On the past and present geographical distribution of large mammals of South Africa. *Proceedings of the Zoological Society* 1:277–293.

Bulpin, T.V. 1950. *Lost Trails on the Lowveld.* Howard B. Timmins, Cape Town.

Bulpin, T.V. 1954. *The Ivory Trail.* Howard B. Timmins, Cape Town.

Bulpin, T.V. & Mulkler, P. 1968. *Low Veld Trails.* Books of Africa, Cape Town.

Bulpin, T.V. 1983. *Lost Trails of the Transvaal.* Books of Africa, Cape Town.

Cattrick, A. 1959. *Spoor of Blood.* Howard B. Timmins, Cape Town.

Chapman, A. 1900. Draft of scheme for preserving and re-establishing the large game and wild animals in the Transvaal. Transvaal Archives, Acquisition C.S 2:211.

Clancey, P.A. 1971. A handlist of the birds of Southern Mozambique. *Memorias do Instituto de Investicâo Cientifica de Mocambique* 10 Serie A 1969–1970:149.

Cronwright-Schreiner, I.S.C. 1925. *The Migratory Springbok in South Africa.* T. Fischer & Unwin, London.

Cumming, R. Gordon. 1840. *The Lion Hunter in South Africa.* John Murray & Co., London.

Cuthbertson, M.B. 1949. African game 400 years ago. *African Wild Life* 3(2):148–150.

Das Neves, D.F. 1879. *A Hunting Expedition to the Transvaal.* George Bell & Sons, London.

Delius, P. 1986. Abel Erasmus: Power and profit in the Eastern Transvaal. In: V. Beimert et al. (eds). *Putting a Plough to the Ground.* Raven Press, Johannesburg.

De Vaal, J.B. 1948. Schoemansdal die verlate Voortrekkerdorp in die Zoutpansberg. *Die Naweek* 19 June.

De Vaal, J.B. 1948. Olifantjag in die dae van Schoemansdal. *Die Naweek,* 10 June & 17 June.

De Vaal, J.B. 1948. Ivoorsmouse in Schoemansdal. *Die Naweek* 24 June.

De Vaal, J.B. 1948. Bernard Francois Lotrie, die baas-olifantjagter van Schoemansdal. *Die Volkstem* 8, 15, 22 & 29 October & 5 November.

De Vaal, J.B. 1953. Die Rol van João Albasini in die geskiedenis van Transvaal. *Archives Year Book for South African History* 16(1):1–166.

De Vaal, J.B. 1978. João Albasini. Unpublished manuscript.

Du Preez, M.H.C. 1993. Die wedervaringe van die Voortrekker Pieter du Preez (1807–1889). *Yearbook of the Africana Society.*

Dicke, B.H. 1923. Tsetse in the Letaba-Shingwedzi Basin, North-East Transvaal. *Union of South Africa, Department of Agriculture Entomology Memoir* 1:60–63.

Dicke, B.H. 1936. *The Bush Speaks: Border Life in Old Transvaal.* Shuter & Shooter, Pietermaritzburg.

Erasmus, J.A. 1898. Letter dated 23 March 1898, from the Native Commissioner, Lydenburg to obtain permission to hunt protected game for the Museum. Transvaal Archives Acquisition S.S.2031 (R8009/89, R464/98:41).

Erskine, St. V.W. 1875. Journey of exploration to the mouth of the River Limpopo. *Journal of the Royal Geographical Society* 45:45–128.

Ferreira, O.J.O. 1978. *Stormvoël van die Noorde.* Makroboeke, Pretoria.

Ferreira, O.J.O. 2002. *Montanha in Zoutpansberg. 'n Portugese handelsending van Inhambane se besoek aan Schoemansdal, 1855–1856.* Protea Book House, Pretoria.

FitzPatrick, J.P. 1907. *Jock of the Bushveld.* Longmans Green, London.

Gibbons, A. St. H. 1898. *Exploration and Hunting in Central Africa.* Methuen & Co., London.

Glynn, H.T. 1929. *Game and Gold.* The Dolman Printing Co. Ltd, England.

Godfrey, P. 1963. Epitaph for a hunter. *Wide World* April:265–267.

Grobler, H.J. 1919. Storie van 'n vergane Voortrekkerdorp. *Die Brandwag* 24 December:217–222.

Grobler, H.J. 1938. 'n Voortrekkerdorp wat verrys en verdwyn het. *Die Volkstem* 9 December:14.

Gunning H.L. 1898. Administrative memo (29 March 1898) from the Director of the State Museum, in which he opposed the issuing of a hunting licence to Abel Erasmus and in which he proposed steps to enhance game preservation. Transvaal Archives Acquisition S.S.2081 (R8009/89, R464/98):41.

Hall, H.L. 1937. *I have Reaped my Mealies.* Betteridge &. Donaldson, Johannesburg.

Hall-Martin, A. 1985. Mofhartbees na jare op pad terug na Suid-Afrika. *Custos* 14(3):4–5.

Hofmeyr, S. 1890. *Twintig jaren in Zoutpansberg.* J.H. Rose & Co., Cape Town.

Huet, P. 1869. *Het Lot der Zwarten in Transvaal.* J.H. van Pearson, Utrecht.

Isaacs, N. S.d. *Travels and Adventures in Southern Africa.* Van Riebeeck Society, Cape Town.

Ivy, R.H. 1983. Wildlife observations in the African bush. Unpublished manuscript. Skukuza Archives.

Kuit, A. 1945. *Transvaalse terugblikke.* Van Schaik, Pretoria.

Labuschagne, R.J. 1958. *60 Years Kruger Park.* National Parks Board, Pretoria.

Labuschagne, R.J. 1968. *The Kruger Park and Other National Parks.* Da Gama Publishers, Johannesburg.

Lategan, F.V. 1974. *Die Boer se Roer.* Tafelberg, Cape Town.

Le Roux, S. 1939. *Pioneers and Sportsmen in Southern Africa, 1760–1890.* Published by the author & Art Printing Works, n.p.

Lombard, S.A., Donovan, M.A.C., Beckett, C.F., Kock, P.J. & Stevenson-Hamilton, J. 1945. *Report of Inquiry into Game Preservation 1945.* Government Printer, Pretoria.

Manners, H. 1980. *Kambaku.* Ernest Stanton, Johannesburg.

Milstein, P. le S. 1986. Historical occurrence of Lichtenstein's Hartebeest, *Alcelapus lichtensteini* in the Transvaal and Natal. Unpublished manuscript.

Moerschell, C.J. 1912. *Der Wilde Lotrie.* H. Stürtz, Würzburg.

Möller-Malan, D. 1953. *Dina Fourie.* Voortrekkerpers, Johannesburg.

Murray, A.K. 1964. Beschrijving der Transvaalsche Republiek. *De Zuid-Afrikaan* 18 January.

Nathan, M. 1941. *Paul Kruger – His Life and Times.* Knox, Durban.

Nichols, J.A. & Eglington, W. 1892. *The Sportsman in South Africa.* The British & Colonial Publications Co., London.

Orpen, J.M. 1964. *Reminiscences of Life in South Africa from 1846 to the Present Day.* C. Struik, Cape Town.

Paynter, D. & Nussey, W. 1986. *Kruger – Portrait of a National Park.* Macmillan S.A. Pty. Ltd, Johannesburg.

Penzhorn, B.L. 1985. An old reference to 'hartebeest' in the Transvaal Lowveld. *Koedoe* 28:69–71.

Pienaar, U. de V. 1985. Indicators of progressive desiccation of the Transvaal Lowveld over the past 100 years, and implications for the water stabilisation programme in the Kruger National Park. *Koedoe* 28:93–166.

Pienaar, U. de V. & Mostert, M.C. 1983. Op die spoor van die ou transportryers deur die Krugerwildtuin. *Custos* 12(10):12–19.

Preller, G.S. 1930. Herrinneringe van Karel Trichardt. *Die Huisgenoot* 24 October.

Preller, G.S. 1940. Stigting en verwoesting van Schoemansdal. *Die Vaderland* 13 December.

Pringle, J.A. 1982. *The Conservationists and the Killers.* Books of Africa, Cape Town.

Punt, W.H.J. 1953. *Louis Trichardt se laaste sk*of. Van Schaik, Pretoria.

Punt, W.H.J. 1958. Voortrekkerdogter was wit vrou van Zimbabwe. *Fleur* (Supplement of *Dagbreek*) 24 August.

Rabie, A. 1976. *South African Environmental Legislation.* Institute of Foreign and Comparative Law – University of South Africa, Pretoria.

Rendall, P. 1895. Field-notes on the antelopes of the Transvaal. *Proceedings of the Zoological Society* 1895:385–362.

Rendall, P. 1899. Lichtenstein's hartebeest (*Bubalis lichtensteinii*) in East Africa. In: H.A. Bryden. (ed.). *Great and Small Game of Africa.* Rowland Ward, London.

Reitz, D. 1929. *Commando.* Faber & Faber, London.

Roos, J. de V. 1930. Herinneringe van Carel Trichardt. *Die Huisgenoot* 24 October.

Roos, J. de V. 1938. Die ou Ivoorsentrum van Transvaal. *Die Huisgenoot* 3 June.

Sandeman, E.F. 1880. *Eight Months in an Ox-waggon.* Griffith & Farran, London.

Schoeman, A.E. 1938. *Coenraad de Buys, the First Transvaler.* De Bussy, Pretoria.

Scully, W.C. 1907. *By Veldt and Kopje.* T. Fisher & Unwin, London.

Scully, W.C. 1913. *Further Reminiscences of a South African Pioneer.* T. Fisher & Unwin, London.

Skirving, J.B. 1899. Letter (8 August) from the State Prosecutor, Barberton to the Attorney-General, including a letter from Acting Commandant M.J. Lombard. Transvaal Archives Acquisition P.214 (SPR 7145/99).

Stevenson-Hamilton, J. 1944. *The Lowveld: Its Wildlife and Its People.* Van Schaik, Pretoria.

Stevenson-Hamilton, J. 1974. *South African Eden.* Collins, London.

Struben, H.W. 1920. *Recollections of Adventures, Pioneering and Development in South Africa 1850–1911.* Maskew Miller, Cape Town.

Strydom, C.J.S. 1966. *Wilde roeping: die verhaal van beroemde grootwildjagters in Suider-Afrika.* Tafelberg, Cape Town.

Stutterheim, C.J. &. Brooke, R.K. 1981. Past and present ecological distribution of the yellowbilled oxpecker in South Africa. *South African Journal of Zoology* 16(1):44–49.

Style, G. 1965. Barnard the elephant poacher. *Outpost* May:19–21.

Suid-Afrikaanse Biografiese Woordeboek 2:165–167. 1972. Insert on: De Buys, Coenraad Tafelberg, Cape Town & Johannesburg.

Suid-Afrikaanse Biografiese Woordeboek 2:5–6. 1972. Insert on: João Albasini.

Suid-Afrikaanse Biografiese Woordeboek III. 1977. Entry on: Lotrie, Francois Bernard Rudolph. Tafelberg, Cape Town.

Swart, N.J.R. & McDonald, W. 1874. Contract to transport military equipment and other goods from Delagoa Bay to Lydenburg for the Gold Commissioner, New-Caledonia Goldfields. Transvaal Archives Acquisition S.S. 18:60–67.

Van der Merwe, A.P. 1975. Verkenners en pioniers. In: C. Barnard (ed.). *Die Laeveld – kamee van 'n kontrei.* Tafelberg, Cape Town.

Van Oordt, G.A. 1980. *Striving and Hoping to the Bitter End: The Life of Herman Frederik van Oordt (1862–1907).* G.A. van Oordt, Fishhoek.

Van Zijl, P.H.S. 1947. Kommandant 'Rooi Regering'. *Die Brandwag* 10 January:4.

Van Zijl, P.H.S. 1948. *Waar en trou.* Afrikaanse Pers, Johannesburg.

Vaughan-Kirby, F. 1896. *In Haunts of Wild Game.* William Blackwood and Sons, London.

Viljoen, B.J. 1902. *Mijne herinneringen uit den Angloboerenoorlog.* W. Versluys, Amsterdam.

Von Moltke, J. 1945. *Jagkonings.* Nasionale Pers, Cape Town.

Von Wielligh, G.R. 1890. Letter (27 August) from the Surveyor-General, Pretoria, in which he laments the excessive extirpation of game. Transvaal Archives Acquisition S.S. 1875:17 (R11730/90).

Wheelwright, F.W. 1913. Letter (12 June) from the Magistrate of Pietersburg to Dr. Austin Roberts, Transvaal Museum. Transvaal Museum Archives.

Willis, P.W. 1950. Letter to Col. J.A.B. Sandenbergh. Skukuza Archives.

Wolhuter, H. 1948. *Memories* of *a Game Ranger.* Wild Life Protection Society of South Africa, Johannesburg.

Wongtschowski, B.E.H. 1987. *Between Woodbush and Wolkberg.* Review Printers, Pietersburg.

Chapter 8
On the trail of the explorers
Dr U. de V. Pienaar and others

In 1725, Francois de Cuiper, an officer of the VOC stationed at Delagoa Bay, became the first European to enter the Transvaal Lowveld. A hundred years later the Cape pioneer and renegade Coenraad de Buys managed to do the same. In 1820–21 De Buys followed a route through the interior to Botswana, the Soutpansberg and from there, it is said, to Sofala.

Hendrik Scholtz, a farmer from the eastern frontier of the Cape Colony, scouted out a route through what is now the Free State and Gauteng up to the Soutpansberg in 1834. In the Soutpansberg he met the Buys people, who had settled there.

Shortly afterwards, in 1836, Lang Hans van Rensburg and Louis Trichardt led their Voortrekker parties to the Lowveld. In the same year the Voortrekker leader Andries Potgieter and J.G. Bronkhorst undertook an expedition from the Free State, where they left the rest of their trek, through the Lowveld to Sofala. They also accompanied the Trichardt search party in an attempt to determine the fate of the Van Rensburg group. Between 1837 and 1850 Karel Trichardt undertook several journeys and expeditions through the Lowveld and up the east coast. The Portuguese trader and entrepreneur João Albasini established a trading post at Magashulaskraal next to the Sabie River in 1846. From there he traded with Delagoa Bay and the interior until 1848.

The journeys and experiences of these people have been discussed in previous chapters. After the establishment of the Voortrekker towns Ohrigstad (1845), Krugerspost (1848), Schoemansdal (1848) and Lydenburg (1850) a new era in the exploration of the Lowveld was heralded by hunters, traders, missionaries and explorers, each with their own objectives.

The most important of these journeys through what was to become the Kruger National Park, were: the diplomatic mission of the Portuguese priest Father Joaquim de Santa Rita Montanha in 1855–1856 from the port of Inhambane on the east coast to Schoemansdal and back; the trade expedition of the Portuguese trader Diocleciano Fernandes das Neves in 1860 and 1861 from Delagoa Bay to Schoemansdal and back; and the explorations of the English surveyor St Vincent W. Erskine in 1868 and 1871 through large parts of the southern Lowveld and Mozambique to find the Limpopo River mouth. Then there were the expeditions the famous German geologist and explorer Carl Mauch made through the southern Lowveld to the Limpopo and Mashonaland from 1868 to 1871 and the river voyage and hike of Captain Frederick Elton down the Limpopo, from the Tati settlement to the confluence of the Limpopo and Olifants rivers, in July and August 1870.

The German geologist E. Cohen undertook a scientific expedition from the Lydenburg goldfields through the southern Lowveld to Delagoa Bay in 1873, while Henri Berthoud, a Swiss missionary, undertook several explorations between 1880 and 1904 from the mission at Valdézia in the Soutpansberg, through the northern Lowveld and Mozambique. Another explorer, William Napier, walked from the goldfields at Lydenburg to Delagoa Bay in 1875.

The missionaries Piet Huet[1] and Stefanus Hofmeyr,[2] who worked among the black tribes of the Soutpansberg from 1857 to 1877, did not follow Berthoud's example and undertook extended journeys, but both authored books with valuable information on conditions in the Soutpansberg area during their stay there.

Their books contain a lot that is of interest, including notes on the climate, environmental conditions, diseases and illnesses, botany, zoology, geography, and the customs and traditions of the different tribes of the area. For instance, Huet (1869) mentioned that game was so abundant that a hunter could go out in the morning and return with as many as 11 hartebeest (possibly Lichtenstein's hartebeest).

The experiences of the pioneers mentioned above during their explorations will be described in chronological order.

---
1 Huet, P. 1869. *Het lot der zwarten in Transvaal*. J.H. Pearson, Utrecht.
2 Hofmeyr, S. 1890. *Twintig jaren in Zoutpansberg*. J.H. Rose R. Co., Cape Town.

8.1 Father Joaquim de Santa Rita Montanha

Dr J.B. de Vaal

In 1854, João Albasini helped Commandant General Piet Potgieter draft a letter in Portuguese to the governor of Inhambane. The letter was carried by two black couriers and arrived at the port on 18 May 1855. In it, the Boers asked the governor to send some of his Portuguese subjects back with the couriers to negotiate trade deals with them.

Two days after he had received the letter, Governor Jacinto Henriques de Oliveira called together a few of the most important inhabitants of the town. He showed them the document and asked them to each make a contribution so they could delegate someone to undertake the mission, as the state could not afford such a step. Everybody agreed and signed promissory notes. Captain Francisco Antonio Rangel was nominated as the representative and Father Joaquim de Santa Rita Montanha, a Roman Catholic priest, as his travelling companion. Three days after his nomination, Captain Rangel refused to go, so Second Lieutenant Antonio de Souza Teixeira was appointed in his place. On the same day the priest, who would now lead the delegation, received his instructions from De Oliveira, as well as a letter addressed to the "Governor of the Boers".

His instructions were the following:

The delegation would consist of the honourable vicar Father Joaquim de Santa Rita Montanha and his adjutant and second lieutenant Antonio de Souza Teixeira. In consultation with each other they had to plan their expedition carefully so as not to arouse the suspicion of the tribes through whose territories they would have to travel. Following their departure from Inhambane they had to keep a diary, in which they had to make notes on the lay of the land, the routes, towns, rivers and minerals. This journal would serve as a report on their mission.

After their arrival at the Boer settlement in the Soutpansberg, they had to contact the Portuguese citizen João Albasini and arrange with him to be introduced to "the Right Honourable Governor of the Settlement", so that they could hand him the letter. Thus he would get his wish of discussing trade relations with representatives of the Portuguese governor and they would be able to discuss with him the best way of arranging trade with Inhambane. They had to determine the best and shortest routes that avoided tsetse fly-infested areas and had to point out to the "Boer Governor" that the tribes in Manukosi's[3] sphere of influence had to be instilled with respect for him and his horsemen, as only this would enable them to travel through the area without being robbed or killed. The route through Manukosi's area was the shortest and therefore the best. The delegation also had to convince the head of the Boer state of the benefits of trading with the Inhambane district. Inhambane possessed a harbour that was independent of Mozambique, suitable for both national and international trade. The Boers could also route all their overseas mail through Inhambane. The two communities would in future be on friendly terms with each other, fulfil each other's needs and so stand to gain from a well organised trade agreement.

After the delegation had discussed all the advantages of mutual trade with the Boer leader, he could compile a letter covering the suggested opening of a trade route to Inhambane. The delegation was also

A section of the Voortrekker town Schoemansdal in 1865, according to a sketch by Alexander Struben about 10 years after the visit of De Santa Rita Montanha.
(ACKNOWLEDGEMENT: THE EDITOR, *CONTREE*)

---

3   Another name for the Changana-Tsonga high chief Sochangana.

given permission to immediately conclude a provisional commercial treaty and do everything else they deemed necessary.

Montanha was very diligent regarding the order to maintain a diary. A little more than a month after his return to Inhambane he handed it to the governor. While doing so, he declared: "I only describe the naked truth as I have experienced and heard it." This document is therefore of great cultural and historical value.

The group that left Inhambane on 25 May 1855 by boat (for the first two days) and then travelled by foot to Schoemansdal consisted of 29 people. They were Montanha, Teixeira, a corporal, four foot soldiers of the Inhambane infantry, a black chief who would organise bearers along the way, the Boers' two messengers (Bangalasse and his partner) and 19 bearers (17 slaves owned by Montanha and two owned by Teixeira) of whom ten had to carry the rifles. In order to trade for food on their journey, they took along boxes of glass beads, some lengths of Indian cotton and striped linen and other trade goods. After a few days they were joined by the Arab trader Mamoed Amad Saiboe and some of his bearers, who accompanied them for part of the way.

On their way to Schoemansdal they followed the route from Inhambane to Chicuque, Mongo, Morrumbene, Corré, Fervella, Mocumba, Guione's (Homoine) kraal, Cambi's kraal, Mulamula's kraal, Lake Murraba, Lake Nitemalla, Maconduene, the territories of headmen Ingoana, Macuacua and Maziva, Lake Inhametenga, the Luize or Changane River, Lake Manhambongo, Chiqueta, Madiacune, a drift through the Limpopo, Chivandana, Mudiacumue's kraal and Machambo's kraal, until the group reached the Shingwedzi Poort through the Lebombo Mountains (not mentioned by name) on 22 June 1865. Along the way they saw large herds of game, including elephant, buffalo, giraffe, rhino, blue wildebeest and zebra. It is clear from the journal that Montanha and his group did not shoot at the large numbers of animals. They had just passed through Shingwedzi Poort when they twice encountered black rhinos in what is now the Dzombo area of the Kruger National Park. They shot at these animals on both occasions, without success. But on their return journey to Inhambane in July 1856, when they were accompanied by two Boers from Schoemansdal, G.H. Marnitz and J. Wessels, the group repeatedly shot at game (especially hippo, wildebeest and zebra). Near the kraal of Chief Guloane, a son of Ingoane, they also shot and killed a rhino.

Montanha noted the many waterbirds at the various inland lakes and pans they passed, and they also saw ostriches several times.[4] All along the route they encountered black people who kept livestock and cultivated the land. He described in detail the topography, plant life, climate and population of the areas through which they travelled, along with the names of the most important chiefs and headmen.

Early on the morning of 3 July 1855, after they had passed through Shingwedzi Poort, they met an Indian, Cassim Camal, who had come from the Boers to guide them the rest of the way. He greeted them, inquired after their health and had an ox slaughtered. Then a messenger was sent to the Boers to inform them of the delegation's imminent arrival. Two days later they met two well-dressed men on horseback – João Albasini and his brother-in-law Hendrik van Rensburg. The group proceeded to Albasini's farm near Piesang Kop. About this part of the journey the priest wrote:

> He (Albasini) offered us the horses to ride on and the lieutenant and I took turns. We eventually arrived at a place that they called the farm of Livufo, which belonged to another white man by the name of August, who had died.
>
> There was a vehicle with four wheels, which was drawn by ten or 12 oxen, that they called a wagon. As it was already dark, we pitched camp and stayed the night. We slaughtered three sheep. Albazine (sic), whose wagon it was, had flour, butter, coffee, sugar, tobacco and snuff in it. We sat talking around the fire the whole evening, had a meal of barbecued meat with bread and butter, and drank coffee. He offered us some snuff and filled my box, gave tobacco to the soldiers and the bearers, and because it

---

4   Montanha took these birds to be emus (a large, flightless Australian bird species). But Montanha's description of 16 July 1856 of how Wessels and Teixeira, on horseback, had followed two "emus", but that they had *run* faster than the horses and escaped, confirms that they had seen ostriches. The area through which they travelled was then and still is the home of ostriches, and one cannot accept the suggestion by Ferreira (2002) that they had seen marabou storks.

was very late, near midnight, we thought of going to bed. My partner, the lieutenant, and I slept in the wagon and the rest around the fire.

Early the next morning they drank coffee, as was customary among the Boers. The blacks loaded their luggage onto the wagon and made the bed in it, in case one of them should want to sleep during the journey. The two passengers climbed into the wagon, while the driver, an English boy of 15 (probably Tom Kelly, the Albasinis' foster child), sat in front to drive the oxen. Albasini, the Boer, the Indian and a black man mounted the horses and rode ahead of the wagon the whole way, while the rest followed behind the wagon. When they were halfway, at about one in the afternoon, they stopped next to the water and outspanned. There they met a man, a woman and their children at another outspanned wagon. These people were on their way to town to seek safety due to the war with rebellious Venda chiefs. After a meal of coffee and bread with butter they continued their journey, accompanied by the other wagon.

A bit further on, a group of horsemen came to meet them and they fired welcoming shots. The soldiers responded with 20 shots. The Portuguese climbed from the wagon and they all greeted each other, while the Boers expressed their satisfaction and delight with the visit. Everyone was Afrikaans speaking and the priest was struck with their open, honest faces. After a while they climbed back onto the wagon while the Boers mounted their horses and slowly rode ahead.

The group arrived in the Boer town by sunset, after a journey of 42 days. They were welcomed by the inhabitants with another salvo of gunfire and cannon shots from the "fort". There were many people in the streets along which they rode, especially in front of Albasini's house, where they outspanned. After they had got down from the wagon, they were received by Mrs Albasini and other Boer women in the lounge. Shortly afterwards one of the servants, dressed in a long shirt, came in and brought them water so they could wash. Every time he entered or left the room he bowed. The men also came in and shook everyone by the hand. It looked as if it would never end. Afterwards they sat down at a laid table. The main dish was beef, and was followed by coffee.

Albasini ordered that the soldiers and blacks be given a meal as well and after that he showed the two leaders and their guards the two small houses where they could stay during their time in the Soutpansberg.

That night they sat talking in Albasini's house until ten o'clock, mainly answering the Boers' questions. The next morning they visited the most important people in town.

At four o'clock in the afternoon there was a meeting at Albasini's house under the chairmanship of Magistrate Duvenage to officially welcome the two delegates. For this occasion Montanha was dressed in his habit and Teixeira in his uniform. After the welcome they presented their letters of appointment, which Albasini read aloud and translated into Dutch. In the absence of Stephanus Schoeman, who had succeeded the late Commandant General Piet Potgieter, the priest handed a letter that had been addressed to Potgieter, to Duvenage. Those present discussed the contents of the letter and questioned the delegation about the route, the population of Inhambane, if there was game and if one could trade there. After that the meeting was adjourned.

That night, Montanha suffered from tremors (malaria), for which he took a healthy dose of gin. Fortunately he was able to shake off the symptoms after a few days in bed.

Albasini immediately translated the letter the late Commandant General Piet Potgieter had written to the governor in Inhambane, as well as the letter from Inhambane to Potgieter, and sent them to Commandant General Schoeman. He was particularly enthusiastic about the trading possibilities and added the following:

> We have been given the best opportunity in the world to fulfil the much wished-for desire to establish and put into work a route to Inhambane.
>
> I have received several other letters from private persons in Inhambane, all with more or less the same contents. I will submit them to you on your arrival here. The Reverend de Santa Rita Montanha told me that many of the most important moneyed people in Portugal are busy founding a trading company there and aim to improve the different Portuguese colonies along the coast, raising them to an equal footing. This would be very advanta-

geous to us, but our speedy and immediate cooperation is of course therefore essential. If we could reach a proper understanding with that nuisance Manakosa, or get rid of him, we should soon have an extended, profitable and opportune trade with not only Inhambane, but also with Sofala, which is no further from us than Inhambane. Then there is Manika, Senna and Tete, which are further, but which each produce large amounts of gold of which the blacks have up to now reaped the most benefit.

After an absence of two weeks, the commando, under the leadership of Commandant General Schoeman, arrived back in Schoemansdal on 22 July, a Sunday afternoon. On arrival they fired a few shots, which the townsfolk answered with rifle and cannon shots. This continued until the Commandant General dismounted and entered his house. The Portuguese soldiers participated in this welcoming. They had been instructed to wear their uniforms with white trousers and were to fire three salvos using 15 loads of gunpowder that Albasini had given them.

It seems Albasini did his best to keep the delegation and the rest of the party happy. Apart from the hospitality that the leaders enjoyed at his home, he also gave each of the soldiers a pair of shoes, a pair of blue trousers and a shirt from his shop.

Schoeman convened a public meeting on 30 July to which the two delegates were invited. After reading certain legal articles, they discussed the correspondence between Soutpansberg and the Portuguese at Inhambane. Subsequently they questioned the two representatives, whose answers can be summarised as follows:

Since Manukosi was a danger, it would be wise to travel with 50 horsemen to Inhambane to open the route. The most suitable route was via Chicualacuala, through the territories of Maziva and Macuacua. If Manukosi, or any of the other high chiefs, caused trouble, the Portuguese would come to their aid. The lieutenant, who had been to Manukosi's kraal, provided the Boers with useful information about him. According to him, the high chief had many subjects and their kraals were situated close together. There were only about 200 "Vatuas" (Zulus), while the rest were conquered subjects. His main kraal was close to the Limpopo River, on the northern side. During their journey they saw that some areas were thinly populated, and sometimes they travelled for as long as four days without encountering any black people.

Montanha did not have an answer to the question of whether the Boers would be allowed to trade from the Soutpansberg with Inhambane and directly with foreign ships. This was an issue about which his government had to decide, and for this reason he could not conclude a treaty. He was, however, convinced that if a treaty was signed, they would be able to trade with both the Portuguese and foreign ships. If they settled in Inhambane and would be prepared to pay customs duties, he had no doubt that they would be free to trade with incoming ships.

The Commandant General for his part informed them that winter was almost over and that it was too late for the Boers to accompany the Portuguese mission on its return journey to Inhambane. Montanha and his party were greatly distressed at this development and he expressed his dissatisfaction to the Commandant General in a letter. But Schoeman stuck to his guns and the result was that they had to wait until the next winter before they could leave. This gave Montanha the opportunity to make a great many observations, which he recorded in his diary without fail. He also attended the wedding of Commandant General Schoeman to General Piet Potgieter's widow in January 1856 in the Waterberg district, recording that the wedding feast lasted for three days.

The original plan was to have a strong patrol accompany the delegation on their way back. When Manukosi sent an envoy to the Boers in the Soutpansberg with the message that no more than five extra whites would be allowed to accompany the delegation, the escort members turned back one by one. Only two Boers from the Soutpansberg, George Hendrik Marnitz and Johannes Wessels,[5] travelled all the way to Inhambane.

Montanha considered the reluctance of the Boer delegates and especially that of the leaders of the pa-

---

5   Pater Montanha had problems with the spelling of Afrikaans proper names and spelled them mostly phonetically. For example, he wrote De Vinary for Duvenage, G. Merenits for G. Marnitz and Jannes Werbes for Johannes Wessels. The same goes for place names: he wrote Moirefier for Mooi River, Pissam Kop for Piesangkop, Ouris Cidade for Ohrigstad, etc. For some reason Montanha was under the impression that Johannes Wessels was Russian, and therefore repeatedly referred to him as the Russian Jannes Werbes.

trol, Abraham Duvenage and C.J. Rabé, to accompany him to Inhambane, as a motion of no-confidence in the whole undertaking. He expressed his disappointment in no uncertain terms in a letter to Commandant General Schoeman. It was clear that his biggest regret was the fact that his compatriot Albasini had decided go elephant hunting rather than accompany him, and he maintained that this had been why the rest of the patrol were so unwilling to accompany him all the way back to Inhambane.

The small group left Albasini's farm on 25 June. Because of Manukosi's hostile attitude they decided to follow a route further north than during their forward journey. Initially they followed the first section of the ancient trade route to Sofala and travelled via Mashao, Shikundu, Ka-Mhinga, Punda Maria and Klopperfontein to Pafuri and Chicuala-cuala next to the Limpopo.

In the vicinity of Ka-Mhinga (just outside the border of the current Kruger National Park), Johannes Wessels shot an animal to which Montanha referred as a "colhú" in his journal. Judging by the Shona name "Inkulonondo", it is quite possible that he was referring to a Lichtenstein's hartebeest – a further indication that this species had once occurred in the northern parts of the Kruger National Park.

Near Punda Maria the party threw blankets over their horses as they had been warned by the guides that they could expect to encounter tsetse flies from there on eastwards. Their route apparently passed the Shantangalani Hills near Klopperfontein, as Montanha wrote: "At sunset we found water among rocks in round holes and pitched camp to overnight." (These holes are a feature of this particular sandstone outcrop.)

One cannot help but notice that Montanha often mentioned the vast amounts of water they encountered, in lakes, pans and rivers, and this during the dry season. He described well-known rivers such as the Shingwedzi (which he did not mention by name), the Klein Letaba, Luvuvhu, Limpopo (Bembe), Nuanedzi, Luize (Changane) rivers and others as flowing so strongly that they were forded with difficulty. This corresponds with the remarks of other explorers such as J.G.S. Bronkhorst, St Vincent Erskine, Carl Mauch, Dr Cohen, Captain Elton and Reverend Berthoud, who travelled through the Lowveld in the nineteenth century. It confirms the theory that the area experienced a long-term climatic cycle (80–100 years) of above average rainfall during the nineteenth century. At the same time it also explains the conviction of J. Stevenson-Hamilton and other Lowvelders that this area had been experiencing a process of progressive desiccation since the beginning of the twentieth century. Even the Dorps River, which provided the people of Schoemansdal with water, is nothing more than a dry donga today. During the 1850s and 1860s it was such a strong perennial stream that a water mill could be erected in town (J.B. de Vaal, 1948; J.A. Bruyns, 1932).

Montanha's repeated references to the large herds of elephants they encountered all along the route in both Mozambique and the Soutpansberg are enlightening. It is not surprising that the ivory trade flourished in the days of Schoemansdal's existence and that the Boer hunters went deep into Mozambique in their search for big ivory carriers. He also mentioned a certain Piet du Preez as having caused trouble in the territory of a certain chief that was six days' journey east of Chicuala-cuala's kraal along the Nuanedzi.

From Chief Chicuala-cuala's kraal Montanha's party followed a route in a southeasterly direction across the Luize River and the swamps and lakes of Ingoana to Mocoduene and Morrumbene at the bay of Inhambane. With the aid of maps of Mozambique of that time, such as the one by the Portuguese explorer Major Serpa Pinto (1881), Montanha's routes can be traced with reasonable accuracy. During their return journey they encountered Manukosi's warriors on several occasions, but fortunately managed to avoid a confrontation.

They finally arrived back in Inhambane on 1 August 1856, after a journey of a month and nine days. Wessels stayed with Teixeira and Marnitz with Montanha.

One of the priest's orders had been to give advice on trade possibilities. In this regard he reported that the Boers did not conduct trade in the same manner as the Portuguese, but more according to the English manner, since they had been dealing with the English for many years.

To avoid misunderstandings they would have to standardise their measurements and weights in consultation with the Boers. Frequent trading depots would have to be built between Inhambane and the Soutpansberg to avoid the accumulation of trade goods. This suggestion was in accordance with the practice of black and other traders, who through the ages always had hidden depots as resting places along the old trade routes between the coastal harbours and the interior. He also told them that not all the Boers were elephant hunters and that many would not be able to bring their products to the coast. It was therefore necessary that the two governments determine the locations of the depots and that these be protected by the governments. This would guarantee the safety of the money traders invested in the industry. The depots would make the distance between the Soutpansberg and Inhambane significantly easier to cover.

After the two Boers had spent a month in Inhambane, they were sent back to the Soutpansberg with a military guard. In light of Montanha's haughty treatment at the hands of the Schoemansdallers, this was a praiseworthy gesture.

A prosperous trade between the Boers and Inhambane was never realised, however, as the Soutpansberg had too many internal problems that took precedence over international affairs. Apart from that, Manukosi was too hostile and the tsetse belt was impenetrable to ox-wagons.

But Montanha's journey was not in vain, as valuable information on geography, ethnology and nature, as well as cultural-historical data regarding the Voortrekkers, would not have been preserved if he had not made note of it in his diary. A few years after he had handed in his report, it was consulted by geographers who knew very little about the hinterland of Inhambane, and in 1862 a shortened version was published in the *Journal of the Royal Geographical Society*. In 2002 the historian O.J.O. Ferreira published Montanha's full description of his journey, in Afrikaans, with extensive references and explanatory notes (see Ferreira, 2002 in References). Together with Louis Trichardt's diary, this is one of the most important sources of information about two different, yet successive eras in the history of the Voortrekkers of the northern Transvaal and eastern Lowveld.

Professor L. Changuion, a historian at the University of the North in Polokwane (Pietersburg), decided to follow in Montanha's footsteps 147 years later and take more or less the same route to Schoemansdal. He started this journey on 1 August 2002 in Inhambane and finished on 31 August in Schoemansdal to great acclamation.

References

De Vaal, J.B. 1953. Die rol van João Albasini in die geskiedenis van die Transvaal. *Archives Yearbook for South African History* 16(1):1–155.

De Waal, D. 2002. Sendeling besoek Schoemansdal ná 147 jaar. *Die Burger* Supplement 12 October.

Ferreira, O.J.O. 2002. *Montanha in Zoutpansberg. 'n Portugese handelsending van Inhambane se besoek aan Schoemansdal, 1855–1856*. Protea Book House, Pretoria.

Liesegang, G. 1990. Três autores sobre Inhambane: vida e obra Joaquim de Santa Rita Montanha (1806–1870), Aron S. Mukhombo (ca. 1855–1940) e Elias S. Mucambe (1906–1969). *Arquivo: Boletim do Arquivo Histórica de Mocambique*, 8 October.

Macqueen, J. 1862. Journey from Inhambane to Zoutpansberg by Joaquim de Santa Rita Montanha. *Journal of the Royal Geographical Society* 32:63–68.

Pinto, S. 1881. *How I crossed Africa*. 2 vols. Sampson, Low, Merston, Searle & Rivington, London.

Transvaal Archives (TA) Acquisition A 81. *De Santa Rita Montanha se reisverslag*.

Van Bart, M. 2002. Boekstawing van Montanha se epiese staptog aanwins. *Die Burger* Supplement 11 October.

8.2 Diocleciano Fernandes das Neves

Dr J.B. de Vaal

D.F. das Neves has already been mentioned in chapter 2.5 (on the subject of the earliest trade routes from the coast to the interior) as he was one of the traders and hunters who travelled along these routes to the Soutpansberg. Das Neves was in fact much more than just a hunter and ivory trader, and proof of this is his engrossing travelogue *A Hunting Expedition to the Transvaal* and a paper written in 1861 titled "Report on the State of the African Boers, its commerce, agriculture and industry". Since parts of his account of his journey are given in chapter 6.1 (about the history of Schoemansdal), some additional information about him is provided to fill the gaps.

Das Neves arrived in Lourenço Marques on 5 October 1855 at the age of 25 and returned to Portugal 13 years later. He was thus only 30 when he undertook his journey to the Soutpansberg, accompanied by 253 Africans. The fact that he could handle such a large number of hunters and bearers by himself proves that he was an efficient leader and organiser. He was also an extremely attentive and intelligent observer, and nothing worth mentioning escaped his watchful eyes. As a result anthropologists, historians, geologists, ornithologists, medical practitioners, zoologists, hunters, agriculturalists and theologians will all find something of relevance to their disciplines in this wealth of information about the Transvaal Lowveld and Soutpansberg of a century and a half ago. He described in detail how he had recruited his bearers and hunters, what he paid them, how he cast bullets and how the hunters had to be protected against wild animals by bone-throwing sangomas. The large party set off from Lourenço Marques (Delagoa Bay) on 3 September 1860 on their expedition to the then northern Transvaal (now Mpumalanga) through the area that is now the Kruger National Park.

In addition to the many bundles of trade goods, rifles and ammunition, his own luggage consisted of a mattress, pillow, blanket, a large tin of sugar and another of American biscuits, a trunk with his clothes, a box of tea, sugar, two cups and saucers, a teapot, two bowls, eight packets of candles and a tin containing about three kilograms of salt. An indispensable part of his kit was a medicine chest in which he kept, among others, a supply of quinine-sulphate pills – and this in as early as 1860. With these he managed to save the life of one of his bearers who contracted malaria along the way. He also mentioned that the traditional healers of the time could cure the "fever" with an emetic that was even bitterer than quinine and made from a mixture of different roots.

Das Neves even described his attire. On hunting trips he usually wore a bowler hat with large black and white ostrich feathers, a shirt and waistcoat, trousers and shoes, but no jacket. He also wore a leather belt to which a bullet pouch and powder horn were attached. He carried a rifle over his shoulder and a walking stick in his hand.

When Das Neves's journey is carefully mapped according to distances they covered in a day from recognisable places along his route, other landmarks in the current Kruger National Park can be identified as well. After he had left the Sabie River near Sabiepoort, he noted the following:

When we proceeded for about four hours on our
journey, we came to a valley through which coursed
a stream of excellent water [the Mlondozi Spruit].

A Portuguese cross which had probably been carved by a Portuguese explorer into the trunk of a large leadwood tree during the nineteenth century. This tree grows on the southern bank of the Letaba River along the riverbank between the current rest camp and the high-water bridge.
(01.08.1985)

[Compare this reference to that of R. Kelsey Loveday, who referred to this spruit on his map of 1883 as "Sterk Spruit". The original Swazi meaning of "Mlondozi" is "a perennial river that provides water to people, animals and for irrigation".] This valley was very picturesque, a long dense forest of large trees, whose branches were knit together, extending along the south of the valley [a place fitting this description can still be found along the Mlondozi today, on the northern flank of Muntshe Hill].

Here one of Das Neves's hunters shot either a kudu or a waterbuck ("a large stag"). They continued their journey and at about seven o'clock that evening they arrived at the kraal of a local Valôi chief, "Maximbajándhlofu" (Elephant Dung). This could have been in the vicinity of the current Kumana or even a little further north. The next morning Das Neves sent his hunters out to collect meat for the bearers and the local chief. His hunters, Chanâna, Maxotil and Manova each shot a buffalo, while Mabana shot a rhino and Mandissa a giraffe. That night there were several hyenas around the camp.

The next day they set off and camped next to a spruit with many wild palm trees (this could have been the Sweni or the upper reaches of the Nwanedzi River). This was their last camp before they reached the large, relatively dry plain north of the current Satara.

He wrote the following about the Ngotso Spruit and the Timbavati River:

> In the early morning we continued our journey for a very long distance, walking without halting for several hours, reaching a thickly wooded part through which flowed a stream of clear water [the Ngotso Spruit]. Here we all bathed, and after breakfast continued our journey until we came to a large tract of level ground, through which flowed an affluent [the Timbavati] of the great river Imbélúle [Olifants River], where we camped under majestic trees. The following morning we reached the great river Imbélúle, called by the Dutch 'Elephant river'. I crossed this river on the shoulders of two men, the water reaching up to their chests.

After they had forded the Olifants River, he shot a klipspringer in the mountainous area west of the Timbavati mouth. He managed this from a great distance with a fluke shot, which caused a sensation among his hunters and their followers.

In the area of Beja and Palauri (Phalaborwa), which now falls within the Kruger National Park, one of the locals brought him a white object that resembled a glove. According to a member of his party it was the nest of an African penduline tit (*Anthoscopus caroli*). Their Afrikaans name (kapokvoël, or cotton bird in English) refers to the nests they weave from the snow-white strands of the wild cotton plant (*Gossypium herbaceum*). He traded the nest for some missanga beads and another bunch of beads and val-

Gumbandebvu, east of Punda Maria, where black medicine men used to execute their rain-making rituals during ancient times. (26.04.1984)

The site near the northwestern flank of Muntshe Hill where the Portuguese explorer, D. Fernandes das Neves (1860), Surveyor G.R. von Wielligh (1890), Steinaecker's Horse (1901–1902) and Maj. J. Stevenson-Hamilton (1913) all camped. (04.02.1987)

ued it so much that he would gladly have paid double the price. He wrote:

> The bird is very similar to the wagtail [which is an exaggeration of course], and the nest resembles a stuffed glove, such as is used in boxing matches. The nest is closed all around, with the exception of two narrow apertures or tube-like openings similar to fingers of gloves, one on either side, these openings meeting in the centre of the nest [also a slightly inaccurate description]. One of these tube-like openings, which is smaller than the other, is for the male bird to retire at night, and the opening on the opposite side, which is larger, for the hen bird and her eggs. The nest is suspended from the branches of the cottontree by a few bits of cotton fibre, the apertures for the birds inclining downwards, hence no rain can enter the nest, and the two birds remain perfectly sheltered from the greatest storm. It was a perfect marvel of delicate workmanship, rivalling the finest cotton weaving.

According to Dr U. de V. Pienaar, this is the first description of a bird species and its nest in the area that is now the Kruger National Park.

Das Neves did not personally visit the Lobedus' rain queen Modjadji, but he travelled for several days up the Great Letaba River to the Tsonga headman Shiluvane, from where they travelled through Modjadji's territory back to the main route that led to Albasini. Thanks to the first-hand information he gathered about her, he can be regarded as the first white person to give an extremely interesting description of her and her priests' rain-making ceremonies. Her forecasts were frequently correct – an ability he explained as follows:

> The ministers of Queen Mojaju, who besides being learned priests and venerable in their profession, were also clever astrologers, who knew by this sign, that a great storm was brooding, which undoubtedly would be accompanied by great rains.

After several weeks the group reached a hill near Headman Macia's kraal. A short distance from the kraal stood a small thatched-roofed house whose walls had been constructed of reeds, poles and clay. This was the home of a native of Goa, a clerk working for "Senhor Albazini" who traded ivory. The clerk invited the traveller to join him for lunch and Das Neves also met a Boer sent by Albasini to escort him to Goedewensch. Because Das Neves did not understand Dutch, his host had to act as interpreter. After the meal they saddled their horses and rode to Albasini's house, while the rest of his party remained at Macia for the night.

> I was kindly received by Senhor Albazini, who introduced me to his lovely young wife and four children, and begged me to sup with him. I much enjoyed this meal, because, during the twenty-four days which had elapsed since I left Lourenço Marques, I had fared the same as the negroes, with the sole exception of a few biscuits and tea.

That night, a number of Boers had supper with them. They were very surprised to see the stranger and wanted to know whether he was Asian or European, because the only Portuguese people they knew were all from Goa. Albasini had already informed them that he was not Portuguese, but Italian. Only after Das Neves's arrival in the Soutpansberg and the launch of a regular postal service between Goedewensch and Delagoa Bay, manned by Portuguese soldiers, did they realise that the Portuguese were in fact Europeans.

> My people arrived about noon the next day, and entered the place singing. Here they met relations and friends, who, many years previously, had left Lourenço Marques with Senhor Albazini, and had remained in his service ever since. After their arrival, some of the Dutch came to see what goods they brought.

The first to attract the Boers' attention were the two six-gallon demijohns of Brazilian brandy. Das Neves was prepared to sell a few bottles at six shillings each. After that the powder, lead and tin that he had brought for trading purposes, aroused their interest. The Boers could not do without these items as they needed them to hunt elephants. Das Neves noted that he had sold his wares for a very good price.

Albasini translated the local hunting laws for him

since they applied to Portuguese hunters as well. He then sent his black hunters with three of Albasini's people as guides to the territories of chiefs Chiquaraquara (Chicuala-cuala) and Chinguene to hunt elephants and trade ivory. Das Neves only followed them 25 days later. In the meantime he made an inventory of the supplies the Boers needed the most, keeping future trade relations in mind.

On 2 November he left Goedewensch with some trade goods to join his hunters in the Limpopo Valley. The next day he first visited his friend Muzila, a fugitive from his brother Mawewe who lived with his followers under Albasini's protection on the current farm Madrid.

Their relationship was an interesting one. In 1857 Das Neves was shipwrecked after the ship on which he had been sailing foundered off the coast of Inhambane. He proceeded on his journey to Lourenço Marques through the interior. While travelling through the territory of Manukosi, who was still alive at the time, he met Muzila. Das Neves was the first European that Muzila had encountered, and he treated him in a hospitable manner.

When Mawewe succeeded Manukosi, he started killing Boer and Portuguese hunters in his area, taking their ivory, rifles and ammunition, and effectively bringing all traffic to Lourenço Marques to a halt. Now the two friends had long and confidential talks during which Muzila revealed his plans for taking action against his brother to Das Neves, who in turn passed them on to the Portuguese government.

> Mosila was tall, his form was exceedingly well-developed, and his features regular. He was listening to me with an expression of calm sorrow on his countenance, his chin resting on his left hand, and his elbow on his knee, while his right hand hung down listlessly. Then in very sweet, low accents, and that softness of speech peculiar to Vatuas, or Zulus, when speaking confidently to a friend, he replied 'Yes good white friend, thank you very much'.

Das Neves spent 11 days with Chief Chicuala-cuala, of which eight were occupied with trading a tusk of about 50 kg. Because his ammunition supplies were low, he sent a message to his hunters to rejoin him. By that time they had shot 55 elephants, and he sent this ivory and the ivory they had traded ahead to Albasini.

At 11 o'clock on the morning of 6 December 1860 a heavy thunderstorm was brewing. At the same time, Das Neves received a message from one of Albasini's three guides that a band of Mawewe's warriors was on its way to murder him and his party. Within minutes the whole group decamped from Chiquaraquara's (Chicuala-cuala) kraal amid the crash of thunder and torrential rain. They reached the Limpopo only a quarter of an hour later and forded it with all their equipment. The rising water was already chest deep. Half an hour later it came down in such a flood that the river was impassable.

At three o'clock that afternoon the sun broke through the clouds and shone brightly. From their hiding place behind the reeds they could clearly see about 800 warriors sitting on the far bank of the river with their faces turned towards the direction of Chicuala-cuala. Obviously under the impression that the hunters were still on that side of the river, they were patiently waiting for them to arrive so that they could murder them. On the other side, the fugitives were ready to fire at anybody trying to swim across. Fifteen minutes later, four of Mawewe's messengers came running up to the band and shouted: "Mafambacheka has escaped us, go to the other side!", to which the warriors answered: "The river is full."

A little while later the whole band returned to

Baobab trees in the sandveld near the Mozambican border with Malonga Fountain in the background. The Portuguese explorer, Diocleciano Fernandes das Neves, had travelled past here in December 1860 during his return journey from Chicuala-cuala in Mozambique.
(MAY 1984)

Chicuala-cuala, leaving Das Neves's party free to leave their hiding place and light a fire to dry their clothes. The next day the group of 19 trekked as fast as they could along the ancient trade route through the Nyandu Bush to Malonga. He described the exhausting journey as follows:

> Our next day's march was the longest we had performed during our expedition, for we walked nearly eleven leagues [53 kilometres], arriving about six in the evening, at the base of a granite mountain. From the centre sprang a torrent of water that fell in a crystal cataract into a deep pool of excellent water, and then coursed into a running stream. We encamped under the trees growing in this luxuriant spot.

Today all that is left of this fountain is a spring that has dwindled to a trickle, which is getting weaker and weaker during dry seasons and can only provide a limited number of animals with water.

At the fountain Das Neves hit two klipspringers with one shot and the fresh meat managed to considerably boost the morale of his hungry, footsore party.

After resting for a day at Malonga Fountain, they resumed their journey the next morning.

> Soon after seven we started with renewed vigour, reaching, about six in the evening, a huge mountain, at the foot of which there was much foliage. Here we encamped; the men erecting shelters against the expected rain, because the wind had changed to the south. Their anticipations proved true, for, towards one in the morning, we had a good downpour of rain. The water we found here was most excellent.

This second campsite with the fountain at the foot of the mountain was Dzundwini, which is a notable landmark in the Kruger National Park and a well-known watering hole of the old Soutpansberg hunters who travelled along the old trade route via Malonga Fountain to Chicuala-cuala.

After their return to Goedewensch, Das Neves sent some of his bearers with 318 kg of ivory to Lourenço Marques. They had to hand this to an Arab trader from India, from whom he obtained goods on credit. The rest he wanted to sell in Schoemansdal to visiting English traders from Natal and Grahamstown, so he left for Schoemansdal with Antonio de Paiva Rapoza, a rich merchant from Delagoa Bay. He wrote the following about his visit:

> I was very kindly received by Senhor Casimiro Simões, a native of Goa, who lodged me in his own house. This gentleman, who was one of the wealthiest merchants in the Republic, had three establishments, well supplied with piece-goods and other merchandise; one in Soutpansberg, another in Rhenoster-Port [Rhenosterspoort], and a third in Lydenburg. He was a very estimable person, and an intelligent merchant.

Das Neves made particular mention of the fact that the whole Soutpansberg area was teeming with game. Besides elephants there were large herds of buffalo, blue wildebeest, zebra and numerous antelope species he could not identify by name.

Das Neves must have had a very good knowledge of agriculture, because he discussed it extensively.

> These Boers … hardly grow enough to feed their families and there are many who are obliged by the middle of the year to buy wheat and mealies having finished their own. The Boers of Reynosterpoort grow wheat on a large scale, in order to sell it to many parts of the republic.

He observed the agricultural potential of the ZAR with a prophetic eye and remarked the following:

> And were the Dutch to approach the mountains of Lobombo, the boundary which separates the Transvaal Republic from Lourenço Marques, they would find, on the river sides of the Incómate and the Sáve, lands of marvellous fertility, superior to the Brazils for the cultivation of coffee, sugar-cane, and many of the Indian products.

He was not under the impression that the Boers were lazy, because if they were to plant sugar, coffee and cotton they would not be able to sell their crops due to the long distances to Port Natal (Durban) and Port Elizabeth and the poor conditions of the roads. This is why he considered the construction of a good

The Malonga Fountain near the ancient trade route to the east coast as it appears today. Fernandes das Neves described them in 1860 as "a gushing stream of crystal clear water". (21.06.1984)

road between the Republic and Lourenço Marques of the utmost importance. He also predicted that the road would be followed by a railway within a few years. Das Neves further reported to the governor of Lourenço Marques that the Boers were all Protestants who lived religious lives and kept the Sabbath.

He added that the elephant hunters lived in Lydenburg, Makapane Poort, Rhenosterspoort and the Soutpansberg. Their livelihoods depended on elephant hunts and they left their homes in June, returning in September and October. The more affluent Boers wore ready-made clothes bought in Pietermaritzburg, an English town in the interior, four leagues [about 20 kilometres] from Port Natal and three times further [from the Soutpansberg?] than Mozambique. The Boers usually wore moleskin clothing in different colours made by the women. Shoes were manufactured locally.

Das Neves also visited Pretoria and tried to sell some of his ivory there. Because the English traders were not prepared to pay his prices he took the ivory back to the Soutpansberg and sold it for a reasonable price along the way. As a result he could return to Lourenço Marques sooner than planned.

During the return journey he avoided the route through Sabiepoort as he feared Mawewe's warriors, who were still on the warpath. He therefore crossed the Sabie much further west, probably in the vicinity of the current farm Belfast, and then crossed the Crocodile River. (They made for the mountains of "Massuate", or Swaziland.) From there they probably followed the centuries-old trade route through Mathala Poort to Matola and Lourenço Marques (Delagoa Bay). He noted that he had shot a sable antelope between the Crocodile and Komati rivers. This confirms that the species had indeed occurred south of the Crocodile River (most experts have always regarded the Crocodile River as the southernmost boundary of the distribution range of sable antelope in the eastern Transvaal).

On the return journey he shot a hippo in the Great Letaba while they were looking for a suitable drift at which to cross the river. They also saw the carcasses of a blue wildebeest and two lions at the same place. The one lion had apparently died in a fight, while the other had died of suffocation caused by a large bone lodged in his throat. Das Neves noted that they made porridge from maize they had bought from the blacks and flavoured the porridge with the tasty oil made from pressed marula nuts.

By the Sabie one of his hunters shot a bushbuck ewe (he called it a "fallow deer"). While looking for a place to cross the Crocodile River, they saw a female leopard with a cub. The leopard had caught an impala, which his hunters immediately confiscated to eat themselves. That night three lions kept them awake with their deafening roars and 20 buffalo came to drink near their camp. The next day they shot a few impala ("gazelles") and two buffalo along the Crocodile River before resuming their journey towards the Lebombos. He arrived back in Delagoa Bay on 9 August 1861, eleven months and six days after they had left the harbour. It had taken them a full 30 days to cover the distance between Schoemansdal and Lourenço Marques.

This visit once again emphasised the possibility of trade relations between the Republic and the Portuguese at the east coast. Close relationships had been established and Das Neves also recorded valuable information about the interior and its inhabitants, which otherwise would have been lost to future generations.

References

Das Neves, D.F. 1879. *A Hunting Expedition to the Transvaal.* George Bell and Sons, London.

Pienaar, U. de V. 1985. Indicators of progressive desiccation of the Transvaal Lowveld over the past 100 years, and implications for the water stabilisation programme in the Kruger National Park. *Koedoe* 28:93–167.

Transvaal Archives (TA) Acquisition A.81. *Report on the State of the African Boers, its commerce, agriculture and industry* by D.F. das Neves, Lourenço Marques, 1 March 1861.

8.3 St Vincent Whitshed Erskine's travels through the Lowveld and to the mouth of the Limpopo (1868 and 1871)

Dr U. de V. Pienaar

St Vincent Whitshed Erskine was the son of Major David Erskine, a former colonial secretary of Natal. This surveyor and explorer left Pietermaritzburg on 6 May 1868 to accompany the German geologist Carl Mauch on a leg of his expedition to Matabeleland. They travelled by ox-wagon to Potchefstroom and from there to Pretoria and Lydenburg.

At Lydenburg the two friends went their separate ways and Erskine and his party continued by themselves. He first visited Ohrigstad, the abandoned Voortrekker town, and then went to the farm of a certain Mr Schoeman and to Karel Trichardt's farm (probably Branddraai along the Ohrigstad River). After that he crossed the Drakensberg and travelled down the Blyde River with the so-called "Giants' Staircase". In his expedition journal he called the Blyde River "Umchlasi". A few hours' walk from this river brought him to the kraal of a black chief next to a waterhole, which he called "Imperani". There is reason to believe that it was the kraal of Chief Moletele of the Lepulane tribe, who lived near the foot of Mariepskop in those days. On 9 July he came across his first tsetse flies. The blacks in the area kept a number of scruffy mongrels that apparently were unaffected by the flies. They looked like jackals – mostly light brown with long, unkempt fur. He thought it would be impossible to hunt there on horseback on account of the tsetse flies.

From Imperani Erskine struck off in an easterly direction and crossed the Klaserie, which he called "Umtasiti" or "Umtaseera". He also mentioned that this was the first running water he had encountered since leaving the Blyde.

> Passed the 'Umtaseera' river [Klaserie] ... It was a fine stream, as clear as crystal, flowing over a sandy bed. The country from here [Klaserie] to the Sorgobiti river [Nhlaralumi] swarms with game consisting of giraffe, eland, buffalo, koodoo, zebra, brindled gnu, bastard hartebeest[6], pigs and other kinds. We arrived at the banks of the Imbabati river [Timbavati], passing three or four waterholes on the way [probably waterholes in the Nyameni Spruit]. As is usual with these streams it had a few pools of water here and there – the rest was dazzling sand. On starting in the morning, I crossed a stream of water [the Hlangeni Spruit near the current Houtboschrand ranger post], running into the Imbabati river, which immediately lost itself in the sand.
>
> From here we emerged in 'open front', a distinct thing from 'open bush', and consisting of large trees, between the trunks of which you could see a great distance, the ground being destitute of undergrowth [the Lebombo Plains in the vicinity of Shitsalaleni Pan]. I also met here the 'Zenondo' called by the Dutch 'Bastard hartebeest' – a scarce antelope.

It is almost certain that Erskine and his party followed the ancient trade route between Delagoa Bay, Phalaborwa and Mashonaland (Zimbabwe) from the Timbavati to the Lebombo (De Vaal, 1984). He described the route across the Lebombo Plains as follows:

> The country continues to improve, the trees are more scattered, and the grass grows higher and thicker. I have left the usual low plain and ascended a rise [the Lebombos] which appeared flat on top. The country is thickly inhabited, and is well cultivated [the Pumbe Plateau, which was found in a similar condition by surveyors G.R. von Wielligh and W. Vos in 1890].
>
> After proceeding some distance, I came upon a spring which the caffres informed me was only allowed to be drunk out of by 'inkosi' (chiefs), and that 'Abafokazan' (poor men) were killed if they drunk from it! Next morning I went to visit the chief Mnbondune by a path through impenetrable scrub for about 2 miles [similar dense sandveld shrub as that occurring in the Nyandu Bush in the north of the current Kruger National Park].

From Mnbondune's kraal on the Lebombo, Erskine continued to the confluence of the Olifants (Lipalule) and Limpopo rivers (Bembe or Ouro), where he received help from Chief Macigamana. Chief Manjoba, however, was not so obliging and led him

---

6 According to Milstein (1968) it is possible that Erskine was referring to Lichtenstein's hartebeest, seeing that he referred specifically to tsessebi elsewhere.

astray when he attempted to reach the mouth of the Limpopo. After a gruelling journey through the wilderness, he eventually succeeded in reaching the Limpopo mouth at Inhampura on 30 July 1868.

At the end of his 1868 expedition, when Erskine made his return journey from the Limpopo mouth, he walked 225 kilometres in a southwesterly direction from Manjoba's kraal to accompany his friend Robert du Bois and his party on their journey to Natal. Some members of the party contracted malaria, and one, a certain Mr Wood, died on the way in August 1868. His friends laid him to rest in a lonely grave in Sabiepoort (probably on the Mozambican side). From Sabiepoort they travelled southwards along the Lebombos to Swaziland. They found their wagons by the Umbuluzi River in Swaziland and trekked via Utrecht to Natal.

In 1870, not long after his return to Natal, the Nguni high chief Muzila (Umgila) of Gazaland, the son of Sochangana (Manukosi), asked that the Natal authorities send a delegation to meet him. Erskine was told by Lieutenant General R. Keate of Natal to go back to the Gaza Empire. He left Durban on board the *Congune*, but the Portuguese authorities in Delagoa Bay would not allow him to continue his journey into the interior. As a result he left Delagoa Bay and sailed to Inhambane in July 1871 and instead entered Muzila's territory from there. Afterwards he undertook extended journeys along the Limpopo and into the interior of Mozambique and in the process happened on the murder scene of the Van Rensburg Voortrekker party. He was the first European to visit the terrain after the expeditions by Karel Trichardt and Jan Pretorius. He accurately indicated its location on his map as being near Matsamo's kraal and about 60 to 80 kilometres north of the confluence of the Olifants and Limpopo rivers. He described the Limpopo at this spot as being a stream of about 100 metres wide and 45 centimetres deep, with a sandy bed.

Erskine was an excellent observer of natural phenomena and supplied particularly accurate descriptions of tree species, vegetation regions, geological formations, edible wild fruit, animal life, birds and even invertebrates and fish species he encountered on his journeys through the Lowveld and Mozambique. He noted, for example, that he saw samango monkeys in the riverine forest along the Limpopo. This explains the sudden appearance of these monkeys in the riverine forest at Pafuri more than 85 years later in 1957, during an exceptionally wet year.

He encountered the first cattle he had seen since leaving Lydenburg at Chief Matonse's kraal, about 50 kilometres south of the confluence of the Olifants and Limpopo rivers. Erskine's travels in Mozambique in 1871 took him across the Banyini plain north of the Limpopo to the Great Save River and from there eastwards to Sofala[7] on the coast. From here he journeyed in a southwesterly direction down the coast, past Inhambane and Zavora, and once again savoured the satisfaction of seeing the mouth of the Limpopo at Inhampura. During these travels he also visited the impressive Chimanimani Mountains on the border between Mozambique and Mashonaland.

Erskine undertook a third expedition to the Gaza Empire in the years 1873–74. This time he travelled up the Sabie River (Save) for a while and from there to Muzila's territory. On the way he crossed the upper reaches of the Búzi. On his return journey he travelled northwards along the coast of Chiluane to investigate the mouth of the Gorongoza River. From there he travelled southwards to Inhambane via Cabo Sào Sebastiào. After a short interval he returned to the area south of the Save in November 1974 and occupied himself until June 1875 with hunting trips and expeditions (Leverton, 1972).

St Vincent Whitshed Erskine's greatest legacy is the information he collected and his maps of previously uncharted territories.

References

Erskine, St V. W. 1875. Journey of exploration to the mouth of the Limpopo River. *Journal of the Royal Geographical Society* 45:45–128.

De Vaal, J.B. 1984. Ou handelsvoetpaaie en wapaaie in Oos- en Noord-Transvaal. *Contree* 16:5–15.

Fuller, C. 1923. Tsetse in the Transvaal and surrounding territories. An historical review. *Union of SA Department of Agriculture Entomology Memoir* 1:6–68.

Milstein, P. le S. 1986. Historical occurrence of Lichtenstein's hartebeest, *Alcephalus lichtensteini* in the Transvaal and Natal. Unpublished manuscript.

Suid-Afrikaanse biografiese woordeboek part 2:285–286. 1972. Insert on: Erskine, St. Vincent Whitshed. Tafelberg, Cape Town & Johannesburg.

---

7 Now Beira.

8.4 Carl Mauch's expeditions through the southern Lowveld and to Mashonaland (1868 and 1870)

Dr U. de V. Pienaar

In 1868 Carl Gottlieb Mauch, the famous German scientist and explorer, undertook an expedition from Lydenburg northwards across the Olifants and Letaba rivers. He reached the confluence of the Luvuvhu and Limpopo rivers and travelled from there to Inyathi in Matabeleland (now Zimbabwe).

During the first leg of his journey, from Pretoria to Lydenburg, he was accompanied by the English surveyor St Vincent W. Erskine, of Pietermaritzburg (see chapter 8.3). He left Lydenburg with donkeys and pack oxen, but the oxen were more of a hindrance than a help. They were bitten by tsetse flies and the last one died of nagana before Mauch had even reached the Limpopo, despite the fact that he regularly treated them with ammonia salt diluted in water!

During the journey he noted that all the tributaries of the Olifants River that had their origin in the Drakensberg, had a strong flow. On the other hand, those that originated in the foothills of the Drakensberg (Kahlamba) had a slower flow and were characterised by an abundance of pools. He described the Olifants River (Lepelle or Balule) as a beautiful stream of about 80 metres wide, with banks edged with dense stands of reed and large trees, including wild figs.

He noted that he crossed the Olifants at 24°22' south and 31°35' east. Along the banks of the Letaba (Lehlaba) and Klein Letaba (Lehlabane) he also found dense stands of common reed. Both rivers were flowing strongly at the time, but it was difficult to see the water because of the dense reeds.

Mauch stated that the entire area between the Olifants and Luvuvhu rivers was sparsely populated and that the locals lived in small kraals close to the perennial rivers, or on rocky outcrops near permanent fountains. These were very similar to the settlements at Phalaborwa (where the people made ornaments from copper they had smelted themselves), Matsête (Mashetse) along the Klein Letaba, and Sebolene along the Nsama (Mosana), a tributary of the Klein Letaba. Apart from dogs, the local inhabitants had no domestic animals because of the presence of the deadly tsetse flies. He therefore considered the area unsuitable for settlement by whites, although the soil appeared very fertile.

He found game to be plentiful in the vicinity of the Olifants River and saw giraffe, buffalo, kudu, waterbuck and zebra. There were many hippos in the river, but they were shy and kept out of view.

En route to the Luvuvhu (Limvubu) he also crossed the Shingwedzi River, which he described as being strong flowing. The original name of the river, Molototsi, was noted by Mauch as "Mositotsi" and he was under the mistaken impression that it was a tributary of the Luvuvhu. He considered the Luvu-

Carl Mauch (1837–1875), the famous German scientist and explorer, circa 1870.
(ACKNOWLEDGEMENT: STATE LIBRARY, PRETORIA)

vhu an important river, with quite a few running tributaries (such as the Mutale) originating in the Soutpansberg catchment area to the west.

The grass grew luxuriantly along the banks of the Luvuvhu and he said the riverine vegetation resembled a tropical paradise complete with gigantic trees, lianas and palms. Baobab trees were a typical feature of the landscape (as they are today). Along the upper reaches of the Luvuvhu he found wild bananas with ripe fruit and orchids in large trees. He described the Venda-speaking subjects of Chief Ramabulana as being very inquisitive and constantly begging, especially for lead and gunpowder for their heavy elephant rifles. He found them to have pronounced Semitic features and wondered to what this could be ascribed.

Mauch crossed the Limpopo on 31 August 1868 and found the river bed here (22°13' south, 31°15' east) to be 1250 paces wide, but the water itself was only 150 paces wide and knee-deep. A broad strip of riverine forest with colossal trees and fan palms grew along both banks.

He now found himself in the land of uMzilikazi (Moselikatse) and after his travels in Matabeleland he arrived back in Pretoria in October 1868.

In 1870 the Volksraad of the ZAR (South African Republic) asked Mauch to safely escort one Lieutenant Da Costa Leal from Potchefstroom to Delagoa Bay. Leal was a member of a diplomatic delegation under Baron Carlos Barahona e Costa, the governor of Quelimane. They left Potchefstroom on 18 May 1870 and travelled via Losberg and Wonderfontein to Pretoria. From here they traversed the farms of one Vermaak and Dreyer to the upper reaches of the Komati and Lake Chrissie, an area christened New Scotland by Alexander McCorkindale as he planned to settle 300 Scottish families here. This community would establish contact with the east coast via the Usutu River, on which McCorkindale planned to establish a ferry system. Little came of his schemes and he could only recruit 50 Scottish settlers to settle in the vicinity of Lake Chrissie. McCorkindale died in 1871 on Inhaca Island on the Mozambican coast while searching for a suitable port for the settlement.

Mauch and his party trekked from New Scotland through Swaziland and stayed over for a while on the farm Derby of his old friend St Vincent Erskine, who had done some surveying work for McCorkindale. After crossing the Lebombos at Ingwavuma in Swaziland they travelled northwards to Delagoa Bay and arrived on 8 August, to the great relief of Mauch. (Apparently he did not get on too well with his travelling companion, Lieutenant Leal!)

En route Mauch noted in his journal that thousands of springbok and blesbok near Lake Chrissie fled at their arrival. He also mentioned the large stands of cycads (*Encephalarctos* sp.) in the Lebombo Mountains in Swaziland.

During his stay in Delagoa Bay he met Surveyor G.M. Moodie, who was investigating ways of improving communication between the Transvaal and Delagoa Bay. Moodie later built the road between White

Carl Mauch and his assistant, Fernando da Costa Leal, 17 May 1870.
(ACKNOWLEDGEMENT: STATE LIBRARY, PRETORIA)

River and Burger's Hall, and Moodieskloof in Legogote was named after him.

Mauch's return journey to Pretoria via Lydenburg started on 29 August 1870 along De Oude Wagenweg (The Old Wagon Route) between Delagoa Bay and Komatipoort, after which he trekked through the southern part of the current Kruger National Park to Lydenburg.

On 3 September the party arrived at the Komati River where it flows through the Lebombo Poort and they crossed the river there. Mauch shot two hippos, to the great delight of the locals, who immediately started feasting on the abundant fresh meat.

On 5 September they travelled through the poort and beyond the confluence of the Crocodile (Ngwenya) and Komati rivers. On the way they encountered a lioness feeding on the carcass of a buck under a bush. Further up the Crocodile River, in the vicinity of the current Crocodile Bridge, they discovered that there had been a veld fire. Mauch mentioned the red soil typical of this area and the large tree savannah across which they travelled. The area was teeming with wildlife and he noted that there were thousands of zebra and blue wildebeest. He found the game in the area to be very tame and apparently not used to firearms. These animals were observed close to where Francois de Cuiper's party crossed the Crocodile River in 1725. Interestingly, De Cuiper did not make any mention of having seen game here.

Before crossing the watershed between the Sabie and Crocodile rivers, they crossed a spruit (possibly the Vurhami, or the Bumi's upper reaches) "with numerous small pools and rhinoceros spoor", and found a solitary giraffe near three granite hills (Randspruit?). Along the upper reaches of the Nwatimhiri Spruit and in the vicinity of Renosterkoppies they came across large and annoying swarms of tsetse flies.

They followed a small, dry spruit to the Sabie. The further they went, the fewer tsetse flies they encountered. The grass next to the river was thickly overgrown and this hampered their movement. That night, they slept along the riverbank.

On 9 September they resumed their journey along the Sabie and, after having walked for about four hours, the party reached the ruins of Albasini's shop and home at Magashulaskraal. A tree about five metres high, with a 15 centimetre-thick trunk, was already growing inside the walls of the ruins and nothing could be seen of the once cultivated fields. On a nearby hill he found several rock piles marking the graves of blacks who had lived there before.

There was no sign of the buffalo and rhino he had been told of along this part of the Sabie, while other game also seemed to be scarce.

After having been unable to shoot anything for the pot for three days, Mauch had a malaria relapse. He decided not to trek to the Soutpansberg, as had been the plan, but to find white people as quickly as possible. For this reason, he changed direction and set off towards Lydenburg.

When he forded the Sabie River in the vicinity of the current Sabie River Bungalows, he found a disused wagon trail along the northern bank of the river. It is possible that the route he then followed was the old Klipkraal or Neethling road to Sabie. Along the way he became so feverish that he collapsed. After regaining consciousness it was only with a great amount of effort that he managed to climb the mountain, to the upper reaches of the Blyde River. He followed the river downstream to Khobeng's kraal (after whom Kowyns Pass was named). Eventually he arrived at Jan Muller's farm Doornhoek, exhausted and delirious. The Mullers realised the gravity of his condition, gave him some brandy and transported him by ox-wagon to Pieter de Villiers (General Schalk Burger's father-in-law) at Krugerspost.

He only woke the next morning – in the wagon house – and was then sent by horse and cart to the missionary Reverend During at Lydenburg. Here he was nursed back to health and recovered to such an extent that he could resume his journey at the end of September.

In honour of the repeated visits this celebrated explorer had made to Lydenburg, a mountain peak between Sabie and Lydenburg was named Mauchsberg. Cohen (1875) already makes mention of Mauchsberg in the description of his journey from Lydenburg to Delagoa Bay.

Mauch undertook another journey to the northern Transvaal in 1871, travelling via Botshabelo, Zebediela, Marabastad, Matala and Schoemansdal to Albasini's farm Goedewensch. From here he followed the centuries-old trade route to Mashonaland and officially discovered the Zimbabwe Ruins in May

1871. In the journals of his travels through Africa from 1869–1872 he gave a very detailed description of the ruins. Little did he know that Karel Trichardt had already visited these mysterious ruins in November 1838 during one of his journeys into the interior from the east coast harbour Sofala (Preller, 1920).

Upon his arrival at the Zimbabwe Ruins, Mauch also met the hunter Adam Renders and his wife Elsie.[8] Renders and his wife had come from the Transvaal and in 1868 settled temporarily close to where the Morgenzon Mission was later established (Punt, 1958). Mauch was thus the official discoverer, but not the first white to see the Zimbabwe Ruins.

During his journeys through the Transvaal Lowveld, Matabeleland and Mashonaland, Mauch made meticulous notes on the geology, vegetation, water conditions and human activities in the areas through which he travelled. He repeatedly remarked on signs in the geological formations of the Lowveld that indicated the presence of gold and other minerals and he declared that this part of southern Africa possessed enormous mineral riches. He never got to experience the exploitation of these riches, however, as he died a poor man at the youthful age of 37.

References

Bernard, E. & Bernard, F.O. 1969. *The Journals of Carl Mauch 1869–1872.* National Archives of Rhodesia, Salisbury.

Cohen, E. 1875. *Erlaütende Bemerkungen zu der Routenkarte einer Reise von Lydenburg nach den Goldfeldern und von Lydenburg nach der Delagoa Bai in östlichen Süd-Afrika.* L. Friedrichsen & Co., Hamburg.

De Vaal, J.B. 1984. Ou handelsvoetpaaie en wapaaie in Oos- en Noord-Transvaal. *Contree* 16:5–15.

Fuller, C. 1923. Tsetse in the Transvaal and surrounding territories: an historical review. *Union of S.A. Department of Agriculture Entomology Memoir* 1:6–68.

Lombaard, B.V. 1970. Mauch op Lydenburg in 1870. *Die Lydenburg Nuus* 2 October.

Mager, C. 1895. *Karl Mauch: Lebensbild eines Afrika-Reisenden.* W. Kohlhammer, Stuttgart.

Petermann, A. 1870. *Karl Mauch's Reisen in Inneren von Süd-Afrika.* Justus Perthes, Gotha.

Preller, G. 1920. *Voortrekkermense II. (Herinneringe van Karel Trichardt).* Nasionale Pers, Cape Town.

Punt, W.H.J. 1958. Die wit vrou van Zimbabwe. *Fleur* (Supplement to *Dagbreek*) 24 August.

---

[8] Some sources give her name as Elize Renders (Punt, 1958).

8.5 Captain Frederick Elton's voyage down the Limpopo
Dr U. de V. Pienaar

Looking at the Limpopo River during the winter months today, all one can see is a wide, dry, sandy bed. It is hard to believe that somebody managed to travel 70 kilometres from the mouth of the Shashi in Matabeleland (now Zimbabwe) down the river to the Tolo Azime Falls, about five kilometres west of the current Beit Bridge.

This amazing journey was undertaken in a flat-bottomed boat by Captain Frederick Elton in July and August 1870. In his report on the journey he expressed his conviction that the river could be navigated with flat-bottomed steamboats between its confluence with the Olifants and its confluence with the Nuanedzi (Chicuala-cuala), even in dry spells.

According to him it would also be possible to establish regular trade routes from the last-mentioned point to serve the white settlements in the Soutpansberg and along the Luvuvhu by ox-wagon.

Elton started his journey on 6 July 1870 at the Tati goldfields in Matabeleland. His four metre-long flat-bottomed boat, complete with mast, sails and oars, was loaded onto a wagon with the luggage and, with three pack oxen as a rearguard, they departed for the site from where they would launch the boat.

To reach the mouth of the dry Shashi River he had to trek 300 kilometres through the wilderness, an area with little water but an abundance of lions and game. He travelled through the tribal territory of the Makalakas, who possessed no cattle because of the presence of tsetse flies. They did, however, farm with sheep and goats and cultivated the land. According to him they were lily-livered fighters and the subjects of the Matabele.

Despite this they were hard workers who grew millet (*Panicum miliaceum*), pumpkin, tobacco and hemp. Among the game in the area were rhino, buffalo, blue wildebeest, kudu, zebra, giraffe and elephant. The elephants were wild and irascible as a result of ongoing hunting.

At the confluence of the Shashi and the Limpopo, the Limpopo was about 200 paces wide. The banks were hemmed with massive trees, while several crocodiles were sunning themselves on a sandbank. Soon after daybreak on 1 August 1870, Elton pushed his boat into the water and set off on his adventure. Crocodiles silently slipped into the water as the boat sailed past and at one place a large herd of buffalo stormed out of the reeds when they saw the strange object on the water. Waterbirds were common. Apart from fish eagles and Egyptian geese, they also saw many reed cormorants, egrets, grey herons and hornbills. Gigantic baobabs towered above the vegetation on the sandstone hills along the river, while wild figs and other shady trees grew along the banks.

On 2 August a stream dragged Elton's boat under the branches of a fallen tree and it capsized. In the process he lost his blankets, groundsheet, cooking utensils and coat, while the supply of tobacco and sugar was also ruined. They managed to save the vessel, however, and on 3 August they got hold of the pack oxen, which had been taken from the mouth of the Shashi to Chief Mafelagure's kraal. The river had been easy to navigate from the Shashi mouth to this point and the only obstacle had been a number of smaller rapids. The broad channel of water, be-

The Tolo Azime Falls in the Limpopo, just east of the Umzingwane mouth and west of Messina (Musina). Capt. Frederick Elton had lost his boat here during his river journey in August 1870. (01.05.1986)

tween one and three metres deep, could easily be navigated by boat. Elton first heard about the Limpopo Falls' "wall of water" from Chief Mafelagure's people. This was an inhospitable area where large numbers of lions made settlement impossible, while the river teemed with hippos and crocodiles.

About 14 kilometres east of Mafelagure's kraal the party sailed past a group of steep, pointed hills on the right bank. Rapids became more frequent and increasingly formidable until they reached a large tributary of the Limpopo from the north, the "Mzinyani" (the Umzingwani). Elton described the river here as follows:

> Here the Limpopo, stretching out to more than a mile in width, rushes in a dozen different channels over large boulders, in seething and foaming rapids, interrupted by circling eddies and deep dark silent pools, the abode of hippopotami, who feed on the long waving grass of the thickly wooded islands. At a distance of five miles the river culminates in the cataracts of Tolo Azime.

The speed of the water through the rapids made it difficult for Elton to steer the boat, but at the same time the far-off thunder of water made it essential that they reach the bank. At last they managed to get the boat to the side and landed safely. After their landing they explored the bank further down on foot and, to their surprise, came across a roaring waterfall barely twenty paces away. This fall would certainly have smashed their boat and that would have meant not only the end of their expedition, but probably their lives as well. It looked as if the whole landscape had dropped a step from the higher-lying plateau in the west to the lowlands of the east.

To their dismay they discovered that they had actually been washed onto an island and not the real riverbank. Thus they would have to cross another stream to their left to get out of the river. In due course they managed to achieve this with the aid of a fallen tree trunk higher up. They struggled for two days to get their boat, the *Freeman*, on to safe land, but in the end their efforts were in vain as the river ripped it away and it went over the waterfall, smashing onto the large rocks below and breaking into pieces. This was a great loss to the valiant little group.

Despite their misfortune, Elton was overcome with emotion at the sight of the majestic Tolo Azime Falls, although they bear no comparison with the Victoria Falls in the Zambezi or the Augrabies Falls in the Orange River (according to Van Riet [1966] the Tolo Azime Falls are only about seven metres high). Elton was nevertheless of the opinion that the unique contrasts in the environment made it worthwhile to prominently indicate this waterfall on future maps of Africa. According to him no hunter had ever been to these falls and they had also never been indicated on any maps. There were many hippos both above and below the falls, and they spotted kudus, baboons, monkeys, otters and a few buffalo on the northern bank of the river. The southern bank was rocky with very little vegetation or other forms of life.

Elton and his four black companions continued their journey on foot with only the three pack oxen to carry their supplies. They left the falls on 8 August and pitched their ninth camp opposite one of the northwestern foothills of the Soutpansberg. The party walked along the right-hand bank of the river, but because of the rough, rocky terrain they were forced to wade chest-deep through the river on 12 August, as it would be easier to walk along the left bank. On the way they encountered a pride of lions that was so busy fighting over the carcass of a zebra that they ignored the travellers.

On 13 August they reached Amabaga's kraal. These were the first people they had seen since leaving Mafelagure's kraal. They were cordially received and could rest a while after their difficult journey the previous day. During the next stretch of their journey they found that the river meandered through almost vertical cliffs (the Mabiligwe Hills) and on 15 August they spent almost the entire day in the water, as they had to cross the river no fewer than nine times to make any progress! Shortly afterwards they met Knobnoses (Magwamba) for the first time. These people led a miserable life in the bush as a result of raids by the Nguni warriors of Muzila (Umzila). Once Elton and his party also encountered a group of Muzila's men, but they were fortunately left in peace. A day or so later they arrived at the large kraal of the Maloios tribe – Tsonga-speaking subjects of Muzila. In this region there were many kraals, lush vegetation and fertile soil. They noticed sandstone hills to the south, and at this stage the explorers probably found themselves on the Limpopo side of what would later become known as Crooks Corner, between the Luvuvhu and Limpopo rivers. They managed to recruit a guide who accompanied them to the confluence of the Nuanedzi and Limpopo rivers, which they reached on 19 August. The rest of their journey – to the confluence of the Olifants and Limpopo rivers – took another ten days to complete.

Elton noticed that the soil on both sides of the Limpopo was rich and fertile and cultivated by the large black settlements, especially in the vicinity of its confluence with the Olifants, where large numbers of high chief Madumelane's people lived. They were Tsonga-speaking and also subjects of Muzila, who ruled over the whole area between the Inkomati and Buzi rivers. They planted peanuts, tobacco, hemp, millet, holcus (a type of grass) and other agricultural crops. The animals they saw along this stretch of the Limpopo included elephant, buffalo, rhino, giraffe, blue wildebeest, zebra, eland, kudu, bush pig, brown hyena, wild dog, leopard and lion.

They encountered some tsetse flies on 19 and 20 August and although Elton's oxen were bitten, they showed no negative symptoms. He was therefore convinced that the danger these flies posed was somewhat exaggerated.

Because St Vincent Erskine had already documented in detail the Limpopo between its confluence with the Olifants and the sea, Elton decided to head straight for the Komati (King George River) and from there to Lourenço Marques. On reaching the Komati they had to bribe the ferryman with their last valuables to take them across the river. At last, on 7 September, they arrived in Lourenço Marques; footsore but satisfied with their successful expedition.

After completing the journey, he claimed in his report that if given six months to prepare, he would be able to voyage up the Limpopo to the Nuanedzi mouth in 15 days with flat-bottomed steamboats. From there he would be able to reach the Tati goldfields in Matabeleland via the Soutpansberg within 15 days – thus 30 days in total from the coast. He was convinced that the unhealthy conditions along the east coast and in the Limpopo Valley were greatly exaggerated. Of course nothing came of Elton's grandiose plans, which must have been an enormous disappointment to him.

References

Baines, T. 1854. The Limpopo, its origin, course and tributaries. *Journal of the Royal Geographical Society* 24:288–291.

Birkby, C. 1939. *Limpopo Journey.* Frederick Miller, London.

Elton, F. 1872. Journal of exploration of the Limpopo River. *Proceedings of the Royal Geographical Society* 16:89–101.

Erskine, St. V.W. 1875. Journey of exploration to the mouth of the Limpopo river. *Journal of the Royal Geographical Society* 45:45–128.

Van Riet, W. 1966. *Stroomaf in my kano.* Tafelberg, Cape Town.

8.6 Doctor E. Cohen's journey from Lydenburg to Delagoa Bay, 1873

Dr U. de V. Pienaar

In June and July 1873 the German geologist Dr E. Cohen walked from Lydenburg via the Mac-Mac goldfields to Delagoa Bay. He followed the route of what would soon be a wagon trail, completed in December of the same year by George Compton and his contractors, Shires and Hampton, on the instructions of the Volksraad and President Thomas Burgers. It started at the Lydenburg goldfields and ran north of Lozieskop (Legogote) via Pretoriuskop, Skipberg, Nellmapius and Furley's drifts in the Crocodile and Komati rivers respectively, into Mozambique through Matalha Poort in the Lebombos to Pessene and ultimately Delagoa Bay.

As could be expected, his detailed report on this journey contains a large amount of topographical and geological information (including the first topographical map of the southern part of the Kruger National Park). It also contains a lot of information on the black tribes, the distribution of tsetse flies, and the plant and animal life in the area through which he walked. He noted that the western boundary of the tsetse fly-infested area south of the Sabie

Newu Hill close to the old Delagoa Bay trade route. Dr E. Cohen, the German geologist and explorer, moved past here in July 1873.
(08.06.1984)

River was in the vicinity of Newu Hill (Taba Neu), southeast of Pretoriuskop and placed the eastern boundary of the tsetse belt about 20 kilometres east of the Lebombo Mountains and about 50 kilometres northwest of Lourenço Marques.

He noted that there were no tsetse flies in the rocky parts of the Lebombos where the game paths ended. He also reported that the blacks living south of Delagoa Bay owned cattle, but not those in the area that was to become the Kruger National Park. At Nellmapius Drift he encountered large swarms of tsetse flies, which bit him mercilessly. They were also bitten repeatedly while carrying water to their camp in the area between the Komati River and the Lebombo Mountains after they had crossed Furley's Drift. But by sunset the flies had disappeared completely. Cohen noticed that, depending on the denseness of the bush, the flies occurred in patches; and there were many more in thickets than in open, sunny areas. He also mentioned that a certain MacDonald once lost 100 of his 130 oxen to nagana while travelling along the tsetse-infected route from Delagoa Bay to the Lydenburg goldfields.

In the vicinity of Setigalanga Hill (Sithungwane) he found an abundance of game, mainly zebra and blue wildebeest, while he encountered large herds of zebra, blue wildebeest, kudu, waterbuck, reedbuck and sable antelope between Sithungwane and Newu. The party never actually saw lions, but often heard them roaring after dark. At night they stacked thorn branches around their camp and kept large fires going to keep the lions away after they had found the skeletal remains of a black man who was killed and eaten by lions inside his enclosure.

On 28 June they pitched camp at the foot of Skipberg near a marsh. Here they found standing water between large granite slabs, as well as a running stream (possibly the Samarhole Spruit). Close to Skipberg they encountered giraffes in a knob-thorn thicket and when they reached Newu Hill on 30 June they found the spoor of a large herd of buffalo.

The area between Newu Hill and Makhuthwanini Hill (Taba Umlutschue) teemed with game, especially blue wildebeest and zebra, but water was scarce.

They did, however, find enough water in a pool in a dry watercourse to allow them to pitch camp for the night (possibly the Ntomeni pools). For supper he ate mountain tortoise and found it delicious – apparently it tasted like frog legs! It rained that night, and the next morning they reached Makhuthwanini Hill.

Cohen described the olivine gabbro dyke east of Makhuthwanini in great detail and further mentioned that he found many large pools in the upper reaches of Mtlhowa Spruit, which he crossed. Lastly he also described the open savannah and the game (including giraffe) along the final leg of the route (the upper reaches of the Lwakahle) before they reached Nellmapius Drift in the Crocodile River (Ngwenya) on 1 July.

Here they found large herds of impala, and the knob-thorn thicket through which they travelled before reaching the river (Gomondwane) impressed Cohen to such an extent that he described it as true woodland, in contrast to his descriptions of the other bushveld types they had moved through. He also found the Crocodile River impressive, about 200 paces wide with the water level varying between knee-deep and one meter deep.

On the right bank of the Crocodile River they found the cultivated fields of Chief Umgagane's people. The next day they continued their journey to Furley's Drift in the Komati River and ultimately Delagoa Bay.

References

Cohen, E. 1875. *Erläuternde Bemerkungen zu der Routenkarte einer Reise von Lydenburg nach den Goldfeldern und von Lydenburg nach der Delagoa Bai in östlichen Süd-Afrika.* L. Friedrichsen & Co., Hamburg.

Fuller, C. 1923. Tsetse in the Transvaal and surrounding territories: an historical review. *Union of S.A. Department of Agriculture Entomology Memoir* 1:6–68.

Punt, W.H.J. 1950. Verslag i.s. ondersoek van ou paaie en roetes in die suidelike Wildtuin. Unpublished report to the National Parks Board.

8.7 William Napier's hike from the Mac-Mac goldfields to Delagoa Bay, 1875

Dr U. de V. Pienaar

William Napier, an adventurous, middle-aged digger, walked from the Mac-Mac goldfields via Pretoriuskop to Delagoa Bay in December 1875 and January 1876. In a letter to his friend, the author D.M. Dunbar, he gave a comprehensive account of this hike through the area that today represents the south-western corner of the Kruger National Park.

Napier's journey started on 27 December at the Mac-Mac goldfields. From there he followed the footpath to Mr Pettigrew's "Welcome Farm" instead of the wagon trail from Pretoriuskop to Mac-Mac, which was 30 kilometres further. This Pettigrew was probably the same person who later built the Pettigrew road running from the Barberton goldfields, along the Kaap and Crocodile rivers, to the transport road between Lydenburg and Delagoa Bay, joining it at Nellmapius Drift. This road was, however, hardly ever used as it became notorious for the hordes of tsetse flies along it that pestered the transport riders' animals. Sir Percy FitzPatrick of *Jock of the Bushveld*-fame used this road in the late 1880s and to his chagrin lost almost all his draught animals in the process.

At Pettigrew's farm Napier got a black guide to lead him to Pretoriuskop. The footpath he followed ran along the watershed between the Sabie and Mac-Mac rivers, which the wagon trail from the Mariti Spruit also followed. He crossed the Mac-Mac River about three kilometres west of Pettigrew's farm. Although the river was in flood, they managed to wade through safely just before sunset. They pitched camp later that night, after they had been thoroughly drenched in a thunderstorm. At this stage they were still on the northern side of the Sabie River. The next morning they covered the about 24 kilometres between Mac-Mac drift and the Sabie River. The Sabie was also in flood and they weren't able to cross it, so they spent the night at the drift.

Napier, who had an injured shoulder and arm, struggled for two days before he managed to cross the still flooded Sabie River with the help of a rope. Almost all of their belongings got soaking wet during the crossing and they had little food left. Eventually they were all safely on the river's southern bank on the evening of 30 December.

On New Year's Eve 1875, Napier and his two companions were on their way to Pretoriuskop. After walking for about 16 kilometres through beautiful, open and green country they reached the North Sand River. He described it as "fine clear water running rapidly into the Sabie". He was very impressed with the vegetation in the area, which included wild fruit such as wild plums, spiny monkey-orange (*Strychnos spinosa*) and marula. He also mentioned that the grazing along the whole route so far would be excellent for cattle. Fortunately drongos led them to three beehives and the honey assuaged the worst of their hunger. They spent the night in a deep donga about five kilometres east of the drift through the North Sand River. Napier shared what little food he had left with his companions.

The next morning his guide led him on a direct route to Pretoriuskop. They walked about 12 kilometres in the morning and arrived at the overnight station[9] of Alois Hugo Nellmapius's company (the Lourenço Marques & South African Republic Transport Service) exhausted and very hungry. Here they were cordially welcomed by the Danish station master, Feldschau, who was working in his fields. After washing he was offered a meal of beans and coffee. Napier thought it would be difficult to find an area more suitable to farming than that between the Sabie River and Pretoriuskop. The soil, climate, water and topography were all suited to just about any form of farming – a view substantiated in more recent times by the flourishing farming communities at Kiepersol, Hazyview and Burger's Hall. Napier thought that the Pretoriuskop area in particular would be well suited to cattle farming.

It is thus no wonder that Albasini pastured his

---

9   In July 1985 it was determined with a reasonable amount of certainty that Feldschau had built his overnight station and storage facility on the same spot as the outpost of Albasini's headman Manungu – about 500 metres north of Manungu Hill.

cattle here from 1846 to 1848 under the watchful eye of Chief Manungu.

From Pretoriuskop goods were transported by ox-wagon to Lydenburg and the goldfields, but eastwards only bearers were used, who carried the heavy bundles on their heads. For this they received 20 shillings per 100 pounds (or ten shillings for a bundle of 50 pounds). If they were prepared to carry their loads to Pilgrim's Rest, they received a further five shillings for every 50 pounds. The cost of transporting 100 pounds of goods from Delagoa Bay to Pilgrim's Rest was thus £1.10.00. Napier did not consider this to be a profitable business, mostly because the logistics of providing the bearers with food and shelter along the route were lacking. He found almost no food for the manager of the post or the bearers at Pretoriuskop and would later encounter the same situation at the rest of the stations further east. At that stage no flour, tea, coffee, sugar or any other essentials were transported along the route – only brandy, wine, gin and other luxury items. In his opinion the whole organisation left much to be desired and Napier thought that if Nellmapius and his partner failed to take drastic action to improve the situation, the whole enterprise would end in a fiasco.

At dusk he went to bathe in a pool full of catfish (waterholes in the Manungu Spruit just west of the current Pretoriuskop rest camp). (The remains of Feldschau's home and overnight station were found here in 1985 by senior ranger Thys Mostert.[10])

After breakfast on 1 January 1876 the small party left for the next overnight post along the Nellmapius route – Joubertshoop, situated about 25 kilometres southeast of Pretoriuskop. Napier found the soil drier and the vegetation less varied as they walked further east, but still described the area as beautiful grassland which, in his opinion, would be suitable for sheep farming. At the foot of Skipberg the soil was heavier and more clayey and here he found large trees (knob-thorns). He described the area as parkland, attractively dotted with shrubs and beautiful trees. This is also where they encountered the first herds of game and, with the animals, the first tsetse flies. The Joubertshoop overnight post was situated on the western bank of a spruit (Josekhulu),[11] in a small depression, but still high enough not to be unhealthy during the rainy season. (At the time malaria was associated with poisonous swamp gases rising from low-lying areas!)

Thomas Hart, the station commander at Joubertshoop, was a congenial and cheerful young man, who warmly received the travellers in his tent. (His wattle-and-daub house had not been completed yet.) Hart had about 15 labourers working for him and he showed Napier a maize field he was busy laying out. He also bemoaned the fact that he had almost nothing to eat except for the venison he or his black workers had to provide themselves. As a result the bearers also often had to go without food.

Because Napier considered Hart to be an enthusiastic and studious young man, he thought that he should be supplied with books on botany, forestry and mechanics, especially when taking into account the valuable wood to be found in the area (kiaat, Cape beech and others).

That evening Napier again had a bath in a waterhole in the Josekhulu Spruit close to Hart's tent, where he noticed some fish. Their supper consisted of zebra meat and ham and he had a good night's sleep.

The next morning the explorers set off on the next leg of their journey – a distance of about 27 kilometres to the next overnight post at Ludwichslust[12] next to the Crocodile River. The route still headed in a southeastern direction and took them past Newu Hill and the Makhuthwanini Hill. The vegetation there was not as lush as further west, but the grass was nevertheless sweet and of excellent quality for cattle. Tsetse flies were, however, a big problem. Napier believed it necessary to build proper cattle kraals and to plant maize between the different posts for the trek oxen, which the Nellmapius firm in Delagoa Bay was busy breaking in. He also thought ferry boats should be used to carry wagons safely across the Crocodile and Komati rivers. This would place the transport business on a firm footing to make it both practical and profitable.

The landscape became flatter as they neared the

---

10 Rowland G. Atcherley (1879) described in his book, *A Trip to Boerland*, how he and his party spent a night in Feldschau's old store at Manungu Hill, while en route between Delagoa Bay, Spitzkop and Lydenburg. This was shortly before the Sekhukhune rebellion of 1876. At that stage the door and half the roof were already dilapidated.

11 Also spelled Josikhulu in certain sources.

12 Some sources spell it Lodwichslust or Ludwigslust.

Crocodile River and the shrubs fewer and lower. Large, isolated trees decorated the area.

Upon their arrival at the Crocodile River they were pleasantly surprised to see the attractive, neat Ludwichslust overnight post on the southern bank of the river. There was even a road to the river that had been fenced off with branches. Their call from across the river was soon answered and "doctor" J. Birch,[13] the station master, sent his boat to bring them across. Napier thought a good wagon trail could easily be built and maintained between Joubertshoop, the North Sand and the drift through the Crocodile River. This station was only five months old, but Birch and his black helpers had done wonders – constructing a neat home and overnight facilities for the blacks and laying out neatly fenced fields and vegetable gardens. Despite the relatively dry conditions, a thriving crop of maize was standing in the fields. After a good night's sleep, "doctor" Birch served Napier an excellent breakfast and even lent him a donkey on which he could continue his journey to the Komati River. The distance between the two rivers along this route is about 16 kilometres. The soil was more fertile and the vegetation more lush than at any other section of the route, except for the stretch between Sabie and Pretoriuskop.

The overnight post at Coopersdal next to the Komati was also on the southern bank and a boat was once again sent to fetch them. "Doctor" H. Pearce, the station master, was in Lourenço Marques on business, but Napier and his black companions were cordially received by the young Peacock. He also met W. Boby,[14] the station master of the next overnight post at Castilhopolis in the Lebombo Mountains, who had malaria and was on his way to the Pretoriuskop station to recuperate in the healthier climate there.

Very few improvements had been made at the Coopersdal post, which had the same appalling lack of food and other supplies. Peacock had no meal or provisions other than a few chickens of uncertain breed.

Because the next overnight post at Castilhopolis was only 12 kilometres away, Napier decided to push on. After a refreshing dip in the Komati, he once again mounted his faithful donkey and set off towards the Lebombos. Peacock lent him a blanket since Boby's absence from his station meant few provisions would be available.

They reached Castilhopolis station after dark and the black man in charge let them through the gate. There was a fair bit of activity at the post as a number of bearers had just arrived and were sitting around fires, chatting away.

All they found to eat at this isolated post was bread and coffee. There was a heavy thunderstorm that night and they were extremely grateful for the shelter the hut afforded them.

The next morning he realised that the Lebombo overnight post had been built in a very picturesque spot with a magnificent view. But according to Napier it was very unhealthy as it had been built close to a marsh, and he blamed Boby's malaria on this proximity. The 24 kilometres to the next station, Progresso de Guedes, were inside Mozambique and the halfway stop was situated along the eastern foothills of the Lebombos. That day the route turned more to the south and Napier noted that the area was very rocky with little top soil. The waterholes and rocky areas they encountered appeared somewhat "unhealthy" to him, so they pushed on as quickly as possible.

They found the Progresso de Guedes post on the southern slope of a low hill. His attention was first caught by the crowing of a cockerel, and soon afterwards they reached the neat home of the station master, Lempke. The house had been built on the left bank of a small spruit. In the absence of Lempke, who had gone to Lourenço Marques to fetch supplies, they were attended to by his Indian cook, who served them a dish of ham and rice. Lempke had been away for six days and his black helpers were going hungry. Once again Napier doubted the sustainability of a transport business that was so badly organised, and questioned Nellmapius's ability to look after his employees' needs.

On 5 January Napier left the station and noted that tsetse flies were still a nuisance here because of the presence of large herds of game. There were also

---

13 Atcherley (1879) claimed that "doctor" Birch settled near Henry Glynn's shop at Krugerspost after the Sekhukhune Rebellion of 1876, and that he became known as a musician of note using his real name, Louis Julien.
14 Some sources give his name as W. Boly.

a lot of baboons in the area and they caused havoc in Lempke's maize field. En route he ran into a large party of bearers, as well as the "absent" station masters Lempke and Pearce, who were on their way back to their stations with supplies from Lourenço Marques. The last overnight post, Campos de Corvo (which would be run by Nellmapius's partner, Albertos Carlos de Paiva Rapoza), was not completed yet, so they took a route that brought them to the kraal of a black chief (in his report he called it Grey's Camp) about 24 kilometres further on. Here he met a French party, Les Crapauds, and after bathing in a nearby stream Napier decided to move on. They walked about six kilometres in the moonlight before deciding to pitch camp. That night they were pestered by mosquitoes (with fatal results, as would become apparent).

At dawn on 6 January 1876 they set off again and crossed a deep, swampy river (the Matola). The dark, almost black water reached his waist. The last section of their journey led across fertile, cultivated fields. After few hills they at last arrived in Lourenço Marques and he could book into the Royal Hotel, which was under the excellent management of a hospitable Mrs Fernandez. As a last remark in his report, he mentioned that the entire journey from Mac-Mac to Lourenço Marques had taken only eight and a half days. This included his forced wait next to the flooded Sabie and he considered it excellent for a man of his years. The governor of Lourenço Marques and the local merchants received him in a friendly manner and the governor was so impressed with his positive remarks about the fertility of the area that he gave him 2000 acres of land. Napier did not make use of the generous offer as he considered the harbour city, despite its picturesque setting, to be unhealthy because of the nearby malaria-carrying swamps. He therefore decided to return to Lydenburg, but at Pretoriuskop he fell ill with malaria, which he had contracted on the way. From there he was transported to Mac-Mac on a litter in an unconscious state. Soon after his arrival he died, without having regained consciousness.

Napier's valuable report would have been lost had it not been for his good friend, the author D.M. Dunbar, who published it in his book *The Transvaal in 1876*.

References

Atcherley, R.G. 1879. *A trip to Boer Land*. Richard Bentley, London.

Compton, W.G. 1873. Letter to President Thomas Burgers after completion of the Nellmapius transport road. Transvaal Archives. Acquisition TA 1023/128.

Dunbar, D.M. 1881. *The Transvaal in 1876*. Richards, Slater & Co., Grahamstown.

Pienaar, U. de V. & Mostert, M.C. 1983. On the spoor of the old transport-riders. *Custos* 12(9):29–33.

Sandeman, E.F. 1880. *Eight Months in an Ox-waggon*. Griffith & Farran, London.

8.8 A missionary's travels through the Transvaal Lowveld, 1883–1898

P.H. Cuénod, F.S. Cuénod and Y.A. Courtin

Reverend Henri Berthoud, born in 1855 in Switzerland, came to South Africa in 1880. Through his missionary work and his expeditions to Mozambique he became very familiar with the region that is now the Kruger National Park (specifically the area north of the Olifants River).

When he first came to South Africa he settled at Valdézia, the first Swiss mission in the far north of the Transvaal. Valdézia was built on the banks of the Luvuvhu River (Rivubye), in the picturesque area between the later Levubu settlement and the Albasini Dam east of Louis Trichardt (Makhado). Berthoud's calling was the ministry of the Gospel, to look after the spiritual wellbeing of the illiterate blacks and convert them to Christianity.

Preaching and proselytising the Christian Gospel is the first aim of any missionary. Henri Berthoud took his mission seriously and carried it out conscientiously. He had a sharp mind and soon mastered Tsonga, the language of the Magwamba among whom he worked. He also had an insatiable need to know everything about the customs and tribal practices of these people and the land in which they lived. His expeditions to the current Kruger National Park and Mozambique were attempts to establish the eastern boundary of the Magwamba's territory and to find a route linking Spelonken and Lourenço Marques (now Maputo), where the Swiss Missionary Society had another important station.

For the purpose of this work it is necessary to take note of Henri Berthoud's expeditions through the northeastern Lowveld. He normally walked, sometimes accompanied by a European colleague, but always with a number of black guides with oxen or pack-donkeys to transport camping equipment and other supplies. During these expeditions, Berthoud carefully noted down distances and times, and by using his compass he could determine the positions of hills, mountains and other landmarks with remarkable accuracy. A sextant was handy to determine height above sea level. He also used a pedometer to accurately determine distances.

With the extensive notes he collected he was later able to draw up an invaluable map of the northern Transvaal (see Berthoud, H. [1903] – *Map of Zoutpansberg*).

He undertook his first expedition in 1883. During this trip he compiled a complete list of all the Tsonga chiefs and the locations of their kraals between Spelonken and the confluence of the Luvuvhu and Limpopo rivers (the current Pafuri area), and further down the Limpopo to its confluence with the Olifants River. To his surprise he found that the entire area that is currently the northern part of the Kruger National Park was sparsely populated. The Ngumbin region (Shilowa) was inhabited by Pedi (who were later massacred by Gungunyane's warriors), while a single Tsonga chief lived on the Tsende (Tsendze) Plain (which he called the Ngotse Wilderness). The fertile banks of the Great Letaba and Olifants rivers were relatively densely populated, however. Because of malaria, the intense heat and the presence of tsetse flies, this part of the Lowveld was considered to be unsuitable for settlement, especially during the rainy season. The area was also known for its great variations in temperature. One day Berthoud measured a temperature of 0 °C at sunrise, only to find that it had risen to 32 °C within a couple of hours. As a result of these factors, Berthoud always

The Valdézia Mission Station of Rev. Henri Berthoud between Morgenzon and Levubu, circa 1890.
(ACKNOWLEDGEMENT: P. CUÉNOD, ELIM)

tried to restrict his expeditions to August, as it was considered the most temperate and "healthiest" month in the Lowveld.

During his first expedition, Berthoud was particularly impressed with one specific route. They travelled southeast from Valdézia to Ysterberg (Yingwekulu), where they had to decide between a seldom-used route that led directly east and an old hunting road of the Boers that went past the Birthday gold mines. The party picked the last road and could move at a faster pace as it was meant for wagons. It led further east, more or less along the Tropic of Capricorn in the direction of Ngumbin (Shilowa). From here, five hours' hard slog brought them to large river that had not been indicated or named on their map. They had to follow the course of this river for another four hours before they reached the first black settlement. There the local chief informed them that it was the Great Shingwedzi River. By following the course of the Shingwedzi deep into Mozambique Berthoud learned much about the topography and other attributes of this region, as well as of the people who lived there.

The destination of Berthoud's second expedition, which he undertook in 1885, was once again Mozambique. This time he wanted to find a suitable place to establish an outpost, and the most logical spot was Magude's kraal, as he was the senior chief in the area. He was also a Tsonga chief who had succeeded in maintaining a measure of autonomy despite his subjugation to the Nguni high chief Muzila.[15]

This time their route followed another wagon trail of the Boers near Ha-Magoro and Ribanga Hill, and for three days they travelled in a southeasterly direction. Eleven days later they reached the Olifants River after they had walked past Kasteelkoppies and Phalaborwa and crossed the Letaba River. Finding a suitable drift through the Olifants River proved to be a nightmare as it was deep and fast flowing.

On their return journey Berthoud and his companion, one Thomas, followed different routes to determine which was best. While Thomas travelled through a mountainous, arid region that was too steep for an ox-wagon, Berthoud's route ran between the old Boer wagon trail and a road from Marabastad to the Murchison goldfields. It was hardly better than the other as it went through dense thorn-veld, waterless plains and numerous dry river beds and dongas.

On 11 June 1889 Berthoud was once again on his way to Mozambique. This time he was accompanied by H.E. Schlaefli, a young missionary who had just arrived from Switzerland and who studied everything that went on around him with great enthusiasm. He left an interesting report on their journey and in it one finds many titbits his young, unpractised eye found fascinating and exciting. Their continual contact with the Tsongas also provided him with insight into their culture. Schlaefli described their route as follows:

> It (the route) follows a line eastwards from Thabina and follows the Murchison Mountains for a distance of about 30 kilometres to Spitzkop, northeast of Leydsdorp. From here we turned right and moved southwards. Two and a half hours' walking brought us to the Selati River. After we had crossed it, we deviated from the existing routes and footpaths by following the river to its confluence with the Olifants River. We decided to follow the last-mentioned river because it would ensure us of a reliable source of water. The area between the Selati and Olifants rivers is nothing but a number of quarts-strewn hills that go on endlessly, one after the other. This inhospitable world stretches to the Olifants River and the local blacks have a very important name for it … 'Makoarrrrrrr!' [which

Henri Berthoud (1855–1904), Swiss missionary, explorer and founder of the Valdézia Mission Station. Circa 1885.
(WITH KIND PERMISSION OF MR P. CUÉNOD, ELIM)

Kaleka-, Tsale- and Ngodzi hills along the western boundary of the Tsende (Tsendze) Plain, west of Shilowa Poort. Rev. Henri Berthoud travelled past here in 1883 after his visit to the confluence of the Limpopo and Olifants rivers.
(17.07.1984)

---

15   Also called Umzila or Mzila in other sources.

8. ON THE TRAIL OF THE EXPLORERS

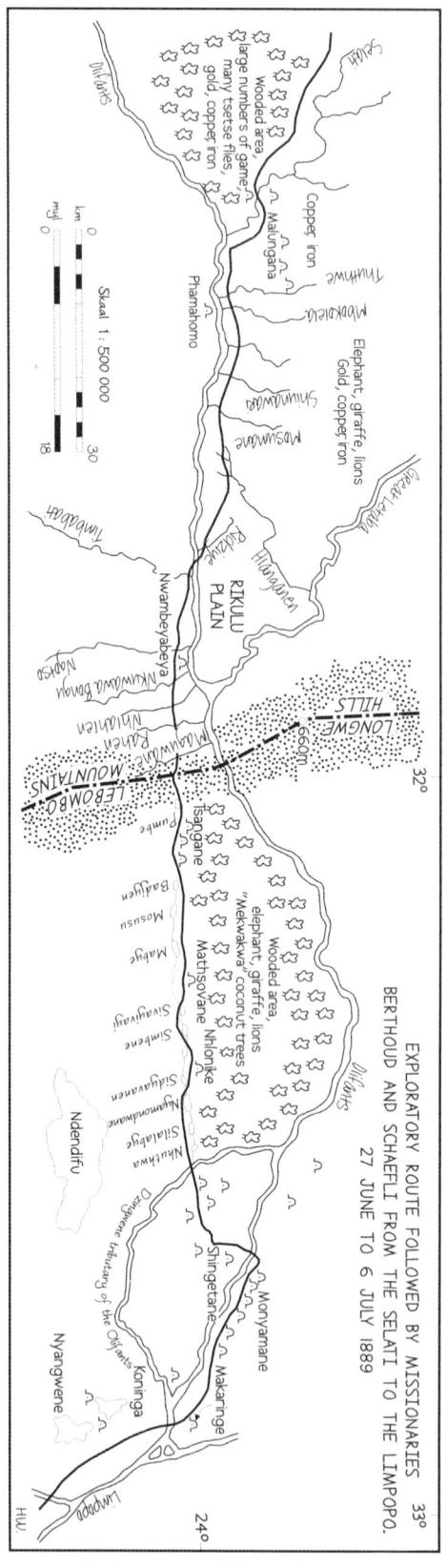

Exploratory route followed by missionaries Berthoud and Schaefli.

means 'land of rocks and stones']. The grass cover is sparse and distributed in small tussocks. There are large, almost impenetrable stands of tree pincushions and woodland landscapes with distorted and stunted little trees [mopane]. Here and there along the river are groups of majestic trees. No people lived in this area, but monkeys, baboons and other game were common. This is the home of the tsetse fly. Whole swarms settled on us and our donkeys like horseflies. When they land, they make a sound like a bee on a flower.

In the Selati Valley we experienced great temperature variations. On 24 June the temperature was 0 °C at sunrise. By the afternoon it had risen to 29 °C and at sunset it was 22 °C. On 25 June the temperatures were 0 °C at sunrise and 27 °C in the afternoon. The nights are bitterly cold. We got very cold at night if we did not keep a large fire going. This is a region that is rich in copper and the local blacks know how to smelt and work it. We found the remains of an old furnace in which metal had once been smelted. Iron ore is also common …

We continued on our journey and found wild grapes and a small shrub that the blacks call 'nandhihane' or 'nandzika' [meaning 'tasty'] growing in the area. It is probably the so-called liquorice wood or something similar. [In fact, Schlaefli was probably referring to *Indigofera bainsii*.] The blacks eat the bark and roots of this plant and say that it is very tasty – hence the name.

As one nears the Olifants River, the hilly landscape changes. It becomes very rocky and broken and makes it impossible for us to follow the western bank of the Selati any further. We are forced to cross this river about three to four kilometres above its confluence with the Olifants. We reach the Olifants by following a rocky path, strewn with iron ore. This footpath meanders between rough hills that look as if the large boulders had been stacked by the giant Titan.

At last we reach the Olifants River, but where we expected to find a drift is only a deep pool – the home of a herd of hippo. As a result we are obliged to follow the northern bank of the Olifants, but before our departure we quickly measure the circumference of a majestic wild fig tree. The trunk's circumference was 4½ arm spans [a single arm span of mine was later determined to be 184 centimetres]. The rest of the tree is proportionally large. It is not possible to follow the river's many twists, so we take a short cut away from it and try not to deviate from a direct easterly line.

While following the bed of the Olifants River, we were surprised to notice black stripes in the white river sand. Closer inspection revealed it to be fine iron ore deposits.

Upon arrival at Phamahomo [near the mouth of the Tshutshi Spruit], we find a suitable drift for ox-wagons. But before one will be able to take a wagon across this granite rift, which stretches right through the river, it will have to be evened out with stones, which will take a lot of labour and money. We have neither the time nor the money, so we trek on to find another drift.

From Phalaborwa the landscape changes and becomes more wooded … a real woodland. Our route turns northeast and again we enter an area abounding with big game – especially elephants. Despite our desire to admire these pachyderms in their natural state, we see only their spoor and dung heaps.

We gradually cross four tributaries of the Olifants River; they have wide sandy beds but little water. The two types of mountain palm found here give the landscape an exotic look. Eventually we reach the old trade route to Mpalaora [Phalaborwa]. We spend the night along the upper reaches of the Hlangane River [Nhlanganini] and cross it the next morning. Then we come to the Rikulu Plain [Bulweni], where we find limestone and tsetse flies. Reverend Berthoud and Thomas also travelled through this area with their ox-wagon four years ago and did not come across a single fly.

After we crossed the Ndzyo [Ndziyo], a spruit in a deep gorge, we again reach the Olifants River. The river here is noticeably wider than at Phamahomo and near the Selati mouth – 100 metres water and another 150 metres of sandy bed.

For the first time on our journey we hear a lion roaring – it will also be the last time. The king of the bushveld was far from us, however, or maybe it was out of fear for our rifles that he uttered his 'Shivindhi mayo' so softly. The words 'Shivindhi mayo' is the interpretation of a lion's roar as translated by the Tsonga. It simply means "a piece of liver for me" and can be explained by the fact that lions always eat the liver of their prey first, before anything else.

After we had waded through the Timbavati, we left the Olifants River behind and set off into a southeasterly direction and ended up in an area that was reasonably bushy, but with trees of more or less the same size. The 'Makoarrrrr' [shrub mopane] has disappeared. We crossed the Ngotso Spruit, as well as the Nkuwa-wa-Bango [Shipembane], the Hlahleni, the Raken [Bangu] and the Maguwane [Ntsumaneni], all tributaries of the Olifants River. They are all running, except for the last-mentioned. We spend the night next to this spruit and have to go to bed hungry as we cannot cook.[16]

The next day, after having walked for an hour and a half, we reached the Lebombo Mountains … The plateau of this mountain range drops imperceptibly to the lowlands of the Olifants and Limpopo rivers. The terrain on top of the mountain differs vastly from that through which we travelled so far. It is sandy and we would travel through sandveld up to the mouth of the Limpopo and from there along the Indian Ocean to Delagoa Bay.

Rather than to pay too much attention to Schlaefli's description of their journey through Mozambique to the coast, it is more important to take note of the trees and shrubs that he listed on their journey to the Mozambican border. He mentions the following, among others:

Our doctor draws our attention to a smallish tree with a thin trunk and small, thick leaves with rounded tips. This is the 'salt tree' [*Salvadora angustifolia* var. *australis*]. We taste the leaves and they indeed have a salty taste. The blacks burn the leaves to ash, wash the remaining ash and then let the water evaporate. This provides them with spotlessly white salt, which they cast into small cone shapes …

---

16 Along this part of the route that Reverend Berthoud and Schlaefli followed, Parks Board officials found two small stone cairns (isiVivans) next to the Ntsumaneni Spruit in 1986, about a kilometre west of the Briscoe baobab tree – an indication that this route was also often used by black travellers.

Two other trees also draw our attention, thanks to their peculiar fruit. One looks very much like a large, leafy oak, but the fruit resemble the sausages made in Bologna [*Kigelia africana*]. The other tree is much smaller and the fruit, which are inedible, are shaped like pears – some become as big as a baby's head … [possibly *Rothmannia fischeri*, *Strychnos madagascariensis* or *Monodora junodii*].

A third type of tree belongs to a totally different species. It is a type of thorny mimosa that reminds the black people of the whites – not because of the white thorns, but as a result of the colour of the trunk. Nothing appears stranger and more mysterious than a thicket of these 'mekelenga' trees [*Acacia xanthophloea*]. At closer inspection the trees show a smooth bark with a soft green colour, covered in a fine, pollen-like dust that feels like talcum powder.

With this report, Schlaefli shed new light on an area that had until their journey been largely unknown and badly documented.

Ribye-ra-Khubyane potholes in the Shingwedzi River, a holy place of the Tsonga people which Rev. Henri Berthoud visited in 1889. (24.04.1984)

In July 1890 Berthoud spent a whole month travelling to the area between the Luvuvhu and Limpopo rivers. He studied the local population as well as the features of this unique region with the aim of establishing an outpost here. He selected a high hill called Hutuene, the furthermost northeastern foothill of the Soutpansberg range, that towers over the Limpopo and Luvuvhu valleys.

It was thanks to his meticulous notes and annotations on the topography of this area that Berthoud was of invaluable help to Fred Jeppe, the official cartographer of the ZAR (South African Republic) and President Kruger. He could provide drawings of the topography with exact indications of landmarks and place names and could therefore also shed light on the points of dispute about the border between Mozambique and the ZAR (Jeppe 1890–1895).

To place the activities of the Swiss mission in Mozambique on a stronger footing, Henri Berthoud in 1981 visited the powerful high chief of the Tsongas, Gungunyane (the son of Mazila), whose main residence was then at Mondlagazi. This kraal was on the left bank of the Limpopo River, a fair distance north of its confluence with the Olifants. This potentate had so much influence that even the Tsonga chiefs and high chiefs in the Transvaal obeyed his every wish and command. He was a feared yet respected man, with tremendous status among the Tsonga of the Lowveld.

Upon his arrival at the kraal of this influential leader, Berthoud was allowed to pitch his tent on the outskirts of the kraal. He was also given an ox to provide him and his people with food while waiting for an interview with the high chief. The long wait was worth his while as Gungunyane gave the missionary a friendly reception and let it be known that he was not opposed to allowing missionaries among his subjects. After this exhausting journey, Henri Berthoud's health deteriorated noticeably and he was unable to undertake any more long journeys on foot. There was, however, one exception. In 1898 he took to the road again to see the place known among blacks as Ribye-ra-Gudzani (also Ribye-ra-Khubyane, now known as the Red Rocks in the Kruger). He had been hearing about this place for years and how black travellers would not go past it without making an offer to their god, Khubyane – even if was only a pinch of snuff, a piece of meat or a length of cloth. Then they would pray:

> We call to you, Khubyane, master of the universe, creator of the trees, the rivers, the grass, the mountains, the clouds that give us rain, the animals and the people. We offer these things to you so that you can guide and protect us along the way; give us meat, the food of the wilderness; give us joy, a good road and the sleep that we need to live.

If they had been lucky enough to kill an antelope on the way, they would cut off a choice piece of meat and sacrifice it to Khubyane with songs of praise, placing it either at the foot of a nkanyi (marula tree) or under a rock (altar).

On his way eastwards to this holy place, Berthoud was struck by the frequent changes in vegetation along the different rivers and spruits. So the Tshange (Tshanga) Spruit was covered in a beautiful reed bed through which a pleasant, crystal-clear stream flowed. But the Nkokodzi Spruit, only a couple of kilometres further east, was overgrown with dense stands of mlala palms (*Hyphaene natalensis*) and high, tree-like ncindzu palms or wild date palms, as they are known today (*Phoenix reclinata*). The spruit was not running. Like many of the smaller rivers in the Lowveld it did not flow during the winter months, but had numerous pools with standing water.

After having been on the road for 12 days, Henri Berthoud reached the remarkable Ribye-ra-Gudzani – a beautiful, weathered outcrop of red sandstone in the Shingwedzi river bed, with deep pot-holes worn away by the water over the centuries. On the banks of this river were large trees that provided the weary travellers with welcome shade.

To reach this place Berthoud had to cross a number of spruits and tributaries: the Marivane, which flows into the Nsama, later the Matsinyane and then the Madzimbane (Maguweni), the Shongololo (Nkayini) and the Tshange (Tshanga). The daytime temperatures during this journey were pleasantly cool. After supper the travellers sat talking around a fire and it was during these peaceful hours that Berthoud learned much about the local blacks' customs, language and traditions.

Berthoud did not get to experience the further development of the Lowveld. The Anglo-Boer War left the area ravaged and plunged into misery. The Swiss missionaries had to obtain passes from the British occupational forces at Fort Edward if they wanted to travel anywhere. (In the Skukuza Archives there is a print of such a pass, which Berthoud had to obtain on 3 June 1901 to travel from Valdézia to Pietersburg [Polokwane], signed by the infamous Captain A. [Bulala] Taylor).

Repeated bouts of malaria caused his health to decline and weakened him physically. He nevertheless devotedly carried on with his missionary work until his death at Valdézia in December 1904. As he was no longer able to undertake long journeys, he dedicated his time to preaching the gospel, teaching the locals to read and write, doing translations and drawing up a new Tsonga grammar. He also collected words for the compilation of a Tsonga dictionary.

References

Berthoud, H. 1903. Map of Zoutpansberg. H. Kümmerly & Frey, Berne.

Jeppe, F. 1890–1895. Series of letters to Missionary H. Berthoud at Valdézia. (References by P. Cuénod.) Personal documents and notes of H. Berthoud with kind permission of P. Cuénod.

Schlaefli, H.E. 1889. Route followed by H. Berthoud and H.E. Schlaefli from the Selati to the Limpopo River. (Unpublished manuscript, with permission of P. Cuénod.)

Wongtschowski, B.E.H. 1987. *Between Woodbush and Wolkberg*. Review Printers, Pietersburg.

Yates, C.A. 1944. 'Jawawa' In: *Varia*. Unpublished manuscript. Skukuza Archives.

Chapter 9
Prospectors and delving
JOHAN VERHOEF

The southern part of the African continent has posed an almost mystical challenge to explorers and adventurers since the earliest times. When the sea route to the East around the Cape of Good Hope was found, the never-ending rumours of the mighty Kingdom of Monomotapa and its gold, exported via Sofala,[1] could finally be investigated.

What compelled Europeans to undertake long, dangerous journeys into the unknown? It was mostly a craving for adventure, awakened by the discovery of riches, that led to those historic journeys. At first the sea routes were of primary importance, but gradually explorers started taking an interest in these strange continents and their uncharted interiors.

It took a long time before Europeans reached the northeastern part of the Transvaal.[2] The Cape was first occupied by Europeans in 1652, who gradually moved into the interior. This was followed by permanent colonisation. Yet, during this time some journeys were undertaken with the sole aim of exploring the unknown, recording observations and returning with astonishing tales of adventure, discovery and the weird and wonderful. The reasons why Westerners took so long to get a foothold in the Lowveld of the former Transvaal, were deadly, if unusual: malaria, transmitted by mosquitoes (a fact not known at first); and nagana, transmitted by the tsetse fly. People succumbed to the fever, which exacted a very high toll, while nagana killed off their animals. This is how the first pioneers and explorers got to know the Lowveld – as an enchantingly beautiful region, an open bushveld rich in game, but largely protected by its own traps and endemic diseases.

Apart from these early explorers and adventurers, two groups of Europeans eventually entered the Lowveld and surrounding areas, driven by different needs and motivations. There were the Afrikaner Voortrekkers who wanted to make it their own and settle there permanently, and then there were those who came to hunt, to prospect and to mine – all in the pursuit of wealth. It is this last group whose fortunes and misfortunes, successes and travels across the Escarpment and the Lowveld are discussed here and whose activities were a major contribution to the history of this part of the Transvaal.

When the Dutch East India Company in 1721 took Delagoa Bay, or Fort Rio de Lagoa,[3] from the Portuguese, these occupiers had already come into contact with gold and iron from the interior. Native inhabitants had shown them gold dust and ostrich feathers and used these as trade goods, and had iron objects that came from the interior. Rumours of an "iron mountain" must have stirred the imagination, and it wasn't long before an expedition departed for the interior in August 1723 in search of the sources of the gold and iron ore.

This expedition was led by Jan Christoffel Steffler, lieutenant to Jan van de Capelle, who was second in

Entrance of an archaic mine shaft slightly south of Shilowa Poort.
(OCTOBER 1983)

1 Now Beira.
2 This now forms part of the Mpumalanga & Limpopo provinces.
3 Later Lourenço Marques, now Maputo.

command at Fort Rio de Lagoa. The group consisted of 20 men and apparently managed to reach the Lebombo Mountains between Goba and Lomahasha in Swaziland. Here they were attacked by hostile warriors and Jan Steffler and Eldret Stigling were killed in the skirmish. The expedition returned without having achieved its mission, but it was nevertheless able to ascertain that there was coal in the Lebombos and that the important "Iron Mountain" was nine days' journey on foot from the Bay, in the northern Transvaal. The gold of which the Dutch had heard at the Bay could have come from what later became known as the Kaapsche Hoop fields, where nineteenth-century miners found signs of old workings. Similarly, the iron could have come from the Malelane area, as the first Europeans also found signs of earlier mining activities here.

The De Cuiper expedition from Delagoa Bay to the Lowveld in June and July 1725 had similar aims – to explore and find the sources of precious metals. It is known that this expedition managed to reach the Gomondwane bush before locals forced them to beat a hasty retreat. The honour of being the first Europeans to set foot in the Transvaal therefore belongs to them. This expedition is discussed in detail in chapter 3.

To this day slag, ore and clay pipes can be found in the southwestern part of the Kruger National Park, indicating the presence of earlier iron smelting operations. A variety of remains were found at the well-known Mangake Hill in the Stolsnek ranger section, as well as terraces where people had lived. Signs that the area around Skukuza was once inhabited by people speaking African languages have recently been uncovered, along with iron slag and ore. Indications of primitive mining activities have also been found further north in the Park and it is now known that both iron and copper ore had been mined and worked in the Phalaborwa hills, as well as further north in Venda. The early inhabitants of the current Kruger National Park mined materials such as red and yellow ochre at places such as Ngodzi Hill and along the Shishangani Spruit near the Nwanedzi ranger post. They used the pigments to paint and decorate their clay pots.

These first explorers and pioneers found signs and remnants of prehistoric cultures and even encountered black tribes, but it was the prospectors who came later that had both the knowledge and the inclination to search for the source of the riches. The story of Mapungubwe is a good example of this: Only fairly recently did archaeologists ascertain (after gold objects had been dug op during excavations there) that this northern Transvaal tribe had access to gold.

In the middle of the nineteenth century the eastern Transvaal (now Mpumalanga) was still relatively unexplored by Europeans. By 1830 there was a Portuguese trading post and settlement at Delagoa Bay, surrounded by swamps and plagued by malaria. A

Ranger A. Labuschagne at the entrance of the antique ochre mine on the southern bank of the Shishangani Spruit near Nwanetsi ranger post.
(11.11.1986)

President Thomas Francois Burgers, who headed the ZAR (South African Republic) government between 1871 and 1877, the years during which both the First Sekhukhune War as well as the discovery of gold at Pilgrim's Rest had taken place.
(ACKNOWLEDGEMENT: TRANSVAAL ARCHIVES)

9. PROSPECTORS AND DELVING

Pioneer prospector Tom McLachlan, one of the first people to discover gold in the Eastern Transvaal, 1892.
(ACKNOWLEDGEMENT: TRANSVAAL PROVINCIAL MUSEUM SERVICES)

The market plain of the diggings at Mac-Mac during President T.F. Burgers's second visit in 1874.
(ACKNOWLEDGEMENT: TRANSVAAL ARCHIVES)

few years later the Voortrekkers entered the Lowveld in their quest for a route to the east coast – with disastrous results (as has been described elsewhere).

When President T.F. Burgers was elected president of the ZAR in 1872, this region was still sparsely populated. White settlements were mainly restricted to higher-lying areas. The Lowveld was relatively untouched, uninhabited by whites and protected by malaria and tsetse flies. When President Burgers assumed office, the ZAR was beset with internal turmoil and state coffers were empty, but the discovery of gold would soon play a decisive role in restoring the ZAR's fortunes. This in turn would lead to imperialist action by the mighty British Empire and even more problems.

The time preceding the discovery of gold in the Transvaal was one of intensive prospecting. Because rumours and signs of the presence of gold had already been circulating for more than a century, it was to be expected that "experts" would eventually appear on the scene. In 1852 the first so-called geologist to visit the Transvaal was John Henry Davis, who disappeared from the scene with few recorded facts about his finds. According to one version he had discovered gold, but was asked to leave out of fear that England would annex the region. Others had it that he had prospected for gold and precious stones, but found nothing. The next year Pieter J. Marais obtained permission to prospect for gold. He found signs of this metal near what is now Johannesburg, as well as alluvial gold deposits in the Crocodile and Jukskei rivers. Marais reported his finds to the Volksraad, which handled the matter with the utmost secrecy. Marais prospected in the Transvaal for another two years (still secretly), but everybody knew what was happening. He reported to the Volksraad again and claimed that there was more gold in the Transvaal than in America and Australia combined, but because of his lack of geological expertise he was unable to proffer the necessary proof. His prediction would eventually become true, but it would take another 30 years before the Witwatersrand would reveal its incredible treasures and 20 years before the eastern Transvaal would yield the first really meaningful quantities of alluvial gold.

In the meantime, the well-known German geologist and explorer Carl Mauch had started his expeditions into the northern interior. Armed with his considerable knowledge of geology, botany and zoology, he traversed large parts of southern and eastern Africa, spurred on by his inquisitive nature. Although he left behind comprehensive records and reports, he also failed to find the goldfields. Nevertheless, today he is considered a pioneering geologist and even the "father" of the gold mine industry, mainly because of his predictions about the mineral riches of South Africa.

Mauch and Henry Hartley explored the area to the north and south of the Limpopo River. He discovered gold in the Tati River in 1867, as well as old workings north of the Limpopo. In May 1871 he became the first European to document the mysterious Zimbabwe Ruins, although he was not the first to see them. Mauch returned to the Transvaal with a theory that would later prove to be true; namely that everything pointed to the fact that those parts of Africa once had a prosperous gold industry and that this would again be the case in the future. Of greater importance, however, is that the rumours of gold, which until then had been rather vague and unreal, now became significant. More and more prospectors and adventurers entered the Transvaal to search for the northern goldfields after the findings of Mauch and Harley became known. Hartley soon went back there, but Mauch returned to the eastern Transvaal. Mauch would come very close to realising his dream but it was not to be, as he died at the age of 37.

By 1869 the Transvaal was overrun by prospectors. With ox-wagon, donkey and even on foot, they ventured into the unknown. At the time interest was centred on the presumed goldfields north of the Limpopo, but one small group, namely Edward Button, Parsons and Sutherland (the last two were known only by their surnames), focused on the Lydenburg area. They had more faith in Mauch's opinions of the area and shortly after Tom McLachlan joined them they got their reward, probably in the vicinity of the Spekboom River.

Afterwards the group split into two. Button and Sutherland moved northwards to the area that would later become known as Murchison Range (and much later as the famed Selati goldfields) while McLachlan and Parsons investigated the streams and creeks in the current Graskop area. In 1870 Button found reef gold on the farm Eersteling and word spread. He informed the magistrate in Lydenburg of his discovery in January 1871 and also mentioned the names of McLachlan and Sutherland. Button became the first person to obtain a concession from the Volksraad to mine gold ore and established the first mining company, the Transvaal Gold Mining Company, with capital from England. He could not get that much gold out of this first gold mine in the Transvaal with his imported twelve-stamper battery, but it was still enough to cause a gold rush in this and the surrounding areas. The Volksraad hurriedly had to adapt its mining and prospecting laws and regulations. In the meantime President Pretorius was succeeded by T.F. Burgers.

On 5 February 1873 the magistrate of Lydenburg could tell the Executive Council in Pretoria that he had seen enough evidence to prove that McLachlan, Parsons and Valentine had indeed found alluvial gold east of the town. They wanted to register their discovery to be considered for the compensation described in the relevant law. The die was cast. Rich alluvial gold deposits were found on Hendriksdal, south of where Sabie is currently situated, with an even richer deposit on the slopes of Spitzkop – the later Malieveld Mine. Magistrate Jansen went to investigate and returned with three or four ounces of gold, mainly nuggets. He asked the government to take immediate steps before the anticipated rush started. He also reported that the farm Graskop had been bought by Abel Erasmus (the same Abel Erasmus who would later become such a well-known figure in the history of the Lowveld). On 14 May 1873 the area was proclaimed as a goldfield and the first true gold rush started.

Initially it was very chaotic and Jansen had his hands full with the necessary administration. Claims had to be registered, levies collected, and law and order maintained, apart from dozens of other problems. After he had appointed one Major Macdonald as gold commissioner and the diggers could start working, relative peace and a short-lived prosperity prevailed.

In the midst of this rush, during which diggers from across the world converged on the goldfield, the gold pioneer Tom McLachlan bought the farm Geelhoutboom with even richer gold deposits than his first. He later also became justice of the peace at the goldfield.

Although Magistrate Jansen probably found these developments rather overwhelming, it is common knowledge that he acquitted himself quite well. He even succeeded in convincing President Burgers to visit the goldfields, with the result that the president arrived at Spitzkop in August 1873. The hotchpotch of nationalities and characters viewed the president with a certain amount of scepticism and suspicion, but after two days his fluent English and pleasant personality had won them over. He also visited Major Macdonald's camp, where he encountered so many surnames starting with "Mac" that he named the place Mac-Mac forthwith. Although this particular area was initially registered as the New Caledonia Gold Field, the name Mac-Mac stuck – and it is still in use today.

President Burgers returned to Pretoria determined to buy the farm Graskop from Abel Erasmus and to lay out a central miners' town there, but subsequent developments put paid to his plans.

In September that year another of the celebrated characters of those days, Alec Patterson, decided to pack up, abandon his exhausted claim and leave the overpopulated and busy goldfields next to the Mauchsberg road. He set off over the nearest hill in a northeasterly direction, probably followed a game path and eventually reached a valley in which he found peach trees, interestingly enough. The stream flowing in the valley eventually flowed into the Blyde River further down, although he was unaware of this. He instinctively started mining and even with his first pan Wheelbarrow Alec realised that his biggest dream had come true.

Shortly afterwards, another digger, William Trafford, left Mac-Mac and happened on the same valley. After finding his Eldorado, according to legend, he called out: "Pilgrim's at rest …" He probably first heard these words from a group of Natal diggers who had arrived at Mac-Mac and referred to themselves as pilgrims – a description later adopted by all the diggers. This was followed by a local gold rush, but when it became known that the new goldfield was not only rich in alluvial gold but also produced nugget gold, diggers hurried there from the far ends of the earth. About a third came from Kimberley's diamond fields and some even from as far as America and Australia. Long and eventful journeys were undertaken from Cape Town and the interior, as well as from Delagoa Bay through the Lowveld, to reach the new goldfield, stake claims and start becoming rich as quickly as possible.

The diggers spread out across the valley where Pilgrim's Rest is today, over the farms Ponieskrantz, Driekop and Grootfontein. The commissioner moved his office there because Pilgrim's Rest now had the biggest concentration of diggers and was already a busy little town by 1874. President Burgers visited the goldfields for a second time that year in order to sort out matters relating to the legal owners of the farms and their rights, as well as to explain the newly promulgated mining laws. As with his previous visit to Mac-Mac, he had to deal with a fairly aggressive group of miners, bad roads and full rivers.

He eventually succeeded in restoring law and order and legend has it that the diggers treated him on his last evening there to a banquet prepared by a chef called the Bosun. Alcohol flowed freely and the president left as a friend to one and all.

Digging at "Middle Camp", Pilgrim's Rest, circa 1874.
(ACKNOWLEDGEMENT: TRANSVAAL ARCHIVES)

Halfway House claim, Pilgrim's Rest goldfields, circa 1874.
(ACKNOWLEDGEMENT: NATIONAL CULTURAL HISTORICAL OPEN-AIR MUSEUM)

The "golden" years for diggers at the Pilgrim's Rest goldfields were 1874 and 1875. There was alluvial gold in abundance, and the hard-working prospectors exhausted one claim after the other. It is said that most of the hundreds of diggers made enough money to live comfortably for the rest of their lives, but eventually no more than about 20 left the goldfields rich men. The rest would all die in poverty, either here or on some other goldfield. Every name became a legend and this adventurous, romantic era is remembered for characters such as the Bosun, French Bob, Sailor Harry, German George, Charlie the Reefer, Wally the Soldier, Black Sam and many more. There were soldiers, sailors, bank clerks and even a dentist – all searching for gold and a better life. Many of the diggers were of Welsh, Scottish or Irish origin and the old cemetery at Pilgrim's Rest is proof of the varied origins of this group of incurable optimists. Some of the diggers would later become famous: the young Percy FitzPatrick arrived at the diggings only in 1884, but he would become known thanks to his famous book *Jock of the Bushveld* and his status as a public figure along with Cecil John Rhodes, Barney Barnato, Alfred Beit, Lionel Phillips and others. He also played a role in the founding of Johannesburg. The Hungarian Alois Hugo Nellmapius was also a digger – one of the few to use his profits wisely. He was the first industrialist in the Transvaal and on his own initiative he had the old route between Lydenburg and Delagoa Bay cleared to make it more passable to ox-wagons. The drift through the Crocodile River in the south of the Kruger National Park, where the old trade route crossed the river, still carries his name. Tom McLachlan[4] played a leading role at the diggings from the beginning and was a respected figure.

The Glynn family was also well known in the area, not as diggers but as gold buyers. They owned land on the other side of the mountain and regularly provided the diggers with supplies. Glynn senior's son, Henry, became well known in the Escarpment and Lowveld areas, especially after he had bought the farm Grootfontein and the family moved to Sabie. They founded Glynn's Lydenburg Company and it is estimated that gold to the value of eight million pounds (then the equivalent of R16 million) was mined here over a period of 60 years. It is also estimated that the Pilgrim's Rest Valley produced gold to the value of £1,5 million and that 1500 diggers had gathered here at the end of 1874. Gold nuggets were also found, but represented a much smaller percentage of the total gold finds. Yet some remarkable nuggets were found: Hugo Nellmapius found one weighing 123 ounces, and then there were the Lilley nugget of 119 ounces and the Breda nugget weighing 214 ounces. There is also a legend of a "monster" nugget, found under a rock at the bottom of a spruit, which apparently weighed 11 kilograms. After a few years the best claims were exhausted, the diggers had to move to higher-lying claims and production costs rose drastically. As a result most were forced to move to other goldfields – just as poor as when they had first arrived.

After the allocation of mining concessions in 1881 by the ZAR government and the founding of mining companies, quartz reef mining started in the hills above the rivers and creeks. Some of these small mines were active until recently. Interesting names such as Theta, Jubilee, Clewer, Peach Tree, Beta, Brown's Hill and Chi were given to these mines and in the history of Pilgrim's Rest they represent a later, but equally interesting period.

Portuguese Reef north of Sabie, in the vicinity of Devil's Knuckles (Long Tom Pass), got its name because the prospectors found signs of old workings there and assumed that the Portuguese had mined there. The possibility exists, however, that these were prehistoric mines that can be linked to those at Kaapsche Hoop.

Pilgrim's Rest during the days of the gold rush, circa 1880.
(ACKNOWLEDGEMENT: TRANSVAAL ARCHIVES, PRETORIA)

---

4   Some sources spell his name MacLachlin.

9. PROSPECTORS AND DELVING

The diggers lived hard but adventurous lives. The first few months after the rush, first at Mac-Mac and then Pilgrim's Creek on the farm Ponieskrantz, were quite disorganised. The first thing they did, of course, was to stake and register their claims. Without exception, the diggers who arrived there were so poor that their food – in most cases no more than tea and coarse flour – had to be bought with the proceeds from their first finds. After that, they had to provide themselves with some form of shelter, which consisted mostly of humble little tents. Old photographs show that there were hundreds of tents on the slopes along the river.

The wealthier prospectors lived in wattle-and-daub huts. Only after the first few years more permanent structures of wood and zinc were erected in Pilgrim's Rest by traders, bar and hotel owners and others. By 1884 it started to resemble a town, complete with streets and erven. The gold commissioner and his diggers' committee maintained law and order. Funnily enough, although there was a conglomeration of rough characters, a few basic principles were always respected. The first and most important was the rule against theft, which was swiftly and mercilessly enforced. A gold thief faced immediate corporal punishment, his hair and beard were shaved on one side and he was banned from the diggings.

In the history of Pilgrim's Rest one such banned gold thief tried to return, but he was sentenced to death after being caught out. His nameless grave can still be seen in the cemetery – it is the only one that is orientated north-south amid all the others facing either east or west. The road between Pilgrim's Rest and Ohrigstad is known today as Rowerspas (Robbers' Pass) after a coach carrying gold to the value of £14 000 (the then equivalent of R28 000) was robbed. The Zulu driver of the coach apparently ran to the house of a Mr Muller and, with a couple of Boers in tow, unsuccessfully searched for the robbers. Two years later the driver recognised one of the criminals, who was arrested and sent to jail, but the gold was never found. In 1876 a second robbery was planned, but it is said the robbers lost their way in the fog.

Nobody was allowed to work after sunset or on Sundays, however much they may have wanted to – especially those who had profitable claims. The tent community was divided into three sections: up-, middle and downtown, and apart from the regular digger tents, bar tents were the most common. Digging was hard work – they worked up to ten hours a day – and in the evenings the bars did good business. In 1875 there were no fewer than 11 canteens and competition was rife among the owners to make a living.

An interesting fact is that gold was the accepted currency. Every trader and canteen owner possessed a good scale and gold dust was tendered. The gold was measured very carefully and often a digger would give them his entire gold supply as credit for alcohol and provisions. This meant that some diggers did not work for long periods, until their credit was spent and they were forced to return to work. The gold weight measure in use was the so-called pennyweight (24 grains), or 24/1000 pounds Troy weight, and was worth between 3s 6d and 3s 9d. according to the then gold price of £3.10s (then the equivalent of R7) an ounce. This meant that a bottle of square-face gin cost two pennyweights of gold, while a bottle of Cape brandy was a little more expensive. Food was the costliest at £3 a bag of maize meal, £4 a bag of coarse flour and 2/6 for a tin of milk. A box of matches was 6d. A prospector therefore had to find at least one ounce of gold (£3 10s) a week to even come close to making a living. The Australian diggers, who worked

The nameless north-south lying grave of a robber in the old Pilgrim's Rest cemetery bears witness to a romantic era in the history of the Transvaal.
(06.09.1986)

the hardest and were the most skilled, could at times maintain a daily yield of up to 8 ounces.

The high prices charged for essentials were due to the high transport costs. Usually a digger carried a few rations with him, but after the establishment of Pilgrim's Rest a plan had to be made to get the necessary provisions there. Two routes served the area. The first ran from Pretoria via Lydenburg and a second from Delagoa Bay across the inhospitable and dangerous Lowveld, past Pretoriuskop to the diggings (i.e. the Nellmapius road, laid out on the old Albasini/Karel Trichardt trade route).

Another road to Lydenburg came from Durban (Port Natal) past Lake Chrissie. It is therefore understandable that supplies were not only expensive, but also hard to come by. Although the route through the Lowveld was by far the shortest, it was a nightmare thanks to malaria and tsetse flies. Several attempts were made to maintain the trade route and one of the better-off and more enterprising diggers, Nellmapius, obtained a concession from the government to open a route and settle station masters at the outspans. This was an attempt to improve the procurement of goldfield supplies. His road followed the routes used by Trichardt and Albasini and the stations along the way were Campos de Corvo, Progresso de Guedes, Castilhopolis, Furley's Drift, Nellmapius Drift, Hart's Shop, Pretoriuskop, Burger's Hall and others, as described elsewhere. President Burgers was sympathetic towards the diggers' cause and tried his best to find a solution to the transport problem. In as early as 1875 he started to investigate the possibility of a railway connection with Lourenço Marques, but in the meantime the diggers had to cope on their own. It is no secret that the ZAR was almost bankrupt in those days, and President Burgers was confident that the discovery of gold would solve their problems. But the whereabouts of all the gold remains one of the best kept secrets of all times, since it definitely did not go to the government of the time. The diggers used their gold as currency, and the merchants must have sold it somewhere else, usually to gold buyers. Apparently the banks in Natal benefited the most as there was not much confidence in the currency of the ZAR and it had little value. It can therefore be accepted that most of the gold left the country.

During his visit to the goldfields in 1874, President Burgers was advised by a Mr Perrin, a Swiss assayer, to start a mint in order to keep the gold in the Republic. Burgers sent 22.25 pounds of gold (10 kilograms) to England to be minted – the result of this was 837 coins, the so-called Burgers pounds. Because the project was not very enthusiastically received by the Volksraad, it never went any further.

Gold from Pilgrim's Rest was also used for the

The main street of Pilgrim's Rest in the 1880s.
(ACKNOWLEDGEMENT: INFORMATION AND MARKETING SERVICES, TRANSVAAL PROVINCIAL ADMINISTRATION)

The well-known Joubert Bridge across the Blyde River just east of Pilgrim's Rest which had been built in 1896/1897 and almost washed away during a large flood in January 1909.
(06.09.1986)

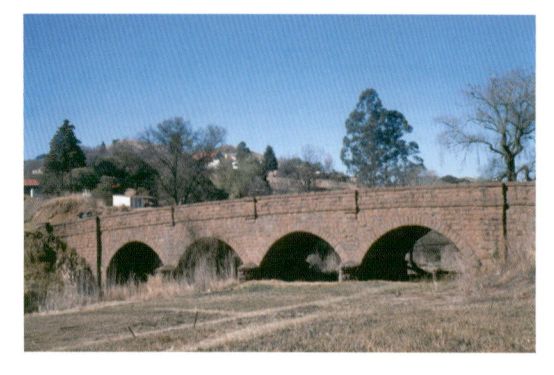

production of Burger Crosses, exceedingly valuable decorations, two of which were handed to a couple of pioneer women: Mrs Tom McLachlan, for unselfishly ministering to the sick on the diggings, and Mrs D. Austin, who had nursed the wounded during the Sekhukhune War of 1876.

By 1877, the year of the British annexation of the Transvaal by Shepstone, alluvial gold was becoming more difficult to find and the diggers had to work much harder and dig deeper to work claims productively. More and more packed up and left for other goldfields. For a while they hoped that the new British government would do more for diggers, such as improving the transport system and getting provisions there more cheaply, but Shepstone himself informed them that the railway would never be built.

For those in search of alluvial gold, the writing was on the wall. It became clear that to mine the alluvial and quartz reef gold of the area, specialised mining techniques and heavy machinery were essential. This heralded the era of the mining companies – also known as the concession period.

In the meantime, the Republic had regained its freedom in 1881 after the Transvaal War of Independence. The new president, Paul Kruger, who was initially elected from the triumvirate government, immediately took steps to improve the economy. D.H. Benjamin, a financier from London, was awarded a concession to develop the Pilgrim's Rest goldfields, with the proviso that all claim owners be bought out and the government compensated. It took Benjamin two years to achieve this. The Transvaal Gold Exploration and Land Company was founded in 1882 and mined its first gold in 1883. In 1894 this company produced gold to the value of £69 000 sterling.

In 1886 the Witwatersrand goldfields were discovered, and what with the rich Kimberley diamond fields and the Barberton goldfields, which had attracted hundreds of diggers in 1882, the time was right for some new names in the mining industry to step to the fore.

Lionel Phillips (the partner of diamond magnate Alfred Beit) negotiated with Benjamin in 1889, obtained the rights to the Pilgrim's Rest goldfields and founded the Lydenburg Mining Estates, which would become the well-known Transvaal Gold Mining Es-

Ox-wagons in Pilgrim's Rest's main street after the Anglo-Boer War.
(ACKNOWLEDGEMENT: INFORMATION AND MARKETING SERVICES, TRANSVAAL PROVINCIAL ADMINISTRATION)

tates with properties and mining rights in large parts of the eastern Transvaal.[5]

This was followed by the Jameson raid and the failed attempt by influential persons (among others Percy FitzPatrick) to topple the Kruger government – the precursor to the Anglo-Boer War.

Phillips and two other directors of the Transvaal Gold Mining Estates (TGME), Percy FitzPatrick and Abe Baily, were taken into custody, fined and banned from the Transvaal. This slowed down the activities of the company somewhat, but with new directors it was business as usual and the mines in the vicinities of Sabie and Pilgrim's Rest could continue to produce gold. The Beta and Theta mines became famous as some of the richest "small" mines. Beta produced gold until 1971, a production period of 85 years!

Before the outbreak of the Anglo-Boer War in 1899, Pilgrim's Rest had undergone many changes and it was no longer a typical prospector's town but a thriving little mining town. The local engineer, Wertheman, had replaced all the old and sometimes dilapidated stamper batteries scattered over the area with a central reduction works and a modern 60-stamper battery. These buildings can still be seen in this museum town. The Belvedere hydroelectrical power station was erected in the lower Blyde River opposite the well-known Bourke's Luck potholes and provided electricity for a cocopan (tram) network to all the mines. All the necessary machinery was transported by ox-wagon, and when one keeps in mind the fact that the nearest station was at Machadodorp, it was clearly no mean feat.

During the Anglo-Boer War all mining activities came to a halt, with only two caretakers and a number of maintenance personnel looking after the interests of TGME. Even by September 1900, with the British forces at Middelburg, and much later still, Pilgrim's Rest remained under the control of the Boer forces. After the fall of Pretoria the State Mint was hastily moved to Pilgrim's Rest. With the gold at Pilgrim's Rest and gold bars from Pretoria, 986 known "Veld Ponde" (Field Pounds) were produced by the "Staatsmunt te Velde" (State Mint in the Field). The "Staatsdrukker te Velde" (State Printer in the Field) was also at Pilgrim's Rest and printed the necessary paper money. Because there was no suitable paper they used school exercise books and TGME ledger paper to print ten, five and one pound notes.

Mining activities resumed after the war, new developments took place and two new mines, Vaalhoek and Elandsdrift (1905), were opened. Today, Pilgrim's Rest is still bustling, but as a museum town under the care of the Mpumalanga department of public works. All the historic buildings have been reconstructed and restored; monuments to a romantic era in the nineteenth century that played a decisive role in the development of the Lowveld.

The discovery of gold in the Pilgrim's Rest area was the most important event of the time and for years played a leading role in the history of the eastern Transvaal. The discovery of gold by Button in 1871 at Eersteling near Marabastad in the Pietersburg (Polokwane) district was the beginning of the so-called Zoutpansberg goldfields. This discovery led to the proclamation of the first gold law (Number 1 of 1871).

The exploration and prospecting of this area is interwoven with the peregrinations of, among others, Mauch, who discovered gold along the Olifants River in 1868, and the travellers who had visited the northern Transvaal before him. One Logegary discovered gold in as early as 1858 and reported this to the government. Button was, however, the first to explore the area north of the Olifants River up to the Limpopo and the most successful prospector. He

The well-known Royal Hotel in the main street of Pilgrim's Rest had been the hub of social activities in the small town even before the Anglo-Boer War. (06.09.1986)

---

5 Today the Mpumalanga area.

The well-known character and prospector, August Robert, better known as French Bob. He was one of the first miners after gold had been discovered by Moodie in 1882.
(ACKNOWLEDGEMENT: TRANSVAAL PROVINCIAL MUSEUM SERVICES)

found gold in the hills along the Klein Letaba River and called these hills the Sutherland Series. When he found more gold south of there, the Murchison Hills got their name. In 1870, while in Lydenburg, he reported this in detail to Dr Sutherland in Natal. In this report he also mentioned that he had found copper deposits in the Phalaborwa hills.

The mining activities at Eersteling were unique in many ways: because quartz reef was being mined, a large round boulder was used to first crush the ore. A pole was fastened to the boulder and two labourers rode see-saw on it to break the ore underneath. After that heavy machinery was imported from Europe at great cost, but there was not enough water to successfully drive the battery and the deposits were not very good. A large smeltery with a tall chimney was later erected and is still standing today. With the outbreak of the Anglo-Boer War, all the equipment was apparently disabled by the Boers.

During its heydays (1888 to 1889) the Soutpansberg goldfields included a number of separate goldfields. The Houtboschberg Delverijen was proclaimed in 1887, followed by the Marabastad goldfields and the well-known Selati goldfields (northeast of the Olifants/Selati corner), which included the Murchison Hills. The Selati railway line, which led from the Eastern Line near Komatipoort northwestwards through the Lowveld, was named after this goldfield. Several mining concessions were awarded by the government, including eight to the Shingwedzi Prospecting and Exploration Company. Some of the smaller mining companies eventually amalgamated under the Letaba Gold Mining Company, which then held concessions over a large area. These concessions were probably why the section of the Kruger National Park between the Letaba and Olifants rivers was the last to be proclaimed as part of the Park; with the eastern half incorporated in 1914 and the rest, up to the current western border, in 1923.

Quartz reef mining was in its infancy at the time, techniques were hampered by a shortage of machinery and water played a major role. As a result gold mining in the Soutpansberg was both time consuming and expensive. By 1892 the De Kaap and the new Witwatersrand mines were already producing more gold.

Apart from gold, the district also had rich iron, copper, lead, mercury, chrome, silver and tin deposits. South of the Alkants River mica could be found. The first and even some of the later prospectors had seen how the native tribes smelted iron and copper in a primitive yet effective manner and worked these metals into hoes, weapons such as assegai blades, and copper ornaments. Famous smiths (the northeastern Sotho or BaPhalaborwa) lived in the vicinity of the Phalaborwa hills and practised primitive mining to obtain ore. Furnaces were also found further north and it is now known that the first iron smelting in South Africa had taken place in about AD 250 on the farm Silver Leaves in the Tzaneen area.

The Soutpansberg was also fine combed by diamond prospectors, as rumours about the presence of diamonds were always circulating. After the discovery of the Kimberley riches, prospectors started moving further north, but very few diamonds were found and the only ones worth mentioning came from Soutpan in the far northwest. Prospectors also searched for diamonds further east, but without success.

The lives of the first prospectors at the Soutpansberg goldfields were made miserable by the harsh climate and endemic diseases. The unhealthy summers in particular, along with the accompanying malaria that made the lower-lying areas so infamous, hampered development. Consequently the Witwatersrand drew much more attention as everything was so much easier there.

The discovery of the De Kaap goldfields in the southern Lowveld only happened after the alluvial gold at Pilgrim's Rest had been exhausted and the diggers had moved elsewhere. They prospected mainly north- and southwards along the Drakensberg and in 1882 indications of primitive workings at Duivelskantoor (Devil's Office) at Kaapsche Hoop were

A view of the De Kaap goldfields near Barberton in the early 1890s.
(ACKNOWLEDGEMENT: TRANSVAAL ARCHIVES)

investigated (Hromník, 1986). Tom McLachlan, the acclaimed prospector and digger who had become so well known as a result of his finds at Eersteling, Mac-Mac and Pilgrim's Rest, changed the Valley of Death into the Valley of Hope. In the same year B. Chomse discovered gold on the farm Berlin, the later Barrett's Berlin, near Kaapsche Hoop. McLachlan was of course familiar with the De Kaap area – he had already investigated it in 1874, but with no success worth mentioning. In that year he wrote a letter to President Burgers informing him that he had found gold on state land, but interestingly enough indicating the Moodie farms as non gold-bearing on his sketch plan. G.P. Moodie, the surveyor-general of the ZAR, had concessions on certain farms in the Lydenburg district. A year after McLachlan had discovered alluvial gold at Duivelskantoor and the gold rush started, in 1883, Auguste Robert, or French Bob as he was known, discovered the first gold reefs in the De Kaap Valley – on Moodie's land! According to legend, French Bob had already found Pioneer Reef the previous year, 1882, but kept it hush hush. His secret was only discovered after he had started running water to his mine. This later became the Agnes Mine (southeast of Barberton) and is still producing gold today.

Because Tom McLachlan's alluvial finds were not as rich as those at Pilgrim's Rest, the discovery at first did not create much enthusiasm among prospectors and adventurers. After the Transvaal War of Independence (1880–1881) things soon changed, however. The recession that followed the war, coupled with news of the finds of Chomse, Charlie the Reefer and others in the De Kaap Valley, as well as wild stories of streets covered in gold, led to huge interest in these areas. It was not long before the usual rush of diggers, adventurers and all kinds of characters from near and far started descending on the De Kaap Valley.

As a result of these discoveries and the disorder left in their wake, the ZAR government in 1882 appointed a commission consisting of General P.J. Joubert, M.W. Pretorius, C.J. Joubert and C.F. Hoolboom, the gold commissioner at Pilgrim's Rest, to investigate the scope of the goldfields and the state of affairs there. The commission found that between 170 and 180 diggers were already working on the so-called "Kodwaan Plate" (Ngodwana Plateau). Most of them

Alluvial gold mining with trough and pan at Waterval, Duivelskantoor (Devil's office), circa 1890s.
(ACKNOWLEDGEMENT: TRANSVAAL ARCHIVES)

Primitive mining using the shaft method in the vicinity of Barberton, circa 1890s.
(ACKNOWLEDGEMENT: TRANSVAAL ARCHIVES)

Graham Hoare Barber, the founder of Barberton, who wrote a letter on 21 June 1884 to the State Secretary in Pretoria in which he announced the discovery of gold at Barberton.
(ACKNOWLEDGEMENT: TRANSVAAL PROVINCIAL MUSEUM SERVICES)

F.H. Barber – another Barber from Barberton, 1889.
(ACKNOWLEDGEMENT: TRANSVAAL PROVINCIAL MUSEUM SERVICES)

Henry Mitford Barber, one of the founders of Barberton, circa 1885.
(ACKNOWLEDGEMENT: TRANSVAAL PROVINCIAL MUSEUM SERVICES)

were Afrikaans speaking and there was already a small digger's town called Duivelskantoor (Devil's Office) – the later Kaapsche Hoop and the centre of all activities at the new goldfields. The diggers had appointed their own diggers' committee to look after their interests, but the commission decided to appoint J.P. Ziervogel, chairperson of the diggers' committee, as justice of the peace and declared the De Kaap goldfields public diggings. The official proclamation only followed in 1885 – more than a year after French Bob's discovery of the Pionieer Ridge in 1883.

Ziervogel was not only responsible for maintaining law and order on the goldfields of the De Kaap Valley, but also had to fulfil his duties as gold commissioner. He had his hands full with the tricks, quarrels and transgressions of the recalcitrant group of diggers that were fast becoming a cosmopolitan community.

Furthermore, there was no building that could serve as a jail – a fact the diggers took full advantage of with uproarious drinking parties, fights, claim jumping, complaints about mining concessions and the wild firing of pistols.

The town Kaapsche Hoop rose overnight on a piece of state land next to the farm Berlin. It was laid out between massive, solitary sandstone boulders on a plateau. These boulders reminded people of the walls of a room and the thick mist that would cover them in the early morning lent the place a certain eeriness – hence the original name Duivelskantoor.

Kaapsche Hoop initially consisted of wagons and tents. Even the government office was housed in a tent and there was no sign of proper streets. But before long the first rickety clay huts, as well as wood-and-corrugated iron buildings, started rising pell-mell in the general disorder.

With the official proclamation of the area as a public goldfield in August-September 1885, the area entered its most active time and experienced a short period of prosperity.

By 1886 there were a couple of thousand inhabitants and quite a few permanent buildings at Kaapsche Hoop. These included the government building and a powder-house, both of which had been completed in 1885. This boom did not last long because of the discovery of the much richer goldfields at Barberton and the Witwatersrand, which were declared public goldfields in 1886. However, this did not mean that the diggers' existence at Kaapsche Hoop came to an end. The little town was, especially in the 1890s and shortly after the turn of the century, still a lively place with two hotels, one of which later burnt down, several bars, a small school, several general dealers and a number of corrugated iron, clay and even brick houses.

In the meantime a jail had also been built and it is said that there was even a tennis and croquet court in town. Kaapsche Hoop's big new post office, built in 1896, was designed by the state architect, S.W. Wierda. This building became the hub of the town's social life and according to one of the older inhabitants one could buy almost anything, from drinks to pancakes, for threepence each at the so-called "tie-kiedraai-aande" (threepence evenings).

Today only memories and a few remnants testify to Kaapsche Hoop's heyday. All that remain are the ruins of the government building and the magistrate's home, the occasional shrub or twisted rose bush on Commissaris Square (the old town square), the Lambourn family's old pear orchard, the cattle dip, desolate homes and shops, and streets bearing the names of the town's first inhabitants and officials. Only the post office is still doing business and the cemetery is being maintained for the last few inhabitants of this once bustling town.

While chaos reigned at the Kaapsche Hoop goldfields, Moodie awarded mining rights to the renowned pioneering miner French Bob. Other diggers had to pay high licensing fees, which led to much discontent. In 1884 Moodie appointed Henry Nourse to run his business and on his way to the goldfields he met Fred and Henry Barber at Lake Chrissie. They accompanied Nourse to the diggings and helped him settle the dispute with the miners. In the meantime Moodie obtained legal mining concessions to his land in 1885 and these 12 farms later became part of the Zuidkaap Goldfields.

While Nourse was trying to sort out the diggers' problems, the Barber brothers were prospecting for gold. In June 1884 they found paying deposits on a farm called Brommers (Blue flies), just above the current town Barberton. A gold rush followed and soon the new Barbers' Camp was a hive of activity. A diggers' committee was appointed, gold commissioner Wilson did the necessary and soon the town Barberton (named after the Barber brothers) was founded.

As was the case with Pilgrim's Rest, Barberton soon had its own character with even more interesting personalities and adventurers. Because the area contained only reef gold, Barberton's goldfields heralded the beginning of a new era for the country's mining industry. Dynamite, cocopans (trams) and elevated rails, as well as water-driven stamper batteries, appeared for the first time on the goldfields. By the middle of 1884 there were already between 400 and 500 houses of all kinds in town. The first of these were the modest wattle-and-daub houses, followed by wood-and-corrugated iron ones. Town planning only took place in 1886–87, when more elegant Victorian houses and buildings were transported piece by piece from Natal. It is estimated that by 1887 the valley and district housed 10 000 people.

Vast amounts of money were pumped into Barberton and it was hopelessly overcapitalised. Only a few mines were productive, while others were nothing more than fictitious dreams on lovely pamphlets and brochures – marketed by imaginative swindlers. President Kruger visited the goldfields in 1886 and met the diggers under a fig tree in Barberton to listen to their grievances and complaints. This fig tree is still standing next to the town's rugby field. After the meeting a dinner was held in the Phoenix Hotel and it is said that the president delivered one of his few speeches in English in answer to a toast.

Although the shallow reef the Barbers had discovered yielded reasonably well at first, it eventually became unprofitable and was incorporated into the

Barberton in 1885.

Sheba Mine. Many other lucrative little mines contributed to the development of the area and the mid-1880s is considered to have been Barberton's heyday. As placer gold became scarcer and organised mines were established, more and more diggers moved to the Witwatersrand, where the richest gold deposits in the country were eventually found.

By that time Barberton had the first gold and stock exchange in the southern hemisphere (the Transvaal Share and Claim Exchange Trust and Agency Company Ltd.), but unfortunately it burned down. In 1887 it was replaced by a permanent building in which symphony concerts and other stage productions, dances and meetings were held. Only its façade remains, and it is one of Barberton's national monuments today.

As was the case at the other goldfields, supplies were a real problem, especially in this mountainous region. Two supply routes were used: the longer but safer ox-wagon route from the end of the railway line at Ladysmith in Natal, and the shorter but dangerous route from Delagoa Bay. Most of the fresh food was produced in the vicinity, but equipment and scarcer supplies had to be brought in by ox-wagon via the named routes. The route through the Lowveld was of course exposed to tsetse flies and malaria, as well as flooded rivers in the summer months. Eventually trading posts or stations were established along the route – Furley's store at Komati Drift on the farm Coopersdal, Alf Roberts's store at Nellmapius Drift and Compton and Hampton on the Lebombo watershed. The Eastern railway line was built between 1888 and 1894 and linked Delagoa Bay and Pretoria via Komatipoort and Nelspruit.

Barberton's colourful boom will always be remembered as being unique of its kind. After the discovery of gold had become known, merchants, hotel owners, barmen and bar girls, doctors, pharmacists, ministers, businessmen and an assortment of unusual characters streamed in and soon the town had almost 5000 inhabitants. Some of the most famous people in South Africa settled in Barberton – Abe Baily, Al-

The well-known Sheba Gold Mine at Barberton, circa 1890.
(ACKNOWLEDGEMENT: TRANSVAAL ARCHIVES)

Die old market plain in Barberton, 1891.
(ACKNOWLEDGEMENT: TRANSVAAL ARCHIVES)

fred Beit, Sammy Marks and Percy FitzPatrick were only a few of the names to capture the limelight for a few years. FitzPatrick apparently often camped under a large thorn tree just outside town and there are murals of him and his famous dog Jock in the Impala Hotel. These were painted by one Genal, who also painted scenes in the Noord-Kaap Hotel, including one of the old Zeederberg coaches that served the area. Legend has it that Cockney Liz was the best-known bar girl and cabaret dancer in a town that at one stage had 200 hotels and bars. It is said that she would be auctioned off to the highest bidder on Saturday nights. Yankee Moore was the first shop owner; Ikey Sonnenberg an interesting gambler/broker; Stafford Parker a businessman, auctioneer, agent, assizer and adviser; Jimmy Winter owned the first bakery; and, last but not least, there was George Thorncroft, an amateur botanist who made the Barberton daisy (*Gerbera*) world renowned.

Adventurers flocked to the area and small communities sprang up all around Barberton. The Sheba reef was discovered by Edwin Bray, who called it Golden Quarry. It would become the most famous and richest mine in the world. In December 1885 J. Sherwood opened a butchery and hotel on a hill between the Sheba and Fairview mines and this little settlement got the romantic name of Eureka City. In this hotel Sherwood's wife was known as the Queen of Sheba, although she was certainly not the most attractive woman in the area. Sherwood's hotel was also named The Queen of Sheba. Eureka City soon reached its peak – in 1886 – and by then already had three shops, three hotels, a bakery, chemist, music hall, several bars and even a race track. Sheba is currently the oldest productive gold mine in the world.

Today one mining group, Metorex, owns and works the mines in the area.

Despite all the prospecting and digging activities in those years, from the northeastern Transvaal[6] down to the south of Swaziland, the lower-lying eastern parts remained untouched, mainly as a result of tsetse flies and malaria. While the area was healthier and more tolerable in winter, people generally avoided it during the summer months. Many prospectors did venture lower down, but they had to pay a high price – often their lives.

C.F. Osborne was one of the better-known prospectors who visited the Lowveld. Bulpin reported that he had found gold in the vicinity of Pretoriuskop in as early as 1865. It is also known that Osborne investigated the route between Pretoria and Delagoa Bay in 1872, and found traces of gold at several places. He did not, however, have any prospecting equipment with him and constantly fell ill with malaria. His report that worthwhile amounts of gold occurred on the escarpment would later prove to be true.

At the time of the establishment of a diggers' community at Pilgrim's Rest and the systematic dispersion of prospectors and diggers across the escarpment, some fortune seekers took their chances in the lower-lying areas. According to records no fewer than 35 prospectors braved the Lowveld in 1873, of

The notorious barmaid "Cockney Liz" of Barberton and Eureka City in the 1880s. She was apparently auctioned to the highest bidder in the bar by Stafford Parker on Saturday evenings.
(ACKNOWLEDGEMENT: TRANSVAAL PROVINCIAL MUSEUM SERVICES)

---

6   Limpopo Province.

9. PROSPECTORS AND DELVING

Prospecting pits on the sandveld plateau south of the Mutale mouth, where prospectors such as Geo Hickey had prospected for minerals including copper before 1912.
(8.10.1984)

The grave of an African chief with many relics from the time, including a well-preserved plough, near the terrain where prospecting had taken place along the Salitji Road east of the Nwatindlopfu Mouth.
(18.07.1984)

C.J. Gerber's prospecting shaft near Malonga Fountain. It is still more than 12 metres deep today.
(30.07.1984)

whom 27 died of malaria. After this the area was regarded with great respect.

On the northern bank of the Sabie River, about two kilometres east of the Nwatindlopfu Mouth, signs have been found that prospecting for gold and/or diamonds on a fairly large scale took place there from 1870 to 1880. Numerous relics dating back to this time have been found on the terrain. Like so many others in the Lowveld, this project was apparently a pipe dream and, judging by the graves in the area, more than a few black workers had to pay with their lives for these castles in the air.

Over the years sporadic prospecting took place within the boundaries of the present Kruger National Park, especially in the north. Geo Hickey prospected in the area between the Mutale mouth and Pafuri (probably for copper) and prospecting pits can still be seen south of the Mutale mouth and at Thulamela at Pafuri. There are also claim stakes near Spokonyolo Pan along the Limpopo, but the origin of these is unknown.

Bill Lusk, or Texas Jack as he was also known, was one of the better-known characters of Crooks Corner. He panned for alluvial gold at the Ribye-ra-Gudzani (also Ribye-ra-Khubyane) potholes in the Shingwedzi River. This terrain is better known today as Red Rocks because of the light red Karoo sandstone ridge that crosses the Shingwedzi River here.

A geologist, Charles J. Hamilton, and one Thomsen unsuccessfully prospected for diamonds in the area south of the current Letaba rest camp from August to September 1917.

J.C. (Chris) Gerber and his brother Jan (of Louis Trichardt[7]) also prospected for diamonds in the early 1920s at Malonga Fountain. Between 1923 and 1926 Gerber and other prospectors, namely Van Deventer, Elton, Jerome, Coetzer, Pienaar and Wright, from Louis Trichardt and elsewhere, prospected for diamonds and other minerals in the area between the Olifants and Letaba rivers. Ranger Ledeboer, stationed at Letaba section, had quite a bit of trouble with these prospectors in the former Shingwedzi Reserve until prospecting and mining in national parks was prohibited by the new National Parks Act of 1926. It was noted in old Parks Board documents that Gerber's prospecting rights were withdrawn on 31 December 1926.

---

7 Now Makhado.

It would seem others must also have prospected fruitlessly elsewhere in the former Sabie and Shingwedzi reserves. Ranger Lynn van Rooyen and research officer Ian Whyte noticed a foreign object in the veld near the artesian borehole Mahlati, in the Shingwedzi section, during an aerial census in 1991. Closer inspection revealed it to be a diamond washing machine, still in good order, with several smallish piles of washed gravel in the vicinity.

After 1926 the National Parks Board still received sporadic requests from prospectors. In 1946 a Mr Van der Westhuizen obtained permission to remove 300 tons of corundum, which had been mined along the Olifants River, from the Park, paying five per cent of his profit to the Board as compensation. He had, of course, acted illegally, but maintained that he had been unaware of the fact that he was working inside the Park.

Even after that every now and again someone would try to prospect and mine illegally. One such case was Mrs Cheeky Stamp, who had a small mica mine next to the Olifants River near the western border of the Park. She was one of many interesting characters in an era when the Kruger National Park was proclaimed and prospecting activities were banned by the new National Parks Act. Cheeky was married to Thomas (Tommy) Harding Stamp. They had moved into the Lowveld via The Downs in 1922 on a donkey cart and prospected for gold in the Blyde River area. After a few months they crossed the Olifants River and prospected at Mica, but could not find any paying deposits.

From there they moved northwards to where Phalaborwa is today, encountered a fair number of big game and shot some lions, among others, using an 8 x 60 mm Mauser. En route they pegged their claims over an area stretching from Mica to Phalaborwa (Lulekop). After that they moved into what would become the Kruger National Park, where they found profitable mica deposits at Madzumbelani next to the Olifants River, where they started mining. They lived at the site in a clay house with a reed roof and transported the mica by donkey cart to the small town of Mica.

With the proclamation of the Kruger National Park in 1926 they moved to the farm Grietjie near Phalaborwa, where Tommy and Cheeky apparently became estranged. Former ranger G.C.S Crous met Cheeky at Letaba Ranch while he was looking after some cattle for somebody from Louis Trichardt.

Tommy joined the British Army, but they still used donkeys to fetch mica in secret from their mine in the park and transport it to Grietjie. It is said that Tommy Stamp and a Commandant Muller tried in vain to draw up a petition protesting the proclamation of the Park. They bought their supplies at Leydsdorp and often camped at the baobab tree there. Cheeky later moved back to Mica and the two separated for good in 1938.

The Stamps were among the first whites to live in this part of the Lowveld and were well-known prospectors. They had two children, Patsy and Stuart, who literally grew up in the bush. Governesses from England taught them at the farm. Their mother Cheeky was apparently a determined and strong-minded woman and it is said that she continued to work the mine years after she and her husband had separated. She mined the mica by hand and transported it to Mica Station by donkey cart, later buying a truck. Her daughter Patsy later married Charles Malan, who was the shop manager at Skukuza for many years and then the camp manager at Lower Sabie.

Cheeky died in 1975 and Tommy in 1976. The tools they used at the Madzumbelani mine are still

The diamond washing machine "Dommie" which was found in 1991 in the vicinity of the artesian Mahlati borehole on the Shingwedzi ranger section.

9. PROSPECTORS AND DELVING

Mrs Cheeky Stamp's mica mine at Madzumboleni along the Olifants River. It was mined until the early 1930s.
(18.06.1984)

Mrs Johanna Susanna Hilda (Cheeky) Stamp, who illegally worked a small mica mine at Madzumboleni, just east of the Tsutsi mouth along the Olifants River during the early 1930s, circa 1931.
(WITH KIND PERMISSION: MR C. MALAN, TSHIPISE)

there. A hundred of their claims near Phalaborwa were sold to Dr Hans Merensky in about 1940 for £100 (the then equivalent of R200). Lulekop, with its incredibly rich phosphate deposits, apparently also belonged to Mrs Cheeky Stamp.

According to the minutes of a Parks Board meeting held in November 1942, two geologists of the Department of Mines were allowed to prospect for cinnabar in the Park, but the result of their efforts is unknown.

People have been prospecting for coal in the Park since the beginning of the previous century. In the 1903 correspondence of the first warden, J. Stevenson-Hamilton, it was mentioned that the railway administration was interested in prospecting for coal in the Sabie Reserve, which subsequently took place. A Mr Fry found coal seams on the banks of the Crocodile River at Komatipoort and the Department of Mines started negotiations to obtain mining rights. This idea was eventually abandoned after long negotiations between Stevenson-Hamilton and several government departments, and would lead to the proclamation of the National Parks Act of 1926, which finally prohibited any prospecting and mining in national parks. Some of the prospecting stakes left in the 1920s in the search for coal in the area north and south of the Crocodile River can still be seen in the Crocodile Bridge section, near the old sandstone quarries west of the rest camp.

In 1907 one W. Depledge asked to prospect for copper in the north of the reserve. This request was also declined after negotiations.

Several other applications were received, even after 1926, and it is understandable why Stevenson-Hamilton had taken such a strong stand against it from early on. He soon realised that this disturbance in the Park would be in direct conflict with the principles of nature conservation.

Towards the end of 1938 C.W.T. Pittendrigh and C.D.C. Geyer, of Carolina, prospected for gold next to the Shingwedzi River at Shangoni. Stevenson-Hamilton suspected that they were working mainly within the Park's boundaries and ranger Harry Kirkman was rapped over the knuckles for allowing it. It was later determined that most of their prospecting pits were in fact outside the Park, except for one shaft in the shale layers of the Swaziland System north of the Shangoni ranger post, where they probably prospected for emeralds. These pits are also still visible.

With the discovery of rich phosphate deposits at Phalaborwa and the economic possibilities it held for the region, that part of the Park, including Lulekop, was exchanged for the farm Peru on the southwestern border of the Park.

This chapter is concluded with a relatively recent incident, namely the prospecting for coking coal in the Punda Maria area in the 1970s. After the size of the coalfield had been determined and confirmed by the Geological Survey Section on behalf of Iscor, history repeated itself as representations were again made to mine the coal inside the Park. Again the Parks Board had to take a strong stand, based on conservation principles. The Parks Board managed to resist the pressure and it was eventually decided to start mining outside the Park in Venda. With this, the modern approach to nature conservation came to the fore, namely the wider understanding of both nature conservation and quality, as well as the sensible utilisation of natural resources. The Parks Board came to the realisation that it was inevitable that the Park's integrity would sooner or later come under the same kind of pressure. This served once again to confirm their responsibility to preserve the Kruger National Park as an untouched conservation area for future generations.

A prospecting pit near the current Shangoni ranger post where Pittendrigh and Dreyer prospected for emeralds in 1938.
(12.11.1986)

Dr Hans Merensky, the famous geologist who discovered the Bushveld Igneous Complex, as well as copper and phosphate deposits at Phalaborwa.
(ACKNOWLEDGEMENT: TRANSVAAL PROVINCIAL MUSEUM SERVICES)

References

Anon. *Vallei van die dood tot vallei van goud.* Publication of the Transvaal Provincial Library and Museum Services, Pretoria.

Atcherley, R.J. 1879. *A Trip to Boer Land.* Richard Bentley, London.

Aylword, A. 1878. *The Transvaal of Today.* William Blackwood & Sons, London.

Baines, T. 1877. *The Gold Regions of South Eastern Africa.* Edward Stanford, London.

Bornman, H. 1975. Goud in die berge. In: Barnard, C. (ed.). *Die Transvaalse Laeveld – kamee van 'n kontrei.* Tafelberg, Cape Town.

Bornman, H. 1984. *Fotogeskiedenis van Barberton 1884–1984.* Town Council of Barberton, Barberton.

Bulpin, T.V. 1950. *Lost Trails on the Lowveld.* Howard B. Timmins, Cape Town.

Bulpin, T.V. 1954. *The Ivory Trail.* Howard B. Timmins, Cape Town.

Bulpin, T.V. 1978. *Illustrated Guide to Southern Africa.* Readers' Digest Association of S.A. (Pty) Ltd., Cape Town.

Bulpin, T.V. 1983. *Lost Trails of the Transvaal.* Books of Africa, Cape Town.

Cartwright, A.P. 1960. *The Gold Miners.* Purnell, Cape Town.

Cartwright, A.P. 1961. *Valley of Gold.* Howard Timmins, Cape Town.

Cartwright, A.P. 1971. *The First South African.* Purnell, Cape Town.

Cartwright, A.P. 1972. *Phalaborwa. A Mining Success Story.* Phosphate Development Corporation.

Cartwright, A.P. 1974. *By the Waters of the Letaba.* Purnell, Cape Town.

De Witt, A. 1984. *The Pilgrim's Rest Story.* Transvaal Provincial Museum Services, Pretoria.

Edwards, D. 1890. *The Gold Fields of South Africa.* D. Edwards & Co., Cape Town.

Elton, G. 1872. Trip from the Tati settlement to Delagoa Bay. *Journal of the Royal Geographical Society* 42:1872.

Ferreira, O.J.O. 2002. *Montanha in Zoutpansberg:'n Portugese handelsending van Inhambane se besoek aan Schoemansdal, 1855–1856.* Protea Book House, Pretoria.

FitzPatrick, J.P. 1979. *South African Memories.* Ad Donker Ltd., London.

Glynn, H.T. 1926. *Game and Gold.* The Dolman Printing Co., London.

Grewar, J.F. 1980. *The Vanished Saddleback – The Story of Phalaborwa.* Purnell, Johannesburg.

Herring, G. 1948. The Pilgrim diggers of the 70s – a short history of the origin of Pilgrimsrest (1973–1881). Unpublished manuscript.

Hromník, C.A. 1986. Gold and God in Komatiland. Unpublished manuscript.

Jeppe, F. 1893. The Zoutpansberg Goldfields in the SA Republic. *The Geographical Journal* 11:213–378.

Jones, G.R. Pilgrim's Rest in the early days. Unpublished manuscript.

Klein, H. 1971. *Hans Merensky: Visionary.* Brochure compiled for the Hans Merensky Foundation. Hortors, Johannesburg.

Klein, H. 1972. *Valley of the Mist.* Howard Timmins, Cape Town.

Letcher, O. 1936. *Gold Mines of South Africa.* Author & Waterlow & Son, London.

Malan, C. 1984. Personal communications.

Mathers, E.P. 1970. *The Goldfields Revisited.* State Library, Pretoria.

Merensky, A. 1889. *Erinnerrungen aus dem Missions-Leben.* Berliner Evangelische Missions Gesellschaft, Berlin.

Mitford-Barberton, I. 1934. *Barbers of the Peak.* Oxford University Press, Oxford.

Nathan, M. 1941. *Paul Kruger – His Life and Times.* Knox, Durban.

Peacock, R. 1950. *Geskiedenis van die Lydenburgse goudvelde tot 1881.* M.Sc. thesis, University of Pretoria, Pretoria.

Raddatz, H. 1886. Das Kaffernland des Unteren Olifants. *Pettermann's Mitteilungen Heft 2.*

Rehman, A. 1883. Des Transvaal Gebiet des südl. Afrika, in phys.-geogr. Beziehung. *Mitteilunger der k.k. Geographische Geselschaft in Wien* 26(8):377–378.

Richards, R. 1884. *The Truth About the New Goldfields.* London, John Walker & J. Pearce, London & Pietermaritzburg.

Sawyer, A.R. 1892. *Geological and General Guide to the Murchison Range.* J. Heywood, London.

Scrymgeour, A. 1916. Pilgrim's Rest goldfields. *Mines and Claim Holders Quarterly* 1(2):1–44.

Scully, W. 1913. *Further Reminiscences of a South African Pioneer.* T. Fisher & Unwin, London.

Slingsby, P. 1987. *Kaapsche Hoop Voetslaanpad.* Map compiled for the National Hiking Trail Board.

Verhoef, J. 1986. Notes on archaeology and prehistoric mining in the Kruger National Park. *Koedoe* 29:149–156.

Von Wielligh, G.R. 1923. *Langs die Lebombo.* Van Schaik, Pretoria.

Webb, H.S. 1954. The goldfields of Sabie and Pilgrim's Rest and transport development, In: *A Survey of the Resources and Development of the Southern Region of the Eastern Transvaal Lowveld.* Publication of the Lowveld Regional Development Association. *Cape Times*, Cape Town.

Webb, H.S. 1954. The penetration of the region by Europeans. In: *A Survey of the Resources and Development of the Southern Region of the Eastern Transvaal Lowveld.* Publication of the Lowveld Regional Development Association. Cape Times Ltd, Cape Town.

Webb, H.S. 1954. The goldfields of De Kaap and the founding of Barberton. In: *A Survey of the Resources and Development of the Southern Region of the Eastern Transvaal Lowveld.* Publication of the Lowveld Regional Development Association.

Wynne K.F. 1934. Gold in the Eastern Transvaal. *Journal of the Royal Geographical Society* 47.

Zeederberg, H. 1971. *Veld Express.* Howard Timmins, Cape Town.

Chapter 10
Transport roads, transport riders and the tsetse fly
10.1 The "Commission" and "Old Wagon" roads
Dr J.B. de Vaal

The Voortrekkers' ideals were "freedom and the sea". In their quest to reach the sea they undertook several expeditions from different places in the old Transvaal. In December 1843 a commission set off with 16 wagons on a journey to Delagoa Bay to find a suitable harbour as well as somewhere closer to live. The commission consisted of 80 people from Winburg, Potchefstroom and the Magaliesberg under commandants G. Mocke and A.H. Potgieter and Magistrate J. Vermeulen, with Karel Trichardt as their guide. When they had to wait for 14 days next to a flooded Olifants River some of them contracted malaria while their animals were stung by tsetse flies, with the result that they had to turn back without having accomplished anything.

At the end of May 1844 a second commission consisting of 50 men, again under the leadership of Potgieter, Mocke and Vermeulen but this time without Trichardt, left for Delagoa Bay. This time it was winter, so flooded rivers and malaria did not pose any problems. They passed west of the place where the little town of Andries-Ohrigstad would be laid out the next year, down the later Ohrigstad River, along the Kaspersnek River and then over Kaspersnek. This crossing or mountain pass over the Drakensberg was discovered by and named after Casper Kruger, the father of the later State President Paul Kruger.

About 24 kilometres on they travelled across the farm later known as Ledouphine and outspanned next to a river. Because the terrain was barely passable they left the wagons and about half the men behind while the rest continued on horseback. The riders managed to reach Delagoa Bay and, with the Dutch merchant Smellekamp as translator, concluded a treaty with the Portuguese governor on 22 July 1844. The treaty stipulated that the Boers could settle in the interior west of the harbour between 10° and 26° south and four days' journey from Delagoa Bay.

Because the members of the party who had stayed behind feared that the Potgieter party had been murdered, they decided to turn back and call the river where they had waited the Treur (Sorrow) River. To their delight the party returning from Delagoa Bay caught up with them 16 kilometres further, at another river. Their joy at seeing their friends was so great that they called this river the Blyde (Joyful) River. Potgieter went back to fetch his trek at Potchefstroom and by the end of July 1845 the new town, Andries-Ohrigstad, had been founded. Karel Trichardt and his family had undertaken the move with Potgieter and settled on the farm Branddraai along the Ohrigstad River.

Unfortunately no information remains about the route the Potgieter commission followed from the

The centenarian, Ngutu Sambo. He led oxen during Karel Trichardt's 1848 expedition from Ohrigstad to Delagoa Bay. It was during this journey that Willem Pretorius had died and was buried near Pretoriuskop.
(PHOTOGRAPH BY: J. PRETORIUS, 1935)

10. TRANSPORT ROADS, TRANSPORT RIDERS AND THE TSETSE FLY

The chop marks in the stem of this large leadwood tree near Kwaggaspan windmill, probably indicated the direction of Karel Trichardt's "Oude Wagenweg" (Old Wagon Route) of 1845.
(08.05.1984)

Treur River to Delagoa Bay. But there was another way of uncovering this information. During a meeting of the "Raad der Representanten van het volk" (the Council of Representatives) in Ohrigstad on 20 August 1845, Potgieter proposed that some of the land across the "Delgoas Berg" (Drakensberg) should be demarcated into farms. It was determined that the first farms to be apportioned would be in the area stretching from the mountain down to "de Eerste revier aan de Commissie drift" (the drift in the Sabie River through which the Potgieter commission went) and from there across the river westwards to the mountain. A commission would be appointed to build a road down the mountain and at the same time land for the demarcation of farms would be assigned. This proposal was enacted on 27 August.

On 8 October 1845 it was decided that the commission would leave for Delagoa Bay on 14 October and that everybody who wanted to go along had to be at the commandant's laager by noon that day. Twenty-six people arrived and a great ruckus was caused when Adolph Coqui said he wanted to join the party. Karel Trichardt and Johannes Steyn would have none of this as Coqui was a foreigner and they wanted no foreigners on the expedition. The rest of the group agreed. Coqui, born in Germany, was a trader from Natal who had married a Boer girl, a Miss Van Breda. He set off for Delagoa Bay under his own steam, leaving on 4 March 1846 with two ox-wagons. He was accompanied by five Europeans, including his father-in-law, and three Africans, but more is told about this expedition and its tragic fate in chapter 18.

On 17 October the council secretary, J.J. Burger, informed Karel Trichardt that he had been appointed as field cornet of the commission that was to undertake the journey to Delagoa Bay. He was delegated by the council to demarcate farms on the eastern side of the mountain on its behalf and on his return he had to hand his demarcation notes to the book-keeper.

Although 26 volunteers had turned up, according to the book *Herinneringe van Karel Trichardt* (Karel Trichardt's Memoirs) only 18 burghers took part in the expedition. Some of them had probably been on the previous expedition with Potgieter to Delagoa Bay, so the direction of the route or "The Commission Road" would not have posed any problems to them. Potgieter's party had travelled with pack oxen and horses, while Trichardt and his group had to find a route and construct a road along which ivory and skins could be transported to the coast by ox-wagon. Near the Kwaggaspan windmill a group of Kruger National Park researchers found particularly deep chop marks on very old leadwood trees, which points to the exploration of this route (Pienaar & Mostert, 1983). On 27 October 1845 Potgieter could inform the magistrate and heemraad (county court) of Potchefstroom by letter that the party had already descended the mountain on the way to Delagoa Bay and that they had found an easy route.

The commission reached Delagoa Bay, but all their draught animals had died of nagana and their horses of African horse sickness. It is quite probable that Trichardt's old friend, João Albasini, had not yet left on his first journey from Delagoa Bay via Magashulaskraal to Ohrigstad and that he had either taken them along or provided them with trek oxen. He also had many black bearers who could have been of assistance. Whatever the case may be, on 10 December 1845 the Trichardt commission was back in Ohrigstad and reported to the Council of Representatives the same day.

For the sake of convenience, the farms that had been selected were as close as possible to the road, so the name of each farm and its position indicated the direction of the route. Examples are Bokkraal (which later became Klipkraal) along the commission road; Eylandsdrief (Elandsdrif) next to the first big river;

10.1 THE "COMMISSION" AND "OLD WAGON" ROADS

Kruysfontein (Kruisfontein) situated where the wagon road goes down the mountain on the other side of the Eerste River; Eerste Kafferskraal below the mountain where the laager stood (that of Potgieter in 1844); Grootfantein (Grootfontein) on the Delgoasberg (Drakensberg), south of the wagon route and south of the first big river (the farm on which Sabie would later be laid out); Nooidgedacht (Nooitgedacht) on the Delgoasberg, south of the wagon road and south of the first big river; and Wetkop (Witklip) next to the third spruit under the "Delgoosberg" on the right-hand side of the road.

By examining the positions of the different farms, the late A.P. van der Merwe was able to determine the route the Potgieter and Trichardt commissions had followed from Ohrigstad to Delagoa Bay. It went northwards along the Ohrigstad River to the Kaspersnek River, along this river over Kaspersnek in a southeastern direction, over the Blyde River and across the farm Ledouphine, along the southern course of the Treur River and Watervalspruit, over the farm Klipkraal and through the Sabie River where Sabie was later laid out. From there it crossed the farms Grootfontein, Spitzkop, Elandsdrif, Kruisfontein, Zwartfontein, Klipkopje, Peebles, Burger's Hall and went to Pretoriuskop. From Pretoriuskop the Trichardt commission explored a route between Shabeni and Pretoriuskop. It went south of Shithlave Hill over the upper reaches of the Mbyamiti and past Kwaggaspan, south of Renosterkoppies and Siyalu Hill, over Godleni Neck and the Lebombo Mountains, north of Pessene to Delagoa Bay. This route was indicated on old maps as the "oudste weg naar Delagoabaai" (the oldest road to Delagoa Bay) and is also known as "The Old Wagon Route", but was not very popular, for various reasons. It was easily traversable but impractical as there was too little perennial water along long sections, especially during the healthy winter months, when people travelled and hunted. As a result it soon fell into disuse.

In January 1847 João Albasini, who had in the meantime settled at Magashulaskraal next to the Sabie, appeared before the Volksraad in Ohrigstad and informed them that he had found a route to Delagoa Bay that was relatively free of tsetse flies and had more than enough water.

After many delays and much bickering, an expedition under J. van Rensburg was sent on 12 July 1847 to explore Albasini's proposed trade route and report back. Karel Trichardt again played a prominent role in what would be a successful expedition, but the matter was not completely finalised and the route had to be demarcated in more detail. Therefore Karel Trichardt, J.J. Burger and a Joubert (probably the J.B. Joubert employed by Albasini and Trichardt) found themselves in Delagoa Bay in July 1848 after having explored the route further. It is probably during this journey that one of the expedition members, one Willem Pretorius, died of malaria and was buried under a marula tree near the current Pretoriuskop by Albasini. The new route had to be made passable and less than a year after their second expedition, on 7 March 1849, Field Cornet Jan de Beer was ordered by the Volksraad to start improving the road through Zoasland (Umswaas' territory) to Delagoa Bay in May the same year. He could use the help of blacks and open collection lists for public donations.

At the same meeting W.H. Neethling was instructed to start from the other side and build a road down the Drakensberg over the upper reaches of the Treur River to find a route that was closer to both Delagoa Bay and Inhambane. It is probably because this "nadre weg" (closer road) went down the Dra-

Mr Jan Pretorius at Voortrekker Willem Pretorius's grave near Pretoriuskop in 1935.
(ACKNOWLEDGEMENT: TRANSVAAL ARCHIVES)

Panorama, with Manungu Hill (left) and Pretoriuskop as seen from the east.
(08.06.1984)

kensberg over Klipkraal and from there eastwards along the Sabie River over the farms Frankfort, Rietspruit, Lunsklip, Richmond and Magashulaskraal, that João Albasini gave the biggest donation on one of the collection lists, namely 12 *muids* (one *muid* being the equivalent of three bushels) of maize, six pick-axes, a crow-bar and an ox.

At a commission board meeting six years later in Lydenburg on 4 July 1855, it was decided to appoint a three-man commission consisting of Counsellor C. Potgieter, Magistrate J. de Clerq and Field Cornet F.J. Combrinck, with Jacob Middel as interpreter and secretary. They had to take a wagon and ten oxen via "Macaxülle" to Delagoa Bay on the "closer" road, but would still be taking the longer commission road over the abandoned Ohrigstad and Kaspersnek and along the Treur River. Karel Trichardt's newly explored route of 1848 to Delagoa Bay was passable, because in June 1849 Field Cornet De Beer had already completed the part for which he was responsible. With the aid of Swazi labourers he had constructed a passable road to Eb en Vloed, about half a day's journey from Delagoa Bay. The Portuguese governor had undertaken to complete it from there through Portuguese territory to the coast.

The new route and later wagon road turned off from the Old Wagon Route in a southeasterly direction, close to Willem Pretorius's grave. It went between Pretoriuskop and Manungu Hill, north of Skipberg over a sandy hill, to avoid the clay soil near the mountain, over the Crocodile and Komati rivers, through Matalha Poort in the Lebombo Mountains and from there straight to Delagoa Bay. This road, which was very difficult to travel by ox-wagon for long sections, only came into its own 20 years later, after the discovery of the Lydenburg goldfields in 1869, when Mac-Mac was used as the mine commissioner's headquarters.

References

Breytenbach, J.H. & Pretorius, H.S. 1844–1850. *Suid-Afrikaanse argiefstukke, Transvaal no. 1: Notule van die Volksraad van die Zuid-Afrikaansche Republiek* I. Cape Town.

Breytenbach, J.H. 1854–1858. *Suid-Afrikaanse Argiefstukke, Transvaal no. 3: Notule van die Volksraad van die Zuid-Afrikaansche Republiek* III. Parow.

Cachet, F. Lion. 1883. *De worstelstryd der Transvalers.* Jaques Dusseau & Co. Uitgevers, Amsterdam & Cape Town.

De Vaal, J.B. 1964. Ou Handelsvoetpaaie en wapaaie in Oos- en Noord-Transvaal. *Contree* 16 July:5–15.

Jeppe, F. 1879. *Map of the Transvaal and the Surrounding Territories.* Unknown publisher, Pretoria.

Lombaard, B.V. 1979. Die bergpas oos van Sabie. *Forestry Museum Pamphlet* no. 27. 19 October, Sabie.

Pelzer, A.N. 1950. *Geskiedenis van die Suid-Afrikaanse Republiek.* 2 vols. A.A. Balkema, Cape Town.

Pienaar, U. de V. & Mostert M.C. 1983. Nuwe feite bekend oor romantiese transport-era. *Custos* 12(9):28.

Preller, G.S. 1920. Herinneringe van Karel Trichardt. In: *Voortrekkermense II.* Nasionale Pers, Cape Town.

Pretorius, H.S. & Kruger, D.W. 1937. Voortrekker archive articles.1829–1849. Pretoria.

Suid-Afrikaanse Biografiese Woordeboek part 3. 1977. Insert on: Coqui, Theodor Julius Adolph. Tafelberg, Cape Town.

Transvaal Archives (TA). 1/148 – Bechtle and Marais. *Schetskaart van de Lebombovlakte tuschen Olifants- en Krokodilrivier,* 1891.

Van der Merwe. A.P. 1950. Map titled: Paaie en roetes (1836–1860). In: *Lydenburg Eeufeesgedenkboek.* Unknown publisher, Pretoria.

Van der Merwe, A.P. 1961. Die eerste pad van Ohrigstad na Lourenço Marques. *Historia* 6(4):291–294.

Van der Merwe, A.P. 1975. Verkenners en pioniers. In: C. Barnard (ed.).*Die Transvaalse Laeveld – kamee van 'n kontrei.* Tafelberg, Cape Town.

10.2 Alois Hugo Nellmapius's transport route

M. LIEBENBERG

Alois Hugo Nellmapius was born in Budapest, Hungary in 1847 and died in Irene in 1893. His mother was Jewish and his father Hungarian. Not much is known about his youth, except that he trained as a civil engineer in the Netherlands.

He was an excellent linguist and apart from his first language, which was German, he was fluent in Hungarian, French, Italian and Dutch. During his voyage to South Africa he also learned to speak English (Kaye, 1978). He was an energetic man with strong convictions. He believed in himself and although his own interests always came first, he never doubted the possibilities and future of the Transvaal. In 1874, after his appointment as Vice Consul of Portugal, he gave President Burgers the assurance that he had adopted the Transvaal as his own and that he would do everything in his power to further the interests of the country. These were not idle promises, as he had the courage to help found industries when the industrial development of the country was still in its infancy. He always tackled ventures with fervour and on a grand scale. With his strong entrepreneurial spirit and persistence, Nellmapius rose above several setbacks, including two bankruptcies.

His arrival in Pilgrim's Rest in 1873 coincided with the discovery of gold and he was swept along in the rush. The fact that he had no knowledge of mining alluvial gold was no deterrent to Nellmapius. He soon increased the amount of alluvial gold he managed to find by not only washing soil in large quantities, but also instituting a night shift. While everybody else was living in tents, he built himself a comfortable home and cultivated vegetables to augment his income (Kaye, 1978).

Nellmapius's initiative and personality soon contributed to him becoming one of the leading figures at the goldfields. In 1874 he acted as the miners' representative when he submitted their grievances about mining regulations to President Burgers. In December 1874 he was appointed as Vice Consul of Portugal and in 1879 as justice of the peace for the goldfields in the Lydenburg district.

The Hungarian entrepreneur and financier, Alois Hugo Nellmapius, circa 1880.
(ACKNOWLEDGEMENT: STATE ARCHIVES PRETORIA)

Even though Nellmapius's activities at the goldfield turned him into a wealthy man – his largest find was a gold nugget weighing 123 ounces, which he called "Die Voortrekker" (*De Volkstem*, 11.05.1875) – he left the goldfields a poor man. The sudden collapse of the Lourenço Marques and South African Transport Service, combined with less income from his claims, led to his reduced circumstances (Kaye, 1978).

At the beginning of the 1880s Nellmapius settled on the farm Heatherley outside Pretoria and in 1882 he was appointed as justice of peace for Pretoria. The same year he married Johanna Corlydia (Lily) Hoffman of Durban. They had four children (Kaye, 1978).

In 1889 Nellmapius bought two-thirds of the farm Doornkloof from the Erasmus family. He had the farm subdivided into three sections and after naming the middle one Irene after his daughter, he turned it into a first-class farm. In January 1896 the newspaper *Land and Volk* called it the finest developed property in South Africa. Arnold Theiler, who

General Piet Joubert, commander of the Boer forces during the War of Independence, January 1881.
(ACKNOWLEDGEMENT: TRANSVAAL ARCHIVES)

later became famous for his work at Onderstepoort, laid the foundations for Nellmapius's cattle herds in 1891. He used the services of the English horticulturist Richard Wills Adlam, and later that of the German horticulturist Hans Fuchs, to lay out Irene's orchards and beautiful gardens (Helme, 1960). It cost him more than £200 000 and the services of 400 workers to cultivate the farm. He also built himself a stylish home in Pretoria on the corner of the current Vermeulen and Bosman streets, which he named Albert House, after one of his sons (Helme, 1960).

During Nellmapius's 20-year stay in the Transvaal he made a success of many undertakings and as such made a valuable contribution to the industrial development of the Transvaal. At his death Mrs Gezina Kruger, the wife of the president, declared that every Boer in the country should raise his hat in respect upon hearing the news (*The Press*, 31.3.1893).

The Voortrekkers and later the Republicans' one great ideal was to obtain an independent harbour. With the annexation of Natal in 1843 it became clear that the British government had no intention of allowing this. But the Boers were determined and set their sights on Delagoa Bay.

During his rule (1872–1877) President Burgers did everything within his power to make a road to Delagoa Bay a reality, since the control the British had over the Republic's supply of weapons, in particular, was a constant threat to the continued existence of the ZAR. He was particularly worried that when the Republic became involved in a war with one of the black tribes, the governor of the Cape would cut off the supply of ammunition and gunpowder. This fear eventually became a reality in the Sekhukhune War of 1876 (Appelgryn, 1979).

Nellmapius's contact with the Portuguese government contributed to his view that a road between the ZAR and Lourenço Marques was imperative.

On 29 January 1875, and again on 26 March 1875, Nellmapius submitted proposals to President Burgers and his government from Pilgrim's Rest for the construction of a road between Lourenço Marques and the goldfields at Pilgrim's Rest. He would transport goods from Lourenço Marques by boat, probably up the Komati River to the foot of the Lebombo Mountains, where the goods would be stored in sheds. From there the merchandise would be transported to Pretoriuskop by relays of bearers stationed at convenient spots, and from there by ox-wagon to Mac-Mac. He would build bridges over impassable rivers and streams, where tolls would be charged.

He also undertook to maintain the route in good order and to build overnight facilities, which would be essential, every 24 kilometres along the road. For this purpose he requested land from the government sizeable enough to provide grazing and allow for the cultivation of food for the travellers, bearers and workers. He further undertook to transport the governmental post weekly to and from Lourenço Marques. Nellmapius pointed out to the government that such an organised route would improve trade as there were already three shipping lines that regularly visited Lourenço Marques, namely the Union Line, The British India Steam Navigation Company and Donald Currie's Castle Shipping Line.

A significant amount of assistance from both the ZAR and the Portuguese government, in the form of benefits and privileges, was essential to the success of the endeavour. Since the Portuguese government had already declared its willingness to contribute, he now requested the cooperation of the ZAR.

His request was discussed by the Executive Council, which referred it with recommendation to the

Volksraad on 10 May 1875. The Volksraad, in turn, discussed the matter in detail. Acting President Joubert was in favour of the plan because he saw this road as a precursor of the planned railway line. The criticism levelled against the application was that the government of the ZAR had already spent £1000 on a road between Lourenço Marques and the goldfields.

While some feared that Nellmapius's road would create a monopoly that would be detrimental to the interests of the ZAR, others maintained that the existing road, because of its impassibility, was of no financial benefit to the Republic. And regarding the possible monopoly, an alternative harbour could always be used (*De Volkstem*, 22.05.1875). Accordingly, the Volksraad decided on 11 May 1875 to approve Nellmapius's request. The next step was the drawing up of a contract. This contract consisted of 13 articles and was signed on 18 May 1875 in Pretoria by Acting President P.J. Joubert, Acting State Secretary H. Stiemens and A.H. Nellmapius. In the contract Nellmapius was awarded a concession to build a road from the goldfields at Lydenburg via Pretoriuskop to Lourenço Marques. The road had to be maintained until it was replaced with a railway line. The government provided Nellmapius with eight 3000 morgen (2508 ha) farms, for the establishment of way stations. Six of these farms – all of them about 24 kilometres from each other and situated between Lydenburg and the Lebombo Mountains in the Lydenburg district – were transferred to Nellmapius on 9 May 1876. They were: Burger's Hall, Joubertshoop, Ludwichslust, Coopersdal, Pretoriuskop and Castilhopolis (*De Volkstem*, 23.10.1875). The other two farms, Parsons 310 and Excellence 309 in the Soutpansberg district, were transferred to his name on 7 April 1877. If any gold deposits were found on one of the farms or if one of them were needed for the building of a railway line, the farm involved would be expropriated by the government while Nellmapius would be compensated with another government farm nearby. If Nellmapius failed to adhere to the conditions of the contract before the railway was completed the government would have the right to repossess all eight farms, except if he, as stipulated in article 10 of the contract, was prevented from doing so by a war or another unforeseeable circumstance.

A contract was also drawn up in Lourenço Marques between Nellmapius and the Governor General of the Province of Mozambique, José Guedes de Carvalho Minezes, as the representative of the Portuguese government. The contract was signed on 7 July 1875 in the St Paulo palace in the presence of Secretary General J. Francisco de Salles Lencastre and the commissioner of crown land in the district, Stuart Lucio da Gama. These negotiations led to the founding of a transport company.

The contract stipulated, among others, that a regular weekly transport system between the Lydenburg goldfields and the Lourenço Marques area had to be implemented at own risk and cost. The company further undertook to build its own storage facilities and accommodation and to transport goods and passengers to and from the ZAR. Five overnight stations where food for both passengers and workers would be cultivated were to be built between Lourenço Marques and the border with the ZAR, all of them about 24 kilometres apart. The governor of Mozambique would provide land for the following purposes – one hectare in Lourenço Marques near the custom offices where storage facilities would be built, another two hectares for the building of the company's offices and employees' houses, and 5000 hectares uncultivated land for the building of the five overnight stations. The Portuguese government would supply the wood for the construction of the stations.

The construction of the stations had to start within six months from the signing of the contract and the work had to be completed within two years.

If the company did not adhere to these conditions or if construction was interrupted for a period of more than three months, the government would declare the contract null and void, except in circumstances beyond human control. If the undertaking failed, no compensation would be paid by the Portuguese government.

To honour the conditions of the contract with the Portuguese government, Nellmapius sealed a partnership with Albertos Carlos de Paiva Rapoza, who lived in Lourenço Marques and had experience in road construction. This partnership, with its head office in Pilgrim's Rest, was sealed on 19 July 1875 in the presence of Magistrate H. Cooper of Lydenburg.

10. TRANSPORT ROADS, TRANSPORT RIDERS AND THE TSETSE FLY

Six thousand pounds sterling had to be paid into the partnership, with Rapoza giving the first £3000, after which each partner would pay in £1500. Land supplied by both governments to the undertaking would become the property of both associates in a ratio of two-thirds to Nellmapius and one third to Rapoza. Profits and losses would be shared equally, while costs would be covered by the company.

On 17 August 1875 Nellmapius, as general manager of the company mentioned, sent a letter to Acting President Joubert to inform him that the transport concession awarded him by the two governments would be known as the Lourenço Marques and South African Transport Service. He gave the assurance that the company's services would always be at the disposal of the government. Because the road between Lydenburg and the Spitzkop goldfields formed part of the main connecting road, Nellmapius asked that the government repair this road. From Spitzkop to Delagoa Bay, road maintenance would be the responsibility of the transport company.

On 22 November 1875, when construction of the road was at an advanced stage, Nellmapius informed Acting President Joubert that several travellers and hunters had already used the completed stations. Immigration was aided in that overnight tariffs were very low and some of the poorer travellers received free accommodation.

The company wanted to prevent stations from becoming gambling dens and therefore Nellmapius asked permission to forbid transport riders to use the facilities at the stations. He also asked that the company be exempted from paying customs duties and licensing fees for the stations, to allow it to maintain this high standard of service. In a decision taken by the Volksraad on 15 June 1877 it was recommended that the company should receive as much help as needed and therefore be allowed to pay only 50 per cent of the customs duties on imported goods.

In an advertisement in *De Volkstem* of 23 October 1875, the company named the stations that had been completed. A variety of fresh produce, as well as skins, were made available to traders.

A comprehensive description of the route between the goldfields at Lydenburg and Lourenço Marques was given by William Napier, who had completed the journey in ten days on foot (Dunbar, 1881).

The road went from the Lydenburg goldfields via Spitzkop to the first station, Burger's Hall, which was 187 kilometres from Lourenço Marques.

About 21 kilometres from Burger's Hall was the second station, Pretoriuskop, run by station master J. Feldshaw[1] (Dunbar, 1881). Joubertshoop was 26 kilometres southeast of Pretoriuskop and 141 kilometres from Lourenço Marques. Thomas Hart was the station master here. Ludwichslust, the next station, was 27 kilometres from Pretoriuskop and 114 kilometres from Lourenço Marques on the right bank of the Crocodile River. Dr J. Birch supplied a medical service here. The ford through this river later became known as Nellmapius Drift (Bulpin, 1957). Coopersdal next to the Komati was 88 kilometres from Lourenço Marques – the distance between the

The site at Nellmapius Drift through the Crocodile River. The junction of the old transport road with the current tourist road is clearly visible in the background. (12.06.1984)

Archaeological excavations at Thomas Hart's overnight post at Josekhulu Spruit, southeast of Ship Mountain.

---

1 The correct spelling, as elsewhere in the book, is Feldschau.

rivers was about 26 kilometres. Medical services were available here as well, with Dr H. Pearce the station master.

Along the Lebombo Mountains, 13 kilometres from Coopersdal, was Castilhopolis with W. Boby in charge. R. Lemke was the station master at the next station, Progresso de Guedes (or de Guides), about 24 kilometres from Castilhopolis. The last station, Campos de Corvo, was 26 kilometres from Lourenço Marques, but had not been completed yet. Carlos de Paiva Rapoza, the manager of the company, was in charge here.

The company employed mainly black bearers to transport goods and luggage. From Lourenço Marques to Pretoriuskop a black bearer could earn 20 shillings for carrying a load weighing 100 pounds. If the destination of the goods was Pilgrim's Rest, he was paid another five shillings for every 50 pounds in weight that he carried. Therefore, one bearer's total pay, food included, was £1.10.0 for every 100 pounds in weight carried from Lourenço Marques to Pilgrim's Rest (Dunbar, 1881). On 12 February 1876 the first goods from Lourenço Marques arrived at the goldfields (Bulpin, 1957).

During the Volksraad sitting in May 1882 it was pointed out that the connection with Lourenço Marques was of great benefit to the ZAR because the import duties on goods coming from Natal was ten per cent, while the levy in Delagoa Bay was only three and a half per cent. The wares coming from Delagoa Bay could therefore be sold just as cheaply as in Cape Town (*De Volkstem*, 10.05.1882).

In May 1876 Pedi high chief Sekhukhune became so aggressive that the ZAR was forced to act against him. His warriors hid in the mountainous areas of Lydenburg and because they had guns, multiple attacks became common (Muller, 1975). The Lourenço Marques and South African Transport Service did not escape these attacks and under these circumstances Nellmapius wanted to halt its activities. However, the transport of weapons was essential to the ZAR as Sir Henry Barkly, the governor of the Cape, had placed an embargo on the selling of weapons to the ZAR. President Burgers made a personal appeal to Nellmapius to carry on regardless, especially with the transport of weapons. He gave the assurance that all losses incurred by the company would by covered by

Sir Garnet Wolseley, commander of the joint British, Boer and Swazi forces who finally defeated Pedi High Chief Sekhukhune in November 1879.
(ACKNOWLEDGEMENT: TRANSVAAL ARCHIVES)

Pedi High Chief Sekhukhune, shortly after he had been captured in November 1879 by Sir Garnet Wolseley's punitive expedition force.
(ACKNOWLEDGEMENT: TRANSVAAL ARCHIVES)

government. President Burgers would also provide them with an armed guard of 40 men under the leadership of Field Cornet Van Niekerk.

Nothing came of the promised commando, with the result that the company's stations were plundered and robbed and its employees murdered. On 11 September 1876, W. Barter, the bailiff and police inspector in Pilgrim's Rest, notified the state secretary in Pretoria of these robberies and attacks. He informed him that he had received news on 16 August about the murder of Thomas Hart, the station master at Joubertshoop. During an investigation on 20 August he found Joubertshoop abandoned, burnt down and pillaged, and Hart's remains mutilated by wild animals. After having buried Hart's skull and remains, he also found Pretoriuskop deserted. He brought back 20 rifles, 544 kilograms of lead and six cases of percussion caps. Hart's murder on 10 August 1876 by a marauding band of Chief Maripe's warriors is described in more detail in chapter 7 (Sandeman, 1880). Dr J. Birch, the station master at Ludwichslust next to the Crocodile River, had also evacuated his station because it was not protected. Barter and his volunteers placed this station's weapons and ammunition in the care of a local chief and later had it brought to safety. Barter brought it to the state secretary's attention that the road could only be maintained with the necessary protection.

On 18 September 1876, John Scoble, the gold commissioner at Pilgrim's Rest, informed the state secretary in Pretoria of the murder of Albertos Carlos de Paiva Rapoza, the manager of the company and station master at Campos de Corvo. As the stations had already ceased to function, this meant the end of the company.

The total losses to the company as a result of the Sekhukhune War amounted to £33 000, but on 12 April 1877, before Nellmapius could seek compensation from the government, the ZAR was annexed by Sir Theophilus Shepstone. The Sekhukhune rebellion was finally quelled by Sir Garnet Wolseley in 1879, supported by a Boer commando and a Swazi regiment under Chief Matafin.

Because his claims, which were first directed at Sir Bartle Frere and later at Sir Owen Lanyon, were unsuccessful, Nellmapius directed a request at the Transvaal government after the state had regained its freedom in 1881. In his request of 13 February 1881 he maintained that because he had honoured the contract without having received any compensation, he felt that he was entitled to the eight farms that had been transferred to his name in 1875.

The Executive Council agreed that this request was reasonable. On 14 February 1883 it was decided to ask the Registrar of Acts to treat the contract of May 1975 as having been honoured and to register the eight farms in Nellmapius's name.

After the Transvaal War of Independence (1880–1881), money was scarce and the financial burden on the state large. One way to fill state coffers was to protect industries by means of concessions (Ferreira, 1984). Countries such as England and the USA had proven that monopolies and concessions contributed towards the establishment of industries (Van der Walt, 1944). President Kruger wanted to follow the same route and declared in 1883 that no private institution had the financial wherewithal at its disposal to establish an industry. His aim was to grant concessions so the country's minerals could be processed inside the country, since he considered it a mistake to depend on others for basic consumer goods (Gordon, 1970).

Nellmapius supported this policy. In September 1881 he brought it to the attention of the Triumvirate[2] that the large labour market of the ZAR was being used to profit "strangers" in Europe, without the ZAR receiving any of the benefits.

Trade goods such as maize, coffee, leather ware, soap, cheese and clothing were all imported while they could be produced and manufactured in the Republic. He emphasised that political independence without economic independence was not enough and suggested that the government should encourage entrepreneurship by way of monopolies, patents and certain benefits.

The years 1881–1882 were busy and meaningful to Nellmapius because this was when he contributed to the establishment of important industries in the Transvaal. In September 1881 he applied for a concession to produce liquor and sugar. He was granted the concession and obtained the sole mandate for 15 years.

---

2 Paul Kruger, M.W. Pretorius and P.J. Joubert.

On 7 June 1882 he founded a company with Sammy Marks and Isaac and Barnett Lewis as partners (Ferreira, 1984). Factories were built on the farm Heatherley along the Pienaars River near Pretoria and these were known as the "De Eerste Fabrieken in de Zuid-Afrikaansche Republiek" (The First Factories in the South African Republic). These factories were officially opened on 6 June 1883 by President Kruger and were named "De Volkshoop" (Hope of the Nation) (*De Volkstem,* 09.06.1885).

While Nellmapius was busy building the liquor and sugar factories, he applied for a concession to manufacture all types of gunpowder meant for various purposes. He was granted certain benefits for a period of 13 years, as well as the exclusive right to manufacture bullets during this period. A predetermined amount of gunpowder would be bought every year by the government to assure Nellmapius of a market. This concession was ceded on 8 July and transferred to The South African Pioneer Powder Factory (Ltd.) (Eerste Kruitfabriek in Zuid-Afrika Beperk). This company was also established in Europe with offices in London (Van der Walt, 1944). Some documents refer to this company as "De Voortrekker Kruitfabrick in Zuid-Afrika Beperkt". Lewis and Marks executed this concession and built the first factories of this kind in the Transvaal on the farm Baviaanspoort, outside Pretoria. Nellmapius and State Secretary W.E. Bok formed a committee of directors, who managed the company in Pretoria.

On 28 June 1882 Nellmapius applied to the Executive Council for support in connection with building iron foundries. He wanted to build these foundries on one or more farms where rich iron ore occurred. He also asked that the government institute a levy of two pennies for every pound of iron that was imported, so the locally produced iron could compete with imports. His request was granted for a period of 20 years. During this period £20 500 had to be paid into state coffers and all iron needed for the state artillery had to be provided free of charge. On 29 March this concession was ceded and granted to J.C.A. Anderson.

The negotiations around the granting of the various concessions necessitated close contact between Nellmapius and several leaders of the ZAR. Discussions regarding the building of the road to Lourenço Marques, for instance, led to Nellmapius winning the trust of both President Burgers and Acting President Joubert. This position of trust in turn led to fast bonds of friendship (Lewson, 1960). Nellmapius was invited to Acting President Joubert's home in as early as 1874. Although Joubert and Nellmapius had different political viewpoints, Joubert, as commandant general during the Mapoch War in 1883, nevertheless used Nellmapius's services as a dynamite expert (*De Volkstem*, 10.02.1883). His success during the war did not go unnoticed. On 30 June 1883 the state secretary wrote a letter to Nellmapius in which he thanked him for his contributions towards the success of the war.

After Nellmapius had settled in Pretoria in 1881, his friendships with high-ranking officials strengthened, especially with Paul Kruger. It was Nellmapius who organised a farewell for President Paul Kruger and his deputation to London in 1883.

Upon their return in 1884 they were welcomed back with an elegant party at Sesmyl Spruit (Krüger, 1963). In 1885 a financial commission was appointed to discuss the budget. Nellmapius was not a member of the commission, but President Kruger had so much faith in his judgement that he insisted that Nellmapius be present during the negotiations with Standard Bank.

On 29 September 1886 Nellmapius was tried by Judge Brand in a court case that lasted nine days. He was found guilty of the embezzlement of funds

The signing of the peace accord after the end of the War of Independence at O'Neil's farmhouse at the foot of Amajuba. (21.03.881). (ACKNOWLEDGEMENT: TRANSVAAL ARCHIVES)

and sentenced to 18 months' imprisonment. At the request of Nellmapius's advocate, one Hollard, six points against this sentence by Judge Brand were reserved for the adjudication of the full bench of the court. According to article 83 of the Constitution, the President and the Executive Council had the right, after having obtained advice from the bench, to pardon a person under sentence or to reduce the sentence (Bell, 1887). With the approval of the Executive Council President Kruger pardoned Nellmapius, without awaiting the further course of events and without the approval of Judge Brand, which in turn led to the judge's resignation. The whole incident created the impression that the president had tried to obstruct justice in a dictatorial manner in the interests of one of his friends (Bell, 1887).

Nellmapius's friendship with the president came in for a lot of criticism. The newspaper *Land en Volk* in particular took a strong stance against the president. Eugène Marais wrote in an article in this paper that Paul Kruger had surrounded himself with Jews and the Dutch: With Nellmapius to his right, the German Eduard Lippert to his left and Hermann Ekstein, Sammy Marks and a whole series of Jews behind him, access to the president was scarcely possible.

With the death of Nellmapius on 28 July 1893 flags in Pretoria were flown at half mast. President Kruger, Vice President N.J. Smit, Commandant General Joubert and other members of the government were present to pay their last respects to Nellmapius (Gordon, 1970).

References

Appelgryn, M.S. 1979. *Thomas Francois Burgers, Staatspresident 1872–1877.* HAUM, Pretoria.

Bell, W.H.S. (ed.). 1887. *The Cape Law Journal, Under Superintendence and Control of the Incorporated Law Society of the Cape of Good Hope* vol. IV. Josiah Slater, Grahamstown.

Bulpin, T.V. 1957. *Lost Trails of the Transvaal.* H. Timmins, Cape Town.

Dunbar, D.M. 1881. *The Transvaal in 1876. Extracts from a diary of the late Hon'ble William Napier while travelling between Lydenburg Goldfields and Delagoa Bay.* Richards, Grahamstown.

Ferreira, O.J.O. 1984. Eerste Fabrieken: Die beginpunt van nywerheidsontwikkeling in die Zuid-Afrikaansche Republiek. *Contree* 15, 15 January.

Gordon, C.T. 1970. *The Growth of Boer Opposition to Kruger 1890–1895.* Oxford University Press, London.

Helme, N. 1960. *Irene. A Pictorial History of a Transvaal Farm and Village.* Unknown publisher, Pretoria.

Kaye, H. 1978. *The Tycoon and the President.* Macmillan S.A. (Pty) Ltd., Johannesburg.

Kotze, J.G. & Barber, S.H. 1903. *Reports of cases decided in the Supreme Court of the South African Republic (Transvaal), January 1885 to December 1888* vol. 11. Juta, Cape Town.

Krüger, D.W. 1963. *Paul Kruger, 1883–1904* part II. Dagbreek, Johannesburg).

Lewsen, P. (ed.). 1960. *Merriman, John Xavier, Selections from Correspondence 1887–1890.* Van Riebeeck Society, Cape Town.

Muller, C.F. (ed.). 1975. *Vyfhonderd jaar Suid-Afrikaanse geskiedenis.* Academica, Pretoria.

Sandeman, E.F. 1880. *Eight Months in an Ox-waggon.* Griffiths & Farran, London.

Van der Walt, N. 1944. *Die dinamiet monopolie: Sy ontstaan en ontwikkeling vanaf 1881–1892.* Unpublished M.A. thesis, University of Pretoria, Pretoria.

Primary sources

De Volksstem: 11 May 1875; 22 May 1875; 23 October 1875; 10 May 1882; 9 June 1983; 10 March 1883.

The Press: 31 July 1893; 31 July 1893.

Transvaal Archives (TA), Pretoria. Archives of the State Secretary, Division Home Affairs.

(a) Incoming documents indicated as SS 1875–1893: SS176, R1480/74: A.H. Nellmapius to T.F. Burgers, Pilgrim's Rest, 3 October 1875.

SS1127, R3233/86 added to R5688/85: L. Ogden to S.J.P. Kruger, London, 10 July 1886.

SS178, R1869/74: A.H. Nellmapius to T.F. Burgers, Pilgrim's Rest, 7 December 1874.

SS8624, BB1553/74: N.J.R. Swart to A.H. Nellmapius, Pretoria, 3 December 1874.

SS353, R2490/79: Magistrate of Lydenburg to State Secretary, Lydenburg, 14 July 1879.

SS8657, R4304/82: W.E. Bok to A.H. Nellmapius, Pretoria, 5 August 1882.

SS8624, BB1553/74: N.J.R. Swart to A.H. Nellmapius, Pretoria, 3 December 1874.

SS184, R294/75: A.H. Nellmapius to T.F. Burgers. Pilgrim's Rest, 29 January 1875.

SS184, R703/75: A.H. Nellmapius to P.J. Joubert, Pilgrim's Rest, 26 March 1875.

SS202, Supl. R6/75: Contract between A.H. Nellmapius and the ZAR Government, Pretoria, 18 May 1875.

SS334, ad. R936/79: Contract between A.H. Nellmapius and the ZAR Government, Pretoria, 18 May 1885.

SS334, ad. R4489/79 added to R936/79: Colonial Secretary to Registrar of Deeds, 06 April 1886. These eight farms would become the property of Nellmapius after he had fulfilled the requirements of the contract. These farms were only transported to his name in February 1883.

SS334, R4489/79 added to R936/79: Contract between Nellmapius and the Portuguese Government, Lourenço Marques, 7 June 1875.

SS146. R950/72: C. de Paiva Rapoza to T.F. Burgers. Lourenço Marques, 17 June 1872.

SS334, R2145/79 added to R936/79: Contract between Nellmapius and C. de P. Rapoza, Pilgrim's Rest. 19 July 1875.

SS192, R1714/75: A.H. Nellmapius to P.J. Joubert, 17 August 1875.

SS198, R2671/75: A.H. Nellmapius to P.J. Joubert, 22 November 1875.

SS334, ad. R236/77: N.J.R. Swart to A.H. Nellmapius, Pretoria, 29 September 1876.

SS215, R2442/76: J. Scoble to N.J.R. Swart, Pilgrim's Rest, 18 September 1876.

SS215, R2359/76: W. Barter to N.J.R. Swart, Pilgrim's Rest, 11 September 1876.

SS215, R2443/76: J. Scoble to N.J.R. Swart, Pilgrim's Rest, 18 September 1876.

SS558, R3613/81: A.H. Nellmapius to Triumvirate, Pretoria, September 1881.

SS558, R4109/81: Secretary of the Volksraad to W.E. Bok, Pretoria, 07 October 1881.

SS825, Petition added to R2881/83: Nellmapius and Sammy Marks to S.J.P. Kruger, Pretoria, 16 June 1883.

SS580, R4992/81: A.H. Nellmapius to Executive Council, Pretoria, 11 November 1881.

SS585, R3312/82: Contract between Government and Nellmapius, Pretoria, 8 December 1881.

SS837, Supl. added to EU653/83: Supplements to the original agreement, Pretoria, 8 July 1882.

SS837, R3225/53 added to R3653/83: A.H. Nellmapius and W.E. Bok to S.J.P. Kruger, Pretoria, 9 July 1883.

SS867, R547/83: The South African Pioneer Powder Factory Directors, London, 12 October 1883.

SS1236, R3706/82 added to R3015/86: A.H. Nellmapius to Executive Council, Pretoria, 11 October 1881.

SS1236, R3015/86: Contract between Nellmapius and Government, Pretoria, 7 July 1882.

SS1236, R1109/97 added to R3015/86: Contract transferred to J.C.A. Henderson, Pretoria, 29 March 1889.

SS188, R1019/75: A.H. Nellmapius to P.J. Joubert, Pretoria, 13 May 1875.

SS8659, BB220/83: W.E. Bok to A.H. Nellmapius, Pretoria, 30 January 1883.

(b) Supplementary document, band SS202:
SS202, Supl. R6/75: Contract between A.H. Nellmapius and P.J. Joubert, Pretoria, 18 May 1875.

(c) Letter Books indicated as BB 30 January 1883; 3 December 1874.

Archives of the Secretary of the Volksraad

(a) Incoming documents indicated as EVR, VRR:
EVR. 232, VRR209/83: VRB, 14 July 1883, art. 770.
EVR. 232, VRR209/93: Report of commission, VRB, 14 July 1883, art. 770
EVR. 250, VRR 314/85: Minutes Financial Commission, Pretoria, 23 July1885.

(c) Executive Council indicated as URB:
UR7, URB, 7 October 1886, art. 225 UR7;
URB, 6 October 1886, art. 257

10.3 Further development of the transport road between Lydenburg and Delagoa Bay, the Old Harbour Road, Long Tom Pass and other transport roads in the Lowveld

Dr U. de V. Pienaar

After the initial logistical and maintenance problems that Napier had described so well (Dunbar, 1881), Nellmapius's trade route between the Lydenburg goldfields and Delagoa Bay had just started to function properly when the first Sekhukhune Rebellion broke out in May 1876. The station master at Joubertshoop, Thomas Hart, was one of the first victims when he was murdered at his station on 11 August 1876[1] by a marauding band of warriors under Chief Maripe, a subordinate ally of Sekhukhune.

The supervisor of the Campos de Corvo outpost, Albertos Rapoza, was also murdered and the other station masters withdrew to Lydenburg and Pilgrim's Rest for their own safety.

Captain W. Barter, the bailiff and police inspector, and 11 volunteers were sent from Pilgrim's Rest to collect the supplies at the stations, especially rifles, gunpowder and lead. On 20 August they buried Hart's mutilated remains close to the burnt-out ruin of his wattle-and-daub home.

The First Sekhukhune War ended in January 1877 through the mediation of Reverend Merensky of the Botshabelo Mission, but President Burgers's half-hearted attempts to subjugate the Pedi chief soon led to further problems with the black tribes in the area (Aylward, 1879). All traffic between the Lydenburg goldfields and Delagoa Bay was brought to a halt. On 7 April 1877 Sir Theophilus Shepstone annexed the Transvaal and on 4 March 1879 Sir Owen Lanyon was appointed as administrator of the Transvaal. By this time it was clear that huge problems were once again developing with the Pedi high chief Sekhukhune and his headmen, including Maripe, Mapoch and Umsutu. Lanyon (who soon became known as *Ou*

Field Cornet Ferreira, who together with Capt. Conrad Hans van Schlickman and Capt. Mostyn Owen, played a leading role during the Sekhukhune War of 1876–1879.
(ACKNOWLEDGEMENT: TRANSVAAL ARCHIVES)

President Marthinus Wessel Pretorius in about 1860.
(ACKNOWLEDGEMENT: STATE ARCHIVES)

Sir Theophilus Shepstone (middle front) and his personnel after the annexation of the Transvaal by Great Brittain in 1877.
(ACKNOWLEDGEMENT: TRANSVAAL ARCHIVES)

---

1 According to Sandeman (1880) the date was 10 August 1876.

Langjan [Old Longjohn] among the Boers), also failed to control the situation and did not get on with the Boers.

After the Anglo-Zulu wars and the subjugation of Cetshwayo, the Zulu high chief in Natal, British troops under the experienced Sir Garnet Wolseley could be sent to suppress the second Sekhukhune rebellion once and for all. With the help of 8000 Swazi warriors under the leadership of Mataffin, as well as a Boer commando, they conquered Sekhukhune's mountain stronghold on 27 November 1879. Sekhukhune and his indunas were captured three days later by captains Mostyn Owen and Ignatius Ferreira and sent to prison in Pretoria.

The Boers were unhappy with the British occupation of the Transvaal and regained their freedom through great victories over the British troops. The main battles took place at Bronkhorstspruit (20 December 1880), Laingsnek (27 January 1881), Ngogo (8 February 1881) and Majuba, where General George Pomeroy Colley's troops were beaten decisively on 27 February 1881. Colley and 92 of his men were killed in this battle. On 23 March 1881 Sir Evelyn Wood, through the mediation of President Brand of the Free State, signed a peace treaty with General Piet Joubert, and the ZAR (South African Republic) once again became a self-governing state. A royal commission was appointed to draw up a convention that would spell out the terms of independence, and this Pretoria Convention was accepted by the Boer leaders on 3 August 1881. Five days later the Triumvirate – Paul Kruger, Piet Joubert and M.W. Pretorius – took control of the Transvaal and ruled until Paul Kruger's election as president in 1883.

Only then could the traffic in trade goods between Transvaal and Delagoa Bay resume in an organised manner. Nellmapius's Lourenço Marques and South African Transport Service had been driven to bankruptcy by the wars, and therefore the onus to improve and make the old transport routes passable for wagon traffic was on transport riders themselves. They also had to develop a trek procedure to protect their draught animals and horses against tsetse flies in the affected areas (Hall, 1937). The resourcefulness of this hardened group of pioneers was limitless and they soon overcame all these problems. Along the transport route through the area that is now the Kruger National Park they chose new outspans where there was enough water for both man and beast. These new outspans included Pretoriuskop; Fihlamanzi ("the hidden water"), a fountain close to the intersection of the tourist roads between Skukuza and Malelane and between Pretoriuskop and Malelane, and a day's travel from Nellmapius Drift; the Ntomeni waterholes close to the current Voortrekker windmill; and the waterholes in the Samarhole Spruit

General Sir George Pomeroy Colley, commander of the British forces, who died on 27 February 1881 during the battle of Amajuba.
(ACKNOWLEDGEMENT: SOUTH AFRICAN DEFENCE FORCE ARCHIVES, PRETORIA)

An artist's depiction of Gen. George Pomeroy Colley leading his men on Amajuba (27.02.1881).
(ACKNOWLEDGEMENT: SOUTH AFRICAN DEFENCE FORCE ARCHIVES, PRETORIA)

Lt Gen. Sir H. Evelyn Wood who signed the peace accord after the War of Independence on 21 March 1881 on behalf of the British Crown.
(ACKNOWLEDGEMENT: TRANSVAAL ARCHIVES)

Ranger S.H. Trollope's father-in-law, Mr John Glen-Leary and field ranger Follok at Fihlamanzi fountain.
(ACKNOWLEDGEMENT: C.A. YATES AND STEVENSON-HAMILTON-COLLECTION)

Far right: Fihlaminzi fountain ("hidden water") – a well-known outspan along the old Delagoa Bay transport road.
(11.05.1984)

Jock of the Bushveld guarding a kudu which his master, Sir Percy FitzPatrick, had shot.
(ACKNOWLEDGEMENT: E. CALDWELL AND MRS C. MACKIE-NIVEN)

on the western flank of Skipberg. The road itself was improved and drifts through the streams were made more passable. They even found an alternative route through the Lebombos at Matibaskom, just north of Mathala Poort, and this was also improved. Fortune seekers, adventurers, diggers, gamblers, traders, transport riders and other odd characters and hangers-on were still rushing to the goldfields at Pilgrim's Rest and Mac-Mac, with the result that the old Nellmapius transport route was a hive of activity in the 1880s.

Convoys of transport wagons, with long teams of oxen – up to 24 a wagon – conveyed trade goods, mining equipment and other provisions from Delagoa Bay to the interior.

The transport business became very lucrative. Many well-known figures in the Lowveld such as H.L. Hall, Donald Mackay, Org Basson, Charlie Roberts, the Sanderson brothers, Mostyn Owen, Abel Erasmus, James Fullerton, Ted Sievewright and others were in the transport business for periods of varying duration.

One person who participated in this romantic trade and who later became famous through his books *Outspan*, *Jock of the Bushveld* and others, was Sir Percy FitzPatrick. As a young man he went to the Lydenburg goldfields, but soon joined a group of transport riders and undertook numerous journeys from Lydenburg via Krugerspost, Pilgrim's Rest and Pretoriuskop to Delagoa Bay and back. It was during one of these journeys that his famous dog, Jock, was born at a camping site on the eastern bank of the Samarhole Spruit. One of Percy's partners, Ted, had a pregnant bull terrier bitch named Jess. One day in May 1885 she gave birth to a litter of six in the grass next to the camp, and Jock was the runt. Because Percy was the youngest among the group of transport riders he got the ugliest puppy, but it would later prove to be the pick of the bunch. This boerbull-bull terrier cross would have many adventures with his owner on the transport routes through the Lowveld – events that Percy later immortalised in his book.

From Nellmapius Drift through the Crocodile River, the old transport road went in a northwesterly direction across the sodic plain along the upper reaches of the Lwakahle Spruit to Fihlamanzi Fountain (just southwest of the current Skukuza-Malelane-Pretoriuskop junction). From there it passed between a half to one kilometre south of the current tourist road between Pretoriuskop and Malelane, going up to the Ntomeni Drift. Then it followed more or less the course of the current tourist road between Pretoriuskop and Malelane to the vicinity of the Komapiti windmill. After that the old transport route swerved to the north to avoid the impassable gabbro flats on the northern flank of Skipberg (or Langkop, as it was known to the transport riders). Then the route ran between half and a kilometre north of the current tourist road in the Park up to the spot where it crosses the old road between Pretoriuskop and the rhino camp. After that it followed the old road to the current Pretoriuskop rest camp, past the Faai waterholes (the location of the Danish Feldschau's camp, at the foot of Manungu Hill). From there it ran between Manungu Hill and Pretoriuskop, passing close by Willem Pretorius's grave, to Burger's Hall. The old road left the Park area just south of the current Numbi Gate and then followed the route described elsewhere in this book (also see the map titled: "Old transport and wagon roads, southern Lowveld and Escarpment").

Because the old transport road swerved north of the current tourist road in the vicinity of Skipberg, Jock's real birthplace could not be found by former rangers Harry and Henry Wolhuter and Stevenson-Hamilton, nor by the members of the Percy Fitz-Patrick Trust, who researched the old transport routes through the Lowveld in 1948. FitzPatrick described the site as follows in *Jock of the Bushveld*: "We had rested through the heat of the day under a big tree on the bank of a little stream; it was the tree (near Ship Mountain [Skipberg]) under which Soltké prayed and died." The grave of the German Soltké was therefore the key to locating Jock's birthplace. FitzPatrick described the drama of Adolf Soltké's death in *The Outspan*, which was published in 1897. He wrote:

We had passed the Crocodile River and the belt of 'tsetse fly' country. We had passed Josikulus where Hart was murdered by the niggers, and we told Soltké the story of the dead man's sentry-go. We passed Ship Mountain,[2] and pointed out the bush that hid the haunted cave, and told him the weird tradition of the old witch-doctor imprisoned by the rock slide, handling still as a skeleton the implements of magic he used in life … It was on a Sunday morning that we came in sight of Pretoriuskop – a solitary sugarloaf hill – and we lay by as usual during the hours of daylight. We knew it was Sunday, because Soltké had said so, and because we saw him in the early morning kneeling in the shadow of a big tree a few yards from the wagons, prayer-book in hand, absorbedly following the prayers of the Mass. He was a Roman Catholic …

The northwestern flank of Ship Mountain (Long Kop). It is here where the traditional healer's cave collapsed.
(16.06.1984)

Ship Mountain, or Long Kop, as it was known among the transport riders (1876–1892).
(08.06.1984)

2 Also called Langkop, Saddleback or Mhukweni.

Adolf Soltké, a young German of about 23, was found at the side of the road just outside Delagoa Bay by a group of transport riders, with FitzPatrick a member of the party. Apparently Soltké was an ignorant young man who came to find his fortune on the Lydenburg goldfields. He wore a suit and a bowler hat, with an umbrella to protect him from the Lowveld's scorching sun. The transport riders took pity on the young man with his broken English and he joined them on their journey to the interior. Apparently the group constantly poked fun at him, something which he accepted without complaint. His good nature soon won them over and they all became very fond of him. No one could have guessed what tragic fate lay in store for him.

The northeastern corner beacon of the farm Pretoriuskop, as determined in 1875 by Surveyor P.W. de Villiers.
(07.08.1985)

The site of Adolf Soltké's grave and Jock of the Bushveld's birthplace along the Samarhole Spruit, northwest of Ship Mountain as seen from the south.
(16.06.1984)

At the outspan described above, a roller flew past the camp. Soltké, who in the meantime had learnt to shoot, was collecting birds to show his parents in Germany. In his excitement at the sight of this beautiful specimen he jumped onto the wagon and grabbed a double-barrelled shotgun. While jumping to the ground, both barrels of the shotgun went off by accident, and shattered the young man's right leg below the knee.

The nearest doctor was more than 100 kilometres away at the Mac-Mac goldfields and although one member of the group, Donald Mackay, set off immediately to find help, the only assistance that arrived after two days was a drunken chemist called "Doc" Munroe. He was so inebriated that he could do nothing for the injured man.

Another two days dragged by before a qualified doctor arrived from Mac-Mac, after a journey of 110 kilometres on horseback. It was too late, however, and although he amputated Soltké's leg that very night, the young German died early the next morning of gangrene. His friends buried him under the same tree where he had prayed a few days before. An unknown party later stacked stones on the grave.

Jock was born a few months later at this very spot, during a journey between Lydenburg and Delagoa Bay.

Thanks to the careful tracking by senior ranger Thys Mostert and his team of trackers from Pretoriuskop – Charlie Nkuna, July Mona, Ngwela Mangane, Salmão Mongoe and Million Zitha – the entire route of the old transport road between the western border of the Park and Fihlamanzi Fountain was redetermined and demarcated. At the spot where the old road crossed the Samarhole Spruit, within sight of both Pretoriuskop and Skipberg, they found a white person's grave on the eastern bank of the spruit.

An examination of the remains in the grave confirmed that it was the grave of Adolf Soltké, who had been resting there for over a 100 years. Pieces of his boots and shirt and a belt buckle were found, as well as two iron rings and a rope that had been used to lower the body into the grave.[3] This find and the examination of the contents of the grave confirmed it as being Jock's birthplace.[4] The grave was disturbed as little as possible and was afterwards carefully restored to its original condition.

3   The iron rings could also have been part of his belt.

4   It is claimed in certain circles (Gray, 1986 and others), that Jock of the Bushveld never existed and that the whole story was a young man's flight of imagination. He is said to have personified everything he had not achieved yet in the heroic deeds of his dog. The finding of Adolf Soltké's grave at the place he had described in his book *Outspan* will certainly give these skeptics food for thought.

10.3 FURTHER DEVELOPMENT OF THE TRANSPORT ROAD IN THE LOWVELD

The grave of another white person was found about two kilometres northeast of that of Soltké, on the right-hand bank of the Mbyamiti Spruit (Pretorius River, or Veldkraal Spruit in the days of the transport riders). It was probably that of a hunter or transport rider who had died of malaria, or had been killed by a wounded animal such as a buffalo.

Other sites of historical importance that could be identified, or reidentified, by rangers Thys Mostert, Mike English and the author, were: Willem Pretorius's grave between Numbi Gate and Pretoriuskop; the outpost of Manungu (Albasini's induna and cattle herder) on the eastern flank of Manungu Hill; the Danish Feldschau's overnight station just north of the Pretoriuskop-Manungu road; and the boundary beacons (piles of stones) of the farms Pretoriuskop and Joubertshoop, which had been given to Nellmapius in 1875 in reward for the construction of a road from Pilgrim's Rest to Delagoa Bay. There was also a large spekboom (*Portulacaria afra*) stand on the upper reaches of the Mbyamiti Spruit, west of the Môrester windmill, which had probably been planted by the transport riders (these trees do not occur anywhere else in this part of the Park); a camping place on the western bank of a tributary of the Mbyamiti that had probably been used during the Anglo-Boer War by Steinaecker's Horse (judged by the many military bullets, buttons – including one from a uniform of the 24th Warwickshire Infantry Regiment, which had fought against the Zulu at the battle of Rorke's Drift – and other relics); and the well-known outspan at the Ntomeni waterholes where one can find, apart from the many relics dating back to those days, a leadwood tree that was used for target practice by the transport riders. They shot at it with their Martini-Henry and Westley-Richards rifles, and some lead bullets can still be seen in the trunk. The ruins of Thomas Hart's overnight station and his grave were found on the western bank of the Josekhulu Spruit; the pile of stones marking the spot near the upper reaches of the Hlambanyati Spruit (the Shipalatsene) where John Glen-Leary, the father-in-law of ranger S.H. Trollope, was attacked by a leopard on 3 August; the camping site where one Dartnell apparently ran a smithy for a while in the 1880s, where the Barberton transport road turns off the Lydenburg-Delagoa Bay road (one of these hills is probably the

A large stand of spekboom along the old Delagoa Bay transport road, northwest of Ship Mountain and west of Môrester windmill. (12.11.1985)

The remains of the foundations of Alf Roberts's shop, Tengamanzi, at Nellmapius Drift through the Crocodile River. (27.07.1984)

Martini-Henry bullet holes in the trunk of a leadwood tree along the Ntomeni Spruit at Voortrekker windmill. This had been a popular outspan of transport riders during the nineteenth century. (16.06.1984)

10. TRANSPORT ROADS, TRANSPORT RIDERS AND THE TSETSE FLY

Cigarette Kopje of which FitzPatrick spoke when telling how he had got lost while following a wounded kudu bull); the outspan at Fihlamanzi; and the ruins of Alf Roberts's store Tengamanzi on the northern bank of the Crocodile River at Nellmapius Drift.

This store was the logical outcome of the decade 1882–1892, when the transport road between Lydenburg and Delagoa Bay was at its busiest. A whole series of overnight stations and stores were built at strategic places along the route and delivered a much valued service to the transport riders. The best-known of these amenities were Tom Barnett's store at Piscene in Mozambique, the Indian trader Dorabjee Dhunjbhoy's store at the Matibiskom border post, Julius Furley's "hotel" on the southern bank of the Komati River at Furley's Drift, Alf Roberts's store Tengamanzi on the northern bank of the Crocodile River at Nellmapius Drift, and Nolan's Lekkerlag hotel at Spitzkop near the turn-off, about 13 kilometres southeast of Sabie. At Nolan's hotel a branch of the road turned off to the Mac-Mac goldfields and McLachlan's Creek.

Tom Barnett's store became well known as this was where Jock of the Bushveld died. He was left there by Percy FitzPatrick in the care of his friend Barnett after he had moved to the goldfields at Barberton. Tom accidentally shot Jock one night, under the impression that there was a stray dog in his chicken pen.

Near Jock's grave is the grave of his arch-enemy, Field Cornet Seedling of Krugerspost, who once tried to have Jock savaged by his tame male baboon (with fatal results to the baboon). In revenge Seedling set the village of Jim Makokela (FitzPatrick's wagon driver) on fire and fled on horseback to Mozambique. He was followed, but all they found was his body – by the side of the road in Mozambique where he had died of heatstroke. He was buried at Tom Barnett's store.[5]

Legend has it that there were a lot of crocodiles at Furley's Drift through the Komati. Apparently Furley fed these crocodiles, the idea being that they would frighten travellers into using his ferry to cross the river. Later policemen shot his tame crocodiles, which was a great setback to him. Wild crocodiles soon arrived to replace them and once again created a dangerous situation at the drift. Once, when the well-known adventurer Mostyn Owen (also known as "Mad Mostyn") was travelling along this road, he crossed the Komati at Furley's Drift on horseback and his horse was attacked by a crocodile. He hit the crocodile on the head with his stirrup until it let go of the horse, but this fine animal was badly mutilated.

During the scuffle Owen lost his rifle in the river. Outraged, he walked to Furley's hotel and asked for a rifle to shoot the crocodile. Nobody would help him, but eventually a travelling Jewish tailor named Haller offered his assistance. When the two reached the drift, Owen said: "Now listen. Here's the rifle.

Porcelain insulators on the trunk of a large wild fig tree on the northern embankment of the Crocodile River. These are the only remains of the telegraph line to Tengamanzi Shop.
(27.07.1984)

Sir Percy FitzPatrick's daughter, Mrs C. Mackie-Niven and assistants at the place where Jock of the Bushveld was buried at Pessene in Mozambique in 1949.
(WITH THE KIND PERMISSION OF MRS C. MACKIE-NIVEN)

---

[5] According to W.H.J. Punt, who visited Piscene (or Pessene) on 31 March 1953, there were other graves in the vicinity of Tom Barnett's store, about 300 metres southwest of the railway station. These were the graves of Johannes Gideon Slabbert (born 13 Nov. 1846, died on 28 July 1918) and his son Johannes Gideon Slabbert (born on 7 Dec. 1891, died on 24 July 1929).

I'll jump into the river and when the crocodile comes for me, shoot it!"

Haller immediately protested. "But I can't shoot. What if I hit you!"

"Fine," said Owen. "Give me the rifle and you jump in!" Haller took off his clothes, folded them neatly, whispered a prayer and jumped in. Owen got his crocodile, but it is said that Haller lost all of his hair soon afterwards – according to him it was because of delayed fright! (Bulpin, 1950). Alf Roberts's store at Tengamanzi next to the Crocodile River was a very popular outspan – attesting to this fact are the many broken bottles, gin crocks (demijohns), rusty bully beef tins, harnesses, a rusted wagon brake and other wagon parts that can still be found. The place even had a telegraph line from Hectorspruit. The line's porcelain insulators can still be seen today on the trunk of a large wild fig tree on the river's northern bank.

Alf Roberts had another store at Malelane where he, according to W.H.J Punt, was murdered in 1941 by an unknown assailant, or assailants. The remains of Birch's overnight station on the southern bank of the river have not been found yet.

In sharp contrast with the inns that provided transport riders with shelter and meals and the shops where they could buy necessities, there were also wooden and corrugated iron shops erected in the southeastern part of the Kruger National Park with the express purpose of robbing the blacks returning from the mines on the Witwatersrand of their hard-earned money. Alcoholic concoctions of dubious origin as well as wares of inferior quality were sold at exorbitant prices to unsuspecting customers. One of these dishonest shopkeepers was Sardelli the Greek – a rough and brutal man and as cunning as a crocodile. He and his partner Charlie Woodlands, a miserable layabout who always had a pipe clamped in his slobbering mouth, built a string of these "shops" along the footpaths used by the returning mine workers. The blacks referred to Sardelli as "Mfishane" (the short little man). His "shop" was at Gomondwane, about 10 kilometres northwest of Crocodile Bridge, where until recently there was a stand of eucalyptus trees they had planted. A second shop was situated on the Nhlanganzwani firebreak, about 7,5 kilometres south of the dam at the foot of Ndzengendzenge Hill in the Lebombos near the Sabie River. There was also a third one next to the Crocodile River just west of the current Nkongoma outpost. It was in this little corrugated-iron building near the Sabie River that Bill Sanderson was interned for several months by Steinaecker's Horse during the Anglo-Boer War because he would not fight the Boers.

After the surveyor George Moodie had completed the road from White River to Burger's Hall in later years, a certain Logie built a boarding house at the bottom of Moodieskloof at the foot of Legogote. The ruins of this building can still be seen and this was apparently the origin of the old name for Legogote: Lozieskop (or Logieskop) and of the Nsikazi River (Lozies or Logies River).

After the discovery of the Kaapsche Hoop and Barberton goldfields in 1882 and the foundation of Barberton in 1884, a shorter route than the one via Pretoriuskop and Burger's Hall to the new digger community became essential.

A branch route for wagons was built at about this time (1883) from a point near the Makhuthwanini hills, southwards past Rooikop and Stolsnek to the Nsikazi Drift (or Logies River Drift), near (north

Eucalyptus trees on the terrain at Gomondwane where the shop of the Greek, Sardelli and Charlie Woodlands used to be. (12.06.1984)

of) the Mncumeni outpost and the Luketseni mouth. From here the route went via Karino to Barberton.

Recent investigations inside the Park have revealed that the original transport road to Barberton had swerved westwards towards Barberton near Fihlamanzi fountain and followed a mostly westerly route from there. It followed the watershed between the Mlambane and Mbyamiti streams, went south of Newu Dam to the vicinity of the drift through the Mangake Spruit near the current Stolsnek ranger post, where it joined the later Makhuthwanini-Stolsnek route to Barberton (see map titled "Old transport and wagon roads, southern Lowveld and Escarpment).

To avoid the tsetse belt to the east, this road was built from Karino and Crocodile Poort through Salvation Valley to Barberton. The road followed the course of the Amen Spruit, a tributary of the Crocodile River, and on its banks Edward Hardie Gould and his wife Florence built the Resurrection Hotel. Legend has it that they cultivated excellent tobacco here (Bulpin, 1965). In those years hunters also constructed a road past the southern flank of Skipberg. Senior rangers M.C. Mostert and M. English redetermined the course of this road and in the process found signs of a hunting camp between Skipberg and Inhliziyo's village.

In later years, first Pettigrew (1886) and then the well-known French delver and prospector Auguste Robert, also known as French Bob (1888), tried to build shorter roads from Barberton and Eureka City to the old Lydenburg-Delagoa Bay road. Pettigrew constructed his road along the Kaap River and from there along the Crocodile River to Nellmapius Drift, where it joined the main transport route to the Bay. This road was passable, but it went through one of the worse tsetse belts in the Lowveld and was therefore not very popular among the transport riders. Percy FitzPatrick used the road once and in the process lost almost all of his trek animals to nagana. French Bob's road from Sheba Mine was built further south, closer to the Swaziland border, and from there it followed the northern bank of the Mlumati River to Furley's Drift through the Komati, but this was also a difficult road because of the steep, mountainous terrain.

Members of the Lowveld Diggers and Transport Riders' Association, especially B.P. Simmons and D. Coetzee, made an intensive study of the transport routes in the southern Lowveld outside the boundaries of the Park and in 1985 published a map that shows all the known routes. A road that was constructed in as early as 1889 by the well-known hunter and native commissioner of the ZAR, Abel Erasmus of Krugerspost, went from Magashulaskraal, where Albasini had his trading post from 1846–48, through Mambatine's Drift in the Sabie to his farm Orinoco, where he kept his prized Afrikaner cattle. From here the road was later extended to Leydsdorp.

One of the old transport riders who could clearly remember this part of the transport road through the Kruger National Park was Org Basson, or Uncle Org as he was better known. Uncle Org visited the Park in 1950 with Dr W. Punt. They were accompanied by Izak Holtzhausen of Barberton (more about him in chapter 16.1), who had been a Zarp (member of the ZAR Police) stationed at Komatipoort. During the visit Uncle Org recalled some of his experiences during his transport riding days. He also confirmed that the Lowveld as a whole had been much wetter during the nineteenth century than today and that they never had any trouble finding water for their large teams of trek-oxen along the transport routes.

Org Basson rode transport on the Lydenburg-Delagoa Bay road, which crossed the southwestern corner of the Park, between 1883 and 1885. According to him, the road opened up by Nellmapius in 1875–76 followed more or less the same route as the

Transport riders at Poort City, Crocodile Poort during 1890.
(ACKNOWLEDGEMENT: ALBERT SCHATZ COLLECTION AND MR H. BORNMAN)

current tourist road between Pretoriuskop and Malelane. He was, however, of the opinion that the old road had been between one and three kilometres north of the tourist road in the vicinity of Skipberg. This was later confirmed by senior ranger Thys Mostert's research.

The following remarks made by Org Basson are also important (see Punt, 1950):

> They saw eland and oribi at Shabeni Hill in the years 1883–85. Izak Holtzhausen confirmed this.
>
> There were a lot of lions in the vicinity of Pretoriuskop and he and the Stols brothers often hunted them. One of the Stols brothers died of a heart attack after one of these lion hunts and was buried at the foot of Manungu Hill. The best lion hunting area, according to him, used to be the area between the Crocodile and Komati rivers.
>
> The vegetation on the slopes of Pretoriuskop was much denser at that time than in 1950. He also found the grass cover in general much sparser, with the scarcity of buffalo grass particularly glaring. His first remark was that the Crocodile River had been an open stream with sandbanks 67 years ago, while it was overgrown with reeds now. (The same phenomenon can also be seen on old photographs of the Sabie River at Skukuza and can probably be ascribed to veld fires being less frequent in modern times and the fact that people who had lived on the river banks would chop open the riverine forest to keep lions from their livestock.)
>
> Basson and Holtzhausen found the whole area between Barberton and the Lebombo Mountains much drier than 70 years before. At Pretoriuskop, between the hill and the current rest camp terrain, are waterholes in which bulrushes grow, and these waterholes held much more water 70 years ago. This used to be a popular outspan with the transport riders, as were the Ntomeni waterholes, which also used to be a favourite drinking place of buffalo herds.
>
> Next to the Josekhulu Spruit,[6] at the same spot as where Hart managed one of Nellmapius's overnight stations, had been one of Albasini's trading posts managed by one of his indunas, Josekhulu (Big Josef). Thus Punt found confirmation that the transport road between Lydenburg and Delagoa Bay had been built on the route of the centuries' old trade route from Delagoa Bay, via Josekhulu and Pretoriuskop, to the interior. The accuracy of this deduction was confirmed in July 1988 when it could be determined with reasonable certainty that Feldschau's overnight station and shed had been built on the site of the outpost of another one of Albasini's indunas, Manungu.

A transport wagon on its way near Pilgrim's Rest in the 1880s.
(ACKNOWLEDGEMENT: LIBRARY AND MUSEUM SERVICES, TRANSVAAL PROVINCIAL ADMINISTRATION)

After Ohrigstad and the surrounding area had to be abandoned due to the unhealthy climate and accompanying malaria threat, Krugerspost became the seat of the Volksraad until the new town, Lydenburg, was built on higher-lying ground in 1850. As the town grew in size and activity, there arose a need for a shorter wagon road to Delagoa Bay than the old one via Krugerspost, Ohrigstad and Kaspersnek to the upper reaches of the Sabie (Grootfontein) and down the mountain via Spitzkop and Kruisfontein.

The people of Lydenburg had been a headstrong and independent bunch since the early days. When Commandant General A.H. Potgieter moved to the Soutpansberg in 1848, he remarked with a great deal of despondency that there was no hope of maintaining contact with Lourenço Marques by wagon. The obstacles would have been the tsetse belt, malaria, steep and impassable mountains, the lack of water in certain areas in winter and the delays caused by swamps and drifts through the bigger rivers in summer, not to mention the marauding bands of robbers in the Lowveld.

---

6 In some sources also spelled Josikhulu.

However, the people of Lydenburg would not be discouraged in their efforts to establish trade with the Netherlands via Delagoa Bay, and had the Dutch wholesaler Ohrig and the shipping magnate and businessman J.A. Smellekamp mediate. When this failed, it was the people of Lydenburg who tried time and again to establish a suitable route connecting them with the non-British port of Delagoa Bay. On their own initiative and after their short-lived declaration of independence in March 1857, several delegations were sent to the harbour city to establish permanent trade relations with the Portuguese.

After their reunion with the rest of the Transvaal into the ZAR, President M.W. Pretorius decided that Delagoa Bay had to be used as the land-bound republic's port since he, like his father Andries Pretorius, could not persuade Natal to release its stranglehold on import duties.

In 1864[7] he visited Lydenburg and travelled on horseback with his party to the vicinity of Pretoriuskop – an event which convinced some historians that that was why the name Pretoriuskop had been given to this landmark in the Kruger National Park (Lombaard, 1969). It is said that the well-known German explorer Carl Mauch was asked by the Lydenburgers during one of his visits in 1868 or 1870 to survey a pass over the mountain. Unfortunately nothing about this appears in any of his documents. After President Thomas Burgers had visited the goldfields at Pilgrim's Rest at the end of August 1873, he made renewed efforts to have the road from Sabie to Lydenburg, and the one between Sabie and Delagoa Bay, completed. The construction of the latter was a half-hearted affair at the time. He obtained £3000 for this purpose from the Volksraad.

The president's road building project to Delagoa Bay gained momentum in May 1873 when two would-be prospectors, Walter George Compton and his brother-in-law James Atherton Hampson, arrived from Durban at the Mac-Mac goldfields.

At first the two prospected for gold near the home of the pioneering prospector Tom McLachlan, but without much success. Then, in June 1873, the two decided to scout out a passable transport route from the Mac-Mac goldfields to Delagoa Bay. On 18 June 1873 they set off on their adventurous journey to Delagoa Bay via Spitzkop, Kruisfontein, Pretoriuskop and the route that had been opened up by Field Cornet De Beer in June 1849. After an eventful journey, in which several lions played a part, they finally arrived in the harbour city on 8 July 1873.

James Hampson kept a diary of their journey, which is now housed in the Killie-Campbell Africana Library in Durban. It has recently been reworked into manuscript form by B.P. Simmons of Karino (Simmons, 1987).

Hampton's diary contains many interesting references to game and environmental conditions in the area that is now the Kruger National Park. He noted, for instance, that they had tried to hunt eland, giraffe and zebra, without success, in the area around Pretoriuskop. In the area just east of Skipberg they found the first tsetse fly on Compton's horse and shot a tsessebe for the pot.

After the two pathfinders had returned to Mac-Mac via Swaziland, the diggers there were visited by President Thomas Burgers on 27 August 1873. The diggers cordially entertained the president and he renamed their camp Mc Mc (Mac-Mac) because of the numerous diggers with Scottish surnames there.

The president heard about Compton and Hampson's expedition to Delagoa Bay and on 31 August he held talks with them and Mine Commissioner

Annotated drawing of a *kakebeenwa* (jawbone wagon) as used by the Voortrekkers.
(WITH KIND PERMISSION OF THE CULTURAL HISTORICAL MUSEUM, PRETORIA)

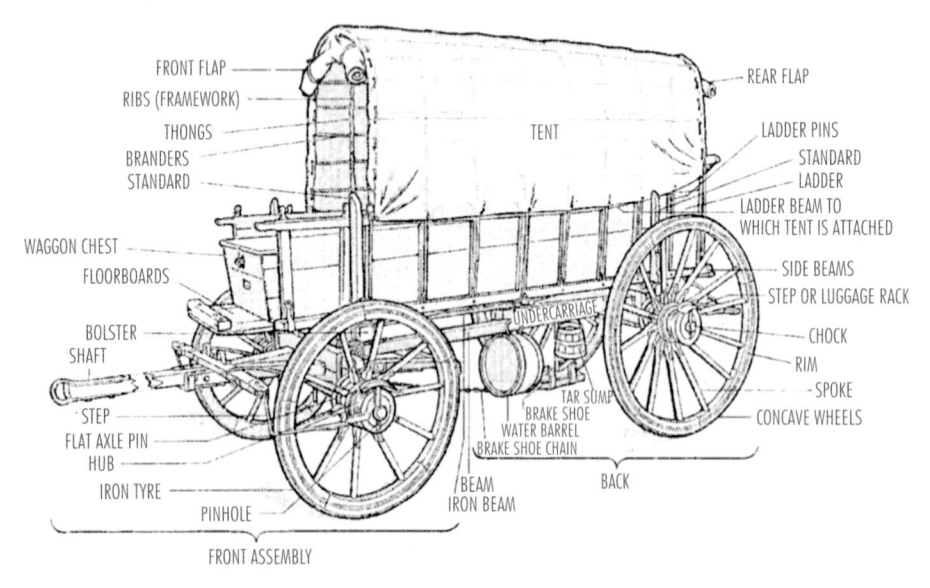

---

7 Lombaard (1969) quoted General Schalk Burger (1916) as saying that the year of President Pretorius's visit was 1865, but doubted this himself.

MacDonald to discuss the possibility of building a road to Delagoa Bay.

It was decided to appoint a road commission consisting of Gold Commissioner MacDonald, Tom McLachlan, Geo Compton, James Hampson, Cameron, Wainwright and Muller to ensure that the road building project started as soon as possible. The project was put out to tender and on 19 September Compton and Hampson's tender of £1000 was officially accepted by the road commission. Work started on 29 September 1873 and it was decided to tackle the project in three sections. Geo Compton and Tom McLachlan would construct the section from the Crocodile River to the pass through the Lebombos (the road east of the Lebombos to Delagoa Bay was already usable). Hampson, with his base at Pretoriuskop, would build the road eastward to the drift through the Crocodile River and westwards to the Kruisfontein Pass. Joseph Shires, Tom McLachlan's brother-in-law, would be responsible for the western mountain passes to Spitzkop and the goldfields.

Despite small delays and problems with the climate, lions that attacked their animals, African horse sickness, nagana, malaria and a narrow escape from an enraged buffalo, these pioneer road builders made remarkable progress. On 4 December 1873 commission member Walter George Compton was able to report in writing to President Burgers that the road from the Lebombos to Mac-Mac would be completed by 15 December.

In fact, Tom McLachlan and commission member Cameron left on 10 December to inspect the completed road. Both contracted malaria on this trip but nevertheless returned safely on 30 December with the good news that the road had been approved for use by transport riders.

Compton and Hampson were paid in full on 2 January 1874 for their difficult job and the two, by now completely cured of their gold fever, returned to Natal via Lydenburg on 6 January. Later (1875–76) Alois Nellmapius's transport service would follow more or less the same route to and from the goldfields at Pilgrim's Rest as the one constructed by the road-building pioneers Compton and Hampson in 1873. He shortened the route by going from Pretoriuskop via Burger's Hall to Spitzkop, instead of via Peebles, Klipkopje and Kruisfontein.

Now only the direct route between the goldfields at Mac-Mac and Pilgrim's Rest via Spitzkop to Lydenburg remained in use.

In 1874 Abraham Espag began the enormous task of building a mountain pass over Mount Anderson. The work that black workers with pickaxes and crowbars couldn't do was accomplished by blasting, with gunpowder loads serving as a blasting agent. This information was published by Jan Coetser, who had known Abraham Espag well, in the *Lydenburg News* in 1933.

During the winter of 1874 the first wagons with loads of rifles and ammunition and a few cannons arrived in Lydenburg from Delagoa Bay. They had trekked from Pretoriuskop along the original wagon road between Ohrigstad and Delagoa Bay to Spitzkop, from where they turned off onto the new road that went westwards over Devil's Knuckles and Whisky Creek. At last Commandant General Potgieter's fondest dream – a direct route from Lydenburg to the sea – became true. This explains the signboards for the Old Harbour Road along the current Long Tom Pass (see map titled: "Old transport and wagon roads, southern Lowveld and Escarpment").

The ingenuity Abraham Espag and his workers had shown in building the Old Harbour Road is even more commendable when one takes into account the fact that the current Long Tom Pass over Mount Anderson follows for the greater part the original course of the old road. Today this pass can be regarded as a monument to the vision and dogged perseverance of the Lydenburg Voortrekkers, their descendants, presidents Pretorius and Burgers, and

Spitzkop, as the area appeared in July 1902.
(ACKNOWLEDGEMENT: STEVENSON-HAMILTON COLLECTION, SKUKUZA ARCHIVES)

The old road to the coast (Harbour Road), still clearly visible between the Devil's Knuckles and Blystaanhoogte.
(22.06.1986)

Wagon spoor clearly engraved into the shale deposits of the old Harbour Road against the ascent of The Devils's Staircase in the Long Tom Pass.
(22.06.1986)

Abraham Espag and his nameless black workers.

Today it seems almost impossible that wagons with their heavy loads could have ascended this steep mountain pass. Even in the days of the hardened and experienced transport riders, who could work wonders with a team of oxen, this part of the journey was a nightmare to both animals and drivers. Napier (1881) gave a contemporary description of what a Boer could achieve with his wagon and oxen:

> There is no denying that for sagacity, courage, and perseverance as road makers they (the Boers) are unequalled. With his vrouw and children and household goods stowed away in a wagon of his own construction, drawn by oxen of his own breaking-in, there is no distance, no difficulty, no danger, the Boer will not successfully encounter. The mantle of courage has fallen from the shoulders of his ancestors on those of worthy descendants – a high quality accomplishing great deeds in this wilderness of a continent where a man, under Providence, can rely only on himself ...

Gold Commissioner McDonald of Pilgrim's Rest was in charge of the first group of eight wagons that brought rifles and gunpowder from Delagoa Bay to Lydenburg. He achieved this with the few remnants of his 16 teams of oxen that had managed to survive the onslaught of the tsetse flies.

Two years later robbers plundered six wagon loads belonging to J.C. Jankowitz after he had been forced to leave his wagons in the Lowveld as all his draught animals were dead of nagana. The stretch of the road built by Geo Compton and his helpers Shires and Hampson between the goldfields at Pilgrim's Rest and the Lebombos soon became impassable to ox-wagons, mainly due to inadequate maintenance and the danger of tsetse flies. This was why the Volksraad granted a concession to the Hungarian Alois Hugo Nellmapius's Lourenço Marques and South African Republic Transport Service to transport goods from Delagoa Bay to the goldfields. The company used black bearers to transport trade goods to Pretoriuskop, from where the goods were taken up the mountain by ox-wagon. Ox-wagons were also used between Delagoa Bay and the Lebombo Mountains.

Hunter and author F. Sandeman had the following to say about his journey between Sabie and Lydenburg in 1877:

> Our second trek took us over the most dreaded part of the road, called the 'Devil's Knuckles'. These infernal knuckles consist of four steep hills. When on top of one of these points the wagon looked as if it were stuck on the point of a sugar loaf, and that any attempt at descent must result in a headlong roll down many hundred feet.

It is said that a certain Atcherley went over the ledge here in 1879, "wagon and all".

On a somewhat level area between Devil's Knuckles and Blystaanhoogte, just ahead of the steepest section of the pass, there was an outspan and trading post on the farm Rhenosterhoek, which had apparently belonged to João Albasini or one of his partners. This old trading post has another interesting historical name: April se Stellasie (April's Scaffold). A few Boers had apparently been hunting in the area and left their biltong hanging on a scaffold in the care of April. April seized the moment and ate the lot, after which the scaffold burned down "accidentally"!

No wonder many transport riders (among others Percy FitzPatrick and his companions) took the longer route via Pilgrim's Rest, Kaspersnek and Krugerspost to Lydenburg, especially in the busy 1880s. They usually rested their animals for a while on top of the mountain near the current Graskop at their Paradise Camp, before setting off on the last shift to Krugerspost and Lydenburg.

To avoid the Devil's Knuckles another pass was built between Visierskerf and Sabie about 40 years after the completion of the first road. Although it was wider and did not have such steep sections, it was nevertheless a feared stretch of road and was replaced 40 years later by the Long Tom Pass. This meant a return to the Devil's Knuckles, but this time their sting was taken out with the aid of modern road engineering design, bulldozers and a tarred surface.

The Long Tom Pass was officially opened and named on 22 July 1953 by the Administrator of the Transvaal William Nicol. Its name recalls a historical incident on 9 September 1900 when the Boer army retreated via the Old Wagon Route with two of their four Creusot (Long Toms) cannons.

These two cannons were set up below Devil's Knuckles, where they were used in a shoot-out with the English cannons of General Buller in an effort to protect the Boers' rearguard and to give their front line an advantage. The cannons were later removed from their mountings in a great hurry and in a misty rain the burghers themselves had to be the breaks – holding on to the cannons with hide straps to prevent them from tumbling over the precipice. A veteran later said that even Commandant General Louis Botha jumped in to help with the back-breaking work. Travellers can still see some of the craters made by the cannon balls fired from these famous big guns along the Long Tom Pass. In 1984 the National Heritage Council erected an exact replica of a Long Tom cannon next to the pass. It was unveiled by the ad-

The old road to the coast is still clearly visible on the farm Rhenosterhoek (Misty Mountain) between Devil's Knuckles and Blystaanhoogte (22.06.1986). Apparently there are remains of an ancient mine shaft which had been dug by early visitors from the East Coast (Indonesians or Arabs) just west of here.

A replica of a Long Tom canon which was erected in October 1984 along the Long Tom Pass at Devil's Knuckles, where these famous Boer big guns had been used for the last time during the Anglo-Boer War. (22.06.1984)

This sandstone school building at Krugerspos dates back to the nineteenth century.

ministrator of the Transvaal, W.A. Cruywagen, on 10 October 1984.

It might be appropriate to dwell for a while on this breathtakingly beautiful region with its hills and dales, paying respect to the many known and forgotten pioneers of the decades from 1870 to 1890 – those who made it possible for future generations to live in and use this fertile part of South Africa.

Of course, the old transport routes in the Transvaal Lowveld and to the northern Transvaal Bushveld were not used only by the transport riders with their ox-wagons.

From the earliest days, after the first whites had settled in these areas, there was a need for a passenger service between far-flung places and established communities. In 1862 two transport riders, Hendricus Buykes and Daniël Brink, started conveying goods from Potchefstroom to settlements as far north as Schoemansdal. Then they began to accept paying passengers and the demand soon became so great that they acquired a light carriage (a four-in-hand), similar to the ones used by the Mormons in America. The journey through the wild Bushveld of the northern Transvaal was inevitably filled with danger. Brink drove the carriage himself, with two coloured men as armed guards. One incident took place in February 1863 while the carriage was going through a thicket of dense brushwood near the Dwars River. Suddenly a volley of shots rang out and the guard in front, fatally wounded, toppled over between the horses. Brink immediately handed the reins to one of the passengers, jumped down and freed the guard's body from the horses' hooves, just as about 20 black warriors came charging from the bush. Brink laid into the bloodthirsty gang with his whip despite more shots being fired – luckily these passed harmlessly over their heads. Only later did Brink realise that there were bullet holes in the crown of his hat and the sleeve of his jacket, and that another bullet had nicked his arm.

Although this service was exposed to a certain amount of danger, Brink nevertheless found it very lucrative and it was only terminated when Schoemansdal was evacuated in 1867 after repeated Venda uprisings.

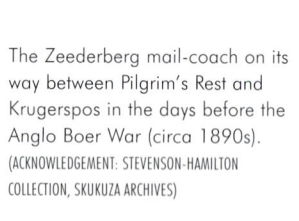

A Long Tom shell hole along the Long Tom Pass on the northern side of Blystaanhoogte. (22.06.1986)

The Zeederberg mail-coach on its way between Pilgrim's Rest and Krugerspos in the days before the Anglo Boer War (circa 1890s). (ACKNOWLEDGEMENT: STEVENSON-HAMILTON COLLECTION, SKUKUZA ARCHIVES)

General Munnik next to the Zeederberg mail-coach at Louis Trichardt (Makhado) in the 1890s.
(ACKNOWLEDGEMENT: AFRIKAANSE PERS AND MR P. DE VAAL, LOUIS TRICHARDT)

This was the last passenger service in the Transvaal for many years. Then gold was discovered in 1872 at Eersteling and a year later at Lydenburg and Pilgrim's Rest.

Thousands of prospectors, diggers, fortune seekers and other followers were drawn to these areas by the discoveries. A coach service between the goldfields and the more established communities in Transvaal and the Natal and Cape harbours became a necessity. There were a number of private vehicles at Pilgrim's Rest that undertook infrequent trips to Lydenburg, but the owners were not eager to transport passengers as they had little taste for the subsequent burden of responsibility.

A certain Pearce of Pilgrim's Rest, who would later work for Nellmapius, suggested to William Leathern that a light coach be bought and a weekly passenger service between Lydenburg and Pilgrim's Rest be instituted. Pearce would provide the funds while Leathern had to get the service up and running. This mode of transport got a lot of support from the miners and others.

A year later the Chevalier Oscar Wilhelm Alric Forssman, a consular representative for Portugal and a Potchefstroom businessman, announced that he wanted to start a passenger coach service between Potchefstroom and the diamond fields at Kimberley and the Vaal River that would be called the Transvaal and Goldfields Extension Transport Company. In time he also expanded the business to provide a passenger service to the goldfields at Lydenburg via Pretoria (Zeederberg, 1971). Oscar Forssman was a younger brother of the Chevalier Magnus Johan Frederick Forssman, the surveyor-general of the ZAR.

Forssman's company used light, four-wheeled buggies that were closed against wind and rain. The journey from Potchefstroom to Lydenburg via Pretoria took just over 100 hours to complete.

In the meantime, the goldfields at Pilgrim's Rest and Lydenburg attracted more and more prospectors and diggers, but the Sekhukhune uprisings and the first confrontations between the Boers and Great Britain brought passenger services to and from the goldfields to a halt. Forssman had his company liquidated and it was taken over by Erasmus and Green.

By 1882 the economy of the Transvaal had improved to such an extent that coach services to the eastern Transvaal goldfields once again became lucrative and could be reinstated.

The Gibson brothers built up a flourishing stagecoach service – the Red Star Line – and extended their activities to Pilgrim's Rest and Lydenburg. An interesting fact about the Red Star Line was that their coaches were named after steamboats of the Castle Steamship line and the conductors got the rank of captain.

Major gold finds in the northern and eastern Transvaal in 1884, including those at Barberton, Kaapsche Hoop and Duivelskantoor, led to the extension of the stagecoach service of Geo Heys & Co., as well as that of the Gibson brothers' Red Star Line.

At this time two of the well-known Zeederberg brothers, Louw and Piet, decided to extend their transport services operating south of the Vaal River to the north to include the Transvaal and Rhodesia.[8] With six coaches in their fleet, they called on their brother Abraham (Dolf), who had been forced by ill-health to end his career as a pharmacist, to lend them a hand.

After his arrival in Kimberley, Dolf's brothers gave him two wagons and sent him to transport badly needed equipment and provisions to miners' settlements as far north as the Limpopo, while they started a transport service to Potchefstroom, Pretoria and the eastern Transvaal. This undertaking was so successful that they had ten wagons on the road by the end of that year.

In 1885, the Zeederbergs summoned their youngest brother, Christiaan (Doel), to help them expand their business even further. Doel was convinced that the future lay in passenger transport and pointed to the success of Heys & Co. and the Red Star Line's stagecoach services.

Apart from the ever-present danger of being attacked and robbed by highwaymen, the passengers en route to Barberton and other isolated areas had many adventures in those days. The passengers of the Red Star Line, for instance, had an experience one June that they would remember for a long time. Lions were often a nuisance along the coach routes and sometimes would delay a coach for hours. One of the most dangerous places in this regard was the Elandsdrif near Machadodorp (Zeederberg, 1971). The coach drivers usually gave the surrounding bush here a careful once-over before moving on. During

The Zeederbergs' team of zebra with two donkeys as rearguards ready to depart from Pietersburg (Polokwane) to Leydsdorp (1881).
(ACKNOWLEDGEMENT: H. EXTON COLLECTION)

---

10 Now Zimbabwe.

this particular incident the coach was moving forward very slowly when one of the passengers on the coach suddenly noticed a pride of lions below the drift. The horses caught the scent of the lions and while the driver was trying to bring them under control his assistant, with great presence of mind, blew hard on the coach bugle. This had the desired effect and the lions, which had been lazily sunning themselves in the winter sun, jumped up and took to their heels in bewilderment.

Coaches on their way to the towns that had gone up like mushrooms around the eastern Transvaal goldfields, often also transported female passengers. There was a constant stream of bar girls, such as the famed Cockney Liz of Eureka City, stage girls and the diggers' wives. Yet female passengers had nothing to fear in those days – especially not from their rough fellow passengers. Woe betided the man who failed to treat these ladies with the necessary courtesy and respect!

The scarcity of horses and mules gave the Zeederbergs the idea to tame zebras and use them to draw their coaches. Piet Roets, a prominent farmer in the Soutpansberg, actually gave Doel Zeederberg the idea. The first real attempt to tame zebras took place in 1881 when Jacob Erasmus succeeded in capturing several zebras and bred them on his farm near Leydsdorp. Roets, who farmed near Haenertsburg, and Izak van der Merwe, of the Pietersburg (now Polokwane) district, followed his example. They succeeded in taming zebras, bred them and even managed to get them to draw a plough. Roets offered some of his tame zebras to Doel Zeederberg and, to his surprise, Doel accepted his offer and bought 20 animals from him for £200.

Zebras were still common in the northern Transvaal in those days and until the 1890s herds of more than 500 could be found (Zeederberg, 1971). Over short distances they were faster than horses or mules, but they lacked stamina and a good horse would run them off their feet within ten minutes. Doel offered £1.0.0 for each zebra young farmers could catch and within a few weeks there were 150 to 200 zebras in camps near Haenertsburg.

The task of breaking in the zebras and training them to work in a team fell to Jock Bannatyne and Frank Klopper. They had little experience but an abundance of enthusiasm and within three months they had three to four teams with mules as team leaders, ready for action.

A direct stagecoach service was established between Pietersburg (Polokwane) and Leydsdorp and coaches stopped at several important mining centres, including Haenertsburg, New Agatha and Thabina.

In July 1881 the first coach trip with a full team of zebras took place between Haenertsburg and Leydsdorp on one of the worst roads in the whole of the Transvaal. Their performance made history and surprised everyone. The mountainous section of the pass was almost too much for the zebras, but they were nevertheless much more sure-footed on the rough terrain than horses, mules or oxen. On the 40 kilometre level stretch between Gravelotte and Leydsdorp they maintained a speed of 12 to 16 kilometres an hour. Their arrival at the Leydsdorp hotel caused a sensation and within moments 400 miners had gathered to gape at the spectacle.

Terry King, an agent for the Mosenthal brothers who owned the hotel, provided a barrel of beer and the Zeederbergs' breakthrough could be celebrated in a manner befitting the occasion. "What will such a team of zebras cost me?" King asked Pieter Zeederberg. "Oh, I think I can let you have a team for about £500!" "Excellent," was King's gratified acceptance of the offer and two weeks later he received his team, which he immediately sold at a sizeable profit to local farmers and prospectors. Although the zebra teams had not fared badly on their first trip in front of a coach, they were temperamental and clearly did not possess the necessary stamina to replace horses or mules as draught animals. This eventually forced the Zeederbergs to accept that their interesting experiment was too impractical to succeed.

By the late 1890s highwaymen were less active than before, but one day in December a coach was taken from the Zeederberg stables in Pilgrim's Rest to the offices of the Transvaal Gold Mining Estates and secretly loaded with crates of gold bars. The crates, destined for Lydenburg, were hidden under the driver's seat. The driver was one Piet du Plessis and the value of the gold was estimated at £10 065 at the time. Wilson was the assistant driver and the coach with its valuable load was soon climbing Pilgrims Hill. After that, the route went over Ponieskrantz and Rosenkrantz, Karel Trichardt's old farm, but they never reached the farm as they were waylaid by two masked robbers, one on each side of the road, and ordered off the coach. The driver, his assistant and the passengers were tied to the coach wheels after which the robbers filled their saddlebags with their booty and loaded these onto two of the coach's mules. After chasing the other mules into the bush they disappeared into thin air. The captives were set free by a passer-by soon afterwards. Although an armed commando searched the area for days on end, they found no sign of either the robbers or their spoils. Three years later Doel Zeederberg received a letter from a highly respected person in Lourenço Marques thanking him for his welcome contribution towards getting his business off the ground. He had the gall to offer to tell Doel the whole "interesting story" of their escape should he take the trouble to visit him in Lourenço Marques. Doel never took him up on his offer and never revealed the identity of this impertinent highwayman.

The year 1898 was an unfortunate one for the Zeederbergs. A whole series of mishaps kicked off with a serious accident in January involving one of their coaches on its way to Lydenburg. The coach with 20 passengers was creeping up the steep Long Tom Pass when the link connecting the beam to the front axle snapped. The coach ran backwards and came to a stop against a steep bank. The coach was badly damaged, but the shocked passengers managed to escape without any serious injuries. This accident and another one only a few weeks later at Bührmanns Drift cost the firm £5000 in repairs and compensation to the passengers. The outbreak of the Anglo-Boer War in October 1898 hit the Zeederbergs very hard and all coach services were temporarily discontinued.

After the war travellers using the well-known Jock of the Bushveld road between Pilgrim's Rest and Lydenburg had no trouble with highwaymen for a number of years. In June 1912 this situation changed dramatically, however. Andries Johannes du Plessis, the driver of a coach en route from Lydenburg to Pilgrim's Rest, heard an order from the side of the road: "Stick 'em up!" The masked robber with gun in hand ordered the frightened coach driver to unload the chest of "gold" from the coach. Du Plessis meekly obliged, after which he was ordered to carry on with his journey. Once he reached the top of the hill he set off on a headlong rush to Pilgrim's Rest to report the heist. In the meantime the robber removed the "gold" from the chest, put it into his saddlebags and cleared off with a grin under his mask. A while later, in the privacy of an abandoned farm house, he opened the saddlebags only to discover that the bags of "gold" only contained a heap of silver coins that amounted to a measly £358.

The following day the local constable, Cook, was informed that a servant of the 43-year-old barber, Thomas Dennison, had paid his account with John Her, a businessman, with a pillowcase full of silver coins. This aroused Her's suspicions and Dennison was arrested forthwith for the robbery on the coach.

He was found guilty and sent to jail for three years. Having served his time, he returned to Pilgrim's Rest and resumed his former occupation among the diggers. Dennison smoked the most expensive cigars in the whole town and it was often wondered how he could afford these on his meagre income as a barber. One day Barry, the manager of the Transvaal Goldmining Estates, arrived at his shop and pointed an accusing finger at Dennison. "Dennison, you must stop melting my gold and casting buttons from it – I know what you're up to!"

Dennison slowly removed his cigar and answered politely: "Mister Barry, if that is the case, then you

Memorial tablet erected at the site where the highway robber, Thomas Dennison, had ambushed the mail-coach between Pilgrim's Rest and Lydenburg on 7 June 1912 and robbed it of £129 in silver coins.
(06.09.1986)

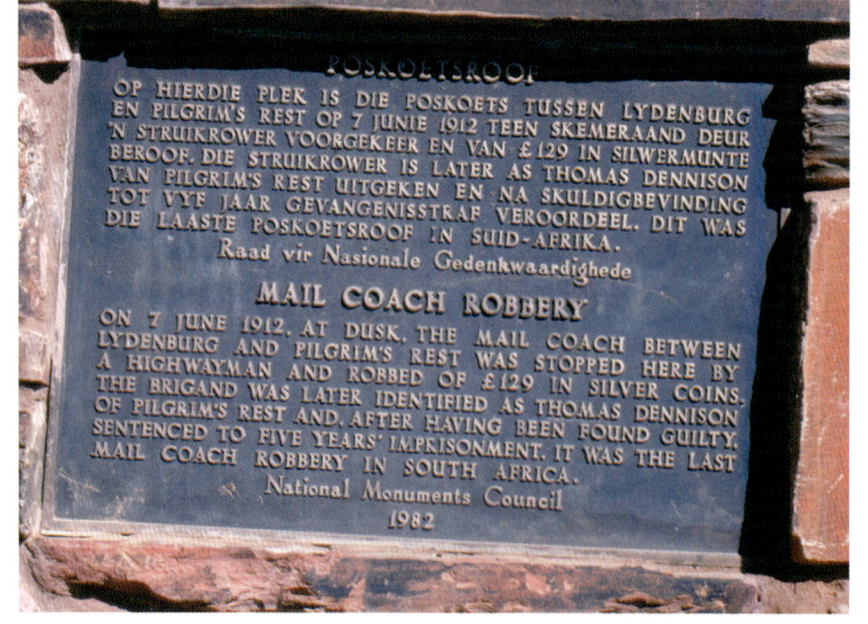

have to see to it that your men don't sell your gold to me!" (Zeederberg, 1971)

He was to be the last of the many characters active in the pioneer days at Pilgrim's Rest – an era that came to an end with the outbreak of the First World War.

In the Lowveld the old transport routes stretching in all directions were still in frequent use until 1892, when the Eastern railway line between Delagoa Bay and Nelspruit was completed. Two years later the whole line was completed all the way to Pretoria, followed by branches to Barberton, Sabie, Graskop and White River.

After that the old transport routes were used especially in winter by a few transport riders and more often by hunters and sheep farmers, who moved with their livestock to the Lowveld, but in time they fell into disuse and oblivion. After the Anglo-Boer War, in the infancy years of the Sabie Reserve, part of the old Lydenburg-Delagoa Bay transport road was used by the first rangers as a communication route between Malelane and Pretoriuskop. After the proclamation of the Kruger National Park in 1926 certain sections were also incorporated into the tourist road between Pretoriuskop and Malelane.

Only the shadows of a forgotten and romantic era in the history of South Africa still drift along the rest of the old roads today. If you should stand at a drift where wagons used to pass and close your eyes, you can once again hear the cracking of long whips, the crunch of wagon wheels over rough roads, the voices of the transport riders and the bellowing of the trek animals that trampled the transport roads through the Lowveld.

The mail-coach arrives at the Pilgrim's Rest post office in the 1880s.
(ACKNOWLEDGEMENT: INFORMATION AND MARKETING SERVICES, TRANSVAAL PROVINCIAL ADMINISTRATION)

References

Adendorff, G. 1984. *Wild Company*. Books of Africa, Cape Town.

Anon. 1924. Bushveld days. Why there can be no equal to Jock. *The Star* 17 October.

Anon. 1937. History as you drive. *The Star* 26 June.

Anon. 1949. The last resting place of Jock of the Bushveld. *Sunday Times* 27 February 1949.

Anon. 1984. *The Historical Associations of Sabie: The Long Tom Pass over the Drakensberg* pamphlet no. 14. Forestry Museum, Sabie.

Anon. 1987. Pettigrew's Nek herbenaam. *The Lowvelder* 10 November.

Atcherly, R.G. 1879. *A trip to Boer Land*. Ritchard Bentley & Son, London.

Axelson, E. 1940. *South-East Africa 1488–1530*. Longmans Southern Africa, Cape Town.

Axelson, E. 1969. *Portuguese in South-East Africa 1600–1700*. Witwatersrand University Press, Johannesburg.

Aylward, A. 1879. *The Transvaal of Today*. William Blackwood & Sons Ltd., London.

Baines, T. 1854. The Limpopo, its origin, course and tributaries. *Journal of the Royal Geographical Society* 24:290–291.

Baines, T. 1864. *Journal of Residence in Africa (1850–1853)* vol. 2:177 *et seq.* Kennedy, R.F. (ed.). Van Riebeeck Society, Cape Town.

Bechtle, A. & Marais, C. 1891. Schetskaart van de Lebombovlakte tusschen Olifants- en Krokodilrivier. TAB,1/148.

Bellairs, Lady. 1885. *The Transvaal War 1880–1881*. Unknown publisher, Edinburgh. Faxcimilee reprint 1972, C. Struik, Cape Town.

Berthoud, H. 1903. Sketch map of the N.E. Zoutpansberg, ZAR 1:333 000, Berne.

Beyers, C. 1938. Report on the origin of the Jock of the Bushveld road to Col. J. Stevenson-Hamilton. Unpublished report.

Bulpin, T.V. 1950. *Lost Trails on the Lowveld*. Howard B. Timmins, Cape Town.

Bulpin, T.V. 1983. *Lost Trails of the Transvaal*. Books of Africa, Cape Town.

Bulpin, T.V. & Miller, P. 1968. *Lowveld Trails*. Books of Africa, Cape Town.

Burger, S. 1916. Reminiscences of General Schalk Burger. *Pelgrim's Rest Goldfields, Mines and Claimholder's quarterly*.

Cameron, T. & Spies, S.B. 1986. *An Illustrated History of South Africa*. Jonathan Ball, Johannesburg.

Cartwright, A.P. 1971. *The First South Africa*. Purnell, Cape Town.

Cartwright, A.P. 1978. *Transvaal Lowveld*. Purnell, Cape Town.

Cohen, E. 1875. *Erläuternde Bemerkungen zu der Routenkarte einer Reise von Lydenburg nach der Delagoa Bai in ostlichen Süd-Afrika*. L. Friedrichsen & Co., Hamburg.

Compton, W.G. 1873. Letter to President Thos. Burgers after completion of the Pilgrim's Rest-Delagoa Bay road. T.A. 2013/128.

De Vaal, J.B. 1953. Die rol van João Albasini in die geskiedenis van die Transvaal. *Archives Year Book for South African History* 16(1):1–155.

De Vaal, J.B. 1982. João Albasini (1813–1888). *Camoês Annual Lectures* no. 3. Ernest Oppenheimer Institute for Portuguese Studies.

De Vaal, J.B. 1984. Ou handelsvoetpaaie en wapaaie in Oos- en Noord-Transvaal, *Contree* 16:5–15.

De Vaal, J.B. 1984. Handel langs die vroegste roetes. *Contree* 17:5–14.

De Villiers, P.W. 1875. Inspectie Rapport van Pretoriuskop. TR.2402.

Dicke, B.H. 1941. The Northern Transvaal Voortrekkers. *Archives Year Book for South African History* 1:67–170.

Dunbar, D.M. 1881. *The Transvaal in 1876*. Richards Slater & Co., Grahamstown.

Erskine, St. V.W. 1887. Map of the Kaap Gold Fields, S.A. Republic.

FitzPatrick, J.P. 1897. *The Outspan*. William Heineman, London.

FitzPatrick, J.P. 1907. *Jock of the Bushveld*. Longmans Green & Co., London.

Garson, N.G. 1957. The Swaziland question and a road to the sea. *Archives Year Book for South African History* 2.

Glynn, H.T. 1926. *Game and Gold*. Dolman Printing Co. Ltd., London.

Gray, S. 1986. Jock of the Bushveld. *African Wild Life* 40(5):167–173.

Hall, H.L. 1937. *I Have Reaped my Mealies*. Batteridge &. Donaldson (Pty) Ltd., Johannesburg.

Jeppe, F. 1879. Map of the Transvaal and the surrounding territories.

Jeppe, F. 1892. *Die Neue Grenze zwischen der Süd-afrikanischen Republik und den Portugiesischen Besitzungen*.

Krüger, D.W. 1938. Die weg na die see. *Archives Year Book for South African History* 1:31–233.

Leibnitz, H.M.C. 1951. *A brief historical sketch of the Eastern Lowveld area below Kowyns Pass*. Manuscript compiled for the Drakensberg Farmers' Society.

Lion Cachet, F. 1898. *De Worstelstrijd der Transvalers*. Jaques Dusseau &, Co. Uitgevers, Amsterdam & Cape Town.

Lombaard, B.V. 1964. Long Tom-pas en die pad na die Baai. Unpublished manuscript for the Regional Officer, Roads Department, Lydenburg.

Lombaard, B.V. 1969. Herkoms van die naam Pretoriuskop. *Koedoe* 12:53–7.

Lombaard, B.V. 1978. Ou pad van Ohrigstad noordwaarts na die Laeveld. *Lydenburg Nuus 28* September.

Lombaard, B.V. 1979. Die bergpas oos van Sabie. *Lydenburg Nuus* 28 September.

Loveday, R.K. 1883. *Map of the Lydenburg Gold Fields, S.A. Republic* (Transvaal). Pretoria, (TA1/267).

Mackie-Niven, C. 1968. *Jock & Fitz*. Longmans, Cape Town.

Merensky, A. 1875. Original map of the Transvaal or South African Republic including the gold and diamond fields. 1:1 850 000. Berlin.

Meyer, A. 1986. *'n Kultuurhistoriese interpretasie van die Ystertydperk in die Krugerwildtuin.* D. Phil. thesis, University of Pretoria.

Peacock, R. 1950. *Geskiedenis van die Lydenburgse goudvelde tot 1881.* M.A. thesis, University of Pretoria, Pretoria.

Peterman, A. 1870. Originalkarte von C. Mauch's Reisen im Innern von Süd-Afrika. 1:2 000 000. Gotha.

Pienaar, U. de V. & Mostert, M.C. 1983. Nuwe feite bekend oor romantiese transport-era. *Custos* 12(9):27–29.

Pienaar, U. de V. & Mostert, M.C. 1983. Kleurryke karakters deel van Wildtuingeskiedenis. *Custos* 12(10): 12–14.

Pienaar, U. de V. 1984. Geskiedkundige fontein in Krugerwildtuin opgespoor. *Custos* 13(7):4–7.

Preller, G.S. 1920. *Voortrekkermense II. Herinneringe van Karel Trichardt.* De Nationale Pers, Cape Town.

Prinsloo, H. 1980. Die ou Delagoabaaipad. Unpublished manuscript.

Pretorius, H.S. 1936. Report to Col J. Stevenson-Hamilton about the role of A.H. Nellmapius in the construction of the Lydenburg-Delagoa Bay road. Unpublished manuscript.

Pretorius, H.S., Krüger, D.W. & Beyers, C. 1937. Voortrekkerargiefstukke 1829–1849. Government Printers, Pretoria.

Punt, W.H.J. 1950. Verslag i/s ondersoek van ou paaie en roetes in die suidelike Wildtuin. Unpublished manuscript to National Parks Board.

Punt, W.H.J. 1950. Unpublished notes. Skukuza Archives.

Punt, W.H.J. 1962. 'n Beknopte oorsig van die historiese navorsing in die Nasionale Krugerwildtuin. *Koedoe* 5:123–127.

Punt, W.H.J. 1975. The relationship between the pioneer routes in the Transvaal and ancient trade routes. *Occasional Paper of the South African Archaeological Society.* Johannesburg.

Read, F. 1978. *Jock – the Art of Edmund Caldwell.* Frank Read Press, Mazoe.

Sandberg, C.G.S. 1943. *Twintig jaren onder Krugers Boeren in voor en tegenspoed,* Amsterdamsche Keurkamer, Amsterdam.

Sandeman, E.F. 1880. *Eight Months in an Ox-waggon.* Griffith & Farran, London.

Scully, W.C. 1937. The Lowveld: First wagon journey from the coast. *The Star* 25 June.

Simmons, B.P.& Coetzee, D. 1983–84. Personal communication.

Simmons, B.P. & Coetzee, D. 1985. *The Old 'Transport' or Jock of the Bushveld Roads.* Map by Promedia.

Simmons, B.P. 1987. Pioneers of the Delagoa Road: James Atherton Hampson 1846–1903 and Walter George Compton. Unpublished manuscript.

Stevenson-Hamilton, J. 1974. *South African Eden.* Collins. London.

Van der Merwe, A.P. 1947. Professor I.D. Bosman in die Krugerwildtuin. *Historiese studies* 8(1):28–32.

Van der Merwe, A.P. 1950. Beskawing kom na Oos-Transvaal. In: *Lydenburgse Eeufeesgedenkboek.*

Van der Merwe, A.P. 1951. Waar kom die naam Pretoriuskop vandaan? Beroemde ruskamp en 'n eensame graf. *Die Transvaler* 11 July.

Van der Merwe, A.P. 1961. Die eerste pad van Ohrigstad na Lourenço Marques. *Historia* 6(4):291–293.

Van Jaarsveld, F.A. 1947. *Historiese Studies* 8(1):51–52.

Vaughan-Kirby, F. 1896. *Map of Kahlamba and Libombo.* William Blackwood & Sons, Edinburgh & London.

Voight, J.C. 1969. *Fifty Years of History of the Republic of South Africa (1795–1845).* Unknown publisher, Cape Town.

Wolhuter, C.A. 1948. *Memories of a Game Ranger.* Wildlife Protection Society of South Africa, Johannesburg.

Yates, C.A 1935. *The Kruger National Park.* George Allen & Unwin Ltd., London.

Zeederberg, H. 1971. *Veld Express.* Howard Timmins, Cape Town.

10.4 The influence of tsetse flies, malaria and the rinderpest epidemic of 1896–97 on the development of the Lowveld

Dr U. de V. Pienaar

Among the factors that played a significant role in the colonisation and development of the Lowveld through the ages were the endemic tsetse fly and malaria and the great rinderpest epidemic (panzootic) of 1896–97.

Tsetse fly and malaria are associated with large parts of the Lowveld; to such an extent that after the arrival of the first white settlers this area soon became known as "the white man's grave". However, the rinderpest epidemic of 1886 was a cataclysmic event originating in tropical Africa that spread across the country like wildfire, decimating livestock and certain game species. It had a far-reaching influence on the economy of the entire country, as well as the ecology of the Lowveld.

Summers (1967), Dicke (1932), Fuller (1923) and others deduced that the tsetse fly had occurred in the lower-lying regions of the Transvaal and the adjacent coastal areas of Mozambique and Natal since before the beginning of the Common Era.

According to Fuller (1923), the only nagana-transmitting species that ever occurred in the Transvaal was *Glossina morsitans Westwood*. The type specimen of this biting fly was first collected by Oswell and Vardon during an expedition in 1846 at Siloquana Hill along the middle Limpopo, just east of 30° longitude. Summers (1967), on the other hand, believed that the east coast species *Glossina pallidipes* Austen, which occurred mainly in Mozambique and what was then Natal (formerly Zululand), could have penetrated the eastern Limpopo valley. It is also possible that another east coast species, *Glossina brevipalpis* Newstead of Natal, could have occurred in the southern-eastern part of the Transvaal Lowveld. It is quite probable though that *Glossina morsitans* was the only tsetse fly species of any importance in the Transvaal Lowveld.

As the name "morsitans" indicates, this biting insect is the carrier of deadly diseases, including nagana in livestock and sleeping sickness in humans. Nagana is caused by the trypanosome *Trypanosoma brucei*[1] and sleeping sickness by the trypanosome *Trypanosoma rhodesiense*. Humans and their tame livestock, both highly susceptible to these illnesses, are only the secondary hosts of tsetse flies, as their primary hosts are wild ungulates immune to the effects of these micro-organisms in their blood. The most important primary hosts of *Glossina morsitans* are forest dwellers such as buffalo, kudu, bushbuck and warthog.

The famous hunter F.C. Selous observed in as early as 1908 that there was a distinct correlation between the distribution of tsetse fly and of buffalo. The flies prefer fairly wooded areas such as riverbanks and other shady areas and are seldom found in open, sunny places.

Before 1850 tsetse flies occurred in a broad band along both banks of the Limpopo, from its headwaters in the vicinity of Vliegpoort (just north of Brits where Cornwallis Harris encountered tsetse flies for the first time in 1836), north of the Soutpansberg and throughout the Lowveld down into Swaziland and northern Natal.

Dicke (1932) speculated that this tsetse fly-infested area had a significant impact on the colonisation of the Transvaal by black tribes from the north in the era before AD 1500.

Archaic mines, interesting ruins and irrigation works that have been discovered in the interior south of the Zambezi indicate that the area had been mined by foreign colonists long before the Portuguese settled along the east coast. Neither the identity of these early adventurers and miners (Arab, Goanese, Persian or Phoenician) nor their mysterious disappearance has been explained satisfactorily. Dicke (1932) provided a possible explanation. The black settlers in the northern Transvaal, who were relatively late arrivals from the north, held the belief that a mystical

---

1 Named after Major David Bruce AMS, who discovered the organism in 1895 in a hut in Ubombo, Zululand, while carrying out research. His discovery led to the realisation that tsetse flies are the carriers of nagana and not the cause.

power called "madumadungana" was responsible for the disappearance of the earlier settlers. They cannot explain the meaning of the word or what the phenomenon represents, but it may be significant that the word ends with the suffix "ngana", so similar to "nagana", the name most commonly used for the feared tsetse fly-transmitted livestock disease.

It is commonly accepted these days that the summer rainfall regions of southern Africa, like other regions in the world, are subjected to long-term climate cycles of higher and lower rainfall periods (Hall 1976, Summers 1967, Tyson 1977, Pienaar 1986 and others). Such climate cycles in turn influence the development of vegetation, animal life, natural fire regimes and human colonisation of these areas. Naturally such climate cycles will also create conditions favourable or unfavourable to the survival of the tsetse fly, causing an ebb and flow pattern in their long-term distribution. Available historical data clearly indicate that the distribution of tsetse flies in the neighbouring Zimbabwe was not static and that significant changes occurred over the centuries (Fuller 1923, Dicke 1932, Summers 1967).

Short-term fluctuations in the distribution of tsetse flies in a particular region also took place, with these fluctuations probably linked to the accepted short-term rainfall cycles in the summer rainfall areas of southern Africa.

Fiedler and Kluge (1954) had another theory. They believed that these fluctuations could be ascribed to the numbers and distribution of a parasitic wasp that is a natural enemy of tsetse flies. Many parasitic wasp species attack flies, but this specific one, *Thyrindanthrax*, limits its parasitism to tsetse flies. The wasp lays its eggs in the pupae of tsetse flies, where the larvae develop and feed on the pupae. It is quite possible that the population growth curve of these parasites is also coupled to climatic conditions.

Dicke (1932) speculated that the mysterious early inhabitants of the Transvaal and Zimbabwe had settled here at a time when environmental conditions did not favour tsetse flies. Later the wider distribution range of the tsetse fly and malaria either wiped them out or forced them to leave the area.

Shortly after the end of the Anglo-Boer War, old Portuguese graves were discovered at the foot of the Blouberg on the farm Leipzig in the northwestern Transvaal. Older local black inhabitants still tell stories of these Portuguese that were handed down to them by their forefathers. Legend has it that these people came from the north to exploit the mineral richness of the area, and died of malaria. Their livestock also died after being stung by tsetse flies.

Dicke (1932) also believed that the tsetse belt along the Limpopo River had been responsible for the fact that the colonisation of the Transvaal and areas further south, from the northern tribal areas, took place along the coastal plains and through what is now Botswana (which was relatively free of tsetse). This was mainly as a result of the deadly obstacle tsetse flies on both sides of the Limpopo River posed to large migrations of people with livestock. This of course did not prevent smaller groups of people (without susceptible livestock) from moving into the hinterland of the Transvaal and settling there long before AD 1500 (compare the iron and copper smelters at Phalaborwa and elsewhere in the current Kruger National Park).

It is, however, accepted that the large-scale immigration of tribes from the north down the east coast reached Delagoa Bay by about AD 1500. From here splinter groups crossed the Lebombo Mountains and settled in the area that is now the Kruger National Park, mainly as a result of tribal conflicts and wars.

It is not clear what the situation was with regard

A tsetse fly, *Glossina morsitans Westwood*, the carrier of the deadly nagana and human sleeping-sickness.
(ACKNOWLEDGEMENT: ONDERSTEPOORT VETERINARY RESEARCH INSTITUTE)

to the tsetse fly, but it is known that, when Francois de Cuiper undertook his landmark expedition from Delagoa Bay to Gomondwane (in the southern part of the current Kruger National Park) in 1725, there were no tsetse flies in the region. Looking at his descriptions of environmental conditions, rivers, etc., it would seem that the journey was undertaken during a generally wetter climate cycle. He used pack oxen and the party also encountered cattle at many of the black settlements they visited. De Cuiper also mentioned sizeable cultivated lands and large settlements and villages with hundreds and even thousands of inhabitants. The lack of game was also noticeable – a phenomenon that can be ascribed to the amount of people that hunted them with the aid of pitfalls, game drives, etc., as well as the intensive cultivation of the land. Regular man-made fires to benefit the livestock were also likely.

All of these factors inhibited the occurrence of tsetse fly and the distribution range was as such much smaller than in later years. These conditions probably prevailed until the start of the great wars of conquest and annihilation among the various tribes at the beginning of the nineteenth century. The establishment of the mighty Zulu empire under Shaka and his successor, Dingane, and their reign of terror, caused the widespread dispersion and extirpation of neighbouring tribes, especially from 1818 to 1835. This time of devastation is known in the Nguni languages as "Mfecane", while in Sotho it is called "Difaqane" (Berg & Berg 1984).

The story of the Mfecane/Difaqane is one of human slaughter and the eradication of tribes on an unprecedented scale, of turning to cannibalism to survive, of famine and tribes fleeing to mountainous areas and isolated regions. The clashes that provoked this chain reaction of violence among the black tribes of southern Africa had already occurred before Shaka's rule with the power struggle between the Mthethwa of Dingiswayo and the Ndwandwe of Zwide Nqumayo. After the death of Dingiswayo it was continued by his successor, Shaka. After Shaka's decisive defeat of Zwide's army in 1818, the Nwandwe army splintered and various groups under the leadership of Sochangana, Zwangendaba and Nxaba fled to southern Mozambique. Here they clashed with the indigenous Tsonga tribes as well as the Portuguese settlers. In 1821 another one of Shaka's commanders, uMzilikazi, and his followers fled the Zulu high chief and established the AmaNdebele empire in the southern and southwestern Transvaal. From here he conquered all the surrounding tribes and those to the north, as far as the Limpopo, and to the east, as far as the Drakensberg.

South of the Vaal River the Sotho-speaking Tlokwa under Queen-Regent Mmanthatisi (and her successor Sekonyela) themselves maintained a reign of terror and either subjected or massacred the smaller tribes in the vicinity.

These wars of annihilation by successive tribal high chiefs of neighbouring power blocs had far-reaching consequences for the Transvaal Lowveld. While the established Tsonga- and Sotho-speaking tribes had lived in relative peace among one another, they were eventually also drawn into the whirlpool of violence. One by one they were subjected to the campaigns of Sochangana (Manukosi) and, after him, those of Mawewe, Muzila and Gungunyane's Nguni warriors from the east, uMzilikazi's AmaNdebele hordes from the west and Sobhuza's Swazi soldiers from the south. Later they were further harassed by Sekwati's Pedis from their mountain strongholds in the Drakensberg. As a result the previously densely populated Lowveld was almost completely depopulated and those that remained had to make a precarious living hiding in rocky ravines or isolated and sheltered corners in the bush. Because of this depopulation the amount of game increased. Soon the entire area between the Lebombos and the Drakensberg was teeming with game, but with this amount of animals in the area, the tsetse fly soon penetrated large parts of the Lowveld.

Bergh & Bergh (1984) pointed out that, apart from the socio-political changes and migrations caused by the Mfecane/Difaqane, it also led to large, fertile parts of Natal and the interior becoming uninhabited. The tribes in the interior had not even had time to reorganise themselves after the wars when the Voortrekkers made their appearance in 1836. In certain instances some of these displaced tribes were unable to return to their old territories because Voortrekkers had already settled there. Only the Tswana tribes living at the edge of the Kalahari and the Venda in the Soutpansberg managed to stay in their original settlement areas.

When the Voortrekker parties led by Lang Hans van Rensburg and Louis Trichardt reached Soutpan on the northwestern flank of the Soutpansberg in 1836, they had encountered hardly any settled black tribes along the route. In the Soutpansberg Trichardt met the descendants of the adventurer and renegade Coenraad de Buys, and the Venda-speaking tribes of this mountainous area. On 12 March 1837, Trichardt first mentioned the presence of tsetse flies, which had apparently stung the oxen of Jan Pretorius as he was leading a search party to find the Van Rensburg trek. The main trek encountered tsetse flies near the Brak River (a tributary of the Sand River, north of Soutpan).

During their catastrophic journey to Delagoa Bay in 1837 the Trichardt trek once again encountered tsetse flies along the Olifants River close to its confluence with the Steelpoort River. After they had set off down the Drakensberg mountains and reached what is now the Kruger National Park, they encountered tsetse flies in the vicinity of Vutomi Spruit (northwest of Tshokwane). When they spoke to the local chief Halowaan, he told them his people had once been able to keep a few cattle and showed them a couple of old kraals.

Stevenson-Hamilton (1947) mentioned that the current Kruger National Park had several well-known tsetse fly-infested areas until 1896, namely: (i) a 16 kilometre-wide strip along the northern bank of the Crocodile River up to Kaapmuiden; (ii) along the southern bank of the Sabie River between the hippo pool (Skurukwane) and Lower Sabie (the dense *Acacia welwitchii* veld west and east of the present Skukuza); (iii) a small area west of Tshokwane (again *Acacia welwitchii* thickets between Leeupan and the area just north of Kumana); (iv) the Punda Maria sandveld and a belt along the Luvuvhu down to the eastern border of the park. He was apparently not aware of the fact that, in the years preceding the rinderpest epidemic in the nineteenth century, the tsetse fly could be found in the Nyandu Bush, the vicinity of Shingwedzi Poort (Vos 1890), along the Olifants River east and west of its confluence with the Selati and near Timbavati (Erskine 1875 and Schlaefli 1889), as well as at Renosterkoppies (Mauch 1870). Erskine (1875), Sandeman (1880) and Vaughan-Kirby (1896) described the tsetse fly-infested areas of the Lowveld, and of these Vaughan-Kirby's map of the southern Lowveld is probably the most reliable.

A rinderpest patrol post along the Limpopo on the border between the former ZAR (South African Republic) and Rhodesia (Zimbabwe), 1896.
(ACKNOWLEDGEMENT: ONDERSTEPOORT VETERINARY RESEARCH INSTITUTE)

South of the Crocodile River, the tsetse fly occurred in almost the entire area east of Crocodile Poort to the mountains of Swaziland, and formed an effective barrier to ox-wagon traffic between the interior and Delagoa Bay. Apart from malaria, the tsetse fly was the main reason why the whole area that is now the Kruger National Park was unsuitable for white settlement and was safe only in winter. This is also why the game populations in the area could survive the regular onslaught of hunters during the winter months until the proclamation of the old Sabie Game Reserve by President Kruger in 1898.

After the establishment of the Voortrekker towns Ohrigstad (1845), Schoemansdal (1848) and Lydenburg (1850), the combination of the tsetse fly and malaria mosquitoes to a great extent managed to scupper several efforts to establish a continuous communication and a trade route to the sea. The "Oude Wagenweg" (Old Wagon Route) between Ohrigstad and Delagoa Bay, which Carel Trichardt had explored in 1845, was a failure because of the lack of water for draught animals along the route in winter. The later route (1848) followed – more or less – the centuries-old trade route from the interior past Pretoriuskop to Delagoa Bay. In 1873, after the discovery of the Lydenburg goldfields, this route was improved by Nellmapius, but it was only used as a permanent transport road after the Sekhukhune wars of 1879.

Even then the tsetse fly was a significant stumbling block and special measures had to be taken to limit the contact between oxen and tsetse flies to a minimum. Hall (1937) said that they trekked early in the morning after entering the tsetse belt east of Skipberg. At about nine or ten in the morning, when the tsetse flies became active, they outspanned in an open area until evening, when they completed another shift before setting up camp for the night.

After the discovery of the Barberton goldfields, several attempts were made to build a road from there to Delagoa Bay. These included the Barberton-Crocodile Poort-Delagoa Bay road along Salvation Valley in 1883, Pettigrew's road from Barberton along the Kaap and Crocodile rivers (1886) and French Bob's road further south to the Swaziland border and Furley's Drift (1888). All of these roads had the same tsetse fly problem as the old Lydenburg-Delagoa Bay transport road, and this was one of the main considerations that led to President Paul Kruger's decision to push for the construction of a railway line between Pretoria and Delagoa Bay. The Eastern Line was completed in 1894 and at last provided the ZAR (South African Republic) with permanent access to the sea (Krüger 1923). The settlement of whites in the far western and northwestern parts of the Transvaal and the progressive eradication of game, especially buffalo, resulted in the gradual disappearance of the tsetse fly from the western part of the old Limpopo tsetse belt to the area east of the confluence of the Limpopo and the Sand River (shortly before 1896) (Fuller 1923).

After the rinderpest epidemic of 1896 had decimated both livestock and susceptible game, it was noticed with some surprise that the tsetse fly had vanished from the rest of the Lowveld as well – despite the fact that several species of game that had not been affected by rinderpest still occurred in reasonably large numbers.

According to Stevenson-Hamilton (1947), old hunters told him that they had noticed, as they were leaving their hunting grounds in September 1896, that kudu were dying of rinderpest. Tsetse flies were then still very active in all the known tsetse belts, but when they visited the Lowveld again in May 1897 they were immediately struck by the fact that the tsetse flies had completely disappeared from the entire area.

Sir Arnold Theiler (1867–1936), famous veterinarian and first director of Onderstepoort.
(ACKNOWLEDGEMENT: VETERINARY MEDICAL LIBRARY, ONDERSTEPOORT)

10.4 THE INFLUENCE OF TSETSE FLIES, MALARIA AND THE RINDERPEST EPIDEMIC

What ultimately caused the total eradication of tsetse flies in the Lowveld is still a matter of speculation, but it can be assumed with a reasonable amount of certainty that the tsetse fly could not hold its own in an area where the numbers of its preferred hosts had suddenly dropped to a minimum, despite the continued presence of certain secondary hosts.

It took an epidemic on the scale of the rinderpest panzootic of 1896 to rid the Transvaal Lowveld of one of its ancient plagues and to make it suitable for farming with livestock, especially cattle. But it must be remembered that the tsetse fly infestation of the Lowveld until 1896, as well as the lack of effective malaria prophylactics, indirectly contributed just as much to the preservation of the Lowveld's unique wild mammal composition as the actions of the conservation-minded President Paul Kruger and Messrs Loveday, Glynn, Chapman and others.

While the ZAR government in Pretoria was still struggling to cope with the upheaval and political recriminations left in the wake of the Jameson raid early in 1896, rumours of a feared livestock disease started circulating. This cattle plague had spread southwards from north Africa and had already killed countless antelope and cattle in Uganda. On 22 February 1896 the disease claimed its first victims in the vicinity of Bulawayo. C.E. Gray, then the acting veterinary officer of the Chartered Company in charge of Rhodesia (now Zimbabwe), diagnosed the disease on 18 March that year – rinderpest! Rinderpest is a highly infectious viral disease, endemic among certain game populations in North Africa and deadly to cattle.

On 4 March, Dr Hutcheon, the senior state veterinarian of the Cape Colony, made an urgent appeal to the Cape's administration to restrict the movement of cattle from the infected area and to euthanise animals showing any symptoms of the disease. By then it was already too late to save the Bechuanaland Protectorate. The resident commissioner there, Francis Newton, telegraphed the high commissioner in the Cape to isolate Matabeleland and set up a strong border patrol. At that stage widespread fatalities among cattle and game had already occurred in the land of King Khama. The Tswanas' livestock died off on a large scale.

The Cape high commissioner, Robinson, in turn sent a cablegram to Colonial Secretary Joseph Chamberlain in London, informing him of the scale of the disaster that was about to overwhelm the entire

In certain areas of the Transvaal up to 95 per cent of all cattle herds died during the rinderpest epidemic in 1896.
(ACKNOWLEDGEMENT: ONDERSTEPOORT VETERINARY RESEARCH INSTITUTE)

subcontinent unless immediate and drastic measures were taken. Hutcheon, who had to witness the deaths of 500 000 head of cattle in 1865 during a similar epidemic in Britain, was convinced the only way of stopping the spread of the disease was large-scale quarantine measures and the killing of infected animals (the stamping-out policy). In the meantime Robinson also informed President Kruger of the danger. The president reacted immediately and on 11 March 1896 promulgated a proclamation that prohibited the movement of cattle across the borders of Bechuanaland, Matabeleland and Mashonaland. Guards patrolled the borders to make sure no cattle were smuggled across. Acting State Secretary C. van Boeschoten contacted Arnold Theiler (later the founder of Onderstepoort), who was then doing research and farming pigs in Les Marais, a suburb of Pretoria. He was asked to visit Matabeleland to determine the extent and nature of the epidemic so the Republic could act on professional advice.

It took Theiler ten days by mule-drawn coach to reach Bulawayo. On 19 March, while en route, he telegraphed that the disease was indeed rinderpest and that the veld was strewn with hundreds of cattle carcasses.

Upon his arrival in Bulawayo he met his colleague C.E. Gray and together they carried out several post-mortems in the veld amid an overwhelming sense of impotence. Theiler sent the following message: "One might as well have tried to stop a rising tide on the sea-shore, as prevent this dreadful disease from travelling steadily down the main roads, leaving nothing but rotting carcasses and ruined men behind it." (Gutsche, 1979)

By now the disease had already spread to the Transvaal – despite the border guards. The first cases were reported at Nylstroom (now Modimolle) in the north and the Crocodile River (Zeerust district) in the west. Apart from the telegraph system, all communication with the north was cut off due to a ban on road traffic. The only news that trickled through was the desperate cries for help by the afflicted farmers and livestock owners. The catastrophe caused the AmaNdebele to rebel and they killed someone. Theiler reconfirmed his diagnosis by telephone on 25 March and insisted on the strictest security measures. All horses in Rhodesia were commandeered for military purposes and rinderpest patrols came to a halt. Theiler managed to escape the beleaguered Bulawayo via the Tati goldfields. On his journey south he discovered to his shock just how quickly the disease was spreading – at a rate of 130 to 160 kilometres a week!

Many contemporary observers, including Selous (who had temporarily been appointed as livestock inspector), Melton Prior and others, made notes of the conditions in Matabeleland, but Theiler himself said nobody who had not seen it themselves could have any idea of the heat, the stench and the piteous bellowing of the dying animals. By this time, the columns that were supposed to relieve Bulawayo were already on their way from Mafeking (now Mafikeng). One of their officers reported as follows about events:

> All the way, the road was inches deep in dust and this, disturbed by the hooves of 60 horses, kept everyone enveloped in continued cloud. When trotting, it was impossible to distinguish any trace of the man immediately in front of me. Added to this was the terrible stench from the decaying carcasses of dead oxen – victims of rinderpest – which lined the roadside. It was a common occurrence to see the remains of whole spans, twenty or thirty, lying about within a radius of a hundred yards. The air was never entirely free from this pestilential taint. Now and again, wagons were met – derelicts of the veld – laden with timber, furniture and cases of all kinds of merchandise, drawn up in the bush just off the road and left to look after themselves. All the trek oxen had succumbed and the transport riders had no alternative but to abandon their loads. There was wholesale looting …

To make matters worse, horses and mules started dying of African horse sickness. Only donkeys survived the scourge and were the only remaining draught animals in these difficult times. Theiler reached Palachwe on 30 March, speechless at the dreadful disaster. Hutcheon had in the meantime arrived from the Cape and urged everybody to destroy infected herds. He was set on preventing the epidemic from reaching the Cape Colony – a "cordon sanitaire" had to be established before the Molopo border. Thousands of head of cattle belonging to the Baralong were shot and, like the AmaNdebele, they were about to rebel.

Khama's followers had already lost an estimated 90 000 cattle and faced total ruin. Theiler eventually reached Mafeking, from where he travelled southwards by train via Kimberley to De Aar, and from there northwards through Bloemfontein to Pretoria. On 7 April 1896 he arrived in Pretoria and that night he conveyed the seriousness of the circumstances to the Volksraad. President Kruger immediately comprehended the gravity of the situation, as well as the fact that science was their only hope of salvation.

Theiler was delegated to attend a conference in Mafeking on 17 April 1896 where representatives of the Cape, the Free State and the Transvaal would discuss measures to prevent the further spread of the rinderpest epidemic. They decided on a campaign of eradication and all the states had to pledge their support. Hutcheon, seconded by Lloyd of Natal, suggested that a double fence be erected along the borders of the Transvaal, as far as deemed necessary. His proposal was accepted. At this stage there was still no question of immunisation against the disease, simply because there wasn't an effective vaccine. President Kruger issued a proclamation according to which each district of the Transvaal had to assemble a rinderpest committee, which would receive its orders from the state veterinarian in Pretoria – a post to which Theiler had not even been appointed yet!

The control measures had no effect. Despite the strictest measures possible and the destruction of thousands of cattle, the epidemic spread through the Transvaal like a runaway fire. In its wake it left a trail of death and misery that may have been even worse than that of the Anglo-Boer War three years later. Many remedies and drugs were tested, including Cooper's powder, salt and paraffin, sodium permanganate, quinine, copper sulphate, the "soup" made by cooking a sick animal, and other secret cures. All was in vain and the devastated Transvaal buckled under the combined effects of a drought, a locust plague and the rinderpest. While Theiler was busy testing a cure, 15 000 head of cattle died in the Waterberg district and 2000 carcasses were rotting near Rustenburg. The State spent £30 000 on buying mules so essential transport services could be maintained.

By now the whole country had come to the terrible realisation that nothing could stop the catastrophic panzootic. Hutcheon's river corridor and fences along the Molopo River were a complete failure. Rinderpest penetrated the northern Cape and rapidly spread south- and eastwards.

In the Waterberg district Theiler witnessed the epidemic at its worst and saw cattle and susceptible antelope species dying by the score. At that time (August 1896) newspapers started running articles about a "vehicle that can walk", developed by the firm Daimler-Benz and imported by J.P. Hess and Co. of Pretoria. "Just think – one can now travel without horses, harnesses, grooms and grain!" one editorial read. When the odd-looking vehicle arrived in Pretoria, President Kruger inspected it, tapped it with his walking stick and refused to set foot in it!

A second two-day rinderpest conference took place in Vryburg at the end of August 1896 and was attended by representatives of the Cape, Transvaal, Natal, Orange Free State, Basutoland, Bechuanaland and German West Africa.[2] Once again methods to

Dr J.W.B. Gunning, Director of the National Zoo in Pretoria during the rinderpest epidemic of 1896. (ACKNOWLEDGEMENT: STEVENSON HAMILTON COLLECTION, SKUKUZA ARCHIVES)

2   Today Lesotho, Botswana and Namibia.

contain the disease were discussed, but the pessimistic conclusion was that very little could be done. Cattle and game could be contained in camps, but people themselves were the main carrier of the disease. There were even reports that unscrupulous individuals were maliciously spreading the virus.

Theiler and Watkins-Pitchford pointed out that the only hope of treating the disease successfully was effective immunisation by means of an antiserum.

The epidemic was still spreading rapidly through the rest of the Transvaal, the Free State and the northern Cape. There was great bitterness among the people of the Cape Colony and the prime minister, Sir Gordon Sprigg, was accused of dereliction of duty. Farmers demanded his resignation and petitions for the appointment of the dynamic John X. Merriman were doing the rounds. By the time Theiler and his colleague Watkins-Pitchford arrived in the Marico district to set up an experimental station for the development of a vaccine, 16 808 out of a total of about 30 000 head of cattle had already died of rinderpest, while 4027 had been shot. Only 6766 eventually survived.

In the Cape Colony Dr Edington started conducting experiments similar to those of Theiler and Watkins-Pitchford – injecting cattle with the serum of animals that had recovered from the disease. He was working in the Taung district of the northern Cape. In these times of despair the Cape parliament approached Baron Von Schuckman in German West Africa with the request that the famous Dr Robert Koch be asked to help. This was clearly a motion of no-confidence in their Dr Edington.

In the meantime, his scientists' lack of success caused President Kruger to lose his trust in science. He circulated his own complex "remedy" among the farmers in the Transvaal that was just as ineffectual as the hundreds of other quackish cures!

By mid-December Theiler paid a visit to Pretoria and reported to President Kruger that he had managed to develop a promising-looking vaccine from the blood of cattle that had survived the disease. However, a lot of work still had to be done to improve the serum and conduct clinical tests. Rinderpest had already reached Pretoria and environs and caused mass fatalities among cattle and game.

Robert Koch disembarked in Cape Town on 1 December 1896, accompanied by his wife and assistants. He was met by the Cape director of health, George Turner. The De Beers diamond company had set up laboratory facilities for Dr Koch in Kimberley.

Just as his own experiments to create a vaccine started showing promising results, Theiler was recalled to Pretoria in January 1897 to assist two French veterinarians, M. Jean Danysz and Jules Bordet, with further research to develop a cure for rinderpest. The epidemic had in the meantime reached its climax and had spread through almost the entire Transvaal to the northernmost bushveld regions.

Hunters had seen kudu dying of the disease in September, and in the following months thousands of kudu, nyala, bushbuck, eland, warthog, bush pig and almost all of the buffalo succumbed (according to J. Stevenson-Hamilton only between 15 and 20 buffalo survived in the Nwatimhiri bush of the current Kruger National Park). Smaller numbers of less susceptible game such as blue wildebeest, impala, sable and roan antelope probably also succumbed to the disease, but fewer than in the case of highly susceptible animals (incidentally also the primary hosts of the tsetse fly). As was the case elsewhere in the Transvaal and other infected areas, there were large-scale fatalities among cattle in the Lowveld.

H.F. van Oordt, the warden of the Pongola Reserve, mentioned in his 1896–1897 annual report to the provincial secretary that rinderpest had not yet affected the animals in the reserve, but that it was expected. (The area was hit late in 1897.)

Van Oordt's observations about rinderpest drew the attention of the authorities. J.W.B. Gunning, the director of the State Museum in Pretoria, was asked to contact Van Oordt in connection with the disease and particularly to get information about the relative susceptibility of the different species. In March 1898 Van Oordt sent Gunning a report with the relevant information.

Hardly any cattle survived the epidemic in the Soutpansberg, while African horse sickness decimated the horses. Tribes were hit the hardest as their crops had been destroyed by drought and the locust plague. Famine and plunder were the order of the day.

Thousands of people died of starvation, while the resulting shortage of labour had a crippling effect on the economy as a whole. The whites in the Transvaal, especially the farmers, suffered just as much. Of a cattle stock amounting to almost a million, more than two-thirds had died. Many farmers were reduced to beggary and became bywoners (tenant farmers) of those who still had something left. Game, which used to be a supplementary source of food, had been decimated to such an extent that hunting could not provide significant relief.

Government bought thousands of mules and donkeys at the exorbitant price of £10 to £12 a head to prevent transport services from collapsing. Many farmers eked out an existence by riding transport with donkeys or mules, while the president and the churches created a disaster fund to relieve the worst suffering.

In the meantime, Koch and Theiler were working tirelessly to create lasting immunity in cattle by dosing them with a wide range of experimental vaccines. Robert Koch was the first to achieve a breakthrough on 10 February 1897. He obtained active immunity by injecting healthy animals with a serum prepared from a diseased animal's gall. Theiler and his helpers, Danysz and Bordet, independently developed a similar vaccine from the blood, serum and gall of diseased animals at their experimental station Waterval north of Pretoria. Koch, however, stole a march on them when he announced his results to the world. This caused Theiler a great deal of frustration, but later he improved on Koch's method with the aid of his colleague Watkins-Pitchford, since it emerged soon after Koch's departure for India that his "serum" did not provide permanent immunity. Theiler eventually perfected an immunisation method that was 85 per cent effective – a considerable improvement on Koch's treatment. These results were soon published and replaced the Koch immunisation method throughout the country.

Because of a shortage of trained personnel and apparatus (almost no hypodermic syringes were available in the Transvaal and they had to borrow 35 from the Cape!), as well as a fierce debate about the efficacy of the two alternative immunisation methods, rinderpest was still raging on. By August 1897 it had spread to Natal and down to the coast in the Cape. A rinderpest conference was organised in Pretoria under the chairmanship of Magistrate C.E. Schutte, and the different methods of immunisation were endlessly discussed and debated. Eventually they came to the conclusion that the immunisation method developed by Theiler and Watkins-Pitchford and subsequently improved by two researchers at the Pasteur Institute, Danysz and Bordet, was in fact the safest and best way of fighting rinderpest. It would therefore receive further attention.

Both the Koch and the Theiler-Danysz-Bordet methods were subsequently used in efforts to contain the epidemic. A significant degree of success was achieved only in the eastern Cape. In the end, the epidemic cost the Cape £1 196 360 and 35 per cent of its cattle, while Natal suffered a loss of over £200 000. The Transvaal and the Free State suffered the most, however. When taking into account the fact that the Anglo-Boer War broke out a mere three years after the epidemic had ravaged the unfortunate country, it becomes a lasting symbol of the people's grit in the face of adversity, as well as their courage, faith and trust in the future.

Just as the famine in Ireland at the end of the nineteenth century had left the British public and Queen Victoria cold, the devastation caused by the rinderpest epidemic of 1896–1897 was never properly evaluated in ecological terms. Apart from the severe blow it dealt the country's economy, it had other direct and indirect consequences. As a result of the general shortage of meat, an interest in the use of refrigerating machines developed so meat could be stored when available. Because of the shortage of draught animals such as oxen (as a result of the rinderpest), horses and mules (as a result of African horse sickness) and donkeys (as a result of foot rot), the Volksraad of the Transvaal urgently had to pay attention to the development of alternative modes of transport. The completion of the Pretoria-Delagoa Bay railway line in 1894 was a far-sighted step and other lines soon followed. Interest in motor vehicles and steam tractors rocketed and in 1894 Sammy Marks imported two steam tractors to plough his lands. The bicycle business flourished and electricity was used to run trams in cities. The donkeys imported by the government to supplement the oxen shortage created their own problems in that their numbers increased to such an extent that they eventually became a plague that had to be eradicated at considerable cost.

The price of salted cattle became exorbitant – up to £60 was paid for an ox that had recovered from rinderpest, whereas the original price for a good trek ox had been only £6.

In the Lowveld, certain game species such as buffalo and eland almost died out as a result of the disease and it took many years before their numbers recovered.

According to an interview W.H.J. Punt had conducted with H.F.A. Schiel, of the old museum in Pretoria, the only kudus to survive the rinderpest epidemic of 1896 in the area currently the Kruger National Park could be found on a hill south of Klopperfontein – probably Dzundwini.

One bonus that came about as a result of the epidemic was the disappearance of the tsetse fly from the Transvaal.

Malaria, like the nagana in cattle and horses spread by tsetse fly, is also transmitted by a blood-sucking insect, namely the *Anopheles* mosquito, and for centuries played an important role in the human colonisation of the Transvaal Lowveld. The difference, however, was that this deadly disease, which killed people in their thousands, was never associated with an insect bite. It is one of the oldest known diseases and was described by Hippocrates in as early as the fifth century AD. He also distinguished between malaria that resulted in fever attacks every day, on alternative days, or on every third day. Malaria, as can be deduced from the name, was associated with bad air rising from swamps. Only in 1894 did the British pioneering researcher in tropical medicine, Sir Patrick Manson, point to the mosquito as a probable carrier of the disease. In 1898 Sir Ronald Ross, a British medical doctor, claimed that malaria in birds was transmitted by the *Culex* mosquito in India and described the life cycle of the blood parasite in mosquitoes. A year before he found pigmented cysts in an *Anopheles* mosquito that had sucked the blood of human malaria victims. In 1898 the Italian researchers A. Bignami, G.B. Grassi and G. Bastianelli succeeded for the first time in transferring malaria to humans through mosquitoes. They described the life cycle of the malaria parasite in humans and placed it on record that human malaria was transmitted by *Anopheles* mosquitoes. The first person to notice the parasite in human blood and describe it was the Frenchman A. Laveran in 1880 in Algeria. An Italian, Camillo Golgi, indicated in 1886 that fever attacks in malaria sufferers coincided with the segmentation and release of masses of parasites (merozoites) into the bloodstream by the disintegration of infected red blood cells.

A treatment for the fever attacks had been known since the 1630s, when the Countess of Chinchón was apparently cured of a malaria attack by drinking a decoction of bark from a tree that grows in the jungles of Peru. These trees were named after her and chinchona or kina trees, *Chinchona ledgeriana* and *C. succirubra,* were later exported by the Dutch to Java, improved and cultivated on a large scale. Until the outbreak of the First World War the bark of these trees was the raw material for the world's most important anti-malaria remedy, quinine sulphate. It is known that some of the earliest explorers in the Lowveld, including D. Fernandes das Neves, who travelled from Delagoa Bay to Zoutpansbergdorp (Schoemansdal) in 1860, carried a supply of quinine sulphate for use against "marsh fever". This bit of foresight enabled Das Neves to save the life of one of his bearers.

Over the centuries the indigenous black population had developed a degree of immunity to the disease, but it remained a major cause of mortality among younger children in particular, mainly at the end of high rainfall seasons when mosquito populations increased to maximum levels.

This disease was fatal to Europeans right from the start, and after 1721, when the Dutch settled in Delagoa Bay, deaths in the Lowveld were so common that they were accepted with resignation. Experience showed that the disease was contracted mostly during the wet summer months in low-lying areas and these areas were thus avoided at these times. The high death toll as a result of malaria was the main reason why the settlers at Ohrigstad had to abandon the town in 1849 to establish a new town, Lydenburg, in 1850 in a higher-lying, healthier environment.

Like the tsetse fly, malaria showed a clear connection to climate cycles. It wasn't always equally bad everywhere and the incidence varied according to above- or below-average rainfall seasons. While Louis Trichardt's laager was at Soutpan, the members of his party did not contract malaria, but when they started moving eastwards he first referred to the feared "yellow fever" in his diary of January 1837. That month Jan Pretorius wrote a letter to Trichardt from the Lowveld, saying that Gert Scheepers and three of his children, two of Hendrik Botha's children, Louw and Izak Albach, as well as his own family, were all ill with fever. Jan Pretorius and the rest of the search party were on their way back from the Lowveld after their fruitless search for the Van Rensburg trekker party. Gert Scheepers succumbed to the disease shortly afterwards. After the Scheepers family had returned to Trichardt's camp, their daughter Anna died on 12 March 1837 of "de koorts siekte" (the fever sickness).

Shortly afterwards, on 21 April, Alida Strijdom died while Jeremiah Trichardt and another five members of the Strijdom family fell ill. After April 1837, until their departure for Delagoa Bay in August 1837, Trichardt did not mention any more cases of malaria, but only six days after they had set off for the Bay there was another death. This time it was Louis Trichardt's young daughter. During their arduous journey to the coast no more malaria cases occurred, but shortly after their arrival in the unhealthy surroundings of the east coast harbour city on 13 April 1838, the members of Trichardt's group started dying. Eventually, out of a total of 53 adults and children, only 25 women and children and three servants were brought to the safety of Port Natal by the sailing ship *Mazeppa*.

Delagoa Bay itself, ever since the arrival of the first settlers of the Dutch East India Company, was known as an area deadly to Europeans. Records show that during their short stay of nine years (1721–1730) there, no fewer than 1000 Dutch people died of malaria. This was the main reason why the VOC's settlement at Delagoa Bay failed (Punt, 1975). It is not surprising therefore that the Dutch named the later Fort Rio de Lagoa of the Portuguese, Fort Lijdsaamheid (Fort of Suffering).

Schoemansdal was also plagued by malaria before its evacuation in 1867 – a fact attested to by the more than 120 piles of stone in the cemetery.

The terribly high death toll malaria sometimes exacted among individual families travelling through the Lowveld is described in chapters 9.3 (Dina Fourie, 1859) and 18 (A. Coqui, 1848).

It is estimated that one worker died for every kilometre of railway line during the construction of the Eastern and Selati railway lines. Once again, it is the numerous piles of soil or stone in lonely cemeteries that bear silent witness to this.

Somebody came up with the "original" suggestion of planting eucalyptus trees in lower-lying areas to "force" malaria out of these areas, as the aromatic oil in the leaves had medicinal attributes. The ZAR's Railway Company laid down official guidelines instructing construction workers working on the Eastern railway line to plant eucalyptus trees wherever they could. This is probably how eucalyptus trees came to be planted in what is now the Kruger National Park, among others at Skukuza and at the site of Sardelli the Greek's shop at Gomondwane.

Malaria was also a factor that had to be considered by the fighting forces in the Lowveld during the Anglo-Boer War. Many a soldier passed away after being bitten by a mosquito.

After the war, malaria put a huge damper on the rebuilding of the devastated region and made life difficult for the first warden and rangers of the old Sabie Reserve. Stevenson-Hamilton (1944) wrote about it in his book, *The Lowveld – Its Wildlife and Its People*. Until the outbreak of the Second World War almost all of his annual reports referred to malaria attacks among his staff. Fortunately malaria treatment has made great strides over the years and effective prophylactics are now available. Highly suc-

Dr Siegfried Annecke (middle front) and personnel of the Institute for Tropical Diseases, Tzaneen in 1938.
(ACKNOWLEDGEMENT: MR G.J. VAN EEDEN, TZANEEN)

cessful campaigns to eradicate mosquitoes were also initiated in the early 1930s by the then Department of Health and Welfare.

The person responsible for this was the acclaimed Dawid Siegfried Annecke, who founded and expanded the National Institute for Tropical Diseases in Tzaneen. He did pioneering work of inestimable value with regard to adult mosquitoes and the control of their larvae. During his 25 years of service, until his death in August 1955, he tirelessly devoted himself to his life's work of making the fertile Lowveld habitable and safe for settlement and farming. Although his main goal was the eradication of malaria, he spent some time doing research on bilharzia as well. The blood parasite that causes the most deadly form of malaria – cerebral malaria and blackwater fever – is *Plasmodium falciparum*. This is the most common malaria parasite in the Lowveld and is transmitted through the bite of infected mosquitoes of the genus *Anopheles*, namely *A. funestus* and *A. gambiae*.

Annecke's malaria control campaigns were mainly aimed at: the extermination of both mosquito species in their adult form through the spraying of houses with DDT and other chlorinated hydrocarbon compounds; the control of larvae by spraying breeding sites with available compounds; and preventing people from being bitten by mosquitoes by supplying households with mosquito netting, administering prophylactics and running continual information campaigns among all classes. It is evident that he and his diligent workers succeeded to a great extent as malaria is under control in the Transvaal Lowveld and only sporadically shows flare-ups during climate conditions (good rainfall years) particularly favourable to mosquitoes.

Whereas the Lowveld was considered the "white man's grave" at the beginning of the twentieth century, it is now known as "South Africa's pantry". For this, the current generation should honour one of the great sons of South Africa, the "Man of Ironwood" Siegfried Annecke, and his qualified assistants and award him a place of honour in the annals of scientific achievement.

References

Annecke, D.H.S. & Pitchford, R.J. 1954. Malaria, bilharzia and health services. In: Webb, H.S. (ed.). *A Survey of the Southern Region of the Eastern Transvaal Lowveld*. Lowveld Regional Development Association.

Austen, E.E. 1903. *Monograph of the Tsetseflies*. British Museum, London.

Baines, T. 1854. The Limpopo, its origin, course and tributaries. *Journal of the Royal Geographical Society* 24:288–291.

Baines, T. 1864. *Journal of Residence in Africa (1850–1853)* vol 2:177 *et. seq*. Kennedy, R.F. (ed.). Van Riebeeck Society, Cape Town.

Baines, T. 1877. *The Gold Regions of South-Eastern Africa*. Edward Stanford & J.C.W. Mackay, London & Port Elizabeth.

Bergh, J.S. & Bergh, A.P. 1984. *Stamme en ryke*. Don Nelson, Cape Town.

Bernard, E. & Bernard, F.O. 1969. *The Journals of Carl Mauch 1869–1872*. National Archives of Rhodesia, Salisbury.

Buxton, P.E. 1955. *The Natural History of Tsetse-flies*. Lewis, London.

Cohen, E. 1875. *Erlaüternde Bemerkungen zu der Routenkarte einer Reise von Lydenburg nach den Goldfeldern und von Lydenburg nach der Delagoa Bai in östliches Süd-Afrika*. L. Friederichsen & Co., Hamburg.

Cumming, A.G. 1850. *Five Years of a Hunter's Life in the Far Interior of South Africa*. Vol.1 & 2. John Murray, London.

Das Neves, D.F. 1879. *A Hunting Expedition to the Transvaal*. George Bell &. Sons, London.

Dicke, B.H. 1932. The tsetse fly's influence on South African history. *South African Journal of Science* 29:792–796.

Dunbar, D.M. 1881. *The Transvaal in 1876*. Richards, Slater & Co., Grahamstown.

Elton, F. 1872. Journal of an exploration of the Limpopo river. *Journal of the Royal Geographical Society* 42:1–49.

Erskine, St. V.W. 1870. The tsetse fly. *Natal Mercury* 1 September.

Erskine, St. V.W. 1875. Journey of exploration to the mouth of the Limpopo river. *Journal of the Royal Geographical Society* 39:45–125.

Estes, R.D. 1984. Rinderpest in Sub-Sahelian Africa. *Gnusletter* May:1–4.

Estes, R.D. 1984. Rinderpest in Africa. *IUCN Survival Service Commission Newsletter* 4:8.

Fiedler, O.G.H. & Kluge, E.B. 1954. The influence of the Tsetsefly eradication campaign on the breeding activity of Glossinae and their parasites in Zululand. *Onderstepoort Journal of Veterinary Research* 26(3):389–97.

Fiedler, O.G.H. & Kluge, E.B. 1954. The parasites of Tsetse flies in Zululand with special reference to the influence of the hosts upon them. *Onderstepoort Journal of Veterinary Research* 26(3):399–404.

FitzPatrick, J.P. 1907. *Jock of the Bushveld.* Longmans Green & Co., London.

Ford, J. 1971. *The Role of Trypanosomiasis in African Ecology: A Study of the Tsetse Fly Problem.* Clarendon Press, Oxford.

Fuller, C. 1923. Tsetse in the Transvaal and surrounding territories: A historical review. *Union of S.A. Department of Agriculture Entomology Memoir* 1:6–68.

Glynn, H.T. 1926. *Game and Gold.* The Dolman Printing Co. Ltd., London.

Gutsche, T. 1979. *There was a Man – the Life and Times of Sir Arnold Theiler.* Howard B. Timmins, Cape Town.

Hall, H.L. 1937. *I Have Reaped My Mealies.* Betteridge & Donaldson, Johannesburg.

Hall, M. 1976. Dendroclimatology, rainfall and human adaptation in the Later Iron age of Natal and Zululand. *Annals of the Natal Museum* 22(3):693–703.

Harris, W.C. 1839. *The Wild Sports of Southern Africa.* John Murray, London.

Hofmeyr, S. 1890. *Twintig jaren in Zoutpansberg.* J.H. Rose &. Co., Cape Town.

Krüger, D.W. 1938. Die weg van die see. *Archives Year Book for South African History* 1:31–233.

Lapage, G. 1956. *Mönnig's Veterinary Helminthology and Entomology.* Baillière, Tindall and Cox, London.

Leipoldt, C.L.1937. *Bushveld Doctor.* Jonathan Cape Ltd., London.

MacQueen, J. 1962. Journey from Inhambane to Zoutpansberg by Joaquim de Santa Rita Montanha. *Journal of the Royal Geographical Society* 32:63–68.

Mager, C. 1895. *Karl Mauch: Lebensbild eines Afrika-Reisenden.* W. Kohlhammer, Stuttgart.

Meyer, A. 1986. *'n Kultuurhistoriese interpretasie van die Ystertydperk in die Krugerwildtuin.* D.Phil. thesis, University of Pretoria.

Mitchell, B.L. 1963. A new aspect of the effect of bush fires in connection with the tsetse, *Glossina morsitans. Kirkia* 3:26–9.

Möller-Malan, D. 1953. *Dina Fourie.* Voortrekkerpers, Johannesburg.

Naude, J.F. 1956. Dr. Dawid Siegfried Annecke – so onthou ek hom. Radio talk on SABC. 22 March.

Oswell, W.E. 1900. *William Cotton Oswell, Hunter and Explorer.* Volumes 1 & 2. William Heinemann, London.

Petterman, A. 1870. *Mitteilungen aus Justus Perthe's Geographischer Anstalt über wichtige neue Erforsungen auf den Gesamentgebiete der Geographic.* Justus Perthes, Gotha.

Pienaar, U. de V. 1985. Indicators of progressive desiccation of the Transvaal Lowveld over the past 100 years, and implications for the water stabilisation programme in the Kruger National Park. *Koedoe* 28:93–166.

Preller, G.S. 1917. *Dagboek van Louis Trighardt.* Het Volksblad Drukkerij, Bloemfontein.

Punt, W.H.J. 1953. *Louis Trichardt se laaste skof.* Van Schaik, Pretoria.

Punt, W.H.J. 1975. *The First Europeans in the Kruger National Park.* National Parks Board, Pretoria.

Richard, D. 1974. *Man van Ysterhout.* Tafelberg, Cape Town.

Sandeman, E.F. 1880. *Eight Months in an Ox-waggon.* Griffith & Farran, London.

Schlaefli, H.E. 1889. Route followed by H. Berthoud and H.E. Schlaefli from the Selati to the Limpopo. Unpublished manuscript with kind permission of P. Cuènod.

Selous, F.C. 1908. *African Nature Notes and Reminiscences.* Macmillan, London.

Stevenson-Hamilton, J. 1939. The health of wild animals. *Journal of South African Veterinary Medical Association* 10(2):56–64.

Stevenson-Hamilton, J. 1944. *The Lowveld – Its Wildlife and Its People.* Van Schaik, Pretoria.

Stevenson-Hamilton, J. 1945. Tsetse-flies in the Lowveld. Unpublished manuscript Skukuza Archives.

Stevenson-Hamilton, J. 1947. Tsetse Fly menace to the Kruger National Park. *The Star* 9 August.

Stevenson-Hamilton, J. 1957. Tsetse fly and the rinderpest epidemic of 1896. *South African Journal of Science* 53(8):216–218.

Summers, R. 1968. Archaeological distributions and a tentative history of tsetse infestation in Rhodesia and the Northern Transvaal. *Arnoldia* 3(13):1–18.

Tyson, P.D. 1977. The enigma of changing world climates. *South African Geographical Journal* 59:77–116.

Van Onselen, C. 1972. Reactions to the rinderpest in southern Africa, 1896–1897. *Journal of African History* 13(3):473–488.

Van Oordt, H.F. 1897. Van Oordt's third annual report on the Pongola Reserve. T.A. SS 2031. RS009/89.

Van Oordt, H.F. 1898. Report by Van Oordt to Dr. Gunning about rinderpest in the Pongola Reserve. T.A. SS 4975a. R8748/95.

Vaughan-Kirby, F. 1896. *In Haunts of Wild Game.* William Blackwood & Sons, London.

Vaughan-Kirby, F. 1916. Tsetse-flies in Zululand. *South African Journal of Science* 12:375–396.

Vos, W. 1890. Dagboek van opmeting van oosgrens tussen ZAR en Mosambiek. National Parks Board Archives.

Webb, H.S. 1954. The earlier part of the twentieth century and the dawn of development. In: Webb, H.S. (ed.). *A Survey of the Resources and Development of the Southern Region of the Eastern Transvaal Lowveld.* Publication of the Lowveld Regional Development Society.

Chapter 11
The Eastern and Selati railway lines
A. Minnaar and Dr U. de V. Pienaar

Much has already been written and told about the origins and development of the historic Eastern and Selati railway lines. This documentation ranges from summaries and matter-of-fact references to comprehensive articles and even books, while the contents of these documents range from cold historical facts to humorous anecdotes. It can become difficult to distinguish between reality and fantasy.

Many yarns and even legends are coupled to both these lines, but few of these could be included in this chapter. Interested readers are therefore referred to Bulpin's well-known *Lost Trails of the Transvaal* and Buchan's *The Chronicles of a Contractor*, which tells the story of the famous railway contractor George Pauling.

Although both railways have important connections with the Kruger National Park, the Selati Line in particular played a key role in the development and management of this conservation area.

What makes these two railway lines relevant in the context of this work? There are essentially two aspects that are of importance. One is the colourful and romantic history of both infrastructures and the other the fact that sections of both run through the Kruger National Park, or are adjacent to it. In a sense, the origin of these lines seems an anomaly. On the one hand one has the unspoilt old Sabie Reserve and later Kruger National Park and on the other the train as symbol of man's destruction of Creation through technological advances. These two man-made structures represent a tale of idealism, pride, perseverance, suffering, patriotism and hard labour, but also of disloyalty, deceit, robbery, murder, humiliation, disappointment and death.

In one aspect the Eastern Line is unique in the history of the South African railway system. Wildenboer (1975–76) points out that it is the only complete main line in South Africa built between a city in the interior and a seaport. The rest all started as short railway lines at ports and were afterwards gradually lengthened to reach inland cities and towns.

The Eastern Line roughly follows a line from west to east, from Pretoria via Middelburg, Machadodorp, Waterval-Boven and Waterval-Onder, Nelspruit and Kaapmuiden, and then along the southern border of the Kruger National Park through Komatipoort to Maputo (originally Delagoa Bay, later Lourenço Marques).

The Eastern Line had its forerunners in the old trade routes from the interior to the ports along the east coast and the later transport roads from Ohrigstad, Lydenburg and Barberton to Delagoa Bay. In as early as the 1840s the Portuguese Lowveld pioneer João Albasini had visions of steam-driven wagons transporting goods from the Portuguese ports on the east coast to the interior and back. The first Voortrekkers who settled in the Transvaal soon realised the importance of a passable route to the sea for trade purposes and untiringly worked towards the establishment of a regular communication channel with the Portuguese. The completion of the Eastern Line in 1896 was the realisation of the Transvaal settlers' cherished ideal – a way to the sea!

Louis Trichardt was the first Voortrekker leader who managed to reach Delagoa Bay from the inte-

The Eastern railway line through Crocodile Poort shortly after its completion in 1895.
(ACKNOWLEDGEMENT: A. SCHWARTZ COLLECTION AND H. BORNMAN)

rior of the Transvaal after his epic journey of 1837–38 by ox-wagon. Unfortunately his death put paid to his dream of establishing a permanent trade route between the port and Transvaal. It was therefore eminently suitable that the remaining link to this pioneering era – the 84-year-old Karel Trichardt – and some of his family members were on the train that made the inaugural trip from Pretoria to Lourenço Marques. Tragically, the Eastern Line was also the way by which President Paul Kruger, who had the line built during his term of office, left his fatherland on 10 September 1901 to find freedom in exile.

Commandant General Andries Hendrik Potgieter continued with Louis Trichardt's efforts to establish a trade route to Delagoa Bay from 1844 onwards. In 1848 he had the Oude Wagenweg (Old Wagon Route) between Ohrigstad and Delagoa Bay established with the help of Karel Trichardt. This route was unsuitable for regular ox-wagon traffic because of a lack of water along the way during the healthier winter months and gradually fell into disuse, especially after Potgieter's move to Schoemansdal in 1849.

After the discovery of gold at Pilgrim's Rest in 1873 and at Barberton shortly afterwards, a road connecting these two areas with Delagoa Bay became a matter of urgency. In 1875 President Thomas Francois Burgers instructed the Hungarian entrepreneur Alois Hugo Nellmapius by way of the Volksraad to build a suitable road between the goldfields and Lourenço Marques.

Nellmapius achieved his aim and in 1876 the first goods were transported by long files of bearers from Mozambique to the mines. Ox-wagons could travel only part-way along this road and the arrival of goods at their destinations was too erratic. As a result President Burgers started contemplating the construction of a railway line between Lydenburg and Lourenço Marques, but several of his attempts to obtain funds in the form of loans in Europe failed. Of the £300 000 required, he could only raise a loan of £83 838. With this he had rails, sleepers and other material bought in Belgium. These were duly delivered in Lourenço Marques in 1877, but in the meantime war had broken out with the Pedi high chief Sekhukhune and on 12 April 1877 the Transvaal was annexed by the British under Shepstone.

At that stage it was impossible to build the railway line and President Burgers's rails and other material gathered dust and rust in the Lourenço Marques harbour.

After the Sekhukhune rebellion had been quelled by Sir Garnet Wolseley, assisted by a Boer commando and a Swazi army under Chief Mataffin, the ZAR (South African Republic) burghers became dissatisfied with the British occupation. In 1881 the Republic regained its independence through the Transvaal War of Independence and Stephanus Johannes Paulus Kruger was elected as State President in the first election after the war, in April 1883. He obtained 3431 votes against the 1171 for General Piet Joubert.

During the British annexation a half-hearted attempt was made to investigate the possibility of a railway line to Lourenço Marques. In May 1879 an engineer named Farrell was sent to investigate the suggested route through Swaziland, but that is as far as it got. With the railway lines that slowly but surely crept into the interior from Cape Town and Durban, it was inopportune in political terms to support the building of a third line from an east coast harbour.

After Paul Kruger's election as president of the once-again independent Transvaal Republic, the Eastern Line project was resurrected. By then the Kaap goldfields were also in full production and a railway connection was now a matter of the utmost

The docks and mooring place of the Nederlandsche Zuid-Afrikaansche Spoorweg-Maatschappij in Lourenço Marques, circa 1890s.
(ACKNOWLEDGEMENT: NATIONAL CULTURAL HISTORICAL OPEN-AIR MUSEUM)

urgency. Both the Portuguese authorities and the Volksraad gave the project their full support.

Early in 1883 the Portuguese sent Major Joachim José Machado to Pretoria to discuss the scheme with the Republic's Railway Commission, which had been appointed by President Kruger. During his trip to Pretoria, Machado became aware of a completely new route along the southern bank of the Crocodile and then the Elands River to the Highveld and from there via Belfast and Middelburg to Pretoria.

As a result of Machado's visit, President Kruger and a delegation again left for Europe to negotiate financial assistance for the Transvaal part of the project. After the usual bickering, a group of Dutch financiers took on the project. On 16 April 1884 the Volksraad awarded them the concession to start construction of the Transvaal section of the line.

In the meantime the Portuguese had allocated the concession for their part of the line to an American financier, E. McMurdo, on 14 December 1883. President Kruger wanted the line completed as soon as possible, for both political and economic reasons, but another couple of years went by as a result of the sponsoring company dragging its feet in obtaining the necessary funding. In March 1887 McMurdo finally established The Delagoa Bay and East African Railway Company in London and the actual construction work could begin.

Lourenço Marques, which had been a dreary little harbour town, was suddenly given a new lease on life as a result of the railway project. The American Hotel was opened by Otto Berg, canteens and businesses flourished and a multitude of newly appointed construction workers and mountains of material collected in the vicinity of the harbour.

Sir Thomas Tancred was appointed as the line's chief construction engineer and more than 200 white and 3000 black workers were recruited. The labourers were divided into teams, each with a specific task, and rivalry and even fights between gangs were common. Some teams gave themselves colourful names such as the Salvation Army and the Irish Brigade. The latter consisted mainly of British deserters and other desperadoes. Their escapades under their leader, George Hutchinson (or Captain Moonlight), would become legendary in the Lowveld. The usual practice was that the workers were taken back to Lourenço Marques by train every weekend, where they would become roaring drunk and paint the town red.

Despite all of these antics, construction proceeded rapidly. On 2 June 1888 the first section was inaugurated with great ceremony. A colourful group of Portuguese, Goanese, Arabs and Africans went along and Padre Annibal sprinkled the little steam locomotive with "holy water" and christened it Don Luiz Filipe.

Sir Thomas Tancred loved ceremonies and often

Komatipoort Station with the Lebombo Mountains in the background. 8 November 1912.

organised banquets after a section of the line was completed successfully. The Portuguese part of the railway line was completed in December 1888 up to the point regarded as the border with the Transvaal. It was inaugurated on 14 December with a cheerful fiesta attended by the Portuguese governor. The customary speeches and formalities took place at the end of the line, after which they awaited the special train that would deliver the food and drinks to celebrate the occasion. When the train arrived three hours late, Sir Tancred heard to his dismay that the Irish Brigade had hijacked the train and seized all the food and alcohol, with the "consolatory" message that they would drink to Sir Tancred's health!

On Christmas Day 1888, Sir Tancred fulfilled his part of the contract by constructing the railway line up to the Transvaal border, but Machado would soon discover that the border was in reality 9,6 kilometres further west. A great hullabaloo followed and the Portuguese made the most of the opportunity by confiscating the entire railway line in June 1889, on the grounds that the contract had not been completed in the required time. The matter was referred to three Swiss jurists for arbitration. After a long squabble the question was finally settled in March 1900 with the allocation of the railway line to the Portuguese government and a payment of £941 511 to the construction company.

After the termination of their contract, the Irish Brigade ran amok in Lourenço Marques for one last time – in the process nearly sinking a Portuguese gunboat and stripping it of valuables. This was their swan song, after which they drifted apart to resume their nefarious activities elsewhere.

On the Transvaal side, the Nederlandsche Zuid-Afrikaansche Spoorweg-Maatschappij (Dutch South African Railway Company) was waiting patiently for the 80 kilometres of the Portuguese section of the line to be completed. This company, which was registered on 21 June 1887, had already started recruiting labourers and finalising preparations for its part of the project from the border to Pretoria. Sub-contractors such as Van Hattum & Co. were appointed to start clearing two kilometre-long sections of bush and to prepare and level the ground. They were paid fixed tariffs of ten shillings a metre for rock work, six shillings for shale and two shillings for groundwork.

The grave of railway engineer J.E. Ysendyk in the old Komatipoort cemetery. He died of malaria on 10 February 1888 during the construction of the Eastern railway line.
(30.01.1986)

Hectorspruit Station along the Eastern Line to Delagoa Bay, circa 1895.
(ACKNOWLEDGEMENT: TRANSVAAL ARCHIVES)

Construction of the Eastern railway line through Crocodile Poort, 1891.
(WITH KIND PERMISSION: C. BARNARD AND H. BORNMAN)

11. THE EASTERN AND SELATI RAILWAY LINES

At Komatipoort a whole town mushroomed. Julius Furley moved his hotel from the old Komati Drift along the old transport route to the new town and the well-known Indian, Dorabjee Dhunjbhoy, built a shop.

Inevitably all kinds of shady characters streamed to the scene of such feverish activities and as a result there was a great deal of illegal recruitment of black labour, robbery, theft and boozing.

George Pauling, the famous railway contractor and builder of the Eastern railway line (1892–1894), as well as a section of the Selati Line (1909–1912).
(ACKNOWLEDGEMENT: CONSTABLE & CO., LONDON)

Construction work on the railway bridge across the Komati River, circa 1890.
(WITH KIND PERMISSION: MR H. BORNMAN)

Among the subcontractors were men of various nationalities, but mainly British and Italian. The Dutch construction company, of course, employed mainly Dutch workers and groups of 500 were brought at a time from the Netherlands to complete the work in shifts.

Malaria immediately began claiming victims among the workers. On the Portuguese side of the border more than 200 workers died, and matters weren't much better on the Transvaal side. Especially the Dutch workers were killed off at a terrifying tempo while they were working in the Lowveld. Two of the first engineers who had arrived from Holland as overseers, J.E. van Ysendijk and G.H. van der Meulen, died in Komatipoort within a month of each other, in January and February 1888 (Kuit, 1942). The chief engineer, Snethlage, resigned as a result of the malaria threat and W. Verwey was appointed in his place.

The engineers' first big project was the construction of the strategic railway bridge across the Komati River. General Smit laid the foundation stone on 27 October 1890 and after that engineer Byderwellen pushed construction as much as possible.

Many factors hampered the project. Not only were malaria and the Lowveld heat great obstacles, but all types of swindlers and crooks tried to benefit from the project. One subcontractor convinced the company that no stone in South Africa was suitable for construction of bridges or crossings. As a result all the dressed stone was imported from the Netherlands at great cost, until the deception was discovered.

The Komati Bridge was completed at the end of May 1891 and on 12 July that year the line was opened up to Komatipoort Station. Komatipoort was a very unhealthy place for the construction workers, and 50 whites and many Basuto from the Highveld found their final resting place here. From Komatipoort the line advanced rapidly across the flat bush country to Hectorspruit, a hamlet surveyor Servaas de Kock had named after his dog.

By December 1891 the line was completed up to Malelane and three months later Kaapmuiden was reached. It was at this point that the branch line along the Kaap River to Barberton joined the main line in April 1897. At the end of the Anglo-Boer War a troop train taking the Hampshire Regiment back to England derailed on a narrow bridge across a tributary of the Kaap River on this branch line, with great loss of life.

Before the construction of this branch line, contact between Barberton and the Eastern Line had been established by way of the coach road running along a valley west of the infamous tsetse fly-infected Pettigrew's road (along the Kaap River) to Crocodile Poort.

This route was discovered by the canteen owner Tom Gould. Because it eliminated the danger of nagana among their trek animals, the grateful coach operators and transport riders named it Gould's Salvation Valley. Crocodile Poort is just outside Komatipoort in a particularly hot area of the Lowveld, and the rugged, mountainous terrain here brought an end to the relative easy construction work from Komatipoort. One of the most difficult construction terrains along the whole track lay ahead. They had to blast a road along the steep cliffs on both sides of the poort through which the Crocodile River had carved its own path through the ages.

Originally Machado's proposed route avoided Crocodile Poort because of the inhospitable terrain and passed north of it, but because it would shorten the line by 20 kilometres it was decided to go through the poort. P.H. Boutons, one of the engineers, described the problems they had to face as follows: "What we housed were thousands of baboons, blue monkeys, hippos, snakes, crocodiles and panthers [leopards]."

The contract for this part of the railway line was awarded to George Pauling, a well-known railway contractor with international experience. He erected his construction camp on the farm Madeliefie near the current Crocodile Poort Station and called it Poort City. It was a cheerful and rowdy place, despite its being plagued by malaria.

Many workers died at Poort City and some renamed Crocodile Poort the Valley of Death. Pauling established his own hospital here and appointed a full-time doctor, Simões. However, there was very little the doctor could do to improve the condition of those in his care. The temperatures of the patients in the 24 beds were taken every afternoon and the one with the highest reading won a prize!

George Pauling noted in his books that the daytime temperature at Poort City often rose to above 120 ºF (almost 50 ºC) and that of his sick contractors to 108 ºF (42 ºC), at which stage they usually became delirious.

In the ten months that Poort City, with its billiard saloons, canteens and bar girls, existed, more than 127 workers died of malaria.

It is said that for every 1000 white workers who worked on the line an average of 135 died of malaria, and that almost nobody managed to escape the disease.

Physicians were still very ill informed about the causes of this disease. In 1895 the theory was that lush vegetation growing next to rivers that rotted in the damp conditions polluted the air (malaria is French for "bad air") and so caused malaria. The solution they came up with was planting eucalyptus trees at each station. The SASSAR of January 1976 reported claims that this tree was an effective antidote and had led to a slight improvement at Komatipoort once the trees reached a certain height. It could have been this belief that prompted Sardelli to plant the stand of eucalyptus trees at his store at Gomondwane and that led to Stevenson-Hamilton planting some at Skukuza (Sabie Bridge).

Pauling used a lot of draught animals during this stage of the construction. Diseases and other factors also exacted a high toll among them. Tsetse flies made merry among the donkeys, while lions often came to catch their share. During his contract period, which lasted 12 months, Pauling lost no fewer than 500 donkeys.

On the other side of the Crocodile Poort lay the Upper Crocodile River valley. This was a relatively

A steam locomotive of the South African Republic (ZAR) on the Delagoa Bay railway line shortly after its opening in 1895.
(ACKNOWLEDGEMENT: TRANSVAAL ARCHIVES)

Nelspruit Station along the Eastern railway line, August 1902.
(ACKNOWLEDGEMENT: STEVENSON-HAMILTON COLLECTION, SKUKUZA ARCHIVES)

unpopulated, wooded valley with a bad reputation for fever. A few sheep farmers from the Highveld, including the Nel brothers Gert, Andries and Louis, came here every year with their sheep for winter grazing. Later, H.L. Hall, a transport rider and the son of President Burgers's railway engineer, settled on the farm Riverside, along the Crocodile River, which he had rented at £54 per annum during a farm auction in 1890. At the same auction the Nel brothers obtained the farm through which the Nels River flows. Along with Hall and B.M. Bester, they were the first farmers to settle in the Upper Crocodile River valley.

The Eastern Line reached what is now Nelspruit and the station built there on 20 June 1892. The line was now 200 kilometres long and had climbed to a height of 716 metres above sea level. Nelspruit would remain the end of the railway line for another 12 months as the construction company had run out of money. During this time the town developed rapidly and a number of shops, a hotel and a police station were built. In later years branch lines were constructed from here to Sabie and Graskop (1911) and to White River and Plaston (the farm of the old foreman Willem van der Plas) in 1926.

On 1 June 1893 another 17 kilometres of line westwards to Alkmaar was finally opened to traffic. This part of the line ran through an area that in later years would become one of the richest agricultural regions in the country. A station was built on H.L. Hall's farm Riverside. He named the station Tomango, but it was later renamed Mataffin, after the Swazi chief who had settled there in the 1870s after an altercation with Swazi King Mbandzeni (Umbandine).

From Alkmaar the Eastern Line gradually snaked upwards through the Elands River valley. This lovely valley was still uninhabited and had been visited by only a few hunters and prospectors, including Hartley and Keene. In the 67 kilometres to Waterval-Onder the line climbs by no fewer than 507 metres. This section of the track was inaugurated on 20 January 1894. The most difficult part of the railway line, from a construction point of view, started at Waterval-Onder. The original railway line (it was later diverted) followed a gradient of one in twenty and had to be provided with a cog-wheeled middle line (rack railway) over a distance of four kilometres to prevent locomotives from sliding backwards down the steep line. A tunnel of 230 metres that was completed in September 1893 by the engineering firm E. Warren & W. Royce gave access to more level ground on the farm Doornhoek, where the locomotive depot and marshalling yard Waterval-Boven was later built. This part of the railway line was completed on 20 June 1894, and brings us almost to the end of this story. Here the line left the Lowveld, and work progressed rapidly to join the section under construction across the Highveld. A station was built on Cornelius Potgieter's farm Geluk, 313 kilometres from Lourenço Marques. It was named Machadodorp in honour of Major Machado, who had surveyed the original route of the railway line.

On 20 October 1894 the eastern and western sections of the line met at Balmoral on the farm Elandsfontein, 90 kilometres from Pretoria. President Kruger travelled especially to Wilge River on 2 November 1894 to tighten the last bolt, a fact mentioned in the book commemorating the occasion.

The completion of this railway line was both a historic moment and the realisation of a dream. On 1 January 1895 a regular train service was instituted between Pretoria and Lourenço Marques and, according to *The Star*, one Parfitt was the first driver. The railway line was officially opened on 8 July 1895, when a special train transported the Portuguese governor from Lourenço Marques to Pretoria with much revelry and celebration. When the train steamed into

Pretoria Station at 11:00 the proud President Kruger declared the line open. Then the festivities started. After the official speeches and receptions, President Kruger and a number of invited guests (including the elderly Karel Trichardt) travelled the whole length of the line.

This railway line would later have a lion's share in the development of the Lowveld and the Kruger National Park. But there would also be tragic accidents, daring robberies and even vicious murders. The first serious accident occurred on 15 September 1894 when a goods train derailed and both the driver and stoker died.

Incidents with lions were commonplace. Apparently, on 10 December 1897, four lions attacked a group of railway workers on the line between Malelane and Kaapmuiden. The ensuing skirmish lasted two hours, ending only with the fortuitous arrival of a train. One of the workers was badly mutilated, but they managed to kill two of the lions.

It is not known exactly how many people died during the construction of the line, but it must have been hundreds – most of whom will remain unknown forever. Only the product of their labours still stands as a monument to their perseverance.

Bulpin (1983) described it aptly:

> Each train that rides the Eastern line is a memorial, in its way, to the workers of the line. The long, dark, clanging goods trains, the passenger trains with their twinkling strings of lights swinging in around some bend, and then vanishing again into the mystery of the night. The rhythm of their wheels on the rails seems to sing a hymn in praise of those whose spirits, I sometimes think, still sit in the shadows beneath the acacias and watch the trains go passing by.

When gold was discovered near Gravelotte, Leydsdorp, Rubbervale and Trichardtsdal, the Eastern Line was not finished yet, and in 1886 there was also a rush in Murchison Ridge. At that stage there was already a need for a railway line to connect the northeastern Transvaal with the central markets of the ZAR. The discovery of the Selati goldfields removed any further hesitation. Particularly President Kruger, who had always applied himself to the development of railways, realised the importance of such a railway. He therefore convinced the Volksraad in July 1890 to approve in principle the construction of a railway line connecting the Soutpansberg and the Selati goldfields with the main line to Lourenço Marques. The area through which this railway line would run was inhospitable and uninhabited. The large region between Komatipoort and Soekmekaar belonged solely to nature.

With this decision the Volksraad started one of the most absorbing episodes imaginable in the history of railway development. The then Member of Parliament for Soutpansberg, B.J. Vorster, armed

President Paul Kruger at his Landauer coach at Pretoria Station during the opening of the railway line to Delagoa Bay in 1895.
(ACKNOWLEDGEMENT: TRANSVAAL ARCHIVES)

11. THE EASTERN AND SELATI RAILWAY LINES

The old quarries near Crocodile Bridge where the sandstone blocks to build the train bridges across the Sabie and Crocodile rivers were taken out.
(12.06.1988)

with 1132 signatures of locals who wanted the railway line, competed for the concession to construct the line. At the same time, a charming 23-year-old Frenchman, Baron Eugène Oppenheim, appeared on the scene with a whole bag of rather sly tricks. He lived the high life, constantly entertaining influential politicians and other figures, and was a swindler of no little ability. Even before the Volksraad could reach a decision about awarding the concession, he had convinced Vorster to sell him his concession for £40 000. Vorster and his partners, L.A. Porcheron and W. Stephenson, then had to admit that the concession had not been officially awarded to them yet, but Oppenheim did not consider this to be a problem.

In an underhand manner, with flattery, lavish receptions and gifts, he managed to win the support of the decision makers. One of these gifts, for example, was a painting valued at £300 given to President Kruger to hang in the council chamber. This failed to impress the president – what did impress him was the fact that Oppenheim was French and would therefore not use British capital to construct the railway line. This would prevent Great Britain from again becoming involved in the matters of the Transvaal.

With a little insight into human nature and by studying history it becomes quite clear that Oppenheim was not interested in the actual construction of the railway line at all. Like Vorster, he wanted to make money from the concession – lots of money. But in the process he had to adhere to certain conditions, of which the most important was the creation of a company with paid capital of £500 000. He hoped to raise the money in France, but failed to do so. It later transpired that he was regarded with suspicion in the highest financial circles in France.

In the meantime, the matter became increasingly complicated. Vorster and his associates realised that they had made a grave error by selling their concession to Oppenheim for next to nothing. A lot more than £40 000 could be made from this transaction, so they moved heaven and earth to try to convince the government that they were the rightful concession holders, and not Oppenheim.

After nine months the stipulated company was still not founded. The matter also acquired a diplomatic flavour when the Transvaal's representative in the Netherlands, Beelaerts van Blokland, was appointed as special railway commissioner to oversee the railway concession on behalf of the ZAR.

Eventually it was Porcheron who managed to secure the money to found a company, but the government still regarded Oppenheim as the legal concessionaire. Beelaerts van Blokland, who had to act as middleman and arbitrator, was of the opinion that it was only a matter of monetary rivalry and jealousy between the two parties. As can be expected, he was impressed with Oppenheim and eventually supported the cunning young man.

In the meantime the government's patience had run out and it demanded that the company be founded immediately. At this stage Oppenheim had not been able to raise a cent yet and his only option was to pretend that he had managed to do so.

Oppenheim and Porcheron came to an agreement and the company was registered in Brussels, without any capital. There it was easy to exploit loopholes in company law; something that would have been just about impossible in France. The ink was scarcely dry when the ZAR was asked to pay out the first £500 000 that it had guaranteed in shares. While negotiations about this were still taking place the company awarded the contract for the construction of the railway to one Louis Warnant. Once again, this was just another scam, because just two days later Warnant handed over the contract to an English engineering firm, Westwood and Winby of London.

President Kruger, who specifically wanted to prevent any British involvement, had been duped by the sly baron – Warnant's brother was one of the directors in Oppenheim's company. Oppenheim did not want to openly transfer the contract to the British, as one of the stipulations of the contract with the ZAR was that there would be no British involvement. That is why he used Warnant as a middleman.

The government was not aware of the transfer of the construction contract and only found out later. Oppenheim's company demanded the rest of the remaining £1 000 000 from the government, and it decided to pay £500 000 to the company with the imposing name of La Compagnie Franco-Belge du Chemin-de-Fer du Nord de la Republique Sud-Africaine.

The first plans for the railway line were approved in May 1893 by the railway commissioner J.S. Smit. The Selati Line would be 307 kilometres long and completed in three years at an estimated cost of £6000 per kilometre. To construct this remarkable railway line, Oppenheim's company would supply the first £500 000 and the ZAR government the remaining £1 500 000, with four per cent interest on the investment capital.

Surveying of the line was completed by November 1892 by surveyors Bechtle and Marais, working for the construction company Westwood and Winby. Groundwork started early in 1893, and by July 1893 40 kilometres had been completed. When the company then demanded that a further £500 000 be paid by the government, railway commissioner Smit, who was also the magistrate of Pretoria, smelled a rat and ordered an investigation of the books. According to the contract, the books had to be kept in Pretoria and written in Dutch, but this was not done. The books were in fact kept in the company's offices in France, in French.

In August 1894 Oppenheim's entire swindle was exposed. By that time, 120 kilometres of the railway line had been completed from Komatipoort to Newington, a temporary bridge had been built across the Sabie River, and the railway bridge across the Crocodile River had been finished. The government halted all work on the line and the whole project was in limbo.

This was a great scandal and an utter fiasco. Chaos ruled. Empty liquor bottles ("medicine against the fever"), tools such as spades and picks, and bits of building material were scattered about in the bush. There was also some looting, as it was said that the material belonged to anybody who could be bothered to fetch it.

Malaria also took a toll among the mainly Irish and British workers during this first phase in the construction of the Selati Line. In the old cemetery in Komatipoort are the gravestones of the following people who found their last resting place there between 1893 and 1895: Horatio Offord (15 May 1893), Patrick O'Connor (11 June 1893), Aubrey Drury (24 June 1895), Frank Wilson (18 September 1893), George Charles Bovy (30 November 1893) and John Frederick Farrall (21 August 1894). Several unmarked mounds of soil bear further witness to the high cost in human lives that this railway line exacted.

The workers on the railway line ate mainly venison, from game shot by teams of hunters in the vicinity of the line. Game numbers were soon drastically reduced. If it weren't for the later proclamation of the area as a game reserve, it is doubtful whether the Kruger National Park would ever have been realised.

After much agitation and many court cases, both here and overseas, the concessions awarded to Oppenheim were declared null and void. The ZAR government was very embarrassed by the whole affair.

Oorsprong halt along the Eastern railway line (Delagoa Bay railway line), circa 1895.
(ACKNOWLEDGEMENT: TRANSVAAL ARCHIVES)

Steinaecker's Horse's blockhouse at Sabie Bridge which was temporarily used as quarters by Maj. J. Stevenson-Hamilton in 1904.
(ACKNOWLEDGEMENT: STEVENSON-HAMILTON COLLECTION, SKUKUZA ARCHIVES)

Accusations of corruption and bribery streamed in against those members of parliament that had received kickbacks while Oppenheim, the real culprit, was safe and snug in Europe. The shareholders had lost all their money in the process while the government of the ZAR was left with a useless, half-finished railway line and a lot of unpaid interest on borrowed capital.

One of the big shareholders was one Tolmay, who had no intention of losing everything. He therefore decided to confiscate the line, all equipment, the locomotive and carriages until he was certain that his money would be paid back. Jules Diespecker of Komatipoort was appointed as his agent to keep an eye on his interests. Tolmay also instructed Diespecker to drive the locomotive to Sabie Bridge from time to time, to check if everything was in order. Diespecker was "annoyingly" diligent with regards to this instruction, until the outbreak of the Anglo-Boer War. A small regiment under the command of Prussian "Baron" Ludwig Steinaecker received orders from the British forces to guard the border between Mozambique and the Transvaal after the occupation of the Lowveld. Steinaecker used his position and commandeered the railway line, as well as Diespecker's home, for military use. He also erected guard posts all along the border from Swaziland to north of the Letaba River, manning each with an officer and a few soldiers. Two of these guard posts were of importance in the later history of the Kruger National Park, namely the one at Gomondwane and the other on

The railway trolley which was used to commute between Sabie Bridge (Skukuza) and Komatipoort, circa 1910.
(ACKNOWLEDGEMENT: STEVENSON-HAMILTON COLLECTION, SKUKUZA ARCHIVES)

the southern bank of the Sabie River at Sabie Bridge. Stevenson-Hamilton recruited his first ranger at Gomondwane – E.G. (Gaza) Grey – while the blockhouse at Sabie Bridge became Stevenson-Hamilton's first home here, in 1902. The old Sabie Bridge post later developed into the current Skukuza. Steinaecker used the train to transport his provisions and men between Komatipoort and Sabie Bridge.

Diespecker did not take kindly to these events. As far as he was concerned Steinaecker was an ugly fly in the ointment, especially after the war, when Steinaecker tried to convince the authorities that the presence of his regiment would be essential for the maintenance of peace in this border region. At that stage the area between the Sabie and Crocodile rivers was reproclaimed as a game reserve and the appointed warden, Stevenson-Hamilton, was impatient to assume his duties. Steinaecker only agreed to withdraw his troops and hand back the train to Diespecker six months after the conclusion of peace in November. Stevenson-Hamilton immediately moved into the Sabie Bridge blockhouse, while Diespecker himself gladly used the opportunity to evict the pompous little "baron" from his home by force.

Stevenson-Hamilton convinced Diespecker to allow him to use the railway line. It is also known that Stevenson-Hamilton used a significant amount of the useless railway building material and equipment lying around for the development of the Sabie Reserve. Initially reluctant, Diespecker stipulated that Stevenson-Hamilton had to undertake to maintain and look after the line, the equipment and the workers' huts. Stevenson-Hamilton fabricated a trolley that consisted mainly of a platform with wheels. As a vehicle it was uncomfortable and difficult to handle, as the workers had to push it up slopes and then, when it reached the top, quickly jump aboard while the whole contraption sped downhill pell-mell. A trip to Komatipoort took about five hours, because it was downhill almost all the way, while the return journey took up to ten hours or longer. It was, however, a much faster mode of transport than a horse or wagon. Stevenson-Hamilton later obtained a more "luxurious" trolley, which used pump action, with seats on the platform and a roof to keep out the sun. To ride the trolley the lever had to be moved up and down, which was a tiring process in itself, but much easier than walking and pushing ... in Stevenson-Hamilton's own words: "I called this the passenger and the old one the goods train."

Stevenson-Hamilton was in Pretoria in 1909 when he heard that the Selati Line was going to be dismantled. He had mixed feelings about this, because it meant that they would again have to make use of horses and ox-wagons as means of transport. He was hardly back in Skukuza when a group of surveyors arrived, asking permission to travel through

Constable Henderson on the railway trolley at Sabie Bridge (Skukuza), May 1905.
(ACKNOWLEDGEMENT: STEVENSON-HAMILTON COLLECTION, SKUKUZA ARCHIVES)

Construction work on the railway bridge over the Sabie River at Skukuza, December 1909.
(ACKNOWLEDGEMENT: STEVENSON-HAMILTON COLLECTION, SKUKUZA ARCHIVES)

the reserve to do surveys, as the railway line would be extended to Tzaneen and the Sabie Bridge completed. He also had conflicting emotions about this. On the one hand it would mean comfortable transport, but on the other there were many negatives, including veld fires, uncontrolled construction workers and the large-scale disturbance of the environment.

It was only after unification in 1910, when the different railway companies in South Africa were consolidated into the South African Railways, that surplus funding became available and work on the railway line could resume. The contract to complete the work was awarded to the respected George Pauling & Co. and in 1912 the line was completed up to Tzaneen. Work on the train bridge across the Sabie River had already started by the end of 1909 and was completed in 1910. Pauling's main camp was at Newington, an uninhabited state farm about 40 kilometres northwest of Sabie Bridge. In time a small town developed here with station buildings, workshops, storerooms and magazines, a shop and even a hospital with a resident doctor and nurses. Some of Pauling's construction workers on the Selati Line became legends in their own right. There was John Orcheson, for instance, who imbibed so heavily one night that he imagined himself to be a mule. At a dare he pushed his head deep into a bag of maize, started eating and suffocated. They made him a simple coffin and buried him next to the railway line. Legend has it that his spirit is still sitting next to the line, watching passing trains!

It cost £200 000 to finish the construction of the remaining 150 miles of the Selati Line, compared to the £1 000 000 spent in 1893 on the abortive construction attempts of the original 75 miles. The official opening took place on 25 October 1912[1] with a splendid ceremony. Festivities started at Komatipoort

Construction work on the Selati Line north of Sabie Bridge (Skukuza), 1912.
(ACKNOWLEDGEMENT: STEVENSON-HAMILTON COLLECTION, SKUKUZA ARCHIVES)

The completed railway bridge across the Sabie River at Skukuza, 1912.
(ACKNOWLEDGEMENT: STEVENSON-HAMILTON COLLECTION, SKUKUZA ARCHIVES)

---

1   Rocher (1912) gives this date as 9 November and the date of completion of the train bridge as 22 April 1910.

and reached their climax at Reserve (Sabie Bridge, later Skukuza) when the banner- and ribbon-bedecked train, filled with dignitaries, steamed slowly over the Sabie Bridge.

At that stage, the Selati goldfields had almost been worked out and gold digging would soon become a less profitable proposition. Now the economic value of the railway line came under serious discussion again. In 1921 concessions were granted to prospect for coal along the Selati Line and until 1922 prospecting beacons regularly made their appearance in the southern part of the Sabie Reserve. Some of these old beacons can still be seen today between Bume and Crocodile Bridge. There was also a proposal to open up the land along the Selati Line to farming, including the old Sabie Reserve. Opposition to this scheme came from within the ranks, among others the artist and advertising official of the SAR, Harry Stratford Caldecott. He supported the idea of using the railway line for tourism and promoted the Sabie Reserve's value as a tourist destination. Stevenson-Hamilton once wrote of Caldecott's untiring advertising work in the interests of nature conservation: "There was no single man in South Africa who worked more strenuously and successfully for the cause of wildlife preservation on the subcontinent."

To achieve something concrete, the SAR came up with the idea of instituting a tourist train service through the eastern and northeastern Transvaal. This soon became known as the "Round in nine", as it took nine days to complete. The first of these journeys took place in 1923, with the trip starting and ending in Johannesburg. All the most interesting sights were visited, including Lourenço Marques (Maputo). The highlight of the trip was undoubtedly the leg through the game reserve. The Kruger National Park was proclaimed in 1926, and as a result this service became increasingly popular with tourists. The trip also included a campfire gathering at Huhla Station close to Reserve (the later Skukuza). This was a real experience. The trips became so popular that a game ranger travelled with the train and took the tourists on short walks through the bush at every stop.

In 1925 the Prince of Wales made an official visit to South Africa, and used the opportunity to travel by train through the area. The honour of escorting the heir to the British throne from Komatipoort to Sabie Bridge fell to Stevenson-Hamilton. During the journey he told the prince about the reserve and its planned development.

Stevenson-Hamilton was well disposed towards the train drivers, while being aware of the fact that they were poaching game. This he could not tolerate. Although he suspected one particular driver and his stoker, he could never catch them red-handed. One night they shot a kudu bull next to the line between Skukuza and Komatipoort and loaded it onto the footplate of the train. The police suspected something was up and awaited the train in Komati-

Group photograph of guests who accompanied the inaugural journey of the Selati Line. Komatipoort, 8 November 1912.
(ACKNOWLEDGEMENT: STEVENSON-HAMILTON COLLECTION, SKUKUZA ARCHIVES)

The official opening of the Selati Line and the inauguration of the bridge across the Sabie River on 25 October 1912.
(ACKNOWLEDGEMENT: STEVENSON-HAMILTON COLLECTION, SKUKUZA ARCHIVES)

poort, but could find no sign of a carcass. This was simply because a friend of the driver had warned them before they reached Komatipoort, so they burned the kudu to ashes in the firebox of the locomotive!

Another of the stories coupled to the Selati Line can definitely be taken with a pinch of salt. Apparently one of the drivers hit upon the idea of spreading birdlime on the branches of one of the trees along the line. He had noticed that this particular tree was always covered in guineafowl at sunset. What could be easier than "picking" a few birds from the branches, like fruit from a tree, on the way back? He was very excited to see that the tree was indeed full of birds, but then his eyes nearly popped with surprise. The birds, frightened by the noise of the locomotive, all tried to fly away, and in the process pulled the whole tree, roots and all, out of the ground and took it with them!

In 1929 it was decided to organise vehicles from Nelspruit to take tourists on sightseeing trips in coordination with the train service. A trip in one of these vehicles from Crocodile Bridge was apparently an unforgettable experience in many ways, although it was anything but comfortable. Roads were underdeveloped and in a bad condition. There were no luxuries and something like air conditioning was a fantasy. Most days were scorchers. It is said that one day, during a heavy rainstorm, some overseas visitors got stuck in a muddy drift and faced the threat of being washed away when the spruit came down. They had no choice but to abandon their vehicle in the pelting rain. Fate had even more for them in store, because soon lions started roaring nearby. The sodden bunch of tourists took refuge in some nearby thorn trees, where they had to wait until they were eventually released from their misery by ranger McDonald of Crocodile Bridge. But the worst was yet to come a few days later, in the form of malaria.

As the years passed it became increasingly difficult and expensive to maintain the Selati track. Even the telephone lines had to be strung tremendously high so giraffes could not reach them or animals be injured by them. Despite their best efforts, the trains caused numerous veld fires and it was not always possible to avoid hitting animals, especially at night. Impalas, giraffes, kudus, buffalos, lions and even elephants were among the main victims. During the night of 14 January 1970 a train smashed into an elephant bull near the eighteenth-mile post and on 10 July 1970 another train ran into a herd of buffalo and killed or maimed 19.

On New Year's Day 1968 the bloodiest train accident in the history of the reserve took place four kilometres from Randspruit. A goods train, heavily loaded with phosphate ore and drawn by two diesel locomotives on its way to Crocodile Bridge, crashed into the back of a stationary passenger train. The diesel locomotives and several of the passenger coaches were derailed and crushed. P. Breedt, the driver, and 13 passengers died in the accident, while another 38 sustained serious injuries. The Park's personnel provided valuable help at the scene of the disaster, but it took days before the wreckage could be removed and the line reopened. This and other incidences contributed towards this line being dubbed the "track of death".

The extensive developments in the Lowveld region (especially the phosphate and copper mines at Phalaborwa), coupled with the subsequent influx of people to the area, led to the Selati Line becoming very busy. The authorities realised that the electrification of the line was essential. With all of these factors known to the SAR, it was decided to reroute the

The ganger's home at the former Klipspringer (Duba) Halt along the Selati Line.
(16.07.1984)

railway along the outside of the western border of the Kruger National Park.

During the construction of the new line a train one night drove into a herd of elephant bulls. One was killed on impact while the other two seriously injured animals stumbled into the bush.

A new railway line was built in 1968–69 from Kaapmuiden in the south, west of the Nsikazi River. It connected with the old Selati Line at Metsi, a couple of kilometres north of Newington. After the completion of the new line, one train a day still transported goods from Komatipoort to Skukuza from April 1971 onwards, but after the completion of the tar road between Skukuza and Hazyview this service was finally halted in 1973. This also brought to an end the unrestricted transport of livestock and even living wild animals in sealed cattle wagons through the Park to Komatipoort Station and elsewhere.

The Parks Board took over the historic train bridges across the Sabie and Crocodile rivers from the SA Transport Services. A locomotive (No. 3638), one of the hundred class 24 locomotives that had been used by the SAR, was donated to the National Parks Board as a souvenir of the old Selati Line. This locomotive, which really ran on the Selati Line, is a suitable monument to commemorate the close relationship that had developed through the years between the SAR and the Kruger National Park. The official handing-over ceremony took place on 23 October 1978 and the locomotive became part of the popular Selati Train Restaurant and Museum complex at Skukuza.

The Selati Line has long since been in disuse, and the tracks will eventually be reclaimed by nature and disappear. However, the memories and nostalgia will remain, and the role this railway line played in the early development of the Park should not be underestimated. This role was so important, in fact, that one can state with certainty that had it not been for this line, the establishment and development of the Sabie Reserve may not have been possible.

Both the Eastern and the Selati Line played a key role as a communication and transport channel during the early years of the Sabie Reserve and the later Kruger National Park. These two lines also determined the placing of the first ranger posts. Sabie Bridge (later Skukuza), Crocodile Bridge, Kaapmuiden, Malelane, Msutlu and Rolle were all posts that owed their establishment to the vicinity of the Eastern and Selati lines. Equipment for the building and maintenance of the fledgling reserve's infrastructure was almost entirely brought in by train. The poor transport and wagon roads that had to serve the area were no longer a major problem. The railway lines brought order and direction to the transport system and the reserve was planned and laid out around this. The railway, more than anything else, contributed towards making the reserve accessible to the public. This helped them to become aware of its beauty, peacefulness and tranquillity, especially

Ruins of the old ganger's home at Kemp's Cottage along the Selati Line.
(16.07.1984)

The last steam locomotive that ran on the Selati Line. It was donated to National Parks Board on 23 October 1978 by the Department of Transport and is currently part of the Selati Train Restaurant complex at Skukuza.
(07.05.1984)

11. THE EASTERN AND SELATI RAILWAY LINES

The railway bridge across the Crocodile River at Crocodile Bridge.
(12.06.1984)

during a time when there was a great deal of pressure to get rid of the reserve. It is in this context that we should remember these two railway lines. Then it becomes possible to see these man-made structures in perspective and appreciate their worth. The relationship between the Eastern and Selati lines is very much like the camaraderie between the two leaders of a team of oxen opening up a new route to a faraway, inaccessible area. Like the two draught animals that bend under the yoke at the front of a team, aware of the strength and support of the other, the Eastern and Selati lines unlocked the untamed Lowveld of the Transvaal and made an invaluable contribution to the creation and growth of our young country's pride and natural inheritance – the Kruger National Park.

References

Books and other sources

Barnard, C. 1975. Vyf-en-veertig plekname. In: Barnard, C. (ed.). *Die Transvaalse Laeveld – Kamee van 'n Kontrei.* Tafelberg, Cape Town.

Boonzaaier, W.A. 1953. Transvaal se grootste skandaal. *Die Huisgenoot,* 21 August.

Bornman, H. 1979. *Nelspruit 75 in '80.* City Council, Nelspruit.

Bouten, P.H. 1941. *De aanleg van't Oosterspoor.* Unknown publisher, Pretoria.

Buchan, D. 1926. *The Chronicles of a Contractor.* Constable & Co., London.

Bulpin, T.V. 1950. *Lost Trails on the Lowveld.* Howard B. Timmins, Cape Town.

Bulpin, T.V. 1963. *Lost Trails of the Transvaal.* Books of Africa, Cape Town.

Bulpin, T.V. & Miller, P. 1968. *Lowveld Trails.* Books of Africa, Cape Town.

Caldecott, S. 1929. *The Tourist Train.* The Cape Times, Cape Town.

Coetzee, J. 1975. Die Oosterspoor. In: Barnard, C. (ed.). *Die Transvaalse Laeveld – Kamee van 'n Kontrei.* Tafelberg, Cape Town.

De Jong, B.C., Van der Waal, G.M. & Heydenrych, D.H. 1988. *NZASM 100. The Buildings, Steam Engines and Structures of the Netherlands South African Railway Company.* Chris van Rensburg, Pretoria.

Engelenburg, F.V. 1894. *De Delagoabaai-spoorweg, een terugblik.* Volksstemdrukkerij, Cape Town.

Hickey, W.A. 1971. Die Selatiskandaal. *SASSAR,* February.

Kaye, H. 1978. *The Tycoon and the President.* Macmillan S.A., Johannesburg.

Kuit, A. 1942. *Transvaalse gister.* Van Schaik, Pretoria.

Loubser, J.H.G. 1978. Speach delivered at the handing over of locomotive no. 3638 to the National Parks Board on 23 October 1978. Skukuza Archives.

Meiring, P. 1982. *Agter die skerms in die Krugerwildtuin.* Perskor, Pretoria.

N.Z.A.S.N. 1895. *Gedenkboek Uitgegeven ter Gelegenheid der Feestelijke Opening van den Delagoabaai Spoorweg.* N.Z.A.S.N., Amsterdam.

Pringle, J.A. 1982. *The Conservationists and the Killers.* T.V. Bulpin & Books of Africa, Cape Town.

Prinsloo, H. 1978. *The Conservationists and the Killers.* T.V. Bulpin & Books of Africa (Pty.) Ltd., Cape Town.

Rocher C.G.C. 1910–1912. Personal diary. Skukuza Archives

Stevenson-Hamilton, J. 1974. *South African Eden.* Collins, London.

Wildenboer, L.A. 1975–76. Stasiename op die Oosterlyn. *SASSAR* Nov. 1975/Jan. 1976.

Webb, H.S. 1954. The coming of the railways and the eighteen-nineties. In: *The South Eastern Transvaal Lowveld.* Publication of the Lowveld Regional Development Association.

Wolhuter, H. 1958. *Memories of a Game Ranger.* The Wildlife Protection Society of South Africa, Johannesburg.

Yates, C.A. 1935. *The Kruger National Park.* George Allen & Unwin Ltd., London.

Magazines and newspapers

Custos. August 1973. Die Selati-treinspoor deur die Krugerwildtuin.
Die Volksstem. 23 October 1878.
Gold Fields News. November 1912.
Nuus uit ons Parke 5(3). March 1965.
The Star. 11 August 1970.

Chapter 12
The demarcation of the border between Mozambique and the ZAR
Dr U. de V. Pienaar

Since 1843, when the first contact between the Voortrekkers and the Portuguese authorities took place in Delagoa Bay, both parties had recognised the need to demarcate the territory over which each would have jurisdiction.

Several delegations under the leadership of General Andries Hendrik Potgieter, with Karel Trichardt as guide, visited the Portuguese harbour city in 1844 and again in 1845. Their mission, apart from establishing trade relations with the Portuguese, was to find a suitable trade route between Ohrigstad and the coast. In July 1848 yet another expedition under the leadership of Karel Trichardt was sent to find a route with more water for the trek animals and where the tsetse fly would be less of a threat. It would seem they picked the ancient trade route past Pretoriuskop and Skipberg for this purpose, on João Albasini's advice.

It was during this trip that Willem Pretorius fell ill with malaria near Skipberg. He was sent back by Trichardt, but did not get very far; dying under a marula tree slightly to the west of Pretoriuskop, where Albasini buried him. Afterwards the nearby hill became known as Pretoriuskop and the river that rises here became the Pretorius River (now the Mbyamiti). Preller (1920) noted that during this visit Karel Trichardt had discussions with the Portuguese governor about the determination of the border between the Portuguese area and the land the Voortrekkers had bartered from King Sobhuza I of Swaziland. The Portuguese governor favoured the idea that the Drakensberg should be the border between the young Boer republic and the Portuguese territory. Karel Trichardt, however, insisted that the border should be four days' trek from the Drakensberg; along the Lebombo Mountains. Krüger (1938) mentioned that the newly appointed Portuguese governor, Joachim Carlos D'Andrade, started corresponding with the Volksraad in Lydenburg at the beginning of 1850 to discuss a trade agreement as well as the border issue.

In answer, the Volksraad sent a delegation consisting of A.F. Spies, C. Potgieter, P.J. Coetzer and N.T. Bührmann, with Karel Trichardt acting as guide and interpreter, to Delagoa Bay to continue negotiations. It is claimed that it was during this visit that Trichardt managed to convince Governor D'Andrade to accept his suggestion that the Lebombos should be the international border. Unwittingly, Trichardt therefore was instrumental in having the entire Kruger National Park area fall within the borders of the ZAR (South African Republic).

Trichardt's arguments in favour of the Lebombos being the legal border rested on practical considerations and carried enough weight to convince D'Andrade. In order to establish meaningful trade relations between the communities at Lydenburg and Ohrigstad and Delagoa Bay, the Boers at least had to manage a significant part of the trade route, as this was a wild, tsetse fly and malaria-infested area with the added danger of predators and hostile black tribes. The Portuguese would not be able to provide transport riders with the necessary protection since their influence stretched no further than the immediate vicinity of the east coast harbours they had annexed. The delegation managed to reach a written agreement with D'Andrade on 30 July 1850 in which it was agreed that the final determination of the border would be undertaken by a Portuguese officer.

On 24 August 1850, Governor D'Andrade – accompanied by two officers from the garrison at Lourenço Marques, F.J. de Mattos and G.G. De Castellão – attended a commission board meeting at Ohrigstad where the agreement that had been reached in Delagoa Bay was ratified and extended. They also agreed that each side would appoint a delegation to demarcate the border (Preller, 1920).

However, D'Andrade's agreement was declared null and void by the Governor General of Mozambique in 1854. According to him, D'Andrade did not have his government's mandate to reach an agreement with the Boers. Despite this setback, the negotiations and agreements reached in 1850 with Governor D'Andrade would nevertheless serve as a foundation for later negotiations and the border demarcation between the Lydenburg Boer Republic (as well as the later ZAR) and the Portuguese authorities.

João Albasini was very keen that the dragging border issue between the Boers and his Portuguese compatriots be finalised. After he had settled at Goedewensch in the Soutpansberg, he donated the land along the Sabie River that he had bartered from Chief Magashule to the Portuguese government with the proviso that it should become part of the Portuguese government's possessions abroad and that he be put in charge of it. He even named the piece of land "São Luiz" in honour of the Portuguese king, which is how the area was indicated on old maps. Without Albasini's knowledge, the Portuguese Consul-General Alfredo du Prat negotiated with the government of the ZAR in 1869 (ten years after Albasini had promoted the possibility of a Portuguese colony in the interior). In "An accord of friendship, trade and border demarcation between Portugal and the Transvaal Republic", Albasini's claim on the São Luiz area along the Sabie was nullified. The Executive Council's viewpoint was that they had already legally obtained the land from King Sobhuza I of Swaziland and that Magashule, a lower-ranking captain, had no right to barter the land to Albasini. Therefore none of Albasini's grandiose plans to establish a Portuguese colony in the interior close to the Boer communities came to fruition. A map that indicated the suggested borderline between the Transvaal and Mozambique was drawn up by Surveyor-General Magnus Forssman of the ZAR and signed by President M.W. Pretorius and Consul-General Alfredo du Prat of Mozambique on 29 July 1869.

Article 23(ii) of the treaty determined that the Portuguese kingdom's sphere of influence would extend over the area "with a southern border of 26° 30' S in a line running westwards from the coast to a point on top of the Lebombos, from where it would run to the 'Comatie (Komati) River' Poort through the mountain range 'Le Bombo'". It would then head north-north-east up the mountain and from there to Pokioneskop, just north of the Olifants River, turn north-north-west to the closest point on the Chicundu Mountains (Sierra Chicundo), along which the river "Umbovo" flowed, and from there in a straight line to the confluence of the "Pafori" and Limpopo.

This vague description of the borderline, especially north of the Olifants River, later resulted in major disputes between members of the official border commissions that had been appointed to determine the border beacons here.

The British government disputed the Portuguese's claim to an area that far south of Delagoa Bay, as one Captain Owen had apparently already claimed it in 1823. The French president, MacMahon, who had been called in as arbiter, confirmed the Portuguese claim, however, and the treaty was acknowledged by Britain in Lisbon on 11 June 1891.

At the London convention of 27 February 1884 between the ZAR and Great Britain, the borderline between the Republic and Mozambique was once again described according to the original treaty of 1869:

> … Mananga, a point in the Libombo range; thence to the nearest point in the Portuguese frontier on the Libombo range, thence along the summits of the Libombo range to the middle of the poort where the Komati River passes through it, called the lowest Komati Poort; thence in a north by easterly direction to Pokiones Kop, situated on the north side of the Olifant's River, where it passes through the

Gideon Retief von Wielligh (1859–1932), surveyor general of the ZAR who surveyed and erected beacons along a large section of the border between Mozambique and the Transvaal in 1890 (eastern border beacons A–T).
(ACKNOWLEDGEMENT: STATE ARCHIVES, PRETORIA)

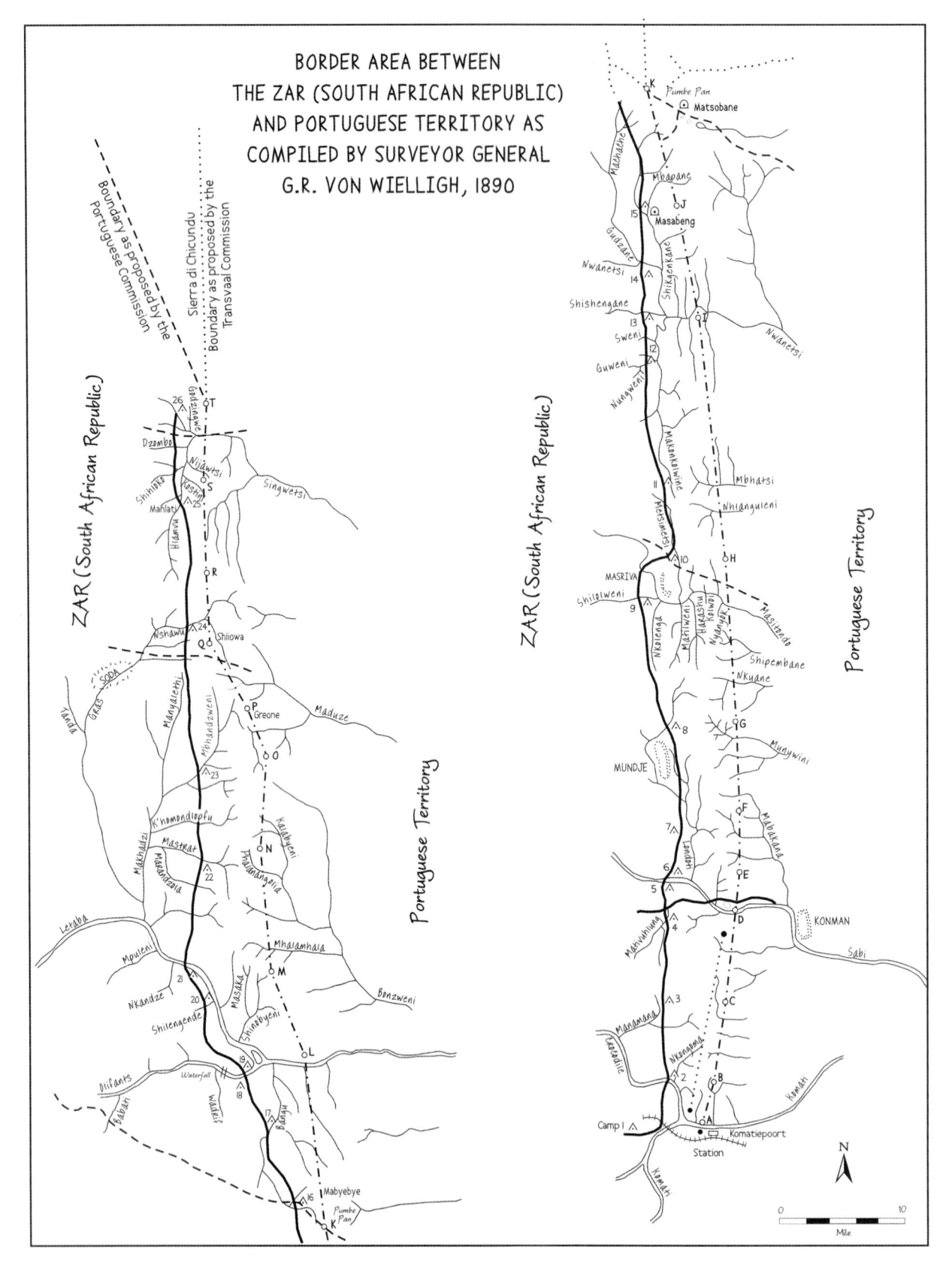

12. THE DEMARCATION OF THE BORDER BETWEEN MOZAMBIQUE AND THE ZAR

Border between the ZAR (South African Republic) and Portuguese East Africa.

ridges; thence about north north-west to the nearest point of Sierra di Chicundu; and thence to the junction of the Pafori River with the Limpopo or Crocodile River …

This unclear description was accepted by both parties, until the railway line between Komatipoort and Lourenço Marques had to be built and certain difficulties arose regarding the exact position of the border.

In 1887 a commission demarcated the border as running from the Matingatinga River on the Swaziland border with Mozambique to the Komati River at Komatipoort. Seeing as the railway crisis had been solved by then, they terminated the project at Komatipoort.

Nothing more was done about the border until 1890, when a joint border commission was appointed, on the insistence of President Kruger, to determine the exact location of the border between the ZAR and Mozambique from Komatipoort in the south to the Limpopo in the north.

Gideon Retief von Wielligh, the surveyor-general of the ZAR, was nominated as commissioner of the ZAR's delegation. The other members of his team were surveyor M.C. Vos, who would assist with the astronomical observations and measurements; J.H. Luttig, who would act as secretary; Native Commissioner Jacobus Abel Erasmus of Krugerspost, who would be in charge of the black bearers and policing; and a servant, Cornelis, would be tasked with general services. One Mr Hiron would transport the group's heavier equipment by ox-wagon.

The members left Barberton on 26 May on a hired ox-wagon and on 29 May they pitched camp on the northern side of Furley's Drift in the Komati River. Here they were cordially received by the owner of the shop, Captain "Mal" ("Mad") Mostyn Owen, who got his nickname because of his fearlessness during the Sekhukhune wars, and another incident in which he hit a crocodile on the head with his stirrup when it tried to catch his horse while Owen was crossing the drift. The unexpected turn of events convinced the crocodile to seek his lunch elsewhere. Julius Furley, the owner of the hotel, and 'T Hoofd, the Transvaal tax collector, lived on the southern side of the Komati Drift and Von Wielligh and his group spent the night there. The next day they joined Abel Erasmus at Komatipoort and found that he had shot a four metre-long crocodile, as well as two hartebeest[1] and a kudu, as provisions for the bearers. The native commissioner Davy and his assistant Van Niekerk had recruited about 100 bearers, mainly from the Mambayi tribe, but they needed to recruit many more to carry all the equipment.

Gen. Sir Joachim José Machado, the renowned Portuguese surveyor, administrator and patriot who was involved with the determination of the border line between the ZAR and Mozambique, as well as with the building of the Eastern railway line.

Ressano Garcia, the small town near the Mozambican/South African border, photographed from eastern border beacon A. (15.06.1984)

---

1   Milstein (1986) speculated that the hartebeest to which Vos referred in his diary could have been Lichtenstein's hartebeest. However, judging from current distributional data, as well as the testimony of Izak Holtzhausen, the first ranger in the eastern part of the Sabie Reserve before the Anglo-Boer War, it appears these animals could just as easily have been tsessebe.

On 1 June Von Wielligh's group received a visit from Dr Dekena, one of the engineers busy with the groundwork for the railway line from Lourenço Marques to Pretoria. He informed them that malaria was rife in the region during the summer months, but did not cause many problems in winter. African horse sickness, on the other hand, was a problem throughout the year.

On 2 June Von Wielligh and his assistants met the representatives of the Portuguese commission under the leadership of Colonel Joachim José Machado. The other members of the Portuguese commission were Captain Alfredo Augusto d'Andrade, Major Alfredo Augusto Caldis Xavier, Captain José Antonio Matheo Serrano and an Italian engineer, E. Mazzena. Captain Alfredo Augusto D'Andrade would replace Machado as leader of the delegation when the latter was appointed Governor General of Mozambique. They were the only members of the delegation who could speak English and D'Andrade would eventually become Minister of Foreign Affairs in Lisbon.

During their first meeting, which was conducted in English, the joint commission agreed to build a pyramid-shaped border beacon – known as the Von Wielligh beacon – in the Komati Poort (south of the river) with dressed stone. On the Transvaal side the inscription would be chiselled into the stone in Dutch, with Portuguese on the Portuguese side. The building costs would be borne by the two governments and it would be built by the Portuguese, after the decision had been confirmed by both governments.

The first border beacon A, north of the Komati River, was positioned near a wooded Lebombo cliff. The next day they had another meeting, at which Colonel Machado submitted a map issued by the Portuguese Department of Foreign Affairs (Ultra marino). He suggested that this map be used as guideline for the determination of the border. Surveyor Von Wielligh objected, however, because the ZAR had no part in drawing up the map, and instead suggested that the map that had accompanied the treaty of 1869 be used. The amendment was generally accepted and Colonel Machado withdrew his suggestion. Von Wielligh's proposal made it possible for the ZAR commission to negotiate reasonably freely, as the map of 1869 had been drawn up very inaccurately and vaguely. A further suggestion by Von Wielligh that was accepted was that further beacons be erected on the peaks of the Lebombo in a generally northern direction from the Komati Poort beacon. If it should appear that one of the parties was losing land in the process, on arrival at marked points on the map of 1869, including Pokiones Kop and Sierra Chicundo, land would be demarcated on a give-and-take basis to serve as compensation. Meetings would be held at different points along the route to be demarcated where the work done so far would be reviewed and jointly approved. Colonel Machado suggested that his team erect beacon A, while Von Wielligh's group would be responsible for building beacon B north of it, as their camp was closer to this point. (The ZAR group's camp number 2 was on the northern bank of the Crocodile River near Nkongoma.)

Eastern border beacon A in the Lebombo just north of the confluence of the Crocodile and Komati rivers.
(15.06.1984)

Eastern border beacon B in the Lebombo Mountain northeast of Ressano Garcia.
(12.06.1984)

Von Wielligh and his assistants erected beacon B about six kilometres northeast of beacon A. When the Portuguese commission found that the beacon was not directly north of beacon A they complained in earnest and claimed that the ZAR would gain a large chunk of territory over a distance of 350 kilometres. But Von Wielligh managed to convince them that they would probably gain similar pieces of land as a result of the uneven way in which the beacons had to be erected. This satisfied the Portuguese and is exactly how matters eventually turned out.

Colonel Machado left the joint commission at this stage and handed over control to Captain Alfredo d'Andrade. The ZAR's representatives had 183 black bearers to transport their equipment along the border, while the Portuguese used three wagons and oxen. They also sent for guns and ammunition as they had heard that the native population in the vicinity of the Olifants River was rebelling.

Surveyor M.C. Vos kept an accurate diary of subsequent events during the demarcation of the eastern border from Komati Poort to Shingwedzi Poort. His notes make for very interesting reading about conditions along this part of the country's border in 1890, especially if studied in conjunction with Von Wielligh's remarkably accurate map (archival acquisition R5879/87 and R11709/90) of the eastern border strip, the rivers and spruits, topography and positions of the border beacons, as well as their base camps.

As a result of the size of the workforce that had been hired to transport the ZAR team's equipment and supplies, large amounts of meat had to be provided and hunters were sent out almost daily. Near the Crocodile River camp they shot eight kudu and an impala. The bearers overindulged to such an extent that they each had to be dosed with Epsom salts!

On 10 June they moved their camp further north to the vicinity of Panamana and erected beacon C on the crest of a long eastern ridge of the Lebombos. Vos described the adjacent Lebombo flats as an excellent place to hunt – an open marula savannah with small knob thorn trees (it still looks like this today). By that time, the Portuguese had already lost two of their five horses to African horse sickness. The ZAR delegation, on the other hand, had salted horses, which were still in a good condition. Near beacon C and on the plain they hunted again and shot four zebra, two impala, and a mountain reedbuck (in the Lebombos). Vos wounded a reedbuck on the flats – apparently he was a rather poor shot!

On 11 June they again moved camp, this time closer to the Sabie River, and on the way hunted on the flats, with mixed results. They shot six blue wildebeest in a large herd, a zebra and a steenbok, among others, and pitched camp next to a spruit (probably the Mativuhlungu) about two kilometres south of the Sabie. This was their fourth camp. The Sabie River is described as being about 200 paces wide

Mabakana beacon (F) along the eastern border as seen from the air.
(18.07.1984)

A group of surveyors from the office of the surveyor-general at Mabakana beacon (F) on the eastern border (1924).
(ACKNOWLEDGEMENT: OFFICE OF THE SURVEYOR-GENERAL, PRETORIA)

here with low banks and a sandy bottom, densely overgrown with common reed. Large trees, mainly waterberry and wild fig, grew along the riverbanks. The water level was low and the flow not very strong. The Sabie's tributaries near the Lebombos were all dry.

They found a shrub in the mountains with a fleshy stem and thorns, which must have been a kudu lily (*Pachypodium saundersii*). They also found a lot of *Achatina* land snail shells here. In the meantime the Portuguese lost another horse.

While crossing the Sabie River from their fifth camp just west of Mativuhlungu mouth, one of Hiron's wagons became stuck and had to be pushed by 100 of the bearers. Each of them got a tot of rum for his efforts after they had reached the far bank. Here they shot impala and waterbuck to supplement their rations. On 14 June they crossed the Sabie on horseback and erected beacon D just north of the river in Sabiepoort. After that they erected beacon E on Matukanyana Hill, five kilometres further north. Here Von Wielligh's horse fell ill and died the same evening. Vos mentioned the large euphorbias growing in the Lebombos on both sides of the poort and an island in the river close to the poort on which large trees grew. The group pitched their sixth camp just north of the Sabie River, across from the Mativuhlungu mouth.

On 19 June the commission again moved camp to the Mlondozi River, which Vos called the "Nondoote River".[2] Camp number 7 was slightly west of Ntente near the Matimakule waterhole in the Mlondozi. They found a gigantic python and killed it, after which they spent the rest of the day hunting. Vos noted that they had great success with hartebeest and had killed four.

The next day they built beacon F on Mabakana Hill, a lone hill in the Lebombos. Abel Erasmus shot two waterbuck and the Portuguese group, led by the Italian engineer Mazzena, returned from an expedition to the confluence of the Sabie and Komati rivers. On 21 June they moved camp again and trekked northwards along the eastern flank of Muntshe to about 1,5 kilometres north of the mountain, where they pitched camp number 8 next to the Mlondozi Spruit. This is also where Steinaecker's Horse later had an outpost during the Anglo-Boer War. According to Vos, the spruit did not have much flowing water, but it had many deep pools covered in water lilies, while the banks were bordered by large wild date palms and beautiful fever trees. (R. Kelsey Loveday called the Mlondozi "Sterk Spruit" on his 1883 map of the Lowveld. "Mlondozi" in Swazi means "a perennial river that provides water to people, animals and for irrigation".) The trees on the flats were mainly knob thorn, marula and red bush willow. He further mentioned the black, clayey soil and the stretches of grass growing a metre high. The soil was fertile and the locals grew three kinds of beans, maize, manioc and millet. The hunters shot a sable antelope, two zebra, a klipspringer and a few warthog in this area, and wounded a hartebeest.

They sent some of the bearers ahead in search of water and on their return they reported that there was water about four kilometres further north (the Shilolweni or Nkelenga Spruit).

On 23 June they erected beacon G north of Munweni Poort and shot a hartebeest. The previous day a hunting party had wounded a large giraffe.

They moved camp again on 24 June and encountered a large herd of sable antelope near a village by the Nkuwana. They shot a large bull and wounded a few others. According to Von Wielligh's map, the ninth camp was situated next to the Shilolweni Spruit, west of its confluence with the Nwaswitsontso and south of the current Tshokwane.

The group reached the Nwaswitsontso on 25 June. On the way there Abel Erasmus shot an impala. They pitched camp next to the Metsi-Metsi on the northern side of the Nwaswitsontso after they had trouble getting their wagons through the river. Camp number 10 was more or less where the current firebreak to Shingedzene crosses the Metsi-Metsi. Captain D'Andrade went hunting and almost shot one of the natives' goats by mistake before Mazzena stopped him.

Beacon H was erected on the Nhlanguleni ridge on 26 June. Vos mentioned that the Lebombos were barely 121 metres high here and that they found it difficult to pinpoint the beacon's position.

The next day they travelled further northwards

---

2   In Von Wielligh's official map of the 1890 border expedition, the name was more correctly spelled "Londoti".

and Vos noted that it was the first time they had not encountered any game along the route. On 28 June they trekked along a footpath that led in the direction of Chief Magebain's village. They reached a flowing spruit (the Makongolweni) and pitched camp for the night – their eleventh camp. The next day they trekked further and found another spruit with a strong flow (the Ngunwini) that had not been marked on their map. Here Abel Erasmus and Von Wielligh shot two giraffes and wounded a third. They reached the Sweni River on 29 June, which also had a strong flow. They set up their camp number 12 close to the confluence of the Sweni, Guweni and Nungwini spruits. That night they feasted on giraffe meat and marrow – a dish everyone relished. They also received a letter from the hunter Henry Glynn, who said that he had crossed the Lebombo Mountains and had already shot more than 60 head of game. The next day they reached the Shishangani, just west of the Nwanedzi confluence, and while crossing the spruit one of Hiron's wagons broke. They set up camp number 13, just south of the Shishangani. The flowing Shishangani and Nwanetsi spruits both had large hippo pools and the bearers reported that a hippo was hiding in one of the pools. They used dynamite to chase the animal out of the water, after which Luttig shot it. They were all surprised to see such a scenic river (the Nwanedzi) there, since it had also not been indicated on their map. On 1 July they built beacon I on the southern bank of the Nwanedzi in the poort. Luttig collected some seeds from Rhodesian teak pods in the area. That night they heard lions roaring in the distance. On 2 July they moved camp to the confluence of the Gudzani and Nwanedzi spruits (camp 14).

After having trouble with a broken shaft they reached Chief Magebain's village on 3 July. Here they pitched camp number 15 in a thicket along the Mbhandzweni Spruit. Abel Erasmus went hunting and shot three kudu, while they encountered large flocks of guineafowl feeding in the locals' cultivated lands.

On 4 July the delegation ascended the Lebombos on foot to erect beacon J near the upper reaches of Ntomeni Spruit. Von Wielligh explored the area further northwest on the Lebombo plateau and came across some elephant spoor. They also found large black settlements in the area. The people lived in thickets and their kraals could only be reached via footpaths that had been cut through the undergrowth. Guineafowl were particularly easy to come by and the locals were friendly and helpful.

That night the Portuguese contingent treated the ZAR delegation to a seven-course meal and copious amounts of wine.

The commission left Magebain's village on 5 July, travelling in a north-northwesterly direction. After about six kilometres they found water in a spruit (the upper reaches of the Mbhandzweni), from where they travelled for four and a half hours before finding water again (the upper reaches of the Bangu near Mabyebye).

Here they found the footpath that led from Mujaji's village over the Lebombos to Delagoa Bay – a centuries-old trade route that was also followed by missionary Henri Berthoud. Locals told them that there were hunters along the Timbavati that had shot quite a few hippos. The commission set up camp 16 alongside the ancient trade route to wait for the approaching wagons and shot two waterbuck nearby.

On 7 July they ascended the Lebombos via the trade route and built beacon K at a small pan (Pumbe Pan). They found many black settlements on the sandy Lebombo plateau, an area covered in shrubs that to Vos's eye resembled granaatbos (*Rhigozum obovatum*), but were in fact large false mopane and other sandveld shrubs. The locals here were astonished at the group's horses, as they had never seen any before. Abel Erasmus and Von Wielligh shot a large giraffe bull and wounded another two – they had to fire ten shots before the bull went down. They broke up camp on 8 July to move further north to the Olifants River. They left too late and could not reach the river, so they pitched camp number 17 next to a spruit at the foot of the Lebombos (Bangu) from where they explored the mountain. They found the whole mountain plateau covered in a dense forest and thought that the trees (msimbit) looked like olive trees. The blacks living in this area had all hastily abandoned their kraals as a rumour had reached them that the border commission was indiscriminately shooting blacks on the way.

The day after that, they finally reached the Olifants River at a point near the site of the old Gorge

The historical baobab tree near the confluence of the Olifants and Letaba rivers, where surveyor-general G.R. von Wielligh's party pitched their camp number 19 on 10 Julie 1890, circa 1930s.
(ACKNOWLEDGEMENT: STEVENSON-HAMILTON COLLECTION, SKUKUZA ARCHIVES)

Ironwood pegs which could possibly have been hit into the trunk of the baobab tree at the confluence of the Olifants and Letaba rivers by G.R. von Wielligh's party. (Compare to the diary of Surveyor M.C. Vos).
(18.06.1984)

tourist camp and pitched camp number 18 on the southern bank. The black guides who had led them there bemoaned the fact that they were kept awake by a pride of lions that had killed a kudu. They found that the Olifants River cleaves through the Lebombos in a narrow poort, with the high banks almost vertical in some places. On every sandbank they saw either crocodile spoor or the reptiles themselves. Davy set off to investigate the lion kill and surprised another lion that had caught a zebra. He shot the lion in the neck, wounding it. Elsewhere he found a hyena that had been bitten almost in half by lions, and later he encountered a pack of wild dogs hunting a waterbuck. The scenery at the confluence of the Letaba and Olifants rivers was particularly impressive and the poort through which the river flowed was a rough yet magnificent natural site.

On 10 July 1890 the group forded the Olifants River on foot. The river had come down the previous night and the current was therefore very strong. They made their nineteenth camp under an enormous baobab tree on the northern bank. This giant measured 53 feet (more than 15 metres) in circumference (this tree was again measured on 24 June 1985 and its circumference was then 18 metres [60 feet]). Von Wieligh claimed in his book *Langs die Lebombo* that the circumference of the tree was 54 feet (16 metres). In the years since 1890 it has therefore grown only about two metres in circumference. From this information its age can be deduced as being between 600 and 700 years. It was a camping site popular with hunters and many initials and dates were carved into the bark, including E.G., M.E., H.A.G. (Henry and Arthur Glynn?), H.F.F., 1896, M.E.S., 1919, C.V. 19.9.26, A.W.F., T.V.D., C.F. 27.10.26, C.G. (Charles Gerber?), J.E.G. 5.12.16, N. 1893, L.T. On the east-facing side of the tree a series of iron spikes have been driven into the wood so one can climb to the top of the main trunk. The bearers found two elephant tusks and brought them to the camp. They found numerous elephant tracks in the area and there was apparently no shortage of lions either – two ashen-faced workers made their terrified appearance at the camp after having been chased by lions.

The next morning they received news that there was a hippo on the northern bank of the Letaba River. After crossing the river, which was flowing strongly, with difficulty, they found the hippo at the edge of a pan next to the northern bank and shot it. They cut the skin into strips and removed the best cuts, as well as the fat, and gave the rest of the carcass to the workers. They also found the carcass of an elephant, which had apparently been wounded by hunters months before and had died here. The animal's impressive tusks had not deteriorated and they removed these. They returned to their camp on the southern bank of the Letaba and waited for the workers to carry the Portuguese party's equipment across the Olifants River.

On 12 July, Von Wielligh and Vos, accompanied by Abel Erasmus, Luttig and the Portuguese members of the commission, went to investigate the northern bank of the Olifants River in search of Pokioneskop (indicated on the old border map of 1869). There was, however, no sign of such a prominent hill and after much bickering they agreed that beacon L should be erected on an elevated point on the northern bank of the Olifants River, near the eastern end of the poort. D'Andrade stayed behind to take astronomical readings in an effort to determine the precise position of the beacon.

On the way back to camp they shot a Sharpe's grysbok. There they drove wooden pegs into the trunk of the large baobab tree, climbed up and at the top found a hollow, about three metres in diameter, filled with water. (The pegs are still there.) The next day they shot a few impala and took some provisions to D'Andrade.

On the morning of 14 July they received a note from D'Andrade in which he asked that a couple of workers take him back to camp as he was very ill. They shot seven impala during the day.

After D'Andrade's return they struck camp and travelled along the southern bank of the Letaba to another large baobab about 12 kilometres from the confluence (camp number 20). The following year, in 1891, Von Wielligh again camped under this tree and carved his initials and the year into the trunk. This tree can still be seen today, just west of the road between the Letaba and Olifants rivers. Elephants have, however, damaged the tree and in the process deleted Von Wielligh's graffiti. In the vicinity of the Olifants River Poort, on the southern bank of the river and just east of the old Gorge rest camp, there is another baobab tree with the name and date "H.M. Borter, July 1890" carved into the trunk. Further south, in the Ntsumaneni Poort (a tributary of the Bangu Spruit), there is yet another baobab tree with the name and date "Briscoe, 1890" carved into its trunk in a particularly neat fashion. Both these people could have been members of the hunting party that had been shooting hippo at the Timbavati mouth, according to Vos's diary.

The Letaba's banks, as well as parts of the river bed, were overgrown with common reed and there were beautiful trees with dark leaves (Natal mahogany) growing along the banks. Hiron's wagons progressed with difficulty on the broken terrain along the river and the Portuguese decided to follow the next day.

On 16 July their camp was nearly destroyed by a veld fire and had it not been for the help of the black workers they would have lost everything. Abel Erasmus shot a bushbuck and also brought a few baobab fruits back to camp. The Portuguese contingent arrived at the camp by sunset.

The next day, Von Wielligh and Erasmus rode into the Lebombos to erect the first beacon north of the Olifants River – beacon M, a few kilometres northwest of Ramiti Pan. The previous day's veld fire flared up and again threatened the camp, this time from the opposite direction. Pitch-black, burnt grass surrounded the camp in a great swathe. Thunderclouds started gathering and there was a heavy rainstorm during the night. At daybreak it was still raining and this prevented them from moving camp early. They then travelled further along the Letaba and found that the locals still had green maize and tobacco on their lands thanks to the tropical climate. These people were subjects of Umzila's son Gungunyane, and the surveyors expected trouble from them. To their surprise they encountered a black man who could speak Dutch.

They found a passable, sandy drift through the Letaba east of the Nkandze mouth and pitched camp number 21 close by. They crossed the river on 19 July. Near the drift they found a boat made of marula tree bark. The group trekked along the Letaba for a while before turning north. At one stage they feared that the expedition would have to pitch a dry camp

The large baobab tree along the Letaba River where surveyor G.R. von Wielligh's group pitched their camp No. 20 on 15 July 1890. (18.06.1984)

Surveyor G.R. von Wielligh returned in 1891 to the tree above and carved his name and the date, 1891, into its trunk.
(WITH KIND PERMISSION: MRS P. ADENDORFF, LOUIS TRICHARDT)

Surveyors and their assistants at eastern border beacon M (Mala Mala) during the re-surveying of its position in 1924.
(ACKNOWLEDGEMENT: OFFICE OF THE SURVEYOR-GENERAL, PRETORIA)

12. THE DEMARCATION OF THE BORDER BETWEEN MOZAMBIQUE AND THE ZAR

(camp 22), but they found water in a tributary of the Letaba (the Mastrat tributary of the Makhadzi). An exploration of the surrounding area revealed the topography to be very flat. To the west they could clearly see two small hills – over a distance of more than 30 kilometres (possibly Masorini and Shikumbu, or Ngodzi and Tsale).

They failed to shoot any game and an ox was slaughtered – the first time during the expedition that they had to resort to such a measure. The trees in the area were mostly mopane (called "turpentine" trees by Vos).

During the morning of 20 July some of the workers went hunting and shot a few lions. Later that day Abel Erasmus and Davy followed four giraffes and killed two, wounding the rest. D'Andrade and the other Portuguese arrived at sunset and told them that they had had enormous problems getting their wagons through the Letaba River. One donkey drowned in the process.

On the morning of 21 July, Abel Erasmus and Von Wielligh went up the mountain to build the next beacon (N) at Tshavandlopfu. The workers found lions with cubs and caught one of about two months old, which they sold to Vos for one pound. Locals arrived at the camp and offered dagga for sale, and one of Hiron's donkeys was caught by lions.

On 22 July they moved camp again. After about two and a half hours' travelling they found a small waterhole (in the Mbandweni tributary of the Makhadzi), where they pitched camp number 23. On the way they crossed a large footpath running in the direction of the Lebombos. They were told that it led to the Limpopo (Vembe) (probably one of the old trade routes to Inhambane).

On 23 July they again ascended the Lebombos on horseback (D'Andrade had lent his horse to Vos) to erect beacons O (Ncindweni) and P (Greone). One of the members of the Portuguese delegation left them at this point to travel across the Lebombos to the confluence of the Olifants and Limpopo rivers and from there all along the Limpopo back to Delagoa Bay.

A difficult trek without any water awaited the group on 24 July. They eventually found water in a valley near the Lebombos after travelling for six exhausting hours. (This was the Shawu Valley, which Von Wielligh called the Soda or Grass River on his map.) They failed to shoot any large game and had to make do with a few guineafowl. The expedition members were tired and footsore and their spirits low. In the vicinity of their camp (number 24) there must once have been a large black settlement. They found the remains of many huts and skeletons – probably those of people attacked and murdered by Gungunyane's warriors.

After a thick mist had cleared on the morning of 26 July, they erected beacon Q on Shilowa Hill.

The next day they trekked further north to erect beacon R (Shibyatsangela) on the Lebombos, about eight kilometres north of the Shilowa beacon. In this area the mountain range is no more than 30 metres high and small shrub mopane, interspersed with knob thorn trees, cover the flats. The whole area had been

The Tshavandlopfu beacon (N) along the eastern border, north of the Letaba River.
(17.07.1984)

A Portuguese cross carved into the trunk of a wooden-banana tree at eastern border beacon N (Tshavandlopfu). It could possibly have been done by the Portuguese surveyors in 1890.
(17.07.1984)

burnt in veld fires and game was scarce. They often heard lions at night. The soil formation also changed – Vos described it as "granite (sic) with layers of lime and blue rock resembling Kimberley blue". (This is probably why prospectors later ceaselessly searched for diamonds in this part of the Lebombos). According to Von Wielligh's map, they pitched camp number 25 next to the Hlamvu tributary of the Dzombo, in the vicinity of the current Mahlati borehole.

On 27 July they managed to shoot a few head of game to augment their rations. D'Andrade threw dynamite into a couple of waterholes (probably the Nyawutsi waterholes) and killed a lot of fish. The members of the group were starting to get impatient after so many days in the veld and Abel Erasmus threatened to pack up and go home.

On 28 July they decided to push through to the Shingwedzi and erected beacon S at Kostini on top of the Lebombo ridge. The environment started taking on a tropical hue and Vos mentioned that there were many lala palms – some reaching heights of six metres and more. There was plenty of water in the spruits flowing north (Dzombo, Shihloka and Nyawutsi) towards the Shingwedzi, but they found it to be very brackish.

The group reached the Shingwedzi River on 29 July and pitched camp number 25 on its northern bank near the Gadzingwe mouth. For the first time Vos made mention of tsetse flies, saying that this area was infested with them. The Shingwedzi was a broad river with large stands of common reed, but little running water. They found the trail of an enormous python in the sandy river bed.

On 30 July the commission members had one last meeting. D'Andrade wanted them to find the mountain marked "Sierra Chicundo" on the 1869 treaty's map. He also claimed that the Shingwedzi had been incorrectly identified as the Umvobo on the old map. The locals were not aware of such a mountain in the area, but said that there was a hill further northwest that they called Zundo (Dzundwini), next to a river (the Shisha). D'Andrade wanted to investigate this possibility, but Von Wielligh objected and pointed out that the word "Sierra" indicated a mountain range and not a single mountain or hill. He was convinced that the "Sierra" referred to on the map was the Lebombo range north of Shingwedzi Poort and that the "Umvobo River" on the map was indeed the Shingwedzi.

After much bickering Von Wielligh managed to convince the Portuguese that it would be wisest to erect beacon T on Gondegonde, north of Shingwedzi Poort. D'Andrade agreed, but made it clear that he was not satisfied with the position of the beacon. He wanted to further investigate the precise location of Sierra Chicundo before the rest of the borderline up to the confluence of the Limpopo and Pafuri rivers was determined. He once again pointed to Zundo (Dzundwini) as possibly the correct Sierra Chicundo.

It was just as well that the members of the two delegations did not know that there was a mountain called Shikundu west of the current Punda Maria, as it would almost certainly have led to further arguments, as well as the possible loss to the ZAR of a large tract of land. They eventually agreed that the members of the ZAR commission would return to beacon T. The Portuguese, under the command of D'Andrade, would travel northwards in the direction of Zundo (Dzundwini) and from there to the Limpopo to satisfy themselves that Zundo was not the Sierra Chicundo indicated on the 1869 map.

In numerous letters sent by Surveyor Fred Jeppe to the Swiss missionary Henri Berthoud at Valdézia during 1891–95 (Berthoud had helped him draw up maps of the northern Transvaal) he expressed his concern that the existence of a hill named Shikundu, west of Mhinga, would lead to great problems in determining the rest of the eastern border.

The ruins of a former Pedi settlement at Shilowa Poort where G.R. von Wielligh's party found skeletal remains of the former inhabitants in 1890, who had been murdered by Gungunyane's warriors. (28.04.1984)

Surveyors W.P. Murray and R.E. Antrobus (on horseback) (above) and the observation towers which were built in the Nyandu bush during the surveying of the eastern border between Shingwedzi Poort and the confluence of the Limpopo and Luvuvhu rivers in 1924–1925.
(ACKNOWLEDGEMENT: OFFICE OF THE SURVEYOR-GENERAL, PRETORIA).

Von Wielligh's party started their return to civilisation on 2 August and travelled via the Shawu Valley and the Tsende (Tsendze) River to Ngodzi, Tsale and Kaleka hills. They crossed the Letaba on 4 August and arrived in Phalaborwa on 5 August. On 8 August they reached the Olifants River, where they hunted hippo one last time, and on 10 August the Blyde River. During their journey southwards they encountered several hunting parties, including one in the Shawu Valley and one by the Letaba. They reached Lydenburg on 15 August. The ZAR members of the border commission held their last meeting at Krugerspost, where they also compiled their final report.

The Portuguese delegation encountered severe problems on their way northwards. Almost all their trek animals contracted nagana and died. When they finally managed to reach the Limpopo after many trials and tribulations, they had to send their wagons on rafts down the Limpopo to the sea. They continued on their journey with one wagon, which eventually had to be drawn by blacks as all their trek animals had died.

After the Anglo-Boer War, in 1903, G.R. von Wielligh, who had played such an important role in the demarcation of the eastern border, visited Lourenço Marques accompanied by his brother, P.A. von Wielligh, to collect surveying instruments that they had ordered. There was a carnival in the streets of the harbour city and the two brothers stood closer to have a better view. A paper bag containing lime rather than flour was thrown in their direction and exploded in Von Wielligh's face. A local doctor washed his eyes with a diluted silver nitrate solution, causing further damage. Years later, a Dr Wood of Cape Town treated him and he regained partial sight in his left eye, but he was permanently blind in his right eye. After the tragic accident he could not carry on with his surveying work and settled on his brother Marthinus's farm in the Hendrina district. Here he wrote several books, including *Jacob Platjie, Langs die Lebombo, Nimrod Seeling, Gwennie Barneveld, Dierestories* and *Boesmanstories*. He played an important role in the Afrikaans language movement until his death in 1932. (He was a member of the Genootskap van Regte Afrikaners [Movement for True Afrikaners] and an honorary member of the Akademie vir Taal, Lettere en Kuns [Academy for Language, Literature and Art]).

The dispute between the Portuguese and Union governments over the northern part of the eastern border between the Shingwedzi and the confluence of the Limpopo and Luvuvhu rivers dragged on for a long time. Another border commission was appointed in 1894 to resolve the matter, but they were unable to start their work as a result of the war between the Portuguese authorities in Mozambique and Gungunyane.

With the proclamation of the Transvaal boundaries in 1903, this part of the eastern border was described as follows: "... to Impundwine a point in the Lebombo range; thence northwards along the border of the Portuguese East Africa Territory to the confluence of the Pafuri and Limpopo rivers".

The entire border had now been demarcated, except for the stretch between the Shingwedzi and Limpopo rivers. This part of the border remained a point of dispute between the two parties and despite negotiations and the efforts of more than one commission, a final agreement between the governments of Mozambique and South Africa was only reached in 1923.

This agreement stipulated that surveyor Von Wielligh's view of the border, that is to say a straight line between beacon T, just north of the Shingwedzi River, and the confluence of the Limpopo and Luvuvhu (Pafuri) rivers, was correct. This was confirmed in a treaty signed by both governments on 18 February 1926 in Pretoria. The agreement was finally registered with the former League of Nations on 25 September 1928. The treaty of 1923 stipulated that this section of the border had to be surveyed and beacons erected by a border commission comprising representatives of both governments. The Portuguese nominated Commander Viera da Rocha and captains Semedo and Moura Braz as commissioners.

The representatives of the Union government were surveyors W.P. Murray (commissioner) and R.E. Antrobus. The two commissions first met in July 1924 at Pafuri near the northern end of the borderline, where the borders of Mozambique, Rhodesia (Zimbabwe) and the Union (now the Republic of South Africa) meet. Paul Neergaard, district manager of the Witwatersrand Native Labour Association (WNLA) at Soekmekaar, and his agents McCurdy and MacKay, at Pafuri, were very cooperative and even helped to transport some of the equipment. The commissioners visited both WNLA posts at Pafuri and in his report Murray mentioned the unique baobab avenue leading to the Portuguese WNLA post at Pafuri. They agreed to erect a series of beacons between the confluence of the Luvuvhu and Limpopo rivers and the last surveyed beacon of 1890 north of the Shingwedzi (beacon T). The positions of the beacons would be determined with the aid of trigonometrical surveys, using elevated landmarks on both the Transvaal and Mozambican sides of the border. The Portuguese commissioner insisted that at least half of the landmarks should be in Mozambique, but after three months of hard work it was clear that there were no suitable places in Mozambique from where bearings could be taken. Only then did the Portuguese commissioner agree that beacons on the Transvaal side of the border could be used for the purpose of determining a series of numbered beacons (1–14) along this part of the border. Unfortunately the rainy season had started by this time and all further work had to be suspended until June 1925.

The landmarks selected for the trigonometrical surveys were Dwaal, Luck, Shikokololo, Shikundu, Dzundwini, XXII, Ponda, Shingwedzi and Long Edge. Everything went according to plan, until the surveying group encountered the almost impenetrable Nyandu Bush, about 50 kilometres north of beacon T. This dense botanical region occurs on an even sandveld plateau about 30 to 60 metres higher than the surrounding Lebombo flats. It was so overgrown with shrubs (mainly *Baphia massaiensis* and *Guibourtia conjugata*) of between three and five metres high that it was impossible for them to continue without cutting open a path with axes. The local blacks told the surveying team that nobody had ever managed to cross this dense bush as no landmarks were visible from it and one could only navigate by the sun. Almost no grass grew inside the bush and game and water were scarce. Crested guineafowl, a variety of other birds and field mice were common and here and there they found shallow pans.

Eventually they had to clear a nine metre-wide swathe through the 15 kilometre-wide bush. To accurately determine the positions of the beacons they also had to build a high lookout tower inside the bush. This "Nyandu Station" was built with more than seven metre-long poles they had felled outside the bush and then dragged inside with an ox-wagon's rear axle. On top of the poles they affixed a sturdy platform and lookout tower for the instruments. A few lookout towers from where surveying could be done were also erected in the crowns of tall Rhodesian teak trees. These towers were of enormous help and work could be completed during the winter of 1925. Fortunately only beacons 8 and 9 were inside the Nyandu Bush. Beacons 1 to 7 were erected south of the bush from Gondegonde to Mashumbela and beacon 10 just north of Shilahlandonga Poort. Beacon 11 was on the edge of the Pafuri scarp, 12 on the elevation just south of the WNLA post at Pafuri, 13 between the WNLA post and the river and 14, the northernmost beacon, on the southern bank of the Luvuvhu at its confluence with the Limpopo on the international border. This point does not coincide with the description of the international border between South Africa and Zimbabwe (formerly Southern Rhodesia) as being in the middle of the Limpopo River (on a large island in the middle of the Limpopo). After the completion of the border expedition Antrobus contracted malaria and nearly died.

The precise coordinates of this series of numbered eastern border beacons between the Kruger National Park and Mozambique were determined in 1924–25 by Murray and Antrobus. The coordinates of the line of letter beacons erected by the 1890 border commission were determined by surveyors Gilfillan (beacons A–J) in 1898 and Fripp (beacons K–T) in 1935. Surveyors Le Roux and Van Bergen further verified these coordinates during the 1930s and 40s.

In time the eastern border beacons became overgrown and could only be found with great effort by Surveyor Strydom in 1975–76, when the eastern border between the Kruger National Park and Mozambique was clearly demarcated by a solid fence. The fence was erected eight metres west of the actual international border line to enable teams to do maintenance work on the eastern side without interference.

The series of letter beacons erected in 1890 are all stacked piles of stone, some capped with a typical

One of the numbered beacons and a pod mahogany in the Nyandu bush. The tree was changed into an observation tower during Murray and Antrobus's survey of the eastern border between Shingwedzi Poort and the Limpopo/Luvuvhu confluence (1924–1925).
(ACKNOWLEDGEMENT: OFFICE OF THE SURVEYOR-GENERAL, PRETORIA)

"surveyors' crown" consisting of an iron peg with a crossed plate at the point on which the letter of the beacon is indicated. The series of number beacons between beacon T and the confluence of the Limpopo and Luvuvhu rivers were made of concrete, square at the bottom and tapered upwards, on which Roman numerals were stencilled on the wet concrete.

The fact that all the beacons along the eastern border are known today and that a passable road exists along the entire length of the international border of the Park emphasises the enormous – and sometimes almost insurmountable – problems with which the survey teams had to cope since 1890 in their attempts to demarcate an internationally recognisable borderline. No wonder it took almost 35 years to complete this gruelling task.

A large leadwood tree, of which some branches have been chopped off, along the eastern edge of Shirhombe Pan in the Nyandu bush. This tree was probably one of those which had been used as observation towers by surveyors W.P. Murray and R.E. Antrobus in 1924–1925 during their survey of the eastern border north of Shingwedzi Poort
(PHOTOGRAPH: P.J. VAN STADEN)

WNLA Officer H.H. Mockford of Pafuri at eastern border beacon 14 near the confluence of the Limpopo and Luvuvhu rivers, circa 1940.
(ACKNOWLEDGEMENT: STEVENSON-HAMILTON COLLECTION, SKUKUZA ARCHIVES)

References

Anon. 1959. *Geskiedenis van die totstandkoming van die Transvaal-Mosambiekgrens.* Internal report to the Surveyor-General. Files no. 4628/1 & 4628/4).

Anon. 1963. *South Africa's International Boundaries.* Internal Report of the Deparment of Foreign Affairs.

Antrobus, R.E. 1926. Note on observation stations. Transvaal-Portuguese Border Survey. *South African Survey Journal* July:228–229.

Beyers, C. 1938. Reports to Col. J. Stevenson-Hamilton in connection with the old Delagoa Bay road. Unpublished report.

De Vaal, J.B. 1985. João Albasini 1913–1888. Unpublished report to the National Parks Board.

De Villiers, J. & De Villiers, R. 2004. Uit verlange se dae – Die Oosterspoor. *Marloth Indaba* 2(2):8-9.

Forssman, M. 1869. Schetskaart aantonende de grenslijnen in Zuid Afrika tusschen de gronden behoorende aan het Koninglijke Portugeesche Gouvernement en de Zuid Afrikaansche Republiek overeenkomstig het traktaat dd. 29 Julie 1869. Witnessed by M.W. Pretorius (State President) and Alfredo du Prat (Consul-General).

Jeppe, F. 1890–1895. Series of letters to missionary Rev. H. Berthoud of Valdezia. Owned by P. Cuènod, Valdezia.

Jeppe, F. 1893. The Zoutpansberg goldfields in the S.A. Republic. *The Geographical Journal London* 11:213–237.

Krüger, D.W. 1938. Die weg na die see. *Archives Year Book for South African History* 1(1):31–233.

List of Coordinates: Transvaal-Mosambican border. Internal document, Office of the Surveyor-General.

Milstein, P. le S. 1986. Historical occurrence of Lichtenstein's hartebeest, *Alcelaphus lichtensteini* in the Transvaal and Natal. Unpublished manuscript.

Mockford, H.H. 1985. Report about the settlement of the WNLA posts in the Kruger National Park. Unpublished manuscript.

Murray, W.P. 1926. Definition of the northern section of the Transvaal-Portuguese Boundary line. *South African Survey Journal* July:125–132.

Preller, G.S. 1920. *Voortrekkermense II. (Herinneringe van Karel Trichardt).* Nasionale Pers, Cape Town.

Prinsloo, H. 1980. *Von Wielligh se boom.* Unpublished report to the National Parks Board.

Stevenson-Hamilton, J. 1927–1940. Warden's reports to the National Parks Board. Unpublished reports.

Strydom, J.A. 1975. Opmetings Oostelike Grensbakens: Nasionale Krugerwildtuin. Internal report to the Surveyor-General.

Thomas, P.W. 1967. Boundary Beacon: Pafuri–Limpopo. Internal report to the Surveyor-General.

Von Wielligh, G.R. 1890. Section of diary, 29 Mei–15 August 1890 about surveying the boundary between Portuguese East Africa and the Z.A. Republic. Unpublished report. T.A. A249(III).

Von Wielligh, G.R. 1890. Plan aantoonende N.O. Grenzen der Z.A. Republiek tusschen Komatiepoort en samenloop van Pafuri en Limpopo. Archive Reference. R5879/87 and R11709/90.

Von Wielligh, G.R. 1895. Rapport van den Landmetergeneraal over het jaar 1894. Pretoria No./15.

Von Wielligh, G.R. 1925. *Langs die Lebombo.* Van Schaik, Pretoria.

Vos, M.C. 1890. Diary kept while surveying the border between the Transvaal and Portuguese East Africa. Unpublished report.

Chapter 13

Game conservation in the Transvaal 1846–1898 and the proclamation of the Sabie Game Reserve

Dr E.J. Carruthers & Dr U. de V. Pienaar

In conservation circles worldwide the latter half of the nineteenth century is known as the "period of extinction". In Europe the industrialisation process quickly engulfed rural areas, and in Australia and New Zealand white settlers moved into the interior with their sheep and cattle and to a large extent killed off the indigenous fauna.

At the same time, large numbers of people moved westwards in North America and the construction of new railway lines put wilderness areas and wildlife under great pressure. Thanks to the use of sophisticated weapons it did not take the new colonists long to decimate the enormous herds of bison (*Bison bison*) and hunt the massive flocks of passenger pigeons, *Ectopistes migratorius*, to extinction. In the 1870s flights of passenger pigeons 45 kilometres long and five to six kilometres wide could still be seen in Wisconsin, but by 1914 they had disappeared completely.

Man, the ultimate predator, started playing a decisive role in the extermination of entire species and disrupting complicated ecosystems, often in a reckless and irrevocable manner.

Fortunately this "century of extinction" also heralded the "century of conservation". People everywhere became aware of the fact that all life forms on earth are interdependent and have a right to both survival and conservation.

As a result of this widespread damage to natural landscapes, as well as to plant and animal communities, the realisation dawned that man's power over the environment is so absolute that its uncontrolled abuse will eventually lead to humanity's own downfall.

This change in attitude that could be detected across the world at the end of the nineteenth century resulted in governments starting to place restrictions on the utilisation of natural resources. They started putting areas aside for the purpose of preserving beautiful or unique landscapes and/or specific types of wildlife. Structures and legislation in this regard were put into execution in many countries and expanded in the twentieth century. These formed the basis of the conservation ethic, organisations and actions we have today.

In terms of its natural heritage, South Africa went the same route and dealt with the same problems as the rest of the world. Until about 1830, white settlement was mainly restricted to the southern and eastern Cape, where the amount of game gradually dwindled.

As the colonists moved northwards, they were astonished at the abundance and variety of wildlife they encountered. In the areas that would later become the ZAR (South African Republic, or the Transvaal) and the Free State, there were herds of springbok, blesbok, black wildebeest, quagga, Burchell's zebra and even buffalo that numbered in the thousands (and sometimes even hundreds of thousands, in the case of springbok).

The black tribes had hunted to supplement their rural existence from time immemorial. The immense numbers of game, their primitive weapons and the lack of a market for game products (except ivory) prevented them from killing game on a large scale. In the northern areas of Transvaal and Natal, the presence of tsetse flies and malaria mosquitoes further protected wild animals.

Southern Africa is home to the greatest variety of large mammals of any zoogeographical region in the world. A lot of species are endemic to this region, while many are particularly spectacular and make sought-after trophies. It is therefore understandable that the unbelievable abundance of game in the Transvaal would attract a great deal of attention.

The first Europeans to set foot in the area were explorers, traders and missionaries. Although most of these pioneers left no record of their experiences, some did keep notes of their observations. Thanks to these journals and reports people can still share their wonder and respect for a region that, in contrast with Europe and the Cape, was still teeming with game. A naturalist was the first to document an expedition to the Transvaal. This was Dr Andrew Smith, the director of the South African Museum in Cape Town, who led an expedition to the Transvaal from 1834 to 1836 and recorded geographical, natural and ethnological findings. The next European visitor to the area north of the Vaal River to leave a written record was the British "sportsman" and soldier from India, William Cornwallis Harris, who undertook an expedition to this hunter's paradise in 1836. Harris travelled far and wide in the Transvaal and killed hundreds of animals – probably wounding even more in the process. Many carcasses were simply left in the veld to decompose. Other hunters, including Roualeyn Gordon Cumming, followed his example. These "fearless" pioneers entertained the reading public in England and the rest of Europe with exciting accounts of the mass slaughter of game, which they described as "sport" and "adventure". Game numbers in the Transvaal and Free State were such that these people never even dreamt that it would all come to an end. No consideration was therefore given to the possibility of game reserves or any other kind of protection of the rich fauna of these areas.

However, foreign visitors are not the only ones to blame for this large-scale extermination of wildlife. From 1836 onwards groups of Voortrekkers left the Cape Colony and moved into the Transvaal because of political considerations and the constant frontier wars with the Xhosa. During the Great Trek, the Voortrekker settlers formed informal and even rival political groups. The town of Potchefstroom, which was founded in 1838 by General A.H. Potgieter and his people, was later moved to its current site and became an independent republic in 1844. The following year Andries Ohrigstad became a formal settlement, but was evacuated in 1848 because of the unhealthy climate. Some of the Lowveld pioneers, led by Potgieter, moved to Zoutpansbergdorp (Schoemansdal), while the remaining people of Ohrigstad founded a new town, Lydenburg, in 1849. In the period 1849 to 1852, attempts were made to form a united Volksraad (Legislative Assembly) to govern the whole Transvaal. In 1853 the name Zuid-Afrikaansche Republiek (South African Republic or ZAR) was given to the area north of the Vaal River. Despite this step and their freedom from British rule, confirmed by the Sand River Convention in 1852, the Voortrekker communities in the Transvaal remained factional. This repeatedly led to confrontation among the leaders, who time and again hovered on the brink of civil war.

Between 1856 and 1860 Lydenburg declared itself independent of the rest of the ZAR. It was only in 1860 that the whole Transvaal was united and a comprehensive constitution could be drawn up. Even then differences had not been settled satisfactorily and problems between the factions and even armed conflict between the burghers of Schoemansdal and those of Potchefstroom and Rustenburg kept flaring up between 1860 and 1864.

Initially the Voortrekkers used the game stock of the Transvaal as their main means of survival. They shot animals for food while the skins and other by-products were worked into straps, strops, shoes and other essential items. In years to follow, the economy of the Transvaal settlements increasingly depended on hunting and its by-products. Animals weren't hunted solely for survival or personal use any more. Hunting trips aimed at trade now became important and vast amounts of ivory and skins were transported from the Transvaal to the east coast harbours and to markets in the interior such as Grahamstown. In 1864 alone, the value of the ivory that left the area was R240 000, that of ostrich feathers R50 000 and leather products R10 000 – enormous incomes at a time when the salary of a young official was R30 a year! In 1872 the value of the exported skins amounted to R180 000, representing about 125 000 animals. During the winter months of the 1860s and 1870s all the able-bodied men of a district would be hunting, most of them accompanied by their families and black workers known as "swart skuts" (black shots).

Game was a national asset. Without the income from game products the settlements in the far northern and western Transvaal would have gone under, as these areas were unsuited to the conventional agricultural methods of the time.

One of the oldest methods of preserving game is to institute laws or ordinances that prohibit hunting, or to place restrictions on the hunting of certain kinds of game or the periods during which hunting is allowed.

From the early years of the Dutch East India Company at the Cape certain measures were implemented to protect natural resources. By the middle of the nineteenth century, legislation regulating hunting had taken on exhaustive proportions. All of these laws and regulations barely made a dent in keeping the systematic extermination of game in certain areas in check. The bluebuck, *Hippotragus leucophaeus*, and the quagga, *Equus quagga*, had already been driven to extinction, while the numbers of most other species had decreased dangerously.

Coming from the Cape with its existing prohibitions on hunting, the Voortrekkers were well versed with this legislation and soon promulgated their own hunting laws and regulations. Taking into consideration the disputes between the various factions, and the fact that the Transvaal had not been completely annexed by the Voortrekkers and large areas were still under the rule of high chiefs and headmen, it is understandable that these hunting laws did not have much effect in the Transvaal. It must also be kept in mind that these communities were egalitarian. Unlike Europeans they did not have a tradition of a rural aristocracy who, thanks to their personal power and wealth, created parks for the protection of deer and curtailed poaching by appointing gamekeepers.

The first law in the Transvaal aimed at controlling hunting was promulgated by the Volksraad of Andries Ohrigstad in January 1846. No restrictions were placed on the hunting of elephants as this was the district's main source of income. As far as other game was concerned, the law stipulated that families were only allowed to shoot enough for their own use and that each hunter could collect only one wagonload of carcasses. The law also stipulated that only Transvaal burghers, thus Voortrekker settlers, were allowed to hunt. No foreigners could hunt in the Ohrigstad district and any burgher who allowed a foreigner to hunt there was guilty of an offence. The restriction on hunting by foreigners was aimed especially at hunters and traders from the south who deprived the Transvalers of their rightful share of game products, and at people who could not even claim to be legal traders and mostly were fugitives or other disreputable and unwanted elements.

The 1846 law was therefore specifically proclaimed to prevent anybody not living in the district from benefiting from an important resource. Scarcely three months later Joseph McCabe, an English hunter and trader, was found guilty of breaking the law and sentenced to pay a maximum fine.

During the nineteenth century hunting was an important part of life at the outposts in the Transvaal. Hunting parties often stayed in the hunting fields for months at a time, accompanied by their families and large numbers of black hunters – often as many as 200 to 300 in total.

Occasionally there was conflict between blacks and whites over the hunting of certain species or the use of certain areas, but at the same time it created opportunities for cooperation between the different races, as all profited. Sometimes the black hunters were sent into the hunting areas by themselves, to return weeks or months later to hand over the trophies to their sponsors. This system later led to serious problems with especially the Venda hunters. They refused to hand over all the ivory they had collected and later also absconded with their employers' rifles. This led to conflict and drawn-out wars with the Venda and the evacuation of Schoemansdal.

Between the years 1856 to 1859 the Lydenburg Volksraad hired six black hunters as a result of a serious shortage of money. It provided them with rifles and ammunition and ordered them to hunt elephant along the foothills of the Drakensberg. They had to carry on hunting right through the unhealthy summer months. It was hoped that the income obtained from selling the ivory they brought back would help fill the empty coffers. This project was not a great success and only 1000 rix-dollars (about R2500) could be added to the Lydenburg coffers after the ivory, horns and skins had been sold at auction. Most of the inhabitants of Lydenburg then submitted petitions to the Volksraad to be allowed to hire black hunters themselves. The Volksraad refused on the grounds that it would not be worth their while and that arming blacks without proper supervision was dangerous.

By the late 1850s game numbers in the Transvaal, even in remote areas, had declined to such an extent that more encompassing laws to protect the remainder became essential.

In March 1857 a public meeting was held in the Soutpansberg. As a result of the recommendations submitted, a more comprehensive hunting law was put into place the following year. The main aim of this law was to ensure the continued existence of the hunting community in the Soutpansberg.

By that time elephants had been hunted to such a degree that they could only be found in the more remote and unhealthy parts of the Lowveld. Every year scores of people succumbed to the malaria they contracted while on hunting trips. A decision was made that nobody would be allowed to hunt elephant in the malaria-infested Lowveld in summer. As a result of the diminishing game numbers, the old Ohrigstad hunting clause, which stipulated that nobody was allowed to hunt more than needed for own use and that no more than a single wagonload of carcasses could be collected on a hunting trip, was included in the Soutpansberg hunting legislation.

Contraventions of these regulations would be fined. Game species were divided into three categories – large, medium and small – so the correct fine for the contravention could be imposed. The law was only applicable to edible game and excluded rhino, for instance. The purpose of the legislation was mainly to provide as many of the inhabitants of the Soutpansberg district as possible with the opportunity to benefit from the by-products of hunting by either selling it or using it themselves. It was also aimed at preventing a small group of wealthy hunters from obtaining a monopoly over hunting in the area. The law also decreed that no white hunter could hire more than two black hunters and that no black hunter could hunt on his own without being accompanied by his employer or having been registered with the magistrate. They also tried to use this law to prevent the trading of arms and ammunition among and to blacks.

The record of contraventions of the hunting law of 1858 is incomplete, but in one case recorded in 1863, the brothers Gert and Kornelis Bezuidenhout were accused of wasting game in a "scandalous manner". They were only after skins and left scores of buffalo, giraffe and eland carcasses in the veld for the vultures and scavengers.

When they were confronted by four eyewitnesses and Field Cornet De Klerk tried to confiscate the skins, the Bezuidenhouts assaulted him and the row turned into a free-for-all.

Unfortunately the 1858 hunting law was never widely enforced in the Soutpansberg district. When an attempt was made to do so in the mid-1860s, the burghers objected that it was be unfair and would threaten their existence.

It is obvious that the 1858 law was completely ineffective in halting the large-scale extermination of game when one looks at the growing number of agenda points at Volksraad meetings that dealt solely with game affairs, as well as the official correspondence of the Transvaal at the time. More and more influential people in the Transvaal began to express concern at the fast disappearing herds of game. The magistrate of Lydenburg, for instance, in 1860 officially objected to the decimation of game in his district.

When people started realising how fast the wildlife was disappearing, there was often a search for a scapegoat. A finger was readily pointed at the effortless way in which blacks could obtain rifles and ammunition, but people rarely looked closer to home.

The 1858 Zoutpansberg hunting law was again published in the *Staats Courant* (Government Gazette) in 1869. An admonition by President Pretorius was added in which he warned that the time had come to adhere to the provisions of the law, as he constantly received complaints from field cornets about the reckless eradication of game.

During the 1860s a number of private landowners published notices in the *Staats Courant* prohibiting people from hunting on their property. By this time the amount of game had decreased to such an extent that the governments of the Free State, Transvaal and Natal were accusing each other of allowing their respective subjects to hunt across their borders in neighbouring states.

Eventually so little game was left – with survivors mostly in the tsetse belts – that professional hunting as an occupation started dying out. The inhabitants of the Transvaal had to turn to livestock farming to make a living. Hunting expeditions now became social events on land where game were given a measure of protection by the owners. In the 1870s the Transvaal game laws were adapted to the changing circumstances. The first of these changes took place in 1870, when the game law was consolidated. The draft bill was published in the *Staats Courant* on 17 May for the public's consideration on 28 June, as was the norm in those days. It was then discussed by the ZAR's Volksraad. Basically the new law incorporated most of the regulations of the old one: A close season on hunting certain kinds of game in certain areas was reconfirmed; limitations were placed on hunting by blacks; and the fines with regard to the contravention of restrictions on the waste of game or game products and the collection of more than a single wagonload of game carcasses per person, were reconfirmed.

The law also contained three completely new principles. It made provision for the appointment of "jacht-opzieners" (rangers) at the request of the burghers of a district. These rangers would be responsible for the enforcement of the game and hunting laws and would be entitled to half of all money raised from fines. Anybody who was aware of the contravention of a hunting law and who failed to report it to the magistrate or local ranger was also guilty of a contravention and could be fined. The second new principle was a close season on hunting ostriches and collecting their feathers. Thirdly, the use of traps or snares to capture game was banned. With only one nay among its 20 members the Volksraad almost unanimously supported and accepted the new law.

At this early stage there still was no clear policy direction to back game legislation in the Transvaal. Laws were made on an ad hoc basis and were aimed at preserving wildlife as a natural resource for the inhabitants of the Transvaal and providing everyone with equal access to it. As yet there was no aesthetic or sentimental basis for the legislation – it was only there to ensure that a valuable commodity was not destroyed or wasted.

It must be emphasised that the members of the ZAR Volksraad were agreed, as in few other cases, that the area's wildlife had to be protected.

However, the legislative assembly had to adapt to circumstances and the wishes of the burghers. In 1871, scarcely a year after the new legislation had been implemented, hunting in the tsetse fly belt north of the Soutpansberg was allowed during the close season at the request of the Soutpansbergers, who had great difficulty in tracking down game to hunt. The ban on the hunting of ostriches and the collecting of their feathers was also lifted. These regulations were only implemented again four years later.

In 1874 an important step forward was taken. In answer to a petition from certain voting districts, the Volksraad decreed a close season from 15 September to 15 January every year. During this period nobody was allowed to hunt unless a licence had been obtained from the local district ranger – if there was one. The licence was free and issued at the discretion of the ranger, who also determined the number of animals every person could shoot. This decision was not unanimously accepted by the Volksraad – 15 members voted for and five against the proposal – but it was nevertheless accepted that a close season during calving season was necessary. This law was implemented in 1875.

Although the introduction of a close season was an important improvement on existing legislation, the formulation of new laws was but one aspect of game preservation. A law means very little if not enforced, something which was difficult at the time.

There were no rangers in most districts and farming practices were very primitive. For want of an economic infrastructure many of the border farmers were dependent on hunting – not only for survival, but also to ensure an income.

The burghers did not want to be kept in check as far as hunting was concerned and paid little attention to the legislation. The close season, for instance, did not apply to landowners who hunted on their own properties.

It is therefore no surprise that at the time both visitors to and inhabitants of the Transvaal could testify to the failure of the hunting laws and the continuous drastic decline in game in all the districts of the country.

On 12 April 1877 the Transvaal was annexed by Great Britain and once again petitions for the appointment of rangers were submitted. Although the British administration acknowledged the need for stricter hunting laws, it had more pressing problems that needed urgent attention. The only law in connection with wildlife conservation accepted during this time was the prohibition on killing fish by using dynamite or other explosives.

In those years public opinion in the Transvaal was expressed by way of petitions. Most of the laws in

President Paul Kruger as a young man. The photograph shows his left hand without a thumb. It was crushed during a hunting accident when his gun exploded.
(FROM: *DIE HERINNERINGE VAN PAUL KRUGER*, VOL. 1. 1902)

connection with hunting and game preservation had their origins in reactions to petitions submitted by voters to either the Executive Council or the Volksraad.

The petitions that dealt with hunting and other wildlife matters usually only received attention after a long time because there were always political and external matters that were more urgent. By 1884, the correspondence regarding hunting and wildlife preservation was already considerable. In 1882, for instance, the acting magistrate of the Soutpansberg requested further restrictions on hunting because almost all the wildlife had disappeared from his district. In April that year about 29 burghers of the Bethal district asked for the appointment of more rangers to enforce the hunting laws more stringently. In 1883 the Zeerust voting district warned that people were hunting recklessly in the district to make biltong, which they sold at the Kimberley diamond fields. The magistrate there asked that the amount of game that could be shot in one day be determined and that proper licences be acquired first.

In the same year complaints also came from Lichtenburg about people who crossed the border to hunt in the district. Biltong loads of 20 to 30 bags were carted away. Lydenburg's magistrate also reported his concern about the lack of game in his district. In contrast to these calls for stricter conservation measures, some burghers from Standerton and Middelburg appealed for the abolishment of the hunting laws.

The reaction of the secretary of state to these conflicting petitions was to send out a general circular in July 1883 asking all the burghers for their opinion on the matter so the Volksraad could look into it. By 1884 numerous replies had been received. Most magistrates asked for a longer close season, heavier fines, better control over hunting by blacks and a restriction on the amount of game that could be shot. But a lot of people still earned a living from hunting and the legislation would have to be carefully formulated. The reaction to the circular confirmed the terrible extent to which game had been slaughtered. The magistrates of Ermelo and Heidelberg noted that there was almost no game left in their districts.

The thorough review of game and hunting laws did not take place in 1884 as expected. During the Volksraad session that year only the 1882 petition from the Bethal burghers to appoint more rangers was put forward for discussion and approval.

During the session they also debated the effectiveness of the 1870 hunting law. According to J. Du Plessis de Beer, the member for Waterberg, the law failed to prevent the extensive damage caused by "skin" hunters. The member for Standerton, D.P. Taljaard, proposed that rangers be appointed throughout the Transvaal. President Kruger declared that the government could hardly appoint rangers everywhere, as they were not needed everywhere. In places where they could fulfil an important function, they would be appointed on request.

The plea from Bethal that a more encompassing hunting law with greater restrictions be implemented received substantial support. But the proposal by the member for Middelburg, N.J. Smit, for the implementation of a close season that would also apply to landowners, received heavy opposition. Few were prepared to accept restrictions on their own land and the issue of the rights of landowners would become an even bigger headache in years to come. Better control over hunting by blacks and a restriction on the number of animals that could be hunted for own use also came up for discussion.

The Volksraad, president and Executive Council unanimously decided that the 1870 law still provided ample protection to the country's game, and so let an opportunity slip by to implement stricter rules on hunting.

There is no reference in the 1884 minutes of the Volksraad to President Kruger declaring himself in favour of the preservation of wildlife. The desirability of putting an area aside for this purpose was also not mentioned (see Stevenson-Hamilton, 1937; Behrens, 1947; Pringle, 1982 and others).

Tattersall (1975) even claims – probably on the basis of unreliable biographies by Fisher and Marjorie Juta – that the president said it would be useless to only issue hunting ordinances and decree close hunting seasons. Of greater necessity was putting aside an area for animals where nature could remain untouched as the Creator had intended. Game was an asset and should not be wasted.

It would seem that 1884 came to be seen as the year the idea of nature conservation took hold in Transvaal because of a misprint in Stevenson-Hamilton's book *South African Eden* (1937). In his diary of 20 May 1935 he clearly stated that Dr Beyers, the chief archivist at the Transvaal Archives, in a conversation had pointed to 1889 as the year the first reference was made to the idea of setting aside a game reserve in the Transvaal.

In his book he wrote of 1884 in reference to this occasion without it having been mentioned previously in any of the available literature. Later authors time and again repeated this mistake in their documents without confirming the date in archival sources. The ZAR was on the verge of bankruptcy at the time and the president and the Volksraad had to pay attention to more important matters than wildlife preservation.

The Republic's wildlife and hunting laws were not discussed in the Volksraad for the next five years, although many burghers, as well as immigrant mine workers on the goldfields of Pilgrim's Rest and Barberton, had written to the government to express their concern about game conditions.

When wildlife affairs came up for discussion again in 1889, members were asked to take a far-reaching decision. On 31 July 1889 the Executive Council took a very important step – probably in reaction to a proposal from Magistrate J.C. Krogh of Wakkerstroom.

In a letter to the state secretary dated 13 March 1889, he asked that part of the state land adjacent to the Lebombos in the Piet Retief district be closed to hunting for a few years to give game numbers an opportunity to recover.

The proposal struck a chord and the Executive Council accepted resolution 482. The Volksraad was asked to mandate a ban on hunting on certain tracts of state land and to formulate regulations and penalty clauses with regard to transgressions of this ban. In essence, this decision of the Executive Council made provision for the establishment of game reserves on state land, even if it was only temporarily.

Two days later the matter came up for discussion in the Volksraad. Two members, De Beer and Taljaard, immediately opposed the draft legislation. Unlike Taljaard, De Beer had no objection in principle to a ban on hunting in certain areas, but he wanted them to first define the exact boundaries rather than give the government a free hand in determining the areas where the ban would be imposed.

It was then that President Kruger gave his historic response to the two men, which clearly illustrates the sentiments of the dignified and wise old statesman, who had once been a renowned hunter and had even lost his left thumb when his gun exploded after he had wounded a rhino near the Steelpoort River in 1845. He told Taljaard and De Beer that it would be impossible at that stage to say which areas would become game reserves. He was in fact in favour of all state land being closed to hunting. In some areas little more than a few antelopes had survived and he said one area the government was seriously considering was in the southeastern Transvaal bordering on the New Republic, Swaziland and parts of Zululand, where the Pongola River flows through the Lebombo Mountains.

The president revealed his intimate knowledge of game and wildlife conditions by pointing out to the members that there was still a large variety of game, including buffalo, elephant and giraffe, in the area in question. He thought it was essential that steps be taken to preserve this remaining wilderness area in the Transvaal for following generations. Game move back into areas where they were not constantly hunted and that was where they should be preserved "voor de verre toekomst" (in other words, for ever-

Herman Frederik van Oordt, the first warden of the Pongola Reserve (1894–1899), which was proclaimed on 3 June 1894 by President Paul Kruger.
(ACKNOWLEDGEMENT: L.C. AND G.A. VAN OORDT AND DR. J.A. PRINGLE)

more). There were also other areas in the Transvaal where similar reserves could be established, and here he specifically mentioned the area north of the Soutpansberg up to the Portuguese (Mozambican) border.

The president probably also had all the unsurveyed state land between the Limpopo and Olifants river in mind, south of where quite a few private farms had already been allocated and surveyed. This area would later become the Shingwedzi Reserve, which is currently the northern part of the Kruger National Park.

The first time this area was mentioned as a possible game reserve was in President Kruger's speech in the Volksraad in 1889. In later years his wishes would come true and eventually his name would be coupled with an ideal he had supported at this early stage. On this particular day in the Volksraad, Jan de Beer, the member for Waterberg, and Daantjie Taljaard of Standerton were not satisfied with the president's explanation. De Beer wanted the assurance that blacks would be subjected to the same regulations as whites. Taljaard claimed that state land did not belong to the government but to the nation, and that the nation therefore had the right to use it as it saw fit.

Again the president explained that they were wrong in assuming that members of the public could do as they wished on state land.

No Volksraad members other than De Beer and Taljaard objected to the Executive Council's recommendation. Support for the decision came from member J.P.L. Lombard, who pointed out that there were already farms in the Cape Colony on which hunting was prohibited and that it was high time that the Transvaal took similar steps.

The ensuing discussion was fairly short and when they voted on the matter on 2 August 1890, only De Beer and Taljaard voted against Article 1244, which authorised game reserves in the Transvaal.

The Pongola Game Reserve only became a reality in 1894, almost five years after it had first been proposed in the Volksraad by President Kruger and his supporters. It was then that the boundaries of the area were determined satisfactorily (after Lucas Meyer, former president of what had been the "New Republic", had been consulted), and a warden was appointed for this first "gouvernements" (government's) game reserve in the Transvaal. The area was small and consisted of only seven farms with a combined surface of 17 400 hectares. In time it became obvious that the reserve was too small to serve its purpose as it was situated on a game migration route between Swaziland and Zululand. Despite this, the reserve was proclaimed in the *Staats Courant* on 13 June 1894 – a historic event and a test for similar legislation.

For administrative purposes the Pongola Game Reserve was placed under the control of the superintendent of education, with Dr J.W.B. Gunning, the director of the State Museum in Pretoria, as the administrative head and H.F. van Oordt as the first warden. Although there later were claims of irregularities during his term of service, Van Oordt was in more ways than one a wise choice as official in charge of the reserve. He fulfilled his administrative duties diligently and made a valuable contribution by reporting in detail on the habits, numbers and movements of game in the reserve.

Herman Frederik van Oordt was therefore the first real game ranger in Africa. He had emigrated from Holland as a young man and arrived in Cape Town in 1881, first settling in German West Africa (now Namibia) and later moving to the Transvaal, where he was a civil servant. In 1889 he and J.J. Ferreira were sent to the southeastern outpost of the ZAR as resident agent and administrative official for Swazi Chief Sambana[1], who lived with his subjects on the summit of the Lebombo Mountains. Van Oordt settled at Wolfshoek on the Lebombo plateau. After his appointment as warden of the Pongola Reserve in 1894, he served in a double capacity, as the eastern border of the reserve was also the boundary of Sambana's territory.

Van Oordt handed in his first report as warden of the Pongola Game Reserve in 1895. He mentioned seeing kudu, waterbuck, hartebeest (Lichtenstein's hartebeest?), tsessebe, blue wildebeest, zebra, impala, mountain reedbuck, bushbuck, grey duiker, steenbok, klipspringer, red duiker, nyala (very rare), warthog and bush pig in the reserve. He also mentioned that a few black and white rhino still frequented the area, as well as some lion and hippo. Elephant and eland

---

1   Also referred to as Zambaan or Zambane in other sources.

had already disappeared from the area. Leopards were so numerous that he set traps for them, but hyenas, wild dogs and jackals were scarce. There were many bird species and he described a curious-looking bird with a black, feathered crest – probably a crested guineafowl. Van Oordt reported that he had killed all the stray dogs in the area and that is was noticeable that the animals were becoming tamer and that their numbers were on the increase. His staff consisted of four black rangers and later two armed constables were seconded to the reserve to help with patrols and controlling poaching.

He arrested many black poachers and summarily fined them. In 1896, 1897 and 1898 he again submitted reports that contained valuable information about the rinderpest epidemic. At the outbreak of the Anglo-Boer War his reports dried up when he joined the Piet Retief commando. He served under General Louis Botha, but was captured and deported to St Helena. He returned after the war, but was not reappointed to his former position. (The appointment of Major A.A. Fraser as warden of the reserve is discussed in chapter 16.2.) Van Oordt's later years were difficult and lonely. He tried to make a living as a trader in the northern Cape and South West Africa (Namibia), but without much success. His wife and children left him and he started drinking heavily. He died in 1907, alone and penniless (G.A. van Oordt, 1980).

The 1890s were characterised by continued interest in matters relating to game, and there were two intensive discussions of the game laws in the Volksraad. This not only reflected a growing acknowledgement of the seriousness of the matter, but also concern at the drastic and worsening drop in game numbers. It is clear that the burghers of the Transvaal, the government and public representatives started fearing the total eradication of a valuable natural resource. They now wanted to implement practical measures to save what remained. Other natural resources of the republic such as indigenous forests and natural pastures also started showing signs of overuse and therefore also came under the spotlight.

The Volksraad's role in attempts to protect the dwindling game population was of particular importance, especially if taken into account that this body was empowered to draft legislation.

In 1891, after a delay of nine years, the often discussed amendments to the Jagwet (Hunting Law) 10 of 1870 again appeared on the Volksraad's agenda. By that time the Volksraad had been divided into two chambers – a First and a Second Volksraad. The Second Volksraad gave a voice to the mining communities, who were controlled mainly by foreigners.

The First Volksraad remained the most important legislative body in the Republic. Once again written representations or petitions formed the basis of discussions in the Volksraad and, as always, widely divergent opinions were aired. For the first time the differing sentiments of rural and urban communities became apparent. Petitions asking for more restrictions on hunting were received from Pretoria and Johannesburg, similar to those in Natal. In contrast, the burghers of Ermelo asked that hunting licences be made cheaper.

Some farmers from Potchefstroom and Heidelberg worried that the close season would be made applicable to privately owned land, and that their crops and livestock would suffer damage if they could not eradicate destructive game, birds and predators during the close season.

Two petitions were also received from the Second Volksraad. Because the other petitions were relevant to the existing hunting laws, the "Petitions Commission" advised that they be consulted during discussions on proposed amendments to the law.

In June 1891 the First Volksraad debated the petitions and the draft amendments. The petitions and appeals formed the basis of discussions and the old

Members of the First Volksraad of the ZAR (South African Republic) after the War of Independence. R.K. Loveday (second from left, back) and J.L. van Wijk (fifth from left in the middle row) were the two members who actively promoted President Kruger's proclamation of the Sabie Game Reserve on 26 March 1898.
(ACKNOWLEDGEMENT: TRANSVAAL ARCHIVES)

Richard Kelsey Loveday (1854–1910), surveyor, MP for Barberton and great campaigner for the creation of nature reserves. He had submitted the motion which led to the proclamation of the Sabie Reserve in 1898.
(ACKNOWLEDGEMENT: STATE ARCHIVES, PRETORIA)

matter of limits on the rights of individual landowners to hunt on their land during any season once again resulted in heated debate. The same fate befell the question of whether game belonged to the state, or to the landowner on whose farm it grazed, or if it *res nullius* belonged to nobody.

After two days of fairly fruitless discussions the entire matter was referred to a committee consisting of J.M. Malan (Rustenburg), P.L. Bezuidenhout (Potchefstroom) and L. de Jager (Lydenburg). The commission handed in its report later that month and the details of each of the 21 clauses in the proposed bill on hunting were carefully studied by each member of the Volksraad and discussed in light of their implications for voters in their specific districts.

The majority of Volksraad members were in favour of greater control over hunting, but were uncertain if the new law would be the most effective way to put it into practice. Certain clauses of the draft bill were accepted or reconfirmed without any problems, but others elicited strong emotions and confrontations. They had particular problems with the issuing of licences to landowners to hunt on their own land, the cost of licences, the length of the close season, the amount of game that could be hunted, the number of black assistants allowed to hunt on behalf of their employers and the right of blacks to hunt in their own territories. While some Volksraad members made legitimate objections and experienced fewer problems, quite a number seemed set on trying to sabotage the bill's stricter regulations. They tried the patience of President Kruger and their colleagues to the limit. Clause 2 of the new bill prescribed the issuing of licences for hunting particular species. Member Loveday, supported by President Kruger, proposed that the hunting season be shortened, running from 1 April to 15 August, rather than 1 February to 15 September. This clause was accepted by 20 votes to three.

Matters relating to the conservation of natural resources will inevitably contain an element of "you may not", and there will always be those who object to any restrictions to their real or imagined individual rights or freedom. This last category was amply represented by members for the border districts, as game was still relatively abundant and an economic resource in these areas. The inhabitants of the border areas were also generally less well off than their farming or urban compatriots, and regarded the law as a real threat to their way of life and even to their chances of survival.

The third reading of the bill took place on 1 July 1891, but it was only promulgated on 1 January 1892. (R.K. Loveday, the representative for Barberton who later became one of the greatest champions of game and conservation legislation, voted against the new law since he believed that it would fail to protect game in practice. With this view he was ahead of his colleagues.)

Many of the new law's clauses were only repetitions of those of Law 10 of 1870. The main omissions were the many measures referring to black hunters. The most important new provisions related to an increase in the species that enjoyed complete protection (such as rhino and elephant) and the species that could be hunted only during the "open" season; licences to hunt on state and private land and for which certain fees would be levied (although landowners did not need a licence to hunt during the open season); and better control over the trade in game products.

At the next session of the Volksraad in June 1892 a law declaring the secretary bird a protected species was put through as a result of a petition from Pretoria. At the same time the Volksraad rejected two other petitions that asked for an even stricter hunting law. The member for Barberton, R.K. Loveday, asked that his objection to this decision by the Volksraad be recorded in the minutes. In coming years his voice would be heard more and more on wildlife and hunting matters and he often aired his dissatisfaction with measures that he believed were not strict enough.

It was also suggested that the term "for own use" be better defined in the legislation by stipulating exactly how many head of game each hunter was allowed to shoot. Many members were in favour of this, even Jan de Beer, but the president felt that the provisions of this clause should not be too circumscribed.

During the next session in June 1893, the urban community of Pretoria once again prompted the Volksraad to act. No fewer than nine petitions were submitted demanding stricter control over hunting. During discussions it came out that these petitions represented the efforts of the newly established Wildlife Protection Society in Pretoria. Other petitions again demanded a loosening of hunting restrictions.

The "Petitions Commission" found it difficult to reconcile these conflicting viewpoints and wanted to appoint a separate commission to investigate the matter. Arguments in the Volksraad once again became heated as many members felt that the existing law was strict enough and more than comprehensive, but the matter was eventually referred to a commission.

A month later the chairperson of the Game Commission, J.A. Malan of Rustenburg, suggested that a number of other animals (including buffalo, eland, giraffe and rhino) be added to the list of those that enjoyed specific protection. He also took the other petitioners' request into account by suggesting that the close season be shortened by a month.

Once again opposition came from the same quarters as before, but the recommendations of the commission were eventually accepted with a larger majority than the amendments to the Act of 1892 were.

Reference has already been made to the fact that individual farmers had been trying to preserve the game on their own land since 1867 by publishing notices in the *Staats Courant* that prohibited people from hunting on their land. In total 200 owners protected about 300 farms in this manner between 1867 and 1881. The farmers of the Lydenburg district played a leading role here by prohibiting hunting by outsiders on 62 farms. The same was done at a further 57 farms in the Waterberg district, as well as at 43 farms in the Pretoria district and 23 in the Wakkerstroom district. One of these farms in the Wakkerstroom district belonged to Alexander Marsh Robertson (*Staats Courant* 476 of 3 June 1873, p. 3). V.L. Robertson (1947), the son of this pioneer, later said that his father, after many years' peregrinations as a trader and hunter, decided in 1867 to settle in the Wakkerstroom district where there were still large numbers of game. He bought two adjacent farms, Rolfontein and Elandsberg, which together extended over 7600 morgen of sweetveld with perennial water. He settled on Rolfontein and obtained another farm, Indlovudwalile ("where the elephant got lost"), along the Swaziland border for winter grazing.

Fencing material became available during this time and Robertson sent his wagons to Pietermaritzburg to fetch a load of barbed wire. Using leadwood poles from his Lowveld farm he fenced about 500 morgen with two rows of barbed wire to prevent his horses from wandering off – this was probably one of the first fenced camps in the eastern Transvaal. The double fence soon proved effective in both restraining his horses and providing protection to ground game, which crawled in underneath the fence and soon realised they were fairly safe inside the camp. Game in the camp soon increased and Robertson regarded it as a significant if unexpected success in his efforts to protect the game on his farm.

Blesbok and springbok were by far the most common game in the district, but there were also healthy populations of mountain rhebok, grey rhebok, oribi, steenbok, duiker, wildebeest, zebra and eland. The eland were a problem because they damaged the fences. Many of the farmers in the district decided to halt all unnecessary hunting. Among them were Fritz Roch, Org (George) Bezuidenhout, Jan Greyling and Paul Fick. Albertus Stoop was the Volksraad member for Wakkerstroom at the time and his daughter was engaged to President Kruger's son Tjaart, whom she later married. Stoop, a great supporter of wildlife conservation, now had a direct link to the president.

The president had the habit of touring rural areas to visit his people. He had heard about Robertson's efforts to fence in game and wrote to Stoop that he would like to see the experiment at first hand and spend a night at Rolfontein. His trip was organised and the president showed great interest in the Wakkerstroom farmers' efforts to preserve game and encouraged them to persevere.

In 1892 the president visited Rolfontein again and he was very pleased with the farmers' efforts to secure a refuge for wild animals.

The night after his visit to the game camps, at a barbecue held in his honour, he apparently said that small private reserves like Rolfontein played an important role in protecting smaller game, but that his main concern was the big game of the Transvaal – such as wildebeest, zebra and eland – which needed a much larger habitat to thrive under protection. He then explained to those present that he planned to

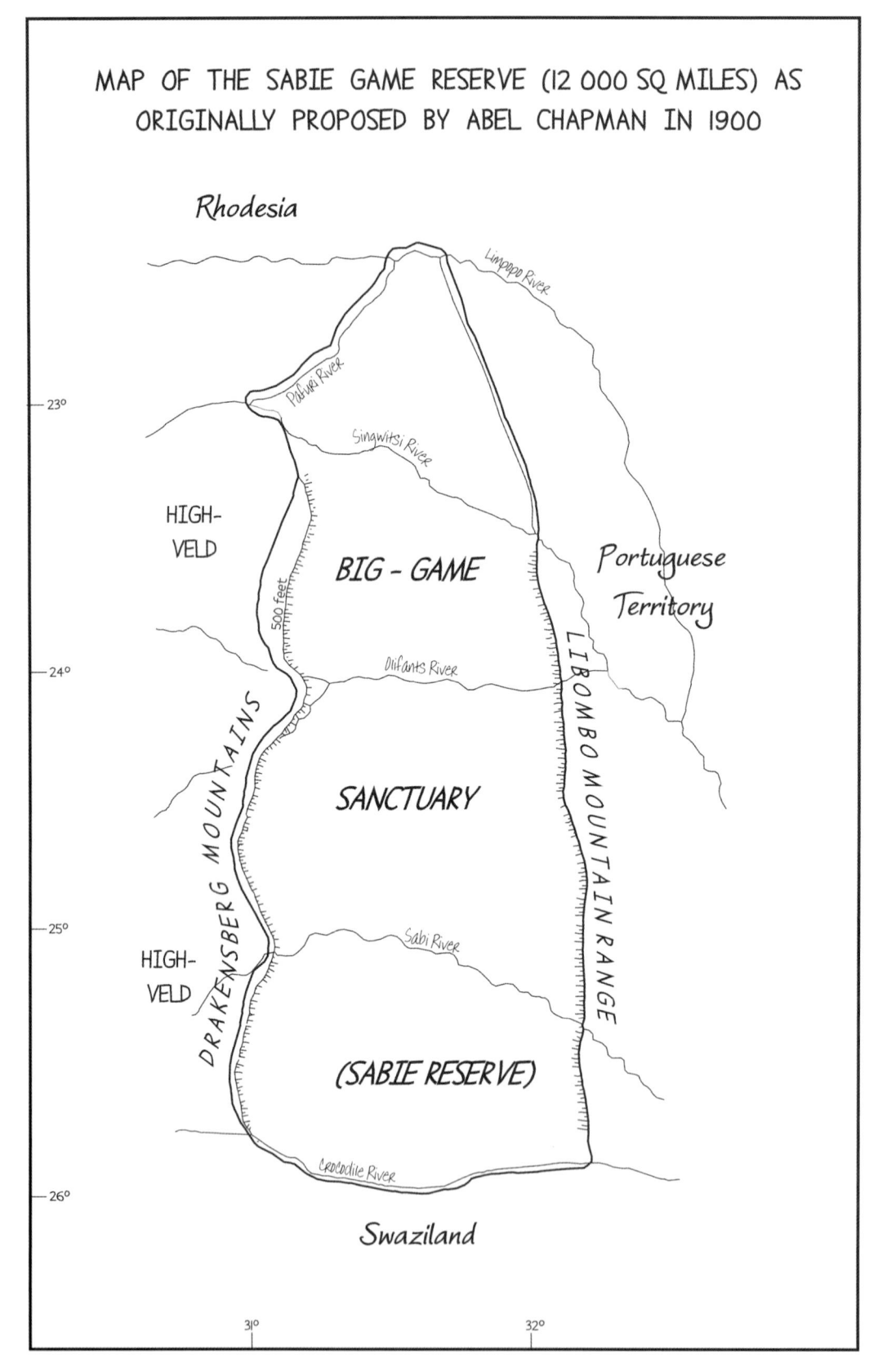

13. GAME CONSERVATION IN THE TRANSVAAL AND THE PROCLAMATION OF THE SABIE GAME RESERVE

Sketch-map of the Sabie Game Reserve as originally suggested by Abel Chapman in 1900.

establish a large game reserve along the Sabie River where hunting would be prohibited and where the Republic's big game would be protected for generations to come.

It is clear, if Robertson's account is genuine, that President Kruger may already have had his eye on a large game reserve in the Lowveld – an ideal that would gain momentum in later years after many petitions from the Lowvelders themselves, including the pioneers Henry Glynn, J.C. Ingle and F. Vaughan-Kirby, Native Commissioner Abel Erasmus and especially the Volksraad member for Barberton, R.K. Loveday.

The popular author and poet C. Louis Leipoldt (1937) also wrote about an interview he had had with Charles Marais, one of the ZAR's first surveyors. Marais apparently told him that he had informed President Kruger of the large-scale eradication of game in the Lowveld after returning from a trip to the Sabie area.

The president queried him at length and at Marais's insistence that something be done, the president answered that he would see what he could do, if things were as bad as Marais claimed. According to Marais, President Kruger stuck to his promise by proclaiming the Sabie Game Reserve.

In the years after 1892 matters also took a positive turn in the Volksraad. By 1894 there were so many petitions and submissions about wildlife matters that the Executive Council asked the Volksraad to consider another amendment to the existing game and hunting legislation. When the matter came up in the Volksraad there was much less discussion than in the past, as members were already well versed in both the legislation and the arguments for stricter wildlife protection measures. A fact that furthered the cause was the gradual disappearance of game, even in the far-off frontier districts. The biggest obstacle remained the ownership of game on private property. In an effort to bring an end to this argument, C.J. Tosen, the member for Piet Retief, suggested that a ban should be placed on hunting on all state land, but that no restrictions be placed on private landowners. This is the same suggestion President Kruger had made in 1889.

Jan de Beer once again claimed that many of the inhabitants of the Waterberg district could not afford to buy land (even at the rock-bottom prices of those years) and that they would have to beg for a living if they were to be prohibited from hunting on state land. J.H. de la Rey, of Lichtenburg, was also concerned that the suggested ban on hunting on state land, as well as the high licensing fees, would benefit the rich while the livelihoods of poorer burghers would be under serious threat.

Evidence obtained by S.W. Burger of Lydenburg showed that the existing laws were being completely ignored. He submitted proof that people in his district were still hunting with the aid of 50 to 100 black hunters only to collect skins. After a few days' debate, during which they voted on every clause, all the suggested amendments were accepted.

The 1894 Act at last stipulated that the term "for own use" meant no more than 15 to 20 head of game. The law also said that all field cornets and native commissioners would be ex officio rangers, and that they had to report on game conditions in their areas of control on a regular basis – a measure suggested in 1892 by E.P.A. Meintjies, a prominent Pretoria resident.

Apart from that, the power vested in the Executive Council in 1889 to proclaim game reserves was extended by the right to prohibit the hunting of certain rare species in the Transvaal by way of proclamation. Before 1889 this regulation only applied to oribi.

The Volksraad was, however, not consistent in accepting stricter legislation in the 1890s. In June 1895 a petition by 50 inhabitants that hunting be prohibited in the Waterberg district for a period of five years, was rejected. The petition had been sent without the knowledge of the Volksraad member for the district, Jan de Beer, and his arguments about the poverty of the burghers in the district carried much weight in Volksraad. The petitioners were told to go straight to the government with their request, because it could be put into practice only by way of proclamation and did not need the Volksraad's approval. At the same time, a petition by 52 farmers from the Potchefstroom district was considered. This petition asked that landowners only be allowed to hunt game on their own land after it was determined that the animals had been there for at least a month. The request was aimed at preventing neighbours

President Paul Kruger, circa 1883, which was the time when he started getting involved with the cause of nature conservation.
(ACKNOWLEDGEMENT: STATE ARCHIVES)

from luring animals onto their land to hunt them. This petition was also rejected.

When it became apparent that stricter legislation alone would not prevent the wholesale slaughter of the country's wildlife, the alternative – creating game reserves – started enjoying increasing prominence.

Their exact purpose and role in the country's economy still had to be determined.

In as early as 1888 the Transvaal Lowveld was singled out as an area worth preserving. One Williams, a Bloemfontein farmer who knew the Lowveld well and had a nose for business, suggested that a game reserve should be established in the Barberton district and then be hired out to British sport hunters at great expense. Others had similar ideas, and in November 1890 two well-known republican officials made a serious attempt to have part of the eastern Transvaal set aside as a game reserve. One of them was Commandant G.J. Louw, a special officer of the peace in the Komatipoort ward. In his petition to the government that a game reserve be proclaimed between the Crocodile and Sabie rivers, he pointed to the success of such conservation areas in Europe and the Cape Colony. In the same month (November 1890), the native commissioner and Voortrekker descendant Abel Erasmus, who grew up in the Ohrigstad district, suggested closing more or less the same area to hunting. No game should be hunted here except with the president's permission.

The Volksraad did not react to either of these two proposals and did not even answer their letters. The whole issue of wildlife conservation dragged on for another three years until November 1893, when G.P.J. Lottering of Bethal repeated the request that a part of the eastern Transvaal Lowveld along the Portuguese border be closed to hunting and that he be appointed as warden of the reserve. Two years later, in 1895, the state secretary got a letter in a similar vein from H.T. Glynn, a gold mine owner and hunter from Sabie, and Frank Streeter, the customs official at Komatipoort. The government at last paid attention to this series of requests and asked Abel Erasmus and H.F. van Oordt, the warden of the Pongola Game Reserve, for their opinions about these recommendations. Both men wholeheartedly supported the idea of a second game reserve in the Transvaal.

In January 1895, H.P. van der Walt asked that a game reserve be proclaimed on a number of state-owned farms in the Waterberg district. After a delay of three years – and to the consternation of Jan de Beer, the member for Waterberg, when he realised what his constituents had achieved without his knowledge – the government reacted to Van der Walt's request. In a proclamation issued in April 1898 hunting was banned on 12 farms in the Waterberg district. A game ranger was even appointed to keep an eye on the area. Jan de Beer was successful in his counter-proposals that, at the discretion of the ranger, poor people should be allowed to shoot one antelope a week in the reserve for own use.

The petition for the proclamation of a game reserve in the Waterberg was not the only one received in 1895.

The most crucial decision the Volksraad took with regard to wildlife conservation at the end of the nineteenth century was Article 482 of 1889 of the Executive Council and the subsequent Article 1244 of the Volksraad, which laid down the principles for the establishment of game reserves. This decision by the Volksraad provided the Executive Council with the power to proclaim game reserves on state land in the Transvaal without having to refer such decisions to the Volksraad for confirmation.

With these powers the Executive Council issued Proclamation R8009/89 of 13 June 1894, making the Pongola Game Reserve the first official game reserve in the Transvaal. In January 1895 the Executive Council also took note of the request to establish a game reserve in the Waterberg. After a petition by A.D.W. Wolmarans in August 1894, hunting was prohibited for three years on the commons of Pretoria and the farm Groenkloof, (on which the Fountains Valley is situated and where the current head office of the South African National Parks is located) by means of a proclamation in 1895. In 1898 this prohibition was extended for another three years. None of these proclamations had been preceded by a discussion in the Volksraad as the Law of 1889 had made it unnecessary.

The following events, which led to the tabling of a motion in the Volksraad through which the Executive Council would be requested to establish a game reserve in the Lowveld, are therefore not very clear.

Two members of the Volksraad – J.L. van Wijk,

Gen. Koos de la Rey, member of parliament for Lichtenburg, who voted against certain aspects of President Kruger's conservation legislation in Parliament.
(ACKNOWLEDGEMENT: TRANSVAAL ARCHIVES)

for Krugersdorp, and R.K. Loveday, for Barberton – gave notice during a session of the Volksraad on 6 September 1895 that they would table a motion asking the Executive Council to proclaim a game reserve in the Lydenburg district.

The area they had in mind was between the Crocodile and Sabie rivers, bordered in the west by a line running north-south along the Nsikazi River past Pretoriuskop and in the east by the international border on the Lebombo Mountains. This was the same area mentioned by Kruger in 1892 to his friend Robertson, and part of an area he had pointed out in 1889 as being ideal for a large game reserve.

On 9 September the motion was submitted by Van Wijk (seconded by Loveday) and it was decided that the matter would be discussed the following week. This agreement was reached with only a slight majority in favour (12 votes against 11) – until then the closest vote in the Volksraad about wildlife matters.

It can be assumed that President Kruger supported Van Wijk and Loveday in their submission and that the three must have felt rather uncertain about the final verdict on their proposal.[2] The reserve could have been proclaimed by the Executive Council without interference from the Volksraad, but it is possible that the president was uncertain about his support in this body and preferred to manoeuvre the matter in a round-about way through the Volksraad. Although the president was a veteran politician and had already been elected as president for three consecutive terms, the race between him and General Piet Joubert had been very close in the 1893 elections. The president's stature and popularity among voters of the Transvaal was greatly boosted by his handling of the Jameson Raid in 1896, and then just went from strength to strength. It may be that 1895 represented a low point in his career and that he met with opposition in putting his plans and ideas into action.

As the case may be, the motion by Van Wijk and Loveday was raised in the Volksraad on 17 September 1895. The president, Van Wijk and Loveday had probably launched a successful caucus campaign among the other members, because it is surprising that there was no debate about the matter and that no counter-proposals were submitted, especially considering that on 9 September there was only a slight majority in favour of discussing the motion. In the final submission to the Volksraad Loveday proposed the motion, seconded by Van Wijk, and it was adopted with enthusiasm ("bij acclamatie werd aangenomen"). No vote count was minuted, but it is clear it was not unanimous since three members asked that their dissentient votes be noted. One of them was J.M. Malan, the member for Rustenburg, who in the past had supported most legislation in connection with wildlife and hunting matters and who was also the chairperson of several game commissions. He gave no reason for his opposition. The other dissenters were A.D.W. Wolmarans, of Pretoria, and, as expected, the old Waterberg warhorse J. du P. de Beer. Both argued that it would be unfair to curtail the hunting activities of white burghers while there was no control over hunting by blacks in their own territories.

This decision by the Volksraad led to the immediate establishment of the Sabie Game Reserve (as it would become known). The Volksraad only asked the Executive Council to proclaim a game reserve in the Lydenburg district. Another few years would pass before it became a reality, as had been the case with Pongola and Waterberg.

In the two years after the decision to proclaim the Sabie Game Reserve, game matters were pushed to the background by two separate incidences. The first was the Jameson Raid in the first week of January 1896 and the second the rinderpest epidemic, which spread like wildfire through the country from April 1896 to February 1897 and killed hundreds of thousands of cattle and susceptible game species such as buffalo, kudu and eland. The Executive Council even asked the Volksraad to lift certain hunting bans for three years to ensure that the impoverished population had a food source. The request was granted.

When, after two years, nothing had been done about the proclamation, R.K. Loveday confronted the Executive Council during the Volksraad session in November 1897. He wanted to know why no proclamation had yet been promulgated. The fol-

---

2   Tattersall (1975) and other sources claim that President Kruger stated in the original debate: "If we do not close this small part of the Lowveld, our grandchildren will never know what a kudu or a lion or an eland look like." No reference to this statement of the president could, however, be found in the official minutes of the Volksraad.

Copies of the *Staats Courant* (Government Gazette) of 1898 in which the official proclamation of the Sabie Game Reserve by President Paul Kruger was published on 26 March 1898.

(ACKNOWLEDGEMENT: STATE ARCHIVES, PRETORIA)

lowing month the Executive Council extensively discussed this, the many recommendations it had received, and the verbal communications from the surveyor-general and the mine commissioner of Barberton.

At long last, the historic proclamation R8748/95 was promulgated by the president on 26 March 1898. It was published in the *Staats Courant* on 13 April 1898. Dr J.W.B. Gunning, the director of the State Museum in Pretoria, was placed in charge of the running of the new reserve. In April 1898 he wrote to the superintendent of education and recommended that the German Paul Krantz, the museum's taxidermist, be appointed as the first warden of the reserve because of his intimate knowledge of game and of the area itself.

The government was not in a position to appoint somebody, and once again Loveday had to ask the Volksraad why a warden for the Sabie Game Reserve had not been appointed yet. The inhabitants of the eastern part of the Transvaal were also becoming impatient about the matter, especially since poaching by Mozambicans was on the increase in the region. In September 1898 the Executive Council agreed to appoint a warden, but because of a lack of funds the tasks of patrolling the area and enforcing the hunting laws fell to two ZAR police force members (Zarps). They were Izak Holtzhausen, stationed at Komatipoort, and Paul Bester, of Nelspruit. (See Chapter 16.2 for more information about these two). It should be mentioned that the fine for poaching in the Sabie Game Reserve was staggering for the time: £150 or a 12-month jail sentence!

While steps were being taken to establish the Sabie Game Reserve, other measures were also implemented to strengthen and expand game conservation in the Transvaal.

In the 1890s several public bodies and societies were established aimed at ensuring wildlife and nature conservation. The Wildlife Protection Society has already been mentioned, but there were also the Natural History Society (1891), a branch of the Society for the Protection of Animals (SPCA) that had been established in 1895, and the vibrant State Museum in Pretoria. Game and conservation matters also started attracting more publicity in the press.

The largely improved legislation and greater public involvement and awareness contributed greatly to a remarkable change in attitude towards game and nature. In 1898 the Executive Council, at the request of the burghers themselves, forbade hunting on the commons of Nylstroom (now Modimolle), Belfast, Marthinus Wesselstroom (Wakkerstroom) and all state land in the wards Hex River (Waterberg district), Elands River and Zwagershoek (Waterberg dis-

> R.2742/95.
>
> PROCLAMATIE door ZHEd.den STAATSPRESIDENT.
>
> Ik, Stephanus Johannes Paulus Kruger, Staatspresident der Zuid-Afrikaansche Republiek, met advies en consent van den Uitvoerenden Raad, blijkens art. 1011 zijner besluiten dd. 29 December 1897, uit kracht van Volksraadsbesluit art. 1244 van 1889, waarbij aan de Regeering machtiging werd verleend om sommige plekken Gouvernementsgrond, het jagen op wild geheel te beletten en daaromtrent de noodige regeling en strafbepalingen vast te stellen, en ingevolge opdracht aan de Regeering gedaan bij Eerste-Volksraadbesluit art. 1230 dd. 17 September 1895, een stuk Gouvernementsgrond tusschen Krokodil- en Sabi-rivier te proclameeren als wildtuin, onder de voorzieningen en machtiging verleend bij bovengenoemd Volksraadbesluit art. 1244 van 1889, maak bekend en proclameer mits dezen als
>
> GOUVERNEMENTS-WILDTUIN
>
> de Gouvernementsgronden gelegen tusschen de Krokodil- en Sabi-rivier, welke gronden begrensd zijn als volgt: Van het punt in Komati-Poort waar de Komati-rivier de grenslijn snijdt met de grens tusschen de Zuid-Afrikaansche Republiek en de Portugeesche bezittingen tot aan de Sabi-rivier, vandaar met de Sabi-rivier op tot waar de Noordelijke Zandrivier in de Sabi-rivier loopt; vandaar met eene rechte lijn naar Pretorius-Kop, en vandaar met de spruit af tot aan de Logies- of Sikasi-rivier, dan met deze rivier af tot waar die in de Krokodil-rivier loopt en vandaar met Krokodil- en Komati-rivier af tot aan de Portugeesche grens, het eerstgenoemde punt in Komatipoort.
>
> En voorts proclameer en maak ik mits dezen bekend, dat het een ieder verboden is, in dezen wildtuin zooals boven aangegeven, eenig wild of gevogelte te jagen en te schieten, op te zoeken of op eenige manier te bemachtigen, te verjagen, te verdrijven of op eenige manier te verontrusten.
>
> En ten slotte proclameer en maak ik bekend, dat overtreders hiervan zullen gestraft worden met inbeslagneming van alle bij hen gevonden wild en schiet- en jachtgereedschap en eene boete van ten hoogste Een Honderd en Vijftig pond sterling, of bij wanbetaling met gevangenisstraf van 12 maanden niet te bovengaande, met of zonder harden arbeid, zullende overtredingen in deze vervolgd en uitgewezen worden in termen van art. 19 van de jachtwet no. 5 van 1894.
>
> GOD BEHOEDE LAND en VOLK
>
> Gegeven onder mijne hand ten Gouvernementskantore op dezen dag der maand Maart van het jaar achttien honderd acht en negentig.
>
> Staatspresident
>
> Staatssecretaris.

trict) for five years. Big game in the Marico district was protected for the same period. In 1899 Middelburg, Potchefstroom and Bloemhof also asked that their commons be closed to hunting. Hunting was also banned for five years in Magatosland, north of the current Makhado (Louis Trichardt) in the Soutpansberg.

"Jachtopzieners" (rangers) were appointed in most districts and they handed in complete reports on game censuses and contraventions of the hunting laws in their wards. Licences were issued to hunters, but it is improbable that the few remaining records accurately reflect the number of hunters. In Barberton, for instance, a town with a large population as a result of the goldfields, the most licences the magistrate issued in one year, from 1892 to 1900, was 85.

It is clear there was very little game left to hunt, a fact amply demonstrated by one of the last conservation measures implemented by the government of the ZAR before the outbreak of the Anglo-Boer War in October 1899.

In June 1899 the magistrate of Vryheid reported that seven hippos had moved into the Pongola Game Reserve and settled at the confluence of the Pongola and Baviaans rivers. He believed that hunters posed a serious threat to their continued existence. With the help of warden H.F. van Oordt, J.S. Vermaak was employed the following month at a salary of £12 (then about R24) a month to guard the small herd and to ensure that they did not leave the reserve. The fact that the teeming hordes of game in the Transvaal had been thinned to such a pathetic extent that the few that remained had to be protected by a special guard, is an indication of the unbelievable destruction left in the wake of the "century of extinction".

On the other hand, the ZAR, underlying motives aside, had implemented several conservation measures by the end of the century that brought it in line with the global zeitgeist, and it deserves the undying gratitude of posterity.

The "century of conservation" had dawned! The United States of America had proclaimed Yellowstone as its first national park in 1872 and a variety of nature reserves had been set aside by Australia (1879), New Zealand (1855), Canada (1887) and Natal (1897). There were also plans for many more.

The burghers of the ZAR, from the highest government official down to the humblest farmer, had implemented timeous measures to preserve the small country's rich natural diversity for future generations.

The Volksraad members Loveday and Van Wijk, and of course President Kruger, undoubtedly had the lion's share in the struggle for and formative process of nature conservation.

In two recent publications, Carruthers (1994 and 1995) tries to depict President Kruger's role in the establishment of the Sabie Game Reserve as insignificant. All the honour goes to the president's political opponents – R.K. Loveday and J.L. van Wijk. There is no question that these two had indeed played a very important role in realising the vision of a conservation area in the Transvaal Lowveld, and they should be honoured for submitting the motion at the Volksraad session on 6 September 1895 that set the ball rolling. One can also not deny the effect of Loveday's continued pressure on the Executive Council, leading to the reserve's proclamation on 26 March 1898.

President Kruger's attitude to conservation, and to the establishment of sanctuaries for the ZAR's dwindling game resources, has been well known ever since his noteworthy address to parliament on 2 August 1889. His continued support of a scientific basis for the maintenance of both livestock and indigenous fauna is less well known. In 1892 the president, on the advice of Dr W.J. Leyds, gave permission for the establishment of a State Museum in Pretoria, along with the National Zoological Gardens, with the groundwork for the former already laid in 1897, with Dr J.W.B. Gunning as the first director and P.A. Krantz as his taxidermist. The cornerstone of this building, which would eventually become the Natural Cultural History Museum in Pretoria, was laid in 1899, but the building was only officially opened after the war, in 1904. Dr Arnold Theiler, a Swiss veterinarian, arrived in South Africa in 1891. He was appointed as the ZAR's state veterinarian in 1893 and would in time become one of South Africa's first-class pioneering scientists. He played a leading role in the containment of the rinderpest epidemic of 1896–97 and in the establishment of the Onderstepoort Veterinary Research Institute. The Geological Society of South Africa was founded in 1895 by David Draper, a ZAR burgher. In 1896 N.T. Nichols, another ZAR burgher, founded the Transvaal Agricultural Union and foresaw its eventual expansion to other areas.

It would therefore be naive to claim that opposition members in the Volksraad could have carried a motion "met acclamatie" if the president and the majority of his supporters in the Volksraad had opposed the concept or science of conservation. In fact, it was the president who repeatedly had to reprimand or call to order the opponents of wildlife preservation in the Transvaal, including Jan de Beer of Waterberg and Daantjie Taljaard of Standerton.

One also has to remember that even an esteemed member of the Volksraad such as General Koos de la Rey of Lichtenburg voted against some of the president's game laws. It is therefore not far-fetched to consider the possibility that, in a clever tactical move, the president used two of his political opponents to submit a motion about a concept that was still considered eccentric and for which general support in the Volksraad was uncertain. The fact that there was no debate about the matter and that the motion was carried with enthusiasm, increases the possibility of canvassing ahead of the vote and the full support of the president himself.

The Executive Council's long delay in announcing the proclamation of the Sabie Game Reserve can be ascribed to a chain of events that had a very negative impact on the ZAR, including the Jameson Raid in January 1896, the rinderpest epidemic of 1896–97, and the ZAR's preparations for possible resistance against the mighty British Empire's continued policy of conquest and its imperialist urges.

As war clouds gathered over the country in October 1899, there was very little indication of the leading role the Transvaal would come to play in the area of nature conservation in southern Africa, and indeed the rest of the world.

References

Archival sources

Balliol College, Oxford: *Bodley papers*.
Kruger National Park, Skukuza: Pienaar, U. de V. The Kruger Park Saga, 1898–1981.
Transvaal Archives:
    Barberton Landdrost
    Eerste Volksraad (EVR) (First Volksraad)
    Commandant-General (CG)
    Education Department (ED)
    State Prosecutor (SP)
    Secretary of State (SS)
    Executive Council (EC)

Published works

Anon. 1881. The Boers at home. *Blackwoods Edenburgh Magazine*.

Atcherly, R.J. 1879. *A Trip to Boer Land*. Richard Bentley, London.

Aylward, A. 1878. *The Transvaal of Today*. Blackwood, Edinburgh.

Bigalke, R.C.H. 1966. South Africa's first game reserve. *Fauna and Flora* 17:13–18.

Bozzoli, B. (ed.). 1983. *Town and Countryside in the Transvaal*. Ravan, Johannesburg.

Bryden, H.A. 1894. The extermination of great game in South Africa. *Fortnightly Review* October.

Cachet, F. Lion 1882. *De worstelstryd der Transvalers aan het volk van Nederland verhaald*. J. Dusseau Uitgevers, Amsterdam.

Carruthers, E.J. 1985. The Pongola Game Reserve: an eco-political study. *Koedoe* 28:1–16.

Carruthers, E.J. 1985. Attitudes to game conservation in the Transvaal, 1900–1910. Seminar submitted to History Department, University of Cape Town.

Carruthers, E.J. 1988. Game protection in the Transvaal, 1900–1910. *South African Historical Journal* 20:35–56.

Carruthers, E.J. 1988. *Game protection in the Transvaal 1926–1946*. Ph.D. thesis, University of Cape Town.

Chapman, A. 1928. *Reminiscences and Impressions of a Hunter-naturalist in Three Continents, 1851–1928*. Gurney & Jackson, London.

Chase, J.C. 1843. *The Natal Papers*. C. Struik, Cape Town. Facsimile reprint 1968.

Churchill, J. 1979. *A Merchant Family in Early Natal: Diaries and Letters of Joseph and Marianne Churchill, 1850–1880*. A.A. Balkema, Cape Town.

Clements, K.A. 1984. Herbert Hoover and conservation, 1921–1933. *American Historical Review* 89(1):67–88.

Clepper, H. (ed.). 1966. *Origins of American Conservation*. John Wiley & Sons, New York.

Curson, H.H & Hugo, J.M. 1924. Preservation of game in South Africa. *South African Journal of Science* 21:400–424.

Delius, P. 1983. *The Land Belongs to Us*. Ravan, Johannesburg.

Fisher, J. 1974. *Paul Kruger*. Secker & Warburg, London.

Fuggle, R.F. & Rabie, M.A. (eds.). 1983. *Environmental Concerns in South Africa*. Juta, Cape Town.

Gillet, F. 1908. Game Reserves. *Journal of the Society for the Preservation of the Wild Fauna of the Empire* 4:42–45.

Glynn, H.T. 1926. *Game and Gold*. Dolman Printing Co., London.

Gunning, J.W.B. 1898. Letter (1 April 1898) from the Director of the State Museum to the Superintendent of Education, Pretoria, in which Paul A. Krantz was suggested as Ranger of the Sabie Reserve. Transvaal Archives E.D. 217 (O.R. 4296/98):44–47.

Gutshe, T. 1979. *There Was a Man: The Life and Times of Sir Arnold Theiler K.G.M.G. of Onderstepoort*. Howard B. Timmins, Cape Town.

Hofmeyr, S. 1890. *Twintig Jaren in Zoutpansberg*. Rose & Co., Cape Town.

Huet, P. 1869. *Het lot der Zwarten in Transvaal*. J.H. van Pearson, Utrecht.

Jeppe, F. 1868. "Die Transvaalsche oder Süd-Afrikanische Republik". Mittheilungen aus Justus Perthe's geographischen Anstalt über wichtige neue Erforschungen auf dem Gesammtgebiete der Geographie von dr. A. Petermann, 24 September.

Kleynhans, W.A. 1966. *Volksregering in die Zuid-Afrikaansche Republiek: die rol van die memories*. Van Schaik, Pretoria.

Kruger, D.W. 1961 & 1963. *Paul Kruger*, 2 vols. Afrikaanse Pers, Johannesburg.

Kruger, P. 1902. *The Memoirs of Paul Kruger,* 2 vols. T. Fisher & Unwin, London.

Kruger, S.J.P. 1947. *Gedenkskrifte van S.J.P. Kruger.* Bredell, H.C. & Grobler, P. (eds.). Van Schaik, Pretoria.

MacQueen, J. 1862. Journey from Inhambane to Zoutpansberg by Joaquim de Santa Rita Montanha. *Journal of the Royal Geographical Society* 32:63-68.

Marks, S. & Atmore, A. (ed.). 1980. *Economy and Society in Pre-industrial South Africa.* Longman, London.

Naude, P. 1954. *Boerdery in die Suid-Afrikaanse Republiek 1858–1899.* D. Litt. thesis University of South Africa, Pretoria.

Pelzer, A.N. 1950. *Geskiedenis van die Suid-Afrikaanse Republiek. Deel I: Wordingsjare.* A.A. Balkema, Cape Town.

Potgieter, F.J. 1958. Die vestiging van die blanke in Transvaal (1837–1886) met spesiale verwysing na die verhouding tussen die mens en die omgewing. *Archives Year Book for South African History* 21(2).

Pringle, J.A. 1982. *The Conservationists and the Killers.* Bulpin, Cape Town.

Rabie, A. 1976. *South African Environmental Legislation.* Unisa, Pretoria.

Roche, H.A. 1878. *On Trek in the Transvaal.* Sampson, Low. Marston, Searle & Rivington, London.

Sanderson, J. 1860. *Memoranda of a Trading Trip into the Orange River (Sovereignty) Free State and the Country of the Transvaal Boers.* State Library Reprint no. 91, Pretoria.

Silver, S.W. & Co. 1978. *Handbook to the Transvaal British South Africa.* S.W. Silver & Co., London.

Smit, F.P. 1951. *Die staatsopvattinge van Paul Kruger.* Van Schaik, Pretoria.

Staats Courant. 1863–1900.

Stevenson-Hamilton, J. 1907. Notes on the Sabi Game Reserve. *Transvaal Agricultural Journal* 5(20):866-871.

Stevenson-Hamilton, J. 1937. *South African Eden.* Cassell, London.

Stuart, J. 1854. *De Hollandsche Afrikanen en hunne Republiek in Zuid-Afrika.* G.W. Tielkemeijer, Amsterdam.

Tattersall, D. 1975. Wildbewaring in die Laeveld. In: Barnard, C. (ed.). *Die Transvaalse Laeveld: kamee van 'n kontrei.* Tafelberg, Cape Town.

Van Oordt, G.A. 1980. *Striving and Hoping to the Bitter End: The life of Herman Frederik Van Oordt, 1862–1907.* G.A. van Oordt, Cape Town.

Viljoen, P.J. 1925. *Die lewe van Paul Kruger 1825–1904.* De Bussy & HAUM, Pretoria & Cape Town.

The text in this chapter is largely the same as it was published in 1990 and has not been updated. Since 1990 there have been many different viewpoints on the subject. The reader is referred to the following sources for further information.

Behrens, H.P.H. 1946. Oom Paul's great fight to preserve game. *African Wild Life* 1(1):12-22.

Behrens, H.P.H. 1951. Paul Kruger – wildbeskermer, aspek van president se lewe wat selfs sy biograwe vergeet. *Huisgenoot* 12 October 37(1542):6-7.

Carruthers, E.J. 1994. Dissecting the myth: Paul Kruger and the Kruger National Park. *Journal of South African Studies* 20(2):263–283.

Carruthers, E.J. 1995. *The Kruger National Park. A social and political history.* University of Natal Press, Pietermaritzburg.

Carruthers, E.J. 2001. *Wildlife and Warfare. The Life of James Stevenson-Hamilton.* University of Natal Press, Pietermaritzburg.

Grobler, J.C.H. 1988. Natuurlik het Kruger 'n rol gespeel. *Beeld* 8 October.

Grobler, J.C.H. 1988. *Politieke leier of meeloper – Die lewe van Piet Grobler (1873-1942).* Scripta Africana, Johannesburg.

Grobler, J.C.H. Kruger hét ver gekyk. *Die Burger* 6 September:10.

Leipoldt, C.L. 1937. *Bushveld Doctor.* Jonathan Bell Cape Ltd, S.I.

Robertson, V.L. 1947. Early attempts to preserve wildlife. *African Wildlife* 1(2):21-23.

Van Oordt, H.F. 1898. *Paul Kruger en de Opkomst der Zuid-Afrikaansche Republiek.* Jacques Dusseau, Amsterdam.

Chapter 14
War clouds loom over the Lowveld 1899–1902
Dr U. de V. Pienaar

On 11 October 1899, President Paul Kruger declared war on the mighty British Empire after it had rejected his 9 October 1899 ultimatum to withdraw its troops massed on the borders of the ZAR (South African Republic). He was supported by President Marthinus Theunis Steyn of the Republic of the Orange Free State.

The Boer army invaded Natal and the Cape and besieged the British military bases at Mafeking (Mafikeng), Kimberley and later Ladysmith.

Despite the resounding successes the Boers achieved at first at battles such as Nicholson's Nek, Modderspruit, Chieveley, Willow Grange, Colenso, Spioenkop and Vaalkrantz in Natal, as well as Belmont, Graspan, Stormberg, Sannaspos and Magersfontein in the Northern Cape, the Boers failed to capitalise on their gains. This gave the British forces an opportunity to consolidate and put large numbers of troops ashore in Cape Town and Durban. Lord Roberts took over command of the British forces from General Buller. With his superior number of troops and weaponry he soon halted the advance of the Boer forces and then forced them into retreat. In February 1900 Kimberley was relieved by General French and Ladysmith by General Buller soon afterwards.

On 13 March Roberts's troops entered Bloemfontein. Twelve days later, on 27 March, General Piet Joubert, the Commander-in-Chief of the Boer army, died and General Louis Botha was placed in command. Mafeking was relieved on 17 May by Mahon and Plumer and on 31 May 1900 Roberts's troops marched into Johannesburg. Pretoria, the

Gen Sir Redvers Buller, Commander-in-Chief of the British forces during the battle of Dalmanutha, August 1900.
(ACKNOWLEDGEMENT: TRANSVAAL ARCHIVES)

Burghers crowd in front of the Royal Hotel in Pilgrim's Rest with the start of the Anglo-Boer War, 1899.
(ACKNOWLEDGEMENT: INFORMATION AND PROMOTIONAL SERVICES, TRANSVAAL PROVINCIAL ADMINISTRATION)

14. WAR CLOUDS LOOM OVER THE LOWVELD 1899–1902

President Paul Kruger in his railway carriage at Hectorspruit Station in 1900 during his journey to Lourenço Marques and exile.
(ACKNOWLEDGEMENT: PROVINCIAL ARCHIVES, PRETORIA AND MR H. BORNMAN)

Boer troops in their positions during the battle of Spioenkop, 23–25 January 1900.
(ACKNOWLEDGEMENT: SOUTH AFRICAN NATIONAL MUSEUM OF MILITARY HISTORY)

capital of the ZAR, fell into British hands on 5 June 1900. By that time, President Kruger and his staff had already moved to Waterval-Onder in the Lowveld.

On 21 July Roberts started marching to Komatipoort, while the Boer forces, after the demoralising surrender of General Cronjé at Paardeberg and General Prinsloo at the Brandwater Basin, desperately tried to regroup for the last great artillery battle of the war at Berg-en-dal (Dalmanutha) – the portal to the Lowveld.

After one of the heaviest artillery bombardments of the war on the key position Berg-en-dal, which was heroically defended by a section of the ZAR Police from Johannesburg, the Boer forces of generals Botha, Kemp and Viljoen succumbed once again before the superior British numbers. They were forced to retreat to the Lowveld via Lydenburg and the Eastern railway line.

On 7 September 1900 the British forces were deployed in a line from Carolina to Belfast, Waterval-Onder, Helvetia and Lydenburg, while the Boer commandos were regrouping at several different places, from Barberton to nearby Lydenburg. Lord Roberts thought the end of the war was near after trapping the Boer commandos against the Mozambican border near Komatipoort and forcing them to surrender, but he got a rude awakening thanks to the military acumen of the Boer generals, and the war would drag on for another two years.

President Kruger and his officials had in the meantime moved to Nelspruit, while General Botha and the larger commandos moved to Lydenburg and further north, in an attempt to escape Buller's encircling manoeuvre.

On 10 September 1900, the Executive Council of the Transvaal granted President Kruger six months' leave (which was later extended indefinitely), so he could travel to Europe to plead the cause of the Boer republics with sympathetic European nations. Due to his advanced age it was impossible for him to accompany the commandos. On 11 September 1900, the president said farewell to his men with a heavy heart. Accompanied by Piet Grobler (who could speak German, French and Italian) and a small following, he set off by train for Lourenço Marques.

One can only wonder at his thoughts as the beautiful landscape of the Sabie Game Reserve flashed by the train window – an area he had proclaimed on 26 March 1898 for future generations as a "Goewernements Wildtuin" after years of hard work.

After more than a month's stay in Lourenço Marques, the president and his entourage boarded the Dutch cruiser *De Gelderland,* sent for the purpose by Queen Wilhelmina, and soon they were on their way to the French port Marseille – and exile.

The departure of President Kruger also heralded a new phase in the struggle between Boer and Brit: the guerrilla war. The Lowveld was almost immediately drawn into the whirlpool of violence.

After the battle of Berg-en-dal (Dalmanutha), generals Kemp and Ben Viljoen, with commandos from Krugersdorp and Johannesburg respectively, moved in the direction of Waterval-Onder to cover the commandos' retreat along the railway line. When they reached the Godwana River they were met by President Steyn, who addressed and encouraged them with his inspirational enthusiasm.

Unfortunately President Steyn's heartening words could do little to alleviate the chaos at railway stations such as Nelspruit, Crocodile Poort, Kaapmuiden and Hectorspruit caused by fleeing people, hundreds of trainloads of war provisions and goods, horses, draught animals, etc. General Ben Viljoen grippingly described the confusion of people, rolling stock and goods in his war memoirs. He and his commando had been ordered by the ill General Botha to set off forthwith for Hectorspruit and meet him there.

All along the way, through Nelspruit, Crocodile Poort Station and Kaapmuiden, he encountered

Lord Roberts of Kandahar, Commander-in-Chief of the British forces during the first phase of the Anglo-Boer War.
(ACKNOWLEDGEMENT: TRANSVAAL ARCHIVES)

Gen. Ben Viljoen, Commander of the Boer forces in the southern Lowveld between August 1900 and his capture in January 1902.
(ACKNOWLEDGEMENT: TRANSVAAL ARCHIVES)

hordes of people rushing around aimlessly, with some getting as drunk as possible and others trying to steal goods from the loaded trains – goods they eventually had to bring back as there was no way to transport it somewhere else. Loosely translated from the Dutch, this is how he described the situation:

> These sad incidents are unsettling and contribute nothing towards the state of affairs, or the spirit of the burghers. The main cause was not the fact that the officers had little control over the people; it was as if the devil had been set loose and was doing his best to entertain himself among us.
>
> Hectorspruit is the second last station before reaching the Portuguese border and about 70 miles from Ressano Garcia. The commandos here planned to move northwards and then westwards and across the mountains near Lydenburg. On the very day I arrived at Hectorspruit, President Steyn left along this route, accompanied by an escort of 100 men.
>
> On 16 September 1900 at Hectorspruit one could observe an incident that is difficult to describe. Hectorspruit is one of the many rather unattractive stations along the Delagoa Bay railway line, situated between the Great Crocodile River and the sombre, black hills or foothills of the "Kaapsche" Mountains and the Lowveld, better known as the Bushveld. This area is overgrown with black thorn bushes, where one, if in the middle of the bush, can see only the sky and the place on earth where one is standing ...
>
> The station was filled with train carriages, loaded with all types of goods and an unknown number of passenger coaches. The platform and its vicinity swarmed with people, all in a great hurry. The one packed in and the other carted out, one stood sniffing in the doorway while the other was looting. Most, however, wandered around aimlessly. They did not know what to do, what was going on, or what should be done. Only the officers were keeping themselves busy handing out food, ammunition, etc.

In the meantime, General Viljoen had received an order from General Botha to call the burghers together and tell those who no longer had a horse (about 1800 men) that they would be taken to Komatipoort by train. There they would prepare for a final stand against the advancing columns of General Pole-Carew. Those with horses would be led by General Viljoen through the southern Lowveld to Leydsdorp, which would temporarily serve as the seat of government.

General Viljoen's order was a difficult one and was greeted with bitterness and sorrow by those doomed to travel to Komatipoort with only one Long Tom (Schneider siege gun) and two Armstrong cannons taken from the British forces at Sannaspos. General Coetser and a few hundred burghers resisted the order and escaped northwards along the Lebombos.

Before setting off into the Lowveld, General Viljoen blew up the remaining heavy artillery (about ten Krupp and other cannons) on the banks of the Crocodile River at Hectorspruit. When he later wrote about this he also mentioned that they had buried the rest of their equipment, "which will probably only be rediscovered by later generations".

These were prophetic words, because on 20 June 1970, former ranger Jan de Kock of the Malelane section, found the chassis of a cannon (which over the years had lost its barrel) in the bed of the Crocodile River about three kilometres west of the old Nellmapius Drift. He also found an axle with brass hubs, a wheel hoop and part of a shaft, all of which were sent to Skukuza for safekeeping. At a later stage, while the river was very low (on 30 August 1983), another attempt was made to find the remains of General Viljoen's heavy artillery. Rangers J. van Graan and E.A. Whitfield were lucky enough to find a whole and remarkably well-preserved cannon wheel in the river bed about three kilometres west of the former Lwakahle ranger outpost. A memorial tablet was erected next to the tourist road between Malelane and Crocodile Bridge at the closest point to this site, marking the spot where this dramatic event took place.

Penning (1904) described the fate of the Boer contingent that had travelled to Komatipoort by rail. On 18 September this demoralised Boer army held a council of war led by General Coetser, Commandants Pienaar, Schreuder and Ricchiardi, and Justice of the Peace J.H. Janson. The council decided that the available forces had to be organised immediately to offer the maximum amount of resistance to the approaching British army.

The wheel of a Krupp or Armstrong canon of the Boer forces which was blown up on 16 September 1900 on the banks of the Crocodile River opposite Hectorspruit to prevent it from falling into the hands of enemy forces. (AUG. 1983)

14. WAR CLOUDS LOOM OVER THE LOWVELD 1899–1902

The three cannons, including the remaining Long Tom, would be set up on the far side of the Crocodile River while the men would be instructed to fight to the last, but it would seem the Long Tom was too heavy to take across the river.

According to J.J. Claassen, an eyewitness, the Long Tom was set up between the bridge and the Portuguese railway station at Ressano Garcia, right against the Lebombos at a distance of about 800 paces from the railway line. (Information obtained from a letter in *Die Huisgenoot* 1937.)

They did try to transport the two remaining cannons (the Armstrong and a Nordenfelt-maxim) across the Crocodile River, but these became hopelessly stuck and were left there until the following morning.

During the night, General Coetser, accompanied by the remaining 150 mounted men and a number of wagons, left the Boer emplacements. Some of the other commandants and their men surrendered to the Portuguese troops that had gathered east of the border. Before General Pole-Carew's forces reached Komatipoort, where he captured the remaining burghers without meeting any resistance, the command had been given that the cannons and 200 railway carriages containing provisions and ammunition should be either blown up or burned. It took 20 cases of dynamite to destroy the cannons.

According to Claassen (1937) of Hopetown, the Long Tom was first buried with a load of dynamite at its emplacement on the southern side of the river and then blown to pieces. (But his date of 5 August 1900 is incorrect – this took place on 19 September 1900.)

In his classic narrative on the fate of the Long Toms during the Anglo-Boer War, *Silence of the Guns*, Changuion (2001) described the last resting place of this Long Tom as follows (loosely translated):

The wheels of a Nordenfeldt-maxim canon which the Boer forces blew up on 19 September 1900 in a drift through the Crocodile River opposite Komatipoort Station.
(JULY 1986)

A Schneider siege gun (Long Tom) of the Boer forces in action during the siege of Mafeking (now Mafikeng), October 1899.
(ACKNOWLEDGEMENT: TRANSVAAL ARCHIVES)

14. WAR CLOUDS LOOM OVER THE LOWVELD 1899–1902

A live 6-inch Long Tom grape-shot with two Mauser rifles and a Martini-Henry bandoleer. Former ranger A. de Clerck found three similar shells in the bed of the Crocodile River close to its confluence with the Komati River in August 1965.
(ACKNOWLEDGEMENT: SANDF ARCHIVES, PRETORIA)

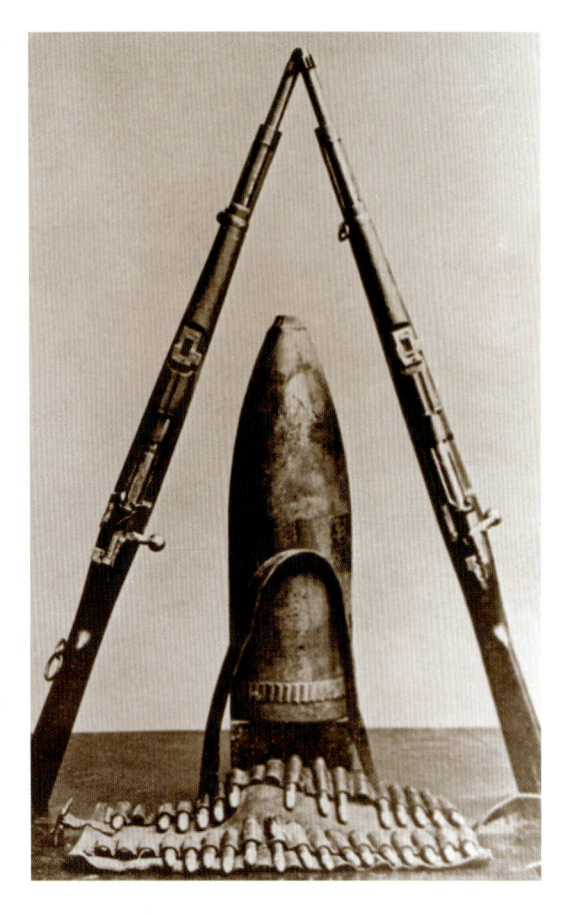

The location in the Crocodile River between Nkongoma and Bob, where former ranger A. de Clerck found a number of Long Tom and Armstrong shells in the water in August 1965.
(27.07.1984)

Wonderboom Tom and a few other guns from Barberton, not via the main line through Hectorspruit, arrived at Komatipoort early in September 1900. General Francois Pienaar and all those burghers who had decided to leave the country were to keep the British occupied here for long enough to give the commandos time to make good their escape into the Lowveld. The initial decision had been that the burghers of General Pienaar would dig in, and the three pieces of artillery – the Long Tom, an Armstrong and a Nordenfelt-Maxim – were assigned to their chosen emplacements. The Long Tom was off-loaded from the train, and because it was too heavy to be dragged over the rough terrain and through the river, it was placed at the bottom of a nearby foothill of the Lebombos, south of the railway line and about 800 paces from the confluence of the Crocodile and Komati rivers.

Many Boers, however, left Komatipoort in the next two days, making the intended defence of the terrain futile – especially seeing as the British forces under the command of General Pole-Carew were already at Hectorspruit.

Before the Boers finally evacuated Komatipoort on 22 September, they decided to destroy the weapons and ammunition they could not take with them – this included the Wonderboom Tom. Schreuder, one of the Boer commandants, had brought dynamite for this purpose from Barberton and gunner Rode was instructed to organise the destruction of the cannons. Many of the foreign volunteers who had fought on the Boer side were also still with these commandos at Komatipoort.

One of the toughest of these men was an American named John N. King, who was also known as 'Dynamite Dick' because of his knowledge of explosives and his extensive experience in blowing up bridges and trains. His last job for the Boers was now to deliver the *coup de grâce* to the Long Tom. And so the first of the four Long Toms came to its end.

Two days later Pole-Carew's British forces arrived in Komatipoort and found the remains of the long Tom and the other cannons, including 40 shells for the Long Tom, in the railway yard. The British were disappointed at having failed to capture a Long Tom, and sent a telegram to London dated 25 September that summed up the situation: 'Official: one Long

Tom – one Creusot gun – found destroyed at Komatipoort.'

The war correspondent L. Penning, of *De Telegraaf*, wrote that under these circumstances the Republic greatly appreciated the services of the Dutch, who had acquitted themselves well in those dangerous hours. He made mention of F.C. Keuzekamp, Vierdag, Paré van Egmond, Voogt and Kerminga, but said there had been many others that deserved being mentioned.

It seems that the Armstrong cannons and the Nordenfelt-maxim were blown up in the drift through the Crocodile River. Marks on the rocks in the drift indicate that a big explosion took place here. When pieces of the cannons were later found in the drift through the Crocodile River at Nkongoma it was confirmed as the final resting place of the remaining Boer artillery.

In August 1965, when the water level in the Crocodile River was particularly low, former ranger A. de Clerck found three big cannon shells, and a smaller shell, about one kilometre upstream of its confluence with the Komati River. The larger shells were undoubtedly those of the Long Tom and the smaller one that of an Armstrong. Unfortunately this valuable and historic find was handed to the local police, who destroyed it.

The water level in the Crocodile River was very low again in July 1986. This time the wheels of an Armstrong and Nordenfelt-maxim and other iron remnants of the blown-up cannons were found in the shallow water, just below a natural rocky drift through the river east of the old Nkongoma post, opposite Komatipoort Station. Signs of the use of explosives were also found on the slab of rock, and this, along with the remains of the Boers' artillery in the river bed of the Crocodile River, provides ample proof that this is where the last of the Boer cannons (except the Long Tom) were destroyed on 19 September 1900.

The Long Toms, of which the Boers had four, were 155 mm Le Creusot cannons of French manufacture. They had originally been imported to protect the forts around Pretoria and had even longer ranges than some of the large ship's cannons of the British. They became celebrated in the early part of the war in Natal and many a battle was won by the Boers because of their accurate bombardment. They were also used during the siege of Ladysmith. Three of these famous big guns were used together for the last time at the battle of Dalmanutha and afterwards to try to stop the advance of the British forces.

In the opinion of the British, and even Lord Roberts himself, the war was finished after Komatipoort had fallen. Little did they realise that this was only the beginning of an exhausting guerrilla war that would last another 20 months and claim thousands of lives on both sides.

General Ben Viljoen, who would take over command of the scattered Boer commandos in the southern part of the Transvaal Lowveld after leaving Hectorspruit, was a particularly ingenious and daring military leader. He would remain a thorn in the side of the British forces until his capture in January 1902. Viljoen was a feared and respected opponent, particularly popular with his own burghers and even the common "Tommies" thanks to his sense of humour. For example, after he had taken the Lady Roberts, an enormous ship's cannon, from the British at Helvetia, he wrote to General Smith-Dorrien at Belfast (with a jab at the recent British move to banish women from Pretoria as they were considered "undesirables"):

I have been obliged to expel the 'Lady Roberts' from Helvetia, this lady being an 'undesirable' inhabitant of that place. I am glad to inform you that she seems quite at home in her new surroundings, and pleased with the change of company.

Ben Viljoen was also a good judge of human nature who did not close his eyes to the failings of his compatriots. During their journey from Hectorspruit through the arid Lowveld, food was scarce and he wrote as follows (loosely translated):

Game is abundant in this region. One can see wildebeest, hartebeest,[1] impala, etc. in groups of five to

---

1   According to Milstein (1986) General Viljoen could have been referring to Lichtenstein's hartebeest. These had been seen in the vicinity of the Sand River, near the farm Orinoco of Abel Erasmus or just adjacent to it, where taxidermist Paul A. Krantz had collected two Lichtenstein's hartebeest for the State Museum in Pretoria, according to indications in the Transvaal Museum records, while on an expedition to Abel Erasmus's farm in 1894.

20 and at night the roaring of lions and the cries of wolves [hyenas] can be heard. Some have even had encounters with lions at daytime. One of the failings of the Afrikaners, and definitely their worst weakness, is that they cannot help shooting when they see game and they couldn't care less if it is legal or not. Each commando appointed hunters to shoot for the pot, but many others went in secret so the dense bush was crawling with hundreds of hunters. As a result it sounds as if a real battle is going on when a bunch of burghers come across game.

An encounter with a lion near the Blyde River during one of these unauthorised hunting expeditions resulted in one burgher getting the fright of his life. Viljoen's description of this small drama is a priceless bit of comedy in such a miserable time:

One afternoon a burgher, whose name I won't mention, arrived with his clothes in tatters and without his hat and rifle. I had just finished bathing in the drift and was busy getting dressed. I could hear some burghers laughing at him from a distance, so I yelled at him: 'Man, what on earth did you see? It looks as if a devil in disguise ambushed you!'

His hair stood on end and his voice trembled badly when he answered: 'Lordy, General, I nearly died. I walked into the veld this morning to shoot some game and I wounded a blue wildebeest, which ran into a bush. I could see the animal was dying so I followed it, but the bush was so dense that I could only see right in front of me and no further. I struggled on like this, following the blood spoor, and saw the wildebeest where it was lying. When I climbed over a fallen tree trunk I put my rifle down, and just as I was lifting my leg over the trunk I saw a huge lion standing with one front paw on the wildebeest. Oh lord, I thought, I have seen my last day, because he looked me straight in the eyes and gave a small little roar. I swear General, if it had been an Englishman I would have surrendered. I whipped round and hotfooted it out of there. I left my rifle, hat and everything else behind and I wouldn't go back for all the money in England. The General may as well punish me for losing the rifle, because I'm not going back to fetch it.'

I asked him what the lion did next, but he didn't hang around long enough to see 'next'. One burgher who was standing nearby remarked: 'I think this guy saw a dog or a meerkat; he ran past me first and he was so scared that his own mother wouldn't have recognised him. I think that if I had asked him then, he wouldn't have known his own name.'

The terrified man was by now so indignant that he immediately offered to go and show us where the lion had been, or at least find its spoor. There was no time, however, and the blockhead had a difficult time of it as everybody poked fun at him. He was immediately christened 'Vaalverschrikt' (White-with-fright).

Deneys Reitz (1929), in his book *Commando*, also described conditions in the Lowveld in 1900 when about 5000 Boer riders, after their regrouping at Hectorspruit, moved in the direction of Leydsdorp:

Our road ran through the Sabi low country teeming with big game of all descriptions. By day great herds of zebra, wildebeest and sable stood fearlessly gazing at us, and at night lions prowled around our camps. Of hunting we had our fill and to me the journey through a strange and remote region was full of fascination, for we were passing through country as untouched as that upon which the old pioneers had looked when first they came north in the days of the Great Trek.

Shortly after the outbreak of the Anglo-Boer War, when the Boer forces were still in control of the Transvaal, the British high command came up with a plan to blow up the railway line between Pretoria and Delagoa Bay, cutting off the ZAR from the rest of the world. Bulpin (1950) writes of quite a few failed attempts. In December 1899 a valiant effort was made by a group of British subjects from Lourenço Marques. A small force under the command of Captain David Forbes junior also tried and failed in April 1900. As a result of these failed attempts the Boer commanders reinforced the troops guarding the railway line.

A more ambitious effort by British strategists followed. At the end of May 1900, 600 men of Strathcona's Horse were transported by ship to Durban, where two companies landed before marching overland to Swaziland. A third company would land at Kosi Bay, from where they would prepare the way for the larger force, with David Forbes as guide.

But the British high command under Lord Roberts received news that the railway bridge at Komatipoort was guarded by a strong force of 300 Boers. The warship *Widgeon* waited in vain at Kosi Bay for the appearance of Forbes and his friend Brooks. On 5 June 1900, Roberts sent a telegraph to Lord Milner saying that the expedition had been called off. Forbes and his friend later arrived at Eshowe in Zululand, close to dead from malaria and hunger.

At about the same time General Redvers Buller, Commander-in-Chief of the British forces in Natal, became interested in a certain bombastic German-Bulgarian adventurer – Sergeant Francis[2] Christian Ludwig Steinaecker from Port Shepstone, born in Berlin in 1854. He had served in the Prussian army from 1871 to 1879. After living in Bulgaria (at the time of Prince Alexander von Battenberg) and German West Africa (Namibia), he arrived in Port Shepstone in Natal in 1890 and joined the Colonial Scouts when the war broke out. Steinaecker was always boasting of his own and his German compatriots' ability to use explosives effectively, which so impressed Buller that he ordered Steinaecker to try to blow up the Eastern Line. Accompanied by six men and a mule laden with dynamite, Steinaecker (now a lieutenant) travelled through Swaziland to the

A group of Steinaecker's Horse officers at Komatipoort, 1902. The person on the far left is Captain E.G. (Gaza) Gray, who later became a ranger in the Sabie Reserve. Colonel L. von Steinaecker is the man in the middle without a horse. The horse, second from right was Tommy, which Harry Wolhuter later sold for £40.
(ACKNOWLEDGEMENT: STEVENSON-HAMILTON COLLECTION, SKUKUZA ARCHIVES)

Steinaecker's Horse's blockhouse at Komatipoort (1900–1902). Major Stevenson-Hamilton took the photograph in October 1905.
(ACKNOWLEDGEMENT: THE STEVENSON-HAMILTON COLLECTION, SKUKUZA ARCHIVES)

---

2   His father had been a colonel in the German forces and he was christened Franz Christian Ludwig. It would seem that he took the name Francis after becoming a British citizen in 1897.

14. WAR CLOUDS LOOM OVER THE LOWVELD 1899–1902

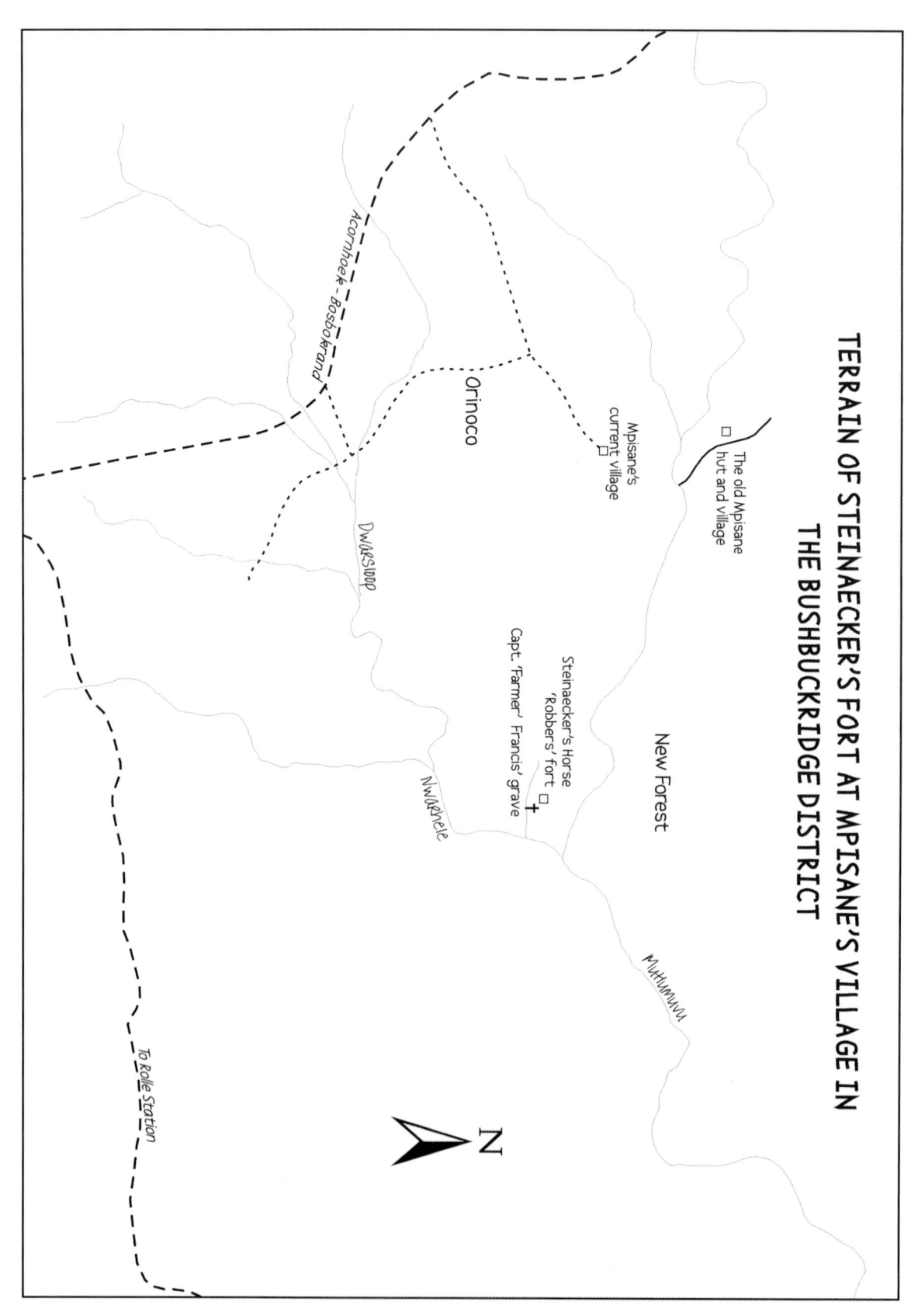

TERRAIN OF STEINAECKER'S FORT AT MPISANE'S VILLAGE IN THE BUSHBUCKRIDGE DISTRICT

Acornhoek - Bosbokrand

Orinoco

Mpisane's current village

The old Mpisane hut and village

Dwarsloop

Nwanhele

Capt. 'Farmer' Francis' grave
Steinaecker's Horse 'Robbers' fort

New Forest

Muhumuvu

To Rolle Station

N

Terrain of Steinaecker's fort at Mpisane's village in the Bushbuckridge district.

Crocodile River Valley. They found that the bridge over the Komati River close to the Mozambican border was too heavily guarded, but he succeeded in blowing up a smaller bridge over a spruit between Malelane and Kaapmuiden on 17 June 1900. Because of this a goods train was damaged and traffic to Lourenço Marques was delayed for 14 days, until the bridge was repaired.

Steinaecker then fled to Swaziland. Here, in April 1901, he lured a small section of Boers into an ambush at Chief Mbudla's kraal, killed a few and captured Commandant Van Dam of the Johannesburg ZAR Police.

As a result of his successes, Steinaecker was promoted to the rank of major and allowed to recruit an armed force of 450 men that had to make life difficult for the Boers in Swaziland and the adjacent Transvaal Lowveld. This recruitment drive led to the establishment of the notorious Steinaecker's Horse. He made his headquarters in Swaziland, on the Mozambican border at Nomahasha ("the place of horses"), and built a small fort, complete with a heliograph, on top of the 800 metre-high Stegi Hill, from where he could see just about the whole of Swaziland. In March 1901 he raided Bremersdorp (now Manzini). He found a few Republicans and British citizens still living there and captured eight men. Steinaecker was very proud of being in charge of the town and systematically stripped it of anything useful.

After this, Steinaecker and his unit started carrying out raids across the Republic's border and seized herds of cattle. He also captured a couple of small groups of Boers who tried to cross the border to Mozambique and managed to make an enemy of the Swazi queen and her subjects.

The Boers bided their time, and on 24 July 1901 General Tobias Smuts surrounded Bremersdorp and attacked Steinaecker's Horse with about 150 men of the Ermelo Commando. But Steinaecker had been warned of the danger the previous night and slipped out of town to "summon reinforcements from Nomahasha".

The Boers took the town after a short firefight. Four of Steinaecker's men were killed, four wounded and 17 captured. Captain A.W. Greenhill-Gardyne, Steinaecker's adjutant who was later placed in charge of the garrison at Sabie Bridge (now Skukuza), managed to escape with the rest of the men.

After the conquest of the Lowveld by Lord Roberts's forces under the command of General Pole-Carew, Steinaecker moved his headquarters to

Gen. Reginald Pole-Carew, Commander of the British forces along the Eastern railway line in the southern Lowveld after the battle of Dalmanutha.
(ACKNOWLEDGEMENT: SANDF ARCHIVES, PRETORIA)

Gen. Ben Viljoen (in the middle) with Capt. Herklaas Malan to his left and Capt. Blignaut to his right. Note the hairy Burgher (left, front) with his short-barrelled Martini-Henry!
(ACKNOWLEDGEMENT: SANDF ARCHIVES, PRETORIA)

Steinaecker's Horse's blockhouse at Sabie Bridge (Skukuza), April 1909. Col. J. Stevenson-Hamilton used it as living quarters for a while.
(ACKNOWLEDGEMENT: STEVENSON-HAMILTON COLLECTION, SKUKUZA ARCHIVES)

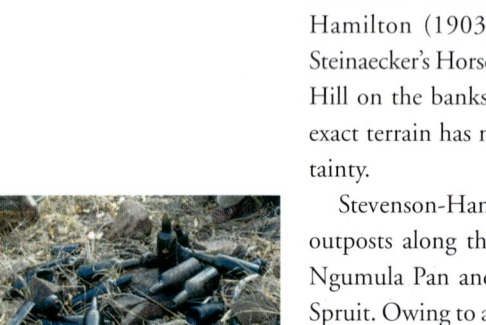

A heap of rum bottles found north of the Crocodile River near Komatipoort.

Komatipoort. Here he built a blockhouse and commandeered the quarters of the inspector of the Selati Line, Jules Diespecker, for his personal use. Diespecker was, understandably, unimpressed. Steinaecker then received an order to erect a series of border posts along the border with Mozambique to prevent any traffic between the two countries. Apart from the previously mentioned posts in Swaziland, Steinaecker set up bases and/or observation posts on the northern bank of the Crocodile River near its confluence with the Komati; at Gomondwane; at Sabie Bridge (where he also built a blockhouse next to the unfinished railway bridge on the southern side of the river); on the northern bank of the Nwaswitsontso just west of the Metsi-Metsi mouth; and next to the Makhadzi Spruit north of the Letaba River, about a kilometre north of the Zombe waterhole.[3] According to an old map of Stevenson-Hamilton (1903) there was another outpost of Steinaecker's Horse on the northern flank of Muntshe Hill on the banks of the Mlondozi Spruit, but the exact terrain has not yet been determined with certainty.

Stevenson-Hamilton's map also showed military outposts along the Sweni Spruit in the vicinity of Ngumula Pan and at Malihane by the Mavumbye Spruit. Owing to a faulty map reading by Stevenson-Hamilton, it is possible that the Sweni Spruit outpost was near the current Satara rest camp instead.

The terrains of all the other old military posts have been found and, apart from interesting military relics of the time, it is clear there was no shortage of liquid fortification, judging from the large number of rum bottles found at the observation post next to the Crocodile River. It does not take much guesswork to imagine what effect the regular rum rations, in the hot Lowveld climate, had on the morale and general wellbeing of the troops of this notorious unit!

In 1984 ranger Ben Lamprecht, then of Letaba section, discovered a British military outpost next to the Makhadzi Spruit. Although there is no reference in archival material to an outpost of Steinaecker's Horse this far north, this terrain must nevertheless be regarded as their northernmost outpost as the Bushveld Carbineers never established or manned outposts along the Lebombo Mountains.

During a visit to the terrain on 17 July 1984 and judging from the artefacts strewn about, I came to the conclusion that Steinaecker's Horse had in fact maintained a border post there between 1900 and 1902, manned by both white members of the unit and black members of the Native Police or Border Guards. According to the remains found at other outposts, including the one near the Metsi-Metsi mouth, these had been manned exclusively by black Border Guards.

During my visit to the terrain the following were found, among others: a number of Martini-Henry, 12-bore shotgun and Lee-Metford rifle cartridges and even live ammunition; numerous lids of bully beef tins and other tinned foods; pieces of earthenware; the glass lids of Worcestershire sauce bottles; medicine bottles; forks; spoons; various kinds of buckles; nails; a pair of scissors, part of a mouth organ; and empty rum bottles (similar to those found along the Crocodile River; see photograph).

During later, systematic archaeological digs by a team of archaeologists led by Dr A.C. van Vollenhoven of the National Cultural History Museum in Pretoria and later by the private archaeological consultants Archaetnos, a great number of military and other artefacts were found in the rubbish heaps and elsewhere on the terrain, which confirm that the post was manned by a unit of Steinaecker's Horse. A collection of these artefacts are currently exhibited at a museum at the site as well as at the reception area of the nearby Mopani rest camp.

One of the important finds by Van Vollenhoven

---

3   Interesting archaeological excavations have been conducted since 1996 by the cultural and cultural resources consultants Archaetnos under the leadership of Dr A.C. van Vollenhoven at this and other military terrains in the Kruger National Park. Several papers have already been published (see References). There is an exhibition of the artefacts found at the Makhadzi outpost of Steinaecker's Horse at Mopani rest camp in the Kruger National Park. Also see the colour photograph on p.418.

and his team is a "Maxim" bullet, stamped Maxim 1897. This .557/.450 Martini-Henry calibre bullet was made for a machine gun manufactured by Maxim Arms Co. in London. It confirms that the unit at Makhadzi was armed with a machine gun, apart from other weapons such as Lee-Metford and Martini-Henry rifles, 12-bore shotguns, and .380 and .350 Wembley revolvers.

Van Vollenhoven's team also found indications that more permanent accommodation than tents had been erected on the terrain, and that these huts and the terrain itself were used at a later stage (circa the 1920s) by whites.

The officers and non-commissioned officers of the unit stationed here probably built more comfortable accommodation in the form of traditional mud-and-daub huts with the help of Chief Makuba's subjects, whose village was a little north of the outpost. This would explain the presence on the terrain of clay pots and other artefacts associated with black habitation. The artefacts pointing to a later era of utilisation (the 1920s) probably indicate the presence of ranger Ledeboer of the Letaba section, who often patrolled along the Makhadzi Spruit between 1920 and 1929 and would certainly have used the hut(s) to sleep in, if they still existed. It could also be that Major A.A. Fraser, who manned the Malunzane outpost west of the Tsende (Tsendze) River between 1904 and 1920, used this terrain during patrols to the eastern border. The small carbon-dioxide "bombs" one could use to make soda water found in 2001 by Van Vollenhoven and his team at the Makhadzi terrain, are similar to those I found on a rubbish dump at Major Fraser's Malunzane quarters.

The pioneers of the Lowveld were in the habit of using existing accommodation or campsites. In this way, the nice campsite underneath high thorn trees on the northwestern flank of Muntshe Hill was used in turn by the Portuguese trader and explorer D. Fernandes das Neves (1860), the surveyor G.R. von Wielligh (1890), Steinaecker's Horse (1901–1902) and Major Stevenson-Hamilton (1913). The little store of Sardelli the Greek at Gomondwane was also commandeered for use as an outpost by Steinaecker's Horse, with Captain E.G. (Gaza) Gray in command.

Even a patrol hut built by ranger H.E. (Bert) Tomlinson in 1934 at Shingomeni, near the eastern boundary of Shingwedzi section, was used by a series of rangers for the same purpose and is still there today!

Three Scottish brothers – Bill, Bob and Tom Sanderson – lived near Legogote (then Lozieskop), where the farm Peebles is today, when the war broke out. They had started farming here long before the war and made a name for themselves among the locals as dedicated hunters and experts on this part of the Lowveld. Like many others before them, they did not have much luck on the alluvial diggings at Pilgrim's Rest and Spitzkop.

Bob, who farmed about six miles from Klipkoppies, also rode transport for a while. Bill was an eccentric character and very proud of his reputation as a crack shot. One day ranger Harry Wolhuter visited him and wanted to test a new rifle, so they set up an empty tin. Wolhuter, a champion shot himself, hit the target in the centre. He then handed the firearm to Bill to give it a go. After Bill had fired, they walked over to the target and still found only one hole in the tin. Without turning a hair Bill said: "Jist as I was thinking, Harry – clean through the same hole as your yin!"

The Sandersons were on good terms with the Boers. At the outbreak of war they decided to remain neutral as they had friends on both sides.

Bill Sanderson and his son at Legogote, 1902.
(ACKNOWLEDGEMENT: STEVENSON-HAMILTON COLLECTION, SKUKUZA ARCHIVES)

14. WAR CLOUDS LOOM OVER THE LOWVELD 1899–1902

Steinaecker's Horse warrant-officers at Komatipoort, 1902.
(ACKNOWLEDGEMENT: STEVENSON-HAMILTON COLLECTION, SKUKUZA ARCHIVES)

Remains of trade articles found at the site of Sardelli and Charlie Woodland's shop at Ndzengendzenge near Sabie Poort, where Bill Sanderson was interned for a while by Steinaecker's Horse in 1901.
(OCTOBER 1986)

Thanks to mediation by their friend Commissioner Abel Erasmus, the Boers did not commandeer them and they were allowed to stay on their farms. Bob, however, decided that it would be safer to take their cattle and horses across the Portuguese border and he and Harry Wolhuter went to Mozambique together. After a month Bob received a letter from Bill telling him to return immediately with his cattle and horses as the Boers were making things difficult for him. Bob later fell ill and died in Barberton Hospital. (One source claimed that he had been hit by lightning.)

After the British forces had taken over the Lowveld, they heard of an English-speaking farmer who lived among the "enemy" in the White River district. Major Steinaecker sent a message to Bill saying that he would soon be "relieved". His answer was that he was perfectly happy and did not want to be "relieved". Steinaecker started to suspect that Bill was a traitor in collusion with the Boers and a whole contingent of Steinaecker's Horse was soon dispatched via the old Delagoa Bay road to capture Bill. (In 1984 ranger Mostert discovered a spot along the old transport road, northwest of Skipberg, where the group had probably camped, and collected many bullets and military items here.)

Steinaecker's mounted force soon arrived at Sanderson's home and would not accept his explanation that he was neutral. He was summarily ordered to start packing and accompany them. Just as they were about to leave a small Boer commando arrived and the two groups started shooting at each other. Steinaecker's group, with Bill in their midst, beat a hasty retreat and the Boers confiscated all Bill's cattle and other belongings, "seeing that the old crook had now joined the British!"

In the meantime the unfortunate Bill had a difficult time of it as he was "interned" in one of Sardelli the Greek's little corrugated iron stores on the Lebombo flats near Sabie Poort. Here he nearly died of the heat and malaria before being allowed to return to his farm – but not before Steinaecker commandeered his best horse and rifles. Back at his farm he found that everything had been destroyed or looted and he had to start from scratch again.

While on his farm he met Stevenson-Hamilton for the first time when the latter arrived in the Lowveld in July 1902. They eventually became fast friends and Stevenson-Hamilton would learn a lot about game and conditions in the Lowveld from him.

The ruin of the shop where Bill Sanderson had been "interned" has since been rediscovered next to the Nhlanganzwani firebreak, about half a kilometre south of the Sabie River.

Charlie Sanderson, Bill's son with one of his handful of black wives, later made a name for himself as an artisan in the Kruger National Park. Many of the first tourist amenities built in the 1930s, as well as a few ranger houses, were his handiwork.

After seeing what had happened to Bill, Harry Wolhuter was in no mood to receive similar treatment from Steinaecker's Horse and joined the unit instead.

Steinaecker was not very particular when it came to recruiting. There were nevertheless a few people of integrity in the unit, such as his adjutant Major A. Greenhill-Gardyne, previously of the Gordon Highlanders (he was a nature-lover, kept poaching in check and left a remarkable report on game and suggestions for its preservation in the Lowveld to Stevenson-Hamilton). Other good men included Major R. Robertson, who was in charge of the blockhouse and unit at Sabie Bridge, the well-known hunter H.F. (Farmer) Francis, the brothers P.W. and B.C. Willis, A. Mockford, and the future rangers Harry Wolhuter, E.G. (Gaza) Gray, Harold Trollope and others. The majority, however, was a collection of unsavoury characters such as the Greek D. Sardelli, Tom Paulin, Charlie Woodlands, Ben Harper and others, flotsam from the Lowveld community who joined this unit because they viewed it as an opportunity for personal gain. These opportunities presented themselves in the form of raids on Boer farms and the looting of military equipment issued to the unit. The unit soon enraged the Boer forces in the veld because it armed blacks and used them as spies or to help during the raids on the vulnerable Boer farms. At the time several murders were blamed on them, including those of the four Stols brothers (family members of the acclaimed hunters), who were killed in September 1901 near White River.

Harry Wolhuter (1948) wrote the following in his memoirs about his former commander:

> Major Steinaecker was a pompous little cocksparrow of a man, standing some 5'3" in his boots, spare and wiry-looking in figure, and of possibly 120 lbs fighting weight. The most striking feature of his lean cadaverous face from which, under bushy eyebrows, gleamed two truculent black eyes, was a vast and remarkable moustache which, well waxed and turned up at the ends, extended to some nine inches on either side of an aggressive jaw, but failed to conceal a mouth from which the front teeth, all but a few yellow and broken fangs, had vanished. He took immense pride in his various self-invented uniforms. His feet and legs were usually encased in soft brown Wellington boots, the tight-fitting overalls severely drawn down by straps fastened under the insteps. The heels of the boots were decorated with huge silver box spurs, which flashed and clinked as he walked. When in full regalia, his formidable tasselled sword trailing from his hip, wearing his heavily laced cap, corsetted, and clad in a long wasp-waisted semi-naval frock coat with enormous heavily fringed epaulettes, he reminded me, as he strutted about, of a peacock showing off his plumage! He was no doubt a true pattern of the traditional swashbuckler.

There was very little discipline to be found in the unit and the men thought nothing of openly poking fun at their commander. Wolhuter wrote that Steinaecker once addressed them as follows while on parade:

> 'Men! I will have you understand that this is not vun dam picneek: that it is vun military organisation, and I vill have you obey me! If you won't I'll damned well break you.' At this point somebody shouted, 'Good old Stinky!' (This being his affectionate nickname among us) and the Major's long moustaches fairly bristled as he screamed, at the top of his voice, 'Step forward, that man! I'll have you court-martialled and shot!'
>
> Needless to say nobody stirred so the thoroughly infuriated Major promptly threatened to shoot the lot of us, whereupon some other bold spirit shouted: 'You can't do that', and so on. Such is a sample of the prevailing discipline. Eventually he told the S.M. to dismiss the parade: told us we were a lot of 'Pluddy ruffians' – perhaps he was right!

It is clear that General Ben Viljoen and the Boers had the same opinion of him and his men (loosely translated):

> The so-called Steinaecker's Horse was a corps of all the desperadoes that could be gathered. It consisted of disreputable shopkeepers, hunters, smugglers and all the dregs of society that could be found in the wilderness of the North. They also had a black unit, consisting of several hundred blacks under the command of a man named Francis, a well-known hunter. The corps had its headquarters at Komatipoort and the commander was Major Steynaecker. Who or what this "shining light" was, I could never

Maj. R. Robertson, commander of the Steinaecker's Horse garrison at Sabie Bridge (the later Skukuza), 1902.
(ACKNOWLEDGEMENT: STEVENSON-HAMILTON COLLECTION, SKUKUZA ARCHIVES)

Colonel Ludwig von Steinaecker, commander of the British unit Steinaecker's Horse in the Lowveld, 1900–1902.
(ACKNOWLEDGEMENT: SOUTH AFRICAN NATIONAL MUSEUM OF MILITARY HISTORY)

14. WAR CLOUDS LOOM OVER THE LOWVELD 1899–1902

The remains of Col. L. von Steinaecker's Fort Mpisane on the farm New Forest, northeast of Bushbuckridge.
(11.11.1986)

find out. Judging by his men and the work they did, he could be fairly closely related to Rinaldo Rinaldini, the notorious leader of the Italian bandits. The guarding of the Portuguese border was probably delegated to Mr Steinaecker, while he had to have had *carte blanche* as his power seemed boundless. His main aim was obviously to accumulate as much loot as possible …

The booty consisted mainly of cattle and other goods taken during raids on the isolated Boer farms. For this purpose Steinaecker built a fort on the farm New Forest near Chief Mpisane's village. About 30 soldiers were stationed here and placed under the command of H.F. (Farmer) Francis. They then armed Mpisane and his followers and, by avoiding contact with the Boer forces, waged a reign of terror in this part of the Lowveld. The looted livestock were transported to Sabie Bridge under supervision and from there to Komatipoort, where the commander disposed of it to the benefit of himself and his men. For the Boers it was the last straw when Mpisane's raiders stole Abel Erasmus's entire herd from his farm Orinoco near the fort. Abel Erasmus was a legendary figure in the Lowveld, much respected by the blacks. They called him "Dubula duze" – the one who shoots from nearby (more about him in chapter 7). Over the years Erasmus had bred a prize herd of cattle that he called "bruin geelbekke" (brown yellow-muzzles) and it was this herd that fell prey to Steinaecker's looters. When General Viljoen heard of this, he decided to teach the raiders a lesson once and for all. Harry Wolhuter happened to be among those who herded the cattle to Sabie Bridge, and so escaped the subsequent events at Fort Mpisane.[4]

General Viljoen ordered a strong division of his commando under the command of Commandant Piet Moll to attack and destroy the "raiders' fort". An eyewitness account of this battle (the last big battle in the Lowveld) on 7 August 1901[5] by Commandant Andries Francois du Toit, better known as "Old Red Government", was recorded by P.H.S. van Zyl (1948), who wrote the following (loosely translated):

> General Viljoen and Commandant David Schoeman were not there. Commandant Piet Moll had 45 men there, while I had 55; and Field Cornet Jan de Beer, from Lydenburg, also arrived with 20 men to help out. Commandant Moll and I were in charge of the 120 men. Captain Herklaas Malan was second in command under Commandant Moll, and Lieutenant Jan Mellema, who was in charge of General Ben Viljoen's scouts, would also have been there with Moll's men.
>
> Before the attack I had my two best spies – the old hunters Marthinus Koekemoer and Gabriël Stols [probably a son of the Stolse buried at Pretoriuskop in the Park] – reconnoitre the area. We trekked through the night to a wooded ravine. The next morning Commandant Moll, Koekemoer, one other man and I went on a reconnaissance mission. The road between the camp of Steinaecker's Horse at Sabie Bridge and Fort Mpisane was three miles north

---

4   Stevenson-Hamilton (1974) claimed that Wolhuter had been on his way to Sabie Bridge with reports from Captain Francis and thus managed to escape the attack on the fort. Apparently they had passed close to the small hill where the Boer force was hiding, but were left unhindered by the Boers out of fear that they would reveal their position.
5   General Ben Viljoen gave the date as 6 August 1901.

of our base in the kloof. When we reached a rise I climbed into a tree next to the road.

I ordered two men, one black and one coloured, armed with assegais, to surrender when they passed underneath the tree, and Moll and the others captured them. We took them to our camp and questioned them, and they drew the position of the fort and the tents. That night at about ten o'clock we went to the road; with guards in front and behind them the two captives had to lead us along the road. At about three o'clock in the morning we crossed a small river and Fort Mpisane was now only about a mile ahead. We left a guard of about ten men with the horses and proceeded towards the fort on foot early that morning. Mpisane's Fort was round, about 100 paces in diameter, with a trench around it. There were three gates with small bridges across the trench, and in the trench and under the bridges there were tunnels from the middle of the fort. The walls on the inside were at least six feet high and neatly built. On the outside the wall went up at an angle from the edge of the trench, which was about four or five feet deep with right-angled walls. If you stood inside the trench, you were neatly in line with the loopholes in the fort's wall. These were at different heights and concealed with grass and other plants on the outside. If a shadow fell over a loophole, a shot immediately flashed from the inside.

We first attacked at the large gate on the eastern side. There Commandant Moll's jaw was shot off with a dum-dum bullet. My men and I attacked the tents and small camp south of the fort and then the big gate on the southwestern side, while Captain Malan moved around the fort to attack the gate on the northwestern side. We jumped into the trench. By doing this we gained a huge advantage on the defenders; we were too quick for them otherwise they definitely would have wanted to take up position in the trench themselves.

We helped Gert van den Heever up the wall and he was the first at the top, but I told him not to stick his head out and wait for everyone else, for they would certainly blow it off. He said: 'Commandant, there is one, he is naked; he is only wearing a shirt; he wants to mount his horse.' (The horses were inside the fort.) He raised his head and took aim. I shouted: 'Gert, what are you doing!' With

Cmdt A.F. du Toit (in the middle) an eyewitness of and reporter on the attack of the Boer forces on Fort Mpisane, with Willie du Toit (left) and Japie Fouché (right).
(ACKNOWLEDGEMENT: SANDF ARCHIVES, PRETORIA)

Relics collected at the site of Steinaecker's Horse's northernmost outpost along the Makhadzi Spruit. (17.07.1984)

this his shot rang out and he fell backwards, almost on top of me and his brother Danie, whom I was busy helping up the wall. Gert was at least six feet tall, a big, well-built Afrikaner, and his heart was just as big as his body. There were five Van den Heevers with me – four sons and Uncle Danie van den Heever, and their cousin Giep – all brave heroes. With a thousand men like these I would have charged any British camp. The man Gert killed was Captain Francis, who had jumped out of bed wearing only a shirt. He was killed at almost the exact same moment.

We helped each other up and then I quickly ran east around the fort to Captain Malan. I saw Com-

Cmdt Piet Moll (in the middle), whose jawbone had been shot off on 7 August 1901 during the attack on Fort Mpisane. Standing to his right is Capt. J. McKenny and to his left Adj. Fleisch.
(ACKNOWLEDGEMENT: SANDF ARCHIVES, PRETORIA)

mandant Moll walking away from the trench with his jaw hanging – too terrible. I said to Malan: 'Hurry. Help your people up, and when I blow the whistle, you blow yours and we jump over the wall.' It happened like that. When the whistles blew, the men jumped up and fired into the fort.

The men inside the fort surrendered immediately, but while they were standing in the fort with their hands in the air, a black commando came out of the bush on the western side. I told Malan to disarm the British and the blacks inside the fort; he entered through the gate with his men and they surrendered. Ten of my men stayed on the wall to help him. With the other 40 I took up position inside the trench.

During the attack on the fort, Commandant Moll and I had sent Lieutenant Potgieter, Corporal Maritz, my adjutant Chris Joubert and a few other men to the side of the bush on the western side, as we feared that reinforcements would arrive from there under cover of the bush. There was a bare expanse around the fort. Before they realised what was going on, our spies saw some blacks descending on them from the direction of the bush. They shot and fell back immediately. Chris Joubert came running towards us, screaming: 'There are blacks around us!' When Maritz and the others reached us, the blacks were no more than 20 paces from us. We could not shoot earlier out of fear of hitting our own men. Then we opened fire.

The next minute all we could see were the blacks charging and throwing assegais. Each had a throwing assegai and a shorter one to jab with, with a short Lee-Metfords in the left hand as well. I saw one of them throwing an assegai over Maritz's shoulder that landed with such force that it was quivering in the ground. When we started shooting they ran off to one side rather than straight back into the bush, which made for the easiest killing in the world. A few turned back and some of my men followed them for a while on the British horses that had been inside the fort.

It was an act of providence that the blacks outside had not charged a few minutes earlier. In the end there were two blacks about three hundred paces away with my uncle Andries Welman in pursuit – a brave old soldier and a good runner. He loosed off a few shots on the fly, but missed. Then old Field Cornet De Beer said to me: 'Wait, let me show you how to fire a good shot.' I was still saying: 'Wait! Don't shoot; you'll hit my uncle!' when De Beer felled one of the blacks in front of my uncle. My uncle stopped and looked around in confusion; there had still been one man ahead of him and now he was gone.

After the battle Captain Malan and I and one other man walked around the fort. I noticed a tunnel and peered inside. A shot went off in my face; I nearly fell on my back, but it missed me. I immediately backed off and called a few men to shoot into the tunnel. Then I called their sergeant major and said: 'Look, if one of our men is killed, we'll shoot all of you.' He called to them to come out and 28 blacks, each also armed with two assegais and a short Lee-Metford rifle, came out of the three tunnels underneath the fort.

We tied them up in twos, one Englishman and one black man together, and drove them on. The next morning we held a council of war at Bushbuckridge, five or six miles from Fort Mpisane. The council consisted of Captain Herklaas Malan, Field Cornet Jan de Beer, Lieutenant Potgieter and Lieutenant Swart (I don't know if Lieutenant Mellema was present as well) and me. The captured blacks were sentence to death for armed treason. Captain Malan insisted that the whites should also be shot, but I objected and De Beer supported me. The blacks were lined up beside a prospecting pit, shot and buried in the pit.

The total loss to the enemy was: Captain H.F. Francis killed and another two whites wounded, including Rifleman J.E.D. Travers, while thirty British soldiers were captured at the fort. On our way back from the fort we captured another six Brits next to a wagon. The several hundred blacks killed at the fort were buried in a long trench southeast of the fort. Our booty consisted of a considerable number of cattle and some food. Lieutenant Mellema and I shared Captain Francis's uniforms.

Our losses were 17 (sic) dead and wounded. The names of the 13 (sic) dead, all reburied [1907] in the municipal cemetery at Lydenburg [and reburied at the Berg-en-dal Burgher Monument in 1970] were: G.D. van den Heever, P.E. Muller, Jan Nel, C.J. van der Merwe, Jacobus Nel, Jan de Villiers, L.C. Schoeman, J.C. Pelzer, S. Walters, Egbert Meintjies, H.W. Dykstra and S. Swart. Some of them later died of their wounds. The other wounded, apart from Commandant Moll, were C.A. Geerdts and A.F. Lombaard.

I sent Commandant Moll with the wounded British soldiers on a trolley to their camp at Sabie Bridge. He was later sent to Ceylon[6] as a prisoner of war, but escaped on board a Russian ship. Like Willie Steyn and others he eventually reached President Kruger in the Netherlands. He stayed there and his jaw was very competently reattached with a small silver plate. After the war he turned up at my home in Belfast with this silver jaw; and shortly afterwards, both he and Captain Herklaas Malan emigrated to Argentina.

My report to General Viljoen about the battle at Fort Mpisane was sent to President Kruger in the Netherlands from Pilgrim's Rest by an American courier named Adams, who hid it in the hollow heels and soles of his boots. The main purpose of the report was to prove in Europe that the British had armed blacks and used them against us.

General Viljoen was not sure what he should do with the British prisoners of war from Fort Mpisane as he was not convinced that they were legitimately part of the British army. At first he was going to have them court-martialled as common highwaymen, but then wrote to Lord Kitchener to see what he thought of the men's status. Kitchener confirmed that they were attested members of the British forces and asked for their extradition. This correspondence between General Viljoen and the British Commander-in-Chief led to a meeting between him and General Bindon Blood on 27 August 1901 in Lydenburg. After the meeting 24 prisoners were handed over to the British, after they had solemnly promised not to revert to their barbarous ways.

Very little remains of Mpisane's Fort on the farm New Forest close to the confluence of the Mutlumuvi and Nwarhele spruits, about 11 kilometres west-south-west of Rolle Station. Locals have long since carried off the stones of the fort's walls. All that remains visible is the trench around the fort, and a ring of very large trees grow in the loosened soil. The military cross that had been erected on Captain Francis's grave southeast of the fort after the war has been removed by somebody and only a small pile of stones indicates his final resting place. Close to it is another grave – that of R. Luxford, who had died of malaria – but its exact position is unknown. A great-grandson of Captain Francis, George A. Francis, worked in the Kruger National Park from 1 June 1970 to 7 September 1983 as assistant-superintendent of the National Parks Board's by-product depot.

Interestingly, the people living close to the old fort have a feast every September/October in honour of those who had fallen during the battle of Mpisane. They call the place "Shikanikiso" (the place where there was fought to the death).

Chief Mpisane Nxumalo, whose village was on the farm adjacent to Abel Erasmus' Orinoco, died in September 1927. His hut, about nine kilometres northwest of the old fort, is still kept in its original condition by his descendants.

As is common in times of war, most of the injuries sustained by the members of Steinaecker's Horse did not occur on the battlefield. It has been recorded that only 11 of the 114 fallen were killed in action. Twenty-seven died of other causes, mainly malaria,

---

6   Now the island Sri-Lanka.

14. WAR CLOUDS LOOM OVER THE LOWVELD 1899–1902

Senior ranger Mike English at the grave of Capt. H.F. (Farmer) Francis who had been killed on 6 August 1901 during the attack of the Boer forces under Gen. Ben Viljoen on Fort Mpisane.
(11.11.1986)

The grave of Capt. H.S. (Farmer) Francis who was shot by Boer forces under Cmdt Moll on 6 August 1901 at Fort Mpisane, on the farm New Forest, in the Lowveld. This photograph dates back to the 1930s.
(ACKNOWLEDGEMENT: INFORMATION AND PROMOTIONAL SERVICES, TRANSVAAL PROVINCIAL ADMINISTRATION)

27 were wounded and the other 49 were captured or simply disappeared.

Other interesting information found in the South African list of casualties between 1899 and 1902 is the following: Rifleman S. Smart was killed by a lion at Sabie Bridge on 4 October 1900; Rifleman R. Chambers broke his neck after being flung from his horse on 24 November 1900 at Nomahasha; Rifleman O. Engerstroom committed suicide at Josanes in Swaziland on 13 June 1901; Rifleman G. Gaines was caught by a crocodile on 30 November 1901 in the Usutu River in Swaziland; and Rifleman W. Halpin died as a result of an epileptic fit at Komatipoort on 21 February 1902.

In his memoirs Harry Wolhuter gave a fairly accurate description of the life of a member of this infamous unit and also related what happened to some of his comrades after the war. He, for example, often had to take wages or rations on horseback to far-off outposts as far north as the Olifants River. On more than one occasion he found the entire garrison at the outpost sick with malaria and he first had to nurse them and dose them with quinine pills. Once they trekked through an unfamiliar area near the Olifants River, but managed to convince a local man, Ngotso, to show them the rest of the way. When they later reached a fairly sizeable tributary of the Olifants River and learned that it did not have a name, they straightaway named it Ngotso, in honour of their guide. This name is still in use today.

After roaming through Rhodesia (Zimbabwe), Wolhuter's great friend, Perry the Canadian, bought a farm along the Sabie River, close to the main road between White River and Bushbuckridge. Perry's Farm became a well-known landmark and the Sabie River Bungalows were built on his land. Tom Boyd, who had been the driver of the locomotive that served the railway line between Sabie Bridge and Komatipoort during the war, drank himself to death. His stoker, Bertram Churchill Willis, or "Clinkers" as he was known to friends, took over from him. His brother Percy was in charge of the pump that supplied water to the barracks at Komatipoort and so got the nickname "Pump". The brothers Willis became reputable traders at Acornhoek after the war. D. Sardelli, the Greek who had absconded during the battle at Fort Mpisane, drifted off and sold rifles he had stolen from the unit to locals. Later he became a cattle herder at Mica for a salary of £3 and a bag of maize meal a month. He eventually died in a mental institution.

Major R. Robertson, a man deep in his fifties, joined a South African regiment as a common rifleman after the war and was sent to Somalia to fight the mad "Mullah". He died a year or so later in South Africa.

Steinaecker was promoted to lieutenant colonel at the end of the war and tried his utmost to have his unit turned into a permanent border guard. When Stevenson-Hamilton was appointed as the first warden of the reproclaimed Sabie Reserve there was a lot of friction between him and Steinaecker, who still thought he was in charge.

Steinaecker, who received the DSO on 8 April 1902, was ordered to disband his unit in February 1903. He was highly indignant at this "insult", but the people of Komatipoort were overjoyed, and nobody more so than Jules Diespecker, who immediately reclaimed his home after, or so the story goes, giving Steinaecker a black eye into the bargain.

For some time Steinaecker tried to get a permanent commission rank in the British army. When he did not succeed, he wrote long letters to British newspapers in which he demanded a pension. Eventually he farmed tobacco and cotton on the farm London near Bushbuckridge, but these efforts came to nothing and he had to leave the farm. He then lived for a while with Griffiths, the Native Commissioner of

Graskop, until his wife gave him an ultimatum – it was either her or Steinaecker, but one of them had to go. John E. Travers of the farm Champagne, himself a former member of Steinaecker's Horse who had been seriously wounded at the battle at Fort Mpisane, took pity on the colonel. Steinaecker did odd jobs on the farm (close to his old fort near Mpisane's village) in exchange for pocket money and a home. Here he stayed until the outbreak of the First World War, when he became aggressive and started bragging about Germany's imminent victory. In April 1917 Travers gave him notice to leave the farm, but he refused. Travers asked the police at Bushbuckridge to remove him, but when they arrived at the farm on 30 April 1917 he had taken a dose of strychnine[7] and died a painful death at the age of 66 (Bulpin, 1950). Today he rests in a lonely grave under a small pile of stones in the bushveld. Nobody mourned his death.

While the members of Steinaecker's Horse were busy in the southern Lowveld, another irregular British unit settled in the northern Transvaal and soon gained an even worse reputation than the notorious Steinaecker's Horse.

This unit, who called themselves the "Bushveldt Carbineers", settled in July 1901 in the Spelonken area of the Louis Trichardt (Makhado) district. Its members had been recruited mainly in Australia and consisted of an even more disreputable collection of characters than Steinaecker's Horse. Their headquarters were at Fort Hendrina (which they renamed Fort Edward), near the Swiss mission at Elim. At first they were under the command of Captain James Huntley Robertson, a young Australian officer with flashy manners and no clue about commanding.

There was soon a complete lack of discipline among the men and many instances of drunkenness, thievery and even rape occurred. Robertson's leadership defects were underlined by the negative influence a subordinate, Captain Alfred Taylor, had on him.

It is as well to know more about this Captain Taylor, since he was a shadowy figure whose influ-

John Edmund Delacoer Travers on whose farm, Champagne, Col. L. von Steinaecker was buried.
(ACKNOWLEDGEMENT: *SOME LOWVELD PIONEERS*)

Gen. Ben Viljoen being transported by mule wagon after he had been captured near Lydenburg during the night of 25 January 1902 by a mounted section of British under command of Maj. Orr.
(ACKNOWLEDGEMENT: TRANSVAAL ARCHIVES)

---

7 According to Alexander Mathews, an old Lowvelder who had founded the Citrus Research Station at Nelspruit and later became the owner of the farm Fleur de Lis, John Travers told him that Steinaecker had shot himself with his service revolver. However, his official death certificate, obtained from the Registrar of Births and Deaths, confirms that he committed suicide by drinking poison.

Possible site of Col. L. von Steinaecker's grave near the ruins of John Travers's former home on the farm Champagne in the Bushbuckridge district.

The Boers' steel fort in the Soutpansberg which was named Fort Hendrina in honour of Cmdt Gen. Piet Joubert's wife. It fell in the hands of a British unit, the Bushveldt Carbineers, during the Anglo-Boer War and they renamed it Fort Edward. It can now be seen in Louis Trichardt (Makhado).
(ACKNOWLEDGEMENT: SANDF ARCHIVES, PRETORIA)

ence could be detected in all the unit's atrocities, and he left a trail of blood in the Soutpansberg of which the effects can be felt to this day.

According to Lee (1972), Lord Kitchener himself appointed Captain Alfred James Taylor (1862–1941) as a special intelligence officer to control the thousands of blacks in the district. He was an Irishman born in 1862 in Dublin, and had immigrated to South Africa in 1881 at a youthful age. After he had settled in the Transvaal in 1884, he was commandeered by the Boers to fight Chief Masjou in the Schweizer-Reinecke district. In 1886 he went with Rhodes's pioneering group from Kimberley to Matabeleland to erect mining machinery. Here he managed to win Lobengula's trust and was present when the Rudd concession was signed with Lobengula on behalf of Cecil John Rhodes. In 1890 he was a dispatch rider for F. Selous and the pioneer columns to Fort Tuli and Rhodesia (Zimbabwe). He participated in the Matabele wars of 1893 and 1896 and was wounded in the leg in the Matopo Hills. He was also a member of the British forces that besieged Fort Tuli in October 1899 and wanted to enter the Transvaal at Pont Drift. After that he helped during the relief of Mafeking (now Mafikeng). Mphefu, the fleeing Bavenda chief, had secretly been visited by Taylor before the war of 1899 to incite him against the ZAR. He was also advised to request British annexation. After the outbreak of the war, Mphefu ordered his subjects to help the British and rise up against the Boers. As a result many white inhabitants of Louis Trichardt (Makhado) and vicinity had to flee into the unhealthy Lowveld with their families.

During the Anglo-Boer War, Taylor showed himself to be a bloodthirsty and merciless man. The black nickname he got as early as the Matabele campaign of 1896, "Bulala", meant "kill", and the Shangaan referred to him as "Bulala zonke Matshangane" (killer of all the Shangaan). His headquarters were on Bristow's farm Sweetwaters, close to Elim's mission hospital. This was near the farm Lovedale Park and Fort Hendrina (named after General Piet Joubert's wife).

Captain F.R. de Bertodano, the intelligence officer appointed by the British high command to investigate Taylor's activities in the Soutpansberg, wrote the following about him: "I disliked him instinctively; a hard face, light grey eyes and never looked one in the face … He had also married a native or half-caste woman." He was convinced that Taylor was the real instigator of the troubles in the area and recommended that he be relieved of his duties.

After the war Captain Taylor moved to Rhodesia (Zimbabwe). He took part in the First World War (1914–1918) and received the DSO in France. Later he became a member of the Rhodesian legislative assembly. He died in 1941 at Que-Que and took all his dark secrets with him to the grave.

The commanding officer of the British unit at Pietersburg was Major Robert William Lenehan, who also did little to maintain order in the unit. The local commander at Fort Edward, Captain Robertson, was replaced by Captain Percy Frederick Hunt on 16 April 1901. At the same time, the soon-to-be notorious Lieutenant Harry Harbord Morant of Queensland joined the unit. He was a champion horse rider and tamer and as a result became known as "Breaker".

Other members of the Bushveld Carbineers were: Lieutenant Peter Joseph Handcock, a veterinary lieutenant from New South Wales, where he had been a blacksmith; Lieutenant George Ramsdale Witton, another Australian with no previous experience as an officer who later became an artillerist in the Victorian Permanent Artillery in Australia; Lieutenant Henry Picton, a British drifter and mercenary; Lieutenant Leonard Henry Ledeboer, an intelligence officer of Dutch origin, born in India, who later became a labour recruiter at the infamous Crooks Corner and eventually served as a ranger in the Kruger National Park at Letaba and Satara (1921–1946); and Corporal Herbert Sharpe. Among the remainder were 15 soldiers and officers who sent a petition on 4 October 1901 to higher ranks about the cruelties committed by Captain Taylor, lieutenants Morant, Handcock and Witton and other members of the unit; and one Van Buuren, who had absconded from the Boers to join the unit.

As background to the incidents described here, it must be remembered that the Boer forces were reorganised into guerrilla units after the siege of Pretoria on 5 June 1900. These units were under the command of generals Botha, De Wet, De la Rey, Smuts, Viljoen, Kemp, Beyers and others. The commandos were usually divided into smaller, mobile units of 100 men or fewer and they terrorised the British forces for another 18 months with no let-up. Lord Kitchener, who took over as Commander-in-Chief from Lord Roberts, came up with the dishonourable strategy of placing the Boer women and children in concentration camps and destroying the Boers' homes and crops. The Soutpansberg was the last unoccupied area of Boer resistance where small groups fought on persistently. Assistant Commandant General C.F. Beyers and his commando, in particular, made the area unpleasant for the British. Two other Boer commanders who were constant thorns in the sides of the British were Captain Jack Hindon, the specialist Boer spy, and Field Cornet (later Commandant) Tom Kelly.

The Bushveld Carbineers were ordered to clear the district of Boer resistance fighters using the same guerrilla tactics as the Boers. "Bulala" Taylor had to control the black people of the Soutpansberg area since Kitchener feared that João Albasini II (a grandson of the well-known João I, white high chief of the Magwamba), who was fighting with General Beyers's commando, could convince the Shangaan to fight on the Boers' side.

The Medingen Mission, with Reverent Reuter of the Berlin Missionary Society in charge, was about 100 kilometres southeast of Fort Edward. At the end

Capt. O.J. (Jack) Hindon, who became a thorn in the side of the British forces in the Northern Transvaal, where he blew up several trains on the line between Pretoria and Pietersburg during the second half of the Anglo-Boer War. (ACKNOWLEDGEMENT: SANDF ARCHIVES, PRETORIA)

A group of Bushveldt Carbineers officers in 1901. From left to right: Lt P.J. Handcock, Lt H.H. Morant, Dr Johnston, Capt. P.F. Hunt, Capt. A.J. (Bulala) Taylor and Lt H. Picton.
(ACKNOWLEDGEMENT: ANGUS & ROBERTSON, LONDON)

of July 1901 Reverend Reuter visited Fort Edward and told Captain Hunt that Field Cornet W.H. Viljoen and his two sons were living in Duiwelskloof and causing havoc near the farm of Sergeant Frank Eland, a member of the Bushveld Carbineers. After one of his men confirmed this, Captain Hunt decided to attack Barend Viljoen's farmhouse, accompanied by a large group of rain queen Modjadji's subjects armed with assegais, knobkerries, pistols and rifles. Kitchener did not officially approve the use of armed blacks, but the Bushveld Carbineers, like Steinaecker's Horse, made their own rules. There were 15 Boers inside the house and another 40 on one Botha's adjacent farm.

During a skirmish in the night of 5 August 1901, both Captain Hunt and Sergeant Eland were shot dead. Two Boers – J.J. Viljoen and G. Hartsenberg – also died. The next morning Hunt's men discovered that the Boers had escaped and that the body of Captain Hunt had been mutilated. It later came to light that Viljoen's body had suffered the same fate. Boer settlers in the area were convinced that both bodies had been mutilated by muti doctors. The next day Lieutenant Morant arrived with 70 men, decided that his friend had been killed in cold blood and vowed to seek revenge. With this, he and other officers of the Bushveld Carbineers began a reign of terror in the district and ordered (claiming that the order had come from Lord Kitchener himself) the execution of all Boer prisoners of war.

This systematic murder of Boer prisoners only ended once the British high command, after many reports of atrocities had reached them, disbanded the unit and court-martialled the officers.

During the court-martial it came out that the practice of shooting Boer prisoners of war had started before Captain Hunt's death, when Captain Robertson was still in charge and on the orders of Captain Bulala Taylor. One day a message reached Fort Edward that a number of Boers were on their way to surrender. There were six men in the group, including a seriously ill man on a wagon carrying a white flag. A Sergeant Oldham carried out Taylor's instructions and shot them all, including the invalid. This particular patrol included Rifleman Van Buuren, the "joiner", who objected to the shooting of helpless prisoners. But in doing this he signed his own death warrant. On Taylor's orders Lieutenant Handcock took Van Buuren on a patrol, during which the two of them "conveniently" wandered off into the dense bush. Here Handcock shot Van Buuren in cold blood. Later he told the others that they had been ambushed by the Boers, who had shot Van Buuren.

Shortly after the fight in which Captain Hunt was killed, Lieutenant Morant attacked a Boer camp and captured one Josef Visser. The captive was wearing part of a British uniform, which Morant immediately "recognised" as belonging to the slain Captain Hunt. (Later it was proved that Morant had claimed all Hunt's uniforms for himself.) Contrary to all military regulations and customs, Morant organised an impromptu "war tribunal" and sentenced the poor Visser to death. Ledeboer told him of the sentence and Picton shot him in the head with a pistol. Morant reported the execution to Major Lenehan.

A week or two later another incident took place in which eight or nine Boers surrendered to intelligence officer Ledeboer. On the way to Ford Edward they were met by Morant, Handcock and Witton. Morant sent the rest of the patrol back to Ford Edwards while he stayed behind with the other officers and three riflemen, including the "joiner" Theunis Botha. They shot all the prisoners by the side of the road and the bodies were still warm when a young German missionary, Reverend C.A.D. Heese, arrived at the scene in a buggy. Daniël Heese saw the bodies and was allowed to carry on in the direction of Bandolierkop, after having been warned by Morant to be "careful" of the Boers. Handcock and Morant then had a private conversation and Handcock was seen to follow the missionary. The bodies of Reverend Heese and the coloured boy who had accompanied him were found next to the road soon afterwards, and a local witness later confirmed that he had seen a horse similar to Handcock's next to the murdered missionary's cart.

And the slaughter didn't end here. The court also heard that Lieutenant Hannam had shot at a group of Boers with children, killing two little boys and wounding a girl in the neck, but there was not enough evidence to secure a conviction. What could be proved was that three Boer prisoners, including a 12-year-old boy, had been shot dead by a patrol led by Morant, Handcock and Witton.

On 16 September 1901, Morant and Witton left Fort Edward with a 30-man patrol and moved eastwards into the bushveld. They had received information that Commandant Tom Kelly, the Boer commander who had caused them so much trouble, was near the Portuguese border at Banniella's kraal by the Dzombo Spruit (in the current Kruger National Park). A week later they reached the site of Kelly's camp. Early the next morning they attacked and captured the Boers (Kelly and another nine men). They had heard that Kelly had cannons with him and were very disappointed not to find any. Tom Kelly was a big man of over six feet (1,8 metres) and about 52 at the time of his capture. His father, John Thomas Kelly, was Irish and his mother, Engela Botes, an Afrikaans girl. They were both buried in the Schoemansdal cemetery after dying of malaria. Tom Kelly had married Mieta du Plooy and they had four sons and two daughters.

Over the years the Kellys had proven themselves to be excellent hunters. It is said that Tom Kelly even hunted elephant during the war in the area between the confluence of the Olifants and Letaba rivers. Early on in the war, on 2 November 1899, he had also been part of the group under General Van Rensburg that launched a successful attack on two British forts in the Tuli Block (Penning, 1904). He had been involved in blowing up armoured trains on the Pietersburg (Polokwane) railway line, as well as with an attack on Fort Edward itself.

After his capture by Morant he was taken via the Birthday Mine to Fort Edward. Commandant Kelly had his wife and daughters with him at the time. It is said that every time Morant ordered him to move away from the wagon, his wife and daughters went with him. This probably saved him from the same fate as the other Boer prisoners of war and he was handed over to the authorities at Pietersburg (Polokwane).

Commandant Kelly returned from the prisoner of war camp in Ceylon (now Sri Lanka) to the Soutpansberg district after the war and died on 27 March 1932 on the farm Boskoppies.

At about the same time that Tom Kelly was captured, Captain Bulala Taylor apparently heard that there were Boers with quite a bit of gold in their possession at the small Louis Moore and Birthday mines.

Cmdt Tom Kelly, who was captured by the Bushveldt Carbineers led by Lt Morant in September 1901 within the area which is now the Kruger National Park. He was a well-known hunter during his time.
(ACKNOWLEDGEMENT: MR P. ELOFF)

A section of *Zuid-Afrikaansche Politie* (Zarps) at one of the steel forts, Fort Schutte, which had been erected in the years preceding the Anglo-Boer War in the vicinity of Louis Trichardt (Makhado). These forts were initially built as protection against hostile Venda tribes. One of them, Fort Hendrina, southeast of Louis Trichardt (Makhado), had been occupied by the Bushveldt Carbineers during the Anglo-Boer War and renamed Fort Edward.
(ACKNOWLEDGEMENT: MR P. DE VAAL)

He decided that the gold had to be seized and set off with a division of Carbineers to capture them. But the Boers had received news of their imminent arrival and, knowing Taylor's reputation, loaded their gold and other possessions onto a wagon and fled into the area that is now the Kruger National Park. Their route apparently followed the Bububu River in the direction of the Phugwane. At the Phugwane some blacks told them that Taylor and his band were very close, so they decided to push the wagon and everything on it into a deep waterhole (possibly the Dili). After this they fled. However, Taylor's mounted patrol soon caught up with them and they were captured. When the men refused to reveal where they had buried the gold, they were allegedly shot on the spot, along with the little black boy who led their oxen. The women and children were left behind in the veld and Taylor and his men returned to Fort Edward, having accomplished nothing.

These murders were not mentioned during the court proceedings, but circumstantial evidence indicates that the two neat graves with headstones that were found by ranger Mike English just east of the Zari-Phugwane confluence in 1964 are in fact the graves of the two men. Friendly locals had helped the women and children back to their homes, where they reported the murders to General Beyers.

Interestingly, a tributary of the Phugwane west of the murder scene is called Mashadya, which means "crying children". This may refer to the sad return journey of the murdered men's wives and children.

One of the incomprehensible aspects of the British court martial's findings is that there was "insufficient evidence" to find Captain Taylor guilty, and he was acquitted. It is also strange that he, a special intelligence officer, could not be court-martialled despite the fact that he had shot a certain Van den Bergh on his farm Paardekraal, behind Hangklip at Louis Trichardt (Makhado), in August 1901 in front of his wife and children. Schulenburg (1981) mentioned several other murders for which no one was charged because of a lack of evidence, and supplied a complete list of the Boer prisoners of war who had been murdered. Joe Albasini (1971) also mentioned the grave of four Boers who had been shot by Taylor and which he only recently discovered. According to Albasini, Taylor had also shot many blacks (especially Shangaan) in cold blood on the pretext that they would support the grandson (João II) of their former chief, João Albasini, fighting with General Beyers's commando.

After the war, the graves of the Boers murdered by Taylor, Morant and the others were tended by the Burgher Grave Committee. A simple gravestone was erected with the following wording: "Helden van Zuid-Afrika vir Vaderland en Vryheid 1899–1902" (Heroes of South Africa for Fatherland and Freedom 1899–1902), followed by a list of names of the deceased.

The courts martial, which heard the murder cases separately, started their proceedings on 16 January 1902 in Pietersburg (Polokwane), with Lieutenant Colonel H.D. Denny as president and six members. A Major J.F. Thomas of the New South Wales Mounted Rifles acted for the defence.

After hearing exhaustive evidence (some of which was damning, such as that of Corporal Sharpe, himself no blameless character), Lieutenants Morant, Handcock and Witton were found guilty of several charges of murder and sentenced to death. Corporal Sharpe then announced that he would gladly walk

The graves along the Phugwane River close to Zari mouth, which could possibly be those of two Boer prospectors who had presumably been murdered by Capt. Bulala Taylor of the Bushveldt Carbineers during the Anglo-Boer (28.04.1984)

the 100 kilometres to Pietersburg (Polokwane) to be a member of the firing squad that would execute Morant and the others! Lieutenant Picton received a dishonourable discharge and Major Lenehan was reduced to the ranks and severely reprimanded.

Taylor was tried as a civilian by the military court and acquitted, but was immediately reappointed to his military rank. He fled to Rhodesia (Zimbabwe) immediately after the war out of fear of retribution. Lieutenant Witton's death sentence was commuted to life imprisonment by Lord Kitchener, and he only spent a few years in jails at Gosport, Lewes and Portland in England. After his release in 1904 he returned to Australia, where he settled in Gippsland, Victoria. Here he wrote a book, *Scapegoats of the Empire*, about this unsavoury corner of British military history.

Lieutenants Morant and Handcock were taken to the Central Prison in Pretoria and executed at six o'clock the morning on 27 February 1902 by 18 men of the Cameron Highlanders under command of Lieutenant Thompson. Morant smoked a last cigarette, refused to be blindfolded and said in a typical Australian manner: "Shoot straight, you bastards, don't make a mess of it!" Other Australian soldiers took the bodies and buried them in a single grave in the Church Street cemetery (Schulenburg, 1981). Just a few days after the execution of Morant and Handcock, Theunis Botha was shot and killed while on a white horse in front of the Central Prison in Pretoria, by an unknown person, probably a Boer who considered Botha a despised traitor.

Only later did the Australian public hear about the courts martial and executions, and there was an outcry at the manner in which their countrymen had been executed. There have never been adequate answers to the questions of why Witton's death sentence was commuted and Taylor acquitted. This is probably how Breaker Morant became a folk hero in his own country – especially after the publication of the book *The Breaker*, authored by fellow Australian Kit Denton (1979), and a subsequent eponymous film on the topic of a persecuted soldier. To the distress of the Australian public, Denton himself later rejected his own romanticisation of Morant in another book, *Closed File* (1983).

Fourteen years after he had turned Breaker Morant into a symbolic victim of injustice and hypocrisy, Denton said that he had studied new information on the matter in depth and the man he had made out to be a folk hero was nothing less than a cold-blooded murderer. He added that Morant was unable to distinguish between right and wrong and was a cruel and violent man who would kill just for the experience. With regard to the courts martial themselves, Denton (1983) wrote that there had definitely been irregularities during the hearings of the various officers of the Bushveld Carbineers. He also stated that there were many unanswered questions in connection with Kitchener and the orders of his legal advisers at the time.

These aspects of the case were not, however, enough to change Denton's opinion that the trial had been as proper and fair as circumstances allowed – especially if taken into account that the people who had to hand down the verdict had no legal training, often had to listen to conflicting evidence and must have had their own personal and military preconceptions.

There are still many people in Australia who do

After their execution on 27 February 1902, Henry Morant and P. Handcock were buried in the same grave in the old cemetery in Church Street west, Pretoria.

Gen. Christiaan Frederick Beyers, commander of the Boer forces in the Northern Transvaal during the second half of the Anglo-Boer War. (ACKNOWLEDGEMENT: SANDF ARCHIVES, PRETORIA)

not accept Denton's (1983) explanation of the Morant saga. This is clear from a recent book by Nick Bleszynski (2002), *Shoot Straight You Bastards!*, in which he tries to prove that Morant and Handcock were executed wrongly because they, according to the evidence he presents, only carried out Kitchener's orders. He also tried to convince the Australian government to reopen the investigation into the execution, and believed that such an investigation would prove that the two accused, like Witton, did not deserve the death penalty, and that their names should in fact be placed on the Australian military roll of honour!

Bleszynski certainly has the makings of a strong case holding Kitchener (as Commander-in-Chief of the British forces in South Africa during the guerrilla stage of the Anglo-Boer War) ultimately responsible for the murders in the northern Transvaal. That is, if it can be proven that he had given the order that the Bushveld Carbineers should take no prisoners of war. Lord Kitchener was honoured by his countrymen as the personification of British military and aristocratic excellence, but history has shown him to be a cruel and merciless warlord who fought any opposition to the conquering mania of the British Empire with all the might available to him, even with barbarous methods.

Today his "scorched earth" policy and the horror of the concentration camps for the wives and children of the Boers of the two republics would certainly be considered war crimes of the worst kind by any international war tribunal, and Kitchener himself would probably have been called before such a court to account for his deeds.

It is therefore quite possible that he had personally given the order to the commander of the Bushveld Carbineers that they should not take any prisoners of war, in order to break the morale of the Boers in the northern Transvaal and thereby ending the war as soon as possible. It also seems fairly certain that he personally intervened to have Captain A. Taylor exonerated of all blame – especially since Joseph Chamberlain, the British Colonial Secretary, had judged that "two executions ought to be sufficient!". Therefore the Australian public should certainly ask questions about Kitchener's direct or indirect connection with the inhumane acts of the Bushveld Carbineers, but there can be no mitigating circumstances when it comes to the cold-blooded murders of the missionary Daniël Heese and his assistant by Morant and Handcock. Arthur Davey's (1987) sober analysis of the incident should carry the most weight; namely that, except in the case of "Bulala" Taylor, the members of the British court martial had meted out the correct punishment to the accused before them.

The one person who really knew what had happened in the northern Transvaal bushveld in those days was the pardoned George Witton. In 1929 he wrote a letter to Major Thomas, the counsel for the accused, in which he said the following:

> Personally I think the attitude you take with regard to Morant and Handcock in the Heese case is not the right one ... I saw Handcock shortly afterwards and asked him about the Heese business. He said, 'Why, weren't you standing beside Morant when he asked me if I was game to follow the Missionary and wipe him out?' I believe Morant got Handcock to deny the previous statement in which he had made a clean breast of everything and they got to work to frame up an alibi which you know was successful and the means of their acquittal (on this charge).

And so, with his declaration Witton undermined the premise of his book: that the accused, himself included, had been the scapegoats of the British Empire. It is clear that, with the exception of Taylor, justice was served.

By the time the court proceedings in Pietersburg (Polokwane) had come to an end, the war was entering its last phase. Representatives of the different Boer commandos were sent to Vereeniging in May to negotiate peace. But on 1 October 1901 General Beyers still attacked a British camp on the farm Pruizen, south of Potgietersrust (Makopane), and forced the British to surrender.

From Pruizen General Beyers went to Maleps Poort south of Pietersburg (Polokwane), where he found two small commandos under Commandant Daantjie de Villiers. From here he went with 200 men into the unhealthy Lowveld to help where he could. From Maleps Poort he went to Houtbosdorp and from here to Altenroxel (near Tzaneen), Leydsdorp and to the district of the current Phalaborwa. From here he travelled to the confluence of the Klein and Great Letaba, eastwards to the Mozambican border and then northwards along the border to a point about 30 miles (48 kilometres) from the confluence of the Limpopo and Luvuvhu rivers. Here he turned southwards and travelled to the Shingwedzi. He then trekked westwards along the Shingwedzi, a route that brought him to the Spelonken area, where he found the neglected camping sites of the Boer guerrilla fighters. Here and there along the way they captured black spies.

For a time General Beyers's commandos stayed in the vicinity of Bristow's farm Sweetwaters, and here he was nearly captured by the British. He managed to escape, however, despite the fact that he was suffering from a serious bout of malaria.

While court proceedings against Morant, Handcock, Taylor and the other accused were in progress, General Beyers attacked the Pietersburg concentration camp on 23 January 1902 and freed 150 burghers. In the dying moments of the war, in March–April 1902, he attacked and seized Ford Edward, but had to abandon the fort before an onslaught by the British forces under the command of Colonel Colenbrander.

Commandant A.F. du Toit described how he had fired what was probably the last shot of the war (loosely translated):

> After General C.H. Muller and Field General Wynand Viljoen had left for the peace negotiations in Vereeniging, I was Acting-General, while Jac Pienaar, later Administrator of the Transvaal, was one of my best commandants. Field Cornet Andrew Pienaar (later chairperson of the Public Service Commission), was another energetic officer at my side. We wanted to attack a joiner camp with 4000 cattle between Nelspruit and Barberton. Then the British commander at Lydenburg wrote that he had orders to shoot if we crossed the front line.
>
> We had a telephone line from Sabie to Pilgrim's Rest – a line the British had laid, but for which we had the apparatus. It was already 5 June 1902 (five days after the peace accord had been signed), and we still had heard nothing about the result of the negotiations. Commandant Giel Joubert of White River had an order from me to shoot if the British stepped over the front line; and we obstructed the road between Sabie and Lydenburg with stones. Commandant Joubert reported to me that Steinaecker's Horse (who had their camp at Sabie Bridge) was moving across the front line. He stopped them and they told him peace had been declared. He asked me what to do, so I sent an order that he should give them until three o'clock to lay down their weapons, then he had to shoot – I knew nothing about peace. I sent the same message to Major Steinaecker. Joubert and Steinaecker were sitting talking to each other when my two messages reached them.
>
> Steinaecker told Joubert that he would be back to inform him of their decision about what they intended doing. Then he went to his men and they galloped away along the old road down Legogote Hill. Joubert said to his men: 'Shoot, but aim high; maybe it is peace!' He chased after them for a bit. That night we received the message through Sabie that peace had been declared.

On 31 May 1902 the two Boer republics lost their independence at Vereeniging. After three years of blood, bitterness and misery, they could start rebuilding the Lowveld, but the earth was wild and bleak and pillars of smoke rose from the plains ...

Boer officers photographed by Maj. J. Stevenson-Hamilton during peace negotiations at Vereeniging in May 1902. Gen. C.R. de Wet is standing second from the right. (ACKNOWLEDGEMENT: STEVENSON-HAMILTON COLLECTION, SKUKUZA ARCHIVES)

References

Albasini, J. 1971. Personal communication to J.B. de Vaal.

Amery, L.S. 1900–1909. *The Times History of the War in South Africa,* vol. 5. Sampson, Low, Marston, London.

Anon. 1937. Tales of Col. Steinaecker. *Barberton & Nelspruit News,* April.

Badenhorst, S., Plug, I., Pelser, A.J. & Van Vollenhoven, A.C. 2002. Fauna/analysis from Steinaecker's Horse, the northernmost British Military outpost in the Kruger National Park during the South African War. *Annals of the Transvaal Museum* 39:57–63.

Beyers, C. 1944. Laaste lewensjare en heengaan van President Kruger. *Archives Year Book for South African History* l:19–67.

Bleszynski, N. 2002. *Shoot Straight You Bastards!* Random House, Milsons Point.

Breytenbach, J.H. 1969–1983. *Die geskiedenis van die Tweede Vryheidsoorlog in Suid-Afrika,* 5 vols. Government Printers, Pretoria.

Buchan, J. 1903. *The African Colony: Studies in the Reconstruction.* William Blackwood & Sons, London.

Bulpin, T.V. 1950. *Lost Trails on the Lowveld.* Howard B. Timmins, Cape Town.

Carnegie, M. & Shields, F. 1979. *In Search of Breaker Morant.* H.H. Stephenson, Armadale.

Cattrick, A. 1959. *Spoor of Blood.* Howard B. Timmins, Cape Town.

Changuion, L. 1986. *Pietersburg: die eerste eeu (1886–1986).* Pietersburg City Council, Pretoria.

Changuion, L. 2001. *Silence of the Guns. The History of the Long Toms in the Anglo Boer War.* Protea Book House, Pretoria.

Claassen, J.J.H. 1937. Die Vier "Long Toms". Letter to the Editor, *Die Huisgenoot* 28 June.

Conway, O.P.M. 1978. Baron von Steinaecker, D.S.O. Steinaecker's Horse. *Journal of the Medal Society* 1(2):3.

Davey, A.M. 1986–1987. Personal communication.

Davey, A.M. 1987. *Breaker Morant and the Bushveldt Carbineers.* Van Riebeeck Society Publication 18.

De Clerck, A. 1965. 'n Interessante ontdekking. *Nuus uit ons Parke* August:3.

Denton, K. 1979. *The Breaker.* Angus & Robertson Publishers, London.

Denton, K. 1983. *Closed File.* Bok Books International, Durban.

Diespecker, D. 1996. The naming of Steinaecker's Horse. *Military History Journal* 3(10):97–101.

Fitzgerald, P. 1939. Personal communication to G.J. van Niekerk, Soekmekaar.

Forsyth, D.R. 1972. Steinaecker's Horse. *Journal of the Orders and Medals Research Society* December:21–23.

Forsyth, D.R. 1972. Steinaecker's Horse. *Journal of the Orders and Medals Research Society* December:21–23.

Forsyth, D.R. 1975. Lt Col Francis Christiaan Ludwig Steinaecker. D.S.O. *Newsletter* no. 6 July, of the Military Medal Society of South Africa.

Francis, H.F. 1901. Letter from H.F. Francis to Lieutenant Pasement 02.08.1901. Skukuza Archives.

Gardyne, A.D.G. 1949. Colonel Ludwig Steinaecker of Steinaecker's Horse. Unpublished manuscript. Skukuza Archives.

Gensichen, M. 1902. *Bilder von unserem Missionsfelde in Süd- und Deutsch-Ost Afrika.* Berliner Evangelischer Missiongesellschaft, Berlin.

Greenhill-Gardyne, Maj. A. 1902. *Report on game preservation.* Skukuza Archives.

Heese, J.A. 1959. *Grossvater und Grossmutter Heese 1859–1959.* Paarlse Drukpers (Pty) Ltd, Paarl.

Jones, H.J. 1994. The Delagoa Bay railway line and the origin of Steinaecker's Horse. *Military History Journal* 6(9):102–109.

Jordaan, G. 1944. *So het hulle gesterf.* HAUM & De Bussy, Cape Town & Pretoria.

Kemp, J.C.G. 1941. *Vir vryheid en vir reg.* Nasionale Pers, Cape Town.

Kruger, P.R. 1973. M'pisana's Fort. *Historia* September:200–201.

Lee, S.J. 1973. *Krygsverrigtinge in Soutpansberg en Waterberg gedurende die Anglo-Boereoorlog 1899–1902.* Typed manuscripts issued in book form by the Human Research Council Library.

Lombaard, B.V. 1984. Die Longtomkanonne na die slag van Dalmanutha. *Militaria* 14(4):21–26.

Loock, J.C. 2001. *The ammunition from the northern outpost of Steinaecker's Horse.* Unpublished manuscript.

Louw, A.M. 1968. *Die groot gryse.* Tafelberg, Cape Town.

Magnus, P, 1980. *Economy and Society in Pre-industrial South Africa.* Penguin. Harmondsworth.

Malan, A. 1999. *Oorwinnings sonder roem. 'n Foto-album van herinneringe aan die Anglo-Boereoorlog 1899–1902.* J.P. van der Walt, Pretoria.

Matsebula, J.B.M. 1972. *A History of Swaziland.* Longmans, London.

Matthews, C.H.S. 1914. *Bill, a Bushman.* Edward Arnold, London.

Milstein, P. le S. 1986. *Historical occurrence of Lichtenstein's hartebeest, Alcephalus lictensteini in the Transvaal and Natal.* Unpublished manuscript.

Moerschell, C.J. 1910. *An der Grenze der Zivilisation.* H. Stürz, Würzburg.

Möller-Malan, D. 1966. *Oos van Doornrivier.* Tafelberg, Cape Town.

National Archives Depot. Master of the High Court MHG 32062. Death notice and estate of the late F.L. Steinaecker, 1918.

National Archives Depot – Transvaal Public Department, TPD 540, G1013/06: Transvaal Deptartment of Agriculture.

Naudé J.F. 1903. *Vechten en vluchten van Beyers en Kemp bokant de Wet.* Nijgh R. Van Ditmar, Rotterdam.

Paynter, D. 1986. Bismarck of the Bushveld. In: *Kruger – Portrait of a National Park.* Macmillan SA (Pty) Ltd, Pretoria.

Pakenham, T. 1979. *The Boer War.* Jonathan Ball, Johannesburg.

Penning, L. 1904. *De oorlog in Zuid-Afrika* vol. III. D.A. Daamen, Rotterdam.

Pelser, A. 1999. The Boer attack on Fort Mpisane, 7 August 1901. *Research by the National Cultural History Museum* 8:50–59.

Pelser, A & Van Vollenhoven, A 1998. The daily activities at the northernmost outpost of Steinaecker's Horse as reconstructed through archaeology. *Research by the National Cultural History Museum* 7:27–46.

Preller, G.S. 1916. *Kaptein Hindon – Oorlogsavonture van 'n baasverkenner.* Nasionale Pers, Cape Town.

Preller, G.S. 1923. *Oorlogsoormag en ander sketse en verhale.* Nasionale Pers, Cape Town.

Preller, G.S. 1937. Die Vier Longtoms. *Die Huisgenoot* 28 May.

Pretorius, F. 1985. *Die Anglo-Boereoorlog 1899–1902.* Don Nelson, Cape Town.

Pretorius, F. 2001. *Verskroeide Aarde.* Human & Rousseau, Pretoria.

Pretorius, J.L. 1910. Ons Suidafrikaanse Militaire Tradisie vols 1–4.
Die Brandwag 28 June:69–72
Die Brandwag 8 July.
Die Brandwag 15 July:104–106
Die Brandwag 1 August:143–145
Die Brandwag 1 September:205–209

Reitz, D. 1929. *Commando.* Faber & Faber, London.

Renar, F. 1902. *Bushman and Buccaneer. Harry Morant: His Ventures and Verses.* H.T. Dunn, Sydney.

Renfrew, B. 1984. Breaker Morant a ruthless killer. *Rand Daily Mail,* 19 June.

Scholtemeyer, J. 1987. Personal communication.

Scholtz, G.D. 1914. *Generaal Christiaan Frederik Beyers, 1869–1914.* Voortrekkerpers, Johannesburg, Pretoria & Potchefstroom.

Schulenburg, C.A.R. 1981. Die "Bushveldt Carbineers" – 'n greep uit die Anglo-Boereoorlog. *Historia* 26(l):37–58.

Spies, S.B. 1977. *Methods of Barbarism? Roberts and Kitchener and Civilians in the Boer Republics.* Human & Rousseau, Cape Town & Pretoria.

Stevenson-Hamilton, J. 1974. *South African Eden.* Collins, London.

Steyn, P. 1984. Lokalisering van Steinaecker's Horse se fort by Mpisane se kraal, New Forest. Personal report.

Stirling, J. 1907. *The Colonials in South Africa.* W. Blackwood, London.

Trichardt, S.P.E. – *Geschiedenis, werken en streven van S.P.E. Trichardt, Lt. Kol. der vroegere Staatsartillerie ZAR.* Human Sciences Research Council, Pretoria.

Van Vollenhoven, A.C., Pelser, A.J. & Teichert, F.E. 2003. A historical archaeological investigation of the northernmost outpost of Steinaecker's Horse, Letaba district, Kruger National Park (Report no. III). Report no. AE 302:63pp.

Van Vollenhoven, A.C., Van den Bos, J.W. & Pelser, A.J. 1996. 'n Eerstefase kultuurhulpbron-ondersoek na enkele militêr-historiese terreine in die Nasionale Krugerwildtuin. Unpublished manuscript. National Cultural History Museum, Pretoria.

Van Vollenhoven, A.C., Van den Bos, J.W. & Pelser, A.J. 1997. 'n Kultuurhulpbronstudie van die Britse blokhuisstelsel van die Tweede Anglo-Boereoorlog (1899–1902) in die voormalige Transvaal (Zuid-Afrikaansche Republiek). Unpublished manuscript. National Cultural History Museum, Pretoria.

Van Vollenhoven, A.C., Van den Bos, J.W. & Pelser, A.J. 1998. Historical-archaeological investigation of the northernmost outpost of Steinaecker's Horse, Letaba district, Kruger National Park. Unpublished manuscript. National Cultural History Museum, Pretoria.

Van Vollenhoven, A.C., Van den Bos, J.W. & Pelser, A.J. 1998. A historical-archaeological investigation of an Anglo-Boer War British outpost in the Kruger National Park. Unpublished manuscript. National Cultural History Museum, Pretoria.

Van Vollenhoven, A.C., Van den Bos, J.W. & Pelser, A.J. 1998. A historical-archaeological investigation of an Anglo-Boer War British outpost in the Kruger National Park. *Koedoe* 41(2):113–121.

Van Vollenhoven, A.C., Van den Bos, J.W. & Pelser, A.J. 2001. A historical-archaeological investigation of the northernmost outpost of Steinaecker's Horse, Letaba district, Kruger National Park II. Unpublished manuscript. Archaetnos. Wonderboompoort, 57 pp.

Van Zyl, P.H.S. 1946. Baasverkenner – kaptein Jan Mellema se avonture. *Die Brandwag,* 27 December.

Van Zyl, P.H.S. 1947. Kommandant "Rooi Regering". *Die Brandwag* 10 January..

Van Zyl, P.H.S. 1948. *Waar en trou.* Afrikaanse Pers, Johannesburg.

Viljoen, B.J. 1902. *Mijne herinneringen uit den Angloboeren oorlog.* W. Versluys, Amsterdam.

Wallace, R.L. 1976. *The Australians at the Boer War.* Australian War Memorial and the Australian Government Publishing Service, Canberra.

Wederpohl, J. 1921. *Sy herinneringe oor Eerw. Daniël Heese.* Unknown publisher, Pietersburg.

Wiid, J.A. 1959. Oorlogsmisdade in Suid-Afrika. *Die Huisgenoot,* 11 September.

Wilson, H.W. 1902. *After Pretoria: the Guerilla War.* The Amalgamated Press, London.

Witton, G.R. 1907. *Scapegoats of the Empire.* Angus & Robertson (U.K.) Ltd, London.

Wolhuter, H. 1948. *Memories of a Game Ranger.* Wild Life Protection Society of South Africa.

Wongtschowski, B.E.W. – *Between Woodbush and Wolkberg.* Review Printers, Pietersburg.

Chapter 15
Reproclamation of the Sabie Game Reserve and later border changes
Dr U. de V. Pienaar

During the last few months of the Anglo-Boer War the British administration, headed by Governor Lord Alfred Milner, decided to reproclaim President Kruger's Sabie Game Reserve in its original format, mainly as a result of the representations of prominent residents of the Lowveld such as Henry Glynn of Sabie, J.C. Ingle of Bushbuckridge, the naturalist and author Abel Chapman, Sir Alfred E. Pease and others. In as early as June 1902 this was accomplished under Administrative Proclamation No. 11 of the Lieutenant General of the Transvaal. The original Dutch text was used in the description of the boundaries.

A year later, on 28 August 1903, Lieutenant Governor Arthur Lawley, on the authority of Acting Colonial Secretary P. Duncan, issued Administrative Proclamation No.38 in which the boundaries of the Sabie Game Reserve were described in English. A large tract of land (both state land and land belonging to the Transvaal Consolidated Land and Exploration Company) between the Sabie and Olifants rivers was added to the original Sabie Game Reserve of the ZAR (South African Republic). The complete proclamation reads as follows:

Proclamation

By his Excellency the Lieutenant Governor

Whereas by the Section of Ordinance No. 29 of 1902 entitled 'The Game Preservation Ordinance 1902' it is provided that Reserves within which it shall not be lawful to kill game without the special written permit of the Colonial Secretary may from time to time be established by the Lieutenant Governor by Proclamation in the Gazette defining such Reserves:

Now therefore under and by virtue of the powers thus vested in me I do proclaim and make known that the area within the following boundaries, i.e.,

North – The Olifants River;
East – The Portuguese Boundary;
South – The Crocodile River;
West – The watershed running from the Crocodile Poort Railway Station to Logies Kop, thence following the North Sand River to where it joins the Sabi River, thence following the Sabi River to Mambatene's Kraals thence northwards by the old waggon road to where it touches the Selati Railway Extension, thence following the Selati Railway Extension to the Olifants River,

shall be and is hereby established as a Reserve within which it shall not be lawful to kill game without the special written permit of the Colonial Secretary and that any person who shall take, kill, pursue, hunt, shoot or destroy by any means whatever any game within the said reserve without the special written permit of the Colonial Secretary shall on con-

Lord Alfred Milner, British High Commissioner in South Africa from 1897 and Governor of the Orange Free State and Transvaal from 1900 and 1902 respectively. He resigned in April 1905 and was succeeded by Lord Selborne.
(ACKNOWLEDGEMENT: STATE ARCHIVES, PRETORIA)

viction be liable to a penalty not exceeding one hundred pounds and in default of payment thereof imprisonment with or without hard labour for any period not exceeding six months as provided by the aforesaid Ordinance.

God save the King.

Given under my Hand and Seal at Pretoria, this 28th day of August, One thousand Nine hundred and three.

Arthur Lawley
Lieutenant Governor
By order of His Excellency the
Lieutenant Governor
P. Duncan
Acting Colonial Secretary

In a letter to the assistant colonial secretary dated 22 April 1903, the secretary of native affairs mentioned that the commissioner of native affairs believed that a game reserve should be proclaimed in the Soutpansberg district. The boundaries would be the same as those set out in a letter dated 19 December 1902 from L.H. Ledeboer to the magistrate of Pietersburg (Polokwane). This reserve was proclaimed without much further ado by the Lieutenant Governor under Administrative Proclamation No. 19 of 1903. The border of the new Shingwedzi Reserve went from the confluence of the Great and Klein Letaba rivers in a straight line northwards to Shikumdu Hill and the Luvuvhu River, and then northeast along the Luvuvhu River to its confluence with the Limpopo River, from where it followed the Transvaal-Mozambique border to the Olifants River. It then followed the Olifants River to its confluence with the Great Letaba and then went along the Great Letaba westwards to its confluence with the Klein Letaba.

The border of this reserve remained more or less unchanged until 1913, when the Mhinga and Shikumdu locations (No. 284 and 285) were cut out of the reserve.

In 1906, Administrative Proclamation No. 31 of 6 April 1906 added a considerable tract of land to the old Sabie Game Reserve. The addition consisted of the old Kaap blocks E & F, which were state land and excellent game country and only used by sheep farmers from the Lydenburg, Carolina and Ermelo districts as winter grazing areas.

After that the boundaries of the Sabie Game Reserve read as follows:

From the southwestern beacon of Lot No. 333, Section 'E', Kaap Block (near Crocodile Poort Station), generally northwards along the boundaries of and including Lots Nos. 333, 332, 331, 327, 317, 328, 310, 227, 274, 192, 194, 146, 145, 196, 198, 199 and 200, Section 'E', Kaap Block, to the northwestern beacon of the last mentioned lot, continuing with the boundaries of and including the farm Engelbrechtshoop No. 112 to its northwestern beacon, Lozies or Logies Kop; thence generally northwards along the boundaries of and including Lots Nos. 134 and 135, Section 'F', Kaap Block, to where the western boundary of Lot No. 135 cuts the North Zand River; thence northwards down the middle of the North Zand River to its junction with the Sabi River; thence eastwards down the middle of the Sabi River to where it is crossed by the old waggon road near Mambatene's Kraals; thence northwards along the said road to where it touches the Selati Railway; thence following the eastern side of the Selati Railway to the middle of the Mklaseri River; thence down the middle of the

Sir Arthur Lawley (second from the right, back) with Lord and Lady Chamberlain.
(ACKNOWLEDGEMENT: TRANSVAAL ARCHIVES)

15. REPROCLAMATION OF THE SABIE GAME RESERVE AND LATER BORDER CHANGES

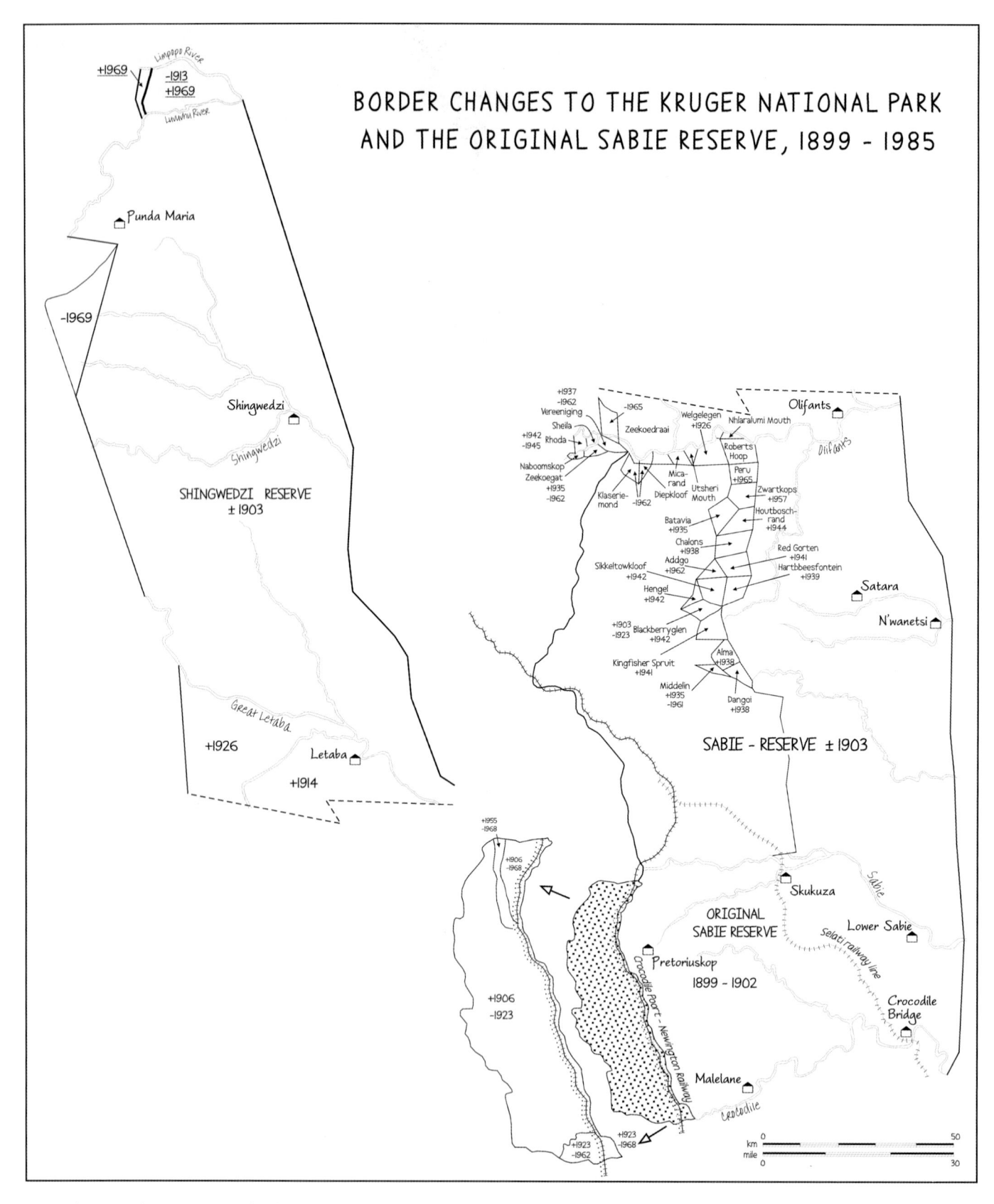

Border changes to the Kruger National Park and the original Sabie Game Reserve.

Mklaseri River to its junction with the Olifants River; thence eastwards down the middle of the Olifants River to the provincial Boundary; thence southwards along the provincial Boundary to where it crosses the Komati River, thence westwards up the middle of the Komati River to its junction with the Crocodile River, thence up the middle of the Crocodile River to the southwestern beacon of Lot No. 333, Section 'E', Kaap Block.

In 1911 the farm Burger's Hall was added to the Sabie Game Reserve.

In 1913 the Administration also found it necessary make excisions from and additions to the Sabie Game Reserve, apart from the adjustments to the boundaries of the Shingwedzi Reserve. Along the southwestern border of the Sabie Game Reserve the farms Logies Farm, Engelbrechtshoop and Burger's Hall were cut out of the reserve as they were already settled by whites. A significant piece of land between the Sabie and Shingwedzi reserves, which until then had fallen under the Department of Mines, was added to the protected area (with the department retaining mineral and prospecting rights). As a result the Sabie and Shingwedzi reserves were combined into one big reserve. Administrator Johann Rissik proclaimed the additional area as a game reserve under Administrator's Proclamation No. 48 of 1 December 1914, and it was described as follows:

From the junction of the Manedze [Nwanedzi] spruit with the Groot-Letaba River, generally south-eastwards down the middle of the Groot-Letaba river to its junction with the Olifants River; thence generally westwards of the middle of the Olifants River to the junction of the Malalani [Mulalani] spruit; thence generally north-westwards up the middle of the Malalani spruit to its source; thence north-eastwards to Shishene beacon, continuing north-eastwards in a straight line to the source of the Manedze spruit, thence generally north-eastwards down the middle of the Manedze spruit to the place of origin.

In 1916 the Sabie and Shingwedzi reserves were consolidated into the Transvaal Game Reserve, managed by the provincial secretary.

In March 1916 the then Transvaal Administration decided to appoint a commission under the chairmanship of Advocate J. Ludorf (later the chairperson of the Parks Board), with S.H. Coetzee, A. Grant, F.A.W. Lucas, H. de Waal, G. Hartog and C. Wade as members. Their mandate was to investigate the extent, adequacy and administration of the Transvaal Game Reserve and make recommendations. The commission's report was submitted in 1918 and in it several recommendations were made in connection with the Transvaal Game Reserve (the former Sabie and Shingwedzi reserves) that would

Member of the Executive Council Adv. J.F. Ludorf, chairperson of the provincial commission who had to report about the right of existence of the Sabie and Shingwedzi reserves in 1918. He was also a member of the National Parks Board between 1930 and 1952 and chairperson of the Board in 1951.

The Reform Committee after their arrest during the Jameson Raid in December 1895. Sir Arthur Lawley, who became lieutenant governor of the Transvaal after the Anglo-Boer War (middle, front) with the roll of documents in his hand.
(ACKNOWLEDGEMENT: TRANSVAAL ARCHIVES)

have far-reaching implications for this conservation area.

The commission's first recommendation was that, apart from a small area between the confluence of the Nsikazi and Crocodile rivers (the so-called Sigambu Block), the rest of the old Kaap blocks E and F, which had been added to the Sabie Game Reserve in 1906, should be cut out of the reserve again and made available as winter grazing to sheep and other farmers. A big area between the Sabie and Olifants rivers, on both sides of the Selati railway line, would also be cut out of the reserve. In this way, the whole area that is currently the Sabie Sand and Timbavati reserves was cut out of the Transvaal Game Reserve.

This decision would be regretted in later years, especially when it became clear that the excised areas represented the entire habitat suitable to oribi and red duiker, as well as the best habitat for mountain reedbuck. Some of the best habitat for sable and roan antelope was also lost.

A positive recommendation was that more land should be added to the part of the reserve between the Olifants and Letaba rivers to establish a more meaningful connection between the former Sabie and Shingwedzi reserves. A draft border description for the newly proposed reserve was drawn up, and it was recommended that the new area be proclaimed a national park as soon as possible.

When the entire area was proclaimed as the Kruger National Park on 31 May 1926, the draft border description suggested by the Ludorf Commission was used in its entirety as the boundaries of South Africa's first national park. These boundaries were stipulated as follows (translated from the original Afrikaans):

> From the confluence of the Pafuri and Limpopo rivers in a southeastern and southern direction along the eastern border of the Transvaal Province to where it crosses the Komati River; thence in a western direction along the right bank of the Komati River to its confluence with the Crocodile River; thence in a western direction along the right bank of the Crocodile River to the extension of the western boundary of Lot 375, Section E of the Kaap Block, District Barberton; thence in a northeastern direction along the boundaries of the following lots of the previously mentioned Section E, so they fall within the described area, namely: Nos 375, 374, 373, 369 and 368 to the northeastern corner, on the Logies River, of the last-mentioned lot; thence in a northwestern direction along the right bank of the Logies River to the communal beacon of lots Nos 173 and 174 in Section F, Kaap Block, District Barberton; thence in a northern direction along the eastern boundary of the mentioned Section F to the left bank of the Sabie River; thence in an eastern direction along the left bank of the Sabie River to the southeastern corner of the farm Kingstown No. 318, District Potgietersrust; thence in a northern direction along the boundaries of the following farms in District Pilgrim's Rest, so those farms fall within the described area, namely: Sabie Bridge No. 63, Prairie No. 341, Albumen No. 102, Congconia No. 393, Brandybal No. 388, Kelvin No. 334, Atherstone No. 390, Helena No. 335, Gowrie No. 348, Imlay No. 345, Tadema No. 404, Rabelais No. 451, Kameelkop No. 116, Balla Harra No. 499, Roodewal No. 147, Doornspruit No. 149, 's Gravenhage No. 399, Grootlaagte No. 131 to the northwestern beacon of the last-mentioned farm; thence in a western direction along the southern boundaries of the following farms in District Pilgrim's Rest, so these farms fall within the defined area, namely: Robertshoop No. 169, Welgelegen No. 157, Mica-rand No. 124, Zeekoeidraai No. 92, Diepkloof No. 91, Klaserimond No. 90 to the northwestern corner, on

The old western boundary of the Kruger National Park from Masala Hill northwards, before Mrs Eileen Orpen bought seven farms and donated them to the Kruger National Park.
(17.07.1984)

the Olifants River, of the last-mentioned farm; thence northwards along the right bank of the Olifants River to its confluence with the Selati River; thence along the right bank of the Selati River until it reaches the southern boundary of the farm Loole No. 1199 in the District Pietersburg; thence in a straight line eastwards to the southeastern corner of the mentioned farm; thence northwards along the eastern boundaries of the following farms in the District Pietersburg, so these farms fall outside the described area, namely: Loole No. 1199, Laaste No. 1198, Silonque No. 1172, Genoeg No. 1171, Ophou No. 1144, Letaba Ranch No. 1143 to the northeastern corner of the last-mentioned farm on the Great Letaba River; thence eastwards along the left bank of the Great Letaba River to its confluence with the Klein Letaba River; thence northwards along the right bank of the Klein Letaba River to the northwestern corner of the farm Mariakootch No. 2103, Pietersburg district; thence north-westwards in a straight line to the southeastern corner of the farm Alten No. 1237, District Zoutpansberg; thence north-westwards along the eastern boundary of the following farms in the District Zoutpansberg, so these farms fall outside the defined area, namely: Alten No. 1237, Plange No. 1238, Lombaard No. 1239, Kelder No. 1240, Schutte No. 1242, Kluster No. 1244, Pagelee No. 1252, Johnstone No. 1253 to the left bank of the northernmost Shingwedzi River (Mphongolo); thence eastwards along the left bank of the Shingwedzi River (Mphongolo) to the southwestern corner of Tshikundus Location's extension No. 1250, Zoutpansberg district; thence in a northern direction along the eastern boundary of the mentioned Tshikundus Location's extension No. 1250, Mhinga's Location's extension No. 1251 and the northern boundary of Mhinga's Location No. 185, all in the District Zoutpansberg, so they fall outside the defined area, to the Pafuri River; thence in a northeastern direction along the left bank of the Pafuri River to the exit.

Notable about this boundary description is the fact that the riverbank, rather than the middle of the river, is now used to indicate the border of the protected area.

Before the proclamation of the Kruger National Park in 1926, 26 state-owned farms known as the Alexandra Block, with Newington as the centre, were added to the Transvaal Game Reserve in February 1925. During the extensive negotiations between Minister Piet Grobler and the Transvaal Consolidated Land and Exploration Company (TCL) (which owned many farms between the Sabie and Olifants rivers) before the proclamation of the Kruger National Park could become a reality, the farms making up the Alexandra Block were offered to the TCL in exchange for the land it would have to surrender to the reserve. The offer was accepted and proclamation could go ahead.

After the excision in 1923 of the old Kaap Blocks E and F from the Transvaal Game Reserve, warden J. Stevenson-Hamilton made representations to the government that the areas be placed under his control again due to the large-scale eradication of game by blacks who had settled there. The Park's rangers were given permission in 1927 to combat poaching in these areas. This policy was a success until the outbreak of the Second World War, after which the concession was withdrawn and the areas became part of the former KaNgwane homeland.

The Law on National Parks was promulgated on 31 May 1926, and after the establishment of the Kruger National Park under Proclamation No. 197 in the *Government Gazette* of 10 September 1926 the boundaries of the area underwent considerable changes (see map on page 436), until it was fenced in the late 1970s.

The first change took place in 1933 when the warden of the Kruger National Park was put in charge of game preservation in the adjacent Pafuri Reserve – the Crooks Corner area between the Luvuvhu and Limpopo rivers. This arrangement lasted until June 1954, when the Parks Board itself discontinued it.

In April 1954 the state proclaimed the farms Zeekoegat No. 4, Knaboomkop No. 5 next to the Olifants River, Middelin No. 106, Johniesdale No. 355 and Batavia No. 298 as part of the Park. At the same time, the farm Chalons No. 229 of 11 099,8 acres, which had been bought by Eileen Orpen, was given to the state to be added to the Park. In February 1936 the farm Johniesdale was exchanged for the farm Vereeniging No. 3 (private land) along the

15. REPROCLAMATION OF THE SABIE GAME RESERVE AND LATER BORDER CHANGES

Mrs Eileen Akerman Orpen (1882–1954), greatest benefactor of the Kruger National Park. She and her husband donated seven farms comprising 24 530 hectares along the western boundary of the Central District between 1934 and 1944.
(ACKNOWLEDGEMENT: DR E. GREGOR, RONDEBOSCH)

The former Rabelais entrance gate to the Kruger National Park, before the donation of seven farms by Mrs E. Orpen, which necessitated the movement of the gate eight kilometres westwards. The new Orpen Gate was opened in May 1954.
(17.07.1984)

Olifants River. At the same time the Board asked that the farm Sumatra No. 299 (state land) also be added to the Park, but this request was denied by the Minister of Lands in May 1936. This farm was added to the Park in 1937. In January 1938 the state also bought the farms D'Argai No. 403 and Alma No. 453 and added them to the Kruger National Park. In February 1938 an attempt was made to have the old farm Joubertshoop, between Pretoriuskop and Malelane, subdivided into units of 50 morgen (41,8 ha), and sell these units to members of the public, giving owners the right to hunt! Fortunately the Board managed to scotch this wild scheme.

In September 1938 a Mr Heroldt, the owner of the farm Kingfisherspruit, offered the Park the small piece of his farm on which Rabelais Gate was situated, with the proviso that the Board carry the transport costs. Mrs Eileen Orpen bought the following farms and donated them to the state to be added to the Park: Kingfisherspruit No. 406 (10 341 acres) and Red Gorten No. 126 (9 943,6 acres) in February 1941; Blackberry Glen No. 154 (8 765 acres), Hengel No. 121 (2 056,9 acres) and Sikkeltowkloof (10 007,9 acres) in June 1942; and Houtboschrand No. 87 (8 398 acres) in February 1944. Eileen Orpen thus donated seven farms to the Kruger National Park, with a total surface of 60 611 acres (24 281 ha). In current land values this gift would be worth millions of rands today. No greater gift could have been made to the South African nation!

In November 1938 the Parks Board decided to approach the government about the incorporation of the farms Rhoda and Sheila (state land) in the Park.

In January 1942 the farms Rhoda and Sheila, on the northern bank of the Olifants River and west of Phalaborwa, were placed under the care of the Board with a view to protecting the game there. In 1945 black cattle farmers were settled on Rhoda and Sheila and the game protection rights of the Board were withdrawn. In February 1943, Secretary of Mining V.H. Osborn asked the Board to act in a supervisory capacity on the farms Wegsteek No. 494 and Loole No. 199, but it was later denied that such an agreement had ever been reached.

In July 1946 Eileen Orpen bought the farm Newington and donated it to the Board, but as there was no way of connecting this land to the Kruger National Park, the Board regretfully had to hand back the title deeds.

In as early as the 1930s the Board realised that it had accidentally moved outside the proclaimed boundaries of the Park during the development of Numbi Gate and its facilities, as well as during the placement of certain ranger outposts, including Metsel and Mkiweni, on the western border north of Numbi Gate. To resolve the problem and to determine the western border more precisely, the firm of surveyors J.M. Rabie & Co was appointed in 1949 to investigate the situation and to report back to the Board. On 24 August 1949 it reported that several of the beacons had disappeared, but that there was no doubt that the Park had intruded far into the "native" area.

To rectify the situation, Parks Board offered land to the SA Native Trust in exchange for the about 2500 morgen of Trust land the Park had "swallowed". After long and complicated negotiations the matter was finally settled to everybody's satisfaction in the mid-1950s. In the swap with the SA Native Trust, the Board relinquished about 1500 morgen of the illegally claimed land near Numbi Gate, as well as the farm Daannell 33 NU (No. 212) southwest of Numbi Gate, while the farms Numbi No. 211 and Rooiduiker No. 210 north of Numbi Gate were added to the

Park. The tracts of land under discussion here were respectively proclaimed and deproclaimed on 12 April 1957. The farm Rooiduiker, which had been added to the Park, included Numbi Hill. In this manner the Park regained a suitable habitat for red duiker.

In 1956 a stock-proof fence was erected between the Kruger National Park and the adjacent trust land from Numbi Hill to the Sabie River.

During 1961–1962 the western border of the Kruger National Park was fenced to control the spread of foot-and-mouth disease. Due to practical considerations and to keep the fence as straight as possible, the western part of the old Sigambu Block, just west of the confluence of the Nsikazi and Crocodile rivers, was cut out, as were the farm Middelin No. 106 and sections of the farms Klaserimond No. 90 and Diepkloof No. 91. In this process the western border fence between Numbi Hill and the Sabie River was erected slightly to the east of the 1956 fence. In exchange for the land the Park had lost, the state bought the eastern half of the farm Addger No. 974 and added it to the Kruger. In 1965 Phalaborwa Phosphate Mine had difficulties finding suitable land for its silt dams and dumps and asked the Parks Board to make an area between Phalaborwa Gate and the Klaseri Mouth (north of the Olifants River and including Mahulule Hill) available.

In exchange, the Phalaborwa Mining Company

The buildings on the farm Kingfisherspruit which Mrs Eileen Orpen had bought in 1941 and then donated to the Board. This is where the Kingfisherspruit ranger post is currently situated. (17.07.1984)

bought the farm Peru No. 208, just north of the farm Zwartkops, for inclusion in the Kruger National Park. This transaction was concluded in 1965.

The construction of the new railway line between Kaapmuiden and Metsi and the diversion of this part of the old Selati Line, which had run through the Park, to the western border of the conservation area, resulted in the Park losing the rest of the old Sigambu Block between the veterinary fence and the new fence along the Nsikazi River. The Park also lost the following land: the farm Rooiduiker No. 210, including Numbi Hill and the farms Vaalribbok No. 547 JU, northeast of Numbi Hill, and Ingrid No 591 JU further north, which included the whole section between the old western border, the Phabeni Spruit and the Sabie River boundary of the farm Calcutta. It also meant that the last suitable red duiker habitat and an eventual settlement area for both grey rhebok and oribi were lost. Despite urgent appeals by the Parks Board's research personnel that the farms Rooiduiker, Vaalribbok and Ingrid west of the new railway line should remain part of the Park, to be used as a breeding camp for rare antelope species (red duiker, grey rhebok, oribi and roan antelope), the Board decided otherwise and it was agreed to let those sections go. To compensate, certain areas between the new railway line (the new western border for this section) and the Nsikazi River south of Numbi Gate were added to the Park, but in the process it lost more than 1000 ha. The land exchanged between the Board and the SA Native Trust has still not been proclaimed or deproclaimed by the government departments involved, which has resulted in chaos – the farm Vaalribbok was illegally occupied by squatters and a whole town, Nyongane, sprang up; while a luxury holiday farm, the Nyala Lodge, was built on the farm Ingrid, with plans for an airstrip. All of this on land that is still legally part of the Kruger National Park. It was certainly one of the factors to blame for the fact that the land claim by the Ntimane community submitted to the Land Claims Court included the farms Rooiduiker, Vaalribbok and Ingrid!

In the meantime, the Parks Board was still negotiating to receive the land owed it – a 1000 ha corridor from Hazyview, along the southern bank of the Sabie River to the Park's new Phabeni Gate, close to the Albasini Ruins.

After years of negotiations, which had started in the 1930s, a final exchange took place in 1972 in which the old Pafuri Reserve (including Makuleke's location), a total of 19 842,6 ha, was exchanged for the Mthlaveni area, with a surface of 20 344,3 ha, south of the Punda Maria Gate and north of Shangoni ranger post. This exchange happened thanks to the severe drought in the 1960s. The relocation of people from the Makuleke area to Mthlaveni was completed in August 1969. The land in compensation west of the Mutale mouth that would make the two areas equal in size was surveyed in November 1968, from a point about one and a half kilometres west of the Mutale mouth, north-westwards for about five kilometres and then north-eastwards to the Limpopo River across from Mabyeni Hill. Only in 1972 was this new area officially proclaimed as part of the Park.

In 1998 the Makuleke community succeeded in their land claim, on political grounds, to get back their traditional territory between the Limpopo and Luvuvhu rivers. According to the agreement that was eventually reached the Parks Board would continue to administer and manage the area on a "contractual basis", while all income generated from tourism or otherwise would go to the Makuleke community. They would also be able to stay on in the Mthlaveni area, southwest of Punda Maria, which had been exchanged in 1969 for the Makuleke area on a hectare for hectare basis, without any compensatory land being added to the Park. With this ruling the Land Claims Court set a precedent for a whole series of land claims against South African National Parks (formerly South African National Parks Board).

After the completion of the fencing of the eastern border in 1975–1976 and the northern border along the Limpopo River in 1980, the whole Kruger National Park was finally fenced. At the time, it was believed that there was very little possibility of additions to the Park or changes to the existing boundaries. At that stage the Kruger National Park consisted of 1 948 528 hectares and was already one of the largest national parks in the world.

In 1993 large sections of the original foot-and-mouth fence between the Sabie and Olifants rivers and the adjacent private game reserves were removed, with the permission of Veterinary Field Services, and a new disease control fence was erected further west.

Negotiations on the establishment of the Great Limpopo Transfrontier Park, which will consist of the Kruger National Park, the old Coutada 16 in Mozambique (now the Limpopo National Park) between the Olifants and Limpopo rivers and the Ghona-re-Zhou National Park in Zimbabwe, create new horizons. An experimental section of the old eastern border between the Kruger National Park and Mozambique has already been removed, but the expectations created will only come to fruit if the area is managed with both skill and circumspection.

Aerial photograph of the triangular eastern part of the farm Addger that was added to the Kruger National Park in 1962.

References

Administrative Proclamations of the Lieutenant Governor of Transvaal no. 11 of 1902; no. 19 of 1903; no. 35 of 1903; no 31 of 1906; no. 20 of 1909.

Administrative Proclamations of the Lieutenant Governor of Transvaal no. 48 of 1914; no. 71 of 1923.

Catrick, A. 1959. *Spoor of Blood.* Howard B. Timmins, Cape Town.

Changuion, L. 2001. *Silence of the Guns. The history of the Long Toms of the Anglo-Boer War.* Protea Book House, Pretoria.

Chapman, A. 1928. *Retrospect, Reminiscences and Impressions of a Hunter-naturalist in Three Continents 1881–1928.* Gurney and Jackson, London.

De la Rey, H. 1998. *The ware generaal Koos de la Rey.* Protea Book House, Pretoria.

Groenewald, C. 1992. *Banneling oor die oseaan. Boerekrygsgevangenis 1899–1902.* J.P. van der Walt, Pretoria.

Annual Reports of the Warden of the Kruger National Park to the Board 1902–1972.

Jooste, G. & Oosthuizen, A. 1998. *So het hulle gesterf. Gedenkboek van teregstellings van Kaapse rebelle en Republikeinse burgers tydens die Anglo-Boereoorlog, 1899–1902.* J.P. van der Walt, Pretoria.

Ludorf, J. 1918. *Report of the Game Reserves Commission.* Government Printers, Pretoria.

Marais, P. 1999. *Die vrou in die Anglo-Boereoorlog.* J.P. van der Walt, Pretoria.

Meijer, J.W. 2000. *Generaal Ben Viljoen 1869–1917.* Protea Book House, Pretoria.

Minutes of meetings of the National Parks Board of Trustees. 1927–1972.

Pretorius, F. 1999. *Life on Commando during the Anglo-Boer War 1899–1902.* Human & Rousseau, Cape Town.

Van der Merwe, C.N. & Rice, M. 1999. *A Century of Boer War Stories.* Johanthan Ball, Johannesburg.

Woolmore, W. 2002. *The Bushveldt Carbineers and the Pietersburg Light Horse.* Slouch Hat Publications, McCrae.

Chapter 16
Early years: The Sabie and Shingwedzi reserves and the Kruger National Park 1900–1946

16.1 The early evolution of the Sabie Game Reserve (1900–1926) and the proclamation of the Kruger National Park

Dr E.J. Carruthers and Dr U. de V. Pienaar

Although the Sabie Game Reserve was already proclaimed in March 1898, it would take an entire generation for it to become a fully fledged national park, and in this process of metamorphosis it was subjected to many vicissitudes and tumultuous happenings.

The signing of the Treaty of Vereeniging on 31 May 1902 after the Anglo-Boer War was a watershed – not only did it bring an end to hostilities, but it also meant the end of the Zuid-Afrikaansche Republiek (South African Republic) and brought about the closure of the Transvaal borders. Effective imperialism was implemented and with it came suggestions regarding social and economic "progress". The Transvaal economy would be reconstructed by the mining industry, and British immigrants (former soldiers and others) would be encouraged to settle in rural areas to "modernise" the farming industry. This new development would sound the death knell for two deeply ingrained views of the environment that still existed in the Transvaal: that the natural resources of the area were inexhaustible (with regards to both plant and animal life) and that wildlife represented an economically sustainable resource. To the British the wild animals of the Transvaal meant much more, and were regarded as a source of inspiration and "sport".

The Transvaal pioneers had to make use of the country's wildlife to survive, and later the game also became a source of real material gain. This was not the case with the new British settlers here.

There was nevertheless some measure of continuity in thinking about wildlife during the pre- and post-war eras, in the sense that the preservation and conservation of fauna was considered important because of its value to people and not for its own sake. The Republicans hunted for food and income; the British colonists, on the other hand, hunted mainly for sport and pleasure. In the developing countries of the world, hunting for sport came to the fore when hunting for survival was no longer necessary or possible. As a result people living in countries where game had almost been eradicated often went to great lengths to protect wild animals and birds, and even bred them so they could be hunted for pleasure. England itself was a good example of such a country, with a hunting system that had developed over centuries.

Game and conservation affairs were of great importance to the new British administration in the Transvaal in the light of the prevailing view on sport hunting in England.

The British governors of the Transvaal after the war had seen game in their own country being decimated and were keen to prevent the same from happening in a colonial situation under their control.

Sir Godfrey Yeatman Lagden, private secretary of Sir Owen Lanyon and secretary of the Executive Council. He was chairperson and High Commissioner of the Intercolonial South African Native Affairs Commission (1902–1904).
(ACKNOWLEDGEMENT: NATIONAL CULTURAL HISTORICAL AND OPEN-AIR MUSEUM, PRETORIA)

16.1 THE EARLY EVOLUTION OF THE SABIE GAME RESERVE (1900–1926) AND THE KRUGER NATIONAL PARK

While the Anglo-Boer War was in full swing, the British played a key role in preparations for the first international conference in London in 1900 on the protection of wildlife in Africa. The conference led to the London Convention on Nature Conservation, which was unfortunately never ratified.

Helped and motivated by this British tradition, sentiment and sport rather than the maintenance of a food source became the main reasons for and aims of game preservation in the first years of the twentieth century.

That does not mean that there were no sentimentalists or sportsmen in the pre-war Transvaal. Those living in the ZAR had important victories in the interest of nature conservation, but the post-war era provided these first champions of nature conservation with many allies, especially in the influential Civil Service.

As an indication of how strongly the British occupational forces felt about the conservation of South African fauna, it is illuminating to know that they gave attention to nature conservation even before the end of the war. It may have been a typically Victorian characteristic that time could be spent on such a relative unimportant matter in the midst of a very expensive and destructive war.

The British forces occupied Pretoria in the first week of June 1900. At the end of September, barely three months later, the first notice regarding restrictions on hunting – a close season, based on the old Republican Act 5 of 1894 – appeared in the *Government Gazette*. It seems that this step was taken because British soldiers were shooting game around Pretoria to augment their rations. The English "Tommies" obviously did not share the enthusiasm of the conquered territory's administrators for nature conservation!

On the other hand, in the far-off Lowveld, Major A.W. Greenhill-Gardyne, an adjutant of the infamous ragbag of British border guards Steinaecker's Horse, did his utmost to check uncontrolled hunting by his unit at Sabie Bridge (later Skukuza). At the end of the war this officer sent a comprehensive report on wildlife conditions in the southern Lowveld to the head of the Department of Native Affairs, Sir Godfrey Lagden. He also made suggestions regarding the steps that should be taken to protect what little game remained. Major Greenhill-Gardyne's report would eventually prove to be of great help to the first official warden of the Sabie Game Reserve, Major J. Stevenson-Hamilton.

Apart from a convenient law that could enforce a closed season, the British administration also inherited a number of Republican era game reserves. In some of these areas a complete ban on hunting was in force, with the first category being the commons of many Transvaal towns on which so little game was left after the war that it was not worthwhile to have them reproclaimed.

Then there were the bigger areas where hunting had been prohibited for a period of five years shortly before the outbreak of war. These areas were in the Marico, Magatosland (north of the Soutpansberg), Elands River and Zwagershoek wards. Even before the war game conservation – despite the restrictions that had been imposed – was almost impossible here as a result of the rinderpest epidemic and the poverty of the inhabitants, both black and white.

The three proclaimed reserves – the Sabie Game Reserve in the eastern Transvaal Lowveld, the Pongola Game Reserve in the southeastern border area, and the Waterberg Reserve, northeast of Nylstroom (now Modimolle) – were in the third category.

Mr J.C. Ingle of Sabie who had acted as guide for author-explorer Abel Chapman in June to August 1899 during his journeys through the Lowveld. September 1902.
(ACKNOWLEDGEMENT: STEVENSON-HAMILTON COLLECTION, SKUKUZA ARCHIVES)

The Sabie Game Reserve was the first find the British made in their spoils of war. Even before the British occupation of the eastern Transvaal, the Foreign Office in London had received an application for the post of warden of the reserve. Arthur H. Glynn, a farmer from Lydenburg, well-known hunter and brother of the better-known Henry Glynn of Sabie, was in England at the time. In a letter to the Foreign Office in London in October 1900 he asked to be considered for the post of warden if the new administration reproclaimed the Sabie Game Reserve. He added that competition for the post would be fierce and that he was aware of at least two other applicants (of whom one was F. Vaughan-Kirby, the well-known hunter and author, who would later become the first warden of the Zululand reserves), but that he regarded himself to be the superior candidate. Nothing came of Glynn's application, as the Lowveld was still controlled by the Boer forces.

It was probably inevitable that the Sabie Game Reserve would be reproclaimed, but the matter was definitely expedited after the former Volksraad member for Barberton, R.K. Loveday, directed a request in this regard to Lord Alfred Milner, the British Governor of the Orange Free State and Transvaal. The Pongola Game Reserve was also reproclaimed soon after the British administration had become aware of it.

An important motivation for the reproclamation of the Sabie Game Reserve was undoubtedly the British aristocracy's interest in protecting wildlife. Another important reason was that the appointment of a warden or ranger in this part of the Lowveld would be one way of re-establishing a civil administration here. Other factors that contributed to the speedy reproclamation of the area in June 1902 by Lord Milner (in the original Dutch text of the 1898 proclamation) were the positive reports that had been received from various people about its potential. They included Major A.W. Greenhill-Gardyne; Henry Glynn of Sabie; J.C. Ingle; the explorer and author Abel Chapman, who had toured the Lowveld for three months between June and August 1899; Tom Casement, the assistant mine commissioner of Barberton; and Sir Alfred Pease, the magistrate of Barberton.

However, the deciding factor was probably the fact that, in contrast with the Waterberg Reserve, which was considered to be suitable for agricultural development, the Sabie and Pongola game reserves were thought to be unsuitable for human habitation as both were situated in endemic malaria areas.

Although the motives for the reproclamation of the Sabie Game Reserve were never formally set out, it would seem that the colonial government was keen to develop the reserve as an area suitable for sport hunting as soon as the numbers of the different species made it possible.

For this reason only the herbivores in the reserve were protected. Predators such as lions, leopards, cheetahs and wild dogs were declared vermin and shot as a matter of course. This concept differed radically from the later ideal that the area should be transformed into a national park in which all forms of life, and the environment in which they occurred, were protected as an ecological unit.

By 1901 the British occupational forces were in firm control of the greater portion of the southern Lowveld and they had already appointed civil officials to most towns. The mining commissioner of Barberton, Tom Casement, would play an important role in the preservation of the Sabie Game Reserve. He was an ardent nature lover and a keen conservationist. At the beginning of 1901 he wrote to the authorities in Pretoria to object to the issuing of permits to military personnel at Barberton to shoot game without valid hunting licences. He also objected to the fact that similar transgressions were taking place inside the Sabie Game Reserve.

The local magistrate, J.B. Skirving, was asked to take steps to stop this practice, but he said that the situation was uncontrollable since he had only a few policemen at his disposal. Casement's complaints were viewed in a serious light by the administration in Pretoria, and in May 1901 instructions were issued to all British units forbidding the hunting of game. Apparently most British army units adhered to this instruction, but in the Lowveld the particularly undisciplined Steinaecker's Horse did not take much note of official orders of this nature. This unit was deployed along the Transvaal/Mozambican border at Komatipoort, Sabie Bridge (Skukuza) and other places in the Lowveld, where they amused themselves raiding Boer farms and black villages, "guarding" the

16.1 THE EARLY EVOLUTION OF THE SABIE GAME RESERVE (1900–1926) AND THE KRUGER NATIONAL PARK

border and hunting where and whenever the opportunity presented itself (despite the efforts of some of their officers, including Major Greenhill-Gardyne).

They not only hunted for pleasure and to augment their rations, but also to sell trophies to other British units to take back to England.

One of Colonel Steinaecker's officers, Captain H.F. Francis, brought this state of affairs to Casement's attention and was rewarded with the appointment as honorary game inspector of the Sabie Reserve with effect from 4 July 1901.

Captain Francis's period of service was short-lived, however, as he was killed on 7 August 1901 during a Boer attack on Fort Mpisane, where he was the commander.

The only step he had taken before his death was to request that a number of black rangers be allocated to him to assist him in the task. After his death, another officer of the same unit, Lieutenant (later Captain) E.G. (Gaza) Gray, wrote a letter to Casement in which he asked to be appointed to Francis's position. His application was unsuccessful, however, as Casement had received several reports of illegal hunting by officers of Steinaecker's Horse and feared that appointing Gray would only encourage them.

Casement therefore suggested a more "reliable" candidate to the authorities in the person of William M. Walker. He had been a prospector at the Moodie diggings and joined a British cavalry unit at the outbreak of war, later becoming a member of the intelligence service. His candidature was supported by the fact that he knew this part of the Lowveld intimately and was fluent in Afrikaans as well as the local black languages.

Walker was officially appointed as the first postwar ranger of the Sabie Game Reserve on 24 October 1901, a position in which he did not acquit himself well. Because of the dangers posed by deserters (from British and Boer sides both) and wild animals during his visits to the Sabie Reserve, he was paid the unusually high salary (for the time) of £40 (then R80) a month.

Walker did very little to earn this generous salary, as he continued living in Barberton, visiting the Sabie Game Reserve only once – and that by travelling by train to Komatipoort. Along the way he warned everybody he encountered that he was the ranger and that they had to stop hunting immediately if they did not want to get into trouble with him. Not surprisingly, his superiors were unimpressed with the way in which he executed his duties as official in charge of the Sabie Game Reserve, and he was asked to resign at the end of January 1902.

After Walker's dismissal, another warden had to be found for the Sabie Game Reserve. The preferred candidate was Major A. St Hill Gibbons, an experienced and intrepid explorer who had, among others, led an expedition in 1898 and 1899 up the Zambezi to unknown parts of Barotseland. He was not interested in the position, however. For some reason the application of another suitable candidate, the hunter and author Frederick Vaughan-Kirby, was overlooked.

Major James Stevenson-Hamilton, who had accompanied Major Gibbons on his Barotseland expedition, was eventually chosen for the position.

There can be no doubt that he was the right choice, and Stevenson-Hamilton was destined to play a crucial role in nature conservation in South Africa during the first half of the nineteenth century. At the same time it was suggested to the administration that W.G. Treadwell and George Roy (of Lydenburg) should be appointed as ranger and senior ranger re-

Major James Stevenson-Hamilton as a young man during the time of his appointment as warden of the Sabie Reserve in 1902.
(ACKNOWLEDGEMENT: STEVENSON-HAMILTON COLLECTION, SKUKUZA ARCHIVES)

The coat of arms and motto of the Stevenson-Hamilton family.
(ACKNOWLEDGEMENT: STEVENSON-HAMILTON COLLECTION, SKUKUZA ARCHIVES)

Major J. Stevenson-Hamilton (left) with his unit, the 6th Inniskilling Dragoons at Schmidt's Drift in the Cape Colony, circa March 1902.
(ACKNOWLEDGEMENT: STEVENSON-HAMILTON COLLECTION, SKUKUZA ARCHIVES)

spectively under Major Stevenson-Hamilton. Nothing came of these proposals and the major later appointed his own white and black rangers.

Stevenson-Hamilton was of Scottish descent and heir to a large estate at Fairholm in Lanarkshire, Scotland. He was born on 2 October 1867 in Dublin, Ireland, where his father, James Stevenson, served as an officer in the 12th Royal Lancers Regiment. His mother was a Hamilton from Fairholm, Lanarkshire. The estates Fairholm and Kirkton in Scotland had been in the possession of the Hamilton family for 400 years, but when the old laird Hamilton died in the 1870s there was no male heir and his daughter Elizabeth (Hamilton) Stevenson inherited the estate. It was unthinkable that the name Hamilton should die out, so the two families agreed that their names and coats of arms would be combined. In this manner Captain Stevenson became James Stevenson-Hamilton and their only son, James, the heir to the title and estate.

The young James was only eight years old when his mother died and he became the laird of Fairholm. James attended the famous English public school Rugby and then entered the Royal Sandhurst Military Academy to prepare for a career in the military. After leaving the academy, he became a member of the Royal Archers Scottish Bodyguard before joining the 6th Inniskilling Dragoon Guards, an Irish cavalry regiment in which he became an officer. James came to South Africa for the first time in 1888. As a young man of 21, he and his regiment were sent to South Africa to help quell the Zulu rebellion in Natal. In August 1888 he had his first experience of active service in an area between the Black and White Umfolozi rivers, where his regiment encountered one of Dinizulu's impis. Shortly afterwards he saw his first South African antelope – a reedbuck family, which he could observe closely. The impression this first meeting left on him was indelible.

Shortly afterwards his regiment returned to Pietermaritzburg, where he made a name for himself as a horseman in races and on the polo field. During this time he also visited Johannesburg, and in his diaries he gave a lively and fascinating account of his adventures in the pioneer Gold City of those days (Carruthers, 1986).

Two years later, he and Captain (later Field Marshal) Allenby planned an expedition to the far-off Nyasaland. Captain Allenby was a keen amateur zoologist and taught the young Stevenson-Hamilton much about South Africa's wildlife.

Unfortunately nothing came of their planned expedition, as his regiment was suddenly recalled to England.

The time he had spent in South Africa took hold of his imagination and the mundanity of life in England made him long for Africa's unspoiled nature and wildlife. In 1898 he took a year's leave from his regiment to join Major A. St H. Gibbons's expedition in search of the source of the Zambezi River in Barotseland.

While they were having adventures and exploring the Caprivi and Barotseland they were unaware of the fact that the Anglo-Boer War had broken out, and that Lieutenant Stevenson-Hamilton had, on 31 December 1899, been promoted to captain.

Stevenson-Hamilton contracted malaria in Barotseland, and when he eventually returned to civilisation at the end of 1899 he heard about the war between Boer and Brit in the Free State and Transvaal. After recovering from his illness, he hurried to join the British forces in their fight against the Boer republics, boarding the *Herzog* at Chinde on the Mozambican coast in December 1899. The ship, carrying among others French and Belgian nurses off to help the Boer forces, was on its way to Delagoa Bay. Near their destination they were intercepted by the British battle ship *Thetis* and the captain of the *Herzog* was ordered to sail to Durban, where Stevenson-Hamilton would report for active duty in February 1900.

At first he served as officer in charge of a small section of Colonial Scouts, which was attached to General Clements's columns and was stationed at Barberton for a while. He was soon able, however, to join his own regiment, the 6th Inniskilling Dragoons, where he served until the end of the war.

His unit was especially active in the Free State and northwestern Cape. On his birthday in 1901 they were involved in a skirmish with a small Boer force near Cornelia in the northeastern Free State. Two Boers were killed and a number were taken hostage. They also seized several wagons, tumbrils and 1700 head of cattle.

Eventually the Treaty of Vereeniging was signed in May 1902. Stevenson-Hamilton was mentioned in dispatches and promoted to major.

Maj. James Stevenson-Hamilton on his horse Charlie, ready to leave for a patrol in the Sabie Reserve, 1909.
(ACKNOWLEDGEMENT: STEVENSON-HAMILTON COLLECTION, SKUKUZA ARCHIVES)

16. EARLY YEARS: THE SABIE AND SHINGWEDZI RESERVES AND THE KRUGER NATIONAL PARK 1900–1946

Stevenson-Hamilton did not want to return to England after the war and was looking for a suitable position in the colonial administration. On 2 June 1902, while in Bloemfontein, he wrote to the Commissioner of Native Affairs in Transvaal, Sir Godfrey Lagden, and applied for a suitable position. He had his eye on the area north of the Limpopo (more specifically Barotseland), but to his surprise and joy he was offered the position as "head ranger" of the Sabie Game Reserve on the recommendation of Major Gibbons.

Despite the warnings of the Rand Club's worthy members that he would not last long in the Lowveld, he accepted the job. On 1 July 1902 Stevenson-Hamilton was appointed as warden of the neglected game reserve at a salary of £500 a year, with an additional annual allowance of £180. He would get a raise of a further £100 a year in 1903.

His immediate superior was the Native Commissioner of Lydenburg, E.G. Hogge, but Sir Godfrey Lagden was the senior administrator. Stevenson-Hamilton was delighted with his new position – according to Sir Godfrey a "very interesting and sporting job" – but acknowledged that he had little idea what was expected of him in this remote part of the Transvaal. When he asked Sir Godfrey for instructions he was nonchalantly ordered "to go down there and make yourself thoroughly disagreeable to everyone!"

Stevenson-Hamilton was a dedicated, almost obsessive writer of diaries and reports, for which the current generation of nature conservation officers are eternally grateful. From these diaries and journals it is clear that he did not enjoy cities or crowds, preferred a simple, solitary existence and was never happier than when alone or with friends or family in the wilderness. He loved travelling and visiting far-off places, and when he was on leave he went on hunting and photographic safaris to the remotest corners of Africa. He was a keen amateur photographer and left behind a wealth of well-documented photographs, which give a vivid impression not only of his lifestyle, but also of the evolution of the Sabie Game Reserve and the later Kruger National Park from 1902 until his retirement in April 1946.

Although he lived a simple life, it is clear that he gradually surrounded himself with the accoutrements of a colonial gentleman of the time – spotless table linen, a silver tea service, a library, study, etc.

For recreation and relaxation he rode (he was an excellent horseman), angled in the Sabie and other rivers, and played tennis with his co-workers at Sabie Bridge and later his neighbours, the Crosbys of the adjacent farm Toulon, neatly outfitted in trousers, shirt and tie. He built the first tennis court at Sabie Bridge (Skukuza) in 1912 for this purpose.

Stevenson-Hamilton was short in stature and vigorous and restless in nature. He once admitted to a good friend, the artist Strat Caldecott, that any opposition impelled him to take effective action, and he certainly had no shortage of opposition during his years in service of conservation. What began as a

Col James Stevenson-Hamilton in a typical patrol outfit, circa late 1920s.
(ACKNOWLEDGEMENT: STEVENSON-HAMILTON COLLECTION, SKUKUZA ARCHIVES)

mere adventure for which he took two years' leave from his regiment soon turned into a career as warden of the Sabie Game Reserve and the Kruger National Park that for 44 years kept him in close contact with his great love and interest – nature conservation. During this long and illustrious career, through keen observation and study, he obtained knowledge of the wildlife of the Transvaal Lowveld that was unequalled at the time. Eventually he wrote four authoritative works: *Animal Life in South Africa* (1912), *The Lowveld: Its Wild Life and Its People* (1927), *South African Eden* (1937) and *Wild Life in South Africa* (1947). Apart from the books, he also penned many brochures and more than 30 articles in scientific and other magazines.

As a result of his chosen career and lifestyle, Stevenson-Hamilton stayed a bachelor until his sixties. Then, during his year-long return to England from May 1928 to June 1929 due to health problems, he met the 27-year-old Hilda Cholmondeley (pronounced Chumley). This accomplished artist, born in Australia in 1901, captured his heart. After Stevenson-Hamilton's return to South Africa to resume his duties as warden of the Kruger National Park, Hilda followed him. At first she stayed with an old school friend, Mrs Graham of Plaston, but on 13 August 1930 she married the 62-year-old bachelor and moved in with him at Skukuza.

Hilda Stevenson-Hamilton enjoyed the peaceful and unfettered lifestyle as much as her husband and developed an active interest in the park and its fauna and flora. Her sketches and paintings, many of which are exhibited in the Stevenson-Hamilton Memorial Library in Skukuza, had a unique quality and were so fresh and true-to-life that some of her themes are still being used to decorate table linen and curtains in the Park. The tall and stately woman was also a gifted sculptor, and after her husband's death on 10 December 1957 at the age of 90 on his estate Gibraltar near White River, she cast a life-size bronze statue of him, which is now exhibited in the entrance hall of the Stevenson-Hamilton Memorial Library. Hilda also designed a beautiful stained glass window for the small church near the Stevenson-Hamilton estate in Lanarkshire in memory of her husband. It depicts Saint Francis of Assisi, surrounded with wild South African animals and birds.

The Stevenson-Hamiltons had three children: Margaret (1931), Jamie (1933) and Anne (1935). Sadly, Margaret died of meningitis before her fourth birthday. Hilda herself passed away on 10 January 1979.

It is therefore only just that the ashes of this remarkable couple, who had played such a decisive and remarkable role in the history of the largest and most important national park in South Africa, were strewn in the wind at Shirimantanga by their daughter, Mrs Anne Doyle of England, on 10 April 1979, a lovely autumn day. This hill had been one of Stevenson-Hamilton's favourite haunts near Sku-

Stained glass panel built into the little chapel at Fairholm, Lanarkshire, Scotland. The panel had been designed and installed by Hilda in honour of her husband Col J. Stevenson-Hamilton (in life the first warden of the Kruger National Park). The coat of arms is that of the Stevenson-Hamiltons of Fairholm and the inscription reads as follows: "To the Glory of God and in memory of James Stevenson-Hamilton of Fairholm, soldier, explorer and guardian of wildlife. Born 2nd October 1867. Died in South Africa 10th December 1957. This memorial was designed by his widow." The text from the Scripture is Isaiah 11: "They shall not hurt nor destroy in all my Holy Mountain for the earth shall be full of the knowledge of the Lord and the waters cover the sea."
(ACKNOWLEDGEMENT: DR SUE HART, JOHANNESBURG)

Mrs Hilda Stevenson-Hamilton on her husband's farm Gibraltar, circa 1950s.

Anne and Jamie Stevenson-Hamilton at Skukuza, 1940.
(ACKNOWLEDGEMENT: STEVENSON-HAMILTON COLLECTION, SKUKUZA ARCHIVES)

The Count of Clarendon during a visit to Sabie Bridge (Skukuza) in 1932.
(ACKNOWLEDGEMENT: STEVENSON-HAMILTON COLLECTION, SKUKUZA ARCHIVES)

The Count of Athlone during his visit to Sabie Bridge (Skukuza) in 1930.
(ACKNOWLEDGEMENT: STEVENSON-HAMILTON COLLECTION, SKUKUZA ARCHIVES)

Members of the Executive Council of the Transvaal Provincial Administration during a visit to Sabie Bridge (Skukuza) on 17 June 1926. From left to right: F. Brown, J.H. Hofmeyer (administrator of the Transvaal), E. de Souza, Ben Pienaar, Col J. Stevenson-Hamilton, A.E. Charter and T.C. Stoffberg.
(ACKNOWLEDGEMENT: STEVENSON-HAMILTON COLLECTION, SKUKUZA ARCHIVES)

kuza, and on a clear day one can see the whole of the southern Lowveld from its summit. The warden had already, with the Board's permission, blasted a grave for himself out of the solid granite in 1930. A simple yet striking bronze memorial plaque on the side of a gigantic rock at the foot of the hill indicates where these founders of the Kruger National Park found their final resting place.

Although Stevenson-Hamilton was appointed to serve in one of the loneliest and remotest areas of southern Africa, he never lost contact with the outside world or allowed himself to stagnate as a result. He regularly undertook long journeys overseas and to the untouched corners of Africa, read widely, subscribed to books and periodicals and regularly corresponded with colleagues in the nature conservation field. He regularly saw his superiors in Pretoria to attend meetings, which gave him an opportunity to meet fellow government officials. In this way he even met ministers and got to know other high-ranking people while maintaining contact with friends and acquaintances. The contacts he made over the years in these circles would eventually bear fruit. As could be expected in the circumstances, he received important guests at his headquarters at Sabie Bridge from the start, including General J.C. Smuts (then the Minister of Defence), Julius Jeppe (the Surveyor General), Johann Rissik (the Administrator of the Transvaal) in 1912, the Prince of Wales in July 1925, the Executive Committee of the Transvaal Administration in 1926, Lord Athlone in 1930, Count Clarendon and his wife in 1932 and Sir Patrick Duncan (the Governor General of the Union of South Africa) in 1936, to name a few. After his retirement in April 1946 he received a special request to act as guide and host to the royal party, King George VI of England and his family, during their visit to the Kruger National Park in 1947.

In the early years of the Sabie Game Reserve Stevenson-Hamilton regularly returned to England to visit his family and estate. While in London he often gave lectures to societies and people in high circles, including Dr W.T. Hornaday, director of the New York Zoo, US President Theodore Roosevelt and others. He also became a member of the Society for the Protection of the Fauna of the Empire, the Royal Zoological Society of London, the Royal Geographical Society and others.

In 1904 he resigned his commission in his regiment when he was called up for duty, in order to continue with his chosen career. To do this he had to return to England and in November 1904 he sailed aboard the *SS Africa* up the east coast of Africa, calling at Beira, Kilindini, Mombasa, Aden, Port Said and the Suez Canal on the way.

From July to August 1908 he accompanied the British consul for Mozambique, R.C.F. Maugham, as cartographer on an expedition from northern Mozambique to Fort Maguire at Lake Nyassa (Malawi), and in 1909 a report about this journey was published in the journal of the Royal Geographical Society.

Between August 1909 and January 1910 he also undertook a long journey through Kenya, Uganda and the upper Nile to Khartoum in the Sudan and from there to Cairo, Athens, Constantinople (now Istanbul) and England. In 1910 he was asked to assemble a representative collection of African fauna for the British monarch.

Stevenson-Hamilton also undertook repeated journeys to Mozambique to discuss border problems, poaching, etc. with officials there, and in 1903 he visited Swaziland and the Pongola Game Reserve, where Major A.A. Fraser was in charge. In March 1907 he visited Northern Rhodesia (Zambia) and used the opportunity to visit his old friend Major Gibbons on his farm Pumba. In June the same year he visited the Waterberg and the Belgian Block to assess the wildlife situation there. He even undertook an expedition to the Kalahari in April–May 1912, and in April 1913 he once again set off to explore the Bechuanaland Protectorate (now Botswana). After his return from active duty in the First World War, he undertook another hunting trip to Uganda and the Sudan in December 1921.

But not all his journeys were pleasurable. At the outbreak of the First World War he immediately enlisted and left the running of the Sabie Game Reserve in the capable hands of ranger C.R. de Laporte. He left Cape Town on board the *SS Garth Castle* and travelled to Southampton via St Helena, Sierra Leone and the Cape Verde Islands. At the beginning of the war, his unit (the Yeomanry Brigade) served at Ramleh near Alexandria and in Cairo, Egypt. He requested a transfer and his new unit, the Scottish Lowland Division, was entrenched from July 1915 to January 1916 at Cape Helles in the Gallipoli peninsula. In 1916 he was stationed in Ismailia, but he later returned to Cairo as provost marshal with the rank of lieutenant colonel.

From 1917 until the end of his career in the British army at the end of 1919, he was seconded to the Egyptian army and served at different localities in the Sudan, including Mongalla, Rejaf, Kojokaji, Lau-Nuer and Duk. He decided against accepting an administrative position in Egypt or Palestine and returned to South Africa, where he resumed his responsibilities as warden of the Sabie Game Reserve in April 1920. During the war years the reserve was badly neglected by Major A.A. Fraser, who had taken over from De Laporte as acting warden in 1917.

Lt Col James Stevenson-Hamilton in his military uniform during the First World War, 1918.
(ACKNOWLEDGEMENT: STEVENSON-HAMILTON COLLECTION, SKUKUZA ARCHIVES)

Maj. James Stevenson-Hamilton during a hunting trip in the Nile Province of Uganda. Christmas 1909.
(ACKNOWLEDGEMENT: STEVENSON-HAMILTON COLLECTION, SKUKUZA ARCHIVES)

At the end of his long career in the Sabie Game Reserve and the Kruger National Park, Stevenson-Hamilton had already received international recognition for his contribution to nature conservation in general and for the conservation of the fauna and flora of the Transvaal Lowveld in particular.

For these services he was awarded a silver medal by the Zoological Society of London and two honorary doctorates in law – one in March 1935 by the University of the Witwatersrand and the other in November 1945 by the University of Cape Town. The National Parks Board also honoured the first warden of the Kruger National Park by appointing him warden emeritus of the Park on 27 June 1950, four years after his retirement.

Soon after his arrival at Sabie Bridge, local blacks nicknamed Stevenson-Hamilton Skukuza,[1] which means "he who sweeps clean". The name apparently referred to his large-scale campaign to resettle blacks living within the boundaries of the reserve in areas outside the reserve. This, along with all his other efforts to protect the fauna and flora of the area from further destruction, confirmed that his appointment in July 1902 had been a wise and admirable choice.

The characteristics the Administration wanted in a warden for the Sabie Game Reserve were those accepted at the time as the qualifications of a British head ranger. In contrast with the East African rangers of the time, to whom wildlife conservation meant convenient hunting for themselves and others, Stevenson-Hamilton started off with the right attitude. He was a painstaking administrator with military discipline, physically in a good condition and inured to hardships, highly intelligent, well-spoken, a keen observer and a natural leader.

Another qualification that counted in his favour was the fact that he had a private income. His starting salary, measured at today's standards, was a meagre £500 a year plus a horse allowance of £180 a year.

The game reserves in the Transvaal were placed under the supervision of Sir Godfrey Lagden, the Transvaal Commissioner of Native Affairs. He was a keen sportsman who helped Stevenson-Hamilton in word and deed with his difficult task during the first years of the Sabie Game Reserve.

The precedent set before the war in the Pongola Game Reserve determined that the warden of a Transvaal game reserve was a commander plenipotentiary in the area, as well as an ex officio justice of the peace. Stevenson-Hamilton made full use of his powers in this regard.

The establishment of the game reserve and the game ordinances that protected it got extensive publicity. After staying a few months in a corrugated-iron shack belonging to Gaza Gray at Crocodile Bridge, Stevenson-Hamilton erected his headquarters at Sabie Bridge and made the old blockhouse of Steinaecker's Horse at the bridge his temporary quarters. He then set about stamping his authority on the reserve, explored it properly and appointed his first rangers, both black and white, to assist him in his task.

Col James Stevenson-Hamilton on patrol in the Kruger National Park, 1930s.
(ACKNOWLEDGEMENT: STEVENSON-HAMILTON COLLECTION, SKUKUZA ARCHIVES)

---

1   Sometimes wrongly interpreted as meaning "the person who changes everything". Skukuza is also the name that was given to the first headquarters of the reserve, Sabie Bridge, in September 1932, at the insistence of Gustav Preller, one of the Board members, in acknowledgement of the long, excellent service of the park's first warden, Colonel James Stevenson-Hamilton.

The reserve was not open to the public and people wanting to visit it for whatever reason first had to obtain a permit. This ensured that the warden always knew who was in the reserve. Apart from being the local justice of the peace, he was also appointed as the resident native commissioner. In 1903 he arranged that the South African Constabulary withdraw from the reserve and, apart from a small police force at Sabie Bridge, he and his rangers took over the policing duties in the reserve.

That year the police would realise exactly how serious Stevenson-Hamilton was about his responsibilities as warden, when he was responsible for the arrest and trial of two very senior police officers for poaching in the reserve. Their conviction sent a clear warning to everyone else with similar aspirations.

In March 1903 Stevenson-Hamilton succeeded in convincing the Regional Commissioner of Native Affairs for the Eastern Transvaal, Mr Hogge, to ban the movement of cattle through the reserve. This ensured that officials of the newly established Department of Agriculture would not travel through the reserve. In July that year he also managed to frustrate the plans of the railways to mine coal along the Crocodile River, and successfully agitated for a halt to all prospecting in the reserve. He also got the cooperation of the Portuguese authorities to stop people from freely crossing the international border and to protect game on their side of the border. With these measures the Sabie Game Reserve was effectively managed under paramilitary guidelines.

In the Transvaal colony new hunting or conservation laws were promulgated almost every year between 1902 and unification in 1910. Ordinance 29 of 1902 revoked the old Republican legislation[2] and replaced it with new and expanded legislation.

The next year Ordinance 30 would give the Administration the right to promulgate game reserve regulations that had been left out of the 1902 law. In 1904, Ordinance 46 strengthened the restrictions on hunting even further, forbade angling from May to September every year, and formalised trout fishing. Ordinance 6 of 1905 consolidated and extended previous legislation on game. Slight concessions were made for those who had to hunt or fish for survival and it allowed the fishing of certain species throughout the year.

In 1907 Britain granted the former republics home rule, and the Legislative Assembly adopted three laws with regards to hunting. The first, Act 13, was aimed at the prevention and curtailment of the sale of biltong, in an effort to put a stop to formal hunting of any kind. The second, Act 23, restricted the keeping of hunting dogs by blacks and decreed that each dog had to be licensed by the owner for £5 annually. A third law, Act 30, prohibited the export of ostriches or their eggs. Lastly, Act 11 of 1909 gave the government greater scope in drawing up lists of game that should be protected or could be hunted, and stipulated that game in certain areas could be protected through regulations rather than specific laws. Discussions in the Legislative Assembly regarding wildlife matters were very one-sided when the Transvaal was a British colony and differences of opinion were rare. Prominent members of the Transvaal Game Protection Society, including R.K. Loveday, A. Woolls-Sampson[3] and E. Rooth, were all members of the Legislative Board and As-

Sir Patrick Duncan, Colonial Secretary just after the Anglo-Boer War and later Governor General of the Union of South Africa. He visited Skukuza during the Christmas holidays of 1937.
(ACKNOWLEDGEMENT: STEVENSON-HAMILTON COLLECTION, SKUKUZA ARCHIVES)

---

2   Ordinance 29 of 1902 not only maintained the bans on the hunting of elephant, hippo, buffalo, eland, giraffe, rhino and ostriches in the Republican Act 5 of 1894, but another 11 species were added to the list, including Lichtenstein's hartebeest, sable antelope, roan antelope and tsessebe.
3   A. Woolls-Sampson was the first to mention the idea of a national park in the Legislative Assembly of the Transvaal in 1907.

sembly and it would seem their views were shared by their fellow members. Their only opponent was usually W. Hosken, the member for Von Brandis, who regularly castigated his colleagues about the strict laws they had introduced that, according to him, were archaic compared with legislation in other countries. He was also concerned about the waste of food sources encouraged by the substantial restrictions on the sale of game products. His appeals had little effect and the pro-conservation members of the assemblies were particularly influential at the time.

Officials of the Department of Native Affairs, under whose jurisdiction the Transvaal game reserves fell until September 1905, and the department of the Colonial Secretary Sir Patrick Duncan, which controlled them until Unification in 1910, gave their continuous support to conservation in these areas. Other government departments were, however, less enthusiastic about the principle of conservation. While the land was worthless, it was safe, but two departments – Agriculture and Mining – realised it had economic potential.

Stevenson-Hamilton's successful petitions in 1902 to prohibit mining and prospecting in the Sabie Game Reserve were lamented by the Secretary of Mining, who thought this decision effectively "locked away" the mining potential of the eastern Transvaal. This opinion gained support in 1908 from the Department of Lands, but this caucus group was not influential enough to have the legislation changed. All applications to prospect in the Sabie Game Reserve were refused between 1900 and 1910, although permits were issued to a number of individuals or syndicates in search of the "Kruger Millions", which Stevenson-Hamilton and his personnel bitterly resented.

Relations with the Department of Agriculture were less sharply defined. Stevenson-Hamilton wrote a number of articles on the topic of wildlife preservation and the Sabie Game Reserve for the newly established *Transvaal Agricultural Journal* – an indication of the good relationship between the warden of the Sabie Game Reserve and this department. The editor of the journal also advised farmers to join the Game Protection Society. The reserve's officials co-operated with the Department of Agriculture in matters such as vermin control and the spread of animal diseases. The view that wildlife in the Transvaal was being conserved mainly for sport hunting led to several agricultural specialists commenting that sport should not stand in the way of agricultural development.

The dispute over the underutilisation of crown lands steadily increased in intensity and reached a climax in the mid-1920s.

As long as the climate and diseases like malaria discouraged human habitation, the continued existence of wildlife and the game reserves was ensured, but in as early as 1904 it was claimed that the Lowveld was becoming healthier (for settlement) and more valuable.

Stevenson-Hamilton probably expected this. Shortly after his appointment and with the support of Sir Godfrey Lagden, he started working towards extending his jurisdiction and ensuring that his control over the whole eastern border of the Transvaal was formalised.

After the appointment of the first rangers – E.G. (Gaza) Gray, Rupert Atmore and Harry Wolhuter in 1902, and C.R. de Laporte, T. Duke and A.A. Fraser (all former officers of the British forces during the war) shortly afterwards – Stevenson-Hamilton was confident that he would be able to exert control over larger areas. Because the areas he had in mind were then considered just as "useless" for settlement as the Sabie Game Reserve, he encountered very little opposition in his attempts to extend the reserve's boundaries. After only three months on the job he proposed that the Sabie Game Reserve's boundaries be extended both north- and westwards to protect more and a larger variety of game. In August 1903 he obtained control over game in the area north of the Sabie River to the Olifants River, and in 1906 the reserve was extended westwards to include the old Kaap Blocks E and F between the Nsikazi River and the upper reaches of the North Sand River, as well as the foothills of the Drakensberg.

In May 1903 a big new reserve was proclaimed – the Shingwedzi Reserve, situated between the Letaba River in the south and the Luvuvhu and Limpopo rivers in the north, bordered in the east by Mozambique. It would seem Stevenson-Hamilton had a hand in the decision to proclaim the enormous new reserve, but inhabitants of the northeastern Transvaal

Col and Mrs J. Stevenson-Hamilton and former ranger Harry Wolhuter in the garden of the Stevenson-Hamiltons' farm, Gibraltar, circa 1956.

had also previously sent petitions to the colonial secretary regarding the matter. One of the most prominent petitioners had been L.H. Ledeboer, who was later employed as a ranger by Stevenson-Hamilton.

In 1903 the land between the Sabie and Shingwedzi reserves (the area between the Sabie and the Olifants rivers) belonged to private land companies, and Stevenson-Hamilton offered to protect and administrate the game on their behalf for a period of five years. The landowners accepted his offer enthusiastically and even gave up their hunting rights for the period. This additional responsibility increased the area under Stevenson-Hamilton's control to over two million hectares.

The Sabie and Shingwedzi reserves were inhabited by small groups of blacks of different tribes, some of whom had been living there for many years. Although Stevenson-Hamilton resettled most of these people outside the boundaries of the reserve, he allowed some to stay, providing that they did not possess rifles or dogs. He used these people as informants about poaching in the more remote areas and recruited most of his field rangers and labour force from their ranks. The taxes they had to pay were also a welcome source of income – it had been decided in May 1905 that black settlers on state land had to pay an annual squatter tax. This income was later used to help finance the reserve's development, such as the roads built mainly by convicts.

Stevenson-Hamilton was determined that the game reserves under his control would not be a drain on the Transvaal treasury and he initiated measures to have the reserves largely finance themselves. Collecting taxes from squatters was one method. As has been stated elsewhere, the eventual purpose of the game reserves was make them accessible to hunters, and in those early years Stevenson-Hamilton's attempts were mainly aimed at getting game numbers to increase to such an extent that hunters would pay for the privilege of hunting there. Between 1900 and 1910 this was not possible, however, as numbers had not increased to a level where they could be culled.

In 1906 it was suggested that Stevenson-Hamilton should catch ostrich chicks and sell them to the burgeoning ostrich industry, but he pointed out that such a project would be impossible to implement due to the terrain and the lack of staff.

Another scheme to settle rare and even exotic game in the reserves, allowing them to breed and selling the surplus, also failed to get off the ground despite a special allocation of £500 in the budget. At the end of the decade Stevenson-Hamilton tried to catch a number of young lambs and calves of different game species, rear them by hand and then sell them to the National Zoological Gardens in Pretoria. This project also failed and only caused bad blood between the reserves' staff and those of the zoo.

During the early years of the twentieth century, wildlife management and ecology were not yet acknowledged as two different, clearly differentiated sciences. "Scientific" was a label mainly reserved for botanists, zoologists and veterinarians. The relationship between the staff of the Sabie and Shingwedzi reserves and the scientific fraternity was not particularly harmonious, although the differences in opinion about matters relating to wildlife were not as acrimonious as in later years.

The chief enemy of conservationists in those days was the Veterinary Services division in the Department of Agriculture. These officials increasingly expressed concern about the relation between wildlife and livestock diseases and broadcast the view that game were carriers of rinderpest, foot-and-mouth disease and other dangerous threats to the country's livestock and therefore should be exterminated. This was, of course, a view Stevenson-Hamilton and his supporters vehemently opposed and, as in many other clashes, they won this battle – to the benefit of posterity.

Although reliable censuses of the game populations were not possible in those years, it soon became clear that the Sabie and Shingwedzi reserves had succeeded in their main purpose, seeing as game populations were on the increase only in these two areas. Elsewhere in the Transvaal the early twentieth century was characterised by a decrease in the numbers of all kinds of wildlife. Although hunting licences were issued sparingly and in most years no licences were issued to hunt rare species (royal game), the existing hunting legislation was not sufficient to stop the eradication of game in the face of hunting for survival, sport and professional reasons, the human population increase in rural areas and the development of agriculture. As game was becoming increasingly scarce

Maj. James Stevenson-Hamilton fishing in the Sabie River near Sabie Bridge (Skukuza). January 1914.
(ACKNOWLEDGEMENT: STEVENSON-HAMILTON COLLECTION, SKUKUZA ARCHIVES)

16.1 THE EARLY EVOLUTION OF THE SABIE GAME RESERVE (1900–1926) AND THE KRUGER NATIONAL PARK

Col James Stevenson-Hamilton – the father of nature conservation in the Lowveld.
(1955)

everywhere in the Transvaal, new game reserves came up for discussion every now and again in the first decade of the previous century.

However, none would ever match the Sabie or Shingwedzi reserves in terms of size or the variety of wildlife that could be protected. The Pongola Game Reserve still existed, but it was too small and remote to be viable and game there had been depleted to such an extent that even Stevenson-Hamilton showed no interest in its continuation. After ranger A.A. Fraser's transfer from this reserve to the Shingwedzi Game Reserve in 1904 this first conservation area of the Transvaal fell into even greater neglect. Sporadic attempts were made to restore it, but these were so half-hearted that the reserve was deproclaimed in January 1921. Other reserves were suggested, including one near Nylstroom (now Modimolle), in the Waterberg, Messina (Musina) and northwest of Rustenburg, but in the end only the last area, which had the active support of General Jan Smuts, the colonial secretary, was proclaimed (in 1909). This reserve was not placed under the control of either Stevenson-Hamilton or the administration that ran the eastern reserves and never enjoyed the same measure of protection as the Sabie and Shingwedzi reserves. It is therefore no surprise that the lifespan of the Rustenburg Reserve was fairly short (1909–1914). Its chances of survival were further hampered by an inefficient warden, P.J. Rieckerdt, problems with private landowners, increasingly capitalistic farming practices and a lack of interest from Stevenson-Hamilton himself.

Although he was consulted about the desirability of the proposed reserves, he seldom became involved as he considered the maintenance and financial management of the Sabie and Shingwedzi reserves his main responsibility.

When Stevenson-Hamilton took on the job as warden in 1902, the future of the Sabie Reserve was by no means guaranteed and his initial period of service would in any case only be two years. By 1910 Stevenson-Hamilton's belief in the long-term prospects of both the Sabie and Shingwedzi reserves started to rest on a solid foundation. Both reserves were by then precious areas that effectively protected most kinds of wildlife in the Transvaal. Opinions regarding these conservation areas had also changed and there was no more talk of opening the reserves to controlled hunting.

After 1905 Stevenson-Hamilton became inspired with the ideal of having his reserves declared a national park that could be used by visitors, like the Yellowstone National Park in the United States. He could have been encouraged in this through his association with William T. Hornaday, director of the New York Zoo; Theodore Roosevelt, later president of the US; and others. In 1910 he prepared a memorandum on the subject and took it to Pretoria to submit it to the colonial secretary, General Jan Smuts. Unfortunately the general had already left for Cape Town to attend meetings that would lead to Unification and Stevenson-Hamilton had to return without having accomplished anything.

The prospect of both reserves becoming national institutions or assets and being developed for this express purpose had already been tentatively put forward in the Legislative Assembly in 1907. The suggestion came from the member for Parktown, A. Woolls-Sampson, who compared the Sabie and Shingwedzi reserves to the Yellowstone National Park in the United States.

But the time was not ripe yet to transform the reserves into a national park. According to the contemporary international definition a national park was a relatively extensive area proclaimed by the country's highest authority that enjoyed statutory protection and where the plant and animal life and environment, representing one or more ecosystems, were protected in such a manner by the country's most important conservation body as to remain unaltered for the continued enjoyment of, inspiration to, education of and use by current and future generations.

After Unification, when the Transvaal reserves were placed under the jurisdiction of Administrator Johann Rissik, Stevenson-Hamilton once again had to explain his idea of upgrading the reserves to national park status to him, as he had done before during a visit to Sabie Bridge in 1912.

The Sabie Game Reserve was originally established by way of an inferior legislative procedure, in other words, through a proclamation by the Executive Council and not a law formally adopted by the Volksraad. Proclamations could be changed or revoked as easily as they were promulgated – both the Rustenburg and Pongola reserves were abolished by means of a simple proclamation of the Executive Council and the voting public had no say in the process.

The future of the Sabie and Shingwedzi reserves started looking bleaker after South Africa became a Union in 1910. The Transvaal provincial authorities were not nearly as enthusiastic about wildlife conservation as their colonial predecessors.

Economic development was of much greater importance to the province at that stage, and all possibilities of boosting it were examined.

One of the first threats to the game reserves was the completion of the Selati railway line, which connected the Murchison goldfields and northern Transvaal to Delagoa Bay. Transvaal was the only province with a surplus in its coffers at the time of Unification and built the Selati Line with a portion of these funds, rather than transferring the funds to the coffers of the central government. On top of that, land suit-

16.1 THE EARLY EVOLUTION OF THE SABIE GAME RESERVE (1900–1926) AND THE KRUGER NATIONAL PARK

able for large-scale agriculture was becoming increasingly difficult to obtain, and in 1913 the big land companies started to consider repossessing the land that had belonged to them before being added to the Sabie Reserve.

Even legislators started investigating avenues that would make the reserves more economically viable.

Barely a year after Unification a petition was submitted to the Transvaal Provincial Council with a request that the Sabie Game Reserve be made smaller and that farmers be allowed to let their animals graze in this area. The Provincial Council referred the petition to the administrator, but nothing was done about it for several years. The farmers were, however, allowed to use certain areas of the reserve west of Pretoriuskop as winter grazing, and in February 1913 8800 sheep flocked into the Sabie Game Reserve, temporarily transforming it into a sheep camp.

In 1914 a large tract of land between the Letaba and Olifants rivers was added to the Sabie Game Reserve that connected it to the Shingwedzi Reserve. This addition was a major breakthrough in the sense that it consolidated the Transvaal reserves into one unit, but the proclamation that had made it possible was just another piece of paper. The land that had been added included almost 250 000 hectares belonging to private owners or companies.

In June 1913 the member for Soutpansberg, T.J. Kleinenberg, introduced an important motion in the Provincial Council in which he stated that the time had arrived for the "Sabie and Singwetsi [sic] reserves be nationalised, and that the Union government had to be encouraged to take the necessary steps".

Kleinenberg's motion also included the request that the central government should investigate the possibility of establishing other "National Reserves" elsewhere in the country, and that a commission of inquiry should be appointed to propose suitable areas for this purpose as well as the best methods to ensure the successful conservation of South Africa's fauna. These proposals were almost verbatim reiterations of resolutions passed by the Transvaal Wildlife Protection Society at their annual meeting of 1912.

Mr R. Knobel, director of the National Parks Board visiting Col J. Stevenson-Hamilton on his 90th birthday on 2 October 1957 on his farm, Gibraltar, near White River.

The mule wagon of the 1917 Game Commission near Pretoriuskop.
(ACKNOWLEDGEMENT: A.K. HAAGNER)

The Provincial Council did not show much interest in Kleinenberg's proposals and after a short debate the agenda point was held over for discussion the next day. The point would appear on the agenda every day, but it was never discussed, and when the session was adjourned it was struck from the agenda.

The pro-conservationists let a valuable opportunity slip through their fingers, and when the Transvaal reserves came up for discussion again in the Provincial Council, in 1916, it was to take a unanimous decision as requested by the administrator to insist that the Union government make the Sabie reserve smaller. Why the central government had to be involved is not clear.

In the end it was decided to appoint a provincial commission under the chairmanship of Member of the Provincial Council J.F. Ludorf to investigate the status of the Transvaal reserves and make recommendations on the amount of land needed for conservation and the amount of land that could be ceded to human use. The commission visited the reserves in 1916 and 1917, during the absence of Stevenson-Hamilton, who had departed in 1914 for active duty.

The commission issued its report in August 1918, recommending that the boundaries should be maintained, except for an area west of the Nsikazi River and Pretoriuskop (old Kaap Blocks E and F, which had been added to the reserve in 1906), which could be expropriated and made available to sheep farmers as winter grazing (this decision was executed in 1923). The report included a pronouncement that would turn out to be of great importance:

In the course of our investigation we were not a little struck by the uselessness of having these magnificent Reserves merely for the preservation of the fauna – in an area practically unknown and, by the effect of a somewhat stringent policy, made to a great extent inaccessible to the bulk of people – a policy which it will be increasingly difficult to maintain as applied to so large an area. For these and other reasons we recommend that the policy of the Administration should be directed toward the creation of the area ultimately as a great national park where the natural and prehistoric conditions of our country can be reserved for all time.

This proposal had already been submitted in 1912 by the Transvaal Wildlife Protection Society and referred to the Provincial Council in 1913, and was an ideal of Stevenson-Hamilton since 1910.

In August 1914 the First World War broke out and Stevenson Hamilton enlisted. Ranger C.R. de Laporte agreed to temporarily take over as warden (1914–1916), after which Major A.A. Fraser took over from him (1917–1919).

16.1 THE EARLY EVOLUTION OF THE SABIE GAME RESERVE (1900–1926) AND THE KRUGER NATIONAL PARK

After his involvement with civil administration in the Sudan after the war, Stevenson-Hamilton returned to the Transvaal in 1920, only to find that the reserves he had developed with such care were sadly neglected. Fraser's administration had been chaotic and poachers operated freely as a result of the reduction in staff. His disappointed reaction was as follows: "On the whole the impression I received is that there has been a general retrogression, bringing the state of things now obtaining back to about the position occupied in 1904." The war had also stimulated industry, mining and agriculture and land for these purposes became increasingly scarce and valuable.

The abolishment of the Pongola Game Reserve in 1921 clearly demonstrated the fragile and vulnerable legislative foundation of the Transvaal reserves.

Two opposing standpoints became apparent in connection with the reserves – some wanted to have the areas conserved because of their intrinsic value and others championed the abolishment of the reserves so the land could be used for more economically viable purposes.

After the game commission's report was made public in 1918, relative peace prevailed for a year or two, but in 1921 dark and menacing clouds started gathering that threatened to engulf the reserves. Certain mining companies wanted to mine coal deposits in the southern Lowveld, and started prospecting in the area between Crocodile Bridge and Bume; the large landowners and companies between the Sabie and Olifants rivers refused to make further concessions; calls for land for returning soldiers became insistent; farmers south of the Crocodile River and west of the Nsikazi agitated for more land and the right to shoot predators; and the Selati Line ran at a loss, leading to requests by South African Railways that the adjacent areas, including the Sabie Game Reserve, should be developed as a farming area in order to make the line profitable.

Stevenson-Hamilton was putting out fires everywhere and did his utmost to propagate the value of the conservation area at the highest legislative level. In 1923 he was summoned to the office of the administrator, where all interested parties would reflect on the continued existence of the reserves. During the meeting, Secretary of Lands Somerville announced that the Department of Lands, on the insistence of the Land Board, was planning to subdivide all farms on crown land, including the Sabie Game Reserve.

Stevenson-Hamilton feared all was lost, but with Somerville's mediation he secured an interview with the Minister of Lands, Colonel Deneys Reitz. During their discussions, Stevenson-Hamilton once again issued a plea to Sir Patrick Duncan, General Smuts and Johann Rissik that the Sabie and Shingwedzi reserves be nationalised and opened to the public.

On Stevenson-Hamilton's invitation, Colonel Deneys Reitz visited the reserves in August 1923. When the minister realised the enormous tourism potential of the areas he immediately became a staunch supporter of the idea of national parks. But an end to the problems was not yet in sight. The Transvaal Consolidated Land and Exploration Company started farming cattle on a large scale on the farm Toulon, inside the protected area just north of the Sabie River.

When the company ordered that a wildebeest be shot because it was eating the cattle's grass, there was a test case in court. After drawn-out legal arguments over the definition of grass and the status of game on private land, the court eventually ruled against the land companies. This thwarted their development plans.

His Hon. Col Deneys Reitz, Minister of Lands in the Smuts government before 1924, and afterwards member of the National Parks Board of Trustees (1926–1942).

In the midst of all this turmoil, South African Railways initiated a very popular fortnightly service in the winter of 1923 called the "Round in nine" and included the Sabie Game Reserve in the itinerary. It soon became apparent that the Sabie Game Reserve was the main draw and hundreds of tourists undertook the journey so they could spend a night around the campfire at the Huhla siding just north of Sabie Bridge, listening to the roar of lions and the other mysterious and romantic sounds of the wilderness.

In August 1923 Colonel Deneys Reitz, accompanied by members of the Provincial Council, undertook an inspection tour through the Sabie Game Reserve. By the time the group left they all had a much clearer view of the viability of national parks.

The Wildlife Protection Society and other conservation bodies also made their voices heard and started campaigning for a national park for South Africa, while the South African Railways started to give the Sabie Game Reserve great prominence in their publicity campaigns.

For this purpose they called in the help of the well-known artist/journalist Harry Stratford Caldecott, a personal friend of Stevenson-Hamilton. After a visit to the reserve to get a "feel" for the place, he wrote a never-ending stream of articles in the daily press in which he promoted the reserve as a potential national park. He also delivered a series of lectures to opinion-makers on the subject and as such played an important role in creating the right climate for action. (See chapter 21 for more about individuals and organisations that played a critical role in propagating the concept of a national park.)

The main obstacle to the proclamation of a national park remained the private ownership of large tracts of land between the Sabie and Olifants rivers, about which vehement polemics grew. As a compromise, a western border was determined that would leave about 70 privately owned farms within the boundaries of the proposed park, and this land would be expropriated. The remaining farms, which covered a continuous block between the confluence of the North Sand and the Sabie rivers in the south and the Klaserie mouth in the north, would be cut out of the reserve and remain in the hands of private owners. The excision, along with that of the old Kaap Blocks E and F in the south just west of the Nsikazi, came into effect on 4 December 1923 under Proclamation 71 of the administrator, A.G. Robertson. Unfortunately this meant that the best roan and sable antelope habitat and almost all of the oribi and red duiker habitat in the south were cut out of the future national park. However, the proposed national park benefited from the addition of state land between the Olifants and Letaba rivers.

Minister Oswald Pirow, minister of defence in the Hertzog government from 1924 to 1939. He served as member of the National Parks Board from 1926 to 1940. He introduced Col Stevenson-Hamilton to Minister of Lands, Piet Grobler, in 1924. Grobler became the father of the National Parks Act.

Harry Stratford Caldecott, major publicity agent and campaigner for the Kruger National Park at Salitje picket, September 1925.
(ACKNOWLEDGEMENT: STEVENSON-HAMILTON COLLECTION, SKUKUZA ARCHIVES)

16.1 THE EARLY EVOLUTION OF THE SABIE GAME RESERVE (1900–1926) AND THE KRUGER NATIONAL PARK

But before a draft national parks bill could be drawn up, there was a change of government after the election in 1924, in which the National Party led by Hertzog came into power. Fortunately many members of the National Party were in favour of conservation, but matters were delayed while they got the new government's legislative programme going.

Oswald Pirow, then the Minister of Defence and a great nature lover, organised a meeting between Stevenson-Hamilton and the newly appointed Minister of Lands, P.W. Grobler, a grand-nephew of President Kruger. Grobler was very taken with the idea of a national park when Stevenson-Hamilton explained it to him and he was also convinced of the merits of such a project. Piet Grobler was a conservation-minded person in his own right, with great respect for his country's natural beauty and fauna and flora, and he assured Stevenson-Hamilton that he would do his best to promote and realise the idea of a national park.

During this final waiting period several incidents took place that would influence public opinion. Great swarms of brown locusts had descended on the country and caused immense damage to crops and grazing, including in the Sabie and Shingwedzi reserves. Farmers along the southern Crocodile River also complained loudly about lions killing their livestock.

Nature conservation enthusiasts closed ranks and threw their weight behind the effort to promote the idea of a national park. Well-known photographers such as Colonel F.R.G. Hoare, Paul Selby and Herbert Lang took excellent photographs of the rich wildlife of the two reserves that were published and circulated among parliamentarians. Charles Astley-Maberly, the artist and naturalist, wrote letters to the press and figures of influence, and his beautiful sketches and paintings of mammals and birds gained wide interest and support. Harry Caldecott's unceasing marketing campaign also ensured that many articles on and posters and paintings of the reserves' unique natural assets were published. In some sources he is given the honour, along with Stevenson-Hamilton, of propagating the name Kruger National Park for the proposed national park.

However, it is now accepted that *Die Burger*, an Afrikaans newspaper, was the first to officially propose the name Kruger National Park in its editorial of 14 December 1925. This suggestion got strong support in the media and from the National Monuments Commission.

The idea was inspired and eminently suitable, as it not only honoured a great statesman who had been the founder of the country's first conservation areas but also gained the support of many people on both sides of the political spectrum. It stimulated an awareness of a joint heritage – the magnificent fauna and flora of South Africa – and laid the foundation for reconciliation between English- and Afrikaans-speaking South Africans, who were still deeply and bitterly divided in the 1920s. The chosen name for South Africa's first national park was symbolic of this process and gave South Africans of all races a common goal to strive towards.

Opponents of conservation launched a final attack. Sheep farmers sent a petition to parliament in which they condemned the plan to create a national park. Individual landowners tried to pull strings. State veterinarians protested loudly against the creation of a large conservation area for game and met trouble halfway by predicting a renewed tsetse fly invasion of the Lowveld, with catastrophic results for the country's livestock!

To counteract this there was a flood of articles and speeches by Drs Gill, Haagner, Warren and other prominent scientists in South Africa that ensured a great deal of support for the proposed national park.

People who were involved with the proclamation of the Kruger National Park, from left to right: Dr A.A. Schoch, government's chief legal advisor; Dr A.K. Haagner, director of the National Zoological Gardens and chairperson of the Transvaal Game Protection Society; and Col J. Stevenson-Hamilton, warden of the Sabie Reserve, circa 1926.

465

16. EARLY YEARS: THE SABIE AND SHINGWEDZI RESERVES AND THE KRUGER NATIONAL PARK 1900–1946

His Hon. P.G.W. Grobler, Minister of Lands in the Hertzog government (1924–1938), who tabled the National Parks Act on 31 May 1926 in parliament. Following this, the Kruger National Park had its origin. It also led to the proclamation of the Kalahari Gemsbok National Park, the Addo Elephant National Park and the Bontebok National Park (in 1931), and the Mountain Zebra National Park (in 1937).

Members of the Wildlife Protection Society such as H.B. Papenfus, Dr A. Haagner, Dr A. Schoch, H.S. Caldecott, Colonel F.R.G. Hoare and D.F. Gilfillan were all actively campaigning behind the scenes to sway influential figures.

On 9 December 1925 Minister Piet Grobler met representatives of the big landowning companies and their chairperson, P. Greathead, at Corner House in Johannesburg. Extremely diplomatic at first, he made it increasingly clear that the scheme had his full support and that it was a matter of national interest. Afterwards he listened to objections and reiterated the state's offer of an exchange of land. He would not compromise on this matter and informed the group that those landowners who refused to negotiate would have their land expropriated.

The companies conceded and about 100 000 hectares of land was exchanged. This left the consolidated Sabie and Shingwedzi reserves with an eventual area of about 1 900 000 hectares, with only slight changes to the western and northern boundaries in later years.

Everything was now ready and officials of the Department of Lands increased their efforts to com-

Session of the first National Parks Board of Trustees in Pretoria on 16 September 1926. Front (left to right): His Hon. Sen. W.J.C. Brebner (chairperson), His Hon. Col D. Reitz. Back (left to right): R.A. Hockley, A.E. Charter, W.A. Campbell, Dr G.S. Preller, C.R. de Laporte (acting warden, KNP), Dr A.K. Haagner, J.A. de Ridder (secr.). Absent: Sir Abe Bailey, O. Pirow and B. Papenfus.

plete the documentation for the eventual proclamation. The draft national parks bill, drawn up by the state's legal adviser Arnold A. Schoch, was scrutinised by Minister Grobler himself, who made an important amendment that made provision for the creation of additional national parks elsewhere. In the interviews he granted he time and again referred to other areas where wildlife was under threat, making specific mention of the white and black rhinos that should be protected on a national level.

The big day eventually arrived. On 31 May 1926 Piet Grobler, the Minister of Lands, tabled the National Parks Bill (Act 56 of 1926) in parliament. He held it up as the realisation of a great ideal of one of the country's foremost statesmen, President Paul Kruger. Everybody was deeply touched at the realisation that the proposed legislation would entail the nationalisation of a portion of South Africa's natural assets.

There was loud applause from both sides of the house when Grobler announced that the consolidated Transvaal game reserves would in future be known as the Kruger National Park of South Africa.

The motion was seconded by General Smuts, leader of the opposition. He delivered an impressive plea for the statutory protection of the many other unique natural wonders of South Africa. After other speakers such as H.B. Papenfus, who was a great proponent of the bill and of nature conservation in general, had a chance to express their support, former Minister of Lands Deneys Reitz got up and gave an historical overview of how far the country had come in the battle for the recognition of one of the country's most precious natural assets.

After the Speaker closed the debate, the National Parks Act and the proclamation of South Africa's first national park, the Kruger National Park, was accepted unanimously and to loud applause. It was confirmed by the senate in June 1926 and announced in banner headlines in all the newspapers.

There was praise for the far-sightedness of President Kruger and the people of the Transvaal in setting aside conservation areas where fauna and fauna could be protected, for Stevenson-Hamilton's dedicated service and almost fanatical efforts to save the area for posterity, for the public's whole-hearted support of the idea, and for the politicians and commercial interests that had made it possible to bring the project to fruition.

The battle was won! South Africa got its first national conservation area only a year after the first national park in Africa, the Albert National Park in the former Belgian Congo (Democratic Republic of Congo), was proclaimed in 1925.

A new dawn had broken in the history of nature conservation in South Africa.

References

Unpublished sources
Cape Archives:
    Strat Caldecott Collection
Central Archives Depot:
    Department of Justice
    Department of Lands
Johannesburg Public Library:
    Transvaal Landowners' Society
Skukuza Archives, Kruger National Park:
    Various files
    The Stevenson-Hamilton Collection
Transvaal Archives:
    Colonial Secretary
    Clerk of the Legislative Assembly
    Executive Committee: Transvaal
    Governor of the Transvaal and Orange River Colony
    Legislative Assembly: Original Annexures
    Lieutenant Governor of the Transvaal
    Military Governor Pretoria
    Prime Minister
    Secretary of Native Affairs
    Secretary of Mines
    Transvaal Provincial Secretary
    Transvaal Department of Agriculture
    J.S. Smit Collection

Carruthers, J. 1985. Attitudes to game conservation in the Transvaal, 1900–1910. Unpublished report.
Pienaar, U. de V. 1981. The Kruger Park Saga. Unpublished manuscript, Skukuza Archives.
Stevenson-Hamilton, J. 1902–1946. Annual Reports to the National Parks Board. Skukuza Archives.

Published sources

Anon. 1958. The late Lieutenant-Colonel James Stevenson-Hamilton. *African Wild Life* 12(1):5–9.

Anon. 1965. *Some Lowveld Pioneers.* Publication of the Lowveld 1820 Settlers Society.

Bannister, A. & Gordon, R. 1983. *The National Parks of South Africa.* C. Struik, Cape Town.

Behrens, H.P.H. 1947. His name is Skukuza. *African Wild Life* 1(2):46–66.

Bulpin, T.V. 1978. *Illustrated Guide to Southern Africa.* The Reader's Digest of South Africa (Pty) Ltd, Cape Town.

Burger, Die. 1925. 14 December.

Caldecott, S. 1925. Create a national park. *South African Nation* 2(85)21 November:7–8.

Caldecott, S. 1926. Sabi Bushmen paintings. *South African Nation* 3(113) June:17.

Caldwell, K., 1924. Game preservation: its aims and objects. *Journal of the Society for the Preservation of the Fauna of the Empire* Part 4:45–56.

Cape Argus. 1926.

Cape Times. 1926.

Carruthers, J. 1986. Journal of a visit to the Gold fields in 1889 by J. Stevenson-Hamilton. *Africana Notes and News* 27(3):91–104.

Carruthers, E.J. 1989. Creating a national park, 1910 to 1926. *Journal of Southern Afrcan Studies* 7(2):188–216.

Carruthers, E.J. 1993. "Police boys" and poachers: Africans, wildlife protection and national parks, the Transvaal 1902 to 1950. *Koedoe* 36(2):11–22.

Carruthers, E.J. 1995. *The Kruger National Park: A social and political history.* University of Natal Press, Pietermaritzburg.

Carruthers, E.J. 2001. *Wildlife and Warfare. The Life of James Stevenson-Hamilton.* University of Natal Press, Pietermaritzburg.

Carson, G. 1972. *Men, Beasts and Gods: A History of Cruelty and Kindness to Animals.* Scribner's, New York.

Chapman, A. 1928. *Retrospect: Reminiscences of a Hunter-naturalist in Three Continents.* Gurney and Jackson, London.

Chapman, A. 1930. *Memories of Fourscore Years Less Two, 1851–1929.* Gurney and Jackson, London.

Fitter, R. & Scott, P. 1978. *The Penitent Butchers: the Fauna Preservation Society, 1903–1978.* Fauna Preservation Society, London.

Grobler, J. S.d. Minister Piet Grobler. In: *Soos ek onthou.* Publication of the SABC.

Haagner, A.J.1925. The conservation of wild life in South Africa. *South African Journal of Industries* December:761–775.

Hingston, R.W.G. 1930. The only way of saving African fauna. *The illustrated London News* 13 Desember:1062.

Hornaday, W.T. & Haagner, A.K. 1922. The vanishing game of Southern Africa. *Permanent Wildlife Protection Fund Bulletin* 10 September.

Hyam, R. 1976. *Britain's Imperial Century 1815–1914: A Study of Empire and Expansion.* Batsford, London.

Kloppers, H. 1968. *Game Ranger.* Juta, Cape Town.

Labuschagne, R.J. 1958. *60 Years Kruger National Park.* National Parks Board, Pretoria.

Labuschagne, R.J. 1968. *The Kruger Park and Other National Parks.* Da Gama, Johannesburg.

Meiring, P. 1976. *Die Wildtuin is hul woning.* Koöperasiepers Ltd, Johannesburg.

Ortega y Gasset, J. 1972. *Meditations on Hunting.* Scribner's, New York.

Paynter, D. & Nussey, W. 1986. *Kruger – Portrait of a National Park.* Macmillan S.A. (Pty) Ltd, Johannesburg.

Pease, Sir A. 1908. Game and reserves in the Transvaal. *Journal of the Society for the Preservation of the Wild Fauna of the Empire* 4:29–34.

Pringle, J.A. 1902. *The Conservationists and the Killers.* Books of Africa, Cape Town.

Rand Daily Mail. 1926.

Saturday Review. 1907. The dying fauna of an Empire. *Journal of the Society for the Preservation of the Wild Fauna of the Empire* 3.

Scholtz, J. du P. 1970. *Strat Caldecott 1886–1929.* A.A. Balkema. Cape Town.

Star, The. 1926.

Stevenson-Hamilton, J. 1903. Game preservation in the Eastern Transvaal. *The Field* 14 March.

Stevenson-Hamilton, J. 1905. Game preservation in the Transvaal. *Journal of the Society for the Preservation of the Wild Fauna of the Empire* 2:20–45.

Stevenson-Hamilton, J. 1906. Aantekeningen omtrent de wild reserve aan de Sabi. *Het Transvaalsche Landbouw Journaal* 4:636–650.

Stevenson-Hamilton, J. 1906. Game preservation in the Transvaal. *Blackwood Magazine* March:407–411.

Stevenson-Hamilton, J. 1907. Opposition to game reserves. *Journal of the Society for the Preservation of the Wild Fauna of the Empire* 3:53–60.

Stevenson-Hamilton, J. 1907. Notes on the Sabi Game Reserve, Part I. *Transvaal Agricultural Journal* 5(19): 603–617.

Stevenson-Hamilton, J. 1907. Notes on the Sabi Game Reserve, Part II. *Transvaal Agricultural Journal* 5(20): 866–871.

Stevenson-Hamilton, J. 1909. Notes on a journey through Portuguese East Africa from Ibo to Lake Nyassa. *Geographical Journal* November:414–529.

Stevenson-Hamilton, J. 1921. The preservation of the African elephant. *Journal of the Society for the Preservation of the Fauna of the Empire* 1:34–42.

Stevenson-Hamilton, J. 1922. Empire fauna in 1922. *Journal of the Society for the Preservation of the Fauna of the Empire* 2:403–405.

Stevenson-Hamilton, J. 1924. The Transvaal game reserve. *Journal of the Society for the Preservation of the Fauna of the Empire* 4:35–44.

Stevenson-Hamilton, J. 1926. The Transvaal game sanctuary. *Journal of the African Society* 99:211–228.

Stevenson-Hamilton, J. 1927. The great game of South Africa. *South African Railways and Harbours Magazine* December:2023–2032.

Stevenson-Hamilton, J. 1937. *South African Eden.* Cassell, London.

Stokes, C.S. 1942. *Sanctuary.* Cape Times Ltd, Cape Town.

Tattersall, D.H. 1972. *Skukuza.* Tafelberg, Cape Town.

Tattersall, D.H. 1975. Wildbewaring in die Laeveld. In: Barnard, C. (ed.). *Die Transvaalse Laeveld – kamee van 'n kontrei.* Tafelberg, Cape Town.

Transvaal Colony. 1903–1906. *Debates of the Legislative Council.*

Transvaal Colony. 1902–1906. *Minutes of the Legislative Council.*

Transvaal Colony. 1907–1910. *Votes and Proceedings of the Legislative Assembly.*

Transvaal Colony. 1907–1910. *Debates of the Legislative Assembly.*

Transvaal Colony. 1901–1907. *Minutes of the Executive Council.*

Transvaal Police Magazine. 1911. News from Bushbuck Ridge District.

Transvaal Province. 1910–1926. *Votes and Proceedings of the Provincial Council.*

Transvaal Province. 1918. *Report of the Game Reserves Commission.* TP 5-'18, Pretoria.

Union of South Africa. 1926. *House of Assembly Debates.* 3rd Session, 5th Parliament, 1926.

Union of South Africa. 1926. *The Senate of South Africa: Debates.* 3rd Session, 5th Parliament (2nd Senate).

Wallis, J.P.R. (ed.). 1953. *The Barotseland Journals of James Stevenson-Hamilton, 1898–1899.* Chatto and Windus, London.

Wolhuter, H. 1976. *Memories of a Game Ranger.* Wild Life Protection Society of South Africa, Johannesburg.

Yates, C.A. 1935. *The Kruger National Park.* George Allen & Unwin Ltd, London.

Chapter 16
16.2 The first rangers
Dr U. de V. Pienaar

Nature conservation legislation is of little use if not enforced in practice by the relevant inspectorate. Since medieval times the application of game preservation laws, promulgations and regulations has been the task of the ranger or gamekeeper, who has to ensure that unauthorised people do not poach his landlord's game or game birds. These functions have been gradually extended to include other aspects of nature conservation. It is clear, however, that the first nature conservation officials in the Sabie Game Reserve were mostly left to their own devices when it came to formulating and applying nature conservation practices. When the appointed warden of the Sabie Game Reserve asked his senior, Sir Godfrey Lagden, what was expected of him when it came to conservation, he answered: "Just go down there and make yourself thoroughly disagreeable to everyone!"

The first rangers of the Sabie and Shingwedzi reserves were therefore mainly intent on bringing some order to a region of the Lowveld that was still inhabited by people who viewed wild animals as a source of free meat if one could shoot them or catch them in snares. Gradually, Major Stevenson-Hamilton and his rangers and field rangers, despite their meagre resources, managed to make their influence felt. Their mere presence was a great provocation to the local population, black as well as white, who saw their actions as a threat to their own survival. In time the warden and his rangers won this battle and most of the black settlements were moved to areas outside the reserves, which was how Major Stevenson-Hamilton got the nickname "Skukuza", meaning "the one who sweeps clean" – "Sikhukuza".

As they managed to bring home the purpose of their presence and to establish permanent stations, they could pay greater attention to a more comprehensive conservation strategy, mainly based on the functions of the traditional English gamekeeper. These functions included curtailing poaching, effective patrolling of the area of responsibility, the control (where possible) of damaging veld fires and the control of predatory "vermin" that preyed on the game in their care.

In the beginning, the battle to survive under the most difficult circumstances imaginable was part of their daily routine. There were almost no means of communication other than the Selati Line and later a telephone to "Reserve", the headquarters of the Sabie Game Reserve (now Skukuza). Transport was limited to ox and mule wagons over almost non-existent roads and until 1910 their accommodation consisted of traditional wattle-and-daub huts with thatched roofs. As a result of the absence of medical services, people were often laid low by malaria, flu, pneumonia and other infections. Their provisions were mainly the bare essentials needed for survival and all the rangers' staple food was venison and maize meal porridge. The Lowveld of those years was thought to be no place for women and children, and only after their accommodation and communication improved did the white rangers even consider getting married.

Today a ranger's work is highly specialised and requires formal training. The functions of nature conservation personnel entail a formidable list of duties relating to the maintenance of order and the monitoring of scientifically-based wildlife management strategies. In contrast with past practices it does not focus only on the conservation of the larger mammals and certain bird species, but on the ecosystem as a whole. The equipment and techniques available today are sophisticated and among the best in the world.

Still, the pioneering work of the first warden of the Kruger National Park and his small corps of rangers should not be disregarded. With the knowledge and resources available to them they managed to do pioneering work of vital importance. Without their ceaseless work and dedication there would not be a Kruger National Park today. We should therefore salute the achievements of this brave little troop of men, who often encountered and overcame problems that would seem insurmountable today. Taking into consideration their primitive living conditions, often inadequate equipment, meagre salaries and the lack of medical services, they executed a task of which posterity can be proud.

Major James Stevenson-Hamilton was appointed in 1902 at the princely salary of £500 a year and an extra allowance of £180 a year, and ranger L.B. Steyn was appointed in 1929 at a salary of £250 plus an allowance of £184 a year. In 1946, after having served for 17 years, he still only received £57.19.20 a month. Major Stevenson-Hamilton's successor, Colonel J.A.B. Sandenbergh, was appointed in 1946 at an all-inclusive £86.18.11 a month! Between 1898 and 1926 the handful of black and white rangers were the only workforce in the Sabie and Shingwedzi game reserves and they had to undertake all projects themselves, even road-building and construction work.

After the proclamation of the Kruger National Park in 1926 and the opening of the conservation area to the public, the rangers had to build roads and rest camps and serve as gate officials, apart from their normal duties. Before the first camps were completed, they were expected to welcome the tourists into their own modest little homes and see to their comfort! Then they had to serve as permit officials at the gates, rest camp officials and supervisors, and even petrol attendants. The field personnel stoically endured the extraordinary pressure until the first rest camp supervisors and gate officials were appointed in 1931.

Because their main duty in those days was law enforcement, the first rangers were recruited from either the police or the army – men with the necessary discipline and, where possible, a basic knowledge of the veld and black languages, which most hunters and Lowvelders at the time possessed.

Dr J.W.B. Gunning, director of the old State Museum in Pretoria, submitted an extensive report to the Superintendent of Education in the Transvaal in as early as 1897 in which he made suggestions regarding the proclamation and boundaries of the proposed Sabie Game Reserve. He also put forward Paul Krantz, the taxidermist of the State Museum, as a candidate for the position of the first warden of the envisaged reserve because of Krantz's intimate knowledge of the specific area and its wildlife. Because of financial and other constraints, Gunning's proposal could not be implemented.

From the proclamation of the Sabie Game Reserve on 26 March 1898 until the outbreak of the Anglo-Boer War on 11 October 1899, control of this third reserve in the Transvaal was entrusted to two members of the old "Zuid-Afrikaansche Republiek Politie" (South African Republic Police, or Zarps), Izak Holtzhausen and Paul Bester, who were stationed at Komatipoort and Nelspruit respectively.[1]

Izak Cornelis Holtzhausen (1873–1966)

Izak Cornelis Henry Oliver Holtzhausen was born on 13 February 1873 as the thirteenth of sixteen children. His parents were from the Cape Colony and had moved to Barberton by ox-wagon. They were of German extraction and his grandfather was H. von Holtzhausen, but the family decided to drop the "von".

Izak attended the Barberton Government School. After the discovery of gold at Kaapsche Hoop, Izak once saw a digger by the name of Retief find a nugget weighing 216 ounces. A coach driver was fortunate enough to discover a nugget that weighed 13.5 pounds – in those days gold was worth £9.17.06 an ounce.

After leaving school Izak went to work in a shop in town where he befriended the brother of Lilian Cubitt, who later married Percy FitzPatrick. Later he worked in a shop in Warmbaths (now Bela-Bela) for a while and served President Paul Kruger when he visited the hot-water springs.

Izak Cornelis Holtzhausen (1873–1966), member of the ZARP and ex officio ranger of the old Sabie Reserve before the Anglo Boer War (1898–1899). He was stationed at Komatipoort.
(ACKNOWLEDGEMENT: BARBERTON MUSEUM)

---

1 The historian Jane Carruthers (1995) recently expressed doubt about who had a supervising responsibility in the conservation area between the Sabie and Crocodile rivers after its proclamation on 26 March 1898. She questioned the view held by South African National Parks that the two ZAR policemen, Izak Holtzhausen and Paul Bester, had received this responsibility as part of their duties, "as there are no official records of their appointments". There is, however, enough direct evidence in the archival documents of the South African National Parks that these two took over control of the reserve from 1898 until the outbreak of the Anglo-Boer War from the field cornets of these wards, namely N.H. Versfeld of the Komati ward and D.J. Schoeman of the Crocodile River ward. They can therefore be regarded as the first *ex officio* rangers of the Sabie Reserve. The historian and expert on the history of the Lowveld, Dr W. Punt, invited Izak Holtzhausen and the former transport rider Org Basson in 1950 on an inspection tour of the old transport route to Delagoa Bay to obtain more information on the wildlife and veld conditions in the park area during the transport-riding era of the 1880s. Dr Punt's report specifically mentioned Holtzhausen's knowledge of game conditions in the old Sabie Reserve, which he surely would not have obtained had he not patrolled the area on a regular basis to curb poaching activities. It is as a result of Punt's report that a former Parks Board official visited Izak Holtzhausen at Barberton to record more of his observations on tape. In the Skukuza Archives there is also a comprehensive report by Paul Bester's daughter, Mrs Alice Briedenhann of Nelspruit, in which she gives an extensive account of her father's experiences as policeman/ranger in the Sabie Reserve before the outbreak of the Anglo-Boer War. She also made photographs of her father's patrol hut at Sabie Bridge (later Skukuza) available. See page 546.

Izak Cornelis Holtzhausen, one of the first two rangers of the Sabie Reserve in the days of the ZAR (South African Republic), photographed in his South African Police uniform in later years.
(ACKNOWLEDGEMENT: BARBERTON MUSEUM)

In 1880 he returned to Barberton and travelled to Komatipoort via Louw's Creek and Pettigrew's road. On his arrival at Komatipoort he found that construction work on the Eastern Line between Lourenço Marques (now Maputo) and Pretoria was already in full swing. Here he worked as a construction worker on the railway bridge over the Komati River, before joining the Railway Police.

While walking from Komatipoort to Malelane one day, Izak came across two black men pushing a railway trolley. On the trolley was a money trunk with the wages of the construction workers at Kaapmuiden. The two Dutch men who were supposed to accompany the trolley had apparently imbibed a little too heavily and were lying next to the side of the road somewhere. Holtzhausen delivered the money safely at Kaapmuiden and received a £25.0.0 reward from the railway engineer.

Izak was stationed at Machadodorp for quite a while and rose to the rank of sergeant-major in the ZAR Military Police. In 1895 he accompanied Bhunu, the son of the Swazi king Mbandzeni (Umbandine), who lived at Legogote and wanted to offer his services to President Kruger. In the Anglo-Boer War Izak Holtzhausen was a member of the Boer forces under the command of General Piet Cronjé that crushed the Jameson Raid of 1896. Afterwards he accompanied the captives to Pretoria. Holtzhausen was also a member of the President's Guard for a while, after which he was stationed at Komatipoort. At this time that he was given the added responsibility of supervising the eastern part of the newly-proclaimed Sabie Game Reserve.

In later years, Uncle Izak, as he was known then, had much to say about this interesting time in interviews with Mr R.J. Labuschagne, the former liaison officer of the National Parks Board, and Dr W. Punt, the well-known historian.

He said he had been responsible for the strict enforcement of game legislation in the area between the Komati and Olifants rivers from 1892 onwards as a ZAR policeman. According to him, there were blue wildebeest, zebra, buffalo, giraffe and numerous tsessebe on the Lebombo flats north of Komatipoort. This was confirmed by surveyor M.C. Vos's report (1890), in which he said that they had hunted tsessebe on the flats between the Sabie and Crocodile rivers. Holtzhausen also had a contract to shoot game for the construction workers working on the Selati railway line. He denied that large groups of hunters used the Selati Line to recklessly shoot game in its vicinity.

Predators caused havoc among the livestock, while hippos, elephants and a few rhinos damaged fences and crops.

In 1890 he encountered tsetse flies in the vicinity of Satara as well as in a thorn thicket a few kilometres northeast of the current Skukuza.

Izak Holtzhausen was also familiar with the transport routes of the time and in 1950 he and the old transport rider Org Basson showed some of these routes through the current Kruger National Park to Dr W. Punt. Both Holtzhausen and Basson said that the part of the Park through which the Delagoa Bay transport route ran used to be much wetter.

Holtzhausen told Dr Punt that Nellmapius Drift used to be an open stream with sandbanks and not as overgrown as in the 1950s. The vegetation 16 kilometres southeast of Josekhulu Spruit was the same knob-thorn savannah as now, but Pretoriuskop and its surrounds, now covered in dense shrubs, was almost unrecognisable – in the 1880s and 1890s this used to be an open grassland in which oribi could be found near Shabeni Hill and eland were common as far away as Newu Hill.

They never encountered any rhinos but lions were plentiful, especially near Pretoriuskop and along the Crocodile and Komati rivers. Elephants had already been eradicated in the southern Lowveld and the Stols brothers had to travel to the Limpopo to hunt these giants. The Ntomeni pools were a favourite outspan of the transport riders and a waterhole for herds of buffalo.

For a while during the Anglo-Boer War Izak Holtzhausen was in charge of the 450 English prisoners of war who had surrendered at Roodewal near Standerton, and he befriended some of them. Later one of these friends helped him to escape from the prisoner of war camp at Barberton, after he had been captured by the British forces.

In 1896 Izak married a Miss Jones, whose parents also lived in Barberton. After the war he settled in Barberton again and in 1905 he bought the 2000 morgen farm Sunnymeade near Barberton, where he farmed for 21 years and became a pillar of society. One day while quenching his thirst at a small creek, the Tutusi near the Swaziland border, he discovered an asbestos outcrop, collected some samples and pegged off a few claims. Eventually these claims were taken over by Turner and Newall and this was the beginning of the Havelock Asbestos Mine.

In 1914 Izak started a branch of the National Party at Barberton. After his wife's death in 1952 he remarried and the staunch old pioneer spent his last years living a peaceful life in Barberton and died at the age of 93.

Paul Machiel Bester (1870–1952)

Paul Machiel Bester was born on 6 March 1870 on the farm Waaihoek in the Ladysmith district of Natal. He was one of 12 children of Adriaan Willem Petrus Bester and Maria Gregory. His parents later moved to the Lowveld and he first settled on the farm Hilltop between Nelspruit and Barberton, and later moved to the farm Bester's Last (the site of the Rob Ferreira Hospital today) near Nelspruit. Here he and his brother Attie farmed dairy cows, fruit and vegetables. They also lived with their parents on the farm Retreat for a while, and there their youngest sister, Phyllis Natalie, was born. The Besters were friends with the well-known Native Commissioner Abel Erasmus and the two families regularly visited each other.

In the years before the Anglo-Boer War, Paul Bester was a transport rider between Lourenço Marques (now Maputo), Natal and Pretoria. After the construction of the railway line between Delagoa Bay and Pretoria had started in 1888, Paul and Attie signed a contract with the NZASM to provide wood for sleepers and stayed at Charlie Bernds's hotel in Machadodorp with a group of Dutch contractors. Apparently Paul was something of a practical joker and loved pulling people's legs. According to Johnny Goodman, a friend of his from Lydenburg, his nickname was "Poker Paul", which says a lot.

Once, Paul and the Dutch were having a discussion in the hotel bar about Machadodorp's height above sea-level and Paul added a few extra thousand feet to test their knowledge. "No, Mr Bester," one said, "if that were true, we would now be sitting at the table of the Lord!"

Paul Machiel Bester (1870–1952), member of the ZARP and *ex officio* ranger of the Sabie Reserve before the Anglo Boer War (1898–1899). He was stationed in Nelspruit.
(ACKNOWLEDGEMENT: MRS P. VAN ROOYEN, LYNNWOOD)

On another occasion, while he and a few friends were standing on the platform of the Machadodorp station, a train came steaming in. In one of the train's windows they saw a beautiful girl with her mother, sisters and brothers. Paul immediately bet his friends that he would kiss the girl. He casually stepped onto the train and introduced himself to the surprised mother as a long-lost relative and without so much as a pause kissed all of them in welcome – the pretty one included. When the train left he claimed his money and hotfooted it out of there.

After his contract with the NZASM had expired, Paul joined the ZAR Police (Zarps) at Barberton. He was stationed at Nelspruit and after the proclamation of the Sabie Game Reserve by Paul Kruger on 26 March 1898, Paul was ordered to keep an eye on illegal hunting in the western half of the reserve. Thus he and Izak Holtzhausen became the first two rangers of this remarkable conservation area.

Paul patrolled the area on horseback and had an outpost at the unfinished Sabie Bridge, which was manned by a black constable. He sometimes lived in the hut for extended periods of time while keeping an eye out for poachers. The hut's door was made of wildebeest skin and at night hyenas would often come to pull at it – until Paul would fire a few shots in discouragement. One of the black constables yearned to own a pair of boots, which Paul then gave to him. One day, the constable failed to return from a patrol and a search only turned up one bloody boot at the place where he had been caught and eaten by lions.

While roaming the reserve he came across an albino kudu in the area between Skukuza and Pretoriuskop. A photograph of Paul Bester in front of his patrol hut at Sabie Bridge was donated to the Skukuza Archives by his daughter, Mrs Alice Briedenhann, in the 1960s.

At the outbreak of the Anglo-Boer War, Paul joined the Boer forces (probably General Viljoen's commando). There was chaos in the Lowveld after the Boers had been beaten at the battle of Berg-endal in September 1900. Paul and his brother Attie, like Harry Wolhuter, Bob Sanderson and others, decided to take a herd of stray cows they had gathered and slip across the Mozambican border, rather than fall into the hands of the British. It was probably at this time that he befriended Harry Wolhuter – a friendship that would last until his dying day. Bob Sanderson and Harry Wolhuter returned to the Lowveld after a few months, but Paul and Attie stayed on in Mozambique.

While Paul and Attie were looking after their herd near Ressano Garcia, Attie contracted malaria. Members of Steinaecker's Horse heard of the two Boers with their great cattle herd and sent some armed blacks to capture the two and rustle the cattle. They managed to capture Attie and the cattle, but Paul tricked them by running down the ditch in which he was supposed to be shot. This border violation was brought to the attention of the Portuguese authorities in Lourenço Marques. The Portuguese governor laid an official complaint with Colonel Steinaecker and Attie was released. They also had to pay him and his brother compensation for the stolen cattle. As "punishment", however, the two were exiled to the Portuguese territory for three years and were as such not allowed to return to the Transvaal immediately after the war. The two settled in Lourenço Marques in a small house next to the Polana Hotel, where they made a living transporting provisions to the military outposts for the Portuguese authorities. They also became fluent in Portuguese. It may have been at this time that Paul met his future wife, Lilian Maud Kestell, a niece of Father J.D. Kestell. Her father, Cornelis Benjamin Kestell, owned a smithy at Nomahasha on the Swaziland border, but before the war he had lived in Bremersdorp. The British consul in Lourenço Marques married Paul Bester and Lilian on 19 July 1904 – Paul was 34 at the time – and their marriage was blessed with two daughters and a son. After their marriage the couple went to farm with his mother and brother Attie on Bester's Last.

The farm was not big enough to sustain two farmers, so Paul bought the farm Woodhouse, where the Kruger Mpumalanga Airport (Nelspruit) is situated today, and started farming fruit and vegetables. In 1914 he received a letter from General Louis Botha that warned him not to be influenced by the political ideas of Mr Hull, who aspired to win the Nelspruit parliamentary seat. He also bought the farms Moria and Ross in the Lowveld and the family often went hunting there in winter. He later sold Woodhouse to H.L. Hall and bought Sterkspruit, across from Cairns Station. He then sold this farm to the state and a certain Mr Strauss and moved to Friedenheim between Nelspruit and White River. At that stage there were still impala, bushbuck and duiker in the area. Paul made Friedenheim his home and became a prominent and conservation-minded farmer until bad health eventually forced him to move to White River. He died there on 14 June 1952 and was buried in the old Nelspruit cemetery.

After reproclamation
At the end of the Anglo-Boer War, the British Administration under Lord Milner reproclaimed the Sabie Game Reserve, mainly at the insistence of certain prominent Lowvelders such as Henry Glynn, J.C. Ingle, Sir Alfred Pease, the author and naturalist Abel Chapman, Major Greenhill-Gardyne of Steinaecker's Horse and Assistant Mine Commissioner Tom Casement of Barberton. Reproclamation took place in June 1902 under Administrative Proclamation No. 11 of the Lieutenant General of the Transvaal. The original Dutch text was used.

Even before the reproclamation there had been efforts to regain a measure of control over the former conservation area. In May 1901, Tom Casement, the assistant mine commissioner in Barberton, received an application from Captain H.F. Francis to be appointed as ranger in control of the old Sabie Game Reserve. This application was forwarded to the Provincial Secretary in Pretoria and on 4 July Casement was empowered to appoint Francis as a ranger to curb poaching in the area. However, Francis was killed shortly afterwards, on 7 August 1901, when a Boer force attacked Fort Mpisane of Steinaecker's Horse. Once again Tom Casement started lobbying to get a warden appointed to the reserve. On his insistence one W. Walker of Barberton was appointed on 24 October 1901 by the district commissioner.

Walker's feeble efforts and the subsequent appointment of Major Stevenson-Hamilton as warden of the Sabie Game Reserve on 1 June 1902 are discussed in detail in chapter 16.1. Because Stevenson-Hamilton was a professional soldier himself, it is not surprising that the first rangers he appointed came from the ranks of either the British forces that had served during the Anglo-Boer War or the South African Police. He implemented this policy until the mid-1920s and of the 29 rangers he appointed during his 44 years as warden, 15 were either former soldiers or policemen.

The first ranger Stevenson-Hamilton appointed after his arrival on 29 July 1902 in the Sabie Game Reserve was Edward George (Gaza) Gray.

Captain Edward George (Gaza) Gray
Captain Gaza Gray was a seasoned Lowvelder and about 38 when he first met Stevenson-Hamilton at his camp at Lower Sabie. At that stage he was still the officer in charge of the Steinaecker's Horse outpost at Gomondwane, but before the war he had been a recruiting officer for the Rand gold mines in the adjacent Portuguese East Africa (Mozambique). After the establishment of Steinaecker's Horse during the Anglo-Boer War, he joined the unit on 20 August 1900. On 8 January 1901 he was promoted to lieutenant and he left the unit on 31 August 1902 at Nomahasha in Swaziland with the rank of captain. Gray enjoyed hunting and visited Gazaland, Mashonaland (now Zimbabwe) and central Africa for the purpose. After the death of Captain H.F. Francis at Mpisane on 14 August 1901, he applied to the Civil Commissioner in Pretoria to be considered for the position of warden of the Sabie Game Reserve, but without success.

Originally from the eastern Cape, he had been living in the Lowveld for many years and was fluent in Tsonga. He was also an expert on the local Tsongas'

A group of Steinaecker's Horse officers at Komatipoort in 1902. The person to the left is Capt. E.G. (Gaza) Gray, the first rangers to be appointed in the newly reproclaimed Sabie Reserve. (ACKNOWLEDGEMENT: STEVENSON-HAMILTON COLLECTION, SKUKUZA ARCHIVES)

customs and traditions, to such an extent that they believed he had supernatural powers. His reputation as an "mthakati" was common knowledge and led to a superstitious fear among many locals, whose nickname for him was "Mastutele" ("the quiet one"). Gray built himself a comfortable camp at Lower Sabie that consisted of half a dozen rondavels surrounded by a reed fence. He also had a large herd of cattle. (Some of his cattle kraals were at an outpost about five kilometres north of the present Nhlanganzwani Dam.)

Stevenson-Hamilton was particularly impressed with Gray's knowledge of the area and the locals and offered him a position as acting ranger, which he accepted. He informed his new boss that there was still plenty of game in the Lebombo Mountains, but that many blacks had settled in the area between the Crocodile and Sabie rivers in order to be closer to the game.

According to an agreement that was reached with Gray, he would be allowed to keep his cattle in the reserve, but in exchange he had to serve as acting or honorary ranger in the area east of the Selati Line, between the Sabie and Crocodile rivers. He had to tell new arrivals to move out of the area before they started ploughing and in general make people aware of the area's new status as a game reserve in which no hunting was allowed.

Gray offered an old corrugated-iron shack of his at Crocodile Bridge to Stevenson-Hamilton as a temporary residence. It had originally been a shop similar to those at Gomondwane and Nkongoma, and was six kilometres west of Sabie Poort (where Bill Sanderson had been interned during the war). Stevenson-Hamilton made the cottage at Crocodile Bridge his headquarters for a few months, until November 1902, while establishing contacts at Komatipoort.

Gray did not waste any time and by the middle of November he had already told chiefs Mjonjela and Mahalani that they had to move and informed Captain Kongwazi, who lived opposite Komatipoort, that she had to settle elsewhere.

Steinaecker's Horse was dissolved on 7 February 1903, despite the best efforts of the commanding officer, and Gray was free to give his full attention to his duties as ranger.

Stevenson-Hamilton had moved his headquarters to Sabie Bridge on 20 December 1902. He started apprehending poachers and arrested some of Chief Mjonjela's subjects for illegally hunting near the Mlondozi River.

On 5 May 1903 the head of the Witwatersrand Native Labour Association, a Mr Pickard, arrived at Stevenson-Hamilton's headquarters and informed him that he had instructed Gray to return to Portuguese East Africa to resume his duties as recruitment officer. He was surprised to hear that Gray had not informed the warden of his absence and promised to rectify the matter.

Thomas Duke, appointed by the warden to take over Gray's post at Lower Sabie, then received a visit from a very arrogant and abusive Gray, who told Duke that his cattle were too sick to leave the reserve and that he refused to move them.

On 26 May Stevenson-Hamilton officially discharged Gray and two days later Gray wrote a reply in which he agreed to leave the reserve with his cattle and expressed his indignation at his summary discharge. And thus ended the short, if tumultuous, career of Gaza Gray as a ranger in the Sabie Game Reserve – on a much sourer note than it had begun.

Rupert Maurice Atmore (1878–1966)

Just a few days after Stevenson-Hamilton had appointed Gaza Gray as acting ranger at Lower Sabie, he was in Komatipoort to discuss certain matters with the Commanding Officer of Steinaecker's Horse, L. von Steinaecker, and to recruit a few more rangers

Maj. James Stevenson-Hamilton's temporary living quarters, a corrugated-iron building at Crocodile Bridge, which belonged to Capt. E.G. (Gaza) Gray of Steinaecker's Horse. August 1902.
(ACKNOWLEDGEMENT: STEVENSON-HAMILTON COLLECTION, SKUKUZA ARCHIVES)

for the newly proclaimed reserve. Here he met Rupert Atmore, who was working for the British Intelligence Service in the Barberton district. Atmore had got to know the southern part of the reserve well during hunting trips. According to Tattersall (1972) he was appointed on 15 August 1902 as the first permanent ranger in the Sabie Game Reserve.

Stevenson-Hamilton wrote that he appointed Atmore at more or less the same time as Harry Wolhuter. (According to the official Civil Service List of 1924, Wolhuter's date of appointment was 17 August 1902.) It therefore would appear that Atmore, and not Wolhuter, was the first permanent ranger in the Sabie Game Reserve.

Rupert Maurice Atmore, whose Swazi name was "Mhlati" ("the man with the large jaw that throws a shadow") was born on 8 September 1878 at Katberg in the eastern Cape and came to the Lowveld in 1884.

After his appointment as ranger, he was stationed at Boulders on the northern bank of the Crocodile River opposite Kaapmuiden Station and the southwestern part of the Sabie Game Reserve was his responsibility. He had to patrol eastwards to Crocodile Bridge and northwards along the Nsikazi to Pretoriuskop. To make life easier he had two rondavels built at every outpost – one for him and one for his horses – as well as accommodation for his assistants.

One day, two men crossed the Crocodile River on horseback at Kaapmuiden, after which they rode downstream in the direction of Malelane. Atmore could not catch up with them, so at noon he unsaddled his horse, hobbled it and sat down to eat his lunch. Suddenly he heard a shot and, when he went to investigate, discovered to his shock that the poachers had shot his horse and run away.

Atmore asked the warden to replace his horse, but Stevenson-Hamilton told him that the loss was due to his own negligence and that he had to replace it from his own horse allowance of £8.00.00 a month. He then bought another horse, but soon afterwards it died of African horse sickness.

Atmore then informed the warden that he would quit if the second horse was not replaced. Stevenson-Hamilton told him that he did not see how he could replace the horse and that Atmore should resign if that was how he felt, which he then did in December 1902. His short career as a ranger did not frighten off Rupert Atmore, who started farming just south of the Crocodile River on Malelane Estates in partnership with Captain G.L. Elphick.

Rupert Maurice Atmore, one of the first rangers to be appointed by Maj. Stevenson-Hamilton in August 1902. He served for only a few months and was stationed at Boulders opposite Kaapmuiden Station.
(ACKNOWLEDGEMENT: SOME LOWVELD PIONEERS)

In 1912 he started farming on his own with vegetables, papayas and tomatoes on land north of Malelane Station along the river (Riverside Estates). He later became the "papaya king" of the Lowveld and was a well-to-do man in every respect. In 1914 he joined General Botha's forces, which attacked German South West Africa (Namibia), and left with the rank of captain at the end of the campaign. In 1916 he married Gertrude Meintjies and after her death in 1931 he married Janetta van der Walt of Middelburg in the Cape in 1937. Atmore brought the first motorised vehicle, a Trojan with solid tyres, into the Lowveld and later served on the Road Transport Commission.

In 1932 ranger H.E. Tomlinson caught Atmore's workers cutting reeds in the bed of the Crocodile River and laid a charge against him. Atmore defended his case in the High Court, but Judge Tindall ruled in favour of the National Parks Board. This brought an end to a dispute of many years about the legal southern border of the Park, which is the high-water mark or the inside bank of the Crocodile River along its southern side. Despite his failed career as a ranger, Atmore was a very successful farmer and true pioneer of the vegetable and citrus industries in the Crocodile River Valley. He died on 18 August 1966.

Henry Charles Christoffel Wolhuter (1876–1964)

In contrast to the first two rangers Stevenson-Hamilton had appointed and who each only served a few months, the next person Stevenson-Hamilton met, after the mediation of Major Greenhill-Gardyne at Komatipoort, carefully considered his offer for three days before making a decision. Henry Charles Christoffel Wolhuter, or Harry as he was commonly known, was a fit young man of 26 and a sergeant in the British border guard unit Steinaecker's Horse. He was on the verge of resigning from the army.

After he had accepted the appointment on 17 August 1902,[2] a bond would develop between him and his new head that would last for 55 years – of which they would spend 44 years in the big new conservation area in the Lowveld.

According to Harry (1948), he was born on 14 February 1877 in Beaufort West in the Karoo.[3] As a young boy he spent his happiest days in the veld, observing and hunting the Karoo's wildlife.

He was only 13 when his parents moved to the Transvaal and he got his first job in a store in Maraisburg. Later he accepted a position in a billiard saloon and canteen, the Black Cat, but he dearly wanted to follow his parents, who had since settled on a farm in the White River district near Legogote. His chance came when he got an opportunity to lead the oxen of a transport wagon on its way to Barberton. After his arrival in the Lowveld he went to live with his parents on their farm and helped out in their shop.

Harry went on his first real hunting trip when he accompanied an acquaintance of his parents on a visit to the current Satara area. His parents later moved to Belfast to give the younger children a better education, but Harry stayed behind on the farm, where he built up a considerable herd of cattle. He issued loincloths to his black workers and in turn they gave him the nickname "Lindanda" (meaning loincloth).

His nearest neighbour was the legendary hunter and pioneer Bill Sanderson, who farmed on Peebles, and in time they forged a close friendship. Bill's brother Bob farmed a few kilometres away on another farm.

The first rangers of the old Sabie and Shingwedzi reserves. October 1904. Front: T. Duke, Maj. J. Stevenson-Hamilton and Maj. A.A. Fraser. Back: C.R. de Laporte and H.C. Wolhuter.
(ACKNOWLEDGEMENT: STEVENSON-HAMILTON COLLECTION, SKUKUZA ARCHIVES)

One of Maj. Stevenson-Hamilton's first rangers, Harry Wolhuter, as a young man in 1905. He served as ranger of the Pretoriuskop section from 1902 to 1946.
(ACKNOWLEDGEMENT: STEVENSON-HAMILTON COLLECTION, SKUKUZA ARCHIVES)

---

2   Harry Wolhuter always used to give his date of appointment on official reports as 1 July 1902, which is incorrect.
3   His official birth certificate, obtained from Christ Church, Beaufort West, gives his date of birth as 14 February 1876 and the date of his christening as 15 March 1877.

Harry lost almost all his cattle in the rinderpest epidemic of 1896–1897 and he was forced to work for H.L. Hall on his farm Riverside (later Mataffin). While there, he was commandeered to help the Boers quell the Makhado rebellion. After returning from this campaign he decided to undertake an extensive hunting expedition through the area that would become the Kruger National Park, as far north as beyond the Olifants River. On his return he learned to his dismay that the Anglo-Boer War had broken out. His friend Bob Sanderson decided to move with a herd of his cattle to Portuguese East Africa (Mozambique) and Harry followed suit. When they finally returned the two were in such a haggard state that "Pump" Willis, another friend of theirs, claimed that they looked like a band of wild men that would cut your throat for half a crown!

By that time the legendary and later notorious Colonel Steinaecker was busy raising a border guard corps at Nomahasha in Swaziland. Harry and many other English-speaking volunteers from the Lowveld decided to join Steinaecker's Horse. Harry served in this irregular British unit for the entire war and rose to the rank of sergeant. While he was stationed at the unit's Sabie Bridge outpost he helped to build the blockhouse at the northeastern end of the Selati Line. His wartime exploits are discussed in chapter 14.

At the end of the war Harry was stationed at the headquarters of Steinaecker's Horse in Komatipoort and it was here that he met Stevenson-Hamilton in July 1902 and accepted the position as ranger. His only condition was that he be allowed to settle permanently at Mtimba on the western border of the Sabie Game Reserve. Stevenson-Hamilton readily agreed and Harry carried out his duties as ranger from Mtimba for 36 years.

After the area west of the Nsikazi River, including Mtimba, was excised from the reserve in 1923 on the suggestion of the 1918 Game Reserve Commission and after the development of the Pretoriuskop rest camp in the early 1930s, it became difficult for Harry to carry out his duties in and around Pretoriuskop from his base at Mtimba. In 1938 a rondavel camp was built for him near the rest camp and in 1940 it was replaced with a comfortable brick house. Harry earned the distinction of having manned a single section, namely Pretoriuskop, for the longest period of time (44 years). In the old days, when the only means of transporting provisions was an ox-wagon, it was a real exodus to visit the reserve's headquarters at Sabie Bridge. As a result he built a comfortable overnight post at Doispane, the kraal of the black staff member Doispane (Holozane) Mongwe, on the west-

"Spanplek" – ranger Harry Wolhuter's outspan between Doispane and Skukuza, with the Old Wagon Route in the background. (08.06.1984)

The magnificent northern flank of Mangake Hill with its rare stand of *Euphorbia evansii*, which had probably been discovered by ranger Harry Wolhuter. (08.06.1984)

ern bank of the Mtshawu Spruit. From here they trekked to a place called Spanplek, about midway to Sabie Bridge, where they outspanned and ate. Then they made for Sabie Bridge, which they usually reached late in the afternoon. The Old Wagon Route at Spanplek is still clearly visible from the air.

In 1903, during a patrol, Harry Wolhuter and his brother-in-law C.A. Yates, a teacher and scout officer from Barberton, discovered the ruins of João Albasini's house at Magashulaskraal along the Sabie River near the Phabeni Mouth. They were therefore the first whites to see this lonely outpost of civilisation since Carl Mauch on 9 September 1870 and the hunter and pioneer Henry Glynn in the 1880s. Yates was an accomplished photographer and took a good photograph of the ruins as it appeared at the time.

During the Sabie Game Reserve's early years, white supervision was sporadic and Harry had to patrol vast areas, up to the Olifants River in the north. During one of these patrols in August 1903 he had his greatest adventure – one that would make him a living legend.

On 26 August 1903 he was on his way back from a long horseback patrol to the Olifants River. As evening fell the small party was still a distance from its destination, so he rode ahead with his wire-haired dog Bull to find a suitable camping site near water. His four black field rangers followed at a distance with the three pack donkeys.

It was already dusk when he saw a movement in the long grass along the footpath. He was under the impression that it was a couple of reedbucks, but Bull dashed ahead and started barking. Harry called the dog back, but the next moment he saw a large male lion to his right that was about to jump. There was no time to lift his rifle, a .400 Express, but he pulled his horse to the left and dug his spurs into its flanks. The horse bucked, with the result that the lion only managed to give his prey a swipe across the hindquarters. In the horse's crazed attempts to get away from the lion Harry was thrown from the saddle. At that moment he saw a second lion storming towards him from the opposite side. He landed right in front of this lion and the animal immediately grabbed him while its mate set off in pursuit of the fleeing horse.

When Harry came to, he was being dragged by the lion along a game path. He was on his back under the belly of the lion, which was dragging him by his right shoulder and arm. It kept stepping on Harry and his thighs and trousers were torn to shreds. Harry had of course lost his rifle during the attack and the lion, presumably in anticipation of the awaiting meal, made a soft purring noise much like a cat that had caught a mouse.

Because he was on his back, his spurs would catch in the bushes and the lion would give a yank that caused excruciating pain in the arm and shoulder of its prey. Harry tried to prepare himself for the worst, but then he remembered his hunting knife, which he

This old cattle dip at Phabeni along the Sabie River fell into misuse after 1938. (11.11.1986)

Ranger Harry Wolhuter on his horse, ready for a patrol, circa 1903. The knife in the sheath on his side would later save his life when he was attacked by a lion. (ACKNOWLEDGEMENT: STEVENSON-HAMILTON COLLECTION, SKUKUZA ARCHIVES)

wore in a sheath on his right side. Because the lion had a hold on his right shoulder, he had to work his left hand underneath his back and feel for his knife, and it seemed like an eternity before he managed to get a proper grip on it. It was a sharp knife with an 11.5 cm long blade manufactured by the firm T. Williams in London. Harry had only one thought and that was to survive! His entire concentration was focused on using the weapon as effectively as possible. He carefully felt with his left hand for the spot where he thought the lion's heart should be and then gave it two stabs in quick succession behind the right shoulder. The lion immediately dropped him but still stood over him, growling. Harry desperately struck again and apparently hit a major artery because he was drenched in the animal's blood. On receiving the last stab the lion sprang away and backed off into the

A small bronze statue by Sculptor B. Lawrence, which depicts Harry Wolhuter's death-struggle with a lion in 1903. Next to it is part of the trunk of the tree into which he had climbed after the incident. These items can be seen in the Stevenson-Hamilton Information Centrum, Skukuza.

An artist's impression of the dramatic incident when ranger Harry Wolhuter was attacked by a lion in August 1903 and succeeded to kill it with his hunting knife. This illustration appeared shortly after the incident in a Barberton newspaper. (ACKNOWLEDGEMENT: STEVENSON-HAMILTON COLLECTION, SKUKUZA ARCHIVES)

16. EARLY YEARS: THE SABIE AND SHINGWEDZI RESERVES AND THE KRUGER NATIONAL PARK 1900–1946

dark. Harry rose painfully and the two adversaries stood facing each other with only two or three feet between them. The ranger was convinced the growling lion would attack him again and came up with a plan that would save his life – he started yelling and swearing at the lion at the top of his voice. This had the desired effect as the fatally wounded animal turned around and disappeared in the darkness. Harry could hear the growls turn into groans, which became fainter and fainter and finally ceased.

As quickly as his wounded shoulder allowed Harry climbed into at young false-thorn tree, because the other lion was still nearby. He was barely in the tree when the other male lion turned up.

Fortunately Bull distracted the lion with his constant barking. The lion charged the dog a few times with no success, and when it discovered that it could not reach the wounded man in the tree or get rid of the annoying dog, it walked off in the same direction as its mate. Harry, by this time faint with the loss of blood, tied himself to the tree with his belt.[4]

After a while he luckily heard the voices of his field rangers and when he started calling they quickly came to help. The badly injured ranger was in great pain and terribly thirsty, but he had to walk another seven kilometres with his colleagues to one of his old outposts during his days with Steinaecker's Horse. They carried burning branches to keep the lions away. After what felt like an eternity they eventually arrived at the camp and he could quench his thirst with a little muddy water that the men had to fetch from quite a distance. They also tried to wash his wounds, but he found it too painful and there were no bandages for the wounds. The next morning he had a fever. The field rangers found his horse, which was unharmed except for the wound to its hindquarters. They also found his rifle and the dead lion, close to the scene of the attack the previous evening. The old male's canines show little wear, but his stomach was empty and he must have been terribly hungry. They skinned the lion and carried Harry in a litter all the way to Komatipoort. Only here did he receive medical treatment, after which he went to Barberton Hospital by train, accompanied by his friend Dickson. He arrived at the hospital a full six days after the attack. The very septic wounds would heal completely, but Harry's stiff right shoulder and upper arm was a constant reminder of the gruesome experience.

The lion skin and the knife that saved his life can be seen in the Stevenson-Hamilton Memorial Library and Information Centre at Skukuza.

The Lindanda Memorial, which indicates where ranger Harry Wolhuter was attacked by a lion in August 1903.
(18.07.1984)

The skin of the large black-maned lion that was killed by ranger Harry Wolhuter with his sheath knife (in the photograph) after he had been attacked by the lion and knocked out of his saddle in August 1903. The lion skin, sheath and knife are exhibited in the Stevenson-Hamilton Information Centre at Skukuza.
(1984)

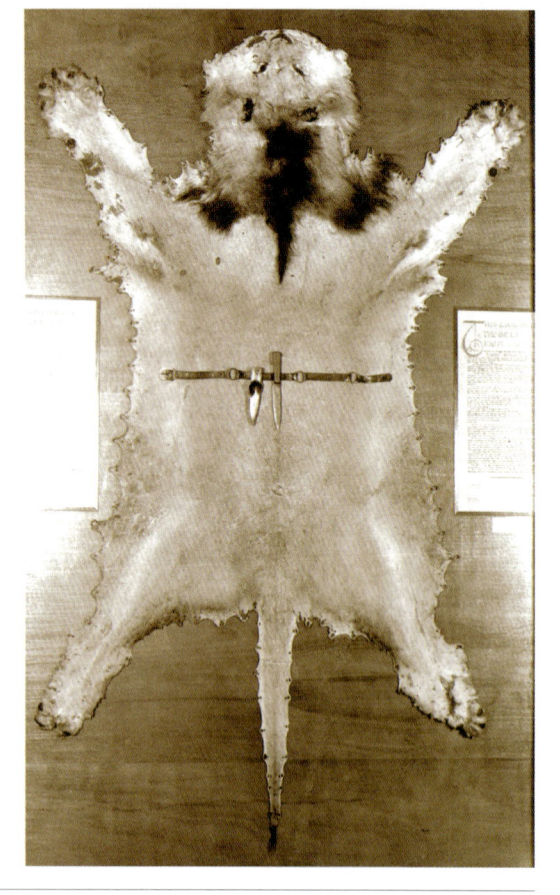

4   In an earlier report (1904) Wolhuter said that he had used his handkerchief and neckcloth.

In November 1937 Harry showed Stevenson-Hamilton the place where the dramatic incident had taken place. In 1938 three memorial plaques were erected at the site, financed by a donation of R100 by Mr Vinnie Robertson of Amersfoort.

The tree in which Harry found refuge has long since disappeared, but a log from it is displayed with the skin and the knife in the Stevenson-Hamilton Memorial Library. The site of the attack was named Lindanda[5] in honour of the bravery of a sinewy young ranger who refused to be the meal of two hungry lions.

Harry's next close shave happened a few years later while on his way from Pretoriuskop to Skukuza, when he took his horse to the Sabie River for a drink of water. As he knelt to have a drink he found himself eye to eye with a massive crocodile, which would surely have caught him had the water not been so clear. Harry said that it was the quickest somersault he had ever executed!

Amid all of this excitement Harry still had to carry out the many daily duties of a ranger, which often took him away from home for days or even weeks at a time. Poaching was a big problem and a lot of his time was taken up with predator control, fighting veld fires and road construction. Fortunately there was also time to study and admire the rich variety of wildlife. He even managed to tame a bush pig that would achieve him great fame locally.

Harry married Alice M. Webster, of Gourock, Scotland, on 14 February 1919. Their only son, Henry Charles, who would follow in his father's footsteps, was born on 28 June 1920.

All his life, Harry loved good horses and good dogs, and his friends and acquaintances readily acquired the horses or dogs he had trained. On his farm De Rust, which he bought in 1910, he kept a big team of dogs and Arabian horses. The horses were all salted animals and were trained not to shy when a rider fired a rifle from their backs.

Like most rangers of his time (and today), Harry was not very keen on administrative duties and although he was the senior ranger, Stevenson-Hamilton usually named ranger C.R. de Laporte the acting warden when he went away. The honour nevertheless fell to Harry in 1931, 1936 and again in 1939. On 1 January 1940 he was appointed as acting warden in the warden's absence. In 1941 he had the opportunity to visit the Kalahari Gemsbok National Park in the company of several board members.

After the excision of the section west of the Nsikazi River from the Sabie Reserve, Harry was shocked and upset about the large-scale slaughter of game that he had so carefully helped to preserve. After many representations, Stevenson-Hamilton obtained the right to protect wildlife in this area again.

To the great chagrin of the people who had settled in the excised area of the old Kaap E and F blocks, Harry regularly arrested poachers, confiscated unlicensed firearms and killed many of the emaciated hounds.

His illustrious career in the Kruger National Park came to an end on 30 April 1946, when he retired along with his friend, Colonel J. Stevenson-Hamilton. He was already 70 years old then and settled on his farm Lindandene just outside White River. His position as ranger of Pretoriuskop section was taken over by his son, Henry. At Lindandene, Harry wrote a book about his adventurous life and work called *Memories of a Game Ranger*. The book, which was illustrated by the well-known wildlife artist C.T. Astley-Maberly of Duiwelskloof, was published in

Col James Stevenson-Hamilton with the Wolhuter couple, Harry and Alice, and their son Henry in 1928.
(ACKNOWLEDGEMENT: STEVENSON-HAMILTON COLLECTION, SKUKUZA ARCHIVES)

---

5   Harry Wolhuter's Changana name, which according to him means "impressive man".

16. EARLY YEARS: THE SABIE AND SHINGWEDZI RESERVES AND THE KRUGER NATIONAL PARK 1900–1946

Former ranger Harry Wolhuter and Col J. Stevenson-Hamilton.
(SEPTEMBER 1956)

Former ranger Harry Wolhuter and author Joy Adamson at Lindandene, White River, shortly before his death in January 1964.

1948 and became very popular. Harry received honorary membership of the Wildlife Protection Society of South Africa in 1947.

Harry received many friends and prominent visitors in his old age at Lindandene, including Joy Adamson of "Elsa" fame, before he died on 30 January 1964 at the age of 88.

Thomas Duke (1860–1934)
After Rupert Atmore's resignation, Thomas Duke was appointed by Stevenson-Hamilton on 5 December 1902. He was temporarily stationed at Kaapmuiden. After Stevenson-Hamilton had moved from Robert Gray's corrugated-iron hut at Crocodile Bridge to the blockhouse at Skukuza in December 1902, Duke moved to Crocodile Bridge. He had to move again soon afterwards because he was transferred to Lower Sabie after Gray's resignation in May 1903. Here he was stationed until his retirement in the middle of 1923 after a dedicated career of more than 20 years.

During the First World War he manned the Rolle ranger post next to the Selati Line for a while, but at the end of the war he returned to Lower Sabie.

Thomas Duke was born in Ardee, Louth County, Ireland and in 1860 came to South Africa with his parents while still a baby. They settled in the eastern Cape and the young Duke learned to speak Xhosa fluently, which led to him being called "M'Xhosa" by the Shangaan during his years as ranger in the Sabie Game Reserve. His father was a successful farmer, but during the Xhosa rebellion of the 1870s their farmhouse was burnt down. The 12-year-old Tom was captured, tied up and abused, but was later freed by a loyal servant, who helped him to escape.

He served in the Cape Mounted Police for 16 years and received a Distinguished Conduct Medal, as well as medals awarded after service in the Tembu and Basuto wars of 1880. During this time he rose to the rank of sergeant. Afterwards he went to the Free State where he joined the Criminal Investigation Department, in which he served until the outbreak of the Anglo-Boer War.

During the Anglo-Boer War he served in Rimington's Guides and he and Stevenson-Hamilton fought together under General Mike Rimington for a while. He was also an intelligence officer with Le Gallais's columns, which nearly succeeded in trapping and capturing General Christiaan de Wet, the famous Boer commander. During the battle that followed the encircling manoeuvre, Le Gallais, the commander of the Eighth Hussars, was fatally wounded.

During the time they had spent together in the war, Stevenson-Hamilton became impressed with Thomas Duke's qualities as a person and especially his proficiency in Xhosa, his ability to communicate with blacks and his knowledge of their customs and beliefs.

He therefore had no hesitation in appointing his former comrade-in-arms as ranger in the Sabie Game Reserve. Through the years Duke's service was of such a high standard that Stevenson-Hamilton never regretted his choice.

In 1937 he wrote the following about his recruit:

Duke, an old Cape Mounted Policeman, and for a time a detective in the C.I.D., possessed an almost uncanny knowledge of the native mind. When some animal had been illegally killed and no traces of the offenders were discernible, Duke was like a bloodhound on the trail. Following devious ways, it was seldom he failed ultimately to land the right man in court.

At the time of his appointment, Duke was already an experienced and seasoned man of about 40, afraid of nothing. An example of his tough nature was when he broke his leg in 1908 falling from a train trolley between Sabie Bridge and Komatipoort. The plaster cast was hardly off when he fashioned a crutch from a straight branch and marched away along the Selati Line in search of poachers. Like his colleague Dick de Laporte, he loved wearing a pith helmet, of the kind that was very popular with those who went "on safari" in Africa.

He was a man of few words and even less inclined to do administrative work than Harry Wolhuter. Today there is barely any record of official documents he had sent to his head at Sabie Bridge.

One of the few documents that could be unearthed in the Skukuza Archives was a report to Stevenson-Hamilton about an incident with a wounded lion. In his neat handwriting he wrote how he had stalked and shot the lion (one of three) from behind an anthill. Unfortunately it was not a mortal wound and the wounded animal charged at him in a rage. There was no time for a second shot and he had to take to his heels. His corporal, Mpampuni Ubizi, saved his life that day as he pulled him behind a pile of rocks, from where he could deal the lion a death blow. Years later, on 15 October 1935, Corporal Mpampuni was killed by a lion while on patrol near Crocodile Bridge – one of the many black rangers who gave their lives to nature conservation in the Kruger National Park and elsewhere.

After the rinderpest epidemic of 1896–97, in which buffalo had almost completely disappeared from the Lowveld, Duke's reports of rising buffalo numbers were very encouraging to the warden. On 10 September 1913 he reported that he had seen a herd of 15 buffalo between Lower Sabie and Crocodile Bridge.

During his first few years as a ranger, Thomas Duke lived in a very primitive wattle-and-daub hut at Lower Sabie. In 1910 a neat wood and corrugated-iron house was built for him there, similar to the ones built in 1909–1910 for the warden at Sabie Bridge, Harry Wolhuter at Mtimba, C.R. de Laporte at Kaapmuiden and G.R. (Tim) Healy at Nwanetsi

(now Satara). Here Tom Duke lived and carried out his duties as a ranger without disturbance until his retirement in 1923.

He then got the Board's permission to run a small store for the locals at Sabie Bridge. In 1931 he became the gate official and permit officer at Crocodile Bridge. He was succeeded in 1932 by his son, T.R. Duke, who nursed his aged father until his death in March 1934. In 1976 a waterhole, supplied with water by two boreholes, by the Shimangwana Spruit, about ten kilometres south of Lower Sabie, was named after him.

Tom Duke junior also had a close encounter with a male lion. The lion had knocked down one of his trackers and, in an effort to get a clear shot through the dense vegetation, he started climbing a tree after handing his rifle to another tracker who was already in the tree. But the noise made by all of this frantic activity caught the lion's attention and he jumped up, grabbed Duke and pulled him from the tree. Duke played dead, but after a while the lion picked him up by the right leg and shook him like a dog would a rat. The man in the tree could not shoot and Duke was helpless in the lion's grip. After a while the lion gave him a final bite on the buttock, walked away and found a comfortable spot under a tree about 120 paces away. Duke's wife Sylvia, who had in the meantime been summoned to the scene of the incident, came to her husband's rescue with their car and took him to hospital. Duke had to spend two months in hospital before his wounds healed to such a degree that he could be released (C.P. de L. Beyers, 1970). During his stay at Crocodile Bridge, Tom Duke junior once shot a giant crocodile that had terrorised both man and beast. The crocodile measured 16 feet (4.88 m) long and was so heavy that it had to be pulled from the water by a couple of oxen.

Cecil Richard de Laporte

After Gaza Gray's resignation in May 1903, Stevenson-Hamilton was a ranger short. On 13 May 1903 he filled the vacancy by appointing Cecil Richard de Laporte, formerly a lieutenant in charge of intelligence with General Rimington's columns during the Anglo-Boer War. Dick de Laporte was an energetic young man of about 28 at the time of his appointment and, as would become apparent later, one of

Three rangers from the early days – from left to right: C.R. de Laporte, G.R. Healy and T. Duke. 1906.
(ACKNOWLEDGEMENT: MR T.R. DUKE, SOMERSET WEST)

the best appointments Stevenson-Hamilton made in the Sabie Game Reserve's early years. He was an extremely capable, dedicated and tactful official. He was also very enthusiastic about his new job and his administrative abilities soon earmarked him as the most suitable candidate for the position of acting warden in Stevenson-Hamilton's absence – a task he carried out often and for long periods, with great distinction.

In contrast to the other pioneering rangers, he often moved station and at short notice had to fill posts where particular problems developed. He was initially stationed at the Kaapmuiden ranger post in 1903. He made himself as comfortable as possible in the small wattle-and-daub house there, which he tongue-in-cheek referred to as Bleak House. (Only in 1910 was a proper wood and corrugated-iron house built for the ranger on the northern bank of the Crocodile River, opposite Kaapmuiden Station.)

Within a few months he was asked to man the post at Lower Sabie in Duke's absence. While stationed there in August and September 1903 he undertook an extended patrol to the Olifants River during which he apprehended two senior police officials who had illegally shot a giraffe and other game in the Bangu area. Despite the obstacles placed in their way at even the highest levels in the Administration, De Laporte, backed by his warden, charged the officers and they were found guilty in the Pietersburg (now Polokwane) magistrate court. This was a major breakthrough and a great morale booster for the warden and his rangers at a time when the concept of nature conservation was still new and foreign.

In 1903 Dick de Laporte went to meet the warden with an ox-wagon at the Olifants River after Stevenson-Hamilton's return from a long patrol in the newly proclaimed Shingwedzi Reserve.

De Laporte soon returned to his post at Kaapmuiden, where he carried out his duties conscientiously and with dedication. Like many of his contemporaries he was fond of wearing a pith helmet, which he always kept in an immaculate condition. His fondness for puttees soon gave rise to the nickname "Makaose" among the locals – a name he kept for the rest of his career. He loved animals and his horses and dogs were among the best in the reserve – to such an extent that his colleagues often tried to obtain puppies from him. Quite a few of his favourite dogs, such as Dansch, were buried in the dog cemetery at Skukuza.

In September 1909 Stevenson-Hamilton accompanied R.C.F. Maugham, the British Consul to Mozambique, on an expedition as cartographer and only returned early in 1910 after visiting Dutch and Brit-

Ranger C.R. de Laporte in his British military uniform, 1902.
(ACKNOWLEDGEMENT: STEVENSON-HAMILTON COLLECTION, SKUKUZA ARCHIVES)

Ranger Cecil Richard de Laporte, ranger at Kaapmuiden, Rolle and Crocodile Bridge (1903–1928) and acting warden in 1909, 1914–1916, 1919 and 1926. Photograph taken at Skukuza, 1926.
(ACKNOWLEDGEMENT: STEVENSON-HAMILTON COLLECTION, SKUKUZA ARCHIVES)

The headstone which ranger C.R. de Laporte erected in 1915 for his favourite dog, Dansch, in the dog cemetery at Skukuza.
(07.05.1984)

ish East Africa and the Sudan. De Laporte served as acting warden for the first time and was very effective in halting a serious poaching problem along the eastern boundary of ranger Healy's section.

In addition to his extensive knowledge of wildlife, De Laporte also had experience as a prospector and had studied geological formations and minerals. In the early 1920s he wrote the first report on geological formations in the Park, while in 1914 he had written: "With a not inconsiderable knowledge of minerals and prospecting, I have never yet found anything of value (in the mineral field), during the whole of my 14½ years wandering in the reserves." Little did he know that large reserves of coking coal in the northern part of the former Shingwedzi Reserve would later attract the attention of exploiters!

With the outbreak of the First World War, Stevenson-Hamilton reported for duty in the Allied forces' campaign in the Middle East (where he served at Cape Helles, Gallipoli, Cairo and the Sudan). Once again, De Laporte was chosen to act as warden at Sabie Bridge (Skukuza). In June 1916 he accompanied the Commission of Inquiry into Game Reserves and Nature Conservation under the chairmanship of J.F. Ludorf on a tour through the Sabie Game Reserve.

He carried out his responsibilities at Sabie Bridge in a particularly competent manner, until he felt compelled to enlist in the South African contingent. Major A.A. Fraser then assumed the mantle of acting warden until the end of the war.

Within a few months De Laporte was promoted to officer in the famous South African 9th Division and earned his spurs in battles in France in 1917–1918. He was seriously wounded in a skirmish on the day before Armistice in November 1918.

After his return from Europe he once again served as acting warden until Stevenson-Hamilton, who had since been promoted to lieutenant colonel, was later that year released from his military responsibilities in the Middle East.

Mrs De Laporte had stayed on at the ranger post at Kaapmuiden (Boulders) during the first few years of the war. While looking after their property, which included a herd of cattle, she was told that a pride of six lions had caught an antelope nearby. Fearing that the lions would get into their herd of cattle, she grabbed her husband's .303 Lee-Metford and, accompanied by two field rangers, hurried to the site where the lions had been spotted. When they got there they found the lions still with their prey, about five kilometres from the ranger post. Mrs De Laporte started shooting at the lions and killed three. A fourth charged them, but she got it with a well-aimed bullet through the head when it was barely eight paces away. This incident caused a sensation in the Lowveld and Mrs De Laporte was toasted as a very brave woman. A trained nurse, she joined the Allied medical services in 1915 and served in France and elsewhere.

After Stevenson-Hamilton's return in 1919, De Laporte was sent to Rolle outpost for a while as a serious poaching problem had developed there during the war years when there were inevitably fewer ranger patrols. During that time the new ranger, W.W. Lloyd, relieved him at Kaapmuiden. De Laporte returned to Kaapmuiden later in 1919 and stayed on until 1920, when he was transferred to Crocodile Bridge. Here he moved into a temporary residence (which he called the Rat Pit). A proper brick house was built for him in 1923 and he was the first person in the Sabie Game Reserve to buy a brand new Model-T Ford. He was paid a transport allowance of £1 a month to use the vehicle in an official capacity. Stevenson-Hamilton only bought himself a second-hand "Tin Lizzie" Ford in 1927.

Because none of the roads in the reserve was suitable for motor vehicles, both cars had to be transported there by train. De Laporte built a special circular road from Crocodile Bridge to Lower Sabie so he could use his car. This later became the first tourist road in the Park.

After the official proclamation of the Kruger National Park on 31 May 1926, Stevenson-Hamilton decided to take extended leave to get matters in order at his estate in Scotland.

De Laporte was once again asked to take over as acting warden and he stayed on in this position until Stevenson-Hamilton's return in November 1926. During this period he had the honour of attending the first official meeting of the National Parks Board on 16 September 1926 in his capacity as warden. De Laporte gave such a good account of himself that the Board decided to pay him an *ex gratia* award of £135. At the end of 1926 he moved back to Crocodile Bridge, where he served until 23 December 1928. He then took extended leave and handed the section to the newly appointed ranger H. McDonald. He officially retired on 1 April 1929 and settled on the farm Glenshiel near Haenertsburg on the advice of an old acquaintance, Dr Louis Thompson. The childless couple stayed there for about three years and then moved to Halfway House between Johannesburg and Pretoria, where they finally disappeared from public life.

With his retirement, the Kruger National Park lost one of its best rangers and a man who had played a very important role in the early history of this big conservation area. In 1948 the Wildlife Protection Society of South Africa awarded him honorary membership for life. Next to the Napi road to Pretoriuskop, about seven kilometres from Skukuza, is a windmill that was named after him in later years.

Major Affleck Alexander Fraser (1855–1929)
After President Paul Kruger proclaimed the Pongola Game Reserve, situated in the corner between Swaziland, the ZAR and Natal, on 13 June 1894, N.F. van Oordt was appointed as warden of the Transvaal's first official conservation area.

When the Anglo-Boer War ended, the Pongola Game Reserve was reproclaimed by the British Administration in 1903 and placed under the care of the warden of the Sabie Game Reserve, Major J. Stevenson-Hamilton. He was instructed to appoint a senior ranger as soon as possible to take over the conservation duties in the area. On the recommendation of Secretary of Native Affairs W. Windham, he offered the position to the Scottish-born professional soldier Major A.A. Fraser.

Affleck Alexander Fraser received his first commissioned rank in the British army on 21 September 1875. On 26 February 1896 he was promoted to the rank of vice major and on 21 March 1900 to second in command of the First Bedfordshire Regiment at Mooltan in India.

After that he was seconded to the Second Bedfordshire Regiment and was second in charge of the unit when it was involved in the battle of Sannaspos near Bloemfontein on 1 January 1901. At the end of 1901 he left the Bedfordshire Regiment and returned to Scotland, where he received the offer to become the Pongola Game Reserve's ranger.

The 48-year-old Fraser was an enthusiastic hunter and fisherman, as well as a keen observer of game and nature, a crack shot and a phenomenal consumer of his national tipple – whisky!

He was a giant of a man, over six feet in height and broad shouldered, with a luxuriant red beard and moustache.

On 12 June 1903 he reported for duty to Stevenson-Hamilton at Sabie Bridge and spent a week or so there before leaving for Pongola. He had his own horses but had to stop at the Repatriation Department in Wakkerstroom to get donkeys to take his camping equipment and supplies to Pongola. There were no donkeys to be had at either Wakkerstroom or Volksrust and he eventually managed to find some at Piet Retief.

When he arrived at Pongola he moved into the ruins of Van Oordt's home on the Lebombo Mountains and found that the cellar and a few of the outbuildings were still intact.

His neighbour was a black missionary called Moses Zinyawo (or "Moses with the large feet"). This particular cleric apparently devoted himself to the "spiritual" wellbeing of especially young women in the area and gathered a large following at his home – to the ire of the authorities, who had to fend off complaints by the women's husbands because Moses refused to send them home.

Fraser had a solution to the problem and, despite Moses' protests, sent him and his followers packing.

After his first reconnaissance patrols of the newly

reproclaimed reserve, Fraser noticed that game was relatively scarce and the grazing of fairly poor quality. Poaching was rife and it was very difficult to catch poachers because of the mountainous terrain, bordered on three sides by foreign territories. But he did try to the best of his ability to protect what little game remained.

Meanwhile, in September 1903 Stevenson-Hamilton visited the newly proclaimed Shingwedzi Reserve north of the Letaba River, which was also under his control. He first wanted to get to know this enormous area, find out how much game was left and make it clear to the black chiefs and village headmen that hunting, the possession of firearms, dogs or snares, poaching and starting veld fires were now illegal. At a place called Malunzane on the northern bank of the Shongololo Spruit, a tributary of the Tsende (Tsendze) River, he found an abandoned but serviceable hut of the labour recruiting agency of the Rand gold mines, the Witwatersrand Native Labour Association. Stevenson-Hamilton decided that this particular spot would make a very good ranger post in the Shingwedzi Reserve.

In December that year, Stevenson-Hamilton made his first visit to Fraser and the Pongola Game Reserve. When he heard Fraser's report on the scarcity of game, the inhospitality of the area, the extent of poaching and the distance from the nearest shop and post office, Stevenson-Hamilton realised that a senior ranger was wasted there. It was then decided that Fraser would train two reliable black rangers to take over from him and that he would be transferred to the Shingwedzi Reserve to serve as first section ranger of this vast area.

On 3 March 1904 Fraser arrived at Komatipoort with his horses and donkeys. His other possessions were sent by railway to Pietersburg, where he had to collect them before he could settle at Malunzane. He finally moved into his new residence on 3 May 1904.

After his arrival he did not let the grass grow under his feet and when Stevenson-Hamilton arrived there in the middle of June, during his annual visit to the north, he had already built a second hut, a kitchen and a bathroom for himself, a hut for his dogs, another one for his field rangers and a stable for his riding and draught animals – all neatly white-washed with leadwood ash and the surrounding area cleared of vegetation.

Today the Shongololo is a dry watercourse at Malunzane, but in those days there were large perennial waterholes in which Fraser caught many a fat bream or mackerel for breakfast. There was little game in the vicinity, but enough to stimulate Fraser's keen eye.

In 1908 he reported seeing the first eland, which according to him had entered the reserve from Portuguese East Africa (Mozambique). In 1914 he reported that all kinds of game were increasing significantly, especially waterbuck and kudu. He had seen elephant 40 miles (64 kilometres) west of the Lebombos and was convinced that they would soon move back into the reserve on a permanent basis. The year before, 1913, he reported that waterbuck had died in great numbers of a mysterious disease, with symptoms similar to those of the illness that had caused the death of many kudus in the Soutpansberg. It was a particularly dry year – so dry in fact that Fraser temporarily had to move residence as the water at Malunzane had dried up. Taking into account the climatic conditions and the symptoms of the animals that had died, this was probably the first anthrax epidemic documented in the official annals of the Kruger National Park.

Fraser, despite his excellent capabilities, had certain personal oddities that often amused his superior and colleagues, but could also drive them nearly insane.

At the end of 1904 Stevenson-Hamilton had to ask Fraser to act as warden at Sabie Bridge while he was on a long voyage on the *SS Africa* to the Sudan, Aden and Somalia. Fraser was not at all keen and in his own words "did not come here to be a dammed clerk" – and Stevenson-Hamilton found that this statement was all too true when he had to bring some form of order to the administration at Sabie Bridge on his return in June 1905.

In August 1906 the warden ordered Fraser to move into Chief Malihane's village next to the Mavumbye Spruit, north of Satara, where he had to attend to the constant poaching problem. Fraser apparently resolved the issue to his head's satisfaction and went back to Malunzane, where he resumed his own unusual daily routine. He divided his working hours in quite a unique way. Shortly before daybreak he would call together his field rangers and assigned them their day's duties – all of this in English, since he refused to learn any other language. One or two of the rangers would be told to accompany him and he would go fishing, or patrol the veld on horseback to observe game numbers, grazing and water conditions. When he needed meat, he would shoot something for the pot, usually a warthog, which he considered vermin.

By ten o'clock in the morning he was usually back at camp and after a hearty breakfast he would take a nap until evening, when the field rangers had to line up to report about their day. And woe betide the man that dared to bother "Manjoro" when he was resting with anything less than a matter of life and death. Yet he still managed to establish a special relationship with his workers and in the years after his retirement the older black workers would refer to "Manjoro" with respect and nostalgia.

At about six in the evening he ate his supper, accompanied by a healthy tot of whisky and soda (judging by the numerous small soda water bombs that can still be found on the terrain of his old Malunzane post). After supper he cleaned his rifles and tinkered with his fishing equipment, or he would read his stack of *Field* magazines, which he had ordered especially from England, over and over. At two or three in the morning he would go back to bed for an hour or so until dawn. Despite this odd routine he knew more about what was happening in his huge section than some of his colleagues with far smaller sections.

The letters he wrote to the warden and colleagues show that Major Fraser was a learned and well-educated man, but he had an aversion to any form of administration or the keeping of monthly returns and statements.

When Stevenson-Hamilton and De Laporte left for active duty in the First World War, Fraser had no choice but to take over the responsibilities of the warden at Sabie Bridge from 1917 to the middle of 1919. Harry Wolhuter had refused point-blank to take on this responsibility. The departments of Native Affairs, Justice and Lands sent letters and official inquiries to Fraser, but to no avail – all of their correspondence ended up in a waste-paper basket. Finally a policeman was appointed at Sabie Bridge to act as clerk and public prosecutor. At the same time he could handle the acting warden's correspondence. This arrangement suited Fraser to a tee, but the poor man that had been appointed first had to wade through the spiderwebs in the office to reach the files and desk! He tried his best, but eventually had to throw in the towel due to the warden's abhorrence of paperwork.

The end of the war arrived just in time to prevent the complete disintegration of communication between the Sabie and Shingwedzi reserves and the government. After his return at the end of 1919, Stevenson-Hamilton disappointedly reported:

> I find that the system of control, carefully built up since 1902, has been seriously impaired since I left in 1914 ... On the whole the impression I received is that there has been a general retrogression, bringing the state of things now obtaining back to about the position in 1904.

Both Fraser and his head were greatly relieved that the former could revert to his eccentric ways at Malunzane!

Once a year Fraser travelled to Pietersburg to deposit the fines he had collected in the area north of Letaba as justice of the peace. On one occasion he walked into the magistrate's office, dropped a small bag full of coins before the surprised official and turned around to leave. When the magistrate asked him what was in the bag, Fraser duly explained, and when the magistrate wanted to know where the receipts and other documentation were Fraser lost his cool.

"Documents! Receipts!" he snapped. "Sir, if you think my word is not good enough then say so. The money is all there. I didn't even count it. You can jolly well do so yourself and then make out your own damned receipts!"

This colourful character's aversion to administrative procedures was so great that he would often pay his staff out of his own pocket and even buy their rations, rather than write out requisition forms. He had no kith or kin, lived simply and had a heart of gold. He preferred spending his money in this manner to dealing with irritating receipts or forms.

While he was acting warden at Sabie Bridge a couple of acquaintances from Komatipoort turned up for a visit one day. They all sat on the veranda of the warden's house, chatting and drinking whisky. The mosquito netting around the veranda had been ripped full of holes, so his many dogs could come and go unhindered. The garden had gone to rack and ruin, as he preferred the unspoiled bush to fruit trees and vegetable beds. Fraser entertained his guests late into the night with hunting adventures from his "Shikari days" in India. Every time he put down his glass, somebody would secretly top it up again. By three o'clock the next morning all the guests had passed out. The host helped them to bed one by one and then sat down on the veranda to finish smoking his pipe. At the crack of dawn he woke up his hungover friends by dousing them with a jug of icy water and invited them to go fishing with him!

"Manjoro" loved his dogs and at one stage had 26 animals of questionable pedigree and no discernible breed. On one occasion Harry Wolhuter visited Fraser at Sabie Bridge during his stint as acting warden. Wolhuter was forced to overnight and was offered the spare bedroom, a pillow and two blankets. Late that night it became very cold and Harry went to ask his host for another blanket. In Fraser's room one candle was burning weakly and in one corner Wolhuter noticed a dark heap from which an unearthly snoring sound came. Closer inspection revealed Fraser, fully clothed and lying flat on his back right at the bottom, with the 26 mongrels draped over and around him! When Harry mentioned this the next morning, Fraser explained that he had given all his blankets to his guest and that he often slept like that since dogs could keep you warmer than any blanket.

At the end of April 1920 this rare old pioneer of the Shingwedzi Reserve had to retire and, aged 65, he handed over his position and equipment at Malunzane to ranger Coetser of Punda Maria. It took him several months to get his business in order and he only left the reserve on 9 November 1920. In the meantime he had built himself a comfortable little stone cottage near Hlomela's village by the Klein Letaba, about eight kilometres northwest of the Sautine outpost. He named his home "Magowbey" and here the old bachelor spent his last days peacefully with his dogs, rifles and fishing tackle. On 16

Maj. A.A. Fraser and his large pack of dogs at his quarters at Malunzane, in the former Shingwedzi Reserve, 1911.
(ACKNOWLEDGEMENT: STEVENSON-HAMILTON COLLECTION, SKUKUZA ARCHIVES)

January 1929 Harry Wolhuter received a letter from a mutual friend, E.G. Ireland, in which he informed him that Fraser had died a few days earlier of liver cancer in a nursing institute in Johannesburg. With the help of a Colonel Thackeray six officers were commandeered to act as bearers and the old soldier was buried.

Today only the ruins of Fraser's homes at Malunzane and Hlomela remain, but near his post next to the Shongololo is a windmill that was named Frasersrus in honour of this stalwart old pioneer ranger of the north.

Guy Rambont Healy
Because of his various duties relating to the maintenance of law and order, James Stevenson-Hamilton was appointed as justice of the peace shortly after his appointment as warden of the Sabie Game Reserve in 1902. To help him fulfil his duties with the required dignity and precision, he was provided with a public prosecutor and court clerk. This post was filled from the beginning by a junior corporal of the South African Constabulary.

After several misfits, including one who drowned in the Crocodile River after a week of non-stop drinking, the authorities sent the 21-year-old Guy Rambont Healy to "see if he would do".

Healy was born in Ireland and studied medicine in Dublin before coming to South Africa during the Anglo-Boer War as a second lieutenant in an Irish militia regiment. He received his original commission in the Fourth Battalion of the Royal Munster Fusiliers. After the war he resigned from the army and joined the South African Constabulary, where he soon made corporal.

Tim Healy, as he was known to friends, was a tall, loose-limbed young man with blond hair and innocent blue eyes, which often gave the uninformed an erroneous impression. Later he would become one of the most energetic and capable of the old guard of rangers.

On the day of his arrival at Sabie Bridge in 1904 there was almost a tragedy. Stevenson-Hamilton tried to cross the flooded Sabie River with his horse Pompey to fetch some urgently needed supplies on the northern bank. But the current was too strong and swept rider and horse downriver. After a few hundred metres they came close to the right bank where Healy and a crowd of black workers, who had raced along the bank keeping up with them, were waiting and managed to drag him ashore safely. Pompey reached the bank on his own steam.

Early in 1907 the first general election since the war took place in the Transvaal and Sabie Bridge was named a subdistrict of the greater Barberton constituency. There were exactly two voters on Sabie Bridge's election roll: Major Stevenson-Hamilton and his clerk, Corporal Tim Healy.

Stevenson-Hamilton was the official in charge of proceedings during the election and Healy had to guard the ballot box and later deliver it safely in Barberton. On the day of the election Healy, suitably attired in his police uniform, arrived at the office of the justice of the peace with the ballot box, which was carried by two field rangers. The ballot box was placed on a table and Healy duly mounted guard to ensure that nobody tampered with his precious cargo. Stevenson-Hamilton decided to cast his vote later that day in a manner suited to his dignity, but unfortunately a Mozambican spitting cobra got him in the eye and he had to stay in bed for the rest of the day with bandaged eyes. Healy thus had to

Ranger G.R. (Tim) Healy with a lion which he had shot at Sabie Bridge (Skukuza) in 1907.
(ACKNOWLEDGEMENT: STEVENSON-HAMILTON COLLECTION, SKUKUZA ARCHIVES)

cast his single vote by himself and take his "precious cargo" of one vote all the way to Komatipoort with the railway trolley. At one point the tracks ran downhill and the trolley hit an anthill on the track at high speed. Healy, his helpers and the ballot box were all flung from the trolley. In the process the ballot box broke and, to add insult to injury, it took them so long to get the trolley back on the track that Healy missed the train from Komatipoort to Barberton. In Barberton the chief election officer was almost apoplectic with rage as votes could not be counted until the tally of ballot boxes was complete. But eventually the ballot box reached its destination and Mr R.K. Loveday was elected with a considerable majority as the Transvaal Volksraad member for Barberton.

In August 1903 a large tract of land between the Olifants and Sabie rivers was added to the Sabie Game Reserve under a proclamation by the Lieutenant General of the Transvaal, Arthur Lawley. Stevenson-Hamilton urgently needed a senior ranger that could take over this enormous area. Because of his enthusiasm and other excellent qualities, the warden decided on Tim Healy for this position and left no stone unturned to get him appointed. However, he had to wait until 1908 to get permission to appoint Healy and then on the proviso that he pay Healy's salary out of savings! This Stevenson-Hamilton managed by cutting his own salary and not filling a number of field ranger vacancies.

In the meantime Healy did extremely well as prosecutor and clerk of the court at Sabie Bridge, shot his first lions and entertained everybody with his genial personality and his attempts to play sangoma (traditional healer or diviner). He was soon nicknamed "Mthepe" by the locals thanks to his weakness for caps.

Healy loved dogs and his favourite pet was a greyhound bitch named Mary. On a sweltering day in November 1907 he went on a patrol west of Skukuza, accompanied by his dog. However, Mary got heatstroke and collapsed at a spruit about ten kilometres west of Sabie Bridge along the Doispane road. Healy tried to revive her and carried her all the way back to Skukuza in his arms, but despite all his efforts she died later that day.

The inconsolable Healy buried Mary in the dog cemetery near the old blockhouse at Sabie Bridge and erected a neat sandstone tombstone. This tombstone, along with others at the site that were found intact, was moved to a spot near the Nature Conservation offices at Skukuza when the terrain along the old Selati railway line was needed for the enlargement of the administrative service area. The little spruit where Healy's dog fell ill is still known as "Inja-ka-Mtepe" (Healy's dog).

After his appointment as ranger in 1908 Healy still served as relief ranger at Sabie Bridge, but in 1910 a wood and corrugated-iron house was built for him next to a tributary of the Nwanedzi River (where Satara rest camp is today).

Inja-ka-Mthephe Hill along the Doispane road, where ranger G.R. Healy's dog Mary contracted heatstroke in November 1907.

The headstone which ranger G.R. Healy erected in 1907 for his beloved dog Mary in the dog cemetery at Sabie Bridge. (07.05.1984)

Healy and his corps of 14 black field rangers could devote all of their attention to their nature conservation duties in the vast area between the Sabie and Olifants rivers. Because poaching took place in both the western and eastern parts of his section, he had to build a suitable outpost near the western border. The spot he picked was on a hill close to Rolle Station on the Selati railway line, southeast of Acornhoek. He stayed there for long periods and even planted fruit trees. A gnarly old mulberry tree can still be found on the site of the old Rolle Outpost (in the former Gazankulu homeland). Close nearby is a big old marula tree scarred with bullet holes, a sign that Healy and his successors used the tree trunk to adjust their rifles (Pienaar, 1986).

In 1914 the young Healy married a 19-year-old girl from Meath County, Ireland, in Cape Town. This could have been the same Miss Monk who had visited him in 1910 at his Nwanedzi post. Healy proudly told everybody later that the famous Rudyard Kipling had delivered a speech at the wedding reception. The young Mrs Healy was passionate about horse riding and showjumping – to such an extent that he built her a gymkhana field at the old Rolle outpost so she could practise for the next big showjumping competition in Johannesburg.

Healy built a cart drawn by Stevenson-Hamilton's old pony, Jantjie, so he could travel the 64 kilometres between the outposts at Nwanedzi and Rolle in comfort.

Stevenson-Hamilton often mentioned Tim Healy and his field rangers in his annual reports and in 1911 he wrote: "Ranger Healy is a hard-working and conscientious ranger."

In 1910, while the warden was on extended leave and C.R. de Laporte the acting warden, Tim Healy played an important role in the resettling of Chief Mnyamane and his gang of poacher-subjects from the Nwanedzi area in an area outside the western border.

After the outbreak of the First World War, Healy decided to join the Allied forces and fought in the East African campaign under General Van Deventer. On 25 May 1915 he received his commissioned rank in the 3rd Battalion of the King's African Rifles, but in 1916 his colleagues received the tragic news that Lieutenant Guy Rambont Healy had fallen on 27 March 1916 during a battle in Tanganyika (now Tanzania).

Thus the promising career of one of the best young rangers of the old Sabie Game Reserve ended on a tragic note. In 1928 Stevenson-Hamilton sent a testimonial to the honorary secretary of the Officers' Family Fund in Mayfair, London, so that they could provide the widow Healy with assistance for the education of her son, John.

Paul Siewert
The appointments and periods of service of the next few rangers in the old Sabie and Shingwedzi reserves are all related to the period just before and during the First World War.

In 1910 Paul Siewert was appointed as junior ranger by Stevenson-Hamilton to help him with a scheme to capture young antelope of different species and raise them by hand. The plan was to use the National Zoo in Pretoria to sell these animals to zoos overseas, thereby providing the reserves – which always struggled to obtain sufficient funds from the authorities – with an income. Although they were successful with the capture of several young bucks, including kudu, bushbuck and others, and later even kept a small herd of tamed eland at Sabie Bridge (Skukuza), the scheme came to nothing because of losses both during and after capture.

Siewert's temporary appointment was a huge success in other ways. There was a real need for a section ranger on the northern side of the Sabie River because the river made it almost impossible to curtail poaching on the northern side during the rainy season and before the completion of the railway bridge and the railway line between the bridge and Tzaneen in 1912.

Because he was an ardent nature lover, a bachelor and someone who preferred a solitary existence, the warden settled him at the Msutlu outpost on the northern bank of the Sabie River, about five kilometres northwest of Sabie Bridge. Here he soon organised a comfortable camp for himself. Shortly after his move there, he and two black field rangers were involved in a shoot-out with a big band of armed poachers near the eastern border. About 50 shots were fired, but fortunately without any ill effects to the rangers.

Siewert had been a shopkeeper before his appointment and was a nephew and employee of Mr Hanneman, the well-known German shopkeeper at Komatipoort. He was also of German birth, but had been adopted and raised by American parents. It is clear from his reports and the surviving correspondence in the Board's old files that Paul Siewert was a conscientious and hard-working young man. Because he was unmarried and had a youthful lust for adventure he was often used to relieve older rangers when they went on leave or had to fulfil other duties. His first stint as a relief ranger was in 1910 at the Kaapmuiden ranger post when ranger C. de Laporte had to stand in for Major Stevenson-Hamilton. He was stationed at Kaapmuiden from September 1910 until December 1911, but after De Laporte's return in January 1912 he went back to his Msutlu base camp by the Sabie. In 1913 he did another stint at Kaapmuiden and on 26 April 1913 he got the welcome news that his permanent appointment as ranger had been confirmed by the authorities at a salary of £210 a year!

From 1913 until the outbreak of the First World War he was back at Msutlu and of particular use with Stevenson-Hamilton's ill-fated game capture scheme. By that time he had already earned two nicknames from the black rangers – "Mavhusngwenya" ("somebody in whom a crocodile can be awakened") or "Mafekete" ("don't try to bluff me"). In his annual report of 1912, Stevenson-Hamilton mentioned that two of the remaining hippos in the Sabie River, a cow and her calf, had taken up residence in a pool opposite ranger Siewert's camp at Msutlu.

After ranger Healy's departure in 1914 for active service in the East African Campaign, the central district of the Sabie Game Reserve was left without a ranger and Paul Siewert was transferred to the Rolle ranger post along the Selati Line. Here he also did a sterling job, but ranger Thomas Duke wrote on 8 September 1915 in a confidential letter to acting warden De Laporte that the Portuguese authorities were on the lookout for Paul Siewert, who they claimed had shot a black poacher on the Mozambican side of the eastern border in the Lebombos. Nothing came of this accusation and Paul Siewert stayed on at Rolle until his discharge from the Civil Service in 1916.

When Stevenson-Hamilton departed in 1914 to add his weight to the war effort of the Allied forces, he went to say goodbye to his old friend Hanneman at Komatipoort. The German shopkeeper waved him goodbye with tears in his eyes and said: "Oh, Major, I wish I never lived to see this happen."

The poor man would not see the end of the war. His shop was boycotted by local "patriots" and although he had been a prominent figure in Komatipoort for many years, his former friends avoided him. His loneliness aggravated an illness he suffered from and he died in 1916.

His daughter committed suicide by drinking poison and the poor, friendly Mrs Hanneman ended up in a mental institution.

Because of his German origins, Paul Siewert also fell victim to persecution by over-zealous "patriots", whose petitions resulted in his dismissal from the Civil Service in 1916. This shock was too much for the young Siewert, who started drinking too much and shortly afterwards fatally shot himself – a victim of man's unreasonable aggression and vindictiveness at times of war.

Frank Streeter

After Major Fraser had taken over as acting warden from C.R. de Laporte early in 1917, he asked an acquaintance of his, Frank Streeter, to man the ranger post at Malunzane in his absence and keep an eye on the Shingwedzi Reserve.

Frank Streeter was a seasoned Lowvelder of British origin with extensive knowledge of the veld. Before the Anglo-Boer War he was a customs officer at Komatipoort and had written to the state secretary on 24 September 1895 suggesting that the state should proclaim a game reserve between the Sabie and Crocodile rivers. He was known for his comprehensive collection of birds' eggs, in which even the late Dr Austen Roberts, the curator of the Transvaal Museum, was interested. He was also a qualified taxidermist and collected and stuffed small mammals and birds as a hobby.

Not much is known about his service period at Malunzane between 1917 and 1919. It seems, however, that he managed to exert at least a measure of control in this vast area. After Fraser's return, Frank Streeter settled in a rondavel on Huntley Pohl's farm

Baviaanskloof near Mooketsi. In 1923, Streeter accompanied Eric Pohl, a certain Prince and a Botha on a hunting trip to Mozambique. Before the expedition he wrote to ask ranger De Laporte about the breed of dog they should take along if they wanted to hunt lions. De Laporte's answer was "fox terriers".

During the trip the party had a lot of trouble with field mice in their camp and Streeter set several traps for the pesky rodents, including one under his bed. One night he was woken by the trap snapping shut. He stuck his hand under the bed to remove the trap and its prey, only to find he had trapped a snake, which promptly bit him on the finger. Streeter was unfazed – he calmly walked over to the butcher's block and lopped off the finger with a hatchet!

They had planned a three-month trip, but because of trouble with their workers they ended up staying for five months. Streeter fell seriously ill with malaria and the rest of the group returned without him, after leaving him at Malvernia to recuperate. Beyond all expectations he returned, after three months, by train via Rhodesia (Zimbabwe) and Mafeking (Mafikeng) to Pretoria and again settled on the farm at Mooketsi. Some of the older inhabitants of the area, including Messrs Bands and Vincent Pohl, remembered him as a friendly old man with a prominent white moustache.

In 1937 Frank Streeter once again contracted malaria and died a few days after being admitted to Duiwelskloof Hospital. He was buried in the Duiwelskloof cemetery.

Andrew James Wolhuter
Because three of the Sabie Game Reserve's rangers – the warden and rangers De Laporte and Healy – had been on active service since the beginning of 1917, acting warden A.A. Fraser received permission in 1918 to appoint Harry Wolhuter's younger brother, Andrew James Wolhuter, as acting ranger. He was stationed at De Laporte's post at Kaapmuiden.

According to a report by Fraser, Andrew Wolhuter acquitted himself very well of his task and as a result of his excellent judgement in disputes he became known as "Ngea-ngea" ("the arbiter") among the local blacks. This third member of the famous Wolhuter ranger corps fell prey to the Spanish flu epidemic of 1918 like so many others and died of this feared illness on 14 November 1918 in his brother's house at Mtimba. All the other rangers also contracted the flu, but luckily they survived it – although Harry Wolhuter and Frank Streeter nearly died.

Andrew Wolhuter was buried near the homestead and his grave, which was covered with a cement block, can still be seen in the backyard of a home in KaNgwane, built on the site of the former Mtimba ranger post. The bronze plaque with an inscription has long since disappeared, however.

Captain J.J. Coetser (1872–1935)
Long before Major Fraser, the only section ranger in the Shingwedzi Reserve for many years (1904–1919), retired in 1920, the authorities had realised that he could not possibly control such a large area alone. A senior ranger was needed in the far-northern section of the Shingwedzi Reserve to stop the illegal activities of hunters, smugglers, recruiting agents for the mines and other undesirables in the notorious Crooks Corner between the Luvuvhu and Limpopo rivers.

Captain J.J. (Kat) Coetser was chosen for this purpose and he accepted his appointment as ranger in this far-off area on 1 May 1919. He was the first Afrikaans-speaking ranger to be appointed in the

Ranger Harry Wolhuter (left) and his younger brother, Andrew James, who was temporarily appointed in 1918 as ranger at Kaapmuiden.
(ACKNOWLEDGEMENT: STEVENSON-HAMILTON COLLECTION, SKUKUZA ARCHIVES)

Sabie and Shingwedzi reserves. On 6 May 1919 another Afrikaans-speaking man, Colonel P.J. de Jager, was appointed as a ranger in the Sabie Game Reserve. An account of ranger Coetser's life and work in the Shingwedzi Reserve and later Kruger National Park also appears in chapter 19.

Capt. J.J. (Kat) Coetser, ranger at Punda Maria and Satara (1919–1928), in his military uniform during the East African Campaign (1916).
(ACKNOWLEDGEMENT: MRS W.H. ROGERS, PRETORIA)

Mrs Maria Coetser, ranger J.J. Coetser, Col J. Stevenson-Hamilton and Mr Paul Neergaard, manager of WNLA at Soekmekaar, relaxing at Punda Maria. October 1922.
(ACKNOWLEDGEMENT: STEVENSON-HAMILTON COLLECTION, SKUKUZA ARCHIVES)

Johannes Jacobus Coetser was born on 3 May 1872 in Lydenburg. He married Maria Johanna de Beer and they had 14 children, of whom four died young. He served as bailiff in Standerton for a while, but later joined the then Department of Native Affairs as he could speak several African languages. He then became a member of the State Artillery and was promoted to captain at the outbreak of the Anglo-Boer War. Coetser was in charge of the Johannesburg Fort until Judge Krause took over from him and he left for active duty. He was eventually captured and interned in the concentration camp at Green Point. After the war he joined the army and became an intelligence officer. With the outbreak of the First World War he joined General Van Deventer's scouts and served throughout the whole African campaign, during which he mastered Swahili as well.

After his return he and Colonel Piet de Jager were appointed as rangers even before Stevenson-Hamilton's return from active service.

Coetser built his first camp at the fountain below Dzundwini Hill, but soon moved to a more suitable terrain at the strong Shikokololo Fountain at the foot of Dimbo Hill. Here he built a comfortable camp (some of the huts are still standing) and named it Punda Maria. According to the former native commissioner of Sibasa, C.L. Harries, in a letter to Stevenson-Hamilton in 1933, Coetser explained to him why he had chosen such an odd name for his post. Apparently it had to do with the fact that the first animals he had seen on his arrival in the Shingwedzi Reserve were zebras. The Swahili name for a zebra is "punda milia" ("striped donkey") – from there the Punda. His wife's name was Maria and she loved wearing white and black striped dresses. He therefore thought that Punda Maria would be a very suitable name for the ranger post.

On a later occasion Coetser confirmed the origin of this name to the well-known artist and author Charles Astley-Maberly.

In 1959 the Board decided to change the name of the section and camp to Punda Milia under the misapprehension that Coetser had misspelt the Swahili name. The correct facts were, however, brought to the Board's attention in 1981 and the original name was reinstated.

Coetser became known as "Gungunyane" among

the locals and built a hut over the fountain at Punda Maria to keep it clean. Legend has it that there used to be a couple of tame catfish in the fountain that would quickly devour any insects that fell into the water. In those days it took Stevenson-Hamilton three weeks to reach Punda Maria from Sabie Bridge and he could therefore not visit the post more than once a year.

Paul Neergaard, the WNLA manager at Soekmekaar, often visited Coetser at Punda Maria and so managed to have post and supplies delivered to Pafuri via the later WNLA bus service. At one stage Neergaard heard that Coetser was ill with blackwater fever and rushed to Punda Maria to help him in Mrs Coetser's absence.

Stevenson-Hamilton was not completely satisfied with Coetser's services at Punda Maria and transferred him to Satara in July 1924, where he stayed until his resignation on 15 December 1928. In 1925 he built the first earthen dam in the Park in the Shitsakana Spruit. It later broke during a big flood and was swept away. In 1928 the Board complimented him on the roads he had built in his section.

After leaving the Park, Coetser joined the Department of Agriculture – technical services. In 1934, while the warden was on sick leave, he entered the Park in the company of Colonel Van Deventer, the senior locust control official of the Pilgrim's Rest district, and issued unauthorised instructions to the staff. This resulted in his transferral to an area further away from the Park.

On 21 November 1935 he was serving as a foot-and-mouth disease control officer along the Limpopo River east of Messina (Musina) when he was charged by an enraged elephant. His gun-bearer ran away and the poor Coetser was caught by the elephant and trampled to death (Godfrey, 1963). He was buried in Pretoria.

Lieutenant Colonel Petrus Lafras de Jager

Petrus Lafras de Jager was appointed as ranger in the Sabie Game Reserve only days after his colleague Coetser, on 6 May 1919. That was towards the end of Major Fraser's term as acting warden at Sabie Bridge and De Jager was instructed to temporarily man ranger Healy's old post at Rolle. He took over from Thomas Duke, who had also been temporarily

Field rangers J.J. Coetser and Six at Punda Maria in October 1922.
(ACKNOWLEDGEMENT: STEVENSON HAMILTON COLLECTION, SKUKUZA ARCHIVES)

Col Petrus Lafras (Piet) de Jager, who served as ranger at Rolle, Kaapmuiden and Punda Maria (1919–1929) in his military uniform at Rolle 1920.
(ACKNOWLEDGEMENT: MR P. DE J. MOUTON, BEDFORDVIEW)

stationed here, and stayed on until 1920. He was then transferred to the Kaapmuiden (Boulders) section after ranger C.R. de Laporte's move to Crocodile Bridge.

Piet de Jager was a professional soldier with an excellent record at the time of his appointment as ranger in the Sabie Game Reserve. He had been a commandant in the Boer forces during the Anglo-Boer War and a colonel in the mounted corps under the command of General. J.L. van Deventer during the East African campaign of 1916. He distinguished himself as soldier in both wars and received the DSO in the East African campaign. He also received several other decorations and honorary awards, including a Russian or Serbian medal that was awarded to him through the personal mediation of General Smuts. He was already 50 years old at the time of his appointment, but was still fit for his age and a tall, well-built man with aristocratic features. De Jager was an excellent shot and horseman, from where his name among the locals, "Madubula" ("the one that shoots well"). Others referred to him as "Mhlepuna" ("the one who was wounded"). At Rolle ranger post he often used to practise his target shooting using a large marula tree close to the house and the bullet holes can still be seen in the tree.

According to his daughter, Mrs L.E. Kruger of Hoedspruit, they attended school in Piet Retief at the time, but after his transfer to Kaapmuiden they went to school in Barberton. The ranger's quarters at Rolle consisted of two rooms, two rondavels and a kitchen – all of wattle and daub and a real pioneer's home typical of the time. The house at Kaapmuiden, or Boulders as it was also called, was a more comfortable wood and corrugated-iron structure with a veranda that was enclosed with mosquito netting and built on a series of concrete pillars, which are still standing. The quarters were on the northern bank of the Crocodile River opposite Kaapmuiden Station, which was the main source of supplies and news of the outside world.

While De Jager was stationed at Kaapmuiden, it became customary that a group of Boy Scouts from Barberton would camp in the reserve every year during the winter holidays under the leadership of their officer, C.A. Yates, Harry Wolhuter's brother-in-law. De Jager fetched them from Kaapmuiden Station and rowed them across the river, after which he took them to a suitable camping site with his ox-wagon. Here they would camp under his supervision for a week or so and spend an adventurous and educational holiday in the wilderness. One of these fortunate children, H.V. Hodge, later recounted some of his experiences in the Board's popular *Custos* magazine.

After ranger Coetser's transfer to Satara in July 1924, De Jager was asked to man the Punda Maria section. He went to this far-off post without his wife and children and found the isolation just too overwhelming for a man of his years. He also had serious relapses of malaria, which he had contracted in East Africa. In July 1927 he asked to be transferred back to Malelane section, but this request could unfortunately not be granted.

Despite all his problems he got to know his section well, and it was thanks to him that the occurrence of nyala in the Pafuri area could be confirmed without a doubt. There had been talk of nyala in the area during Coetser's time, but De Jager sent Stevenson-Hamilton the horns of a bull that had been killed by lions. The presence of these rare antelope in the Shingwedzi Reserve caused quite a stir among zoologists at the time.

A new brick house was built for De Jager at Punda Maria in 1926, but it was an unpractical Highveld design without a veranda to provide some much-needed shade. This was added later.

De Jager missed his family and farm and left on extended leave on 30 November 1928, before he retired in January 1929 at the age of 61.

The old ranger post at Kaapmuiden (Boulders) was demolished in 1925 and the section's headquarters moved to Malelane, while the terrain itself was cut out of the Park by the foot-and-mouth fence along the western border of the Park. In 1985 a new private camp, sponsored by the firm Murray & Roberts, was built close to the Phalaborwa-Mooiplaas tar road about nine kilometres southwest of the Mooiplaas ranger post. The camp was called Boulders in honour of the old ranger post.

William Walter Lloyd
The deaths of rangers Healy and A. Wolhuter in 1916 and 1918 respectively made the appointment of another ranger in 1919 imperative. The successful

candidate was the 53-year-old William Walter Lloyd. Before his appointment he had farmed on the farm White Waters near Kiepersol.

Lloyd was of Welsh extraction and had emigrated to South Africa just before the start of the Anglo-Boer War. He settled in the eastern Cape and was a crack shot at an early age. This is proven by the fact that he won a trophy at a target shooting competition as a member of the East London Rifle Association in 1894/5. This trophy is currently in the possession of his granddaughter, Mrs D.B. Glanville of Lyttelton (now Centurion).

He started his career as a ranger at Kaapmuiden (Boulders) next to the Crocodile River. During his time there he was visited twice by C.A. Yates, the Boy Scout officer of Barberton and brother-in-law of Harry Wolhuter. After the return of ranger De Laporte at the end of 1919 he was transferred to Rolle, where he took over from Piet de Jager. In 1920 Stevenson-Hamilton transferred him to ranger Healy's old post at Nwanedzi (Satara), where he served until his tragic death in November 1922.

Despite his advanced age, Lloyd was a very energetic and busy ranger. He adored horses and often undertook extended patrols on horseback. The floor of a neatly constructed stable that could house five horses has since been discovered at Pumbe, on the eastern border with Mozambique. A revolver cartridge found at the site could be an indication that the stable had been built and used by Lloyd.

While on patrol Lloyd often slept outside, without food and in wet clothes that consisted of a self-made pair of lion skin trousers, a light shirt and a wide-brimmed hat. Locals called him "Kukuzele" ("the man who walks crookedly") and "Nwashibeyilana" ("the one who carries a small axe"). After a long patrol on a sweltering day in November 1922 Lloyd arrived home and, soaked in perspiration, sat down on the veranda to cool down – not surprisingly he contracted pneumonia. His young wife did the best she could with the little medicine at her disposal, but it was to no avail and he died on 15 November 1922.

The next night Stevenson-Hamilton was woken by a black messenger with a note from Mrs Lloyd informing him of her husband's death the previous night. The messenger had covered the 88 kilometres between Satara and Sabie Bridge in less than 24 hours!

Before daybreak Stevenson-Hamilton left for Satara in his buggy to help the young widow and her three small children (a girl and two small boys). He covered the distance in a day, but on arrival found that Mrs Lloyd had already buried her husband with the help of the field rangers under an umbrella thorn tree just north of the residence.

The warden helped her to pack her belongings and when his ox-wagon arrived after two days they loaded everything onto the wagon and Lloyd's donkey cart and left for Sabie Bridge, from where Mrs Lloyd would travel by train to her family.

At Tshokwane the company was caught in a tremendous thunderstorm and they were forced to spend the night in the patrol huts there. That night two lions caught one of the donkeys next to the hut, but the next morning they tracked down one of the lions that had been wounded during the night with the help of ranger Lloyd's mongrel Irish terrier Bles.

The family later erected a decent headstone on ranger W.W. Lloyd's grave. Apart from Stevenson-Hamilton himself (his ashes and his wife's were scattered on Shirimantanga Hill on 10 April 1979 by

Ranger W.W. Lloyd on his horse at Satara in 1920. During his career he had consecutively been stationed at Rolle, Kaapmuiden and Satara where he died (1919–1922).
(ACKNOWLEDGEMENT: MRS F.J. DEN BAKKER, DURBAN)

Floor of an old horse stable near the former Pumbe picket, which had probably been built by ranger W.W. Lloyd, circa 1929. (11.08.1985)

their daughter, Ann Doyle), Lloyd was the only ranger buried in the Kruger National Park.

The elderly Mrs F.I. den Bakker (the little girl who had accompanied her mother and two brothers on the sad journey from Satara to Tshokwane) donated the only known photograph of her father to the Skukuza Archives in 1977.

Leonard Henry Ledeboer (1868–1959)

After Major A.A. Fraser's departure from the Shingwedzi Reserve in November 1920, a new ranger was appointed to fill the vacancy. Leonard Henry Ledeboer accepted the position on 1 April 1921. He was not stationed at Fraser's old camp at Malunzane, however, as there was no perennial water left. Instead, the warden asked him to man a new post at Hatane, at the mouth of the Makhadzi Spruit. From here he would be able to keep a better eye on Chief Makuba's people living along the Makhadzi and Tsende (Tsendze) rivers, among whom there were known poachers. As a result of the unusually heavy late summer rains, Ledeboer struggled to get his wagon and oxen from Malunzane to Hatane and only arrived there on 18 April with his wife and baby daughter. Shortly afterwards Ledeboer also developed pneumonia, but he was more fortunate than his colleague William Lloyd. Over a period of two very difficult months in a ramshackle mud and grass hut, his wife managed to nurse him and their baby, who had also fallen ill, back to health.

Of Dutch extraction, Ledeboer was born on 28 September 1868 in India. In 1888, at the age of 21, he came to South Africa and made a living as a transport rider and hunter. From 1891 to 1893 he hunted mainly in what was then Rhodesia (now Zimbabwe) between the Limpopo and Sabie rivers, as well as in Portuguese East Africa (Mozambique). In 1895–1896 he hunted in the area between the Olifants and Limpopo rivers in what is now the Kruger National Park and from 1908 to 1914 he hunted without a break in Rhodesia (Zimbabwe) and Mozambique with two permanent camps along the Sabie and Limpopo rivers. During this period he apparently met all the well-known hunters of the time, including Selous, Jackson, Finnaughty, Van Rooyen and Goosen.

During the Anglo-Boer War he served in the ZARP unit, but after his capture in May 1901 by Captain Bulala Taylor he defected to the British forces and was appointed as an intelligence officer in the notorious Bushveldt Carbineers regiment, which was stationed at Fort Edward near Elim in the northern Transvaal.

In January 1902 he testified in the Pietersburg court martial of several officers of the unit who were accused of having murdered Boer prisoners of war.

After the war, on 19 December 1902, Ledeboer wrote to the magistrate at Pietersburg and recommended that the area between the Limpopo and Letaba rivers be proclaimed as a game reserve. The matter was supported by a letter on 22 April 1903 from the secretary of Native Affairs to the assistant colonial secretary. The Shingwedzi Reserve was proclaimed under Proclamation No. 19 of 1903 of the Lieutenant Governor shortly afterwards, with the same boundaries as those proposed by Ledeboer. In this manner, Ledeboer had probably made his most significant contribution to nature conservation even before being appointed as a ranger.

After the war Ledeboer worked in the office of the native commissioner at Sibasa for a while, but later moved to Crooks Corner between the Luvuvhu and Limpopo rivers to work as a recruiting agent for the mines. His sojourn and experiences here are described in chapter 19. This is also where he married Constance Elizabeth Ash, sister of the well-known shopkeeper Morty Ash. They had three children, one of whom was Frederick Leonard (Freddy), who in later years was twice attacked by lions and managed to survive both attacks. In the second incident, on Letaba Ranch, he lost his left arm (Minnaar, 1985). After the death of his first wife Ledeboer married his children's governess, Anna Maria Christina Bindemann, in 1915. She and their baby daughter accompanied him when he started working as a ranger at Hatane in the Shingwedzi Reserve.

On 10 August 1921 Anna unexpectedly died of a heart attack and he buried her just west of the confluence of the Makhadzi with the Letaba River.

As a result of the tragedy and other reasons, Ledeboer got the warden's permission to move his quarters to a better location on a low ridge about one kilometre northeast of the current Letaba rest camp. He named this post Mondzwani and lived here from December 1921 until November 1927. It would seem his baby daughter also died during this time, judging from the sad little pile of stones near the ruins of the old Mondzwani homestead.

Chief Makuba and his followers left the Shingwedzi Reserve on 3 November 1923 and Ledeboer could then focus on his primary task. In those days Ledeboer had endless problems with prospectors searching for gold, mica and especially diamonds in the area between the Letaba and Olifants rivers. (Prospecting was banned only in 1926, by the National Parks Act.) Many prospectors, including Van Deventer, Elton, Jerome, Coetzer, Pienaar, Wright and the brothers Gerber of Louis Trichardt (J.C. Gerber later also dug the prospecting hole at Malonga in search of diamonds), believed great riches could be unearthed in the area near the confluence of the Letaba and Olifants rivers. They were constantly turning up with prospecting permits, only to leave empty-handed and return yet again the next winter. They were only successful in poaching game in the areas where they prospected, which placed an added burden on Ledeboer and his field rangers.

Judging from the piles of washed pebbles near the old Mondzwani outpost, the prospecting bug also bit Ledeboer and periodically he would try to conjure up a few shiny pebbles from the ground. Maybe it was these activities that led to him being called "Mthakati" ("sorcerer"). Otherwise he was known as "Mabalane" ("the secretary" or "writer") among the locals.

In April 1922 he married his third wife, Kate Häsler (Evans) in Johannesburg. They had a daughter, Esmé (1923), and a son, Leonard Cecil (1929).

On 2 April 1924, while at Mondzwani, Ledeboer saw a meteor fall about 19 to 23 kilometres from his home near the Hlanganini Spruit's upper reaches. The meteor has never been found, but will probably be discovered some day. In the winter of 1923, after his return from a relief stint of six months at Satara following Lloyd's death, he discovered that scores of waterbuck and kudu had died of a mysterious disease on the Lebombo flats north of Letaba. This happened again the next winter and the symptoms he described point unmistakably to anthrax – thus the second recorded outbreak of this disease in the history of the Shingwedzi Reserve.

During the spring and summer months of 1924/25 large swarms of brown locusts descended on the Lowveld and denuded the Sabie and Shingwedzi reserves of vegetation. All the personnel had their hands full spraying the locusts (with lead arsenate!). These conditions repeated themselves in the period 1932 to 1934 and especially in 1934 the senior personnel, assisted by more than 300 black workers, were kept busy round the clock exterminating thousands of swarms of locusts in the hopper stage (in 1934 alone more than 40 000 swarms were killed!). However, these were red locusts, as in 1906.

Leonard Henry Ledeboer, ranger at Letaba and Satara sections (1921–1946). This photograph was taken during the 1930s at Satara.
(ACKNOWLEDGEMENT: STEVENSON-HAMILTON COLLECTION, SKUKUZA ARCHIVES)

In 1934 Ledeboer reported that he had noticed eland spoor on the Lebombo flats and had encountered small groups of roan antelope. In 1925 he saw giraffe spoor for the first time on the way to Shingwedzi and came across a small herd of buffalo along the Tsende (Tsendze).

Elephant numbers were also increasing, and a breeding herd had also settled in the area between the Letaba and Olifants rivers. In 1928 he reported discovering for the third time the remains of a sable antelope bull and a lion that had killed each other in a fight. In May 1924 Ledeboer reported that he had found significant salt deposits about eight kilometres from the Ngwenyeni Spruit on the way to Makushane's village. Locals went there in the rainy season to collect large amounts of salt from the clayey soil. During this period he also waged a merciless war on wild dogs and killed them indiscriminately (this was policy at the time). He even dug the puppies from their holes and poisoned them.

In December 1923 he was told that his section would be extended westwards to the confluence of the Olifants and the Selati and from there to the confluence of the Great and Klein Letaba rivers. On 16 September 1926 he received the news that the Sabie and Shingwedzi reserves had officially been proclaimed as the Kruger National Park.

In one of his reports he also mentioned that the Mbaula Ranch had been proclaimed as a private nature reserve in 1929. At the time it belonged to Joe Albasini, a great-grandson of the famous João Albasini.

Every now and again Ledeboer would write to the warden to complain about Mondzwani's remoteness and eventually it was decided that he should settle south of the river to prevent floods from cutting him off from the outside world on such a regular basis.

In November 1927 a temporary residence was built for him on the southern bank of the Letaba River, just west of the current rest camp, but it was destroyed by a hurricane on 23 October 1928 and had to be rebuilt. In September 1929 Ledeboer was transferred to the Satara section, where he was stationed until his retirement on 30 April 1946.

Ledeboer was prone to exaggeration and the warden often had to take his reports with a pinch of salt. So, for example, in answer to an official request from the Board in August 1936, he insisted that there were more than 1500 lions in the Satara section and unless their numbers were drastically reduced there would soon be no game left in his section.

While living at Satara his third wife died on 23 November 1937. He later married Martha Maria Meyer, who outlived him. Ledeboer received honorary membership for life from the Wildlife Protection Society of South Africa in 1947.

After his retirement in 1946 he settled on his farm Radoo in the Letsitele district, where he also managed a small store, but he later moved to White River, where he died on 25 February 1959 at the age of 91.

Satara ranger quarters in 1930.
(ACKNOWLEDGEMENT: STEVENSON-HAMILTON COLLECTION, SKUKUZA ARCHIVES)

Jack R. Brent

After the permanent transfer of ranger W.W. Lloyd from the old Rolle outpost to Satara in 1920, a new ranger was appointed to fill the vacancy. Jack Brent was a young British bachelor who, before coming to South Africa, had served for three and a half years in the Royal Flying Club and Royal Air Force, where he received a commissioned rank during operations in France in the First World War.

Little is known about his time of service at this remote ranger post, but according to a report by Major Stevenson-Hamilton he was a pleasant young man who carried out his duties with care and thoroughness.

The western part of the central district of the Sabie Reserve was an enormous area consisting of a hotchpotch of state land and farms belonging to the Transvaal Estates & Development Company and the Transvaal Land Company (TLC).

Black people lived on parts of the TLC farms and every year they had to pay taxes and rent to the owners. The person appointed to collect the money was J.E.D. Travers of the farm Champagne in the Bushbuckridge district.

John Travers had been a member of Steinaecker's Horse during the Anglo-Boer War and was wounded in the shoulder in a Boer attack on Fort Mpisane. During the First World War his former commanding officer, Colonel Ludwig von Steinaecker, found asylum on his farm, became mentally unstable and committed suicide by drinking strychnine. He was buried on Travers's farm. Since he was the rent collector Travers soon became known as "Maticket" among the locals. During ranger Brent's time at Rolle he and Travers were involved in an ongoing feud because the young ranger would arrest his leaseholders for poaching.

However, in his correspondence with Travers, Stevenson-Hamilton would time and again come to Jack Brent's defence, pointing out that he was only doing his job by arresting trespassers and summoning them to court.

Brent regularly patrolled his section on horseback up to its northern boundary, the Olifants River. After one such a patrol in August 1922 he mentioned in a report that he had seen very little game on the TLC farms Jersey, Guernsey and Winchester, but many signs of poaching by blacks with dogs and snares. He did, however, see a few tsessebe and a sable on the farms Nederland, Sumatra and Roodekrans near the Olifants River.

In March 1923 Brent wrote to the warden that Pietersburg's district commandant of police, Captain Coxwell, and Sergeant H. D'Arcy had arrived with several mounted constables at his post to help one Brinley Pritchard look for a hidden treasure in the vicinity of Rolle. After six days' fruitless search over a distance of 140 miles they threw in the towel.

In April 1923 Brent reported that he had shot a lion caught in a trap at the Rij Hills, but that the grass was too thick and tall to hunt lions on foot.

A month later he received instructions from the warden to meet a special train with tourists at Newington Station at the end of July and to accompany them on a trip through the reserve. Although his reports show that he still diligently carried out his duties, after three years the isolation at Rolle became too much for Brent. After a visit to Rolle, De Laporte told Colonel Stevenson-Hamilton on 8 May 1923: "Young Brent told my wife that he would go mad if he remained in Rolle much longer."

His boss reacted on this information and after the retirement of Thomas Duke in 1923 he decided to transfer Jack Brent to Lower Sabie. But he found this post equally lonely and in February 1924, shortly after his arrival, he resigned. It's not clear what happened to him afterwards, but in May 1927 he wrote to Stevenson-Hamilton from an address on the East Rand that he had tried his luck on the diggings, without success, and asked for a testimonial about his years of service in the Sabie Game Reserve.

Harry Brake

After Colonel Piet de Jager's transfer from the Kaapmuiden (Boulders) section to Punda Maria in July 1924, a new ranger called Harry Brake was appointed to man the post.

Very little is known about him, but it would appear that he was not at all qualified for the various tasks expected of a ranger and resigned in January 1925 as a result of a nervous breakdown.

After the excising of the old Kaap E and F blocks from the Sabie Game Reserve in December 1923 there was no reason to have a post at Kaapmuiden.

Ranger S.H. (Harold) Trollope who served as ranger of Malelane section between 1925 and 1929.
(ACKNOWLEDGEMENT: MR GLEN TROLLOPE, THABAZIMBI)

Construction of ranger S.H. Trollope's living quarters at Malelane by contractor Durie, 1925.
(ACKNOWLEDGEMENT: MR GLEN TROLLOPE, THABAZIMBI)

Therefore Stevenson-Hamilton decided to demolish the ranger homestead there and rebuild it at Malelane. This was a much more central position for the new ranger and would in future be the headquarters of the southwestern section of the old Sabie Game Reserve.

Contractors would be responsible for moving the ranger's quarters, but Harry Brake was told to build a temporary rondavel at Malelane in the meantime. He obviously did not feel up to the task and wrote to Stevenson-Hamilton asking that his neighbours, the Elphicks, on the southern side of the Crocodile River, help him. This request was summarily denied and he was told to start with the job immediately and supervise a team of 20 labourers that was busy exterminating locust hoppers in his section.

According to a report from ranger De Laporte to Stevenson-Hamilton, Harry Brake could not carry out any of these orders satisfactorily, and this, together with other problems, led to his resignation. After his departure, Mr Harold Webb of the Thornhill Trading Co. at Hectorspruit inquired about £1.0.0's worth of maize meal Brake had ordered from him, but never paid for.

T.F. (Tom) Elphick, the son of C.W. Elphick and nephew of the well-known Captain George "Mkonto" Elphick, applied for the Malelane ranger post, but Harold Trollope was appointed in 1925.

Stephen Harold Trollope (1881–1949)

After ranger Harry Brake's brief stay at Malelane, Stevenson-Hamilton appointed Stephen Harold Trollope as ranger to serve in this new post.

At the time of Trollope's appointment there was considerable unhappiness among farmers along the Crocodile River over the cattle losses they suffered because of predation by lions from the reserve. The pressure soon started taking on a political flavour and because Stevenson-Hamilton was busy at that stage with sensitive negotiations to get the Sabie and Shingwedzi reserves declared a national park, he could not afford any negative publicity. He therefore needed an experienced hunter to solve the lion problem along the Crocodile River and found the ideal man for the task in Harold Trollope.

Stephen Harold Trollope was born on 7 July 1881, the second son of a former magistrate in the Cape and Transvaal. He married Irene Glen-Leary, whose father, John Glen-Leary, would later die a tragic death in the Park. Harold and Irene had three daughters and a son.

During the Anglo-Boer War Harold and his older brother, Wesley Booth Trollope, a well-known Afrikaner cattle breeder known as "Oom Wessels" in the Bushveld, fought on the Boers' side. Wesley was wounded by the British. After the war, Harold made a name for himself as a hunter and became an excellent shot. He loved hunting on horseback and was utterly fearless. He usually shot lions from the back of a horse and would wait until the lion charged before felling it with a well-aimed shot. Once he allowed this habit to go a little too far and his brother Wesley was charged by a lion from a distance of 15 paces. It came within a whisker of getting Wesley, but Harold managed to kill it just in time.

Shortly after his arrival at Malelane Harold was saved from being savaged by a lion when Corporal Nombolo Mdluli shot the wounded animal, one of a group of four Trollope had shot at, just as it was about to pounce. The dead lion collapsed on Trollope's legs. This incident took place on the banks of the Hlambanyathi Spruit. Trollope never forgot his corporal's bravery and long after he had left the Park he still sent him £2.0.0 every year.

Trollope took his order to cull lions in the Malelane section so seriously that he was often on their spoor by daybreak and even burned sections of grazing to track them down. His habit of getting up very early earned him the nickname "Mavukane" ("the one who rises early").

After a year or so of intense activity, Harold Trollope could report to Stevenson-Hamilton that the lions in his section had been almost completely eradicated, and his colleagues in neighbouring sections ruefully reported that there were no predators left in their sections either.

Trollope's obsession with predator control caused his father-in-law John Glen-Leary's fatal injury by a wounded leopard on 3 August 1926, next to the old Delagoa Bay transport road, along the upper reaches of the Hlambanyathi Spruit. (This episode is described in detail in chapter 18.) A stone cairn and a memorial plaque were erected at the spot the tragedy took place and the leopard's skin and skull were later donated to the Stevenson-Hamilton Information Centre at Skukuza.

Trollope also confirmed that mountain reedbuck were common in the mountainous part of Malelane section.

He resigned in January 1928 and moved to the eastern Cape. Here he became the first warden of the Addo Elephant National Park on 25 August 1931, where he made a valuable contribution to protecting the last remaining group of eastern Cape elephants until his retirement on 30 April 1935. He was also the first ranger to have served in two national parks. Trollope died on 15 May 1949 and in 1952 his family made a donation to the Board with which a comfortable hut was built at Malelane in memory of Stephen Harold Trollope.

Herbert Ernest Tomlinson

Herbert Ernest Tomlinson, a bachelor, was appointed on 1 September 1927 as a ranger in the Kruger National Park – the first appointment in nature conservation ranks after the area was proclaimed a national park in 1926. During his first two months as ranger he relieved ranger De Laporte at Crocodile Bridge while the latter was on extended leave. After Harold Trollope's resignation in January 1928, Bert Tomlinson was transferred to the Malelane section, where he would serve for the next five years.

Herbert Ernest (Bert) Tomlinson was born in New

Ranger S.H. Trollope ready to depart on a lion hunt at Malelane, 1926.
(ACKNOWLEDGEMENT: STEVENSON-HAMILTON COLLECTION, SKUKUZA ARCHIVES)

Ranger S.H. Trollope's camp along the Matjulu, Malelane section, which he used when travelling to Skukuza by ox-wagon. 1924.
(ACKNOWLEDGEMENT: MR GLEN TROLLOPE, THABAZIMBI)

Ranger S.H. Trollope's cattle kraal at Malelane, 1925.
(ACKNOWLEDGEMENT: MR GLEN TROLLOPE, THABAZIMBI)

16. EARLY YEARS: THE SABIE AND SHINGWEDZI RESERVES AND THE KRUGER NATIONAL PARK 1900–1946

Zealand and came to South Africa as a young man just after the Anglo-Boer War. Being a farmer through and through, he joined the Transvaal Consolidated Land Company as farm manager and worked first on the Springbok flats, then in the Waterberg and after 1924 on the TLC's large cattle farm at Toulon, just outside the western border of the Kruger National Park and close to Sabie Bridge. When Mr Crosby, the manager at Toulon, resigned, TLC sent Bert Tomlinson with 250 cattle from Bloubank in the Waterberg to Toulon in the Lowveld. Here he cleared about 100 acres of bush and tried to plant maize and sorghum, but the game ate everything.

Lions also drove him up the wall as they regarded his company's cattle as a welcome variation to their menu. In the four years he tried to farm cattle there he had to shoot more than 100 lions and two of his workers either caught or shot another 100. Bert Tomlinson was a calm and deadly shot and thought nothing of following a wounded lion and delivering the mortal shot while it was charging at him. After his appointment as a ranger in the Kruger National Park a good friend of his from his Bloubank days, Harry Kirkman, became manager of Toulon and would also eventually leave his mark on the Park.

Like many strong and fearless men of his calibre, he was mild-mannered with a calm voice and a quiet, pleasant disposition. He rarely lost his temper or ran into problems with either his colleagues or the public. In fact, "Uncle Bert", as he was commonly known, was a well-loved personality and a man's man. His nickname among locals was "Rivalo" ("the one who inspires respect").

During his period of service at Malelane he built himself an outpost next to the Mahula Spruit to keep an eye on the poaching by the eccentric old foreman Kemp of Kemp's Cottage and his labourers. The ruins of this comfortable outpost are still visible, including a deep concrete bath he had built for himself.

In 1929 he was acquitted in the Barberton court on a charge of assaulting one of border farmer Tom Elphick's black labourers. In March 1931 Bert Tomlinson took a group of Boy Scouts of the St. Andrews High School in Bloemfontein on a tour through his section on ox-wagon – an unforgettable experience for the boys.

In 1932 the warden decided that the central part of the northern district, the area along the Shingwedzi River, did not get the attention it deserved and that a

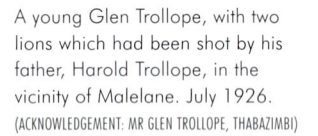

A young Glen Trollope, with two lions which had been shot by his father, Harold Trollope, in the vicinity of Malelane. July 1926.
(ACKNOWLEDGEMENT: MR GLEN TROLLOPE, THABAZIMBI)

ranger post should be established there, with the added objective of developing the area for tourism.

Because he did not want to appoint a new ranger to this responsible position, Stevenson-Hamilton decided to transfer Bert Tomlinson there in March 1933. Bert was very unhappy with this state of affairs as he thought it was a vote of no confidence in his services at Malelane. However, the chairperson of the Board assured him in writing that he had been chosen because of his excellent service at Malelane. This satisfied Tomlinson and he threw his full weight behind the new venture. But before his departure he had one last adventure with lions near the Bumi mouth.

This time he planned to trap poachers crossing the river and pitched camp in a clearing among the trees. He made a fire close to his truck and after his field rangers had collected enough wood, he gave them permission to go and sleep at a nearby outpost. He leaned his rifle against the truck's radiator and poured a cup of coffee. For want of a teaspoon he reached for his knife, which he had left on the bonnet of the truck. As he put his arm out he became aware of a dark shape barely four paces in front of him. His first thought was that it was a bundle of firewood, but with a cold chill creeping up his spine he realised that it was a lion preparing to jump. He grabbed his rifle and tried to push a bullet into the barrel, but the mechanism failed. Helpless, with a rifle in his hand, he yelled at the top of his voice at the lion, and the surprised male did an about-turn. Tomlinson grabbed the chance to jump into his truck and switch on its lights. To his amazement he saw 12 other lions lying in a semicircle around the fire. This time he managed to load his rifle properly and shot three before the other nine disappeared into the dark. After this incident Bert never mistook a lion for a bundle of wood again!

His first task in his new section was to select a suitable terrain along the Tsende (Tsendze) River and construct a tented camp, in time for the tourism season of 1933. He chose an open terrain near fresh water, about one kilometre northwest of the current Grysbok windmill. This site was later named Mabohlelene because of the number of bottles in the rubbish hole. Tomlinson was ready for the tourist season and a Mr Wyckerd acted as camp manager. A few remnants of this old tented camp can still be seen at the site today, including a lane of *Combretum mossambicensis* shrubs that had been planted there. The camp even had a mobile petrol pump.

With the supervisor's permission, Bert Tomlinson moved the Tsende (Tsendze) tented camp to a more suitable terrain in 1934. The new site was along the Shingwedzi River, where the Shingwedzi camp is now. A comfortable home was built for him on a hill to the west of the old main dirt road from Letaba (currently the mechanic's quarters). The builder was Charlie Sanderson, the son of Bill Sanderson of Peebles. During 1934 only a tented camp could be

Ranger Herbert Ernest (Bert) Tomlinson had served as ranger between 1927–1949 at Malelane and Shingwedzi.
(ACKNOWLEDGEMENT: THE LATE D. WOLFF, PRETORIA)

The original ranger quarters at Shingwedzi, circa 1934. It later became the home of the resident mechanic.
(ACKNOWLEDGEMENT: STEVENSON-HAMILTON COLLECTION, SKUKUZA ARCHIVES)

16. EARLY YEARS: THE SABIE AND SHINGWEDZI RESERVES AND THE KRUGER NATIONAL PARK 1900–1946

Ranger Nombolo Mdluli at the carcass of an elephant that had been shot by ranger Bert Tomlinson along the Crocodile River in 1932.
(ACKNOWLEDGEMENT: MR D. TOMLINSON, CAPE TOWN)

The patrol hut that was built by ranger H.E. Tomlinson at Shingomeni in the Shingwedzi section, circa 1934.
(21.06.1984)

The Matiyovila hot-water spring near Mafayini Fountain which was discovered by ranger H.E. Tomlinson.
(28.04.1984)

erected for the tourist season, but the experiment was very successful and the beautiful environment was greatly enjoyed by the visitors.

In 1935 Bert Tomlinson started building more comfortable and permanent accommodation for visitors at Shingwedzi, assisted by the tourist officer, George Meade. He managed to complete three huts, with three rooms each, as well as a dining room and quarters for the camp manager. The building material used was mainly scrap material that he had transported piece by piece from Skukuza and other places. The poles of Lebombo ironwood (msimbit) used for the walls and rafters had been collected at Punda Maria and later also came from the adjacent Mozambique. The camp's water came from a well at the foot of the hill on which Tomlinson's home had been built. To beautify the camp they planted shrubs, especially impala lilies (*Adenium obesum* var. *multiflorum*) – today these are a remarkable feature of the camp.

The camp was completed bit by bit over the next four years as funding and material became available. After the locust campaigns of 1933 to 1935 a large surplus of locust poison (arsenic) had been left over and Tomlinson dipped the ironwood poles in this before planting them. No wonder then that when the old huts (all built in the period 1935 to 1939 by

Bert Tomlinson, George Meade and their black assistants), were renovated 50 years later, many of the ironwood poles were practically undamaged.

In the meantime Tomlinson found the time to expand and improve the road network and to build two earthen dams – one in the Mooiplaas valley (1937) and another in the upper reaches of the Dzombo (1938). Both structures were still standing by the late 1980s and especially the former was an effective reservoir after good summer rains.

Tomlinson built himself a comfortable patrol hut, in the same style as those in the Shingwedzi camp, at Shingomeni to be able to better control poaching along the eastern border. This hut is still standing and was renovated in the eighties (Zway, 1986).

The epizootic foot-and-mouth disease of 1938–39
At the end of 1937 delegates to a veterinary conference in Lourenço Marques (Maputo) reported incidences of foot-and-mouth disease among cattle at Moamba near Komatipoort. Soon afterwards the disease was diagnosed among cattle on the Transvaal side of the border as well. The outbreak of this feared economic disease had far-reaching effects, not only in the Kruger National Park but in the whole Transvaal Lowveld. To begin with all the infected cattle herds between the Mozambican border and Hectorspruit were destroyed. This brought about a lull in the spread of the disease, but it also resulted in the sharply intensified inspections of cattle herds by Veterinary Services in the whole Lowveld region, including the Kruger National Park.

At that time blacks living along the western border of the Park owned large herds of cattle. Inside the Park there were only a few isolated herds of 10–20 animals, while the section rangers each kept a few milk cows. Along the Sand River, about seven kilometres north of Skukuza, there was a herd of 30 oxen and breeding stock that belonged to the Board, as well as three stabled Friesian cows belonging to Stevenson-Hamilton.

For a few months there were no reports of sick animals in the Kruger National Park, but one day the stock inspector stationed at Satara discovered a sick animal with foot-and-mouth lesions among Ledeboer's milk cows. The whole herd was immediately destroyed and the wheels and feet of vehicles and people leaving Satara first had to be sterilised with carbonate of soda.

Things were quiet for a while, but then the storm hit in November 1938. Foot-and-mouth disease was discovered among cattle belonging to people living on the farm Toulon close to the Park's border at Skukuza. Just a day or two later the warden heard to his great disappointment that his prized milk cow, Rhoda, was also ill. In the meantime the disease had spread quickly outside the Park's borders and all cattle and other cloven-hoofed livestock inside the ever-widening cordons were systematically destroyed.

In December 1938 the final order came to destroy all domestic cloven-hoofed animals within the borders of the Park as a preventative measure against the spread of the epizootic. The carcasses were buried en masse in dongas dug by bulldozers. However, the summer of 1938–1939 was one of the wettest in years, with the result that many of the decomposed carcasses were washed open and some even washed into the rivers!

About 3000 cattle, as well as all the sheep, goats and pigs belonging to the black people living inside the Park, were killed, as were the herds of milk cows, oxen and other cloven-hoofed animals belonging to the section rangers. Even poor Bert Tomlinson's herd at the isolated Shingwedzi post were second to last on the list. The Department of Veterinary Services also ordered that all blacks who came into the Park

Ranger Bert Tomlinson and his bride, Martha, on their wedding day at Elim in February 1940.
(ACKNOWLEDGEMENT: D. TOMLINSON, CAPE TOWN)

16. EARLY YEARS: THE SABIE AND SHINGWEDZI RESERVES AND THE KRUGER NATIONAL PARK 1900–1946

The dip at Dipene near Shingwedzi Poort, which had been erected in 1938 during a foot-and-mouth disease outbreak to sterilise the feet of illegal immigrants from Mozambique.
(26.04.1984)

from Mozambique had to wash their feet with disinfectant at Shingwedzi Poort and other entrances. Tomlinson built a trough for this purpose, which can still be seen there today, and the place has been known as "Dipene" ever since.

The last of the livestock in the Park, that of ranger Kirkman, was destroyed at Shangoni in March 1939. To prevent renewed outbreaks of foot-and-mouth disease, Veterinary Services determined that in future no domestic livestock could be kept within the borders of the Kruger National Park.

Although the government compensated the owners of destroyed livestock, the amount was inadequate when all the implications of the campaign were taken into account. Especially to the black inhabitants of the Park, the loss of all their livestock, which had in most cases represented the sum total of their earthly possessions, was an absolute disaster.

Most of their descendants still regard the destruction of their animals with rancour and harbour great bitterness against the authorities of the time. Because of the stipulation that they could keep no livestock other than donkeys, mules and horses within the borders of the Park, many of the settled black inhabitants decided to move westwards with their families, outside of the boundaries of the foot-and-mouth cordon. Those who remained had to adapt to the rule by keeping only a few donkeys and some poultry. The disasters hitting the Lowveld and the Park's people were not at an end yet, however, because in September 1939 the Second World War broke out in Europe.

Captain Elliot Howe
(Acting warden 23 May 1928 – 1 June 1929)
The promulgation of the National Parks Act on 31 May 1926 led to the consolidation of the former Sabie and Shingwedzi reserves and the proclamation of the Kruger National Park on 16 September 1926. Shortly after the promulgation, Colonel Stevenson-Hamilton left for his estate in Scotland to get his affairs in order there, while C.R. de Laporte acted as warden during his absence. Stevenson-Hamilton returned from Scotland in November 1926 and informed the Board that he would not be able to stay on as warden much longer because of his failing health and that they had to find a suitable successor.

He recommended Captain Elliot Howe of the South African Police for the position as he was convinced that he possessed the necessary administrative and management capabilities to fill the post successfully. Captain Howe was the brother of the well-known author and journalist Ethelreda Lewis, an acquaintance of Stevenson-Hamilton's and a champion in the press for the Kruger National Park and nature conservation in general. H.V.B. Potter's name was mentioned as an alternative candidate should the SAP refuse Howe's transfer.

Howe had joined the South African Constabulary (SAC) in May 1901 during the Anglo-Boer War. He served at the divisional headquarters of the SAC

Capt. Elliot Howe, acting warden of the Kruger National Park between 23 May 1928 and 1929.
(ACKNOWLEDGEMENT: STEVENSON-HAMILTON COLLECTION, SKUKUZA ARCHIVES)

in Bloemfontein from January 1902 until April 1904 and was promoted to staff sergeant.

From April 1904 until April 1913 he served in various districts of the Orange Free State and during the last four years of this period he was superintendent of police in the Orange Free State. At the establishment of the South African Police Force in April 1913 he was transferred to Cape Town and appointed to divisional personnel.

In the years 1914 to 1916 he did relief work in different districts of the western Cape and in 1915 he was promoted to subinspector. Between 1917 and 1918 he was provost marshal of Cape Town harbour. In November 1920 he was promoted to inspector and until August 1923 he served in the secretarial office of the SAP headquarters in Pretoria. In 1923 he was appointed as district commandant of the SAP at Rustenburg, where he earned a salary of £642 per annum.

After extended correspondence between the chairperson of the Board, Senator Jack Brebner, the Board's secretary, F.S. Potgieter, and both Stevenson-Hamilton and the Commissioner of Police, it was at last decided that Captain Elliot Howe would be seconded to Skukuza as acting warden for a period of one year, after which the situation would be reviewed. His temporary appointment in this capacity was confirmed by the secretary of the Civil Service Commission on 11 April 1928.

Howe officially assumed command at Skukuza on 23 May 1928, but he had already arrived in the Kruger by the end of March to be properly informed by Stevenson-Hamilton about his numerous duties. Stevenson-Hamilton went overseas in the second week of May and Howe held the position until 23 May 1929. The Commissioner of Police was not prepared to lengthen his period of service as acting warden beyond that and if the Board wanted to appoint him as warden, he would have had to resign from the Police. He was not prepared to do this, however, and, running out of options, the Board wrote to Stevenson-Hamilton and asked him to consider returning to his former position.

In the meantime, an application had been received from Lieutenant Colonel H.F. Trew, the Deputy Commissioner of Police in the western Cape, to be considered for the position. Stevenson-Hamilton cabled that he would return for a period of six months in June 1929 to enable the Board to appoint a new warden in his place. The representative of the Wild Life Protection Society serving on the Board, Paul Selby, was also interested in the post, but after his return Stevenson-Hamilton's health improved to such an extent that he resumed full responsibility after the six months' transition period had expired. He eventually managed the Kruger National Park until his retirement on 30 April 1946. Paul Selby withdrew his application and resigned from the Board on 6 August 1930.

Elliot Howe had been an able and dedicated acting warden. He managed the Kruger National Park well and attended several Board meetings during the time. Howe habitually travelled with his dogs in his car and always wore a pith helmet and a tobacco pouch on his belt. As a result of his police training he was a hard taskmaster. It was his unpleasant duty to give ranger Coetser six months' notice as a result of unsatisfactory service. The locals named him "Mkonto" ("spear") – certainly not for nothing. During his time as acting warden Howe appointed two new rangers, J.J. Botha and H. McDonald, while Piet de Jager retired in November 1929. For some or other reason Howe and Harry Wolhuter never got on very well, probably because Wolhuter questioned his knowledge of the veld.

During Howe's period of service, 424 kilometres of new tourist roads were constructed, the building of tourist rondavels at Satara and Gorge started and the total annual income from tourism was £179.10.6. In contrast with the dozen or so vehicles that had visited the Park in 1927, the number in 1928 was a promising 183.

Howe reported that a black rhino had been encountered by a ranger patrol that year in the Nwatimhiri bush at Mile 44 on the Selati Line. He also mentioned a fair-sized herd of roan antelope about 16 kilometres outside the borders of the Park on the Acornhoek-Rabelais road. Roan had apparently been common on the Mlondozi flats of Tshokwane section at the time.

He pointed out that the type of roads that the rangers could build would not be able to carry the envisaged heavier tourist traffic in future, and expressed the conviction that the roads would have to be built up and provided with permanent surfaces. Prophetic words indeed!

After his return to the SAP, Howe was promoted to divisional inspector in the Free State in November 1929. He retired with pension on 31 October 1930 and offered his services to the Board in any suitable position. But because of his age and other factors it was not to be his destiny to work in the Park again.

Gerhardus Christoffel Snyman Crous
Long before 1928 Stevenson-Hamilton had built a comfortable outpost on the former farm Strasburg at Tshokwane, which he and some of the other rangers often used during their patrols to the northern sections. After the resignation of Harold Trollope in January 1928 and a number of internal transfers, the warden saw the opportunity to settle a permanent ranger post at Tshokwane with the appointment of ranger G.C.S. Crous.

Gerhardus Christoffel Snyman Crous started his career in the Park on 1 January 1928 and was immediately stationed at the outpost at Tshokwane, while ranger H.F. Tomlinson filled the Malelane post.

Before his appointment as ranger, Crouse had served as station master at Komatipoort since 1922. It was here that Stevenson-Hamilton got to know him and earmarked him as a future ranger. Before his employment by the South African Railways, this bachelor had a particularly colourful and adventurous career. He went to America in 1912, where he was taught Morse code at the Marconi School. Later he joined the American merchant fleet and visited the Barbary Coast, Hong Kong and Shanghai, among others. With the outbreak of the First World War he tried to join the US marines, but fell ill and returned to South Africa, where he became station foreman until his eventual promotion to station master. During 1918–1919 he went hunting behind the Soutpansberg and in 1924 he hunted and toured in Swaziland.

He remained a bachelor throughout his life, but it is said that during his service period at Letaba in the 1930s, he had his eye on a divorced woman, Mrs Johanna Susanna Hilda "Cheeky" Stamp, who ran a small mica mine next to the Olifants River a few kilometres east of its confluence with the Tsutsi. He often undertook patrols in this direction and made sure that the patrol road along the Olifants River, which had been opened up by his predecessor, Ledeboer, stayed in good order. There were even claims that he had invested in Cheeky Stamp's mine. When the romance soured, however, Mrs Stamp (the mother-in-law of a later shop manager at Skukuza, Charles Malan) left the Park and ranger Crous washed his hands of the female sex.

As is often the case with bachelors, ranger Crous was reserved and not particularly talkative, while extremely precise and neat in everything he did. Even his tea and mealtimes were precisely scheduled and if guests arrived at any other time, they had to do without. His firearms were always spotless and well oiled and he allowed nobody to touch them "with their sweaty hands". His desk was painfully neat with nothing out of place and his monthly returns were always sent promptly and faultlessly to Skukuza. The patrols taken by his black rangers were painstakingly planned. His only company was a tame baboon that went on patrols with him and warned him of any potential danger.

His favourite dishes were fish and the cane rat (which he called "rietvarkies" [cane piglets]) stews he cooked. The one luxury he allowed himself was a two-seater Packard that he bought in 1936. While on leave he criss-crossed the country in it. Crous was a dedicated ranger in all respects and took his work very seriously. It is no wonder he declared war on the armed bands of poachers from Mozambique that decimated the game along the eastern border of the Satara and Tshokwane sections during the 1920s.

The home of rangers Ledeboer and Crous at Letaba. (18.06.1984)

During a skirmish in April 1929 he shot and killed one of them. To his shock and surprise the police charged him with murder, but in the resulting trial on 20 August in the circuit court at Pilgrim's Rest he was exonerated of all blame and acquitted.

In February 1929 building work on the ranger's house at Tshokwane started, but the contractor absconded and work could only resume in October when a Swiss contractor, Bergen, took on the project, which he completed in April 1930.

While stationed at Tshokwane, Crous shot a white lion, but he was apparently very reticent about showing the skin to outsiders. This was the first white lion ever recorded in the Kruger National Park.

In June 1929 he was transferred to Satara where he did duty for four months after new ranger I.J. Botha's relief period, which lasted from January to June 1929, had come to an end.

On 1 October 1929 he took up duty at Letaba and ranger Ledeboer was transferred from there to Satara. Crous served here uninterruptedly from 1929 until his retirement on 31 December 1950. During this time he went on long leave a number of times. In 1935–1936 he was relieved by ranger Harry Kirkman for several months, in 1939 by a clerk, D.H. Swarts, and in 1949 by ranger W. Lamont. During his service period Crous distinguished himself in particular by the good tourist roads he built and maintained north and south of his post at Letaba. He was naturally also involved with the construction of tourist facilities at Letaba rest camp. A new house was built for him on the site of the current

Ranger G.C.S. Crous (with rifle) and friends at an elephant carcass during the early 1930s. He served at Tshokwane and Letaba sections from 1928 to 1950.
(ACKNOWLEDGEMENT: MR A. COURSE, PRETORIA)

Matilolo Dam, which was built by ranger Crous near the old Malopeni-Letaba Drift junction. (18.06.1984)

ranger's home in 1932 and his old quarters were turned into stables and storerooms. Crous also built the Matilolo Dam on the sandstone ridge near the turnoff of the Shingwedzi road from the old Malopeni road. His nickname among the locals was "Mashangana" as he could speak the language fluently.

In 1936 Crous estimated the number of elephants in the Letaba Section at 150, and eland, roan, sable and tsessebe at 50 each. According to him, there were only 80 hippos in both the Letaba and Olifants rivers. Ranger Crous managed his section remarkably well until his retirement on 31 December 1950, after which he settled in the hotel at White River. From here he often visited the Park in his new DKW car and gave advice to the younger rangers. He died in 1975 and was buried in White River.

Hector McDonald

With the retirement of Piet de Jager and J.J. Coetser in November 1928 and December 1928 respectively, there were two vacancies for rangers.

Acting warden Howe recommended that, out of 555 applicants, Hector McDonald and Izak Johannes Botha be appointed. Hector McDonald officially entered service on 1 January 1929 at the Crocodile Bridge section, but he had already taken over from ranger C.R. de Laporte on 28 December, when the latter left on extended leave before his retirement on 31 March 1929.

Hector McDonald was born on 8 April 1886, but very little is known about his history before his appointment as Park ranger. He was, however, a well-known inhabitant of the Lowveld and had hunted everywhere in the area since 1910, especially on the Lebombo flats and in the northern part of Swaziland.

Shortly after his appointment in 1929, in March that year, when tourism in the Kruger National Park was still in its infancy, a large group of American tourists arrived at Crocodile Bridge in a luxurious train. Using the train as their base, the group entered the Park on two large trucks for a tour of the area. They were caught in a terrible thunderstorm and the two trucks got hopelessly stuck. The visitors, wearing thin summer clothes, were thoroughly drenched and by nightfall they all scrambled into trees when lions started roaring nearby. Fortunately Hector McDonald and a few of his field rangers arrived with food and blankets for the soaked Americans. They had to spend the night inside the two cramped outpost huts at Gomondwane. The next day the visitors viewed the whole event as a great adventure, but several of them contracted malaria. The resulting negative publicity led to the closure of the Park to visitors in summer months for many years.

Poaching across the eastern border was still a major problem during the early years of Hector McDonald's service period. He and his field rangers were once involved in a mini battle with armed black poachers under the leadership of a few whites from Mozambique. The poachers fled and none of them could be arrested.

The thing that most probably scared them away was the heavy thundering of McDonald's double-barreled .470 Rigby elephant gun, for which he received 50 free rounds of ammunition every year instead of the 200 rounds of .303 bullets that were annually provided to the other section rangers.

Hector McDonald was an extrovert who loved to extol his abilities as a man of the veld and as a marksman. It is said that he once won a bet in the bar at Komatipoort by firing six shots through the same hole in the corrugated-iron roof – to the great admiration of the onlookers. Later he gave away the show – the first shot was a live bullet while the rest were blank cartridges. His nickname among his black colleagues was "Ndlovane" ("little elephant"), possibly because of his figure.

As the stream of Afrikaans-speaking visitors gradually increased through the years, Hector McDonald and his colleague, Tom James, were instructed by the Board to improve their Afrikaans as they were the only two officials who could not speak the second official language at all.

On 15 October Hector McDonald lost his faithful and extremely efficient corporal, Mpampuni Ubizi, in a confrontation with a lion.

Mpampuni was one of the oldest serving black rangers and someone with a world of field knowledge and experience, which he had gained during his years in the Park. This fateful afternoon he was standing outside his hut watching a group of vultures circling nearby. Armed with only his knobkerrie, he walked to the lion kill and came upon a lioness and her cubs with their catch. The lioness growled threateningly and Mpampuni wisely decided to back off, but as he turned around, the lioness charged him. A woman working in a nearby field shouted a warning. Instead of turning around and confronting the lioness directly, which would probably have made her turn around, Mpampuni took to his heels and headed home.

Despite his lead, she quickly caught up with him and grabbed him by the thigh. She severely wounded his groin and with his penknife Mpampuni slashed across her nose, at which she let him go and went back to her cubs. Mpampuni stumbled back to his hut where his wife tried to dress the wounds. Ranger McDonald arrived soon afterwards and took him to Komatipoort in his car, after which he was taken to Barberton Hospital by train. His groin wounds were too serious, however, and he died a few days later in hospital as a result of shock and blood poisoning.

In a report to the Board in 1936, McDonald mentioned the fact that the biggest number of black rhino he had ever found on his section was three. He also mentioned that the number of ostriches had decreased by 90 per cent and roan and sable by 50 per cent. Although there are still a few ostriches on the Lebombo flats and a small herd of sable in the vicinity of Randspruit, roan has since completely disappeared from the Crocodile Bridge section. It is not clear if successive droughts and predator pressure were responsible for this.

A drilling programme was started in 1933 in the southern district of the Park to augment the scanty natural water resources during this great drought cycle. Five holes were drilled successfully: at Komapiti, Muhlambamadvube, Randspruit, Randspruit North and Gomondwane. At the last borehole Hector McDonald built a square concrete drinking trough, just east of the large wild fig tree by Sardelli's eucalyptus trees. The structure is still standing and had been mistakenly identified as the foundations of Sardelli's shop.

On 30 May 1937, Hector McDonald, warden Stevenson-Hamilton and his colleague Harry Wolhuter participated in a radio programme about the

The old water trough at Gomondwane which was erected in about 1934 by ranger H. McDonald, just east of where the Greek Sardelli's shop had been.
(15.06.1984)

Ranger Hector and Mrs McDonald at Crocodile Bridge during the early 1930s. He was ranger of Crocodile Bridge section between 1929 and 1951.
(ACKNOWLEDGEMENT: STEVENSON-HAMILTON COLLECTION, SKUKUZA ARCHIVES)

16. EARLY YEARS: THE SABIE AND SHINGWEDZI RESERVES AND THE KRUGER NATIONAL PARK 1900–1946

Kruger National Park broadcast worldwide from Skukuza to millions of listeners. The program was organised by James McClury of the Cape Town studios of the South African Broadcasting Corporation.

Before his retirement, McDonald had the pleasant task of serving as a guide to Miss South Africa and two foreign beauties, Miss Great Britain and Miss America, on 15 March 1948. McDonald often boasted to his colleagues that he had specifically been chosen for this important task.

He finally retired on 31 October 1951 after an eventful career of almost 23 years in the Kruger National Park.

Izak Johannes Botha

Izak Botha was the second ranger whom acting warden E. Howe appointed in January 1929 to fill the vacancies that opened up after rangers De Jager and Coetser had retired. He was one of 555 applicants for the two positions.

Botha was born and grew up on a farm in the Cape and was interested in nature from a very young age. As a boy he had the opportunity to hunt jackal, baboon and springbok on his father's farm and on those of friends. Before his appointment as ranger, he had served as station master at Malelane.

After his appointment on 5 January 1929, he was immediately stationed at Satara as relief ranger after the departure of ranger J.J Coetser on 15 December 1928.

Then he was transferred to the headquarters at Skukuza after ranger G.C.S. Crous had temporarily taken over the Satara section in June 1929.

Botha was stationed at Skukuza until April 1930. In the middle of that month he was transferred to Punda Maria to take over from ranger T.L. James, who had served there from February 1929 to April 1930.

While stationed at Skukuza, Botha suggested in December 1929 that the Board take tourists, especially schoolchildren, on ox-wagon rides through the Park. The warden welcomed the idea, but felt that the time was not ripe yet.

After his move to Punda Maria, he initially reported that game was very scarce in the northern part of the Park, but even then he was fortunate enough to see three nyala at Pafuri on 12 June 1930. With this he confirmed the sightings of this species that had been reported by rangers De Jager and James.

In 1934 Botha calculated the amount of game in his section and came to the following estimation: 1000 waterbuck, 400 kudu, 150 eland, 60 nyala, 10 elephants, 30 buffalo, 650 zebra, 200 sable antelope, 200 roan antelope, 150 tsessebe, 200 blue wildebeest, 250 impala, 25 hippo, 75 ostriches and 100 reedbuck and bushbuck. Since then, the relative composition of the game, as well as the numbers of the various species in the area, has changed drastically as a result of a wide range of factors.

In the winter of 1931 tourists stayed in a tented camp in Punda Maria for the first time. After the end of the season, ranger Botha and a team of labourers started building the traditional pole and mud thatched rondavels that are still there after so many years, though in a slightly renovated form. This series of huts was completed in July 1931 and an ablution block was added in 1933. By then Botha had also constructed the network of roads north of Punda Maria rest camp to Dongadziva, Shidzavane, Mago-

The baobab tree which ranger I.J. Botha planted on 4 February 1931 at the Punda Maria quarters. (27.04.1984)

Izak Johannes Botha, who consecutively served as ranger at Satara, Skukuza and Punda Maria (1929–1938).
(ACKNOWLEDGEMENT: STEVENSON-HAMILTON COLLECTION, SKUKUZA ARCHIVES)

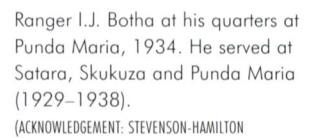

Ranger I.J. Botha at his quarters at Punda Maria, 1934. He served at Satara, Skukuza and Punda Maria (1929–1938).
(ACKNOWLEDGEMENT: STEVENSON-HAMILTON COLLECTION, SKUKUZA ARCHIVES)

vani and Klopperfontein. Because these roads were quite sandy, every now and then visitors travelling on them had to get out to push their vehicles. While working on the road to Magovani, Botha camped along the Luvuvhu under a large baobab tree and carved his name into the trunk of this old bush giant. Many years later, in 1954, ranger Gus Adendorff came across Botha's handiwork.

Botha planted a baobab seedling in the garden of the Punda Maria ranger quarters on 4 February 1931. Fifty-five years later it had grown into an impressive young baobab with a trunk circumference of 5.67 metres and a crown height of 18.9 metres.

During his patrols in the vicinity of Klopperfontein Botha discovered the interesting potholes in the small sandstone ridges at Shantangalani. For some reason he did not get along with his black colleagues and as a result they gave him the somewhat unflattering nickname "Mawawa" ("the dominating one").

In 1935 he obtained permission from the Board to build an earthen dam just south of the rest camp. This provided water to animals in the area for many years. Botha got married on 26 June 1936.

In 1937 he constructed a number of byroads branching off the main road between Punda Maria and Shingwedzi and so opened the interesting routes along the Shisha and Mphongolo rivers to the public. Izak Botha resigned on 1 May 1938 and left the service of the Board after eight productive years as ranger at Punda Maria. Some of his handiwork can still be seen here today.

The unique baobab forest (northern stand) on the sandveld plateau east of Madzaringwe Ridge, which had probably been discovered by ranger I.J. Botha, during the early 1930s.
(30.07.1984)

The newly completed ranger's quarters at Punda Maria in 1926.
(ACKNOWLEDGEMENT: STEVENSON-HAMILTON COLLECTION, SKUKUZA ARCHIVES)

Thomas Llewellyn James (1886–1971)
Thomas Llewellyn James was appointed as a relief ranger on 1 February 1929, but was sent to the isolated Punda Maria section from the get-go to take over from Piet de Jager, who had just left the service. He was to stay there until April 1930, after which the warden recalled him to headquarters in Skukuza to assist with administrative and relief duties.

Tom James was born on 22 January 1886 at Te'Ore, near Auckland in New Zealand, and came to South Africa after the Anglo-Boer War. From 1910 to 1920 he farmed in the Waterberg district and also worked in the Zaaiplaats tin mines. He met his future colleague Bert Tomlinson while living in the Waterberg. There he also frequently hunted on farms and acquired an intimate knowledge of the wildlife. He was married in 1913 at the age of 27.

During his period of service at Punda Maria, Tom James became only the second ranger after Piet de Jager to see the shy and rare nyala in the flesh. His sighting of six animals (four bulls and two ewes) took place during a patrol in October 1929 along the banks of the Luvuvhu River. During this time he was also instructed by the Board to improve his Afrikaans, a language which he never managed to master.

While in Scotland, Stevenson-Hamilton wrote a letter to the Board with the following request:

Were I to die while Warden of the Game Reserve

or within ten years of cessation of employment therein, it would be my wish to be buried on the top of the highest of the rocky kopjes standing west of the Selati line near mile 43 [old kilo 67]. A small granite pyramidal pillar with inscription on a concrete stand, facing the Sabie river. A small sum would be set aside from the Estate for the erection and upkeep.

The purpose of the warden's somewhat unusual request was that his earthly remains would help to control poaching in that part of the Park even after his death, since superstitious locals would believe that "Skukuza" was watching them from the top of Shirimantanga Hill if they were to transgress within sight of it!

The Board approved Stevenson-Hamilton's request and the task of dynamiting a grave out of the solid granite fell to Tom James. He had a blasting certificate and completed the job successfully. When Stevenson-Hamilton was questioned about the hole on top of the hill in later years, he referred to it as a "drinking place he had made for the klipspringers".

After Stevenson-Hamilton's death his daughter, Mrs Ann Doyle of England, scattered his ashes and that of his wife Hilda on top of Shirimantanga Hill on 10 April 1979.

Ranger Thomas Llewellyn (Pop) James at Malelane, circa 1940. He had served from 1929 until 1950 in the Punda Maria, Skukuza and Malelane sections.
(ACKNOWLEDGEMENT: MRS D. VAN DER VEEN, WHITE RIVER)

Tom James became known among his black colleagues as "Mabalane" ("the secretary" or "writer") because he fulfilled administrative as well as secretarial duties. This name followed him even after his transfer to Malelane section in April 1933, where he took over from ranger Bert Tomlinson.

After settling at his new post, ranger James soon encountered the curse of this particular section – unfriendly neighbours on the other side of the Crocodile River and the organised poaching by members of the Mambayi tribe from the adjacent Nsikazi trust area. The gangs of armed poachers entered the Park with donkeys and then stayed in caves and rock shelters in the rough, mountainous area northwest of the Malelane ranger post. Catching these wily poachers, who had extensive knowledge of the veld, before they could do permanent damage to the game in the area was almost a full-time job. The poachers often fired at ranger James and his field rangers from their mountain shelters and he fought a never-ending battle against these enemies of the Park.

James was an ingenious man. One day he came upon a poachers' camp after the culprits had taken to their heels, with the meat of the animals they had shot still hanging in the trees. Because he was convinced that they would come back during the night to fetch the meat, he and his field rangers set a trap. He was right, but during the ensuing rumpus the poachers managed to escape again, leaving behind the four donkeys they had planned to use to transport the meat. Ranger James noticed that one of the donkeys was lactating and decided to give her free rein, certain that she would return to her foal. That is exactly what she did, leading the rangers straight to the poachers' village, where the necessary arrests could be made.

In 1937 George Meade, who had been camp manager at Malelane for several years, developed gangrene in his leg while at Shingwedzi. Meade had originally owned a bar in Potgietersrust (Makopane) and later at Naboomspruit (now Mookgophong). Ranger Bert Tomlinson brought him to Malelane in 1931.

As there were no medical facilities at Shingwedzi, Tomlinson asked his colleague, who also happened to be a friend of Meade's, if he would look after the old man at Malelane. Tom James readily agreed, but despite his best efforts Meade's condition deteriorated and he died. He was buried close to the section ranger's house. In the 1980s a memorial plaque was erected at the grave of one of the first and most popular tourism officials in the Kruger National Park.

Uncle "Pop" James, as he would be known in later years, sent his children to school in Johannesburg. Arriving at Malelane for their holidays, the children had to be rowed across the river in a small boat to reach their home. The Park's bachelors soon started taking notice of his eldest daughter, Daphne, and a romance ensued between her and the warden's clerk, H.C. van der Veen. They were married on 1 July 1939. Mr "Van" or "Makuloskop", as he was called by the blacks, would later become the efficient and respected manager of tourism of the Kruger National Park.

Rumour has it that Sergeant Helfas Nkuna of Pretoriuskop section once found a rusted tin filled with gold pounds under a hollow cliff south of the Mlambane Spruit, in James's section. The treasure was apparently divided among several people, but James only heard about it much later via the bush telegraph. From that day onwards relations between him and his colleague Harry Wolhuter became somewhat strained and he threatened to arrest any field ranger from another section who trespassed in his area!

In 1945 "Pop" James obtained permission from the Board to build a concrete dam in the Mlambane Spruit near the infamous "Treasure Hill". Work was completed the same year. This dam, which had provided water to many a thirsty animal, later silted up and now the only way game can get water here is to dig holes in the sand.

On 21 August 1951, shortly before Pop James's retirement, his home was struck by lighting at 3 a.m. and burned to the ground. The main building (built from wood and zinc) was completely destroyed, but the adjacent brick kitchen and office could be saved. The James family's loss amounted to £1 144.7.0 and this was paid out to him by the Board.

T.L. (Pop) James finally retired at the end of February 1952. He first settled on his small farm near Malelane, but then moved to Ladysmith and after that to Port Elizabeth, where he spent his last years with his daughter. He often visited his other daughter and son-in-law, the Van der Veens of Skukuza, until his death on 16 December 1971.

Remains of the ranger's house at Malelane after it had been destroyed by a fire on 22 August 1951.

This ravine in which tree aloes occur was only discovered in 1976 during an aerial census on the southern slope of Khandizwe Hill, west of Malelane.
(08.06.1984)

L.B. Steyn, ranger of Tshokwane section (1929–1954) and warden of the Kruger National Park between 1 May 1954 and 30 April 1961, circa 1955.

Louis Botha Steyn (1901–1962)

The Board was informed on 12 April 1929 that ranger S.H. Trollope was not interested in being re-appointed as ranger after the resignation of ranger C.R. de Laporte. Shortly before Stevenson-Hamilton's year-long stay in Scotland came to an end, it was decided to appoint L.B. Steyn as ranger. He started his career on 3 May 1929 and after a few months' training at Skukuza he was posted to the old ranger post at Tshokwane (Strasburg).

Louis Botha de Wet de la Rey Steyn was born on 23 April 1901 in Pretoria. He attended Eendracht Primary School and completed his high school education at Boys' High School. After matric he taught for a year at Marico and then returned to Pretoria, where he worked in the office of the auditor general for seven years.

A friend who later became his brother-in-law, Fritz Wichman, invited him on a hunting trip to Kenya in 1927 where they hunted big game for three months.

After his return from East Africa he desperately wanted to work in the outdoors and he was lucky enough to be appointed as ranger in the Kruger National Park shortly afterwards. In the long run he would serve as ranger for 25 years and another seven years as warden, making a total of 32 years' service in the Kruger National Park.

The new ranger moved into the old rondavels of the old outpost at Tshokwane for the first few months because the Swedish building contractor, Bergen, only finished his new quarters in April 1930. It is said that the city boy at first became a bit nervous when lions roared outside the huts at night. He came up with the plan to cement a kudu horn into the hut's wall to accommodate his nightly needs!

Lou Steyn, as he became widely known later on, was an obstinate, abrupt, almost aggressive character. This soon earned him the nickname "Mafunyane" ("the one who wants to rule over everything"). He often bumped heads with colleagues, his bosses and even members of the public. In addition, he had very firm convictions and in later years vehemently fought scientifically-based proposals to amend existing management practices such as predator control and veld fires.

Despite his very human shortcomings, Lou Steyn fulfilled certain official duties with great diligence. An official population census took place in September 1960 and Lou Steyn was appointed as chief census official for the Kruger National Park.

At the time of the census, his son, Louis junior, was at home on holiday and was forthwith enlisted by his father. On the morning of the designated day he had to report to his father's office to be sworn in as a census registration officer. After a knock on his father's office door he was told to enter. While sternly glaring at him over his pince-nez, the communication more or less went like this: "Ahem! Are you Louis Botha de Wet de la Rey Steyn?" "Ja, Pa!" came the quick reply. "Listen here, I'm not Pa, I'm the Commissioner of Oaths. Say, 'So help me God!'"

In his younger years Lou Steyn was very interested in the history of the Voortrekkers, especially the mystery that surrounded the Van Rensburg murder site, as well as the route Louis Trichardt had taken through the Lowveld and the Kruger National Park. During the winter of 1939 he undertook an extended journey along the Limpopo River in Mozambique to the terrain where, according to him, the Van Rensburg murders had taken place. His findings were published in a series of articles in *The Star* of 18–20 September 1939. These findings were, however, proven to be incorrect by the well-known historian Dr W. Punt during a later expedition in 1959, as were his indications of Louis Trichardt's route through the Lowveld. Lou Steyn could not accept this and initiated a drawn-out and futile correspondence in newspapers and with historians.

He had more success with the location of the eastern border beacons. In 1937–1938 he was of great help to the State Surveyor Commission under Mr Le Roux, who had been appointed to find and survey these beacons. They so valued his contribution that they presented him with a theodolite, an instrument of which he was extremely proud. It proved very useful in later years when he surveyed and built several earthen dams in his section, namely Kumana, Mazithi and Salitje. He also enlarged the well-known Leeu Pan and dug contour furrows so it would fill up more easily during the rainy season.

The roads in the vicinity of Tshokwane were also his handiwork, and a remarkable road he helped to construct was the one to the lookout point on top of Nwamuriwa Hill. During the construction they discovered some fossil bones, which he sent to Dr Robert Broom, the curator of the Transvaal Museum. These bones were identified as belonging to a fossil hyena that was named after Steyn – an honour of which he was quite rightly proud.

Lou Steyn was married in 1933 to Chrissie and they had a son Louis and a daughter Marianne. The young Louis later became a well-known batik artist and Marianne married George F. Barkhuizen, who had been a research assistant and traffic officer in the Kruger National Park. Barkhuizen later became park warden of Golden Gate National Park and deputy director of nature conservation in the Free State. He would eventually become director of nature conservation in this province, a position he held until his death.

On 20 November 1937 Lou Steyn accompanied Harry Wolhuter, the warden and Vinnie Robertson to determine the site where Harry had nearly been killed by a lion in 1903, before killing it with his hunting knife. They managed to find the place and Robertson, a farmer from Amersfoort, donated an amount of money that was used for the erection of three memorial plaques there.

Bronze memorial plaques on rock faces between Tshokwane and Skukuza were unveiled during Steyn's time in the Park, one in honour of President Paul Kruger and Minister P.W. Grobler on 22 September 1933 and one in honour of Eileen Orpen on 4 October 1944.

Lou Steyn, like most of his colleagues, had his share of run-ins with lions. Late one night in July 1940 he wounded a lion that had caught several of his squatters' donkeys with his 7.9 mm Mauser. When they had to follow the wounded animal the next morning he took along his heavier 9.3 mm Mauser. The wounded lioness had found shelter in a dense stand of reeds. She charged when her pursuers were about eight paces from her. There was only time for a single shot and the heavy 9.3 mm bullet struck the lioness in the mouth, but her momentum carried her on top of Lou Steyn and her front paws tore deep gashes across his shoulders. When he fell backwards, he pushed his right upper arm into her crushed mouth and grabbed her tongue. His corporal shot and killed the badly wounded animal on top of him. His wife and a Dr Jacobs from Germiston treated his wounds, which healed quite well.

In 1937 he nearly died when the plane, in which he flew to obtain a B-licence, crashed and he was in hospital for several months. Apart from the other fractures and cuts, he was always reminded of the accident when looking at his broken nose in a mirror, or the amputated index finger of his left hand.

In 1938 he again had to go Salitje squatter camp – a man-eating lion had attacked and mutilated a woman. It had also carried away and eaten a young cattle herder. This time he luckily had no trouble killing the guilty animal.

When Stevenson-Hamilton retired in April 1946, Lou Steyn unsuccessfully applied for the position of warden, even though he was the senior ranger. An Air Force officer, Colonel J.A.B. Sandenbergh, was appointed in the position instead. Lou Steyn considered this a slap in the face and he and his new head soon clashed. Despite the best efforts of the Board's secretary, Mr Van Graan, as well as the chairperson himself, the two could never see eye to eye and often argued about even the most trifling matters. After the distasteful political echoes left by the Hoek Investigation into the administration and management of the Board's national parks in September

1952, the dissention between the warden and senior ranger widened even further. Once, when Sandenberg asked Lou Steyn to help Harry Kirkman shoot a couple of elephants that had been destroying farmers' crops, Steyn immediately asked: "Is it an order or a request?" "It's an order!" the warden answered. Steyn replied: "Well, if it is an order, the Board will have to provide me with a 10 ton tank for protection – elephants are dangerous animals!" – and that was the end of it.

Sandenbergh found the situation intolerable and for this, as well as other reasons, resigned on 1 February 1954. As expected, Lou was appointed as acting warden from 1 February 1954 to 30 April 1954, after which he officially became warden of the Kruger National Park on 1 May 1954. He served in this position until his retirement on 30 April 1961. Unfortunately, Lou Steyn became increasingly short-tempered as he aged, especially during his tenure as warden at Skukuza. He was not popular among his equals and the terror of the younger rangers. Even outsiders sometimes came under fire.

One day he had an argument with a senior police officer in his office. After the policeman left the office without having achieved anything, Hannes Kloppers, who worked in the office next door, offered the colonel a cup of coffee. It was accepted with gratitude, and during the ensuing conversation the colonel asked Hannes: "Tell me, Mr Kloppers, I hear you people do game counts in the Park these days, what is the latest estimate for the number of buffalo?" Innocently Kloppers answered: "Colonel, to the best of my knowledge I reckon there are about 10 000 in the Park." "No, Mr Kloppers," the colonel answered, "you have made a mistake – the correct number should be 10 001!"

In the meantime Lou Steyn kept himself occupied by organising general game censuses in the Park and surveying earthen dams at several places with his theodolite and then having them built (Nhlanganzwani, Panamana, Jones and Mafunyane dams). The last-named dam was not a huge success and once when senior ranger Colonel Rowland-Jones asked the Shangoni ranger, Jan Meyer, "How is the work progressing with the old man's dam down there, Jan?" Jan answered in his unique Shangoni English: "I don't want to make *kommentaar*, Colonel, but they are making a *bleddie gemors* down there!" ("I don't want to comment, Colonel, but they're making a bloody mess down there!")

On 30 April 1961 Lou had to leave the Park and all it had held for him, both good and bad. He settled in Pretoria, but the city life was too much for the staunch old nature lover and he died on 7 October 1962.

Walter Henry (Harry) Kirkman (1899–1989)
Walter Henry Kirkman was born on 31 March 1899 on the farm Grootvlei in the Steytlerville district, at the foot of the Winterhoek Mountains. There this son of the Karoo enjoyed an idyllic life. Sometimes he heard travellers talking about the great plains north of the Vaal River and of the fertile Lowveld with its game, fever and extreme heat – all of which captured his young imagination. His parents were of Scottish extraction and farmed angora goats and merino sheep.

At the outbreak of the First World War on 28 July 1914 Harry was only 15 years old. He wanted to join the Allied forces, like his brothers John and Joseph, but his father dug in his heels – he was too young.

However, Harry kept harping on the subject and

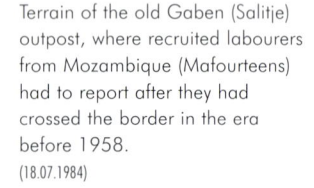

Terrain of the old Gaben (Salitje) outpost, where recruited labourers from Mozambique (Mafourteens) had to report after they had crossed the border in the era before 1958.
(18.07.1984)

his father, at his wits' end, sent him to the magistrate, a Mr Lawrence, where he gave his age as 18. This enabled him to join the South African Brigade of General J.L. van Deventer, which would face the Germans under the command of General Paul von Lettow-Vorbeck in German East Africa (Tanzania). He first went to Potchefstroom for basic training and then became a member of the Sixth Infantry Regiment.

In 1915 he sailed from Durban to Mombasa along the east coast of Kenya, and from there he travelled by train to Voi where the Allied headquarters were.

At Salita he and his friend Louis Rossouw had their baptism of fire. Another clash took place at Latimanek and the Germans offered fierce resistance. They advanced to the Himo River. Here Harry and many of his comrades contracted malaria and were sent back to the Union to recuperate. After a while he was allowed to return home to recover properly. In 1918 he returned to Potchefstroom to resume active service – this time he was trained as an artillerist. While he was receiving training, the Spanish flu epidemic spread through the country like wildfire. The soldiers on the home front, including Harry, all had to muck in and help with emergency services and nursing. He went from Potchefstroom to Kimberley where he worked until the war ended on 11 November 1918. Afterwards he returned to the farm in the Karoo.

Farming conditions in the region deteriorated to such an extent that the Kirkmans decided to move to the northern Transvaal bushveld. To the 20-year-old Harry, the 1600-kilometres trek was much more than a move – it was pure adventure.

The family settled on the farm Haakdoringdraai in the Waterberg district and it was here that Harry met ranger Bert Tomlinson. Although Bert was a fair bit older than him, the two eventually became very close friends.

Bert Tomlinson was the manager of a large estate belonging to the TCL (Transvaal Consolidated Land and Exploration Co. Ltd) along the Sand River in the southern Lowveld. "You become attached to it, Harry," Bert often said. "If you have lived in the Lowveld once, you can't get it out of your blood again!" Bert advised Harry to learn Tsonga, the language of the Shangaans, and within a year he could speak it fluently.

Ranger Walter Henry Kirkman, who had been the last surviving pioneer ranger by the 1980s. He served as ranger at Skukuza and Shangoni sections between 1933 and 1950 and as roads foreman at Skukuza from 1950 until 1958.

One day Harry received a letter from Bert in which he informed him that he had accepted a post in the Kruger National Park, which meant that his post at the TCL's Toulon Estate was vacant – if Harry was interested.

After an interview with the manager, Percy Greathead, the matter was settled. Harry was appointed as farm manager of the Toulon Block or Sabie Ranch, where they had been farming cattle on a large scale for a number of years. He started there in January 1927.

The estate consisted of ten farms – Belfast, Lisbon, Kingstown, Toulon, Shaws, Dudley, Charleston, Flockfield, Malamala and Marthly – and covered an area of 30 835 hectares. Harry moved to Toulon to look after a massive herd of 2500 cattle. Strong knob-thorn kraals were built to protect the cattle against predators, but despite all their efforts lions exacted a heavy toll. Harry waged a constant battle to protect the valuable Sussex and Hereford cattle against the big cats.

He had a lot of adventures and many lions were killed. Many were caught in traps and killed. During his three years as manager of the estate, Tomlinson killed 100 lions but, according to Leo Lownds (1931), Harry Kirkman killed 400 in the following six years by either trapping or shooting them.

Despite their concern over their cattle, the landowners were not negatively inclined towards the game on their farms. In 1928 they initiated a game protection scheme by appointing two black rangers to assist Harry Kirkman to stop poaching. Kirkman met Stevenson-Hamilton at Skukuza during a courtesy visit and questioned him at length about conditions inside the Park, including the problems and duties of a game ranger. The answers he received immediately awakened in him an interest in the ranger profession.

The great drought of 1930 to 1933, coupled with a locust plague, made cattle farming difficult. The TCL increasingly questioned the wisdom of its undertaking and sold some of the farms to other owners, with the result that Harry feared for his future.

Early in 1933 Stevenson-Hamilton let Harry know that there would be a vacancy in the Park from 1 April. Harry needed no convincing to accept the coveted position, and on 1 April 1933 he moved into his office at Skukuza as clerk / relief ranger. He lived in a simple rondavel with a small veranda.

Harry Kirkman was in his element now. His restless and energetic personality soon led to him being called "M'lilwane" ("the small fire") among the black workers.

In the spring of 1934 he married Ruby Ross, the stepdaughter of Albert Cass, the shop manager at Skukuza.

From December 1935 onwards he and his young bride had to relieve ranger Crous at Letaba section for six months while Crous was on extended leave. Somehow he contracted plant poisoning and had to be taken across the Olifants River by the pontoon to be treated at Nelspruit. He stayed on at Letaba until May 1936 and then returned to Skukuza as section ranger. H.C. van der Veen became the warden's full-time clerk and this released him from his administrative duties. Harry Kirkman particularly endeavoured to determine the status of black rhino in the park at that stage and went on extended patrols in the thick Nwatimhiri bush in search of these rare animals.

On 22 July 1936 he found the spoor of a black rhino along the Nwatimwambu Spruit and a week later even fresher ones in the vicinity. He could also determine that the animals had browsed on tamboti, sickle bush and a certain type of creeper.

On 6 October he again found rhino spoor near the confluence of the Mhlupheka and the Nwatimhiri. On 15 October 1936 Harry Kirkman and his black rangers were the last people to see a live black rhino cow in the Kruger National Park until a few of these animals, caught in Natal, were released on 17 May 1971 along the Nwaswitshaka Spruit.

In 1937 Harry helped Bert Tomlinson to arrest some white poachers at Mahlambandlophu Pan. In January 1938 he found himself at Punda Maria where he relieved ranger Rowland-Jones while he was on long leave. There he undertook a patrol along the Luvuvhu River to Pafuri. Two weeks later he became very ill with malaria. As sick as he was, though, he got up to destroy a man-eating lion that had caught and eaten a boy in the Mhinga area outside the Park.

At the end of May 1938 he was sent to Shangoni to supervise the building of a new ranger post. Hunt, Leuchars & Hepburn Ltd. of Pretoria were the contractors and the total cost of the project amounted to £806.0.0. Due to the remoteness of the area, work progressed very slowly and Harry could only move into his new home on 1 October 1938. Shortly after his arrival at Shangoni, a certain C.W.T. Pittendrigh and C.D.C. Geyer of Carolina arrived to prospect for gold along the Shingwedzi River close to the residence. The warden was under the impression that

Ranger Harry Kirkman at the spoor of a black rhino bull which he and two field rangers found on 22 July 1936 along the Nwatimwambu Spruit in the Nwatimhiri forest. This had been the last sign of black rhino in the Kruger National Park until they were re-introduced from Natal on 17 May 1971.

they were active inside the borders of the Park and reprimanded Kirkman, but it would later become apparent that most of the prospecting pits had been dug outside the Park's western border.

One day in May 1939, while Harry was patrolling on horseback along the Mphongolo River, he found a large pool. He rode closer, his mind on the possibility of crocodiles, when a lion suddenly charged out of the long grass, jumped up against his horse's neck and grabbed it behind the head. Harry could not use his rifle from the horse's back and jumped off. Although the spurs on his boots tripped him, he still managed to get upright in time to shoot the lion in the shoulder. The lion released the horse and disappeared into the thickets on the riverbank. Harry fired another two shots at the lion and then followed the dying animal's spoor. They soon found it and, as it tried to get up, field ranger George inflicted the mortal wound. Harry immediately returned to Shangoni to treat his injured horse, but the wounds were so severe that the animal died. The Board paid him £18.11.2 in compensation.

In 1940, after the outbreak of the Second World War in 1939, Harry applied for leave to go on active service. Because of the already thinned-out ranger corps at that stage, the Board turned down his application.

In May 1941 Harry was asked by the warden to help former ranger Harold Trollope chase a number of elephants that had wandered across the western border of the Park, back into the Park. They managed to do so, but in the process Harry and Trollope had to shoot quite a few uncooperative bulls.

1942 was a very difficult year for the Kruger National Park. As a result of fuel rationing only Pretoriuskop and Skukuza could be kept open during the tourist season. The severe drought caused a famine in the adjacent black homelands and poaching along the borders increased drastically. In the absence of Rowland-Jones, Kirkman had to run Punda Maria section as well. During patrols through this enormous area he saw the interesting Nyandu Sandveld and Malonga Fountain for the first time.

He again applied to join the Allied forces in 1942 and this time the Board approved his request. He left on 15 October 1942 and Bill Lamont took over his duties at Shangoni. In contrast with his baptism of fire in the First World War, the battlefield where Harry's unit was sent in October 1943 was the desert and endless sand flats of northern Africa. During the day they had to cope with the scorching sun and the roaring of the German case-shot and at night with the cold and grey desolation. After the end of the war in 1945 Harry resumed his duties at Shangoni. Within two months of his return Stevenson-Hamilton retired and the Kruger National Park personnel had to bid their beloved "Skukuza" farewell.

In March 1947 Harry Kirkman was transferred from Shangoni back to Skukuza by the new warden, Colonel J.A.B. Sandenbergh. In 1950 he was put in charge of the Park's road construction and maintenance section, with A. d'Assonville as his assistant.

While mainly concerned with the construction of new roads and the maintenance of existing tourist roads and firebreaks, Harry also tackled soil erosion works, including the building of a series of four earthen walls in a deep donga along the eastern flank of Shaben Hill. Assisted by two builders, D.J. Roets and G. Jordaan, he also built a series of five concrete weirs in the Mahula Spruit, south of Renosterkoppies, in 1949–1950.

In the autumn of 1948 former ranger Harold Trollope was once again called in by Sandenbergh to help drive elephants back into the Park that had crossed the Crocodile River and ran amuck in the

Shangoni ranger post which was built in 1938 for ranger Harry Kirkman.
(30.07.1984)

In 1942 ranger Harry Kirkman undertook a foot patrol along the Luvuvhu River and became one of the first white officials to have ever seen the impressive Makahane Poort.
(21.06.1984)

The mysterious and seemingly endless Nyandu Sandveld along the eastern boundary of the old Punda Maria section. This area was difficult to enter for rangers such as M. Rowland-Jones and Harry Kirkman.
(30.07.1984)

Malahlapanga waterhole on Shangoni section – another one of the unique natural phenomena which ranger Harry Kirkman discovered on Shangoni section while on patrol.
(28.04.1984)

citrus orchards. Trollope, who had other obligations, recommended that Kirkman be tasked with this project. When the raids reached a climax in September 1950, Kirkman, accompanied by the warden and ranger Eric Mingard, began chasing about 44 elephant bulls back into the Park. Every elephant that would not return, had to be destroyed. During this dangerous operation they managed to drive back 32 elephants, while only seven had to be killed.

Harry Kirkman, together with his colleague M. Rowland-Jones, retired on 31 December 1958. He was only due to leave on 31 March 1959, but his accumulated leave allowed him an earlier retirement date.

After his retirement he became a ranger in the Sabie Sand private nature reserve and was once again stationed at Toulon, where he had started his career in the Lowveld 32 years before.

While working in Sabie Sand, Harry Kirkman had another nasty experience with a lion in June 1964 that nearly cost him his life. Two very young men accompanied the old ranger on the spoor of a wounded lion in the reed thickets of the Sand River. They found fresh blood in a very dense spot, which indicated that the lion was nearby. Scarcely two paces further the male charged them with a deafening roar. When Harry turned around, disaster struck as his rifle sling got caught on a shrub's root. He jerked it loose, but it was too late – the lion was already on top of him. In an effort to protect his throat he threw up his left arm, only to hear it crack as the lion's powerful jaws tore through flesh and bone. Harry realised that he had to play dead and softly whispered to the young man closest to him: "Michael! Shoot him, Michael!" His words had the desired effect and the deafening report of a rifle echoed through the reed thickets. Although it was not a fatal wound, the lion released its prey and disappeared into the reeds. They found the wounded lion about 50 paces further and Harry gave a mercy shot to the only lion that had gotten the better of him in his long career. After that he still had to walk two miles to get home before he could go to hospital.

Fortunately, thanks to his hardy constitution, he recovered from his wounds and carried on with his duties as warden of this private reserve until he finally retired at the age of 70 in July 1969.

After his retirement, he settled on Irene Chalkley's Folly Farm near White River, where his wife died in October 1983. He died on 12 June 1989 in an old-age home in Eshowe.

Mike Rattray, a later owner of Toulon, turned the old house on the estate into a tourist camp and named it Kirkman's Camp in tribute to Harry. Another installation on the part of the farm that had belonged to Don Roche, close to the western border of the Park, was also turned into a small tourist camp and is known as Harry's Huts.

Harry was even honoured in Swaziland, when the first warden of the original National Park of Swaziland, Mr T. Reilly, named the park "M'lilwane" in honour of his old mentor at Sabie Sand.

In the Kruger National Park itself, a borehole close to the series of weirs he had built at the foot of Shaben Hill was named Kirkman.

Hannes Kloppers recorded the details of Harry Kirkman's life and times in *Veldwagter* ("Game Ranger"), which was published in 1967.

Lieutenant Colonel Maurice Rowland-Jones (1899–1959)

Maurice Rowland-Jones was born on 8 March 1899 in the town of Misterton in Nottinghamshire, England. After school he joined the British army. His unit was sent to India, where he was awarded a commissioned rank, and eventually he was promoted to captain. In the late 1920s he resigned from the army and came to South Africa. In 1931 he accepted a post in the Kruger National Park as supervising gate official at Numbi Gate near Pretoriuskop. Before then, entrance permits to visit the Park had been issued by agents at White River, Acornhoek, Gravelotte and Punda Maria (until 1928) and in the southern part of the Park at A. Moodie's shop and tearoom at the foot of Legogote (from 1929).

Rowland-Jones knew something of building and working with concrete, and outside the tourism season the warden put him to work completing certain construction works. In 1931 he was ordered to build the first concrete dam in the Kruger National Park in the Ntomeni Spruit, close to the old Delagoa Bay transport road, between Pretoriuskop and Nellmapius Drift. He finished the project in 1932, complete with a large wheel sluice. Unfortunately he had failed to anchor the foundation of the wall properly and the first flood tore a large hole under the wall. The break was fixed, but the dam never held water well. It later silted up, but during the rainy season enough water is stored in the sand to maintain an almost permanent waterhole under the wall.

In 1932 the warden asked him to build a permanent concrete low-water bridge across the Sand River at Jafuta. This terrain may have been named after Sergeant Jafuta Shithlave, one of Stevenson-Hamilton's stalwart field rangers during the pioneering years of the former Sabie Reserve. Before construction started there had only been a pole bridge over the Sand River, which caused endless problems and washed away almost every year. Rowland-Jones worked on this bridge until May 1932, when he had to resume his duties as gate official at Numbi. Bricklayer P.J. Joubert completed the project.

Legend has it that the barrels of a large number of destroyed poachers' rifles were used in the reinforcement of the bridge.

After the end of the tourist season in 1932, Rowland-Jones built the concrete reservoir of 12 000

The hollow baobab tree along the upper reaches of the Zari Spruit, Shangoni section, a well-known shelter for poachers. (30.07.1984)

Lt Col Maurice Rowland-Jones, ranger of Punda Maria section (1938–1954), circa 1944.
(ACKNOWLEDGEMENT: MRS V.M. ROWLAND-JONES, LOUIS TRICHARDT)

Lt Col Maurice Rowland-Jones, senior ranger at Skukuza (26 June 1954–31 December 1958), circa 1957.

gallons at Skukuza. This structure is still standing in the garden of the Nature Conservation terrain.

In 1933 Rowland-Jones tore down the old windmill at Skukuza and re-erected it on a well at Letaba rest camp to provide water for the camp. He also built an ablution block at Satara. The walls were of concrete and when it had to be demolished later to allow new extensions, it had to be dynamited.

Rowland-Jones only became a permanent personnel member in 1937, after he had already served for almost seven years as gate official at Numbi.

Ranger I.J. Botha of Punda Maria resigned on 31 May 1938 and his position in section 8 was filled by the newly appointed ranger, Rowland-Jones, on 1 June 1938. Shortly afterwards he was also appointed special justice of the peace for the Punda Maria region.

After the outbreak of the Second World War, Rowland-Jones was called up by his unit, the Natal Carbineers, on 21 May 1940. In 1942 he spent some time in Madagascar, but his unit was later stationed in the Middle East, where he served until the end of the war. He was eventually promoted to the rank of Lieutenant Colonel and became the commander of Unit 133. While he was away on active service, his wife Elsie (née Roper) obtained permission from the Board to remain in the ranger quarters at Punda Maria. The section was run by ranger H. Kirkman of Shangoni and later by ranger Bert Tomlinson of Shingwedzi.

In 1942 Mrs Jones fell seriously ill and Harold Mockford, the WNLA official of Pafuri, had to take her to hospital in Pietersburg (now Polokwane). She was later transferred to a hospital in Pretoria where she died on 4 March 1942. She was buried in Pretoria.

Rowland-Jones returned from the war to the Park with an OBE medal and on 1 January 1940 he resumed his duties as ranger at Punda Maria. After Stevenson-Hamilton's retirement, he unsuccessfully applied for the position of warden of the Kruger National Park. The widowed ranger later married a widow, Violet Myrtle Syme (née Thomas), commonly known as "Carrots" because of her flaming red hair.

While serving at Punda Maria, Rowland-Jones built a comfortable patrol hut on top of Hapi Hill at Pafuri. In later years a big lion with a wire snare around its neck moved into the hut.

Rowland-Jones was a tall, bony man who loved to wear a Baden-Powell slouch hat. Because of his prominent Adam's apple he was known as "Malokoloko" among the locals.

After Sandenbergh's retirement in February 1954, Rowland-Jones was promoted to senior ranger on 26 June 1954 and transferred to Skukuza. He served there under the new warden, L.B. Steyn, until his retirement on 31 December 1958. His period of service at Skukuza was typified by his regular visits with his blue Chev Imp to junior rangers and his sergeant-major-like behaviour toward all of those who, in his opinion, had not acquitted themselves according to expectations. He had a keen interest in the birds of the Kruger National Park and compiled the first official bird list of the area in 1957. This list contained 299 species, but the number has since almost doubled to 491 species, with 33 species in doubt.

After Maurice's retirement, the Jones couple settled in Haenertsburg and he was appointed by Tom Carruthers, the manager of WNLA at Soekmekaar, as acting relief official. In August 1959 he relieved Harold Mockford, the WNLA official at Pafuri, while he was on leave. On 15 August 1959 Carruthers received a radio message from Mrs Jones saying that her husband had suddenly died of a heart attack that morning. He was buried at White River and his widow moved to Louis Trichardt. Later she moved to Pietersburg (Polokwane), where she died on 31 August 1982.

William Henry Lamont (1884–1974)

William Henry Lamont was born on 24 August 1884 in Greenock, Renfrewshire, Scotland. He became a sailor while in his teens, then went back to Scotland to farm, but could not resist the call of the sea. Shortly afterwards he joined the Scottish Horse Regiment and was called up for service during the Anglo-Boer War.

By the time their ship reached Las Palmas the war had ended, with the result that Bill Lamont entered South Africa not as a soldier, but as an unemployed young man. He worked on the mines for a while and then tried his hand at prospecting, without much success. Later he farmed maize in the Standerton district, but soon became bored and moved to the northern Transvaal bushveld, where he farmed in the Potgietersrust (Mokopane) district for 16 years.

In 1937 he was appointed as tourism officer at the new Shingwedzi camp on the recommendation of Bert Tomlinson when the appointed tourism officer, George Meade, was unable to continue working due to bad health.

Lamont had acquired some knowledge of explosives and well-digging during his stint as mine worker. For this reason he was appointed as technical official in charge of dam-building and the digging of wells in the northern district of the Kruger National Park after the tourist season of 1938.

He started with this task in 1939 and by 1941 had completed a series of 24 wells at Shipandani, Nwamba, Mashagadzi, Ndlopfini, Tsumani, Dzombo West, Dzombyana, Nkokodzi, Shingwedzi quarters, Nwarihlangari South, Nkulumbeni North, Shingomeni, Babalala, Stangene, Nwashitsumbe North, Shidzavane, Shashanga (Salane), Mahembane (Siafa), Mgambine (between Tsumenene and Klopperfontein), Klopperfontein, Mashikiri, Mazanje, Nkovakulu and Nkovakulu North. Of this series of water points, the wells at Nwarihlangari South, Babalala, Stangene, Nwaswitsumbe North, Mashikiri, Mazanje and Nkovakulu are still in use.

It was during his well-digging period on the Lebombo Flats that Bill Lamont was named "Siafa" by the black workers. He always used this expression when they struggled with the work and it literally means "we are dying", or "we will die" (of the hard work). It was also Bill Lamont who discovered the hot spring Tshalungwa along a tributary of the Madzaringwe Spruit (probably while digging the well at Klopperfontein). It was called "Siafa" Fountain for a while.

In 1939 the warden instructed him to build a tented camp next to the Luvuvhu River at Pafuri. The terrain that was selected is where the Pafuri picnic site is today. Some of the Lebombo ironwood fence poles that he had used can still be seen there. Lamont had to provide the camp with a sturdy fence, shower complex and toilets.

One day while they were working Lamont poked a stick into the hollow trunk of a large jackalberry on the construction terrain. To his surprise he pulled a tobacco pouch from the hidden depths, and when he shook the contents of the bag into his hand, he found several gold coins! He later showed these to Harold Mockford, the WNLA official at Pafuri. The tented camp at Pafuri was only used for a few years and was finally closed in 1948 because of health reasons and flood damage.

In 1940 Bill Lamont had to temporarily bring his well-digging operations to a halt as he was needed as tourism officer at Punda Maria, where the appointed official had resigned shortly after his appointment.

When Harry Kirkman left for active duty in October 1942, Bill Lamont was sent to Shangoni as acting ranger. He also did relief work in Punda Maria section during this time.

Ranger William Henry (Bill) Lamont, who served as ranger at Shingwedzi and Shangoni sections between 1942 and 1950.
(ACKNOWLEDGEMENT: MR J. LAMONT, NEWCLARE)

The awe-inspiring Nkovakulu Poort carved out by the Luvuvhu River – one of the unique scenes which ranger Harry Kirkman witnessed during his foot patrol in 1942.
(21.06.1984)

The Egyptian fruit bat cave in Lanner Gorge, west of the Pafuri floodplain, had escaped the attention of all the pioneers until its discovery in 1983 by Dr Leo Braack.
(21.06.1984)

The grave of former ranger Bill Lamont at Serondella along the Chobe River in Botswana.
(ACKNOWLEDGEMENT: COL. J.A.B. SANDENBERGH)

Tshalungwa hot-water spring which was discovered by ranger W. Lamont during the early 1940s along a tributary of the Madzaringwe Spruit on Punda Maria section.
(27.04.1984)

Stangene well and windmill. It is one of the wells which had been dug by ranger W. Lamont during the early 1940s.
(30.07.1984)

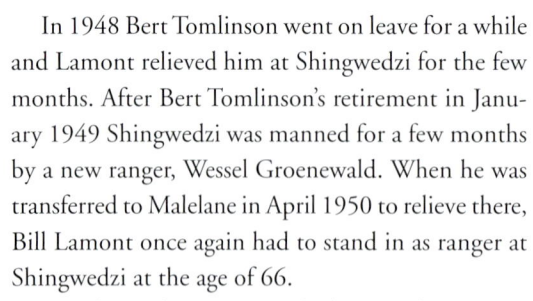

In 1948 Bert Tomlinson went on leave for a while and Lamont relieved him at Shingwedzi for the few months. After Bert Tomlinson's retirement in January 1949 Shingwedzi was manned for a few months by a new ranger, Wessel Groenewald. When he was transferred to Malelane in April 1950 to relieve there, Bill Lamont once again had to stand in as ranger at Shingwedzi at the age of 66.

Board member B.A. Key died in April 1949 and after the Board had given permission to his family to scatter his ashes near his favourite camp, Shingwedzi, Bill Lamont was instructed to create a simple, yet suitable sandstone memorial. This memorial, with the wording: "Bran Key lives on here 17–4–1949" can still be seen close to the northern flank of the Kanniedood dam wall.

On 30 June 1950 Lamont finally retired and settled on a piece of land along the Chobe River near Kasane in Botswana. This piece of land in the Chobe Game Reserve had been donated to him by a former minister of lands of Botswana, Mr Haskins. Lamont hunted crocodiles and pottered around until he finally started living a more peaceful life at the age of 70.

This colourful old figure, like so many of the other pioneer rangers, lived a long life and reached the age of 90. He was buried in 1974 at Serondella by the Chobe River in Botswana.

A windmill situated five kilometres northwest of Shingwedzi rest camp next to the Mphongolo River was named after Bill (Pop) Lamont in remembrance of his contribution towards the water provisioning programme of the Kruger National Park.

Dawid Hercules Swarts (1904–1970)

Dawid Hercules Swarts was born on 28 November 1904 in Parys in the Free State. After his education at Vredefort and in Johannesburg, he joined the police force and served as mounted constable at Louw's Creek, Barberton and Squamans, between Komatipoort and the Swaziland border.

According to him, there was still a small herd of roan antelope roaming over this part of the Lebombo flats during the early 1930s.

Dawie Swarts was promoted to sergeant and was seconded to Skukuza in 1936 as clerk to the special justice of the peace, J. Stevenson-Hamilton. In April 1938 he resigned from the police and accepted a position as clerk and storeman in the warden's office at Skukuza.

In December 1939 ranger Crous went on extended leave and Dawie Swarts was sent to Letaba as acting ranger. He stayed there until June 1940, when Crous returned to his section.

Dawid Swarts married Sannie Coetzee in October 1940 in Nelspruit. Sannie worked in the post office at Skukuza and they had one son, young Dawid.

After ranger Ledeboer's retirement in March 1946, Swarts was appointed as the new ranger of Satara section on 1 April 1946. He soon made significant improvements to the residence and gardens and also planted a range of indigenous trees (including a number of baobab and mopane trees) in the rest camp and in front of the field rangers' quarters. He loved gardening and laid out a vegetable garden and fruit orchard close to the homestead, which soon became known as the pick of the gardens in the Kruger National Park.

Shortly after the war, he and warden Sandenbergh were involved in capturing a number of giraffes for the Portuguese authorities in Mozambique. This operation was not a great success and the last surviving giraffe, which had been destined for the Gorongoza reserve, was released near Klopperfontein.

Dawie Swarts was a short man, but he cut an imposing figure in his white pith helmet and thin moustache. He was a particularly neat person who completed his official correspondence and returns in good time and with great accuracy. He officially declared all documents "correct" and then signed each stylishly: Dawid H. Swarts.

In 1954 Swarts was promoted to district ranger of the central district of the Park – the area between the Sabie and Olifants rivers – and he was a good teacher to the junior rangers at Kingfisherspruit and Tshokwane. During his service period he had a whole series of important patrol roads opened up along the Nwanetsi, Sweni and Timbavati rivers, along the western border from Orpen to Naboomkop, from the old Gorge rest camp to the waterfalls in the Olifants River (where the Olifants River hiking camp is today), to Bangupoort, the Pumbe plateau and to interesting places like the Ngirivane hills and Mshatu and Ngumula pans. Some of his old patrol roads, like those along the Timbavati, Nwanetsi and Sweni rivers, later became extremely popular tourist roads. At the Pumbe outpost on the sandveld plateau on top of the Lebombo Mountains, next to the Mozambican border, he also built himself a comfortable overnight hut near a large Rhodesian mahogany. For many years this facility was used by officials who had to work in the area. He built a similar outpost at Tseri along the Olifants River. At all of the outposts he erected sturdy kraals for his pack donkeys and horses using knob-thorn poles. He, Harry Wolhuter and a few others were among the last group of rangers who still patrolled using horses and donkeys.

Dawid Swarts ruled his field rangers and other workers with an iron fist, which certainly led to him being nicknamed "Mafelele" ("the cruel one", or "the one who keeps everything for himself").

From 8 to 13 November 1958 he and Dr N.J. van der Merwe, the liaison officer of the Board, accompanied the historian Dr W.H.J. Punt and Mr C. Trichardt, a great-grandson of the Voortrekker leader Louis Trichardt, on an expedition to Mozambique. Their aim was to determine the place where the Van Rensburg murders had taken place. Unfortunately heavy rains and inaccessible roads prevented them from finding the exact spot, but they collected enough reliable information to allow Dr Punt to determine the murder site with certainty the next spring.

When ranger Rowland-Jones retired in March 1959, Swarts was his obvious successor as senior ranger. Unfortunately certain irregularities brought an abrupt halt to his career in the Park and he left the service on 31 May 1959. He settled on his small farm Hilltop between Nelspruit and Barberton, where he died in 1970.

Henry Charles Wolhuter junior (1920–1964)

Henry Charles Wolhuter was born on 29 June 1920 in Barberton Hospital. He was the only son of the famous Harry (Henry senior) Wolhuter and his wife Alice. He spent his childhood years at Mtimba, which

Dawid Hercules Swarts, ranger of Satara section between 1946 and 1959.

was part of the old Sabie Reserve until 1923 and where his parents lived until 1939 before they moved to their new residence at Pretoriuskop.

Henry was the first white baby in the Park. He received his primary education at Miss Fuller's school in White River, which later became Uplands School, where he was one of the first seven pupils. After completing his primary education he went to St John's College in Johannesburg where he matriculated shortly before the outbreak of the Second World War.

After school he worked at the Chamber of Mines for six months and then joined the army. The greater part of his active service was spent in the Middle East, where he was a member of the Animal Transport Unit.

After the end of the war he worked at the Chamber of Mines for another six months, but when his father retired at the end of April 1946, Henry jnr. was appointed as ranger of Pretoriuskop section on 1 May 1946.

He married Rosemary Dyke-Wells on 19 April 1948 and they had two daughters. The couple was later divorced. In April 1954 Henry was promoted to district ranger in charge of the southern district of the Park (the area between the Sabie and Crocodile rivers), a position which he filled with distinction. On 19 May 1956 he married Joan Merriman and the couple had three sons, Geoffrey, the eldest, and twins, Henry and Kim. On 1 April 1961 he was promoted to senior ranger and transferred to Skukuza. Once again he excelled in this position, until his death at the early age of 44.

Henry was a son of the Lowveld and had shot his

The spectacular waterfall in Bangu Poort, which was discovered by ranger D.H. Swarts in the late 1940s.
(18.06.1984)

Henry Charles Wolhuter as a young boy on Pretoriuskop section, standing at the carcass of a wildebeest bull which he had shot as rations, circa 1930.
(ACKNOWLEDGEMENT: STEVENSON-HAMILTON COLLECTION, SKUKUZA ARCHIVES)

first lion before his twelfth birthday. He dedicated nearly 19 years of his life to the Park and nature conservation. Like most nature lovers he was quiet and reserved, but at the same time big-hearted and a man of integrity.

His love of nature found its expression not only in his excellent knowledge of the plant and animal life, but also in his familiarity with the black people of the area, their language, culture and habits.

He spoke Swazi since a child and his habit of starting to work before daybreak gave him the nickname "Kwezi", which is a variant of the Swazi word "likhweti", which means "morning star".

He knew the southern part of the Park, and especially the Pretoriuskop section where he worked for the greater part of his career, like the back of his hand. With his faithful field rangers Simon Mangane, Helfas Nkuna and Charlie Nkuna, he apprehended many a poacher and experienced countless adventures during patrols and in the execution of the then predator control policy. He applied himself to stabilise the natural water in his section and in 1944 built the small concrete dam in the upper reaches of the Komapiti Spruit. Later he also built the stone dam on the rock slab at Tlapa-la-Mokwena, as well as the two earthen dams close to Napi Hill.

The translocation of white rhino from Natal to the southern part of the Kruger National Park, which started in October 1961, was a great pleasure to and challenge for Henry. He played an important role in this translocation programme and often went on the long and difficult journey from the Natal (KwaZulu-Natal) parks through Swaziland to the Kruger National Park with a precious load of white rhino.

In December 1963 he developed serious emphysema for the first time and this debilitating lung disease led to his being hospitalised repeatedly until May 1964. By May 1964 his condition improved to such an extent that he could work at head office in Pretoria and by August he could resume his duties in the Park. Everybody was happy and thankful for his remarkable recovery. On 10 December the emphysema monster pounced again and this time he lost the battle. After a sickbed of only a few days, he passed away on 14 December 1964 in the park warden's home. He was buried at White River.

Henry's sudden death was a great shock to his colleagues and everybody who had known him. With his death the Wolhuter tradition that had been built up in the Park over a period of 62 years came to an end, but the legacy "Kwezi" and his father "Lindanda" left will not easily be erased.

The small Tlapa-la-Mokwena Dam which was constructed in 1954 by ranger Henry Wolhuter on a rocky face near the Mbyamiti. (08.06.1984)

Henry Charles Wolhuter, ranger of Pretoriuskop section between 1946 and 1959 and senior ranger at Skukuza from 1 April 1959 until his death on 14 December 1964.

Remarkable black field rangers

It is understandable that the names of quite a few white leader figures come to the fore in a discussion of the early years of the Sabie and Shingwedzi reserves and later Kruger National Park. Under Stevenson-Hamilton they laid a foundation on which later generations could build a conservation and development strategy for the Kruger National Park of which South Africa and the rest of the world can justifiably be proud.

The stories with which the black field rangers were associated and which normally took place behind the scenes, are seldom highlighted unless something interesting or heroic can be associated with it. These people constantly had to face danger, whether it came in the shape of lions and other wild animals, bands of armed poachers, difficult living conditions and the lack of even the most basic medicines, inadequate transport available or malaria, diphtheria and other epidemic diseases. Then there were the natural disasters like epidemics among the game (especially anthrax), devastating veld fires, widespread locust plagues, long droughts and flooded and washed-away communication roads. These are only a few of the occurrences that called on the conservation staff to show their true mettle.

Colonel Stevenson-Hamilton selected his white rangers with care, and with one or two exceptions they overcame the challenges of their occupations in a remarkable manner. It must, however, be said, without any fear of opposition, that all their efforts would have been futile if it had not been for the continuous and faithful support of the many black field rangers and other helpers. Their contribution can never be praised enough and it must be acknowledged that the Kruger National Park would never have developed to its full potential without them.

Not only did the black rangers play a vital role, but so did the labourers, supervisors, artisans, drivers, road workers, rest camp staff and indunas and all the other black workers. They also made many sacrifices, often the highest – their own lives.

Early on Stevenson-Hamilton appointed several field rangers, labourers and other assistants who lightened his load considerably. They were mostly locals who knew the area, the veld and the history of the area very well. Some had been habitual poachers, but with the right leadership and motivation they eventually became excellent rangers. Before Stevenson-Hamilton moved most of the inhabitants to areas outside the reserves, he had used informants to find out about illegal activities. In the first few years, the

Field rangers Kukise Khoza, Fasha, Joe Nyalungwa, Office Ubisi, Mahanje, George, Jafuta Shithlave (sergeant), Cement Mathlali and Corporal Njinja Ndlovu (in front) on parade at Sabie Bridge (Skukuza). February 1921.
(ACKNOWLEDGEMENT: STEVENSON-HAMILTON COLLECTION, SKUKUZA ARCHIVES)

only weapons field rangers were allowed to carry were assegais and hatchets. Of course these provided very little protection against poachers (especially those from Mozambique), who had firearms and did not hesitate to use them.

It says a lot about the dedication and courage of Stevenson-Hamilton's field rangers that, despite the lack of effective weapons, they still managed to arrest countless poachers. In 1905, one of the first field rangers to die in a confrontation with poachers from Mozambique was a man called Mehlwana.

In the days before, during and just after the First World War, when patrols had to be curtailed as a result of belt-tightening measures and weapons were easily obtainable in Mozambique, armed poaching increased. Quite a few field rangers, including Cement Mathlali and Mfitshane Koza, were overpowered by gangs of poachers and seriously injured. Some of these gangs were led by former field rangers who had been dismissed from the Park due to misconduct. Among the best-known gang leaders were Mashashe Gumbani, Mashashe Mathebula and one Mataffin, who operated out of the Matukanyane area.

Mfitshane Koza was a small man with the heart of a lion. Once, he and a companion came upon a gang of poachers who had just killed a waterbuck near the eastern border. The poachers immediately attacked the two rangers with assegais and axes. Mfitshane's colleague ran away to find help, but Mfitshane defended himself with such determination that his attackers eventually decided to retreat. The brave constable wanted to arrest at least one of his attackers, so he overtook the slowest running one and grabbed him by the collar. Unfortunately he lost his balance and fell, upon which his "prisoner" turned around and plunged his assegai into Mfitshane's lungs.

Mfitshane was left for dead, but eventually he recuperated to such an extent that he could resume his duties as field ranger. One day, however, while once again chasing after some poachers up the rocky slopes of the Lebombo, a blood vessel in his lungs burst, close to the old wound, and he bled to death.

In 1927 Corporal Cement Mathlali and his colleague Stephanus were attacked by a group of more than 50 armed poachers from the Mapulanguene area in Mozambique. Cement was hit over the head with a knobkerrie and when he came to, he found that the poachers had left and Stephanus was lying dead next to him.

By that time, Stevenson-Hamilton had already made representations to the Union Army to allow his field rangers to carry and use rifles in self-defence. This request was only approved in 1927 and initially 24 old Martini-Henrys were issued to field rangers who patrolled the most dangerous zones. These were later replaced with more modern Lee-Metford or Lee-Enfield rifles, which made a huge difference to the tracking and arresting of armed poachers. In the meantime, the Portuguese authorities had, in the interests of neighbourliness, withdrawn the rifles using bullets in the possession of black people in Mozambique. After that they were only allowed to use shotguns or antique old muzzle-loaders for hunting purposes. This action sounded the death knell for poaching out of Mozambique and for many years poaching from across the border was hardly an issue.

Along the western and southern park borders it was another matter, however. Nyokane Shilobane was shot and killed by poachers on 19 November 1950 near Malelane. Johannes Maluleke and his partner were attacked by three poachers on 22 September 1958 in the vicinity of Matukwala Dam and Maluleke received a fatal axe wound to the head. All three culprits were tracked down, arrested and found guilty of murder, thanks to efficient action by the police and the field rangers of Punda Maria.

One of the first black rangers whom Stevenson-Hamilton appointed was "Toothless Jack", a character who features in Sir Percy FitzPatrick's *Jock of the Bushveld*, as well as Henry Glynn's book, *Game and Gold*.

The first field rangers he appointed included Baptine Ubisi, Basket Mandhlasi, Bob Llamula, Doispane Mongwe (after whom the Doispane road was named and whose kraal was close to the Mtshawu Drift along this road), Sergeant Fifteen Ubizi, Sam Nkuna (Sikomela's father), Jack Mathebula, Indaba Khoza, Jack Nyalungwa, Manthai Zitha, Jase Makukula, Mahashi Ngwenya, Sergeant Jafuta Shitlave, English Shabangu, Rice Ubisi, Matches Ngomane, Jim Mbambo, Ntambo Ngwambu, Jonas Mkanzi, Breakfast Xihamba Nkuna, Makapani Nkumalo, Masodja Ndlovu, Matilwane Nkuna, Mfitshane

16. EARLY YEARS: THE SABIE AND SHINGWEDZI RESERVES AND THE KRUGER NATIONAL PARK 1900–1946

Sergeant Roberts of Steinaecker's Horse and "Toothless Jack" saddling their horses in the Sabie Reserve, September 1902.
(ACKNOWLEDGEMENT: STEVENSON-HAMILTON COLLECTION, SKUKUZA ARCHIVES)

Ranger Sergeant Jafuta Shithlave, one of the most outstanding field rangers to have served in the Kruger National Park through the years, standing ready for inspection of equipment. 1913.
(ACKNOWLEDGEMENT: STEVENSON-HAMILTON COLLECTION, SKUKUZA ARCHIVES)

Koza, Corporal Mpampuni Ubizi, Night Mdluli, Piet Mdluli, Matches Ntivane, Ngungwa Mkabe, Corporal Office Ubisi Seboma, Piccanin Mkari, Willem Sifunda and Corporal Sakabona Nkabela.

Of these men, Sergeant Jafuta (Masabane) Shithlave was undoubtedly the most remarkable personality. He was stationed at Sabie Bridge and responsible for the training of new rangers. He is still remembered with great respect by the old guard for his dedication and strong sense of discipline – a man among men!

Ranger Harry Wolhuter's right-hand man in those years was Corporal Forage Ngomane, a highly intelligent and very capable man who was also a good builder and bricklayer.

Then there was Band Nyalungwa (Shongwane), who served for 22 years (1905–1927). His son, Sinias Nyalungwa (Shongwane), saved a woman and child from drowning in the flooded Crocodile River in 1940, 400 paces below the railway bridge where they had fallen in. He was in service of ranger Hector McDonald of Crocodile Bridge at the time. In 1940 he received a bronze medal for bravery from the Royal Life Saving and Royal Humane societies, as well as a reward of £5.0.0 from the Board!

A later group of excellent field rangers who served from 1920 to about 1946, included Sergeant Njinja (Ginger) Ndlovu, Corporal Kukise Khoza (respectively Stevenson-Hamilton's senior black ranger at Skukuza and his regular translator), Matches Khoza, Mbambali Mdluli, Ngwelane Makabela and Joe Nyalungwa, all of whom served at Skukuza. At Malelane there were Corporal Jim Mbazine (Nombolo Mdluli's uncle), Jotai Ubisi, Watch Moyane, Newone Mathebula, Amos Sibuye, Folok Mashave and Moses Ngomane.

The excellent people at Pretoriuskop section included Corporal Charlie Nkuna, Sergeant Helfas Nkuna, Sikomela Nkuna, Merriman Nkuna and George Lubambu. Watch Masingi, Sinias Nyalungwa (Shongwane), while Kabana Ngomane and Waistcoat Nkuna were stationed at Crocodile Bridge. At Tshokwane were Corporal Maplank Ngomezulu, Madubula Sibuyi (David Sibuyi's uncle), Keppies Mzimba, Mistele Mdaga and Klaas Sihlangu. Satara's best men included Corporal (later Sergeant) Jim Zwane, Dick Mdluli and Mapipi Mnisi. Corporal Sokkies Mongwe, Mapoisa Dyambukeni and Malunzane Mhlongo served at Letaba and Corporal Nombolo Mdluli, Watch Moyani, Kostine Mbawana, Manzi Mack Nyalungu and Jack Nyalangwa at Shingwedzi. At Punda Maria were Corporal Jack Mabaso, Nduma Mathebula (the son of Chief Babalala Mathebula), Guija Sivalo and Jim Sibuye and at Shangoni were Corporal Bonyoti Mhlanga, Pencil Mhlongo, George Hlongwane and Samuel Sivalo.

By the mid-1940s many of the old stalwarts were still active, with service periods ranging between 25

16.2 THE FIRST RANGERS

TRADITIONAL SITES OF FIELD RANGER OUTPOSTS, 1926-1984

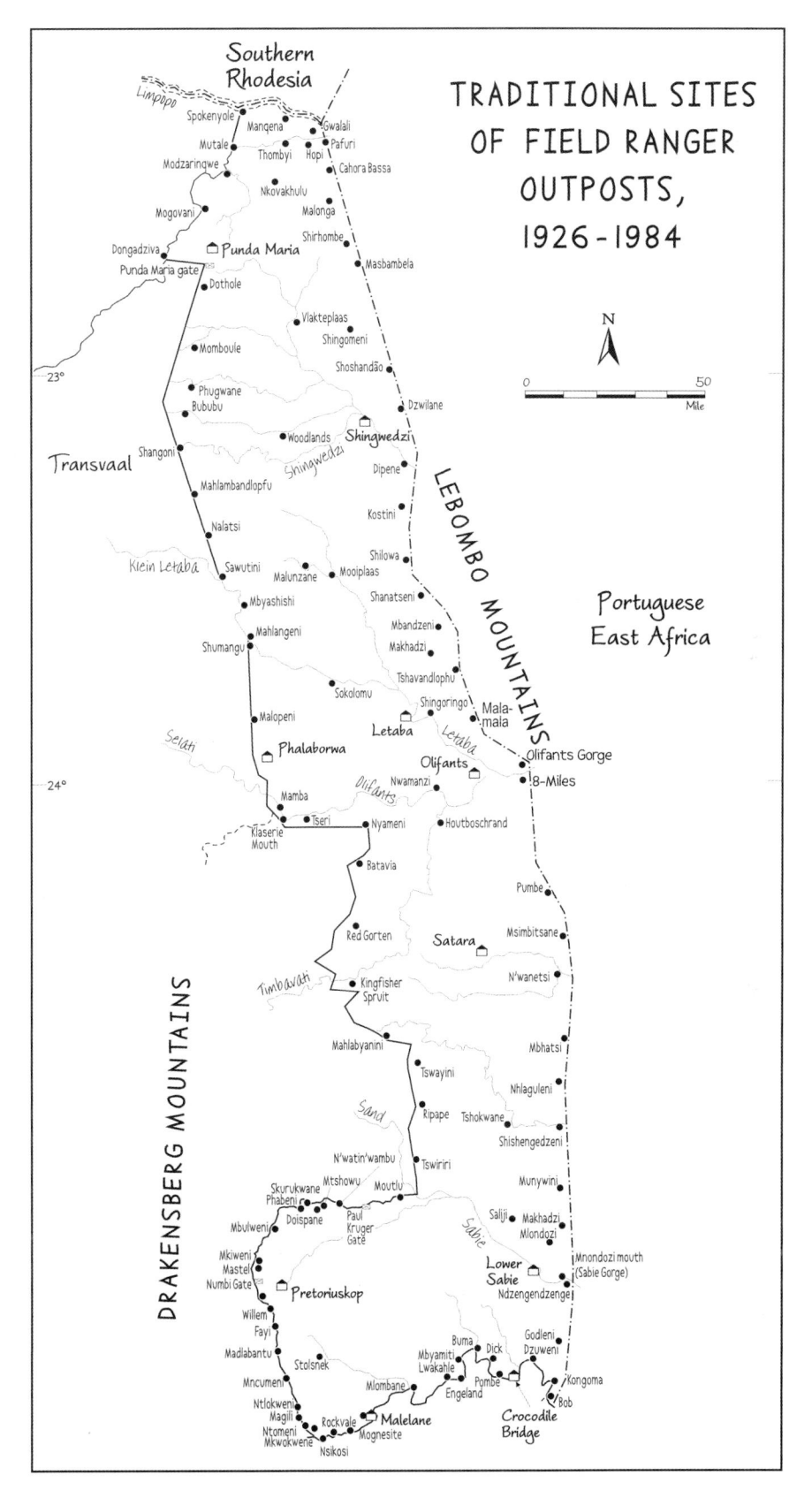

Map of the Kruger National Park indicating the localities of the former pickets (outposts), 1926–1984.

Field ranger Sergeant Helfas Nkuna, the right-hand man of rangers Harry and Henry Wolhuter and their successors on Pretoriuskop section. He passed away on 27 March 1975 after 46 years of faithful service. Circa 1970.
(ACKNOWLEDGEMENT: MR C.J. VAN DER WALT)

and 35 years. Others succeeded those who had already retired and filled their positions with distinction.

In 1944 Njinja Ndlovu was still the chief induna at Skukuza, but in the meantime Mistele Mdaka, Jim Mebu, Bambali Mdluli and Seventeen Nyambi also joined. Helfas and Merriman Nkuna were still serving at Pretoriuskop, but the corporal was Charlie Sifundsa. Added to their numbers were Mapoyisa Sifundsa, Basket Mandhlasi, Maswazi Njunda, George Lubambo and Mbango Mkabela.

Jim Mbazima took control of Malelane section. Still serving were Jotai Ubisi (even though he was injured in 1945 while following a white poacher who had been poaching game in the Park), Amos Sibuye and Watch Matonsi. New additions were Japie Mathebula, Solomon Sibusa and Frans Sibusa.

At Crocodile Bridge, headman Ubisi was the corporal by that time. Sinias Nyalungwa was still there, but new additions were Maxim Ubisi, Johannes Sibuyi, Fifteen Ubisi and Seventeen Ngonyama.

At that time Tshokwane had two corporals, Keppies Mzima and Maplank Mgomezula. Other field rangers there were Jim Zwane, Frans Sibuye and Johan and Solomon Ubisi.

Jim Madonga was the corporal at Satara with some of his reliable colleagues Frans Manyike, Makwapin Koza, Philemon Ngomane and Dick Mtunga.

At Letaba old Malunzane Mhlongo, who would later save ranger Geldenhuys from a sure mauling by a wounded leopard, was already the corporal, with a crew consisting of Amos Mashele, Mack Ndlovu, Pencil Ndlovu and Minute Hlatswayo.

Nombolo Mdluli, the brave ranger who in 1926 had shot a lion standing over ranger Trollope along the Hlambamyathi River and who had served under no fewer than six section rangers since 1919, was still ranger Tomlinson's corporal at Shingwedzi. Here he single-handedly shot a man-eating lion and Mrs Stevenson-Hamilton made him a cap from the skin.

Some of his men were Kostine Mbawana, Manzi Mack Nyalungu, Jack Nyalungu, Jim Sibuye, Manuel Maluleke and Fifteen Mbombi.

Sergeant Jan Hatlane, after whom Jan-se-Pan south of the airstrip was named, was the senior field ranger at Punda Maria, with Nduma Mathebula, Office Guija, Guija Sivalo, Cement Mashava, Andries Ramashiya and Samson Mathebula serving under him.

Pencil Mathebula was the corporal at Shangoni. Serving under him were Elias Mabasa, George Hlongwane, Dick Hlongwane and Kukise Mgogo.

The original black workers of the Sabie and Shingwedzi reserves in general were extremely diligent and loyal people and many of them grew old in the service. Njinja Ndlovu, Colonel Stevenson-Ham-

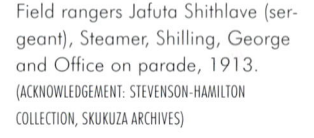

Field rangers Jafuta Shithlave (sergeant), Steamer, Shilling, George and Office on parade, 1913.
(ACKNOWLEDGEMENT: STEVENSON-HAMILTON COLLECTION, SKUKUZA ARCHIVES)

ilton's loyal right hand man, died in 1945 of tuberculosis after 35 years of excellent service.

By 1947 there were already quite a few pensioners who had served in the Park for more than 15 years. Among them were Joe Nyalungwa (28 years), Kostine Mbawane (17 years), Newone Mathebula (34 years), Jack Mbambalas (31 years), former sergeant Jan Hatlane (30 years), George Dumela Hlungwane (30 years), Sokkies Masinga (20 years), Watch Matonsi (23 years) and George Libambo (18 years). Later several others would join this group of veterans, including Jerry Sibuye, who started his career in 1910 as a cattle herder for Stevenson-Hamilton and retired as a cook in 1972 after 62 years' service. Samuel Maluleke, who was a familiar figure at Pafuri picnic spot for almost 30 years, retired in the 1970s after 54 years' service.

Mapoisa Shifundza, who was appointed at Pretoriuskop in 1939, retired in 1963 after 23 years' service. Jotai Lubisi of Malelane retired in 1968 after 30 years' service, while Joël Zima had more than 47 years' service behind him when he retired on 20 January 1985 as supervisor of Nwanetsi rest camp.

After 50 years of service, Judas Mashele was still the gate guard at Numbi and probably one of the most photographed men in the country, if not the world.

Sergeant James Chauke of Crocodile Bridge, and after that Phalaborwa, also had more than 30 years' service by the mid-1980s.

Black employees who left their mark after the Stevenson-Hamilton era or while he still was in the Park, include the following: Sakabona Mkabela, who had an altercation with a lion at Lower Sabie, Corporal Mpampuni Ubisi and Mafuta Shabangu, who were killed by lions at Malelane and near Kemp's Cottage respectively, Sergeant Fifteen Ubizi, Sergeant Malunzane Mhlongo (the father of Piet Mhlongo), Gidya Shivalo, Sergeant James Maluleke, Sergeant Hlangomo Hlengane Maluleke, Gamba Maluleke, Corporal Manuél Kubayi, Corporal Aaron Nyukere, Sergeant Elmon Masiye, Sergeant Judas Ndlovu, Sergeant Nombolo Mdluli (who retired in January 1959), Sergeant Helfas Nkuna, Sergeant Charlie Nkuna (the son of Merriman Nkuna), Sergeant John Mnisi, Sergeant Simeon Mangane, Sergeant Amos Mafukula Sibuye, who was killed by a tractor driven

Field ranger Sergeant Nombolo Mdluli at Malelane.
(CIRCA 1930)

Field ranger Judas Mashele who daily stood on guard at Numbi Gate for more than 50 years.
(16.06.1984)

16. EARLY YEARS: THE SABIE AND SHINGWEDZI RESERVES AND THE KRUGER NATIONAL PARK 1900–1946

Field rangers Faskoti Nkuna and Mestele Mdaka at Skukuza in the 1930s.
(ACKNOWLEDGEMENT: STEVENSON-HAMILTON COLLECTION, SKUKUZA ARCHIVES)

Field ranger Solly Kubayi at the former Ntlokweni picket on Stolsnek section.
(08.05.1984)

by a malicious person on 26 August 1967 near Malelane, Corporal Piet Mhlongo, Corporal Aaron Mathebula, Corporal Jan Mdluli, Corporal Pios Hlonwane, Corporal Titus Mkabela, Ngwela Mkabela, Joe Rangane, Jim Ngomane, July Mona, Ngwela Sambo, Kai-Kai Mavundla, Jim Baloyi, William Mahundla, Jack Sibuye, Dawid Sibuye, Bob Ngomane, Sikomela Nkuna, Office Lubisi, Office Chauke, Nambrete Sithole, Radzingane Matumba, Gamba Maluleke, Andries Ramashiya, Solly Kubayi, Philemon Nkuna, Maxim Singwane and Philemon Chauke.

Mpampuni's fatal run-in with a lion near Crocodile Bridge was described earlier in this book. In 1926 a similar fate befell field ranger Mafuta Shabangu. The warden had received a message from the ranger at Kemp's Cottage that lions were terrorising him and his workers at night and decided to send Mafuta, an old and reliable ranger and a good shot, to investigate. On the ill-fated day Mafuta, who was on his own and armed with his Martini-Henry rifle, a hatchet and a hunting knife, set off on the spoor of the pride of lions. He found them where they were resting in the shade of a tree and took aim at and shot a big lioness. The wounded lion did not have the strength to keep up with the rest of the pride, but she still managed to stay ahead of Mafuta. After a while she lay down in a dense patch of shrubs and awaited her pursuer. She charged the moment she spotted him, and he could only manage one shot and scramble into the lower branches of a small tree nearby. The lioness followed him and dragged him out of the tree, after which they both fell to the ground. Mafuta managed to kill the lion with his pocketknife, but by then she had already mangled him with her claws and severed his femoral artery with a bite to his thigh. Mafuta dressed the wound with his puttees and then started for home. He had hardly covered 400 metres before he became too weak and collapsed under a tree, where he died an hour or so later of loss of blood. The other lions returned later and dragged his body around a bit, but did not disturb it further.

One of the most gruesome, if rarest, misfortunes to befall a field ranger in the veld happened on 25 November 1939. Peculiarly, it had a marked similarity to what had nearly been the fate of his employer 36 years before. One scorching summer day, on the instructions of his supervisor Harry Wolhuter, field ranger Joseph (Mankoti) Nkuna was on a foot patrol near the confluence of the Phabeni Spruit and the Sabie River. Although there were no eyewitnesses to the rest of this tragic story, as Mankoti was on his own, several capable trackers would later reassemble the pieces of this macabre puzzle.

While walking down above the high embankment of the spruit, Mankoti was unexpectedly attacked by a lion – the reason could never be determined. The distance of the charge was short and he had only a second or two to pull his service rifle from his shoulder to fire a single shot in defence. Under the circumstances it was not a badly placed shot, but it was not fatal and the lion pulled the man to the ground. During the fall he lost his rifle, but Mankoti immediately pulled out his hunting knife and tried to fight off the lion. He stabbed it three times behind the shoulder and twice in the chest, after which it collapsed and died. The badly mauled man managed to get to his hands and knees and crawled 20 metres to the riverbed.

Here Mankoti used the last of his strength to scratch a hollow in the sand and drink some of the seepage, after which he died. The next day Mankoti's own father, not knowing of the tragedy that had befallen his son the day before, came upon his body as the vultures were starting to move in. Only 14 days later two lions caught a field ranger between Lower Sabie and Gomondwane, tearing him apart and eating the body in front of his three terrified colleagues perching in a tree.

And so the good earth that is today the Kruger National Park was baptised with the blood of its keepers. The Board erected a suitable memorial plaque at Paul Kruger Gate, the main entrance gate to the Park, in acknowledgement of those who had paid with their lives for the ideal of preserving nature.

Labor omnia vincit!

Sergeant Judas Ndlovu with a captured armed poacher. 1957.

References

Ackerman, D.J. 1951. Nombolo the brave. *Custos* 10(6): 17.

Ackerman, D.J. 1981. Veteran game ranger. *Custos* 10(1): 17.

Adendorff, G. 1984. *Wild Company.* Books of Africa, Cape Town.

Anon. 1914. A ducal ranger. Written for: *The Gold Fields News.*

Anon. 1931. History of the Kruger National Park – some reflections and comparisons. *The Star* 7 October.

Anon. 1937. The man who made the game – fascinating story of 35 adventurous years. *The Star* 9 June.

Anon. 1954. Nombolo se lewensgeskiedenis in die Wildtuin. Unpublished manuscript Skukuza Archives.

Anon. 1958. The late Lt. Col. James Stevenson-Hamilton. *African Wildlife* 12(1):57.

Anon. 1965. *Some Lowveld Pioneers.* Publication of the Lowveld 1820 Settler's Society.

Anon. 1975. Phabeni se gryskop begrawe. *Custos* 4(6):13–19.

Agendas and minutes of National Parks Board of Trustees meetings. 1926–1946.

Bands, A.C. 1986. Personal communication (Ranger. F. Streeter).

Bannister, A. & Gordon, R. 1983. *The National Parks of South Africa.* C. Struik, Cape Town.

Barnard, C. 1975. *Die Transvaalse Laeveld – kamee van 'n kontrei.* Tafelberg, Cape Town.

Behrens, H.P.H. 1947. Hunter turns protector – The story of Henry Wolhuter. *African Wild Life* 1(3):36–50.

Birkby, C. 1937. Amazing story of the bearded giant of the Bushveld. *Cape Argus* 24 April.

Braack, L.E.O. 1983. *Kruger National Park.* C. Struik, Cape Town.

Bulpin, T.V. 1950. *Lost Trails on the Lowveld.* Howard B. Timmins, Cape Town.

Bulpin, T.V. 1954. *The Ivory Trail.* Howard B. Timmins, Cape Town.

Bulpin, T.V. 1983. *Lost Trails of the Transvaal.* Books of Africa Ltd, Cape Town.

Bulpin, T.V. & Miller, P. 1968. *Lowveld Trails.* Books of Africa Ltd, Cape Town.

Carruthers, E.J. 1985. The Pongola Game Reserve: An eco-political study. *Koedoe* 28:1–16.

Carruthers, E.J. 1988. *Game protection in the Transvaal 1846–1926.* Ph.D thesis, University of Cape Town.

Cattrick, A. 1959. *Spoor of Blood.* Howard B. Timmins, Cape Town.

Coetser, E. 1981. Personal communication (Captain J.J. Coetser).

Crous, A. 1977. Personal communication (Ranger G.C.S. Crous).

Crous, G.C.S. 1929–1944. Diaries of the ranger at Letaba. Skukuza Archives.

D'Arcy-Herman, P. 1986. Personal communication (Sergeant H. D'Arcy).

De Laporte, C.R. 1923–1928. Diaries of the ranger at Crocodile Bridge. Skukuza Archives.

De Leeuw Beyers, C.P. 1965. *Kruger Park Adventures.* Living World Adventures. Living World Films, Sandton.

De Leeuw Beyers, C.P, 1970. *Lurking Danger in the Kruger Park.* C.P. de L. Beyers, Pretoria.

Den Bakker, F.I. 1977. Personal communication (Ranger W.W. Lloyd).

De Villiers, R.M. 1972. *Better Than They Knew.* Purnell, Cape Town.

Doran, M. 1977. Personal communication (Captain J.J. Coetser).

Dreyer, G. 1981. Personal communication (Captain J.J. Coetser).

Duke, T.R. 1977. Personal communication (Ranger T. Duke).

Fourie, P.F. 1985. *Kruger National Park – Questions and Answers.* S.A. Country Life, Nelspruit.

Glen-Leary, H. 1984. Glen Leary's death – new facts. *Custos* 13(2):26–27.

Godfrey, P. 1963. Epitaph for a hunter. *Wide World* April:265–267.

Gordon, D.T. 1977. *The Trollopes of South Africa.* Shuter & Shuter, Pietermaritzburg.

Gunning, J.W.B. 1897. Letter (27 July 1897) to the Board of Trustees (Curatorium) of the State Museum, Pretoria. Transvaalse Archives Document: O.D. 166 (OR 9538/97).

Gunning, J.W.B. 1898. Letter (1 April 1898) from the Director, State Museum, to the Superintendent of Education, Pretoria. Transvaal Archives O.D. 217 (OR 4296/98):44–47.

Harries, C.L. 1933. Letter to Col. J. Stevenson-Hamilton about the origin of the name Punda Maria. Skukuza Archives.

Hodge, H.V. 1986. Memories of 1927: the Boy Scouts go to Malelane. *Custos* 14(11):13–14.

Ireland, E.G. 1929. Letter to Harry Wolhuter in connection with the death of Maj. Fraser. Skukuza Archives.

James, T.L. 1933–1936. Diaries of the ranger at Malelane. Skukuza Archives.

Kingshott, M.J. 1977. Personal communication (Majoor A.A. Fraser and Ranger G.R. Healy).

Kirkman, W.H. 1985. Personal communication.

Klein, H. 1952. *Land of the Silver Mist.* Howard B. Timmins, Cape Town.

Kloppers, J.J. 1967. *Veldwagter!* Janssonius & Heyns, Johannesburg.

Kloppers, J.J. 1972. *My vader wat is 'n held?* Janssonius & Heyns, Johannesburg.

Kloppers, J.J. 1968. *Game Ranger.* Juta, Johannesburg.

Kruger, L.E. 1977. Personal communication (Colonel P. de Jager).

Lamont, J. 1980. Personal communication (Ranger W. Lamont).

Lamont, W. 1946. Dam making and well sinking. Unpublished manuscript, Skukuza Archives.

Ledeboer, L.H. 1921. Letter to Col. J. Stevenson-

Hamilton. Skukuza Archives.

Ledeboer, L.H. 1923–1929. Diaries of the ranger at Mondzwani and Letaba. Skukuza Archives.

Ledeboer, L.H. 1936–1946. Diaries of the ranger at Satara. Skukuza Archives.

Lloyd, W.W. 1977. Personal communication (Ranger W.W. Lloyd).

Lownds, L.E.O. 1931. Toulon. Unpublished manuscript.

Marais, M. 1954. Eensame graf in die Wildtuin. *Die Vaderland* 21 August.

McDonald, H. 1928–1929. Diaries of the ranger at Crocodile Bridge. Skukuza Archives.

Meiring, P. 1976. *Die Wildtuin is hul woning.* Koöperasiepers, Pretoria.

Meiring, P. 1982. *Agter die skerms in die Krugerwildtuin.* Perskor, Johannesburg.

Minnaar, A. 1985. Twee maal in die kloue van 'n leeu. *Custos* 14(8):33–34.

Minnaar, A. 1985. Ledeboer vermink in bloedige geveg. *Custos* 14(9):34–35.

Minnaar, A. 1988. Vergeet ons nie. Unpublished manuscript.

Nel, P. 1985. Personal communication.

Paynter, D. & Nussey, W. 1986. *Kruger: Portrait of a National Park.* Macmillan S.A. (Pty.) Ltd, Johannesburg.

Pease, Sir A.E. 1913. *The Book of the Lion.* John Murray, London.

Personal files of Col J. Stevenson-Hamilton, Capt. E. Howe, H.E. Tomlinson, T.I. James, W.H. Kirkman, Col M Rowland-Jones, H.C. Wolhuter sr, H.C. Wolhuter jr. L.B. Steyn, S.H. Trollope,. H.MacDonald, L.H. Ledeboer, I.J. Botha, G.C.S. Crous and W. Lamont. Skukuza Archives.

Pienaar, U. de V. 1981. The Kruger Park Saga 1898–1981. Unpublished manuscript. Skukuza Archives.

Pienaar, U. de V. 1983. Geheim van die vergete graf. *Custos* 11(10):13–16.

Pienaar, U. de V. 1984. Fihlamanzi se geheim ontrafel. *Custos* 13(7):4–10.

Pienaar, U. de V. 1986. Vergete veldwagters van die ou Sabie- en Shingwedzireservate. *Custos* 15(1):36–37.

Pohl, V. 1986. Personal communication (Ranger F. Streeter).

Pringle, J.A. 1982. *The Conservationists and the Killers.* Books of Africa, Cape Town.

Prinsloo, A.A. 1980. Majoor Fraser van Malunzane. Unpublished manuscript.

Rattray, G. 1986. *To Everything its Season. Malamala. The Story of a Game Reserve.* Jonathan Ball, Johannesburg.

Rogers, W. 1981. Personal communication (Captain J.J. Coetser).

Stevenson-Hamilton, J. 1902. Personal diary of J. Stevenson-Hamilton 2 September 1902–28 June 1903.

Stevenson-Hamilton documents in trust. Skukuza Archives.

Stevenson-Hamilton, J. 1902–1946. Die Stevenson-Hamilton documents in trust. Skukuza Archives.

Stevenson-Hamilton, J. 1907–08. Aantekeningen omtrent die wild reserve aan de Sabi. *Het Transvaalsche Landbouw Joernaal* 4:636–650.

Stevenson-Hamilton, J. 1907. Opposition to game reserves. *The Wild Fauna of the Empire* 3:53–59.

Stevenson-Hamilton, J. 1909. Transvaal Government reserves and regulations. *The Wild Fauna of the Empire* 5:61–68.

Stevenson-Hamilton, J. 1902–1946. Official annual reports of the Warden, Kruger National Park to the South African National Parks Board of Trustees. Skukuza Archives.

Stevenson-Hamilton, J. 1928. Correspondence with rangers A.A. Fraser, H.C. Wolhuter, T. Duke, C.R. de Laporte, and others. Stevenson-Hamilton documents, Skukuza Archives.

Stevenson-Hamilton, J. 1937. *South African Eden.* Collins, London.

Stevenson-Hamilton, J. 1944. *Die Laeveld – sy natuurlewe en mense.* Van Schaik, Pretoria.

Stokes, C.S. 1942. *Sanctuary.* Cape Times Ltd, Cape Town.

Swarts, D.H. 1946–1959. Diaries of the ranger at Satara. Skukuza Archives.

Tattersall, D. 1972. *Skukuza.* Tafelberg, Cape Town,

Thompson, E.A. 1986. Personal communication (Ranger C.R. de Laporte).

Tomlinson, H.E. 1928–1933. Diaries of the ranger at Malelane. Skukuza Archives.

Trollope, S.H. 1925–1928. Diaries of the ranger at Malelane. Skukuza Archives.

Trapido, S. 1984. Poachers, proletarians and gentry in the early twentieth century Transvaal. African Studies Seminar. Paper read on 12 March 1984, University of the Witwatersrand.

Van der Veen, H.C. 1985. Personal communication.

Wallis, J.P.R. 1953. *The Barotseland Journal of James Stevenson-Hamilton 1898–1899.* Chatto & Windus Ltd, London.

Wolhuter, H.C. 1904. In the lion's jaws. Sworn affidavit by the resident magistrate of Barberton, Alfred E. Pease. *Journal of the Society for the Protection of the Fauna of the Empire* Vol. 1.

Wolhuter, H.C. 1948. *Memories of a Game Ranger.* Wild Life Protection Society of SA., Johannesburg.

Yates, C.A. 1925. Brave Dogs of Gameland. *The Cape Argus* 23 May.

Yates, C.A. 1935. *The Kruger National Park.* George Allen & Unwin Ltd, London.

Zway, P. 1986. Early days at Shingomeni. *Custos* 15(5):22.

Chapter 16
16.3 Forgotten ranger posts
Dr U. de V. Pienaar

By the mid-1980s, nature conservation responsibilities in the Kruger National Park fell under a Head: Nature Conservation, assisted by a chief ranger with a corps of 22 section rangers, more than 200 field rangers, nine trails rangers, three traffic officers, two instrument makers, a pollution control officer and a labour liaison officer. In contrast to this, the first warden of the Sabie and Shingwedzi reserves had to control these enormous areas from August 1902 assisted by only four section rangers and 48 field rangers.

During the period from the proclamation of the Sabie Reserve by President Kruger on 26 March 1898 until the outbreak of the Anglo-Boer War on 11 October 1899, control of the area was entrusted to two members of the *Zuid Afrikaansche Republiek Politie* (ZARP), constables Izak C. Holtzhausen of Komatipoort and Paul Bester of Nelspruit. Thus they can be regarded as being the first two rangers of the Sabie Reserve, with Komatipoort and Nelspruit the first ranger posts.

Keeping the inaccessibility of the area and the lack of roads or proper communication routes in mind, one can see why Stevenson-Hamilton had to settle his first white rangers at places where there already was some form of housing available. These places also had to be accessible by existing wagon roads or railway connections, including the Eastern and Selati railway lines. Eventually some of these posts were moved to other locations, while border changes excised some and others became unoccupied for other reasons.

After his arrival in the Sabie Reserve, Stevenson-Hamilton had to move into E.G. (Gaza) Gray's old rat-infested little corrugated-iron room at Crocodile Bridge, if only for a few months.

After Steinaecker's Horse evacuated their base at Sabie Bridge and finally disbanded in February 1903, Stevenson-Hamilton moved there in November 1902. He used the blockhouse at the unfinished bridge over the Sabie just east of the railway line as his new central headquarters. Soon afterwards he built

Paul Bester next to his patrol hut at Sabie Bridge (later Skukuza), 1899.
(ACKNOWLEDGEMENT: MRS ALICE BRIEDENHANN, NELSPRUIT)

16.3 FORGOTTEN RANGER POSTS

an office, complete with a Union Jack up a long, rough pole, just west of the railway line, as well as a compound for his black workers east of the railway line in the vicinity of the current transport depot.

From here Stevenson-Hamilton managed the reserves under his supervision until 1909, when a new building was erected on the later site of the military radio mast.

His first section ranger was the recruiting officer E.G. (Gaza) Gray, whom he had first met at Steinaecker Horse's base at Gomondwane. Stevenson-Hamilton appointed Gray on a temporary basis on 12 August 1902. Gaza Gray moved into an outpost that consisted of six rondavels and a kraal for his horses and donkeys on the site of the current Lower Sabie rest camp. From here he patrolled and supervised the southeastern part of the Sabie Reserve.

On 5 May 1903 Stevenson-Hamilton was informed that Mr Pickard, the head of the Witwatersrand Native Labour Association (WNLA), had transferred Gaza Gray back to Mozambique.

This forced the warden to transfer Thomas Duke to Lower Sabie. Duke had been appointed on 20 December 1902 and was temporarily stationed at Crocodile Bridge. He manned the post at Lower Sabie until his retirement in 1924. The post was then occupied by a ranger Brent for a few months until his resignation in February 1925, after which the post was closed until October 1977, when new quarters

Ranger Thomas Duke's quarters at Lower Sabie, January 1907.
(ACKNOWLEDGEMENT: STEVENSON-HAMILTON COLLECTION, SKUKUZA ARCHIVES)

Ranger C.R. de Laporte and his wife at their original "Bleak House" at Kaapmuiden. October 1907.
(ACKNOWLEDGEMENT: STEVENSON-HAMILTON COLLECTION, SKUKUZA ARCHIVES)

Ranger Harry Wolhuter's quarters at Mtimba in the Sabie Reserve, 1914. It had been built in 1910 and he lived in it until 1940, when he moved into the new quarters at Pretoriuskop.
(ACKNOWLEDGEMENT: STEVENSON-HAMILTON COLLECTION, SKUKUZA ARCHIVES)

16. EARLY YEARS: THE SABIE AND SHINGWEDZI RESERVES AND THE KRUGER NATIONAL PARK 1900–1946

Ranger M.C. Mostert and Sikomela Nkuna at the remains of what had been ranger Harry Wolhuter's post at Mtimba (in the former KaNgwane).
(08.06.1984)

Ranger M.C. Mostert and his wife, Eileen, at the site of former ranger Harry Wolhuther's picket at Doispane.
(08.06.1984)

were built. The first ranger of this "new" post was M.C. Maritz.

Shortly after the temporary appointment of Gaza Gray in 1902, Stevenson-Hamilton appointed the first white permanent officials, namely H.C. (Harry) Wolhuter and Rupert Atmore. According to official records, both were appointed on 17 August 1902, although Tattersall (1972) claims that Atmore was recruited a few days before Wolhuter.

Rupert Atmore was stationed on the northern bank of the Crocodile River across from Kaapmuiden Station and this post was known as the Nsikazi, Kaapmuiden, or Boulders ranger post. It was strategically well situated to control poaching from the White River area, the area adjacent to the Eastern Line and by locals from nearby settlements. Kaapmuiden Station was handy for the sending of mail or to obtain supplies.

Atmore could not adapt to Stevenson-Hamilton's strict discipline and after a dispute about the care of his horses he resigned on 5 September 1902. C.R. de Laporte replaced him and he served at this post from 13 May 1903 until the outbreak of the First World War in 1914. De Laporte moved around quite a bit initially and in 1903 he also served at the outposts at Lower Sabie and the Olifants River.

During his time of service, ranger De Laporte often had to take over from Major Stevenson-Hamilton as acting warden when the warden had to leave Sabie Bridge (Skukuza) for long periods due to vari-

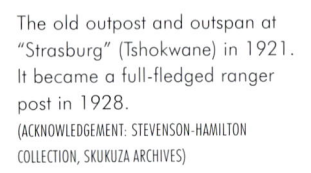

The old outpost and outspan at "Strasburg" (Tshokwane) in 1921. It became a full-fledged ranger post in 1928.
(ACKNOWLEDGEMENT: STEVENSON-HAMILTON COLLECTION, SKUKUZA ARCHIVES)

ous reasons. During his absences from the Kaapmuiden ranger post he was first relieved by Paul Siewert, a new ranger who had been appointed in 1910. Siewert served at Kaapmuiden between September 1910 and December 1911 and again in 1913.

In the meantime, Harry Wolhuter spent a few months at his family's farm near Legogote while getting the permanent post at Mtimba organised. Mtimba was about 16 kilometres west of the current Pretoriuskop camp and ranger post. This post was strategically well positioned after a large section of land west of the Nsikazi River (sections E and F of the Old Kaap blocks) had been added to the Sabie Reserve on 6 April 1906. Because this post was so far from Skukuza and an ox-wagon was the only transport available, Harry Wolhuter built himself outposts at Pretoriuskop and between Pretoriuskop and Skukuza. The latter post was at the kraal of one of his black rangers, Doispane (Dustbin) Nyathi, just west of the Doispane road's current drift through the Mtshawu Spruit.

After a while he built a comfortable home at Doispane of which only part of the foundation is still visible. During his trips to Skukuza he usually outspanned and had lunch at Spanplek, which is about ten kilometres east of Doispane and 16 kilometres west of Skukuza. After the western border was moved back to the Nsikazi River in 1923 and the loss of Kaap blocks E and F, Mtimba became inconvenient as a permanent ranger post for the Pretoriuskop section and an alternative site had to be found. However, Harry Wolhuter's outpost at Pretoriuskop was only improved in 1938, while a permanent home was built for him in 1940.

All that remains of the old Mtimba post today is a few of the foundation stones in the middle of a busy street in a town in the former KaNgwane homeland. A few of the large mango trees that had been planted by the Wolhuters can still be seen, as can the neglected and forgotten grave of Andrew James Wolhuter, Harry's younger brother, in one backyard. In 1918 he had served for a few months at the Kaapmuiden post, but fell victim to the great flu epidemic the same year and was buried there.

During the early years of his service period, Harry Wolhuter had to patrol the whole western and northern part of the Sabie Reserve up to the Olifants River. These patrols were sometimes undertaken by ox-wagon, but more often on horseback with pack donkeys to transport provisions and camping equipment.

Because of the great distances several outposts were established, especially in the area north of the Sabie River, where rangers could safely overnight during these lengthy patrols. Some of these outposts were at Tshokwane (where the tearoom is now), Kumana, Nwanetsi (Satara), Malihane (on the Mavumbye Spruit), Bangu and on the Olifants River (Balule). Of these, Nwanetsi (Satara) became a fully fledged ranger post in 1908 and Tshokwane[1] in 1928. It was during one of these patrols in 1903 that Harry Wolhuter had his famous encounter with two hungry lions at Lindanda. (See chapter 16.2)

After the proclamation of the Shingwedzi Reserve in 1903, Stevenson-Hamilton undertook an extended journey by ox-wagon through this enormous new conservation area. On this journey he discovered a few huts that had been built in 1902 by a

The terrain of the former Boulders ranger post in the Sabie Reserve opposite Kaapmuiden. It was used between 1903 and 1925. (27.07.1984)

---

1   Originally named "Strasburg" after the old farm.

recruiting agent for the mines, situated on top of a hill on the northern bank of the Shongololo Spruit, about three kilometres west of its confluence with the Tsende (Tsendze) River. This place was known as Malunzane and Stevenson-Hamilton considered it a suitable and convenient site for the first ranger to be stationed at in the Shingwedzi Reserve. He already had somebody in mind in the person of Major A.A. Fraser, who was then the ranger in charge of the Pongola Reserve between Swaziland and Natal.

Fraser, a retired soldier and bachelor, was transferred to the Sabie Reserve in February 1904 and manned the Malihane outpost north of Satara for a few months until June 1904. He and his large pack of dogs then moved into his new quarters at Malunzane, and for many years he was the most isolated ranger in the two game reserves. In 1905 and again from 1917 until mid-1919 Fraser acted as warden at Skukuza in the absence of Stevenson-Hamilton, who first went on extended leave and then left for active service during the First World War. He retired in 1920 and left the Malunzane ranger post on 9 November 1920. He spent his last years in his cottage near Hlomela's village, near the eastern bank of the Klein Letaba (about five and a half kilometres southwest of the Nalatsi outpost). He died in January 1929 and was buried at Pietersburg.

During Fraser's absence as acting warden at Skukuza between 1917 and 1919, the Malunzane outpost was manned by another Lowvelder, Frank Streeter.

After Fraser's retirement the Malunzane ranger post was closed. The new ranger, Henry Leonard Ledeboer, was appointed on 1 April 1921 and was stationed at Hatane at the confluence of the Makhadzi Spruit and the Letaba River. Here he and his second wife Maria Christina and their young child led a precarious existence. The reason for his placement at Hatane was mainly to keep an eye on Chief Makuba and his subjects who lived further north along the Makhadzi on the northern flank of the Lebombo Mountains and who were notorious poachers. There was also a possibility that some of Makuba's people had been responsible for the murder of a white shop owner who had had a trading post there. Another reason why the Malunzane post was not manned again was that the perennial waterhole there had dried up during a number of consecutive droughts.

Ledeboer's wife died of a heart attack on 11 August 1921 at Hatane and was buried close by. Because of the dreadful living conditions he and his field rangers had to endure at Hatane, it was decided that they should all move to a new site on the northern bank of the Letaba River, about one kilometre northeast of the current Letaba rest camp. This new post

Maj. A.A. Fraser's quarters at Malunzane along the Shongololo River in the Shingwedzi Reserve, June 1904.
(ACKNOWLEDGEMENT: STEVENSON-HAMILTON COLLECTION, SKUKUZA ARCHIVES)

Ruins of Maj. A.A. Fraser's quarters at Malunzane, along the Shongololo, Shingwedzi Reserve.
(10.06.1959)

Buildings and implements at Maj. A.A. Fraser's ranger post at Malunzane in the Shingwedzi Reserve, 1911.
(ACKNOWLEDGEMENT: STEVENSON-HAMILTON COLLECTION, SKUKUZA ARCHIVES)

16.3 FORGOTTEN RANGER POSTS

A grave, maybe that of ranger L.H. Ledeboer's little girl at Mondzwani rangers post along the Letaba River.
(17.07.1984)

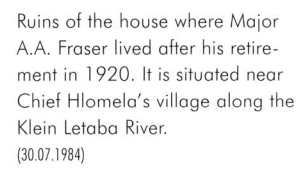

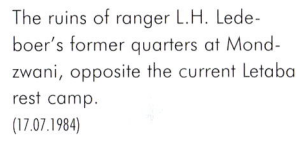

Ruins of the house where Major A.A. Fraser lived after his retirement in 1920. It is situated near Chief Hlomela's village along the Klein Letaba River.
(30.07.1984)

The ruins of ranger L.H. Ledeboer's former quarters at Mondzwani, opposite the current Letaba rest camp.
(17.07.1984)

Ranger L.H. Ledeboer and his second wife, Maria, in front of the door of their dilapidated home at Hatane, at Makhadzi Mouth in the Shingwedzi Reserve, 1921. Mrs Ledeboer had died here on 11 August 1921 and was buried close to the old quarters.
(ACKNOWLEDGEMENT: STEVENSON-HAMILTON COLLECTION, SKUKUZA ARCHIVES)

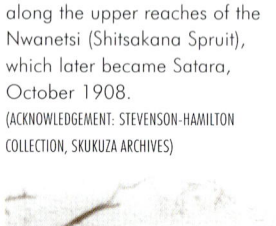

Ranger G.R. (Tim) Healy's camp along the upper reaches of the Nwanetsi (Shitsakana Spruit), which later became Satara, October 1908.
(ACKNOWLEDGEMENT: STEVENSON-HAMILTON COLLECTION, SKUKUZA ARCHIVES)

was named Mondzwani and Ledeboer lived here until November 1927, when he moved into better housing that had been built for him just south of the river (the current storerooms at Letaba ranger post).

Judging by the small grave close to the ruins of the house at Mondzwani, Ledeboer's young daughter died and was buried there. While living at Mondzwani he met and married his third wife, Kate Häsler (Evans). The current Letaba section headquarters were only built in 1932 and were first occupied by ranger G.C.S. Crous.

The first ranger to be stationed in the area between the Sabie and Olifants rivers was G.R. (Tim) Healy. Healy had been a member of the South African Constabulary stationed at Skukuza. Early in 1908 Stevenson-Hamilton managed to get this pleasant and energetic young man appointed as ranger. That year Healy was stationed in temporary accommodation along the Shitsakana Spruit, a tributary of the Nwanetsi. This so-called Nwanetsi ranger post later developed into Satara rest camp and the current ranger post.

The original living quarters of the older rangers were nothing more than thatched clay and mud huts with whitewashed walls and only the barest amenities. In 1909 and 1910, houses of corrugated iron and wood, each resting on a series of metre-high concrete pillars, were built at five of the ranger posts. Wide, netted verandas kept the sun from the corrugated-iron walls and the buildings were surprisingly cool, even during the worst of the summer heat. These new quarters were built at Skukuza (for the warden), Lower Sabie, Nwanetsi (Satara), Kaapmuiden and Mtimba. Stevenson-Hamilton wrote the following about Healy's new home:

> The building of Healy's new house at Nwanetsi (later known as Satara) by two young German contractors, had been a long and arduous proceeding. It was 55 miles from the nearest point on the Selati line at Sabie Bridge, with no connecting road or track. All the material had to be conveyed in the warden's light wagon drawn by six oxen.

However, these buildings would serve their purpose for many years and the Skukuza quarters served as the warden's home until 1956. The house at Satara was only replaced in February 1962 and sections of it are still standing.

Ranger Healy had an enormous district to patrol before the excision of a large area between the Sabie and Olifants rivers in 1923. For this reason he built himself a fairly comfortable outpost about 700 metres southeast of Rolle Station on the Selati Line in 1910. This post gradually became an important ranger post and was often inhabited for longer periods than the better quarters at Nwanetsi (Satara). This was mainly as a result of the convenience afforded by the proximity of the train, which could deliver post and provisions literally to the back door of the ranger, as well as the telegraphic contact with Sabie Bridge (Skukuza).

The Rolle ranger post was originally manned by ranger G.R. Healy (1910–1914) – afterwards he joined the Allied Forces and was killed in action on 26 March 1916 in German East Africa (Tanzania). He was succeeded at Rolle by rangers Paul Siewert (1914–1918), Thomas Duke and C.R. de Laporte (1919), Colonel P.L. de Jager (1919–1920), W.W. Lloyd (1920) and Jack Brent (1920–1923).

After the excision in 1923 of the area between the North Sand and Klaseri rivers east of the current western border of the reserve, Rolle post was closed and ranger W.W. Lloyd settled at Nwanetsi (Satara) on a permanent basis. The daughter of Colonel Piet de Jager, the ranger stationed at Rolle from 1919 to 1920, described the house itself as consisting of two rooms, two rondavels and a kitchen built of mud and reeds – a typical wattle-and-daub house! Nothing is left of this building. Close to the terrain is a Lowveld mangosteen (*Garcinia livingstonei*) that had been almost completely destroyed by local medicinal healers. There is also a big marula tree with many bullet holes in the trunk. "Madubula" (the name which the locals had given De Jager) and some of the other rangers had certainly used this particular tree to set their rifles. In the Skukuza Archives is a photograph of Mrs Healy – an enthusiastic rider – practising gymkhana on a course laid out for her by her husband close to the quarters at Rolle. In 1923 ranger Brent still helped Sergeant D'Arcy of the SAP to keep an eye on one Pritchard, who wanted to search for hidden treasures in the vicinity of Rolle.

Another ranger post in the area between the Sabie and Olifants rivers that was manned during the early years of the Sabie Reserve's existence but has since been abandoned, is the Msutlu ranger post. This post, situated on the northern bank of the Sabie River

The ranger quarters along the upper reaches of the Nwanetsi (Shitsakana) – now Satara, 1920.
(ACKNOWLEDGEMENT: STEVENSON-HAMILTON COLLECTION, SKUKUZA ARCHIVES)

Rolle ranger post in the Sabie Reserve, circa 1920. It had successively been managed by rangers G.R. (Tim) Healy, P. Siewert, T. Duke, C.R. de Laporte, P.L. de Jager and W.W. Lloyd, since 1910 until its closure in 1923.
(ACKNOWLEDGEMENT: MR P. DE J. MOUTON, BEDFORDVIEW)

16. EARLY YEARS: THE SABIE AND SHINGWEDZI RESERVES AND THE KRUGER NATIONAL PARK 1900–1946

A large marula tree with bullet holes in its trunk on the terrain of the former Rolle quarters, Sabie Reserve.
(04.06.1984)

Mrs G.R. Healy on her horse at the former Rolle ranger post in the Sabie Reserve, 1914.
(ACKNOWLEDGEMENT: STEVENSON-HAMILTON COLLECTION, SKUKUZA ARCHIVES)

Ranger Paul Siewert's quarters at Msutlu along the northern side of the Sabie River, circa 1913.
(ACKNOWLEDGEMENT: STEVENSON-HAMILTON COLLECTION, SKUKUZA ARCHIVES)

about five kilometres northwest of Skukuza, was settled in 1911. Until 1914 ranger Paul Siewert was stationed here to control the area between the Sabie and Olifants rivers. Msutlu was created because this area was almost impossible to reach from Skukuza in especially the summer months when the Sabie River was in flood (until the completion of the railway bridge). With the outbreak of the First World War Siewert was transferred to Rolle ranger post, as ranger Healy had joined the South African forces and had been sent to the front line in East Africa. Msutlu was never manned by a section ranger again, although field rangers maintained it as an outpost until the early 1960s.

Kaapmuiden (Boulders) ranger post was temporarily evacuated in 1914 by ranger C.R. de Laporte while he acted as warden at Skukuza until 1916, when he also left for active duty. It was, however, temporarily manned by acting ranger Andrew James Wolhuter in 1918, until he fell ill with Spanish flu and died in his brother Harry's house at Mtimba.

Afterwards ranger W.W. Lloyd manned the post for a while in 1919 until ranger De Laporte could resume his duties there. De Laporte had returned to Skukuza from the war in 1918, but again acted as warden in Colonel Stevenson-Hamilton's stead at Skukuza until Stevenson-Hamilton completed his military duties at the end of 1919. He could then resume his duties as warden of the Sabie and Shingwedzi reserves. De Laporte was permanently transferred to Crocodile Bridge in 1920, after he had also relieved at Rolle for a while when Colonel Piet de Jager was transferred from Rolle to Kaapmuiden section. He was stationed there from 1920 until 1924, after which he was transferred to Punda Maria. Ranger Harry Brake was stationed at Kaapmuiden from July 1924 to January 1925. In 1923, the Old Kaap Blocks E and F in the Nsikazi area west of Kaapmuiden were cut out of the Sabie Reserve. After that, there was no point in keeping the Kaapmuiden post going. It had already been decided in 1924 to tear down the building and move it to Malelane (where the current ranger post is). Contractors completed this task in 1925. This was the beginning of Malelane ranger post, with ranger Harold Trollope as the first incumbent. He was appointed on 27 May 1925.

Another interesting temporary outpost in the southern part of the Kruger National Park was the Mahula outpost of ranger Bert Tomlinson, who served at Malelane from 1928 until 1933. He built this post in 1930 on the northern bank of the Mahula Spruit, about 700 metres east of the old Skukuza-Malelane road. The purpose of this outpost was to keep an eye on the activities of the eccentric foreman Kemp of Kemp's Cottage and his workers, who were known not to object to the disappearance of a buck or two from time to time. The remains of this comfortable post, the well and the cement bath that Tomlinson had built for himself, can still be seen.

Shortly before Fraser's retirement as the only section ranger in the old Shingwedzi Reserve, Stevenson-Hamilton decided to create another ranger post in

The remains of ranger Paul Siewert's quarters at Msutlu along the Sabie River, northwest of Skukuza.
(16.07.1984)

The foundation pillars of the ranger quarters at Boulders, opposite Kaapmuiden. The rest of the building was moved to Malelane in 1925.
(27.07.1984)

The ranger's quarters at Malelane, 1925. The section had successively been managed by rangers S.H. Trollope, H.E. Tomlinson and T.L. James until August 1951, when it burnt down.
(ACKNOWLEDGEMENT: STEVENSON-HAMILTON COLLECTION, SKUKUZA ARCHIVES)

Ranger H.E. Tomlinson's camping site along the Mahula Spruit near Kemp's Cottage. This site was utilised during the early 1930s. (16.07.1984)

The site of ranger J.J. Coetser's original camping spot at the fountains at the foot of Dzundwini Hill. He moved from here to Punda Maria in 1919. (27.04.1984)

Ranger J.J. Coetser's quarters at Punda Maria, October 1922.
(ACKNOWLEDGEMENT: STEVENSON-HAMILTON COLLECTION, SKUKUZA ARCHIVES)

the far northern part of this area. The main purpose of the new post would be to curtail illegal hunting and the recruitment of black workers for the mines in the notorious Crooks Corner, as well as to establish better control over this remote area in general.

The appointed ranger, Captain J.J. Coetser, whose story is told in chapter 16.2, assumed his duties on 1 May 1919. As there were no beaten roads, he could not reach his station by vehicle and a plan had to be devised. In the end his personal belongings were carried along the ancient trade route that went past Dzundwini Hill in the direction of Mozambique. Coetser set up a temporary camp at the Dzundwini Fountain, under the giant sausage tree that is still there today. He stayed there for a while, but as winter progressed the water in the fountains decreased and became so brackish that it was impossible to drink. In the meantime he had heard about a strong fountain in the vicinity of Chief Shikokololo's village, about 18 kilometres northwest of his camp, so he moved there. He built a permanent camp at this picturesque setting at the foot of Dimbo Hill, close to the bubbling fountain, and then his family joined him at the post he named Punda Maria.

The origin of this unusual name is explained in chapter 16.2. All that remains of Coetser's original campsite at Dzundwini Fountain is a few bullet holes in the trunk of the large sausage tree. These bullets bear silent witness to the first phase of the settlement of a new ranger post that would become famous in later years for more than one reason.

In 1964 the departments of Customs and Excise and Justice made an amount of R69 000 available for the erection of a number of border posts at strategic points along the international borders of the Kruger and Kalahari national parks. After careful consideration it was decided to build two posts, one at Nwanetsi and the other at Mhalamhala, just north of Ramiti Beacon on the eastern border between the Olifants and Shilowa poorts. There was a footpath at the latter point that had been used for many years by people living in the Massingire district in Mozambique to cross the Park's border and enter South Africa in search of work. The creation and manning of these two border posts would control the to and fro of the illegal traffic across the Mozambican border much more efficiently than had been the case until then. Both border posts were completed in 1964 and the new ranger post at Mhalamhala was named Klipkoppies. The first ranger to man the post was S.C.J. Joubert (later warden of the Kruger National Park). He served here from 1 November 1964 until January 1967, after which he left on study leave to Pretoria. After that the post was manned consecutively by rangers S.G. Botha (1 February–31 December 1969), E.A. Whitfield (1 April–30 June 1970) and T.B. Mollentze (1 July–31 October 1973). The post was then closed as traffic across the border at this particular point had not been allowed for years, the only legal entry points from Mozambique being Ressano Garcia, Nwanetsi and Pafuri. The security situation along the Mozambican border had also been deteriorating steadily before the official Frelimo takeover in 1975 and a remote border post so close to the border became a security risk. Aside from these factors, the entry road to the post also gave them a lot of problems, while during the rains of the 1970s a fountain started bubbling from underneath the house's foundations that eventually damaged the house so badly that it became unsafe for human habitation. All these reasons contributed towards the closure of the post. The living quarters were demolished, but the outbuildings were left to be used during border patrols.

As an alternative to Klipkoppies, a new ranger post was built on a small hill in the Mooiplaas area in 1973–1974. Ranger B.R. Bryden moved into these quarters on 30 April 1974 as the first section ranger to be stationed at the new post. The Nwanetsi border post was retained and, after the Portuguese had ceded Mozambique to Frelimo in 1975, it became the only legal passageway between the Park and Mozambique.

Although the old ranger posts do not exist anymore, each of them played an honourable role in the pioneering history of the Kruger National Park.

Ranger J.J. Coetzer's camp at Punda Maria. October 1922.
(ACKNOWLEDGEMENT: STEVENSON-HAMILTON COLLECTION, SKUKUZA ARCHIVES)

The ruins of Klipkoppies ranger post which was used between 1964 and 1973.
(18.06.1984)

References

Ackerman, D.J. 1981. Veteran game ranger. *Custos* 10(1): 17.

Ackerman, D.J. 1981. Nombolo the brave. *Custos* 10(6): 17–19.

Adendorff, G.M. 1984. *Wild Company.* Books of Africa, Cape Town.

Coetser, E. 1981. Personal correspondence.

De Leeuw Beyers, C.P. 1965. *Kruger Park Adventures.* Living World Films, Sandton.

De Leeuw Beyers, C.P. 1970. *Lurking Danger in the Kruger Park.* C.P. de Leeuw Beyers, Brooklyn, Pretoria.

Kirkman, H. 1984. Personal correspondence.

Kruger, L.E. 1977. Personal correspondence.

Pienaar, U. de V. 1986. Vergete veldwagters van die ou Sabie- en Shingwedzireservate. *Custos* 15(1).

Prinsloo, H. 1980. Majoor Fraser van Malunzane. Unpublished manuscript, Skukuza Archives.

Rogers, W.H. 1981. Personal correspondence.

Stevenson-Hamilton, J. 1902. Diary of J. Stevenson-Hamilton, 2 Sept. 1902 – June 1903. Skukuza Archives.

Stevenson-Hamilton, J. 1902–1925. Annual reports of the Warden of the Kruger National Park to the National Parks Board of Trustees. Skukuza Archives.

Stevenson-Hamilton, J. 1974. *South African Eden.* Collins, London.

Tattersall, D. 1972. *Skukuza.* Tafelberg, Cape Town.

Chapter 16
16.4 The early developmental history of the Sabie and Shingwedzi reserves and the Kruger National Park, 1898 to 1946

Dr S.C.J. Joubert

The history of the Lowveld, especially of the area that is now the Kruger National Park, provides a backdrop against which the early development of the Sabie and Shingwedzi reserves, as well as the later Kruger National Park, can be measured. This history has been discussed in chapters 1 to 15.

With the proclamation of the Sabie and Shingwedzi reserves at the beginning of the twentieth century, the concept of nature conservation was provided with a tangible foundation, although the Pongola Reserve had already been proclaimed a few years earlier in 1894 and the Hluhluwe and Umfolozi reserves in 1897. However, the establishment of these two large reserves in the Transvaal Lowveld created many problems: there were still several black settlements inside the areas occupied by these reserves; they were the preferred hunting areas of many white hunters who had a very vague understanding of nature conservation (and some were downright hostile); there was no administration to control the areas; and, to crown it all, the concept of nature conservation was still in its infancy – to such an extent that no distinct policy or direction could be coupled to it. Add to that the Lowveld's sweltering climate, the danger of malaria, the lack of accommodation and navigable roads, and it becomes clear that the establishment of the Sabie and Shingwedzi reserves was an almost unreachable dream that could be achieved only by the most devoted and determined supporters of the newly-born nature conservation concept!

Fortunately the opportunities also brought men of action to the fore. The first few years were very difficult. Between 1898 and 1902 it was unsuccessfully attempted to at least discourage poaching and hunting, while the development and application of a more encompassing nature conservation policy only happened after the appointment of Major J. Stevenson-Hamilton in 1902.

The beginning ...

To properly understand the early development of the nature conservation philosophy, management practices and general administration, it is also necessary to form a picture of prevailing conditions at the time

Maj. Stevenson-Hamilton and his pack donkeys wading through the Letaba River during a patrol in August 1911.
(ACKNOWLEDGEMENT: STEVENSON-HAMILTON COLLECTION, SKUKUZA ARCHIVES)

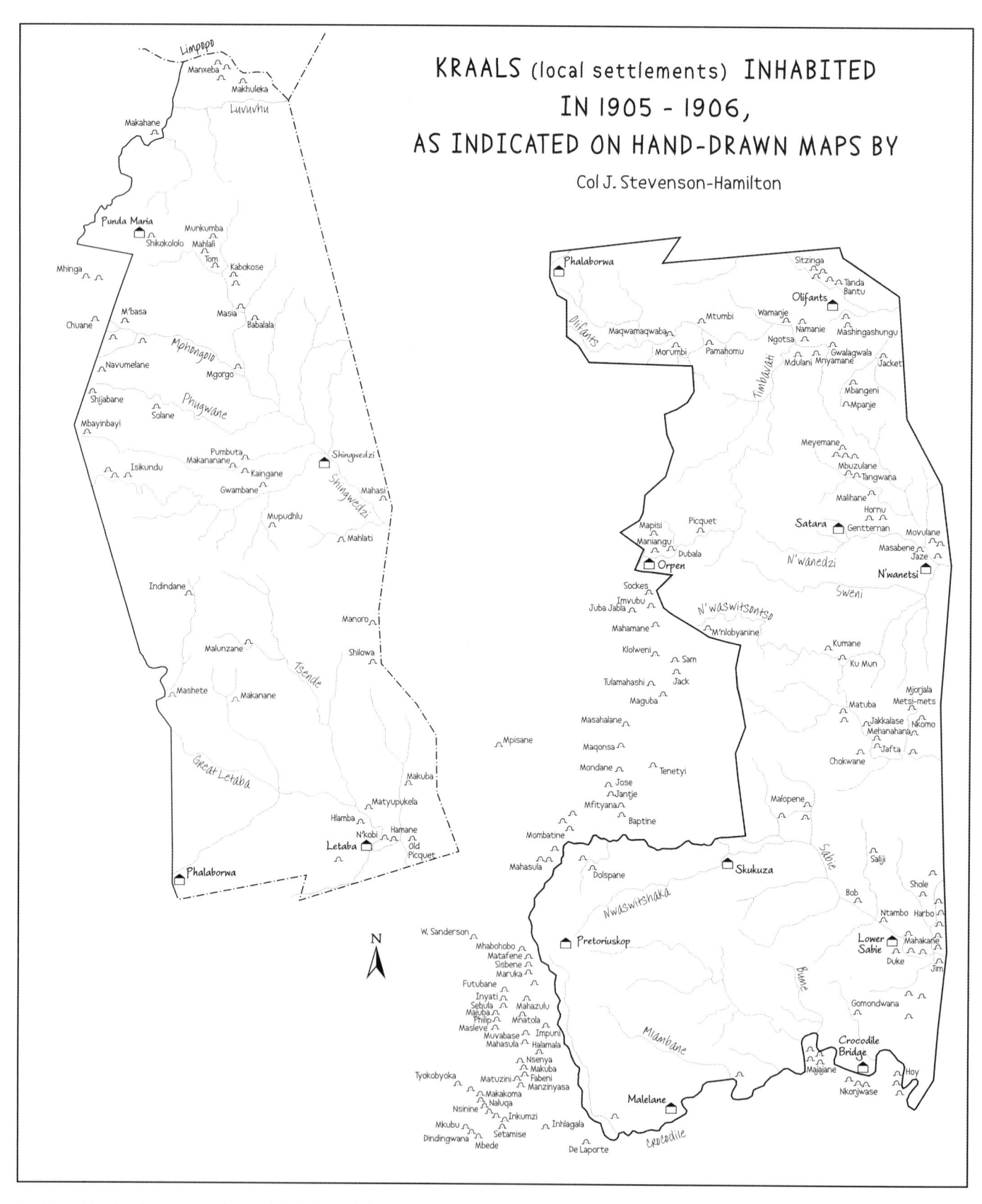

16. EARLY YEARS: THE SABIE AND SHINGWEDZI RESERVES AND THE KRUGER NATIONAL PARK 1900–1946

Localities of local settlements in and around the Sabie and Shingwedzi reserves 1905–1906.

16.4 THE EARLY DEVELOPMENT OF THE SABIE AND SHINGWEDZI RESERVES AND THE KRUGER NATIONAL PARK

of the proclamation of the Sabie and Shingwedzi reserves.

In 1903 Stevenson-Hamilton differentiated between the areas west and east of the Selati railway line. West of the line the rainfall was higher and more reliable and the soil fertile. East of the railway line the climate was considerably drier, there was little perennial water and the agricultural potential therefore much lower. These conditions led to a high density of people living on a permanent basis in the first area. In the eastern area the population density was considerably lower and the area was utilised mainly during the summer months. These seasonal movements were mostly opportunistic in nature – both the Portuguese and the ZAR (South African Republic) tax collectors could be avoided in this manner, while there were ample opportunities to hunt. Agricultural activities played a minor role and the inhabitants only tried to provide for the winter months. This nomadic lifestyle of the black population was summed up by Stevenson-Hamilton (1903) as follows:

> None of these people are, nor ever have been, in any sense a settled population, but have merely wandered about after the game and lived entirely by hunting, but for the attractions of which no natives would have been disposed to live in this inhospitable and unhealthy country. Few have ever paid taxes to either Portuguese or Boer Government in the past;

they possess neither livestock, nor do they attempt to cultivate except for their immediate wants, and in all probability when tax-collecting time comes, many will cross into Portuguese territory hoping to return when the danger had passed.

During his first visit to the Shingwedzi Reserve in 1903, Stevenson-Hamilton found many black settlements along the Letaba and Luvuvhu rivers, as well as along the Tsende (Tsendze) and Shingwedzi rivers.

One thing that really annoyed Stevenson-Hamilton in the early years was the "professional" hunters. Every year hordes of hunting parties descended on the Lowveld from the higher-lying adjacent areas. These hunters did not hunt only to provide for their own needs, but also earned an income from

Maj. Stevenson-Hamilton's office and living quarters at Sabie Bridge (Skukuza). August 1905.
(ACKNOWLEDGEMENT: STEVENSON-HAMILTON COLLECTION, SKUKUZA ARCHIVES)

An ox-wagon in the drift at Sabie Bridge (Skukuza) during the winter months of 1904.
(ACKNOWLEDGEMENT: STEVENSON-HAMILTON COLLECTION, SKUKUZA ARCHIVES)

the sale of skins, horns and biltong. As a result they were unscrupulous, showing little regard for the animal's species, age or sex.

Halting the hunting activities of both whites and blacks was an almost impossible task, as Stevenson-Hamilton (1903) himself said:

> The difficulties of protecting game in the Low Veldt are serious, the country is so immense, so thickly covered with bush, and so wild, that to altogether put a stop to individual poaching – especially with the Portuguese border so near at hand – is practically impossible; all we can do is catch offenders now and then, and by getting exemplary punishment inflicted, rely on putting a salutary fear of a similar fate into the minds of would-be law breakers.

At the time of the proclamation of the two reserves, the Selati Line was already completed up to Newington, although there was no bridge over the Sabie River until 1912. This railway line was the only entry route to the Sabie Reserve. Although there were ancient trade routes that crossed the reserve, none of these were suitable for vehicles.

When the "Gouvernement Wildtuin" ("Government Game Reserve") was proclaimed in 1898, there was no formal accommodation other than the traditional thatched huts used by the black inhabitants. The living quarters of the first Lowveld pioneers such as Joubert and Albasini, as well as the trading posts along the trade routes, had long since been evacuated and were in ruins.

During the Anglo-Boer War the military unit known as Steinaecker's Horse built a blockhouse near the uncompleted train bridge across the Sabie River at Sabie Bridge (Skukuza). One of its officers, Captain Gaza Gray, also built a simple wood and corrugated-iron hut at Crocodile Bridge, as well as a camp consisting of a number of comfortable thatched rondavels at Lower Sabie. When Stevenson-Hamilton arrived here in 1902, the area was truly wild!

Certain deductions about the environmental conditions at the beginning of the century can be made from the early reports of people such as Stevenson-Hamilton and Captain Greenhill-Gardyne of Steinaecker's Horse. Water sources, especially perennial water, were relatively scarce. In as early as 1903 Stevenson-Hamilton reported that even the biggest seasonal spruits, like the Nwanetsi and Nwaswitsontso, had a run-off only after heavy rain.

Apart from the bigger rivers it was difficult to find perennial water sources. In 1906 Stevenson-Hamilton gave the following overview of the water supplies:

> Water is scarce, a few large perennial streams water the country – flowing from west to east, and being, generally, direct or indirect affluents of the Limpopo. There are, besides, a multitude of watercourses varying from the size of considerable rivers

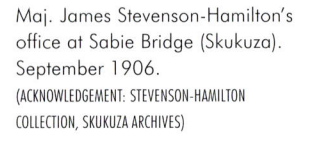

Maj. James Stevenson-Hamilton's office at Sabie Bridge (Skukuza). September 1906.
(ACKNOWLEDGEMENT: STEVENSON-HAMILTON COLLECTION, SKUKUZA ARCHIVES)

Panorama with Manungu, Pretoriuskop and Shanbeni Hill in the background. October 1902. (ACKNOWLEDGEMENT: STEVENSON-HAMILTON COLLECTION, SKUKUZA ARCHIVES)

to that of small drains, all only holding water during the rainy season, and then only for a few days after each deluge of rain, and in the larger of which, during the remainder of the year, water is only found with difficulty, in stagnant pools or by digging in the sandy beds …

There are no perennial springs known in the low country, though these are not infrequently to be met with amongst the foot hills (of the Drakensberg).

In his annual report for 1912 he also described the Lowveld as a region "where water is often, even under the most favourable conditions, difficult to find away from the perennial streams". In later annual reports (see the annual reports of 1912, 1913, 1925, 1932 and 1944) he also expressed the conviction that the Lowveld underwent progressive desiccation in the period immediately before and during the 44 years of his service (see also Pienaar, 1985).

Little mention is made of the vegetation in those early years and most descriptions are rather vague. Stevenson-Hamilton (1903) pointed out that there was a dearth of open veld except in the far west and a small strip along the Lebombo Mountains, and that the rest of the area was covered in dense bush.

In reference to the area north of the Sabie River, the trees and shrubs were described as being much sparser than that south of the river, and a man on horseback could ride through it with relative ease. From descriptions of the Pretoriuskop area by the old hunters (see Sandeman, 1880; Vaughan-Kirby, 1896; Glynn, 1929; and others), it is clear that the vegetation of this area was considerably more open than is the case today. Stevenson-Hamilton's reports show that the bush encroachment in these parts had already started in the 1930s. Indications that the vegetation of the Lowveld were more open in earlier years were also discussed in detail by Pienaar (1985).

In the last years of the nineteenth century poaching and hunting activities by both white and black inhabitants had decreased the game numbers considerably. White rhino had disappeared from the area even before the end of the century and several other game species faced the same fate. As a result of the favourable climatic conditions, especially the higher and more reliable rainfall, and the resulting higher population density in the area west of the Selati railway line, the game there suffered greater losses. In the drier eastern parts with nomadic black inhabitants, conditions for the survival of game were better and game could still be found there in greater quantities. However, the game conditions weren't satisfactory anywhere and in 1906 Stevenson-Hamilton described the general condition in his annual report as follows:

… the game was found to be in a far from satisfactory condition; it was scarce, timid, and constantly on the move, a state of things not conducive to an adequate increase of species. Some kinds of animals were upon the verge of disappearance, and it was possible to cover very large tracts of country, especially in the western portions where the winter hunters had been most active and numerous, without seeing a sign of life, nor even a solitary spoor. The bush in such places seemed 'dead'. Moreover the carnivorous animals had not suffered in relation to the game.

It may seem surprising, but it is nevertheless a fact, that it was rather the exception than the rule for white hunters to devote any attention whatever to the predatory animals ... Native hunters practically never wantonly interfered with the larger carnivora; it is usual with natives to destroy only such individuals as have made themselves a nuisance by attacking livestock, a rare occurrence so long as there is game about.

The great flood in the Sabie River on 23 March 1925. The flood had already subsided 3 metres by the time this photograph was taken.
(ACKNOWLEDGEMENT: STEVENSON-HAMILTON COLLECTION, SKUKUZA ARCHIVES)

The newly-completed warden's quarters at Sabie Bridge (Skukuza) 25 November 1909. Note the Stevenson screen for meteorological observations in the foreground.
(ACKNOWLEDGEMENT: STEVENSON-HAMILTON COLLECTION, SKUKUZA ARCHIVES)

The sharp decline in the numbers of the great game herds during the second half of the nineteenth century had also been pointed out by old hunters such as Vaughan-Kirby (1929). At the beginning of Stevenson-Hamilton's period of service, the populations of most of the herbivores had been drastically reduced and as a result he was of the opinion that the predator populations were disproportionately high when compared to the prey populations.

Against this shaky background the Sabie and Shingwedzi reserves took their first uncertain steps to meet the future.

Yet, from these humble beginnings, a great national park would develop. At the end of his career Stevenson-Hamilton could look back with satisfaction and compare the Sabie and Shingwedzi reserves with a Cinderella that had become a princess. However, this metamorphosis took a very long time to complete, with many difficulties to overcome.

Nature conservation: a new concept

In 1902, shortly after the end of the Anglo-Boer War, Sir Godfrey Lagden was appointed Secretary of Native Affairs. In this position the Sabie Reserve became his responsibility. It was also Sir Godfrey, a nature lover, who clinched Stevenson-Hamilton's appointment. At that stage neither he nor Stevenson-Hamilton had any idea what nature conservation entailed, and when Stevenson-Hamilton asked Sir Godfrey what would be expected of him, he could not give a clear answer. The best advice he had to offer was that Stevenson-Hamilton should make himself "generally disagreeable". At best, the concept of nature conservation then only meant that hunting and poaching within the proclaimed area should be stopped. Little did they know what new perspectives would be born of this adventure.

To obtain a better understanding of and appreciation for the development of a nature conservation and management concept, it is important to take a look at the observations made through the years about natural phenomena, and the interpretation thereof. In this instance, the following should be considered:

Climate
At the beginning of the twentieth century there were no weather stations in the Lowveld, nor any infor-

mation that could be used to gain a better insight of the general weather patterns. Admittedly, enough was known about the general attributes of weather patterns in this region, such as a definite summer season with high temperatures and rainfall, compared with a cooler, generally frost-free and dry winter season. However, there was no information available to describe the extent of temperature fluctuations, rainfall patterns or any other climatic attributes. Stevenson-Hamilton started recording rainfall measurements from the beginning. In 1908, a few years after he had settled at Sabie Bridge, the first official rain gauge and Stevenson screen for the recording of temperature fluctuations, humidity, etc. were installed. Eventually rain gauges were also installed at other ranger posts.

From the information available, it appears that the first few years of the Sabie Reserve's existence, until 1916, were exceptionally dry. This dry period was preceded by a period between 1890 and 1895 of exceedingly high rainfall (Sandenbergh, 1947). This variation in the intensity of rainfall convinced Stevenson-Hamilton that the Lowveld was subject to a gradual process of desiccation. This conclusion would later play an important role in the development of the conservation management policy, especially regarding water provision. In his annual report of 1913, Stevenson-Hamilton for the first time hinted that the series of dry years might be part of a rainfall cycle. As the years went by and more information could be gathered about the consecutive high and low rainfall periods, the concept of rainfall cycles took root even further. By the end of his career Stevenson-Hamilton had no doubts about this. He was also able to ascribe similar tendencies in vegetation and animal populations to the varying high and low rainfall cycles and the associated changes in natu-

Leeupan, near Tshokwane – a popular tourist attraction since the earliest days.
(18.07.1984)

Pack-donkeys enjoying a break at a waterhole during a patrol in 1911.
(ACKNOWLEDGEMENT: STEVENSON-HAMILTON COLLECTION, SKUKUZA ARCHIVES)

ral water resources. He elaborated on this view in his annual reports of 1937 and 1944. Seen in the light of the meagre information at his disposal, his identification of climatic cycles and their role in natural systems must serve as an excellent example of Stevenson-Hamilton's acute powers of observation.

Later analyses of the rainfall figures of this period confirmed the occurrence of rainfall cycles. With the available rainfall data for the period 1908 to 1946 (unfortunately a few years' data, between 1902 and 1925, are missing), Gertenbach (1980) could show that a dry (low rainfall) cycle occurred between 1908 and 1913 (it could, however, have lasted longer on either side) and again between 1926 and 1933, while a third dry cycle started in 1943. High rainfall cycles occurred in the periods 1917 to 1924/25 and 1934 to 1942.

Water resources

As in the case with the climate regime, it is a difficult task to obtain a reliable picture of the natural water resources and their permanency in the early years of the twentieth century.

Pienaar (1985) gives a comprehensive summary of all the available information on water distribution and perennial water resources based on the prehistoric and historical distribution of the San (Bushmen) and black tribes in the Lowveld; the trajectories of ancient trade routes; and conditions recorded by early explorers such as F. de Cuiper (1725), the different Voortrekker reconnaissance patrols (1836–1848), Fernandes das Neves (1860), St Vincent W. Erskine (1868), Carl Mauch (1870), E. Cohen (1875), hunters like Henry Glynn, F. Vaughan-Kirby, E.F. Sandeman, B. Lotrie, C. Bernard and others, surveyors G.R. von Wielligh and M.C. Vos (1890) while surveying the eastern border, old transport riders and explorers such as W. Napier, H.L. Hall, O. Basson, I. Holtzhausen and others, the Boer General Ben Viljoen, delvers and prospectors such as B. Lusk, C. Gerber and others, and by Stevenson-Hamilton himself during his period of service from 1902 to 1946.

Pienaar (1985) speculated in depth about statements by Stevenson-Hamilton and other Lowvelders about a progressive desiccation of the Transvaal Lowveld over a period of 100 years, and the possible occurrence of long and short-term climatic cycles of above and below average rainfall.

In preceding chapters more information is given about the occurrence and permanence of naturally-occurring water in the Lowveld, as well as the condition of the vegetation in the nineteenth century, until Stevenson-Hamilton's appointment as warden of the Sabie Reserve in 1902.

Information obtained from the map that had been drawn up by the surveyor-general of the ZAR (South African Republic), G.R. von Wielligh, as well as from the diary that surveyor M.C. Vos kept during the survey of a large part of the eastern border in 1890, indicates that the spruits near the eastern border, between the Crocodile and Sabie rivers, were dry during their trip in June 1890. North of the Sabie River they found several pools in the Mlondozi River, but little running water. Further north in the vicinity of Muntshe, water was scarce and they only found sufficient water for use in the Shilolweni or Nkelenga Spruit. Although the larger spruits were mentioned by name, including the Nwaswitsontso and Metsi-Metsi, Vos did not mention any water in them. Running water was recorded for spruits that, according to Von Wielligh's map (and judging by their camping sites) were the Makongolweni, Nungwini, Shishangani and Nwanetsi. On their trek further northwards, they found water in the upper reaches of the Mbatsana Spruit, after which they travelled through a very dry area before finding water again in the Mabyebye, in the upper reaches of Bangu Spruit.

From Stevenson-Hamilton's early reports it also appears that perennial water, apart from that found in the larger rivers, was very scarce. In fact, in 1903 he already referred to the low rainfall and the resulting scarce water resources in the area that is currently the central region of the Park – in other words, the area between the Sabie and Olifants rivers. Even in the main spruits in the area, for instance the Nwaswitsontso, Timbavati and Nwanetsi, water was limited to a few sparsely distributed waterholes during the winter months. In 1912 he reported as follows about the availability of water:

The lack of rain during the past two years had been responsible for the drying up of many pools usually considered to be permanent, and this, in a country where water is often even under the most favourable conditions, difficult to find away from the perennial streams ...

The relative lower density of game in the former Shingwedzi Reserve was also ascribed to the lack of water in the area, although the Tsende (now Tsendze) and Shingwedzi rivers had perennial waterholes. The area that is currently the southern region of the Park (between the Sabie and Crocodile rivers) was also described as a dry area in which only a few isolated waterholes could be found in the larger spruits during the winter months.

Stevenson-Hamilton was convinced that the Lowveld was subject to a process of gradual desiccation. He even aired the opinion that certain parts were changing into a desert. This opinion was based on the fact that waterholes that had been regarded as perennial, were gradually drying up. In 1913 he pointed out that there were reports every year that more and more waterholes were drying up. He further based his opinion on the evidence of black inhabitants of the Lowveld who maintained that spruits that were dry and overgrown used to hold perennial water in their younger days. In 1925 he even pointed out that "the native name 'Manzendlovu' [Nwatindlopfu] given to a now dry, sand river, would indicate that in former days it was either permanently running, or held pools large and deep enough to satisfy the water-loving elephant".

By the end of his career, Stevenson-Hamilton nevertheless accepted the occurrence of low and high rainfall cycles and the influence thereof on vegetation and animal populations. With reference to the cyclic nature of rainfall and the impressions it could leave, Sandenbergh (1947) explained the situation with regards to the availability and distribution of water resources at the beginning of the century as follows:

No records (rainfall) within the Park go back more than 40 years; but, from old residents and longer established meteorological stations outside the Park, we know that between 1890 and 1895 there occurred a period of extremely high rainfall – which was general over the whole country. In an area with the Park's topography these wet years must have created considerable areas of swamp, vlei and pan. All streams must have been very strong and seasonal rivers must have run for a long time. Perennial rivers certainly reached enormous flood levels and all their pools scoured out to a considerable depth. This must have kept the country in what might, possibly, have been an 'artificial' condition for many years. First impressions are lasting impressions and the paradise for wildlife that the Park was in the early years – possibly directly attributable to this wet period – lives forever in the memories of those who knew the Park in its early years.

Sandenbergh's impressions about the earlier distribution of natural water in the Lowveld are based on only seven years' experience and should be read in conjunction with Pienaar's data (1985).

In light of the available information, or lack thereof, one can speculate about the historical distribution of perennial water resources in the Park. If it is kept in mind that the accessibility of the area was very limited, it stands to reason that the early inhabitants of the Lowveld would not have had a complete view of either the distribution or the permanence of all the water resources. It is clear that there has been a shortage of perennial water resources over large areas for certain periods since the earliest years and that this played an important role in the later evolution of management strategy.

Vegetation
Unfortunately very little information is available about the appearance of the vegetation at the begin-

A typical scene on the Lebombo plains between Shingwedzi and Punda Maria, 1925. These large ilala palms were almost all completely destroyed by a series of intensely hot veld fires in the early 1950s.
(ACKNOWLEDGEMENT: STEVENSON-HAMILTON COLLECTION, SKUKUZA ARCHIVES)

ning of the twentieth century. There are, however, references to specific tree species, especially unusual trees such as wild figs and jackalberry, but contemporary historical documents contain few references to the nature and structure of the vegetation. In a general description of the vegetation of the Sabie Reserve, Stevenson-Hamilton (1903) pointed out that large areas were covered in dense thorn bushes and that there was a shortage of open plains. He also refered to a more open strip of vegetation along the Lebombo Mountains and in the western parts. North of the Sabie River the bush was also not as dense as south of it and a man on horseback could find his way through it with relative ease.

There are also references to the thorn thickets along the Sabie River and in the vicinity of Skukuza and the relatively open Tshokwane area to the east in contrast with the fairly dense thickets southwest of it (also between the Mutlumuvi and Nwatindlopfu spruits). The area between Gudzani and Bangu was described as a very dry, flat region with relatively open vegetation dotted with stunted trees, while the area between Bangu and the Olifants River was typified by relatively dense thorn thickets. The area between the Letaba and the western border was described as being quite wooded.

In correlation with his belief that the Lowveld was undergoing a process of desiccation, Stevenson-Hamilton also believed that the vegetation changed from a primary to a secondary "forest". He attributed this change to the long droughts, deforestation in the earlier days by blacks as well as whites and the occurrence of annual veld fires. These secondary forest conditions were viewed as an intermediary phase leading to a treeless desert, if the general desiccation were to continue. It is clear Stevenson-Hamilton was not taken with this secondary forest, which he described as "not inspiring nor even attractive in appearance, consisting as it does mainly of deciduous and rather undersized trees, seldom exceeding a height of twenty feet and, even in full leaf, possessing no wealth of foliage." In Stevenson-Hamilton's opinion, only two factors prevented the region from turning into a desert – a reasonable groundcover of plant material, which prevented severe erosion, and the pans formed by the game on watersheds.

The dynamic properties of the vegetation were recorded early in the developmental history of the Park. With reference to old photographs taken in the Pretoriuskop area in about 1902–1903, it was remarked that "the general impression conveyed is that of a piece of high or higher middle veld, wide open spaces, with only here and there a bush or tree to be seen. Today it is rapidly assuming the character of ordinary bushveld." The same change was also described around Tshokwane; in fact, Stevenson-Hamilton expressed the opinion (in his annual report of 1943) that bush encroachment occurred widely in the Lowveld. He later withdrew this statement and applied it only to Tshokwane and Pretoriuskop. The increase in woody plants was ascribed to the lack of veld fires, which had been an annual occurrence before 1925. In 1943 he had also predicted that the numbers of short-grass grazers such as blue wildebeest would decrease as a result of bush encroachment, while the numbers of browsers would increase – a prediction that proved to be correct in later years. The only change in vegetation in the northern area that was recorded was the decrease in tall mlala palms on the Tsende (Tsendze) Plain. In 1912 Stevenson-Hamilton ascribed this phenomenon to the large-scale felling of single-trunked mlala palms by locals to brew palm beer.

In 1937 the first official plant list of the Park was published. A total of just over 300 woody plant species were recorded as a result of surveys undertaken by the well-known botanical artist Cythna Letty and the botanist A. Obermeijer in 1930. Photographer and naturalist Herbert Lang added to their list in 1932 (Obermeijer, 1937).

Animal populations
Initially the primary reason for the proclamation of the Sabie and Shingwedzi reserves was to protect the dwindling numbers of the big game species. As has been pointed out in previous chapters, poaching by the locals, and more especially continued hunting by white hunting parties in the second half of the nineteenth century, led to great losses in the numbers of especially the herbivores. These declines were acerbated in the last decade of the nineteenth century by two developments in the Lowveld: the construction of the Selati railway line, which started in 1893, and the decimating effect of the rinderpest

epidemic of 1896 and 1897, which spread through the area from the north. It is well known that the Selati Line's construction workers were dependent on venison, with the result that large amounts of animals were killed to feed them. The rinderpest epidemic also caused great losses among certain species and almost led to the eradication of kudu, eland and buffalo from the Lowveld.

This state of affairs led to several game species being eradicated, or being on the brink of eradication, by the beginning of the twentieth century. White rhino had already disappeared from the Lowveld by the end of the nineteenth century, while black rhino numbers were so depleted that they could not recover and disappeared from the area by 1937. Lichtenstein's hartebeest also could not survive the onslaught and the last survivors were shot before 1920 north of the Soutpansberg. Even elephants were eradicated from the area that is now the Kruger National Park, but fortunately they could later repopulate the area from the adjacent Mozambique. In the early years only one small herd of buffalo was known to frequent the area south of the Sabie River, in the Nwatimhiri bush. Other species managed to survive the onslaught, although in reduced numbers, and in some instances even they disappeared from parts of their former distribution areas.

Predators got off lightly compared to the herbivores, with the result, according to Stevenson-Hamilton at the beginning of the twentieth century, that the proportion of predators to their natural prey was unhealthy. During this period predators survived on alternative prey (the livestock of black and white settlers) and as a result they were regarded as both vermin and a threat to the nature conservation effort. This viewpoint led to important management decisions during the early years of the Sabie and Shingwedzi reserves, as will be discussed later.

Fortunately, in as early as 1912 Stevenson-Hamilton could report that the strict conservation measures that had been instituted were producing results and that the game populations were reacting very favourably. Official estimations of the various game species were made in 1912 and again in 1918 (Table 1). In 1925 he also pointed out that his estimations of 1912 were very conservative and that the total had increased fourfold since then. He also remarked that

... I have been told by hunters who knew the district as long ago as 1880, which was before it had been very seriously exploited for hides and biltong, that game today is at least as numerous as it was then, and that the larger carnivora are much less in evidence (Stevenson-Hamilton, 1925).

From the beginning specific attention was also afforded to the effect of environmental conditions on animal populations. In this regard, it was pointed out that game species that had been eradicated from the Pretoriuskop area before 1902 were starting to return there in 1912.

Another matter that caused concern early on was the inadequate water resources. Especially during the drought of 1912–1913 animals had to travel far to reach water. This led to large concentrations of game at isolated waterholes with the resultant overgrazing there, while big areas with sufficient grazing remained untouched.

These observations led to two important conclusions, namely Stevenson-Hamilton's emphasis on the fact that big areas were imperative for a successful nature conservation strategy, as well as the viewpoint that the creation of additional water resources would bring about a more uniform utilisation of grazing (Stevenson-Hamilton's annual report for 1925).

Donkeys and horses are being prepared for a patrol at the storerooms and drinking trough which were just west of the current park warden's office complex. May 1921.
(ACKNOWLEDGEMENT: STEVENSON-HAMILTON COLLECTION, SKUKUZA ARCHIVES)

The development of nature conservation

The preceding review shows that a number of matters that had their origin in the prehistory of the Sabie and Shingwedzi reserves, had to be rectified. It is also clear that after the proclamation of the reserves certain schools of thought developed that would lead to intervention in natural processes. The attempts that followed represent the first steps in the development of nature conservation practices in South Africa, and included the following:

Control of hunting and poaching

The original proclamation of the Sabie Reserve clearly prohibited all hunting in the area. Stevenson-Hamilton diligently enforced this ban from the get-go and effectively put it into operation with the appointment of a number of section rangers, each with a corps of field rangers under his supervision.

To further prevent poaching, the black inhabitants were told to leave. Most of them then moved out of the reserve (as well as the later Shingwedzi Reserve), and some families who were given permission to stay, were kept under strict surveillance. It was also expected of many of them to periodically work in the area on, for instance, the maintenance of roads, or to serve as informants about poaching. A few members of the families to whom this allowance was originally made, lived in the area until recently. The allowance only applied to the original inhabitants and could not be transferred to their descendants.

Water provision

It has already been pointed out earlier that natural perennial water sources in the reserves were sometimes scarce and considered to be inadequate.

As a result game gathered at a few isolated waterholes during droughts, which in turn caused overgrazing in these areas, while grazing in waterless areas remained largely untouched. It also led to the more nomadic types of game migrating to areas outside the conservation area with better grazing and more water during extended droughts. In addition to this there was the belief that the Lowveld was subject to a long-term process of desiccation and that the already insufficient water sources would continue to decrease. For these reasons, the possibility of creating additional water sources was considered from very early on. This viewpoint is clearly illustrated by the following extract from the Board's annual report for 1928:

ANIMAL TYPE	Sabie 1912	Sabie & Shingwedzi 1918	National Kruger Park 1988
Elephant	25	65	7344
Black rhino	6	6 (plus a few)	±150
White rhino	0	0	1323
Hippopotamus	100	195	2744
Giraffe	200	210	6067
Buffalo	250	250	28 491
Zebra	2000	3500	30 733
Waterbuck	4000	6500	3833
Blue wildebeest	3000	5000	14 309
Sable antelope	1000	3500	2026
Kudu	1000	3000	9080
Tsessebe	700	1000	844
Roan antelope	300	800	183 **
Impala	6000	6800	124 757
Reedbuck	4000	4500	*
Bushbuck	2000	4000	*
Klipspringer	2000	4500	*
Mountain reedbuck	200	750	*

* Not included in census surveys, but plentiful in suitable habitats. Between 1974 and 1976 a total of 370 impalas were transferred from the Bergkwagga National Park to the Kruger National Park and released in the Stolsnek area.

** Actual figures ±350.

> Now there are many river beds and spruits in the Park which flow only in the rainy season. Some of these leave pools behind and others do not. It is therefore recommended that in order to ensure a more even distribution of drinking pools (and in consequence a better distribution of feeding grounds) more use be made of these normally dry watercourses.
>
> Such a scheme would no doubt distribute the game more and make the Park less spectacular in certain parts from the tourist's point of view. But the primary object of the Park being the preservation of its fauna, this is a minor consideration, and in view of the prevailing theory that the country is drying up, it is imperative that if the fauna is to be preserved

the first essential is a good and well distributed water supply.

In as early as 1927 the possibility of sinking a few boreholes to ensure ample water supplies during droughts was considered. In 1929 this was considered again. However, the warden pointed out that engineers had told him that boreholes would not be successful and that the scheme was impractical. The Board accepted his advice and decided to build a series of dams instead. The warden was duly instructed to draw up a list of suitable dam sites.

During the drought of 1926–1928 mention was made several times that game had to migrate across the Park's borders to reach water. The prevention of such migrations became another important consideration for the creation of additional watering holes.

After consulting his rangers, Stevenson-Hamilton in 1930 submitted four alternative possibilities for the creation of water sources to the Board, instead of a list of suitable sites to build dams. These alternatives included the following: cement weirs at suitable places in watercourses; dam walls made of wired netting; the planting of tsamma (*Citrullus lanatus* – a creeper bearing water-retaining fruit that occurs in the Kalahari); and the enlargement of natural pans.

There was some doubt about the construction of concrete weirs, as the cost would be very high, the building would necessitate big labour teams and these weirs would silt up as most of the watercourses had sandy beds. It was thought that cement walls in crevices could work, but not as a general measure. Dam walls made of gabions were more acceptable as such walls could be removed and placed elsewhere if the dam became silted up.

Experiments to cultivate tsamma melons started in about 1930 after 23 kg (50 lbs) of seeds were

Windmill, dam and water trough in the Kruger National Park, erected as part of the water provisioning programme.

Blue wildebeest and zebra drinking at Muwawi waterhole in the Shisha. Photograph by the well-known naturalist and photographer, Herbert Lang. Early 1930s.
(ACKNOWLEDGEMENT: K. SCHWEIKERDT)

Randspruit windmill, one of the first series of boreholes which were sunk in 1933. It is still providing water to game.
(12.06.1984)

The oldest concrete dam in the Kruger National Park. It was erected during 1931-1932 in the Ntomeni Spruit along the Pretoriuskop-Malelane road by Capt. M. Rowland-Jones.
(16.06.1984)

Steam-driven drilling machine in Pretoriuskop rest camp, 1933.
(ACKNOWLEDGEMENT: STEVENSON-HAMILTON COLLECTION, SKUKUZA ARCHIVES)

ordered from the then South West Africa (Namibia) and distributed among the rangers. The original experiment failed, but this was ascribed to bad seeds and the intention was to acquire a few hundred kilos of seed and to continue the experiment. However, nothing came of this.

The most accepted method of stabilising water resources was the enlargement of natural pans, which had clay soil that would retain the water, and the pans would also be protected against silting.

However, no final decision could be reached about this form of water provision and in 1931 the services of an engineer were obtained to further investigate the feasibility of sinking boreholes in the Park. A favourable report was received in which the likelihood of securing water at a depth of approximately 200 feet (65 m) was expressed. The lack of funds effectively hobbled the implementation of a drilling programme. At the same time the warden submitted a list of 45 suitable dam sites to the Board. The first of the dams had to be built in the Ntomeni Spruit, southeast of Pretoriuskop, as a matter of priority. But the board members still disagreed and member Preller maintained that only boreholes with concrete drinking troughs would guarantee permanent watering holes. It was then decided to approach the Minister of Lands for financial support.

In the meantime attempts to build dams continued and the first earthen weir was erected in 1925 in the Shitsakana Spruit at Satara, while a concrete dam was completed in the Ntomeni Spruit in 1931.

The policy to sink boreholes was approved in principle in 1931. As a result the warden insisted that the Board purchase a drilling machine for the reserve, but this proposal was not accepted and it was decided that such work should be undertaken by the Department of Irrigation. A series of six boreholes was approved by the Board. In 1932 it was also proposed that when the pontoons in the rivers were replaced, causeways should be built that would also serve as weirs.

The six boreholes that were approved by the Board in 1931 were sunk in 1933. These were at Pretoriuskop, Komapiti (near Ship Mountain), Muhlambamadvube, Gomondwane, Randspruit and Manzimahle. Apart from this series of boreholes, 1933 was also a year of great significance to the water provision programme, as it was the year during which Bertram F. Jeary of Cape Town launched at his own initiative a "Water for game" fund-raising project.

In support of this fund the Board decided to make any person who contributed 50 guineas (then R105 – a guinea was worth £1.1.0) or more towards the fund a life-long member. In practice this meant that these benefactors could enter any national park free of charge for the rest of their lives. Two years later, in 1935, the "Water for game" fund received R60 000 from the government, of which R20 000 was destined for the creation of water sources.

In 1935 the Department of Irrigation was again approached to assist in determining suitable dam sites. In 1937 a report by this department, compiled by a Mr Lignau, was submitted to the Board. It suggested a total of 40 dam sites. Due to limited funds preference was initially given to ten, and this figure was later scaled down to five, of which four were to be built in the north and one in the south of the reserve. At this time ranger Bill Lamont was appointed to supervise the dam construction.

In the meantime the Department of Irrigation had sunk a few more boreholes in the reserve, but because of its own busy schedule it could not make a drilling machine available to the Park for long periods. Once again it was urged that the Board should acquire its own drilling machine. As before, nothing came of this.

In 1938 tenders were invited for the drilling of ten boreholes in the reserve, but nothing came of this either. The warden then reported that the water table was so close to the surface in the northern areas of the Park that it would be possible to dig wells. Ranger Lamont was duly instructed to dig a series of wells in an attempt to find water.

In the midst of all these frustrations, Board member Rood proposed that the Department of Irrigation be approached regarding the possibility of channelling water from perennial rivers to the seasonal watercourses. This proposal was accepted and the report resulting from this investigation showed that such a scheme would be feasible. Water could be channelled over distances of between 40 and 64 kilometres from the Luvuvhu River to the Shingwedzi, Bububu and Mphongolo rivers. The Hutomi Spruit and a number of the Sabie's tributaries could be fed

Col Stevenson-Hamilton in the veld at Gudzani with his buggy. August 1922.
(ACKNOWLEDGEMENT: STEVENSON-HAMILTON COLLECTION, SKUKUZA ARCHIVES)

Oxen dragging away ranger C.R. de Laporte's horse which had died of horse-sickness at Sabie Bridge. 1904.
(ACKNOWLEDGEMENT: STEVENSON-HAMILTON COLLECTION, SKUKUZA ARCHIVES)

Pitching camp at Kumana during an inspection tour through the Sabie Reserve. November 1908.
(ACKNOWLEDGEMENT: STEVENSON-HAMILTON COLLECTION, SKUKUZA ARCHIVES)

Ranger Harry Wolhuter and a friend, Bowden, at their camp at Pretoriuskop, with Manungu Hill in the background. Photograph: Dr A.K. Haagner, circa 1904.
(ACKNOWLEDGEMENT: STEVENSON-HAMILTON COLLECTION, SKUKUZA ARCHIVES)

by the Sand River while water could be channelled from the Sabie River to the upper reaches of spruits in the Kemp's Cottage and Renosterkoppies area. These proposals never came to fruition either. When it was proposed in 1943 that the annual water-for-game budget should be raised from R400 to R4 000, the warden complained that he would not be able to work off that amount as he did not have enough money to replace his vehicle's worn tyres!

Shortly before his retirement in 1946, Stevenson-Hamilton once again voiced his disapproval of building dams because of the siltation problem. Board member Krige pointed out that dams built in watercourses in the Western Province, where he came from, often resulted in the spruits below the dams holding more water, even if the dams had silted up. The Board accepted his proposal that this might be the case in the reserve as well and instructed the warden to carry on building a dam in the Sweni Spruit.

Despite all these frustrations, differences of opinion and perpetual shortage of funds during the early years of the reserve, much nevertheless was achieved. Apart from the dams and boreholes already mentioned, the following were also accomplished before 1946: in 1933 a project to build a dam in the Kumana Spruit was started; dams at Satara and in the Sweni Spruit were completed; the well-known Leeupan (also known as Ngwandlopfu or Lily Pond) was deepened; in 1934 three more boreholes were sunk, one next to the Makukutswana Spruit near the former Rabelais entrance gate, the others at Bangu and Marheya; in 1935 boreholes were sunk at Gomondwane and Gudzani while the building of Mazithi Dam started; in the Letaba area, boreholes were sunk at Shivhulani, Nwanetsi, Ngwenyeni, Malopeni and Malopeni North in 1935; two dams, Pretoriuskop (Fayi Dam) and one in the Mooiplaas Vlei were completed in 1937; in 1938 a dam was completed in the Dzombo Spruit; and in 1939 and 1940 ranger Lamont dug wells on a full-time basis.

By 1941 there were 18 wells in the northern areas and one at Kingfisherspruit. These 18 wells in the north of the Park were at Mashagadzi, Ndlopfini, Tsumani, Nkokodzi, Nkulumbeni North, Shingomeni, Dzombyana, Nwarihlangari South, Nwashitsumbe North, Babalala, Stangene, Shidzavane, Shashanga, Mashikiri, Klopperfontein, Mazanje, Nkovakulu and Nkovakulu North. Three dams were built in 1941, namely the Rabelais Dam near Kingfisherspruit and the Matukwala and Shashanga dams near Punda Maria; in 1942 ranger Lamont completed a new dam at Pretoriuskop; in 1944 a concrete dam was built in the Komapiti Spruit with funds donated by the Percy FitzPatrick Memorial Fund; and the Orpen Dam in the Nwaswitsontso Spruit east of Tshokwane was completed in 1944 with funds donated by Mrs Eileen Orpen. In 1945 the construction of a concrete dam in the Sweni Spruit started and ranger L.B. Steyn built an earthen dam in the upper reaches of the Salitje Spruit, but it was washed away shortly afterwards.

A view of the Sabie River from a point east of the Nwaswitshaka mouth. July 1926.
(ACKNOWLEDGEMENT: STEVENSON-HAMILTON COLLECTION, SKUKUZA ARCHIVES)

Aerial view of Kumana Dam from the east. It had been one of the first earthen dams in the Kruger National Park (1933) and it washed away several times, but was time and again repaired. (27.07.1984)

16.4 THE EARLY DEVELOPMENT OF THE SABIE AND SHINGWEDZI RESERVES AND THE KRUGER NATIONAL PARK

Veld fires

It used to be common practice among both the black population and later white farmers who used the area as grazing for their herds in the winter to burn the veld every autumn. The construction workers working on the Selati Line did the same. Stevenson-Hamilton was not pleased with this state of affairs and soon after his appointment started trying to control the veld fires. He even applied punitive measures to local squatters found guilty of arson.

Although Stevenson-Hamilton believed that veld fires were to blame for mortalities among several bird species, the destruction of ostrich nests and mortalities among young animals, these were not the underlying reasons for his disapproval of regular veld fires. His condemnation rather stemmed from his interpretation of the influence of these fires on the vegetation. He believed that the vegetation of the Laeveld once consisted of "primary forests" with high trees, and that this vegetation had to make way for "secondary forests" thanks to veld fires.

In as early as 1912 he noted that the "gnarled, twisted, and stunted appearance of practically all the trees is due mainly, if not entirely, to the grass fires". He ascribed this change in the vegetation to the fact that the veld fires first scorched the lower branches and bark, causing the bark to fall off and so exposed the trees to wood-boring insects. As has been mentioned, Stevenson-Hamilton was not impressed with the "secondary forest" and regarded it as merely an intermediate step between the vanished primary forest and arid desert, in an area he believed to be subject to a progressive desiccation.

It is quite understandable that the first step of someone holding this opinion would be to gain control over and prevent the annual fires. On the other hand, he also realised the danger hot fires and the accumulation of combustible material would pose to plants in an area that had not been burnt for a long time. He therefore instituted the first burning programme in 1912 "by burning strips alternately in different places and at the proper season, especially where rank grass exists".

It would seem that these measures were not successful, as every year fires still raged throughout the Lowveld until 1925. In 1926 a general ban was placed on the deliberate burning of grass, and thereafter the

The old water reservoir at Skukuza rest camp shortly after completion, 1932.

Shivhulani windmill, Letaba section. One of the first series of boreholes which were drilled in the Kruger National Park, 1934.
(ACKNOWLEDGEMENT: STEVENSON-HAMILTON COLLECTION, SKUKUZA ARCHIVES)

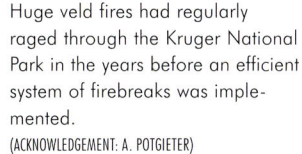

Huge veld fires had regularly raged through the Kruger National Park in the years before an efficient system of firebreaks was implemented.
(ACKNOWLEDGEMENT: A. POTGIETER)

16. EARLY YEARS: THE SABIE AND SHINGWEDZI RESERVES AND THE KRUGER NATIONAL PARK 1900–1946

In 1928 tourists could still camp in the veld at their own chosen spots.
(ACKNOWLEDGEMENT: STEVENSON-HAMILTON COLLECTION, SKUKUZA ARCHIVES)

Lions in the vicinity, but no traffic officer. Early 1930s.

Breakfast at Satara rest camp, June 1932.
(ACKNOWLEDGEMENT: MR T.C. SINCLAIR DAVID, DURBAN)

During the early days tourists could leave their vehicles along the Lower Sabie road. June 1932.
(ACKNOWLEDGEMENT: MR T.C. SINCLAIR DAVID, DURBAN)

veld fires apparently either started outside the Park or were accidentally ignited. But this ban also did not have the desired effect and in 1935 the Board's executive subcommittee decided that veld fires should not be banned, but rather controlled.

On the recommendation of this subcommittee the Board accepted in principle a system of firebreaks. At that stage there could be no suggestion of cutting firebreaks through the bush, and in 1936 the Board instructed the rangers to burn firebreaks. The next year (1937) it was also suggested that flame-throwers should be used to burn the firebreaks.

The Board's chronic shortage of funds at that time meant that this method was considered too expensive in practice. Board member Rood then pointed out that farmers usually burnt the veld every second year, alternating between January and March, thereby preventing hot veld fires and ensuring that too little combustible material remained to make winter fires possible. The Board accepted this method for the Kruger National Park and instructed the warden to apply it along the borders and tourist roads.

This method failed to serve its purpose, and flame-throwers were then tried. They, too, proved unsuccessful in creating effective firebreaks, so the warden suggested that the Board buy lawnmowers to cut open firebreaks. This again held financial implications for the Board, so they decided to rather try a method suggested by Board member Campbell. He told the meeting that the farmers in Natal hauled branches over the green grass in summer, and within a few days the grass at the bottom would be dry enough to be burned. This method was tried, but found to be ineffective. Eventually two new lawnmowers were bought to cut firebreaks. When firebreaks were cut they concentrated on the Park's borders to prevent fires from crossing into the Park. It is not known how successful this method was. It is unlikely that it was used for a considerable period, and it wasn't until the 1950s that they started cutting open a network of firebreaks.

Just as the first measures aimed at controlling fires were successful in containing the annual burning of the veld, they soon realised that a complete ban on burning would also not be acceptable. They soon realised that game tended to avoid overgrown areas, and that the accumulation of plant material kept old grass warm, which resulted in uncontrollable fires. It also became clear that woody shrubs quickly encroached on unburnt areas. These considerations soon led to the acceptance of fire as an important factor in the management of the Park. In as early as 1937 Stevenson-Hamilton maintained that the veld had to be burned every second year to get rid of old grass. In 1947 Stevenson-Hamilton suggested the following burning policy:

> The best policy to pursue, having in view, as the main object, the well being of the wild animals of the Park, would seem to be to burn about half the grass each year, and as early as possible in the year, selecting, if such occurs, a spell of a few weeks of dry hot weather during the months of February, March or April, for the purpose ... if half the grass be left unburnt, it will be available for burning early in the following year, while the other is left untouched and so on annually, in rotation.

Picnickers at the Lower Sabie picnic spot in June 1932.
(ACKNOWLEDGEMENT: DR R.E. KAY, DURBAN)

These days veld fires in the Kruger National Park are controlled with sophisticated fire extinguishing equipment.

16. EARLY YEARS: THE SABIE AND SHINGWEDZI RESERVES AND THE KRUGER NATIONAL PARK 1900–1946

Apart from the aims of this policy that have already been mentioned, the following considerations were also important:

i in the period during which fires had to be set, there was still a fair amount of moisture in the grass and the fires would consequently not be very hot;
ii there usually weren't strong winds at that time of year;
iii late summer rains would provide lush green grazing during the winter months; and
iv the cooler fires would inflict the least damage on the riverine vegetation.

These principles and policies regarding veld fires stayed in place until the end of Stevenson-Hamilton's career in 1946.

Large areas of grazing burnt down in the early years during the dry season.
(ACKNOWLEDGEMENT: A. POTGIETER)

The cattle dip at Sabie Bridge (Skukuza). July 1926.
(ACKNOWLEDGEMENT: STEVENSON-HAMILTON COLLECTION, SKUKUZA ARCHIVES)

Animal populations

The primary concern of conservation in the newly proclaimed Sabie and Shingwedzi reserves initially revolved around resettling the larger herbivores. These were the game populations which had been affected most during the past few years of the nineteenth century, mainly by hunting and the rinderpest. To build up these depleted populations successfully, Stevenson-Hamilton had two primary alternatives: to do everything possible to prevent further losses and to bring species which had become extinct from elsewhere and resettle them. His first efforts were to enforce the new law on hunting and poaching, while at the same time a total onslaught on predators was launched.

As has been pointed out before, most animal populations responded well to the newly enforced stricter control measures. However, these increases were only applicable to those populations of which there had been sufficient numbers left to bring about increases. In some instances, the numbers had been too low for species to recover, while some species had already died out. The reason for the onslaught against predators had been mentioned already, and the policy and means of depletion will be discussed later.

First we will look at the efforts to settle and increase the larger herbivores. Since 1903, Stevenson-Hamilton had anticipated that elephant and rhino would again colonise the Lowveld from the adjacent Mozambique. The same possibility was not expected in the case of hartebeest (probably Lichtenstein's hartebeest) and eland, as it was known that they had already disappeared from the area a long time ago. In reality there still had been a number of isolated herds of Lichtenstein's hartebeest in the far northeastern Transvaal, but these would also die out before 1920.

Within the first few years of the proclamation of the Sabie Game Reserve, Stevenson-Hamilton brought a few young eland from Mozambique to Skukuza where they were kept in kraals. These young animals were to serve two purposes. First they were to be tamed with the aim of replacing the cattle which were still being kept in the reserve, and secondly to settle an eland population in the reserve again. Not one of these objectives could be achieved. When staff had been prohibited from keeping cattle after 1938, the possibility of replacing them with eland and buffalo was mentioned again. The eland would provide milk and the buffalo meat to establish a meat industry. This would have been undertaken in cooperation with the Department of Veterinary Services, but the proposal was rejected by the Board on the grounds that such a project would be better suited to a zoo than a game reserve.

After the establishment of the National Parks Board and the proclamation of the Kruger National Park in 1926, the Board had their first meeting on 16 September 1926. At this meeting, the then Minister of Lands who had been one of the greatest campaigners for national parks, Senator P.G.W. Grobler, pleaded that the reintroduction of white rhino into the reserve should be seen as a matter of high priority. At the next Board meeting in 1927, the Board reiterated the importance of resettling white rhino in the reserve, but at the same time realised that it would virtually be impossible to capture these large, dangerous animals in Natal and to transport them to the reserve. As a more acceptable alternative they then decided to rather propose that the Umfolozi and Hluhluwe reserves be proclaimed as national parks.

In 1928 negotiations with the Zululand Cash Wholesalers took place to capture and deliver 12 white rhino to the Park by train at a total cost of £8000. The Board firstly did not have the finances for this project and secondly were rather sceptical about the success of such an operation. The Natal Provincial Administration held the same opinion as the Board and was of the opinion that if animals were to be moved, then it should be done by experts rather than on a purely commercial basis. The minister was not overly impressed with these failures, but had to resign himself to putting the project on hold.

In the years following the proclamation of the reserve, attention was also afforded to several other projects concerning the reintroduction, or in several cases, only the importation of game. While attention was being afforded to the reintroduction of white rhino, the Board received an offer in 1927 to buy a number of hartebeest (probably red hartebeest). Because of a lack of funding, this offer had to be rejected. Another offer involved the British Prince Arthur of Connaught, who offered two giant sable antelope from Angola to the Board. Some expressed their misgivings about whether the animals were of the same subspecies as the local sable and also if they could ever be settled in the reserve. Ultimately, the funding still lacked for such projects and this offer had to be rejected also. In 1930, Board member W.A. Campbell donated a total of 20 crested guineafowl to be released into the reserve.

In 1930 the Board received an offer of nine lechwe from Zambia. The only cost to the Board would be the import tax, as the animals themselves had been donated. The consideration of this offer led to the first difference of opinion based on principal grounds among the board members. Stevenson-Hamilton pointed out the lack of suitable habitat, which would be swamps, in the reserve, but nevertheless thought that they would be able to adapt to the Punda Maria area. The chairperson of the Executive Subcommittee of the Board, member Selby, thought it to be a

A cattle herd at the windmill on the bank of the Sabie River at Sabie Bridge (Skukuza) in 1913. This borehole provided water to man and animals, as the Sabie River had been contaminated at the time by effluents of the gold mines along the upper reaches of the river. It was situated just east of the current Volkskas guest house.
(ACKNOWLEDGEMENT: STEVENSON-HAMILTON COLLECTION, SKUKUZA ARCHIVES)

good idea, but member O. Pirow differed from him on the grounds that "... the principle was not right, as the Board was concerned with the preservation of the local game". He suggested that the animals rather be offered to the Pretoria Zoo. Member Papenfus thought wrongly that the National Parks Act did make provision for the import of new animal species. As a result, the matter was referred to the Executive Subcommittee of the Board, who in turn accepted the offer. However, before arrangements could be made to fetch the animals, news was received that all the rams had died, upon which it was decided to abandon the project.

In 1931 the Board received an offer of 30 black wildebeest, but they declined the offer as members had been of the opinion that the general conditions in the reserve were not suitable for these animals. Shortly after this offer, an offer involving oribi was received from Mr Evans from Stegi in Swaziland during 1932. The animals would be free of charge, but the Board had to capture and transport the animals. Stevenson-Hamilton saw this offer as a great opportunity to introduce new blood into the dwindling oribi population of the reserve. The Board pointed out that oribi "... did not seem to thrive in the Park and the warden feared that the animals would tend to wander out of the Park into more suitable country". However, in the end it was not this opinion that prevented the operation, but merely the lack of funding.

In 1934, another farmer, Tom Elphick from Magnesite near Malelane, offered the Board a number of red duiker on the same conditions as the oribi deal. This time, the warden pointed out incorrectly that red duiker did not occur in the reserve and he doubted if they would survive because of a lack of suitable hills. He nevertheless considered the project to be worthwhile. However, Board member Hoare disagreed with him and pointed out that "... as a rule attempts to introduce wild animals where they were not indigenous, proved to be failures". Member Papenfus stuck to the idea that any new additions to the animals in the reserve should be welcomed. Attempts to capture some Natal duiker were launched, but failed miserably.

In 1936, member Rudolf suggested that consideration should be given to the release of blesbok and springbok in the vicinity of Pretoriuskop. The warden was against this proposal based on the grounds that none of these two species had ever occurred in the area and that they would be caught by lions soon. Member Hoare in turn, was not so much concerned about the fact that lions would catch the buck, but was against the proposal in principal, as he was of the opinion that only animals found in the area previously should be considered for reintroduction. Once again, nothing came of this proposal. After this there were no further discussions about the import or release of animals until the end of Stevenson-Hamilton's career in 1946.

From the cases above, it is clear that neither Stevenson-Hamilton nor the Board had any clear principles or guidelines for the release of animals in the reserve. Although Board members, including Hoare and Pirow, pleaded for the release of only indigenous species, they were in the minority. Their viewpoints were rejected by other Board members and sometimes also by Stevenson-Hamilton. Today it is accepted policy that only animal species which have been proven without a doubt to have occurred in a specific area be considered for reintroduction into national parks. This policy was finally accepted by the National Parks Board in 1951. Therefore, it appears as if the lack of funding which had prevented these projects should not always be seen as having been a spoke in the wheel of progress, but rather as an ally in the maintenance of the quality of our national parks.

Capturing game

It had already been decided during the earliest years of the Sabie and Shingwedzi reserves to capture animals to sell alive. This decision was taken in the hope that such a project would generate income to augment the puny funding available for the development of the reserves. For this purpose they obtained nets and other equipment and several game capturing operations were launched during 1912 and 1913. Apart from a few adult animals caught, they rather attempted to capture young animals for hand-rearing. Although they managed to capture quite a few animals during these operations, the project never met the expectations – mainly because of the animals being injured during the capturing process and also because of the financial losses incurred. Subsequently, these attempts were stopped in 1914.

After the establishment of the Kruger National Park in 1926, the matter of game capture was only raised again in 1928. In this instance the Pretoria Zoo approached the Board for two impala, but the request was refused as the Board was not equipped to undertake such an operation and because the animals could just as easily be captured outside the reserve. Later in the same year the Pretoria Zoo once again approached the Board and offered an eland bull in exchange for a heifer. In this instance the Board supported the warden's point of view in that the provision of animals to other organisations should only be considered once animals outside the Park had become so scarce that the Park would be the only source.

In 1930 the Pretoria Zoo suggested to the Board that young animals be captured and hand-raised at Skukuza, after which they would be distributed to interested parties. The Board refused again as the project was undesirable and impractical in their opinion. In similar vein a request for animals for a zoological garden in the United States during 1935 was also refused.

It appears as if Stevenson-Hamilton did not only have objections against the capture of game during the early years of the Kruger National Park, but that he was also opposed to the development of game capturing techniques. In as early as 1928 an application was received from a Captain Harris who wanted to test an "automatic hypodermic bullet" which would not kill, but only drug the animals, so that they would be easier to handle. Stevenson-Hamilton maintained that the use of weapons inside the Park, whether lethal or not, would create a dangerous precedent. The Board accepted his viewpoint and refused permission.

By 1946 the Board's attitude had changed somewhat and on hearing of the use of a so-called "mercy bullet" used for game capture in Rhodesia (Zimbabwe), the warden was instructed to obtain more information for possible use in future. It is not clear if this "mercy bullet' was the same as the one which Captain Harris had wanted to use in the reserve, nor if the warden ever managed to obtain the necessary information.

There is also no evidence of such a technique ever being used in the Park during the 1940s.

Young kudu calves are being fed at Sabie Bridge, February 1913. They formed part of a scheme that had been launched to hand-raise young animals to sell to the Pretoria Zoo to generate funds. However, it was not very successful, mostly because of official opposition.
(ACKNOWLEDGEMENT: STEVENSON-HAMILTON COLLECTION, SKUKUZA ARCHIVES)

Predators

It has already been mentioned that by Stevenson-Hamilton's assessment there were too many predators in relation to their prey species in the Sabie and Shingwedzi reserves. This imbalance was largely ascribed to intensive hunting and poaching activities on the larger herbivores by the end of the nineteenth century, while carnivores had come off largely unscathed. Predators were initially considered to be vermin which hampered the increase of prey populations and as such had to be exterminated. Despite the initial policy to kill all predators, it is also clear that Stevenson-Hamilton firmly believed right from the beginning that carnivora had an integral role in maintaining a healthy balance in nature. As early as 1906 he had the following to say:

> Even left to themselves carnivora never increase out of a due proportion to the game on which they prey, as is sometimes ignorantly stated, always supposing that man does not step in, and by destroying the game unduly upset the balance of Nature. The latter, indeed, may be trusted to be a tolerably good regulator of such matters, and if she suffered her laws to be so inefficiently drawn up as to allow the purely flesh eating animals to be equally prolific with the others, surely long before man made his appearance upon the scene at all, the latter would have entirely disappeared.

To Stevenson-Hamilton there was but one way to regain the balance between predators and their prey and that was to drastically reduce the carnivore populations. He found justification for this in that it would at least accelerate the rate in which prey populations would increase. In his own words:

> The reduction of carnivorous animals to, and their retention within proper and moderate limits when carried out on methodical principles, ought not to, and does not, present any insurmountable difficulties. Hence, far from increasing, they ought to decrease in numbers considerably, while the game, on the other hand, multiplies unhindered. With the decrease in predatory animals comes a corresponding decrease in competition, which cannot act otherwise than as an incentive to retain them in their accustomed haunts.

Therefore, it is clear that the aims of culling of predators were two-fold: to give prey species an opportunity to increase to their former numbers and secondly to reduce competition among predators, through which they would be less inclined to cross the boundaries of the protected area, where they caused havoc among neighbouring farmers' livestock.

The early onslaught against these "vermin" included not only hunting, capturing and poisoning the larger carnivores but also the smaller ones such as wild cats, crocodiles, raptors, snakes and even honey badgers. The predator culling programme had been so intense that it was noticed as early as 1913 that leopards had disappeared from certain areas. This did not bother nature conservators of the time much; their only disappointment being that baboons had increased in numbers as a result of the programme. The only benefit ascribed to lions and leopards was that they hunted bushpigs. These had also been considered to be the "most mischievous vermin" and were also killed along with cane rats, porcupines and other rodents.

In 1912 Stevenson-Hamilton stopped the use of poisons (except in the case of wild dogs), because the dying-off of the smaller wild cats led to an unacceptable increase in the rodent population. With that, the killing of smaller cats was also ended "… to keep the balance adjusted as far as possible" (Stevenson-Hamilton in his annual report of 1913).

By 1925 Stevenson-Hamilton's attitude towards predators had changed completely and he emphasised the natural balance which should exist between predators and their prey. However, large influxes of carnivores from adjoining areas took place and to control this situation, Stevenson-Hamilton imposed the following guidelines in 1925:

> … to confine all predatory species within strictly limited bounds as regards numbers, … both from the point of view of the welfare of the herbivorous types, and of scientific and popular interest, any complete extermination of these creatures was a deplorable catastrophe.

At the very first meeting in September 1926 after the proclamation of the Kruger National Park, it was decided to stop the culling of lions "for the time being". The acceptance of this policy change had to take place without the knowledge of the general public. The reason for this would become clear later on. It appears as if this decision of the Board was not strictly enforced. This was confirmed in 1928 after it was claimed that lion numbers were busy decreasing. It was then spelled out clearly that rangers would only be allowed to shoot lions under exceptional circumstances, and then only with the direct approval of the warden. Although the shooting of lions was stopped, the opposite policy was accepted with regards to wild dogs. During a Board meeting in 1928 it was pointed

Major James Stevenson-Hamilton with "trophies" at Sabie Bridge. November 1906.
(ACKNOWLEDGEMENT: STEVENSON-HAMILTON COLLECTION, SKUKUZA ARCHIVES)

out that significant deaths had been occurring among reedbuck. These deaths were attributed to bad environmental conditions due to the drought, rather than as a result of wild dogs. It was nevertheless claimed that wild dogs had been seen again in the vicinity of Skukuza for the first time since 1921 and that they had caught almost the entire crop of impala lambs. It was also believed that the wild dog population had increased since 1924 and it was subsequently decided to make a deliberate effort to get rid of them. To serve as encouragement and as a bonus for the hard work to kill wild dogs, it was decided to pay £1.0.0 for every wild dog killed. This remuneration would be paid out by a magistrate after sufficient evidence had been submitted. To continue with this scheme, the Transvaal Provincial Administration's permission had to be obtained first. However, this was refused and ranger staff had to duly carry on with the eradication of wild dogs without any compensation.

The policy regarding lions and jackals received attention again in 1929. Stevenson-Hamilton argued that lions had become too numerous and could pose a serious danger, especially to tourists, if not reduced. Some Board members differed and felt that lions were a huge tourist attraction and reaffirmed the policy that lions could be shot only in extreme circumstances and only by authority of the warden. The question was also raised whether the time had not arrived to stop shooting jackals as well. However, the warden was of the opinion that certain jackal species caused considerable damage and through this revealed that he was in favour of carrying on with their killing. After consideration, the Board resolved that the shooting of jackals should be stopped and that killing them could also only take place at the discretion of and with permission of the warden.

In April 1930 the warden reported to the Executive Subcommittee that a young black man had been mauled by a lion, while another's life had been threatened by an aggressive lion. Following these incidents the warden felt that lions were losing their fear for humans and subsequently were becoming a danger. To prevent similar situations it had already been decided the previous year that rangers were to hunt down lions on foot only where necessary so as to instil fear of humans! As a result of the incidences mentioned, it was even suggested that tear gas be used in an effort to instil fear of humans. This proposal was never taken further.

During 1930 the Executive Subcommittee received complaints from farmers in the Letaba district that lions had become a danger to their livestock. However, in this instance the warden defended the lions and pointed out that lions were also found outside the Park in that area and had therefore not necessarily been the Park's lions. The Board nevertheless gave instructions that lion numbers be restricted in that area. By the end of 1930 pressure was again exerted on the Board to resume lion and jackal culling, but once again the Board stuck to their guns and their existing policy was reaffirmed.

In 1931 the customary habit that rangers could keep the skins of the lions which they had shot to sell for their own gain came up for discussion. It was argued that this practice could become a lucrative business and result in unprofessional conduct, and should therefore be abolished. The argument was countered by pointing to the dangers involved in shooting lions, but the Board could not come to a decision and this old practice was maintained.

Although strict guidelines had been laid down by the Board in connection with the shooting of lions and jackal, it was not accepted equally enthusiastically by all Board members and officials. In June 1932 Board member Dr Preller reported that during a visit to the Park he had been struck by the

Lion skins in front of the warden's living quarters in Skukuza. The animals were shot as part of the predator control programme, Skukuza, 1926.
(ACKNOWLEDGEMENT: STEVENSON-HAMILTON COLLECTION, SKUKUZA ARCHIVES)

absence (or few) birds along the roads. He was probably referring to ground and predatory birds, including guineafowl, francolin and partridges. This phenomenon was ascribed to the predatory habits of the jackal. In contrast to his viewpoint of two or three years earlier, the warden countered that jackals were not the only predators of birds and that "it has been proven time and again that nature would itself adjust the balance should there be an abnormal increase of any one type of animal over another". The Board decided on a majority vote to reinstate the shooting of jackal, but only in the vicinity of roads.

This decision was not accepted without a fuss by all the Board members and in December 1932 member Hoare informed the Board that steenbok, duiker and groundbirds were common in habitats suitable to them. Because of this, there was therefore no need to kill jackal. The Board resolved to reconsider its policy at a later date and in the meantime the shooting of jackal along the roads would continue. A further attempt to have the resolution to shoot jackals revoked at the March 1933 Board meeting, also failed.

Following a plea by a member of the public, Mr Forbes from Swaziland in 1934, that the lions in the Park should be reduced as they would eventually exterminate the prey species, the Minister of Lands himself intervened and rejected the proposal. He answered that there was no apparent decrease in the number of prey animals and that lions had proven themselves to be a prominent tourist attraction. In 1935 more pressure was applied on the Board to reduce the number of predators. As a result, the Board instructed the warden to carry out a complete census of all game, including predators and birds in the reserve. Later that same year, the warden reported to the Board that such a census would be an impossible task and the best figures that could be obtained would probably have a plus or minus 30 percent margin of error. The census was thus not done.

A request at about the same time by ranger Steyn of Tshokwane to reduce lions as they were killing his livestock elicited a sharp rebuke from the Board. His request was not only declined, but he was also reminded that the Park was a tourist resort!

In his annual report of 1935, Stevenson-Hamilton staunchly supported the beneficial role played by predators in a natural system, despite increasing pressure from several quarters to reinstate predator control. To support his viewpoint, he pointed out that natural ecological adaptations to the relationship between carnivores and prey would not allow one species to eliminate another.

By 1936 the differences of opinion about the desirability of predator control had not come to a rest yet. Some Board members expressed their concern about the general scarcity of game in the reserve. However, member Hoare maintained that the perceived impression could be ascribed to the density of the vegetation and not as a result of predator pressure. Following this, a decision was taken to circulate questionnaires to the rangers to obtain their views. When the matter came up for discussion again at the next meeting, member Pirow proposed that the warden be requested to consider alternatives and to inform the Board about what the best option would be. The three alternative options proposed were: Should there be an experimental lion reduction campaign in areas from which complaints had been received?; Should there be a general predator reduction campaign throughout the Park?; or Should there be no control at all?

Following this suggestion, the Executive Subcommittee proposed early in 1937 that a drastic reduction of predator populations would not be advisable. Two areas, each about 1280 square kilometres in size, were proposed as experimental lion culling areas. The one was along the Crocodile River and the other along the southern bank of the Olifants River. This proposal was accepted and the warden was instructed to have lions shot at his discretion in the two areas and to report periodically to the Board. Even these measures were not sufficient to satisfy the increasing number of supporters of stricter measures against predators. The media also joined the fight and several articles in which the adoption of an encompassing predator policy was pleaded, appeared. The whole matter came to a head in 1939 when the Board deemed it necessary to lay down a definitive policy about predator control and issued a press release.

Strong differences of opinion still existed even among the Board members. Member Gilfillan argued that the natural balance between predators and prey could not be maintained in a confined area, while member Hoare was equally adamant that predators play an important role in the maintenance of sound prey populations. Member Pirow maintained that lion control should be confined to the stretch along the Crocodile River, while member Hockly supported a system of well-defined areas where predators could be shot on sight, with no shooting outside these areas. Amidst these differences of opinion and the public debate which at times became highly emotional, the Board nevertheless managed to adopt a policy providing for the following:

i that the shooting of old, decrepit and dangerous lions be continued at the discretion of the warden;
ii that the eradication of lions along the Crocodile River be continued;
iii that no reduction of lions take place beyond these measures; and
iv that the policy be applied for at least three years.

The adoption of this policy brought about that the public debate died down. From 1939 until Stevenson-Hamilton's retirement in 1946, the whole matter of predator control disappeared into the background. In 1945 the Department of Agriculture appealed to the Board to erect predator-proof fences on the boundaries of the reserve. The Board undertook to oblige where possible.

In the whole debate about predator control, the focus fell increasingly more on lions and in a lesser degree on the wild dogs and jackal, while other predator species received little or no attention. From official reports it seems as if the eradication of leopard, cheetah, wild dog, hyena, jackal, crocodile and "other predators" only continued until 1935. After that no further mention was made about any control of these predators in official reports and returns. Although the reports, especially before 1926, had not always been totally accurate, the following numbers of predators killed between the proclamation of the Sabie and Shingwedzi reserves until 1946 were: 935 lions, 248 leopards, 34 cheetahs, 387 wild dogs, 147 hyenas, 961 jackals, 252 crocodiles and 3090 others. The "others" included caracal, serval, civet, genet, predatory birds, baboons, badgers, python, poisonous snakes, otters, porcupines, bush pigs and giant eagle owls.

In judging the development of the predator policy, as is the case with all other wildlife management policies, it should be borne in mind that the circumstances, conditions and attitudes in the early developmental years of the Kruger National Park were totally different from what the case is today. The rural population was rather sceptical about nature conservation, while predators were generally considered to be vermin which posed a threat to humans and their livestock. The understanding of natural phenomena was subject to the opinions of individuals rather than tested against objective facts. As a result one can therefore have sympathy for the problems with which the first nature conservators had to struggle, and at the same time also show appreciation for the courage they showed to adopt the necessary changes according to their convictions!

The development of a nature conservation philosophy

Seen against the background of current knowledge of nature conservation, there will understandably be differences of opinion about the way in which important matters or principles had been handled by the first wildlife managers. The more sympathetic will find earlier arguments and validations for certain actions interesting and even amusing, while those whose judgement is more critical, will reject certain actions as being without principle. To those who realise the excellent powers of observation and long-term monitoring necessary to understand nature's mysteries, there can only be appreciation for the sagaciousness with which so many of the important laws of nature had been recognised and documented. Some of the outstanding contributions in this regard came from Stevenson-Hamilton himself.

Throughout his 44 years as warden, it is abundantly clear that he had been a firm believer in the functional value of each of the constituent parts of the ecosystem to maintain a healthy balance in nature. This conviction gradually gained strength. Initially the emphasis had been restricted to individual constituents or groups, but in time more and more elements of nature as a whole were added to his list. At first the larger herbivores had been his main concern, but later on he also acknowledged the important role of carnivores. Before his retirement he had not only considered all natural functional elements as an integrated whole, but also described its dynamic attributes.

By 1957 he had gained enough experience to reflect objectively upon varying periods of drought and high rainfall and the effect these had had on vegetation and animal life. From his records he could identify the short-term cyclic nature of natural processes about which he wrote the following:

> In ecological matters we have certainly to allow for the incidence of curves which are continuous with their alternating elevations and depressions, in the life history of all species, and are in major degree quite independent of the assaults of natural enemies … Even in the relatively short space of under 40 years during which I have, within a single area, had the opportunity of studying animals in their wild state at first hand, I believe that several of these phenomena of specific curves have taken place, and been completed, though from inexperience they may at the moment not have been recognized for what they really were.

With reference to these natural cycles he also remarked:

> Although it is of course easier to note phenomena among the larger animals, I have from time to time observed apparent indications of the same curve of alternating increase and decrease among lesser types, notably some of the smaller carnivores, rodents and snakes.
>
> There is no reason to suppose that the same law does not operate in respect of the flora, though very much longer periods of time must be allowed for the completion of the curve. Yet in the case of smaller plants and weeds it is possible sometimes to notice the same working of the process in the great increase over several years of a certain, for example, weed in a particular area, followed by its decrease and even temporary disappearance owing possibly to having over increased and choked itself out.

It is not only an understanding of the degree to which natural ecosystems and their dynamic cyclic nature are integrated – which still form the cornerstones of rational environmental management practices – for which we have to thank Stevenson-Hamilton. During the difficult formative years of the Kruger National Park, he had to create a clear picture for himself of the values of a national park. Values not only based on sentiment alone, but which would also serve to conquer efforts to alienate the land area of the Park. He had to formulate concepts which would validate the right of existence of the Kruger National Park and this he had to do in an era in which tourism only gradually started developing and the associated financial benefits could not yet serve as motivation. But Stevenson-Hamilton believed in his cause and recorded the following about the values of the Kruger National Park:

> Great interest is being taken in the wild life of the country by the general public since they have had the opportunity of seeing it for themselves in the Kruger Park.
>
> But for the actual conditions governing this wild life there is not a very wide knowledge … nor is it perhaps always completely realised what a national park stands for. The ideal embodied in the latter, in addition of course to its popular object of displaying unspoiled nature and giving the public some notion of how Africa appeared before the white man came to it, is to try and find the answers to many scientifically interesting questions of natural history, which can only be sought within an area where all the species indigenous to it are permitted to live their natural lives unhampered by artificial aids or restrictions … we want to see them (animals) following their strictly natural habits and not those imposed on them … We also would like to learn experimentally how Nature, if left entirely alone, manages her own affairs with regard to limitation and increase of various species. Full answers to the above cannot be expected today because a certain amount of artificiality is inevitable, but in the Kruger National Park we have the opportunity of getting nearer the truth than is possible anywhere else in the world today.

This remarkable appraisement of the Kruger National Park was not only applicable to Stevenson-Hamilton's time, but is still valid today.

The development of tourism

As pointed out already, both the Sabie and Shingwedzi reserves had been very poorly developed during the time of their proclamation. With the exception of the Selati railway line, there were no roads suitable for general traffic. With the gaining and establishing of control over the areas and the few personnel whose services could be used, it is no wonder that the possibility to develop these areas for tourism had enjoyed such low priority (even if considered) that it has not been noted anywhere officially.[1]

Tourism as a whole and the creation of a National Park was only considered as a possibility by the authorities in 1916 with the naming of the Game Reserves Commission under chairpersonship of J.F. Ludorf KC. In their official report of 1918 the Commission proposed that the two Transvaal reserves be combined and proclaimed as a national park. They clearly indicated that nature conservation was the primary objective of the two reserves. The development of tourism facilities could be considered as it would not necessarily clash with the primary objective. As motivation for this view they particularly emphasised the educational and research opportunities which the reserves offered and in this regard the opportunity which the general public would be afforded to experience nature in a totally natural environment.

Initially nothing came of these proposals and it was only in 1923 when the South African Railways instituted a "Round in nine" tour to the Lowveld and adjacent Lourenço Marques (Maputo) in Mozambique, that the potential for the reserves as a tourist attraction came up for discussion again. An overnight stop in the Sabie Reserve at Sabie Bridge (Skukuza) had only been included because it was convenient and not because of the possibility that tourists would be interested in the wildlife. Stevenson-Hamilton had his work cut out for him to convince the Railway Commissioner that rides during the day through the Sabie Reserve would improve the popularity of the "Round in nine" railway excursions. Stevenson-Hamilton's pleas led to the "Round in nine" being scheduled so that trains would move during the day from Komatipoort to Sabie Bridge. On his part, Stevenson-Hamilton organised that a ranger accompany the tourists on this leg of the journey and spend the night at Sabie Bridge with them. There were no amenities at Sabie Bridge, so the tourists had to sleep in the train. The ranger would make a large camp fire at night and entertain the tourists with interesting information about animals and the reserve. He even took the tourists on short excursions during the day. This became a great attraction and was particularly popular among the tourists.

When the Kruger National Park was proclaimed in 1926, the concept of tourism had already taken hold. The value of tourism as a source of income was also mentioned at the very first Board meeting on 16 September 1926. To encourage tourism with the aim of obtaining an income from it, it was decided to construct a main road with secondary branch roads for game viewing. After that they would appoint guides who would escort the tourists and for whose services a fee could be charged. It was further decided that fees would be charged for taking photographs. A third source of income would be the writing of articles which would be offered for sale

The main road to the Kruger National Park between Legogote and Numbi Gate. June 1935.
(ACKNOWLEDGEMENT: PROF. M.C. BOTHA, PRETORIA)

The original ZASM tunnel between Waterval-Boven and Waterval-Onder was used by visitors to the Kruger National Park, circa 1935.

---

1   While in London in 1905, Stevenson-Hamilton obtained first-hand information about the American national parks and their success as public attractions. This provided him with a real goal to work for, to develop the areas into tourist attractions (Stevenson-Hamilton, 1937).

16. EARLY YEARS: THE SABIE AND SHINGWEDZI RESERVES AND THE KRUGER NATIONAL PARK 1900–1946

This is the way tourist roads looked in the old days. This photograph was taken in the vicinity of Satara, 1931.
(ACKNOWLEDGEMENT: STEVENSON-HAMILTON COLLECTION, SKUKUZA ARCHIVES)

The original six huts which were erected in 1930 in Balule rest camp are still standing today. (18.06.1984)

Satara rest camp as seen from the east during the late 1930s.
(ACKNOWLEDGEMENT: STEVENSON-HAMILTON COLLECTION, SKUKUZA ARCHIVES)

on the one hand and to attract overseas visitors on the other.

Lack of accommodation for tourists proved to be a huge problem and in January 1927 the South African Railways requested the Board if they could erect a hostel and lease it to them. Nothing came of this scheme, but in the same year an agreement was reached between the Board and the SAR through mediation of Stevenson-Hamilton, whereby the Board and the SAR would work in close collaboration towards a combined strategy for the development of the tourism industry. The Board undertook to build roads, rest houses and other facilities, to provide guides and other protection services and in the meantime not encourage independent traffic. The SAR in turn undertook to provide transport for tourists, both by rail and road, to launch advertising campaigns, to do the catering at rest houses and to pay the Board a percentage of the income obtained through tourism.

For the purpose of this scheme there had initially only been four two-spoor roads available: the one from White River to Pretoriuskop; the one between Crocodile Bridge and Lower Sabie which had been constructed by ranger C.R. de Laporte; the one from Acornhoek via Satara to the Mozambican border; and the one from Gravelotte to Makuba's village near Letaba. The latter two had been constructed by the WNLA.

In August 1927 the Board resolved that the Pretoriuskop area should be opened to tourists on the following conditions: that an entrance permit be obtained for £1 from either the Board Secretary in Pretoria, at White River, from the warden, or the ranger of Pretoriuskop who was stationed at Mtimba. Also that they would be allowed to travel only on a specified route from White River, via Mtimba to Pretoriuskop and back via the same route on the same day, as there were no overnight facilities available. The last stipulation was that no rifles were permitted, only revolvers for personal protection.

The arrangement in connection with the obtaining of permits was confusing to many visitors and they often travelled past Mtimba (ranger Harry Wolhuter's post) without reporting. In 1929 the Board appointed A. Moodie as permit agent in Moodieskloof until the appointment of a permanent gate official, Captain M. Rowland-Jones at Numbi Gate in 1931.

By the end of 1927 several additional proposals had been considered or adopted by the Board to stimulate tourism traffic. A proposal by the SAR to build a hotel at Sabie Bridge was rejected by the Board as "impractical". The SAR also submitted a proposal for suitable motor crossings over the Crocodile River. The Board in turn requested that the SAR open the railway bridges over the Crocodile, Sabie and Olifants rivers to cars, that the train service on the Selati Line be changed to make it more convenient for tourists and officers of the Board and to accept responsibility for the construction of a road from Crocodile Bridge to Satara and Acornhoek.

The construction of tourist accommodation only started in 1928. The first three so-called rest houses were built at Satara, Pretoriuskop and Skukuza (at that time still known as Sabie Bridge or "Reserve"). Six additional huts were planned at the time. These huts, or rest houses, each consisted of a set of rondavels or rooms and a car port. Nothing came of the second set of six huts, but in 1929 two rondavels each with a diameter of six metres and ten with diameters of a little over four metres, were erected at Skukuza, while two additional rondavels were built at Satara. Rest camps of the size of Skukuza were also planned for Pretoriuskop, Satara and Letaba. Two smaller rest camps with six rondavels each were also planned for Balule (then known as Olifants camp) and Olifants Poort (better known as Gorge) near the confluence of the Olifants and Letaba rivers.

Construction of the rest camp at Olifants Poort had already started in 1929. Expansion of tourist facilities gained momentum in 1930. Apart from two additional rondavels at Skukuza, four were built at Pretoriuskop as well to augment the four existing rondavels there. Fifteen huts were also erected at Satara, twelve at Letaba, six at Balule, one at Olifants Poort and four at Malelane. At Lower Sabie, the five-room wood-and-iron guest house which had once served as ranger Tom Duke's living quarters, was renovated to accommodate tourists.

All the rondavels built at the time (and which are still to be seen at Balule rest camp), had been designed by Paul Selby according to the so-called "Selby-style". Paul Selby had been an American mining engineer who served on the Board. He designed the huts with a gap between the wall and the roof and no windows. There was a peep-hole through the "stable" doors, which enabled tourists to peep through to see if there were wild animals outside the hut before they left it. Rest camps were of course not fenced in at the time. The Selby huts were soon criticised as being inadequate as they were too cold at night in winter, too dark due to the lack of windows, while the peep-holes enabled strangers to see into the huts. The most serious criticism, however, was that they provided no protection against mosquitoes! Since 1931 all new huts had windows installed.

During the early 1930s great progress was made towards the provision of additional tourist amenities. The old guest house at Lower Sabie had proven to be unsuitable due to its dilapidated condition and it was decided to use it as the ranger's quarters while

Col J. Stevenson-Hamilton visits the former Gorge rest camp in the 1930s.
(ACKNOWLEDGEMENT: STEVENSON-HAMILTON COLLECTION, SKUKUZA ARCHIVES)

The oldest hut in Pretoriuskop rest camp which was built in 1930. (16.6.1984)

a new camp was built along the Crocodile River. Eight rondavels were completed at Crocodile Bridge in 1931, while the "guest house" at Lower Sabie was demolished in 1932.

In 1931 tents provided additional accommodation for the first time. The tents, each with four beds, were initially put up in Skukuza and afterwards at Satara.

Apart from the rest camps already mentioned, another six were created during this time. The building of the Rabelais entrance gate also commenced in 1931 and in 1932 the first huts of a new rest camp at Punda Maria were built in the traditional wattle-and-daub style, since cement was so expensive at that time.

Construction in 1932 also included a small rest camp at Malopeni. In 1933 a small tented camp was erected along the Tsende (Tsendze) River at Mabodhleleni. This camp was only used for a few months until building of the Shingwedzi rest camp started in 1934. This rest camp had initially also been a tented camp. In 1935 the first units, each consisting of three rooms, were completed. The wooden structures of these huts, as well as others which followed are still in used today.

In 1932 the first ablution block with four baths and four showers, was built in Skukuza. That same year the rest camps were all fenced in for the first time to keep animals out.

In 1935 they experimented with a new hut design and the so-called Knapp huts were erected at Skukuza, Crocodile Bridge and Letaba. These were square units with corrugated-iron roofs. Large hollow concrete blocks were used to build the walls with. These had been rather unsightly and unpopular, with the result that the experiment was soon terminated.

The last two rest camps that opened before 1946 were Lower Sabie and Pafuri. After the closure and later demolition of the guest house at Lower Sabie, it was decided to build a new rest camp there. The first buildings of this new camp, which had been designed by architect Gerhard Moerdyk, were completed in 1936. It consisted of three units with six bedrooms each which were laid out in a U-shape. A small tented camp was opened at Pafuri in 1939 situated along the Luvuvhu River at the terrain of the current picnic site. It was closed only a year later as a result of flood damage, although the mosquito problem also played a large role. It would only be reopened after the war.

The development of the tourism industry in the Kruger National Park was in many ways very similar to that of wildlife management. The Board had to deal with a totally new and unique type of development for which there had been no clear principles or policies. Decisions were originally made at random and in many ways experience had to make fools wise. For instance, with the building of the first little rest houses and rondavels in 1928, they had not given a single thought to the possibility of rest camps at all. Board member Oswald Pirow pointed out in 1929 that the few rondavels and rest houses would be insufficient once visitor numbers started increasing. He suggested that no new huts be built, but rather that rest camps of approximately 100 x 100 metres be fenced off with barbed wire, with an elongated shed with corrugated-iron roof on poles in the centre. Here a receptacle for boiling water should be provided to tourists. He was of the opinion that such a construction would meet the need as most tourists preferred

Traditional hut accommodation in Shingwedzi rest camp, 1957.

Col. J. Stevenson-Hamilton and ranger I.J. Botha at the newly completed huts in Punda Maria, 1933.
(ACKNOWLEDGEMENT: STEVENSON-HAMILTON COLLECTION, SKUKUZA ARCHIVES)

to camp. The Board accepted his proposal, but revoked it in the same year!

The Board's close relationship with the South African Railways in connection with the development of a tourist industry has already been pointed out. In 1930 the Board undertook to build a rest camp for the SAR in the vicinity of Skukuza as soon as its own construction programme had been completed. However, the Board could not stick to their promise and the project was called off in 1931.

Although hot water is nowadays expected as a given in any tourist facility, it had not been so during the early years of the tourism industry in the Park. Shortly after the completion and opening of the road between Letaba and Punda Maria, the Board received recommendations that bathrooms in both these two camps should be provided with hot water. The argument being that the road between the two camps was not only very long, but also very dusty. (This road went over dusty, black clay soils for most of the way, which could not be gravelled during the construction phase.) The chairperson of the Board at the time, Senator Jack Brebner, was not at all impressed with this proposal and it was initially rejected on the grounds that it would be an unnecessary luxury. However, petitions kept on coming in and in 1933 the Board accepted the principle with the provision that tourists pay one shilling per bath!

The matter of hot water for baths and showers came up for discussion in 1936 again after it had already been decided to provide this "luxury" in several other rest camps. This "unnecessary luxury" led to more arguments. The fact that the Board did not have the necessary funding for this in any case, was also mentioned. Once again the matter was put on hold for reconsideration. The Board only finally agreed to install hot water systems during the 1939 works programme, and then only hot and cold showers for men and hot baths for ladies between 17:00 and 21:00 daily!

Despite the rate at which accommodation for tourists had been erected during the early thirties, it could not meet the ever growing demand. In 1934 the SAR and Transvaal Publicity Conference submitted an urgent request to the Board to make more accommodation available. It was even suggested to the SAR to park a number of rail coaches at Skukuza to serve as accommodation. However, this proposal could not be carried out as it would have been illegal.

In 1935 the South African National Publicity Conference, as well as the Pretoria and Johannesburg publicity conferences made representations to government to make a grant of £50 000 available to the Board for the expansion of tourist facilities. The addition of another 150 beds at Pretoriuskop and a new rest camp at Lower Sabie to accommodate 200 people were considered to be the most urgent priorities. Apart from the new expansions, all existing huts also had to be made mosquito proof.

The entrance gate to Skukuza camp with the reception office and post office on either side of the gate and the Papenfus clock tower in the background (originally outside the camp terrain). 1955.

The newly completed Struben guest house in Skukuza camp with Col and Mrs J. Stevenson-Hamilton in the foreground, 1937.
(ACKNOWLEDGEMENT: STEVENSON-HAMILTON COLLECTION, SKUKUZA ARCHIVES)

Blue wildebeest and zebra in a burnt area near Pretoriuskop, with Numbi Hill in the background. Photograph: Herbert Lang. Early 1930s.
(ACKNOWLEDGEMENT: K. SCHWEIKERDT)

16. EARLY YEARS: THE SABIE AND SHINGWEDZI RESERVES AND THE KRUGER NATIONAL PARK 1900–1946

Crocodile Bridge rest camp from the air. 5 August 1948.

Satara rest camp from the air from an east-west direction. August 1948.
(ACKNOWLEDGEMENT: AIRCRAFT OPERATING COMPANY)

Skukuza rest camp from the air. 5 August 1948.

In August 1935 Government announced that it would make a grant of £30 000 (R60 000) available for the development of tourism facilities in the Kruger National Park. The Board decided that only £20 000 (R40 000) of this amount would be spent on tourist facilities while the remaining £10 000 (R20 000) would go towards the establishment of additional water resources for game. In the meantime, the Board had already contracted the architectural firms of Leith and Moerdyk to draft plans for the envisaged expansions. Gordon Leith was to prepare the plans for the new Lower Sabie rest camp, while Gerhard Moerdyk had to design the expansions to Pretoriuskop and Malelane. By the end of 1935, the plans were ready, while a proposal that eight to ten wattle-and-daub huts be erected at Tshokwane was also accepted by the Board, provided that funds were available.

The architects, as well as the Executive Subcommittee of the Board, had not taken into account the opinion of the chairperson of the Board. Judge J. de Wet was not impressed with the architects and considered items such as hot water for showers and baths as unnecessary luxuries and he could also not associate himself with extravagances like septic tanks, trading premises, a dining room, et cetera in rest camps, as he considered those to be superfluous and unwarranted. All that was needed was sleeping accommodation! Therefore, Leith's plans for Lower Sabie were rejected based on the fact that it would be too expensive and the contract was allocated to Moerdyk to draw up new and significantly cheaper plans. The huts planned for Tshokwane were also rejected.

In 1938 the warden expressed concern about the safety of tourists at picnic spots. However, the Board felt that because rest camps were far apart, tourists needed a place to stretch their legs. After consideration it was decided that they would be retained, but that black attendants had to be appointed and that all grass and shrubs be cleared from the areas. Stevenson-Hamilton was still concerned in 1939 when he repeated his warnings. The Board then decided that picnic spots would no longer be indicated on tourist maps and that notice boards with warnings would be erected at the sites.

One of the principle motivations for the initial conception of a tourism industry in the Park, was that it would be a welcome source of income. With the initial opening of the Pretoriuskop area to day visitors, the only money which could be charged was an entry fee of £1.0.0. In 1928 the Board decided that entrance fees of five shillings per person had to be charged at all the other gates, with the single exception of Punda Maria, while a minimum amount of £1.0.0 per vehicle also had to be charged. Entry permits could be obtained from rangers, but also from agents in White River (Legogote) and Rubbervale (Gravelotte). At the same time, visitors on the SAR's "Round in nine" tour had to pay £5.0.0 per day. This fee included the services of the ranger who accompanied them.

To discourage heavy traffic inside the Park, an entry fee of £5.0.0 was charged for each heavy vehicle that entered the Park for trading purposes.

An additional source of income came from the punts which ferried visitors across rivers. Until the end of 1931 tickets could be bought for five shillings. These tickets were valid for seven days and covered any number of crossings with the same punt. In 1932 the rates were increased and motor vehicles had to pay 2/6 every time a punt was used, while trucks had to pay five shillings. Pedestrians were not exempted and had to pay 2/6 for the first five people and an additional sixpence for every person above that number.

For the uninformed it might be difficult today to understand why the Board did not develop its own trading facilities in the Park. However, it must be borne in mind that it was the acute lack of funding which had been hampering the early development and also prevented the Board from launching its own large projects. That was not the only reason. After a request had been received in 1929 to open a café or tea-shop at Pretoriuskop, it was rejected because it would detract from the "natural facilities" of the reserve, whatever that might have meant!

By 1930 it was pointed out to the Board that the rangers at Skukuza and Satara could no longer handle the administration and maintenance of the rest camps together with their normal duties. Hence, it

Malelane rest camp from the air. 5 August 1948.

Lower Sabie rest camp as seen from the air. 5 August 1948.

16. EARLY YEARS: THE SABIE AND SHINGWEDZI RESERVES AND THE KRUGER NATIONAL PARK 1900–1946

Pretoriuskop rest camp in 1930.
(ACKNOWLEDGEMENT: STEVENSON-HAMILTON COLLECTION, SKUKUZA ARCHIVES)

Visitors queuing to book into Pretoriuskop rest camp, 1949.

Lunch in a tent at Pretoriuskop, circa 1955.

Pretoriuskop rest camp with the exotic flamboyant trees which were planted there as adornment and to provide shade. June 1937.
(ACKNOWLEDGEMENT: PROF. M.C. BOTHA, PRETORIA)

was proposed that contractors be appointed who could handle such duties under supervision of the rangers. This proposal was accepted after discussion and in March 1931 a contract for the management of Skukuza and Satara was awarded to P.W. Willis. In September of the same year, Willis was awarded a similar contract for Letaba due to the increasing tourism traffic there.

The contracts awarded to outside contractors handed over total administration of the rest camps to them. This included issuing of permits, all trading and catering, the renting out of bedding and even the erection of their own buildings. In November 1935 the Board decided to appoint its own rest camp supervisors, which meant that the outside contractors were restricted to trading and catering. The first supervisors at rest camps were appointed in 1936 at Pretoriuskop, Skukuza and Lower Sabie.

In the early years, the Board could obviously not make much profit from its rest camps and the income was more or less restricted to the renting out of huts. At that stage the huts were also furnished only with wooden beds with woven leather bases. All the rest, including mattresses, linen, blankets, pillows, towels, etc. had to be hired from the contractors. Initially the renting of a hut had been one shilling per person per night, but when the Board started providing luxuries such as hot water and the services of hut attendants, the daily rate was increased to 2/6 per person.

This rate increase led to great dissatisfaction among the tourists as it was applicable to children under 16 years of age as well. The Board then wisely reduced the rate for children under 16 years to one shilling per night. Visitors who did not use hut accommodation had to pay one shilling per adult and a sixpence per night for children under the age of 16 for a camping site.

In 1933 the Executive Subcommittee expressed concern that the rental of three shillings charged by

The lookout tower Kambani, near Pretoriuskop, circa 1950.

Opening of the swimming pool donated by the Rembrandt Tobacco Company at Pretoriuskop on 6 December 1956.

A notice board which was erected to indicate the position of the western boundary of the Kruger National Park. Circa 1930s.
(ACKNOWLEDGEMENT: STEVENSON-HAMILTON COLLECTION, SKUKUZA ARCHIVES)

the contractor for a mattress and four blankets was excessive. The result was that the pricing structure was adapted in 1934 as such: a mattress would cost a sixpence per night, a blanket and pillows a trippence each per night. They also offered packages and a mattress, blanket, pillow and pillow case could be hired for seven shillings per week or otherwise a complete set of bedding including sheets for ten shillings per week! The Board only started renting out its own bedding in 1940 at the low cost of 2/6 a night for three blankets, two sheets and two pillows per night, nine pence for any additional nights or ten shillings per week! By that time inner-spring mattresses could be rented for two shillings per night.

In 1938 the Board rejected a proposal that towels be rented to tourists – initially only in Pretoriuskop camp – as it would only generate a minimal income.

The outside contractors provided the Board with many problems through the years and the tendency became more and more for the Board to take over their responsibilities, although such steps had initially been taken with great reservations. After repeated complaints, the Board appointed its own rest camp supervisors in Satara, Letaba and Punda Maria. A proposal that the Board take over catering in the rest camps as a result of general dissatisfaction with the contractors, was not accepted. They offered this responsibility to the SAR instead, but they also did not want to handle it.

In 1943 Board member Orpen proposed that a subcommittee of the Board be appointed to thoroughly investigate trade contracts, catering services, control over rest camps and tourism. The Board adhered to his request and a subcommittee was appointed a year later. The subcommittee already suggested in November 1944 that the Board should take over all trading and catering. They also suggested that an official be appointed to accept responsibility for all trading, catering, rest camps and tourism. This person would have to report directly to head office (to the secretary of the Board).

The Board could not execute all the proposals immediately, but it nevertheless led to the appointment of the first tourism manager, H.C. (Van) van der Veen in 1948, as well as an amendment to the National Parks Act in 1946, which would entitle the Board to take over all commercial activities in the Kruger National Park. In 1945 a contract was awarded to a firm called Kruger Park Services until the Board could eventually take over all commercial activities in 1955.

A request in 1946 by the contractor to obtain a liquor licence was summarily refused by the Board. This was to be only the first of many similar requests which fell on deaf ears as far as the Board was concerned, until the organisation took a decision in the 1960s to sell liquor to tourists.

One of the matters to which the Board had to give attention during the early years was the type of accommodation and the layout of rest camps. With the planning of the "new" Pretoriuskop rest camp by architect Moerdyk, they already had aesthetic aspects in mind and wanted to deviate from the barracks-like layout with huts in straight rows. The question of the type of roofing to be used for buildings also resulted in conflict when certain Board members expressed themselves against the traditional thatched roofs as they harboured bugs. Fortunately this matter was settled in favour of thatched roofs.

Another matter which brought on much discussion was the question of hotels. The Board had already rejected a proposal to build a hotel at Skukuza on the grounds that it would be impractical. In 1930 a request was received from Messrs Mostert and Potgieter of Johannesburg to build a hotel in the Park. This request was turned down on the grounds that "at present there is no intention to erect hotels in the Park".

With great pressure on the Board to supply more accommodation urgently, an appeal was made once again that a hotel with about 300 beds be erected. This motion was even supported by *The Star*. The chairperson of the Board, Senator Jack Brebner, was clear in his opposition and pointed out that it would not be profitable, as the Park would only have an "open" season of six weeks per annum.

In 1935 the idea of hotels even received support from some Board members. Member Papenfus pointed out that if the Board should build its own hotels, it would also be in complete control of these hotels. He was supported by member W.A. Campbell, but this made no impression on the chairperson and he rejected the proposal.

In 1939 Mr Lawson, a businessman, requested the Board to state its attitude to the erection of hotels on its boundaries and the possibility of additional entrance gates. In a response which included a proposed hotel at Toulon, the Board stated that it would be prepared to consider granting access to the Park at additional gates, provided that the hotel was well run and reputable, and also provided further that the granting of access caused no losses to the Park through dislocation of existing entrances or in any other way involved the Park in any additional expenditure. As a result of these negotiations it was decided to establish entrance gates at Toulon and along the Nsikazi River to accommodate Lawson.

When Lawson approached the Board in April 1937 with the request that reduced entrance rates be charged for guests at his proposed hotels at Toulon and Plaston for second and subsequent visits to the reserve, the Board summarily refused. The Board's rejection of reduced entry fees made the hotel scheme no longer feasible.

Because of the lack of funding during the early developmental years of the Park, the Board readily accepted donations from benevolent individuals or organisations. Board member W.A. Campbell, who had previously owned the farm Mala Mala on the Park's border, donated the amount of £150 to build a "rest house" in 1929. He made more donations later on and one of the rondavels which had been built with funding obtained from him, was later changed into the Museum Hut in Skukuza camp. Donations to and benefactors of the old Sabie and Shingwedzi reserves and the Kruger National Park are discussed in chapter 21.

Because of the position of the Park and the long distances to be travelled to reach it, the provision of fuel was essential right from the start. The Vacuum Oil Company had already approached the Board in September 1929 with the request to sell Pegasus petrol in the rest camps. The Board was agreeable to the request, and according to an agreement, petrol would be sold at approximately three shillings a gallon (just over 6c per litre!), of which six pennies would go to the Board. Finality with regards to the agreement could not be reached immediately. During this period the warden pointed out that petrol was only required at Satara and Letaba as former ranger T. Duke, who had a trading store at Skukuza, intended to install a petrol pump there. Also that Crocodile Bridge and Pretoriuskop did not need pumps as they were close enough to towns where petrol was available. In the meantime, Shell Oil Company had also submitted an application to sell fuel in the Park.

By August 1930 Pegasus petrol was already available at Satara and Letaba. When Texas, a third oil company, also applied to sell its product in the Park, the Board decided in principle by the end of 1930 that only one company should be allowed to sell petrol in the Park. This matter was referred to the Executive Subcommittee for further consideration. The Board did not stick to its initial decision and in 1931 the Shell Co. was granted permission to sell petrol at Skukuza and Malelane, while Pegasus petrol was sold at Crocodile Bridge, Satara and Letaba and Punda Maria. Apparently Shell had not erected any petrol pumps at Malelane and in 1934 it was noted

The main building of Punda Maria rest camp with the petrol pumps in the front, 1937.
(ACKNOWLEDGEMENT: STEVENSON-HAMILTON COLLECTION, SKUKUZA ARCHIVES)

The reception office at the old Malelane entrance gate, circa 1940s.

that Atlantic petrol was sold there and later also at Pretoriuskop. However, an application of the Atlantic Oil Company to sell its petrol elsewhere was turned down.

When the Park was opened for tourists in the late 1920s, there were very few rules or regulations, except for a moratorium on bringing in firearms. At the time that overnight facilities were created inside the reserve, tourists were not even expected to return to rest camps at night. Therefore, they were free to leave their vehicles at will, build a fire in the bush and even spend the night there. It soon became clear that such a state of affairs would lead to trouble and in November 1930 the first list of regulations, which had been compiled by A.A. Schoch for the Board, was published. Contravention of which these regulations could lead to fines with a maximum of £50. These regulations largely formed the basis of the regulations which are currently valid and included the following: tourists were confined to rest camps at night, driving was confined to the time between half an hour before sunrise and half an hour after sunset, vehicles were restricted to roads, a speed limit of 25 m.p.h. (40 km/h) was imposed, it was not permissible to cut off or damage plants, or to damage or deface any objects in the Park and it was an offence to drop litter.

By 1932 it was pointed out to the Board that regulations had very little meaning if they could not be enforced. As a result, the Board decided to institute motor, or motorcycle patrols from 1933, which would enforce regulations. It was later decided not to pursue the motor patrols for the same old reason – a lack of sufficient funding.

In 1932 the Board approved a request by the Automobile Association (AA) to station a road scout at Skukuza during peak periods to do road patrols. The request was granted, on the condition that he would not interfere with the vehicle repair service of the outside contractor based in Skukuza.

The contractor and the AA could not reach a mutually satisfactory agreement, which prevented the road patrol system to be introduced in 1934. To overcome this problem, the outside contractor's contract was amended in 1934 in such a way, so that the AA could undertake the patrols. The AA's service was eventually introduced during the tourist season of 1935. They undertook not only to assist tourists, but also to help enforce regulations.

The Executive Subcommittee of the Board felt so strongly about enforcing the regulations that they even suggested that the AA approached the Department of Justice to appoint their official as a "special police constable".

The patrol service soon proved to be a huge success and in 1936 the Royal Automobile Club (RAC) applied to introduce a similar service. The request was granted and it was decided that the AA would patrol the area south of the Olifants River and the RAC the area north of it.

Although the first regular aerial service in the Park was only instituted in the late 1960s, the Executive Subcommittee had already considered a request by the Armstrong Siddeley Development Company to airlift tourists to the Park in 1930. The preliminary proposal suitably impressed the Board, but after a more formal application was requested, no further action took place.

In August 1930, Board member Papenfus brought up the matter of an air service again. He informed the Board that he had been in consultation with the Johannesburg Light Plane Club and anticipated that a regular service could be instituted by 1931. However, it was only in 1932 that the Johannesburg Aeronautical Association (formerly known as the Johannesburg Light Plane Club) formally approached the Board to institute a regular air service to the Park.

It was also suggested that the service would be to Satara. However, the Board was worried, following a report in *The Star* newspaper, that flights over the Park would cause panic among the animals. Throughout the negotiations it was emphasised that the flights would not be for sightseeing and it was suggested that the flights be limited to altitudes above 2 000 feet above ground level. The Board considered the request favourably. They did not only see the advantage of such a service as an additional source of revenue, but also its possible use in emergencies.

During the initial negotiations, Malelane had been mentioned as an alternative to Satara, but by June 1932 they decided on Satara and an area 10 kilometres to the north of Satara was indicated as what would become a landing strip. The minimum altitude for flights had been set at 4 000 feet above ground level and the Association gave an undertaking to construct and maintain the landing strip at its own expense. For this the Board was prepared to give the Association an exclusive five-year contract. The Association also undertook to provide vehicles to officials who would take tourists on sightseeing trips. An entrance fee of five shillings per person would be charged and the agreement was finally approved in September 1932.

However, the Civil Aviation Board could not be persuaded to issue a regulation prohibiting flying below 4 000 feet over the Park. It was then decided that if the landing strip was to be utilised by private aeroplanes they would have to pay and entrance fee of £5, although the agreement between the Association and the Board had made provision that private planes would not be allowed to land.

In 1933 the air service was advertised in a tourist guide of the Park as follows:

> In a comfortable two- or three-seater machine (aeroplane), the journey is accomplished over beautiful scenery, to an aerodrome six miles north of Satara, in the very heart of the Kruger National Park.
>
> On arrival the tourists are met at the aerodrome by a roomy five-seater sedan car, and the pilot, who is well acquainted with the KNP, undertakes the dual role of chauffeur and guide.

The club's first aeroplane landed on the Mavumbye airstrip for the first time during the winter season of 1933 and as such made history. Seven aircraft, of which one had been illegal, kept the service going during this first season. An insurmountable problem for the club was the transport of passengers from the airstrip to Satara and this resulted in the service not being successful. After 1933, no further mention could be found of this operation and it is

Sergeant Gillmore of the South African Police at Sabie Bridge (Skukuza), 1927.
(ACKNOWLEDGEMENT: STEVENSON-HAMILTON COLLECTION, SKUKUZA ARCHIVES)

Col J. Stevenson-Hamilton in front of the newly completed SAP offices at Skukuza, 1932.
(ACKNOWLEDGEMENT: STEVENSON-HAMILTON COLLECTION, SKUKUZA ARCHIVES)

16. EARLY YEARS: THE SABIE AND SHINGWEDZI RESERVES AND THE KRUGER NATIONAL PARK 1900–1946

[left] Crossing the Olifants River by pontoon at the former Olifants camp (Balule). Early 1930s.
(ACKNOWLEDGEMENT: STEVENSON-HAMILTON COLLECTION, SKUKUZA ARCHIVES)

[right] The pontoon over the Sabie River at Skukuza, with the restaurant building at the top on the right, 1934.
(ACKNOWLEDGEMENT: MR E.P.H. MEINTJIES, ALKMAAR)

[above] The pontoon crossing the Crocodile River at Crocodile Bridge, 1936.

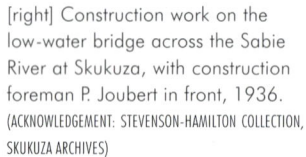

[right] Construction work on the low-water bridge across the Sabie River at Skukuza, with construction foreman P. Joubert in front, 1936.
(ACKNOWLEDGEMENT: STEVENSON-HAMILTON COLLECTION, SKUKUZA ARCHIVES)

[left] The original drift through the Sand River north of Skukuza, 1928.
(ACKNOWLEDGEMENT: STEVENSON-HAMILTON COLLECTION, SKUKUZA ARCHIVES)

[right] The first low-water bridge across the Sabie River at Lower Sabie was built in 1936.
(11.11.1986)

16.4 THE EARLY DEVELOPMENT OF THE SABIE AND SHINGWEDZI RESERVES AND THE KRUGER NATIONAL PARK

[left] A row of tents in Letaba rest camp. July 1954.

[right] Letaba rest camp with eight completed rondavels. June 1932.
(ACKNOWLEDGEMENT: MR T.C. SINCLAIR DAVID, DURBAN)

[left] The well-known umbrella thorn tree in Satara camp, circa 1955.

[right] Different types of accommodation in Letaba rest camp, 1957.

The shop and dining room in Satara rest camp as seen from the north, circa 1930s.

16. EARLY YEARS: THE SABIE AND SHINGWEDZI RESERVES AND THE KRUGER NATIONAL PARK 1900–1946

obvious that the service never really got off the ground and was therefore terminated.

Two aeroplanes of the SA Air Force used the airstrip in 1934 when locust control officials of the Department of Agriculture were flown in to undertake inspection tours of the locust control campaign. An open area, which is all that is left of the former airstrip, can still be seen just north of Mavumbye windmill.

Until the proclamation of the Kruger National Park in 1926, the Selati railway line, ox-wagons, buck-carts, pack-donkeys and horses had been the only modes of transport. There were no motor vehicles or roads for that matter. The first tourist service which had been instituted by the SAR in 1923, was also confined to train transport.

Since 1927, the building, or rather the clearing of roads started in earnest. Naturally, the first roads were mainly connecting roads between the settled ranger posts. The first roads constructed were the ones from White River to Pretoriuskop, from Pretoriuskop via Doispane to Skukuza, from Skukuza to Satara and the Crocodile Bridge and from Crocodile Bridge via Gomondwane to Lower Sabie (which had already been constructed by ranger C.R. de Laporte in the mid-1920s for his own convenience, after he had acquired the first motorised vehicle – a Model-T Ford – in the Park). From Satara a road was constructed to Olifants Poort (Gorge), to Letaba via Olifants rest camp (Balule) and to Acornhoek, from Acornhoek to Nwanetsi and from Letaba to Gravelotte (built by WNLA). Then there was a road from Louis Trichardt to Punda Maria and the Limpopo River (Pafuri) (also built by WNLA) and from Skukuza via Salitje to Tshokwane.

In 1928 construction of the road between Skukuza and Lower Sabie started, but it was only completed in 1931. As a result of the inspired road construction programme, a total of 386 miles (617 km) of tourist roads had already been completed by 1929. During the period 1927 and 1929, three pontoons were put to work. One to take tourists across the Crocodile River (at Crocodile Bridge), one on the Sabie River (at Skukuza) and one on the Olifants River (at Balule). A so-called "Corduroy" drift was constructed across the Sand River at Jafuta. In 1930 a second pontoon service was introduced to take people and vehicles across the Crocodile River at Malelane.

The road building programme carried on uninterruptedly until the mid-1940s and by 1934, 800 miles (1200 km) of road had been completed. This included the important connecting roads, including the ones between Letaba and Shingwedzi (1933) and between Malelane and Crocodile Bridge (1933).

Notice board in Letaba rest camp to warn visitors about the danger of elephants, circa 1940s.
(ACKNOWLEDGEMENT: STEVENSON-HAMILTON COLLECTION, SKUKUZA ARCHIVES)

The first aeroplane of the Johannesburg Aeronautical Association (later Rand Light Plane Club) lands at the Mavumbye airstrip north of Satara, 1933.
(ACKNOWLEDGEMENT: STEVENSON-HAMILTON COLLECTION, SKUKUZA ARCHIVES)

During 1932 a new drift was built across the Sand River to replace the "Corduroy" bridge, while another drift was built across the Letaba River.

Although a number of new tourist roads were constructed between 1935 and 1946, the main efforts centred on maintenance and improvement to the existing road network. This was made possible by the acquisition of a large number of new road-building machines in 1938, including two graders, two bulldozers, a tractor and a trailer. Much attention was also given to replace the pontoons, which had never been a satisfactory option, with drifts. In 1936 a low-water bridge across the Sabie River at Skukuza was completed, as well as one in 1937 across the Olifants River at Balule, while the building of one across the Crocodile River at Malelane was almost completed. In 1938 a low-water bridge was built across the Shingwedzi River, while the completion of a low-water bridge across the Crocodile River at Crocodile Bridge brought about the end of the pontoon service in the Park.

The road network as it looked in 1946 (see the map on page 604) was in itself a great achievement if the ongoing chronic financial position of the Board, the shortage of equipment and manpower, as well as the fact that the Lowveld was as a whole still rather inaccessible to motor vehicles, is taken into account.

The few widely distributed rangers were mainly responsible for the construction of the road network, as well as rest camps. The only two vehicles to their disposal were two Chevrolet trucks which had been bought in 1929 and a single bulldozer bought in 1933. Therefore, it is no wonder that the Board endeavoured in every possible way to get help.

The Board had already approached the SAR and the Minister of Lands in 1927 to obtain financial support. They also requested the Transvaal Provincial Administration (TPA) to help with the road building programme. In 1928 the Transvaal Publicity Conference urged the TPA to expedite work on the access roads and causeways to the Park and that the principle be adopted of using roads to the Park as access roads for the development of the rest of the Lowveld.

The old airstrip which was built by the South African Defence Force in 1940 just south of Klopperfontein. (18.07.1986)

Ranger C.R. de Laporte with his Model-T Ford, the first motorised vehicle in the former Sabie Reserve, which he had bought in 1927. (ACKNOWLEDGEMENT: STEVENSON-HAMILTON COLLECTION, SKUKUZA ARCHIVES)

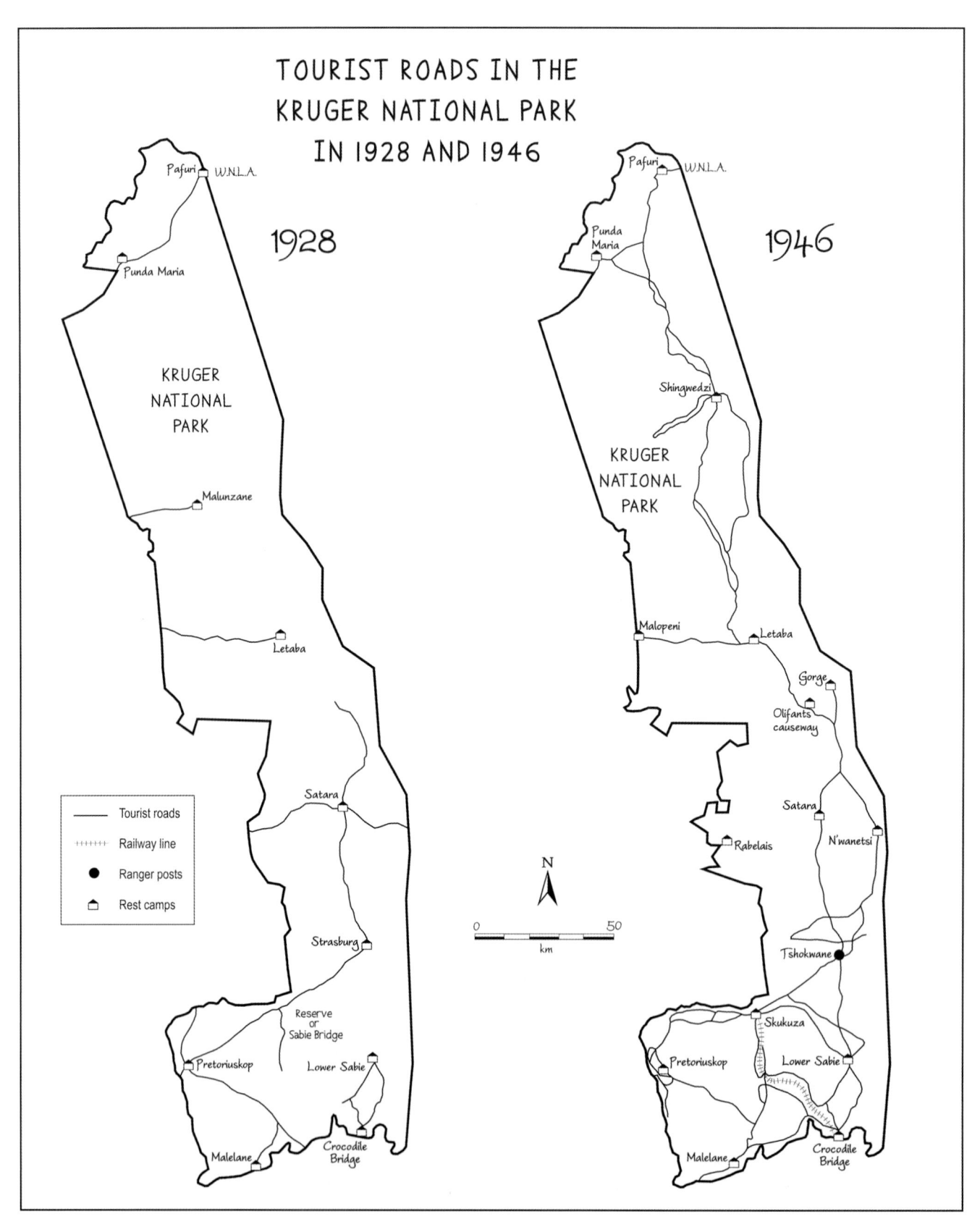

16. EARLY YEARS: THE SABIE AND SHINGWEDZI RESERVES AND THE KRUGER NATIONAL PARK 1900–1946

Tourist roads in the Kruger National Park in 1928 and 1946.

16.4 THE EARLY DEVELOPMENT OF THE SABIE AND SHINGWEDZI RESERVES AND THE KRUGER NATIONAL PARK

Enthusiastic chefs discuss their cooking around a campfire in Gorge rest camp. 1932.
(ACKNOWLEDGEMENT: DR R.E. KAY, DURBAN)

Men washing and shaving in the river at Lower Sabie, 1932.
(ACKNOWLEDGEMENT: DR R.E. KAY, DURBAN)

Visitors around a barbecue fire at Skukuza, 1957.

Prof. Hannes Eloff contemplates his route for the day. Pretoriuskop, July 1938.
(ACKNOWLEDGEMENT: MR A. WEWEGE, NELSPRUIT)

Charlie Sanderson, a son of Bill Sanderson of Legogote, who had been a builder in the Kruger National Park during the 1930s, circa 1932.
(ACKNOWLEDGEMENT: STEVENSON-HAMILTON COLLECTION, SKUKUZA ARCHIVES)

The main road between Skukuza and Malelane in 1947.

Several attempts by the Board to have the railway bridges in the Park opened to motor traffic by the SAR to allow motor traffic during the early 1930s, failed.

In 1935 an encompassing development plan for the Park was drawn up in anticipation of a grant from Government. Provision was even made in the plan for roads consisting of two-spoor tarred strips. However, the grant of £30 000 (R60 000) was not sufficient to meet all the demands, therefore the Board approached the TPA and the National Roads Council for assistance. When it became known in 1935 that the SA Defence Force (now SANDF) needed a road parallel to the eastern border, it was suggested that they be approached to construct a tar road from Punda Maria to Crocodile Bridge.

All attempts to get the roads tarred, failed and in 1937 the secretary of the Executive Subcommittee was instructed to inquire at the University of Pretoria about a tar emulsion for use in Pretoriuskop rest camp to alleviate the dust problem. When Chairperson De Wet heard about these developments, he reacted vehemently to the so-called improvements. "Are we going to maintain Pretoriuskop as a rest camp or a glorified resort?" he wanted to know. Member Papenfus maintained that the dust that accumulated on the vegetation along the roads in general was harmful to animals and strongly urged that a loan be sought to tar the roads. At the Executive Subcommittee meeting in September 1937, the Board adhered to Papenfus's proposal and it was decided to proceed with experimental striptarring on the Pretoriuskop/Doispane and Skukuza/Lower Sabie roads. This decision was never carried out.

Stevenson-Hamilton remained strongly opposed to tarred roads until his retirement. He felt that they would give rise to unnecessary speeding which would result in too many animals being killed. He felt so strongly that he even maintained: "the death toll of animals would increase beyond all bounds". However, he had no objection to tarred roads inside the rest camps.

By the end of 1946 nothing had come of any improvements to the roads. There were in fact very few of the existing dirt roads which had been properly gravelled. The tarring of road surfaces in the Park had to wait until August 1965, when tarring of the

Napi road between Pretoriuskop and Skukuza started. Currently, there are more than 850 kilometres of tarred roads in the Park, apart from 1444 kilometres gravel roads and more than 4200 kilometres firebreaks.

Administration

Until the proclamation of the Kruger National Park in 1926, all the executive and administrative duties had been carried out by the warden and a handful of rangers. Initially their duties were limited to conservation functions and Stevenson-Hamilton expressed the requirements to be a ranger as follows:

> A man must be prepared not only to lead a lonely life, and to travel about the country at all seasons and in all weathers, but must cheerfully incur the certainty of ordinary malarial fever and the possibility of black water. He must also be thoroughly conversant with native languages, customs and idiosyncrasies, interested in game and all pertaining to it, an expert bushman, and possessed of some knowledge of public works, and of methods of working native constables. Above all, he must be thoroughly active, reliable and trustworthy.

With the development of the tourism industry in the Park from 1926 onwards, the rangers themselves had to create the necessary infrastructure, including rest camps and roads. However, it did not stay there and they had to carry out duties including directing tourists to their accommodation, issuing permits and checking and collecting the payments for accommodation. They also had to run the pontoons, provide information and even sell petrol coupons.

To achieve all of that during the pioneering years and later in the Kruger National Park, the handful of white and black rangers and their black assistants used the services of the group of blacks living inside

Orpen camp shortly after its completion, circa 1955.

The Tshokwane picnic site has been very popular since the earliest years, 1957.

the Park. These included the few people who were allowed to remain in certain indicated areas as squatters or land tenants, after most had left the Sabie and Shingwedzi reserves in 1903 on instruction of the warden. To be able to remain in the reserve they had to undertake, according to the conditions of the Squatters' Act of 1895, and especially after the Administrative Proclamation of 6 April 1906, which qualified all blacks living on state land as squatters, not to keep dogs or firearms and adult men had to pay tax amounting to the amount of £1.0.0 per annum to the state. They further had to work for the state for free for three months of the year. They were allowed to keep a maximum of ten head of large livestock per household. Because some of these squatters lived in traditional terrains such as Doispane, Phabeni, Gabeni near Salitje, Nsemani, Crocodile Bridge, Malelane, Makhuba (later Makhadzi Mouth), Shidzavane and Stangene, they were forced to earn a living outside the reserve, especially during drought years. During such times they were compensated for their three-month work periods, but during favourable climatic years, they had to work without pay at places that had been indicated by the warden or his rangers.

In view of the little funding that the Sabie and Shingwedzi reserves received from the colonial government and even after the proclamation of the Kruger National Park in 1926 also from the Union government, all organised labour was dependent on the leasehold system. Soon another source of labour developed through migratory labour from Mozambique, which could be utilised by the rangers to perform larger construction works. These illegal immigrants were mostly young men or boys between 17 and 19-years-old who had left home for the first time to earn their own income. Because they entered the Park illegally, they were originally arrested and then had to work for 14 days, before being issued a permit to go and find work outside the Park's boundaries.

This practice of utilising temporary migratory labour was prohibited in 1923 by the Department of Native Affairs. People who illegally crossed the border after that had to be arrested and handed over to the SAP for repatriation to Mozambique. After the adoption of the Mozambique Convention in 1928, which had organised labour affairs between the Union of South Africa and Mozambique, Mozambicans were once again allowed to go job-hunting in the Union. This was apart from the recruits which had been legally recruited by recruiting companies such as WNLA for labour on the mines. The only provisions were that they had to have the necessary identification documents and pay five shillings for an entry permit at certain approved points in the Park. This is how the "Mafourteen" migratory labour system had its origin and it was used from 1928 until 1958 by the Park authorities as an additional source of labour.

As most of those searching for work did not have the money to pay for the entry permit, they had to work in the Park for 14 days after which they were issued with a work permit. A large proportion of work to create infrastructures in the Kruger National Park during the early developmental years had been carried out with the help of "Mafourteen" labour. These tasks included: the construction of roads; maintenance of drifts; opening up of firebreaks; the building of pickets; control of insect plagues like grasshopper swarms; the digging of wells; erection of dams and windmills; and help with the construction of rest camp facilities by the Board's own technical staff.

The recruits came along footpaths over the Lebombo Mountains at specific points. These were at Pafuri, Shingwedzi Poort and Kostini, Mala Mala, Nwanetsi, Mabakana and via Bob and Nkongoma pickets along the Crocodile River. They were then transported to Punda Maria, Shingwedzi, Letaba, Satara, Salitje and Skukuza, where they were allocated to working teams and supervisors. The "Mafourteens" were in general a pleasant bunch and received the same treatment and allowances as the Board's own labourers. They were certainly not misused as a source of "forced slave labour" as certain publications now claim.

The "Mafourteen" migratory worker system ended in 1958 through the intervention of an SAP officer at Skukuza who believed that the system was against certain later immigration and labour laws of the Union government. As a result of this intervention by the SAP, all thoroughfare of blacks from Mozambique, except for the WNLA recruits was stopped by the warden in April 1960.

Employment of Mozambicans was then also

stopped or limited to the minimum. The Mozambican personnel who had already been in service, were provided with the necessary documents to legalise their presence in the Union. All illegal immigrants from Mozambique were arrested in the years after that and handed over to the SAP to be repatriated.

The situation in which the rangers, their assistants and labourers had to undertake all developmental and managerial actions in the Park, could not continue indefinitely and the first technical and tourism personnel were appointed in the early 1930s, albeit temporarily only, to undertake specialised duties.

The first mason was appointed in 1932 to build the low-water bridges across the Letaba and Sand rivers. In the same year the first mechanic was established at Skukuza. The mechanic initially worked in a private capacity and was only officially appointed in 1933. A second mechanic was appointed in 1938 and in 1939 a house and workshop were built for him at Shingwedzi. He was responsible for the mechanical maintenance of vehicles and equipment in the northern part of the Park.

More rest camp supervisors and tradesmen were appointed temporarily, while the permanent staff in the Park in 1946 still consisted only of the warden, two clerks, a relief ranger, nine section rangers and two mechanics.

In 1928 a Professor Cowley from the USA requested permission from the Board to undertake research on ticks in the Kruger National Park. The Board immediately agreed that the project could be undertaken and that research facilities would be made available. In 1933 it was suggested that the Board should appoint a veterinarian and a botanist, but as it was felt that such appointments would lead to demands from other institutions for similar representation, the proposal was turned down. However, the Board did express itself in favour of scientific research in the Kruger National Park. When it was again suggested in 1938 that a biologist be appointed in the Kruger National Park, the Board reacted negatively though. They maintained that such an appointment would not be necessary as long as good relations could be maintained with Onderstepoort, so that they would provide scientists for projects which had to be undertaken.

As a result of the contact which the warden had up to that stage with the research staff of the Division of Veterinary Services, it is certainly not strange that research officers were regarded as synonymous with veterinarians.

In 1944 the question of appointing a biologist in the Kruger National Park again surfaced. This resulted in a meeting in 1945 between the Board Subcommittee, appointed to investigate the organisation and administration of the Board, and a delegation from Onderstepoort under the chairmanship of the director, Dr P.J. du Toit. At this meeting much of the discussions revolved around the segregation of game and livestock due to the transmission of foot-and-mouth and other diseases. They again discussed the advisability to appoint a biologist. This person could advise the Board on such matters as animal

The site of the former "Malopene" entrance gate to the Park. It was replaced by Phalaborwa entrance gate in 1960.
(18.06.1984)

The official opening of Phalaborwa Gate on 3 June 1960.

migrations, distribution patterns, diseases, breeding seasons, veld burning, et cetera. It was felt that such "scientific knowledge would materially contribute and strengthen the argument for better preservation". The Board was advised to appoint a biologist, but a firm decision about the matter was not taken. Member Orpen was nevertheless instructed to make enquiries about a suitable candidate and an appropriate salary.

By the end of 1945 the Board had interviewed three scientists, including Doctors Pole-Evans, Smit and Curson. These three confirmed to the Board that they did not see the necessity of a biologist for the Board itself, but that they were strongly in favour of the appointment of researchers in the Park. This matter was taken further in 1946 when delegates representing various scientific bodies had a meeting with the Minister of Lands. The minister could not see the necessity of such a step and maintained that "... there was nothing wrong with the present administration of the reserves, and men without scientific training had done great work in the preservation of the Union's wild life".

Instead they agreed to create a National Advisory Board for National Parks and Game Reserves to tend to the research needs of all nature conservation resorts. This advisory board was appointed, but the Board only appointed its own biologist in October 1950.

The Board recognised the necessity of dissemination of information very early, even if at first it would only serve as a source of income and to attract tourists to the Park. As early as in 1927 the Board urged the directors of education to distribute information about wildlife in schools, as well as encourage visits of groups of children to the Park.

In 1928 the first illustrated pamphlet about the Park was produced. The pamphlet came about as a result of a combined effort between the Board and the SAR, titled *Kruger National Park: Handbook and Guide*. This was later followed by *Unspoilt Africa*. In

The former Shawu picnic site between Letaba and Shingwedzi rest camps. July 1955.

The former Gorge camp with Olifants Poort in the background. June 1932.
(ACKNOWLEDGEMENT: MR T.C. SINCLAIR DAVID, DURBAN)

1943 The Wild Life Protection Society of Southern Africa encouraged the Board to compile guide books about the geology, vegetation and animals of the Park.

The need to disseminate information was not only restricted to reading matter and in 1934 the Executive Subcommittee of the Board suggested that an Information Bureau be established, but the Board was not in favour of this and suggested that the distribution of information was actually the responsibility of the rest camp contractors. Despite this, the Information and Registration Bureau was established at Skukuza in 1935. The bureau's first appointment was a Mr Zaayman. No further mention is made of this bureau and it is probable that it had come to nothing after a while.

To improve communication in the Park, a telephone connection between Skukuza and Crocodile Bridge had been established in 1913 already, while the telephone connection between Skukuza and Pretoriuskop was established in 1936.

The ineffective communication between rangers and especially the rest camps caused many problems, as this was the primary obstruction in the way of the implementation of an advanced booking facility.

In 1935 Board member W. A. Campbell donated two radio sets for use in the Park, while the Board appointed a committee to investigate the possibility of radio communication between the Head Office in Pretoria and Skukuza. These two sets were installed at Numbi and Skukuza, but were totally inadequate. Earlier in 1935 it had been reported that radio sets were obtainable at £175 a piece. Two sets were purchased as an experiment. In February 1936 Bartle and Company installed one set in Pretoria and one in Johannesburg as a demonstration to the Board. The trials proved highly successful and four sets were purchased and installed at Skukuza, Numbi Gate, Letaba and Punda Maria.

In 1942 these four sets were replaced with radios provided by the SA Defence Force and a fifth set was installed at Satara. In 1943 the SADF installed radios at all the ranger posts with the single exception of Tshokwane. A radio was eventually installed at Tshokwane in 1945 after which all the permanent posts in the Park had radio communication. However, there still was no radio communication between Pretoria and Skukuza. This only became operational

Traditional canvass notice boards at Molopene (Malopeni) entrance gate, 1949.

The place where the Pafuri tented camp used to be. This camp had been used between 1939 and 1941, but was demolished in 1948 and is currently used as a picnic site.
(26.04.1984)

in 1947 together with an additional radio network which connected Skukuza with Punda Maria, Shingwedzi, Letaba, Satara and Lower Sabie rest camps.

The site of the former tented camp along the Tsende (Tsendze) River, just north of Mabodhleleni, used in 1933. (30.07.1984)

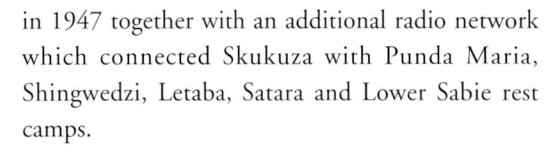

Closure

From the facts in this chapter it is clear that the Sabie and Shingwedzi reserves and later Kruger National Park had a long and colourful genesis. A history which had not been without its ups and downs, but thanks to the belief and perseverance of a small group of dedicated and hardened people, a national park left to current and future generations, which is regarded as one of the shiniest stars among the national parks of the world.

All involved parties should keep on working with dedication to preserve the integrity of the Kruger National Park, because we, as Dr Anton Rupert aptly put it, did not inherit this rich natural heritage from our forefathers, but are simply borrowing it from our children!

References

Agendas and minutes of the South African National Parks Board of Trustees meetings: September 1926 – December 1946.

Annual reports of the Warden of the Sabie Reserve. 1902–1905; 1907–1908; 1911–1918; 1920; 1924–1928.

Annual reports of the Warden of the Shingwedzi Reserve. 1904–1905; 1907–1908; 1911–1918; 1920; 1924–1928.

Annual reports of the Warden: Kruger National Park. 1929–1946.

Annual reports of the National Parks Board of Trustees. 1926–1946.

Anon. 1948. A breach of faith. *African Wild Life* 2(2):1–5.

Bigalke, R.C.H. 1968. *Nature conservation in the Transvaal.* Bulletin no.2. Transvaal Provincial Administration, Nature Conservation Division, Pretoria.

Bigalke, R.C.H. 1939. *A Guide to Some Common Animals of the Kruger National Park.* Van Schaik, Pretoria.

Bigalke, R.C.H. 1939. *National Parks and Their Functions, With Special Reference to South Africa.* Pamphlet no. 10. South African Biological Society, Pretoria.

Bigalke, R.C.H. 1943. Our National Parks: past and future. *South African Journal of Science* 40:248–253.

Bigalke, R.C.H. 1976. Science and nature conservation in the Transvaal. *Fauna and Flora* 27:13–15.

Board meetings: Agendas and minutes September 1926–December 1946.

Brynard, A.M. 1969. Geskiedkundige oorsig van die waterverskaffingsprogram vir wild in die Nasionale Krugerwildtuin. Unpublished Memorandum:1–20.

Carruthers, E.J. 1989. Creating a National Park, 1910 to 1926. *Journal of South African Studies* 15(2):188–216.

Carruthers, E.J. 1990. Towards an environmental history of South Africa. Some Perspectives. *South African Historical Journal* 23:184–195.

Carruthers, E.J. 1993. Police Boys and Poachers: Africans, wildlife protection and National Parks, the Transvaal 1902–1950. *Kleio* 36(2):11–22.

Carruthers, E.J. 1994. Dissecting the myth: Paul Kruger and the Kruger National Park. *Journal of Southern African Studies* 20(2):263–283.

Carruthers, E.J. 1995. *The Kruger National Park. A Social and Political History.* University of Natal Press, Pietermaritzburg.

Davis, D.H.S. (ed.). 1964. *Ecological Studies in South Africa.* Junk, The Hague.

Gertenbach, W.P.D. 1980. Rainfall patterns in the Kruger National Park. *Koedoe* 23:35–43.

Ludorf, J.F. *et al.* 1918. Report of the Game Reserve Commission. Government Printers, Pretoria.

Meiring, P. 1982. *Agter die skerms in die Krugerwildtuin.* Perskor, Johannesburg.

Obermeijer, A.A. 1937. A preliminary list of plants found in the Kruger National Park. *Annuals of the Transvaal Museum* 17:185–227.

Prance, E.L. 1935. *Three weeks in Wonderland: The Kruger National Park.* Juta, Cape Town.

Pienaar, U. de V. 1970. The water resources of the Kruger National Park past and present. *African Wildlife* 21(3):181–191.

Pienaar, U. de V. 1973. Die Nasionale Krugerwildtuin

1946–1973. Unpublished manuscript.

Pienaar, U. de V. 1982. The Kruger Park Saga (1898–1981). In: Bannister, A. & Gordon, R. (eds.). *National Parks of South Africa.* C. Struik, Cape Town.

Pienaar, U. de V. 1984. A history of the Kruger National Park – from the arrival of the first settlers to proclamation in 1926. Paper presented at a Wildlife Symposium: 16–18 March, Berg-en-dal.

Pienaar, U de V. 1985. Indications of progressive dessication of the Transvaal Lowveld over the past 100 years, and the implications for the water stabilization programme in the Kruger National Park. *Koedoe* 28:93–165.

Sandenbergh, J.A.B. 1985. Personal communication.

Schoch, A.A. 1949. Approaches to the Kruger National Park. *African Wild Life* 3(1):41–52.

Speight, A-M. 1972. *Game Reserves and Game Protection in Africa.* University of Cape Town Library, Cape Town.

Stevenson-Hamilton, H. 1962. Preserving wildlife in South Africa. *Optima* 2(5):121–128.

Stevenson-Hamilton, J. 1903. Preservation of game in the Transvaal. Unknown publication, possibly *Journal of the Society for the Preservation of the Fauna of the Empire.* June.

Stevenson-Hamilton, J. 1906. Game preservation in the Transvaal. *Blackwood's Magazine* 1906:407–411.

Stevenson-Hamilton, J. 1906. Notes on the Sabi Game Reserve. *Transvaal Agricultural Journal* 6(4):603–617.

Stevenson-Hamilton, J. 1924. The Transvaal Game Reserve. *Journal of the Society for the preservation of the Fauna of the Empire* New Series 4:35–44.

Stevenson-Hamilton, J. 1928. *The Kruger National Park: The Game Sanctuary of South Africa.* Government Printers, Pretoria.

Stevenson-Hamilton, J. 1930. *The Kruger National Park Handbook and Guide.* Government Printers, Pretoria.

Stevenson-Hamilton, J. 1930. The management of a national park. *Journal of the Society for the Preservation of the Fauna of the Empire* New Series 13:13–20.

Stevenson-Hamilton, J. 1933. A great national park. *The Field* 6 May.

Stevenson-Hamilton, J. 1937. *South African Eden, From Sabi Game Reserve to Kruger National Park.* Cassell & Co, London.

Stevenson-Hamilton, J. 1939. The health of wild animals. *Journal of the South African Veterinary Medical Association* 10(2):56–64.

Stevenson-Hamilton, J. 1941. Wild life ecology in Africa. *Associated Scientific and Technical Societies of South Africa* 5:3–14.

Stevenson-Hamilton, J. 1941–1942. Wild life ecology in Africa. *Associated and Technical Societies of South Africa: Annual Proceedings*:95–106.

Stevenson-Hamilton, J. 1942. The Kruger National Park. *SPCA*:5–9.

Stevenson-Hamilton, J. 1945. General report on the Kruger National Park. Unpublished manuscript: 1–7.

Stevenson-Hamilton, J. 1946. A game warden reflects. *Journal of the Society for the Preservation of the Fauna of the Empire* 54:17–21.

Stevenson-Hamilton, J. 1947. The true approach to wild life preservation. *African Wild Life* 1(2):9–11.

Stevenson-Hamilton, J. 1949. *The Lowveld: Its Wildlife and Its People.* Cassel & Co., London.

Stevenson-Hamilton, J. 1952. The Kruger National Park. *Oryx* 1(6):278–282.

Stokes, C.S. 1953. *Sanctuary.* Maskew Miller, Cape Town.

Struben, F.E.B. 1953. History of the Kruger National Park. *African Wildlife* 7:209–228.

Trapido, S. 1984. Poachers, proletarians and gentry in the early 20th century Transvaal. Paper presented to the African Studies Institute, University of the Witwatersrand, March.

Tyson, P.D. & T.G.T. Dyer, 1978. The predicted above-normal rainfall of the seventies and the likelihood of droughts in the eighties in South Africa. *South African Journal of Science* 74:372–377.

Tyson, P.D. 1986. *Climatic Change and Variability in Southern Africa.* Oxford University Press, Cape Town.

Van Coller, L. 1969. *Die Nasionale Krugerwildtuin: 'n Bibliografie.* University of Stellenbosch, Stellenbosch.

Wilson, J.W.H. 1941. Game reserves in southern Africa. *Ostrich* 11(2):108–111.

Yates, C.A. 1944. *Boy Scouts in Gameland.* In: *Varia.* Unpublished manuscript, Skukuza Archives.

Chapter 17
Quests for buried treasures
A. MINNAAR AND DR U. DE V. PIENAAR

The war-clouds which had hung over the Lowveld from 1899 until 1902 had hardly cleared up when rumours of hidden treasures became more than just a dream of getting rich quickly to many. People searched far and wide and the Sabie and Shingwedzi reserves and later Kruger National Park was not to be excluded. In fact, many of the most interesting and most secretive treasure hunts took place in the Park.

What particularly captured the imagination of most was the so-called Kruger Millions. In his popular book *Memories of a Game Ranger* (1948), Harry Wolhuter recorded details about several search parties who had fruitlessly searched for this alluring treasure. Rumour had it that in 1902 the ZAR (South African Republic) government had buried bar gold to the value of "millions" of pounds somewhere within the Sabie Reserve to protect them against the hazards of war. A multitude of tales about where and when it happened had originated in the meantime. Some claimed that only two men were involved with the matter. They had apparently transported the gold in a Scotch-cart and buried it under at night. Others claimed that a whole commando had transported the gold on an ox-wagon and then buried it. The most interesting account is that some blacks had been employed to dig the necessary hole, after which they were shot. Their bodies were then thrown on top of the gold and buried with it to ensure their silence. Stories about the burying place also varied. Some maintained that the site had been marked by nailing a piece of wood against a tree, while others claimed that the locality was indicated by a cairn. Certainly the most fascinating claim was that no marks had been left, but that the secret was entrusted to an old woman. Nevertheless, the whole affair remains romantic and secretive to this day.

The search for the Kruger Millions started by the middle of 1903 when a certain Captain De Bertodano[1] of the British Headquarters Intelligence Service arrived in the Sabie Reserve. Despite active attempts to keep their activities a secret, it was not long before the story leaked and everybody in the Lowveld knew about it. People were thrilled and excited, especially since it was an official mission. The expedition limited their search to the Lebombo area, but they soon realised that they were busy with an impossible task. Many expeditions followed and all possible attempts were made to find the treasure. One group even consisted of a number of spiritists.

The area which attracted most interest was along the Crocodile River. To the great excitement of everybody, a discovery was made in the area in 1904 by one of the parties digging along the riverbank when they struck a lead of old telegraph wire. They then dug along this wire and after having dug a trench of about 15 metres , they discovered a locked iron safe. Immediately the story spread that the treasure had been found. A message of the finding was telegraphed to Pretoria and soon a whole contingent of armed men arrived to guard the gold, while in transit to a bank's strongroom. The safe was forced open amidst great interest, and then faces turned scarlet as the content was revealed: bundles of old documents connected to the construction of the Eastern Line about 20 years before.

Harry Wolhuter admitted that he had also watched these searches with great interest and admitted that he would not have minded to share in these treasures. He said that one of the blacks who had been working for him once approached him and told him that he had discovered some blacks digging a big hole not far from his camp. Closer examination of the excavation revealed that the hole was in fact a grave, which later turned out to be that of the old hunter and wagon-maker, Gert Stols.

During 1921 an interesting article appeared in *The Star* about the Kruger Millions. According to this, Mr M.M. van Niekerk declared that the so-called Kruger Millions no longer existed. According to him it had been removed from its hiding place years before. Mr Van Niekerk had dedicated almost his en-

---
[1] The same person who investigated the activities of the infamous Captain "Bulala" Taylor of the Bushveldt Carbineers.

tire life to searching for this treasure, but his efforts had until that stage been hampered by the fact that he could not find anybody who had been present when the treasure was buried. Twenty years after the gold had allegedly been buried, he found such a man in the person of John Charlton Denton, a blacksmith from Randfontein. Mr Van Niekerk had heard of Denton by chance and immediately paid him a visit. When Denton was questioned about the Kruger Millions, he was visibly nervous, but eventually agreed to issue a sworn statement. In his affidavit, dated 27 April 1920, Denton declared that he had not been present when the treasure was buried, but arrived shortly afterwards. The exact place, as well as certain landmarks had been indicated to him by a Mr Stoltz, who had passed away in the meantime. In his statement he also undertook to show the site to Mr Van Niekerk. According to Denton, the treasure had consisted of 14 ammunition chests full of gold bars, four chests filled with processed gold and a bag of diamonds.

In August 1920 a group of four men arrived in the Sabie Reserve and reported to the warden, Colonel Stevenson-Hamilton. He referred them to ranger C.L. de Laporte at Rolle. De Laporte did not attach much value to Denton's story and clearly expressed this opinion. However, his skepticism did not put them off and they were determined to carry on with their search.

De Laporte provided them with a team of labourers, as well as a police escort. According to Denton the treasure site was about 63 kilometres from Rolle. The beacons to indicate the site included a number of trees which had been marked in a specific manner. In the trunk of the first marker tree were seven nails. About 60 metres from this tree was a large anthill and 250 metres west of the anthill another tree with a crowfoot carved into its trunk. Close to this were another three trees on which rings had been carved around the trunks, as well as a tree on which an X had been carved. Apart from the marked trees, there was also a cairn.

After searching for several days, they actually managed to find the place. As can be imagined, this caused great excitement among the treasure hunters. Denton was able to indicate the exact position. Great was their disappointment, however, when they discovered a half overgrown, man-made hole measuring about five by one metres in the grass. Everything pointed to the fact that if anything had been buried there, it had been removed years before. Their desperate search which had lasted for days led to nothing but disappointment.

Events which can also be coupled to treasure quests in the Park were recorded in a report of senior ranger Mike English dated 26 April 1984. In this report, ranger English mentioned two graves at Zari in the Shingwedzi area, which had been marked by rough headstones. Ranger English speculated that the graves could be those of former miners. The possibility that their names had been carved on the headstones was further investigated with the aid of infrared photography. According to legend, the graves were those of two Boers who had apparently been murdered in 1901 during the Anglo-Boer War by the cruel war criminal, Bulala Taylor. The two had been too old to join the commandos. They carried on with their gold mining activities at Birthday, or Louis Moore mines. Taylor was stationed at Fort Edward, a few kilometres west of where Elim Hospital is now, along the road between Elim and Bandolierskop.

Field ranger Solly Kubayi in a burrow along the old Barberton transport road near the Nsikazi Drift, where two white men had dug up a "treasure" during the Second World War.
(08.05.1984)

17. QUESTS FOR BURIED TREASURES

Somehow or another, Taylor had received information that these miners were going to deposit an amount of refined gold at the bank in Pietersburg (Polokwane). Taylor regarded this as an ideal opportunity to loot the gold and immediately set off with a contingent of the infamous Bushveldt Carbineers. However, the two Boers had come to hear about their imminent arrival and fled with a wagon and all their possessions over the Bububu River to the Phugwane in what is now the far-northern region of the Park. Near Dhili waterhole, they heard that Taylor and his troops were about to catch up with them, so they pushed the wagon with all their possessions on it over the cliff into a deep waterhole. Then they fled until reaching a black village at Zari mouth where Taylor caught up with them. In an effort to force them to tell where they had buried their gold, Taylor shot their young leader plough-boy in cold blood in front of them. The two aged gentlemen would still not reveal their secret, upon which Taylor had them both shot as well. Their wives buried the two with the kind assistance of the locals and then started walking home. Their children must have been crying terribly on their way back due to shock and exposure as a tributary of the Phugwane is still called Mashadya ("crying children").

The graves were pointed out to ranger English by former field ranger Manuel Kubai. While searching the area, ranger English found significant signs of the events which had taken place, including the remains of some huts, ash heaps and a few potsherds.

In a later report, dated 7 May 1984, ranger English mentioned the so-called Stols grave along the upper reaches of the Stolsnek Spruit, which he had found half dug up in 1977, while stationed in the area. After this discovery, ranger English had the grave repaired. According to legend, the grave had been dug up by blacks searching for hidden treasures before 1958. This is apparently the incident to which Harry Wolhuter referred to above. Former ranger Koos Smit was convinced that it had been locals who were after the treasure and the digging must have taken place in about 1943 or 1944. A year before his death, former ranger Gus Adendorff told his colleague, Mike English, a story linked to this incident. This had in turn been told to ranger Adendorff by Sergeant Helfas Nkuna, a local field ranger at the time. According to this legend, two young men had arrived at Harry Wolhuter's house at Pretoriuskop in 1943 or 1944 and asked if they could search for money or gold, which had been buried in the area by their father or grandfather during the era of the transport riders. According to the two, their father or grandfather's wagon had broken down irreparably inside the area of the current Park. As they had been unable to fix it, they decided to bury their gold or money somewhere nearby. After having done this, they pushed the wagon over the site and set it on fire to disguise the freshly disturbed soil.

As a result of his previous experience with treasure hunters, Harry Wolhuter refused permission for them to carry on with their search. A few days later some of his field rangers arrived breathlessly at his quarters with the news that there were two armed whites with a few black helpers busy digging a deep hole along the bottom reaches of the Nsikazi River. Wolhuter set off in a hurry to the place where the people had illegally entered the Park, but on arrival found only a deep hole and a number of rusted iron hoops which had been used to reinforce wooden

The cliff overhang of "Skatkoppie" (Treasure hill), near Mlambane Dam, where according to legend, ranger Sergeant Helfas Nkuna found a rusted tin full of gold coins during the late 1930s. (11.05.1984)

616

chests (which had been buried here?). Whatever had been hidden inside the hole was gone, while the treasure hunters had disappeared from the scene as well. The place where this treasure had been buried was close to the old transport road from Makhuthwanini (along the Delagoa Bay road) to Barberton and within a few hundred metres from the old drift through the Nsikazi River. Close to the hole there was a pile of stones which had apparently been erected to serve as a marker.

The possible site of this occurrence was later also found by ranger English along the remains of the Old Wagon Route to Barberton. He found glass shards, a rusted screw of the burnt-out wagon and some iron hoops at the site. A more systematic investigation of the terrain would probably shed more light on the thought-provoking secret of what had really happened here.

Not far from this place there is a hill called "Skatkoppie" (Treasure Hill). Not much is known about the history of this particular place, but according to legend a treasure was discovered under a rocky overhang on the western slope of this hill during the 1940s or a little earlier. Skatkoppie is well-known among the black people and there are even rock paintings, which had probably been painted by the San who lived there before.

An interesting legend which could possibly be connected with this particular treasure, centres around the former Swazi Chief Mataffin. This chief's grave lies at the foot of Ship Mountain along the Komapiti Spruit (Imagoroti). By 1887 Mataffin was one of Swazi King Umbandine's (Mbandzeni) advisors and confidants. However, he lost the king's trust when he claimed a young girl on whom the king had an eye himself. When the news reached Umbandine, Mataffin had to flee for his life. The king's warriors followed him up to the Transvaal border, but he and his followers managed to escape and they settled just west of Nelspruit along the Crocodile River. Here he lived in peace and was well-respected by the Boers, as he assisted them during the Sekhukhune wars of 1876 until 1879 with an army of about 2500 Swazi warriors.

In 1891 Mataffin had a disagreement with the well-known native commissioner, Abel Erasmus and decided to seek asylum at Gungunyane, his father-in-law, who lived in Mozambique. He sent his wives and followers ahead and followed later on horseback. Just south of Ship Mountain, he arrived at the village of the Mambayi chief, Inhliziyo, to enquire about the way to the border. This was a bad decision, as this cunning chief hated the Swazis for their cruel raids in the past. He had apparently also heard before that Mataffin had some gold coins in his saddle bags and was therefore devising plans to get this money for himself, so he invited Mataffin to spend the night. The next morning he sent his son and a

Underneath the cliff overhang at "Skatkoppie" near Mlambane. (11.05.1984)

The wooden pole which was used by treasure diggers at "Skatkoppie" near Mlambane. (11.05.1984)

guide to accompany Mataffin. About 800 metres north of the old transport road to Delagoa Bay, Mataffin rode through a small spruit with running water (probably the Komapiti Spruit). Just as he was riding up the bank on the other side, Umgeba shot him between the shoulders. They first took his possessions and then forced his body into an aardvark hole, where it was discovered a week later by another black man by the name of Bob, who had been hunting in the area. He made Mataffin's murder known and had Mataffin's ring as evidence, but Inhliziyo and his followers simply claimed innocence. What became of Mataffin's gold is not clear and the exact locality of Mataffin's grave has also not been discovered yet.

While on patrol during the late 1930s in the vicinity of Mlambane Dam, the famous field ranger, Helfas Nkuna and a few of his men found themselves in the open during a heavy thunderstorm. They found shelter under an overhanging cliff of a nearby hill. While waiting for the storm to abate, Helfas investigated the cave and noticed an old, rusted tin in a crevice. They pulled the heavy tin closer using a forked stick and when they opened the lid, found it filled with gold coins. This finding was immediately reported to ranger Harry Wolhuter, who in turn informed Colonel James Stevenson-Hamilton. It can be accepted that the money was handed over to the authorities, but no mention of it could be found in any official documents. The possibility that this tin of gold coins had been part of Mataffin's gold, hidden in this remote place by Inhliziyo, is quite good if all circumstances are taken into account. News of this discovery soon spread by bush telegraph and was also heard by ranger Tom James, who was in control of the Malelane section where the "treasure" had been found. He was furious because he had neither been informed, nor received a share of the booty. He duly informed his colleague, Wolhuter, to keep his field rangers out of his section in future.

Another interesting character searching for hidden treasures in the Park was Bill Lusk, alias Texas Jack, who had apparently been an excellent revolver and rifle marksman. His colourful career included a stint as pack-mule operator in the American army in the Philippines. He also served in the German army during the Herero War in German West Africa (Namibia). While in German West Africa, he picked up diamonds to the value of £3000 and returned to America with his fortune. However, it did not take him long to spend all the money and he returned penniless to South Africa. Back here, he befriended Buck Buchanan and started prospecting in the Crooks Corner area in 1916. According to Buchanan he had many prospecting licences.

After Buchanan left Crooks Corner, Bill Lusk arrived at his farm one day in 1920 and showed him the best alluvial gold he had ever seen. According to Lusk, he had found his treasure in the Shingwedzi Reserve. Lusk applied for a licence to claim his finding, but it was refused. However, the old prospector was not deterred that easily and every year, during the winter months, he slipped into the reserve by himself and hid his presence so well that nobody could find out where he was. While there, he mined enough

The site along the Larine firebreak where, according to legend, a treasure was buried during the Anglo-Boer War. (08.10.1984)

The anthill with aardvark holes on the Larine treasure terrain where a marker, an old Snider rifle's barrel, had been found during the search of 1957.

gold to provide him with the necessary supplies until the next winter.

One day he arrived at one of the small stations near Mica and after he had unsaddled his donkey, sat down under a tree opposite the local shop. A few days later, somebody remarked that old Lusk had been sitting at the same place for rather a long time. Upon investigation they found that he was dead, with his bags of gold dust lying next to him on the ground. The police attempted to find out where he had found the gold, but this remained a secret until former ranger Harry Kirkman heard from some black people that Bill had mined the gold at Red Rocks (the Gubyane potholes) in the Shingwedzi River.

The theory that the Boers had buried their gold during the Anglo-Boer War out of fear that the British would confiscate it, is quite feasible. According to legend, quite a few Boers actively searched for gold in the vicinity of Fort Hendrina and they were particularly active east of Elim. It was also their habit to sell the gold in Pietersburg (Polokwane) as soon as possible after they had found it. To avoid the British forces, especially Bulala Taylor's murderous band, they followed a detour through the area which is now the Kruger National Park.

On their way to the Lowveld, such a group of Boers usually spent the night on the farm of a Mr Bouwer, a family member of the well-known hunter, Hans Klopper. They were warmly received by the Boer and his wife and left again at daybreak the next morning. According to the legend, they then travelled in the direction of Shingwedzi, and outspanned at a place in the Bububu area, which is now known as Larine. They buried their gold somewhere in this vicinity and marked the spot by placing the barrel of an old Snider-rifle inside a funnel-shaped anthill. They also sawed off the branches of a few marula trees which pointed in the direction of the site. Their aim was to disturb the area as little as possible so that everything would look as natural as possible. After that they returned home. During their return journey, they were once again well received by the Bouwer family. Greyling, one of the Boers, handed his hymn book to Mrs Bouwer, with the request that she should hand it to his wife at Messina (Musina). Apparently there was a sketch in the back of the book indicating the site where the treasure had been buried. Many years later, the Bouwers started searching for the treasure after they had obtained certain information.

A permit which had been issued by the Department of Lands on 8 November 1954, granted Messrs L. Bouwer and S.A. Bouwer permission to search for hidden treasures in the Kruger National Park on behalf of a syndicate. The syndicate consisted of J.L. Bouwer, P.W. Bouwer, B.D. Bouwer, S.A. Bouwer, E. Mingard, H. Fitzgerald, L. Bouwer, H.R.G. Coxwell and R.P.G. Steyn. There is some evidence to indicate that they started their search for the hidden treasure on 4 June 1955. According to the permit, their excavations had to be limited to a radius of 100 metres from the two huts along the Bububu (Larine) River. In the minutes of a National Parks Board of Trustees meeting which took place on 10 December 1957, it was mentioned that according to Board decision No. 74 of 6 December 1965, Mr L. Bouwer and his party, after many failed efforts due to rain, resumed their search for hidden treasures in the Shangoni area of the Kruger National Park on 24 July 1957. The next day they gave up without having achieved anything.

According to former ranger Jan Meyer of Shangoni, who time and again had to accompany the search party, they had found the Snider-barrel on the alleged treasure terrain, but nothing more. According to a personal communication to the authors by the original Bouwers' daughter-in-law, where the two Boers had spent the night, her family had already started searching for the treasure during the early 1930s, often without the necessary permission. All their efforts had also been futile.

Former ranger J.H. Meyer of Shangoni who solved the secret of the hidden treasure at Larine in July 1957.

A marula tree with a sawn-off branch which "points" in the direction of the Larine treasure site. (08.10.1984)

17. QUESTS FOR BURIED TREASURES

Near the crossing of the Larine and Bububu firebreaks is a terrain, which according to all indications in the hymn book, could possibly be the place referred to by Greyling. From this point, about 210 metres in a westerly direction, is a large anthill and a mopane tree. Directly south of this tree is another mopane with five marks on its trunk, thought to be bullet marks. About 103 metres southeast of the tree at the anthill is a large marula tree with a sawn-off branch. In the triangle formed by these three trees, was an evened anthill in which the Snider rifle's barrel was found by ranger Jan Meyer on 25 July 1957. This possibly could be the exact location.

Another treasure legend associated with the Kruger National Park dates back to the era of the Anglo-Boer War. This romantic legend called "Boer hat filled with diamonds" was described by Chilvers in his book *The Seven Lost Trails of Africa* (1930). As so often in the past, this particular treasure resulted in the violent death of three people due to human greed. The resultant drama and murder case made Sir James Rose-Innis, the then chief justice of the Transvaal who heard the case, declare: "It is the most incredible, the most romantic affair that has ever come before me."

This story started at the end of September 1900, when two burghers of General Ben Viljoen's commando were on their way on foot through the southern part of the current Kruger National Park to rejoin their commando. They were part of the group of Boers who General Viljoen had sent to Komatipoort on foot, while the general and the mounted section of the commando crossed the Crocodile River at Hectorspruit on their way to Leydsdorp. To escape the nearing British forces, a large number of ambushed burghers had walked across the border and gave themselves up to the Portuguese authorities. Some of them crossed back over the border and tried to rejoin their commando by themselves, or in small groups. One such group included Philip Swartz and his friend, Pretorius. Swartz had been a short, stocky man, known for his rough manners and unstoppable and violent urges. With the outbreak of the war, there had been a pending case against him in connection with the inexplicable disappearance of a farmer and a team of 16 oxen. However, the case was interrupted by the war. Swartz joined the Boer forces and eventually found himself in the uninhabitable border area between the Transvaal and Mozambique, without a horse or provisions. He and his friend pitched camp in the Lebombo Mountains just north of the confluence of the Crocodile and Komati rivers. The next day they walked into a northwesterly direction in the hope of following their comrades' spoor deeper into the Lowveld.

It was a scorching day and the two footsore travellers were discussing the possibility of finding water in the endless stretch of Lowveld, when they suddenly came upon a gruesome discovery: A human skeleton! The sun-bleached bones had already been stripped clean by vultures and other scavengers, but what immediately caught their attention, was a worn-out leather bag lying to one side. They were amazed and understandably delighted when they opened the bag and found three smaller bags filled with diamonds (apparently enough to almost fill a hat), as well as five small, gold bars.

"We're rich!" was obviously the first thought that went through Swartz's mind. "This stuff means nothing to us if we can't reach the commando and we'll end up exactly like the skeleton," was his mate's sober summary of the situation. "You're a miserable nitwit. We can hide the stuff somewhere and come and fetch it after the war!" Swartz fiercely rebuked.

Venda High Chief Makhado, who had been a thorn in the side of the Boers in the Soutpansberg district for about 30 years (1865–1895), due to land and hunting disputes. His wife was Sotho King Mosheshwe's daughter and President T.F. Burgers was the godfather of their child.
(ACKNOWLEDGEMENT: TRANSVAAL ARCHIVES)

They then divided the treasure and walked on. Eventually they buried their finding east of the Blyde River along the banks of the Brak Spruit at measured distances from the bases of a number of recognisable baobab trees, with the resolve to collect it after the war.

In later years the origin of this treasure was often speculated about and a reasonable assumption was that it had belonged to the Venda high chief Makhado. After the death of Ramabulana, the high chief of the Bavenda in 1864, his youngest son, Makhado, took over from him after bitter tribal conflicts. Makhado had been an experienced elephant hunter, who soon realised the value of firearms in armed conflicts during the golden years of Schoemansdal.

After he and his followers refused to hand back the rifles which had been issued to them by Boer hunters after an elephant hunt, a war situation soon developed in the Soutpansberg. After several failed attempts by the Boers to force the Venda in their mountain strongholds to submission, Schoemansdal was evacuated in July 1867. The once prospering Voortrekker town was burnt down to the ground.

The Venda knew how to mine gold and some of Makhado's followers had also gone to work on the diamond fields at Kimberley. Here, some of them stole diamonds which they then took back to their high chief in the Soutpansberg. Makhado used these diamonds, gold and ivory to buy firearms from certain English traders, as well as from the Portuguese in the eastcoast harbours.

Speculation that it had been one of Makhado's couriers on his way to Delagoa Bay with diamonds and gold to clinch another arms-deal and who died of unknown causes on his way (possibly malaria), is therefore not so far-fetched.

However, Swartz and his companion Pretorius eventually managed to rejoin their commando in the northeastern Lowveld after much roaming about. Swartz was wounded in his thigh during combat with the British, while Pretorius died in action. Swartz recovered from his wound, but was sent to Ceylon as a prisoner of war. He remained there until 1903, but had in the meantime written to some of his lady friends in Johannesburg that his days of poverty would be over as soon as he was back in South Africa. Swartz was in love with one of two lovely sisters who lived with his foster mother. On his arrival back in the country, he learned that she had in the meantime married a Van Dyk, a former member of the ZAR (South African Republic) Police. Her sister had also tied the knot in the meantime with Fanie van Niekerk, a coach driver. Swartz had been highly upset, but apparently soon decided that Mrs Van Niekerk was actually the love of his life and so he started thinking of ways to get rid of her husband. He even went so far as to tell Mrs Van Niekerk that he had seen a vision which revealed to him that she would soon be a widow! In the meantime he also planned the retrieval of his hidden treasure.

He saw an expedition to collect the diamonds and gold which he and Pretorius had buried in the vicinity of the Blyde River as an ideal opportunity to get rid of Van Niekerk. Swartz's friend, Van Dyk, convinced James Colville to finance the expedition from the end of the railway line at Pietersburg (Polokwane), by mule-wagon over Leydsdorp to the Blyde River. Swartz would be accompanied by Messieurs Van Dyk, Van Niekerk, Colville and Donovan.

Swartz was not aware of the fact that Donovan was a member of the Transvaal Criminal Investigation Department. In 1887, while he had been a member of the Australian Police in New South Wales, Donovan single-handedly tracked down a criminal, cornered him in a swamp and arrested him. Donovan first heard about the buried treasure during the train journey on 4 May 1903 to Pietersburg and this put him on guard immediately. Their adventurous journey to the wilderness in the vicinity of the Blyde River with a mule-wagon started the next day. They pitched camp on the bank of the Blyde River on 16 May and were quite relieved to hear from Swartz that they would reach their destination the next day. The expedition set out at dawn and after about 32 kilometres on foot they reached the Brak Spruit without incident at about five o'clock in the afternoon. While the rest of the party pitched their camp, Swartz invited Van Niekerk to see if they could find anything to shoot for the pot. This was the last time that anybody would ever see Van Niekerk alive. While pitching the camp and collecting wood, the other members of the party heard two shots from the southwest, a while later another two shots further east and then it was quiet.

They believed that the two hunters had managed to shoot something for the pot, but waited in vain for their return. It was accepted that the hunters had probably lost their way, but late that night, Donovan heard a noise near the camp and called: "Who's there?" after which Swartz walked into the camp. To their question about the whereabouts of Fanie van Niekerk, he answered that he and Van Niekerk had separated. He had wounded a kudu and followed it, which was why he took so long to return. His explanation about Van Niekerk was that he had probably become lost and was in all likelihood sitting in a tree waiting for dawn. Donovan had an uneasy feeling that something was wrong and decided to keep a careful eye on Swartz.

The rest of the night they repeatedly fired shots to indicate the camp's direction to Van Niekerk, to no avail, and by daybreak there was still no sign of him. After they had left a note to Van Niekerk to indicate the direction into which they were heading, they proceeded to the place where Swartz claimed he had buried the treasure. When they were on top of a rise and could see a far-off hill, Swartz told them to wait right there while he went searching for certain landmarks.

Swartz called to them once and they followed, but could not catch up with him at all as he had disappeared like a needle in a haystack.

The rest of the party was totally demoralised and furious because Swartz had left them in the bush without reason. On Donovan's suggestion they decided to return to their camp along the Blyde River in the hope that they would find Van Niekerk there. Upon their arrival back at the mule-wagon which they had left at the Blyde River, they found a note, apparently written by Van Niekerk, in the spoke of one of the wheels: "Got tired of the fruitless search and have left for Johannesburg today. From there I might go to Bulawayo."

Donovan put the note in his pocket, convinced that it could possibly shed some light on their fruitless and mysterious expedition. An investigation of the mule-wagon revealed that ammunition, blankets, brandy and other items had been removed, but there was no spoor of Van Niekerk, who had owned a distinctive pair of boots, in the vicinity.

Donovan then decided that they should follow the shortest possible route to Leydsdorp to arrive there before Swartz, or Van Niekerk, or both. After they had crossed the Olifants and Selati rivers, they found a black man along the road, who informed them that there was a white man just ahead of them. Shortly afterwards they heard a shot. Donovan commented that it had sounded just like a shot from his own revolver, as it was the only one of its kind in the country. Soon after that, they found Swartz along the river sitting with his feet in the water. He looked tired and bewildered, with his shirt in tatters.

"Where were you?" they chorused. "I told you to follow me, but when you didn't arrive, I found the treasure, tied it into my shirt and buried it again on the riverbank. After that I decided to take the shortest route to Leydsdorp and wait for you there," was Swartz's answer.

After they had flung accusations and insults at each other, Swartz eventually convinced them that they should return to Leydsdorp, as the trek animals were too tired to walk back to the Blyde River. Swartz then started avoiding the rest of the party and spent long periods in solitude. After their arrival at Leydsdorp on 22 May, Donovan immediately reported Van Niekerk's disappearance to the police. The police sergeant thought it quite possible that he had been caught by lions, as there were many in the area. Obviously he was not keen to go searching for the missing man in the bushveld.

Soon after their arrival at Pietersburg (Polokwane), they noticed Swartz exchanging two gold sovereigns for small change. This seemed strange to Donovan, as he knew that Swartz had no money and also that the only member of the group who had had two gold sovereigns had been Van Niekerk. Once on the train back to Johannesburg, Donovan asked Van Dyk to lure Swartz out of the compartment in some or other way. When he had managed to do so, Donovan hastily searched for Swartz's notebook among his possessions. He compared the handwriting in it with the one on the note which had apparently been written by Van Niekerk and left in the wagon's wheel spoke. As he had expected, it was one and the same – Swartz's handwriting!

Upon their arrival back in Johannesburg, Donovan immediately reported his suspicions to the Chief of Police, Howard Chadwick. An official investigation team was sent out and they found Van Niekerk's remains, not far from where their camp had been along the Brak Spruit. The body had already been damaged by vultures, but they were nevertheless able to determine that Van Niekerk had been killed by a bullet from Philip Swartz's rifle. Swartz was arrested, sent to trial and found guilty of premeditated murder and robbery. On 15 February 1904 he paid the highest price for his wrongdoing.

What happened to the treasure, if there ever had been one, still remains one of the many mysteries of the Lowveld bushveld.

These are only a few of the many stories and legends in connection with treasure hunting in the Kruger National Park and its vicinity. There are many more. Read about them in Stevenson-Hamilton's book *South African Eden.* One can also refer to the many applications for permits to look for treasure in the Park, usually with the suggestion that the applicant knew exactly where the treasure was and to Charles Gerber's prospecting hole at Malonga, where he repeatedly searched for diamonds during the winter months of 1926/27. In former ranger H. Ledeboer's diary there are many references to prospectors and treasure hunters, including Gerber, Pienaar, Van Deventer, Coetzer and Fourie, who fruitlessly prospected for diamonds in the vicinity of the confluence of the Olifants and Letaba rivers. This happened while he was stationed at Mondzwani just north of the current Letaba rest camp from 1923 to 1927. Judged by the gravel heaps around the ruins of his former residence, Ledeboer himself also tried to find a few of the elusive gems.

Dr U. de V. Pienaar received a letter from the legendary elephant hunter Bvekenya Barnard's son, Mr Izak Barnard of Geysdorp, in July 1985. In the letter he told Dr Pienaar about his cousin, Leonorah Pullen, who had at that stage been more than 80 years old. Ms Pullen was the daughter of Billy Green, the brother-in-law of Bvekenya Barnard, who had been at Pafuri with him. Ms Pullen had apparently also lived in the area for a year in about 1920. According to her, her father had been the first white person who had camped at Punda Maria. She also mentioned a road which Bvekenya Barnard had built to connect Soekmekaar and Pafuri via Ntoma, Punda Maria and Klopperfontein, and metioned that a few diamonds had been picked up in the vicinity of Klopperfontein. She also said that her father, Billy Green, Bvekenya and another hunting partner, Fred Roux, regularly went hunting in Mozambique where they had a hunting camp. One day the Portuguese visited the camp and arrested Roux. He was in jail for many years, but nevertheless managed to get a message to Bvekenya once. According to the message, Bvekenya had to take 500 stones from an anthill to build a house for Roux's family. Apparently he had meant 500 golden sovereigns, but Bvekenya never managed to discover the anthill at Crooks Corner.

Another interesting story was recorded in an SAP report of 16 March 1923, involving Brinley Pritchard, who managed to convince the police that he had discovered gold in the Park and needed police protection to fetch it. The police accompanied him into the Park. He told Sergeant H. D'Arcy that he and one Ebersohn were employed by a Mr Prins at Mooketsi. One day Ebersohn mentioned to him that he had managed to obtain a permit to search for hidden treasures. During his search he found an old man, Matibela, who lived about 70 kilometres from Rolle in the direction of Pietersburg (Polokwane). When Matibela heard what he was looking for, he told Ebersohn that he knew where there was gold and that he would sell it to him for £25. Ebersohn immediately paid the money and received a number of gold bars of which he buried 75 at a spot which he described to Pritchard. The rest he apparently buried on the farm of an Italian, Beretta. According to Pritchard, Ebersohn had asked him to fetch the gold as he wanted to sell it and then gave him a description of a place near a hill where he had buried the gold. Pritchard declared to the police that he was scared of undertaking the journey on his own, and therefore requested police protection. The police fell for his story and accompanied him into the Park.

The police affidavit described the search in detail and it soon became clear that Pritchard did not have any idea of where the gold was. Sergeant D'Arcy's report ended as follows:

> I do not believe this story ... The Low Country is full of these tales of hidden treasure, and I believe that Pritchard had been listening to them, and came

to the conclusion that with a little assistance he could locate one of the places he had heard referred to, so, to obtain the necessary assistance, concocted the story ... (Unpublished report, Skukuza Archives).

Former ranger W.H. (Bill) Lamont received an instruction in 1939 to prepare a terrain for a tented camp along the Luvuvhu River at Pafuri near Hapi Hill (the terrain of the current picnic spot). While his labourers were fencing in the compound, ranger Lamont scratched with a stick in a hole in the trunk of a jackalberry tree. Great was his surprise when he discovered a tobacco pouch with a number of gold sovereigns and other coins. He later showed his finding to Mr H.H. Mockford, the WNLA official at Pafuri. The two of them speculated that it could have been hidden there by a WNLA recruit on his way back from the gold mines. It is not known what ranger Lamont did with the money.

Colonel James Stevenson-Hamilton mentioned in his book, *South African Eden,* a new profession which had developed at the height of the treasure hunt fever – that of treasure guide. There were not many guides and the "occupation" was also more restricted to a few hardened Lowvelders. Their *modus operandi* was to hang around public places, such as bars, where suggestive hints would be dropped. If somebody made enquiries, the guide would explain that he was the only one alive who knew the exact location where the treasure had been buried. It would simply be a matter of going there and digging it up. He definitely would have gone there himself long ago, but unfortunately lacked the funding. It would not be long before a man of some means agreed to sponsor the expedition. The guide would be paid a fixed remuneration, receive part of the treasure and could live there free of charge. On arrival in the general area where the treasure was supposed to be, the guide would search for certain landmarks, which he had created before. The area where the search took place was usually focused on an isolated hill, like the one just north of the Sand River about 30 kilometres from Sabie Bridge. The search would last until the expedition became discouraged, or had run out of provisions. The search parties always returned empty-handed and disappointed, but then the guide would try his best to convince them that there really was a treasure and that they should try again in future.

That the Kruger National Park or its adjacent areas had probably revealed some of its secrets or some putative hidden treasures, can be deducted (with a pinch of salt) from the remarkable report which ranger J.J. Coetser of Punda Maria sent in May 1922 to Colonel J. Stevenson-Hamilton, at Sabie Bridge.

According to the report, Messieurs Jan Otto and Piet le Roux, accompanied by a black man, George, and a young black boy arrived by car on 13 May 1922 from Pafuri at Punda Maria. They told him that they had become lost and therefore did not report at his camp first. The purpose of their visit apparently was to speak to the Portuguese commandant at Pafuri to obtain a licence to hunt in Mozambique. However, the Portuguese official informed them that they could only obtain a hunting licence through the governor at Delagoa Bay. They would therefore have to go back to Pretoria to obtain the license and return later to go on their hunting trip. They spent the night at his camp. Before he went to bed, he noticed that the two blacks were sleeping in the car and because it was a very cold night, he offered that they could rather sleep in the kitchen. Otto, however, quickly answered that they always slept in the car and declined the offer on their behalf. The next morning the party left for Pretoria.

On 23 May, Coetser was at Klopperfontein accompanying a hunting party consisting of Messrs Albasini (probably Joe Albasini, the great grandson of the famous João Albasini), Van Breda, Beyers and Cilliers, through the reserve to Portuguese East Africa. Just before they were about to depart for Pafuri, Messrs Otto and Le Roux, again accompanied by George, arrived in their car. When Coetser asked them where they were going, Le Roux answered that he had lost his purse with £75 in it during their previous visit, while fixing a flat tyre. Albasini and his group also joined in the discussion and Coetser informed them about Le Roux's problem. Coetser told the two that if they had lost the money in the reserve, he would go with them to find it. He would also ask his black rangers if they had not perhaps picked up the purse during their patrols.

However, Otto called Coetser aside and when they were out of hearing distance he told Coetser that he would like to confide in him. If Coetser would help them, they would pay him ten per cent of whatever they found. Otto said: "Le Roux has lost no money – we are on the spoor of the Kruger Millions, thirteen chests full of gold ingots and specie. It is buried near the Pafuri border in Rhodesia [Zimbabwe} and George knows where the place is." Coetser agreed and followed them in the car to the crocodile waterhole near Makuleka's village. They sent George through the river to determine if the chests with gold were still there. After about two hours, Mr Albasini's party also arrived there. The next day Coetser told Messrs Albasini and Cilliers about the real purpose of Otto and Le Roux's visit to the reserve. Albasini and his party then also crossed the river at Makuleka's village to enter Rhodesia.

About 6 pm that same day, George, accompanied by his wife, arrived back from his mission. Otto informed Coetser later that night that the chests with gold were still at the same place. When Coetser enquired about what Otto was going to do next, he answered that he needed donkeys to fetch the chests, as they were too heavy to carry and that he would like to borrow four donkeys and two horses to carry the load. Coetser agreed to let him have four donkeys, but not the two horses and said that they would have to go back to Punda Maria to fetch them. The next morning Coetser sent field ranger Nganesi with the four donkeys to the crocodile pool at Pafuri, with instructions that he had to wait at Otto and Le Roux's car until they return. Otto had also asked for someone to help with the donkeys and Coetser told him that he could ask Shirilele at the crocodile waterhole to assist him. That night they agreed that they would fetch the chests with gold with the donkeys and then bury it in the reserve. Otto and Le Roux would then take some of it to Pretoria to sell in Johannesburg or Delagoa Bay, after which they would return to collect more. The next morning Otto and Le Roux left for the crocodile pool with George in the car. Coetser gave them a piece of bread as they had run out of food. After their departure, Coetser summoned his corporal, Six, and told him everything he knew. Coetser informed him that as soon as all the chests were buried in the reserve and Otto and Le Roux had left for Pretoria, they would dig out the chests again to bring to Punda Maria. There it would be guarded while the matter was reported to the police at Sibasa.

The next day at about ten in the morning, Otto and Le Roux were back at Punda Maria, in very high spirits. According to them, they had found the place where the chests had been buried, but that George's father had moved the chests to another spot. They were in a great hurry and would not even wait half an hour so that Coetser could offer them something to eat. They then asked Coetser to issue George a pass so that he could visit his father in Rhodesia and determine the new locality of the hiding place. As soon as George had done so, Coetser was to send a telegram to Otto. They would then return to move the chests to the reserve. According to them, the treasure was in a cave, so they wanted to borrow a hunting lamp and a crowbar to open one of the chests to check the contents.

After that Otto and Le Roux hastily left for Pretoria. As soon as they had left, Coetser sent field ranger Babalala with a telegram to Sibasa:

Ranger Coetser,   Commissioner of Police
Punda Maria   Pretoria
Jan Otto, a tram driver from Pretoria, and Piet le Roux, also from Pretoria have left here today 10 a.m. by motor for Pretoria. Please have them closely watched; I think they will arrive Pretoria tomorrow, Tuesday. Letter follows by tomorrow's mail.

George duly came to ask Coetser for a pass to enter Rhodesia in the afternoon and told Coetser the same story as Otto and Le Roux. Coetser then instructed Corporal Six to arrest and handcuff him. He told George that he knew everything about the gold and that he had already sent a telegram to the police in Pretoria to arrest Otto and Le Roux on their arrival. Coetser also informed him that if he told them the truth, he would probably become a state witness and not land into any trouble. About half and hour later, Corporal Six reported that George was willing to tell the whole story. Coetser therefore took him to his office where he made a statement to Coetser, Corporal Six and ranger Jacob.

According to the statement, his father had called him and told him that he wanted to show him something before he dies – something that nobody but himself knew about. He told him that a group of between six and eight Boers arrived at his village during the Anglo-Boer War and had asked him to show them a place where they could bury a large amount of gold. He then showed them a cave in a rock face and they hid two chests there. They also told him not to tell anybody about it, as they would be coming back after the war to fetch it. George's father took him to the cave and showed him the place where the two chests had been hidden. After his father's death George went to Pretoria and worked for many years for Mr Otto on the trams as well as at his house. One night Mr Otto talked to him about the Boer War and then George told him about the two chests of gold which had been buried near his father's village in Rhodesia. Messrs Otto and Le Roux and George travelled to Pafuri shortly afterwards and went to see the Portuguese commandant, so that nobody would become suspicious about their real mission. After that they travelled down the Pafuri River (Luvuvhu) and left the car. They crossed the river into Rhodesia and walked to the cave, then took out one of the chests and because it was so heavy, they had to carry it between them on a pole. After that they returned to the car and hid it under the car's back seat. When they arrived at Punda Maria, Mr Otto told him that he had to sleep on the back seat on top of the chest.

The next day they returned to Pretoria and after their arrival, George was told to get off at Mr Rorke's home. They left him there and he did not know where Messieurs Otto and Le Roux took the chest with gold to. When George saw Mr Otto the next day, he said that they had to go back to the place and collect the other chest as well. They left about a week later and that was when they met Coetser at Klopperfontein. After the arrangement about the donkeys had been made with Coetser, they made Shirelele wait with the donkeys at the foot of the mountain, while the three of them went to carry the chest from the cave to the bottom. They first opened the chest and removed some Kruger pounds, which they wrapped in a shirt to make the chest lighter. They also wrapped the chest in their coats so that Shirelele and Nganesi could not see what was in the bundle which they tied onto the donkey's back. They hid the chest in the car after Nganesi had gone to bed. Mr Otto then told George to walk to Punda Maria and tell Coetser that his father had moved the chests to another location and that he had to get a pass from Coetser to go back to determine where the new hiding place was. They did not want Coetser to find out that the gold had already been found and removed. George had to return to Pretoria later and then they would pay him well. His father died ten years later and he had been his only son.

Coetser further reported that after George had completed his statement, he departed on horseback for Sibasa. Upon his arrival there he found that his messenger had only arrived there after the telephone exchange had already closed. He therefore telegraphed the Commissioner of Police the next morning through the Commandant of Police at Louis Trichardt.

His message (loosely translated) read as follows:

Jan Otto, a former tram driver of Pretoria and Piet le Roux, taxi driver, also of Pretoria, well-known to Mrs Coetser of 25 Brown Street, Pretoria, had left Punda Maria in the Shingwedzi Reserve at 10 am. On 29 May 1922 by car with a chest full of bar gold and Kruger pounds (the chest is about as big as an ammunition chest). It is supposed to be a hidden treasure which they had illegally brought into the country from about 10 miles from the Transvaal border in Rhodesia. Please have their movements carefully monitored. In the meantime I am on my way to the police officer in charge at Louis Trichardt with one eyewitness and all available information. Please also monitor the movements of Mr Rorke, a tram conductor of Pretoria. Otto and Le Roux told me that they would arrive at about 6 pm. on 30 May 1922 in Pretoria. They travel by car.

The next morning at nine, Coetser rode all the way to Louis Trichardt on horseback and arrived there in the evening. He went to Mr Parker, the District Magistrate and also to the police officer in charge. The police commandant organised for George and Coetser to travel to Pretoria by train. After their arrival there, the police took them to a house where he told a Major Roos everything about the incident, after which he returned to Louis Trichardt on the next train. During his absence, Corporal Six was in charge of Punda Maria.

On his arrival back at Punda Maria, Coetser was

told that no hunting parties had crossed into Portuguese territory during his absence and that everything was in order. Then he signed his declaration: "J.J. Coetser, Ranger."

There are no indications in the Parks Board's records of any follow-up by police on the matter, or of Colonel Stevenson-Hamilton's reaction to Coetser's report. What is significant though, is that Coetser was transferred from Punda Maria to Satara shortly afterwards!

There are still many stories about treasures and finding them, most of them ungrounded and farfetched. They are nevertheless very interesting, although hardly any documentary evidence can be provided as proof. General Smuts himself made a remark after the Anglo-Boer War, that the treasure hunting, especially with regards to the Kruger Millions, was an illusion. If there had been a treasure like that, he would have known about it. But the hope and expectations will always be there.

The possibility that may become reality will always allow people to fantasise about great hidden treasures. The Kruger National Park will, with its close historical relation with gold, always live on in people's dreams, but at the same time form an integral part of the mystery and romance of the Lowveld ...

References

Adendorff, G. 1984. *Wild company.* Books of Africa, Cape Town.
Bulpin, T.V. 1984. *Lost Trails of the Transvaal.* Books of Africa, Cape Town.
Carruthers, E.J. 1995. *The Kruger National Park. A Social and Political History.* University of Natal Press, Pietermaritzburg.
Carruthers, E.J. 2001. *Wildlife and Warfare. The Life of James Stevenson-Hamilton.* University of Natal Press, Pietermaritzburg.
Chilvers, H.A. 1930. *The Seven Lost Trails of Africa.* Cassel & Co., London.
Hall, H.L. 1937. *I Have Reaped my Mealies.* Betteridge & Donaldson, Johannesburg.
Schulenburg, C.A.R. 1981. Die Buschveldt Carbineers. *Historia* 26(1):37–58.
Stevenson-Hamilton, J. 1974. *South African Eden.* Collins & Co., London.
Wolhuter, H. 1948. *Memories of a Game Ranger.* The Wild Life Protection Society of South Africa, Johannesburg.
Yates, C.A. 1935. *The Kruger National Park.* George Allen & Unwin, London.

Miscellaneous documentation
Letter to Dr. U. de V. Pienaar: Mr Izak Barnard, 19 July 1985.
Letter to Dr. U. de V. Pienaar: Mrs Pam Pullen, 14 August 1985.
Minutes of a meeting of the National Parks Board of Trustees held at Pretoria on 10 December 1957.

SAP-report
Segeant H. D'Arcy, 16 March 1923, Pietersburg (Polokwane).

Rangers' reports
Senior Ranger M. English, 26 April 1984; 7 May 1984.
Diary: Ranger H. Ledeboer: 1923–1927.

Personal communications
Mrs "Klein" Bouwer, Messina (Musina).
Mr H.H. Mockford, Pafuri.

Chapter 18
Lonely graves in the bushveld
Dr U. de V. Pienaar

There are many graves in the Transvaal Lowveld and the Kruger National Park and these cairns (some with simple headstones) and forgotten heaps of soil bear silent witness of a violent and eventful past.

Through the centuries and after the settlement of the black tribes, a countless number of people were wiped out during bloody campaigns. The first inhabitants of the Lowveld during the Later Stone Age probably had to move out of the area by the end of the Middle Ages to escape the black settlers from the north, or run the risk of being wiped out.

According to Punt (1975) there still were a few remaining Hottentot (Khoi) tribes in the vicinity of Badplaas in the eastern Transvaal by the end of the eighteenth century, but they were all murdered shortly after 1800 by the Swazi. According to legend, the Voortrekkers discovered many skeletons at Badplaas Hill in about 1840.

The never-ending conquering campaigns of the baNgoni warriors of Manukosi (Soshangane), Mawewe, Muzila (Mzila)[1] and Gungunyane from the coastal areas of Mozambique, the Swazi of Somhlolo (Sobhuza I), Mswati (Umswazi) and Umbandine (Mbandzeni) from Swaziland and the Sotho-speaking tribes from the Drakensberg area and escarpment in the west, almost managed to wipe out the original baHlangane and other inhabitants of the Lowveld in the area which is now the Kruger National Park. They were forced to live like animals in the bush, or hide themselves in holes, rocky crevices or caves to escape the aggressive warriors. Some of these cave dwellers died in their hide-outs and were buried there by their tribal members by stacking stones on top of them, such as the grave found in a cave at Bangu Poort.

As a result of the exterminations there were never large black settlements in the Transvaal Lowveld during the eighteenth and nineteenth centuries. As late as 1890, when the eastern border between the South African Republic (ZAR) and Mozambique was demarcated by a commission of the ZAR, surveyors Von Wieligh and Vos discovered the remains of a large number of people at Shilowa Poort who had been murdered by Gungunyane's warriors.

The baHlangane, Kumane-ku-Mhuri, who lived at Kumana in the Sabie Reserve until his death in 1922, gave a life-like description to Col J. Stevenson-Hamilton of one of the Swazi raids under Chief Sihanti on defenseless baHlangane settlements in the bushes along the Sabie River east of Skukuza. This incident must have taken place in about 1860. He told that one cold winter's morning, they were woken up by the barking of a dog. This was followed by a woman's piercing scream. The next moment hundreds of Swazi warriors started shouting and streaming into the village, killing everybody without exception.

He and his widow mother and a few others managed to escape by hiding in the reeds. Later they fled northwards to the Nwatindlopfu hills where he had to leave his mother behind, before fleeing further to the Nwaswitsontso. He and the other survivors settled there and he later became an honoured chief.

João Albasini was probably the first European who lived in the area which is now the Kruger National Park, when he settled along the Sabie River at Magashula's kraal in 1845. Taking the kind of raids described above into account, it is no wonder that he was soon surrounded by many Sotho- and Tsonga-speaking people, who were seeking asylum from the murderous bands of Swazis and Zulu-Shangaan.

After Albasini had moved to the Soutpansberg and settled at Goedewensch along the Luvuvhu River, he again provided protection to the Magwamba tribe, which had fled from Mozambique into the area. They soon acknowledged him as their white chief.

Another reason for the high mortality rate in the Lowveld was the malaria mosquitoes (*Anopheles gambiae* and *A. funestus*). The occurrence of malaria, or blackwater fever was so high in the years before 1930, that the Lowveld soon became known among the whites as the "white man's grave".

During Fort Rio de Lagoa's (renamed to Fort Lijd

---

1 Also spelled "Umzila" (compare Erskine, 1869).

zaamheid, now Maputo) occupation by the Dutch East India Company between 1721 and 1730, more than a 1000 people died of malaria according to calculations. Voortrekker leader, Louis Trichardt, poignantly described the toll malaria took among his people in his diary. Of his original group of 53 whites and seven coloureds, Hans Strydom's wife and four children had already died during their stay in the Soutpansberg. Gert Scheepers had died in January 1837 during the search for the Van Rensburg trekker party and was buried in an unknown grave on the Lebombo Flats. As such he possibly became the first white who found his last resting place in the area which was to become the Kruger National Park.[2]

After a trek of eight months, Louis Trichardt eventually reached Delagoa Bay on 13 April 1838 with his party consisting of 46 whites and seven servants. Within 15 months after his arrival, his wife Martha, himself and another 18 whites and four of the servants were wiped out by malaria. His son Karel (Carolus) was the only adult man to survive. The remaining women and 22 children and three servants were transported to Port Natal in July 1839 with the sailing ship, the *Mazeppa*.

The other Voortrekker leader, Lang Hans van Rensburg and his party of 49 who moved into the Transvaal with Louis Trichardt, had reached the Soutpansberg area before Trichardt. Rather than waiting for Trichardt's party, he decided to continue onwards and reach Inhambane along the Mozambican coast. By the end of July they were attacked by Manukosi's warriors at Sakana's village along the Limpopo. The whole party, excluding a little boy and girl who had been kidnapped, were murdered. Hence, they were the first Voortrekkers to die for the ideal to find a way to the sea in the lowlands of Mozambique.

Malaria mosquitoes also sowed their seeds of death in Ohrigstad and Schoemansdal and both these towns had to be evacuated partly as a result of the unhealthy conditions. Ohrigstad was evacuated in May 1849 and a new town established further south, which the Voortrekkers named Lydenburg to remind them of the suffering and loss of human lives at Ohrigstad. Many people died in Schoemansdal as well and it was finally evacuated in 1867 mainly as a result of the repeated Venda rebellions.

There is a heart-rendering tale of human grief and suffering due to malaria told by a trader, Theodor Julius Adolph Coqui[3] to the explorer Thomas Baines (Baines, 1864). According to Coqui, he and a party comprising five white and three black men, left Ohrigstad on 4 March 1846 with two wagons. They were on their way to Delagoa Bay along the "Oude Wagenweg" (Old Wagon Route) to fetch tradewares. They reached the coast on 2 April. Here they witnessed a ship full of slaves who had been sold to slave traders by Manukosi, being relieved by the British frigate *Cleopatra* and the slaves set free. They also met an emaciated English renegade busy recovering from malaria. His friends had already all died from the disease.

On 7 April they set off on the return journey, totally unaware that they had all already contracted malaria. Seven days later they reached the Komati River after half of their oxen had already died of nagana[4] along the road. The remaining 12 oxen were too weak to pull the wagons through the river and three of the party members were already down with malaria. After another two days, the whole party, except Coqui himself, was seriously ill with malaria. In the meantime they had sent one of the servants with a letter to Ohrigstad to request help. After six days the first black in the party died and had to be buried. Three days later, when they were already expecting help, they heard that their messenger had died along the road. By then all the oxen had died and were lying around the tents, rotting away, while vultures and hyena feasted on their carcasses. In desperation they sent another servant with a letter to Ohrigstad, but just a few days later the first white member of the party, Cobus Snyman, passed away, and Coqui also fell ill. When he finally got better, he found the body of his friend, Van Helsdingen, sitting up on his wagon, with his son sleeping next to him.

The boy died the next day and father and son were buried together. Coqui and his father-in-law were the only two survivors.

---

2 Dr Punt (1953) claimed, however, that Gert Scheepers had died and was buried during January 1837 in the vicinity of the current Karulen Missionary Station, east of Mashao.
3 Possibly the same person after whom Dr Andrew Smith named the coqui francolin in 1843 (*Perdix coqui*).
4 Nagana is a deadly blood disease which is transmitted from wild animals to livestock through the bite of tsetse flies.

In a last desperate attempt to escape the place of death, Coqui hired 20 locals to carry them on a stretcher made from a bed. The hired hands only carried them for a night and a day and then refused to go any further. After an offer of more gifts, they were eventually convinced to carry on until reaching a village, where the two were put in an empty hut. Van Breda died that same afternoon. Too late to achieve much, help finally arrived. Two young Boers, relatives of one of Coqui's group, brought some oxen from Ohrigstad. The next morning they fetched one of the wagons from the Komati River and then trekked to Ohrigstad hastily where they arrived at the beginning of May, with Coqui being the only survivor of the trip. It took him nearly a year to recover from his huge loss and he took a solemn vow never to travel to Delagoa Bay again.

In the epic tale of Dina Fourie, which is related in chapter 6.3, the malaria mosquito also played a primary role.

During the building of the Eastern Line from Pretoria to Lourenço Marques from 1888 to 1894, and the Selati Line from Komatipoort to Pietersburg (Polokwane) from 1893 until 1912, malaria exacted a high toll among the railway workers. The somewhat exaggerated statement is often made that for every four kilometres of railway which run through the Lowveld, at least one worker had paid with his life (Buchan, 1926). However, it is a fact that many people lost their lives, both black and white, but the exact number will never be known (Buchan, 1926). The travel journals and diaries of the earlier explorers, hunters, recruiting agents for the mines, prospectors and missionaries are interwoven with references to malaria, black-water fever and casualties as a result of this dreaded disease. Despite the fact that quinine had already been generally used for the treatment of malaria at the time of the Anglo-Boer War, many Boers and British soldiers died of it.

After the reproclamation of the Sabie Reserve in 1902 and later proclamation of the Shingwedzi Reserve, the first rangers regularly became victims of malaria attacks. Harry Wolhuter mentioned entire families wiped out by malaria or black-water fever.

The Lowveld and the Kruger National Park would today still have held the same terror of death if it had not been for brilliant research efforts. By the end of the nineteenth century, the famous English scientist, Sit Patrick Manson and others, had identified the *Anopheles* mosquito as the carrier of malaria. Dr Siegfried Annecke made a huge contribution during the 1930s to effectively eradicate malaria-carrying mosquitoes, while effective medicines were developed to not only treat the disease, but also prevent it.

In contrast to general expectations, fewer people were killed by wild animals, snakes, crocodiles and other causes in the Kruger National Park, than those who had died of natural causes. However, lions, leopards, elephants, buffalo, crocodiles and poisonous snakes took their toll.

Other causes of death in the Lowveld were as a result of the confrontations between Boers and Brits during the Anglo-Boer War, including those at Mpisane where the raider fort of Steinaecker's Horse was attacked by the Boers, and also war crimes and murders attributed to the Bushveldt Carbineers in the northern areas of the Transvaal (1901 to 1902). Many died as a result of famine, dehydration, heatstroke, lightning and various accidents.

At many of the former habitation areas of blacks in the Kruger National Park there are many graves, often still with the deceased's earthly possessions placed on top, as prescribed by tradition. Of these people, nothing is known, except that they had probably died during the nineteenth century by the hands of bellicose and marauding warriors.

The graves of Hendrik Hartman and William Pye, near Pye's old camping site at Crooks Corner. These graves were covered with cement slabs by Bvekenya Barnard on 14 October 1955. (26.04.1984)

18. LONELY GRAVES IN THE BUSHVELD

Hendrik Hartman's grave. He died of heatstroke on 10 October 1918 near Pye's camp at Crooks Corner. (26.04.1984)

The grave of William Pye, who died of pneumonia. He was buried on 20 November 1918 next to Hendrik Hartman at Crooks Corner. (26.04.1984)

Much more is known about the graves of whites and others in this part of the Lowveld. Former ranger J.J. (Kat) Coetser was trampled by an enraged elephant on 21 November 1935 along the Limpopo River east of Messina (Musina), while manning a foot-and-mouth disease cordon. He was buried in Pretoria. Hendrik Hartman, father of the later chief commissioner of Bantu affairs, Daniël Hartman, went into partnership with legendary elephant hunter, Bvekenya Barnard, to farm sheep behind the Soutpansberg. One day in 1918 he went to Crooks Corner to fetch umbrella thorn branches to build a kraal for his sheep. He surrounded himself with so many thorn branches that he became trapped in his own kraal. It was a scorching day and Hartman soon developed heatstroke. It was already evening when the rest of his party heard his cries for help, but he was already so dehydrated that he died soon after he had been freed. William Pye and Buck Buchanan buried him on 10 October 1918 in Pye's cemetery. Only a few months later, Pye himself developed pneumonia and despite Buchanan and Bristow's careful nursing, he died after a few days. On 20 November 1918, just a few days after Armistice Day at the end of the First World War, he was buried next to Hartman. Bvekenya Barnard returned years later and on 14 October 1955 he covered the graves of his former friends with cement blocks. The place was properly marked later and provided with a historical memorial plaque. Some more characters were buried in Pye's cemetery, including Pat Fye and others, but their graves could no longer be distinguished.

Near the confluence of the Limpopo and Luvuvhu rivers lies the grave of Osorio Terzakis who died here in 1929 of splenic fever. He was a Portuguese official of the administrative centre, Mapai, further down the Limpopo. This young man was about to get married and his fiancée was on her way from Canicado in the south to join him, when he contracted the feared disease. The only place where he could get the necessary treatment, was at the Elim Missionary Hospital near Louis Trichardt (Makhado). While on his way there he went into a coma and died at Pafuri where he was buried.

On the southern bank of the Phugwane River, just east of Zari mouth, are two graves marked with two rough headstones. There are no legible inscriptions on the graves, but according to legend, they are those of two Boers, who had been murdered

The graves of two alleged Boer prospectors who were shot to death near the confluence of the Phugwane and Zari Spruit by Capt. "Bulala" Taylor of the Bushveldt Carbineers. (27.04.1984)

631

18. LONELY GRAVES IN THE BUSHVELD

The memorial stone which was erected along the Shingwedzi, near the wall of Kanniedood Dam, at the place where the ashes of former board member Bran Key had been strewn on 17 April 1949.

during the Anglo-Boer War in 1901 by the cruel British war criminal, Bulala Taylor. The two men were too old to join the commandos, so they carried on with their mining activities at Birthday and Louis Moore mines. Bulala heard from locals that a large amount of gold was available for looting, so he immediately set off with a unit of the infamous Bushveldt Carbineers. The Boers in turn heard that Taylor was on his way and fled into the Park area with all their possessions, over the Bububu River to the Phugwane. Near Dhili Waterhole they learned that Taylor and his unit were very close and decided to push their wagon and all their possessions over the cliff at Dhili into a deep pool. After that they fled further, but were overtaken at Zari Mouth and captured by Taylor. In an attempt to force them to tell where the gold was, Taylor shot the young leader of oxen in cold blood. The two older men still would not talk, so he had them both shot as well. Their wives buried them with the aid of some of the locals and started their journey home on foot. The children must have been crying because of shock and exposure along the way and a tributary of the Phugwane is still called Mashadya (crying children) today.

B.A. (Bran) Key served on the management committee of the Wild Life Protection Society of South Africa for many years. Apart from that, he also served on the National Parks Board of Trustees from 1947 until 1949. After his death in April 1949, his ashes were strewn on the northern bank of the Shingwedzi River, close to the wall of the current Kanniedood Dam. Ranger Bill Lamont carved the epitaph: "Bran Key lives on here 17.4.1949" on a simple sandstone memorial.

Along the upper reaches of the Shipikani Spruit, just east of Stapelkop Dam's wall, is a large sodic area. Here is the grave of one Joubert, who had probably been a young boy from Sabie who came to hunt in the area with a group of hunters and then died of malaria. Others speculate that it could be the grave of João Albasini's partner, Johannes Joubert.

On the western bank of the Nalatsi Spruit, about two kilometres south of Timatoro windmill, another two graves of whites were found.

According to the pot and bottle sherds and Martini-Henry cartridges found on the terrain, these two graves also date back to the 1880s. Just a few hundred metres north of these two graves are quite a few heaps of soil, which could be indicating the graves of blacks from the same period. No reference to these

The grave of Boer hunter, Joubert, who was buried along the Shipikani Spruit near the wall of the current Stapelkop Dam by the end of the nineteenth century. (18.06.1984)

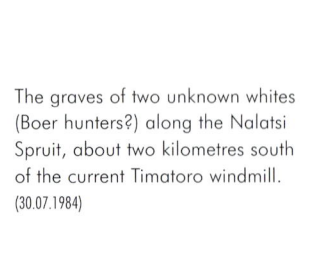

The graves of two unknown whites (Boer hunters?) along the Nalatsi Spruit, about two kilometres south of the current Timatoro windmill. (30.07.1984)

632

18. LONELY GRAVES IN THE BUSHVELD

The graves of several unknown blacks along the Nalatsi Spruit about two kilometres south of the current Timatoro windmill. (30.07.1984)

Ranger Douw Swanepoel at the grave of the notorious poacher, Langone Mbombi, in Shilowa Poort. He had been caught and eaten by a lion here during the 1920s.

graves could be found, but it is possible that a clash took place here during the Anglo-Boer War between General Christiaan Beyers's small nomadic commandos and a group of blacks who had been armed by the Bushveldt Carbineers. If not, it could have been members of an unknown hunting party who had died of malaria.

During the 1920s there was a notorious poacher, Langone Mbombi, who regularly laid in wait for game at waterholes in the Shawu Valley. On one of these occasions, while waiting for some unsuspecting animal to come and drink at Shilowa Poort, he was attacked and killed by a lion. His friends found his remains and piled stones on top of it, where it is still stacked in the shade of the Lebombo Mountains today.

After the retirement of ranger Maj. A.A. Fraser in 1920, Col Stevenson-Hamilton, the warden of the Shingwedzi reserve at the time, decided not to man Fraser's post at Malunzane again, probably because of a water problem. Instead he instructed the new ranger, L.H. Ledeboer, to create a new ranger post at Hatane, close to the confluence of the Makhadzi and Letaba rivers. From here he would be able to keep an eye on Chief Makhuba, who had settled further north along the Makhadzi Spruit.

Ledeboer was already married to his second wife, Anna Maria (née Bindemann), when he settled at Hatane in April 1921. Just a few months later on 10 August 1921, she unexpectedly died of a heart attack. Ledeboer buried her along the northern bank of the Letaba River, close to the Makhadzi Mouth and used a wagon-axle to mark the grave. In the meantime the grave has been restored and provided with a historical memorial plaque. The site is also accessible to tourists by car along the tourist road north of Letaba to Makhadzi Mouth.

William Walter Lloyd was the first ranger to man

The grave of Anna Maria Christina Ledeboer (née Bindemann) at the old Hatane ranger post close to the Makhadzi Mouth. She had died on 10 August 1921 and was buried here by her husband. The incorrect date on the grave was added later on. (28.04.1984)

The grave of ranger W.W. Lloyd at Satara. He died here in November 1922 of pneumonia which he had contracted after becoming soaked during a patrol on horseback. (21.06.74)

633

18. LONELY GRAVES IN THE BUSHVELD

A "tûpi-like" structure – possibly very old and of Indo-Hindustani origin, at Pumbe on top of the Lebombo Mountains east of Satara. One of the ancient trade routes from the interior to the east coast harbours, Inhambane and Delagoa Bay, went past here.
(18.6.1984)

Part of the dome of the "tûpi-like" structure at Pumbe, which was built with small bricks.
(11.7.1985)

Archaeological investigation into the origin of the "tûpi-like" structure at Pumbe, on the Lebombo.
(11.07.1985)

the ranger post at Satara in 1920. Before that, there had been an outpost along the upper reaches of the Nwanetsi, manned by the ranger of Rolle, Tim Healy, from 1910 to 1914.

Lloyd himself, who was called "Kukuzele" ("the one who stalks you") or "Nwashibeyilana" ("the one who carries a small axe") by the locals, manned the ranger posts at Boulders and Rolle for a while during 1919. He had a good hand with horses and loved to undertake long patrols on horseback. This inadvertently led to his death. In November 1922 he got soaked by rain on one of his patrols and developed pneumonia afterwards. There was no radio or other contact between the posts in those days and Mrs Lloyd nursed her husband with her limited knowledge and medicines. Her efforts were in vain and he died on 15 November. With the help of one of the field rangers, the brave woman made a coffin of old planks, dug a grave and buried him just north of the current ranger residence. A black messenger covered the 80 kilometres to Skukuza in a day and a night to ask for help, but by the time Stevenson-Hamilton arrived at Satara in his buggy, there was nothing which he could do but help her pack up her possessions, after which he transported her and her children south.

On top of the Lebombo Mountains is a flat sandveld plateau in the vicinity of the old Pumbe outpost and near the Pumbe pans. It was in these pans where the unique pool-dwelling fish (*Notobranchius* spp.) and lungfish (*Protopterus annectens*) were discovered in later years. One of the centuries-old trade routes from queen Modjadji's fortress to Delagoa Bay passes close to here (just south of boundary beacon K). Scientists A. Potgieter and F. Venter discovered ruins a few hundred metres northwest of the outpost some years ago, which indicated earlier human habitation. These structures, one bigger and one smaller, were apparently dome-shaped originally, but eroded with time and partially collapsed. Both were built with baked bricks measuring 10 x 5 x 3 cm, on a base of round pebbles, which are common in the area. Near one of the structures is a stone floor which indicates a stable for five horses. This structure was probably built during the early 1920s by former ranger W. Lloyd of Satara. A .38 Webley-cartridge case found on the stable floor confirms that the structure had been built recently and does not have any connection to the other older structures.

A Portuguese map of 1893 indicates a place called Mahasane, near Pumbe. This is clearly a word of either Eastern or Arabic origin and might be an indication that Arabic or Goanese traders from the east coast practised their trade here during the nineteenth or eighteenth centuries, or even earlier. According to experts the remains at Pumbe might even have an earlier origin. During the sixteenth century, the Portuguese occupiers of the east coast allowed non Christian Indian traders to trade in the interior for the first time. These Hindu traders could possibly have established a trading post at Pumbe along the centuries-old trade route to Delagoa Bay.

The larger dome-shaped structure at Pumbe is similar to the "tûpi" worship structures in southern India and in other places. If this is so, then these structures are the oldest human-made structures which had remained intact inside the Kruger National Park and as such it should be carefully protected against further decline. Interestingly enough, this trading post had ceased to function by the time that the Von Wieligh border commission trekked past here in 1890. Excavations by the Department of Archaeology of the University of Pretoria during 1985, shed very little light on the secret of the Pumbe domes. The Indo-African expert, Dr Cyril A. Hromník investigated the site in detail on 19 November 1986. His conclusions about the origin of these rare structures can be summarised as follows (Hromník, 1987):

the dome-shaped structure at Pumbe is according to all indications a religious worshiping structure or "tûpi" which had probably been built by a priest of the Dravidian South Indian traders, centuries before the settlement of Europeans in the Lowveld. These people mined gold in the area known as Komatiland during the era which stretched from the third century BC until the tenth century AD. This area lies between the Limpopo River in the north, St. Lucia in the south and the Indian Ocean in the east and the Drakensberg Mountain Range (Quathlamba) in the west. They traded with the indigenous populations (the developing Khoi and later black settlers).

However, Dr Hromník's opinion is questioned by other archaeologists.

Shortly after the arrival of Steinaecker's Horse at Sabie Bridge (Skukuza) by the end of 1900 and the building of the blockhouse to protect the railway line against the Boers, a tragedy played itself out here. The tented camp of the unit was pitched about 150 metres south of where the railway bridge is now, and they had to post guards at night. The black members of the unit refused to leave their tents at night, as lions had already caught quite a few horses in the camp.

One night, while rifleman Sammy Smart was on guard-duty, he was attacked by a lion. He died the next day and was buried there, but his grave was not marked properly and has therefore never been found.

The neglected grave of Captain H.F. ("Farmer") Francis, former officer in charge of Steinaecker's Horse at Fort Mpisane, lies about 12,5 kilometres southwest of Rolle Station, between the Mutlumuvi and Nwarhele tributaries of the Sand River, close to the ruins of Fort Mpisane.

The occupying forces of Fort Mpisane ruled a reign of terror in the area during the Anglo-Boer War and General Ben Viljoen received an order to extirpate the so-called "Pillager Fort". The Boer forces, under the leadership of Commandant Piet Moll attacked the fort during the early morning hours of 7 August 1901. Captain Francis was killed during the short but intense engagement and the fort seized with relative ease.

Maj. J. Stevenson-Hamilton took this photograph of his personal servant, Sharifa Bakhari Ali and his wife, on 5 August 1905. Sharifa originally came from the Comoros.
(ACKNOWLEDGEMENT: STEVENSON-HAMILTON COLLECTION, SKUKUZA ARCHIVES)

18. LONELY GRAVES IN THE BUSHVELD

The grave of Maj. Stevenson-Hamilton's personal servant, Sharifa Bakhari Ali, who had been fatally injured during the night of 5 June 1912 by a railway trolley on the Sabie railway bridge.
(ACKNOWLEDGEMENT: STEVENSON-HAMILTON COLLECTION, SKUKUZA ARCHIVES)

The grave of J. Stevenson-Hamilton's personal servant, Sharifa Bakhari Ali, near Metsi sideline on the former Selati railway line as it appears today.
(ACKNOWLEDGEMENT: STEVENSON-HAMILTON COLLECTION, SKUKUZA ARCHIVES)

The dog cemetery near the park director's office where rangers buried their faithful dogs through the years. This was later moved to a location near the Stevenson-Hamilton Information Centre inside Skukuza rest camp. (28.05.1984)

A number of armed blacks in service of Steinaecker's Horse was summarily put before a firing squad by the Boers and buried in a mass grave near that of Captain Francis. A military cross was erected on Captain Francis's grave after the war, but this had since been removed by vandals and this grave is now only a pile of stones.

Sharifa Bakhari Ali, a native from the Comoros, started working for Col J. Stevenson-Hamilton in 1903. He accompanied Stevenson-Hamilton on his journeys to Portuguese-Nyasaland in 1908 and to British East Africa, Uganda, Sudan and Egypt in 1909. Ali always diligently carried out his housekeeping duties at Sabie Bridge. On the night of 5 June 1912, Ali walked across the railway bridge to deliver medicine to a sick child in a village along the northern bank of the Sabie River. It was a pitch dark night and he apparently did not hear the railway trolley's approach. The trolley, which came from Komatipoort, hit him at high speed and then dragged him along for another 47 metres. Ali was taken to the clinic at Newington, but he died the next morning.

Col Stevenson-Hamilton buried him in the vicinity of Sabie Bridge and later erected a decent headstone. This headstone was recently discovered near one of the warehouses, about one kilometre southeast of the former Metsi Station along the Selati Line. The inscription has become illegible long ago, but the shape of the solid headstone and the marula tree just west of the grave made identification possible.

Colonel Stevenson-Hamilton and the first rangers had the habit of burying their dead dogs in a special graveyard just east of the railway line at Sabie Bridge. Some of the rangers even erected neat sandstone headstones in memory of their pets. Two of the most impressive headstones are those which were erected by ranger Cecil de Laporte for his dog Dansch (23.8.1915) and ranger G.R. (Tim) Healy for his dog Mary (14.11.1907). Mary was on patrol with Healy along the Old Wagon Route to Doispane on a scorching day in November 1907. When she started showing signs of heatstroke, Healy carried her in his arms all the way back to Skukuza. She died despite all his efforts to save her life and was given a hero's funeral. The hill near which she fell ill, was named "Nja-ka-Mthepe" (Healy's dog) by the blacks, with reference to his Healy's Shangaan name "Mthepe" ("the one who wears a cap"). The unique graveyard at Sabie Bridge eventually became threatened by development. The headstones were carefully removed and erected at a suitable site near the then Nature Conservation offices at Skukuza. Several more rangers buried their dogs here to continue the tradition.[5]

Marius Meyer, a young post office official, went fishing with his stepfather, Paul Viljoen, on 18 January 1970 in the Sabie River just west of Skukuza. While walking through the shallow water from the rock where he had been fishing, back to the bank, a gigantic crocodile closed his powerful jaws around his leg and dragged him into deep water. He only called for help once before disappearing under the

---

5 In the meantime, the stones have been moved again and can now be seen next to the Stevenson-Hamilton Memorial Library and Information Centre inside the Skukuza rest camp.

water, never to be seen again, despite a desperate search by police and park staff which lasted for days.

Along the northern bank of the Sabie River, just east of Sabie Poort in Mozambique, is the grave of one Wood who had succumbed to malaria in August 1868. He had been a member of St Vincent Erskine's party who undertook a journey of exploration from Natal to the Limpopo mouth in 1868.

Colonel Stevenson-Hamilton's health left much to be desired in 1927/28. This forced him to take leave of absence on 22 May 1928, while Captain Elliott Howe was appointed as acting warden. On 13 June 1928 Col Stevenson-Hamilton wrote the following to the Board:

> Were I to die while Warden of the Game Reserve or within ten years of cessation of employment therein, it is my wish to be buried on the top of the highest rocky kopjes standing west of the Selati line near mile 43 (old kilo 67). On the grave should be erected a small pyramidical granite pillar with inscription on a concrete stand, facing the Sabi River. The underlying idea is thus that the local natives being still very superstitious, the fact that my being so to say 'in situ' would probably act as a deterrent to native poachers in the vicinity, and as the said kopje is the highest point in the southern areas, and can be seen from many miles around, they would feel 'Skukuza' was watching them, and that what they were doing would certainly bring punishment.

The Board approved his request, but Colonel Stevenson-Hamilton's health improved to such an extent that he could resume his duties as warden in July 1929. He stayed on as warden until his retirement in April 1946 and died on 10 December 1957 at the ripe old age of 90, while his wife, Hilda, passed away on 10 January 1979. Their daughter, Mrs Ann Doyle, strew their ashes in the almost forgotten grave, which had been blasted out of the solid granite rock dome on top of Shirimantanga Hill by ranger T.L. James in 1930.

A simple but impressive bronze memorial plaque against the vertical surface of a giant granite boulder at the foot of the hill was unveiled by his family a few months later.

In the beautiful area of this unique part of our natural heritage, visitors can now reflect and pay homage to this man and woman who spent the most parts of their lives on the development and preservation of the Kruger National Park.

The old trade route between Delagoa Bay and the interior, via Pretoriuskop, also provided its quota of drama and human victims.

After Voortrekker leader Andries Hendrik Potgieter and his people had settled at Ohrigstad, he sent a commission consisting of 26 men under leadership of Karel Trichardt to Delagoa Bay in October 1845 to establish trade relations with the Portuguese. Simultaneously, they had to find a practical route between Ohrigstad and the harbour city. Trichardt

The stone erected by ranger G.R. (Tim) Healy on the grave of his beloved dog, Mary, which died of heatstroke on 14 November 1907.

Col James Stevenson-Hamilton gazes over the Lowveld from the top of Shirimantanga Hill, circa 1927. In 1929 ranger T.L. James had a "grave" blasted for him out of the solid granite of the hill.
(ACKNOWLEDGEMENT: STEVENSON-HAMILTON COLLECTION, SKUKUZA ARCHIVES)

Ranger Wagener standing on roughly the same spot from where Col Stevenson-Hamilton had admired the Lowveld in 1927.

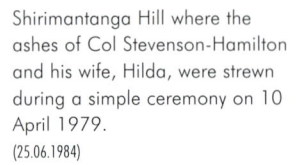

18. LONELY GRAVES IN THE BUSHVELD

The "grave" on top of Shirimantanga Hill which Col James Stevenson-Hamilton had blasted out for him during 1929. His ashes, as well as those of his wife, Hilda, were entrusted to the earth here on 10 April 1976.

Shirimantanga Hill where the ashes of Col Stevenson-Hamilton and his wife, Hilda, were strewn during a simple ceremony on 10 April 1979.
(25.06.1984)

Willem Pretorius's grave between Numbi Gate and Pretoriuskop camp. He was a member of Karel Trichardt's expedition to Delagoa Bay in 1848. He fell ill en route and was buried under a marula tree by João Albasini.
(16.06.1984)

wanted to undertake the journey on foot, but was convinced in the end to take along oxen and horses which later died after being bitten by the tsetse-fly.

It is not known whether the group travelled along the Old Wagon Route, which had been explored by Chief Commandant Potgieter and a previous commission in 1844, or along the more southern ancient trade route that went past Ship Mountain. Because they were using oxen, it is probable that they trekked along the Old Wagon Route. Trichardt's trek oxen died one after the other of nagana during this journey and in the end the group had to be provided with trek-oxen by Albasini to be able to return to Ohrigstad safely.

By the middle of July 1847 and again in July 1848, further commissions under the leadership of J. van Rensburg were sent to Delagoa Bay to investigate and demarcate a better route from Ohrigstad to the Bay with the aid of guides which had been provided by João Albasini. Karel Trichardt played a leading role during both these journeys as guide and interpreter.

One of the members of the expedition, Willem Pretorius, fell ill with malaria across from Ship Mountain. Trichardt decided to send him home with two blacks to accompany and aid him. Just northwest of the current Pretoriuskop, Pretorius lay down under a large marula tree and passed away. His companions immediately went to João Albasini to inform him of the tragedy. This legend was confirmed in 1929 by a very old black man, Ngutu Sambo, who had been Karel Trichardt's leader of oxen in 1848. According to him, Pretorius was buried by Albasini himself at the place where the grave can still be seen today, close to the road between Numbi and Pretoriuskop. It appears as if this piece of history is the reason for the name of the nearby hill, Pretoriuskop. A tributary of the Crocodile River, which has its origin nearby (the Mbyamiti), was also called Pretorius River for a while. According to General Schalk Burger (1916; quoted by Lombaard, 1969), Pretoriuskop was named after President M.W. Pretorius who visited the area in 1865. According to him, the original name of the hill was Mntsobo Hill.

However, the author agrees with Van der Merwe (1951) and De Vaal (1979) who maintained that it would have been contrary to Boer tradition to name an insignificant little hill after their president and that they would much rather have named Shabeni, the more impressive hill nearby, after the president. It would have been far more natural to link the hill where Willem Pretorius had died in 1845 to the incident. It is thus up to the reader to decide which one of the two incidences was the origin of the name Pretoriuskop.

In 1875, President Burger tasked the Hungarian, Alois Nellmapius to improve the trade route between Delagoa Bay and Pilgrim's Rest. After he had cleared the whole route, he erected a number of overnight resting places along the route. These posts were put under the care of station masters and they had to look after the tradewares and the interests of the carriers.

The farms west of the Mozambican border on which the overnight posts were erected, had been allocated to Nellmapius as part of his compensation for construction of the road. These were: Castilhopolis (along the Mozambican border), Coopersdal (along the Komati River), Ludwich's Lust (at Nellmapius Drift through the Crocodile River), Joubertshoop, Pretoriuskop and Burgers Hall (just west of the current Numbi Gate).

On the farm Joubertshoop, where the Josekhulu Spruit[6] crosses the trade route, Nellmapius also erected an overnight post, at the same place where Albasini had an outpost under control of his induna Josekhulu from 1845–1847). This post was placed under the control of a 22-year-old, Thomas Hart. He was a cheerful chap who took his work seriously. He built himself a couple of amenities including a wattle-and-daub house. According to legend, he kept a number of wild birds and animals to keep him company. Unfortunately for Nellmapius and his distant station masters, the Sekhukhune uprisings started in May 1876 and some of these groups moved into the Lowveld. A couple of looters under one of Sekhukhune's deputy chiefs, Maripe, arrived at Thomas Hart's homestead on 10 August 1876 and demanded that he hand over his rifle. He refused, of course. When he walked out of his front door, he was shot in the back of the head by one of the looters who had sneaked round the back of the house. The house was ransacked and burnt down, the animals killed, while Hart's corpse was left to scavengers.

The bailiff and inspector of police at Pilgrim's Rest, William (Billy) Barter, who had been requested by the Secretary of State to visit Nellmapius's overnight stations, accompanied by a group of volunteers to bring the supplies and people to safety, arrived at the ruins of Hart's overnight post on 19 September. By then all that was left of Hart's body, was his skull, which they then buried and covered with stones. This memorial to Hart can still be seen along the tourist road between Pretoriuskop and Malelane.

In the decade before the Anglo-Boer War, Alf Roberts ran a shop at Tengamanzi along the Crocodile River, just north of Nellmapius Drift. After the war, he was killed by unknown attackers at his other shop at Malelane, but was not buried inside the Park.

Just west of Ship Mountain (which the transport riders called Lang Kop), along the eastern bank of the Samarhole Spruit, is the grave of a young German. Adolf Soltké had landed in Delagoa Bay by the end of 1884 and was on his way to make his fortune at the Pilgrim's Rest goldfields. A group of transport riders, which included Percy FitzPatrick and H.L. Hall, took him under their wing and offered him a ride to Pilgrim's Rest. On the way they

Thomas Hart's grave on the bank of the Josekhulu Spruit, along the old Delagoa Bay transport road. He had been one of Alois Nellmapius's station masters and was murdered by a hostile band of Chief Maripe's warriors on 10 August 1876. (09.10.1983)

---

6   Also spelled Josikhulu by certain sources.

18. LONELY GRAVES IN THE BUSHVELD

The grave of an unknown person at "Tengamanzi", Alf Roberts's little shop near Nellmapius Drift. (27.07.1984)

The grave of a young German, Adolf Soltké, who died here in about 1884 of gangrene. This is also the birthplace of Sir Percy FitzPatrick's famous dog, Jock of the Bushveld. (25.09.1984)

often teased him about his lack of knowledge of the bushveld and its inhabitants, but he took it all in his stride and soon became everybody's favourite. They taught him to shoot and he soon started collecting birds which he wanted to take back to show his family in Germany. At their camping site along the Samarhole, Soltké noticed a roller flying past. He was very exited about the beautiful bird and wanted to add it to his collection. When he jumped off the wagon with the loaded double-barreled shotgun in his inexperienced hands, the shot went off and shattered his right leg under the knee. The closest doctor was more than a 100 kilometres away, but one of the members of the group, Donald Mackay, immediately set off to look for help. The only help that arrived after a full two days, was a drunk pharmacist named "Doc Munroe". Such was his state of inebriation, that he could do nothing to help the wounded man. It then took another two days for a qualified doctor to reach them from the Mac-Mac goldfields after a journey of 110 kilometres on horseback. However, it was too late and although Soltké's leg was amputated that same night, he died the next morning of gangrene and trauma. His friends buried him under the same tree where he had kneeled a few days earlier in prayer. Somebody later covered the grave with stones. A few months later, on 4 April 1885, FitzPatrick outspanned at the same place during another trek from Lydenburg to Delagoa Bay. It was right here where the his dog, Jess, gave birth to a litter of puppies of which the famous Jock of the Bushveld was one.[7]

Not far from Soltké's grave, along the southern bank of the Mbyamiti Spruit, another site was also found which to all indications was that of a white person. Who it had been, or when the person was buried here, is not known. It could possibly have been one of the pioneer hunters who had died of malaria or had been mauled by a wounded animal.

On the southeastern side of Ship Mountain, along the Imagoroti Spruit, which is a tributary of the Komapiti, lies the grave of Swazi Chief Mataffin. This elder had been one of Swazi King Umbandine's (Mbandzeni) advisors and confidants in 1887. When Mataffin claimed a girl on which the King had an eye himself, he had to flee for his life. He and his followers managed to escape the wrath of the king

---

7   It was determined by former senior ranger M.C. Mostert in the Grahamstown and Pretoria archives with the help of Sir Percy FitzPatrick's personal documents and official death certificates, that the person to which FitzPatrick referred to in his book *Outspan*, could possibly have been a 23-year-old German, William (Wilhelm) Süpke, formerly from Riversdale in the Cape, who had died after an accident involving a shotgun. According to his death certificate in the Pretoria Archives, he died on 27 October 1884 due to locked jaw, after his right arm (not his right leg) had been amputated as a result of the accident. According to this report, the accident could have taken place along the North Sand River on the farm Burghers Hall, where he was buried with the knowledge of the gold commissioner of Pilgrim's Rest. For some or other reason, FitzPatrick had changed the name William (or Wilhelm) Süpke to Adolf Soltké and weaved a romantic story around the whole incident. That this was one and the same person, was proved by an official inventory of his possessions, including a zither, which had been drawn up by a Master of the Supreme Court and to which FitzPatrick referred to in *Outspan*. Now the question immediately arises of who the person was whose remains had been found in the grave at Samarhole in the Kruger National Park, if it were not those of Süpke, alias Soltké. The description of the incident in *Outspan* indicates a terrain east of Pretoriuskop and west of Ship Mountain. It is of course possible that the accident had happened along the Samarhole and that this spruit was known among the transport riders as Sand River. This possibility is strengthened by the fact that no grave has as yet been found in the vicinity of the transport drift across the Sand River on the farm Burgers Hall. Until further research can shed more light on the subject, it can be accepted that the person who was buried at the Samarhole in the Kruger National Park, had been Süpke (alias Soltké) and that Jock of the Bushveld was born here during FitzPatrick's next journey to Delagoa Bay by the end of March, or beginning of April 1885.

and settled west of Nelspruit along the Crocodile River. Here he lived in peace and managed to gain the respect of the Boers when he helped them during the Sekhukhune wars of 1876 until 1879 with an army of 2500 Swazi warriors.

In 1891, Mataffin had a disagreement with the Native Commissioner of Krugerspos, Abel Erasmus. As a result he decided to go and find asylum at his father-in-law, Gungunyane, who lived in Mozambique. He sent his followers and wives ahead by train from the nearest station along the Eastern Line (probably Malelane) and followed later by horse. Just south of Ship Mountain he reached the village of Mumbai Chief Inhliziyo to ask for the best route to the border.

The cunning Inhliziyo, who hated the Swazi because of their cruel raids during earlier days, invited Mataffin to spend the night. According to legend, Inhliziyo had been aware of the fact that Mataffin carried some gold sovereigns in his saddle bags which he wanted for himself. The next morning, he sent his son and a guide, Umgeba, to accompany Mataffin to show him the way. About half a kilometre east of Inhliziyo's village on the old trade route, Mataffin rode through a spruit with running water (the Imagoroti, a tributary of the Komapiti Spruit). When he rode up the bank on the other side, Umgeba shot the unsuspecting chief in the back. After he had also shot Mataffin's horse, they took the money and then stuffed his body into an aardvark hole on the spruit's bank. About a week later, another black man, Bob, who had been hunting in the area, discovered the body. Bob later became one of Harry Wolhuter's most reliable rangers.

Bob made the murder known and had the chief's ring as evidence, but Inhliziyo and his people simply pleaded innocent. On 12 November 1985, field ranger Sergeant Charlie Nkuna, a cousin of the famous Helfas Nkuna, pointed out the grave along the Imagoroti again. What became of Mataffin's gold is related in chapter 17.

During the era of the transport riders, a witch-doctor lived in a cave against the northern slope of Ship Mountain. One day the cave's roof collapsed on top of him and his skeletal remains are still there, covered by tons of rock. The transport riders believed that the place was haunted and trekked past there as quickly as possible at night. Legend has it that a frightful ghostlike apparition moves up and down in front of the cave on moonlit nights!

Further down the old Delagoa Bay transport route, at the former Piscene in Mozambique, the ruins of Tom Barnett's shop where many a tired transport rider outspanned and found refreshments in the years 1884 until 1892, can still be seen. Under a huge wild fig tree close to the ruins, are two graves. The one is that of the famous dog, Jock of the Bushveld, left in Barnett's care by FitzPatrick. One night he accidentally shot Jock under the impression that it was a stray dog in his chicken coop. The other grave is that of Jock's archenemy, Field Cornet Seedling of Krugerspos, who had provoked his tamed male baboon to tear Jock apart. However, Jock bit and killed the baboon. Seedling was so annoyed that he burnt down FitzPatrick's wagon driver,

Excavations at Adolf Soltké's grave along the Samarhole Spruit to confirm the historical authenticity of the site.
(08.10.1983)

The site of Mambayi Chief Inhliziyo's village south of Ship Mountain.
(25.9.1984)

Jim Makokel's village. After that he fled to Mozambique, but died of thirst and exposure at Piscene. Tom Barnett buried him at the same place where Jock had found his last resting place.

The old transport route witnessed yet another tragedy in later years. Ranger Harold Trollope of Malelane, accompanied by his father-in-law, John Glen-Leary, and photographer C.A. Yates were camping along the old transport route at Fihlamanzi in June 1926. Trollope's mission was to repair the road so that it could be reopened for ox-wagon traffic again. After that they moved camp to the upper reaches of the Hlambanyati Spruit and pitched camp along one of its tributaries, the Shipalatseni Spruit. They stayed there for five days, until 3 August 1926 when they heard a leopard roaring further down the spruit.

Predator control was policy in those days and Trollope immediately set off after the leopard with his pack of hunting dogs, followed by Yates and his father-in-law. Field ranger Folok fired two shots at the leopard, but both missed. When Yates and Glen-Leary appeared on the scene, the dogs had already discovered the leopard and unfortunately started chasing the leopard straight in their direction. When the leopard noticed Glen-Leary, it immediately attacked him. Trollope soon arrived at the scene and shot the leopard standing over his father-in-law. The seriously mauled Leary was carried on a makeshift stretcher to Malelane, while Trollope went ahead to summon a doctor from Barberton. Glen-Leary was admitted to Barberton Hospital and treated, but he died on 6 August 1926 from his wounds.

A cairn and a memorial plaque was erected at the place where the incident took place and the leopard's skin and skull were later donated by the family to the Stevenson-Hamilton Memorial Library and Information Centre at Skukuza.

The Stols brothers had been well-known hunters and wagon-makers in the Lowveld during the two decades before the Anglo-Boer War. In certain publications they are also referred to as the Berg (Mountain) Stolze, but most of them spelled their surname with an "s" and not a "z". Jan Bokwa Stols was a legendary figure, who was so strong that he could lift a wagon with his shoulders from the ground when a wheel had to be changed. Gert Frederik Coenraad Stols was a blacksmith but, like the rest of his family, keen on hunting and they regularly trekked into the Lowveld during the winter months to hunt. Their favourite hunting places were the area around and south of Pretoriuskop. According to the old transport rider, Org Basson, their camping site was about one kilometre southwest of Pretoriuskop.

In August 1886 Gert and his party went hunting again along the upper reaches of the Bukweneni Spruit, a tributary of the Nsikazi, when he contracted malaria. He became another victim of this deadly disease on 21 August 1885. His friends buried him here and the site of the grave later resulted in the

Swazi Chief Mataffin's grave along the western bank of the Imagoroti Spruit (about 1 km east of Inhliziyo's village south of Ship Mountain).

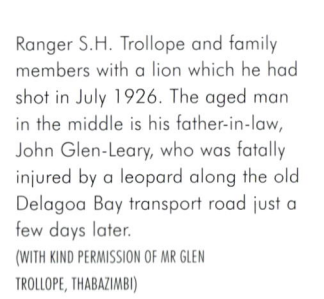

Ranger S.H. Trollope and family members with a lion which he had shot in July 1926. The aged man in the middle is his father-in-law, John Glen-Leary, who was fatally injured by a leopard along the old Delagoa Bay transport road just a few days later.
(WITH KIND PERMISSION OF MR GLEN TROLLOPE, THABAZIMBI)

names of Stolsnek Spruit, Stolsnek Dam and Stolsnek ranger post. According to the late Dr W. Punt, who visited Stols's grave on October 1950, accompanied by ranger Harry Wolhuter, the following inscription was legible on the slate headstone:

> Komt gij allen geslacht en register van Gert Frederik Coenraad Stols. Is geboren in het jaar 1822 den 25 Januarie. Overleden den 21 August 1886.

One of Gert Stols's sons, Gabriël, served as scout under General Ben Viljoen and was a member of the group who attacked Fort Mpisane of Steinaecker's Horse with Commandants Moll and Van Zyl.

According to legend another Stols hunter, incidentally also named Gert, was buried in the vicinity of Pretoriuskop. One day, he had to track down a lion which had been wounded by other hunters (apparently Org Basson, Steele, Van Rooyen and Daniëls) to kill it. He was 86-years-old and died soon after the incident, probably because of a heart attack brought on by the excitement. Apparently Piet Gouws delivered the funeral service. However, this grave has not been found yet.

In September 1901, four members of the Stols family were brutally murdered near White River by a group of blacks armed by Steinaecker's Horse.

Along the upper reaches of the Mnyeleni Spruit is a place where there are many chopped-down tree wisterias. The wood of this tree was apparently very popular during the nineteenth century among wagonmakers for manufacturing certain parts. A white man's grave was found on the southern bank of this spruit, which is probably that of one of the lumberjacks who had died here of malaria.

Ranger Harry Wolhuter lived at Mtimba about seven kilometres northwest of Numbi Gate, for the greater part of his career. Mtimba was in the area which was excised out of the Sabie Reserve in 1923, but he only moved to Pretoriuskop in 1939, after a homestead had been built for him there. During the First World War, Colonel Stevenson-Hamilton and ranger C. de Laporte joined the Allied Forces. This left a shortage of staff among the ranger corpse and the acting warden, Major A.A. Fraser, obtained permission to fill the vacancy at Boulders temporarily.

Mr C.A. Yates and field ranger Folok Mashave at the stone beacon where ranger Harold Trollope's father-in-law, John Glen-Leary, had been fatally injured on 3 August 1926 by a wounded leopard.

The Glen-Leary stone memorial along the Shipalatseni Spruit. (11.05.1984)

The grave of Gert Frederik Coenraad Stols who died of malaria in August 1886 along the upper reaches of the Bukweneni Spruit. (08.05.1984)

The grave of a white (wagon maker?) along the Mnyeleni Spruit. He had probably died here of malaria during the nineteenth century.
(08.05.1984)

The grave of acting ranger Andrew Wolhuter, a brother of Harry, who became a victim of the Great Flu Epidemic on 14 November 1918 at Mtimba.
(25.09.1984)

The memorial plaque against one of the Napi hills which was erected in honour of Mr J.F. (Joe) Ludorf KC. He had been a member of the National Parks Board of Trustees from 1930 until 1951 and chairperson in 1951.
(16.06.1984)

His choice fell on Andrew James Wolhuter, a brother of Harry Wolhuter. Andrew performed his ranger duties diligently and with distinction, but unfortunately he became one of the many victims of the flu epidemic of 1918 and died on 14 November 1918. He was buried at Mtimba and a concrete block is now the only indication of his last resting place in the former Nsikazi Block of the former KaNgwane homeland.

Advocate Joseph Frances (Joe) Ludorf, who served as a member of the commission who made a contribution towards the proclamation of the Kruger National Park in its enlarged form, also served on the National Parks Board of Trustees from 1930 to 1951. In 1951 he was the chairperson of the Board. He passed away on 21 December 1958 and his ashes were strewn to the wind by his family at Napi Hill between Skukuza and Pretoriuskop. A bronze plaque was erected at this place in honour of his memory.

West of Phabeni Spruit, about one kilometre south of the Albasini Ruins, lies the grave of ranger sergeant Helfas Nkuna. Helfas served as field ranger for 46 years and was the right-hand man of Harry Wolhuter, his son Henry, as well as Peet van der Merwe, Koos Smit, Gus Adendorff and Thys Mostert. Helfas feared nothing and nobody and arrested many armed poachers during his career. At the age of 78 he helped kill an injured leopard. The leopard attacked him and he lost his footing, but ranger Mostert killed the beast on top of him. Helfas died of a heart attack on 27 March 1975 at the age of 80. He was buried near his house at Phabeni.

Near the ranger quarters at Malelane, beneath a large umbrella thorn tree, is another lonely grave. This is the last resting place of Uncle George Meade who was one of the first camp managers in the Kruger National Park. Before coming to the Park, he had owned a bar in Potgietersrust (Makopane). Later he lived in Naboomspruit (Mookgophong). Ranger Bert Tomlinson brought him to Malelane in 1931 where he served as camp manager until 1933. After that he went to Shingwedzi with Tomlinson, where he developed gangrene in his leg in 1937. He could no longer continue with his career and was brought back to Malelane on request of former ranger Tom James, where he was cared for until his death later that same year.

On the northern side of the Selati railway line about two kilometres west of Crocodile Bridge, near the deepest quarry and hidden in a dense stand of sickle-bush, lies the grave of A. Hamilton, who was buried here on 26 February 1895. He was of Irish descent and probably one of the stoneworkers who had died of malaria. Somebody engraved a neat clover leaf (shamrock), the Irish national emblem, on the left-hand bottom corner of the headstone.

After the Anglo-Boer War, the tremendous number of human casualties in the Lowveld decreased dramatically, but during the preceding centuries one can totally agree with poet C.L. Leipoldt, 1920:

Dis glad nie reën!
Dis bloed, dis mensebloed!
Dis lewensbloed wat hierdie oord
se bossie-wêreld voed.

Freely translated it means:

Its definitely not rain!
It blood, human blood!
It's life-blood which feeds this region's Bushveld.

Ranger M.C. Mostert at the grave of field ranger Sergeant Helfas Nkuna, who had served a whole series of rangers with distinction during a service period of 46 years. His grave is along the Phabeni Spruit, about 1 km south of the Albasini Ruins.
(08.06.1984)

The grave of camp manager George Meade, near the Malelane ranger quarters. He had died in 1937 as a result of gangrene of the leg and was buried here by ranger T.L. James.
(12.06.1984)

The grave of an Irish stone cutter, A. Hamilton, who died on 26 February 1895 at the quarries near Crocodile Bridge, from where the building stones for the railway bridges across the Crocodile and Sabie Rivers were excavated.
(12.06.1984)

References

Adendorff, G. 1984. *Wild Company.* Books of Africa, Cape Town.

Anon. 1975. Phabeni se Gryskop Begrawe. *Custos* 4(6):13–19.

Baines, T. 1864. *Journal of Residence in Africa.* (1850–1853) volume 2:177 *et seq.* Kennedy, R.F. (ed.). Van Riebeeck Society, Cape Town.

Barter, W. 1876. Report of W. Barter to N.J.R. Swart, Pilgrim's Rest. 11.09.1876. SS 215, R 2359 176. State Archives, Pretoria.

Buchan, D. 1926. *The Chronicles of a Contractor.* Constable & Co. Ltd, London.

Buchanan, D.T. 1955. History of Makuleka. South African National Parks Archives.

Bulpin, T.V. 1950. *Lost Trails on the Lowveld.* Howard B. Timmins, Cape Town.

Bulpin, T.V. 1954. *The Ivory Trail.* Howard B. Timmins, Cape Town.

Bulpin, T.V. 1983. *Lost Trails of the Transvaal.* Books of Africa, Cape Town.

Burger, S. 1916. Reminiscence of General Schalk Burger. *Pilgrim's Rest Goldfields, Mines & Claimholders Quarterly* 1(2):6–10.

Das Neves, D.F. 1879. *A Hunting Expedition to the Transvaal.* George Bell & Sons, London.

De Leeuw Beyers, C.P. 1971. *Lurking Danger in the Kruger Park.* Living World Adventures – Living World Films, Sandton.

De Vaal, J.B. 1978. João Albasini. Unpublished manuscript.

Dunbar, D.M. 1881. *The Transvaal in 1876.* Richard Slater & Co., Grahamstown.

Erskine St. V.W. 1875. Journey of exploration to the mouth of the River Limpopo. *Journal of the Royal Geographical Society.*

FitzPatrick, J.P. 1897. *The Outspan.* William Heinemann, London.

FitzPatrick, J.P. 1907. *Jock of the Bushveld.* Longmans Green & Co., London.

Fraser, A.A. 1918. Warden's report to the Board 1918. Unpublished manuscript.

Glen-Leary, H. 1984. Glen Leary's death new facts. *Custos* 13(2):26–27.

Gray, S. 1986. Jock of the Bushveld. *African Wildlife* 40(5):167–175.

Hofmeyr, S. 1890. *Twintig jaren in Zoutpansberg.* J.H. Rose & Co., Cape Town.

Hromník, C.A.E. 1987. The Tûpi of the Lebombo. Unpublished manuscript.

Krüger, D.W. 1938. Die weg na die see. *Archives Year Book for South African History* 1:31–233.

Ledeboer, L.H. 1921. Personal communication to J. Stevenson-Hamilton.

Lombaard, B.V. 1969. Herkoms van die naam Pretoriuskop. *Koedoe* 12:53–57.

Mackie-Niven, C. 1968. *Jock & Fitz.* Longmans, Cape Town.

Marais, M. 1954. Eensame graf in Wildtuin. *Die Vaderland,* 21 August.

Meyer, J. 1957. Ranger's diary (Shangoni) 11 April.

Mockford, H.H. 1985. Personal communication.

Mostert, M.C. 1986. Personal documentation.

Pienaar, U. de V. 1983. Geheim van die vergete graf. *Custos* 11(10):13–15.

Pienaar, U. de V. 1984. Fihlamanzi se geheim ontrafel. *Custos* 13(7):4–10.

Pienaar, U. de V. 1985. Indications of progressive desiccation of the Transvaal Lowveld over the past 100 years, and implications for the water stabilization programme in the Kruger National Park. *Koedoe* 29:93–166.

Pienaar, U. de V. 1986. Eensame grafte in die bos. *Custos* 15(2):42–46.

Pienaar, U. de V. & Mostert, M.C. 1983. Op die spoor van die ou transportryers deur die Krugerwildtuin. *Custos* 12(9):27–32.

Pienaar, U. de V. & Mostert, M.C. 1983. Kleurryke karakters deel van Wildtuingeskiedenis. *Custos* 12(10):12–14.

Preller, G.S. 1917. *Dagboek van Louis Trichardt.* Het Volksblad Drukkerij, Bloemfontein.

Preller, G.S. 1923. *Oorlogsoormag en ander sketse en verhale.* Nasionale Pers, Cape Town.

Punt, W.H.J. 1953. *Louis Trichardt se laaste skof.* Van Schaik, Pretoria.

Punt, W.H.J. 1958. Die verkenning van die Krugerwildtuin deur die Hollandse Oos-Indiese Kompanjie, 1725. *Koedoe* 1:1–18.

Punt, W.H.J. 1960. Waar die Van Rensburgtrek in 1836 vermoor is. *Koedoe* 3:206–238.

Punt, W.H.J. 1975. *The First Europeans in the Kruger National Park.* National Parks Board of Trustees, Pretoria.

Sandeman, E.F. 1880. *Eight Months in an Ox-waggon.* Griffiths & Farran, London.

Schulenburg, C.A.R. 1981. Die "Bushveldt Carbineers" – 'n Greep uit die Anglo-Boereoorlog. *Historia* 26(1): 37–58.

Stevenson-Hamilton, J. 1944. *The Lowveld: Its Wildlife and its People.* Cassel & Co., London.

Stevenson-Hamilton, J. 1974. *South African Eden.* Collins, London.

Van der Merwe, A.P. 1951. Waar kom die naam Pretoriuskop vandaan? Beroemde ruskamp en eensame graf. *Die Transvaler,* 11 July.

Van der Merwe, A.P. 1961. Die eerste pad van Ohrigstad na Lourenço Marques. *Historia* 26(1):37–58.

Viljoen, B.J. 1902. *Mijne herinneringen uit den Anglo-boerenoorlog.* W. Versluys, Amsterdam.

Vos, W. 1890. Dagboek van opmeting van Oosgrens tussen ZAR en Mosambiek. Skukuza Archives.

Webb, H.S. 1954. The native inhabitants of the southern Lowveld. In: *A Survey of the Resources and Development of the Southern Region of the Eastern Transvaal Lowveld.* Lowveld Regional Development Association.

Witton, G.R. 1907. *Scapegoats of the Empire.* Angus & Robertson (U.K.) Ltd, London.

Wolhuter, H. 1948. *Memories of a Game Ranger.* Wild Life Protection Society of South Africa, Johannesburg.

Wongtschowski, B.E.H. 1987. *Between Woodbush and Wolkberg.* Review Printers, Pietersburg.

Yates, C.A. 1935. *The Kruger National Park.* George Allen & Unwin Ltd, London.

Chapter 19
The wild days of Crooks Corner
Dr U. de V. Pienaar

At the time of the proclamation of the Shingwedzi Reserve after the Anglo-Boer War in 1903, the triangular wedge of land between the Limpopo and Luvuvhu rivers formed part of the new reserve. In 1913 the boundaries of the reserve were changed again and the area, including Makuleke's location, was excised out of the reserve and placed under control of the Department of Native Affairs. This wild, remote piece of land has a common border, the Limpopo River, with Southern Rhodesia (now Zimbabwe) and a point of contact with Mozambique at the confluence of the Luvuvhu and Limpopo rivers. Once it was no longer part of the Shingwedzi Reserve it became a haven for an intermingled assortment of fortune seekers, poachers, smugglers, cutthroats, renegades, expatriates and runaways. It soon gained an unsavoury reputation and with that, the name Crooks Corner. The biggest attractions of the area were the unlimited hunting opportunities, especially elephants, in the adjacent Rhodesia and Mozambique and the recruiting of labourers for the gold mines of the Witwatersrand.

Major A.A. Fraser, who had been appointed in 1904 as the first ranger in the area between the Letaba and Limpopo rivers, was stationed more than 130 kilometres from the Limpopo-Luvuvhu triangle along a tributary of the Tsende (Tsendze) River. Because of the lack of roads and the fact that he could only rely on an ox-wagon or horses for transport, there was very little, if any, supervision over this remote area and the inhabitants could simply carry on with their illegal activities without any interference.

It is almost certain that hunters from Crooks Corner were responsible for shooting the last few remaining Lichtenstein's hartebeest herds on the Lebombo flats south of the Luvuvhu River. Colonel Stevenson-Hamilton still mentioned the occurrence of these animals in his annual report of 1908–09. The rest of the Lichtenstein's hartebeest herds which had still occurred on the Malonga flats and other areas between the Soutpansberg and the Limpopo River until 1916, also fell prey to this lawless community.

The practice to recruit black labourers in Mozambique and Rhodesia (Zimbabwe) who were able to resist the high temperatures in the gold mines, had already started before the Anglo-Boer War. Incidentally, one of the first rangers whom Colonel Stevenson-Hamilton appointed after the war in the reproclaimed Sabie Reserve, was E.G. (Gaza) Gray, a recruiting agent who had his base at Gomondwane near Crocodile Bridge. He had only served at Lower Sabie for a few months until May 1903, after which he resumed his occupation as recruiting agent in Mozambique.

In 1904 Major Fraser moved into an abandoned hut which had been built by another recruiting agent at Malunzane along the Shongololo River in 1902. The mines paid five shillings for every recruit to the agents, in those years, but this price was increased annually until the asking price for an adult recruit was £7.0.0, £4.0.0 for a youngster and £1.0.0 for a

Crooks Corner, the furthest north-eastern corner of the Transvaal (now Limpopo Province) at the confluence of the Limpopo and Luvuvhu rivers.

19. THE WILD DAYS OF CROOKS CORNER

Crooks Corner, where the borders of the RSA, Zimbabwe and Mozambique meet.
(JULY 1952)

The old Bobomeni drift through the Luvuvhu River at Pafuri used to provide entry to Crooks Corner.
(MAY 1984)

Buck Buchanan's baobab tree just east of Manqeba picket on the way to William Pye's old camp in Crooks Corner.
(01.08.1985)

The same baobab tree along the road to William Pye's camp in Crooks Corner, photographed by T.D. Buchanan in about 1914.
(ACKNOWLEDGEMENT: JOHN S. SYKES)

boy. Recruiting normally took place during the winter months. The recruits received £3.0.0 per month and free lodging from the mines for their labour. During the rainy season the workers returned to their homes and then received half of their remuneration at the border.

In 1913 it was determined that no recruiting may be done north of the 22° degree of latitude, as many recruits from the warmer, more tropical areas contracted pneumonia during the cold Highveld winters. However, many of the recruiting agents and their black assistants simply disregarded this stipulation.

Although there probably had been honest people who lured recruits to the mines on a voluntary basis, many irregularities took place. Certain unscrupulous chiefs, who received bribes in the form of money or alcohol from the agents, even forced some of their subjects to join the recruiting teams. Agents coaxed away and even "stole" one another's recruits and this led to many disputes and even fights. In many ways the fate of a team of recruits was not much better than those of the many slaves from the cruelest era in the history of the African continent before 1840.

The main connection route between Crooks Corner and the outside world along which recruits were transported to Soekmekaar Station, was the route which had been used by the hunters before and which was described by Bulpin (1954) as the "Ivory Trail". This road, which was barely passable for a donkey-cart, led from Makuleke's village through the Luvuvhu River at Bobomeni, past Baobab Hill to Klopperfontein (or Shinkhuwa as it had originally been known). From here it went over Dimbye ra

19. THE WILD DAYS OF CROOKS CORNER

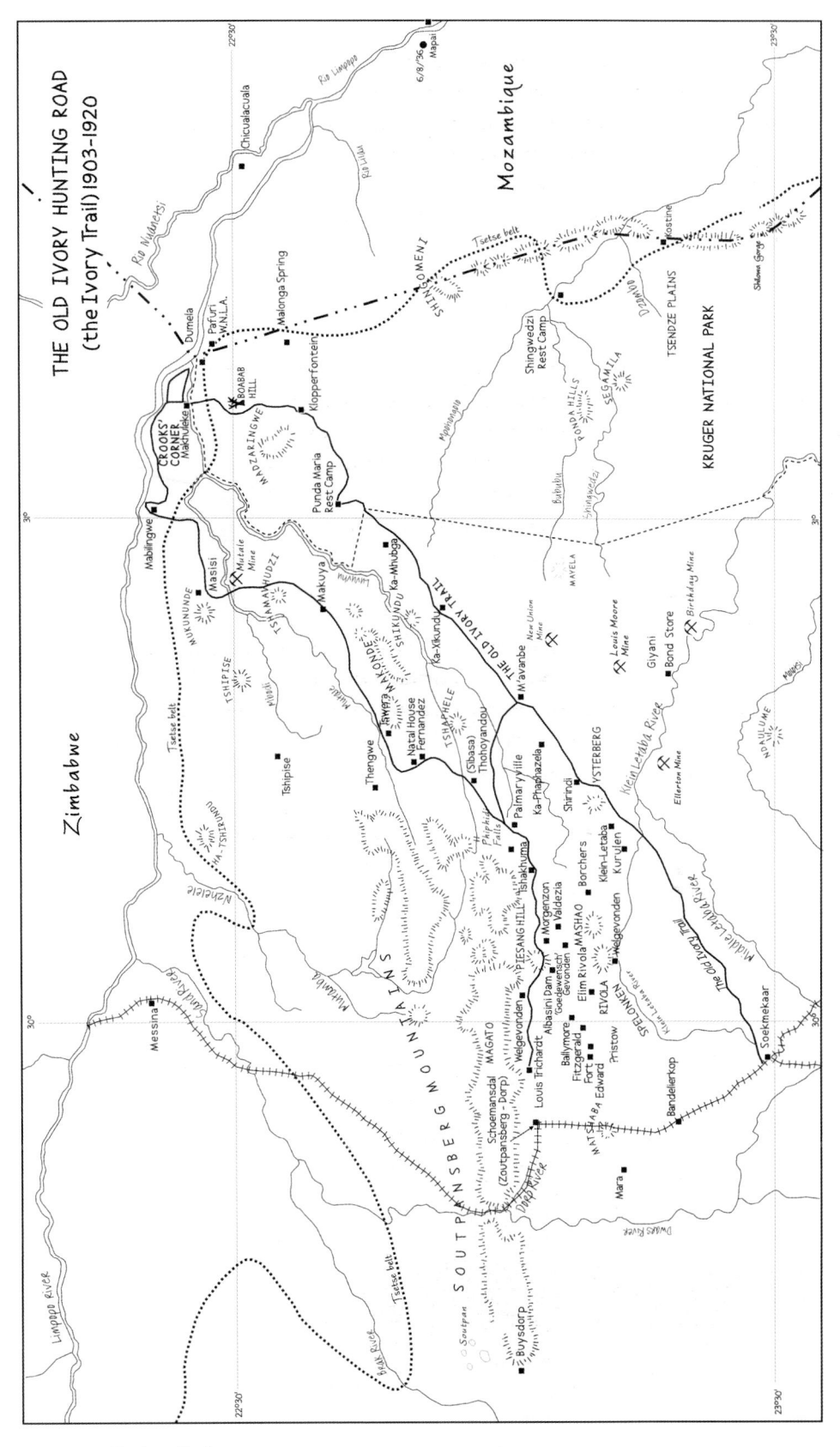

The Ivory Trail.

Shikokololo ("the fertile area of Chief Shikokololo") where there was a strong fountain (later Punda Maria). From Shikokololo it led over Shikumbu, Mavambe, Shirindi, Fonsecas (where a certain De Fonseca ran a kind of inn for travellers between 1910 until 1920), Klein Letaba and Lebjelebore (Lebelle Bowrie) and eventually to Soekmekaar along the Selati railway line.

It was along this route that most of the recruiting agents, hunters and other characters who eventually settled at Crooks Corner first laid eyes on this remote part of the Lowveld. After the First World War, the Witwatersrand Native Labour Association (WNLA) obtained the exclusive concession to recruit mine workers in the neighbouring countries. This organisation gradually improved the route. At first the recruits had been transported from Pafuri to Soekmekaar with donkey wagons, but this function was taken over by Thornycroft buses in 1929.

This route was also the one followed by the best-known figure associated with Crooks Corner, Stephanus Cecil Rutgert Barnard, when he first visited the Lowveld in 1910 with the objective of becoming an elephant hunter. Here he lived a wandering existence as hunter, recruiting agent and later as WNLA official for 19 years only returning to civilisation in 1929 when he got married and settled on his farm Vlakplaas near Geysdorp.

Cecil Barnard had been born on 19 September 1886 in Knysna, but his parents moved to the Western Transvaal where he grew up. After a year or so as a policeman, he succumbed to the call of freedom and his ambition to become a big game hunter. Filled with youthful gumption, but with very little experience, he moved into the wild Lowveld. Here he lived through many adventures, which Bulpin (1954) described in his book *The Ivory Trail*. In time he gradually gained a reputation as being a notorious elephant poacher who had been outlawed by the police of Rhodesia, Mozambique and South Africa, due to his illegal elephant hunting in all three countries. He set up his base camp near the confluence of the Limpopo and Luvuvhu rivers and later told with relish how he had loosened the border beacon between Mozambique and South Africa and moved it from place to place, depending on which country's police wanted to arrest him for illegal hunting.

Barnard got to know the adjacent areas in Southern Rhodesia (Zimbabwe) and Mozambique like the palm of his hand through the years, shot many elephants and also earned good money as a recruiting agent for the mines. One of his dreams was to find the legendary bull elephant, Ndlulamithi ("taller than the trees"), and he spent unlimited time tracking him down without success for many years. One day in 1929 he finally came eye to eye with the magnificent

The wild fig forest galery just west of Bobobmeni drift along the southern bank of the Luvuvhu River.
(19.03.1985)

Cecil (Bvekenya) Barnard, the well-known elephant hunter as a young man, circa 1910.
(ACKNOWLEDGEMENT: IZAK BARNARD AND T.V. BULPIN)

animal after many days of tracking. Squarely in his sights, the great animal with its colossal tusks made such an impression on Barnard that he lifted his rifle and allowed it to go its own way. On 22 August 1967 a large elephant bull was shot by a South African hunter, General Victor Verster, along the Lundi River in the current Gona-re-Zhou reserve in Zimbabwe. Its tusks weighed 62,3 and 48,6 kilograms respectively. It was believed by certain people that this elephant had been Bvekenya Barnard's legendary Ndlulamithi. Mrs Ruth Gillet, formerly of Zimbabwe and now living in Plettenberg Bay, had sent a photograph which had been taken by her husband, Geoff, of this magnificent elephant to the author.

That this elephant could have been Bvekenya's Ndlulamithi is simply not possible, due to the time-lapse of almost 40 years since Barnard's meeting with the animal. Elephant bulls with such large tusks are usually quite old already and Ndlulamithi would probably not have survived much later than the early 1930s, if a hunter had not shot him before the time. In a recent publication, Smillie (2003) suggested that an elephant bull which was photographed in Zimbabwe during the 1920s by the BSA policeman, George Style, in Barnard's former haunts, had been the real Ndlulamithi. He further claimed that a gigantic pair of tusks, weighing 73 and 73,5 kg respectively, which had been offered for sale to R. Balmer, an ivory trader in Lourenço Marques (Maputo), belonged to this famous bull. That this is the most feasible explanation of how Ndlulamithi had come to an end was confirmed by Bvekenya's son Izak.

Like most of the well-known figures in the Lowveld, Cecil Barnard soon acquired a nickname among the local blacks. They named him Bvekenya ("he who swaggers as he walks"), probably because of his excessive self-confidence and way of walking. Many of his adventures are still legendary in the Lowveld, including his escapades to avoid being arrested. His powerfully built body and dominating personality enabled him to demand a large degree of respect and in time he became a kind of folk hero among the other inhabitants of Crooks Corner.

Through the years Bvekenya met all the characters of this early, wild era and built up camaraderie with some which would live on even after his return to civilisation.

The elephant bull, Ndlulamithi, which was later shot by Gen. Victor Verster in the Gona-re-zhou area of the former Southern Rhodesia (now Zimbabwe).
(ACKNOWLEDGEMENT: MRS RUTH GILLET)

One of the most interesting inhabitants of Crooks Corner was Thomas Dorlan (Buck) Buchanan, a Canadian from Montreal. This small, thin man about whose early life very little is known, was born in 1874. When he visited England during Queen Victoria's silver jubilee, a friend asked him to transport 1200 donkeys to Canada. This led to him transporting horses and donkeys between England and Canada for a number of years. After that he lived in Chicago for two years and then in Texas for a while. With the outbreak of the Anglo-Boer War he came to South Africa and joined the Imperial Light Horse

Cecil (Bvekenya) Barnard (left) and Thomas Dorlan (Buck) Buchanan, another one of the characters of Crooks Corner, circa 1953.
(ACKNOWLEDGEMENT: T.V. BULPIN, MUIZENBERG)

19. THE WILD DAYS OF CROOKS CORNER

Bertie Booysen, one of W.F. Borchers's assistants, with a collection of elephant tusks which Borchers had bought from Bvekenya Barnard before the First World War.
(ACKNOWLEDGEMENT: G. BORCHERS, LOUIS TRICHARDT [MAKHADO])

Thomas Dorlan (Buck) Buchanan, who managed Alex Thompson's well-known Makuleke shop on behalf of W. Borchers and Morty Ash between 1912 and 1920, with his hunting dog at Crooks Corner, circa 1914.
(ACKNOWLEDGEMENT: JOHN SYKES ASH)

Brigade. By the end of the war he was in Johannesburg, but moved to the Soutpansberg in 1903 where he tried his hand at farming along the Klein Letaba (Spelonken).

In 1907 he and two friends went to the Motale copper field which had been discovered by Geo Hickey in 1906. He visited the copper claims annually and in 1911 constructed a road to Motale via Natal House, which branched off from the Tengwa road close to one of Alex Thompson's shops at Makonde. This road was later extended over Spokonyole Pan to Makuleke.

In 1912 he was asked to take over and manage the small shop at Makuleke which had been built by Alex Thompson and William Pye in 1910. Thompson had sold the shop to an Australian, Jack Forde, but he returned to Australia after transferring the shop to V.W. Godfrey of Pietersburg (Polokwane). Forde eventually sold the shop to Borchers and Angelbach, two German shopkeepers, who owned a whole series of trading facilities in the Soutpansberg.

Borchers struggled to find a suitable person to run the shop and after he had been disappointed by Bowman and Calder respectively, he asked Buck Buchanan to manage the place on his behalf. It was here where Buchanan met Bvekenya for the first time after Barnard had brought a team of recruits for the mines to the shop. In 1913 he was appointed permanently as shop manager after Gus Cook had not been able to cut the grade.

In 1912 Buchanan married Xalati Tsale Mbokweni at Makuleke. He ran the shop until 1920, when its management was taken over by a Canadian, Bill Milligan, until it was sold to another well-known character of those days, Morty Ash. After this Buchanan left Crooks Corner and went to farm on a piece of land adjacent to that of João Albasini along the Luvuvhu River. A combination of continuous droughts and bad management made him lose the farm in 1931 and he then went to work as farm manager for a Mr Banks along the Klein Letaba. He left Bank's farm in 1937 to manage Grysappel, one of Morty Ash's farms. This farm bordered Waterboom farm, which Morty Ash had bought from the parents of his fourth wife, Caroline Fonseca. Caroline's father managed an inn on this farm along the old road between Klein Letaba and Soekmekaar. Buchanan left Morty Ash's farm in 1944 and again worked as farm manager for a Mrs Kohler along the Luvuvhu. In 1948 he decided to go prospecting with a retired friend, Mr Morgan, who had been postmaster of Rhatomba and later of Klein Letaba. They settled just east of Louis Moore Mine on the bank of the Nsama, which is a tributary of the Klein Letaba. The two spent many years here and lived off the alluvial gold which they managed to mine. Morgan, who had lived a real hermit's life, died in 1952 and was buried at Louis Moore. Buchanan stayed on, until he fell ill in 1956. He was taken to Elim Hospital by his son-in-law, John Ash, where he died at the age of 82. He was buried on the Ash's farm Alverton.

Despite what his drifting life and unsophisticated appearance might suggest, Buchanan had some education and he sent a comprehensive report about the wild days of Crooks Corner to R. Hewitt Ivy in 1955, the agricultural extension officer of the Department of Native Affairs in the Sibasa district at the time, stationed at Ivydale along the Klein Letaba between 1937 until 1955. In 1983 Ivy donated his own notes, as well as those of Buchanan about Crooks Corner, to the Kruger National Park's archives. The contents of this chapter have mainly been derived from these two unpublished manuscripts.

It was interesting to discover from these documents that the area between the Luvuvhu and Limpopo rivers had not been known as Crooks Corner during Thomas Buchanan's stay there (1910–1920). He maintained that the odd assortment of characters there had been referred to as "Alibaba and the forty thieves". Sergeant D'Arcy, who had been in charge of the first police corps to be established at Sibasa in 1910, always enquired in his letters to Buchanan: "How are the other thirty-nine?"

The old hunting route from Soekmekaar to Crooks Corner was long and difficult. The hunters had to stay alert and did not dare take a nap on their horses. Once one of the old hunters by name of Joseph Fourie, was travelling on horseback near Baobab Hill. He was more than half asleep because of the heat and the monotonous singing of the cicadas, when a leopard suddenly came charging out of the grass and grabbed his horse's muzzle. The panic-stricken animal started circling around in an effort to shake off the leopard, but it held on for dear life. By this time Joseph had recovered sufficiently from the sudden shock to drag his rifle from the saddle bag and he managed to kill the leopard with one well-aimed shot. Fourie never dared take a nap along this road again! This same Joseph Fourie apparently took part in the illegal sawing of Rhodesian teak and other hard wood trees in the Shingwedzi Reserve from 1915 to 1920 in the vicinity of Shikokololo (Punda Maria). One Rhodesian teak trunk which Fourie had sawed, was apparently so wide that it filled his biggest wagon from side to side and he needed two teams of oxen to move it.

Bvekenya himself obtained the wood for most of the furniture in his farmhouse from the sandveld around Punda Maria. His fence poles were of Lebombo ironwood (msimbit, which he had also sawn in the vicinity of Punda Maria).

After 1919, with the appointment of J.J. (Kat) Coetser as the first ranger at Punda Maria, these illegal activities were brought more or less under control. Illegal hunting and recruiting of mine workers with Crooks Corner as base also gradually came to an end.

Johannes Jacobus Coetser was born on 3 May 1872 in Lydenburg. He married Maria Johanna de Beer and they had 14 children, of which four died at a young age. He served as bailiff at Standerton for a while, but later joined the Department of Native

The well-known shop which Alex Thompson and William Pye built in 1910 on Sesengambwene Hill near Chief Makuleke's village in Crooks Corner, circa 1915.
(ACKNOWLEDGEMENT: JOHN SYKES ASH)

Some of the merchandise that Buck Buchanan offered for sale in the Makuleke shop, circa 1915.
(ACKNOWLEDGEMENT: JOHN SYKES ASH)

19. THE WILD DAYS OF CROOKS CORNER

The Old Wagon Route against Sesengambwene Hill to Alex Thompson and William Pye's Makuleke shop is still clearly visible.
(01.08.1985)

The site of Alex Thompson and William Pye's well-known Makuleke shop on Sesengambwene Hill, near Chief Makulekes village.
(01.08.1985)

The ruins of Alex Thompson and William Pye's well-known Makuleke shop on Sesengambwene Hill in Crooks Corner.
(24.06.1985)

Affairs as he could speak a number of black languages. Later he became a member of the State Artillery and was promoted to captain with the outbreak of the Anglo-Boer War. At first he was in charge of the Johannesburg Fort until Judge Krause could take over from him, allowing him to go on active duty. However, he was taken prisoner of war and was held in the camp at Green Point. After the war he joined the army and became an intelligence officer. With the outbreak of the First World War, he joined General Van Deventer's scouts and did active duty during the whole East African campaign. While in East Africa, he mastered yet another African language, Swahili.

In 1919 he and Colonel Piet de Jager were selected from a large number of applicants to be appointed as rangers in the Sabie and Shingwedzi reserves. He settled at Shikokololo Fountain and named the post Punda Maria. Under the impression that Coetser had initially misspelled the Swahili name, the Board changed the name to Punda Milia in 1959. In 1981 the correct facts were brought to the Board's attention and the original name was reinstated.

Coetser was named Gungunyane by the blacks. He created an attractive ranger post close to the strong Shikokololo Fountain and also erected a picket at the fountain at the foot of Dzundwini Hill (at the place where he had set up camp before making Punda Maria his permanent post). He built a hut over the fountain at Punda Maria to keep the water clean and apparently there were a number of tamed catfish that caught the insects as soon as they landed in the water.

Coetser was transferred to Satara in 1924 and left the Parks Board on 15 December 1928 to join the Division of Veterinary Field Services. On 21 November 1935, while serving as a foot-and-mouth disease official along the Limpopo River east of Messina (Musina), he was charged by an enraged elephant. His gun-bearer took to his heals and Coetser was trampled to death. He was buried in Pretoria.

Alex Thompson, who had built the original little shop at Makuleke assisted by William Pye in 1910, also sold his other shops in the Soutpansberg district to Borchers and Angelbach. He left his home at Magoli on the road between Natal House and Macdonald's Missionary Station to settle at Louis Trichardt where he eventually built a flourishing business. He died there during the early 1950s.

The man to which W. Borchers sold his shop and recruiting business at Makuleke, had been a character who also left deep tracks in the area. Mortimer Walton (Morty) Ash was born in 1881 on the farm Welgevonden as the son of Frederick Ash, one of three brothers from Cornwall, England who had settled in South Africa. Morty Ash, or Motie as the blacks called him, also tried his hand at elephant hunting during his youth and had, just as Bvekenya Barnard, been outlawed in Rhodesia and Mozambique. He, Barnard and Fred Roux carried on with their illegal hunting activities until the late 1920s. Except for Fred Roux, they could never be arrested by authorities for their illegal hunting, as a result of their effective system of black scouts and their method to warn one another by beating native drums when danger was on its way. They nevertheless had many narrow escapes during their efforts to avoid justice.

Once Morty had been following two elephants in Mozambique when he realised that the animals were moving in the direction of a Portuguese administrative post. He obviously and for understandable reasons did not have any desire to make contact with the officials there, but simultaneously did not want to lose the elephants either, so he sent some of his assistants to the *Chefe do Posto* with the message, that if they wanted to catch Morty Ash, they had to follow his two guides. The Portuguese eagerly followed the two – into the opposite direction. While the Portuguese were on their mission, Ash followed the two elephants and shot them both in their vegetable garden.

He quickly chopped out the tusks and then disappeared back into the bush before the two frustrated, footsore and tired Portuguese officials returned to their post to find two elephant carcasses in the backyard.

As elephant hunting became more and more risky, Morty Ash settled as trader in the Northern Transvaal Lowveld. He eventually owned five shops at Phapazela, Bend Store, Makuleke, Malanye and Medon and became a well-off man. With the outbreak of the Second World War, he sold the shop at Makuleke to Sam Majika, who traded there until the middle of 1950. This historical little building was then demolished by John Fernandez, who used the material to build himself two small shops, one at Gwalala and one just west of the current air strip at Pafuri. These buildings were demolished after the area had become part of the Kruger National Park again in 1969.

Morty Ash had four wives during his lifetime of which only Caroline Fonseca was white. John Sykes Ash, one of his sons with his second wife, Mudlayi, lived in Gazankulu and owned the Gumbani Trading Store. He married one of Buck Buchanan's daughters. Most members of the Ash family were buried on the family farm, Alverton, but Morty was buried on Waterboom farm in 1982, which belonged to his father-in-law, Fonseca.

Another character who frequented Crooks Corner before 1920, was Leonard Henry Ledeboer. Ledeboer was a Dutchman born in India in 1868. In 1888 he came to South Africa at the age of 21 and made a living as hunter and transport rider. During his hunting trips in Rhodesia (Zimbabwe) he met, amongst others, the famous hunter, Frederick Courtenay Selous. During the Anglo-Boer War he served as intelligence officer in the infamous Bushveldt Carbineers at Fort Edward. In 1902 he was a witness at the war tribunal in Pietersburg (Polokwane), where the cases against several officers of the unit were heard who had been accused of killing many Boer prisoners of war.

After the war, Ledeboer worked in the office of the Native Commissioner at Sibasa, but later also became a recruiting officer at Crooks Corner. He was in partnership with Bvekenya Barnard and four others for a long time. While he was recruiting along the Save River in Rhodesia in 1913, a New Zealander by the name of Powdrell also settled at Makuleke as recruiting agent in partnership with a Polish man, known as Adam Thomas, although some claimed that he was the notorious Tom Paulin (alias Polus). One night while Ledeboer was sitting next to a fire while mist settled down over the Save River, the Pole arrived there and told him that Powdrell had been caught by a crocodile. This apparently happened while he was following a wild duck, which he had wounded, into the water. Ledeboer did not believe the story and went to investigate with a few of his assistants the next morning.

They first discovered a pool of blood along the river. Then they followed some drag marks into the reeds where they found Powdrell's body. He had been

Mortimer Walton (Morty) Ash, a well-known shopkeeper and businessman in the days of Crooks Corner. He owned the romantic Makuleke shop for a while, circa 1930s.
(ACKNOWLEDGEMENT: G. BORCHERS, LOUIS TRICHARDT [MAKHADO])

The ruins of John Fernandez's shop close to the current Pafuri airstrip (1954–1969).
(19.04.1985)

shot in the head and his belt with £200 in notes which he had used to bribe chiefs and to pay recruits, was missing.

It was clear that he had been murdered. Ledeboer buried the body and reported the incident to the police. It took them six days to reach the site and by that time the Pole had already made himself scarce.

Ledeboer married Constance Elizabeth Ash, a sister of Morty Ash. They had three children, Frederick Leonard and twin girls, Constance Elizabeth and Jane Marie. His son Freddy was bitten by lions twice during his lifetime, but survived both these attacks, although he lost his left arm on the one occasion (Minnaar, 1985). Ledeboer's first wife passed away and was buried on the Ash's family farm. He then left the children in the care of a governess, Anna Bindemann, whom he married in 1915 before his departure for German East Africa (Tanzania) where he fought under Colonel Wylie. After his return from East Africa he was appointed as ranger in the Kruger National Park in 1921 after the retirement of Major A.A. Fraser. However, Ledeboer was not stationed at Malunzane (major Fraser's old post), propably because of water being a problem there.

Colonel Stevenson-Hamilton requested him to build a new post at Hatlane at the mouth of the Makhadzi Spruit, where he could keep an eye on the locals in the nearby Makuba's village next to the Makhadzi. On 10 August 1921, his second wife Anna Maria Christina (born Bindemann) died of a heart attack at age 55 and he buried her just west of the confluence of the Makhadzi with the Letaba River. According to legend his wife Anna gave birth to a little girl, but it is not known what happened to her. In December 1921 he moved his post to Mondzwani on the northern bank of the Letaba River. Here he lived until 1927 when a new house was built for him on the southern bank of the river, east of the existing WNLA quarters.

In 1922 he married again. He had two children with Kate Häsler (Evans) – Esmé (1923) and Leonard

A group of illegal recruiting agents with a team of recruits for the Witwatersrand gold mines, at a collection point in Crooks Corner, circa 1913.
(ACKNOWLEDGEMENT: JOHN SYKES ASH)

Cecil (1929). In 1929 he was transferred to Satara where he served up to 1946 when he retired. After his retirement, he lived on his farm Radoo near Letsitele where he also managed a small store. His third wife died on 23 November 1937 and he married his fourth wife, Martha Maria Meyer, whose brother Gerry later become a courier for Comair in Skukuza. He later moved to White River, where he died on 25 February 1959 at the age of 91.

William Pye, an Irishman, who built the little shop at Makuleke with Alex Thompson, had originally come from the Catholic section of Ireland. He served in a Hussar regiment under General French during Kimberley's relief. After the war he moved to the Northern Transvaal where he worked for the Greek, Ges Palmary (after whom Palmaryville had been named), as shop manager at the Louis Moore mine. He later left to manage the Natal House Inn for Palmary. Still later he managed the Pepiti shop for Alex Thompson before arriving at Crooks Corner as recruiting officer.

He set up camp near the Limpopo River about two kilometres east of the later Manqeva picket near Dakamila Pan. He was a jovial person and great jokester who loved to mimic others. Apart from that, he served as the official "doctor" and undertaker of Crooks Corner. Those suffering from blackwater fever whom he could not doctor back to health, he buried in the cemetery close to his camp.

Pye founded a recruiting agency with Barnard, Ledeboer, Roux and Diegal. Their opposition was Theo Williams, also called Considine, a Greek and former gambler from Seattle, USA and a Swede, Colson, who worked for Dave and Ticky Erskine of the Native Recruiting Company (NRC). Two other members of their agency were John Dart, a Welshman, and Wieder, a Hungarian.

A Sergeant Pearce of the BSA Police in Rhodesia (Zimbabwe) arrested Pye, Considine and Colson for illegal recruiting and Pye and Considine had to pay fines of £5 each, while Colson had to fork out £85. This was not the end of his troubles, because when Colson arrived back in Crooks Corner, he received a punch from Barnard for allegedly abducting some of his recruits. This was the last to be seen of Colson. He died shortly afterwards and was buried at the Ellerton Mine. Considine died of malaria on his way back from Rhodesia (Zimbabwe) and when Pye returned via Gwanda and Messina (Musina), he had to find out that Buck Buchanan and Baron Cedarström had buried Considine at his old camp along the Limpopo floodplain. During the police investigation afterwards, they found a wig and a bundle of different keys in his hut. His fingerprints were sent to the BSA Police and it was then discovered that he had been convicted of housebreaking in Bulawayo and had spent a number of years in jail. A rumour was spread that he had buried quite a bit of money or gold in his hut and it was dug up from wall to wall – without any success of course!

After Considine's death, a prospector from Natal named Lumley, moved into his camp. He was very deaf and soon became known as Mandlebe ("Ears") among the locals. After his departure in the 1920s, a man by name of Mdoseni, originally from Venda, lived in the house. The ruins of this little building, which dates back to the days of Crooks Corner, were discovered during the 1980s by two nature conservation officials, H. Sleigh and L. Hare. These are about two kilometres west of the Manqeva picket along the perimeter of the Limpopo floodplain.

Hendrik Hartman arrived at Makuleke in about 1918. He met Barnard at the shop and the two decided to start farming sheep together along the Brak River behind the Soutpansberg. Bvekenya collected the sheep during his recruiting journeys in Rhodesia and Hartman had to look after them on the farm. They eventually had 600 sheep and Hartman was to take them to Brak River soon after the arrival of the next group. Barnard duly arrived with another

The Greek Consodine's neglected grave close to where his camp used to be along the Limpopo floodplain, in Crooks Corner. (10.02.1985)

The ruins of the Greek Consodine and the New Zealander Lumley's camp along the Limpopo floodplain about 4,3 kilometres west of Paradys windmill in Crooks Corner. (10.12.1985)

A few of the "crooks" of Crooks Corner, with a collection of their hunting trophies, circa 1914. They have probably played an important role in the extermination of the Lichtenstein's hartebeest north of the Soutpansberg.
(ACKNOWLEDGEMENT: JOHN SYKES ASH)

50 sheep, but there was no sign of Hartman. Their assistants then informed him that Hartman had left there early that morning with an axe to chop down branches to enlarge the sheep's kraal. When he failed to return by evening, they started searching until they eventually heard a faint, far-off call. When reaching Hartman, they discovered him sitting in an umbrella thorn tree. He had been chopping down branches which fell around the tree, creating a kraal from which he could not escape. As a result of spending the whole day in the scorching sun, he contracted heatstroke. They carried him to Pye's camp where everything possible was done to save his life, but he was too far gone and died. On 10 October 1918 Pye and Buchanan buried him in Pye's cemetery.

Hartman's son, Major Daniël Hartman, who later became Chief Commissioner of Native Affairs in Pietersburg (Polokwane), tried for years to determine where his father had been buried. In June 1950, Mr Ivy, the agricultural extension officer of the Sibasa district, organised that Buck Buchanan, whom he had made contact with shortly before, go with him to show him his father's grave. They were accompanied by Colonel M. Rowland-Jones, the ranger of Punda Maria, and Harold Mockford, the WNLA official at Pafuri.

In November 1918 William Pye became a victim of the Great Flu Epidemic. After Barnard had transported him from his camp on a stretcher to Makuleke, he was nursed by Buck Buchanan and Herby Bristow. However, just a few days after Armistice Day at the end of the First World War, he died and was buried on 20 November 1918 next to Hendrik Hartman. Bvekenya Barnard returned many years later and covered the graves of his two friends with cement slabs on 14 October 1955.

An exceptional visitor to Crooks Corner was Bill Lusk, alias "Texas Jack" who had been a member of Texas Jack's popular Wild West Show and an excellent revolver and rifle shot. After that he was a pack-mule operator for the American Army in the Philippines. Still later he served in the German force during the Herero War in German West Africa, where he picked up diamonds to the value of £3000. He returned to America with his fortune, but the money did not last very long and he returned penniless to South Africa. He soon befriended Buck Buchanan and started prospecting in the far northern area of the Transvaal in 1916. According to Buchanan, Lusk had prospecting licences not only throughout the whole Union of South Africa, but also up to the Belgian Congo (Zaire).

After Buchanan had left Crooks Corner, Bill Lusk arrived at his farm one day in 1920 and showed him some of the best alluvial gold which he had ever seen. According to Lusk he had found the gold in the Shingwedzi Reserve. He applied for a license to mine his finding, but this was of course refused.

However, Lusk was not going to let a simple detail like a licence deter him, so every winter he slipped into the reserve by himself and hid his tracks so well that nobody could determine his whereabouts. Then he took out enough gold to provide him with the basic necessities until the next winter. One day he arrived at one of the small stations near Mica and after he had unsaddled his donkeys, sat down underneath a tree across from the local shop. A few days later, somebody noticed that the man had been sitting in the same position for rather a long time. Closer investigation revealed that he was dead. Lying on the ground next to him were a few bags with gold. The police desperately tried to determine where he had mined the gold, but without any success. This remained a secret, until former ranger Harry Kirkman later heard in a roundabout way from blacks that Lusk had taken out the gold at Red Rocks (the Gubyane potholes) in the Shingwedzi River.

Another black sheep of Crooks Corner with an interesting history was the Swede, Baron Cedarström. He was a civil engineer and a brother of the Baron Cedarström who had been married to the famous singer, Adeline Patti. According to Buchanan he had been quite a pleasant fellow, but a little too fond of the bottle. Apparently he was married to a well-off wife who sent him an allowance every three months to keep him out of Sweden! During the Anglo-Boer War he had been an officer in the Scandinavian Brigade of the Boer forces and was wounded and captured at Paardeberg. During his prisoner of war days on St. Helena, he was placed in charge of the waterworks and treated well by the British officers. During his stay at Makuleke as recruiting agent, his son served as aide-de-camp for the Swedish king. After he had left Crooks Corner, he apparently settled at Pietersburg (Polokwane), but nobody knew what became of him after that.

Theo Williams was a Welshman who had managed a hotel in Barberton before settling at Crooks Corner as recruiting agent. He tried to join Buchanan's group with the recommendation that he had a horse which could beat any Portuguese horse in a race! However, Buchanan rejected his proposal so he joined the opposition party of Dave Erskine, who recruited for the NRC using Pietersburg (Polokwane) as base. He had two partners: the Greek Considine and Diegal, who both left him in 1913. He nevertheless carried on with his recruiting activities with the help of an old black man, Solomon, who had previously been a very reliable assistant at the Valdézia Missionary Station. Williams, who was called Makhanda by the locals, left Crooks Corner in 1917 and later died at Umtata in the Transkei.

Martin Jacob Diegal had been born in Cassell, Germany and was a sailor before coming to South Africa. After much roaming about, he eventually arrived at Louis Trichardt. Shortly afterwards he joined the other recruiting agents at Makuleke and became known as Charlie among the blacks. He was a tall, blond man with a light red beard. To evade military duty in Germany, he slipped across the border into the Netherlands and boarded a ship to East Africa where he landed in the Mozambican harbour city, Sofala. Here he fished for a while and made a living by selling dried fish. After his arrival at Makuleke he went into partnership with Jack Lambert, after that with Considine and Williams, but he eventually joined Barnard's syndicate.

Diegal loved hunting, although he had apparently been a rather poor shot. He could, however, plat beautiful whips and shamboks. Once, the Greek left him badly in the lurch with the result that he nearly died of hunger. After that he and Considine became

Charlie Diegal relaxes with his favourite tipple at Crooks Corner, circa 1913.
(ACKNOWLEDGEMENT: JOHN SYKES ASH)

Martin Jacob (Charlie) Diegal, one of the illegal recruiting agents, and his wife at Crooks Corner, circa 1913.
(ACKNOWLEDGEMENT: JOHN SYKES ASH)

arch enemies and their arguments often ended in physical encounters. During one of these fights, Barnard had to intervene. This only infuriated Considine to such an extent that he threatened Barnard with death. Barnard did not take any heed of the Greek's threats, grabbed him and tied him down until his anger had subsided.

After the outbreak of the First World War, Barnard heard that Diegal had been seriously injured by a buffalo in Portuguese East Africa (Mozambique). He had apparently shot the animal with a soft-nosed bullet and only wounded it. The enraged buffalo charged and flung him against the trunk of a large baobab tree. Then it turned around and stormed into the bush, leaving a seriously injured Diegal behind. It had been a difficult time for Barnard. Billy Green, his brother-in-law was with him, seriously ill with blackwater fever. Billy Green had apparently been the first white to have actually lived at Punda Maria. Barnard travelled through the night to reach Diegal. Then he made a stretcher and tried to transport the injured man, but he was in too much pain to be carried on a stretcher, so Barnard made him as comfortable as possible and then returned to attend to his sick brother-in-law. All he could do to get his two seriously ill patients closer together for him to man-age the difficult situation, was to carry Green closer to Diegal until they were about 24 kilometres apart. After that he visited them in turn and cared for them as best as he could. Then they ran out of food. It was the rainy season, the rivers were in flood and Barnard's assistants could not get through with supplies.

Barnard was forced to build a make-shift boat to get back to Makuleke himself to fetch the necessary supplies. Fortunately both his patients were recovering by then. While Barnard was away, the Portuguese heard of Diegal's fate and sent a patrol to arrest him. They had been searching for him for a long time because of his illegal hunting activities and seeing that he was German, he had to be interned. The Portuguese found him where he was recuperating, arrested him and transported him to Beira. He managed to escape, but was soon captured again and taken back to Beira. From there he was sent to Luanda in Angola where he was interned until the end of the war. Some believed that he had died in the internment camp, but apparently Buchanan received a letter from him in 1948 from South West Africa (Namibia) where he was farming under an Afrikaans name.

Jack Forde, the Australian who had originally bought the shop at Makuleke from Alex Thompson, returned from Australia after a while to make his appearance at Crooks Corner again. He was a real troublemaker who caused problems wherever he went, if not with Diegal, then it was with Fred Roux or Buchanan. The BSA police also arrested him once for illegal recruiting and he was taken to Chibi and fined.

Forde was continuously agitating because of one or other grievance. He even wrote to the Portuguese king once, with a drawing of the banks of the Save River where recruiting agents had been shooting many hippos illegally. He also wrote to General Botha and complained because Borchers had not been interned during the First World War. However, Borchers had been a naturalised British subject and after a meeting with Generals Botha and Smuts in Pretoria, he was allowed to carry on with his business activities in the Soutpansberg. His partner, Angelbach, was not so fortunate, however, and he was not only interned, but all his possessions forfeited as well.

Borchers later became a wealthy trader in the district owning about 15 shops throughout the area. His headquarters were at Mashao, not far from João

Albasini's old farmhouse at Goedewensch at Piesangkop. He also loved hunting and often went hunting at Chicuala-cuala and Massangena in Mozambique. The Portuguese authorities in Mozambique even complained to General Smuts about Borchers and Barnard's hunting activities in their country. His shop at Pafuri, a few kilometres west of the WNLA post, was managed by a black manager between 1919 until 1954. He even served visitors to the Kruger National Park in this shop. After Rob Ferreira had transferred his trading concession to the National Parks Board in 1954, Borchers's shop at Pafuri was closed and demolished. However, the foundations remained and can still be seen there.

Barnard had only been captured once by police during his long career at Crooks Corner (1910–1929) and it was not for illegal elephant hunting. This took place early in 1918, while he was visiting William Pye. While sitting in Pye's house one afternoon, chatting about the war, somebody loudly knocked on the door. Thinking that it was Buchanan, Bvekenya opened the door, but there were two Rhodesian BSA policemen outside. "Are you Bvekenya?" one asked. When Bvekenya replied in the affirmative, he was summarily ordered to put his hands in the air, or be shot. Barnard lifted his arms and protested: "But you are on the wrong side of the border!" "It does not matter, just come with us!" they replied. He was handcuffed, lifted onto his mule, Japie, and led northwards towards the Rhodesian border. It took them two weeks to reach the Rhodesian administrative post at Fort Victoria. He was jailed, but released after a local trader paid his £100 bail.

Later he was accused of illegally shooting a hippo

The well-known shopowner and businessman, W.F. Borchers, photographed as a young man in 1892 in New York.
(ACKNOWLEDGEMENT: G. BORCHERS, LOUIS TRICHARDT [MAKHADO])

W.F. Borchers (left) and friends with some duikers which they had shot in the Mhinga area, circa 1910.
(ACKNOWLEDGEMENT: G. BORCHERS, LOUIS TRICHARDT [MAKHADO])

19. THE WILD DAYS OF CROOKS CORNER

in Rhodesia and fined £5. His host lent him the £5 and also enough money to buy a new 7,9 mm Mauser rifle from a local policeman. He then returned home. On the way back he used the opportunity to shoot seven elephants and to recruit a few dozen labourers to compensate him for his "expenses" in Rhodesia!

Back at Makuleke, Barnard had to find out to his great disappointment that large-scale activities were taking place on a hill near the Portuguese border, south of the confluence of the Luvuvhu and Limpopo rivers. Authorities had become increasingly concerned about the extensive illegal activities of the "private" recruiting agents or syndicates in the neighbouring states. As a result it was decided that the concession to recruit black labourers for the mines should fall under central control. It was the Witwatersrand Native Labour Association (WNLA) which acquired sole-rights to do recruiting for the mines in neighbouring states, while the Native Recruiting Company (NRC) obtained the monopoly to recruit in the interior. The buildings rising at Pafuri, were to be the collection point and administrative post of WNLA. Paul Neergaard was appointed as the first official in charge and he laid the foundation here for a flourishing labour recruiting agency in later years.

Neergaard offered Barnard an opportunity to join WNLA, as they realised that they could use his knowledge of the surrounding areas. Barnard considered the offer and visited Mr Gemmill, the official in charge of WNLA in Johannesburg. After it had been arranged that the police would withdraw all pending cases against him, Bvekenya became a WNLA official. He diligently served the organisation for a time. Apart from other accomplishments he was involved in the improvement of the roads from Soekmekaar and Louis Trichardt to Pafuri.

One of his neatly packed stone drifts can still be seen just east of the dam at Klopperfontein. According to him he released a small herd of eland in the vicinity of Klopperfontein, as he had believed that there were none in the Park.

In 1923 the lack of freedom became too much for his restless psyche, so he resigned from WNLA to resume his former career as elephant hunter. He persisted with this until 1929 when he married and settled on his farm, Vlakplaas, near Geysdorp in the Western Transvaal (now North West Province). He farmed there until his death on 2 June 1962. A neat granite headstone with a replica of Astley-Maberley's map in T.V. Bulpin's book, *The Ivory Trail*, engraved on it, indicates the last resting place of this staunch old adventurer.

With Barnard's departure from Crooks Corner things started changing. With the control which R. Hewitt Ivy, the agricultural extension officer of the Department of Native Affairs, and the ranger at Punda Maria under whose control Crooks Corner had been placed, exerted over poaching and other illegal activities, the inhabitants of the area soon left for greener pastures one by one. Barnard's own assistants also eventually left.

Some of Barnard's assistants during his "career" at Crooks Corner included the eccentric De Beer broth-

The ruins of W. Borchers's shop at Pafuri. It was open between 1919 and 1954.
(21.06.1984)

The stone drift which was built by Bvekenya Barnard through a dip at Klopperfontein, circa 1920.
(27.04.1984)

662

ers. Herklaas, Theunis and Jan de Beer left the Waterberg area in 1892 for the Melsetter area in the then Southern Rhodesia. They were born men of the veld who were almost totally dependent on the proceeds of their hunting to survive. Although particularly antisocial, Barnard managed to convince them to enlist some recruits for him while on their hunting trips.

The De Beers had an insatiable capacity to consume meat and usually stayed with a buck's carcase until they had completely finished off the meat. They worked for Barnard until 1914 when Jan died of heatstroke and another friend of theirs, Barend Bezuidenhout, of malaria. Herklaas and Theunis were highly distressed about their brother and friend's deaths, as well as the rumours of war, so they decided to pack up their belongings to move to the Sand River near Pietersburg where they planned to farm.

Barnard, sorry to lose their services, compensated them with £994 and they found this much money so fascinating that they spent days counting and recounting it. On their way to the farm they bought a bull and a few cows from Morty Ash. However, they had not gone very far, before their hunger for meat overcame them, so they slaughtered the bull and ate it and before reaching the farm, they had finished off the cows as well. Gradually they spent all their hard-earned money on meat and had to leave the farm again. Herklaas eventually died in a gravel-slide at the diamond diggings.

Pat Fay had been another odd Irish character who lived at Makuleke for a while. He had been a retired policeman stationed at Sibasa. He had the same outstanding appetite as the De Beer brothers and soon finished off his pension buying food. To earn a little extra, he started smuggling with rifles, but as so many others who had come to this Eldorado, he did not have much success.

One day he returned to his hut after a long trek and was so hungry that he had an ox slaughtered. That night he overate and died as a result of his overindulgence. Many others died as a result of their love for the bottle, but in contrast to those who loved the bottle, Pat Fay's camp could be recognised by the large heap of empty tins near the back door. He was buried at Elim Missionary Station in a coffin made from tinned meat boxes.

Another well-known person who played an important part in the district during the heydays of Crooks Corner and afterwards as well, was Joe Albasini. He was the grandson of the famous João Albasini from the days of Schoemansdal and the first Voortrekker settlers. Joe was a crack shot and as a descendent of the great white chief of the Magwamba, highly respected among the Shangaan. He worked with Hewitt Ivy in the office of the Native Commissioner at Sibasa for many years.

He was also known as Juwawa, the same name which had been given to his renowned grandfather. Ivy in turn was known among the blacks as Gungunyane, which was the name of one of the last great Shangaan chiefs in Portuguese East Africa (Mozambique) where he had succeeded his father, Umzila, in 1884. Umzila was the son of Manukosi (Sochan-

Cecil (Bvekenya) Barnard's grave on his farm Vlakplaas, district Geysdorp.
(ACKNOWLEDGEMENT: MR IZAK BARNARD)

gana), one of Shaka's generals, who fled in 1823 with his followers, members of the Bangoni tribe, from the Zulu potentate's tyranny. They settled near the confluence of the Olifants and Limpopo rivers where Manukosi became the founder of the Shangaan nation. Incidentally, the locals bestowed the name Gungunyane on ranger Coetser of Punda Maria as well.

All these people and many others, had known Barnard and Buchanan before finally leaving this isolated corner of the Lowveld.

After the area between the Luvuvhu and Limpopo rivers had been placed under control of the Department of Native Affairs, most of the whites moved away. The people of Makuleke were born poachers and decimated the game in this area in years to come. They traded the meat to the Venda and Shangaan in the adjacent areas in exchange for grain and other essentials.

A few prospectors still visited the region from time to time in their quests for hidden treasures, but had to turn around without any of them ever finding anything. Some of their prospecting pits can still be seen in the sandstone hills south of the Luvuvhu River at Thulamela and just southeast of the Mutale mouth. In 1944 a "prospector" settled near Mabiligwe along the Limpopo. He smuggled gold, rice and diamonds across the Rhodesian (Zimbabwe) border, but the police was soon hot on his heels, so he hastily had to leave the area.

In 1932 the Parks Board initiated negotiations with the then Department of Native Affairs to exchange the Pafuri area for a piece of land more suitable to agricultural purposes along the western boundary between Punda Maria Gate and the Bububu River (the so-called Mthlaveni area). The Board realised that this ecologically unique piece of land would be an extremely valuable addition to the Kruger National Park. Endless correspondence followed. The Board had to undertake to build a cattle dip in the new area, drill four boreholes and to move Morty Ash's shop there. From the existing correspondence it appears as if the proverbial fly in the ointment, who tried to prevent negotiations from succeeding, had been the assistant native commissioner at Sibasa. Eventually the whole land exchange transaction fell through because the Board could simply not afford to pay the £1250 needed to erect a cattle dip, drill the boreholes and move the shop. In 1932 the Board took note of the fact that the Transvaal Administration had decided to declare the Makuleke area a game reserve (the Pafuri Game Reserve), but that Board officials, together with Mr Ivy, the agricultural extension officer of the Department of Native Affairs, would carry out the necessary conservation actions.

In 1935 all the blacks were forced to move out of the Makuleke area on Ivy's insistence and they settled north of the Limpopo in Southern Rhodesia (Zimbabwe). Only two years later they asked to return to their former home and, to Ivy's frustration, this was allowed by the Board's ranger at Punda Maria. Poaching once again became rife. In June 1954 the National Parks Board made a decision to transfer its conservation functions in the area to the Nature Conservation Division of the Transvaal Provincial Administration.

In June 1965, Pafuri once again reverted to being Crooks Corner. One afternoon, Harold Mockford, the WNLA official at Pafuri, discovered a motor vehicle stuck in the sand at the confluence of the Limpopo and Luvuvhu rivers. On investigation he found that the occupants of the vehicle, a man and a young woman, had walked in the direction of the Rhodesian border. He followed their spoor into the thick bush on the Rhodesian side of the river until it became too dark to carry on. The next morning he discovered that the two had returned to the Transvaal side.

The famous Afrikaans author, Eugène Marais (in the white suit) at Pafuri, September 1934.
(ACKNOWLEDGEMENT: THE PRELLER FAMILY)

At John Fernandez's shop they had hitched a ride on a truck which was on its way to Sibasa. In the meantime, the police had arrived from Sibasa. They set off after the truck and caught up with it about 48 kilometres from its destination. When the truck stopped, the man immediately started shooting at the police with two pistols. The policemen ran for cover, so the outlaws used the opportunity to jump into the police vehicle to speed away. The law enforcement agents shot after their own vehicle with a .303 rifle and apart from destroying it, they managed to hit the man behind the steering wheel. Upon being hit, the man lost control and the vehicle overturned. It was then easy enough to arrest the two. It subsequently appeared that the Hungarian and his Bulgarian girlfriend, had been sought due to several different misdemeanors. The woman, who had been wounded in the arm, and the more seriously injured man were taken to hospital in Pietersburg (Polokwane). Here the Hungarian tried to escape again by attacking a nurse with a pair of scissors, but was quickly overpowered. He later died from his wounds.

One last incident which makes one think back to the wild days of Crooks Corner took place in 1981. This time a young man gassed himself in his car, which he had hidden in a gravel pit near the western boundary along the tar road between Pafuri and Messina (Musina). When the car was eventually spotted during an aerial census of the area, he had already been dead for weeks and the gruesome finding had to be reported to his family.

After renewed negotiations between the Parks Board and the Department of Native Affairs and Development, Crooks Corner was eventually exchanged for the Mthlaveni area south of Punda Maria Gate in 1969. All the Makuleke people were moved to their new home. In 1972 the valuable natural area between the Luvuvhu and Limpopo rivers was officially proclaimed as part of the Kruger National Park to great excitement of conservation conscious people.

John Fernandez and an unknown female companion at Crooks Corner, circa mid-1950s. (ACKNOWLEDGEMENT: G. BORCHERS, LOUIS TRICHARDT [MAKHADO])

References

Ash, J.S. 1985. Personal communication.
Barnard, I.J. 1985. Personal communication.
Barnard, S.C.J. 1962. Taped personal interview with Mr R.J. Labuschagne and Dr N.J. van der Merwe of the National Parks Board. Skukuza Archives.
Borchers, G. 1985. Personal communication.
Buchanan, D.T. 1955. History of Makhuleka. Unpublished manuscript. Skukuza Archives, Skukuza.
Bulpin, T.V. 1950. *Lost Trails on the Lowveld*. Howard B. Timmins, Cape Town.
Bulpin, T.V. 1954. *The Ivory Trail*. Howard B. Timmins, Cape Town.
Bulpin, T.V. 1956. *Lost Trails of the Transvaal*. Howard B. Timmins, Cape Town.
Bulpin, T.V. 1978. *Illustrated Guide to Southern Africa*. Reader's Digest Association of South Africa (Pty.) Ltd, Cape Town.
Bulpin, T.V. & Miller, P. 1968. *Lowveld Trails*. Books of Africa, Cape Town.
Godfrey, P. 1963. Epitaph for a Hunter. *Wide World* April:265–267.
Ivy, R. Hewitt. 1969. Lakes and pans of the Limpopo valley. *African Wild Life* 23(1):60–64.
Ivy, R. Hewitt. 1983. Wildlife observations in the African Bush. In: A place called Crooks Corner. Unpublished manuscript.
Klapwijk, M. 1974. The story of Tzaneen's origin. Unpublished manuscript.
Meyers, R.E. 1986. On horseback in the Kruger: 1920. *Custos* 15(2):6–8.
Minnaar, A. 1985. Twee maal in die kloue van 'n leeu. *Custos* 14(8):33–34.
Paynter, D. & Nussey, W. 1986. *Kruger: Portrait of a National Park*. Macmillan S.A., Johannesburg.
Potgieter, L. 1973. Dhlulamithi die groot olifant. *Landbouweekblad* 27 November:22–23.
Smillie, S. 2003. Who shot Dhlulamithi? *Country Life* June:82–83.
Style, G. 1965. Barnard the elephant poacher. *Outpost* May:19–21.

Chapter 20
The saga of the WNLA in the Kruger National Park
H.H. Mockford & Dr U. de V. Pienaar

Gold was discovered on the Witwatersrand in 1886. This led to the recruitment of black labourers from neighbouring countries including Basutoland (Lesotho), Bechuanaland (Botswana), Mozambique, Malawi and Southern Rhodesia (Zimbabwe). Recruiting started off on an ad hoc basis by self-appointed agents or syndicates, with very little, if any, overhead control. The first ranger, whom Major Stevenson-Hamilton appointed after he had entered office as the first warden of the Sabie Reserve in July 1902, was incidentally one of these recruiting officers for the Rand gold mines. He was Captain E.G. (Gaza) Gray, who had his headquarters at Gomondwane. Gray had served as ranger at Lower Sabie from 12 August 1902 until 20 May 1903, after which he returned to his former career as recruiting agent and moved back to Mozambique.

The Shingwedzi Reserve was proclaimed in 1903 and in 1904 Major A.A. Fraser was appointed as ranger in charge of this enormous reserve. Major Stevenson-Hamilton settled him at Malunzane along the Shongololo River, a tributary of the Tsende (Tsendze). His quarters consisted of a group of vacant huts which had been very conveniently available. These huts had been built by another, but unknown recruiting agent for the mines, but were evacuated during the Anglo-Boer War.

During the heydays of Crooks corner, fortune seekers, poachers, smugglers and other fugitives found a haven in the Lowveld between the Limpopo and Luvuvhu rivers. Here they made an existence by poaching and "black-birding" – the illegal recruitment of black labourers from Southern Rhodesia (Zimbabwe) and Mozambique for the gold mines and later also for the collieries. By the time of his appointment as ranger in 1921, H. Ledeboer had been one of the self-appointed recruiting agents in Crooks Corner.

These uncontrolled recruiting activities eventually took on such great proportions that the authorities had to intervene. After the end of the First World War, the right to recruit labourers in neighbouring countries was awarded to one organisation, the Witwatersrand Native Labour Association (WNLA). The Chamber of Mines also established the Native Recruiting Corporation (NRC) which recruited blacks

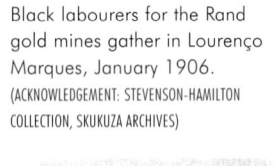

Black labourers for the Rand gold mines gather in Lourenço Marques, January 1906.
(ACKNOWLEDGEMENT: STEVENSON-HAMILTON COLLECTION, SKUKUZA ARCHIVES)

from within South Africa. This step meant the end of illegal recruiting syndicates and also put the stamp of good order on and cooperation with the Transvaal game reserves (the later Kruger National Park). The Witwatersrand Native Labour Association's name had been changed to The Employment Bureau of Africa Ltd. (TEBA) during the 1970s, but this name was never fully recognised as the recruits kept on referring to the organisation as "Wenela".

The WNLA bought 36 morgen of the farm Hasebult from Mr C.J. van der Lith in 1918. This farm was only two kilometres from the small town of Zoekmekaar (later Soekmekaar) along the Selati railway line. The name, Soekmekaar (search for each other), has an interesting origin. The area was often covered with thick mist in which people became lost regularly and then had to be searched for. This is also where the railway line from Messina (Musina) to Johannesburg and the one from Komatipoort to Musina meet.

This WNLA collection point had been well chosen and became the centre through which labourers from the neighbouring countries were channeled. Three main routes were used to bring the recruits to Soekmekaar and the construction of some of these routes involved much planning, sweat and labour.

The initial investigation to determine the most suitable routes was undertaken by William Gemmill and Paul Neergaard. Gemmill was the general manager of the WNLA and Neergaard the regional manager at Soekmekaar. They journeyed by foot, or on horseback into Mozambique and even travelled as far north as the Sabie River.

Camping sites had to be within walking distance of one another, therefore not more than 25 kilometres apart and close to sufficient water. The eventual results of Gemmill and Neergaard's investigations were three routes: (i) Soekmekaar to Pafuri at the confluence of the Limpopo and Luvuvhu rivers; (ii) Soekmekaar via Makhuba (Letaba) to Massingire in Mozambique and (iii) Soekmekaar via Isweni (Nwanetsi) to Mapulanguene in Mozambique.

During this early stage the collection point and depot at Soekmekaar had to be built from locally obtained materials. Despite this drawback, houses, compounds and store rooms were built, a borehole

Mr Paul Neergaard, WNLA official, Mrs M. Coetser, Capt. J.J. Coetser and Col J. Stevenson-Hamilton at Punda Maria, October 1922.
(ACKNOWLEDGEMENT: STEVENSON-HAMILTON COLLECTION, SKUKUZA ARCHIVES)

The homestead of Mr Paul Neergaard, the WNLA official at Cajcino, Inhambane district, Mozambique, December 1905.
(ACKNOWLEDGEMENT: STEVENSON-HAMILTON COLLECTION, SKUKUZA ARCHIVES)

drilled, gardens laid out and trees planted. Paul Neergaard, the first regional manager, was a Norwegian. He was an excellent equestrian, who loved walking and bird-watching.

It was also difficult to obtain fresh vegetables and other essentials during the early years and people were often plagued by malaria, or blackwater fever.

Nine rest camps were laid out along the route between Pafuri and Soekmekaar. These were at Lebellebowrie, Klein-Letaba (Mahonyane) near Morty Ash's shop, Fonsecas, Shirindi, Mawambi, Shingwedzi, Shikundu, Shikokololo (Punda Maria) and Klopperfontein. Because of the fountain at Baobab Hill between Pafuri and Klopperfontein, this site was also utilised later on for this purpose.

Most of the camps were situated on state land and permission had to be obtained from either magistrates, or native commissioners, to erect buildings there. The WNLA had to pay a nominal rent for the privilege. They erected thatched wattle-and-daub buildings, as well as kraals for the trek animals. New recruits and returning workers had to walk the distance to and from their homes, while their luggage was transported by donkey-carts. To protect the donkeys against lions, corrugated-iron kraals were built at Shirindi, Shingwedzi, Shikokololo and Klopperfontein. The donkey enclosure at Pafuri was built with sturdy Lebombo ironwood poles from the area.

The rent for the use of Shikokololo camp had to be paid to the Receiver of Revenue at Louis Trichardt (Makhado). Shikokololo was later to become Punda Maria rest camp in the Kruger National Park.

During those years there had been a strong fountain at Klopperfontein, but it dried up in later years. This problem was solved by drilling boreholes and building a dam. Sometimes, during high rainfall periods such as in the 1970s, the fountain flowed again. The remains of the stone drift which Bvekenya Barnard built through the spruit during the 1920s while working for the WNLA, is still there today. Klopperfontein became even more interesting when rumours that prospectors had picked up one or two diamonds in its vicinity started circulating, but these claims could never be confirmed.

Because of the remoteness of the northern part of the Shingwedzi Reserve, ranger Major A.A. Fraser, could exert very little control over the area and poaching and other illegal activities carried on unhindered until Major Stevenson-Hamilton could settle another ranger, Captain J.J. Coetser, at Punda Maria in 1919. In 1926 the Shingwedzi Reserve became part of the Kruger National Park and with that control over the area improved dramatically.

At one time the WNLA planned to settle their local headquarters at Louis Trichardt (Makhado), but the municipality objected, with the result that Soekmekaar became the district's headquarters. Paul Neergaard was stationed at Pafuri in 1918 where he met the infamous Bvekenya Barnard and offered him a position in the organisation. In 1921 Neergaard became regional manager at Soekmekaar.

Along the route between Soekmekaar and Letaba (Makhuba) were four overnight posts or rest camps: (i) One at Gravelotte along the Selati railway line between Komatipoort and Soekmekaar with Pat Travers as the WNLA agent; (ii) Pioneer Camp, 35 kilometres east of Gravelotte; (iii) M'Quinine in the vicinity of Ngwenyeni Spruit along the Malopeni-Letaba road and (iv) Makhuba camp (where the Letaba rest camp is today). This route stretched across the Mozambican border via Longwe Hill to the Por-

The well-known Baobab Hill, between Pafuri and Punda Maria, where WNLA recruits for the gold mines slept over, during the period when they had still been transported by donkey-cart between Pafuri and Soekmekaar (1919–1927).
(26.04.84)

tuguese administrative post at Massingire.

On the route between Isweni (Nwanetsi) and Soekmekaar were also four main camps: (i) Acornhoek along the Komatipoort-Soekmekaar railway line, with E.H. Whittingstall as the WNLA agent; (ii) Seis Modo on the farm Lilydale; (iii) Spanplek, in the Pelwane (now Phelwana) area and (iv) Zeni or Isweni (now Nwanetsi) close to the Mozambican border in the Nwanetsi Poort. This route was in turn connected to the Portuguese administrative post at Mapulanguene in Mozambique.

The railway line from Komatipoort provided the necessary connection between Acornhoek, Gravelotte and the main base at Soekmekaar.

During the construction phase of the overnight camps and connecting routes, the WNLA experienced many bureaucratic problems with the various government departments, but on the whole the officials were usually well informed and cooperative. Gravelotte, for example, was on state land, Pioneer Camp on land which was under control of the Department of Mines, while Makhuba, M'Quinine, Spanplek and Isweni (Zeni) were all inside the Sabie and Shingwedzi reserves (now the Kruger National Park). The camp terrains were usually half a morgen in size. Once an enclosure to keep out lions was built without obtaining the necessary permission first and the WNLA was duly rapped over the knuckles for the omission! Permission also had to be obtained to open up the routes between the overnight facilities. Land owners who had to be consulted, included the Transvaal Consolidated Land Company, certain government departments, the Transvaal reserves and even private land owners.

A massive amount of correspondence occurred between 1918 and 1926 to solve the WNLA's settlement problems. In those days typewriters were rare and letters had to be written by hand. Carbon paper was also not freely available yet and to make copies of correspondence a special purple ink had to be used. The original letter was placed between two sheets of thin, damp paper and then into in a press for 15 minutes. The original letter became a bit smudged in the process, but the writer could at least file a legible copy of his documents in this way!

During the rainy season the roads became almost

The WNLA headquarters and maintenance depot at Soekmekaar in the Northern Transvaal (now Limpopo), March 1926.
(ACKNOWLEDGEMENT: STEVENSON-HAMILTON COLLECTION, SKUKUZA ARCHIVES)

The WNLA recruiting post at Pafuri during the early 1930s.
(ACKNOWLEDGEMENT G. BORCHERS, LOUIS TRICHARDT [MAKHADO])

impassable and road building teams always had to be ready to carry out maintenance with picks and shovels. Graders were at that stage still a dream!

While all these activities had been taking place in the Transvaal, there were large developments in Mozambique as well, where overnight facilities were even further apart. All transport was by foot or with pack-donkeys. Scotch carts, drawn by two oxen, were used in certain places. The WNLA built several hundred kilometres of road in Mozambique. Here they had to pay annual subsidies to the Portuguese authorities for maintenance of the roads until 1970. Just as on the South African side, wattle-and-daub huts were built along the roads as overnight posts for officials and recruits. H.S. (Mickey) MacKay who had been stationed in Mozambique, once requested funding to broaden a footpath which ran through dense bushes, so that the pack-donkeys could move through with ease.

As if the logistical problems were not bad enough, certain Portuguese officials caused endless trouble. They were poorly paid, far from their superiors in Lourenço Marques and acted as little potentates in their own areas. If the WNLA was not prepared to reward them under the table, they devised all kinds of plans to obstruct or slow down the activities of the organisation. However, the WNLA was not prepared to pay bribes and rather tried different ways to obtain their goodwill and cooperation. When the first Portuguese custom officer arrived at Pafuri, there was nowhere for him to stay. WNLA saved the day and built him wattle-and-daub hut in which he lived for 20 years before his own government built him a brick home.

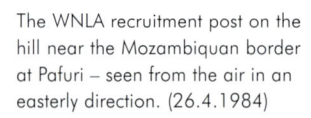

After the WNLA's recruiting campaign had taken off in Mozambique, the stream of recruits to the mines soon grew to such an extent that it became a sustainable enterprise.

Paul Neergaard was succeeded as regional manager of the WNLA at Soekmekaar by Tom Carruthers, who recorded unusual memories dating back to that difficult but romantic era in the history of the WNLA. His superior was W. Gemmill, a strict but fair task master who was highly respected by all his personnel.

Competition was tough in those days as agents were paid by head for the number of labourers which they could recruit. Paul Neergaard founded the recruiting depot at Pafuri and another camp about 95 kilometres downstream along the Limpopo at Mapai. H.S. MacKay originally from Soekmekaar, was first stationed at Pafuri, while McCurdy took over control of the Mapai base camp. Each recruit was issued with two blankets, a loin cloth, a cotton vest and a leather belt before being transported to Johannesburg.

When Wilfred Craw's health had deteriorated to such an extent that he could no longer fulfill his duties at Soekmekaar, Tom Carruthers was transferred there in 1925 as junior clerk. C.L. Impey was transferred there as well during the same year, after C.C. Stubbs had been moved to Louis Trichardt. Personnel transfers often took place as a result of the malaria problem and Soekmekaar became the training centre for agents who were to be placed at the outstations.

These agents also included Portuguese officials, of which Alvaro Montez Ribas, who took over the Massingire recruiting post, was the first. Captain Frank Mutter was transferred to Mapai, and Eric Chapman of Soekmekaar to Pafuri.

In 1928, Tom Carruthers was transferred to the

The WNLA recruitment post on the hill near the Mozambiquan border at Pafuri – seen from the air in an easterly direction. (26.4.1984)

east coast of Mozambique and his position at Soekmekaar was taken over by Major Bluett. After Carruthers had served at Xai-Xai (Vila de João Belo) and Barra do Limpopo, he was placed in charge of the Mapulanguene recruiting post.

At the end of 1929, Carruthers was sent back to Pafuri, because Alvaro Montez Ribas had taken over Terzakis's transport business at Mapai. After that he served at Massingire and Mapai, where he had roads made to Mabote and Vilanculos via Maxaila and Sinkhondlo, 320 kilometres from Mapai.

Gemmill visited Mapai to discuss the possibility of recruiting north of the 22° latitude, as they would be able to withstand the high temperatures in the mines better. On his recommendation, Tom Carruthers and Paul Neergaard carried out a reconnaissance north of the Great Save River from the Espungabera Escarpment up to the Buzi River in the east.

This 1923 expedition took three months. Neergaard suffered a bad case of malaria at Massangena along the Save, but Carruthers managed to safe him with repeated doses of quinine and Atabrin. Despite this, his health still deteriorated to such an extent that he had to retire soon after he had completed his report about the journey. H.S. (Mickey) MacKay succeeded him as regional manager at Soekmekaar. Neergaard's principle of "Recruits from the tropics" was accepted on the grounds of his report, but each recruit had to receive three vaccinations on arrival against typhoid, gastric fever and pneumonia.

Tom Carruthers was sent to Pafuri in 1936 to organise the recruiting of the "tropical" labourers. He was replaced by Harold Mockford on 1 February 1938.

Recruiting came to a halt at Massingire and Mapulanguene during the Second World War and the Isweni (later Nwanetsi) camp of the WNLA was taken over by military authorities.

Soon after the war, Tom Carruthers was sent to Soekmekaar again to take over from MacKay as regional manager and MacKay passed away shortly afterwards. W. Gemmill left the WNLA head office to organise the influx of mine recruits from a head office in Salisbury (now Harare). George Lovett took over his position as general manager of the WNLA and NRC in Johannesburg. After the war, Tom Carruthers reorganised the flow of recruits across the Lebombo to Letaba and Nwanetsi (Isweni) and a new WNLA camp was built on the northern bank of the Nwanetsi River near the confluence of the Nwanetsi and Sweni rivers. Wenela also provided funding to build the dam in Nwanetsi Poort, as well as for the weir in the Nwanetsi just upstream of their new camp. In the meantime, another Gemmill, James, had taken over from Lovett as general manager.

Communication channels had also improved in the form of radio sets. The provision of electricity to the outposts had also been done and improved. Soekmekaar depot had become the maintenance and workshops of the WNLA buses which transported the recruits from the various recruiting posts to Soekmekaar.

Two recruiting posts were built at Pafuri initially, one in the Transvaal on a hill near the confluence of the Limpopo and Luvuvhu rivers and one on the Mozambican side. By the early 1920s the border between the confluence of the Limpopo and Luvuvhu rivers and Shingwedzi Poort had not been properly surveyed yet. The final survey of this portion of the border and the erection of border beacons (1–14) were only completed during 1924 and 1925 by a specially appointed commission. The commission consisted of surveyors W.P. Murray and R.E. Antrobus, representing the Union government and Commander Viera da Rocha and Captains Semedo and Moura Braz who represented Mozambique. Fortunately it was determined during the survey that the international border between the two countries was in fact between the completed camps at Pafuri. Paul Neergaard of Soekmekaar, McCurdy, the agent at Pafuri and his replacement MacKay assisted the international surveying team with the transport of equipment and materials.

The Pafuri route became the most important of the three routes utilised by the WNLA to move black labourers from Mozambique.

In 1919 a brick homestead was built at Pafuri on the Transvaal side. The rest of the buildings were still wattle-and-daub buildings until they began upgrading the buildings in 1938.

The wattle-and-daub buildings were erected with Lebombo ironwood (msimbit) poles and ensured that the well-built ones could serve their purpose

Mr H.H. (Harold) Mockford, who had spent a record period of 47 years between February 1938 until June 1985 in the Kruger National Park in service of one organisation, the WNLA (later TEBA).

A handmill similar to the Colonist mill which had been commonly used in the Lowveld until the middle 1930s to mill maize. (06.09.1986)

The old Mafourteen footpath between Massingire in Mozambique and Letaba, in the vicinity of Longwe. This path had also been used by WNLA recruits to reach their sub-depot at Letaba, from where they were transported to the railway stations at Rubbervale or Gravelotte until 1941. (SEPTEMBER 1984)

for 40 years. The poles were first covered with mud and then whitewashed with lime.

Termites were a problem. Although they could not damage the ironwood poles, they tunneled through the thatch roofs and caused great damage. To solve the problem, a gap was often left between the roof structure and the walls. The poles were then treated with old engine oil to prevent the termites from reaching the roof rafters.

Two specific items dating back to this era which deserve mentioning are the three-legged pot used for cooking purposes and the Colonist mill. The sizes of the cast iron pots varied from a few litres to more than 600 litres. The largest of these were used to cook mealie porridge for large teams of recruits.

Each of the larger camps had a Colonist mill. Each mill had a large wheel with a handle which had to be turned to mill maize between two rollers and the meal had to be crushed a few times consecutively to get it fine enough. Mechanically driven mills eventually replaced the old handmills, but cast iron pots are still very much in use today.

Cattle was kept at several of the camps to provide milk and meat. Because lions often made a meal of the cattle and donkeys, special lion-proof enclosures had to be put up at some of the camps inside the Park. This contract had been awarded to a Mr. Keyter and he was paid £30. While he had to provide his own labour, accommodation and rations, the WNLA provided the corrugated iron as well as 16 gallons of petrol. After motorised transport had been implemented, the lion-proof enclosures were broken down and the material used to build bus ports.

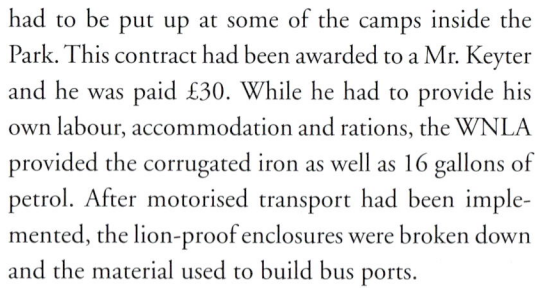

In 1926 the Sabie and Shingwedzi reserves were consolidated and reproclaimed as the Kruger National Park. The Park stretched right up to the confluence of the Luvuvhu and Limpopo rivers in the north and included the land on which the WNLA post at Pafuri had been built. The stretch of the Luvuvhu River after its confluence with the Mutale, was indicated as Pafuri River on maps at the time – from there the name Pafuri recruiting post. Restrictions and control measures which were implemented by the Park's administration on movements to the old Crooks Corner area, demanded that the WNLA officials who worked inside the Park, had to obtain a permit which had to be renewed annually.

By 1926, the WNLA had already built many camps and outposts in Mozambique. White officials were stationed at Pafuri, Mapai, Massingire and Mapulanguene, while black officials manned the outposts at Hassane, Mabalane, Maxaila, Chigubo, Chikondhlo, Sicandavia, Manyarenga, Cubo, Mabele and Mabote. The white officials visited the outposts on foot to inspect food rations and buildings. Water was a problem at many of the outposts and sometimes had to be transported for miles, and then only enough for drinking and cooking purposes. Gardens could not be laid out, except where the posts happened to be close to rivers, for example the Limpopo. Gardeners lived in the gardens on a permanent basis to keep monkeys, baboons and other animals out of the vegetable beds and orchards.

The Portuguese customs official at Pafuri had once laid out a splendid vegetable garden of which he was very proud, with reason. However, one night two elephants paid a visit and completely destroyed the whole lot. Just two nights later the two rogues returned, but this time the official was ready and he wounded one. The injured animal ran back into the Kruger National Park and died there. After the official had come to hear that the elephant's tusks weighed about

30 kilograms each, he offered to exchange them for a much smaller pair which he had in his possession!

Water was often transported in wine barrels in Mozambique. Two thick planks were bolted to the middle of the top and bottom of the each vat. A rope with an eye on each end was then tied to the bolts to make it easy for a single person to drag it. Water barrels were also transported on slays, pulled by oxen, while donkeys were used to transport water in special holders which were carried on their backs like pack-saddles. Once 44 gallon drums became available, they utilised these, and later still, they used old motorcar axles to build special water-carts. To shower, officials had to be satisfied with a bucket of water which was lifted into a tree.

Attempts to recruit labourers were often hampered by Portuguese government officials (*Chefes do Posto*), who stopped recruits and sent them home if they ostensibly did not possess the correct documentation. The recruits then in turn spread the word that the recruiting of mine workers had been terminated. These recurring rumours caused many problems and delays and it always took time to once again collect teams.

Two incidents which were important to WNLA took place in 1928:

i The so-called Mozambique Convention between the Union of South Africa and Mozambique was drawn up. This convention placed black labour affairs, railway transport and trade relations between the two countries on a firm basis.
ii One of the stipulations of the convention was that recruiting agents had to be in possession of a licence which cost £100. Issuing of licenses to people other than Mozambican citizens was to be phased out and by 1934 only Portuguese citizens would be allowed to carry out this function. This forced the WNLA to implement certain personnel movements and even to pay off certain officials.
iii The WNLA started transporting recruits with motorised vehicles instead of utilising animal transport.

Recruits also had to have a passport which cost ten shillings and was valid for 12 months. The passports could be renewed for a further six months by paying an additional five shillings. The Portuguese curator in Johannesburg had to receive one shilling for every recruit who was signed on, a further shilling for each reappointment and two shillings per month per recruit with a total minimum compensation of £1.15. All levies had to be paid in gold to the Mozambique government. By 1933 the total number of recruits for the mines from Mozambique was not to exceed 80 000. The contract period was 18 months, after which a recruit was compelled to return to Mozambique for a minimum period of six months. Deferred compensation was another stipulation of the convention. After a labourer had completed 354 shifts, one shilling had to be put aside per shift. This money was then paid out at their return to Mozambique.

It is understandable that these new regulations caused the WNLA many problems as a result of poorly informed Portuguese officials. There were no quick ways to communicate with Lourenço Marques (Maputo) and it sometimes took weeks to send instructions to the *Chefes do Posto* in the more remote districts. Sometimes they could, or would not understand the instructions and this caused great delays in recruiting activities. Eventually a government official was sent to Lourenço Marques to explain everything to the district officials, but he had hardly left when problems arose once again, because the oral explanations were not followed up in writing. The inconvenience caused to the recruits was never taken into account by the officials and they just had to wait patiently for better days. It was only as a result of the tact and diplomacy of the WNLA officials, that the flow of recruits to the mines gradually became stronger.

After the 1928 convention, recruits who had completed their service periods had to return to Mozambique through Ressano Garcia, Pafuri and Mapulanguene, but not through Massingire. However, referred compensation was only paid at Ressano Garcia and this caused great dissatisfaction, as most recruits were forced to return via this route, as inconvenient as it was to them.

When transport by animals was replaced by motorised transport in 1928 along the Soekmekaar–Pafuri route, the Saunders Transport Company was awarded the contract for this. However, the WNLA soon acquired a number of Thornycroft buses and once again took over the transport of recruits itself. Motorised transport was also initiated along the other Transvaal routes and this was placed under control of J.H. Travers. Motorised transport was gradually also organised in Mozambique and transport of all the WNLA recruits in Mozambique was eventually taken over by the firm Ngala Limitada, including those who returned home via Soekmekaar.

The increase in traffic necessitated a large road construction and maintenance programme. At that stage, the roads in the Northern Transvaal were in a very poor condition, while in Mozambique they were hardly more than two-track roads. The WNLA provided funding to the Transvaal Roads Department on a pound for pound basis to help with the improvement of the roads. In the Kruger National Park itself, the selection of the locations of quite a few of the ranger posts had been the result of the successful maintenance of roads by the WNLA since the early 1920s. Those which were not along or close to the railway lines between either the Johannesburg and Lourenço Marques, or the Selati railway lines (e.g. Kaapmuiden (Boulders), Malelane, Crocodile Bridge, Skukuza (Sabie Bridge) and Rolle), were built along the WNLA's original communication routes (e.g. Punda Maria, Letaba, Satara and later Nwanetsi as well).

During the early developmental years of these posts, the WNLA's vehicles had been just about the only way by which goods and mail could be transported to these remote posts in the Park. The eventual good cooperation and goodwill between the WNLA and Parks Board officials found its origin during those difficult years. The WNLA can also justifiably look back with pride on the role which this organisation had played in the developmental history of the Park.

For instance, each bus which left Pafuri on its way to Soekmekaar had to transport a quota of stones and these were loaded off along the road at bad spots and drifts for later use by road teams. Many tons of stones were transported in this way without any additional cost to the Park.

There were boats at the crossings over the Limpopo and Save in Mozambique so that flooded rivers could not hamper the passage of recruits. These large flat-bottomed boats were built from pine planks which

WNLA recruits for the gold mines at a Thornycroft bus, circa 1928.
(ACKNOWLEDGEMENT G. BORCHERS, LOUIS TRICHARDT ([MAKHADO])

were glued together and painted with pitch. The black boatswains developed a very effective and safe technique to cross the rivers with a load of recruits. They started off by rowing slowly upstream, staying close to the riverbank. Then, at a given point they turned around and started rowing at maximum pace downstream and across to reach the riverbank on the other side always at the right place. During high floods they often rescued locals who had been trapped on islands or in trees from a certain water-death. This was simply part of their daily duties and no bravery medals were ever dealt out!

However, floods brought them slight compensation as they could reach animals which had drowned in the floodwaters with the boats. This meat came in really handy.

There were no bulldozers, mechanical graders or four-wheel drive vehicles available during the pioneering years. There were, however, some ox-drawn graders which worked on the Transvaal roads, but most road maintenance had to be done with picks, spades and shovels. Sometimes thorn branches or a few old tyres were dragged along the roads to improve the surfaces. The Mozambique company controlled the areas between the Limpopo and Save rivers and along the Rhodesian border and permission had to be obtained from them to build new roads in the area. In other areas permission had to be obtained from the Mozambican government. Before the construction of roads in Mozambique by the WNLA, a place like Pafuri had for all intents and purposes been totally cut off from the authorities in Lourenço Marques. A good example of this was the time when a Portuguese official asked the WNLA official at Pafuri to keep an eye over matters while he was away on a journey to Portugal. He travelled through the Transvaal to Durban and boarded a ship there. During his six months' absence, the authorities in Lourenço Marques had been completely unaware of his absence, or the fact that the WNLA official at Pafuri organised matters on their side of the border. It is told that he actually met a senior official in Portugal who wanted to know why he was not at his post in Mozambique, to which he answered that he had never been to Mozambique in his life, let along done duty there, but that he had a twin brother who worked there!

Another WNLA official in Mozambique never returned back to his post on time at Mapai after his annual leave and was usually at least two weeks late. The district manager at Soekmekaar came to hear about this through the bush telegraph and let the involved person know that he would be with him for supper on a specific day. The absentee was two weeks late again as usual, but the manager could also not carry out his promised visit due to unforeseen circumstances. In due course he received a letter from the errant fellow which informed him that he had been expecting the manager for supper for two weeks and wanted to know when he would turn up!

As soon as the bigger rivers started flowing, normally at the end of the rainy season, thoroughfares were built so that vehicles could travel through without trouble during the winter months. In the Limpopo riverbed, two strips of soil were heaped on top of the sand while they packed a thick layer of reeds over the dry area of the riverbed. There was also a wooden bridge, built of mopane poles coupled together by galvanised wire and anchored into the sand with sturdy poles, which could be used to cross the river. During the summer of 1937, the WNLA official H. Mockford of Pafuri departed for Soekmekaar by bus. It took him a full week to travel the 240 kilometres! However, roads in the Northern Transvaal gradually improved especially since tourists started visiting the northern part of the Kruger National Park.

William Gemmill, the general manager in those days, usually visited the Northern Transvaal recruiting officials, as well as those in Mozambique, annually. This enabled him to inspect the condition of the roads at the same time. In 1936 he offered a contract to Mark Spraggon, who had been running a Rhodesian-based transport company, to construct a road in Mozambique along the Mozambican-Rhodesian border from the Limpopo up to the Save River. The road had to be 12 feet wide and all logs had to be removed up to a depth of 18 inches. Spraggon was paid £10 per mile. Large sections of the road had to be made through Lebombo ironwood forests. One section through this dense vegetation was 15 miles long and all shrubs and trees had to be removed by hand! The road surface had to be graded and Spraggon bought a light grader for this purpose which could be drawn by his truck. At one point, about 30 miles from the Save River, a connecting road was constructed to lead to the Portuguese administrative post at Massangena. Spraggon was paid £14 per mile for this stretch of road. Nobody could understand why it had been constructed in the first place as it was not used for many years. However, the general manager had believed that it would become useful in time and in later years, when the transport of recruits along the Rhodesian border came to a virtual halt, Spraggon's almost forgotten road was cleaned and utilised.

It was only through the WNLA's efforts that the roads in Mozambique from the Limpopo as far north as Espungabera were maintained. The Mozambican government contributed nothing. After the Second World War, the firm Ngala Limitada acquired a bulldozer and a few graders and the roads in Mozambique were improved to such an extent that they could be used by large buses as well.

General manager William Gemmill's annual visit to the recruiting camps and outposts was a great occasion. Food supplies, sleeping comforts and tents were transported in two large trucks which arrived several hours before him at a given place. By the time he arrived, everything was ready and even the drinks were waiting! This mobile unit often had to travel into the most remote parts of Mozambique and the agents and recruits had to make intricate preparations so that there would be no delay in travel plans. Everybody feared the little black book which Gemmill carried with him and to have your name noted down in it only meant trouble. The two vehicles which accompanied Gemmill, were extremely well fitted out with built-in cupboards and numbered drawers. Each item was noted in a register and signed off after it had been used, for instance, "Grouse Whisky" and "Bear Brand" milk (which apparently smelled just like fresh Jersey milk) from Scotland. Everything was of the best. The white drivers of the trucks had to see to it that they pitched camp at precisely the right time and place. They often had to break down the general manager's tent if it was not in exactly the right spot according to him. A large amount of spare parts and tools were also taken along.

Since all the routes from Soekmekaar to Mozambique ran through the Kruger National Park, the WNLA had to abide by the Park's rules and regulations. The Isweni (Nwanetsi) to Makhuba (Letaba) route had been under control of the Park's officials, but because of the remoteness of the Pafuri route, it only came under control of park officials after 1919.

Colonel James Stevenson-Hamilton maintained good relations with the WNLA. To reciprocate, the WNLA assisted the Park and its officials with word and deed wherever possible. To begin with, the labour teams of the WNLA constructed and maintained the roads which led to the recruiting posts. Later on, the organisation made an annual financial contribution to Parks Board so that it could take over road maintenance. Cooperation between the WNLA and Park's personnel went even further than the transport of goods and mail. At one stage Colonel Stevenson-Hamilton suspected that the WNLA personnel at Isweni (Nwanetsi) were providing poachers with information about the movements of the Park's rangers. These people were immediately replaced with staff from another area and this had the desired effect. On another occasion, the warden even requested of the WNLA district manager to pay the personnel at Punda Maria while the ranger was on leave.

The three WNLA routes which ran to their destinations through the Kruger National Park in 1926 were:

i Soekmekaar to Pafuri, with camps at Shikokololo (Punda Maria) and Klopperfontein;
ii Gravelotte to Massingire with a camp at Makhuba (Letaba) and
iii Acornhoek to Mapulanguene with a camp at Iseni (Isweni).

The entire length of the last route was constructed by the WNLA.

Despite all pre-emptive efforts, the roads inside the Park usually turned into swamps during the rainy season. Once it took eight days to get a bus which had become stuck along the Pafuri road, out of the mud, despite efforts with a tractor and the help of 100 recruits. The situation in Mozambique was even worse and sometimes recruits had to walk long distances during the rainy season as the roads were totally impassable. In 1929 the WNLA transported all the necessary material to build a pontoon to cross the Olifants River at Balule. Three years later the

WNLA was requested to transport the total supply of petrol needed for the tourist season to Punda Maria – three whole 44-gallon drums. Wenela personnel and recruits often assisted Park's officials to put out veld fires as well. On own initiative, the Wenela also erected road signs outside the Park to indicate distances and directions to visitors travelling to the Park.

By the late 1930s, the routes from Soekmekaar were already quite fixed, but changes were made to the two routes which led to Isweni and Makhuba (Letaba). The Isweni route no longer ran via Acornhoek to Soekmekaar. Both routes now had their railway terminus at Rubbervale near Leydsdorp and entered the Kruger National Park at Malopeni. Improved road building equipment aided the significant improvement of roads in the Transvaal, the Kruger National Park and in the adjacent Mozambique as well. Vehicles were then able to reach even the most remote outposts. Ngala Limitada transported all recruits and returning workers in Mozambique, while Spraggon's Transport fulfilled the same function in Rhodesia (Zimbabwe) along the road which he had constructed on the border between Mozambique and Rhodesia from the Limpopo to the confluence of the Lundi and Sabie rivers. Recruits which were transported by Ngala Limitada were known as "east coast recruits", while Spraggon's recruits were known as "tropical recruits".

During Harold Mockford's 47 years' uninterrupted service as the WNLA recruiting agent and administrative official at Pafuri since 1938, he had the opportunity to traverse the northern Lowveld and even the most remote areas of Mozambique and the former Southern Rhodesia. He declined several promotions and laid out remarkable gardens, planting only indigenous plants on the small hill where the WNLA post was situated. He and his wife, Tiny, were hospitable and endearing people who played a singular role in the conservation history of the Park. They were both avid nature lovers and acquired such an intimate knowledge of the wildlife and environment that they were often referred to as Pafuri's ranger pair. Harold Mockford was deservedly appointed in 1979 as the first lifelong honorary ranger in the Park's history. Among the local blacks he soon became known as "Shitsakise" ("the one who makes others happy"), which was a very apt description of his personality.

To broaden his knowledge of nature, Harold Mockford gradually built up one of the most valuable collections of natural historical Africana in the country. He and his wife investigated the Pafuri area on foot and nothing escaped their sagaciousness or interest. The geology, petrology, San rock paintings and other archaeological phenomena, insects, amphibians, reptiles, birds, mammals and the plant kingdom were all part of nature to them and therefore part of their lives. It was Harold Mockford who discovered the first Zambezi shark in the Park on 4 July 1950 at the confluence of the Luvuvhu and Limpopo rivers. During his long service he also noted the first examples of rare plant, reptile or bird species within the boundaries of the Park. He will be remembered as a great friend of the Park and its people and his memory remains as great inspiration to those who had the privilege to know him.

In 1938 a foot-and-mouth disease epidemic broke out in the Northern Transvaal and the Division of Veterinary Services suspected that the virus had been carried into the Union from Mozambique by recruits. For this reason they wanted all the clothes of the recruits who came into the Park to be destroyed and that new clothes be issued to them at the Makhuba

Mr H.H. (Harold) Mockford, WNLA official and lifelong honorary ranger of the Kruger National Park with his wife, Tiny, in front of the office at their quarters at Pafuri, where they spent 47 productive years in the Park.
(21.06.1984)

An old hut (dating back to the 1930s) of the WNLA in which recruits for the Rand gold mines used to overnight at Pafuri. (26.4.1984)

(Letaba) and Isweni posts. However, the WNLA pointed out that it would be easier to close the two posts until the end of the epidemic. The veterinarians then made the concession that recruits could enter at these points on condition that they wash their hands and feet in a sterilising agent like Jeye's Fluid. These restrictions were soon implemented at the Pafuri post as well, but it was later determined that a washing soda solution should rather be used. It was a great relief to the WNLA officials when these restrictions were lifted, but from then onwards, no recruit was allowed to bring any plant or animal products from Mozambique into South Africa. The recruits who arrived with little bags filled with herbs or traditional healers' medicines were severely upset and refused to hand over these valuable items. If it was discovered that a recruit had a traditional healer's bag with bones, he was summarily sent home.

Smallpox epidemics broke out at different occasions among the local populations in Mozambique and recruits had to be immunised against the disease after their arrival at Pafuri. They also had to undergo a full medical examination. During the later years, the WNLA appointed two full-time medical practitioners at Pafuri. Doctors Adams and Da Souza served in this capacity for many years and performed valuable services. "Tropical recruits" were also immunised against pneumonia, as they were more susceptible to the disease on the cold Highveld. After the Second World War, recruits were also given malaria preventative tablets and still later they also received injections against cholera.

The Isweni and Makhuba routes were both closed with the outbreak of the Second World War in 1939 and the former was taken over by military personnel. Preparations were made to invade Mozambique if it was deemed necessary.

However, the Pafuri route was kept open. A new rule came into effect which stipulated that recruits from Southern Rhodesia (now Zimbabwe) who had been recruited by Spraggon's Transport Company, could not be cleared at the Pafuri post. They were then rerouted to a hastily built post 12 kilometres west of Pafuri, which was known as Bobomeni. The reason for this ruling was that the recruits from the east coast could be properly searched for documents or items which could be utilised by the enemy before making contact with the Rhodesian recruits. A new road had to be made for this purpose from the confluence of the Limpopo and Luvuvhu rivers along the northern bank of the Luvuvhu to Makuleke Drift and Bobomeni. Shortly after the outbreak of the war, Harold Mockford was appointed as temporary customs and migration official and provided with four black border guards. He was also appointed as special constable by the SA Police. Security along the border was generally intensified, but the flow of labourers to the mines was not affected or delayed by this at all.

Improvements to the camps were restricted to the minimum during the war years. Building material was hard to come by and petrol rationing caused trips to be strictly controlled. The medical examinations which had been carried out on recruits at Soekmekaar before 1941, were then done at Pafuri. After motorised transport had been instituted, some of the overnight posts between Soekmekaar and Pafuri also became redundant, while only small changes were made to the outline of the road.

The camps which were not utilised any longer were Lebellebowrie, Klein-Letaba, Fonsecas, Shikokololo (Punda Maria) and Klopperfontein. A new camp, called Mtati, was opened just outside the Park's boundary in the Mhinga area. This became an important camp, as buses could travel all the way from Soekmekaar to Mtati in one day, due to the improved roads. The next morning the buses travelled to Pafuri and by the afternoon they were back at Mtati. The remaining camps at Mawambi, Shingwedzi and

Shikundu were also used less and less and eventually they were also closed. Initially buses from Soekmekaar arrived at Pafuri on Wednesdays, Fridays and Sundays and returned on the same days with a load of recruits to spend the night at Mtati camp. The next day they were back at Soekmekaar. The arrival of buses on Sundays at Pafuri caused much dissatisfaction among government officials and they repeatedly requested that the schedule be changed. However, this schedule suited the officials in Johannesburg and at Soekmekaar very well as it did not intervene with their weekends. Officials at the outposts were therefore informed that they had to be on duty 24 hours a day, seven days a week! In 1942 the trips were reduced to two a week and buses reached Pafuri on Wednesdays and Sundays. The schedule was again changed in 1968 to Wednesdays and Fridays. Now at least one of the complaints of the Portuguese officials was resolved.

Recruiting of labourers for the mines through the Pafuri post carried on throughout the war years without much ill-effect. Strict control was maintained over the articles which recruits brought into South Africa and all letters were censored. The WNLA official at Pafuri, Harold Mockford, was to eventually carry out the duties of a customs and immigration officer for 22 years until the SA Police took over the function in 1965.

After the war, all efforts were made to increase the number of recruits to the mines and larger vehicles were put into service. By then, the firm Ngala Limitada in Mozambique had been taken over by Orlando Paes Mamede, who had left the WNLA's service to do so. Paes Mamede had his headquarters at Mapai and played an important role in the improvement of roads in Mozambique. Wenela in turn, paid subsidies to him for road improvements and also provided him with loans to buy road building equipment and vehicles. This energetic and enterprising Portuguese soon opened up new routes, and before long his buses transported recruits from as far as Espungabera and Mambone.

Mark Spraggon sold his business in Rhodesia (Zimbabwe), which transported recruits to Pafuri, to George Palfrey. This new company conveyed recruits to Pafuri from Southern Rhodesia and Mozambique north of the 22 degree latitude. There were many disputes between the WNLA and Palfrey, but to everybody's surprise he managed to convince general manager Gemmill in 1949 to change the route. Pafuri was passed and recruits from Rhodesia were now brought to a new camp, Mabilikwe, along the Limpopo, about 40 kilometres west of Pafuri. From here they were taken by the WNLA buses to a camp, Munnichshausen, between Messina (Musina) and Beit Bridge and from there by train to Johannesburg. This route was not utilised for long, however, as Palfrey stopped all transport when the WNLA would not grant an increase in transport tariffs. Wenela itself handled the transport in Rhodesia (now Zimbabwe) for a while, until Palfrey complained to a cabinet minister that Wenela was stealing labourers from farmers, with the result that all recruiting licenses were suspended.

From 1952 onwards, "tropical" recruits once again arrived at the Pafuri post, but they were transported by Ngala Limitada along the formerly unused border road which Mark Spraggon had built in 1936 according to the instructions of William Gemmill. A new camp, Mavue, was fitted out along the Save River in Mozambique, just east of its confluence with the Lundi. Paes Mamede's Ngala Limitada was now transporting both "east coast" and "tropical" recruits to Pafuri. After the end of the Second World War, the Letaba (Makhuba) and Isweni routes through the Kruger National Park were utilised again. However, the Letaba route soon became unused again as it was not deemed profitable enough. Isweni, which had been occupied by the military during the war,

Nwanetsi rest camp before it was enlarged and fenced in. November 1954.

was handed over to the Parks Board soon afterwards and inaugurated as Nwanetsi private rest camp. A piece of land along the northern bank of the Nwanetsi Spruit, near the bridge which had been built by the military during the war, was made available to the WNLA and a new camp and overnight post for recruits was built here. The number of recruits rose to expectations, but the Nwanetsi post also had to be closed finally in 1966. This was after a new clause had been added to the Mozambique Convention which stipulated that all recruits for the mines had to travel through either Pafuri or Ressano Garcia.

Since 1938 the wattle-and-daub and other buildings at the WNLA camps had gradually been replaced by more permanent buildings. Hundreds of thousands of bricks were made on the bank of the Luvuvhu and used at the camps at Pafuri, Mtati and Soekmekaar. During 1949 an incidence took place which would change the lives of the WNLA personnel at Pafuri and other outposts drastically. It also changed the WNLA's distant attitude towards their personnel in the bush.

This all happened as a result of a visit from Colonel J.A.B. Sandenbergh, warden of the Kruger National Park at the time, to the WNLA post at Pafuri. To start with, he was delayed for more than ten minutes by a donkey-cart transporting water up the hill. By the time he finally arrived at the post, he was in a thoroughly bad mood and inferred that it was a terrible shame that the Chamber of Mines were still providing such primitive amenities to their staff. He also promised that water would be pumped to the WNLA post from a borehole within three months.

Lovett, WNLA's general manager at the time, approved the project and work to dig a well about a kilometre north of the station near the fever tree forest, could begin. Water was pumped up the hill for the first time early in 1950 and this meant that the post was independent as far as water was concerned for the first time since 1918. The same procedure was followed on the Mozambican side shortly afterwards and the staff there were equally grateful to Colonel Sandenberg for his mediation in the matter.

The digging of the wells was not a simple procedure. After having dug through a thick layer of silt, river sand was reached, which was inclined to cave in. This necessitated the building of a brick wall to stabilise the walls and protect the labourers, a process which had to be carried on as the wells got deeper. Temporary pumps were installed to pump out water to enable the diggers to get deep enough under the water table. Once the well was completed, a pump station was built over it, the pump and engine installed and the pipeline to Pafuri dug. The WNLA agent at Pafuri had to organise and oversee this himself, as it was unheard of that contractors should be appointed for this purpose!

The Portuguese officials apparently had limited technical abilities. A good example of this was when they once came to the South African side to complain that they did not have any water at their post. They were quite certain that there was no leak in the pipeline. A quick inspection revealed that both the engine and the pump were working well. Heavy rains had fallen shortly before and there were large pools in the veld. While walking along the pipeline, the investigator noticed bubbles in one of the pools. This was where the pipeline had broken and where the Portuguese officials' water was disappearing into the sand! It was not an easy task to repair the broken pipeline in the mud, but it had to be done speedily as

This old WNLA guest house at Nwanetsi was donated to the Parks Board after recruiting through the Isweni depot had been terminated in 1941. It was changed into a private rest camp in 1945 and it became very popular in time. (18.06.1986)

the installation often had to work day and night when large teams of recruits arrived.

The next great improvement was when a generator to provide electricity was installed in 1960 at the Pafuri post on the Transvaal side and shortly afterwards on the Portuguese side as well. This was the end of the use of kerosene lanterns. In 1974 an additional 25 KVA generator was installed at the Pafuri side, as well as two air-conditioning units. The district headquarters at Soekmekaar had already been provided with the basic commodities at a much earlier stage. At long last both camps at Pafuri had water and electricity.

Many changes to the buildings at all the camps, as well as at Soekmekaar, took place in the years after the war. At the two Pafuri camps, the original wattle-and-daub buildings were gradually replaced by more solid brick structures. Brick houses were built for the officials, as well as storerooms, kitchens, ablution blocks and an office block. The plans on the Mozambican side had to be approved by government officials first and the completed buildings also had to be inspected. This caused major problems as the officials found fault with everything and their bureaucratic attitude caused delays which lasted years.

The WNLA did not have to do any building at Mapai, as Paes Mamede had already built a well-equipped camp, workshops, etc. there. He built houses with modern amenities for his black officials as well and a clubhouse, using beautiful green stone which he had transported all the way from Espungabera. This lovely base camp was completely destroyed in the late 1970's during the Rhodesian bush war. New bus hangars had to be built at Soekmekaar to shelter the larger buses, while the existing offices and other buildings were improved significantly. As roads became better, the size of the vehicles increased and during the 1970s most of the buses had the capacity to carry 80 passengers.

In 1965 a large South African Police post was built along the border at Pafuri and manned by four white and four black policemen and it was then that after 22 years, they took over the role of customs and immigration from the WNLA representative. In 1962 it became necessary that all recruits had to be in possession of travel documents issued in their country of origin. At their arrival at the post, they had to have passports, as well as contract documents. This meant that their documentation had to be processed on the Mozambican side of the border. A few Rhodesian recruits with passports still reported at the Pafuri post on the Transvaal side, but the trickle soon dried up. On the South African side all the travel documents, contracts and fingerprints were scrutinised by the SAP (now SAPS). During this period the referred compensation was at last also being paid out to returning labourers at Pafuri.

Before this, large numbers of labourers who had initially been recruited through the Pafuri post, had to return to Mozambique via Ressano Garcia. This was not only very inconvenient for the recruits, but Ngala Limitada's buses often had to leave empty from Pafuri, which had huge financial implications to the firm. The firm attempted for many years to organise that labourers could return through the Pafuri post, but the Mozambican agents in Lourenço Marques had personal interests in the transport company which transported recruits from Ressano Garcia to their homes, so that such a ruling would not suit them at all. Therefore, it was a great breakthrough when it was authorised that deferred compensation could be paid out at Pafuri. This also meant that additional transport had to be provided on both sides of the border as well and considerable amounts of money had to be sent from Lourenço Marques to

The WNLA quarters and guest house at Pafuri, as seen from the west. Mr H.H. Mockford was in control here for 47 years. (26.04.1984)

Pafuri. Fortunately there were never any robbery attempts along this lengthy route. The number of recruits who came through the Pafuri post gradually increased and in 1975 it reached the 22 000 mark.

During the early 1960s war-clouds started gathering in northern Mozambique and these gradually moved south. After the death of the Portuguese president, Dr Salazar, a left-inclined government came into power and in 1974 Portugal handed Mozambique over to the guerrilla movement, Frelimo, at a meeting in Lusaka. Frelimo was totally bowled over by this unexpected move and they requested that the transfer should take place over a period of one year. Their request was granted and in 1975 Frelimo took over the governing of Mozambique. News of the Portuguese government's capitulation spread quickly. From then onwards, there were endless problems with the returning workers as well as with the personnel on the Mozambican side at Pafuri. The repatriated workers refused to pay excise levies on their goods and at the referred payment office they demanded that they be paid 50 Escudos for each Rand instead of the 40 Escudos per Rand exchange rate which had been determined by the Frelimo authorities.

Eventually a unit consisting of Portuguese and Frelimo troops had to be called in to assist at the border and the pay office. Similar incidences took place at the Ressano Garcia office and several repatriated labourers were shot during some of the clashes. The Mozambican personnel at Pafuri also continuously demanded higher salaries. When they were informed that they were being paid the maximum wages which the Frelimo government allowed, they demanded bonuses. As soon as they had received a bonus for a given number of recruits, they went on strike again and again demanded higher bonuses.

At that stage there still had been many people who wished to work on the mines and on one occasion there were 1000 recruits in the camp on the Mozambican side. Personnel on the Mozambican side literally had the WNLA in a grip and the situation started getting uncontrollable. The Mozambican clerks were not drinking beer anymore, they had switched to whisky!

WNLA meetings on the Mozambican side were attended by Frelimo guards and usually turned into chaos, or an endless bickering about higher wages. Even building assistants demanded that they be paid the wages of qualified construction artisans. Frelimo often organised their own meetings where they instigated the workers against the white officials. They even maintained that the recruits had been forced in the past to draw wagons over the entire 95 kilometres between Pafuri and Mapai.

It is understandable that the white personnel on the Mozambican side became increasingly nervous. Some Portuguese officials still attempted to work with Frelimo but it soon became clear that lawlessness had taken over and that the AK-47-rifle had taken over the government of the country. The remaining Portuguese started leaving the country in large numbers.

The last team of recruits who came through Pafuri arrived in the Republic on 27 February 1976. In March 1976 Frelimo closed the border and with that all cross-border communication ended. Wenela dismissed all the black employees on the Mozambican side and there are probably some of them who are still reminiscing about the good days of high salaries, bonuses and whisky!

It did not take Frelimo long to move all the citizens who had been living near the border from the area and the three little shops at Pafuri were closed.

In the meantime, the war between Frelimo and Rhodesia (Zimbabwe) had increased in intensity. Quite a number of battles between Rhodesian troop and guerrilla forces from Mozambique could be observed from the Pafuri post on the Transvaal side of the border. There were times when personnel here

Mr H.H. Mockford, WNLA official, photographed at a TEBA (The Employment Bureau of Africa Ltd) notice board at Pafuri, which was the new name of the WNLA (Witwatersrand Native Labour Association). (26.04.1984)

had to duck and dive when automatic rifle bullets hit between the buildings or hit the walls.

Frelimo once fired three 83 mm mortar which landed close to the Wenela officials on the Transvaal side. They also fired at aeroplanes which flew near the border on the Transvaal side and anybody who had been brave enough to cross the border, was arrested and even killed. Frelimo brought to an end all contact across the border.

Paes Mamede's Mapai camp was totally destroyed by the Rhodesian forces and the WNLA representative's house on the Mozambican side was hit by a mortar. Frelimo burnt down the seven large thatched-roofed buildings. They also destroyed all the generators and other equipment out of fear that the Rhodesian forces would come and confiscate them. It was sad to witness how the beautiful camp which had been developed over a period of ten years, was systematically converted into ruins. Those who had been directly involved with the development of this recruiting post can be forgiven if they harbour some bitter feelings about this senseless destruction.

Many of the people who had made great sacrifices in the opening up of large parts of Mozambique in an effort to recruit sufficient labour for the Witwatersrand gold mine industry, have passed away.

Harold Mockford retired at the beginning of 1985 after 47 years' uninterrupted service at Pafuri and settled on his farm in the Haenertsburg district. On 9 June 1985 he died here at the age of 67 years after a long illness. Of all the camps, only the Soekmekaar base and the Pafuri and Nwanetsi camps on the South African side of the border remained. The latter was later occupied by a SA Defense Force unit. There was hope that the Inkomati Accord which was signed on 16 March 1984 by Presidents P.W. Botha of South Africa and Samora Machel of Mozambique would improve relations between the two countries to such an extend that recruits for the mines would come through the Pafuri border post again.

For now it is only suitable that tourists travelling on the tar roads from Musina and Louis Trichardt (Makhado) to the Kruger National Park, should take a moment to think back and acknowledge the pioneering work which had been carried out by the WNLA and its loyal officials to develop and make this untamed wilderness accessible.

References

Carruthers, T. 1980. When Wenela walked through the Kruger National Park. *Teba Times* 15:4–5.
Glynn, W. 1985. Personal communication.
Joubert, S.C.J. 1985. H.H. Mockford – 'n huldeblyk. *Custos* 14(8):39.
Mockford, H.H. 1955–1985. Personal communication.
Mockford, H.H. & FitzSimons, V. 1950. A discovery of great scientific interest. *African Wild Life* 4(3):210–211.
Stevenson-Hamilton, J. 1941. Witwatersrand Native Labour Association. Warden's report to the Board 1941: 6–8. Witwatersrand Native Labour Association.
Wongtschowski, B.E.W. 1987. *Between Woodbush and Wolkberg*. Review Printers, Pietersburg.

Chapter 21
Early supporters and benefactors of the reserves
Dr U. de V. Pienaar

Many people and organisations have supported the Sabie Reserve by both word and deed on its road of development since the earliest years. The role which certain specific people have played in the realisation and proclamation of the first game reserves in the Transvaal, namely the Pongola Reserve (1894), Groenkloof Reserve (1895) and the Sabie Reserve (1898) are discussed in chapter 16.1. Apart from President Paul Kruger, these supporters included Henry Glynn of Sabie, hunter and naturalist F. Vaughan-Kirby, A.M. Robertson of the farm Rolfontein in the Wakkerstroom district, Albert Stoop, R.K. Loveday and J.L. van Wijk (respectively members of the Volksraad for Wakkerstroom, Barberton and Krugersdorp).

After the destruction and misery caused by the hard and bitter Anglo-Boer War, there were fortunately several people who and organisations which endeavoured to get the Sabie Reserve reproclaimed as a conservation area. Following that, there were also those who offered advice and support to the first warden, Colonel James Stevenson-Hamilton. This enabled him to develop the small foundling reserve through the first years of misgivings and opposition into the treasure which it is today.

Amidst all the killing and destruction during the war years there had fortunately been some conservation-concious people as well. Those included Captain H.F. Francis and Major A.W. Greenhill-Gardyne of the British military unit, Steinaecker's Horse, Henry Glynn of Sabie and J.C. Ingle. Another champion for conservation was the explorer-author Abel Chapman. From June until August 1899 Chapman had been travelling through the Lowveld and was impressed by its potential to become a game reserve. Then there was also the Assistant Mine Commissioner of Barberton, Tom Casement. All these people brought the potential and worth of the old Sabie Reserve under the attention of the British administration. Their arguments and advocacy for the reproclamation of the Sabie Reserve carried so much weight, that Lord Milner was convinced to reproclaim the area as early as June 1902 in the original text and language of the Zuid-Afrikaansche Republiek (ZAR).

Arthur H. Glynn, a farmer from the Lydenburg district, had already applied during the war to the British administration to be considered as the warden of the Sabie Reserve, if the administration was considering its reproclamation. He was a brother of the better-known Henry Glynn. In his letter to the British Foreign Office in October 1900, he pointed out to the authorities that his family had delivered valuable services to the occupational forces. However, nothing came of his application.

Captain H.F. Francis had already been selected to become the first warden of the conservation area and was even appointed on 4 July 1901 as acting warden of the area. He was, however, killed soon afterwards on 7 August 1901 in a battle with the Boer forces at Fort Mpisane. After his death, another member of Steinaecker's Horse, Lieutenant E.G. (Gaza) Gray, unsuccessfully applied on 14 August 1901 to the Civil Commissioner in Pretoria, to be appointed in his place. Assistant Mine Commissioner Tom Casement of Barberton suggested that one of the local prospectors at Moodies be appointed as acting warden. Following this, W.M. Walker was appointed as acting warden of the Sabie Reserve on 24 October 1901. Walker did not take his duties very seriously and the closest he ever came to the Sabie Reserve was when he travelled to Komatipoort by train, as he feared both the Boer forces and malaria in the Lowveld.

After the reproclamation of the Sabie Reserve in June 1902, it was wisely decided to appoint a permanent warden for the reserve. There were several candidates for the post, including Major A. St H. Gibbons, a soldier and explorer, and hunter-naturalist F. Vaughan-Kirby. However, the final choice fell on Colonel James Stevenson-Hamilton of the Sixth Inniskilling Dragoons. He was offered the position on 21 June 1902 by Sir Godfrey Lagden, Chief Secretary of Native Affairs. After some consideration he entered upon office on 1 July 1902.

Although his superiors could not provide him with

clear instructions about what was expected of him, Sir Godfrey Lagden simply instructed him to "Go down there and make yourself thoroughly disagreeable to everyone." Despite the lack of guidance, he received much and sometimes unexpected support, advice and help from several different quarters in the Lowveld to ease his difficult task.

Sir Godfrey Lagden himself became a staunch ally and often helped Stevenson-Hamilton to solve problems and ease the way. He also received help and support from his immediate superiors, E. Hogge, the Native Commissioner of Lydenburg and Major L.M. M'Inerney, a magistrate from Barberton.

Major A.D. Greenhill-Gardyne of the Gordon Highlanders and adjutant of Colonel Steinaecker, the commanding officer of Steinaecker's Horse, had still been stationed at Sabie Bridge (now Skukuza) at the time when Stevenson-Hamilton started his career as warden. He was a keen sportsman and lover of nature and drew up a most comprehensive report with respect to game and general conditions in the Lowveld and the steps which aught to be taken for its permanent conservation. This report significantly contributed towards Stevenson-Hamilton's formulation of an initial conservation strategy.

He was also able to learn much of conditions in the area which he would administrate successfully for the next 44 years, from Tom Lawrence, the owner of the only shop at Nelspruit and Bill Sanderson, a pioneer and hunter who farmed in the vicinity of Legogote.

In Komatipoort he received advice and assistance from amongst others Julius Furley, customs officer at the border post, and Jules Diespecker, the manager of the Selati Railway.

The Shingwedzi Reserve was proclaimed in 1903 after Leonard Henry Ledeboer, who later became a ranger in the reserve, had made representations to the magistrate of Pieterburg. The new magistrate at Barberton, Sir Alfred Pease, also served as a great source of support to Stevenson-Hamilton, who suddenly found that the conservation area under his control had more than doubled in size.

After the death of the customs officer at Komatipoort, J. Furley, in 1906, the Director of Customs and Excise in Pretoria organised that Stevenson-Hamilton take over his duties, as well as control over the ten black border guards who had assisted Furley with border patrols before.

After Jules Diespecker had passed away in 1906, Stevenson-Hamilton was appointed as his unpaid successor to maintain the Selati Line between Komatipoort and Sabie Bridge. For this service he obtained the use of the railway trolley which instituted a regular communication service between Komatipoort and his headquarters at Sabie Bridge.

To the great disappointment of Sir Godfrey Lagden, administration of the Sabie and Shingwedzi reserves was transferred from the Department of Native Affairs to the Colonial Secretary in 1905. This department would later develop into the Department of Home Affairs. Sir Patrick Duncan was the Colonial Secretary at the time and he would later become an acclaimed politician and Governor General of the Union of South Africa. From Duncan and his assistant Moor, a soldier from Ceylon, Stevenson-Hamilton also only received good cooperation and assistance.

While in London shortly afterwards, Stevenson-Hamilton met E.N. Buxton, who had founded the Society for the Preservation of the Fauna of the British Empire shortly before and received some valuable pointers from him.

Early in 1907 the first post-war elections were held in the Transvaal. Mr R.K. Loveday, who had been the member for the Barberton district in President Kruger's old Volksraad, was elected with a great majority. Loveday had been one of the most impor-

The living quarters of Mr Jules Diespecker, manager of the Selati railway line, in September 1902. (ACKNOWLEDGEMENT: STEVENSON-HAMILTON COLLECTION, SKUKUZA ARCHIVES)

tant lobbyists to get the Sabie Game Reserve proclaimed in the first place and in him Stevenson-Hamilton would have a trustworthy political ally in the Transvaal parliament.

After the proclamation of the Shingwedzi Reserve, Stevenson-Hamilton soon started dreaming of a large, connected reserve. To promote this ideal, he wanted to call on the owners and managers of the Transvaal Consolidated Land Company (TLC) which owned large sections of the area between the Sabie and the Olifants rivers. Mr Pott of Henderson's Consolidated was a keen hunter and sportsman himself and he was the one who made it possible for Stevenson-Hamilton to meet the other landowners and managers of the TCL. During these meetings he had the opportunity to discuss the possibilities of the transfer of their properties. Unfortunately he did not meet with much success during these first discussions.

In 1910 the Sabie Reserve received a most welcome financial injection of £3000. This came from surplus funds which the Transvaal had available after Unification. Five wood and corrugated-iron houses, provided with mosquito-proof wire gauze to protect the inhabitants against insect invasions, were built for the warden and his rangers. This significantly improved their living conditions.

Many other changes also took place in 1910. The Transvaal nature reserves, as well as the province's whole conservation action were placed under jurisdiction of the Transvaal Provincial Administration. In the new Administrator of the Transvaal, Johan Rissik, the warden of the Transvaal game reserves gained a friend and supporter with firmness of principle. The becoming of a Union seemed to hold out a possibility of bringing the realisation of a national park nearer, but although nationalisation had already been discussed at that stage, it still only remained a pleasant dream.

In 1910 George Pauling & Company, the well-known railway contractors, also completed the iron railway bridge of nine spans which traversed the Sabie River, while the Selati Line was taken over by the South African Railways. Initially a train ran to and from Komatipoort to Sabie Bridge once a week.

During 1910 the warden was also appointed as *ex officio* Native Commissioner of the Sabie and Shingwedzi reserves. This provided him with the authority to move squatters without interference from outsiders. The rough wagon routes between the different ranger posts were also improved and comfortable small permanent camps were built about 32 kilometres apart along these routes, so that officials and their draught animals could sleep over safely. This greatly improved communication between the different ranger posts. Mobility was advanced with the acquisition of more comfortable and faster horse-drawn buckboards.

In 1913 Sabie Bridge became telephonically connected with Komatipoort. This was a great step ahead to enhance communication with the outside world.

In 1912 the warden and Administrator Johan Rissik held talks with Captain Mudge, the manager of the company who owned most of the farms between the Sabie and the Olifants rivers, with a view of exchanging this land for state land further westwards. Although Captain Mudge listened sympathetically, no finality could be reached.

At a meeting of the Witwatersrand branch of the Transvaal Wild Life Protection Society in September 1912, a resolution was adopted that government should be requested to nationalise the Sabie and Shingwedzi reserves. The motion had been proposed by their vice chairperson, Mr J.H. Ryan, and seconded by H.B. Papenfus, who later became a member of the National Parks Board of Trustees. This was the first time that any public body had expressed the

Important visitors to the Sabie Reserve in the warden's office at Sabie Bridge in August 1912. From left to right: Gen. J.C. Smuts (Minister of Defence), Mr Julius Jeppe en Mr Johan Rissik (Administrator of the Transvaal).
(ACKNOWLEDGEMENT: STEVENSON-HAMILTON COLLECTION, SKUKUZA ARCHIVES)

notion of the reserves becoming a national park. One of the founding members of the Transvaal Wild Life Protection Society, C.A.O. Bain, followed this up by writing long articles in newspapers about the sword of expropriation which was hanging over the Transvaal reserves. He further pointed out the necessity of an action programme to ensure their preservation for future generations.

A step backwards during 1912 occurred when the administrator decided to award grazing rights to sheep farmers during the winter months in the old Kaap E and F blocks, as they have had in the past. This is the area west of the Nsikazi River to Legogote. During the first year a full 9000 sheep moved into the area. In 1914 the rent was increased from £5 to £10 per herd. The result of this increase was that nobody utilised the privilege, but in 1917 the rent was lowered again. Furthermore, farmers were allowed to burn the veld and to hunt for the pot. This lured the farmers and their livestock back to the area in their hordes.

In 1916 Government appointed a commission to investigate the whole situation as regards to the size and survival of the Transvaal game reserves. The chairperson of the commission was Mr J.F. Ludorf, who later became a member and still later the chairperson of the National Parks Board. The additional members were Messrs S.H. Coetzee, H. de Waal, G. Hartog, F.A.W. Lucas and C. Wade. They completed and submitted their report to government in 1918.

Their report proposed that certain areas of the old Sabie Reserve, including the Kaap Blocks E and F and a portion between the North Sand and Klaseri rivers and in a north-south line more or less on the 31° 37' latitude be expropriated and excised out of the reserve. This move effectively cut almost all the suitable habitat for oribi and red duiker, as well as the best roan and sable habitat south of the Olifants River, out of the reserve. A very positive recommendation was that the two reserves should be amalgamated by the addition of a piece of land between the Olifants and Letaba rivers and that the status of the whole area should be raised to that of national park.

Apart from nationalisation, the different proposals of the commission were implemented in 1923. Stevenson-Hamilton had already predicted this expropriation of parts of the Sabie Reserve in 1913 and on the insistence of Mr Johann Rissik had also drawn up a plan which would bisect the area between the Sabie and Olifants rivers along a line which would more or less follow the 31° 30' latitude. The result would be that about 400 000 hectares privately owned land and about 200 000 hectares state land would be excised from the reserve, including the winter grazing area for sheep west of Pretoriuskop. The remaining 1 600 000 hectares of the reserve would still include 70 private farms, in total about 200 000 hectares in surface area, which would not be impossible to exchange or buy out. In Stevenson-Hamilton's plan he proposed that two small wedges of land in the southern "grazing" area west of Pretoriuskop should remain part of the reserve, with the specific aim to maintain the ranger posts at Kaapmuiden and Mtimba.

During the excisions of 1923, the first-mentioned area remained part of the reserve, but as a result as a typographical error in the final recommendations, Harry Wolhuter's Mtimba ranger post was excised out of the reserve.

Before the boundary changes of 1923, Stevenson-Hamilton had considered certain options to change the reserve into a financial asset, so that it would not be a burden to the state. One plan he conceived was to capture young animals, tame them and later sell them to zoos all over the world. A small beginning was made with the project at Sabie Bridge. Holding facilities were erected and a few cows acquired to provide milk to the young animals. At one time there were a few waterbuck, a kudu, two sables, a young wildebeest and a number of small buck in captivity, which had all been weaned already.

However, the administrator would not allow the warden to sell the animals directly. Instead they had to be sent to the National Zoological Gardens in Pretoria. They would then sell the animals which they did not want to keep for exhibition purposes on a percentage basis on behalf of the reserve. This effectively doomed the whole undertaking. One cannot help but wonder what the then Division of Veterinary Services would have had to say about the movement of wild animals out of an area where endemic disease conditions occurred, which could be transferred to domestic animals in the vicinity of the distribution points!

In 1922 the Transvaal Consolidated Land Com-

The "Round in nine" train halt at the Skukuza (Reserve) sideline, circa 1925.

Board member H.B. Papenfus (1926–1937) who played a very active role in the early development of the Kruger National Park.

pany sent their general manager, Major Percy Greathead, to investigate the possibility of establishing a cattle farm as close as possible to Sabie Bridge. They did not wait long and by the end of 1922, some 800 hybrid cattle were brought to the farm Toulon, only about six kilometres from Sabie Bridge.

The new manager of Toulon Estate, Mr Crosby, soon befriended the warden and his personnel and before long they were not only visiting each other, but even organised tennis matches between the staff at Sabie Bridge and Toulon! The next step of the company was to instruct the manager to shoot some wild animals to test their rights. By now, Crosby was a firm friend of the reserve. To carry out his instruction, he shot a lone wildebeest bull and immediately afterwards reported his action to the warden. A summons was taken out and the result would be an important test case in the magistrate court.

The defense maintained that the wildebeest had been grazing grass and that grass was "a plant within the meaning of the Act". According to the game laws of the time, land owners were allowed to kill wild animals found in the act of destroying crops or plants. The magistrate could not be convinced that grass was a crop and passed a sentence against the TLC. The company immediately appealed to the High Court, but the judge pointed out that if such an appeal be upheld, it would mean a total rejection of all the game laws. This was a particularly important judgment in favour of nature conservation.

In 1923 the South African Railways launched a scheme to run fortnightly tourist trains during the colder months, to places of interest in the warmer parts of the Transvaal. This service soon became known as the "Round in nine". These tours soon became very popular, as they were affordable, comfortable and efficiently organised. All places which were of interest and the most picturesque were visited during the tours, including the citrus orchards at Nelspruit, the mountain passes at Pilgrim's Rest and Graskop, the forests of Sabie, while the tour ended with a few days at the beach of Lourenço Marques (now Maputo).

When Stevenson-Hamilton became aware of the tour schedule, he was somewhat disappointed to notice that the trains would pass through the Sabie Reserve at night without stopping once. He called the system manager of the SAR and suggested that each train should halt in the reserve at least once, or to at least make the journey during the day.

After long deliberations it was agreed that the train would overnight at Huhla siding just north of Sabie Bridge. Here a campfire would be lit for the passengers and early the next morning they would proceed to Newington, which had still been inside the reserve at the time. To the great surprise of the railway authorities, this part of the tour became the most popular with the tourists, especially the campfire experience at Huhla. Later on it was agreed that the tour should allow more time in the reserve and that a ranger would take the passengers on short bush walks at each halt. This was effectively the beginning of wilderness hiking trails!

The South African Railway administration soon realised that the Sabie Reserve held great publicity value to promote tourism and at once Mr Hoye, the general manager, became an avid supporter of the development of the area as a national park.

The warden had in the meantime had several satisfactory interviews with the Minister of Lands, Colonel Deneys Reitz. The result of these discussions was that the minister promised to visit the reserve during August 1923 to judge conditions for their value on the spot.

During his visit, he was accompanied by Secretary of Lands, Sommerville, and his private secretary, Neser. Members of the Executive Committee of the Provincial Council, including Dr Hjalmar Reitz (a brother of the minister), Messrs Stoffberg and

Kretschmar and Mr Van Velden, the provincial secretary, were the other members of the party. The group also included Dr A. Haagner, the chairperson of the Wild Life Protection Society. This was the first time that the reserve had been visited by government or provincial administration officials since 1917.

By that time, Major Scott, the chairperson of the Land Board and Mr Schoch, the surveyor-general, had already been in the reserve to evaluate the private farms which were still included in the Sabie Reserve. Dr A.A. Schoch, the surveyor-general's brother and government's chief legal advisor, who also happened to be an enthusiastic sportsman and nature lover, had already been busy drawing up a "National Park Ordinance". With a few alterations, this would later become the basis of the National Parks Act of 1926.

The Wild Life Protection Society launched an energetic campaign in favour of a national park, while the SAR administration widely publicised "Visits to our Reserves" in their "Round in nine" tours. Since 1923 the Boy Scouts of Barberton have visited the Reserve on an annual basis during the winter months under their Scoutmaster C.A. Yates, a brother-in-law of Harry Wolhuter.

After deproclamation in 1923 of the land which the Game Reserves Commission had proposed to be excised, Stevenson-Hamilton was fairly optimistic that the Gordian knot had been cut and that proclamation of a consolidated national park would soon follow. He was, as so many times in the past, disappointed yet again. Deneys Reitz summoned Stevenson-Hamilton to Cape Town in January 1924 to advise him about certain technical points in the envisaged legislation. The warden received a severe setback in the form of a telegram from Cape Town, however, which advised him not to come, as the "National Parks Bill" had been put on hold. According to Stevenson-Hamilton himself, it was probably a good thing that the Bill had not come before Parliament then, as "there had not yet been nearly enough propaganda in the country and the public was far from being awakened". There had also not been nearly enough campaigning in political circles.

Early in 1924 there was a general election which the National Party won. Shortly before the election, Stevenson-Hamilton had a meeting with Oswald Pirow, an old acquaintance, in Pretoria. Oswald Pirow was a senior member of the National Party as well as a strong supporter of the nature conservation philosophy. Pirow, who was destined for cabinet status, listened sympathetically to Stevenson-Hamilton and promised that he would do what he could to promote the national park scheme among his colleagues. He was as good as, or even better than his word.

Since the second half of 1925, more and more people campaigned in favour of a national park for South Africa and a persistent information campaign was launched. It was aimed at stimulating authorities at cabinet level, as well as to gain the goodwill and support of the general public. The Wild Life Protection Society did particularly valuable work by publishing several articles and by sending several circulars to its members through its efficient secretary, J.W.H. Wilson. Individual members, including H.B. Papenfus, then a member of parliament, applied himself to promote the national park scheme from public podiums as well as through organised deputations.

Since 1920, Colonel F.R.G. Hoare of the South African Defense Force had been visiting the Sabie Reserve regularly for a few weeks at a time. As an accomplished amateur photographer he managed to take a series of excellent photographs of game in their natural environment with his telephoto lens cam-

Board member, Paul Selby (1927–1930) who represented the Wild Life Protection Society in the National Parks board. He was an accomplished photographer whose outstanding photographs of the animal life in the Sabie Reserve contributed greatly to the nationalisation of the conservation area.

A herd of sable antelope in a burnt area at the foot of Manungu Hill, Pretoriuskop section, photographed by Paul Selby, circa 1924.

Harry Stratford Caldecott, outstanding publicity agent and campaigner for the Kruger National Park, as depicted in a painting by his wife, Florence Zerffi.
(ACKNOWLEDGEMENT: ESMÉ BERMAN, *ART & ARTISTS OF SOUTH AFRICA* AND THE SOUTH AFRICAN ART MUSEUM, CAPE TOWN)

An oribi in a burnt area near Pretoriuskop photographed by Paul Selby, circa 1924.

era. These would eventually be utilised by the Publicity Society of the South African Railways with excellent results. Mr Paul Selby, an American mine engineer, another member of the Wild Life Protection Society, began visiting the Pretoriuskop area in about 1924. With his camera mounted on the back of his Ford truck, he managed to take some startlingly impressive photographs of wildlife and natural scenery. Mr Selby was later requested to supply a set of his photographs to the House of Parliament where it contributed greatly towards making individual members aware of the unique qualities and potential of the Sabie Reserve.

In August 1925, Sir William Hoye, the general manager of the South African Railways, already an assiduous national park supporter, had the insight to appoint artist, H.S. Caldecott as publicity agent to advertise the Sabie Reserve as an enormously valuable potential asset to the country.

Harry Stratford Caldecott was born on 27 May 1886 in Kimberley. He received his education at the South African College School in Cape Town and the St John's College in Johannesburg. In 1912 he left for Paris with a bursary to study at the well-known Académie Julien in preparation for later admittance to the Ecole Nationales des Beaux Arts, where he received advanced art training in the studio of Gabriel Ferrer. With the outbreak of the First World War he served in the British Army, but in 1915 he was demobilised as a result of bad health. From then on he kept himself busy with ameliorative work which was later mentioned in British and French testimonials. After the war he had initially resumed his artistic career in Paris, but returned to Johannesburg in 1923. He was disillusioned by the city, however, and moved to Cape Town in 1924 where he married co-artist Florence Zerffi. He lived in Cape Town until his death at a fairly young age as a result of intestinal cancer.

Caldecott had no previous veld experience, but was still a passionate lover of nature and a member of the Wild Life Protection Society. To qualify him for his task he spent two months with Stevenson-Hamilton at Sabie Bridge where he enthusiastically gathered as much knowledge about the Sabie Reserve and its wild animals as he could. With his return to Cape Town he immediately applied himself ardently to his instruction and became one of the most committed protagonists of the reserve.

On 14 December 1925 the Cape-based newspaper, *Die Burger*, published an editorial which stated that it was time for the proclamation of South Africa's first national park by the Minister of Lands P.G.W. Grobler. The article suggested that the park be named "Kruger-wildpark" (Kruger Wildlife Park) in honour of the great role which President Paul Kruger had played in creating the Sabie Reserve. Caldecott and Stevenson-Hamilton embraced the suggestion as inspiring and lent it support.

Caldecott soon became known throughout the country for his "giraffe poster" which was put up at just about every railway station in the country. He also started publishing an endless stream of articles in newspapers and at the same time motivated reporters and addressed public meetings. According to Stevenson-Hamilton, Caldecott managed to achieve wonders in a very short time. "In fact, until his lamented and premature death in 1929, there was no single man who worked more strenuously and successfully for the cause of wild life preservation in the subcontinent."

Stratford Caldecott's enthusiasm influenced journalists, nature lovers and scientists alike. Many articles and newspaper reports in favour of a first national park for South Africa began appearing all over the country. Some of the people who made contributions include William T. Hornaday, the Director of the New York Zoo and well-known American nature conservator, Leonard Gill (Director of the South African Museum), E. Warren (Director of the Natal Museum), Austen Roberts (Director of the Transvaal Museum, S.H. Skaife of Cape Town, A. Haagner of the National Zoological Gardens, Deneys Reitz, H.B. Papenfus, C.A. Yates, artist C.T. Astley-Maberley, Percy FitzPatrick, Ethelreda Lewis, E. Bell, Ernest Granville, H.S. Webb, Percy Greathead and others.

The Rotarians, the University of Cape Town, The National Society and even the Transvaal Landowners Association also supported the campaign actively.

On the other hand there were of course many influential adversaries who fought the idea of a national park with all their might, with the officials of the Division of Veterinary Services playing a leading role. Claude Fuller, R. du Toit and even the director of Onderstepoort, Sir Arnold Theiler, openly expressed their misgivings about the creation of a large wildlife area on an international border over which wild animals, capable of transmitting a variety of diseases, could move freely.

Their opposition came too late and the battle was almost over, as behind the scenes, A.A. Schoch had already drawn up the final concept Act which would become the National Parks Act to be submitted to Parliament.

On 31 May 1926 Minister P.G.W. Grobler (a grand nephew of the late President Paul Kruger), moved the second reading of the National Parks Act in parliament with acclamation and without any opposition.

Minister of Lands, Pieter Gert Wessel Grobler, was born on the family farm Boekenhoutsfontein in the Rustenburg district. His father was a child of Paul Kruger's sister, while his mother's maiden name had been Du Plessis before her marriage to Wilhelmus Hermanus Cornelius Grobler. Apart from the other great achievements during his illustrious career, Piet Grobler will especially be remembered for his enormous contribution towards nature conservation. Growing up on a farm he not only lived close to nature, but also loved it dearly. As nature conservator he would build on the pioneering work of his famous family member by moving the National Parks Act in 1926.

Through this Act, the former Sabie and Shingwedzi reserves were consolidated into one large national park at last – the Kruger National Park. An important aspect of the National Parks Act was that it made provision for the creation of further national parks in future. Piet Grobler was afterwards directly responsible for the proclamation of the Kalahari Gemsbok, the Bontebok and the Addo Elephant national parks in 1931.

Section 5 of the original National Parks Act of 1926 stipulated that the Governor General should, by notice in the *Government Gazette*, appoint a board of trustees which will be known as the National Parks Board of Trustees, to both control and manage national parks. The Board would consist of ten members (this number was expanded to 12 in 1960) proposed by the Governor General according to the prescription of Section 5. The members will consist of the Administrator of each province in which a national park is situated and one representative of the Wild Life Protection Society. If the society ceased to exist, any society chosen by the Minister of which the aim is wildlife conservation will be entitled to nominate one member to the Board.

The first Board was named by the Governor General in the same year and met on 16 September 1926

Dr W.J. Hornaday, well-known American naturalist, friend of Col J. Stevenson-Hamilton and champion of the national park concept, circa 1926.
(ACKNOWLEDGEMENT: STEVENSON-HAMILTON COLLECTION, SKUKUZA ARCHIVES)

The famous giraffe poster of H. Stratford Caldecott which appeared in 1925 at just about every railway station in the Union. This poster contributed largely towards the publicising of the national park idea. Caldecott had been appointed by Sir William Hoye, the general manager of the SAR, to help promote the Sabie Game Reserve.

Board member G.S. Preller, well-known historian and member of the National Parks board between 1926–1939. He was responsible for the first historical research in the Kruger National Park.

for the first time in Pretoria under chairpersonship of the late Senator W.J.C. (Jack) Brebner. This historical first meeting of the National Parks Board of Trustees was also attended by Minister of Lands P.G.W. Grobler, G.A.W. Schneider of the Department of Lands, as well as the acting warden of the Kruger National Park, ranger C.R. de Laporte. Apart from Senator Brebner who had been elected as chairperson, the other members of the initial Board included Colonel Deneys Reitz, former Minister of Lands, Dr A.K. Haagner, the Curator of the National Zoological Gardens in Pretoria who had been nominated by the Wild Life Protection Society, R.A. Hockly, W.A. Campbell, G.S. Preller, A.E. Charter, nominated by the Administrator of the Transvaal, Sir Abe Bailey, O. Pirow, and H.B. Papenfus. The last-mentioned three members did not attend the first meeting.

The minister opened the meeting and pointed to the fact that there were already many other boards in the country, but that the National Parks Board was the first statutory board and as such could exert full control through Section 12 of the Law. He further declared that it gave him great joy to place the Kruger National Park under the Board's care. He also emphasised the fact that the Kruger National Park represented one of the country's biggest assets and only came into being through the farsightedness of President Kruger and the hard work and sacrifice of its first warden, Colonel James Stevenson-Hamilton.

The minister further expressed the hope that it would be possible to transfer a few of the rare white rhino of which there was only a small population left in Natal, to the Park.

The Board started their activities with an annual financial contribution of £3000 from the Transvaal Provincial Administration and £4000 from Central Government. Minister Grobler suggested to the Board that these funds could possibly be augmented by building roads in the Park and to appoint a courier who could be paid to accompany tourists on excursions. He also recommended that fees be charged for permission to film the Park's fauna.

A suitable tribute to Minister Piet Grobler for the great contribution which he had made to nature conservation in South Africa, is the Groblershoop Dam, which was built in the Timbavati River in the Kruger National Park.

In January 1927, Board member W.A. Campbell of Mount Edgecombe suggested that the Umfolozi, Hluhluwe and Mkuzi reserves in Natal be declared national parks. The Board decided that they would give the matter careful consideration, if the Natal Provincial Administration would carry on administrating the areas. Member A. Haagner resigned and the Wild Life Protection Society elected Paul Selby, an American mining engineer, to take over his position. Cognizance was taken of the income and expenditure statement for 1926 which indicated a total income of £2752, while expenditures totalled £1487.

In March 1927 the Board received an offer from a certain Mr Forder of Natal to sell a herd of hartebeest to them, but the Board declined the offer as they lacked the necessary funding. Member Preller proposed that the possibility to sink a few boreholes in the Kruger National Park to provide water to animals during dry spells should be considered. This was as a result of certain remarks which acting warden, Captain Howe, had made in his annual report.

During a meeting in June 1927, Sir Abe Baily was

dismissed as Board member, as he had not attended four consecutive Board meetings in a row. It was proposed that he be replaced by Colonel F.R.G. Hoare. Notice was served by the Department of Lands in November 1927 that the department was not prepared to contribute £6 per year to maintain the border road which had been opened up by surveyors Murray and Antrobus through the Nyandu Bush. The Board decided that the road would be written off. Later on, Assistant Surveyor General G. McGregor of Pretoria, offered to maintain the road with the proviso that his personal expenses be paid.

During November 1927 the Board also took note with gratitude that Dr W.J. Hornaday, a well-known naturalist in American conservation circles, had made a donation of $1000 to the Board. The Department of Lands informed the Board at the same meeting that the three Natal reserves were in endemic nagana areas and could therefore not be considered for proclamation as national parks.

In June 1928 the Board took the decision to charge entrance fees from visitors to all parts of the Kruger National Park, except the far north (Punda Maria). Entrance fees would be five shillings per person, with a minimum amount of £1 per vehicle. The SA Railways would have to pay £5 per day and night for their round trip through the Park. These tariffs would come into effect in 1929. Entrance for children under ten would be free.

During this time the Board also received an offer from the firm Zululand Cash Wholesalers to capture 12 white rhino in Zululand and to deliver them to the Kruger National Park by railway at the cost of £8000. In view of Minister Piet Grobler's representations that something should be done to save the white rhino from extinction in Natal, the offer was discussed. On recommendation of member W.A. Campbell, the matter was referred to the newly established Natal Parks Board.

In June 1929 member Paul Selby proposed a new plan for the erection of tourist accommodation in the Kruger National Park. The plan which the Board approved, involved the building of rondavels with concrete walls without windows, a concrete floor and a cable roof structure, thatched with either grass or reeds. Member W.A. Campbell kindly offered to cover the costs of an experimental hut at Skukuza (£150).

This particular hut is still standing and has been fitted out as a tourist museum. The Transvaal Wild Life Protection Society and the Orange Free State Game Protection Society each contributed £150 and £120 respectively in the same year towards the building of tourist amenities in Skukuza. In 1930, Messrs Trollope and Pike each donated £12 towards the building of a rondavel at Skukuza.

Stevenson-Hamilton received a request in July 1929 to assist with the capturing of a few young giant sable antelope in Angola, which authorities wanted to present to Prince Arthur of Connaught to settle in the Kruger National Park during his visit to the Union of South Africa. The warden doubted the fact that these animals represented a separate subspecies, so the Board decided to refer the matter to the zoological gardens in Johannesburg, Pretoria or Cape Town to possibly help Angola.

In September 1929 the Board was informed that one McLachlan of the farm Cork had built a weir in the Sabie River without obtaining permission from the Board first.

In June 1930, Mr Werner of Lusaka offered nine tamed lechwe to the Kruger National Park to be settled there. The warden duly informed the Board that there was very little suitable habitat for these animals as they preferred swamps. Despite Member Pirow's objection in principle against the import of

Mr W.A. Campbell had been one of the Parks Board members who served the longest (1926–1950). He had donated money to the Board to build a hut in Skukuza rest camp during 1929, which was later turned into a tourist museum.

21. EARLY SUPPORTERS AND BENEFACTORS OF THE RESERVES

exotic animals, the offer was accepted on recommendation of members Campbell and Papenfus. The Board later received news that all the rams in the herd had died and had to accept with regret that their effort to settle lechwe in the Park had failed.

In that same year, member W.A. Campbell donated ten crested guineafowl to be released in the Park.

During 1930 a pontoon had been constructed to ferry vehicles across the Crocodile River at Malelane and Rupert Atmore of the adjacent farm, Mhlati, agreed that an access road could be constructed across his farm to the point where the pontoon operated from. The Board rejected applications for the building of hotels in the Park.

In September 1930, J.H. Cloete of Clocolan in the Free State donated a windmill to the Park. Board member Selby pleaded that it should be erected in a position where it would be out of sight of tourists, but Mr Cloete later informed the Board that he could unfortunately no longer make the donation.

The Board thanked Dr A.A. Schoch for his large contribution in compiling the new regulations for the management of the Park.

In December 1930, Dr Kind of Johannesburg donated four Basuto ponies to the Board.

In March 1931, the Royal Automobile Club of South Africa (RAC) and the Automobile Association of South Africa (AA) offered to provide notice boards to be put up in the Park without cost. The Board decided that the RAC should serve the area north of the Olifants River, while the AA would function south of the Olifants River.

In August 1931 a proposal was accepted that honorary members and honorary life members be appointed for the Park at a cost of £3.3.0 per annum and a one-off payment of £26.5.0 respectively. These members could enter the Park freely at any time. In 1931 the Automobile Association donated an amount of £30 to the Board as a contribution towards building a hut at the "Reserve" (Skukuza). In 1946 the AA made a further contribution towards building three more huts in the Park.

Mrs Louise Botha offered 30 black wildebeest to the Board in October 1931 at 15 shillings each, but the offer was declined as this species would not be able to adapt to conditions in the Kruger National Park and also because there was no alternative area to release them.

On the suggestion of Mr Paul Sauer it was decided in December 1931 to name a hut in the Park in honour of Harry Stratford Caldecott. This decision was never carried out, or if it had been, then the hut could not be identified later on.

Board member Papenfus suggested that the "Reserve" (Skukuza) rest camp should be renamed in honour of Minister Piet Grobler as a tribute to his great contribution towards moving the National Parks Act through parliament. This proposal was repeatedly modified during later Board meetings. Eventually a decision was made to rename Letaba to Piet Grobler rest camp, but Deneys Reitz was of the opinion that it would be incongruous with the other camps' African names. There was another proposal to name an entrance gate after him. Still later it was

The W.A. Campbell memorial hut in Skukuza which was donated to the Board in 1929 by Board member W.A. Campbell.
(07.05.1984)

The interior of the W.A. Campbell memorial hut which was turned into a tourist museum.
(07.05.1984)

decided to erect a small marble or granite monument with a suitable bronze plaque in honour of Minister Grobler at Skukuza. As a result of the expenditure attached to the project, the proposal gradually became watered down, and it was eventually decided to imbed a few bronze tablets (which were not to exceed the cost of £50) against the side of a granite boulder at the foot of a hill between Tshokwane and Skukuza. Although this frugality will appear strange now, it must be remembered that this happened during the great depression. The wording on the plaques had to pay honour to the contributions which both President Kruger and Minister Piet Grobler had made towards the establishment of the Kruger National Park. It was also decided to invite the minister to unveil the plaques.

The plaques were ordered for the amount of £49 from a Mr Eshen and the unveiling by Minister Grobler finally took place on 22 September 1933.

In August 1932 Board member Gustav Preller proposed that the main camp of the Kruger National Park, formerly known as "Reserve", be changed to "Skukuza", which was the name which the blacks had given Colonel Stevenson-Hamilton. The proposal was accepted and the SA Railways were requested to rename the sideline just north of Sabie Bridge likewise, but it later became known as the Huhla siding. The name Skukuza for the main camp was officially accepted in September 1932.

Rhys Evans of Stegi in Swaziland offered a number of oribi to the Board in September 1932, with the provision that the Board capture the animals. The warden thought that it was a good idea, but too expensive and unpractical, with the result that the project was not approved.

In February 1933 the Royal Automobile Club of South Africa donated £75 to the Board to build a hut, which was eventually erected in Shingwedzi rest camp. During March 1933, Bertram F. Jeary of Muizenberg offered to do a survey of the micro-organisms in the Sabie River to determine the grade of pollution of the water in the river as a result of the small gold mines in its upper reaches. In his report to the warden, Jeary summarised his findings as follows:

> The Sabie River is almost, if not entirely devoid of any form of small aquatic animal life. I can put this down to no other cause than the waste products, which, for many years, have polluted the river from the Graskop mines. The fish have become gradually fewer in numbers as the amount of solid matter carried in suspense is unsuitable to them as well as to the smaller forms of life.

As a result of Jeary's findings, the Board strongly appealed to the Department of Mines to bring an end to the pollution of the Sabie River. All representations fell on deaf ears, however, until the 1940s, when the pollution levels became so bad, that the cattle of the farmers outside the Park started dying when drinking water in the river. Only then were steps taken to eliminate the pollution and in following years, the Sabie River once again developed a very rich water life.

In June 1933 Board member W.A. Campbell donated £200 to the Board with the request that life membership be issued to himself and seven other persons, including Minister Piet Grobler.

The Board also decided that seeing as its own officials had already managed successful campaigns against red and brown locusts in 1906 and from 1923 until 1925, it would not be necessary for officials of the Department of Agriculture to be involved with further locust spraying campaigns in the Park. However, all outbreaks would still have to be reported to the department according to law.

Since 1931 the Park had been ravaged by a severe drought. After discussions with the warden, Mr Bertram Jeary of Muizenberg initiated a publicity campaign to collect funds to sink boreholes in the Park. Jeary approached *The Star* newspaper in Johannesburg and with its help the campaign soon gained momentum. Individuals and organisations from all over the country generously supported the scheme and sent contributions for the Board's drilling programme. In a relatively short time about £5000 was available for the "Water for game" fund and drilling could begin. Three ladies, Mrss E. Orpen, R.A. de Villiers and Armour Hall each subscribed funding for a whole borehole and windmill. All the larger centra also made generous donations, while the State itself also made money available and further agreed to carry the costs of dry holes.

During the drilling programme which started in

Memorial plaque in honour of the original donors of the "Water for game" fund (1933–1935). These were erected at all the entrance gates to the Park.
(16.06.1984)

The Stevenson-Hamilton Memorial Library and Information Centre which was built at Skukuza in 1961 with funds which had been donated by the Wildlife Society of South Africa.
(07.05.1984)

1933 and lasted until 1935, 14 successful holes were sunk. To acknowledge the generous contributions which made it possible, the Board had a number of bronze plaques made with the names of the donors and the name of the borehole which they had donated, on it. These plaques were erected at all the entrance gates to the Park. The Pretoriuskop borehole was donated in 1933 by Armour Hall; Komapiti (Ship Mountain)(1933) by Port Elizabeth and Benoni; Gomondwane (1933) by Pretoria; Randspruit (1932) by Eileen Orpen; Hlambamadvube (1933) by Durban, Manzimahle (1933) by Lourenço Marques; Gudzani (1934) by Johannesburg; Bangu (1934) by the Union Government; Marheya (1934) by the Johannesburg Stock Exchange; Rabelais (1934) by R.A. de Villiers; Malopeni (1935) by the Natal Sugar Mill Association; Ngwenyeni (1935) by the Union Government; Malopeni North and Shivhulani (1935) by the Cape Peninsula.

On 7 February 1934 the chairperson of the Board announced that Mrs Eileen Orpen of Harrismith and donor of Randspruit borehole, had purchased the farm Chalons of 4492 hectares, north of the Rabelais Gate, to be incorporated into the Park. The proclamation of Chalons as part of the Kruger National Park became a reality on 10 April 1935. In later years, Mrs Orpen donated another six farms to the Park, namely Kingfisherspruit (4185 hectares) and Red Gorton (4024 hectares) on 7 February 1941, Blackberry Glen (3547 hectares) on 12 June 1942, Hengel (832 hectares) on 12 June 1942, Sikkeltowkloof (4050 hectares) on 12 June 1942, and Houtboschrand (3399 hectares) on 25 February 1944. In 1946 she also offered the farm Newington (1720 hectares) to the Board for incorporation into the Park, but this offer could not be accepted as the property was too far from the western boundary of the Park.

Through her donations of a total of 24 530 hectares of land along the western boundary of the central section of the Park, Eileen Orpen became the greatest benefactor of the Park since its inception. The land which she had purchased would be worth many millions of Rands today. Although other donors have also made large donations to the Park in more recent times, including the Wildlife Society which donated the Stevenson-Hamilton Memorial Library, the financier Charles Engelhard (the Engelhard Dam in the Letaba River), Goldfields SA Ltd (the Skukuza auditorium), the brothers A. an Y. Naghi, the Dinglers, Baders, Nivens and many other donors of guest houses, family huts and windmills, Eileen Orpen will always be remembered as the most outstanding benefactor in the history of the Park. Apart from the large tracts of land, she repeatedly made other contributions to the Board as well, including the financing of bird baths in four rest camps, as well as the building of the Eileen Orpen Dam in the Nwaswitsontso River near Tshokwane in October 1944.

Eileen Akerman Orpen (née Barry) was a member of the well-known Barry family of Swellendam. During the nineteenth century the Barrys had been so successful as traders and in business that they owned their own coasting-vessel and printed their own banknotes. In 1880, Eileen's father, Thomas Daniel Barry, travelled by ox-wagon from Pietermaritzburg and bought a farm of a full 20 000 hectares in the vicinity of Harrismith. He later became a member of the Cape Parliament.

Eileen passionately loved animals from a very early age. Legend has it that as a little girl, she once tackled a man, who had been mistreating a dog, with a shambok.

Eileen Barry met James Havelock Orpen in Cape Town and they married in 1909 in Johannesburg, where her husband was a surveyor for Geldenhuys Deep Mine. They moved to Rhodesia (Zimbabwe) later on and settled near Bulawayo. In 1914 they moved to the Cape where James had interests in a fishing industry in Kalk Bay. As a result of the bad health of their eldest son, Barry, the family moved to the Free State in 1922 and started to develop the farm Afrikaskop between Harrismith and Bethlehem. This was the last bit of the Barry estate still in their possession. In 1933 Barry married, after which his parents handed the management of the farm over to him. James and Eileen moved to Johannesburg again where he continued his career as surveyor. One of his greatest projects was to open up and survey large sections of the Park's boundaries without charging a cent. Eileen now had the opportunity to live out and intensify her love for nature conservation and she soon became the personification of a real nature lover. Her first donation was a large bird cage to the National Zoological Gardens in Pretoria.

James and Eileen were both members of the Wild Life Protection Society of South Africa and were inspired by the work of the society's chairperson, H.B. Papenfus, who was a member of the National Parks Board at the same time. Their first visit to the Park was on his insistence and from then on it became the all-important interest of their lives. James also planned the outlay of Shingwedzi rest camp and helped to survey quite a number of roads in the vicinity of Pretoriuskop. There was a house on the farm Kingfisherspruit which Eileen had donated to the Park, and this became their base during the winter months.

James Orpen became a member of the National Parks Board in November 1939 and provided excellent services for 14 years. Even before his appointment as a board member, he had already played a particularly active role in the proclamation of the Mountain Zebra National Park in 1937. He also became involved with the fencing of the Addo Elephant National Park, which he had surveyed himself.

As an enthusiastic member of the Wild Life Protection Society, James was elected as president of the organisation in 1943. He served in this capacity until 1949 and made a valuable contribution during this time towards making the general public of South Africa more conscious of conservation. In 1949 the Orpens were deputised to represent South Africa at an international nature conservation conference in America. In May 1954 the South African Government acknowledged Eileen Orpen's generous financial and moral support to nature conservation in South Africa by donating about 11 hectares of land on the edge of the Escarpment near Graskop to her. From here the Orpens had a magnificent view over their beloved Lowveld.

In the meantime, the National Parks Board had also paid tribute to Eileen Orpen. On 4 October 1944 a bronze memorial plaque was fixed to a rock face between Skukuza and Tshokwane. In addition, a windmill along the Shisakashanghondzo Spruit on Chalons was named Eileen to commemorate her contribution to the "Water for game" programme.

Eileen Orpen passed away on 24 May 1954 and James four days later. They were buried next to each other in Bethlehem's cemetery.

A further tribute was made to the Orpens when the Board renamed the Timbavati entrance gate and rest camp, which had been completed in May 1954 along the western boundary of the farm Kingfisherspruit, to Orpen rest camp and entrance gate.

Apart from the Orpens, many individuals and organisations made contributions and offers of help to the Kruger National Park. In 1934, Tom Elphick of Magnesite offered a number of red duikers to the Board. The only proviso was that the Board had to undertake the capturing. A capturing operation was duly launched, but as a result of a lack of the necessary equipment and experience, ranger Tomlinson and his assistants could not capture a single buck.

Mrs Eileen Akerman Orpen, single-most important benefactor in the history of the Kruger National Park, circa 1930s.

Mr James Havelock Orpen, member of the Parks Board (1939–1954) and benefactor of the National Parks of South Africa in various ways, circa 1940s.

21. EARLY SUPPORTERS AND BENEFACTORS OF THE RESERVES

The Transvaal Indian Congress offered to carry the cost of building amenities for Indians, if it could be built at Skukuza. The Board accepted this offer for the 1935 construction programme. There was also a proposal that the camp at Olifants Pontoon (Balule) be reserved for Indians.

As a result of the Board's lack of funding to erect additional tourist facilities, Board member Campbell suggested that an outside organisation be consulted to build a 300-room hotel in the Park. Mr Papenfus supported the proposal, but the chairperson, Senator Jack Brebner, expressed himself strongly against it.

In 1934, W. Borchers of Louis Trichardt (Makhado) wrote a letter to the Board in which he offered to make a financial contribution. The Board answered that a contribution towards the "Water for game" fund would be very welcome, and in March 1935 an amount of £50 was received from him.

In 1935 the Royal Automobile Club donated £150 to build car ports in certain rest camps. In March of the same year the Board gratefully acknowledged a donation of £5000 by the Union Government towards the construction of low-water bridges to replace the pontoons which functioned at the flowing rivers in the Park.

In July 1935 the Board also gratefully heard about W.A. Campbell's offer to pay for the construction of a dining room at Crocodile Bridge, as well as for two radio sets for use in the Park. These two sets were installed at Skukuza and Pretoriuskop and they brought about great improvement in communication! The Board also granted an interview to J. Moe who, along with his brother, owned a farm of 5000 acres between Greytown and New Hanover on which 50 red hartebeest occurred. These had been the only survivors of the Natal subspecies of red hartebeest and Moe believed that the Board should buy the farm as a permanent sanctuary for these and other antelopes. They were prepared to sell the farm for £50 000 or £25 000, provided that they could maintain life interest. It was decided to refer the matter to the Minister of Lands for a decision, but after a month the Minister informed them that he was unfortunately not able to buy the farm. With this decision, what had probably been the last opportunity to save the Natal red hartebeest from extinction, an opportunity was lost.

In March 1936, Board member Ludorf proposed once again that springbok and blesbok be released in the Pretoriuskop area, but the warden fortunately succeeded in convincing the Board not to go ahead with the project.

In May 1936 the South African Railways offered to subsidise a bookstore at Skukuza, while W. Borchers of Louis Trichardt (Makhado) made a contribution of £30 towards maintaining the Pafuri road in October 1936.

In August 1937 the Casa Mia hut was built in Skukuza with funding received from the friends of "Uncle" Bill Huston of Redfern, New South Whales in Australia, and the Casa Mia boarding house in Johannesburg.

In 1936 the warden requested the Board to build a guest house in which he could receive official guests at Skukuza, but the Board did not have the necessary funds at the time to carry out the proposal.

In 1937, Major Frank E.B. Struben of Pretoria offered funds to build the first guest house in the Park. The building was designed by Hilda Stevenson-Hamilton and built just west of where the entrance gate to Skukuza camp used to be. Total building costs amounted to £355.16.00 and it was inaugurated in November 1937. Major Struben received life membership for his contribution.

The famous American naturalist and photographer, Herbert Lang, often visited the Park, especially the northern part during the 1930s, and became world-famous for the quality of animal photographs which he had taken with an old-fashioned wooden

The first guest house in the Park which was built with money donated to the Board in 1937 by Major Frank Struben of Pretoria. It was designed by Mrs Hilda Stevenson-Hamilton and built at Skukuza.
(07.05.1984)

camera. During one of his visits to Punda Maria, he collected an unknown worm-lizard which was described as *Chirindia langi* by Dr V. FitzSimons of the Transvaal Museum in 1939.

Herbert Lang had well-established connections with several American museums, especially the American Museum of Natural History in New York. He visited South Africa in 1926 after a field expedition to Angola and was so impressed with the country that he decided to settle here. He made the Transvaal Museum his headquarters. Shortly after his arrival he encouraged A.S. Vernay to undertake a scientific expedition to the Kalahari and Bechuanaland (now Botswana), during 1930. This expedition delivered extremely successful results.

Lang was a devoted nature conservator and avidly campaigned against the game extirpation project of Veterinary Field Services in Zululand. He also carried out a successful census of white and black rhino in the Zululand nature reserves and in the process managed to obtain excellent photographs of both species.

After his marriage in 1936, Lang found it necessary to bid farewell to his scientific interests and photography. He died in May 1957 after a long illness. His memory is honoured by the unique collection of photographs and engravings which are exhibited at both the Transvaal Museum and in the Stevenson-Hamilton Memorial Library at Skukuza.

Mr V.L. Robertson of Amersfoort donated £50 in September 1937 towards having memorial plaques erected at Lindanda where ranger H.C. Wolhuter had killed a lion while it attacked him, with his hunting knife.

During 1938 the state's surveying team under Mr Le Roux once again spent the winter months in the Park surveying the eastern border. After completion of the project, Le Roux gave ranger L.B. Steyn a theodolite as acknowledgement for his assistance throughout the project.

In 1938 the Board accepted a scheme whereby sponsors, including Lifebuoy Soap, would carry the costs to install mirrors in the tourist huts. The AA and RAC in turn offered £40 each in February 1938 for the building of huts for their patrol officials in the Park.

Also in 1938, J.J. Rose-Innis donated money for

Herbert Lang, a well-known naturalist and photographer during the 1930s.
(ACKNOWLEDGEMENT: TRANSVAAL MUSEUM)

the building of a fish pond and bird bath at the restaurant building in Shingwedzi camp.

In 1938 the Historical Monuments Commission offered to erect a memorial plaque at a suitable place to commemorate Louis Trichardt's epic journey through the Kruger National Park in 1838.

After the death of Board member H.B. Papenfus in 1937, the Board gave permission to his family in August of the same year, to build a sandstone clock tower in front of the Skukuza camp's entrance gate, despite opposition from architect Gerhard Moerdyk. This clock tower was inaugurated by the Judge N.J. de Wet, chairperson of the Board at the time.

The South African Defence Force undertook in 1940 to upgrade and gravel the road from Acornhoek through Rabelais to the Portuguese border at Nwanetsi, so that it could be used in all weather conditions. They also built small bridges of eucalyptus poles and concrete over the Nsemani, Nwanetsi and Sweni spruits. The airstrips at Mavumbye

21. EARLY SUPPORTERS AND BENEFACTORS OF THE RESERVES

(1932), as well as the one which had been constructed in 1938 near Klopperfontein, were also improved during this time.

In November 1941, a hailstorm severely damaged the homestead on the farm Kingfisherspruit. Although Eileen Orpen had been prepared to pay for the damage, the Board nevertheless carried the cost.

The Caltex Oil and Petroleum Company offered in May 1941 to print roadmaps of the Kruger National Park free of charge for distribution among tourists in the Park.

In August 1944, V.L. Robertson of Amersfoort once again made a donation to the Board. This time it was to pay for the construction of a ramp for motor vehicles at Pretoriuskop.

Mrs J.M. Craig of Hastings in England donated an amount of £100 in 1946 towards the "Water for game" project.

As a result of new agreements between the WNLA recruiting company and the Portuguese authorities, from 1941 the organisation no longer allowed recruits to enter via the recruiting posts at Sweni (Nwanetsi) and Letaba. However, activities at Pafuri and Soekmekaar carried on.

The post at Nwanetsi, which included several buildings and a guest house, had originally been built by the WNLA during the 1920s. It was later improved when H.S. MacKay took over the position as regional manager of the organisation at Soekmekaar from Paul Neergaard.

At the end of 1941, the manager of the WNLA handed the keys of the vacated Nwanetsi post over to Stevenson-Hamilton. The guest house section of the camp was upgraded in 1945 to a private camp and made available to visitors.

Mr T.B. Herold offered in November 1940 to carry the building costs for a "Museum hall" at Skukuza. Unfortunately nothing came of this worthy offer.

Former Board member T.B. Bowker made funds available to the Board with which the concrete weir underneath Tshange (Tshanga) outlook in the Shingwedzi section could be constructed in 1952. During the same year, a memorial hut, the "Mavukane hut" was built in Malelane camp in honour of former ranger S.H. Trollope with funding provided by his family.

In July 1946, Board member W.A. Campbell donated £100 to the Board for the purchase of a mosaic aerial photograph to be used by the new warden, Colonel J.A.B. Sandenbergh. The money was used to build a tennis court for the personnel at Skukuza instead. This court used to be where the Stevenson-Hamilton Information Centre was later erected.

Board member Bran Key, who died in 1949, was also honoured in 1956–1957 by the building of a hut in Shingwedzi rest camp with funds which had been made available by his family. During the same year a swimming pool was built at Pretoriuskop with funding which had been made available by the Rembrandt Tobacco Company.

The clock tower in Skukuza, which was donated to the Board in 1937 by the family of Board member H.B. Papenfus, who had suddenly passed away in that year after serving on the Board from 1926–1937.
(07.05.1984)

The small Bowker Dam at the foot of Tshange (Tshanga) outlook which was built in 1952 with money donated by former Board member, T.B. Bowker.
(28.04.1984)

The last two donations during the period under discussion was that of Brigadier C.H. Blaine, a former Board member, who donated money for the construction of a luxury family hut at Skukuza, similar to the one which had been built there for the royal visit, while Messrs George Naylor and F. Herholdt also made funds available in 1947 for the building of two huts in Skukuza.

Pro beneficius semper honorati.

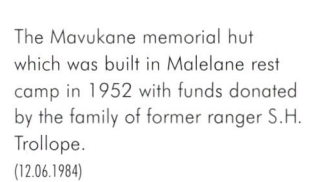

The Mavukane memorial hut which was built in Malelane rest camp in 1952 with funds donated by the family of former ranger S.H. Trollope. (12.06.1984)

References

Agendas and Minutes of National Parks Board of Trustees meetings, 1926–1946. Skukuza Archives.

Bigalke, R. 1950. Wildlife conservation in the Union of South Africa. *Fauna & Flora* 1:1–11.

Chapman, A. 1900. Draft scheme for preserving and re-establishing the large game and wild animals in the Transvaal. Transvaalse Archives Document C.S. 2:211.

Donation Files. 1926–1946. Skukuza Archives.

Francis, H.F. 1901. Letter dated 2 August 1901 to Tom Casement, the assistant mine commissioner, Barberton. Skukuza Archives.

Glynn, H.T. 1929. *Game and Gold*. The Dolman Printing Co. Ltd, England.

Gray, E.G. 1901. Letter to the Civil Commissioner, Pretoria. Skukuza Archives.

Gregor, E.B. 1946. Personal communication.

Grobler, J.C.H. 1988. *Politieke leier of meeloper? Die lewe van Piet Grobler*. Scripta Africana, Johannesburg.

Grobler, J.C.H. 1988. Piet Grobler – vriend van natuurbewaring. *Custos* 17(5):34–37.

Knobel, R. 1971. Oom Paul Kruger die natuurbewaarder. *Nuus uit ons Parke* 11(3):3–6.

Labuschagne, R.J. 1958. *60 Years Kruger Park*. National Parks Board of Trustees, Pretoria.

Labuschagne, R.J. 1968. *The Kruger Park and Other National Parks*. Da Gama, Johannesburg.

Labuschagne, R.J. 1968. Krugerwildtuin 70 jaar. *Nuus uit ons Parke* 8(10):8–16.

Pienaar, U. de V. 1973. The Kruger National Park: 1946–1973. Unpublished manuscript.

Pienaar, U. de V. 1981. The Kruger Park Saga. Unpublished manuscript.

Pienaar, U. de V. 1986. The National Parks Board of Trustees 1926–1986. *Custos* 15(6):12–13 *(et seq)*.

Pringle, J.A. 1982. *The Conservationists and the Killers*. Books of Africa, Cape Town.

Stevenson-Hamilton, J. 1902–1946. Annual reports of the warden of the Sabie and Shingwedzi reserves and the later Kruger National Park. Skukuza Archives.

Stevenson-Hamilton, J. 1974. *South African Eden*. Collins, London.

Suid-Afrikaanse biografiese woordeboek part 3:120–121. 1972. Insert on, Caldecott, Harry Stratford. Tafelberg, Cape Town & Johannesburg.

Suid-Afrikaanse biografiese woordeboek part 2:367–368. 1972. Insert on, Grober, Pieter Gert Wessel. Tafelberg, Cape Town & Johannesburg.

Tattersall, D. 1972. *Skukuza*. Tafelberg, Cape Town.

Wolhuter, H. 1948. *Memories of a Game Ranger*. Wild Life Protection Society of South Africa, Johannesburg.

Chapter 22
Memorials and memorial plaques
Dr U. de V. Pienaar

In Jeremiah 31: 21 of the Old Testament we are instructed to: "Set thee up waymarks, make thee high heaps: set thine heart toward the highway, even the way which thou wentest." Seven years after the proclamation of the Kruger National Park in 1926, the National Parks Board started adhering to this instruction and began erecting and putting up memorials and memorial plaques not only to commemorate certain historical events, donations by benefactors, graves of well-known people and other items of cultural historical importance, but also to bring it under the attention of visitors.

It was only fitting that the first memorial plaques put up by the Board in the Kruger National Park and unveiled by Minister Piet Grobler and Mrs Hilda Stevenson-Hamilton on 22 September 1933 pay tribute to the roles that President Paul Kruger and Minister Piet Grobler had played in the creation of the Kruger National Park.

Since then, about 80 memorials or memorial plaques have been put up and erected at suitable terrains. Many of these are accessible to tourists to the Park. Each place of historical importance will be provided with a memorial plaque and the Park also wants to make those near tourist roads accessible to the public. This is also an excellent way to further enrich visitors' stay in the Park. Apart from the historical memorial plaques, a whole series of plaques have been erected at places where artificial watering points had been created through funding provided by donors. As these plaques are not of historical importance, they are not discussed here.

Following is a list of memorials and memorial plaques in the Kruger National Park and an indication of their location. The number in brackets refers to the number of the historical terrain on the map in the back of the book, titled: "Historical terrains in the Kruger National Park" and in the table with grid references on p. 728.

1. About nine kilometres northwest of the confluence of the Luvuvhu and Limpopo rivers are the graves of William Pye and Hendrik Hartman. They were two of the colourful characters from the days of Crooks Corner (3). The wording on the bilingual inscription on the plaque reads: "This is the grave of William Pye, an Irishman who started the Makuleke store in Crooks Corner in 1910 with Alex Thompson. He died on 20 November 1918 of influenza and was buried by D.T. Buchanan. The grave next to it is that of Hendrik Hartman who died on 10 October 1918 of heat stroke."

2. In August 1986, Izak Barnard, the son of legendary elephant hunter, Cecil Bvekenya Barnard, erected a memorial at Crooks Corner in remembrance of his father. This memorial is eight kilometres northwest of the confluence of the Luvuvhu and Limpopo rivers at Crooks Corner close to where Bvekenya had his camp (248). There are two plaques on the memorial. The English wording on the bilingual plaque reads as follows: "In memory of Bvekenya and his wild companions who followed the Ivory Trail. 19 September 1886–2 June 1862. Unveiled by his son Izak Barnard on 19 September 1986." This

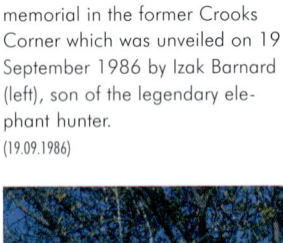

The Bvekenya Barnard Ivory Trail memorial in the former Crooks Corner which was unveiled on 19 September 1986 by Izak Barnard (left), son of the legendary elephant hunter.
(19.09.1986)

site is accessible to visitors.

3. The site where Alex Thompson and William Pye built their later well-known Makuleke shop in 1910 on top of a hill is about one kilometre northwest of the current ranger post at Pafuri (7). This shop used to be the general gathering place during the days of Crooks Corner. The shop became the property of the well-known W. Borchers in 1912 and was managed until 1919 by D.T. 'Buck' Buchanan. After that it was sold to Morty Ash, who had traded there until the outbreak of the Second World War when he sold it to Sam Majika. In the mid-1950s it was taken over by John Fernandez, who demolished it and used the material to build two other little shops, one near the current airstrip and another at Gwalali. The wording on the memorial plaque reads as follows: "The remains of W. Borchers's store, which was established in the days of Crooks Corner and did business from 1919–1954."

4. Along the northwestern flank of the airstrip at Pafuri are the ruins of one of John Fernandez's little stores (8). A bilingual plaque was erected there and the English wording reads: "The ruins of John Fernandez's shop (1954–1969)."

5. Some fence poles and other remains of the former Pafuri tented camp can still be seen at the picnic site at Pafuri (10). The inscription on the plaque here reads as follows: "The site of the old tent camp at Pafuri. This camp was built by ex-ranger W. Lamont and was in use from 1939 to1941. It was closed during the Second World War and finally abandoned in 1948 because of administrative and other problems." The site is accessible to tourists.

6. About two and a half kilometres west of the WNLA post at Pafuri, close to the tourist road on the southern side of the river to the confluence of the Luvuvhu and the Limpopo rivers, are the ruins of the well-known shop W. Borchers that was run here (16). The inscription on the plaque reads: "The remains of W. Borchers's store, which was established in the days of Crooks Corner and did business from 1919 to 1954." The site is accessible to tourists.

7. Baobab Hill with its well-known baobab tree is situated along the road between Klopperfontein and Pafuri. The WNLA established an overnight post for black recruits here many years ago (19). The inscription on the plaque reads: "'Baobab Hill', where black workers recruited in Mozambique for the Rand gold mines by Paul Neergaard of the WNLA organisation during the period 1919 to 1927 and who were transported on foot, by donkey cart, or ox-wagon between Pafuri and the railhead at Soekmekaar, made their first outspan" (19). The plaque is accessible to tourists. On the southern side of the road at this site grows another baobab tree on which the name of an unknown hunter (possibly Ebersohn) had been carved into the trunk.

8. Just east of the confluence of the Madzaringwe and Luvuvhu rivers, three kilometres northwest of the Nyalaland trails camp, are the ruins of the Vha-Lembethu Chief Makahane's village on a high hill right next to the Luvuvhu River (21). The wording on the memorial plaque erected here, reads: "The Makahane Ruin was named after Makahane the Cruel, who was eventually murdered and buried here. He was chief of the Vha-Lembethu, the forerunners of the Vha-Ngona and the Venda in the Northern Transvaal. The ruin with its built-in seat, corresponds with the Dzata, Verdun and Melrose ruins in the Soutpansberg." The site is accessible to Nyalaland trailists.

9. About half a kilometre east of Klopperfontein Dam is an old drift through the Klopperfontein

Memorial plaque against a section of the rebuilt stonewall of the Makahane Ruins at Madzaringwe mouth.
(27.04.1984)

Memorial plaque at the baobab tree that was planted on 4 February 1931 by ranger I.J. Botha at the Punda Maria ranger quarters. (27.04.1984)

valley which was constructed by the legendary Bvekenya Barnard during the time he had been in service of the WNLA (circa 1920) (27). The English inscription on the memorial plaque here reads as follows: "This old drift was constructed by Cecil 'Bvekenya' Barnard of 'Ivory Trail' fame (circa 1920)."

10. About half a kilometre north of the well-known Malonga Fountain is an isiVivan stone cairn (249). The wording on the memorial plaque reads as follows: "An 'isiVivan' cairn along the centuries-old trade route from Soutpansberg to Sofala and Inhambane along the Mozambican coast." Accessible to visitors.

11. Just southwest of Malonga Fountain is a very deep prospecting pit which was dug during the 1920s by a certain C. Gerber from Louis Trichardt while unsuccessfully prospecting for diamonds. The English wording on the bilingual plaque reads as follows: "This old shaft was dug by one C. Gerber, who prospected for diamonds here during the period 1926–27."

12. In the main building of Punda Maria rest camp is a beautiful bronze memorial plaque which was put up during the 1950s by the Historical Monuments Commission to commemorate the expedition of General A.H. Potgieter and J.G.S. Bronkhorst to Sofala during the winter of 1836 (37). The wording on the memorial plaque erected here, reads: "Bronkhorst-Potgieter Journey. In the winter months of 1836 a Voortrekker expedition from the Sand River in the Orange Free State passed this way exploring a road to Sofala (Beira). During the same winter a number of horsemen from the party again passed through this area in search of the trek of Johannes Janse van Rensburg. The participants in these epic expeditions were: General A.H. Potgieter, J.G.S. Bronkhorst, R. Jansen, L. van Vuuren, S. Cilliers, A. Swanepoel, J. Robbertse, A. de Lange, D. Opperman, H. Niewenhuizen, C. Liebenberg, Louis and Carolus Trichardt." Accessible to visitors.

13. Also on the terrain of Punda Maria rest camp is a bronze memorial plaque erected by the Board in honour of the donors of the "Water for game" campaign of 1934–35. Similar plaques were also put up at other entrance gates and camps in the Park, namely Malopene (Malopeni) (later moved to Phalaborwa Gate), Rabelais (later moved to Satara), Numbi Gate, Malelane camp and Crocodile Bridge. The English inscription on the memorial plaque reads: "National Parks Board of Trustees. Boreholes have been provided at the following places with funds donated by:

Pretoriuskop	Mrs Armour Hall
Ship Mountain	Port Elizabeth and Benoni
Gomondwane	Pretoria
Randspruit	Mrs Eileen Orpen
Hlamamaduba	Durban
Manzimahle	Lourenço Marques
Gudzane	Johannesburg
Bangwe	Union Government
Marameyo	Johannesburg Stock Exchange
Rabelais	Mrs R.A. de Villiers
Malopene	Natal Sugar Millers Association
Ngwenyeni	Union Government
Malopene	Cape Peninsula
North Shivhulani	Cape Peninsula"

Accessible to visitors.

14. The late former ranger I.J. Botha planted a baobab seedling in the garden of the ranger quarters at Punda (38). The wording on the memorial plaque at the tree reads: "Baobab planted by the late ranger I.J. Botha as seedling on 4.2.31." Not accessible to visitors.

15. At the foot of Dzundwini Hill along the Dzundwini loop is a fountain where ranger J.J. Coetser had a temporary camp before moving to his permanent post at Punda Maria (41). After that he utilised the site as an outpost. The wording on the historical plaque erected here reads: "The temporary camping site (and subsequent outpost)

of former ranger J.J. Coetser, before he moved to permanent quarters at Punda Maria in 1919." Accessible to the public.

16. According to legend, a treasure had been buried about one and a half kilometres north of the Bububu River and west of the Larine firebreak during the Anglo-Boer War (51). The wording on the plaque erected here reads: "Larine – according to legend a treasure had been buried here during the Anglo-Boer War."

17. About 22 kilometres west of Shingwedzi rest camp, a Karoo sandstone ridge cuts through the Shingwedzi River. The river has scoured out a magnificent number of potholes in the soft sandstone, similar to those at Bourke's Luck in the Blyde River. Apparently some old prospectors discovered alluvial gold in these potholes (57). The wording on the memorial plaque here reads: "'Red Rocks', where the Gubyane sandstone reef is pitted with deep potholes, which were scoured out through the ages by the waters of the Shingwedzi River. According to legend an American prospector from Texas, Bill Lusk, visited this site regularly during the period 1916–1920 to pan for alluvial gold." Accessible to visitors.

18. At the foot of Tshange (Tshanga) Hill outlook, west of Shingwedzi, a small concrete weir was built in the Tshange (Tshanga) Spruit in 1952 with funding donated by former board member T.B. Bowker (58). The wording of the English memorial plaque here, reads as follows: "This dam was donated by Mr. T.B. Bowker, M.P."

19. Along the northern bank of the Shingwedzi River, close to the northern flank of Kanniedood Dam, is a sandstone memorial at the place where the ashes of former Parks Board member Bran Key were strewn in 1949. The memorial stone was chopped out of local sandstone by former ranger W. Lamont (59). The wording on the stone reads as follows: "Bran Kay lives on here. 17.4.49."

20. At the far eastern point of the Shingwedzi River road, about 20 kilometres from Shingwedzi, is a concrete dip tank which was erected during the great foot-and-mouth disease outbreak of 1938. People who crossed the boundary from Mozambique had to dip their feet into a disinfectant in the dip (60). The wording on the memorial plaque reads as follows: "'Dipene.' This old tank was used for several years from 1938 as a footbath to 'disinfect' illegal immigrants from Mozambique ('Mafourteens') and other persons crossing the border against infectious diseases such as foot-and-mouth."

21. Along the Tsende (Tsendze) River, about one and a half kilometres northwest of Grysbok windmill and the mouth of the Mabodlheleni Spruit is the terrain of the former Tsende (Tsendze) tented camp which provided accommodation to tourists during 1933 (64). The English wording on the bilingual memorial plaque is: "The site of the old tent camp along the Tsende River. It was used for only one season during April–October 1933. Mr Wykerd was the camp superintendent and the ranger in charge was H.E. Tomlinson." Accessible to visitors.

22. Along the northern bank of the Shongololo Spruit, a tributary of the Tsende River, about ten kilometres northwest of Pioneer Dam and on top of a hill are the ruins of Major A.A. Fraser's ranger post at Malunzane. It had originally been the post of a WNLA recruiting agent before Major Fraser moved in there in 1904 (68). The wording on the memorial plaque erected here, reads: "The ruins of 'Malunzane' – former ranger Major A.A. Fraser's quarters from 1904 until 1920."

23. The Tropic of Capricorn parallel of latitude where it crosses the main road between Shingwedzi and Letaba about five kilometres northwest of Bowkers Hill (23° 27' 9") (66) is marked by a stone cairn and suitable inscriptions in both Afrikaans and English: "Tropic of Capricorn". Accessible to the public.

The signpost which indicates the position of the Tropic of Capricorn along the main Letaba-Shingwedzi road.
(17.07.1984)

24. Next to a large baobab tree (there were two, but the one had been struck by lightning during the late 1980s and died) at the foot of Bowkerskop is the site of an old hunters camp. This site was used by the Bowker and Barber families in 1888 (69). During the early 1950s, the Bowker family erected a marble memorial with the following English inscription: "Near this Baobab tree a party of hunters from the Rand had their camp in 1888. Their names were Miles Robert Bowker, whose name is carved on this tree, Alec Bowker and Charlie White, his cousins, Fred and Harry Barber, the founders of Barberton, and Miller with his ox-wagon, who provided the transport. M.R. Bowker. July 10, 1888."

25. Just below the wall of Stapelkop Dam and close to the windmill with the same name, lies the grave of a certain Joubert who had been buried here during the second half of the nineteenth century. He was one of the pioneer hunters and possibly an acquaintance of João Albasini. However, according to former ranger J. Meyer of Shangoni, it is the grave of a Joubert boy from Sabie who had camped there before dying of malaria in the years before the Anglo-Boer War (79).

    The wording on the memorial plaque here reads as follows: "Legend had it that a certain Joubert was buried here during the second half of the nineteenth century. He was probably one of the pioneer hunters of the area and may have been an acquaintance of João Albasini."

26. About four kilometres south of Makhadzi windmill and about 300 to 400 metres west of the Makhadzi Spruit, is the site where the infamous British regiment, Steinaecker's Horse, had their northernmost outpost during the Anglo-Boer War (89). The English wording on the bilingual memorial plate reads as follows: "On this site the infamous Steinaecker's Horse had their northernmost outpost during the period 1901–1902 of the Anglo-Boer War."

27. Along the western boundary and about 11 kilometres north of Phalaborwa Gate is the site of the old Malopene (now Malopeni) entrance gate (98). The English wording on the bilingual memorial plaque reads: "The site of the old Malopene entrance gate and camp (1926–1961)."

28. Northeast of Letaba rest camp, about one kilometre from the northern bank of the Letaba River on a gravelly elevation, are the ruins of former ranger L.H. Ledeboer's second ranger post, Mondzwani. Close to the ruins is a small grave which could be that of a child (96). The wording on the memorial plaque erected here reads: "'Mondzwani' – the site of ranger L.H. Ledeboer's ranger post 1921–1927."

29. Just west of Makhadzi Mouth along the northern bank of the Letaba River is the site of former ranger L.H. Ledeboer's first ranger post, Hatane, and his second wife, Anna Maria Christina's grave. She died on 10 August 1921 (97). The English inscription on the memorial plaque here reads: "This is the grave of Anna Maria Christina Ledeboer (née Bindemann). She was the second wife of former ranger L.H. Ledeboer and died here on the 10th August 1921, aged 55." Accessible to visitors.

30. About seven kilometres east of Letaba rest camp is the Engelhard Dam which had been the largest dam in the Park for many years. It was built in 1970 with money donated to the Board by an American financier, Charles Engelhard. A memorial plaque was erected on the southern flank of the dam wall and the wording on the plaque reads: "Engelhard Dam. Charles William Engelhard loved this park and in his concern for the preservation of wild life he donated the funds to build this dam. Dedicated to the memory of his wife and daughters on 2 August 1971." Accessible to visitors.

31. Just west of where the Letaba River road and the road to Olifants camp join, is a large baobab tree on which the surveyor-general of the ZAR (South

Memorial plaque at Engelhard Dam in the Letaba River, east of Letaba rest camp. (27.10.1984)

African Republic), G.R. von Wielligh, carved his name in 1891 (105) (please do not follow his example). The wording on the memorial plaque erected here, reads: "Surveyor and writer G.R. von Wielligh, who surveyed the boundary between the old Transvaal Republic and Mozambique during 1890 with surveyors Vos and Machado, carved his name on the stem of this tree in 1891." Accessible to visitors.

32. About half a kilometre west of the confluence of the Letaba and Olifants rivers, on the southern bank of the Olifants River is the site where Gorge rest camp provided accommodation to visitors from 1930 to 1960 (111). The wording on the memorial plaque erected here, reads: "The site of the Old Gorge rest camp which served as rest camp from 1930 until 1960. Thereafter it was used by visitors as a picnic site until 1979." Accessible to hikers on the Olifants River hiking trail.

33. Between Satara rest camp and the ranger quarters at Satara is the grave of former ranger W.W. Lloyd under a couple of umbrella thorn trees (123). The English epitaph on the head stone of the grave is the following: "W.W. Lloyd. Ranger – Sabie Game Reserve 1919–1922. Died of pneumonia on patrol. Aged 56 years. R.I.P."

34. In 1955 the Wild Life Protection Society erected a bird bath and memorial stone in the bottom part of Orpen rest camp in memory of James and Eileen Orpen (126). The English inscription on the granite slab reads as follows: "Erected by members of the Wild Life Protection Society of South Africa and unveiled by H.H. Gilfillan, president, to commemorate the outstanding work performed by the late Mr and Mrs J.H. Orpen in the interest of wild life preservation. 24 September 1955." Accessible to visitors.

35. Just over eight kilometres from Orpen gate and rest camp along the dirt road to Satara is the site of the former Rabelais entrance gate to the Park. It was used from 1926 until 1954 (128). The English wording on the bilingual memorial plaque erected here, reads as follows: "The site of the old Rabelais entrance gate (1926–1954)." Accessible to visitors.

36. About two kilometres northwest of where the tar road between Tshokwane and Satara joins the upper northern leg of the Vutomi road, grows the southernmost baobab in the Kruger National Park (135). The English wording on the bilingual memorial plaque here reads: "This is the southernmost Baobab tree, in its natural state in the Republic of South Africa." Accessible to visitors.

37. Where the main road between Tshokwane and Satara joins the turnoff to Lindanda and Nwanetsi is a memorial plaque which had been donated to the Park in 1938 by the Historical Monuments Commission to commemorate Voortrekker Louis Trichardt's epic journey through the Park area in March 1838 (137).

The Orpen memorial in Orpen rest camp which has been donated by the Wild Life Protection Society of South Africa.
(17.07.1984)

The southernmost baobab tree in southern Africa about 2 kilometres northwest of the Tshokwane-Vutomi crossroad.
(18.07.1984)

22. MEMORIALS AND MEMORIAL PLAQUES

One of the three memorial plaques at Lindanda, where ranger Harry Wolhuter had his gruesome experience with a lion in August 1903.
(18.07.1984)

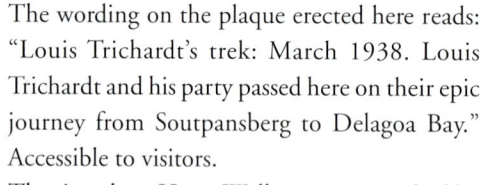

Ranger Harry Wolhuter explains to bystanders where he tied himself up in a fork of the "Lindanda" tree after his confrontation with a lion in August 1903. September 1937.
(ACKNOWLEDGEMENT: STEVENSON-HAMILTON COLLECTION, SKUKUZA ARCHIVES)

Ranger H.C. Wolhuter, Mr Vinnie Robertson, Col J. Stevenson-Hamilton and ranger L.B. Steyn at the "Lindanda" tree.
(ACKNOWLEDGEMENT: STEVENSON-HAMILTON COLLECTION, SKUKUZA ARCHIVES)

The wording on the plaque erected here reads: "Louis Trichardt's trek: March 1938. Louis Trichardt and his party passed here on their epic journey from Soutpansberg to Delagoa Bay." Accessible to visitors.

38. The site where Harry Wolhuter was attacked by a lion in August 1902 and managed to kill the predator with a hunting knife is next to the Lindanda bypass road, about 15 kilometres northeast of Tshokwane picnic site (138). Three memorial plaques were erected here in 1937, with funding which had been donated by V.L. Robertson of Amersfoort. The English wording on the bilingual plaques reads as follows:

    i. "At this spot while alone at night, ranger H.C. Wolhuter, Sabie Game Reserve, was attacked by a lion on 26 August 1903."

    ii. "At this spot ranger Wolhuter, without assistance, killed a lion with a hunting knife after the lion had dragged him here."

    iii. "In this tree ranger H.C. Wolhuter, badly wounded, climbed after having killed the first lion, in order to avoid being attacked by a second close by until assistance arrived."

The tree has perished in the meantime but a piece of it is displayed in the Stevenson-Hamilton Memorial Library. Accessible to tourists.

39. About three kilometres east of the Nwamuriwa outlook and against the eastern flank of the hill, are the trenches and site of the border post which had been manned by Steinaecker's Horse during the last part of the Anglo-Boer War (139). The English wording on the bilingual plaque reads: "At this site the infamous Steinaecker's Horse regiment had their Metsi-Metsi border post and trenches during 1901–1902 of the Anglo-Boer War."

40. About 22 kilometres northeast of Skukuza and immediately west of the tar road between Skukuza and Tshokwane is the massive granite boulder against which the Eileen Orpen memorial plaques were fixed in October 1944 (142). The English wording on the bilingual plaque reads as follows: "In honour of and in gratitude to Mrs Eileen Orpen for her generous donation of the seven farms Chalons, Kingfisherspruit, Hengel, Sikkeltowkloof, Houtboschrand, Red Gorten and Blackberry Glen to the Kruger National Park. This gift will stand witness of her love for nature and the animal world. Erected by the National Parks Board of Trustees. 4 October 1944." Accessible to visitors.

41. About one and a half kilometres northeast of the Orpen memorial plaque, on the eastern side of the tar road between Skukuza and Tshokwane is a large, loose-standing boulder at the foot of a granite hill. Fixed to this boulder are the first memorial plaques which were erected in the Park on 22 September 1933.

These were unveiled by Minister of Lands, Piet Grobler, and Mrs Hilda Stevenson-Hamil-

22. MEMORIALS AND MEMORIAL PLAQUES

The Orpen-tablet "koppie" along the main Skukuza-Tshokwane main road.
(18.07.1984)

The Orpen memorial tablet against the rock face of the Orpen-tablet "koppie".
(25.09.1984)

Board members, officials, rangers and dignitaries at the unveiling of the Eileen Orpen memorial plaque on 4 October 1944. Mr and Mrs J.H. Orpen are standing in the centre at the back.
(ACKNOWLEDGEMENT: STEVENSON-HAMILTON COLLECTION, SKUKUZA ARCHIVES)

709

22. MEMORIALS AND MEMORIAL PLAQUES

ton, to commemorate the proclamation of the Sabie Game Reserve by President Paul Kruger in 1898 and the promulgation of the National Parks Act in 1926 by the same minister. The wording on the two plaques reads as follows:

The terrain where the Kruger-Grobler memorial plaques were embedded in a loose-standing boulder at the foot of a hill between Tshokwane and Skukuza in 1933.
(ACKNOWLEDGEMENT: STEVENSON-HAMILTON COLLECTION, SKUKUZA ARCHIVES)

The unveiling of the Kruger-Grobler memorial plaques on 22 September 1933 by His Honourable Minister P.G.W. Grobler, Senator W.J.C. Brebner (chairperson of the Board) and Mrs Hilda Stevenson-Hamilton.
(ACKNOWLEDGEMENT: STEVENSON-HAMILTON COLLECTION, SKUKUZA ARCHIVES)

> Sabie Game Reserve
> Proclaimed by President Kruger in 1898
> National Parks Act
> Submitted by P.G.W. Grobler, Minister of Lands in 1926.

Accessible to visitors.

42. About five kilometres northwest of Skukuza, along the northern bank of the Sabie River are the ruins of the Msutlu ranger post. This post had been manned by ranger Paul Siewert from 1911 until 1914 (148). The English wording on the bilingual memorial plaque reads: "The site of former ranger Paul Siewert's quarters from 1911–1914 at the old Msutlu outpost."

43. Just less than two kilometres southwest of the Nwatindlopfu drift and north of the Salitji road are indications of old prospecting pits and graves dating back to the period 1879 until 1880 (150). The English inscription on the memorial plaque erected here, reads: "Indications at this site point to the fact that a group of prospectors prospected

22. MEMORIALS AND MEMORIAL PLAQUES

Guests present at the unveiling of the Kruger-Grobler memorial plaques on 22 September 1933.
(ACKNOWLEDGEMENT: STEVENSON-HAMILTON COLLECTION, SKUKUZA ARCHIVES)

The Kruger-Grobler boulder along the Skukuza-Tshokwane road with the memorial plaques. From a historical point of view it is the most important memorial terrain in the Park.
(28.09.1984)

The Kruger-Grobler memorial plaques along the main Skukuza-Tshokwane road are the oldest memorial plaques in the Park.
(28.09.1984)

711

22. MEMORIALS AND MEMORIAL PLAQUES

for gold and/or diamonds in this area during the period 1879–80." Accessible to visitors.

44. About half a kilometre west of the Phabeni mouth and a few hundred metres south of the Sabie River are the ruins of João Albasini's homestead and trading post where he lived between 1846 and 1848. The ruins of this trading post was partly restored in 1979 and an information hut erected. The hut contains information in both Afrikaans and English, as well as some of the artefacts found at the site (155). The site can be reached by vehicle via the Doispane road between Pretoriuskop and Skukuza.

The inscription on the memorial stone at the statue of Paul Kruger at Paul Kruger Gate.
(JULY 1986)

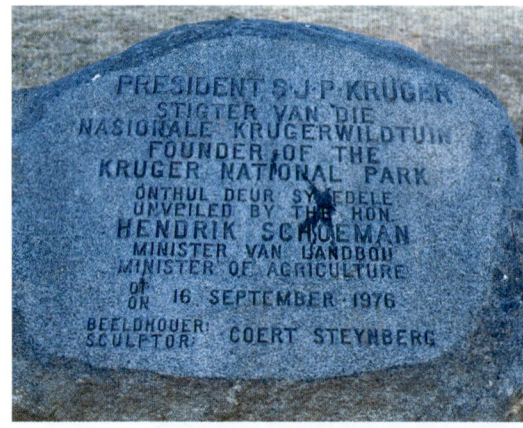

The statue of Paul Kruger which was erected in September 1970 at Paul Kruger Gate.
(07.05.1984)

45. Just west of the entrance road between the bridge over the Sabie River and Paul Kruger Gate is the Paul Kruger monument which was erected in honour of the founder of the Sabie Reserve, later Kruger National Park. The piece of "Douglas Grey" granite out of which the bust was sculptured by celebrated artist, Coert Steynberg, had come from the Park itself. It had originally weighed 36 tons, while the finished work weighed about 24 tons and had to be placed on its pedestal with a crane. Next to the sculpture is a granite block with the following inscription:

President S.J.P. Kruger

Stigter van die Nasionale Krugerwildtuin.
Founder of the Kruger National Park.

Onthul deur Sy Edele Hendrik Schoeman,
Minister van Landbou.
Unveiled by the Hon. Hendrik Schoeman,
Minister of Agriculture

op/on 16 September 1976.
Beeldhouer/Sculptor: Coert Steynberg.

46. A massive leadwood tree trunk was planted on the lawns east of Paul Kruger Gate onto which five memorial plaques, each in a different language, including Afrikaans, English, Tsonga, Swazi and Venda, were fixed. The English plaque reads as follows: "In honour of all those who gave their lives in the cause of nature conservation." Accessible to visitors.

47. When the new main entrance complex in Skukuza was built in 1988, a decision was taken to erect a composite sculpture of life-size impala by sculptor T. Pritchard around a spouting fountain in the open area in front of the buildings. At the same time, a bronze frieze by sculptor Phil Minnaar was fixed against the frontal face of the building's eastern flank to suitably honour the three main founders of the Kruger National Park, President Paul Kruger, Minister Piet Grobler and Colonel James Stevenson-Hamilton. The new building complex and composite sculptures were unveiled by the Chief Director of the National

22. MEMORIALS AND MEMORIAL PLAQUES

Parks Board in March 1989. This frieze is accessible to visitors and the wording on the memorial plaque at the frieze of the park's founders reads as follows:

Founders of the Kruger National Park

(i) Stephanus Johannes Paulus (Paul) Kruger, President of the Zuid-Afrikaansche Republiek. (Transvaal) from 1883–1902. Through his far-sighted statesmanship he was largely instrumental in fostering an awareness of nature conservation in this old Boer republic. He had already proclaimed the country's first conservation area – the old Pongola Reserve in 1894. In 1894 he proclaimed the country's first conservation area, the old Pongola Reserve. At the suggestion of two members of the Volksraad, R.K. Loveday and J.L. van Wijk, he proclaimed the Sabie Game Reserve, the nucleus of the present Kruger National Park, on 26 March 1898. In doing so, he left a living monument and earned himself a place of honour as the founder of nature conservation in the Transvaal.

(ii) Pieter Gert Wessel (Piet) Grobler, Minister of Lands of the Union of South Africa, piloted the National Parks Bill through parliament on 31 May 1926. This simultaneously led to the proclamation of the Kruger National Park, the first national park in South Africa. Under his guidance the national parks system was expanded in 1931 by the proclamation of the Kalahari Gemsbok National Park, the Bontebok National Park and the Addo Elephant National Park.

(iii) Lt Col James Stevenson-Hamilton became warden of the Sabie Reserve on 1 July 1902 after the cessation of the Anglo-Boer War, and later also of the Shingwedzi Reserve. He played a leading role in the preservation and management of both reserves, as well as their consolidation and proclamation as the Kruger National Park in 1926, of which he remained warden until his retirement on 30 April 1946.

48. Just west of the Skukuza clock tower and south of the Stevenson-Hamilton Information Centre and auditorium in Skukuza is the Struben guest house which was built in 1937 with funding donated by Frank E.B. Struben of Pretoria. The building was designed by Hilda Stevenson-Hamilton (163). The English inscription on the bronze plaque erected here reads: "These huts are the gift of Frank E.B. Struben of Pretoria. Designed and built by Hilda Stevenson-Hamilton, 1937."

49. The oldest remaining tourist facility in Skukuza camp is the tourist rondavel which was built in 1929 with funding donated by Board member W.A. Campbell of Mount Edgecombe, Natal. This hut is on the eastern side of the camp next to the Monis guest house. It is currently being used as a tourist museum (164). The inscription on the memorial plaque reads: "Campbell Hut museum 1929." Accessible to visitors.

50. Close to and west of the Campbell tourist museum is the site where a pontoon used to trans-

The memorial at Paul Kruger Gate in remembrance of those who have lost their lives serving the cause of nature conservation.
(JULY 1986)

The memorial plate at the Struben guest house in Skukuza.
(07.05.1984)

port tourists across the Sabie River from 1927 until 1935 (164). A notice board with information on and photographs of the old pontoon was put up here. Accessible to visitors.

51. Among the huts east of the shop in Skukuza is the Casa Mia hut which was built in 1937 with funds donated by friends of 'Uncle' Bill Huston of the Casa Mia guest house in Johannesburg and from New South Wales, Australia. A bronze memorial plaque was erected here with the following inscription: "Casa Mia hut. Erected by friends of 'Uncle' Bill Huston, of Redfern, New South Wales and Johannesburg." Accessible to tourists.

52. About halfway between the main entrance gate and the shop-restaurant complex in Skukuza on the left of the main road, is the Papenfus clock tower which was erected with funding donated by the widow of former Board member H.B. Papenfus K.C. It was unveiled on 16 June 1940 by the then chairperson of the Board, Judge N.J. de Wet (165). The English inscription on the memorial plaque reads: "In memory of Herbert Boshoff Papenfus K.C. 1865–1937. He represented Hospital Hill constituency for 14 years. A friend of all animals and birds. He played an important role in the founding of the Kruger National Park in 1926 and was a member of the Board of Trustees from this date until his death." Accessible to visitors.

53. In the entrance hall of the Stevenson-Hamilton Memorial Library and Stevenson-Hamilton Information Centre is a life-size bronze statue of Colonel J. Stevenson-Hamilton, the first warden of the Sabie and Shingwedzi Reserves and the Kruger National Park. The statue was created and unveiled by his wife, Hilda Stevenson-Hamilton, in 1961. The wording on the memorial plaque at the foot of the statue reads as follows:

> J. Stevenson-Hamilton
> Born 2 October 1867
> Died 10 December 1957
> Warden of the Sabie Game Reserve and
> Kruger National Park 1902–1946
> *Finis coronat opus.*

54. About 17 kilometres from Pretoriuskop on the dirt road past Napi Hill is a memorial plaque against the rock face of the hill in honour of former Board member and chairperson of the Board, J.F. Ludorf MEC. It was put up after his death in December 1958 (172). The wording on the memorial plaque reads (translated): "In memory of Joseph Francis Ludorf, who as offi-

The memorial plaque at the W.A. Campbell memorial hut and museum in Skukuza rest camp. (07.05.1984)

The wording on the memorial plaque of the Papenfus clock tower near the entrance gate to Skukuza rest camp. (07.05.1984)

The information board at the terrain from where the pontoon used to operate in Skukuza camp. (07.05.1984)

cial commission member played a part in the founding of this Park in its extended form and who was in service of the Park as member and also as chairperson of the National Parks Board. His mortal remains were buried here in the soil to which he dedicated his life and strength. Born 14.8.1878. Died 21.12.58." Accessible to visitors.

55. Against the face of a giant granite boulder next to Shirimantanga Hill, about 14 kilometres south of Skukuza, is a bronze memorial in honour of the distinguished role which Colonel James Stevenson-Hamilton and his wife, Hilda, have played in the establishment, preservation and management of the Kruger National Park (174). The ashes of both were strewn on top of Shirimantanga Hill by their daughter Mrs Ann Doyle on 10 April 1979. The English part of the bilingual wording reads as follows:

James Stevenson-Hamilton
Hilda Stevenson-Hamilton 'Skukuza'
2.10.1867–10.12.1957
3.2.1902–11.1.1979

In memory of Lieutenant Colonel Stevenson-Hamilton and his wife, Hilda, whose lives were dedicated to the creation and management of the Kruger National Park, and who chose this koppie as their last resting place. 10 April 1979.

Accessible to visitors.

56. About six kilometres south of where the tar road to Berg-en-dal join the tar road between Skukuza and Pretoriuskop, is a leadwood tree which is about 150-years-old, with deep blaze marks on its trunk. The marks on this tree indicate the Old Wagon Route from Ohrigstad to Delagoa Bay, which was surveyed in 1845 by the Voortrekker, Karel Trichardt (175). The English wording of the bilingual inscription on the plaque is: "One of the old trade routes from the coast to the hinterland, and which was surveyed by the Voortrekker, Carolus Trichardt during October 1845 followed a direction from east to west at this point." Accessible to visitors.

57. About five kilometres from Lower Sabie along the tar road to Crocodile Bridge is the site where the Lowveld pioneer, João Albasini, had a trading post from 1846 until 1848 near Sabie Poort. The remains of this historical site have unfortunately been destroyed during construction of the tar road (178). The English wording of the bilingual inscription reads as follows: "In this area João Albasini, the famous pioneer trader, kept a store during the period 1846–1848, next to the old trade footpath that followed the Sabie River, from his base at Magashula's kraal at Phabeni mouth, eastwards to Delagoa Bay (Maputo). The ruins of this historic little store were unfortunately destroyed in 1969 during the construction of the tarred road between Lower Sabie and Crocodile Bridge."

Bronze statue of Col James Stevenson-Hamilton by his wife, Hilda Stevenson-Hamilton. She also unveiled the statue in the Stevenson-Hamilton Memorial Library in Skukuza rest camp on 15 October 1961.
(SEPTEMBER 1986)

The Stevenson-Hamilton memorial plaque at Shirimantanga Hill where the ashes of both Col James Stevenson-Hamilton, the first warden of the Kruger National Park, and his wife Hilda were strewn on 10 April 1979 by their daughter, Mrs Ann Doyle.

22. MEMORIALS AND MEMORIAL PLAQUES

Bronze statue of two kudu bulls with entangled horns in Skukuza rest camp.
(SEPTEMBER 1986)

The information hut at the Albasini Ruins at Magashulaskraal near the Phabeni mouth.
(16.06.1984)

The bronze sign which commemorates the historical Jock of the Bushveld trek of 1885. Numbi Gate.
(16.06.1984)

58. At Numbi entrance gate to the Park is one of a series of bronze memorial plates which were erected in 1949 along the historical route between Lydenburg and Delagoa Bay by the Percy FitzPatrick Trust. Sir Percy FitzPatrick and his famous dog, "Jock of the Bushveld", followed this route from 1885 until 1888. The bronze plaque has an embossed portrayal of the dog on it with the wording "Jock Trek – 1885" (181). Accessible to visitors.

59. Five kilometres east of Numbi entrance gate and immediately south of the tar road between Numbi Gate and Pretoriuskop is the grave of Willem Pretorius (182). He was one of Karel Trichardt's expedition members from Ohrigstad to Delagoa Bay in 1848. He fell ill with malaria near Ship Mountain and was sent home accompanied by two blacks. At this place his strength gave in and he lay down under a tree where he died. He was buried on the spot by João Albasini. Pretoriuskop was later named after him, although some sources claim that it was named after President M.W. Pretorius. The English part of the bilingual inscription on the memorial plaque which has been erected here, reads: "According to legend Pretoriuskop was named after Willem Pretorius, a member of Karel Trichardt's expedition from Ohrigstad to Delagoa Bay who died here and was buried by João Albasini in 1848."

60. About one and a half kilometres west-northwest of Pretoriuskop rest camp are the ruins of the most western overnight post which the Hungarian entrepreneur, Alois Hugo Nellmapius had erected along the old transport road from the Sabie goldfields to Delagoa Bay. The station master was a Dane, J. Feldschau (184). The English part of the bilingual inscription on the memorial plaque erected here, reads: "This is the site of the most western overnight station erected by Alois Hugo Nellmapius along the old Sabie goldfields – Delagoa Bay transport road." It was managed by a Dane, J. Feldschau during the period 1875–1876."

61. Along the eastern flank of Manungu Hill, about two kilometres west of Pretoriuskop are the ruins of Chief Manungu's trading post and hut. Manungu had been one of João Albasini's indu-

nas and he also watched over Albasini's cattle here during the period 1845 until 1854 (185). The wording on the two memorial plaques erected here, reads:

(i) "Chief Manungu managed a small store (depot) here for João Albasini during the latter's residence at St Luiz (Magashula's kraal) along the Sabie River from 1846 to circa 1853."

(ii) "Here was Manungu's hut. The hole in the floor was probably dug by treasure hunters." Accessible to the public.

62. About nine kilometres southeast of Pretoriuskop rest camp and about one and a half kilometres north of the tourist road from Pretoriuskop to Malelane, is the birthplace (4 April 1885) of Sir Percy FitzPatrick's famous dog, "Jock of the Bushveld", as well as the grave of a young German, Adolf Soltké, who had died following an accident and was buried here in the spring of 1884 (190). The wording on the memorial plaque erected here, reads as follows: "This is the birth place of the famous dog of Sir Percy FitzPatrick – Jock of the Bushveld. The grave at this site is that of a young German, one Adolf Soltké (aged 23 years) who accompanied a group of transport riders in the 1880s. He accidentally shot off his leg with a double-barreled shotgun and died of gangrene despite all efforts to save his life." Accessible to visitors.

63. Along the firebreak from Nhlanganzwani Dam to the Sabie River, about seven kilometres north of the dam and immediately east of the road are the ruins of a store which was run during the 1890s by the Greek, Sardelli, and Charlie Woodlands. The pioneer and hunter, Bill Sanderson, was interned here for a while by Steinaecker's Horse during the Anglo-Boer War (196). The wording on the memorial plaque erected here, reads: "Charlie Woodlands and Sardelli, the Greek, operated a store here in the decade before the Anglo-Boer War. The well-known old hunter, Bill Sanderson, was interned here for a period during the last half of 1901 by Steinaecker's Horse."

64. About four and a half kilometres southwest of Komapiti windmill on the western bank of the Imagoroti Spruit, a tributary of the Komapiti Spruit, is the grave of Swazi Chief Mataffin (200). The wording on the memorial plaque erected here, reads: "Swazi Chief Mataffin's grave. He was murdered here in 1891 by Umgeba, son of the Mambayi Chief, Inhliziyo."

Unveiling of the Jock statue at the Jock of the Bushveld camp by Prof. F. Eloff. 20.09.1985.
(ACKNOWLEDGEMENT: J. VERHOEF)

22. MEMORIALS AND MEMORIAL PLAQUES

65. Along the tourist road from Pretoriuskop to Malelane, about eight kilometres southeast of Ship Mountain, are the ruins of Thomas Hart's overnight post and grave on the western bank of the Josekhule Spruit (202). Thomas Hart was one of Nellmapius's station masters on the old carrier route between the Sabie goldfields and Delagoa Bay. He had been murdered here on 10 August 1876 by a band of black robbers and his remains were buried close to his trading post by Billy Barter, the bailiff and justice of peace of Pilgrim's Rest. The English inscription on the bronze plaque erected here, reads as follows: "A man by the name of Thomas Hart established an overnight station here at Josekhulu Spruit for black carriers who carried goods from Delagoa Bay to Pretoriuskop prior to 1876. He was murdered here on 10 August 1876 by a gang of black robbers and buried near his trading post." Accessible to visitors.

66. About seven and a half kilometres northwest of Afsaal picnic site on the tourist road between Malelane and Pretoriuskop is the site where the first concrete dam in the Park was constructed in 1931–1932 by Captain M. Rowland-Jones in the Ntomeni Spruit (203). The wording on the memorial plaque erected here, reads as follows: "The first concrete dam in the Kruger National Park was built here, in the Ntomeni Spruit by the camp supervisor of Pretoriuskop rest camp, M. Rowland-Jones, during 1931–32." Accessible to visitors.

67. Close to Voortrekker windmill, about six and a half kilometres from Afsaal picnic spot on the tourist road from Malelane to Pretoriuskop, there apparently used to be a popular outspan during the era of the transport riders in the 1880s. The English inscription on the bronze plaque erected here, reads: "This was a favoured outspan site of the old transport riders during the 1880s. They used the large leadwood tree for target practice." Accessible to visitors.

68. Close to where the Randspruit road between Skukuza and Crocodile Bridge joins the Crocodile Bridge and Lower Sabie road at Gomondwane, used to be a number of bluegum trees which had been planted in the 1890s by Sardelli the Greek and Charlie Woodlands at the store which they ran here (206). The English part of the bilingual inscription on the memorial plaque erected here, reads as follows: "These bluegum trees were planted here towards the end of the last century by a certain Charlie Woodlands and a colourful rogue named 'Sardelli the Greek' who established a trading post at this site. The post was abandoned at the outbreak of the Anglo-Boer War, when Sardelli joined the ill-famed Steinaecker's Horse." Accessible to visitors.

69. Along the tar road between Lower Sabie and Crocodile Bridge, about two and a half kilometres northeast of Gomondwane, is the site where François de Cuiper's expedition (the first Europeans to have set foot in the Lowveld), was attacked by Chief Dawano's warriors in June 1725 and forced back across the Lebombo Mountains to the safety of Fort Lijdsaamheid in Delagoa Bay (207). The English inscription on the memorial plaque erected here, reads: "In this area Frans de Cuiper and his 30 men were suddenly attacked on 12 July 1725 by impis of Chief Dawano. This forced the expedition to recross the Lebombo Mountain range and return to the V.O.C. Fort Lÿdsaamheid at Lourenço Marques." Accessible to visitors.

70. During 1870 a memorial plaque was unveiled at the Mpemane borehole along the Vurhami Spruit between Gomondwane and Crocodile Bridge. This borehole and windmill was erected with funds collected by the Technical High School John Vorster and is accessible to visitors. The wording on the memorial plaque erected here, reads:

> Water Year 1970
> This borehole and drinking place has been made possible by a donation from the pupils of the Technical High School John Vorster.
> Pretoria 24.10.70.

71. About three kilometres northwest of where the old Skukuza-Malelane road joins the Malelane-Pretoriuskop road, and a few hundred metres north of the last-mentioned road is the site where John Glen-Leary, the father-in-law of ranger

Harold Trollope, was fatally injured by a wounded leopard on 3 August 1926 (211). The wording on the memorial plaque erected here reads: "Mr John Glen-Leary, father-in-law of former ranger S.H. Trollope, was fatally injured here by a wounded leopard on 3 August 1926." Accessible to tourists.

72. The grave of the hunter and blacksmith Gert Frederik Coenraad Stols, after whom Stolsnek ranger section was named, is about six kilometres west of Stolsnek ranger station and about a half kilometre north of where the Stolsnek firebreak joins the old firebreak to Ship Mountain (212). The wording on the memorial plaque erected here, reads as follows: "The grave of hunter and blacksmith, Gert Frederik Coenraad Stols after whom Stolsnek was named. He died here on 10 August 1886 of malaria. His eldest son, Gabriël Stols, was a scout of General Ben Viljoen and involved in the attack on the ill-famed Col Steinaecker's fort on the farm New Forest in 1901."

73. About 400 metres west of the convergence of the old tourist road between Skukuza and Malelane and the Malelane-Pretoriuskop road is the concealed water point and outspan known as Fihlamanzi. The outspan and fountain had often been used by transport riders during the period 1875 until 1892 and after that by the first rangers of Malelane section (215). The wording on the memorial plaque erected here, reads: "Fihlamanzi ('hidden water'). In the period 1875–1892 this was a well-known outspan used by the transport riders of the time on the route between Lydenburg and Delagoa Bay. Ranger Harold Trollope camped here on 24 July, 1926. Only a few days later on 3 August 1926, his father-in-law – Mr John Glen-Leary, who accompanied him in the veld – was fatally injured by a leopard along a tributary of the Hlambanyati Spruit, barely three kilometres northwest from this site." Accessible to visitors.

74. About a hundred metres north of the ranger quarters at Malelane is the grave of former camp manager, George Meade, who died and was buried here in 1937 (227). The wording on the bronze memorial plaque erected here, reads: "This is the grave of ex-camp superintendent George Meade. He was camp supervisor of the Malelane rest camp from 1931–1934. He was subsequently transferred to Shingwedzi where he served until 1937, when, for medical reasons he was brought back to Malelane. He was cared for by ranger T.L. James up to the time of his death the same year."

75. In 1952 the family of former ranger S.H. Trollope donated funds to erect a memorial hut in Malelane rest camp (228). The English inscription on the memorial plaque erected at the hut reads: "This cottage was erected by the family in memory of Harold Trollope, 'Vukani', first game warden at Malelane. 1925–1929." Accessible to visitors.

76. About three kilometres southwest of Nellmapius Drift, where the old transport road crossed the Crocodile River, is the site where General Ben Viljoen blew up the Boers' heavy guns to prevent it from falling into the hands of the British forces in September 1900 (229). The wording on the memorial plaque erected here reads: "At this point, on the southern bank of the Crocodile River, the Boer forces under command of General Ben Viljoen, spiked and destroyed their canons on 16 September 1900, lest these fell into the hands of the approaching British army." Accessible to visitors.

77. On the northern bank of the Crocodile River close to the old Nellmapius Drift are the ruins of Alf Roberts's store "Tengamanzi". This shop provided valuable service to transport riders during the years 1884 until 1892 (230). The English inscription on the memorial plaque erected here, reads: "This is the site of the trad-

The memorial plaque at the Harold Trollope memorial hut in Malelane rest camp. (27.07.1984)

ing post 'Tengamanzi' which was operated by Alf Roberts for the benefit of the old transport riders during the eighteen eighties." Accessible to visitors.

78. A few hundred metres southeast of the "Tengamanzi" store's ruins is Nellmapius Drift where the old Lydenburg-Delagoa Bay road crossed the Crocodile River in the years 1874 until 1892 (231). The English inscription on the memorial plaque erected here, reads: "Nellmapius Drift. At this point the old transport road from Lydenburg to Delagoa Bay crossed the Crocodile River. The drift was used during the period 1874–1892. After the completion of the Delagoa Bay railway line in 1892 it fell into disuse." Accessible to visitors.

79. Three kilometres west of Crocodile Bridge rest camp on the eastern side of the old Selati railway line, is the site where the sandstone blocks for both the Crocodile Bridge and Skukuza railway bridges were dressed during the period 1894–1912 (232). The English inscription on the bronze plaque erected here, reads: "From this quarry masons obtained the building stone for the railway bridges at Crocodile Bridge and Skukuza during the period 1894–1912. The grave of A. Hamilton, who died in February 26 1895, can also be seen here." The Piet Grobler Dam and memorial plaque in the Timbavati River northwest of Satara. This concrete dam was completed in 1988. The wording on the memorial plaque erected here, reads:

This structure has been erected in honour of Ex-minister of Lands P.G.W. Grobler. Piet Grobler was the grand nephew of President Paul Kruger and founder of the South African National Park System. He piloted the National Parks Bill through parliament on 31 May 1926 and in doing so established the Kruger National Park.

In later years he was also instrumental in proclaiming the following national parks:
Kalahari Gemsbok National Park 1931
Addo Elephant National Park 1931
Bontebok National Park 1931
Mountain Zebra National Park 1937

References

Agendas and Minutes of National Parks Board of Trustees meetings (1926–1946). Skukuza Archives.

Glen-Leary, H. 1984. Glen Leary's death – new facts. *Custos* 13(2):26–27.

Pienaar, U. de V. & Mostert, M.C. 1983. Nuwe feite bekend oor romantiese transport-era. *Custos* 12(9):27–29.

Pienaar, U. de V. & Mostert, M.C. 1983. Kleurryke karakters deel van Wildtuingeskiedenis. *Custos* 12(10):12–14.

Pienaar, U. de V. 1984. Geskiedkundige fontein in Krugerwildtuin opgespoor. *Custos* 13(7):4–10.

Stevenson-Hamilton, J. 1902–1946. Annual reports of the Warden to the Board. Skukuza Archives.

Stevenson-Hamilton, J. 1974. *South African Eden*. Collins, London.

The author's acknowledgements

We, the authors of this book, would first like to sincerely thank the National Parks Board (now South African National Parks – SANP) for the facilities made available to us. These included the Skukuza Archives, the Stevenson-Hamilton Information Centre, as well as other amenities which made it possible to obtain the necessary background material. We would also like to express gratitude to Professor Fritz Eloff, the Chairperson of the Board, who supported the project throughout and agreed to write the foreword, as well as to Mr A.M. Brynard, a former Chief Director of National Parks, and all the other colleagues who supported us through thick and thin, we also say thank you for your moral support.

Many public institutions, organisations and individuals contributed in many ways to the development of this work. Without their support it would never have seen the light. To all who contributed so unselfishly and with enthusiasm, a sincere word of thanks.

Firstly we would like to mention in particular four of the contributors, namely Dr J.B. De Vaal, Dr J. Carruthers, Mrs M. Liebenberg and Mr A. Minnaar. They spent long, fruitful days in state, provincial and cultural-historical archives, the information and publicity sections of the Transvaal Provincial Administration, as well as the archives of the former Department of Native Affairs and the Surveyor-General. An enormous amount of valuable historical material applicable to the prehistory of the Kruger National Park was found and worked into the text.

We would like to extend a word of sincere appreciation to all these organisations for permission to obtain written, as well as photographic material, to copy it and to give permission to refer to the material in this work, or use it otherwise. Specific acknowledgement is given in the text to photographic material which was obtained through these and other sources.

We would also like to thank one of the co-authors, Mr H.H. Mockford posthumously, who contributed by way of the chapter about the role of WNLA in the Park's history. He died on 9 June 1985. It would be difficult to find a more humble or greater devotee of our country's beautiful natural assets.

We are also sincerely grateful to Mr Jamie Stevenson-Hamilton of Fairholm, Lanarkshire in Scotland, son of the celebrated first warden of the Kruger National Park, Col J. Stevenson-Hamilton, for his kind permission to use and refer to his father's personal documents, photo albums and newspaper clippings, of which a large part is kept in trust of the Stevenson-Hamilton Information Centre. This concession was of invaluable benefit and formed the core around which the rest of the work was developed. We would also like to extend our sincere appreciation to his sister, Mrs Anne Doyle of England for her interest and suggestions.

We are also grateful to Mr Willem Punt, chairperson of the Simon van der Stel Foundation, for granting us the right to include his deceased father's previously published work into this comprehensive history of the Kruger National Park. Dr W.H.J. Punt had been an outstanding historian during his time. We are also grateful for permission to reproduce some of his father's valuable historical maps and photographic material.

The same permission was also obtained from Mr T.V. Bulpin in connection with his wide series of published works about the history of the Transvaal Lowveld and also from Mrs Joan Wolhuter with regards to her deceased father-in-law, Harry Wolhuter's book, *Memories of a Game Ranger*. We also thank the firm Angus & Robertson of London, England in allowing certain photographs in Witton's book *Scapegoats of the Empire* to be used.

Mrs P. Nel, wife of the late Dr T.G. Nel, who had been the first ever research officer to be appointed in the Kruger National Park, donated annual reports and other old documents from her late husband's collection to the Skukuza Archives.

Mr R. Hewitt-Ivy, former agricultural officer of Gazankulu and later of Que-Que in Zimbabwe, contributed information in manuscript form about the

ACKNOWLEDGEMENTS

wild days of Crooks Corner. In the same way we are also posthumously indebted to D. Buchanan for his unpublished manuscript about the characters and episodes of Crooks Corner.

Mr W. (Bill) Glynn, manager of TEBA's recruiting office at Soekmekaar made historical manuscripts about the history of WNLA in the Kruger National Park available to us. He also supplied personal information about events of historical importance which took place at Pafuri.

The late Prof. B.V. Lombaard, who possessed an unequalled treasure of knowledge about the history of the Lowveld, worked with great interest and enthusiasm and cooperated with the collection of relevant historical material about the Anglo-Boer War and the old trade routes through the Lowveld. Herewith we would like to pay tribute to his definitive contribution, as well as his donation of valuable historical documents and maps to the Skukuza Archives.

Others to whom we owe thanks for historical pieces, manuscripts, photographs and maps, which were of great value during the compilation of this work are: Col J.A.B. Sandenbergh (former park warden); Mr J.J. Kloppers, former clerk of the warden in Kruger National Park; Mrs Joan Wolhuter of Sandton; Mrs C. Mackie-Niven of "Amanzi", Uitenhage; Dr Sue Hart of Johannesburg; Ms J.J. King of the Lydenburg Museum, Lydenburg; Mr Theo Owen, former MP for Nelspruit; Mr M.J. Steyn, liaison officer of the National Parks Board, Pretoria; Dr Johann Botha, Assistant-Head: Information of the National Parks Board, Pretoria; Mrs Babs Herrmann, photographic assistant, National Parks Board, Pretoria; Mr P.C. de Swardt, National Parks Board, Pretoria; Dr H.P. Junod, "Sisani", Tzaneen; Dr Ina Plug, Dept. Archaeozoology, Transvaal Museum; Mr H. Hoheisen, Stellenbosch; Mr B.P. Simmons, Karino; Mr D. Coetzee, Nelspruit; Mr H. Bornman, Nelspruit; Mr P.F. Fourie, Nelspruit; Mrs Paula van Rooyen, Pretoria; Professor Murray Schoonraad, University of Natal, Pietermaritzburg; Mr A.R. Willcox, Winterton; Mr Bert Woodhouse, Johannesburg; Mr A.B.C. Daneel, Somerset West; Dr J.A. Pringle, KwaZulu-Natal Museum, Pietermaritzburg; Col. A.F. Malan, Valhalla, Pretoria; Mrs C.J.A. of der Merwe, Winterton; Mrs E.A. Thompson, Haenertsburg; Mr D.A. Groenewald, Pietermaritzburg; Mr Ernest Coetser, Port Edward; Mrs L.E. Kruger, Hoedspruit; Mrs M. Doran, Pretoria; Mrs W.H. Rogers, Pretoria; Mrs F.S. den Bakker, Durban; Mr W.W. Lloyd, Wildlife Society of South Africa, Linden; Mr T.R. Duke, Somerset West; Mr J. Lamont, Newclare; Mrs Nora Pullen, Carletonville; Srg. M.J. Kingshott, British Embassy, Pretoria; Mr Vincent Pohl, Duiwelskloof; Mr A.C. Bands, Pretoria; Mr P. D'Arcy-Herrman, Tzaneen; Dr Arthur Davey, Onrusrivier; Mr P. le S. Milstein, Nature Conservation Division, Transvaal; Dr W.G. Winckler (posthumous) and Mrs Cora Hall, Sabie Forestry Museum, Sabie; Dr C. Hromnik, Rondebosch; Mrs J. Grewar, Phalaborwa; Dr E.B. Gregor, Rondebosch; Mr A. Meyer, Dept. Archaeology, University of Pretoria; Mr Menno Klapwijk, "Silver Leaves", Tzaneen; Mr J.G. Stewart, Melrose; Mr H. Stevenson, Alkmaar; Mr P.R. Kruger, Bergvliet; Mr G. Palmer, Historical Monument Commission, Pretoria; Mr J.H. McDonald, "Hope", Phalaborwa; Mr Glen Trollope, Brakvallei, Thabazimbi; Mr Izak Barnard, Geysdorp; Mr C. Malan, Tshipise; Mr P. Cuenod, Elim Hospital, Limpopo, and Mr J. Scholtemeyer, Phalaborwa.

A huge amount of valuable historical information about the Kruger National Park during the period 1930 to 1960 was personally obtained from the late Mr H.C. van der Veen, former tourist manager of the Kruger National Park and from the late Mr Harry Kirkman, former ranger of Shangoni and later roads' foreman at Skukuza.

The following organisations and persons contributed in the form of historical photographic material which are reproduced in this work and which are individually acknowledged with sincere thanks:

The State Archivist, Pretoria; the Provincial Library and Museum Services, Pretoria; the Cultural Historical Museum, Pretoria; the State Library, Pretoria; the Defence Force Archives, Pretoria; the War Museum, Fort Klapperkop, Pretoria; the Museum of War History, Johannesburg; the Office of the Surveyor-General, Pretoria; the Barberton Museum, Barberton; the Sabie Forestry Museum, Sabie; Mr J. Stevenson-Hamilton, Fairholm, Lanarkshire, Scotland; the late Mrs Hilda Stevenson-Hamilton, "Gibraltar", White River; the Onderstepoort Veterinary Medical Institute; Mr H. Bornman, Nelspruit; Mr

ACKNOWLEDGEMENTS

Izak Barnard, Geysdorp; Mr T.V. Bulpin, Muizenberg; Mr Chamba, Dept. Nature Conservation, Mozambique; the late Mr G.K. Gillet, Robertson; Mr Piet de Vaal, Louis Trichardt; Mr Graham Vickers, Nelspruit; Mr J. Wewege, Nelspruit; Mr C.J. van der Walt, Skukuza; Dr F.C. Coetzee, Pretoria; Mr Piet Eloff of Louis Trichardt; Mr Gerd Borchers, Louis Trichardt; Mr P. de J. Mouton, Bedfordview; Mr John Sykes Ash, Gumbani Trading Store, Gazankulu; Mr W.A. Vergeer, Office of the Surveyor-General, Pretoria; Mr T.R. Duke, Somerset West; Mr J. Lamont, Newclare; Mrs W.H. Rogers, Pretoria; Mrs D. van der Veen, White River; Mrs F.J. den Bakker, Durban; Mr D.A. Groenewald, Pietermaritzburg; Mr A. Crous, Pretoria; Mr E.P.H. Meintjies, Alkmaar; Mr Alf Mösz, Pretoria; Mr T.C. Sinclair-David, Durban; Mr G.J. van Eeden, Tzaneen; Mr J. Thompson, Schagen; Mr and Mrs D.S. Mocke, Pretoria West; Dr A. du Leite Cruz, Portuguese Embassy, Mbabane; Mr D. Tomlinson, Cape Town; Col J.A.B. Sandenbergh, Waterval-Boven; Mr Glen Trollope, Brakvallei, Thabazimbi; Dr R.E. Kay, Glenashley, Natal; Geoff Dalgleish, African Bureau, *Rand Daily Mail;* the late Mrs V.M. Rowland-Jones, Louis Trichardt; Mrs P. Adendorff, Louis Trichardt; Mrs M.E. Barkhuizen, Bloemfontein; Mrs B. Meyer, Durban; Dr E.B. Gregor, Rondebosch; Mr W. Pott, Sabie; the Preller family, Pelindaba; the late Mr Dick Wolff, Pretoria; Mr I. Whyte, Skukuza; Mr J. Verhoef, Skukuza; Dr S.C.J. Joubert, Skukuza; Prof M.C. Botha, Pretoria; Mr L. Wagener, Skukuza; Mr M. English, Stolsnek, Kruger National Park; Mrs M. Doran, Pretoria; Mrs L.E. Kruger, Hoedspruit; Mr D. Paynter, Cape Town; Mr J. van Schaik, Pretoria; Mr C. Malan, Tshipise; Mr J. Scholtemeyer, Phalaborwa; Capt. C.M.L. Pretorius, SA Defence Force Archives, Pretoria; Dr Paul van der Reyden, Pretoria; Mrs Paula van Rooyen, Pretoria; and Mrs A. Briedenhann, Nelspruit. Our sincere thanks are also extended to Mrs Babs Herrmann of the SANP Headquarters in Pretoria as well as the SANP photographers, Messrs W. Massyn, W. de Beer, E. Reisinger and Ms L. Stanton, for the reproduction of a large number of historical photographs.

Ms Magda Geringer, cartographer, was responsible for the series of highly accurate and well finished-off maps used as illustrations. A warm word of thanks to her for her continuous enthusiasm and conscientiousness.

Mrs Lollie de Jager had the unenvied task of typing the original voluminous manuscript. The authors would like to commend her for her continuous interest, commitment and exceptionally accurate typing. It was a soul-destroying task that very few people would have been willing to take on.

Mrs M.M. Whyte contributed greatly to the categoration and arrangement of archival material.

Mrs Helena Bryden did an excellent job with the translation. The linguistic care and editing was undertaken by Ms Danél Hanekom and Mrs Carmen Hansen-Kruger – in itself a colossal and time-consuming assignment.

Dr J. Botha and Mr M.J. Steyn were particularly helpful with the technical aspects of the manuscript and also undertook the important liaison with the printers and the publishers.

Ms Hanli Deysel was responsible for the typesetting of the text of the English edition of the book and also designed the cover.

The Head: Wildlife Management, Mr J.J. Kloppers, compiled a complete list of place names in the Kruger National Park, with explanatory notes where possible.

Mr Mike English, senior ranger at Stolsnek, was responsible for the accurate determination and control of all historical sites in the Kruger National Park as well as the allocation of eight-digit grid references on the 1:50 000 military grid map which will be used by generations to come. The authors are sincerely grateful to him for this task, as well as for his pioneering work in localising and mapping all of the known Bushman (San) paintings in the Kruger National Park and the several publications which stemmed from the project. He also made a valuable contribution to the localising and marking of the old trade routes between Delagoa Bay to Lydenburg and from Lydenburg to Barberton, through the Kruger National Park.

The Board's two pilots, Messrs H. van Niekerk and P. Otto, helped in many ways to locate and photograph historical sites. Mr H. Meyer, senior roads inspector of the Park, was responsible for erecting and maintaining the historical plaques along the different tourist roads and elsewhere.

ACKNOWLEDGEMENTS

Lastly it has to be gratefully acknowledged that this book would have been totally incomplete were it not for the diligent and enthusiastic cooperation of all the rangers and other contributors. They helped to relocate the different historical sites (of which many had already sunk into oblivion) sometimes by means of remarkable tracking or historical detective work.

Here special mention should be made of former ranger, the late M. Mostert's redetermination of the old trade route between Delagoa Bay and Lydenburg along its whole route through the Kruger National Park. He also determined historical sites along this route, like Manungu's picket at Manungu Hill, Fieldschau's overnight post at Pretoriuskop, Inhliziyo's village southeast of Ship Mountain, the birthplace of Jock of the Bushveld and Adolf Soltké's grave along the Samarhole Spruit, the outspans along the Komapiti Spruit and Ntomeni and Thomas Hart's overnight post and grave at Josekhulu Spruit.

Then already mentioned is the contribution of senior ranger Mike English, who localised and carefully mapped at least 110 sites with Bushman (San) paintings, especially in the southern part of the Park. He also determined the corner beacons of the historical farms, Pretoriuskop, Joubertshoop and Ludwich's Lust. He determined the old transport routes from the Makuthwanini Hills over Stolsnek to Barberton and discovered the "treasure well" near the Barberton transport drift through the Nsikazi, the grave of a white person along the Mnyelene Spruit, as well as the graves of two Europeans along the Phugwane, at the confluence between the Zari and Pugwane.

Former ranger A. Espag discovered the ruins of the old rangers post directly opposite Kaapmuiden, as well as the remains of Alf Roberts's little shop, Tengamanzi, at Nellmapius Drift and the "treasure hill" near Mlambane Dam.

The late ranger Harry Wolhuter had already rediscovered the ruins of João Albasini's house and shop at Magashulaskraal in 1903. His son Henry, who succeeded him, was responsible for the discovery of the grave of blacksmith and hunter Stols. The late ranger Jan Meyer of Shangoni, found the "treasure site" at Larine as well as the graves of Joubert along the Shipikane as well as that of Mrs C. Ledeboer at the mouth of the Makhadzi.

Late district ranger Gus Adendorff showed a keen interest in the history of the Kruger National Park. Amongst others, he rediscovered the grave of Willem Pretorius near Pretoriuskop, as well as the Von Wielligh and Briscoe baobabs along the Letaba River and Tsumanene Spruit respectively; the hole where C. Gerber prospected at Malonga; the Makahane, Matjigwili, Masorini and Shikumbu ruins, and the isiVivan (cairn) at Matukwala, west of Punda Maria.

The late ranger Peet van der Merwe of Crocodile Bridge, found the ruins of Sardelli's shops at Gomondwane and north of Nhlanganzwani. Together with the late Dr W. Punt he found De Cuiper's Drift through the Crocodile River and the remains of Manungu's picket near Pretoriuskop. Former ranger A. de Clercq found grape-shots of Long Tom and Armstrong cannons in the Crocodile River east of Crocodile Bridge. Former ranger P. van der Hyde discovered the place where the Boer army had blown up the last remains of their big guns in September 1900, in the Crocodile River directly across Komatipoort Station. Former ranger Van Reenen van Vuuren discovered the old quarries and grave of A. Hamilton near the hippo-pool, close to Crocodile Bridge. Former district ranger D. Swarts discovered the grave inside the cave at Bangu Poort, the old wagon drift at Gorge and the small mica mine of Mrs Stamp along the Olifants River (later rediscovered by ranger M.C. Mostert).

Senior ranger L. Wagener of Skukuza discovered the remains of the old Msutlu ranger's post along the Sabie River northwest of Skukuza. The late chief ranger F.D. Lowe determined the locality of Malunzane, which had been the ranger post of Maj. Fraser along the Shongololo, as well as the Bowker plaque at Bowkerkop, the Bran Key-memorial stone and the Dipene dipping-trough near Shingwedzi Poort.

The late ranger Jan de Kock found the chassis of one of General Ben Viljoen's cannons in the Crocodile River directly opposite Hectorspruit.

Ranger P. Wolff found the bulwork (trench) of Steinaecker's Horse at the eastern flank of Nwamuriwa and senior ranger B.P. Pretorius the Borter baobab tree in Olifants Poort. Ranger T.B. Mollentze found a four-pounder elephant gun in a rock crevice along the upper reaches of the Mbanzwe Spruit, as well as caves and ruins in the Lebombo Mountains, where black people had lived. Senior ranger B.P. Lamprecht determined the terrain of the camp of Steinaecker's Horse along the Makhadzi Spruit, as well as the second ranger post of former ranger L.H. Ledeboer at Mondzwani, which is directly across from Letaba rest camp. The location of Maj. A.A. Fraser's retirement house was found by senior ranger J. Krüger outside the Park, along a tributary of the Klein Letaba near Hlomela's village. He also found the graves of two Europeans along the Nalatsi Spruit, the position of the old tent camp along the Tsende (Tsendze) River and the ochre mine at the northern flank of Ngotso Hill. Ranger A.S. Labuschagne found another old ochre mine along the Shishangani Spruit about one kilometre northwest of Nwanetsi ranger post and also Bushman paintings along the upper reaches of the Makahane Spruit.

Regional ranger Flip Nel determined the localities of William Pye and Hendrik Hartman's graves and Buchanan's baobab tree at Crooks Corner. Artisan H. Sleigh found the terrain of Considine's camp and grave near the Limpopo River. Ranger L. Hare discovered an isiVivan close to Malonga fountain, while ranger Paul Zway obtained valuable old photographs and information about Crooks Corner from different sources outside the Park.

Chief ranger B. Bryden discovered, amongst others, several ruins where black people had lived, also primitive mine shafts at Shilowa Poort and Renosterkoppies. The Head of Inland Parks, Dirk Ackerman, discovered the southernmost baobab tree in the Park, north of Vutomi during his stint as ranger at Tshokwane.

Ranger Louis Olivier discovered the remains of the shop which belonged to Sardelli and Charlie Woodlands at Nkongoma, and regional ranger L. van Rooyen and Senior Information Officer J. Verhoef found the site of the Ndzenge-Dzenge shop near Sabie Poort. Trails ranger S. Mostert discovered Bushman engravings and interesting caves in the sandstone hills east of Nyala Trails Camp.

Ranger James Chauke pointed out the terrain of the bulworks (trenches) of Steinaecker's Horse in the Komatipoort, while ranger Sergeant Charlie Nkuna rediscovered Swazi Chief Mataffin's grave and also helped to relocate the old trade routes in the south of the Park. Former ranger Nombolo Ndluli pointed out the locality of the old Rolle ranger post outside the current boundaries of the Park. Skomela Nkuna, a former road machine operator, pointed out the grave of ranger Andrew Wolhuter at Mtimba, as well as the locality of ranger Harry Wolhuter's picket at Doispane. Pollution control officer G. Oelofsen found former ranger Tomlinson's camp along the Mhulu Spruit.

Security officer Nic de Beer and some black helpers found the grave of Col Stevenson-Hamilton's personal servant, Sharifa Bakhari-Ali, near Metsi Station along the old Selati line. Research Officer Ian Whyte found the initials "AWB 1890" engraved on a rock on a hillside along the Sweni and discovered old pitfalls dug by blacks to capture game along the upper reaches of the Nguweni. He also found the ruins of a black settlement along the Letaba near the mouth of the Tsende (Tsendze) River.

The late H.H. Mockford, former WNLA officer at Pafuri, pointed out the sites where the shops of W. Borchers and John Fernandez had been, as well as that of Alex Thompson at Makhuleke. He also discovered the first Bushman paintings in the area between the Luvuvhu and Limpopo rivers. He was also the first person to rediscover the graves of William Pye and Hendrik Hartman near Pye's camp along the Limpopo.

Mr Pieter Scholtz of the Meat Processing Plant at Skukuza, found signs of old diggings and black graves along the Salitje road east of the Nwatindlopfu drift.

Pilot Piet Otto observed and plotted various new ruins of the native population in the Park. Lastly, we are also indebted to former chief agricultural officer Peet Steyn of Gazankulu, who helped to locate the site of Steinaecker's fort close to Mpisane's village (on the farm New Forest) and the grave of Capt. H.F. Francis. Mr J. Scholtemeyer of Phalaborwa helped to determine the locality of Col Ludwig von Steinaecker's grave on the farm Champagne.

The publication of the Afrikaans version of this work was made possible by kind donations from the Trust Fund of the National Parks of South Africa and Saambou Nasionaal Ltd. The Board, as well as the authors, are sincerely thanked for their financial contributions.

Translator's acknowledgements

I would like to express my sincere gratitude to the late Dr U. de V. Pienaar, the principal author and compiler of *Neem uit die verlede,* for the trust he has placed in me to allow me to translate one of his "children". I also wish to thank South African National Parks for financial assistance to complete the work, especially Dr Hector Magome and Kevin Moore who handled the administration.

The Johannesburg Region of the Honorary Rangers of South African National Parks also financially contributed over a period of two years and their assistance is gratefully acknowledged. The Honorary Rangers have played and are still playing an important role in South African National Parks throughout the country. They are all volunteers who offer valuable assistance in the parks in connection with tourism, communication, environmental education, conservation management, professional services, as well as marketing and fund-raising projects. On behalf of all South Africans: Thank you for your unselfish dedication!

I would also like to thank the following Kruger National Park staff who assisted in obtaining some information about the memorial plaques in the Park: Dr Freek Venter, Danie Pienaar and section rangers Karien Keet, Sandra Basson and Neels van Wyk.

Last, but not least, I want to thank Carmen Hart of Knysna for her valuable contribution towards proofreading the manuscript and my husband, Bruce, for his valuable support throughout the project.

Historical terrains – Kruger National Park

Beacons of the farms which had been awarded to A.H. Nellmapius of which the boundaries fall within the Kruger National Park, as well as the subdivisions of the farm Pretoriuskop

Locality and description	8-digit grid map reference	Map number
Original farm beacons of Pretoriuskop		
P/K – A Beacon A = N/W (Shabeni Hill)	22821825	2531 AA
P/K – B Beacon B = N/E Beacon	33421172	2531 AB
P/K – C Beacon C = S/E Beacon	30720505	2531 AD
P/K – D Beacon D = S/W Beacon*	15541167	2531 AA
Subdivision of the farm Pretoriuskop		
(1) "Grootgeluk" (eastern subsection)		
P/K – 1 = N/W Beacon	28751457	2531 AB
P/K – B = N/E Beacon	33421172	2531 AB
P/K – C = S/E Beacon	30720505	2431 AD
P/K – 4 = S/W Beacon	25720720	2531 AB
(2) "Goedgelegen" (central subsection)		
P/K – 2 = N/W Beacon	25551657	2531 AB
P/K – 1 = N/E Beacon	28751457	2531 AB
P/K – 4 = S/E Beacon	25720720	2531 AB
P/K – 3 = S/W Beacon*	20700947	2531 AA
(3) "Pretoriuskop" (western subsection)		
P/K – A = N/W Beacon (Shabeni Hill)	22821825	2531 AA
P/K – 2 = N/E Beacon	25551657	2531 AB
P/K – 3 = S/E Beacon*	20700947	2531 AA
P/K – D = S/W Beacon	15541167	2531 AA
The farm Joubertshoop (No. 108)		
J/H – A = N/W Beacon	41551120	2531 AB
J/H – B = N/E Beacon	51420475	2531 BC
J/H – C = S/E Beacon	48300080	2531 AD
J/H – D = S/W Beacon	40400550	2531 AD
The farm Ludwich's Lust		
L/L – A = N/W Beacon	68269570	2531 BC
L/L – B = N/E Beacon	73009347	2531 BC
L/L – C = S/E Beacon*	72678372	2531 BC
L/L – D = S/W Beacon	65148906	2531 BC

*   These beacons fall outside the Kruger National Park and have not been determined on the ground yet.

Table with grid references of historical terrains

	Locality and description	8-digit grid reference	Map number
1	"Buck" Buchanan's baobab tree	16502410	2231 AC
2	Ruins of Considine and Lumley's camp, circa 1915–1920	12702585	2231 AC
3	Graves of William Pye and Hendrik Hartman	20152495	2231 AD
4	Cave with clay pots on "wooden bier"	06051802	2231 AC
5	Bushman (San) rock paintings	06721782	2231 AC
6	Ruins of black settlement	11351960	2231 AC
7	Site of Alex Thompson's store, Makuleke (1910–1954)	15922284	2231 AC
8	Ruins of John Fernandez's shop (1954–1969)	15702062	2231 AC
9	Bobomeni Drift	15651900	2231 AC
10	Site of the former Pafuri tented camp (1939–1948)	19171975	2231 AC
11	The beacon at Crooks Corner	25801960	2231 AD
12	Old prospecting pits	03301497	2231 AC
13	Old prospecting pits at Thulamela	14301850	2231 AC
14	Ruins of black settlement, Thulamela	14511845	2231 AC
15	Ruins of black settlement, Shaluka	22201750	2231 AD
16	Site of W. Borchers's store, Pafuri (1919–1954)	25351707	2231 AD
17	WNLA post, Pafuri	26151675	2231 AD
18	Ruins of black settlement	*	
19	Baobab Hill	15251178	2231 AC
20	Baobab tree with name Ebersohn on trunk	15381175	2231 AC
21	Makahane Ruins	00451215	2231 AC
22	Cave in Makahane Hill	02521557	2231 AC
23	Shipali Fountain	95050892	2231 CA
24	Tshalungwa hot-water spring	01890545	2231 CA
25	Ruins of black settlement, Matekevele	08309815	2231 CA
26	Ruins of black settlement, Matjigwili	12450025	2231 CA
27	Bvekenya Barnard's Drift, Klopperfontein	12259689	2231 CA
28	Malonga Fountain	28750012	2231 CB
29	Baobab trees with old hunters' names (Bernard Lotrie?)	29259995	2231 CB
30	C. Gerber's diamond prospecting pit and camp (1920s)	28729998	2231 CB
31	Petrified tree trunks, Ntsumanini drift	05879466	2231 CA
32	Bushman (San) shelter and rock paintings, Shantangalani	08429515	2231 CA
33	Old landing strip, Klopperfontein	10589377	2231 CA
34	Shipakweni Fountain	86428922	2231 DB
35	Ruins of black settlement, Dongadzivha	90458631	2230 DB
36	An isiVivan stone cairn along old trade route, Matukwala	90208755	2230 DB
37	Potgieter-Bronkhorst memorial plaque, Punda Maria	96228964	2231 CA
38	Punda Maria fountain, Shikokololo	96558948	2231 CA
39	Baobab tree planted in 1931 by ranger I.J. Botha	96708960	2231 CA
40	Gumbandevu Hill (Nwakama's village)	01058902	2231 CA
41	Dzundwini Fountain, former ranger Coetser's outpost	11608180	2231 CC
42	Ruins of black settlement, Dzundwini	11708167	2231 CC
43	Tshaulu bamboo forest along the old trade route to Sofala	70357875	2230 DD
44	Ruins of black settlement	00577342	2231 CC
45	Stangene well	19407460	2231 CC
46	Malahlapanga waterhole	98986767	2231 CC
47	Shingomeni hut	36276735	2231 CD

TABLE WITH GRID REFERENCES OF HISTORICAL TERRAINS

	Locality and description	8-digit grid reference	Map number
48	Ruins of black settlement, Shingomeni	41876770	2231 CD
49	Well at Nwarihlangari South	29055637	2231 CD
50	a) Mafayini Fountain	18955481	
	b) Matiyovila hot-water spring	19405427	2331 AA
51	"Larini" hidden treasure terrain	03424394	2331 AA
52	Hollow baobab tree shelter used by poachers	05474875	2331 AA
53	Graves of whites, possibly murdered by "Bulala" Taylor	16924815	2331 AA
54	a) Gold prospecting pits, Shangoni	89053792	
	b) Gold prospecting pits, Shangoni	88703735	2330 BB
55	Shangoni ranger quarters	89553663	2330 BB
56	Ruins of black settlement, Phonda Hills	18313877	2331 AA
57	Ribye-ra-Gubyane (Red Rocks), potholes, Shingwedzi River	26503667	2331 BB
58	a) Tshange lookout	19253095	
	b) Bowker Dam, 1952	19403046	2331 AA
59	Bran Key memorial, 1949	42654017	2331 AB
60	"Dipene" dip trough (1938)	51103225	2231 BA
61	White peoples' graves, Nalatsi	01581652	2331 AC
62	Ruins of Maj. A.A. Fraser's homestead, Hlomela, after 1920	96450335	2331 AC
63	"Mondzo-ya-Milwane", leadwood tree	11570751	2331 AC
64	Site of Tsende tented camp, 1933	30760615	2331 AD
65	Shibyatsangela Cave	51200877	2331 BC
66	Tropic of Capricorn	31100620	2331 AD
67	Langone Mbombi's grave, Shilowa Poort	52200480	2331 BC
68	Malunzane, Maj. A.A. Fraser's ranger post (1904–1920)	28100227	2331 AD
69	Bowker Hill & memorial plaque (Miles Robert Bowker, 1888)	36150090	2331 AD
70	Shilowa beacon	53020272	2331 BC
71	Cave sealed with stacked boulders (old mine)	52240430	2331 BC
72	"Jemisi" baobab tree	52300412	2331 BC
73	a) Ruins of black settlement	52570270	
	b) Ruins of black settlement	52650217	2331 BC
74	Ruins of black settlement, hills at Byashishi mouth	08509330	2331 CA
75	Ruins of black settlement, hills at Byashishi mouth	09009200	2331 CA
76	Ruins of black settlement, Kharige Hills	18659178	2331 CA
77	Ruins of black settlement, Milavamisi	08759000	2331 CA
78	Ruins of black settlement, Milavamisi	08929065	2331 CA
79	Joubert's grave (1890's)	22668920	2331 CB
80	Red ochre mine, Ngodzi Hill	37408957	2331 CB
81	Ruins of black settlement, Ngodzi Hill	37168940	2331 CB
82	Former Nshawu tea room	43958865	2331 CB
83	Ruins of black settlement	45949022	2331 CB
84	Giriyondo border beacon	63529122	2331 DA
85	Ruins of black settlement, Mbhandzweni	60408845	2331 DA
86	Ruins of black settlement, Mbhandzweni		
87	Ruins of black settlement, Sokolombe	30527385	2331 CB
88	Old archaeological site, Shipikani mouth	33357265	2331 CD
89	Terrain of Steinaecker's Horse's northernmost outpost	59077561	2331 DA
90	Old causeway through the Nwanetsi	47406760	2331 DC
91	Ruins of black settlement	51656666	2331 DC
92	Leadwood tree with Portuguese cross engraved into trunk	56556535	2331 DC
93	Old Malopeni gate	12346247	2331 CC
94	Ruins of black settlement	*	
95	a) Old WNLA post	54706175	
	b) Letaba ranger quarters	54706150	2331 DC
96	"Mondzwani" – ranger L.H. Ledeboer's ranger post (1921–1927)	56106295	2331 DC

TABLE WITH GRID REFERENCES OF HISTORICAL TERRAINS

	Locality and description	8-digit grid reference	Map number
97	"Hatani" – former ranger post and grave of Mrs Anna Ledeboer, 1921	60606385	2331 DC
98	Old "Klipkoppies" ranger post	73136132	2331 DD
99	Ruins of black settlement, Nwanetsi windmill	38945905	2331 CD
100	Matilolo Dam	50375960	2331 DC
101	Ruins of black settlement, Shikumbu Hill	21905800	2331 CD
102	Ruins of black settlement, near Sable Dam	21455112	2331 CC
103	Ruins of black settlement, Masorini	23505355	2331 CD
104	Ruins of black settlement, Vudogwa	22005070	2331 CD
105	Von Wielligh baobab tree (1891)	69465210	2331 DC
106	Old stone beacon	75324857	2331 DD
107	Old mica mines, Madzumboleni	33503982	2431 AB
108	Old mica mines, Madzumboleni	35524045	2431 AB
109	Baobab tree with hunters' names (1896)	80304590	2331 DD
110	Baobab tree with name H.M. Borter, 1890	81014675	2331 DD
111	Old Gorge rest camp and drift through Olifants River	80754560	2331 DD
112	Grave in cave, Bangu Gorge	83984585	2331 DD
113	Old stone beacon	66124000	2431 BA
114	Former pontoon site, Balule	71103995	2431 BA
115	Baobab tree with name Briscoe, 1890, Ntsumaneni Spruit	84304069	2431 BB
116	Bushman (San) rock shelter, Mbhandzwe Spruit	55602680	2431 BA
117	Holkrans, Shisakashanghondzo	58552665	2431 BA
118	Dome-shaped stone structures of Arabian or Indian origin with religious connotation, Pumbe	90902140	2431 BB
119	Notobranchiuspan, Pumbe	91452140	2431 BB
120	Former airstrip, Mavumbye	76751200	2431 BD
121	Site where Chief Malihane lived	*	2431 BD
122	Old Hartbeesfontein camping site	46000184	2431 AD
123	Grave of ranger W.W. Loyd, Satara (1922)	76000165	2431 BD
124	Ruins of black settlement, Shishangani	*	2431 BD
125	Ruins of black settlement, Shishangani	*	2431 BD
126	Orpen bird bath and memorial	36929250	2431 AD
127	Kingfisherspruit ranger post	42629420	2431 AD
128	Former Rabelais entrance gate	44709372	2431 AD
129	Former WNLA post, Nwanetsi	96749584	2431 BD
130	Old bridge across Nwanetsi Spruit	96709570	2431 BD
131	Bushman (San) rock shelter, Sweni River, R.W.B 91 chiseled into rock face	71659135	2431 BD
132	Pitfalls dug by blacks to capture game	83558612	2431 DB
133	Pitfalls dug by blacks to capture game	87628005	2431 DB
134	Mbhatsi Poort	99008010	2431 DB
135	Southernmost baobab tree in the Kruger National Park	78206875	2431 DB
136	Former Ripape outpost	60306428	2431 DA
137	Louis Trichardt memorial	80756786	2431 DB
138	Lindanda memorial	93156845	2431 DB
139	Trenches dug by Steinaecker's Horse, Nwamuriwa (1901)	89755945	2431 DD
140	Tshokwane, historic sausage tree	84695855	2431 DD / 2432 CC
141	Leeupan	79655290	2431 DD
142	Orpen memorial tablet set into giant boulders	75254815	2431 DD
143	Kruger memorial tablets	76204880	2431 DD
144	Former Rolle ranger post (1913–1923)	22306510	2431 CA
145	Site where Mpisane's village used to be	06006240	2431 CA
146	Steinaecker's fort and the grave of Capt. H.F. Francis (1901)	10056040	2431 CC
147	Sharifa Bakhari-Ali's grave (1912)	32605140	2431 CD

TABLE WITH GRID REFERENCES OF HISTORICAL TERRAINS

	Locality and description	8-digit grid reference	Map number
148	Paul Siewert's former ranger post at Msutlu (1911–1914)	54903917	2431 DC
149	Ruins of black settlement, Mfenhe Hills	75364020	2431 DD
150	Prospecting pits and graves (circa 1870s)	76763714	2431 DD
151	A chief's grave at a Lebombo candelabra tree	77753722	2431 DD
152	Former Salitje picket at Gabeni	86753180	2531 BB
153	Outpost of Steinaecker's Horse (1900–1902)	*	2431 BB
154	Ruins of black settlement, Muntshe	91123380	2531 BB
155	Albasini Ruins, Phabeni (1845–1848)	22823170	2531 AA
156	Stand of Spanish reeds planted by Albasini in 1846	23073165	2531 AA
157	Statue of Paul Kruger, Paul Kruger Gate	46953660	2431 CD
158	Black rangers' memorial, Paul Kruger Gate	47103650	2431 CD
159	Former Huhla Station	57853654	2431 DC
160	Sabie railway bridge and Selati Train Restaurant & Information Centre	58343542	2431 CD
161	Site of original borehole, Skukuza	57703555	2431 DC
162	Rangers' dog cemetery, Skukuza	57723526	2431 DC
163	Struben guest house, Skukuza (1937)	57873525	2431 DC
164	W.A.C. Campbell hut (tourist museum), Skukuza (1929) and the old pontoon site	58403533	2431 DC
165	Papenfus clock tower, Skukuza (1937)	57903565	2431 DC
166	Sergeant Helfas Nkuna's grave (1975)	22353057	2531 AA
167	Harry Wolhuter's outpost at Doispane (used until 1940)	30053242	2531 AB
168	Site of Doispane's village	30203210	2531 AB
169	"Spanplek" – outspan along Old Wagon Route from Doispane to Skukuza	41503315	2531 AB
170	Ruins of black settlement, Matupa	31002715	2531 AB
171	Ruins of black settlement, Mlaleni Hill	40352582	2531 AB
172	J. Ludorf memorial, Napi Hill	36222185	2531 AB
173	Tlapa-la-Mokwena rock	40652047	2531 AB
174	Stevenson-Hamilton memorial and grave, Shirimantanga Hill	58522310	2531 BA
175	Old Wagon Route, blaze marks on leadwood tree, Kwaggaspan (c. 1845–1848)	54741845	2531 BA
176	Bushman (San) shelter, Renosterkoppies	59452080	2531 BA
177	Old brickworks, Lower Sabie	88502342	2531 BB / 2532 AA
178	Possible site of João Albasini's trading post near Sabie Gorge	93601987	2531 BB
179	Matukanyana border beacon	02632032	2532 AA
180	Site of former Mtimba ranger post (1902–1940) and grave of Andrew Wolhuter (1918)	11251857	2531 AA
181	"Jock" memorial plaque, Numbi Gate	18301700	2531 AA
182	Grave of Willem Pretorius (1845)	21771627	2531 AA
183	The Old Wagon Route's crossing with Naphi road	27451750	2531 AB
184	Site of J. Feldschau's overnight station, Delagoa Bay transport road (1976)	23901567	2531 AB
185	Chief Manungu's store and outpost (1846–1848)	23821567	2531 AB
186	Oldest hut in Pretoriuskop rest camp (1930)	25401526	2531 AB
187	Ranger quarters, Pretoriuskop	25631553	2531 AB
188	Group of giant spekboom trees near old transport road	28501400	2531 AB
189	Grave along Mbyamiti River	28891397	2531 AB
190	Soltké's grave and birthplace of Jock of the Bushveld (1885)	31551255	2531 AB
191	Camp site of Steinaecker's Horse (1901), "Outspan"	32901200	2531 AB
192	Komapiti windmill (1933)	38451020	2531 AB
193	Kemp's Cottage (1930s)	60901617	2531 BA
194	Former ranger B. Tomlinson's camp (1930–1933)	61651375	2531 BA
195	Old Randspruit siding	73451472	2531 BA

TABLE WITH GRID REFERENCES OF HISTORICAL TERRAINS

	Locality and description	8-digit grid reference	Map number
196	Possible site of "Gaza" Gray's outpost and cattle kraal (1899–1902)	96501425	2531 BB 2522 AA
197	Wood's grave, Sabie Gorge (Mozambique), 1868	*	
198	Ruins of black settlement, Ship Mountain (Langkop)	35401085	2531 AB
199	Old hunter's camp south of Ship Mountain	33700818	2531 AB
200	Chief Mataffin's grave, 1891	35820612	2531 AD
201	Site of Chief Inhliziyo's village	36050745	2531 AB
202	Thomas Hart's overnight station and grave along the Delagoa Bay transport road, Josekhulu (1876)	43200725	2531 AB
203	First concrete weir in Park, Ntomeni Spruit, 1931	45570670	2531 AD
204	Old outspan along Delagoa Bay transport road at Voortrekker windmill (1884–1892)	45620656	2531 AD
205	Ruins of black settlement, Makhuthwanini	51420475	2531 BC
206	Sardelli's store and bluegum trees, Gomondwane (1892–1898)	83600617	2531 BD
207	Frans de Cuiper memorial plaque, Gomondwane	84450415	2531 BD
208	Godleni Poort	01850435	2531 BD
209	Ruins of black settlement, Newu	40450517	2531 AD
210	Ruins of black settlement, Mangake	34100450	2531 AD
211	J. Glen-Leary memorial, 1926	55859986	2531 BC
212	Grave of Gert Frederik Coenraad Stols, 1886	31849875	2531 AD
213	"Skatkoppie" (Nguleni Hill)	45029497	2531 AD
214	Cave shelter of blacks southeast of "Skatkoppie" (Nguleni Hill)	44749345	2531 AD
215	"Fihlamanzi" outspan, old Delagoa Bay transport road	57239738	2531 BC
216	Old Bume siding along Selati Line	84759725	2531 BD
217	De Cuiper's Drift through the Crocodile River, 1725	90759800	2531 BD
218	Treasure seekers' pit and old drift, Barberton transport road	27279467	2531 AD
219	Old Ntlokweni outpost	29269046	2531 AD
220	Grave of white person, Mnyeleni Spruit	32829057	2531 AD
221	Ruins of black settlement, Maqili (R/F of Hill)	31808650	2531 AD
222	Site of former Boulders ranger station (1914–1924)	31927592	2531 CB
223	Old hunters' camp, Matjulwana Spruit	*	2531 AD
224	Boomalwynkloof	41158166	2531 AD
225	Ruins of black settlement, Khandizwe	42038126	2531 AD
226	Old cattle dip, Malelane	49848165	2531 BC
227	George Meade's grave, Malelane, 1937	49908250	2531 BC
228	H. Trollope memorial plaque, Malelane	50408170	2531 BC
229	Site where Boer cannons of General Ben Viljoen were spiked	65628900	2531 BC
230	"Thengamanzi", Alf Robert's store along the old Delagoa Bay transport road (1884–1892)	67179115	2531 BC
231	Nellmapius Drift through Crocodile River, old Delagoa Bay transport road (1884–1892)	67569105	2531 BC
232	Old stone quarries (for railway bridges) and grave of A. Hamilton, 1895	86109582	2351 BD 2532 AC
233	Bushman (San) painting, Crocodile River hippo-pool	86199435	2531 BD 2532 AC
234	Crocodile Bridge railway bridge	88809492	2531 BD 2532 AC
235	Site of old store circa 1890–1900	973924**	2531 BD 2532 AC
236	Wheel of Boer cannon (1900) found in Crocodile River bed	97009145	2531 BD 2532 AC
237	Border beacon B	00879220	2531 BD 2532 AC

TABLE WITH GRID REFERENCES OF HISTORICAL TERRAINS

	Locality and description	8-digit grid reference	Map number
238	Site where cannon shells were found in Crocodile River	95808780	2531 BD
			2532 AC
239	Trenches and breastworks of Steinaecker's Horse opposite Komatipoort	97108790	2531 BD
			2532 AC
240	Border beacon A	98458733	2531 BD
			2532 AC
241	Rock overhang with Bushman paintings	04401365	2231 AC
242	Hyena Cave	05291406	2231 AC
243	Rock overhang with large clay pot	04601182	2231 AC
244	Cave with Bushman (San) paintings	06101245	2231 AC
245	Ruins of black settlement	07451227	2231 AC
246	Rock arch with Bushman paintings	10250470	2231 CA
247	Cave and ruins of black settlement (Renosterkoppies)	59002125	2531 BA
258	Bvekenya Barnard memorial	20652245	2231 AD
249	isiVivan stone cairn along ancient trade route to Sofala and Inhambane	28850082	2231 CB
250	Some of Chief Makuba's followers lived here	61546337	2331 DC
251	Ruins of old huts (Hatani)	61606337	2331 DC
252	Site where Bill Sanderson was interned (old store), 1901	96251590	2531 BB
			2532 AA
253	Bushman (San) paintings of elephants, Nwatindlopfu	73804115	2431 DD
254	Ancient ochre mine, Shishangani	95709590	2431 BD
255	Old cattle dip, Phabeni	94609600	2431 BD

*   Still has to be determined accurately

List of maps

San rock art sites in the southern region of the Kruger National Park	12
Ethnic distribution of black tribes in the Transvaal Lowveld, circa 1903	63
Old trade routes from the harbours along the east coast through the Lowveld	67*
Frans de Cuiper's land journey, 1725	94
Voortrekker routes through the Kruger National Park	103*
Van Rensburg murder site: July 1836	105
The Trichardt trek route through the Lowveld to the Lebombo Mountains	122
Map of Albasini's fort and surroundings at Goedewensch	159
Exploratory route followed by missionaries Berthoud and Schaefli from the Selati to the Limpopo, 27 June to 6 July 1889	276
Border area between the ZAR (South African Republic) and Portuguese territory as compiled by Surveyor-General G.R. von Wielligh, 1890	370
Map of the Sabie Game Reserve (12 000 sq miles) as originally proposed by Abel Chapman in 1900	394
Terrain of Steinaecker's fort at Mpisane's village in the Bushbuckridge district	412
Border changes to the Kruger National Park and the original Sabie Reserve 1899–1985	436
Traditional sites of field ranger outposts 1926–1984	539
Kraals (local settlements) inhabited in 1905–1906	560
Tourist roads in the Kruger National Park in 1928 and 1946	604
The old ivory hunting road (the Ivory Trail) 1903–1920	649

* *There is also a larger version of this map in the back of the book.*

The following five maps can be found in large format in the back of the book:

Old trade routes from the harbours along the east coast through the Lowveld
Voortrekker routes through the Kruger National Park
On the spoor of the explorers through the Lowveld, 1855–1875
Old transport and wagon roads through the southern Lowveld and escarpment
Historical sites in the Kruger National Park

Register

A

Aberham, Org 173
Acornhoek 65, 127, 243, 422, 495, 513, 529, 588, 589, 602, 669, 676, 677, 699
Acropolis 45, 74
Adams, Dr N. 678
Adamson, Joy **484**, 485
Addger (farm) 441, **443**
Addo Elephant National Park **466**, 507, 691, 697, 713, 720
Adendorff, Gus (ranger) 79, 519, 616, 644
Adlam, Richard Wills 306
Agnes Mine 291
Aitchison, Jasper 177
Albagh, Isak jnr 137
Albagh, Isak 110, 124, 133–134, 137
Albagh, Louw 137
Albasini Dam 67, 115, **165**, 167, 274
Albasini Ruins 81, 146, **147**, 154, 442, 644, **645**, **716**
Albasini Strand 151
Albasini, Antonio Augusto 151
Albasini, Gerrie 211
Albasini, João Augusto I, 166
Albasini, João Augusto II (grandson) 166, 425, 428, 663
Albasini, João Augusto III (Joe, great grandson) 218, 243, 504, 624
Albasini, João 59–60, 67, 68, 69, 70–73, 79, 84, 132, 141, 143–144, 146–167, 175, 179–180, 190–199, 218–220, 224, 242, 243, 246–250, 251, 255–256, 270, 287, 302–303, 322, 327, 350, 368–369, 504, 562, 624, 625, 628, 638, 652, 706, 715–717
Albasini, Susanna Maria Elizabeth 166, 249
Albert House (Vermeulen & Bosman Str.) 306
Ali, Sharifa Bakhari **635**, 636
Alkmaar 356
Allenby, Capt. 448
Alma (farm) 440
Aloe Spruit 96
Alverton (farm) 653, 655
Amabaga (chief) 266
AmaLala tribe 81–82
AmaNdawo tribe 60
AmaNdebele tribe 56, 60, 102, 115–120, 133–136, 143, 170, 204, 210–211, 338, 342–343
American Hotel 352
Anderson, C.J. 132
Anderson, J.C.A. 311
Andries-Ohrigstad (1839) 151, 153–154, 157, 170, 173, 181, 301, 384
Angelbach (Borchers's partner) 652, 660
Anglican Church 224
Anglo-Boer War (1899–1902) 8, 64, 123, 144–145, 160, **207**, 212, 217, 221, 229, 232, 235–238, 242–243, 279, 288–290, 319, 321, **327**, 332–333, 343, 345, 347, 354, 360, **371**, 374, 380, 391, 399, 403–431, 434, **437**, 444–445, 449, **455**, 471–475, 479, 485–486, 489, 493, 496, 498, 500–502, 505, 508, 519, 530, 547, 562, 564, 615, **618**, 619, 620, 626–627, 630–633, 635, 639, 642, 645, 647, 651, 654, 655, 659, 666, 684, 705–708, 717–718
Annecke, Dr Dawid Siegfried **347**, 348, 630
Annibal, Padre 352
Anopheles mosquito. *See* malaria.
Anthrax 241, 490
Antonio (Portuguese guide) 125–126, 204
Antrobus, R.E. **380**, 380–381, 671, 693
April se Stellasie (trading post) 327
Ash, Mortimer Walton (Morty) 241, 502, 652, 653, 655–656, 663, 703
Assangno River. *See* Inkomati River.
Astley-Maberley, C.T. (Charles) 691
Atcherley, Rowland G. 271, 272, 326
Athlone, Count of 453
Atmore, Rupert Maurice (ranger) 456, 476, 477–478, 548, 694
August (farmer) 248
Austin, Henry 179
Austin, Mrs D. 288
Automobile Association of South Africa 598, 694
Aylward, Capt. A. 227

B

Babalala (chief) 31, 538
Babalala (field ranger) 625
Badenhorst brothers 223

Baden-Powell, Col R.S.S. 242
BaHlangane tribe 628
Bailey, Abe 466, 692
Bain, Andrew Hudson 216
Bain, C.A.O., 68
Baines, Thomas 132, 216, 629
Bainsvlei (farm) 216
Baker, Sir Herbert 224
Bakutswe tribe 39
Baldwin, William Charles 218
Balfour, Lord 238
Balmer, R. 651
Balois, Kohtshane 114
Baloyi, Jim (field ranger) 542
Balule River (Tributary of Olifants River). *See* Olifants River.
Balule 36, 98, 125, 261, 549, 588, 589, 602, 603, 676, 698
Bamangwato tribe 204
Bangalasse (guide) 248
Bangoni tribe 664
Bangu Spruit **534**
Banjailand 209
Banks, Mr 652
Bannatyne, Jock 331
Banniella (chief) 427
Baobab Hill 72, 217, 648, 653, **668**, 703
Bapai/Bambhayi tribe 39, 65
Baphalaborwa tribe 38–42, 57–58, 65
Barahona e Costa, Baron Charles 262
Barber, Frederick Hugh 239, **292**, 293
Barber, G.H. **292**
Barber, Henry Mitford (Harry) 239, **292**, 293
Barberton (1892) 146, 217, 226, 236–239, **290**, 291–295, 319–323, 330, 333, 350, 354, 371, 389, 392, 398–399
Barberton goldfields 270, 293, 294, 321, 340
Barberton Herald 237
Barkhuizen, George F. 523
Barkly, Sir Henry 309
Barnard, Billy 224
Barnard, Izak 702
Barnard, Stephanus Cecil Rutgert (Bvekenya) 218, 239–241, 623, **630**, 631, **650**, 650–668, 702–704
Barnato, Barney 285

Barnett, Tom 311, 320, 642
Baronga 61
Barotseland 57, 218, 447, 449, 450
Barry, (Manager of Transvaal Goldmining Estates) 332
Barry, Thomas Daniel 697
Barter, Insp. William (Billy) 234, 310, 314, 639, 718
Basson, J.G. (Org) 155, 316, 322–323, 471, 472, 566, 642
Bastianelli, G. 346
Batavia (farm) 439
Bechtle (surveyor) 359
Bechuanaland (Botswana) 201, 218, 304, 342, 343, 453, 666, 699
Beelaerts, Gerhard Jacob Theodoor 358
Begemann, Rev. A.J. 183
Beit, Sir Alfred 224, 237, 285, 288, 294–295
Beja (farm, now Madrid) 160, 192, 254
Bekker, Pieter 119
Bell, W.D.M. 218
Bellevue (farm) 69, 71, 72
Belvedere hydroelectrical power station 289
Benjamin, D.H. 288, 289
Berg, Otto 352
Bergen (Swedish building contractor) 515, 522
Berg-en-dal (Dalmanutha) 404, 405, 420, 715
Bergfontein (farm) 221
Bergh, O. 32
Bergvliet (farm) 194
Berlin (farm, later Barrett's Berlin) 291–292
Berlin Lutherin Mission (Rooiwal) 78, 151, 172, 425
Berthoud, Henri (missionary) 69, 73, 75, 246, 251, 274–279, **375**, 379
Bester, B.M. 356
Bester, Paul Machiel 398, 471, **473**, 473–474, **546**
Beta Mine 285, 289
Beyers, Dr 389
Beyers, Gen. C.F. (Christian) 425–431, **429**, 624
Bezuidenhout brothers (Gert & Kornelis) 386
Bezuidenhout, Barend 663
Bezuidenhout, George (Org) 393
Bezuidenhout, P.L. 392
Biccard, Mrs M.M. (Maria) 154, 160, 163, 178–180, 196, 219
bilharzia 173, 348
Bilini 59
Birch, Dr J. (Louis) 272, 310, 348

Birthday or Louis Moore Mine 275, 427, 615, 632
Blackberry Glen (farm) 440, 696, 708
black-water fever 630
Blaine, C.H. 701
Blignaut, Capt. **413**
Blood River, Battle of 110, 181
Blood, Gen. Bindon 421
Blouberg 79, 171, 337
Bluebell (farm) 79
Bluett, Maj. 671
Blyde River 126, 157, 227, 259, 263, 283, **288**, 288–289, 303, 380, 410, 621–622, 705
Blystaanhoogte 159, **326**
Bob (hunter) 618, 641
Bobomeni 45, **648**, 678
Boby, W. 272, 326
Boda (ship-broker) 110, 139
Bok, State Sec. W.E. 311
Bolts, William 85
Bomvu Ridge (Swaziland) 30
Bontebok National Park 6, **466**, 691, 713, 720
Boomplaats (farm) 77
Booysen, Bertie **652**
Borchers, W.F. 218, 243, **652**, 652–655, 660, 698, 703
Border Cave (border of Swaziland & KwaZulu–Natal) 18, 32
Bordet, Jules 344–345
Borter, H.M. 217, 377
Bosch, Jan 172
Boskoppies (farm) 242, 427
Botes, Engela 242, 427
Botes, the widow 242
Botha, A.P. 172
Botha, Antjie 137
Botha, Emmi Adrijana 173
Botha, Frederik 124, 129, 132, 133, 134, 137
Botha, Gen. Louis 238, 327, 391, 403, 404–406, 425, 429, 474, 660
Botha, Hendrik 110, 124, 132, 133, 134, 137
Botha, Izak Johannes (ranger) 516, 518–519, 530, **590**, 704
Botha, J.J. 513
Botha, Jan Valenteyn (field cornet) 172–175
Botha, Mrs Louise 694
Botha, Piet 120
Botha, S.G. (ranger) 557
Botha, Theunis 172, 426, 429
Botshabelo Mission Station 227, 327
Boulders. *See* Kaapmuiden.

Bourke's Luck potholes 133, 289, 705
Boutons, P.H. 355
Bouwer, Cornelia 133
Bouwer, S.A. & L. 618–619
Bovy, George Charles 359
Bowker Dam **700**
Bowkerkop 238, **239**
Bowker, Alec 239, 706
Bowker, Miles Robert 218, 238–239, 706
Bowker, Thomas B. 700, 705
Bowlby, Capt. 223
Boyd, Tom 429
Boy Scouts 500, 508, 689
Brak Spruit 80, 111, 135, 621
Brake, Harry (ranger) 505–506
Brakfontein (farm) 203
Brand, Judge 311–312
Brand, Pres. J.H. 217
Branddraai (farm) 143, 144, 259, 301
Brandwagt (farm) 203
Bray, Edwin 295
Brebner, Sen. W.J.C. (Jack) **466**, 513, 591, 597, 692, 698, **710**
Breed, Jan & Maria 176
Breedt, Johannes Augustus (field cornet) 179, 190, 198
Breedt, P. (train driver) 364
Brent, Jack R. (ranger) 505–506, 547, 553
Brink, Daniël 328
Briscoe, Adam William **215**, **217**, 239, 377
Bristow, Herby 658
British India Steam Navigation Company 306
Brommers (farm) 293
Bronkhorst, G. 103, 119
Bronkhorst, Gysbert jnr 102
Bronkhorst, Gysbert, 102
Bronkhorst, J.G.S. 78, 100–103, 106–107, 110, 114–119, 134, 170, 200, 246, 251, 704
Bronkhorst, Lourens 175
Bronkhorst, Sybrand 102
Brooks, Mr 411
Brown, F. **452**
Brown's Hill (mine) 285
Bryden, B.R. (Bruce) (ranger) 557
Bryden, Henry Anderson 218
Bububu River 428, 616, 632, 664, 705
Buchanan, Thomas Dorlan (Buck) 618, 631, **648**, 651–661, 664, **653**, 702, 703
Bucherman, Hendrik 144
Bührmann, N.T. (teacher) 108, 114, 368
Bührmanns Drift 332

Buiskop (Buyskop) 74, 204
Buller, Gen. Redvers H. 223, 327, **403**, 403–405, 411
Bhunu (Mbandzeni's son) 472
Burger, Frederik **171**
Burger, Gen. Schalk Willem 8, 9, 112, **171**, 223, 263, 395, 638
Burger, J.J. 144, 157, 158, **171**, 302
Burger, Willem François 171
Burger's Hall (farm) 76, 263, 287, 303, 307, 308, 317, 321, 325, 437, 639
Burgers, Pres. Thomas François 166, 169, 223, 227, 234, 268, 270, **281**, **282**, 283, 287, 291, 305–306, 309–311, 314, 324, 325, 351, 357, **620**, 639
Bushbuck Ridge 469
Bushman (San) 15–16, 19–21, **29**, **31**, 32
Bushman Rock Shelter (Mpumalanga) 15–17, 19–24
Bushmans River 203
Bushveldt Carbineers 422–425, **423**, **424**, **425**, 427–428, 502, 614, 616, 630–633, **631**, 655
Button, Edward 283, 289
Buxton, E.N. 685
Buykes, Hendricus 238
Buyskop 74, **204**
Buysdorp **204**, **205**
Buys family. *See* De Buys.
Buzi River 139, 671
Byderwellen, Gen. 354

C

Cairo (marble mine) 178
Calcutta (farm) 442
Caldecott, Harry Stratford 363, 450, 464, 464–466, 690, **691**, 694
Caledon River 102, 124, 132
Caltex Oil and Petroleum Co. 700
Camal, Cassim 248
Cameron (commission member) 325
Cameron Highlanders 429
Camitto, Gov. 141
Campbell, W.A. **466**, 577, 579, 597, 611, 692, 693–695, 700, 713, 714
Campos de Corvo 76, 273, 287, 309, 310, 314
Carrington, Gen. Frederick 223
Carruthers, Tom 530, 670–671
Casa Mia hut 698, 714
Casement, Tom 236, 446–447, 475, 684
Cass, Albert 526
Castilhopolis 76, 272, 287, 307, 309, 639

Cedarstrom, Baron 657, 659
Celliers, Jan F.E. 186
Cement (field ranger) **536**, 537, 540
Cetswayo (Zulu paramount chief) 315
Ceylon (farm) 223
Chabane (Tsonga chief) 207–208, 210–212
Chadwick, Howard 622
Chalons (farm) 439, 696, 697, 708
Chamberlain, Lady Mary **435**
Chamberlain, Lord Joseph 341, 430, **435**
Chambers, Joseph (Joof) 209
Chambers, R. 422
Chamberts, Jannie 210
Chamberts, John 180, 190, 206–210, 217
Chamberts, Maria 208, 210
Chamberts, Susan 208, 210
Champagne (farm) 423, **424**, 505
Chanâna (Das Neves's hunter) 254
Changane River 248, 251
Chapman, Abel 243, 434, **445**, 446, 475
Chapman, Eric 670, 684
Chapman, James 132, 218, 341
Charter, A.E. **452**, **466**, 692
Chauke, James (field ranger) 541
Chauke, Office (field ranger) 542
Chauke, Philemon (field ranger) 542
Chi Mine 285
Chicuala-cuala (chief) 208, 251
Chicuala-cuala (Mozambique) 72–74, 117, 192, 208, 217, 251, 256–257, 265, 661. *Also see* Chiquaraquara.
Chikovele (Chabane's son) 207
Chilwane (Mapulana chief) 63
Chinculo 96
Chinguene (chief) 256
Chiquaraquara 255–256. *Also see* Chicuala-cuala.
Chokwane. *See* Tshokwane.
Chomse, B. 291
Chuhwini Pan 232
Churchill, Joseph Fleetwood 181–182
Churchill, Lord Randolph 237
Cilliers, Sarel 115, **119**, 704
Claasen, Johannes 119
Claassen, J.J. 407
Clarendon, The Count of **452**
Clements, Gen. 449
Clensana (chief) 128
Clewaan (chief) 128
Clewer Mine 285
Cloete, J.H. 694

coastal traders 81–86, 117
Cockney Liz (bar maid) **295**, 331
Coetser, Maria 498, 627, **667**
Coetser, S.P. 75
Coetser, Jan 325
Coetser, Johannes Jacobus (Kat) (ranger) 241, 242, 406–407, 492, 497–499, **498**, 503, 556, 623, 624–627, 631, 653–654, 664, **667**, 668, 704, 705
Coetzee, D 322
Coetzee, Rev. M.P.A. 212
Coetzee, Sannie 532
Coetzee, S.H. 437, 687
Coetzer (prospector) 503
Coetzer, Flip 144
Coetzer, P.J. 368
Cohen, Dr E. 233, 246, 251, 263, 268–269
Colenbrander, Col Johan 218, 431
Colley, Gen. George Pomeroy 216, 233, **315**
Colonial Scouts 411, 449
Colonist mill **672**
Colson, Mr 657
Combomune 112, **113**, 114
Combrink, Andries 173
Combrink, F.J. (field cornet) 303
Combrink, Mr 158
Commission Road 302, 304
communication route (Malelane–Pretoriuskop) 229, 333
Compton, Walter George 269, 294, 324–326
conservation. *See* game preservation.
Considine, Mr 657–660
Consort Gold Mine 29
Cook, Const. 332
Cook, Gus 652
Cooper, Magistrate H. 307
Cooper, Wim 147, 149
Coopersdal 97, 272, 294, 307, 308, 309, 639
Coqui, Theodor Julius Adolph 302, 629–630
Cork (farm) 693
Cornelis (servant) 371
Coupanne (chief) 97
Covetti (trader) 84
Coxwell, Capt. H.R.G. 505, 619
Craig, Mrs J.M. 700
Craw, Wilfred 670
Crocodile Bridge 96, 97, **100**, **156**, 263, 298, 321, **358**, 363, 364–366, **365**, 406, 454, 463, **476**, 476–477, 485–488, 500, 507, 516, **517**, 538, 540, 542, 546, 547, 554, 562, 590, **592**, 597, **600**, 603, 604, 606,

608, 611, **645**, 647, 674, 698, 704, 715, 718, 720,
Crocodile Poort (gorge) 322, 340, **350**, **353**, 355, 405, 434, 435
Crocodile River 19, 36, 38, 39, 63, 76, **97**, 98, 99, **100**, 213, 225, 234, 258, 263, 269, 271, 272, 285, 298, **308**, 310, 317, **319**, **320**, 320–323, 325, 339, 340, 342, 366, 371–373, 406–409, 413–414, 434, 437, 438, 455, 463, 465, 477–478, 487, 493, 500, 501, 506, **510**, 520, 527, 538, 548, 584, 585, 589–590, **600**, 602, 603, 608, 614, 617, 620, 638–639, 641, 694, 719–720
Cronjé, Gen. Piet 228, 404, 472
Crooks Corner **239**, 240, 241, 267, 296, 425, 439, 497, 502, 556, 618, 623, **630**, 631, 647–665, **665**, **666**, 672, 703
Crosby, Mr 508, 688
Croucamp, Hendrik 102
Crous, Gerhard Christoffel Snyman (ranger) 297, 514–516, **515**, **516**, 526, 532, 552
Cruywagen, W.A. 200, 328
Cubitt, Elizabeth Lilian 237, 471
Cubitt, John 237
Cumming, Roualeyn Gordon 213, 215, 218, 384
Curson, Dr 610

D

D'Albuquerque, Capt. 61
D'Andrade, Capt. Alfredo Augusto 372–373, 377–379
D'Andrade, Gov. Carlos Joachim 144, 368–369
D'Arcy, Sergt. H. 505, 553, 623, 653
D'Argai (farm) 440
D'Assonville, A. 527
D'Urban, Sir Benjamin 132
Da Costa Leal, Lieut. Fernando **262**
Da Gama, Stuart Lucio 307
Da Purificacão, Maria 151
Da Silveira, Jose Antonio 153, 157
Daannel (farm) 440
Dahl, Capt. D.C.O. (Oscar) **206**
Dales, Col Hartley 224
Dansch (Ranger De Laporte's dog) 487, 488, 636
Danysz, M. Jean 344, 345
Dart, John 657
Dartnell, Mr 319
Das Neves, Dicleliano Fernandes 71, 72, 82, 85, 177–178, 192, 253–258, 346

Davhana (Ramapulana's son) 162, 192–193
Davis, John Henry (geologist) 282
Davy, Comdt 371, 376, 378
Dawano (chief) 97, 718
De Beer (field cornet) 303–304, 418, 420–421
De Beer, Jan du Plessis 388–390, 393, 396–398, 400
De Beer, Maria Johanna 498, 653
De Beers brothers 662–663
De Beers diamond company 344
De Bertodano, Capt. F.R. 424, 614
De Bucquoi, Jacob 92
De Buys, Aletta 204
De Buys, Coenraad 91, 115, 118, 121, 134, 180, 203, **205**, 217, 246
De Buys, Doors 80, 112, 118, 134, 135, 136, 203, 204
De Buys, Dorha 203, 204
De Buys, Elizabeth (uMzilikazi's sister) 204
De Buys, Gabriël 118, 134, 203
De Buys, Georg Frederik 204
De Buys, Jan 204
De Buys, Johannes 204
De Buys, Michael 205
De Buys, Maria Magdalena 204
De Buys, Petrus 204
De Buys, Piet 204
De Carvalho, Augusto 179
De Castellão, G.G. 144, 368
De Clerck, A. (ranger) **408**, 409
De Clerq, Magistrate J. 158, 304
De Couto, Jacob 164
De Cuiper, Francois 39, 70, 83, 94(map), 95–100, 213, 246, 263, 281, 338, 566, 718
De Doorns 135
De Driefonteinen (farm) 203
De Jager family 142
De Jager, Col P.J. (Piet) 498–501, 513, 516, 519, 553, 555, 654
De Jager, Lieut-Col Petrus Lafras (ranger) **499**, 499–504
De Kaap goldfields 81, **290**, 292
De Kaap Valley 291–292
De Klerk (field cornet) 386
De Kock, Jan 406
De Kock, Servaas 354
De la Rey, Gen. J.H. (Koos) 395, **396**, 400, 425
De Lange, Adriaan 115, 704
De Laporte, Cecil Richard (ranger) 27, 453, 456, 462, 466, **478**, 483, 485, **486**, 486–

489, 495–497, 500–501, 505–507, 512, 516, 522, **547**, 548, **553**, 554, 573, 588, 602, 615, 636, 643, 692
De Laporte, Mrs 488
De Mattos, F.J. 144, 368
De Mist, J.A. 203
De Oliveira, Gov. Jacinto Henriques 247
De Ridder, J.A. **466**
De Rust (farm) 483
De Souza, Dr **452**
De Souza, Mariano Luiz 159
De Souza, Pedro Fransisco 153
De Villiers, Daantjie 430
De Villiers, Jan 421
De Villiers, Mrs R.A. 695, 696, 704
De Villiers, P.D. (Pieter) (field cornet) 223, 225, 263
De Villiers, P.R. 9
De Villiers, P.W. **318**
De Villiers, Piet 224
De Volkshoop (*Eerste Fabrieken in de Zuid–Afrikaansche Republiek*) 311
De Vrij, Jan 142
De Waal, H. 437, 687
De Wet, Gen. C.R. 425, **431**, 485
De Wet, Jacobus 102
De Wet, Judge 592, 606, 699, 714
Dekena, Dr 372
Delagoa Bay and East African Railway Company, The 352, 353
Delagoa Bay railway line 161, 166, 345, **355**, **359**, 406, **720**
Delagoa Bay 39, 59–60, **71**, 68–76, 80–87, 91–94, 100, 102, 104, 108, 111, 121, 125–131, 135–144, 150–161, 166, 170–171, 192, 204, 208, 213–214, 224, 225, 233, 234, 237, 246, 255–263, 271, 277, 280–281, 284, 285, 287, 294–295, 301–309, 340, 345–347, 350–351, 368–369, 375, 378, 406, 410, 416, 449, 460, 472, 473, 507, 529, 617, 618, 621, 624–625, 629–630, 634–635, 637–642, 708, 715–720
Delportshoop diamond fields 223
Dennison, Thomas (barber) 332
Denny, Lieut-Col H.C. 428
Denton, John Charlton 615
Depledge, W. 298
Derby (farm) 262
Devil's Knuckles 233, 285, 325
Devil's Office (later Kaapsche Hoop) 290, 292
Dhlulamithi (elephant bull) 665

Dhunjbhoy, Dorabjee 320, 354
Dicke, B.H. 67–72, 80, 84, 108, 336–337
Dickson (Wolhuter's friend) 482
Diegal, Martin Jacob (Charlie) 657, **659**, **660**
Diepkloof (farm) 438, 441
Diespecker, Jules 360–361, 414, 422, 685
Difaqane (Shaka's terror campaign) 213, 220, 338
Dili waterhole 428
Dingane (Zulu king) 59, 108, 141, 338
Dingiswayo (Zulu king) 338
Dipene (now Dipeni) 512, 705
Djindi River 114, 141
Doispane **479**, 494, **548**, 549, 602, 606, 608, 636, 712
Don Luiz Filipe (steam locomotive) 352
Donald Currie Line (shipping firm) 306
Doorn River 135, 136
Doornhoek (farm) 263, 356
Doornkloof (farm) 305
Doornkop (farm) 75, 23
Doornspruit (farm) 135, 240, 243, 438
Dorenvanteyn (farm) 175
Doringbult 172, 181
Dorps River 135, 196, 200, 251
Dos Santos, Vicente Thomas 153
Drakensberg Mountains 25, 39, 62, 65, 71, 75, 76, 81, 104, 116, 118, 124, 125, 133, 136, 142, 144, 146, 148, 154–155, 159, 170–171, 213, 229, 259, 261, 290, 301–303, 338–339, 368, 385, 456, 563, 628
Draper, Daid 400
Driekop (farm) 284
Drury, Aubrey 359
Du Bois, Robert 260
Du Plessis, Andries Johannes 332
Du Plessis, J.H. (field cornet) 162, 192–194, 197
Du Plessis, Jan 198
Du Plessis, Piet 331
Du Prat, Alfredo (Portuguese Consul Gen.) 144, 164, 369
Du Preez, Flip 175–176
Du Preez, G.P. 79
Du Preez, Jan. (field cornet) 190, 199
Du Preez, Piet 216, 217, 251
Du Preez, Pieter 172
Du Toit, Comdt Andries Francois (Old Red Government) 418, **419**, 431
Du Toit, Dr P.J. 609
Du Toit, Johannes 119

Du Toit, R. 691
Du Toit, Willie **419**
Duba 364
Duke, T.R. (Tom, ranger) 456, 476, **478**, 485–487, **486**, 496, 499, 505, 547, 597
Dumbana waterhole 232
Dumeri 208, 209, 211
Dunbar, D.M. 270, 273
Duncan, Sir Patrick 217, 434, 435, 453, **455**, 456, 463, 685
During (missionary) 263
Dutch East India Company (*Vereenigde Oos–Indiese Compagnie* or VOC) 92, 123, 280, 347, 385, 629
Dutch Reformed Church 169, 190
Duvenage, Kosie 180
Duvenage, Magistrate A.C. (Abraham) 174–179, 183, 207, 249–250
Duvenage, Magistrate A.P. 193–195, 197
Dwars River 172, 328
Dyambukeni, Mapoisa (field ranger) 505, 541
Dykstra, H.W. 421
Dzata Ruins (Zimbabwe) 79, 150
Dzombo (Tsonga chief) 65, 248, 511
Dzombo Camp 242, 248,
Dzombo River 379, 427, 531, 574
Dzundwini Hill 73, 346, 379, 498, **556**, 654, 704
Dzundwini waterhole 103, 257

E

E. Warren & W. Royce (engineering firm) 356
Eastern boundary (demarcation of) 274, 438,
Eastern railway line (Pretoria–Delagoa Bay) 228, 294, 333, 347, 350, 353, 354, 356, **359**, **371**, 404, 413
Eastern Tvl. Consolidated Mines 295
Eb en Vloed 304
Ebersohn family 217
Echo Caves (Mpumalanga) 16, 19
Eckstein, Hermann Ludwig 237, 238
Edington, Dr Alexander 344
Edwards, Samuel 175
Eersteling (farm) 283, 289–290, 329
Eiland (salt sheet) 36, 39
Eland, Sergt. Frank 426
Elands River 352, 356, 398, 445
Elandsberg (farm) 74, 393
Elandsdrif (farm) 302, 303, 330
Elandsdrift Mine 289
Elandsfontein (farm) 356

Elim Mission Station 47, 423, 631, 652, 663
Ellerton Mine 657
Eloff, Capt. Sarel 180, 196, 242
Eloff, D.G. (née Grobler) 190
Eloff, Frederik 186
Eloff, P.A. 173
Eloff, Pieter Johannes (Piet) 176, 190, 207, 212, 218
Eloff, Prof. Hannes 200, **606**
Eloff, Zybrand 186
Elphick, C.W. 506
Elphick, Capt. G.L. (George) 477, 506
Elphick, T.F. (Tom) 506, 580, 697
Elton, Capt. James Frederick 233, 246, 251, 265–267, 503
Employment Bureau of Africa Ltd (TEBA), The, 87, 667
Engelbrechtshoop (farm) 437, 439
Engelhard Dam 696, 706
Engelhard, Charles William 696
Engerstroom, O. 422
English, Mike (ranger) 19, 24, 27, 68, 319, **422**, 428, 615
Enslin, J.A. 174
Episcopalian Church 190
Erasmus family 305
Erasmus, G.J. 80
Erasmus, Gert (ranger) 68
Erasmus, Jacobus Abel 218, **227**, 226–229, 283, 316, 330, 331, 371, 374–379, 395–397, 416, 418, 421, 473, 617, 641
Erasmus, Stephanus Petrus 119, 142
Erasmus, Zacharija Geertruida Magdalena (née Kruger) 227
Erskine, Maj. David 259, 657, 659
Erskine, St Vincent Whitshed 70, 113–115, 132, 138, 233, 246, 251, 259–262, 267, 566
Erskine, Ticky 657
Eselsjagt (farm) 203
Eshen, Mr 695
Espag, Abraham 150, 325
Eureka City 295, 322
Excellence (farm) 307

F

Fairholm 448
Fairview Mine 295
Farrall, John Frederick 356
Farrell (engineer) 351
Fay, Pat 663

Fayi waterholes 233, 571
Feldschau, J. 270–271, 308, 716
Fernandes, Antonio 91
Fernandes, Felazio Cornelio 180
Fernandez, John **665**, 703
Fernendez, Mrs 273
Ferreira, Capt. Ignatius Philip (Naas) 243, **314**, 315
Ferreira, J.J. 390
Ferreira, Rob 473, 661
Fick, Antonie H. 158, 179
Fick, Paul 395
Fihlamanzi (now Fihlemanti) fountain 315–322, **316**, 642, 719
Fim Sibuyi (field ranger) 538
Finnaughty, William 218, 502
First War of Independence (1880–1881) 306, 311, 315, 391
First World War 215, 241, 333, 346, 423–424, 453, 462, 485, 488, 491, 495, 496, 498, 505, 514, 524, 527, 537, 548, 550, 554, 631, 643, 650, 652, 654, 658, 660, 690
Fitzgerald, H. 619
Fitzgerald, Terrence William 242
Fitzpatrick, Sir James Percy 218, 236–238, **237**, 270, 285, 289, 295, **316**, 317–322, 327, 471, 537, 574, 639, 640, 641, 691, 716–717
FitzSimons, Dr V.F.M. 699
Fleisch, Adj. **420**
Fleur (Makhado's field-general) 193–194
Folok (field ranger) 538, 642, 643
Foot-and-mouth disease epidemic (1938–1939) 511–512
Forbes, Capt. David jnr 410
Forbes, Capt. David 411
Forbes, Mr 584
Forde, Jack 652, 660
Forder, Mr 692
Forssman, Chevalier Magnus Johan Frederik 329
Forssman, Chevalier Oscar Wilhelm Alric 329
Forssman, Surv.-Gen. Magnus 369
Fort de Rio Lagoa 70, 93, 95, 99, 280, 381
Fort Edward **207**, 242, 279, 423–428, 502, 615, 655
Fort Hendrina (renamed Ford Edward) **207**, 423, **424**, 427, 619
Fort Maguire 453
Fort Mpisane 229, 236, 418, 419–421, **422**, 447, 475, 635, 643, 684
Fort Schutte **427**
Fort Tuli 242
Fouché, Japie **419**
Fourie family 141
Fourie, Dina 206–212, 241, 630
Fourie, Gert 206, 211
Fourie, Hans 194
Fourie, Josef (Dina Fourie's father) 209
Fourie, Josef 206
Fourie, Joseph 241, 653
Fourie, Klein Joof 209, 211
Fourie, Susanna Catharina 206
Francis, Capt. H.F. (farmer) 218, **224**, 229, 421, 447, 475, 684
Francis, George A. 422
Frankfort (farm) 304
Fraser, Maj. Affleck Alexander (ranger) 229, 391, 415, 453, 456, 459, 462, **478**, 488, 489–493, 496–499, 502, 550, **551**, 633, 643, 647, 656, 666, 705
Frasersrus 493
Frelimo War 557, 682–683
French Bob 285, 290, 291, 293, 322
French, Gen. J.D.P. 403, 657
Frere, Sir H.B.E. (Bartle) 310
Friedenheim (farm) 475
Fripp (surveyor) 381
Frisgewaagd (farm) 240
Fry, Mr. 298
Fuchs, J. 306
Fuller, Claude 691
Fullerton, James (transport rider) 316
Funjufunju (also known as Tromp) 192
Furley, Julius 320, 354, 371, 685
Furley's Drift 76, 269, 287, 320, 322, 340, 371
Furley's store (Komati Drift) 294

G
Gaben **524**
Gaines, G. (rifleman) 422
Gamaal, Cassimo 179
game 23, **689**, **690**. *Also see* hunting. Conservation areas and reserves 10, 25, 31, 90, 225–226, 229, 232, 236, 360–361, 384, 389–390, 395–397, 438, 445, 456, 463, 465, 467, 476, 502, 547; conservation ethics 10; endangered species 385; extinction 693, 698; Game Commission 393, 462; game preservation 225, 387–388, 439, 445, 470; hunting licences 387–388, 391; hunting season 155, 179, 191, 392; National Parks Act 296, 297, 298, 464, 466, 467, 503, 512, 580, 596, 689, 691, 694, 710; national parks 30, 239, 396, 340, 438, 444, 446, 464, 579, 580, 691; National Zoological Gardens 400, 458, **465**, 687; nature conservation legislation 455, 688; nature conservation philosophy 559, 585–586, 689; poaching 363, 385, 391, 398, 417, 439, 453, 455, 457, 470, 495, 503, 505, 511, 516, 537, 548, 559, 578, 662, 664, 666; SPCA 398; squatter taxes 457; *Transvaal Agricultural Journal* 465; Transvaal Game Protection Society 455, 465; Transvaal, in 382–401; veld fires 470, 536, **567**, 575–578; white and black rhino 699; game products 216, 383, 384, 385, 387, 392, 456
Gamitto, Antonio Candido Pedrozo 128, 132, 136–137, 141, 153
Gazankulu 61, 422, 495, 655
Geelhoutboom (farm and diggings) 223, 227, 283
Geerdts, C.A. 421
Geldenhuys (ranger) 540
Geldenhuys Deep Mine 697
Geluk (farm) 356
Gemmill, James 662
Gemmill, William 667, 670–671, 675–676, 679
Genal (artist) 295
Geo Heys & Co. (coach service) 330
George (field ranger) 527, **536**, 538, **540**, 624
George Pauling & Co. (railway contractors) 350, 354–355, 362, 686
Gerber, C. 566, 704
Gerber, C.J. 296
Gerber, Charles 376, 623
Gerber, J.C. (Chris) 296, 503
Gerber, Jan 297
Geyser, (field cornet) 172, 190
Geyser, C.D.C. 299
Geyser, Frederik 194, 197
Geyser, Hendrik 173
Ghona-re-Zhou National Park (Zimbabwe) 443
Giants' Staircase 259
Gibbons, Maj. A. St H. 447, 450, 453, 684
Gibraltar (farm) 451
Gibson brothers (John Alexander & James Kidney) 330
Gilfillan (surveyor) 381
Gilfillan, D.F. 466, 584
Gilfillan, H.H. 704
Gill, Dr E.L. 465, 690

Gillmore, Sergt. **599**
Gingelim 71
Glen-Leary, John 316, 719
Glynn, Arthur Henry 375, 446
Glynn, Gerda Green 226
Glynn, Gertrude Gilbertson (née Dales) 224
Glynn, Henry Thomas 218, 222–226, 223, 224, 233, 341, 375, 395–396, 434, 446, 475, 480, 563, 566, 684
Glynn, Henry 222–223
Glynn, Joe (ranger) 223
Glynn's Extension Mine 224
Glynn's Lydenburg Company 224, 285
Goa 180, 255, 257
Godfrey, V.W. 652
Goedehoop (farm) 67, 144
Goedgedacht Mission Station 194, 205
Goedewensch (farm) 60, 67, 69, 70, 72, 149, 151, **158**, 159(m), 160–164, **165**, **166**, 179–180, 192–193, 211, 219–220, 255, 256, 257, 283, 369, 628, 661
Golden Gate Highlands National Park 523
Golgi, Camillo 341
Gomondwane (Dawano's village) 70, 83, 90, 99, 269, 281, 321, 338, 347, 355, 360–361, 414, 415, 475, 476, 516–517, 543, 547, 572, 574, 602, 647, 666, 696, 704, 718
Gomondwane bush 281
Gompies River 76, 172
Gondegonde (now Gondogondo) 379, 381
Goosen (hunter) 502
Gordon Highlanders 417, 685
Gorge rest camp 226, 377, 533, **589**, **605**, 707
Gorongoza National Park 224, 533
Goudmyn (farm) 76
Gould, Edward Hardy 322
Gould, Florence 322
Gould, Tom 355
Gould's Salvation Valley 355
Gouvernement Wildtuin (Government Game Reserve) 562
Gouws, Piet 643
Graaff-Reinet 123, 131, 203
Grant, A. 437
Graskop (farm) 227, 283, 356
Gravelotte 331, 357, 529, 588, 593, 602, 668
Gray, C.E. 341–342
Gray, Capt. Edward George (Gaza) 411, 415, 417, 447, 454, 456, 475–476, **476**, 547, 548, 562, 666, 684
Great Shingwedzi River 70, 275

Greater Limpopo Transfrontier Park 443
Greathead, Maj. Percy 466, 525, 688, 691
Greefswald (farm) 150, 220
Green, Billy 175, 623, 660
Greenhill-Gardyne, Maj. A.W. 413, 417, 445, 446, 447, 475, 478, 562, 684–685
Greone 107, 378
Grey's Camp 273
Greyling (farmer) 619, 620
Greyling, Jan 393
Grietjie (farm) 297
Griffiths (native commissioner) 422
Grobler, Mrs (widow) 185
Grobler, Annie 178
Grobler, Douw Gerbrand 169, 183, 185, 190
Grobler, Hermanus, Jacobus ('Uncle' Koos) 169, **170**, 185, 243
Grobler, Minister Pieter Gert Wessel (Piet) 405, 439, **464**, 465–467, **466**, 523, 579, 690–692, 694–695, 702
Grobler, Niklaas 211
Groblershoop Dam 692
Groblersplaats (farm) 82, 169, 170
Groenewald family 142
Groenewald, Wessel (ranger) 532
Groenkloof (farm) 218, 396
Groenkloof nature reserve 684
Grootfontein (farm) 71, 223, 224
Grootriviersdrift (Masambo Drift) 104
Grootspelonken 169
Grützner, C.H.T. 180
Grysappel (farm) 652
Grysbok windmill 509, 705
Gubyane potholes 619, 659, 705
Gudzani 231, 232, 568, **573**, 574, 696
Gumbandebvu **254**
Gungunyane (Changana chief) 60, **61**, 278, 377, 380, 498, 617, 628, 641, 654, 663–664
Gunning, Dr J.W.B. 232, **343**, 344, 390, 398, 400, 471

H
Haagen, J. (magistrate's clerk) 176
Haagner, Dr A.K. **465**, 466, 573, 689, 691–692
Haakdoringdraai (farm) 525
Haenert, Ferdinand 180,
Haenertsburg 331, 489, 530, 683
Hall, Armour 695, 696, 704
Hall, Hugh Lanion 316, 356, 474, 479, 566, 639

Hall, Richard Thomas 233, 474
Haller (Jewish tailor) 320–321
Halowaan (chief) 128, 339
Halpin, W. (rifleman) 422
Hamilton, A. 645, 721
Hamilton, Charles J. 296, 512, 615, 637
Hampshire Regiment 356
Hampson, James Atherton 324–326
Hampton (Compton's contractor) 268, 294
Handcock, Lieut Peter Joseph **425**, 425–431
Hannam, Lieut 426
Hanneman, Mr (shop owner) 496
Hapi Hill (patrol hut) 530, 624
Harbour Route, old 233
Hare, L. **45**, 657
Harper, Ben 417
Harries, C.L. 498
Harris, Capt. William Cornwallis **56**, 218, 336, 349, 384, 581
Harry's Huts 529
Hart, Thomas 156, 234, 271, 308, 310, 314, 639, 718
Hartley, Henry 181, 207, 216, 218, 224, 282, 356
Hartman, Hendrik **630**, **631**, 657, 658, 702
Hartman, Maj. Daniël 631
Hartog, G. 437, 687
Hartsenberg, G. 426
Hasebult (farm) 667
Hatane 502–503, 550, **552**, 633, 702
Hatlane, Jan (field ranger) 540, 656
Havelock Asbestos Mine 473
Headlands (farm) 150
Healy, Guy Rambont (Tim) (ranger) **486**, 493–495, 497, 500, 552–554, 634, 636, **637**
Healy, Mrs **554**
Heather Mine 224
Heatherley (farm) 305
Hectorspruit 321, **353**, 354, 404, 405, 406, 408–410, 506, 511, 620
Heese, Rev. C.A.D. (Daniël) 426, 430
Heinlein, Johan 212
Helberg, Jan 185
Henderson, Const. 361
Henderson's Consolidated 686
Hendriksdal (farm) 223, 227, 283
Hengel (farm) 440, 696, 708
Her, John 332
Herbst, J.C.J. (Christiaan) (field cornet) 180, 196
Herholdt, F. 701

Herold, T.B. 700
Heuningneskrans 18, 19, 20
Hewitt Ivy, R. 653, 662, 663
Heys & Co. (coach company) 330
Hickey, Geo 296, 652
Hindon, Capt. Oliver John (Jack) 425
Hintsa (Xhosa high chief) 131, 132
hippo pool (Skurukwane) 193, 339
Hiron, Mr 371
Hlangeni Spruit 259
Hlatswayo, Minute (field ranger) 540
Hlomela 493
Hlongwane, Dick (field ranger) 540
Hlongwane, George (field ranger) 538
Hlongwane, Pios (field ranger) 540
Hluhluwe Reserve (1897) 579
Hoare, Col F.R.G. 465, 466, 580, 584–585, 689
Hockly, R.A. 585, 692
Hoek Investigation 523
Hoffman, Johanna Corlydia (Lily) 305
Hoffman, Rev. C. 113
Hofmeyer, J.H. **452**
Hofmeyr, Rev. Stephanus Johan 205, 216, 246
Hogge, Comdt E.G. 174, 450, 455, 685
Hollard, Adv. William Emil 312
Holocene 15, 20, 22
Holtzhausen, Isak Cornelis Henry Oliver 322–323, 398, 471–473, 474, 566
Homo erectus 13, 18
Homo sapiens afer 15
Hoolboom, C.F. 291
Hornaday, Dr W.J. 453, 460, 690, **691**, 693
horse sickness (African) 82, 139, 154, 171, 178, 243, 302, 325, 342, 344, 345, 372, 373, 477, **573**
Hosken, W. 456
Hottentots (Khoikhoi) 32–33, 95–96, 628. *Also see* Bushmen (San).
Hottentots River 95
Houtboschrand (farm) 259, 440, 696, 708
Houtboschberg *Delverijen* 291
Houtrivier (farm) 205
Howe, Capt. Elliot 512–514, 518, 637, 692
Hoye, Mr 688, 690, 691
Hoyi (Ntuyi's son) 60
Hromník, Dr Cyril A. 81–82, 91, 635
Hübner, A. 243
Huet, Rev. D.P.M. (Piet) 179, 185, 216, 246
Hugo, Mrs (widow) 172
Hugo, P.P. (field cornet) 172

Huhla Station 363
Hunt, Capt. Percy Frederick 424, **425**, 426
Hunt, Leuchars & Hepburn Ltd (building contractors) 526
Hunter, J.H. 218, 326
hunting 216, **226**, 649(m); elephant hunting 178, 181, 192, 216, 219, 220, 222, 251, 650, 655, 661; sport hunting 444, 446; well-known hunters 217, 224, 502, 642
Huntington (Glynn's residence) 224, 226
Huruthshe tribe 204
Hutcheon, Dr 341–343
Hutchinson, George (Captain Moonlight) 352
Hutuene Hill 278
Hyaena Cave 24, 27

I
Imbaloele (Balule). *See* Olifants River.
Impala Hotel (Barberton) 295
Imperani waterhole 259
Imperial Light Horse 237–238, 651
Impey, C.I. 670
Inkomati River 81, 95–96, 267. *Also see* Hottentots River.
Indlovudwalile (farm) 393
Ingle, Capt. J.C. 218, 226, 395, 434, **445**, 446, 475, 684
Ingoane (chief) 248
Ingrid (farm) 442
Inhambane 40, 59, 67–73, 84, 91–92, 102–115, 134, 138, 150, 153, 157, 159, 171, 177, 179, 182, 189, 208, 222, 246–252, 256, 260, 303, 378, 629, 634, 667, 704
Inhambpura 260
Inhliziyo (Mambayi chief) 617, 618, 641, 717
Inja-ka-Mthephe **494**
Inkomati Accord (16 March 1984) 683
Inniskilling Dragoons **448**, 449, 684
Inthowelle (Venda chief) 70
Ireland, E.G. 493
Irish Brigade 350–353
Iron Age 11, 24, 26, 29, 30, 35, 65
isiVivan (cairn) **68**, **76**, 704
Isweni Camp 671
ivory trade 93, 153, 161, 181, 191, 251
Ivy, Mr 653

J
J.M. Rabie & Co. (surveyors) 440
Jackson (hunter) 502
Jacob (field ranger) 625

Jacobs, Jan 190
Jacobs, Petrus Jacobus (Piet) 176, 218, 240
Jacobs, S.A. 190
Jacobsen, Jan 142
Jacobs' Fountain 240
Jakkalsbessie River. *See* Moutomme River.
James, Thomas Llewellyn (ranger) 519–520, **520**
Jankowitz, J.C. 326
Janse, Roelof 115
Jansen, Magistrate 283, 406
Jansen, R. 119, 704
Janson, C.B. 163
Janson, Justice of the Peace J.H. 704
Janssens, Gen. 203
Jeary, Bertram F. 573, 695
Jeppe, Friedrich Heinrich (Fred) 70, 233, 278, 370
Jeppe, Sir Gottlieb Ferdinand Julius 453, 689
Jerome (prospector) 296, 503
Jersey (TLC farm) 505
Jock (Fitzpatrick's dog) 225, 237–238, 294, 316, 316–320, 640–641, **716**, **717**, 717–718
Jock of the Bushveld Camp 238
Joel, Woolf 238
Johannes (Sekhukhune's subservient chief) 227
Johniesdale (farm) 439
Johnston, Dr **406**
Jordaan, G. 527
Josekhulu (*Groot Josef*) 156
Josekhulu Spruit 156, 234, 271, 308, 319, 323, 472, 639
Josia (chief) 85
Joubert Bridge **288**
Joubert, Adv. C.J. (Chris) 292, 420
Joubert, Comdt. G.J. (Giel) 78, 431
Joubert, David 158
Joubert, Dawid jnr 112
Joubert, Gen. Petrus Jacobus (Piet) 123, 207, 292, **306**, 306–308, 351, 397, 403
Joubert, Johannes Bernardus 155, 158, 303, 632
Joubert, P. 600
Joubert, S.C.J. (ranger) 557
Joubert, Willem Francois 174
Joubertshoop (farm) 155, 234, 271, 307–309, 319, 440, 639
Juzeh (Simoês's body servant) 180

K

Kaapmuiden (Boulders) 72, 224, 339, 350, 354, 357, 365, 405, 413, 442, 472, 477, 486–487, 496–501, 505, 547, 548–549, 554, **555**
Kaapsche Hoop 281, 285, 290, 291–293, 300, 330, 471
Kaffervoetpad (farm) 76
Kakebeenwa **324**
Kalahari Gemsbok National Park 466, 483, 713, 720
Kaleka Hill **275**
Kalkbank 74, 163,
Kalkfontein 172
Kamanjan 67, 78, 83
KaNgwane homeland 439, 497, 548, 549, 644
Kanniedood dam 532, **632**, 705
Karanga tribe 38, 41, 223
Kaspersnek River 301, 303, 304, 323, 327
Kaspersnek 159
Katse-Katse (*Katlagter*/Khatelage, chief) 163, 192–194, 196, 197
Keate, Lieut-Gov. Robert William 260
Keene, Mr 356
Keiskamma River 203
Kelly, Comdt. Johan Thomas (Tom) 160, 173, 218, 242, 249, 425, 427
Kelly, Engela 173
Kelly, Maria Johanna (Mita, née Du Plooy) 242, 427
Kelly, Piet 242
Kemp, Gen. Jan Christoffel Greyling 404–405, 425
Kemp's Cottage **365**, 508
Kerkbult 173
Key, B.A. (Bran) 532, 632, 700, 705
Keyter (contractor) 672
Khaliphi (amaNdebele commander) 119–120
Khama (Tswana king) 341
Khami 45, 52
Khandizwe Hill (now Khandzalive) **521**
Khobeng 263
Khoisan. *See* Bushman (San).
Khoza, Kukise (field ranger) **536**, 538
Khoza, Matches (field ranger) 538
Khubyane (Tsonga chief) 65
Kimberley diamond fields 164, 288, 388
Kind, Dr 694
King, John N. (Dynamite Dick) 57, 408
King, Terry (agent of Mosenthal brothers) 331
Kingfisherspruit (farm) 31, 440, 441, 533, 574, 696–697, 700, 708

Kirkman, Harry (ranger) 48, 298, 508, 512, 515, 524, 525, 526–531, 619, 659
Kirkman, Walter Henry (ranger) 524–532, **525**
Kirkman's Camp 528
Kitchener, Lord H.H. 408, 421, 424, 425, 426, 429, 430
Klaserie River 127
Klaserimond (farm) 438, 441
Klein Afrika (farm) 75
Klein Letaba River (Lehlabane) 70, 73, 104, 290, 261, 439, 551. *Also see* Letaba River.
Klein Sand River 128. *Also see* Sand River.
Kleinenberg, T.J. 461
Kleinspelonken 67, 151, 159, 194, 219
Klipdam (farm) **172**, 187, 206, 207, 211
Klipkoppies (farm) 235, 415, **557**
Klipkraal (farm, previously Bokkraal) 76, 84, 157, 159, 263, 302, 303, 304
Klipspringer **364**
Klopper, Frank 331
Klopper, J.P.J. (Hans) 218, 240, 242–243, 619
Klopperfontein airstrip (1938) **603**
Klopperfontein **364**, **603**
Kloppers, Hannes 524, 529
Knaboomkop (farm) 439
Knobel, Rocco **461**
Knobnoses 160, 266
Koch, Robert 344–345
Kodwaan Plate (Ngodwana plateau) 291
Koekemoer, Gert 172
Koekemoer, Marthinus 418
Koesterfontein (farm) 169
Kok, K.J. 79
Kololo tribe 56–57
Komapiti windmill 317, 717
Komapitit Spruit 535, 574, 617, 618, 641, 696
Komati Bridge 354
Komati River 59, 60, 68, 71, 81, 93, 99, 263, 269, 272, 306, 320, **354**, 369, 371, 372, **408**, 409, 413, 437–438, 472, 629, 630, 639
Komatipoort (1892) 217, 235–236, **353**, 354–355, 357–359, 363, 396, 409, 418
Kongwane (Ngomane paramount chief) 59
Kongwazi (chief) 476
Kostini 14, 379, 608
Koza, Indaba (field ranger) 537
Koza, Makwapin (field ranger) 540
Koza, Mfitshane (field ranger) 537
Kraalkop (farm) 164
Kranskop 196
Kranspoort Mission Station 205

Krantz, Paul A. 229, 398, 400, 409, 471
Krause, Judge F.E.T. 241, 498, 654
Kremetart Mountain 150
Kretschmar, Mr 689
Krige, (board member) 574
Krogh, magistrate J.C. 389
Kruger Gate 543, 712, 713
Kruger Millions 456, 614–615, 625, 627
Kruger, Casper Jan Hendrik 301
Kruger, Gezina Susanna Frederika Wilhelmina, **8**, 306
Kruger, Jacob 134, 138
Kruger, Karel 119
Kruger National Park: Administration 437, 559, 593, 595, 609, 672; animal population 13, 20–21, 30, 215, 565, 567, 578–580; bird species 255, 391, 470, 580, 677; boundaries 57, 151, 225, 296, 298, 322, 444, 446, 454, 512, 582, 608, 647, 677, 697; entrance fees 593, 599, 693, 693; establishment 389, 439, 454, 581, 695, 715; first Europeans in the area 25, 90, 100; gold at Red Rocks 619, 659; graves 19, 20, 24, 54, 68, 169, 173, 179, 296, 320, 337, 428, 615, 628–633, 641, 702; memorial plates 702–720; naming of 587; predator control 483, 507, 522, 535, 583–585; proclamation 226, 297, 298, 333, 359, 439, 444–467, 471, 489, 512, 582, 602, 607–608, 630, 702, 713; Pumbe domes 635; radio communication 611; rainfall 13, 118, 231, 232, 242, 251, 337, 348, 561, 563–567, 668; regulations for the management of 694; rock art 24, 25–31;

treasure: Bill Lusk's treasure 296, 618, 647, 658, 705; Bvekenya Barnard's treasure 623; Kruger Millions 456, 614–615, 625, 627; Mataffin's treasure 617; Otto and Le Roux's treasure 626; Larine (now Larini) treasure terrain 618–619; Philip Swartz's treasure 620–623; Pritchard & Ebersohn treasure 505, 553, 623; Skatkoppie 616–617; vegetation 13, 15, 28, 50, 224, 231, 260–264, 266, 270, 279, 337, 563, 567–568, 611; water resources 566–567

Kruger, Pres. Stephanus Johannes Paulus 8, **10**, 8–9, 163–164, 166, 176, 190–191, 196–199, **213**, 218, 225, 288, 301, 311, 312, 315, 340–341, 351, 357, **388**, 395, 403, **404**, 467, 471, 474, 489, 684, 690–691, 702, 710–713, 720
Krugerspos 159, 223, 227, 229, **327**, **328**, 641

Kruisfontein (farm) 303, 323, 324, 325
Kubayi, Manuél (field ranger) 541
Kubayi, Solly (field ranger) **542, 615**
Kumana 254, 339, 523, 549, **573**, **574**, 628
Kumane-ku-Mhuri 628
Kurulen Mission Station 106, 111
Kwaggaspan windmill 302, 303

L

La Compagnie Franco-Belge du Cheminde-Fer du North de la Republique Sud-Africain 359
Labuschagne, A. (Riaan, ranger) **281**
Labuschagne, R.J. 147, 240, 472
Lagden, Sir Godfrey Yeatman **444**, 445, 450, 454, 456, 470, 564, 684–685
Lake Chrissie 262, 287, 293
Lake Nyassa 453
Lambert, Jack 659
Lambourn family 292
Lamont, William Henry (Bill, ranger) 515, 527, 530–532, 573–574, 624, 632, 703, 705
Lamprecht, Ben (ranger) 414
Lanarkshire (farm) 448, 451,
Land Company 288, 505, 508, 669, 686
Landman, Karel Pieter 142
Landsberg (German trader) 184
Landsberg, August 179
Lang, Herbert 465, 568, 571, **591**, 698, **699**
Langkop. See Ship Mountain.
Langone Mbombi (poacher) **633**
Lanner Gorge **531**
Lanyon, Sir Owen (*Ou Langjan*) 288, 310, 314, **444**
Laveran, Fransman A. 346
Lawley, Lieut-Gov. Arthur 434, 435, 437, 494
Lawrence, R. 238
Lawrence, Tom 685
Lawson, Mr 597
Leary, John Glen 316, 319, 506, 642–643, 718–719
Leathern, William 329
Lebombo Mountains 38–39, 61, 69, **70**, 71, 76, 85, 93, 106, 127, **128**, 157, 171, 225, 226, 229, 248, 262, 269, 272, 277, 281, 303, 306–307, 309, 323, 326, 337, 368, 375, 389, 390, 397, 476, 550, 563, 569, 603, 620, 633, **634**, 718
Lebombo overnight post 272
Lebombo watershed 294
Ledeboer, Anna Maria Christina (née Bindemann) **633, 706**

Ledeboer, Henry Leonard (ranger) 296, 415, 425–426, 435, 457, 502–504, **503**, 514, 515, 550, **552**, 623, 633, 655– 657, 666, 685
Ledouphine (farm) 301, 303
Leeupan 339, **565**, **574**
Legogote (formerly Lozieskop) **235**, 236, 263, 268, 321, 415, 431, 472, 478, 529, 549, 587, 593, 606, 685, 687
Legrange, Jan 172
Legrange, Thobias 172, 174
Leipoldt, C. Louis 395, 649
Leipzig (farm) 337
Leith, Gordon 592
Lemba tribe 45, 78
Lembethu tribe 38–39, 46, 703
Lempke, R. 272, 273
Lencastre, Sec.-Gen. J. Francisco de Salles 307
Lenehan, Maj. Robert William 424–426, 429
Le Roux (surveyor) 381, 523
Le Roux, Mr 699
Le Roux, Piet 624, 625
Les Crapauds (French group) 273
Letaba Gold Mining Company 290
Letaba Ranch 297, 439, 503
Letaba River 15, 19, 35, 104, 239, 253, 275, 360, 376–378, 456, 490, 503, 550, **559**, 603, 633, 656, 696, 706. *Also see* Klein Letaba River.
Letaba 503, **514**, 516, 526, 530, 532, 538, 550, 551, 552, 589, 602, 694, 700
Letty, Cythna 568
Lewis, Mr 311
Leyds, Dr W.J. 400
Leydsdorp (1890) 217, 229, 297, 322, 331, 357
Libambo, George (field ranger) 541
Lichtenstein, Dr M.H.C. 174, 215
Liebenberg, Barend 119
Liebenberg, Christian 115, 119, 704
Liebenberg, H. jnr 119
Liebenberg, H. 119
Liebenberg, S. 119
Lignau, Mr 573
Limpopo Falls 266
Limpopo National Park 443
Limpopo River 19, 38, 44, 72, 109, 111–113, 118, 220, 246, 250, 265, 278, 282, 337, 381, 435, 442, 499, 522, 602, 631, 635, 647, 654, 657, 675
Limpopo Valley 59
Lindanda 128, 220, 478, 482, 549, 699, 707

Little Muck (farm) 82
Livingstone, David 132, 139
Livufo (farm) 248
Llamula, Bob (field ranger) 537
Lloyd, Mrs 634
Lloyd, William Walter (ranger) 343, 488, 500–502, **501**, 505, **553**, 553–554, 633–634, 707
Lobedu tribe 57–59
Lobengula (chief) 424
Logegary (tourist) 289
Logies Farm (farm) 437
Lombaard, A.F. 143, 324, 421, 439, 638
Lombard, J.P.L. 390
London (farm) 422
London Missionary Society 203
Long Tom canon **327**, 406, **407**. *Also see* Anglo-Boer War.
Long Tom Pass (Devil's Knuckles) 285, 314, **328**, 332
Longford (farm) 74
Losha 54–55
Lotrie, Jacobus (Koos Lotring) 206–209, 216
Lotrie, Cornelis 205, 222
Lotrie, Francois Bernard Rudolph (Frans) 175, 178, 218, **220**, 220–222, **222**, 566
Lotrie, Hannie 206, 208
Lotrie, Helena Beatrix 206
Lottering, G.P.J. 396
Louis Moore Mine 427, 615, 632, 652
Louis Trichardt (town) 144
Lourenço Marques (now Maputo) 71–72, 82, 92, 95, 144, 146, 151–153, 161, 164, 166, 180, 253–256, 269–274, 287, 307–309, 350–352, **351**, 356, 371–372, 380, **404**, 405, 413, 472–474, 587, 651, **666**, 673, 675, 681, 688, 704
Lourenço Marques and South African Transport Service 305, 308, 315, 326
Louw, Comdt G.J. 228, 396
Lovedale Park (farm) 424
Loveday, Richard Kelsey 218, 254, 341, **391**, 392, 395–398, 400, 446, 455, 494, 684, 685, 713,
Lovett, George 671, 680
Lower Sabie 71, 84, 147, **156**, 157, 297, 339, 475, 476, 485–488, 505, 541, 543, **547**, 552, 562, 589, 590–595, **600**, 606, 612, 647, 666, 715
Lowveld 88, 167; bamboo bush 117; chinchona trees 346; climate 414; coffee plantations 167, 219; common reed 147, 173, 219, 261,

374, 379; cultivation & agricultural plants 177–178, 267, 283; eucalyptus trees 321, 347, 355, 517; exploratory journeys to 90, 115, 233. Also see Portuguese exploration & development of the Lowveld.

    fruit trees 154, 160, 167, 177, 219; indigenous trees (Satara) 533; indigenous tribes 11, 35–42, 56–65; mango trees 549; rainfall 13, 20, 26, 118, 231–232, 241, 337, 346, 348, 565–567, 586, 668; tsamma 571;

    vegetation – observations and descriptions 13, 15, 28, 50, 99, 156, 224, 231, 260, 262, 264, 265, 270, 271, 272, 279, 323, 337, 472, 486, 490, 503, 563, 566–568, 575, 578, 584, 586, 606, 611, 675; water resources 517, 566–569, 592

Lowveld Diggers and Transport Riders Association 322
Lubambu, George (field ranger) 538
Lubisi, Jotai (field ranger) 541
Lubisi, Office (field ranger) 542
Lucas, Frank Archibald William 437, 687
Ludorf, Adv. Joseph Frances (Joe) **437**, 438, 462, 488, 587, 644, 687, 698, 714
Ludvonga (Swazi paramount chief) 57
Ludwichslust (farm, also known as Crocodile Station) 272, 307–308, 310
Luize River 74, 251
Lumley (prospector) 657, 658
Lundi River (Zimbabwe) 240, 651
Lunsklip (farm) 224, 304
Lusk, Bill (Texas Jack) 296, 566, 618, 619, 658–659, 705
Luttig, J.H. 371, 276
Luvuvhu River 38, 40, 44–46, 67, 69–70, 72, 136, 167, 193, 240, 274, 435, 519, 526, 528, 531, 573, 590, 624, 628, 647, 648, **650**, 652, 664, 672, 703
Luxford, R. 421
Lwamondo (Venda chief) 163, 194
Lydenburg (1850) 143, 217, 246, 304, 340, 350, 368, 629
Lydenburg goldfields 268, 304, 307
Lydenburg Mining Estates 288
Lydenburg-Delagoa Bay road 285, 319, 322, 720

M

M'Inerney, Maj. L.M. 685
Mabakana Hill 374
Mabasa, Elias (field ranger) 540

Mabaso, Jack 538
Mabiligwe Hills 266
Mabotte, Samaal (chief) 99
Macartney, Gov. George 203
Macatees tribe. See Mantatese tribe.
MacDonald, Mr 269
Machado, Maj. Joachim José 352, 353, 356, 372–373, 707
Machadodorp 39, 289, 330, 350, 356, 472, 474
Machangana (followers of Sochanga) 59
Macia (chief) 255
Macigamana (chief) 259
MacKay (WNLA agent) 363
Mackay, Donald 316, 318, 640
MacKay, H.S. (Mickey) 670, 671, 700
MacKidd, Mrs 205
MacKidd, Rev. Alexander 205
Mackie-Niven, Jack & Cecily 238
Mac-Mac (1874, formerly Geelhoutboom) 223, 224, 227, 268, 270, 273, **282**, 284, 286, 291, 304, 306, 325, 640
Mac-Mac goldfields 268, 270, 318, 324, 324
MacMahon, Pres. 369
Madadzhe (Mashande's son) 47
Madeliefie (farm) 355
Madonga, Jim (field ranger) 540
Madrid (farm) 80, 192, 256
Madumelane (paramount chief) 198, 267
Madzivhandila (chief) 192
Madzumbelani Mine 297
Mafelagure (chief) 265–266
Mafourteen migratory labour system 608
Magaliesberg Mountains 133, 142
Magashula (Kutswe chief) 154, 164
Magashulaskraal 67, 71, 81, 84, 143, **147**, **148**, 154, 155–158, 164, 218, 220, 224, 246, 263, 302–304, 322, 480, 628, 715, 716, 717. Also see Albasini Ruins.
Magebain (chief) 375
Magneetshoogte 76, 175
Magor (chief) 175
Magor Hill 68, 192
Maguba (Pedi chief) 228
Magude (Tshonga chief) 60, 71, 275
Magwamba tribe 85, 151, 159, 160, 266
Mahalani (chief) 476
Mahanje (field ranger) **536**
Mahasane 68, 72, 635
Mahazulu (Tsonga chief) 60
Mahlala (Pedi leader) 63
Mahlati borehole 297

Maholoholo, Battle of 63
Mahon, Mr 403
Mahula Spruit 508, 527, 555, 556
Mahumane (trader) 70, 84
Mahundla, William (field ranger) 542
Majika, Sam 655, 703
Makabela, Ngwelane (field ranger) 538
Makahane (Vha-Lembethu chief) 40, 45–47, 55
Makahane Poort (gorge) 43, **528**
Makahane Ruins 38, 40, **43**, 44, 47, 703
Makakikiaan (Chabane's induna) 207, 210–212
Makalaka tribe 265
Makapaan's Poort 174, 194
Makapan's Caves 176
Makeghula 125
Makhado (Venda paramount chief, son of Ramapulana) 162–163, 192–196, 242, 479, 620–621
Makhadzi military terrain 13, 378, 415
Makhadzi Spruit 13, 414, 415, 419, 502, 550, 633, 656
Makhuba (Tsonga chief) 65, 608, 633
Makkasana (chief) 157
Makoebie (warrior) 193–194
Makokela, Jim 320
Makomo, Tjali (chief) 132
Makonde Mountain 104, 115, 208, 211
Makongolweni Spruit 375, 566
Makuba (chief) 415, 502, 503, 550, 588, 656
Makukula, Jase (field ranger) 537
Makuleke (Tsonga chief) 65, 442, 653, **654**
Makwapa tribe 60
Mala Mala (farm) 377, 597
Malahlapanga Fountain 528
Malan, Capt. Herklaas **413**, 419, 420, 421
Malan, Charles Wynand Marais 297, 514
Malan, J.A. 393
Malan, J.M. 392, 397
malaria 68, 83, 90, 93, 95, 100, 109, 111, 121, 124, 128–129, 134–138, 140–143, 150–151, 157, 171–173, 179, 185, 191, 200, 204, 207–211, 217, 218, 242, 249, 253, 260, 263, 271, 272–274, 279, 280–282, 287, 290, 294, 295, 296, 301, 303, 319, 323, 325, 336–341, 346–348, 354–355, 364, 368, 372, 381, 383, 386, 411, 416, 421, 422, 431, 446, 449, 456, 470, 474, 497, 516, 525, 526, 536, 559, 621, 628–630, 633, 638, 642–645, 657, 663, 668, 670, 671, 678, 684, 706, 719

Malatji, Capt. Brown 41
Malelane 29, 150, 281, 315, 317, 321, 323, 333, 354, 357, 406, 413, 440, 472, 477, 500, **506**, **507**, 506–509, 514, 518, 520–521, 532, 537, 538, 540–541, **555**, 572, 580, 589, 592, 593, 597–599, 602, 603, **606**, 608, 618, 639, 642, 644, **645**, 674, 684, 700, **701**, 704, 717–719
Malieveld Mine 283
Malihane 414, 549, 550
Malitel (Zulu induna) 109
Maloios tribe 266
Malonga fountain 68, 73, 76, 113, 222, **256**, 257, 258, 296, 527, 704
Malopeni entrance gate **611**, **706**
Maluleke, Gamba (field ranger) 541, 542
Maluleke, Hlangomo Hlengane (field ranger) 541
Maluleke, James (field ranger) 541
Maluleke, Johannes, (field ranger) 536, 537
Maluleke, Manuel (field ranger) 540
Maluleke, Samuel (field ranger) 541
Malunzane 415, 490–493, 496, 502, **550**, 633, 649, 656, 666, 705
Mambayi tribe 371, 520, 617, 641, 717
Mambeete (chief) 99
Mambo (chief) 99
Mamoed Amad Saiboe 82, 248
Mampuru (Pedi chief) 63, 228
Mandhlasi, Basket (field ranger) 537, 540
Mangake Hill (granite) 281, **479**
Mangane, Ngwela (field ranger) 318
Mangane, Simon (field ranger) 535, 541
Manika (farm) 250
Manjoba (chief) 259
Mankopane (Mapela, chief) 190
Manqeva outpost 117, 118, 657
Manson, Sir Patrick 346, 630
Mantatese tribe 116, 118
Manukosi (Changana-Tsonga paramount chief, also known as Soshangana) 58, 108–109, 111, 114, 119, 124, 134, 136, 141, 151, 152, 159, 162, 171, 191, 208, 219, 250, 256, 260, 346, 628, 629, 663
Manungu (Albasini's induna) 155–156, 176, 232, 271, 319, 323, 563, 716, 717
Manungu Hill 155, **303**, 317, 319, 323, 573, **689**, 716
Manungu Spruit 271
Manyike, Frans (field ranger) 540
Mapacane 107

Mapai (administrative centre) 110, 631, 670, 675, 681, 682
Maphumbo (chief) 99
Mapog (Ndebele chief, subservient to Sekhukhune) 228
Mapulana tribe 39
Mapulanguene recruiting post 671
Mapungubwe 37–38, 45, 51–52, 81, 83, 85, 150, 220–221, 281
Maputo River 141
Mara (Soutpansberg district) 78, 178, **204**, 205, 221
Maraba (chief) 171
Marabastad 76, 163, 172, 198, 199, 263, 275, 289, 290
Marais (surveyor) 359
Marais, Charles 395
Marais, Eugéne N. 312, 664
Marais, Pieter Jacob 282
Marais, Willem 172
Maré (trader) 179
Maré, Dietlof Siegfried 193, 199
Maré, Paul 195–196, 197, 198
Maré, W.W. 186
Marico 120, 174, 344, 399, 445
Mariepskop 63, 65, 125, 259
Maripe Moshile (chief, subservient of Sekhukhune) 234, 314, 639
Maritz, Corp. M.C. (ranger) 420, 548
Marks, Sammy 295, 311, 312, 345
Marmanella Omsana, chief 126
Marnitz, G.P. 183, 248,
Marnitz, George Hendrik 250–251
Martins, Edward 175
Marula Spruit 125
Mary (Healy's dog) **494**
Masai tribe 26, **28**
Masambo (chief) 104, 108, 114
Mashaba (chief) 171
Mashande (Makahane's son) 47
Mashao (Masjou, Venda chief) 80, 104, 110, 135, 167, 219
Mashao Hill 67, 68, 72–73, 102, 104, 106, 111, 115, 116, 117, **124**, 135
Mashashe (leader of a band of poachers) 537
Mashava, Cement (field ranger) 540
Mashave, Folok (field ranger) 538, 642, **643**
Mashele, Amos Mafukula (field ranger) 540
Mashele, Judas (field ranger) 541
Mashonaland (Zimbabwe) 68–69, 78–79, 84, 138, 216, 218, 222, 246, 259, 260, 261, 263,

264, 342, 475
Masini, Johannes 79
Massingire 240, 556, 667, 669, 670, 673, 676
Masinga, Sokkies (field ranger) 541
Masingi, Watch (field ranger) 538, 540, 541
Masiye, Lemon (field ranger) 541
Massawane (chief) 98
Maswanganyi tribe 156, 159
Matabele tribe 218, 265
Matabele wars 237, 424
Matabeleland 144, 206, 207, 216, 218, 259, 261–265, 267, 341, 342, 424
Mataffin (Matsafene, Swazi chief) 227, 342, 351, 356, 617–618, 640–641, 717
Mataffin Station 356
Mateky (chief) 99
Mathebula, Aaron (field ranger) 542
Mathebula, Jack (field ranger) 537
Mathebula, Japie (field ranger) 540
Mathebula, Nduma (field ranger) 538, 540
Mathebula, Newone (field ranger) 538, 541
Mathebula, Pencil (field ranger) 540
Mathebula, Samson (field ranger) 540
Mathlali, Clement (field ranger) 536, 537
Matibawati River (Timbavati) 98
Matibela (induna) 623
Matibiskom border post 320
Matilolo Dam **516**
Matimakule waterhole 374
Matintonde River (Nwaswitsontso, now N'waswitsontso) 98
Matiyovila hot-water spring 510
Matlala Mission Station 78
Matola River 93, 95, 99, 213, 258, 273
Matola (chief) 99
Matonse (chief) 260
Matonsi, Watch (field ranger) 540, 541
Matsamo (chief) 260
Matšhêma Ruins (Sand River) 78, 81, 150
Matukanyana Hill 374
Matukwala Dam 537
Matumba, Radzingane (field ranger) 542
Mauch, Carl Gottlieb 69, 82, 132, 138, 147, 148, 157, 217, 233, 246, 251, 259, 261–264, 282, 289, 324, 480, 566
Mauchsberg 263, 283
Maugham, R.C.F. 453, 487
Mavukane Hut 700, **701**
Mavumbye airstrip 599, **602**, 699
Mavumbye area 231–232, 414
Mavumbye Spruit 414, 490, 549

Mavundla, Kai-Kai (field ranger) 542
Mawewe (also known as Langapuma, Manukosi's son) 60–61, 161, 166, 169, 192, 208, 210–212, 256, 258, 338, 628
Maynier, Magistrate H.C.D. 203
Mazzena, E. (engineer) 372, 374
Mbambalas, Jack (field ranger) 541
Mbambo, Jim (field ranger) 537
Mbandzeni (Umbandine, Swazi paramount chief) 57, 356, 472, 617, 628, 640
Mbangezithe tribe 159
Mbaula Ranch 243, 504
Mbawane, Kostine (field ranger) 541
Mbazine, Jim (field ranger) 538
Mbhandzweni Spruit 375
Mbhatsi Poort 69, 128
Mbombi, Fifteen (field ranger) 540
Mbudla (chief) 413
Mbyamiti Spruit (also Pretorius River/Veldkraal Spruit) 157, 233, 235, 303, 319, 322, 368, 638, 640
McCabe, Joseph 155, 175, 385
McCorkindale, Alexander 262
McCurdy (WNLA agent) 381, 670, 671
McDonald, Comdt 326
McDonald, Hector (ranger) 364, 489, 516–518, 538
McDonald, Maj. Walter 214
McGregor, G. 693
McKenny, Capt. J. **420**
McLachlan, Mrs 288
McLachlan, Tom (prospector) **282**, 283, 285, **291**, 324–325, 693
McLachlan's Creek 320
McMurdo, E. 352
Mdaka, Mistele (field ranger) 540, **542**
Mdluli, Bambali (field ranger) 540
Mdluli, Dick (field ranger) 538
Mdluli, Jan (field ranger) 542
Mdluli, Mbambali (field ranger) 538
Mdluli, Night (field ranger) 538
Mdluli, Nombolo (field ranger) 506, **510**, 538, 540, **541**
Mdluli, Piet (field ranger) 538
Meade, (Uncle) George 510, 511, 520, 531, 644, 645, 719
Mebu, Jim (field ranger) 540
Medingen Mission Station 425
Mehlwana (field ranger) 537
Meintjies, E.P.A. (Egbert) 395, 425
Meintjies, S.J. 194, 196, 197

Mellema, Lieut Jan 418, 420, 421
Merensky, Alexander (Reverend) 78, 180, 227, 314
Merensky, Dr Hans 298, **299**
Merriman, Glen 147, 224
Merriman, John X 344
Metsi-Metsi River 374, 566, 708
Meyer, Aletta 186
Meyer, Francisca 186
Meyer, J.H. (Jan, ranger) 619, 706
Meyer, Marius 636
Meyer, Pres. I.J. (Lucas) 390
Mgogo, Kukise (field ranger) 540
Mhinga (Tsonga chief) 65, 72, 379, 435, 526, **661**, 678
Mhlanga, Bonyoti (field ranger) 538
Mhlangana (Shaka's half brother) 59
Mhlati (farm) 694
Mhlongo, Malunzane (field ranger) 538, 540, 541
Mhlongo, Pencil (field ranger) 538
Mhlongo, Piet (field ranger) 541, 542
Mica 297, 421, 619, 659
Michel, Jean 83, 90
Middel, Jacob (interpreter) 304
Middelburg 78, 102, 133, 140, 145, 289, 350, 352, 388, 399, 477
Middelfontein (farm) 176
Middelin (farm) 438, 441
Millais, John G. 218
Millar, Exley 224
Miller, Mr 239
Miller, Sidney 48, 53–54, 200, 706
Miller, Stanley 54
Milligan, Bill 652
Milner, Lord Alfred 411, **434**, 446, 475, 684
minerals 68, 92, 296, 310; coal 93, 281, 298, 299, 363, 455, 463, 488; copper 11, 36–39, 42, 52, 53, 68, 69, 74, 75, 77–79, 83, 84, 88, 90, 91, 92, 93, 98, 137, 162, 261, 281, 290, 296, 299, 337, 343, 364, 652; diamonds 290, 296, 379, 503, 615, 620, 621, 623, 664, 668, 704; gold and goldfields 81, 92, 224, 227, 234, 237, 246, 265, 268–271, 275, 281–285, 287–294, 304–308, 314, 316–320, 324, 326, 329–330, 340, 342, 351, 357, 361, 460, 716; iron ore 38, 41–42, 77, 79, 276, 277, 280, 311; mica 118, 290, 297, **298**, 503, 514; mining & prospecting 237, 282, 283, 289, 293, 296–299, 363, 437, 455, 456, 463, 488, 503, 527,

530, 618, 623, 652, 658, 664; ochre 16–17, 24, 29–30, 76, 281; tin 67, 74, 79, 82–83, 91, 93, 118, 290, 519
Minezes, Gov. Gen. Jozé Guedes de Carvalho 307
Mingard, Eric 528, 619
Minnaar, Phil 704
Minnie, Dirk 201
Minnie, Geertruy 203
Mjonjela (chief) 476
Mkabe, Ngungwa (field ranger) 538
Mkabela, Mbango (field ranger) 540
Mkabela, Ngwela (field ranger) 542
Mkabela, Sakabona (field ranger) 541
Mkabela, Titus (field ranger) 542
Mkanzi, Jonas (field ranger) 537
Mkari, Piccanin (field ranger) 538
Mkonde (Madadzhe's daughter) 47
Mkuzi Reserve (KwaZulu-Natal) 692
Mlondozi River (Sterk Spruit) 71, 374, 476, 566
Mmanthatisi (Sotho paramount chief) 338. *Also see* Mantatese.
Mnbondune (chief) 259
Mnisi, John (field ranger) 541
Mnisi, Mapipi (field ranger) 538
Mnyamane (chief) 495
Mocke, Comdt. G. 301
Mockford, A. (ranger) 417
Mockford, H.H. (Harold, WNLA official) 48, **382**, 530, 531, 624, 658, 664, 671, **677**, 678, **682**, 683
Mockford, Tiny **677**
Modderfontein (farm) 67
Modjadji (Lobedo chief) 59, 192
Moe, J. 698
Moerdyk, Gerhard L.P. 590, 592, 593, 596, 699
Moerschell, Carl 221
Mohr, E. 233
Mokopane (AmaNdebele chief) 176, 530
Mokwasele (Tswana chief) 201
Moletele (Lepulane chief) 259
Moletše (chief) 80
Moll, Comdt Piet 418–421, **422**, 635, 643
Mollentze, T.B. (Tom, ranger) 557
Möller, Comdt P.W. (Piet) 82, 84, 102, 218, 243
Mona, July (field ranger) 318, 542
Mondlagazi 278
Mondzwani 503, **551**, 552, 623, 656, 706

Monene (Manukosi's field-general) 155, 156, 159, 162, 192, 197
Monganje River. *See* Crocodile River.
Mongoe, Sâlmao (field ranger) 318
Mongwe, Doispane (field ranger) 479, 537
Mongwe, Sokkies (field ranger) 538
Monna, Johannes 95–97, 100
Monomotapa (chief) 57, 83, 90, 91–93, 95, 280
Montanha, Father Joaquim de Santa Rita 70–74, 177, 183, 189, 217, 246–252
Moodie diggings 447
Moodie, A. 588
Moodie, G.M. 262
Moodie, G.P. (George) 233, 290–293, 321
Moodieskloof 283, 321, 588
Mooi River 129, 142, 250
Mooiplaas 557
Mooiriviersdorp (later Potchefstroom) 170
Moolman, Jacobus 142
Moor (Duncan's assistant) 685
Moorddrif 76, 175, 176
Moore, Louis 427, 615, 632, 652
Moore, Yankee 295
Mopani rest camp 414
Morant, Lieut H.H. (Breaker) 242, 424–431
Môrester windmill 319
Morgan, Mr 652
Morgenzon Mission Station 264, 274
Moria (farm) 474
Moroka (Barolong chief) 142
Mosalie (chief) 126
Mosenthal brothers (Adolph & Joseph) 319
Mostert, Eileen **548**
Mostert, M.C. (Thys, ranger) 68, 147, 156, 271, 318, 319, 316, **548**, 596, 640, 644, **645**,
Motale copper field 652
Motsari, Alari (chief) 97
Mount Anderson 91, 235
Mountain Zebra National Park **466**, 697, 720
Moura Braz, Capt. 380, 671
Moutomme River (Ntomeni) 98
Moyane, Watch (field ranger) 538
Mozambique Commission 228
Mozambique Convention 608, 673, 680
Mozambique 48, 57–64, 70, 83, 91–95, 99, 102–103, 107–109, 113, 117, 121, 128–131, 134–141, 144, 151, 153, 160, 164–166, 178, 179, 207, 216–217, 223 –224, 235, 239–243, 246–247, 251, 256, 258, 260, 268, 272, 274, 275–278, 307, 320, 336, 338, 340, 351, 360, 368–372, 380–381, 413–416, 443, 453, 456, 474, 487, 490, 497, 501–502, 510, 512, 514, 516, 522, 524, 533–537, 547, 556–557, 569, 578–579, 587, 608, 617, 620, 623, 624, 628–629, 641–642, 647, 648, 650, 655, 660, 661, 663, 666–667, 669, 670–683, 703, 705, 707
Mpampuni Ubizi (field ranger) 486, 517, 538, 541
Mpemane borehole 718
Mphaphuli (chief) 192
Mphefu (Bavenda chief) 424
Mpisane Nxumalo (chief) 418, 421
Mpisane (Fort) 229, 236, 418, 419, 420, 421–423, 447, 475, 635, 643, 684
Mshosho Ruins 150
Msutlu 365, 495, 496, 553, 554, 555, 710
Mswati (Umswazi, chief) 57, 109, 198, 628
Mthethwa tribe 58, 338
Mtimba 479, 486, 497, 533, 547, 548–549, 552, 554, 588, 643, 644, 687
Mtunga, Dick (field ranger) 540
Mudge, Capt. 686
Mulder, Sergt. Jan 93, 100
Müller, Gen. C.H. 297, 431
Muller, P.E. 421
Munnik, Gen. 329
Munroe (pharmacist) 318, 640
Muntshe Hill 58, 254, 374, 414, 415, 566
Murchison Hills 275, 290, 357
Murchison Range. *See* Selati goldfields.
Murphy (retailer) 179
Murray, A.K. 191, 220
Murray, Comdt W.P. **380**, 380–381, 671
Murray, Dr A. 173, 183
Musgrow (retailer) 179
Mutter, Capt. Frank 670
Muwawi waterhole **571**
Muzila (Umgila, Nguni paramount chief) 60–61, 162, 191, 192, 198, 256, 260, 266, 267, 275, 338, 628
Mzimba, Keppies (field ranger) 538
Mzinyani River (Umsimgwani River) 266

N

Naba (Shaka's lieutenant) 59
nagana 90, 127, 134, 154, 215, 233, 261, 269, 280, 302, 322, 325, 326, 336–337, 355, 380, 629, 638, 693. *Also see* tsetse flies.
Naghi brothers 696
Napier, William 246, 270–273, 308, 314, 326, 566
Natal Parks Board 693
National Park of Swaziland 529
National Parks Board 27, 100, 114, 168, 239, 240, 297, 365, 382, **437**, 442, 454, **461–466**, 472, 477, 489, 579, 580, 619, 632, 644, 661, 664, 686, 687, 692, 697, 702, 715
National Zoological Gardens 400, 458, **465**, 687, 691, 692, 697
Natural History Society 398
Naylor, George 701
Ncindweni 378
Ndanga Cave 150
Ndengeza tribe 159
Ndlabu (Ngomane vice chief) 60
Ndlovu, Judas (field ranger) 541, **543**
Ndlovu, Mack (field ranger) 540
Ndlovu, Masodja (field ranger) 537
Ndlovu, Njinja (field ranger) **536**, 538, 540
Ndlovu, Pencil (field ranger) 540
Ndlulamithi (elephant bull) 240, 650–651
Ndwandwe tribe 57–59, 338
Ndzenedzenge (Sardelli & Woodlands's shop) 416
Neale, A.L. 224
Nederland (farm) 505
Nederlandsche Zuid-Afrikaansche Spoorweg-Maatschappij **351**, 353
Neergaard, Paul 380, **505**, 499, 662, 667–671, 700, 703
Neethling road 224, 263
Neethling, Rev. Johannes Henoch 173, 183
Neethling, W.H. 109, 303
Nel brothers (Gert, Andries, Louis) 356
Nel, Jacobus 183, 185, 421
Nel, Jan 421
Nel, P.J. (Flip, ranger) 48
Nellmapius company (Lourenço Marques & South African Republic Transport Service) 270
Nellmapius Drift 268–270, 287, 294, **308**, 315, 317, **319**, 320, 322, 406, 472, 529, 639, 640, 719–720
Nellmapius, Alois Hugo 234, 272, 273, 285, 305–312, 314, 319, 351, 639–640, 716
Nellmapius, Irene 305
Nelombe (Makahane's brother) 46–47
Nelspruit 31, 217, 225, 228, 294, 333, 350, 356, 364, 405, 431, **473**, 533, 546, 617, 641, 685

Neser (Somerville's private secretary) 688
New Agatha Mine 331
New Caledonia Gold Field 283
New Forest (farm) 229, 236, 418, 421, **422**, 719
New Scotland 262
New South Wales Mounted Rifles 428
Newington (farm) 359, 362, 439, 562, 688, 698
Newmann, Arthur 218
Newton, Resident-Comdt Francis 341
Newu Hill **268**, 269, 271, 473
Ngala Limitada (transport company) 674–679
Nganesi (field ranger) 625, 626
Ngodzi Hill 279, 281
Ngomane tribe 39, 59–60, 64, 65
Ngomane, Bob (field ranger) 542
Ngomane, Corp. Forage (field ranger) 538
Ngomane, Jim (field ranger) 542
Ngomane, Kapana (field ranger) 538
Ngomane, Matches (field ranger) 537
Ngomane, Moses (field ranger) 538
Ngomane, Philemon (field ranger) 540
Ngomezulu, Maplank 538
Ngoni tribe 56, 58, 61
Ngonyama, Seventeen (field ranger) 540
Ngotse (now Ngodzi) Hill 29, 281
Ngotso 422
Ngqika (Gaika, Xhosa chief) 203
Ngumbin (Shilowa) 274–275
Ngumula Pan 215, 217, 239, 414
Nguni group 56–57, 61, 134, 266, 275, 338
Ngwambu, Ntambo (field ranger) 537
Ngwane tribe (Nguni group) 57
Ngwenya, Mahashi (field ranger) 537
Nhlanganzwani firebreak road 321, 416
Nhlanguleni ridge 374
Niabel 228
Niewenhuizen, H. 704
Njunda, Maswazi (field ranger) 540
Nkabela, Sakabona (field ranger) 538
Nkokodzi Spruit 279
Nkongoma post (now Kongoma) 321, 372, 409, 476
Nkovakulu Poort 531, 574
Nkumalo, Makapani (field ranger) 537
Nkuna, Breakfast Xihamba (field ranger) 537
Nkuna, Charlie (field ranger) 318, 535, 538, 541, 641, 644, 645
Nkuna, Faskoti (field ranger) **542**
Nkuna, Helfas (field ranger) 147, 521, 535, **540**, 541, 616, 618

Nkuna, Joseph (Mankoti, field ranger) 543
Nkuna, Matilwane (field ranger) 537
Nkuna, Merriman (Phutadza, field ranger) 147, 149, 538, 540, 541
Nkuna, Philemon (field ranger) 542
Nkuna, Sam (field ranger) 537
Nkuna, Sikomela (field ranger) 538, 542, 548
Nkuna, Waistcoat (field ranger) 538
Nngwe 55
Nobre, Antonio José 141, 147
Nolan's *Lekkerlag* Hotel (Spitzkop) 320
Nombolo, Mdluli (field ranger) 506, 510, 538, 540, 541
Nooitgedacht (farm) 75, 80, 82, 159, 303
Nourse, Henry 293
Ntimane (Kostini) 152, 155, 442
Ntivane, Matches (field ranger) 538
Ntlokweni outpost 542
Ntomeni waterholes 98, 315, 319, 323
Ntsumaneni Spruit 235, 319, 375, 529, 572, 718
Ntuyi (Ngomane paramount chief) 60
Numbi (farm) 440
Numbi Gate 317, 440, 442, 529, 541, 587, 589, 611, 638, 639, 643, 704
Numbi Hill 441–442, 591
Nunes, L.M. 179
Nwamanungu tribe 159
Nwamuriwa (now N'wamuriwa) Hill outlook 523, 708
Nwanetsi (N'wanetsi) River 533, 562, 634, 671
Nwanetsi (now N'wanetsi, the current Satara) 486, 549, 552–553,
Nwaswitsontso (N'waswitsontso) River 59, 98, 128, 231, 374, 414, 562, 566, 628, 696
Nwatimhiri (N'watimhiri) Bush 225, 263, 344, 513, 569
Nwatindlopfu (N'watindlopfu) Spruit 29, 567, 568
Nxaba (Nguni leader) 57, 59, 338
Nyalunga, Sinias (field ranger) 538, 540
Nyalungu, Jack (field ranger) 540
Nyalungu, Manzi Mack (field ranger) 540
Nyalungwa, Band (Shongwane, field ranger) 538
Nyalungwa, Jack (field ranger) 537, 538, 541
Nyalungwa, Joe (field ranger) 536, 538
Nyambi, Seventeen (field ranger) 540
Nyandu Bush 257, 259, 339, 381, 693
Nyandu Sandveld 527, 528
Nyathi, Doispane (Dustbin) 549
Nyawutsi waterholes 379

Nyongane 442
Nyukere, Aaron (field ranger) 541

O

O'Connor, Patrick 359
O'Neil, Mr **311**
Obermeijer, A. 568
Offord, Horatio 359
Ohrigstad River 71, 143, 144, 157, 259, 301, 303
Ohrigstad (1945) 108, 129, 140, **142**, 142–144, 154–158, 170, 217, 229, 246, 259, 286, 302, 368
Oldham, Sergt. 426
Olifantspoort 81, 217, 224
Olifants River 14, 36, 57, 59, 69, 71, 81, 115, 118, 126, 133, 136, 142, 225, 236, 254, 261, 274, 275–277, 289, 297–298, 301, 339, 369, 373–376, 422, 434, 437, 440, 456, 479, 480, 487, 505, 514, 526, 533, 548, 568, 599, **600**, 603, 676, 687, 694, 707
Onverwacht (farm) 212
Oosthuizen family 142
Oppenheim, Baron Eugène 358–360
Opperman, Daniël Jacobus Elardus (*Rooi* Daniël) 115, 119, 704
Orange Free State Game Protection Society 693
Orange River 16, 204, 266
Orcheson, John 362
Orinoco (farm) 229, 322, 409, 418, 421
Orpen Dam 574, 696
Orpen rest camp & entrance gate 440, 697, 707
Orpen tablet koppie **709**
Orpen, James **697, 709**
Orpen, Mrs Eileen Akerman (née Barry) **438**, 439–441, 523, 574, 695, 696–698, 700, 704–708
Osborn, V.H. 440
Osborne, C.F. 295
Oswell, William Cotton 218, 336
Otto, Jan 624–626
Owen, Capt. Charles Mostyn (Mad Mostyn) 174, 223, 315–316, 320, 321, 369, 371
Owen, Capt. William Fitzwilliam 59
Owen, Rev. Frances 137

P

Paardekraal (farm) 428
Paes Mamede, Orlando 679, 681
Pafuri Reserve 439, 442
Pafuri 14, 48, 70, 72, 73, 81, 104, 108, **118**,

150, 240, 251, 260, 274, 296, 380, 381, 499, 500, 518, 526, 530–531, 557, 590, 623, 624, 626, 631, 655, 662, 664, 665, 667–668, 671–673, 698, 703
Pago (chief) 192, 197
Palachwe 342
Palfrey, George 679
Palmary, Ges 657
Palmietfontein 212
Panda (Zulu king) 141–142
Pandeka (chief) 159
Papenfus clock tower **591**, 699, **714**
Papenfus, Herbert Boshoff **466**, 467, 580, 597, 599, 606, **686**, 689, 691, 692, 694, 697, 698, 699, **700**
Paradise Camp 327
Parfitt (train driver) 356
Parker, Stafford 295
Parsons (prospector) 283
Parsons (farm) 307
Patterson, Alec 283
Paulin, Tom 417, 655
Pauling, George Craig Saunders 350, 354–355, 362, 689
Peach Tree Mine 285
Peacock (Pearce's assistant) 272
Pearce, Dr H. 272–273, 309, 329, 657
Pease, Sir Alfred E. 434, 446, 475, 685
Pedi tribe (North Sotho tribe) 56–58, **62**, 63–64, 65, 116, 118, 227–228, 274, 338, 379
Peebles (farm) 235, 303, 325, 415, 478
Pegado, Señor 210
Pelzer, J.C. 421
Pennefather, Capt. Edward Graham 223
Penning, Louwrens 406, 409
Pereira, Bras Piedade 180
Perrin (Swiss assayer) 180
Perry (Wolhuter's friend) 422
Perry's Farm 422
Peru (farm) 299, 442
Petermann, A. 71
Pettigrew, Mr 270, 322, 340, 350, 472
Pettigrew's road 340, 355, 472
Pfeffer, Daniël 80, 111, 124, 129, 135, 136
Phabeni Gate 442
Phabeni Spruit 146, 149, 154, 218, 442, 543, 644, 645
Phalaborwa Gate 441, **609**, 704, 706
Phalaborwa mines 77, 83, 337, 364, 441
Phalaborwa 36, 98, 118, 254, 275, 277, 380, 431

Phamahomo 277
Phillips, Lionel 224, 285, 288
Phugwane River **428**, 616, 631, 632
Pickard, Mr 476, 547
Pickstone, H.E.V. 238
Picton, Lieut Henry **425**, 426, 429
Pienaar (prospector) 296, 623
Pienaar, Andrew (field cornet) 431
Pienaar, Ben **452**
Pienaar, Comdt 406
Pienaar, Dr U. de V. 27, 35, 71–73, 147, 222, 255, 623
Pienaar, Gen. Francois 408
Pienaar, Gen. Johannes Jacobus (Jac) 431
Pienaarspoort 102
Piesang Kop (Luonde) 159, 248
Pietermaritzburg 137, 179, 180, 183, 216, 233, 258, 259, 393, 448, 697
Pike, Mr 693
Pilgrim's Creek (Ponieskrantz) 286
Pilgrim's Rest goldfields **281**, 285, 288, 289,
Pilgrim's Rest (1873) 217, 223–224, 234–237, 271, 284–293, **284**, **287**, 288, 305–306, 310, 316, 325, 330, 333, 379, 415, 641
Pinto, Maj. Serpa 132, 251
Pionieer Ridge 292
Pirow, Minister Oswald **464**, 465, 466, 580, 585, 590, 689
Piscene (Mozambique) 320, 641, 642
Pistorius, F. 179
Pittendrigh, C.W.T. 298, 299, 526
Plaston (farm) 356, 451, 597
Plumer, Col Herbert Charles Onslow 242, 403
Pokiones Kop 369, 372
Pole-Evans, Dr Illtyd Buller 610
Pole-Carew, Gen. Reginald 406–408, 413
Pongola Reserve (1894) 218, 226, 229, 344, 390–391, 396, 399, 445–446, 453, 454, 459, 460, 463, 489–490, 550, 559, 684, 713
Ponieskrantz (farm) 284, 286, 332,
Ponsonby, Rev. Maurice 224
Pont Drift 424
Poort City (construction rest camp) **322**, 355. Also see Crocodile Poort.
Porcheron, L.A. 358
Portuguese forts 59, 91
Portuguese involvement with the exploration & development of the Lowveld 39, 41, 52–53, 60–64, 83–85, 91–95, 98, 102–108, 111, 124, 126, 128–129, 131–146, 149, 151–166, 170, 179, 180, 192, 203–204, 207, 212, 216, 218, 240, 246–251, 255–256, 258, 260, 280–281, 285, 301, 304–307, 324, 336–338, 350–354, 356, 368–381, 390, 406, 455, 474, 533, 537, 557, 561, 562, 620–627, 631, 635, 637, 655, 659–663, 669–673, 675–676, 680–682, 699; demarcation 370(m)
Portuguese Reef Mine 285
postal service 161, 190, 255
Potchefstroom (1838) 129, 142, 159, 160, 170–175, 189, 259, 262, 301, 328–330, 384, 399, 525
Potgieter, A. (official) 635
Potgieter, Andries (son) 175
Potgieter, C. 303, 368
Potgieter, Comdt-Gen. Andries Hendrik 91, 102–103, 111, 113–120, **133**, 142–143, 159, 167, **169**, 174, 190, 199, 204, 207, 225, 227, 246, 301, 323, 325, 351, 368, 384, 637, 704
Potgieter, Comdt-Gen. Pieter Johannes (Piet) 175–177
Potgieter, Cornelis 144
Potgieter, Dr Evert 142
Potgieter, Elsje-Maria Aletta (née Van Heerden) 176, 189, 212
Potgieter, Frederik 180
Potgieter, Hermanus 115, 169, 176,
Potgieter, Lieut 420
Potgieter, Martha 170
Potgieter, Nikolaas 120
Potgieter, Sec. F.S. 513
Potgieter trekker party 103
Potgieter-Bronkhorst patrol 103, 106, 118
Potgietersrus 118, 176
Pott, Mr 686
Powdrell, Mr 655
Preller family 217
Preller, Gustav Schoeman 113, 115, 117, 127, 146, 169, 454, **466**, 572, 583, **692**, 695
Pretoria 95, 162, 169, 181, 183, 194, 197, 200, 228–229, 258, 263, 283, 294, 341–346, 350–361, 372, 390–398, 403, **429**, 435, 458, 557, 625, 654, 684
Pretoria-Delagoa Bay railway line. *See* Eastern railway line.
Pretorius family 108
Pretorius, Breggie 113, 125, 137–138
Pretorius, Comdt T. 197
Pretorius, Comdt-Gen. A.W.J. (Andries) 129, 137, 169, 172–174, 324

Pretorius, Johannes (Jan) 80, 106–113, 124–125, 132–137
Pretorius, Maj. Phillip Jacobus 218
Pretorius, Pres. Marthinus Wessel 143–145, 157, 163, 165, 167, 185, 189–191, 194–199, 207, 283, 292, 310, 314, 315, 324, 369, 386, 639, 717
Pretorius, Willem 143, 157
Pretoriuskop (farm) 270, 307, 318, 319
Pretoriuskop 76, 143, 147, 157, 215, 218, 224–225, 229, 231, 233–234, 268, 269–272, 287, 301, **303**, 304, 306–310, 317–325, 368, 397, 435, 461, **462**, 472–473, 483, 489, 529, 534, 540, **547**, 549, 563, **572**, 573, 580, **594**, 595–598, **606**, 611, 616, 637, 687, 690, 698, 700, 715–719
Pretorius River (now Mbyamiti River) 157, 319, 368, 638
Prince's Hill (farm) 79–80
Prins, Mr 623
Prinsloo, Gen. A.M. 404
Prinsloo, Klaas 112, 130, 133, 138
Prinsloo, Marthinus 102
Prinsloo, N. 112
Prinsloo, Nellie 175
Prinsloo, Nicolaas (*Klaas Nuweveld*) 102
Prinsloo, Willem 175
Prior, Melton 342
Pritchard, Brinley 505, 623
Pritchard, Mr 553
Pritchard, T. 712
Progresso de Guedes (de Guide, farm) 76, 272, 287, 309
Pruizen (farm) 430
Pumba (farm) 453
Pumbe 272, 375, 501, 502, 533, 634–635
Punda Maria 38, 72–73, 91, 106, 110, 117, 241, 251, **254**, 299, 339, 379, 442, 492, 498–450, 510, **518**, **519**, 518–519, 538, 540, 555, **557**, **567**, **590**, 612, 623–627, 650–665, **667**, 678, 693, 704
Punt, Dr W.H.J. (Willem) 39, 67–74, 82, 100, 102, 104, 108, 114, 147, 150, 155–156, **170**, 320, 471–472, 533, 628–629, 643
Punt, W. jnr **96**
Pye, William **630**, **631**, 652, 657–658, 661, 702–703

Q

Queen of Sheba Hotel (Eureka City) 295
Quelimane 138–139, 153, 262

R

Rabé, C.J. 190, 251
Rabe, C.L. 173
Rabelais Dam 574
Rabelais entrance gate 440, 574, 590, 696, 704
Ramabulana (Venda paramount chief) 162, 171, 192, 262, 621
Ramakok's Kraal 76, 172
Ramashiya, Andries (field ranger) 540, 542
Ramavhoya (Venda paramount chief) 134
Randberg 116
Rand Mines 238
Randspruit 263, 364, 572, 517, 572, 696, 704
Rangel, Capt. Francisco Antonio 247
Rapoza, Albertos Carlos de Paiva 179, 257, 273, 307–309, 314
Rapozo, Ignacio de Paiva 161
Rasethau (Rossetoe, Venda paramount chief) 135–136, 171, 203–204
Rattray, Mike 529
Ravhura (Venda chief) 46
Red Gorten (farm) 440, 708
Red Rocks. *See* Gubyane potholes.
Red Star Line (coach company) 330
Reform Committee 237, **437**
Reformed Church 198
Reilly, T. 529
Reitz, Col. Deneys 410, 463–464, **466**, 467, 688–689, 692, 694
Reitz, Dr Hjalmar 688
Reitz, Francis William 8
Rembrandt Tobacco Company 595, 700
Renders, Adam 138, 217, 264
Renders, Elsie 138, 217, 264
Renosterkoppies 28, 30, 303, 339, 527, 574
Ressano Garcia Station 96, **371**, **372**, 406, 407, 474, 557, 673, 680–682
Resurrection Hotel 322
Retief (prospector) 471
Retief, Piet 172, 226, 389, 391, 395, 489, 500
Reuter, Rev. 425–426
Rhenoster Poort **172**, 176, 181, 257
Rhenoster River 116
Rhenosterhoek (farm) 159, 326
Rhoda (farm) 437, 440
Rhodes, Cecil John 285, 424
Ribeiro, Dionysio Antonio 152
Ribye-ra-Gudzani potholes. *See* Gubyane potholes.
Richmond (farm) 304
Rietfontein (farm) 76

Rietpol (farm) 172
Rietspruit (farm) 126, 224, 304
Rimington, Gen. Michael Frederick (Mike) 485–486
Rinaldini, Rinaldo 418
rinderpest epidemic (1896–1997) 96, 218, 235, 240, 336–345, 391, 397, 400, 445, 479, 486, 569
Rissik, Johann 437, 453, 460, 463, **686**, 687
Riverside (farm) 356, 479
Robbertse, Jan Harm 103, 134, 218, 704
Robert, Auguste (French Bob) 285, 290, 291, 293, 322
Roberts, Alf 321, 639, 720. *Also see* Tengamanzi.
Roberts, Charlie (transport rider) 316
Roberts, Lord 403–404, 405, 411, 425
Robertson, A.G. 464
Robertson, Alexander Marsh 393
Robertson, Maj. R. **417**, 422, 423
Robertson, V.L. 699, 700, 708
Robertson, Vinnie 483, 523, 708
Robinson, Comdt. 341–342
Roch, Fritz 393
Roche, Don 529
Rode (gunner) 408
Roets, D.J. 527
Roets, Piet 331
Rolfontein (farm) 393, 684
Rolle 365, 485, 487, 488, 495–496, 500–505, 552–555, **553**, **554**
Rood, (Dutchman) 188
Rood, Karel 573, 577
Roodekrans (farm) 505
Rooiberg tin mines 67, 68, 76, 82, 91
Rooiduiker (farm) 440, 442
Rooiwal 172, 212
Rooth, Edward Isaac 455
Rorke's Drift, battle of 319
Rose-Innis, Sir J.J. (James) 620, 699
Ross (farm) 474
Rossetoe (chief). 80, 135. *See* Rasethau.
Roux, Fred 623, 655, 660
Rowerspas (Robbers' pass, Pilgrim's Rest-Ohrigstad) 286
Rowland-Jones, Lieut-Col Maurice (ranger) 524, 526–530, **530**, 533, 572, 658, 718
Rowlands, Col 228
Roy, George 447
Royal Automobile Club of South Africa 598, 694, 695
Royal Geographical Society 113, 453

Royal Hotel (Lourenço Marques) 273
Royal Hotel (Pilgrim's Rest) **289, 403**
Royal Zoological Society 232
Rozenkrans (farm) **143**, 144, 158
Rubbervale 357, 593, **672**, 677
Rudd concession 424
Rustenburg nature reserve 459, 460
Rustenburg 174, 189, 193, 343, 384, 397, 691
Rustplaats (farm) 71, 76, 84, 144, 157, 158, 159
Rÿ Hills **220**

S

Sabe River. *See* Sabie River.
Sabie (1895) 8, 207, 217, 223–224, **226**, 289, 303, 356, 431, 602
Sabie Goewernements Wildtuin 8, 424
Sabie River 27, 71, 143, 146, 154, 164, 219, 224, 225, 268, 270, 273, 323, 339, 359, **362**, 363, 373, **458**, 493, 562, **564**, 574, **560**, 686, 710
Sabie River Bungalows 263, 422
Sabie Sand nature reserve 127, 438, 528–529
Sabie Bridge (later Skukuza) **61**, 65, 229, 236, 355, 360–364, 365, 413–414, 417–421, 431, 445, 450, 452–455, 458, 460, 464, 471, 474, 476, 479–480, 485–488, 490–495, 499, 508, 538, 546, 548, 552, **561**, **562**, **564**, 565, 573, 578, **579**, 581, 582, 587, **599**, 624, 635, 685–686, 688, 695
Sabie Reserve 10, 27, 64–65, 146, 150, 218, 226, 229, 232, 236, 298, 333, 347, 350, 361, 363, 365, **392**, **401**, 422, 447, **449**, 461–462, **465**, 471–473, 475, 505, 523, 538, 547–549, 553–555, 562–566, 573, 579, 587, 603, 614–615, 628, 643, 647, 666, 684–685, **686**, 687–690, 712–714; boundary changes 432–443
Sabie Poort **71**, **156**, 416, 476, 637, 715
Saiboe, Mamoed Amad 248
Sakana (Manukosi, chief) 104, 106–113, 134, 138, 629
Salazar, Dr Antonio 682
Salisbury, Lord 238
Salitje 464, 523, **524**
Salvation Valley 340
Samarhole Spruit 269, 315, 316, **318**, 639, **641**
Sambana (Swazi chief) 340
Sambo, Ngutu 301, 638
Sambo, Ngwela (field ranger) 542
Sandeman, E.F. 218, 233–234, 563

Sandenbergh, Col J.A.B. 242, 471, 523–524, 527, 530, 533, 565, 567, 680, 700
Sanderson brothers 218, **235**, 316, 321, **415**, 416, 474, 476, 478, 479, 509, **606**, 685, 717
San Luiz. *See* Magashulaskraal.
Sand River Convention 384
Sand River 74–75, 115, 119, 127, 134, 150, 170, 172, 174, 199, 270, 511, 529, 574, **600**, 602, 603, 624, 635, 663, 704
Sardelli, D. 321, 347, 355, 415–417, 422, 717, 718
Satara 40, 232, 236, 239, 254, 425, 470, 499–450, 502–504, **504**, 511, 513, 515, 518, 530, 532–533, 540, 549–553, 572, 574, **576**, **588**, 589–590, 593, 595–597, 599, 601, **602**, 608, 611, 612, 627, 634, 654, 657, 674, 704, 707
Sauer, Minister Paul Olivier 694
Saunders Transport Company 673
Sautine outpost 492
Schack, W. **96**
Schlaefli, H.E. (missionary) 275–278, 339
Scheepers, Anna 134
Scheepers, Antjie 111, 125
Scheepers, Christina 203
Scheepers, Gert 110, 111, 132, 133, 134, 137, 347, 629
Scheepers, Koot 124
Scheepers family 142, 347
Schiel, Col Adolf Friederich **163**
Schmidt's Drift **448**
Schneider, G.A.W. 692
Schoch, Dr A.A. 465, 467, 598, 689, 691, 694
Schoeman, Comdt David 228, **243**, 418
Schoeman, Comdt-Gen. Stephanus 161, 163–167, 176–179, 189, 189–193, 205, 207–209, 212, 218, 249–251
Schoeman, Elsje-Maria 176, 189, 212
Schoeman, Hendrik 195, 713
Schoeman, L.C. 421
Schoeman, M.J. 193–194
Schoemansdal (1848) 115, 129, 133–135, 157, 169–170, **170**, 179, 180–200, 205, 207, 216, 220, 246, 248, 340, 351, 384–385, 629
Scholtz, Hendrik 217, 246
Schoon, Robert 133
Schoonraad, Prof. Murray 27
Schreuder, Comdt 406, 408
Schutte, Magistrate C.E. 345
Schwartz, Marthinus 218

Scoble, State-Sec. John 310
Scott, Maj. 689
Scully, Magistrate William Charles 214–215
Seboma, Office Ubizi (field ranger) 538
Second War of Independence (1899–1902). *See* Anglo-Boer War.
Second World War 347, 439, 512, 527, 530, 534, 615, 655, 671–679, 703
Sedomi (servant of Josef Fourie) 209
Seedling (field cornet) 238, 320, 641
Sekhukhune (Pedi paramount chief) 63–64, 223, 227, 228, 309
Sekhukhune wars 223, 227, 228, 234, 237, 281, 288, 306, 310, 315, 329, 340, 351, 371, 617, 639
Sekhukhuneland 71, 76, 85, 154
Sekonyela (Mmanthatisi's successor) 338
Sekororo (chief) 81, 124, 125
Sekwati (Pedi paramount chief) 62, 63, 125, 174, 175, 227, 338
Sekwati's Poort 125, 136
Selati goldfields 283, 290, 357
Selati railway line 127, 290, 347, 350, 354, 359–366, 414, 434, 438, 442, 460–463, 470, 476, 479, 485, 494–496, 520, 546, 552, 561–563, 568–569, 575, 587, 589, 602, 630, 636, 645, 650, 667–668, 685–686, 720. *Also see* South African Railway Company.
Selati Train Restaurant and Museum 365
Selati Valley 276
Selby, Paul 463, 513, 579, 589, 689, 690, 693–694
Selous, Frederick Courtenay 218, 336, 342, 424, 502, 655
Semedo, Capt. 380, 671
Sena (Portuguese trading post) 59, 91, 139
Senekal, Dawid 203
Senekal, Jacob 203
Senna (farm) 250
Serita Mine 224
Serrano, Capt. José Antonio Matheo 372
Sesmyl Spruit 311
Setjele (Mokwasele's son) 204
Shabangu, English (field ranger) 537
Shabangu, Mafuta (field ranger) 541, 542
Shabeni Hill 323, 473, 638
Shaka (Zulu king) 56–59, 338
Shanbeni Hill 563
Shangoni 298–299, 442, 512, 524–527, **528**, 529, 538, **619**, 706

752

Shantangalani 29, 73, 242, 251, 519
Sharpe, Corp. Herbert 425, 428
Shashanga Dam 531, 574
Shawu Valley 378, 380, 633
Sheba Gold Mine **294**, 295, 322. *Also see* Barberton goldfields.
Sheila (farm) 440
Shepstone, Sir Theophilus 223, 227, 288, 310, **314**, 351
Sherwood, J. 295
Shibyatsangela 378
Shifundza, Mapoyisa (field ranger) 541
Shikokololo (chief) 650
Shikolokolo camp 72, 73, 117, 381, 650, 653, 668
Shikumbu Hill 41, 42, 650
Shilling (field ranger) **540**
Shilobane, Nyokane (field ranger) 537
Shilowa Hill 378
Shilowa Poort 62, 70, 84, 106–107, 274–275, **280**, 379, 556, 628
Shiluvane (Tsonga chief) 255
Shingomeni 415, 510, 531, 574
Shingwedsi Prospecting and Exploration Company 290
Shingwedzi Poort 70, 72, 107, 248, 339, 373, 379, 380, 382, 512, 608, 671
Shingwedzi Reserve (1903) 242–243, 296, 435, 437, 456, 461, 487–503, 549, **552**, 555, 561, 567, 570, 618, 630, 647, 658, 666, 668, 685, 686, 713
Shingwedzi River 13, 65, 70, 107, 240, 261, 275, 278, 296, 298, 379, 380, 439, 508–509, 526, 603, 619, 632, 659, 705
Ship Mountain (also known as *Langkop*) 155, 157, 214, 215, 308, 317, 318–319, 572, 617, 638, 641, 696, 704
Shipandani Valley 239
Shirelele (field ranger) 626
Shires, J.B. (Joseph) 227, 325, 326
Shirilele (field ranger) 625
Shirimantanga Hill 451, 501, 520, 637, **638**, **715**
Shishangani Spruit 29, 281, 375, 566
Shithlave, Jafuta (field ranger) 529, **536, 538, 540**
Shitsakana Spruit. *See* Satara.
Shivalo, Gidya (field ranger) 541
Shivhulani windmill **575**
Shona tribe (Zimbabwe) 44, 59, 79
Shoshangane (Songana, Zulu chief) 104, 106–113, 134, 138, 629. *Also see* Sakana *and* Manukosi.
Sibusa, Frans (field ranger) 540
Sibusa, Solomon (field ranger) 540
Sibuye, Amos Mafukala (field ranger) 538, 540,
Sibuye, Dawid (field ranger) 542
Sibuye, Frans (field ranger) 540
Sibuye, Jack (field ranger) 542
Sibuye, Jerry (field ranger) 541
Sibuye, Jim (field ranger) 538, 540
Sibuyi, Johannes (field ranger) 540
Sibuyi, Madubula (field ranger) 538
Sierra Chicundo 369, 372, 379
Sievewright, Ted 316
Siewert, Paul (ranger) 495–496, 549, 554, 710
Sifunda, Willem (field ranger) 538
Sifundsa, Charlie (field ranger) 540
Sifundsa, Mapoyisa (field ranger) 540
Sihanti (Swazi chief) 628
Sihlangu, Klaas (field ranger) 538
Sikkeltowkloof (farm) 437, 696, 708
Siloquana Hill 336
Silver Leaves (farm) 290
Simangale 98
Simmons, B.P. 322, 324
Simões, Casimiro (Casmier) 159, 179–181, 187, 257, 355
Simões, Mrs 187
Simonella diamond fields 223
Simpson, W. 94
Simswane (chief) 159
Singwane, Maxim (field ranger) 542
Sithole, Nambrete (field ranger) 542
Sivalo, Guija (field ranger) 538, 540
Sivalo, Samuel (field ranger) 538
Six (field ranger) **499**
Skirving, Magistrate J.B. 446
Skukuza 20, 23, 36, 83, 147, 243, 281, 297, 315, 339, 347, 355, **360**, 360–365, 406, 451, 454, 470, 472, 477, 487, 507, 513, 518–530, 536, 542, 549–550, 552–555, 565, 561, 581, **583**, 589–560, **591**, **592**, 593, 595–596, **599**, **600**, 602, 606–612, 694, 636, 694–695, 698–701, 708, 714–**720**
Sleigh, H. 657
Smart, Sammy 422, 635
Smellekamp, Johan Arnold 143, 301, 324
Smit, Comdt N. 193
Smit, Dr 610
Smit, Gen. 354
Smit, J.S. (railway commissioner) 354, 359
Smit, Koos (ranger) 616, 644
Smit, Vice Pres. Nicolaas Jocobus 312
Smith, Dr (Sir) Andrew 132, 384
Smith, Mrs J.H. 238
Smith, Sir Henry George Wakelyn (Harry) 132
Smith-Dorrien, Gen. (Sir) Horace Lockwood 409
Smits, Josina (née Van Vollenhoven) 183, 185
Smits, Rev. G.W. 183
Smuts, Gen. J.C. 238, 425, 453, 459, 460, 463, 467, 500, 627, 660–661, 686
Smuts, Gen. Tobias 413
Smythe, Capt. 223
Snethlage (chief engineer) 354
Snyman, Cobus 629
Snyman, F. 172
Snyman, Nicolaas 176
Sobhuza I (Swazi king) 57, 59, 141, 164, 368–369, 628
Soekmekaar 357, 381, 650–662, 667, 676–681, 683
Sofala (now Beira) 38, 83, 90, 95, 102, 139, 150, 170, 209–210, 246, 249, 260, 659
Sogana (Zulu chief). *See* Sakana *and* Manukosi.
Solomon (Williams's assistant) 659
Solomon, Sir Richard Prince 238
Solomondale railway station (Kalkfontein) 172
Soltké, Adolf 312, 317–318, 639–640, 717
Solvent (farm) 75
Somerville, sec. William 463
Songandaba (subservient chief of Shaka) 58, 141
Soniya (first wife of Gungunyane) 61
Sonnenberg, Ikey 295
Sonqua or San. *See* Bushman (San).
Sotho tribes 38–39, 57
South African Pioneer Powder Factory (Ltd.) 311
South African Police post 681
South African Railways 362, 463–464, 514, 587–588, 591, 686, 688, 690, 698. "Round in nine" 363, 464, 587, 593, **688**
Soutpan 74, 76, 102–103, 106–107, 110–111, 113, 115–119, 134, 135, 170, 290, 339, 346
Soutpansberg (Zoutpansberg) goldfields 290
Soutpansbergdorp (Zoutpansbergdorp) 115
Soutpansberg 42, 46, 78, 109, 121, 156–157, 161–167, 173–174, 176–179, 181–182, 185, 189–191, 194–196, 195, 199, 204, 246, 257, 290–291, 338, 425, 514, 629
Spanish flu epidemic (1918) 497, 525

Spekboom River 282
Spelonken **158**, 192, 174, 423, 431, 652
Spies, Andreas Theodorus (Andries) 144, 368
Spilsby (farm) 79
Spitzkop 223–224, 233, 275, 283, 303, 308, 320, 323–325, 415
Spraggon, Mark 675–679
Sprigg, Sir John Gordon 344
Spruyt, sec. 197
St Hill Gibbons, Maj. A. 447
St Paulo Palace 307
Stagnene **532**
Stamp, Johanna Hilda (Cheeky) 297–298, 514
Stamp, Patsy 297–298
Stamp, Stuart 297
Stamp, Thomas Harding (Tommy) 297–298, 514
Stanley, Henry Morton 132
Stapelkop Dam 217, 632, 706
Steamer (field ranger) **540**
Steelpoort River 171, 175, 339, 389
Steffler, Jan Christoffel 93, 280–281
Steinaecker, Baron Francis Christiaan Ludwig von 360–361, 411, 413–414, 416–418, **417**, 423, 431, 447, 474, 477, 479, 505, 685
Steinaecker's Horse (British border watch unit) 217, 236, 243, 254, 319, 321, 374, **411**, 413, 414–418, 423, 426, 431, 445–447, 454, 474–475, 476–478, **538**, 546, 562, 630, 635, 643, 684–685, 706, 708, 718
Stephanus (field ranger) 537
Stephenson, W. 358
Sterkspruit (farm) 474
Stevenson-Hamilton, Anne **451**
Stevenson-Hamilton, Hilda (née Cholmondeley) **451**, 457, 637–638, 698, 702, **710**, 713, 714, 715
Stevenson-Hamilton, Col James 10, 146, 235–236, 241, 251, 298–299, 317, 339–340, 447–471, **459**, **461**, **464**, **465**, 471, **476**, **483**, 483–501, **573**, **582**, 618, 624, 633, 636–637, **637**, **638**, 676, 684–701, 712–714, **715**
Stevenson-Hamilton, Jamie **451**
Steyn, Johannes 302
Steyn, Louis Botha de Wet de la Rey (ranger) 471, 521–524, **522**, 530, 574, 584, 699, 708
Steyn, Pres. M.T. 403–406
Steyn, R.G.P. 619
Steynberg, Coert 712

Stiebel, Gus 224
Stiemens, H. 307
Stigling, Sergt. Eldret 281
Stoffberg, Theunis Christoffel **452**
Stols, Gabriël 243, 323, 418, 473
Stols, Gert Frederik Coenraad 218, 243, 323, 473, 614, 642, 719
Stols, Jan Bokwa 217, 642
Stols family (also known as the *Berg Stolse*) 217, 643
Stolsnek 13, 15, 19, 27, 218, 281, 321–322, 542, 642, 719
Stone Age 11–24, **14**, **15**, **20**, **21**, **22**, **23**, 26, 29–30, 35, 42, 64, 628
Stoop, Albertus 393, 684
Strassburg (farm) 514. *Also see* Tshokwane.
Streeter, Frank 396, 496–497, 550
Strijdom, Alida 135, 347
Strijdom, Hans 124, 135, 137, 629
Strijdom, Willem 135, 137
Strijdom Tunnel 18, 229
Struben guest house 591, **698**, 713
Struben, Maj. Frank E.B. 698, 713
Strydom (surveyor) 381
Strydpoort 81, 102, 109, 115, 133, 172
Stubbs, C.C. 670
Style, George 651
Suikerbosrand 115, 118, 133
Sumatra (farm) 440, 505
Summit Route (Mpumalanga) 229
Sunnymeade (farm) 473
Sutherland (prospector) 283
Sutherland, Dr 290
Sutherland, Jim 218
Suzéz (Portuguese slave) 139
Swanepoel, A. 103, 110, 115, 119, 134, 704
Swanepoel, D.G. (Douw, ranger) 633
Swart, Lieut S. 420–421
Swartruggens 175
Swarts, Dawid Hercules (ranger) 515, 532–534, **534**
Swazi tribe 39, 56–59, 63, 109, 146, 180, 227, 234, 309, 338, 617, 628, 641
Swaziland 16, 26, 93, 228, 258–262, 322, 340, 371, 389, 453, 550
Sweetwaters (farm) 424, 431
Swene (chief) 97
Sweni area 231
Sweni River 217, 239, 254, 375, 533, 574, 671
Sweni Spruit outpost 414

T

Tait, Capt. 137
Taljaard, D.P. (Daantjie) 388–390, 400
Tamboekies (Thembus) 203, 204
Tancred, Sir Thomas 352, 353
Tati goldfields (Matabeleland) 265, 267, 342
Tati settlement 246
Taylor, Capt. Alfred James (Bulala) 279, 423–431, **424**, 502, 615–616, 631–632
Teixeira, Antonio de Souza 70, 247–251
Teixeira, Antonio Joaquim 153
Tengamanzi shop (Alf Roberts's shop) 76, **319**, 319–321, 639, **640**, 719
Tengwa road 652
Tengwe (Venda chief) 163
Terzakis, Osorio 631
Tete 139, 251
Thaba Nchu 120
Thabina 72, 275, 321
Thackeray, Col Edward Francis 493
Tharfield (farm) 239
Theiler, Sir Arnold 305, **340**, 341–345, 400, 691
theileriosis (east coast fever) 127
Theta Mine 285, 289
Thomas (Berthoud's travelling companion) 275
Thomas, Adam 655
Thomas, Maj. J.F. 428–430
Thompson, Alex 652, 654, 657, 660, 702, 703
Thompson, Dr Louis 489
Thompson, Lieut 429
Thonga tribe 69, 152, 159
Thorncroft, George 295
Thowelle (Zimbabwe) 98
Thulamela 38, 45, 48–55, **49**, 296, 664
Timbavati nature reserve 438
Timbavati River 13, 72, 98, 254, 277, 375, 533, 566, 692, 720. *Also see* Matibawati River.
Tindall, Judge Benjamin Arthur 477
Tlapa-la-Mokwena Dam **535**
Tlokwa tribe 338
Tolmay, Mr 360
Tolo Azime Falls 265–266
Tomango Station 356
Tomlinson, Herbert Ernst (Bert, ranger) 415, 477, 507–512, **509**, **510**, **511**, 520, 525–526, 530–532, 555, 644, 697, 705
Tomlinson, Martha **511**
Toothless Jack (field ranger) 537, **538**
Tosen, C.J. 395

Toulon (farm) 450, 463, 508, 511, 525, 527, 597, 688
tourism **576**; aerial service 599–602; amenities 416, 591–593, 693; development of 586–587; entrance fees 593–595, 693; fuel provision 598; Knapp huts 590; patrols 598–599; rest camps 50, 471, 589–598, 603, 606–607, 611; "Round in nine". *Also see* South African Railways.
  Selby building style 590; tourist huts 513, 699; tourist roads 315, 317, 323, 333, 406, 513, 515, 527, 533, 577, **588**, 602–603, 702, **705**
trade and trade routes, early 40, 68, 85
Trafford, William 284
transport activities and roads 69(m), 67–78, 79–85, 90–91, 104, 117, 150, 166, 252, 265, 350, 378, 562, 566, 634, 715
Transvaal Consolidated Land and Exploration Co. Ltd. (TCL) 434, 439, 463, 525
Transvaal game reserves (Sabie and Shingwedzi reserves) 667, 686
Transvaal Gold Exploration and Land Company 288
Transvaal Goldfields Extension Transport Company 329
Transvaal Share and Claim Exchange Trust and Agency Company Ltd 294
Travers, J.E.D. (John) 421, **423**, 505
Travers, J.H. 674
Treadwell, W.G. 447
Treur River 157, 302, 303, 304
Trew, Lieut-Col H.F. 513
Trichardt Commission 303
Trichardt graveyard **129**
Trichardt trek 80, 102, 127, 133, 144, 339
Trichardt, Anna Elizabeth (Annie) 137
Trichardt, Cornelia 137
Trichardt, Jeremiah 144, 346
Trichardt, Jerimias Josias 137
Trichardt, Karel (Carolus) 100, 131–145, 140, 149, 153–154, 157–158, 210, 246, 259, 263, 301, 302–304, 340, 351, 357, 368, 629, 638, 704, 716–717
Trichardt, Lieut-Col Stephanus Petrus Erasmus 123, **144**
Trichardt, Louis Gustavus 137
Trichardt, Louis 67, 81, 90, 100, 102–104, 106–111, 115–116, 121–129, **122**, 131–139, 152, 167, 169, 203–204, 217, 243, 246, 260, 338, 346, 351, 522, 628–629, 700, 709; diary 95, 106, 108, 111, 121, 135, 252
Trichardt, Martha (Louis' wife) 137
Trichardt, Petrus Frederik (Pieta) 111, 126, 128, 134, 135, 137, 141
Trichardt, Zacherija Gertuida (Sagriet) (née Erasmus) **140**, 141
Trichardtsdal 357
Trichardtsfontein (farm) 144
Trollope, Glen **508**
Trollope, Stephen Harold (ranger) 319, 417, 506–507, **506**, 514, 522, 527–528, 555, **642**, 693, 719
Trollope, Wesley Booth 506
Tsale Hill **275**
Tsendze Plain 107, 274–275, 568
Tsendze tented camp (1933) 509–510, 705
Tseri (now Tsheri) outpost 533
tsetse flies 82, 90, 99, 108, 110–111, 118, 124–128, 138, 143, 154, 158, 171, 178, 208, 213, 218, 233–234, 247, 251, 261, 265, 267–277, 280, 282, 287, 294, 295, 301, 315, 322–324, 326, 336–340, 346, 355, 368, 379, 383, 387, 465, 472, 638. *Also see* nagana.
Tshalungwa hot-water spring **531**
Tshange (now Tshanga) Spruit 239, 705
Tshange (now Tshanga) 239, 279, 700, 705
Tshaulu 117
Tshavandlopfu 378
Tshimbubfe iron mines 68, 77
Tshivhase (Venda chief) 192, 196
Tshokwane **64**, 339, 501, 502, 513–515, 522–523, 538, 548, 549, 565, 584, 707
Tshwenyane village 204
Tsomo (mountain) 124, 132
Tsonga tribe 38–39, 57–62, **63**, 65, 83, 219, 239, 266, 268, 274–279, 338, 712
Tswana tribe 338
Tuan (Sekwati's son) 176
Turner and Newall 473
Turometzane (Sekhukhune's mother) 228
Tzaneen 36, 59, 71, 290, 348, 362, 495

U

Ubisi, Baptine (field ranger) 537
Ubisi, Captain (field ranger) 540
Ubisi, Fifteen (field ranger) 537, 540, 541
Ubisi, Johan (field ranger) 540
Ubisi, Jotai (field ranger) 538, 540
Ubisi, Maxim (field ranger) 540
Ubisi, Solomon (field ranger) 540
Ubizi, Fifteen (field ranger) 540
Ubizi, Mpampuni (field ranger) 541
Ubizi, Rice (field ranger) 537
Umbandine (Mbandzeni, Swazi king) 57, 356, 472, 617, 628
Umbuluzi River (Swaziland) 93, 260
Umbvesha (Tsonga chief) 60
Umfolozi Reserve (KwaZulu-Natal) 232, 559, 579, 692
Umgagane (chief) 269
Umgeba (Inhliziyo's son) 618, 641, 717
Umlazi River 142
Umsutu (chief, subservient to Sekhukhune) 314
Umzamane (Zulu chief) 112
uMzilikazi (Moselekatse, AmaNdebele paramount chief) **56**, 60, 62, 102, 115, 118–120, 133, 134, 136, 141, 170, 198, 204, 211, 262, 338
Union Line (shipping company) 306
Usutu River 57, 141, 262, 422
Utrecht 246, 260
Uys, Pieter Lafras (Piet) 141

V

Vaal River 56, 102, 119, 124, 129, 133, 159, 170, 329, 330, 384, 524
Vaalhoek Mine 289
Vaalribbok (farm) 442
Valdézia Mission Station 69, 75, 165, 246, **275**, 274–275, 279, 379, 659
Valentine (prospector) 283
Van Bergen (surveyor) 381
Van Boeschoten, Cornelis 185–186, 196, 342
Van Boeschoten family 160
Van Breda, Miss 302
Van Breda, Mr 624, 630
Van Buuren (joiner) 425–426
Van Dam, Comdt-Gen. Gerard Marie Johan 413
Van de Capelle, Jan 83, 93, 99–100, 280
Van den Heever, G.D. 419–421
Van der Hoff, Rev. Dirk 183, 190
Van der Horst, Maria 204
Van der Kemp, Dr J.T. 203–204
Van der Lith, C.J. 667
Van der Merwe, C.J. 421
Van der Merwe, Izak 331
Van der Merwe, N.J. 240, 533
Van der Merwe, P.J. (ranger) 96, 115, 147
Van der Merwe, Peet 644

Van der Merwe, W. 102
Van der Merwe murder 193
Van der Meulen, G.H. 354
Van der Plas, Willem 356
Van der Veen, H.C. 521, 526, 596
Van der Walt, Andries 221
Van der Walt, H.P. 396
Van der Westhuizen (prospector) 297
Van Deventer (prospector) 296
Van Deventer, Gen. (Sir) Jacob Louis 495, 499–450, 525, 623, 654
Van Emmenes, A.J. 173
Van Emmenes, Aletta Magdalena 173
Van Emmenes, G.L. 173
Van Emmenes, Hendrik Josephis 173
Van Emmenes, Lourens 173
Van Graan, J. (Johan, ranger) 406
Van Graan party 221
Van Hattum & Co. 353
Van Heerden (field cornet) 190
Van Helsdingen, Mr 629
Van Niekerk (Davy's assistant) 371
Van Niekerk (field cornet) 310
Van Niekerk, Fanie 622–623
Van Niekerk, M.M. 614–615
Van Nispen, Magistrate Reginald Alphons 160, 194–195, 197
Van Oordt, Herman Frederik 344, **390**, 390–391, 397, 340, 396
Van Oordt, N.F. 489
Vanrenen (acting magistrate) 228
Van Rensburg, Gen. 242, 427
Van Rensburg, Mrs 185
Van Rensburg, Gertina Petronella Maria Janse 158
Van Rensburg, Hendrik Janse 190, 248
Van Rensburg, Johannes Jacobus Janse (*Lang Hans*) 59, 67–68, 73, 91, 102–114, 124, 132–138, 171, 217, 246, 304, 638.
   Also see Van Rensburg-Parks Board expedition (1959) 108, 110, 112, 114; Van Rensburg murder site 260, 522, 533; Van Rensburg search party 338, 339, 346; Van Rensburg trekker party 246, 629, 704
Van Rensburg, Acting Pres. W.C.J. 205
Van Rensburg, Lukas Willem Janse 158
Van Rensburg, Susanna Elizabeth Janse (née Oosthuizen) 158
Van Rooyen (hunter) 502, 643
Van Rooyen family 142
Van Rooyen, Lynn (ranger) 297

Van Taak, Willem 83, 93
Van Velden, Mr 689
Van Vollenhoven, Dr A. 183, 187, 414–415
Van Vuuren, Lourens Janse 115, 704
Van Warmelo, Josina (née Van Vollenhoven) 168, **184**, 187–188
Van Warmelo, Rev. Nicolaas Jacobus 85, **170**, 183–188, 197–198, 205
Van Wijk, J.L. 226, **391**, 396, 397–399, 400, 684, 713
Van Wyk children 102, 109, 138
Van Wyk (politician) 218
Van Wyk, Cornelis 190
Van Wyk, Frederik 102, 109, 138
Van Ysendijk, J.E. 354
Vardon, Mr 336
Vaughan-Kirby, Frederick 218–235, 339, 395, 447, 563–564, 684
vegetation. *See* Lowveld – vegetation.
Vegkop 119, **119**, 136, 170
Venda tribe 38, 39, 44–46, 48, 50, 52–55, 57, 65, 70, 75, 77, 79, 80, 86, 91, 116–117, 134, 159, 160, 204, 262, 621, 338, 339
Venda rebellion 162–163, 191, 207, 242, 249, 328, 385, 427, 621, 629
Venter (deacon) 183
Venter family (Doringbult) 172, 181
Venter, Albert 176
Venter, Comdt S.M. 162, 192, 193, 194, 197
Venter, Dr Freek (SANParks official) 634
Venter, P.J.L. 172
Vercueil, Magistrate Jan 192–194, 197
Verdoorn, G.J. 193
Vermaak, J.S. 399
Vermeulen, Magistrate J. 142, 301
Versamelhoek (farm) 170, 176
Verster, Gen. Victor 651
Versveld, Magistrate 205
Verwey, W. 354
Vhalembethu tribe 38, 39, 44–47, 56, 65
Victoria Falls (Mosi-oa-Tunya) 139, 266
Viera da Rocha, Commander 380, 671
Viljoen, Barend 426
Viljoen, Field Gen. Wynand 431
Viljoen, Gen. B.J. (Ben) 229, 236, 404, **405**, 406, 409, **413**, 417, 418, 421, 422, **423**, 425, 566, 620, 635, 643, 719
Viljoen, J.J. 426
Viljoen, Jan 218
Viljoen, Paul 636
Viljoen, Petrus 102

Viljoen, W.H. (field cornet) 426, 431
Villa des Prierriers 9, 10
Visser, Josef 426
Voetpad station 75
Vogelstruis (farm) 74, 79
Volksraad 8, 108, 114, 142–144, 157, 158, 171, 174, 176, 189, 198, 199, 204, 205, 224–226, 228, 262, 268, 282, 283, 287, 303, 307–309, 323, 324, 326, 343, 345, 351, 352, 357, 358, 368, 384–393, 395–398, 400, 446, 460, 494, 684, 685, 713
Von Lettow-Vorbeck, Gen. Paul Emil 525
Von Ludwig, C. 164
Von Schlickmann, Capt. Conrad Heinrich Hans 227
Von Schuckmann, Baron Bruno Helmut Erich 344
Von Veltheim, Ferdinand Karl Ludwig 238
Von Wielligh beacon 372
Von Wielligh, Marthinus 380
Von Wielligh, P.A. 380
Von Wielligh, Surv. Gen. Gideon Retief 228, 254, 259, **369**, 371–373, 375–379, 566, 707
Voortrekker windmill 315, 319, 718
Vorster, Alie 206, 212
Vorster, Barend Jacobus (Rooi Barend) 172, 195, 206, 212, 357, 358
Vos, M.C. 228, 371, 373–376, 378, 379, 566, 628, 707
Vos, W. 259
Vredeburg 174
Vucha Ruins (Belingwe Reserve) 150
Vudogwa 38, 41
Vurhami River 97, 263, 718
Vutomi Spruit 339

W

Waai-Waai (Magamba guide) 80, 135, 136, 204
Wade, C. 437, 687
Wagenbooms River 203
Wagener, L. (ranger) **637**
Wagon Route, old (Ohrigstad-Delagoa Bay) 143, 171, 263, 302–304, 327, 340, 351, 479, 480, 616, 629, 636, 638, 715
Wainwright, Mr 325
Walker, William M. 447, 475, 684
Walters, S. 421
Wangemann, Dr D. 151
Wangemann, Theodor 78, 165
Warnant, Louis 358, 359

Warren, Dr 465
Warren, E. 232, 356, 691
Waterberg Reserve 445, 446
Waterberg 74, 175, 176, 181, 189, 190, 193, 197, 241, 250, 343, 388, 390, 393, 395, 396, 397, 398, 400, 453, 459, 508, 519, 525, 663
Waterboom (farm) 652, 655
Waterval-Boven (railway depot) 356
Waterval-Onder 350, 356, 404
Watkins, Frank 224
Watkins, Susan (née Glynn) 224
Watkins-Pitchford, Herbert 344, 345
Watt, John 182
Watusi tribe 26, 28
Webb, Harold 506
Welcome Farm 270
Welgevonden (farm) 72, 73, 102, 104, 106–107, 655
Welman, Andries 420
Weltevreden (farm) 69, 172, 199
Wenela (WNLA) 667, 671, 677, 679, 682, 683
Wertheman (engineer) 289
Wessels, Johannes 248, 250, 251
Wessmann, Rev. 47
Westwood and Winby (engineering firm) 358, 359
Wetkop (farm, Witklip) 303
Weynek tin mines 74
White (Sandeman's travelling companion) 233
White River Estate Company 224
White River 31, 217, 224, 225, 238, 243, 321, 333, 356, 416, 417, 422, 431, 451, 461, 475, 478, 483, 484, 504, 516, 529, 530, 534, 535, 548, 588, 593, 602, 643, 657
White, Charlie 239, 706
Whitfield, E.A. (field ranger) 406, 557
Whyte, Dr Ian (SANParks research officer) 297
Wierda, S.W. (state architect) 292
Wildlife Society 696
Wilge River 356
Williams, Theo 396, 481, 657, 659
Willis, P.W. ("Pump") 242, 243, 479, 595
Wilson, Frank 359
Wilson, gold commissioner 293, 331
Wilson, Secretary J.W.H. 689
Winburg 142, 170, 301
Winchester (TLC farm) 505
Windham, W. 489
Winter, Jimmy 295
Witklip 179, 303
Witton, Lt George Ramsdale 425, 426, 428, 430
Witwatersrand goldfields 288, 292, 294
Witwatersrand Mine 290, 321, 647, 650, 656, 683
Witwatersrand Native Labour Association (WNLA, Wenela) 380, 476, 490, 547, 650, 662, 666, 667, 682 *Also see* Wenela.
Wolfshoek (farm) 390
Wolhuter, Alice **483**, 533
Wolhuter, Andrew James (Henry, ranger) 497, **497**, 500, 549, 554, **644**
Wolhuter, Henry Charles Christoffel (Harry, ranger) 146, 218, 235, **236**, 236, 317, 411, 415, 416, 417, 418, 422, 456, **457**, 474, 477, 478, **478**, 479, **480**, 481, 482, **483**, **484**, 485, 486, 491–493, **497**, 501, 513, 517, 521, 523, 533, 540, 543, 548, 549, **573**, 614, 616, 618, 630, 643, 644, 689, 699, 708
Wolhuter, Henry Charles jnr 533, **534**, 535, 540
Wolmarans, Sen. A.D.W. 396, 397
Wolseley, Sir Garnet Joseph 228, **309**, 310, 315, 351
Wood, Dr 380
Wood, Mr 260
Wood, Sir Henry Evelyn 217, **315**
Woodhouse (farm) 474
Woodlands, Charlie 321, 417, 717, 718
Wools-Sampson, A. 455, 460
Wright (prospector) 296, 503
Wykerd, Mr 705
Wylie, Col 656
Wymer, John 30

X

Xavier, Maj. Alfredo Augusto Caldis 372
Xhosa tribe 131–132, 203–204, 384, 485

Y

Yates, C.A. 146, 147, 480, 500–501, 642, **643**, 689, 691
Ysendyk, J.E. (engineer) 353
Ysterberg 69, 72, 74, 77, 79, 82, 93, 104, 106, 117, 118, 275

Z

Zaayman, Mr 611
Zambezi River 56, 57, 59, 81, 95, 98, 139, 153, 266, 336, 447, 449
Zari Spruit 529, 631
Zarps (division of the South African Police Service) 398, 427, 471, 474
Zeederberg brothers 330–332
Zeederburg coach service 295, **328**, **329**, 331
Zeekoegat (farm) 439
Ziedie 111
Ziervogel, J.P. 292
Zietsman, Magistrate Johan Philip 142
Zimbabwe Ruins 69, 74, 81, 91, 138, 150, 217, 263, 264, 282
Zimbabwe 99, 207, 330, 647
Zinyawo, Moses (missionary) 489
Zitha, Million (field ranger) 318
Zuid-Afrikaansche Spoorweg Maatschappij. *See* South African Railway Company.
Zuidkaap goldfields 393
Zulu tribe 42, 56, 59, 108, 109, 114, 132, 141, 163, 170, 173, 182, 250, 256, 315, 319, 448
Zululand Cash Wholesalers 579, 693
Zululand reserves 232, 446
Zwane, Jim (field ranger) 538, 540
Zwangendaba (Nguni chief) 57-59, 141, 338
Zwartfontein (farm) 303
Zwide Nqumayo (leader of Ndwandwe tribe) 57, 58, 338